An Introduction to Nonlinear Analysis and Fixed Point Theory

Hemant Kumar Pathak

An Introduction to Nonlinear Analysis and Fixed Point Theory

 Springer

Hemant Kumar Pathak
School of Studies in Mathematics
Pandit Ravishankar Shukla University
Raipur, Chhattisgarh
India

ISBN 978-981-13-4261-5   ISBN 978-981-10-8866-7   (eBook)
https://doi.org/10.1007/978-981-10-8866-7

Printed on acid-free paper

This Springer imprint is published by the registered company Springer Nature Singapore Pte Ltd. part of Springer Nature
The registered company address is: 152 Beach Road, #21-01/04 Gateway East, Singapore 189721, Singapore

Dedicated to my parents
who assigned me this heavenly body
with a soul to think for humanity,
enlightened my mind with wisdom
of thought to think righteously;
and endowed me with a capacity to
render service for mankind.

Hemant Kumar Pathak

Foreword

The book entitled "An Introduction to Nonlinear Analysis and Fixed Point Theory" by Prof. H. K. Pathak covers both the area of nonlinear analysis and fixed point theory in great detail. The book begins with the fundamentals of nonlinear functional analysis, e.g. geometry of Banach spaces, differential calculus in Banach spaces, monotone operator theory, accretive operators and their variants. The book lays special emphasis on applying techniques of nonlinear analysis to model and to treat nonlinear phenomena with which nature confronts us. Coverage of applications includes many branches of science and technology such as control theory, nonlinear stochastic operator equations, variational methods in Hilbert spaces, degree theory, k-set contraction and condensing operators, variational methods and optimization. Applications of monotone operator theory to ODE, integral equations and solution of nonlinear equations by computational schemes and strong convergence results are presented. Applications of fixed point theorems to geometry of Banach spaces, system of linear equations, control theory, game theory, differential equations, nonlinear integral equations, abstract Volterra integrodifferential equations, surjectivity problems, simultaneous complementarity problems and problems of integral inclusion for multifunctions are thoroughly discussed.

In my opinion, the book should be very useful to mathematics students in their final semester course of master's degree and also for the first semester course for Ph.D. students, enhancing their capability to gain the desired insight into nonlinear analysis and fixed point theory.

Jhusi, Allahabad, India

Prof. Satya Deo, Ph.D., FNASc.
Formerly Vice Chancellor APS University
Rewa, and RD University, Jabalpur NASI Senior
Scientist, Harish-Chandra Research Institute

Preface

Nonlinear analysis is the fascinating emerging field of the twenty-first century characterized by a remarkable mixture of nonlinear functional analysis, nonlinear operator theory, topology, mathematical modelling and applications. Its scope of enquiries not only encompasses the geometric theory of infinite dimensional function spaces and operator-theoretic real-world problems but also widens the range of interdisciplinary fields ranging from engineering to space science, hydromechanics to astrophysics, chemistry to biology, theoretical mechanics to biomechanics and economics to stochastic game theory. The deep-rooted concepts and techniques provide the tools for developing more realistic and accurate models for a variety of phenomena encountered in various applied fields. This gives nonlinear analysis a rather interdisciplinary character. Today, the more theoretically inclined nonmathematician (engineer, economist, geologist, pharmacologist, biologist or chemist) needs a working knowledge of at least a part of the nonlinear analysis in order to be able to conduct a complete qualitative analysis of his models. This supports a high demand for books on nonlinear analysis. Moreover, the subject has become so vast that no single book can cover all its theoretical and applied parts. In this volume, we have focused on those topics of nonlinear analysis which are pertinent to operator-theoretic fixed point results, especially metric, topological and lattice-based fixed point theorems and their applications to control theory, dynamic programming, matrix theory, differential and integral equations, calculus of variations and many real-world problems such as stochastic modelling of physical and biological sciences.

The first half of the twentieth century witnessed an extensive theoretical investigation pertaining to linear functional analysis which deals with infinite dimensional topological vector spaces that provide a suitable pathway to mix in a fruitful way the linear (algebraic) structure with topological one and the linear operators acting between them. This investigation facilitated extending standard results of the linear analysis to an infinite dimensional context paving the way for rigorous treatment of linear mathematical models. Systematic efforts to extend the linear theory to various types of nonlinear operators were started in the early 1960s. This marks the beginning of what is known today as "nonlinear analysis".

However, it is quite interesting to note that well before this period there was a short period during the 1930s when the notions of the compact operator and the extension of Brouwer's degree theory in finite dimension to Leray–Schauder degree theory in infinite dimensional context came into existence.

On the one hand, with the advent of nonlinear analysis, several theories have been developed simultaneously in this respect, and today some of them are well established approaching their limits, while others are still the object of intense research activity. The appearance of set-valued analysis, nonsmooth analysis, differential topology, combinatorics, geometry of manifolds and of computational mathematics all of which were motivated by concrete needs in applied areas such as control theory, optimization, game theory and economics is evidence to this effect. Their development provided nonlinear analysis with new concepts, tools and theories that enriched the subject considerably. On the other hand, fixed point theory has an enormous number of applications in various fields of mathematics. Keeping in view the above facts, it is legitimate to introduce the book with the title "An Introduction to Nonlinear Analysis and Fixed Point Theory". In this book, I tried to present most of the significant results in the field of nonlinear analysis, especially monotone operator, fixed point theory, topological degree theory, variational methods and optimization, and then to present various related applications.

Chapter 1 is an introduction to some fundamental concepts needed for the development of the theory of nonlinear functional analysis. We deal with certain large classes of nonlinear operators which arise often in applications. In particular, we examine the Nemytskii, Hammerstein and Urysohn operators and their continuity properties. All these operators are encountered in almost all problems. Finally, we introduce the concept of Sobolev spaces (the suitable spaces for weak solutions of elliptic equations) needed for the development of differential equations.

Chapter 2 deals with geometrical structures such as convexity and smoothness of Banach spaces and of certain broad classes of nonsmooth functions. This chapter also deals with useful properties of duality mappings that interplay with these geometrical structures of Banach spaces. We show that the subdifferential of norm functional is precisely the duality mapping.

Chapter 3 deals with calculus in real Banach spaces. We start with the Gâteaux and Fréchet derivatives. We discuss the generic differentiability of continuous convex functions. This chapter also deals with an important concept of nonlinear analysis—subdifferential of convex functionals. Properties of the derivative are discussed, and some fundamental theorems of calculus are presented, especially Taylor's theorem, inverse function theorem and implicit function theorem.

Chapter 4 deals with monotone and maximal monotone operators and their properties. We give some results regarding the approximate solvability of operator equations involving monotone operators with the hope this will help develop computer algorithms for the approximate solution of operator equations. Monotone properties of the subdifferential of convex functionals are discussed. We close this chapter by introducing various generalizations of monotonicity concepts—pseudomonotonicity, generalized pseudomonotonicity, etc.

Chapter 5 deals with metric, topological and lattice-based fixed point theory along with common fixed point theorems for a family of commuting mappings. We present some fixed point theorems for multifunctions motivated by their applications to integral inclusions. In this chapter, we also present common fixed point theorems for a family of commuting mappings. We conclude this chapter by giving a discussion on fixed point theorems in ordered Banach spaces. Our treatment is brief and is motivated by their applications to the system of linear equations, matrix theory, control theory, differential and integral equations.

Chapter 6 deals with degree theory, k-set contractions and condensing operators. This chapter is motivated with the fact that topological degree theory has an important advantage over the fixed point theory in the sense that it gives information about the number of distinct solutions, continuous families of solutions and stability of solutions. Leray–Schauder degree theory is presented. In the sequel, we present a generalization of Leray–Schauder degree that extended the concept of the degree to the class of limit-compact operators. Subsequently, this theory was used to discuss k-set contractions and condensing mapping.

Chapter 7 provides an introduction and use of variational methods and optimization in nonlinear analysis.

Chapter 8 discusses integral equations in the most general setting. We conclude the chapter by giving computational scheme for the solvability of nonlinear equations.

Chapter 9 provides applications of fixed point theorems to the system of linear equations, nonlinear matrix equations, control theory, dynamic programming, stochastic game theory and existence theorems for nonlinear differential and integral equations. In most of the illustrated problems, the differential equations are transformed into equivalent operator equations involving integral operators and then appropriate fixed point theorems or degree theoretic methods are invoked to prove the existence of desired solutions by recasting the operator equations into fixed point equations.

Chapter 10 deals with applications of fixed point theorems for multifunction to integral inclusions.

A glimpse of fundamentals, exposition of a rich variety of topics, both theoretical and applied, make nonlinear analysis useful to graduate students and researchers, working in analysis or its applications to control theory, variational inequalities, theoretical mechanics, or dynamical systems. An appendix contains requisite background material needed, and a detailed bibliography facilitates further study.

This book evolved from classes taught by the author at Pt. Ravishankar Shukla University, Raipur, India, in a course of Master of Philosophy entitled "Nonlinear Analysis and Topological Structures". Moreover, the book is self-contained and the presentation is detailed, to avoid irritating readers by frequent references to details in other books. The examples are simple, to make the book teachable. We hope that this book will be extremely useful to students having a background in nonlinear functional analysis, operator theory and topological properties.

I would like to express my profound thanks and gratitude to my friends—Prof. Ravi P. Agarwal, USA; Prof. Billy E. Rhoades, USA; Prof. G. Jungck, USA; Prof. Brian Fisher, England; Prof. Yeol Je Cho, South Korea; Prof. Shin Min Kang, South Korea; Prof. J. K. Kim, South Korea; Prof. J. S. Jung, South Korea; Prof. S. K. Ntouyas, Greece; Prof. Rosana Rodriguez-Lopez, Spain; Prof. Donal O'Regan, Ireland; Prof. S. N. Mishra, South Africa; Prof. V. Popa, Romania; Prof. N. Shahzad, Saudi Arabia; Prof. N. Hussain, Saudi Arabia; Prof. M. S. Khan, Sultanate of Oman; Prof. S. S. Chang, P.R. China; Prof. Zeqing Liu, P. R. China; Prof. T. Suzuki, Japan; Prof. V. Kannan, Prof. Satya Deo Tripathi, Prof. S. L. Singh, Prof. M. Imdad, Prof. Q. H. Ansari, Prof. P. Veeramani, Prof. M. T. Nair, Prof. D. R. Sahu all from India—for their kind encouragement at different occasion during the preparation of this book. I would also like to express my sincere thanks to my colleagues—Prof. B. K. Sharma and Prof. B. S. Thakur—for their constant moral support. I also want to thank my project fellow Ekta Tamrakar for her support in typing some part of the book in LATEX.

The author is indebted to the anonymous reviewers for their precious comments and helpful suggestions towards the improvement of the original draft of the book. Finally, the author is very grateful to Dr. Lynn Brandon from Springer UK, Shamim Ahmad and Shubham Dixit from Springer Nature for their indefatigable cooperation, patience and understanding throughout our communication. It is a pleasure to acknowledge the great help and technical support given to us by N. S. Pandian, Praveenkumar Vijayakumar, Krati Shrivastav and their technical team of Springer Nature in their rapid and meticulous publication of the book.

Raipur, India   Hemant Kumar Pathak
July 2017

Contents

About the Author

Prof. Hemant Kumar Pathak is Head of the School of Studies in Mathematics at Pt. Ravishankar Shukla University. Currently, he is also Director of the Center for Basic Sciences and Director of Human Resource Development Centre at Pt. Ravishankar Shukla University, Raipur, Chattisgarh, India. His areas of expertise include general topology, nonlinear analysis, operator theory, integration theory, fuzzy set theory, cryptography, Banach frames, number theory, approximation theory, convergence analysis and summability theory. He has over 250 research papers in national and international journals. He is the author and co-author of over 50 books for graduates and postgraduate studies of science and engineering students. He is an Associate Editor of Springer's journal—Fixed Point Theory and Applications—and Editorial Member of American Journal of Computational and Applied Mathematics. He is also a Reviewer of Mathematical Reviews of American Mathematical Society.

Acronyms

BVP	Boundary value problem
COP	Constrained optimization problem
deg	Degree of the finite-dimensional mapping
Deg	Degree of the infinite-dimensional mapping
EP	Equilibrium problem
EVP	Ekeland's variational principle
FPP	Fixed point problem
IPS	Inner product space
IVP	Initial value problem
MP	Minimization problem
MPL	Mountain pass lemma
NEP	Nash equilibrium problem
ODE	Ordinary differential equation
OP	Optimization problem
$(PS)_c$	Palais–Smale condition
PDE	Partial Differential Equation
SPP	Saddle point problem
VIP	Variational inequality problem

Glossary of Symbols

$\mathbb{N}, \mathbb{Z}_+$	The set of all natural numbers
$\mathbb{K}$	The set of all real or complex numbers
$\mathbb{R}$	Field of all real numbers
$\mathbb{I}, \mathbb{Z}$	The set of all integers
$\mathbb{Q}$	Field of all rational numbers
$\mathbb{C}$	Field of all complex numbers
X	A Banach space (infinite dimensional)
X^*	First conjugate (dual) space of X
X^{**}	Second conjugate (dual) space of X
$x \in X$	x belongs to X
$x \notin X$	x is not a member of the set X
$\vert, :$	such that
$A \subset B$	A is a subset B
$\cup$	Union
$\cap$	Intersection
$-, \setminus$	Difference
I, Λ	Index set
$\emptyset$	Empty set
$A \times B$	Cartesian product of A and B
$\mathscr{D}(F)$	Domain of $F : X \to Y$
$\mathscr{R}(F)$	Range of $F : X \to Y$
∂C	Boundary of C
$(\cdot, \cdot)$	Inner product in $\mathbb{R}^n$ or $\mathbb{C}^n$
$\langle \cdot, \cdot \rangle$	Inner product in an IPS X
$\vert \cdot \vert$	Standard 2-norm in $\mathbb{R}^n$
$\Vert x \Vert$	Norm of x
μ	Measure
$\forall$	For all
$P \Rightarrow Q$	P implies Q
$x_n \to x$	$\{x_n\}$ converges to x

$x_n \rightharpoonup x$	$\{x_n\}$ converges weakly to x
$f : X \to Y$	f maps from X into Y
$x \mapsto f(x)$	Map assigning $f(x)$ to x
$\mathfrak{F}(T)$	The set of fixed points of $T : X \to X$
$f(A)$	Image of A under f
$f^{-1}(B)$	Inverse image of B under f
i_X	Identity map on X
X^*	First dual space of X
X^{**}	Second dual space of X
$\hat{0}$	Null operator of X^*
$f'(\mathbf{x})$	Derivative of $f(x)$
$\{\mathbf{e}_1, \cdots, \mathbf{e}_n\}$	Standard basis of $\mathbb{R}^n$
A°, int A	Interior of set A
$\overline{A}$	Closure of set A
$[A]$	Matrix A
$\det[A]$	Determinant
$D_j f$	Partial derivative
∇f	Gradient of f
$J_r(x)$	Jacobian
$\wedge$	Wedge product
ω_T	Transform of ω
C^k	Classes of differential functions
$\mathcal{L}(X, Y)$	The vector space of all bounded linear operators from X into Y
$i : B \to A$	Inclusion function
$f\vert_A$	Restriction f to A
$\{x_n\}$	Sequence whose nth term is x_n
Δ^n	n-simplex
2^X	Collection of all subsets of X
(a, b)	Open interval $\{x \vert a < x < b\}$
$[a, b]$	Closed interval $\{x \vert a \leqslant x \leqslant b\}$
f^{-1}	Inverse mapping
$f \circ g$	Composition of mappings f and g
$l.u.b.$	Least upper bound
sup S	Supremum of S
$g.l.b.$	Greatest lower bound
inf S	Infimum of S
$\mathrm{Re}(z)$	Real part of z
$\mathrm{Im}(z)$	Imaginary part of z
$C(X)$	Spaces of continuous functions
$C(X, \mathbb{R})$	Spaces of continuous functions
$CB(X)$	The set of nonempty closed and bounded subsets of X
$\mathcal{K}(X)$	The set of nonempty compact subsets of X
$P_n(x)$	Polynomial of degree n

$\overline{\lim}\, x_n$	Limit superior of $\langle x_n \rangle$
$\underline{\lim}\, x_n$	Limit inferior of $\langle x_n \rangle$
$\lim\sup x_n$	Limit superior of $\{x_n\}$
$\lim\inf x_n$	Limit inferior of $\{x_n\}$
$M \oplus N$	Direct sum of M and N
$(x : P(x))$	Set of all x such that $P(x)$
$\delta_i^j = \begin{cases} 1 & \text{if } i = j \\ 0 & \text{if } i \neq j \end{cases}$	Kronecker delta
$f(x) = \circ(g(x)), x \to a$	$f(x)/g(x) \to 0$ as $x \to a$
$f(x) = O(g(x)), x \to a$	$f(x) \leqslant K\, g(x)$ for all x in the nbd of the point a

Chapter 1
Fundamentals

*I love mathematics not only because it is applicable to
technology but also because it is beautiful.*

Rósza Péter

*Mathematics contains much that will never hurt one if one does
not know it nor help one if one does know it.*

J. B. Mencken (1715)

What is clear and easy to grasp attracts us; complications deter.

David Hilbert

The main objective of this chapter is to familiarize the reader to some basic concepts and fundamental results needed for the development of the theory of nonlinear functional analysis.

In Sect. 1.1, we discuss discrete structure needed in subsequent study. In Sect. 1.2, we introduce mainly metric spaces, topological spaces, linear spaces, normed spaces and inner product spaces, while in Sect. 1.3, we discuss spaces of bounded linear operators.

In Sect. 1.4, we discuss the Hahn–Banach theorem, reflexive spaces, weak topologies and basic properties of weakly convergent sequences in normed spaces, while in Sect. 1.5, we mainly discuss compact spaces.

In Sect. 1.6, we discuss various forms of continuity - complete continuity, weak continuity, demicontinuity, hemicontinuity and compactness which are now commonly used terms in the literature. Various theorems and counterexamples are given which illustrate their interrelationships. In Sect. 1.7, we introduce the notion of measure, measurable function and measure space needed in subsequent chapters.

In Sect. 1.8 by giving definitions of some nonlinear operators, namely the Nemytskiĭ, Hammerstein and Urysohn operators and their continuity properties. These operators are, at present, widely used in applications. In Sect. 1.9, we introduce the concept of the Sobolev spaces (the suitable spaces for weak solutions of elliptic equations) needed for the development of differential equations. In Sect. 1.10, we

.© Springer Nature Singapore Pte Ltd. 2018
H. K. Pathak, *An Introduction to Nonlinear Analysis and Fixed Point Theory*,
https://doi.org/10.1007/978-981-10-8866-7_1

discuss optimal control theory while in Sect. 1.11, we introduce nonlinear stochastic operator equations. Finally, we conclude the chapter in Sect. 1.12 by introducing some fundamental concepts of variational method in the Hilbert space.

In what follows, we denote the set of natural numbers, the set of real numbers and the set of complex numbers by $\mathbb{N}$, $\mathbb{R}$ and $\mathbb{C}$, respectively.

1.1 Discrete Structure

1.1.1 Partially Ordered Set

Definition 1.1 Let X be a nonempty set. A *partial order relation on X* is a relation "$\preceq$" defined on X which satisfies the following properties:

($P1$) **Reflexive**. For each $x \in X$, $x \preceq x$;

($P2$) **Antisymmetric**. If $x, y \in X$, then $x \preceq y$ and $y \preceq x \Longrightarrow x = y$;

($P3$) **Transitive**. If $x, y, z \in X$, then $x \preceq y$ and $y \preceq z \Longrightarrow x \preceq z$.

A nonempty set X in which there is defined a partially order relation is called the *partially ordered set* or simply *poset* and is denoted by $(X, \preceq)$.

Sometimes, it is convenient to express $x \preceq y$ by $y \succeq x$. Let S be a nonempty subset of a partially ordered set X. An element $u \in X$ is called a lower bound of S if $u \preceq x$ for all $x \in S$; and a lower bound g of S is called the greatest lower bound (or g.l.b.) of S if $g \geq u$ for every lower bound u of S. Similarly, an element $v \in X$ is said to be an upper bound of S if $v \geq x$ for all $x \in S$; and an upper bound ℓ of S is called the least upper bound (or l.u.b.) of S if $\ell \preceq v$ for every upper bound v of S. The greatest lower bound of S is usually called its infimum and is written as inf S. The least upper bound of S is called its supremum and is denoted by sup S.

Suppose inf S and sup S both exist and are g and ℓ, respectively. If $g \in S$, then it is called *minimum* and is denoted by min S. Similarly, if $\ell \in S$, then it is called *maximum* and is denoted by max S.

(P4) **Comparable**. Any two elements x, y in a poset $(X, \preceq)$ are said to be *comparable* if either $x \preceq y$ or $y \preceq x$, otherwise they are said to be *incomparable*.

A partially ordered relation with property ($P4$) is called a *total* (or *linear*) *order relation*. A partially ordered set $(X, \preceq)$ whose relation $\preceq$ satisfies condition ($P4$) is called *totally ordered set* or a *linearly ordered set*.

Example 1.1 (1) The relation $\leq$ in the usual sense "less than or equal to" is a partial order relation on the set $\mathbb{R}$ of real numbers.

(2) If S is a set of any collection of sets, then the relation $\subseteq$ read as "is a subset of" is a partial order relation on S.

(3) Let $\mathbb{N}$ be the set of all positive integers. Let $m \leq n$ mean that m divides n. This defines a partial ordering on $\mathbb{N}$.

(4) The relation of divisibility is not partial order relation on the set $\mathbb{Z}$ of integers. In fact, the relation is not antisymmetric because $5 \mid -5$ and $-5 \mid 5$ but $5 \neq -5$.

Chain – Let $(X, \preceq)$ be a partially ordered set. A subset $\mathscr{C}$ of X is called a *chain* if $\mathscr{C}$ is a totally ordered set.

An element x in X is said to be *maximal* if $y \succeq x \Rightarrow y = x$. Similarly, an element x in X is called *minimal* if $y \preceq x \Rightarrow y = x$.

Axiom of choice – Let $\mathfrak{X}$ be a collection of nonempty sets. Then, we can choose a member from each set in that collection. In other words, there exists a function c defined on $\mathfrak{X}$ with the property that, for each set S in the collection, $c(S)$ is a member of S. The function c is then called a choice function.

Using the concepts as given in Definition 1.1, we can now formulate Zorn's lemma, which we regard as an axiom.[1]

Theorem 1.1 (Zorn's lemma) *Let $(X, \preceq)$ be a partially ordered set. If every chain $\mathcal{C}$ in X has an upper bound, then X possesses a maximal element.*

Observation

- It may be observed that Zorn's lemma can be derived from the axiom of choice. Conversely, the axiom of choice follows from Zorn's lemma, so that Zorn's lemma and the axiom of choice can be regarded as equivalent axioms.
- Zorn's lemma occurs in the proofs of several theorems of crucial importance, for instance the Hahn–Banach theorem in functional analysis, the theorem that every vector space has a basis, the Tychonoff theorem in topology stating that every product of compact spaces is compact, and the theorems in abstract algebra that in a ring with identity every proper ideal is contained in a maximal ideal and that every field has an algebraic closure.

Definition 1.2 A system $\mathfrak{A} = \langle A, \leq \rangle$ formed by a nonempty set A and a binary relation $\leq$ called a lattice, if $\leq$ establishes a partial order in A and that for any two elements $a, b \in A$ there is a least upper bound (join) $a \cup b$ and a greatest lower bound (meet) $a \cap b$. The relations $\geq$, $<$, and $>$ are defined in the usual way in terms of $\leq$.

The lattice $\mathfrak{A} = \langle A, \leq \rangle$ is called complete if every subset B of A has a least upper bound $\bigcup B$ and a greatest lower bound $\bigcap B$. Such a lattice has in particular two elements 0 and 1 defined by the formulas

$$0 = \bigcap A \text{ and } 1 = \bigcup A.$$

Given any two elements $a, b \in A$ with $a \leq b$, we denote by $[a, b]$ the interval with the endpoints a and b, that is, the set of all elements $x \in A$ for which $a \leq x \leq b$; in symbols,

$$[a, b] = E_x[x \in A \text{ and } a \leq x \leq b].$$

The system $\langle [a, b], \leq \rangle$ is clearly a lattice; it is a complete if $\mathfrak{A}$ is complete.

---

[1]The term "lemma" is for historical reasons. Zorn's lemma, also known as the Kuratowski–Zorn lemma, after mathematicians Max Zorn and Kazimierz Kuratowski, is a proposition of set theory that states that a partially ordered set containing upper bounds for every chain (i.e. every totally ordered subset) necessarily contains at least one maximal element. This lemma was proved by Kuratowski in 1922 and independently by Zorn in 1935.

1.2 Topological Spaces

The concept of a space is different from the notion of a set. A space is distinguished from a set by possessing attributes not possessed by a mere collection of well-defined objects what we call elements of the set, and a subspace is a subset of a space which is assumed to have inherited such defining attributes. A metric (distance) space suggests that given two points of the space there should be a real number that measures the distance between them. To initiate our discussion on the notion of a "metric", it is natural to begin with a pair (X, d) where X is a set and $d : X \times X \to \mathbb{R}$ is a mapping of the cartesian product $X \times X$ into the reals $\mathbb{R}$. If $d(x, y)$ is thought of as the distance between two points $x, y \in X$ it is natural to assume that d satisfies for each $x, y \in X$:
(SM1) $d(x, y) \geqslant 0$;
(SM2) $d(x, y) = 0 \Leftrightarrow x = y$; and
(SM3) $d(x, y) = d(y, x)$.

A pair (X, d) satisfying the above assumptions is called a semimetric space. These assumptions are in some sense minimal when one thinks of a distance. The semimetric spaces form a subclass of the important class of metric spaces defined below, yet it is doubtful whether semimetric spaces themselves offer sufficient structure to yield very deep results. We can introduce very important concept of "limit" in the setting of semimetric space as follows: a sequence $\{x_n\}$ in a semimetric space X is said to converge to a point $x \in X$ if for each $\varepsilon > 0$ there exists an integer $n_0 \in \mathbb{N}$ such that

$$n \geqslant n_0 \Rightarrow d(x_n, x) < \varepsilon.$$

In this case, x is said to be the limit of the sequence $\{x_n\}$ and we simply write $\lim_{n \to \infty} x_n = x$. However, in general, there is nothing to assure that in a semimetric space the limit of a given sequence is unique. To remedy this defect, we discuss a richer structure than a semimetric space what we call a metric space.[2]

1.2.1 Metric Spaces

Definition 1.3 Let X be a nonempty set. A metric[3] on X is a function $d : X \times X \to \mathbb{R}$ which satisfies the following conditions:
(M1) $d(x, y) \geqslant 0$ for all $x, y \in X$;
(M2) $d(x, y) = 0$ if and only if $x = y$;

---

[2]The notion of a metric space was introduced by M. Fréchet in his thesis Sur Quelques Points Du Calcul Fonctionnel (Rendiconti del Circolo Matematico di Palermo, 22(1906), pp. 1–74) and called by him spaces of class (E). However, Fréchet and his immediate successors did not develop the theory. The term metric space was introduced by F. Hausdorff (Grundzüge der Mengenlehre, Leipzig, 1914).

[3]A landmark construction of the notion of metric introduced in 1906 by the French mathematician Maurice René Fréchet.

(M3) $d(x, y) = d(y, x)$ for all $x, y \in X$;
(M4) $d(x, y) \leqslant d(x, z) + d(z, y)$ for all $x, y, z \in X$.

The value of metric d at a point (x, y) of $X \times X$ is called *distance* between x and y, and the ordered pair (X, d) is called *metric space*.

The condition $(M1)$ is usually called nonnegativity of distance. The condition $(M2)$ states that indistancy implies and implied by equality, while condition $(M3)$ is called symmetry. The condition $(M4)$ is called triangularity.

A function $d : X \times X \to \mathbb{R}$ is called pseudometric on X if it satisfies the following conditions:
$(P\text{-}M1)$ $x = y \Rightarrow d(x, y) = 0$;
$(P\text{-}M2)$ $d(x, y) = d(y, x)$ for all $x, y \in X$;
$(P\text{-}M3)$ $d(x, y) \leqslant d(x, z) + d(z, y)$ for all $x, y, z \in X$.

The set X together with the pseudometric d; i.e., the pair (X, d) is called pseudo-metric space.

Pseudometric differs from metric in the sense that

$$d(x, y) = 0 \ \text{ even if } x \neq y.$$

Thus, for a pseudometric $x = y \Rightarrow d(x, y) = 0$ but not conversely.
Notice that every metric space is a pseudometric space but every pseudometric space is not necessarily a metric space.

Example 1.2 (1) A function $f : \mathbb{R} \times \mathbb{R} \to \mathbb{R}$ defined by $d(x, y) = |x - y|$ for all $x, y \in \mathbb{R}$ is a metric on $\mathbb{R}$ usually called *usual metric*, and the ordered pair $(\mathbb{R}, d)$ is called *usual metric space*.
(2) The real line $\mathbb{R}$ equipped with a function $d : \mathbb{R} \times \mathbb{R} \to \mathbb{R}$ by

$$d(x, y) = \begin{cases} 0 & \text{if } x = y, \\ 1 & \text{if } x \neq y, \end{cases}$$

is a metric space on $\mathbb{R}$ usually called *discrete metric* and the ordered pair $(\mathbb{R}, d)$ is called *discrete metric space*.
(3) Let X be a set of all complex sequences $\{x_i\}_{i=1}^{\infty}$ and let $d : X \times X \to \mathbb{R}^+$ be a function defined by

$$d(x, y) = \sum_{i=1}^{\infty} \frac{1}{2^i} \frac{|x_i - y_i|}{1 + |x_i - y_i|}, \quad x = \{x_i\}, y = \{y_i\} \in X. \tag{1.1}$$

Then, d is a metric called *Fréchat metric* for X.
(4) Let $X = \mathbb{R}$. Define a mapping $d : \mathbb{R} \times \mathbb{R} \to \mathbb{R}$ by $d(x, y) = |x^2 - y^2| \ \forall x, y \in \mathbb{R}$. It is easy to check that d is a pseudometric on $\mathbb{R}$. Evidently, $d(x, y) = 0 \Rightarrow |x^2 - y^2| = 0 \Rightarrow x = \pm y$. Thus, $d(x, y) = 0$ does not necessarily imply $x = y$. Hence $(\mathbb{R}, d)$ is a pseudometric space, but not a metric space.

Given a metric space (X, d), $x \in X$ and $r > 0$, we denote

$B_r(x) = \{y \in X : d(y, x) < r\}$, the *open ball* with center x and radius r;

$\overline{B}_r(x) = \{y \in X : d(y, x) \leqslant r\}$, the *closed ball* with center x and radius r;

$\partial B_r(x) = \{y \in X : d(y, x) = r\}$, the *boundary of ball* with center x and radius r.

For a subset A of X and a point $x \in X$, the distance between x and A is denoted by $d(x, A)$ is defined as the smallest distance from x to elements of A. More precisely,

$$D(x, A) = \inf_{y \in A} d(x, y).$$

The distance $\sup\{d(x, y) : x, y \in A\}$ is defined as the diameter of the set A and is denoted by $\operatorname{diam}(A)$. If $\operatorname{dist}(A)$ is finite, then A is said to be bounded. Otherwise, A is said to be unbounded.

Convergence, Cauchy sequence and completeness in a metric space are defined as follows:

Definition 1.4 Let (X, d) be a metric space, $\{x_n\}$ a sequence in X, and let $x \in X$. Then,

(*a*) The sequence $\{x_n\}$ is said to be convergent in (X, d) and converges to x_0, if for given $\varepsilon > 0$ there exists $n_0 \in \mathbb{N}$ such that $d(x_n, x_0) < \varepsilon$ for all $n \geq n_0$ and this fact is represented by $\lim_{n \to \infty} x_n = x_0$ or $x_n \to x_0$ as $n \to \infty$.

(*b*) The sequence $\{x_n\}$ is said to be Cauchy sequence in (X, d) if for given $\varepsilon > 0$ there exists $n_0 \in \mathbb{N}$ such that $d(x_n, x_{n+p}) < \varepsilon$ for all $n \geq n_0$, $p > 0$ or equivalently, if $\lim_{n \to \infty} d(x_n, x_{n+p}) = 0$ for all $p > 0$.

(*c*) (X, d) is said to be a complete metric space if every Cauchy sequence in X converges to some $x \in X$.

1.2.2 Continuity of Mappings

In this section, we discuss continuity of mappings with their properties.

Definition 1.5 Let T be a mapping from a metric space (X, d) to another metric space (Y, ρ). Then, T is said to be

(*i*) continuous at $x_0 \in X$ if for given $\varepsilon > 0$, there exists $\delta = \delta(\varepsilon, x_0) > 0$ such that

$$\rho(Tx, Tx_0) < \varepsilon \text{ whenever } d(x, x_0) < \delta \text{ for all } x \in X.$$

In general, T is said to be continuous at $x_0 \in X$ if for any sequence $\{x_n\}$ in X such that $x_n \to x_0$ implies $Tx_n \to Tx_0$ in Y.

(ii) uniformly continuous on X if for given $\varepsilon > 0$, there exists $\delta = \delta(\varepsilon) > 0$ such that

$$\rho(Tx, Ty) < \varepsilon \text{ whenever } d(x, y) < \delta \text{ for all } x, y \in X.$$

Example 1.3 (1) Every polynomial function is continuous.
(2) Let X be a discrete metric space and Y be any metric space. Then, any mapping $T : X \to Y$ is uniformly continuous.
(3) Let C be a nonempty subset of a metric space (X, d). Then for each pair $x, y \in X$, we have $|d(x, C) - d(y, C)| \leqslant d(x, y)$. That is,

$$|d(\cdot, C)(x) - d(\cdot, C)(y)| \leqslant d(x, y) \text{ for all } x, y \in X.$$

In particular, we see that the function $x \mapsto d(x, C)$ is uniformly continuous.
(3) No polynomial function of degree greater than 1 is uniformly continuous on the usual metric space $\mathbb{R}$.
(4) Let $X = (0, \infty)$ be with usual metric. Then, one can easily see that any logarithmic function on X is not uniformly continuous.
(5) Let $X = (0, 1)$ be a metric space with the metric induced by the usual metric on $\mathbb{R}$ and $Y = \mathbb{R}$ with usual metric. Then, the mapping $T : X \to Y$ defined by $T(x) = \frac{1}{x}$, for all $x \in X$, is not uniformly continuous. To see it, let $\varepsilon = \frac{1}{2}$ and δ be any positive number. Choose $x = \frac{1}{n+1}$ and $y = \frac{1}{n+2}$, where n is a positive integer such that $n > \frac{1}{\delta}$. Then,

$$|x - y| = \left| \frac{1}{n+1} - \frac{1}{n+2} \right| = \frac{1}{(n+1)(n+2)} < \frac{1}{n} < \delta,$$

but $|T(x) - T(y)| = |(n+1) - (n+2)| = 1 > \varepsilon$. Thus, we see that whatever $\delta > 0$ may be, there exist x and y such that

$$|x - y| < \delta \text{ but } |T(x) - T(y)| > \varepsilon.$$

Observation

- Every uniformly continuous mapping from a metric space X into another metric space Y is continuous at each point of X, but the converse need not be true in general. To see it, consider $X = [-1, 1]$ and $Y = \mathbb{R}$ with usual metric. Let $T : X \to Y$ be a mapping defined by $T(x) = x^2$ for all x in X. Then, for given $\varepsilon > 0$, there exists $\delta = \frac{1}{2}\varepsilon > 0$ such that for any $x, y \in [-1, 1]$, we have

$$|T(x) - T(y)| = |x^2 - y^2| = |x + y||x - y| \leqslant 2|x - y| < 2\delta = \varepsilon$$

whenever $|x - y| < \delta$. Keeping y fixed at x_0, it is easy to see that f is continuous at x_0.
Now, if we consider the same mapping $T(x) = x^2$ defined on $\mathbb{R}$, that is, $T : \mathbb{R} \to \mathbb{R}$ such that $T(x) = x^2$. Then, for any x, x_0 in $\mathbb{R}$, we have

$$|T(x) - T(x_0)| = |x^2 - x_0{}^2| = |x + x_0||x - x_0| < \varepsilon$$

whenever $|x - x_0| < \frac{\varepsilon}{|x+x_0|} = \delta$, where δ depends on ε and x_0. It follows that T is only continuous but not uniformly continuous.

- Every uniformly continuous mapping T from a metric space (X, d) into another metric space (Y, ρ) maps a Cauchy sequence $\{x_n\}$ in X into a Cauchy sequence $\{T x_n\}$ in Y.

Definition 1.6 Let X be a metric space and x_0 a point in X. A function $\varphi : X \to \mathbb{R}$ is said to be

(i) upper semicontinuous at x_0 if $\limsup\limits_{x \to x_0} \varphi(x) \leqslant \varphi(x_0)$;

(ii) lower semicontinuous at x_0 if $\liminf\limits_{x \to x_0} \varphi(x) \geqslant \varphi(x_0)$;

(iii) continuous at x_0 if it is both upper semicontinuous and lower semicontinuous at x_0.

In general, φ is said to be

(i') upper semicontinuous on X if for any sequence $\{x_n\}$ in X such that

$$x_n \to x \quad \text{implies} \quad \limsup_{n \to \infty} \varphi(x_n) \leqslant \varphi(x);$$

(ii') lower semicontinuous on X if for any sequence $\{x_n\}$ in X such that

$$x_n \to x \quad \text{implies} \quad \liminf_{n \to \infty} \varphi(x_n) \geqslant \varphi(x);$$

(iii') continuous on X if it is both upper semicontinuous and lower semicontinuous on X.

Example 1.4 (a) Let $\varphi : \mathbb{R} \to \mathbb{R}$ be a function defined by

$$\varphi(x) = \begin{cases} -1, & \text{if } x < 0 \\ 1, & \text{if } x \geq 0. \end{cases}$$

Then, it is easy to check that φ is upper semicontinuous at $x_0 = 0$ but not lower semicontinuous.

(b) Let $\varphi : \mathbb{R} \to \mathbb{R}$ be a function defined by

$$\varphi(x) = \begin{cases} -1, & \text{if } x \leqslant 0 \\ 1, & \text{if } x > 0. \end{cases}$$

Then, φ is lower semicontinuous at $x_0 = 0$ but not upper semicontinuous (Fig. 1.1).

A mapping T from a metric space (X, d) into another metric space (Y, ρ) is said to satisfy the Lipschitz condition or is said to be Lipschitz continuous on X if there exists a constant $L > 0$ such that

$$\rho(Tx, Ty) \leqslant Ld(x, y) \quad \text{for all} \quad x, y \in X.$$

If L is the least number for which the Lipschitzian condition is satisfied, then L is called Lipschitz constant. In this case, we say that T is an L-Lipschitz mapping or simply a Lipschitzian mapping with Lipschitz constant L. Otherwise, it is called non-Lipschitzian mapping.

Observation

- Every Lipschitz continuous mapping T from a metric space X into another metric space Y is uniformly continuous (and hence continuous) on X. Indeed, for a given $\varepsilon > 0$ there exists $\delta = \frac{\varepsilon}{L} > 0$ such that for all $x, y \in X$,

$$d(x, y) < \delta \implies \rho(Tx, Ty) \leqslant Ld(x, y) < L\delta = \varepsilon.$$

- There exists non-Lipschitzian mapping that is continuous. To this end, consider the mapping $T : [-\frac{1}{\pi}, \frac{1}{\pi}] \to [-\frac{1}{\pi}, \frac{1}{\pi}]$ defined by

$$T(x) = \begin{cases} x \sin\left(\frac{1}{x}\right), & \text{if } x \neq 0 \\ 0, & \text{if } x = 0. \end{cases}$$

Notice that T is continuous, but not Lipschitz continuous.

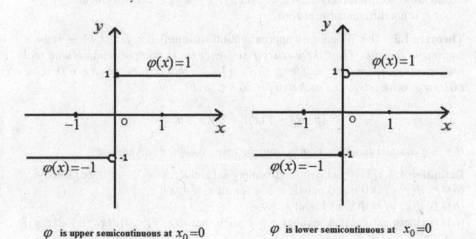

φ **is upper semicontinuous at** $x_0 = 0$ φ **is lower semicontinuous at** $x_0 = 0$

Fig. 1.1 Upper semicontinuity/lower semicontinuity at $x_0 = 0$

Theorem 1.2 (The Weierstrass theorem) *Let C be a nonempty compact subset of a metric space X and $f : C \to \mathbb{R}$ a continuous function. Then, f attains its extremum (maximum and minimum); i.e., there exist $\xi, \eta \in C$ such that*

$$f(\xi) = \inf_{x \in C} f(x) \ and \ f(\eta) = \sup_{x \in C} f(x).$$

The following proposition guarantees the existence of Lipschitzian mappings.

Proposition 1.1 (Criteria for existence of Lipschitzian mappings) *Suppose $T : [a, b] \subset \mathbb{R} \to \mathbb{R}$ is a differential function on (a, b). Suppose T' is continuous on $[a, b]$. Then, T is a Lipschitz continuous function.*

Proof For all $a \leqslant x < y \leqslant b$, we find that
(*i*) T is continuous on $[x, y]$;
(*ii*) T is differentiable in (x, y).
Now applying Lagrange's mean value theorem to T, we have

$$T(y) - T(x) = T'(c)(y - x) \ \text{for some} \ c \in (x, y) \subset [a, b].$$

Because T' is continuous and the interval $[a, b]$ is compact in $\mathbb{R}$, by the Weierstrass theorem, there exists $\eta \in [a, b]$ such that $|T(\eta)| = \sup_{c \in [a,b]} |T'(c)| = L$, say. Thus, $|T(x) - T(y)| \leqslant L|x - y|$ for all $x, y \in [a, b]$, which shows that T is Lipschitz continuous.

Observation

- Lipschitzian mapping need not be differentiable. To see it, consider the mapping $T : \mathbb{R} \to \mathbb{R}$ defined by $Tx = 2|x| + 1$ for all $x \in \mathbb{R}$. Notice that T is Lipschitz continuous with Lipschitz constant $L = 2$; i.e., $|Tx - Ty| \leqslant 2|x - y|$ for all $x, y \in \mathbb{R}$, but T is not differentiable at zero.

Theorem 1.3 (The Weierstrass approximation theorem) *Let $f : [a, b] \to \mathbb{R}$ be a continuous function. Then, there exists a sequence $\{p_n(x)\}$ of polynomials with real coefficients that converges uniformly to f on $[a, b]$; that is, for a given $\varepsilon > 0$, there exists a positive integer n_0 such that for all $t \in [a, b]$*

$$|p_n(x) - f(x)| < \varepsilon \ \forall \ n \geqslant n_0.$$

If f is a complex function, then p_n may be taken complex polynomial.

Definition 1.7 ([31]) Let X be a nonempty set and $d : X \times X \to [0, \infty)$ satisfy:
(*bM1*) $d(x, y) = 0$ if and only if $x = y$ for all $x, y \in X$;
(*bM2*) $d(x, y) = d(y, x)$ for all $x, y \in X$;
(*bM3*) there exist a real number $s \geq 1$ such that $d(x, y) \leqslant s[d(x, z) + d(z, y)]$ for all $x, y, z \in X$.

Then, d is called a b-metric on X and (X, d) is called a b-metric space (in short bMS) with coefficient s.

Notice that the class of b-metric spaces is effectively larger than that of the class of metric spaces. The following examples illustrate the above fact.

Example 1.5 (1) Let $X = \{1, 2, 3\}$. Define $d : X \times X \to \mathbb{R}^+$ by $d(x, y) = d(y, x)$ for all $x, y \in X, d(x, x) = 0$, for all $x \in X$ and $d(1, 2) = 4, d(1, 3) = 2, d(2, 3) = 1$. Then, (X, d) is a b-metric space, but not a metric space because the triangle inequality is not satisfied. Indeed, we have that

$$d(1, 3) + d(3, 2) = 2 + 1 = 3 < 4 = d(1, 2).$$

It is easy to verify that $s = \frac{4}{3}$.
(2) Let (X, ρ) be a metric space and $d(x, y) = (\rho(x, y))^p$, where $p > 1$ is a real number. Then, d is a b-metric with $s = 2^{p-1}$.

Convergence, Cauchy sequence and completeness in b-metric space are defined as follows:

Definition 1.8 ([31]) Let (X, d) be a b-metric space, $\{x_n\}$ be a sequence in X and $x \in X$. Then,

- (a) the sequence $\{x_n\}$ is said to be convergent in (X, d) and converges to x_0, if for given $\varepsilon > 0$ there exists $n_0 \in \mathbb{N}$ such that $d(x_n, x_0) < \varepsilon$ for all $n \geqslant n_0$ and this fact is represented by $\lim_{n \to \infty} x_n = x_0$ or $x_n \to x_0$ as $n \to \infty$.
- (b) the sequence $\{x_n\}$ is said to be Cauchy sequence in (X, d) if for given $\varepsilon > 0$ there exists $n_0 \in \mathbb{N}$ such that $d(x_n, x_{n+p}) < \varepsilon$ for all $n \geqslant n_0, p > 0$ or equivalently, if $\lim_{n \to \infty} d(x_n, x_{n+p}) = 0$ for all $p > 0$.
- (c) (X, d) is said to be a complete b-metric space if every Cauchy sequence in X converges to some $x \in X$.

1.2.3 Topological Spaces

In this section, we give a brief summary of some aspect of topological spaces.[4] In particular, we discuss the notions of open sets, closed sets, interior, closure and boundary of sets. Notions of directed set and net are also introduced.

Definition 1.9 Let X be a nonempty set and $\mathscr{T}$ a collection of subsets of X. Then, $\mathscr{T}$ is said to be a topology[5] on X if the following conditions are satisfied:

---

[4]The word "topology" derives its name from the union of two Greek words, namely topos meaning "surface" and logos meaning "study" or "discourse". Topology thus literary means the study of surfaces and is concerned with the properties of space that are preserved under continuous deformations, such as stretching and bending, but not tearing or gluing. In other words, the study of certain properties that do not change as geometric figures or spaces undergo continuous deformation is called topology. These properties include openness, nearness, connectedness, and continuity.

[5]The term topology was introduced by Johann Benedict Listing in the nineteenth century, although it was not until the first decades of the twentieth century that the idea of a topological space

$(T1)$ $\emptyset, X \in \mathscr{T}$,

$(T2)$ $\mathscr{T}$ is closed under arbitrary unions,

$(T3)$ $\mathscr{T}$ is closed under finite intersections.

The ordered pair $(X, \mathscr{T})$ is called *topological space*. The members of $\mathscr{T}$ are called $\mathscr{T}$-open sets or simply open sets.

Metric topology – The open sets of a metric space (X, d) form a topology, called metric topology or topology induced by the metric d. Thus, a metric space is a topological space with the topology of the metric.

In general, the topology of a topological space $(X, \mathscr{T})$ is not induced by a metric. If it is induced by a metric, it is called a *metrizable space*.

Metrizable space – A topological space $(X, \mathscr{T})$ is said to be a metrizable space if there exists a metric d which generates the same open sets as the open sets of $\mathscr{T}$, i. e., $\mathscr{T}_d = \mathscr{T}$.

Alternatively, a topological space is said to be metrizable if the topology can be obtained from a metric on the underlying space.

Interior of a set – Let G be a subset of a topological space X. Then, the interior of G is the union of all open subsets of G and is denoted by $\mathrm{int}(G)$ (or $G°$). In other words, if $\{O_i : i \in I\}$ are all open subsets of G, then

$$\mathrm{int}(G) = \bigcup_{i \in I} O_i.$$

Observation

- $\mathrm{int}(G)$ is an open set, because it is the union of open sets.
- $\mathrm{int}(G)$ is the largest open set of G.
- If U is an open subset of G, then $U \subset \mathrm{int}(G) \subset G$.

Exterior points and boundary of sets – Let A be a subset of a topological space X. Then, the exterior of A, written by $\mathrm{ext}(A)$, is the interior of the complement of A; i.e., $\mathrm{ext}(A) = \mathrm{int}(X \setminus A)$. The boundary of A is a set of points that do not belong to the interior or the exterior of A. The boundary of set A is denoted by ∂A. Obviously, $\partial A = \overline{A} \cap \overline{(X \setminus A)}$ is a closed set.

Closed set – A subset F of a topological space X is said to be closed if its complement $X - F$ is open.

Theorem 1.4 *Let $\mathscr{F}$ be a family of all closed sets in a topological space $(X, \mathscr{T})$. Then, $\mathscr{F}$ has the following properties:*

$(F1)$ $\emptyset, X \in \mathscr{F}$,

$(F2)$ $\mathscr{F}$ *is closed under arbitrary intersection,*

$(F3)$ $\mathscr{F}$ *is closed under finite union.*

---

was developed. By the middle of the twentieth century, topology had become a major branch of mathematics.

Closure of a set – Let C be a subset of a topological space X. Then, the closure of C is the intersection of all closed supersets of C. The closure of C is denoted by $\overline{C}$. In other words, if $\{F_i : i \in I\}$ is a collection of all closed supersets of C in X, then

$$\overline{C} = \underset{i \in I}{\cap} F_i.$$

Observation

- $\overline{C}$ is a closed set, because it is the intersection of closed sets.
- $\overline{C}$ is the smallest closed superset of C.
- If F is a closed superset of C in X, then $C \subset \overline{C} \subset F$.

Theorem 1.5 *Let C be a subset of a topological space X. Then, C is closed if and only if $C = \overline{C}$.*

Neighborhood – Let $(X, \mathcal{T})$ be a topological space. A subset N of X is said to be a neighborhood of a point $x_0 \in X$ if there exists an open set $G \in \mathcal{T}$ such that $x_0 \in G \subset N$.

Let $(X, \mathcal{T})$ be a topological space. Then, X is said to be

- a T_0- space if for a pair of distinct points in X, there exists an open set that contains one of them, but not the other;
- a T_1- space if x and y are any two distinct points in X, there exists an open set U containing x but not y, and there exists another open set V containing y, but not x;
- a T_2- space or the Hausdorff space if x and y are any two distinct points in X, there exist two open sets U and V such that $x \in U, y \in V$ and $U \cap V = \varnothing$.

Directed set – Let $\mathcal{D}$ be a nonempty set and $\geqslant$ a binary relation on $\mathcal{D}$. Then, the ordered pair $(\mathcal{D}, \geqslant)$ is said to be directed if

(D1)   for all $\alpha, \beta, \gamma \in \mathcal{D}, \alpha \geqslant \beta, \beta \geqslant \gamma \implies \alpha \geqslant \gamma;$ (Transitivity)

(D2)   for all $\alpha \in \mathcal{D}, \alpha \geqslant \alpha;$ (Reflexivity)

(D3)   for $\alpha, \beta \in \mathcal{D}$, there exists $\gamma \in \mathcal{D}$ such that $\gamma \geqslant \alpha, \gamma \geqslant \beta;$ (Existence)

Example 1.6 (1) $(\mathbb{N}, \geqslant)$ is a directed set.
(2) If X is a nonempty set, then $(P(X), \subseteq)$ and $(P(X), \supseteq)$ are directed sets, where $P(X)$ is the power set of X.
(3) If $x \in X$ and $\mathcal{N}_x$ is the family of all neighbourhoods of x, then $(\mathcal{N}_x, \subseteq)$ is a directed set.
(4) Every lattice is a directed set.

Net – A net or a generalized sequence in a set X is a mapping S from a directed set $\mathcal{D}$ into X. The net $\{x_\alpha : \alpha \in \mathcal{D}\}$ is simply written as $\{x_\alpha\}$.

A net $\{x_\alpha : \alpha \in \mathcal{D}\}$ in a topological space X is said to converge to the point $x \in X$, if for every neighborhood U of x, there exists $\alpha_0 \in \mathcal{D}$ such that

$$\forall \alpha \in \mathcal{D}, \alpha \geqslant \alpha_0 \implies x_\alpha \in U.$$

In this case, we write $x_\alpha \to x$, or $\lim_{\alpha} x_\alpha = x$.

1.2.4 Linear Spaces

Definition 1.10 A linear space[6] (or vector space) X over the field $\mathbb{K}$ ($\mathbb{R}$ or $\mathbb{C}$) is a set X equipped with an internal binary operation "+" called *vector addition* carrying (x, y) in $X \times X$ to $x + y$ in X and an external operation"·" called *scalar multiplication* carrying (α, x) in $\mathbb{K} \times X$ to αx in X satisfying the following for all $x, y, x \in X$ and $\alpha, \beta \in \mathbb{K}$:

1. $x + y = y + x$,
2. $(x + y) + z = x + (y + z)$,
3. there exists an element $0 \in X$ called the zero vector of X such that $x + 0 = x$ for all $x \in X$,
4. for each $x \in X$, there exists an element $-x \in X$ called the additive inverse or the negation of x such that $x + (-x) = 0$,
5. $\alpha(x + y) = \alpha x + \alpha y$,
6. $(\alpha + \beta)x = \alpha x + \beta x$,
7. $(\alpha\beta)x = \alpha(\beta x)$,
8. $1 \cdot x = x$ for every $x \in X$ where 1 is the unit element of the field $\mathbb{K}$.

The elements of a vector space X are called *vectors*, and the elements of the field $\mathbb{K}$ are called *scalars*. Instead of saying that X is a vector space over field $\mathbb{K}$ we say that $X(\mathbb{K})$ is a vector space. In what follows, unless otherwise stated, X denotes a linear space over field $\mathbb{R}$.

Example 1.7 (1) $\mathbb{R}$ and $\mathbb{C}$ are linear spaces over $\mathbb{R}$ under usual addition and multiplication as vector addition and scalar multiplication, respectively.
(2) All polynomial over a field $\mathbb{K}$ is a linear space over field $\mathbb{K}$ w.r.t. addition of polynomials and multiplication of polynomials by any scalar.
(3) For any linear differential equation (and linear partial differential equation), the set of all solutions is a linear space.
(4) $X = \{x : x = (x_1, x_2, \ldots)| x_i \in \mathbb{R}\}$ is a linear space over $\mathbb{R}$.

Observation

- A field $\mathbb{K}$ can be regarded as a vector space over any subfield $\mathbb{F}$ of $\mathbb{K}$.
- Because $\mathbb{K}$ is a subfield of itself therefore $\mathbb{K}$ is a vector space over $\mathbb{K}$.
- $\mathbb{R}(\mathbb{Q})$ is a vector space but $\mathbb{Q}(\mathbb{R})$ is not a vector space.

Example 1.8 Let $X = C[a, b]$ $(1 \leqslant p < \infty)$, the set of all continuous scalar functions and let "+" and "·" denote the operations of vector addition and scalar multiplication, respectively, defined by

---

[6]Historically, the first ideas leading to linear spaces (also called vector spaces) can be traced back as far as the seventeenth century's analytic geometry, matrices, systems of linear equations, and Euclidean vectors. The modern, more abstract treatment, first formulated by Giuseppe Peano in 1888, encompasses more general objects than Euclidean space, but much of the theory can be seen as an extension of classical geometric ideas like lines, planes and their higher-dimensional analogs.

$$(f + g)(t) = f(t) + g(t) \text{ for all } f, g \in C[a, b];$$
$$(\alpha f)(t) = \alpha f(t) \text{ for all } f \in C[a, b] \text{ and scalar } \alpha \in \mathbb{K}.$$

Then, $C[a, b]$ is a linear space.

A subset S of a linear space X is a linear subspace (or a subspace) of X if S itself is a linear space. A necessary and sufficient condition for a subset S of a linear space X to be a subspace of X is that S is closed w.r.t. vector addition and scalar multiplication.

Linear span of vectors – If $x_1, x_2, \ldots, x_n$ are given points of a linear space X, then the element

$$x = \alpha_1 x_1 + \alpha_2 x_2 + \cdots + \alpha_n x_n, \quad \alpha_i \in \mathbb{K}$$

is called linear combination of $\{x_1, x_2, \ldots, x_n\}$. The set of all linear combinations of the vectors $x_1, x_2, \ldots, x_n$ is called the *linear span* of these vectors. If S is a nonempty subset of a linear space X, then the linear combination of S is the set of all linear combinations of elements of S. It is subspace called the subspace spanned or generated by these vectors and it is denoted by $[S]$ or $L(S)$.

The vectors $x_1, x_2, \ldots, x_n$ in X are said to be *linearly independent* if the equality $\alpha_1 x_1 + \alpha_2 x_2 + \cdots + \alpha_n x_n = 0$ implies $\alpha_1 = \cdots = \alpha_n = 0$. Otherwise, they are said to be *linearly dependent*.

Basis of a Linear Space – If $\{x_1, x_2, \ldots, x_n\}$ is a linearly independent set of n vectors which spans the whole space X, then $\{x_1, x_2, \ldots, x_n\}$ is said to be a basis (or Hamel basis) for X, and X is said to have dimension n, written dim $X = n$.

Hamel basis – A Hamel basis for a linear space X is a linearly independent set which spans X.

Theorem 1.6 *Every nontrivial linear space X has a Hamel basis.*

Proof Suppose $\mathcal{F}$ denotes the family of all linearly independent subsets of X, partially ordered by set inclusion. For any $x \in X$, $x \neq 0$, the set $\{x\}$ is linearly independent set so $\mathcal{F}$ is nonempty.

Consider a chain, i.e., totally ordered subfamily $\mathcal{C}$ in $\mathcal{F}$. Let $S = \bigcup_{C \in \mathcal{C}} C$. Then, it is clear that S is a linearly independent set in $\mathcal{C}$ and is an upper bound of the totally ordered subfamily $\mathcal{C}$.

By Zorn's lemma $\mathcal{F}$ has a maximal linearly independent set A.

Now for any $x \ni nA$ it follows that $A \cup \{x\}$ is a linearly dependent set in X. Hence, there exist scalers $\{\lambda, \lambda_1, \lambda_2, \ldots, \lambda_n\}$ and a subset $\{e_1, e_2, \ldots, e_n\}$ in A such that

$$\lambda x + \sum_{i=1}^{n} \lambda_i e_i = 0.$$

Notice that $\lambda \neq 0$ because $\{e_1, e_2, \ldots, e_n\}$ is a linearly independent set, so

Fig. 1.2 Illustrative examples of convex and nonconvex sets

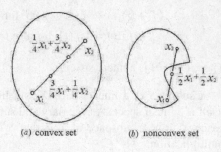

(*a*) convex set   (*b*) nonconvex set

$$x = -\frac{1}{\lambda} \sum_{i=1}^{n} \lambda_i e_i.$$

Thus, we conclude that A spans X and therefore A is a Hamel basis for X.

Dimension of a Linear Space – A linear space X is said to be *finite dimensional* if it has a finite basis; i.e., it is generated by the linear combination of a finite number of elements which are linearly independent. Otherwise, it is *infinite-dimensional*.

Example 1.9 (*i*) The real linear space $\mathbb{R}^n$ is finite dimensional and $\dim \mathbb{R}^n = n$ because $\mathbb{R}^n$ is generated by its n linearly independent elements, namely $e_1 = (1, 0, \ldots, 0)$ and $e_2 = (0, 1, \ldots, 0), \ldots, e_n = (0, 0, \ldots, 1)$.
(*ii*) The complex linear space $\mathbb{C}^n$ is finite dimensional and $B = \{e_1, e_2, \ldots, e_n\}$ forms a basis for it.

Observation

- A Hamel basis, or algebraic basis, of a linear space is a maximal linearly independent set of vectors. Each element of a linear space may be expressed as a unique finite linear combination of elements in a Hamel basis.
- Every linear space has a Hamel basis, and any linearly independent set of vectors may be extended to a Hamel basis by the repeated addition of linearly independent vectors to the set until none are left (a procedure which is formalized by the axiom of choice, or Zorn's lemma, in the case of infinite-dimensional spaces). A Hamel basis of an infinite-dimensional space is frequently very large.
- In a normed space, we have a notion of convergence, and we may therefore consider various types of topological bases in which infinite sums are allowed.

Convex set – Let S be a subset of a linear space X. Then, S is said to be convex if

$$\alpha x + (1 - \alpha)y \in S \text{ for all } x, y \in S \text{ and all scalars } \alpha \in [0, 1].$$

A necessary and sufficient condition for a subset S of a linear space X to be a convex set is stated below (Fig. 1.2).

Proposition 1.2 *Let S be a subset of a linear space X. Then, S is convex if and only if $\alpha_1 x_1 + \alpha_2 x_2 + \cdots + \alpha_n x_n \in S$ for any finite set $\{x_1, x_2, \ldots, x_n\} \subset S$ and any scalars $\alpha_i \geq 0$ with $\alpha_1 + \alpha_2 + \cdots + \alpha_n = 1$.*

Convex hull – Let S be an arbitrary subset (not necessarily convex) of a linear space X. Then, the convex hull of S in X is the intersection of all convex subsets of X containing S and is denoted by $co(S)$. Symbolically, we have

$$co(S) = \cap\{C \subset X : S \subseteq C, C \text{ is convex}\}.$$

Thus, it follows that $co(S)$ is the unique smallest convex set containing S. In other words, $co(S)$ is the set of all finite convex combination of elements of S; that is,

$$co(S) = \left\{\alpha_1 x_1 + \alpha_2 x_2 + \cdots + \alpha_n x_n : x_i \in S, \alpha_i \geq 0 \text{ and } \sum_{i=1}^{n} \alpha_i = 1\right\}.$$

To illustrate the notion of convex hull, let us consider the following: Let X be a linear space. The line segment or interval joining the two points $x, y \in X$ is the set $[x, y] := \{tx + (1 - t)y : 0 \leqslant t \leqslant 1\}$; i.e., $[x, y] = co(\{x, y\})$ is convex hull of x and y.

The closure of convex hull of S is

$$\overline{co(S)} = \left\{\alpha_1 x_1 + \alpha_2 x_2 + \cdots + \alpha_n x_n : x_i \in S, \alpha_i \geq 0 \text{ and } \sum_{i=1}^{n} \alpha_i = 1\right\}.$$

The closed convex hull of S in X is the intersection of all closed convex subsets of X containing A and is denoted by $\overline{co}(S)$. Thus,

$$\overline{co}(S) = \cap\{C \subset X : S \subseteq C, C \text{ is closed and convex}\}.$$

It is easy to observe that closure of convex hull of S is closed convex hull of S; i.e., $\overline{co(S)} = \overline{co}(S)$. Notice that, in general, $\overline{co}(S) \neq co(\overline{S})$.

Observation

- By convention, the empty set $\varnothing$ is convex.
- If A and B are two convex subsets of a linear space X, then
  (i) $A + B$ is convex,
  (ii) λA is convex for any scalar λ,
  (iii) $A \cup B$ need not be a convex set.
- If $\{A_i : i \in I\}$ is any family of convex sets in a linear space X, then $\cap_{i \in I} A_i$ is convex.
- Any translation $A + x_0$ of a convex set A is convex.
- If A is a convex subset of a linear space X, then
  (i) the closure $\overline{C}$ and $\mathrm{int}(C)$ are convex,
  (ii) $co(A) = A$.

1.2.5 Normed Spaces

Definition 1.11 Let X be a linear space over field $\mathbb{K}$ ($\mathbb{K}$ means $\mathbb{R}$ or $\mathbb{C}$) and $\|\cdot\|$: $X \to \mathbb{R}^+$ a function. Then, $\|\cdot\|$ is said to be a *norm* defined on X if the following properties hold:

$(N1)$ $\|x\| = 0$ if and only if $x = 0$;   (strict positivity)

$(N2)$ $\|\lambda x\| = |\lambda| \|x\|$ for all $x \in X$ and $\lambda \in \mathbb{K}$;   (absolute homogeneity)

$(N3)$ $\|x + y\| \leqslant \|x\| + \|y\|$ for all $x, y \in X$. (triangle inequality or subadditivity)

The ordered pair $(X, \|\cdot\|)$ is called a *normed linear space* (abbreviated *nls*) or simply a *normed space*.

Observation

- $\|x\| \geq 0$ and $\| - x\| = \|x\|$ for all x in X.
- $|\|x\| - \|y\|| \leqslant \|x - y\|$ and $|\|x\| - \|y\|| \leqslant \|x + y\|$ for all x, y in X.
- $\|\cdot\|$ is a continuous function; i.e., $\lim\limits_{n \to \infty} x_n = x \implies \lim\limits_{n \to \infty} \|x_n\| = \|x\|$.
- Norm is a convex function; i.e., $\|\alpha x + (1 - \alpha)y\| \leqslant \alpha \|x\| + (1 - \alpha)\|y\|$ for all $x, y \in X$ and $\alpha \in [0, 1]$.
- Every normed space $(X, \|\cdot\|)$ is a metric space with induced metric d defined by

$$d(x, y) = \|x - y\| \text{ for all } x, y \text{ in } X.$$

The induced metric in turn, defines a topology on X called *norm topology*.
- In every linear space, we can define a function $d : X \times X \to \mathbb{R}^+$ by

$$d(x, y) = \begin{cases} 0 \text{ if } x = y, \\ 1 \text{ if } x \neq y, \end{cases}$$

which is a metric space on X. It follows that every linear space (not necessarily a normed space) is always a metric space.

Example 1.10 Let $X = \mathbb{R}^n, n > 1$ be a linear space. Then, the functions defined by $\|\cdot\|_1, \|\cdot\|_p, \|\cdot\|_\infty : X \to \mathbb{R}^+$ defined by

$$\|x\|_1 = \sum_{i=1}^{n} |x_i| \text{ for all } x = (x_1, x_2, \ldots, x_n) \in \mathbb{R}^n;$$

$$\|x\|_p = \left(\sum_{i=1}^{n} |x_i|^p \right)^{1/p} \text{ for all } x = (x_1, x_2, \ldots, x_n) \in \mathbb{R}^n \text{ and } p \in (1, \infty);$$

$$\|x\|_\infty = \max_{1 \leqslant i \leqslant n} |x_i| \text{ for all } x = (x_1, x_2, \ldots, x_n) \in \mathbb{R}^n$$

are norms on $\mathbb{R}^n$.

Remark 1.1 (a) $\mathbb{R}^n$ equipped with the norm defined by $\|x\|_p = \left(\sum_{i=1}^{n} |x_i|^p \right)^{1/p}$ is denoted by ℓ_p^n for all $1 \leqslant p < \infty$.
(b) $\mathbb{R}^n$ equipped with the norm defined by $\|x\|_\infty = \max\limits_{1 \leqslant i \leqslant n} |x_i|$ is denoted by ℓ_∞^n.

Definition 1.12 A sequence $\{x_i\}$ of scalars (real or complex numbers) is said to be *summable* to the sum s if the sequence $\{s_n\}$ of the partial sums of the series $\sum_{i=1}^{\infty} x_i$ converges to s; i.e.,

$$|s_n - s| \to 0 \text{ as } n \to \infty, \text{ or } \left| \sum_{i=1}^{n} x_i - s \right| \to 0 \text{ as } n \to \infty.$$

In such case, we write $s = \sum_{i=1}^{\infty} x_i$.

The sequence $\{x_i\}$ of scalars is said to be *absolutely summable* (or *absolutely convergent*) if

$$\sum_{i=1}^{\infty} |x_i| < \infty.$$

The sequence $\{x_i\}$ of scalars is said to be *p-summable* if

$$\sum_{i=1}^{\infty} |x_i|^p < \infty.$$

Example 1.11 (1) Let $X = \ell_1$, the linear space of all absolutely convergent sequences $\{x_1, x_2, \ldots, x_i, \ldots\}$ of scalars (real or complex numbers), i.e.,

$$\ell_1 = \left\{ x : x = (x_1, x_2, \ldots, x_i, \ldots) \text{ and } \sum_{i=1}^{\infty} |x_i| < \infty \right\}.$$

Then, ℓ_1 is a normed space endowed with the norm $\| \cdot \|_1 : X \to \mathbb{R}^+$ defined by

$$\|x\|_1 = \sum_{i=1}^{\infty} |x_i| \; \forall x = (x_1, x_2, \ldots, x_i, \ldots) \in X.$$

(2) Let $X = \ell_p$ $(1 < p < \infty)$, the linear space of all p-summable sequences $\{x_1, x_2, \ldots, x_i, \ldots\}$ of scalars (real or complex numbers), i.e.,

$$\ell_p = \left\{ x : x = (x_1, x_2, \ldots, x_i, \ldots) \text{ and } \sum_{i=1}^{\infty} |x_i|^p < \infty \right\}.$$

Then, ℓ_p is a normed space with the norm $\|\cdot\|_p : X \to \mathbb{R}^+$ defined by

$$\|x\|_p = \Big(\sum_{i=1}^{\infty} |x_i|^p\Big)^{1/p} \text{ for all } x = (x_1, x_2, \ldots, x_i, \ldots) \in X.$$

(3) Let $X = \ell_\infty$, the linear space of all bounded sequences $\{x_1, x_2, \ldots, x_i, \ldots\}$ of scalars (real or complex numbers), i.e.,

$$\widetilde{\ell_\infty} = \Big\{x : x = (x_1, x_2, \ldots, x_i, \ldots), \text{ and } \exists M > 0 \text{ such that } |x_i| < M \; \forall i \in \mathbb{N}\Big\}.$$

Then, ℓ_∞ is a normed space with the norm $\|\cdot\|_\infty : X \to \mathbb{R}^+$ defined by

$$\|x\|_\infty = \sup_{i \in \mathbb{N}} |x_i| \; \forall x = (x_1, x_2, \ldots, x_i, \ldots) \in X.$$

(4) Let $X = c$, the sequence space of all convergent sequences $\{x_1, x_2, \ldots, x_i, \ldots\}$ of scalars (real or complex numbers), i.e.,

$$c = \Big\{x : x = (x_1, x_2, \ldots, x_i, \ldots), \text{ and } \{x_i\}_{i=1}^{\infty} \text{ is convergent}\Big\}.$$

Then, c is a normed space with the norm $\|\cdot\|_\infty$.
(5) Let $X = c_0$, the sequence space of all convergent sequences $\{x_1, x_2, \ldots, x_i, \ldots\}$ of scalars with limit zero, i.e.,

$$c_0 = \Big\{x : x = (x_1, x_2, \ldots, x_i, \ldots), \text{ and } \{x_i\}_{i=1}^{\infty} \text{ converges to zero}\Big\}.$$

Then, c_0 is a normed space with the norm $\|\cdot\|_\infty$.
(6) Let $X = c_{00}$, the sequence space of all sequences $\{x_i\}_{i=1}^{\infty}$ of scalars having only finite number of nonzero terms, i.e.,

$$c_{00} = \Big\{x = (x_i)_{i=1}^{\infty} \in \ell_\infty : \{x_i\}_{i=1}^{\infty} \text{ has only a finite number of nonzero terms}\Big\}.$$

Then, c_{00} is a normed space with the norm $\|\cdot\|_\infty$.

Observation

It is quite interesting to observe the inclusion relations that hold among sequence spaces $c_{00}, \ell_p, c_0, c, \ell_\infty$ for all $1 \leqslant p < \infty$:

• For all $1 \leqslant p < \infty$, we see that

(i) $x = \{x_i\}_{i=1}^{\infty} \in c_{00} \Rightarrow \{x_i\}_{i=1}^{\infty}$ has only a finite number of nonzero terms

$$\Rightarrow \sum_{i=1}^{\infty} |x_i|^p < \infty$$

$$\Rightarrow x = \{x_i\}_{i=1}^{\infty} \in \ell_p.$$

This shows that $c_{00} \subset \ell_p$.

(ii) $x = \{x_i\}_{i=1}^{\infty} \in \ell_p \Rightarrow \sum_{i=1}^{\infty} |x_i|^p$ is convergent

$$\Rightarrow \{|x_i|^p\}_{i=1}^{\infty} \text{ converges to zero}$$

$$\Rightarrow \{|x_i|\}_{i=1}^{\infty} \text{ converges to zero}$$

$$\Rightarrow \{x_i\}_{i=1}^{\infty} \text{ converges to zero}$$

$$\Rightarrow x = \{x_i\}_{i=1}^{\infty} \in c_0.$$

This shows that $\ell_p \subset c_0$,

(iii) $x = \{x_i\}_{i=1}^{\infty} \in c_0 \Rightarrow (x_i)_{i=1}^{\infty}$ converges to zero

$$\Rightarrow \{x_i\}_{i=1}^{\infty} \text{ is convergent}$$

$$\Rightarrow x = \{x_i\}_{i=1}^{\infty} \in c.$$

This shows that $c_0 \subset c$.

(iv) $x = \{x_i\}_{i=1}^{\infty} \in c \Rightarrow \{x_i\}_{i=1}^{\infty}$ is convergent

$$\Rightarrow \{x_i\}_{i=1}^{\infty} \text{ is bounded}$$

$$\Rightarrow x = \{x_i\}_{i=1}^{\infty} \in \ell_{\infty}.$$

This shows that $c \subset \ell_{\infty}$. Hence we conclude that

$$c_{00} \subset \ell_p \subset c_0 \subset c \subset \ell_{\infty} \text{ for all } 1 \leqslant p < \infty.$$

- For all $1 \leqslant p < q < \infty$, we have

$$x = \{x_i\}_{i=1}^{\infty} \in \ell_p \Rightarrow \sum_{i=1}^{\infty} |x_i|^p < \infty.$$

Now $\sum_{i=1}^{\infty} |x_i|^q = \sum_{i=1}^{\infty} |x_i|^p \cdot |x_i|^{q-p} \leqslant \sum_{i=1}^{\infty} |x_i|^p \cdot \sup_{1 \leqslant i \leqslant \infty} \left(|x_i|^{q-p} \right)$

$$\leqslant M \sum_{i=1}^{\infty} |x_i|^p < \infty$$

$\Rightarrow x = \{x_i\}_{i=1}^{\infty} \in \ell_q$. This shows that $\ell_p \subset \ell_q$. Hence we conclude that

$$c_{00} \subset \ell_p \subset \ell_q \subset c_0 \subset c \subset \ell_{\infty} \text{ for all } 1 \leqslant p < q < \infty.$$

- Let $x = \left\{ (-1)^{i+1} \frac{1}{i} \right\}_{i=1}^{\infty}$. Then, in one hand, we see that

$$\sum_{i=1}^{\infty} |x_i| = \sum_{i=1}^{\infty} \frac{1}{i} = \infty,$$

on the other hand,

$$\sum_{i=1}^{\infty} |x_i|^2 = \sum_{i=1}^{\infty} \frac{1}{i^2} = \frac{\pi^2}{6} < \infty.$$

Note that $x \in \ell_1$, but $x \notin \ell_2$. It follows that the inclusion $\ell_1 \subset \ell_2$ is proper (Fig. 1.3).

Example 1.12 Let $X = C[a, b]$, the linear space of all continuous scalar functions on $[a, b]$. Then, the linear space $C[a, b]$ is a normed space with the norms:

$$\|x\|_p = \left(\int_a^b |x(t)|^p \, dt \right)^{1/p}, \quad 1 \leqslant p < \infty$$
$$\|x\|_\infty = \max_{a \leqslant t \leqslant b} |x(t)|.$$

Example 1.13 Let $X = L_p[a, b]$ ($1 \leqslant p < \infty$), the linear space of all equivalence classes of p-integrable functions on $[a, b]$, i.e., $L_p[a, b]$ is the equivalence classes of Lebesgue measurable functions x on $[a, b]$ such the Lebesgue integral of $|f|^p$ over $[a, b]$ exists and is finite. Then, the linear space $L_p[a, b]$ is a normed space with the norm defined by

$$\|f\|_p = \left(\int_a^b |f(t)|^p dt \right)^{1/p} < \infty.$$

Notice that the elements of $L_p[a, b]$ are equivalent classes of those functions f, where f is equivalent to g if the Lebesgue integral of $|f - g|^p$ over $[a, b]$ is zero.

Example 1.14 Let $X = L_\infty[a, b]$, the linear space of all equivalence classes of essentially bounded functions on $[a, b]$. Then, the linear space $L_\infty[a, b]$ is a normed space with the norm defined by

$$\|f\|_\infty = \text{ess sup} \, |f(t)| < \infty.$$

Fig. 1.3 Inclusion relation of sequence spaces

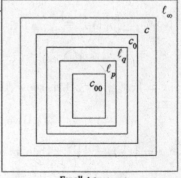

For all $1 \leq p < q < \infty$

For a subset $Y \subset X$, we denote by $\overline{Y}$ the closure of Y, by Y^C (or $X - Y$) the complement of Y, by $span(Y)$ the linear space generated by Y, and by $co(Y)$ the convex hull of Y, that is, the set of all finite convex combinations of elements of Y.

Unit sphere – The sphere with center at the zero vector and radius unity in a normed space X is called *unit sphere*. Unit sphere is often denoted by S_X or simply by S, i.e.,

$$S_X = \{x \in X : \|x\| = 1\}.$$

In $\mathbb{R}^2$, it is easy to observe the shape of the unit closed spheres $\mathbb{S}_p$ as shown in Fig. 1.4 with norms as defined below:

$$\|x\|_1 = |x_1| + |x_2|$$
$$\|x\|_2 = (|x_1|^2 + |x_2|^2)^{1/2}$$
$$\|x\|_4 = (|x_1|^4 + |x_2|^4)^{1/4}$$
$$\|x\|_\infty = \max\{|x_1|, |x_2|\}.$$

Open unit ball – The open ball with center at the zero vector and radius unity in a normed space X is called *open unit ball*. Unit open ball is usually denoted by $B(0, 1)$, i.e.,

$$B(0, 1) = \{x \in X : \|x\| < 1\}.$$

Closed unit ball – The closed ball with center at the zero vector and radius unity in a normed space X is called *closed unit ball*. Unit closed ball is often denoted by $\mathbb{B}_X$ or simply $\mathbb{B}$, i.e.,

$$\mathbb{B}_X = \{x \in X : \|x\| \leqslant 1\}.$$

Definition 1.13 A sequence $\{x_n\}$ in a normed space X is said to be convergent to $x \in X$ if $\lim\limits_{n \to \infty} \|x_n - x\| = 0$, i.e., given an $\varepsilon > 0$, there exists an integer $n_0 \in \mathbb{N}$ such that

$$\|x_n - x\| < \varepsilon, \quad \forall n \geq n_0.$$

Fig. 1.4 Unit spheres for different norms

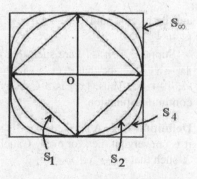

To show that $\{x_n\}$ converges to x, we write $x_n \to x$ or $\lim_{n \to \infty} x_n = x$.

Definition 1.14 (*Cauchy sequence*) A sequence $\{x_n\}$ in a normed space X is a Cauchy sequence if $\lim_{m,n \to \infty} \|x_m - x_n\| = 0$, i.e., given an $\varepsilon > 0$, there exists an integer $n_0 \in \mathbb{N}$ such that

$$\|x_n - x_m\| < \varepsilon, \quad \forall n, m \geq n_0.$$

Theorem 1.7 *Every convergent sequence in a normed space is Cauchy. However, the converse is not true in general.*

Proof Let $\{x_n\}$ be a sequence in a normed space X and let $x_n \to x$. Then, for given $\varepsilon > 0$, there exists $n_0 \in \mathbb{N}$ such that

$$\|x_n - x\| < \frac{\varepsilon}{2} \text{ for all } n \geq n_0.$$

Hence, for $m, n \geq n_0$, we have

$$\| x_m - x_n \| \leqslant \| x_m - x \| + \| x - x_n \| < \frac{\varepsilon}{2} + \frac{\varepsilon}{2} = \varepsilon.$$

This proves that every convergent sequence is Cauchy in a normed space X.

Conversely, suppose $X = C[0, 1]$ is the space of continuous functions on $[0, 1]$ with the norm

$$\|x\|_1 = \int_a^b |x(t)| \, dt.$$

Let $\{x_n\}$ be a sequence in $C[0, 1]$ defined by

$$x_n(t) = \begin{cases} 0, & \text{if } 0 \leqslant t \leqslant \frac{1}{2}, \\ 1, & \text{if } \frac{1}{2} \leqslant t \leqslant 1, \end{cases}$$

and between $\frac{1}{2}$ and $\frac{1}{2} + \frac{1}{n}$, $x_n(t)$ is given by the line joining the points $\left(\frac{1}{2}, 0\right)$ and $\left(\frac{1}{2} + \frac{1}{n}, 1\right)$.

Suppose $m, n \in \mathbb{N}$ are such that $m < n$, then by the induced metric $d_1(x, y) = \|x - y\|_1$ we see that $d_1(x, y)$ is the shaded area in Fig. 1.5b which tends to 0 as $m, n \to \infty$. Thus, $\{x_n\}$ is a Cauchy sequence, but $\{x_n\}$ does not converge to any continuous function.

Definition 1.15 A normed space is said to be *complete* if every Cauchy sequence in it is convergent, i.e., for every Cauchy sequence $\{x_n\}$ in X, there is an element x in X such that $x_n \to x$.

Notice that a normed linear space $(X, \|\cdot\|)$ is a metric space with respect to the metric d derived from its norm, where $d(x, y) = \|x - y\|$.

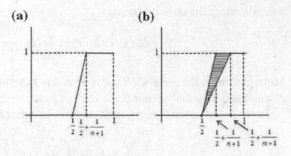

Fig. 1.5 Nonconvergence to any continuous function

Banach space – A Banach space is a normed linear space $(X, \| \cdot \|)$ that is a complete metric space with respect to the metric d derived from its norm.

A Banach space X is called real (complex) Banach space if X is a linear space over the field $\mathbb{R}$ ($\mathbb{C}$).

Example 1.15 (1) The linear spaces $\mathbb{R}$ and $\mathbb{C}$ (the real numbers and the complex numbers) are Banach spaces with the norm $\| \cdot \|$ given by $\|x\| = |x|$, $x \in \mathbb{R}$ (or $\mathbb{C}$).
(2) For $1 \leqslant p < \infty$, we define the p-norm on $\mathbb{R}^n$ (or $\mathbb{C}^n$) by

$$\|(x_1, x_2, \ldots, x_n)\|_p = (|x_1|^p + |x_2|^p + \cdots + |x_n|^p)^{1/p}.$$

For $p = \infty$, we define the ∞, or maximum, norm by

$$\|(x_1, x_2, \ldots, x_n)\|_\infty = \max\{|x_1|, |x_2|, \ldots, |x_n|\}.$$

Then, $\mathbb{R}^n$ equipped with the p-norm is a finite dimensional Banach space for $1 \leqslant p < \infty$.
(3) The linear space ℓ_p, $1 \leqslant p < \infty$ of all p-summable sequences $x = (x_1, x_2, \ldots, x_i, \ldots)$ of scalars is a Banach space with the norm defined by

$$\|x\|_p = \left(\sum_{i=1}^\infty |x_i|^p \right)^{1/p}.$$

(4) The linear space ℓ_∞ of all bounded sequences $x = (x_1, x_2, \ldots, x_i, \ldots)$ of scalars is a Banach space with the norm defined by

$$\|x\|_\infty = \sup_{1 \leqslant i < \infty} |x_i|.$$

(5) The linear space $C[a, b]$ of all continuous functions defined on $[a, b]$ is a Banach space under the norm given by

$$\|x\|_\infty = \max_{a \leqslant x \leqslant b} |x(t)|$$

which induces the uniform metric

$$d_\infty(x, y) = \max_{a \leqslant x \leqslant b} |x(t) - y(t)|.$$

More generally, the space $C(K)$ of continuous functions on a compact metric space K equipped with the sup norm is a Banach space.

In the linear space $C[a, b]$, another norm is defined by

$$\|x\|_1 = \int_a^b |x(t)| dt$$

which induces the metric

$$d_1(x, y) = \int_a^b |x(t) - y(t)| dt.$$

This metric is not complete. Thus, the space of continuous functions $C[a, b]$ with this norm is not a Banach space.

In what follows when we refer $C[a, b]$ as a normed space, we shall always mean the space with norm $\| \cdot \|_\infty$, unless otherwise stated.

(6) The space $P[a, b]$ of all polynomials on $[a, b]$ with the norm $\|x\|_\infty = \max_{a \leqslant x \leqslant b} |x(t)|$ is another example of a normed space which is not complete. Note that this space is an incomplete subspace of the complete space $C[a, b]$.

(7) Consider the subspace $C^1[a, b]$ of $C[a, b]$ consisting of all continuously differentiable functions with the norm $\|x\|_\infty = \max_{a \leqslant x \leqslant b} |x(t)|$. This space is not complete.

However, $C^1[a, b]$ is complete in any of the following norms:

$$\|x\| = \|x\|_\infty + \|x'\|_\infty;$$
$$\|x\| = |x(a)| + \|x'\|_\infty.$$

(8) Let Ω be a domain in $\mathbb{R}^n$, $n \in \mathbb{N}$. Then, the space $C(\Omega)$ of continuous functions on Ω is a Banach space with the norm

$$\|u\|_{C(\Omega)} := \sup_{x \in \Omega} |u(x)|.$$

(9) The spaces $L_p(\Omega)$, $1 \leqslant p < \infty$, of (in the Lebesgue sense) p-integrable functions are Banach spaces with the norms

$$\|u\|_{L_p(\Omega)} := \left(\int_\Omega |u(x)|^p dx \right)^{1/p}.$$

The space $L_\infty(\Omega)$ of essentially bounded functions on Ω is a Banach space with the norm

$$\|u\|_{L_\infty(\Omega)} := \operatorname{ess\,sup}_{x \in \Omega} |u(x)|.$$

Observation

- c_0, c are complete but c_{00} is not so. To see that c_{00} is not complete, let us consider a sequence $\{x_n\}$ in c_{00} where $x_n = \left(1, \frac{1}{2}, \frac{1}{3}, \ldots, \frac{1}{n}, 0, 0, \ldots\right)$. Then, we see that $\{x_n\}$ is a Cauchy sequence in c_{00} in the supremum norm, for $n > m$, $(n, m \in \mathbb{N})$

$$\|x_n - x_m\| = \sup\left\{\frac{1}{m+1}, \frac{1}{m+2}, \ldots, \frac{1}{n}\right\} \to 0 \text{ as } m, n \to \infty.$$

But $\lim_{n \to \infty} x_n = \left(1, \frac{1}{2}, \frac{1}{3}, \ldots\right) \notin c_{00}$. Thus, a convergent sequence in c_{00} need not converge to an element of c_{00}.

- The set of all rational numbers $\mathbb{Q}$ is a normed subspace of $\mathbb{R}$ with norm $\|x\| = |x|$, $x \in \mathbb{Q}$. Notice that $\mathbb{Q}$ is not a Banach space. To see this, let us consider the sequence $\{0.1, 0.101, 0.0.1010010001, \ldots\}$ which is a Cauchy sequence in $\mathbb{Q}$ whose limit is an irrational number.

Product of normed spaces – Let X and T be two normed spaces with norms $\|\cdot\|_X$ and $\|\cdot\|_Y$, respectively. Then, the product $X \times Y$ is a linear space with coordinate-wise addition and scalar multiplication and it can be normed by any one of the following:

$$\|(x, y)\|_p = \left(\|x\|_X^p + \|x\|_Y^p\right)^{\frac{1}{p}}, \quad 1 \leqslant p < \infty$$
$$\|(x, y)\|_\infty = \max(\|x\|_X, \|y\|_Y).$$

The space $X \times Y$ normed by any one of these is called the *product of the normed spaces* X, Y.

In the foregoing discussion we notice that a linear space X has many norms. In general the space may be complete relative to one norm but not complete relative to another norm unless the norms are equivalent.

Equivalent norms – Let X be a linear space over $\mathbb{K}$ and let $\|\cdot\|_1$ and $\|\cdot\|_2$ be two norms on X. Then, $\|\cdot\|_1$ is said to be equivalent to $\|\cdot\|_2$ (written as $\|\cdot\|_1 \sim \|\cdot\|_2$) if there exist two numbers $a, b > 0$ such that

$$a\|x\|_1 \leqslant \|x\|_2 \leqslant b\|x\|_1 \quad \text{for all} \ \ x \in X,$$

or

$$a\|x\|_2 \leqslant \|x\|_1 \leqslant b\|x\|_2 \quad \text{for all} \ \ x \in X.$$

Example 1.16 (1) The following norms on $\mathbb{R}^n$ are equivalent:

(i) $\|x\|_p = \left(\sum_{1 \leqslant i \leqslant n} |x_i|^p\right)^{1/p}, \ 1 \leqslant p < \infty,$

(ii) $\|x\|_\infty = \max\limits_{1 \leqslant i \leqslant n} |x_i|$.

Note that $\|x\|_\infty \leqslant \|x\|_p \leqslant n^{1/p}\|x\|_\infty$. To see this, we observe that

$$\|x\|_\infty = \max_{1 \leqslant i \leqslant n} |x_i| = \Big(\max_{1 \leqslant i \leqslant n} |x_i|^p \Big)^{1/p} \leqslant \Big(\sum_{1 \leqslant i \leqslant n} |x_i|^p \Big)^{1/p} = \|x\|_p$$

and

$$\|x\|_p \leqslant \Big(n \max_{1 \leqslant i \leqslant n} |x_i|^p \Big)^{1/p} = n^{1/p} \max_{1 \leqslant i \leqslant n} \big(|x_i|^p \big)^{1/p} = n^{1/p} \max_{1 \leqslant i \leqslant n} |x_i| = n^{1/p}\|x\|_\infty.$$

(2) In the linear space $C[0, 1]$ of all real continuous functions on $[0, 1]$, the norms $\| \cdot \|_\infty$ and $\| \cdot \|_1$ are not equivalent. To see this, consider the sequence $\{x_n\}$ where $x_n = t^n, n \in \mathbb{N}$. Notice that $\{x_n\}$ converges to 0 with respect to $\| \cdot \|_1$ because

$$\|x_n - 0\|_1 = \int_0^1 t^n \, dt = \frac{1}{n+1} \to 0, \quad \text{as } n \to \infty.$$

But $x_n \nrightarrow 0$ with respect to norm $\| \cdot \|_\infty$. Hence $\| \cdot \|_\infty$ is not equivalent to $\| \cdot \|_1$.

(3) In the product of the normed spaces X, Y it is trivial to see that

$$\| \cdot \|_\infty \leqslant \| \cdot \|_p \leqslant 2^{1/p} \| \cdot \|_\infty.$$

That is, $\| \cdot \|_p$ and $\| \cdot \|_\infty$ are equivalent norms.

Observation

- In a finite dimensional normed space X, all norms are equivalent.
- The relation $\sim$ is an equivalence relation on the set of all norms defined on a linear space X.

A normed space property which holds for a linear space under equivalent norms is said to be a *linear topological invariant property* for normed spaces.

Note that completeness and boundedness of sets are linear topological invariant properties for normed spaces.

Theorem 1.8 *Let $\| \cdot \|$ and $\| \cdot \|'$ be equivalent norms on a linear space X. Then,*
(a) A sequence $\{x_n\}$ is convergent (Cauchy) w.r.t. $\| \cdot \| \leftrightarrow \{x_n\}$ is convergent (Cauchy) w.r.t. $\| \cdot \|'$.
(b) X is complete in $\| \cdot \| \leftrightarrow X$ is complete in $\| \cdot \|'$.
(c) The class of open sets w.r.t. $\| \cdot \|$ is same as the class of open sets w.r.t. $\| \cdot \|'$.
(d) A set is bounded in $(X, \| \cdot \|)$ if and only if it is bounded in $(X, \| \cdot \|')$.

Seminorm – Let X be a linear space over field $\mathbb{K}$ ($\mathbb{R}$ or $\mathbb{C}$). Then, a function $p : X \to \mathbb{R}^+$ is said to be seminorm on X if the following conditions are satisfied:
$(SN1)$ $p(\lambda x) = |\lambda| p(x)$ for all $x \in X$ and $\lambda \in \mathbb{K}$;

(SN2) $p(x + y) \leqslant p(x) + p(y)$ for all $x, y \in X$.

The ordered pair (X, p) is called seminormed space. Note that a seminorm p is a norm if $p(x) = 0 \Rightarrow x = 0$.

Example 1.17 Let $X = \mathbb{R}^3$ and define $p : X \to \mathbb{R}^+$ by

$$p(x) = p(x_1, x_2, x_3) = |x_1| + |x_2|, \quad x = (x_1, x_2, x_3) \in X.$$

Then, p is a seminorm, but not a norm. To see this, we observe that

$$p(x_1, x_2, x_3) = 0 \Rightarrow |x_1| + |x_2| = 0 \Rightarrow |x_1| = 0, |x_2| = 0 \Rightarrow x_1 = 0, x_2 = 0,$$

i.e., $p(x_1, x_2, x_3) = 0$ implies that only first and second components of x are zero.

Definition 1.16 (*Absolutely convergent series*) Let X be normed space and $\{x_n\}$ a sequence of elements of X. If $\sum_{n=1}^{\infty} \|x_n\| < \infty$, then we say that the series $\sum_{n=1}^{\infty} x_n$ is absolutely convergent in X.

Theorem 1.9 (Cantor's intersection theorem) *A normed space X is a Banach space if and only if given any descending sequence $\{C_n\}$ of closed bounded subsets of X,* $\lim_{n \to \infty} diam(C_n) = 0$ *implies* $\bigcap_{n=1}^{\infty} C_n \neq \varnothing$.

We now introduce the notion of topological vector space (also called topological linear space) as follows:

Definition 1.17 A vector space X over field $\mathbb{K}$ is said to be a topological vector space (or, in brief, TVS) if on X, there exists a topology $\mathscr{T}$ such that $X \times X$ and $\mathbb{K} \times \mathbb{K}$ with the product topology have the property that vector addition $+ : X \times X \to X$ and scalar multiplication $\cdot : \mathbb{K} \times X \to X$ are continuous functions. Such a topology $\mathscr{T}$ is called a linear topology on X.

Definition 1.18 A linear topology on a topological vector space X is said to be a locally convex topology if every neighborhood of 0 (the zero vector of X) contains a convex neighborhood of 0. Then, X is called a locally convex topological vector space.

We now have the following interesting result.

Proposition 1.3 *Let X be a locally convex topological vector space over field K. Then, a topology of X is determined by a family of seminorms $\{p_i\}_{i \in I}$.*

1.2.6 Dense Set and Separable Space

Notice that, in general, it is not always possible to find the exact form of Hamel basis in an arbitrary infinite dimensional linear space, although it does exist (Theorem

1.6). As a result, the notion of Hamel basis is not so useful as one might expect for all practical purpose. To overcome this difficulty, another notion of a basis, namely Schauder basis was introduced in Banach spaces.

Schauder basis[7] – A sequence $\{x_n\}$ in a normed space X is said to be a *Schauder basis* of X if each $x \in X$ has a unique representation

$$x = \sum_{n=1}^{\infty} \alpha_n x_n, \tag{1.2}$$

for some scalars $\alpha_1, \alpha_2, \ldots, \alpha_n, \ldots$. Note that the convergence of the series on the right of (1.3.2) is in the norm topology of X, i.e.,

$$\lim_{n \to \infty} \left\| x - \sum_{i=1}^{n} \alpha_i x_i \right\| = 0.$$

Notice that the concept of a Schauder basis is not as straightforward as it appears in (1.2). The Banach spaces that arise in applications typically have Schauder bases, but Enflo showed in 1973 that there exist separable Banach spaces that do not have any Schauder bases. However, we see that this problem does not arise in the Hilbert spaces, which always have an orthonormal basis.

Observation

- The unit vectors $\{e_n\}_{n \in \mathbb{N}}$, where

$$e_n = (0, 0, \ldots, 1, 0, \ldots, 0)$$

$$\uparrow$$

$$n^{th} \text{position}$$

form a basis for spaces c_{00}, c_0 and ℓ_p $(1 \leqslant p < \infty)$.
- The unit vectors $\{e_n\}_{n \in \mathbb{N}}$ is not a Schauder basis for ℓ_∞; rather, the space ℓ_∞ does not possess any basis.
- The sequence $\{e, e_1, e_2, \ldots\}$ is a basis for c, where $e = (1, 1, 1, \ldots)$. To observe this, let us consider an element $x = (x_1, x_2, \ldots) \in c$. Then, $\lim\limits_{n \to \infty} x_n = l \in \mathbb{R}$ and

$$\lim_{n \to \infty} \sum_{i=1}^{n} (x_i - l)e_i = \sum_{i=1}^{\infty} (x_i - l)e_i = (x_l - l, x_2 - l, \ldots) \in c_0.$$

Moreover, $\sum\limits_{i=1}^{\infty} (x_i - l)e_i = x - le$, where $e = (1, 1, \ldots)$. Thus, $x = le + \sum\limits_{i=1}^{\infty} (x_i - l)e_i$.
It follows that the sequence $\{e_i\}$ along with the vector e provides a basis for c.

---

[7] The Schauder basis for the space $C[0, 1]$ was constructed by Schauder himself, while the Haar system is known to be a Schauder basis for $L_p[0, 1]$ with $1 \leqslant p < \infty$.

- The sequence[8] $\{x_n\} \subset C[0, 1]$, where

$$x_0(t) \equiv 1$$
$$x_1(t) = t$$

$$x_{2^n+m}(t) = \begin{cases} 0, & \text{if } t \notin \left(\frac{2m-2}{2^{n+1}}, \frac{2m}{2^{n+1}}\right), \\ 1, & \text{if } t = \frac{2m-1}{2^{n+1}}, \\ \text{linear in } \left[\frac{2m-2}{2^{n+1}}, \frac{2m-1}{2^{n+1}}\right] \text{ and } \left[\frac{2m-1}{2^{n+1}}, \frac{2m}{2^{n+1}}\right] \end{cases}$$

$$(m = 1, 2, \ldots, 2^n; n = 0, 1, 2, \ldots)$$

form a basis for the Banach space $C[0, 1]$.

- The sequence[9] of equivalent classes $\{\tilde{x}_n\}$, where x_n are the functions defined on $[0, 1]$ by

$$x_1(t) \equiv 1$$

$$x_{2^n+m}(t) = \begin{cases} \sqrt{2^n}, & \text{if } t \in \left[\frac{2m-2}{2^{n+1}}, \frac{2m-1}{2^{n+1}}\right), \\ -\sqrt{2^n}, & \text{if } t \in \left[\frac{2m-1}{2^{n+1}}, \frac{2m}{2^{n+1}}\right), \\ 0, & \text{for the other } t \end{cases}$$

$$(m = 1, 2, \ldots, 2^n; n = 0, 1, 2, \ldots)$$

form a basis for the space $L_p[0, 1]$, $1 \leqslant p < \infty$.

Dense set – A subset D of a metric space X is said to be *dense* in X if $\overline{D} = X$. This means that D is dense in X if and only if any one of the following conditions hold:
(1) $D \cap B_r(x) \neq \varnothing$ for all $x \in X$ and $r > 0$;
(2) any open set in X contains a point of D;
(3) for each $x \in X$, there exists a sequence $\{x_n\} \subset D$ such that $x_n \to x$.

Example 1.18 (1) The set of rational numbers $\mathbb{Q}$ is dense in $\mathbb{R}$, i.e., $\overline{\mathbb{Q}} = \mathbb{R}$.
(2) c_{00} is dense in c_0, i.e., $\overline{c_{00}} = c_0$.
(3) The space $C[a, b]$ is dense in $L_p[a, b]$.
(4) The set P of all polynomials is dense in $L_p[a, b]$.

Definition 1.19 A metric space (X, d) is said to be *separable* if it contains a countable dense subset, i.e., there exists a countable set D in X such that $\overline{D} = X$.

Example 1.19 (1) $\mathbb{R}$ is a separable space. To see this, one may observe that the set $\mathbb{Q}$ of rational numbers is countable and dense in $\mathbb{R}$.
(2) $\ell_p^n (1 \leqslant p \leqslant \infty)$ spaces are separable. The set

$$D = \{x \in \ell_p^n : x = (x_1, x_2, \ldots, x_n) \text{ with each } x_i \text{ rational}\}$$

---

[8]The sequence $\{x_n\}$ is called the Schauder system for $C[0, 1]$.
[9]The sequence $\{x_n\}$ is called the Haar system for $L_p[0, 1]$, $1 \leqslant p < \infty$.

is countable and is easily seen to be dense in ℓ_p^n.

(3) $\ell_p(1 \leqslant p < \infty)$ spaces are separable. To see this, consider the set D of all sequences having finitely many nonzero coordinates each of which is rational. Clearly, $D = \bigcup\limits_{i=1}^{\infty} X_i$, where $X_i = \{x : x = \{x_1, x_2, \ldots, x_i, \ldots\}, x_1, x_2, \ldots, x_i \in \mathbb{Q}\}$. Then, it is easy to see that there is a one-to-one correspondence between X_i and $\mathbb{Q}^i$, where $\mathbb{Q}$ is the set of rational numbers and $\mathbb{Q}^i$ is the cartesian product of $\mathbb{Q}$ with itself i times. Thus, X_i is countable and the fact that countable union of countable sets is countable, $\bigcup\limits_{i=1}^{\infty} X_i$ is countable. We now show that A is dense in $\ell_p(1 \leqslant p < \infty)$. Let $x = (x_1, x_2, \ldots) \in \ell_p$ and $\varepsilon > 0$.

Because $x \in \ell_p$, the series $\sum\limits_{i=1}^{\infty} |x_i|^p$ is convergent, so that using Cauchy's criterion there exists an integer $n \in \mathbb{N}$ such that

$$\sum_{i=n+1}^{\infty} |x_i|^p < \frac{\varepsilon^p}{2}.$$

As $\mathbb{Q}$ is dense in $\mathbb{R}$, we can choose n rational numbers $y_1, y_2, \ldots, y_n$ such that

$$\sum_{i=1}^{n} |x_i - y_i|^p < \frac{\varepsilon^p}{2}.$$

Thus, we have $y = (y_1, y_2, \ldots, y_n, 0, 0, \ldots) \in A$ and

$$d_p(x, y) = \left(\sum_{i=1}^{\infty} |x_i - y_i|^p\right)^{1/p} = \left(\sum_{i=1}^{n} |x_i - y_i|^p + \sum_{i=n+1}^{\infty} |x_i|^p\right)^{1/p} < \left(\frac{\varepsilon^p}{2} + \frac{\varepsilon^p}{2}\right)^{1/p} = \varepsilon.$$

(4) ℓ_∞ is not a separable space. To see this, consider $D \subset \ell_\infty$ consisting of sequences of zeros and ones. Clearly, D is uncountable and for any $x, y \in D$, $d_\infty(x, y) = \sup_i |x_i| = 1$. It follows, therefore, that the family $\mathscr{B}$ of open balls of radius $\frac{1}{2}$ centered at each $x \in A$ is an uncountable family of nonintersecting balls. Next, assume that there exists a dense set D in ℓ_∞. Then, each ball in $\mathscr{B}$ must contains at least one point of D which is dense in ℓ_∞. Because the balls of the family B are nonintersecting, this makes D uncountable. Thus, ℓ_∞ does not have a countable dense set. Hence ℓ_∞ is not separable.

(5) $C[a, b]$ is a separable normed space. This follows from the following observations:

(i)   Let P denote the set of all polynomials defined on $[a, b]$. Then, we notice that P is dense in $C[a, b]$. Indeed, for any $x \in C[a, b]$ and $\varepsilon > 0$, by the Weierstrass approximation theorem, we can find a polynomial $p \in C[a, b]$ such that $d(x, p) < \frac{\varepsilon}{2}$.

(ii)   Given a polynomial $p(t)$, there is a polynomial $r(t)$ with rational coefficients such that

$$d(p, r) = \max_{a \leqslant t \leqslant b} |p(t) - r(t)| < \frac{\varepsilon}{2}.$$

It follows, therefore, that

$$d(x, r) \leqslant d(x, p) + d(p, r) < \frac{\varepsilon}{2} + \frac{\varepsilon}{2} = \varepsilon.$$

(*iii*) The set of all polynomials with rational coefficients is countable.

Hence, we conclude that $C[a, b]$ is separable.

(6) The linear space X of all infinite sequences of real numbers with metric d defined by

$$d(x, y) = \sum_{i=1}^{\infty} \frac{1}{2^i} \frac{|x_i - y_i|}{1 + |x_i - y_i|}, \quad x = \{x_i\}, y = \{y_i\} \in X$$

is a separable complete metric space.

Theorem 1.10 *A Banach space with a basis is separable.*

Remark 1.2 In 1932, S. Banach [33] raised the question: Is the converse of Theorem 1.10 true, i.e., Does every separable Banach space possess a basis? This is known as the *Basis Problem*. This problem was settled in 1973 by P. Enflo [223]. As a matter of fact, Enflo constructed a separable Banach space without a basis.

Observation

- If (X, d) is a separable metric space and $C \subset X$ is endowed with induced metric $\tilde{d}$, then $(C, \tilde{d})$ is separable.
- A metric space X is separable if and only if there is a countable family $\{G_i\}_{i=1}^{\infty}$ of open sets such that for any open set $G \subset X$, $G = \bigcup_{G_i \subset G} G_i$.
- Every finite dimensional normed space is separable.
- Every normed space with basis is separable.
- Every subset of a separable normed space is separable.
- $\mathbb{R}^n$, c_{00}, c_0, c, $C[0, 1]$ are separable spaces.
- L_∞ is not a separable space.

1.2.7 Inner Product Spaces

Inner product – Let X be a linear space over field $\mathbb{C}$. An inner product on X is a function $\langle \cdot, \cdot \rangle : X \times X \to \mathbb{C}$ satisfying the following properties:

(*IP1*) $\langle x, x \rangle \geq 0$ for all $x \in X$;

(*IP2*) $\langle x, x \rangle = 0$ if and only if $x = 0$;

(*IP3*) $\langle x, y \rangle = \overline{\langle y, x \rangle}$ where the bar denotes complex conjugation;

(*IP4*) $\langle \alpha x + \beta y, z \rangle = \alpha \langle x, z \rangle + \beta \langle y, z \rangle$ for all $x, y, z \in X$.

The ordered pair $(X, \langle \cdot, \cdot \rangle)$ is called an inner product space. We call $\langle x, y \rangle$ the inner product of two elements $x, y \in X$.

Remark 1.3 The defining properties of an inner product exhibit the following:

(1) Property $(IP4)$ implies that for any $z \in X$ the mapping $\langle \cdot, z \rangle : X \to \mathbb{C}$ is a linear functional on X.

(2) From properties $(IP3)$ and $(IP4)$ it follows that for all $x, y, z \in X$,

$$\langle x, y + z \rangle = \overline{\langle y + z, x \rangle} = \overline{\langle y, x \rangle} + \overline{\langle z, x \rangle}$$
$$= \langle x, y \rangle + \langle x, z \rangle$$

and
$$\langle x, \lambda y \rangle = \overline{\langle \lambda y, x \rangle} = \bar{\lambda} \overline{\langle y, x \rangle}$$
$$= \bar{\lambda} \langle x, y \rangle \text{ for all } \lambda \in \mathbb{C}.$$

So for any $z \in X$, the mapping $\langle z, \cdot \rangle : X \to \mathbb{C}$ is not linear but "conjugate linear".

Example 1.20 Let $X = \mathbb{R}^n$, the set of all n-tuples of real numbers. Then, the function $\langle \cdot, \cdot \rangle : \mathbb{R}^n \times \mathbb{R}^n \to \mathbb{R}$ defined by

$$\langle x, y \rangle = \sum_{i=1}^n x_i y_i \ \forall \, x = (x_1, x_2, \dots, x_n), y = (y_1, y_2, \dots, y_n) \in \mathbb{R}^n$$

is an inner product on $\mathbb{R}^n$ and the ordered pair $(\mathbb{R}^n, \langle \cdot, \cdot \rangle)$ is called real Euclidean n-space.

Example 1.21 Let $X = \mathbb{C}^n$, the set of all n-tuples of complex numbers. Then, the function $\langle \cdot, \cdot \rangle : \mathbb{C}^n \times \mathbb{C}^n \to \mathbb{C}$ defined by

$$\langle x, y \rangle = \sum_{i=1}^n x_i \bar{y}_i \ \text{ for all } x = (x_1, x_2, \dots, x_n), y = (y_1, y_2, \dots, y_n) \in \mathbb{R}^n$$

is an inner product on $\mathbb{C}^n$ and the ordered pair $(\mathbb{C}^n, \langle \cdot, \cdot \rangle)$ is called complex Euclidean n-space or n-unitary space.

Example 1.22 Let $X = \ell_2$, the set of all sequences of complex numbers $x = \{x_i\}_{i=1}^\infty$ with $\sum_{i=1}^\infty |x_i|^2 < \infty$. Then, the function $\langle \cdot, \cdot \rangle : \ell_2 \times \ell_2 \to \mathbb{C}$ defined by

$$\langle x, y \rangle = \sum_{i=1}^\infty x_i \bar{y}_i \ \forall \, x = \{x_i\}_{i=1}^\infty, y = \{y_i\}_{i=1}^\infty \in \ell_2 \tag{1.3}$$

is an inner product on ℓ_2.

Example 1.23 Let $X = C[a, b]$, the linear space of all scalar-valued continuous functions on $[a, b]$. Then, the function $\langle \cdot, \cdot \rangle : C[a, b] \times C[a, b] \to \mathbb{C}$ defined by

$$\langle x, y \rangle = \int_a^b x(t) \overline{y(t)} \, dt \ \forall \, x, y \in C[a, b] \tag{1.4}$$

is an inner product on $C[a, b]$.

Characterizations of Inner Product Spaces

Let $(X, \langle \cdot, \cdot \rangle)$ be an inner product space. Then, the function $\| \cdot \| : X \to \mathbb{R}^+$ defined by

$$\|x\| = \sqrt{\langle x, x \rangle}, \quad x \in X$$

is a norm on X. This norm is commonly known as Hilbertian norm.

Proposition 1.4 (The Cauchy-Schwarz inequality) *Let X be an inner product space. Then,*

$$|\langle x, y \rangle| \leqslant \|x\| \cdot \|y\| \ \text{for all} \ x, y \in X.$$

Proposition 1.5 (The parallelogram law) *Let X be an inner product space. Then, the following holds:*

$$\|x + y\|^2 + \|x - y\|^2 = 2\|x\|^2 + 2\|y\|^2 \ \text{for all} \ x, y \in X.$$

Proposition 1.6 (The polarization identity) *Let X be an inner product space. Then, the following holds:*

$$\langle x, y \rangle = \frac{1}{4} \left\{ \|x + y\|^2 - \|x - y\|^2 + i\|x + iy\|^2 - i\|x - iy\|^2 \right\} \ \text{for all} \ x, y \in X.$$

Orthogonality of vectors – Let x and y be two vectors in an inner product space X. Then, x and y are said to be orthogonal if $\langle x, y \rangle = 0$. If x is orthogonal to y, then we denote this fact by $x \perp y$ and we say that "x is perpendicular to y". Given subset M of X we say that x is orthogonal to M if x is orthogonal to every element of M and we write $x \perp M$.

Remark 1.4 From the defining properties of an inner product, we deduce the following properties about the orthogonality relation.

(1) x is orthogonal to y if and only if y is orthogonal to x; that is, the orthogonality relation is symmetric

(2) From property (IP4), it follows that if x is orthogonal to a subset M, then x is orthogonal to span M and conversely every element of span M is orthogonal to x.

Observation

- The null vector is perpendicular to any $x \in X$, i.e., $0 \perp x$ for all $x \in X$.
- $x \perp x$ if and only if $x = 0$.
- Every inner product space is a normed space with the standard norm $\| x \| = \sqrt{\langle x, x \rangle}$.
- Every normed space is an inner product space if and only if its norm satisfies the parallelogram law.
- If $x, y \in X$ such that $x \perp y$, then $\| x + y \|^2 = \| x \|^2 + \| y \|^2$.

We now show that the orthogonality defined by inner product is actually identical to a "best distance" type of orthogonality defined by the norm. Notice that this notion of orthogonality plays an important role in best approximation problems.

Theorem 1.11 *In an inner product space X, for all $x, y \in X$*

$$\langle x, y \rangle = 0 \text{ if and only if } \|x + \lambda y\| \geqslant \|x\| \text{ for all scalar } \lambda.$$

Proof If $\langle x, y \rangle = 0$ then

$$\|x + \lambda y\| \|x\| \geqslant |\langle x + \lambda y, x \rangle| = \|x\|^2 + \lambda \langle y, x \rangle = \|x\|^2.$$

It follows that $\|x + \lambda y\| \geqslant \|x\|$ for all scalar λ (see, for instance, Fig. 1.6).
 Conversely, if

$$\|x + \lambda y\| \geqslant \|x\| \text{ for all scalar } \lambda,$$

then

$$\|x + \lambda y\|^2 - \|x\| \|x + \lambda y\| \geqslant 0$$

so $\text{Re}\langle x, x + \lambda y \rangle + \text{Re}\lambda \langle y, x + \lambda y \rangle - |\langle x, x + \lambda y \rangle| \geqslant 0$.
Therefore, $\text{Re}\lambda \langle y, x + \lambda y \rangle \geqslant 0$ for all scalar λ.
For real λ, we have
$\text{Re}\langle y, x + \lambda y \rangle \geqslant 0$ for $\lambda \geqslant 0$ and $\text{Re}\langle y, x + \lambda y \rangle \leqslant 0$ for $\lambda \leqslant 0$.
But $\langle y, x + \lambda y \rangle = \langle y, x \rangle + \lambda \|y\|^2 \to \langle y, x \rangle$ as $\lambda \to 0$.
From this, we deduce that $\text{Re}\langle y, x \rangle = 0$.
For $\lambda = i\alpha$ where α is real, we have

$$\text{Re}[\alpha \langle iy, x + \alpha i y \rangle] \geqslant 0 \text{ for all real } \alpha$$

so by the same argument we have $\text{Re}\langle y, x \rangle = 0$ which implies that $\text{Im}\langle y, x \rangle = 0$.
Hence we conclude that $\langle x, y \rangle = 0$.

The Hilbert space – An inner product space $\mathcal{H}$ is called a Hilbert space if it is complete with respect to the induced norm, i.e., if every Cauchy sequence is convergent. Thus, a Hilbert space is an inner product space that is a Banach space with respect to the induced norm.

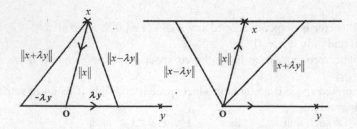

Fig. 1.6 Illustration of "best distance" orthogonality

More precisely, a Hilbert space[10] is aninner product space $(\mathcal{H}, \langle \cdot, \cdot \rangle)$ such that the induced Hilbertian norm is complete.

Example 1.24 For $p \neq 2$, ℓ_p^n is a finite dimensional Banach space but it is not a Hilbert space. Indeed, for $x = (1, -1, 1, 0, \ldots)$ and $y = (-1, 1, 1, 0, \ldots)$, we have $x + y = (0, 0, 2, 0, \ldots)$ and $x - y = (2, -2, 0, 0, \ldots)$. Hence

$$\|x\| = \left(\sum_{i=1}^{n} |x_i|^p \right)^{1/p} = (1^2 + (-1)^2 + 1^2)^{1/p} = 3^{1/p},$$

$$\|y\| = ((-1)^2 + 1^2 + 1^2)^{1/p} = 3^{1/p},$$

$$\|x + y\| = (2^p)^{1/p}, \text{ and } \|x - y\| = (2^p + (-2)^p)^{1/p}.$$

If $p = 2$, then $\|x\| = \sqrt{3}$, $\|y\| = \sqrt{3}$, $\|x + y\| = (2^2)^{1/2} = 2$, $\|x - y\| = (2^2 + (-2)^2)^{1/2} = 2 \cdot \sqrt{2}$ and so the parallellogram law:

$$\|x + y\|^2 + \|x - y\|^2 = 2\|x\|^2 + 2\|y\|^2$$

is satisfied, which shows that ℓ_2^n is a Hilbert space. If $p \neq 2$, then the parallellogram law does not hold good. Therefore, ℓ_p^n is not a Hilbert space for $p \neq 2$.

Observation

- The complex (the real) ℓ_2^n, ℓ_2, $L_2[a, b]$ are Hilbert spaces.
- The complex (the real) ℓ_p^n, ℓ_p, $L_p[a, b]$ $(p \neq 2)$ are not the Hilbert spaces.
- $C[a, b]$ is an inner product space with the inner product defined in (1.4), but not a Hilbert space.

1.3 Spaces of Bounded Linear Operators

An operator[11] maps vectors from one space to vectors in another space. Let X and Y be two vector spaces over the same field $\mathbb{K}$ ($\mathbb{K}$ means $\mathbb{R}$ or $\mathbb{C}$) and $A : X \to Y$ an operator with domain $\mathscr{D}(A)$ and range $\mathscr{R}(A)$. Then, A is said to be a *linear operator* if the following properties hold:

(i) A is additive : $A(x + y) = Ax + Ay$ for all $x, y \in X$;
(ii) A is homogenous : $A(\alpha x) = \alpha A(x)$ for all $x \in X, \alpha \in \mathbb{K}$.

Equivalently, A is a linear operator if and only if

$$A(\alpha x + \beta y) = \alpha A(x) + \beta A(y) \text{ for all } x, y \in X \text{ and } \alpha, \beta \in \mathbb{K}.$$

---

[10]Complete inner product spaces are known as Hilbert spaces, in honor of David Hilbert.

[11]An operator is a function from a vector space to another vector space. Indeed, an operator is a special type of function. Any linear map (i.e. linear function) of vector spaces can be called an operator; this is most common when the map is thought of as "acting on" a vector space.

Otherwise, the operator is called *nonlinear*. If $Y = \mathbb{R}$, then the operator A is called a linear functional.

Example 1.25 (1) Let $X = \mathbb{R}^n$, $Y = \mathbb{R}$, and $A : \mathbb{R}^n \to \mathbb{R}$ an operator defined by

$$Ax = \sum_{i=1}^{n} a_i x_i \text{ for all } x = (x_1, x_2, \ldots, x_n) \in X, a_i \text{ are fixed elements of } \mathbb{R}.$$

Then, A is a linear functional on $\mathbb{R}^n$.

(2) Let $X = Y = \ell_2$ and $A : \ell_2 \to \ell_2$ an operator defined by

$$Ax = \left(x_1, \frac{x_2}{2^2}, \frac{x_3}{3^2}, \ldots, \frac{x_n}{n^2}, \ldots\right) \text{ for all } x = (x_1, x_2, x_3, \ldots, x_n, \ldots) \in \ell_2.$$

Then, A is a linear operator on ℓ_2.

(3) Let $X = L_2[0, 2\pi]$, $Y = \mathbb{R}$ and $A : L_2[0, 2\pi] \to \mathbb{R}$ an operator defined by

$$[Ax](s) = \int_0^{2\pi} x(t)y(t)dt \text{ for all } x \in L_2[0, 2\pi] \text{ and } y \text{ a fixed element in } L_2[0, 2\pi].$$

Then, F is a linear functional on $L_2[0, 2\pi]$.

(4) Let $X = Y = C[a, b]$ and $A : C[a, b] \to C[a, b]$ an operator defined by

$$[Ax](s) = \lambda \int_a^b x(t)dt \text{ for all } x \in C[a, b] \text{ and } \lambda \text{ a fixed real constant.}$$

Then, A is a linear operator on $C[a, b]$.

Proposition 1.7 *Let X and Y be two linear spaces over the same field $\mathbb{K}$ and $A : X \to Y$ a linear operator. Then, we have the following:*

- *(i) $A0 = 0$, 0 being zero element of X and zero element of Y.*
- *(ii) $\mathcal{R}(A) = \{y \in Y : y = Ax \text{ for some } x \in X\}$, the range of A is a linear subspace of Y.*
- *(iii) A is one-one iff $Ax = 0 \Rightarrow x = 0$.*
- *(iv) If $A : X \to Y$ is one-one operator, then A^{-1} exists on $\mathcal{R}(A)$. Moreover, $A^{-1} : \mathcal{R}(A) \to X$ is also a linear operator.*
- *(v)   if $dim(X) = n < \infty$ and A^{-1} exists, then $dim\mathcal{R}(A) = n$.*

Theorem 1.12 *Let X and Y be two normed spaces and $A : X \to Y$ a linear operator. If A is continuous at a single point on X, then A is continuous throughout space X.*

Proof Let the linear operator A be continuous at a point $x_0 \in X$ and $\{x_n\}$ a sequence in X such that $\lim_{n \to \infty} x_n = x \in X$. By linearity of A, we have

$$\|Ax_n - Ax\| = \|Ax_n - Ax + Ax_0 - Ax_0\| = \|A(x_n - x + x_0) - Ax_0\|.$$

Also, $\lim_{n\to\infty}(x_n - x + x_0) = x_0$. Because A is continuous at x_0 and norm is a continuous function, it follows that

$$\lim_{n\to\infty} \|Ax_n - Ax\| = \lim_{n\to\infty} \|A(x_n - x + x_0) - Ax_0\| = \|A(\lim_{n\to\infty}(x_n - x + x_0)) - Ax_0\| = 0.$$

Thus, $x_n \to x \Rightarrow Ax_n \to Ax$, as $n \to \infty$. Hence, A is a continuous operator at an arbitrary point $x \in X$. ∎

Bounded linear operator – Let X and Y be two normed spaces and $A : X \to Y$ is a linear operator. Then, A is said to be *bounded* if there exists a constant $M > 0$ such that

$$\|Ax\| \leqslant M\|x\| \text{ for all } x \in X.$$

According to this definition, the reader may well observe that a bounded operator does not map the whole space X into a bounded set in Y. Rather, it maps a bounded set in X into a bounded set in Y. Moreover, the converse is also true : a linear operator that maps a bounded set a bounded set is bounded.

A linear functional $f : X \to \mathbb{R}$ is called *bounded* if there exists a constant $M > 0$ such that

$$|fx| \leqslant M\|x\| \text{ for all } x \in X.$$

Notice that a linear operator need not be bounded. To see this, consider the following:

Example 1.26 Let c_{00} be the linear space of finitely nonzero real sequences with "sup" norm $\| \cdot \|_\infty$ and $A : c_{00} \to \mathbb{R}$ a functional defined by

$$Ax = \sum_{i=1}^{n} ix_i \text{ for all } x = (x_1, x_2, \ldots, x_n, 0, 0, \ldots) \in c_{00}.$$

Then, A is a linear functional, but it is unbounded.

This example arose a natural question – Under what additional assumption a linear operator would be bounded? The following example shows that such an additional assumption is boundedness of the linear operator.

Theorem 1.13 *Let X and Y be two normed spaces and $A : X \to Y$ a linear operator. Then, A is bounded if and only if it is continuous.*

Proof Suppose the linear operator A is bounded, then there exists a constant $M > 0$ such that

$$\|Ax\| \leqslant M\|x\| \text{ for all } x \in X.$$

Let there be a sequence $\{x_n\} \subset X$ such that $x_n \to 0$. Then, we see that

$$\|Ax_n\| \leqslant M\|x_n\| \to 0 \text{ as } n \to \infty$$

i.e., $$\|Ax_n - A0\| \leqslant M\|x_n - 0\| \to 0 \text{ as } n \to \infty.$$

It follows that A is continuous at zero. By Theorem 1.12, we conclude that A is continuous on X.

Conversely, suppose A is continuous. Then, we have to show that A is bounded. Assume, on the contrary, that A is unbounded. Then, there exists a sequence $\{x_n\}$ in X such that

$$\|Ax_n\| > n\|x_n\| \text{ for all } n \in \mathbb{N}.$$

Because $A0 = 0$, it follows that $x_n \neq 0$. Set $y_n = \frac{x_n}{n\|x_n\|}, n \in \mathbb{N}$. Then, $\|y_n\| = \|\frac{x_n}{n\|x_n\|}\| = \frac{1}{n} \to 0$ as $n \to \infty$ which shows that $\lim_{n\to\infty} y_n = 0$. Next, we observe that

$$\|Ay_n\| = \left\|A\left(\frac{x_n}{n\|x_n\|}\right)\right\| = \frac{Ax_n}{n\|x_n\|} > 1 \text{ for all } n \in \mathbb{N}.$$

This implies that $\{Ay_n\}$ does not converge to zero. Hence A is not continuous at zero, a contradiction to our supposition. Thus, A is bounded. ∎

Notice that in a finite dimensional normed space X, all linear operators $A : X \to Y$ are continuous and hence bounded. In general, a linear operator may be discontinuous in infinite-dimensional normed spaces (see, Example 1.25 above).

1.3.1 Properties of Bounded Linear Operators and the Space $B(X, Y)$

Let X and Y be two normed spaces and let $B(X, Y)$ denote the family of all bounded linear operators from X into Y. Then, we observe that $B(X, Y)$ is a linear space with respect to "addition of operators" and "scalar multiplication of an operator by a scalar", respectively, as defined below:

$$(A_1 + A_2)(x) = A_1x + A_2x \text{ for all } A_1, A_2 \in B(X, Y) \text{ and } x \in X$$
$$(\alpha A)(x) = = \alpha A(x) \text{ for all } A \in B(X, Y) \text{ and } x \in X.$$

The space $B(X, Y)$ becomes a normed space by assigning a norm $\|\cdot\|_B$ as defined below:

$$\|A\|_B = \inf\{M : \|Ax\| \leqslant M\|x\|, \ x \in X\}$$
$$= \sup\left\{\frac{\|Ax\|}{\|x\|} : x \neq 0, \ x \in X\right\}$$
$$= \sup\{\|Ax\| : \|x\| \leqslant 1, \ x \in X\}$$
$$= \sup\{\|Ax\| : \|x\| = 1, \ x \in X\}$$

Notice that the normed space $B(X, Y)$ is a Banach space if Y is a Banach space.

Theorem 1.14 (Uniform boundedness principle) *Let X be a Banach space, Y a normed space, and $\{A_i\}_{i \in I}$ a family of bounded linear operators of X into Y, where I is an index set, such that the set $\{A_i x : i \in I\}$ is bounded for each $x \in X$, i.e., for each $x \in X$, there exists $M_x > 0$ such that*

$$\|A_i x\| \leqslant M_x \text{ for all } i \in I.$$

Then, $\{\|A_i\|\}_{i \in I}$ is a bounded set, i.e., $A_i''s$ are uniformly bounded.

Theorem 1.15 *Let X and Y be two Banach spaces and $\{A_n\}$ a sequence of bounded operators from X into Y. If for each $x \in X$, $\{A_n x\}$ converges to Ax in Y, then*

(a) A is a bounded linear operator, i.e., $A \in B(X, Y)$;
(b) $\|A\|_B \leqslant \liminf\limits_{n \to \infty} \|A_n\|_B$.

Proof (a) Because each A_n is linear, for all $x, y \in X$ and $\alpha, \beta \in \mathbb{K}$ we have

$$A(\alpha x + \beta y) = \lim_{n \to \infty} A_n(\alpha x + \beta y) = \lim_{n \to \infty} A_n(\alpha x) + \lim_{n \to \infty} A_n(\beta y)$$
$$= \alpha \lim_{n \to \infty} A_n x + \beta \lim_{n \to \infty} A_n y = \alpha Ax + \beta Ay.$$

Again, since the norm is continuous,

$$\lim_{n \to \infty} \|A_n x\| = \|\lim_{n \to \infty} A_n x\| = \|Ax\| \text{ for all } x \in X,$$

it follows that for each $x \in X$, $\{A_n x\}$ is a bounded set in Y. By the uniform boundedness principle, there exists a positive constant $M > 0$ such that $\sup_{n \in \mathbb{N}} \|A_n\|_B \leqslant M$. Thus, we have

$$\|A_n x\| \leqslant \|A_n\|_B \|x\| \leqslant \sup_{n \in \mathbb{N}} \|A_n\|_B \|x\| \leqslant M \|x\|.$$

Now taking the limit as $n \to \infty$, we obtain

$$\|Ax\| = \lim_{n \to \infty} \|A_n x\| \leqslant M \|x\|,$$

which shows that A is bounded. Therefore, $A \in B(X, Y)$. ∎

The following is an immediate consequence of Theorem 1.15.

Corollary 1.1 *Let C be a nonempty subset of a Banach space X. For each $f \in X^*$, let $f(C) = \cup_{x \in C}(x, f)$ be a bounded set in $\mathbb{R}$. Then, C is bounded.*

Proof For each $x \in C$, define $T_x(f) = (x, f)$. Then, $T_x \in B(X^*, \mathbb{R})$. For, $a, b \in \mathbb{R}$ and $f, g \in X^*$ we have

$$T_x(af + bg) = (x, af + bg) = a(x, f) + b(x, g) = aT_x(f) + bT_x(g).$$

Because $f(C)$ is bounded, there exists a constant $K > 0$ such that

$$\sup_{x \in C} |T_x(f)| = \sup_{x \in C} |(x, f)| \leqslant K.$$

Further, by the uniform boundedness principle, there exists a constant $M > 0$ such that

$$\|T_x\| \leqslant M \text{ for all } x \in C.$$

1.3.2 Convergence of Sequences in $B(X, Y)$

Let $(X, \|\cdot\|_X)$ and $(Y, \|\cdot\|_Y)$ be two normed spaces, $(B(X, Y), \|\cdot\|_B)$ the space of all bounded linear operators from X into Y and let $\{T_n\}$ be a sequence in $B(X, Y)$. Then,

(i) $\{T_n\}$ is said to be uniformly convergent to $T \in B(X, Y)$ in the norm of $B(X, Y)$ if

$$\|T_n - T\|_B \to 0 \text{ as } n \to \infty,$$

i.e., for given $\varepsilon > 0$, $\exists$ an integer $n_0 \in N$ such that $\sup_{\|x\|_X \leqslant 1} \|T_n x - T x\|_Y < \varepsilon$ for all $x \in X$.

(ii) $\{T_n\}$ is said to be strongly convergent to $T \in B(X, Y)$ if

$$\|T_n x - T x\|_Y \to 0 \text{ as } n \to \infty \text{ for all } x \in X.$$

(iii) $\{T_n\}$ is said to be weakly convergent to $T \in B(X, Y)$ if

$$|f(T_n x) - f(T x)| \to 0 \text{ as } n \to \infty \text{ for all } x \in X \text{ and } f \in Y^*.$$

Observation

- If $\{T_n\}$ is uniformly convergent to $T \in B(X, Y)$, then we notice that

$$\|T_n x - T x\|_Y = \|(T_n - T)(x)\|_Y \leqslant \|T_n - T\|_B \|x\|_X, \quad x \in X$$
$$\to 0 \text{ as } n \to \infty,$$

showing that $\{T_n\}$ is strongly convergent to $T \in B(X, Y)$.
- If $\{T_n\}$ is strongly convergent to $T \in B(X, Y)$, then

$$|f(T_n x) - f(T x)| = |f(T_n x - T x)| \leqslant \|f\| \|T_n x - T x\|_Y, \quad x \in X \text{ and } f \in Y^*$$
$$\to 0 \text{ as } n \to \infty,$$

showing that $\{T_n\}$ is weakly convergent to $T \in B(X, Y)$.

With the above observations, we conclude that the sequence of operators in $B(X, Y)$ follows the following implications

$$\boxed{\text{uniform convergence}} \Rightarrow \boxed{\text{strong convergence}} \Rightarrow \boxed{\text{weak convergence}}$$

However, we notice that the converse implications are not true.

1.3.3 Banach Algebra

In this subsection, first we introduce the concept of the Banach algebra and then prove that the Banach space $B(X)$ is, in fact, a Banach algebra.

Definition 1.20 Let $\mathcal{A}$ be a linear space over the field $\mathbb{K}$. Define the operation of multiplication $\cdot : \mathcal{A} \times \mathcal{A} \to \mathcal{A}$ in such a way that for every ordered pair of elements $x, y \in \mathcal{A}$, there exists a unique product $x \cdot y \in \mathcal{A}$. We usually write $x \cdot y$ as xy. $\mathcal{A}$ is said to be an algebra over $\mathbb{K}$ if the following conditions are satisfied:
(*i*) $x(yz) = (xy)z$
(*ii*) $x(y + z) = xy + xz$
(*iii*) $(x + y)z = xz + yz$
(*iv*) $\alpha(xy) = (\alpha x)y = x(\alpha y)$
for all $x, y, z \in \mathcal{A}$ and $\alpha \in \mathbb{K}$.

An algebra $\mathcal{A}$ is said to be
(*a*) real or complex algebra according as $\mathbb{K}$ is $\mathbb{R}$ or $\mathbb{C}$.
(*b*) commutative (or abelian) algebra if the multiplication in $\mathcal{A}$ is commutative, i.e.,

$$xy = yx, \quad \forall x, y \in \mathcal{A}.$$

(*c*) an algebra with identity if $\mathcal{A}$ contains an element e, called identity element, such that

$$xe = x = ex, \quad \forall x \in \mathcal{A}.$$

Definition 1.21 An algebra $\mathcal{A}$ is said to be a normed algebra if
(*i*) $\mathcal{A}$ is a normed space as a linear space equipped with a norm $\| \cdot \|$; and
(*ii*) $\|xy\| \leqslant \|x\|\|y\|, \quad \forall x, y \in \mathcal{A}.$

Banach algebra – A normed algebra is said to be a Banach algebra, if it is complete as a metric space.

Example 1.27 (1) $\mathbb{R}$ and $\mathbb{C}$ are commutative Banach algebra with identity $e = 1$.
(2) The linear space $C[a, b]$ is a commutative algebra with identity $e = 1$, the product xy being defined as

$$(xy)(t) = x(t)y(t), \quad \forall t \in [a, b].$$

(3) Let $X \neq \{0\}$ be a complex Banach space. Then, the space $B(X)$ of all bounded linear operators on X is a complex Banach algebra. For $S, T \in B(X)$, we define

$$(ST)(x) = S(T(x)), \quad \forall x \in X.$$

1.4   The Hahn–Banach Theorem

The Hahn–Banach theorem is the key ingredient of Functional analysis. Before stating the Hahn–Banach theorem, we need the following definitions:

Sublinear functional – Let X be a linear space and $p : X \to \mathbb{R}$ a functional. Then, p is said to be a sublinear functional on X if

(i) p is subadditive, i.e., $p(x + y) \leqslant p(x) + p(y)$ for all $x, y \in X$,

(ii) p is positive homogeneous $p(\alpha x) = \alpha p(x)$ for all $x \in X$ and $\alpha \geq 0$.

Theorem 1.16 (The Hahn–Banach theorem for real linear space) *Let Y be subspace of a real linear space X, p a sublinear functional on X, and f a linear functional defined on Y satisfying the condition:*

$$f(x) \leqslant p(x) \text{ for all } x \in Y.$$

Then, there exists a linear extension F of f to X such that $F(x) \leqslant p(x)$ for all $x \in X$.

Theorem 1.17 (The Hahn–Banach theorem for normed linear space) *Let X be a real normed linear space and Y a linear subspace of X. If $f \in Y^*$, then there is a $F \in X^*$ such that $F(x) = f(x)$ for all $x \in Y$ and $\|F\|_* = \|f\|_Y$.*

In other words, every bounded linear functional defined on a subspace Y of a real normed linear space X may be extended linearly with preservation of the norm to the whole of X.

The following corollary gives the existence of nontrivial bounded linear functional on an arbitrary normed space.

Corollary 1.2 *Let x be an element of normed space X. Then, there exists a (nonzero) $j \in X^*$ such that*
$$j(x) = \|j\|_* \|x\| \text{ and } \|j\|_* = \|x\|.$$

Corollary 1.3 *Let x be a nonzero element of normed space X. Then, there exists a functional $j \in X^*$ such that*

$$j(x) = \|x\| \text{ and } \|j\|_* = 1.$$

The above Corollary suggests the following:

Remark 1.5 Let $(X, \| \cdot \|)$ be a normed space and $(X^*, \| \cdot \|_*)$ its first dual space. Then, for any $x \in X$,

$$\|x\| = \sup_{\|j\|_* \leqslant 1} \|j(x)\|.$$

Corollary 1.4 *Let X be a normed space, and let $x_0 \in X$ such that $j(x_0) = 0$ for all $j \in X^*$, then $x_0 = 0$.*

Proof Suppose, if possible, $x_0 \neq 0$. By Corollary 1.3, there exists a functional $j \in X^*$ such that

$$j(x_0) = \|x_0\| \text{ and } \|j\|_* = 1$$

which yields $j(x_0) \neq 0$, a contradiction. Hence

$$j(x_0) = 0 \text{ for all } j \in X^* \Rightarrow x_0 = 0.$$

Theorem 1.18 *Let M be a subspace of a normed space X and x_0 an element in X such that $d(x_0, M) = d > 0$. Then, there exists a bounded linear functional $j \in X^*$ with unit norm such that*

$$j(x_0) = d \text{ and } j(x) = 0 \text{ for all } x \in M.$$

We now state complex version of the Hahn–Banach theorem, which is due to Bohnenblust, Sobezyk and Soukhomlinoff.

Theorem 1.19 (The Hahn–Banach theorem for complex linear space) *Let X be a complex linear space and Y a linear subspace of X. If p is a seminorm on X and $f \in Y^*$ such that*

$$|f(x)| \leqslant p(x) \text{ for all } x \in Y.$$

Then, there exists a linear extension F of f such that

$$\|F\|_* \leqslant p(x) \text{ for all } x \in X.$$

1.4.1 Geometric Forms of the Hahn–Banach Theorem

Hyperplane – A subset H of a linear space X is said to be a hyperplane if there exists a linear functional $f \neq 0$ on X such that

$$H = \{x \in X : f(x) = \alpha\}, \alpha \in \mathbb{R}$$

and $f(x) = \alpha$ is called the equation of the hyperplane H.

Example 1.28 Let $X = \mathbb{R}$, $f : X \to \mathbb{R}$ a linear functional defined by $f(x) = \frac{3}{7}x$, and let $\alpha = 4$. Then, the set

$$H = \{x \in X : f(x) = \alpha\} = \left\{x \in X : \frac{3}{7}x = 4\right\} = \left\{\frac{28}{3}\right\}$$

is a hyperplane.

1.4.2 Dual Space (or the Conjugate Space)

Definition 1.22 Let X be a normed linear space. The space $B(X, \mathbb{K})$ of all bounded linear functionals of X into the scalar field $\mathbb{K}$ is called the dual or the conjugate space of X. It is denoted by X^*.

Notice that X^* is a normed space with norm $\| \cdot \|_*$ defined by

$$\|f\|_* = \sup_{x \in \mathbb{S}_X} |f(x)|, \quad f \in X^*.$$

Example 1.29 (1) **The dual of $\mathbb{R}^n$.** The dual of $\mathbb{R}^n$ is $\mathbb{R}^n$ itself. To see this, let $\mathbb{R}^n$ be a normed space with norm $\| \cdot \|_2$ defined by

$$\|x\|_2 = \left(\sum_{i=1}^n x_i^2\right)^{1/2}, x = (x_1, x_2, \ldots, x_n) \in \mathbb{R}^n.$$

Then, for $y = (y_1, y_2, \ldots, y_n) \in \mathbb{R}^n$, any functional $f_y : \mathbb{R}^n \to \mathbb{R}$ of the form

$$f_y(x) = \sum_{i=1}^n x_i y_i, \quad x = (x_1, x_2, \ldots, x_n) \in \mathbb{R}^n$$

is linear. For $\alpha = (\alpha_1, \alpha_2, \ldots, \alpha_n) \in \mathbb{R}^n$, from Cauchy-Schwarz inequality, we obtain

$$|f_\alpha(x)| = \left|\sum_{i=1}^n \alpha_i x_i\right| \leqslant \left(\sum_{i=1}^n \alpha_i^2\right)^{1/2} \left(\sum_{i=1}^n x_i^2\right)^{1/2} = \left(\sum_{i=1}^n \alpha_i^2\right)^{1/2} \|x\|_2.$$

Therefore,

$$\|f_\alpha\|_* = \sup_{x \in \mathbb{S}_X} |f_\alpha(x)| \leqslant \left(\sum_{i=1}^n \alpha_i^2\right)^{1/2},$$

which shows that f_α is bounded. Further, if we take $x = (\alpha_1, \alpha_2, \ldots, \alpha_n)$, then we see that $\|f_\alpha\|_* = \|\alpha\|$.

So, let $j \in B(\mathbb{R}^n, \mathbb{R})$ be any bounded linear functional on $\mathbb{R}^n$. Consider the basis vectors e_i in $\mathbb{R}^n$, where

$$e_i = (0, 0, \ldots, 1, 0, \ldots, 0)$$
$$\uparrow$$
$$i^{th} \text{ position}$$

Suppose $j(e_i) = \alpha_i$. Then, in one hand $\|j\|_* \leqslant \|\alpha\|$ and the other hand $\|\alpha\| \leqslant \|j\|_* \|e_i\| = \|j\|_*$. It follows that $\|j\|_* = \|\alpha\|$. Further, for any $x = (x_1, x_2, \ldots, x_n) \in \mathbb{R}^n$, we have the following representation of x in terms of basis vectors

$$x = \sum_{i=1}^{n} e_i x_i.$$

By the linearity of j, we have

$$j(x) = \sum_{i=1}^{n} j(e_i x_i) = \sum_{i=1}^{n} j(e_i) x_i = \sum_{i=1}^{n} \alpha_i x_i.$$

Thus, we conclude that the dual space X^* of $X = \mathbb{R}^n$ is $\mathbb{R}^n$ itself in the sense that X^* consists of all functionals of the form $f(x) = \sum_{i=1}^{n} \alpha_i x_i$ and the norm on X^* is $\|f\|_* = (\sum_{i=1}^{n} \alpha_i^2)^{1/2} = \|\alpha\|$, where $\alpha = (\alpha_1, \alpha_2, \ldots, \alpha_n) \in \mathbb{R}^n$.

(2) **The dual of ℓ_p, $1 \leqslant p < \infty$.** The dual of ℓ_p, $1 \leqslant p < \infty$ is ℓ_q $(\frac{1}{p} + \frac{1}{q} = 1)$ in the sense that there is a one-one correspondence between elements $\alpha = \{\alpha_n\}_{n=1}^{\infty} \in \ell_q$ and bounded linear functional f_α on ℓ_p, i.e., $f_\alpha \in B(\ell_p, \mathbb{K})$, ($\mathbb{K} = \mathbb{R}$ or $\mathbb{C}$) such that

$$f_\alpha(x) = \sum_{n=1}^{\infty} \alpha_n x_n, \quad x = \{x_n\}_{n=1}^{\infty} \in \ell_p,$$

where

$$\|f_\alpha\|_* = \|\alpha\|_q = \begin{cases} \left(\sum_{n=1}^{\infty} |\alpha_n|^q \right)^{1/q}, & \text{if } 1 < p < \infty, \\ \sup_{n \in \mathbb{N}} |\alpha_n|, & \text{if } p = 1. \end{cases}$$

(3) **The dual of c_0.** Consider the Banach space $(c_0, \|\cdot\|_\infty)$ of all real sequences $x = \{x_i\}_{i=1}^{\infty}$ such that $\lim_{n \to \infty} x_n = 0$ with norm $\|x\|_\infty = \sup_{n \in \mathbb{N}} |x_n|$. The dual of c_0 is ℓ_1 in the sense that there is one-one correspondence between elements $\alpha = \{\alpha_n\}_{n=1}^{\infty} \in \ell_1$ and bounded linear functional f_α on c_0, i.e., $f_\alpha \in B(c_0, \mathbb{R})$ such that

$$f_\alpha(x); = \sum_{n=1}^\infty \alpha_n x_n, \quad x = \{x_n\}_{n=1}^\infty \in c_0,$$

where

$$\|f_\alpha\|_*; = \|\alpha\|_1 = \sum_{n=1}^\infty |\alpha_n|.$$

(4) **The dual of** $L_p[0, 1]$, $1 \leqslant p < \infty$. The dual of $L_p[0, 1]$, $1 \leqslant p < \infty$ is $L_q[0, 1]$ ($\frac{1}{p} + \frac{1}{q} = 1$) in the sense that there is a one-one correspondence between elements $y \in \mathrm{L}_q[0, 1]$ and bounded linear functional f_y on $L_p[0, 1]$, i.e., $f_y \in B(L_p[0, 1], \mathbb{K})$, ($\mathbb{K} = \mathbb{R}$ or $\mathbb{C}$) such that

$$f_y(x) = \int_0^1 x(t)y(t)dt \ \text{ and } \ \|f_y\|_* = \|y\|_q.$$

(5) **The dual of** $C(\Omega)$. The dual of $C(\Omega)$ is the space $\mathcal{M}(\Omega)$ of Radon measures μ with

$$\langle \mu, v \rangle := \int_\Omega v d\mu, \ \ v \in C(\Omega).$$

(6) **The dual of** $L_p(\Omega)$, $1 < p < \infty$. The dual of $L_p(\Omega)$, $1 < p < \infty$, is the space $L_q(\Omega)$ with q being conjugate to p, i.e., $1/p + 1/q = 1$.
(7) **The dual of** $L_1(\Omega)$. The dual of $L_1(\Omega)$ is the space $L_\infty(\Omega)$.
(8) **The dual of** $L_\infty(\Omega)$. The dual of $L_\infty(\Omega)$ is the space of Borel measures.

Definition 1.23 Let X, Y be two normed linear spaces over the same field. A mapping $f : X \to Y$ is called an isometric isomorphism if it is linear, bijective and preserve the norm; that is, if it is linear, one-one, onto and satisfies $\|f(x)\| = \|x\|$ for all $x \in X$. If such a mapping exists, then X and Y are called *isometrically isomorphic*.

Remark 1.6 If X and Y are two normed spaces and are isometrically isomorphic, then we shall mean that they are structurally the same. That is, if two normed spaces are isometrically isomorphic then they are two images of the same abstract object and, as such, one can be identified with the other. Dual spaces are identified in this sense.

Remark 1.7 When we say that the dual of X is Y and write $X^* = Y$, we mean that

(i) there exists an isometric isomorphism from X^* onto Y;
(ii) each functional $f \in X^*$ represented by a $y \in Y$ which is the image of f under the isomorphism; and
(iii) the value $f(x)$ of f at any point $x \in X$ is given by some rule associating y with x.

Observation

- The dual of ℓ_p^n is isometrically isomorphic to ℓ_q^n for all $p, 1 \leqslant p \leqslant \infty$, i.e., $(\ell_p^n)^* = \ell_q^n$.
- The dual of ℓ_p is isometrically isomorphic to ℓ_q for all $p, 1 \leqslant p < \infty$.
- The dual of ℓ_∞ is not ℓ_1. Rather ℓ_1 is the dual of a subspace of ℓ_∞.
- The dual of c is ℓ_1. In fact, c, c_0, c_{00} all the three have the same dual ℓ_1.

1.4.3 Second Dual, Natural Embedding and Reflexivity

The dual of X^*, i.e., the space $(X^*)^*$ of all bounded linear functionals on X^* which is written as X^{**}, is called the second dual of X.

Definition 1.24 (*Duality pairing*) Given a normed space X and its dual X^*, we define the duality pairing as the functional $(\cdot, \cdot) : X \times X^* \to \mathbb{K}$ given by

$$(x, j) = j(x) \ \text{ for all } x \in X \text{ and } j \in X^*.$$

At this juncture, we make an interesting observation. Let $(X, \| \cdot \|)$ be a normed space. Then, $(X^*, \| \cdot \|_*)$ is a Banach space. Let $j \in X^*$. Then, j is regarded as a function on X because it "acts" on any $x \in X$ to give a scalar $j(x)$. We look the other way round. If we fix $x \in X$ and vary j in X^*, we can regard x as a function on X^* because it "acts" on $j \in X^*$ to yield the scalar $j(x)$. Thus, each $x \in X$ can be viewed as a functional on X^* whose value at j is $j(x)$. In a more precise manner, we can say that each $x \in X$ defines a functional J_x on X^* given by

$$J_x(j) = (x, j) \ \ j \in X^*.$$

We show that this functional is linear and bounded, and so $J_x \in X^{**}$. It is linear because

$$J_x(\alpha j_1 + \beta j_2) = (\alpha j_1 + \beta j_2)(x) = \alpha j_1(x) + \beta j_2(x) = \alpha J_x(j_1) + \beta J_x(j_2).$$

Moreover, for $j \in X^*$ we have

$$|J_x(j)| = |j(x)| \leqslant \|j\|_* \|x\|.$$

This shows that J_x is bounded and hence J_x is a bounded linear functional on X^*. This initiate the following definition.

Definition 1.25 (*Natural embedding mapping*) A mapping $\varphi : X \to X^{**}$ defined by $\varphi(x) = J_x$, $x \in X$ is called the natural embedding mapping from X into X^{**}. It has the following properties:

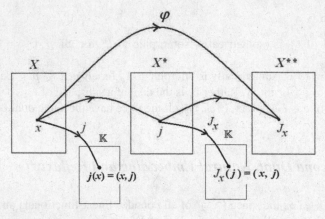

Fig. 1.7 Natural embedding mapping

(*i*) φ is linear: $\varphi(\alpha x + \beta y) = \alpha\varphi(x) + \beta\varphi(y)$ for all $x, y \in X, \alpha, \beta \in K$;
(*ii*) $\varphi(x)$ is isometry: $\|\varphi(x)\| = \|x\|$ for all $x \in X$ (Fig. 1.7).

1.4.4 Multilinear Mappings

In the following $X_1, X_2, \ldots, X_n$ and Y are real Banach spaces.

Definition 1.26 A mapping $A : X_1 \times X_2 \times \cdots \times X_n \to Y$ is said to be multilinear if it is linear in each of the variables separately. For $n = 2$, such a mapping is called a bilinear mapping.

Definition 1.27 The multilinear mapping $A : X_1 \times X_2 \times \cdots \times X_n \to Y$ is said to be bounded if there exists $\mu \geqslant 0$ such that

$$\|A(x_1, x_2, \ldots, x_n)\| \leqslant \mu \|x_1\| \|x_2\| \cdots \|x_n\|.$$

The infimum of such μ is called the *norm* of the operator A. One can show that A is continuous iff A is bounded. $\mathcal{L}(X_1, X_2, \ldots, X_n; Y)$ will denote the Banach space of n-linear continuous mappings of $X_1 \times X_2 \times \cdots \times X_n$ to Y. If $X_1 = X_2 = \cdots = X_n = X$, then we write $\mathcal{L}(X^n, Y)$ for the same.

We have the following theorem immediately from the definition.

Theorem 1.20 *There is an isometric isomorphism between* $\mathcal{L}(X_1, X_2; Y)$ *and* $\mathcal{L}(X_1, \mathcal{L}(X_2, Y))$.

From the above theorem, we inductively have the following relationship:

$$\mathcal{L}(X_1, X_2, X_3; Y) = \mathcal{L}(X_1, \mathcal{L}(X_2, \mathcal{L}(X_3, Y))) \text{ and}$$
$$\mathcal{L}(X^n, Y) = \mathcal{L}(X, \mathcal{L}(X, \ldots, \mathcal{L}(X, Y)) \cdots).$$

1.4.5 Reflexive Spaces

A Banach space X is said to be *reflexive* if the natural embedding $J : X \to X^{**}$ is onto, that is, $J(X) = X^{**}$. In other words, X is *reflexive* if the elements of X represent all the bounded linear functionals on X^* and this amounts to saying that for each $\varphi \in X^{**}$, there is an $x \in X$ such that $\varphi(f) = f(x)$ for each $f \in X^*$. In this case, we write $X \cong X^{**}$ or $X = X^{**}$.

Notice that any reflexive normed space is complete and, hence, is a Banach space. Note also that we speak of reflexivity of Banach spaces only because X^{**} is always complete. Thus, reflexivity implies that the space is Banach.

Example 1.30 (1) ℓ_p^n for $1 \leqslant p \leqslant \infty$ are reflexive spaces.
(2) Every finite dimensional space is reflexive. To see this, suppose $\dim X = n$. Because X is n-dimensional, it follows that X^* is also n-dimensional and so is X^{**}.
(3) Every Hilbert space $\mathcal{H}$ is a reflexive Banach space, i.e., $\mathcal{H}^{**} = \mathcal{H}$.

Observation

- $\mathbb{R}^n$ is reflexive.
- ℓ_p and L_p for $1 < p < \infty$ are reflexive Banach spaces.
- ℓ_1 and ℓ_∞ are not reflexive.
- c and c_0 are not reflexive Banach spaces.
- $C(\Omega)$, $L_1(\Omega)$ and $L_\infty(\Omega)$ are not reflexive.
- A Hilbert space has a countable orthonormal basis if and only if it is separable.

1.4.5.1 Characterization of the Class of Reflexive Banach spaces

Proposition 1.8 (a) *The dual of a reflexive Banach space is reflexive.*
(b) *A closed subspace of a reflexive Banach space is reflexive.*
(c) *The Cartesian product of two reflexive Banach spaces is reflexive.*

Let X be a Banach space and $j \in X^*$. By definition, $\|j\|_*$ is the supremum of the values of j over the closed unit ball $\mathbb{B}_X$. One may notice that j may or may not attain this supremum on $\mathbb{B}_X$. The following theorem shows that if X is reflexive, then j does attain this supremum on $\mathbb{B}_X$.

Theorem 1.21 (James theorem) *If X is a reflexive Banach space, then any $f \in X^*$ attains its norm on the unit ball $\mathbb{B}_X$.*

Proof Let X be a reflexive Banach space and $f \in X^*$. If $f = 0$, then its norm 0 is attained at each vector on the unit ball $\mathbb{B}_X$. So, assume that $f \neq 0$. By the Hahn–Banach theorem (Cor. 1.3) there exists an $F \in X^{**}$ such that $\|F\| = 1$ and $F(f) = \|f\|$. Because X is reflexive, there exists $x \in X$ such that $J_x = F$. Then,

$$\|x\| = \|F\| = 1 \text{ and } f(x) = F(f) = \|f\|.$$

This completes the proof.

Notice that the converse of the above theorem is also true. James has shown that if X is a Banach space and every $f \in X^*$ attain its norm on $\mathbb{B}_X$, then X is reflexive.

Theorem 1.22 *A Banach space X is reflexive if and only if any $f \in X^*$ attain its norm on the unit ball $\mathbb{B}_X$.*

1.4.6 Weak Topologies

Recall the following: If $(X, \| \cdot \|)$ is a normed linear space, then the topology $\mathscr{T}$ induced from the metric $d(x, y) = \|x - y\|$ is called the norm topology, the strong topology, or the topology induced from $\| \cdot \|$.

Let X be a Banach space and X^* its dual. The convergence of a sequence $\{x_n\}$ to x in a Banach space X is usual *norm convergence* or *strong convergence*. That is, $x_n \to x$ if $\lim_{n\to\infty} \|x_n - x\| = 0$. This is concern with the strong topology on X with neighborhood base $B(0, r) = \{x \in X : \|x\| < r\}$, $r > 0$ at the origin.

Notice that there is also a weak topology on X generated by the bounded linear functionals on X. We say that $G \subset X$ is open (or w-open) in the weak topology if and only if for every $x \in G$, there are bounded linear functionals $f_1, f_2, \ldots, f_n$ and positive real numbers $\varepsilon_1, \varepsilon_2, \ldots, \varepsilon_n$ such that

$$\{y \in X : |f_i(x) - f_i(y)| < \varepsilon_i, \ i = 1, 2, \ldots, n\} \subset G.$$

This shows that a subbase σ for the weak topology on X generated by a base of neighborhoods of x_0 in X is the set

$$V(f_1, f_2, \ldots, f_n : \varepsilon) = \{x \in X : |f_i(x) - f_i(x_0)| < \varepsilon, \ \text{for every } i = 1, 2, \ldots, n\}.$$

Notice that, in particular, a sequence $\{x_n\}$ in X converges to $x \in X$ for weak topology $\sigma(X, X^*)$ on X if and only if $(x_n, f) \to (x, f)$ for all $f \in X^*$.
Observation

- The weak topology of a normed space is a Hausdorff topology.
- The weak topology is not metrizable if X is infinite-dimensional.
- The normed space X with the weak topology $\mathscr{T}^*$ is a locally convex topological space.

Weakly convergent – A sequence $\{x_n\}$ in a normed space X is said to converge weakly to $x \in X$ if $f(x_n) \to f(x)$ for all $f \in X^*$. We denote this fact by writing $x_n \rightharpoonup x$ or weak-$\lim_{n\to\infty} x_n = x$.
Weak Cauchy sequence – A sequence $\{x_n\}$ in a normed space X is said to be a weak Cauchy if for each $f \in X^*$, $\{f(x_n)\}$ is a Cauchy sequence in $\mathbb{K}$.
Weakly complete – A normed space X is said to be weakly complete if every weak Cauchy sequence in X converges weakly to some element in X.
Weakly closed – A subset C of a Banach space X is said to be weakly closed if it is closed in the weak topology.

Weakly compact – A subset C of a normed space X is said to be weakly compact if C is compact in the weak topology.

Weakly sequentially compact – A nonempty subset C of a Banach space X is said to be weakly sequentially compact if each sequence $\{x_n\}$ in C has a subsequence $\{x_{n_i}\}$ that converges weakly to a point in C.

1.4.7 Basic Properties of Weakly Convergent Sequences in Normed Spaces

Proposition 1.9 (Uniqueness of weak limit) *Let* $(X, \| \cdot \|)$ *be a normed space and* $\{x_n\}$ *a sequence in X such that* $x_n \rightharpoonup x$ *and* $x_n \rightharpoonup y$. *Then,* $x = y$.

Proof By hypotheses, $x_n \rightharpoonup x$ and $x_n \rightharpoonup y$. Therefore, the sequence of scalars $\{f(x_n)\}$ converge to both $f(x)$ and $f(y)$ for every $f \in X^*$. By uniqueness of limit for a sequence of scalars we must have $f(x) = f(y)$. This implies that $f(x) - f(y) = 0$, i.e., $f(x - y) = 0$. Because $f(x - y) = 0$ for every $f \in X^*$ it follows that $x - y = 0$. Therefore, $x = y$. ∎

Notice that strong convergence always implies weak convergence.

Proposition 1.10 *Let* $(X, \| \cdot \|)$ *be a normed space and* $(X^*, \| \cdot \|_*)$ *its first dual space. Let* $\{x_n\}$ *be a sequence in X such that* $x_n \to x$. *Then,* $x_n \rightharpoonup x$.

Proof By hypotheses, $x_n \to x$, i.e., $\|x_n - x\| \to 0$ as $n \to \infty$. Hence for every $f \in X^*$, we have

$$|f(x_n) - f(x)| = |f(x_n - x)| \leqslant \|f\|_* \|x_n - x\| \to 0.$$

Therefore, $x_n \rightharpoonup x$. ∎

However, the converse of Proposition 1.10 is not true in general. To see this, consider the following example.

Example 1.31 Let $X = \ell_2$ and $\{x_n\}_{n=1}^{\infty}$ be a sequence in ℓ_2 such that

$$x_1 = (1, 0, 0, \ldots, 0, 0, \ldots)$$
$$x_2 = (0, 1, 0, \ldots, 0, 0, \ldots)$$
$$\vdots$$
$$x_n = (0, 0, 0, \ldots, 1, 0, \ldots)$$
$$\uparrow$$
$$n^{th}\text{position}$$
$$\vdots$$

For any $y = (y_1, y_2, \ldots, y_n, \ldots) \in X^* = \ell_2$, we have

$$(x_n, y) = y_n \to 0 \text{ as } n \to \infty.$$

Hence $x_n \rightharpoonup 0$. However, we notice that $\{x_n\}$ does not converge strongly because $\|x_n\| = 1$ for all $n \in \mathbb{N}$. Thus, a weakly convergent sequence need not be convergent in norm.

To see a necessary and sufficient for the weak convergence in ℓ_p spaces, $1 < p < \infty$, let us observe the following proposition.

Proposition 1.11 *For $1 < p < \infty$, let $\{x_n\}$ be a sequence in ℓ_p, where*

$$x_n = (\alpha_1^{(n)}, \alpha_2^{(n)}, \dots, \alpha_i^{(n)}, \dots) \in \ell_p, \quad n \in \mathbb{N}$$

and

$$x = (\alpha_1, \alpha_2, \dots, \alpha_i, \dots) \in \ell_p.$$

Then, $x_n \rightharpoonup x$ if and only if

(i) $\{x_n\}$ is bounded, i.e., $\exists M > 0$ such that $\|x_n\| \leqslant M$ for all $n \in \mathbb{N}$;
(ii) for each i, $\alpha_i^{(n)} \to \alpha_i$ as $n \to \infty$.

Weak* topology – Let $(X, \|\cdot\|)$ be the normed space with norm $\|\cdot\|$ and let $\mathscr{T}$ be the norm topology of X. Let $\mathscr{T}^*$ be the norm topology of X^* generated by the norm $\|\cdot\|_*$ of X^*. Then, the weak topology $\sigma(X, X^*)$ is a subset of the original norm topology $\mathscr{T}$ and there exists a topology denoted by $\sigma(X^*, X)$ on X^* such that $\sigma(X^*, X) \subset \mathscr{T}^*$. The topology $\sigma(X^*, X)$ is called the weak* topology on X^*. The notion of weak* topology on X^* facilitate the concepts of strong neighborhood, strongly closed, strongly bounded, weakly convergence in $(X^*, \|\cdot\|_*)$ and weak* neighborhood, weak*ly closed, weak*ly bounded, weak*ly convergence in $(X^*, \sigma(X^*, X))$, respectively.

Observation

- In a finite dimensional normed space, strong convergence is equivalent to weak convergence.
- Every reflexive normed space is weakly complete.

1.4.8 The Spaces c, c_0, ℓ_1, ℓ_∞ and Schur Property

Notice that the dual spaces of both c and c_0 are isometrically isomorphic to ℓ_1. Moreover, the dual space of ℓ_1 is isometrically isomorphic to ℓ_∞, i.e., $\ell_1^* = \ell_\infty$. These facts lead to three interesting questions.

1. If $\{x_n\}_{n=1}^\infty$ is a bounded sequence in ℓ_1 when does $\{x_n\}_{n=1}^\infty$ converge weakly?
2. When does $\{x_n\}_{n=1}^\infty$ converge in the weak* topology regarding ℓ_1 as the dual of c?

3. When does $\{x_n\}_{n=1}^{\infty}$ converge in the weak* topology regarding ℓ_1 as the dual of c_0?

The answer to these questions help us to sort out subtle differences in the concepts. Moreover, it is quite interesting to note that the answer to the first question is a little surprising. A bounded sequence $\{x_n\}_{n=1}^{\infty}$ in ℓ_1 converges weakly to $x \in \ell_1$ if and only if $\lim_{n\to\infty} \|x_n - x\| = 0$. Spaces which have this property are said to have the Schur property. Thus, the weakly compact subsets of such spaces coincide with the norm compact subsets.

Schur property[12] – A normed space $(X, \|\cdot\|)$ is said to satisfy Schur property if every weakly convergent sequences in X converges in norm; that is, if every sequence $\{x_n\}$ in X such that $x_n \rightharpoonup x$ as $n \to \infty$ implies $\lim_{n\to\infty} \|x_n - x\| = 0$.

To answer the later two questions, we first show that up to isomorphism $c^* = c_0^* = \ell$. To prove this, suppose $y = (y_1, y_2, \ldots) \in \ell_1$ and $x = (x_1, x_2, \ldots) \in c_0$, and define

$$y(x) = \sum_{i=1}^{\infty} x_i y_i.$$

This defines a linear functional. Also,

$$|y(x)| \leqslant \sum_{i=1}^{\infty} |x_i y_i| \leqslant \sup_i |x_i| \sum_{i=1}^{\infty} |y_i| = \|x\|_\infty \|y\|_1.$$

This shows that y is a bounded (continuous) linear functional with $\|y\|_\infty \leqslant \|y\|_1$. Now let $\varepsilon > 0$ and choose N so that

$$\sum_{i=1}^{N} |y_i| > \|y\|_1 - \varepsilon.$$

Let

$$x = (\mathrm{sgn} y_1, \mathrm{sgn} y_2, \ldots, \mathrm{sgn} y_N, 0, \ldots) \in c_0,$$

where "sgn" denotes the function which assigns to a number 1, -1, or 0 according to whether the number, respectively, is positive, negative, or zero. Then,

$$y(x) = \sum_{i=1}^{N} |y_i| > \|y\|_1 - \varepsilon$$

with $\|x\|_\infty = 1$. It follows that $\|y\|_\infty \geqslant \|y\|_1$; thus, $\|y\|_\infty = \|y\|_1$.

---

[12]Schur's property, named after Issai Schur, is the property of normed spaces that is satisfied precisely if weak convergence of sequences entails convergence in norm.

On the other hand, if $f \in c_0^*$ and $x = (x_1, x_2, \ldots) \in c_0$, then since the vectors $\{e_i\}$ form a basis for c_0 we have $x = \lim\limits_{n \to \infty} \sum\limits_{i=1}^{n} x_i e_i$. This yields

$$f(x) = f\left(\lim_{n \to \infty} \sum_{i=1}^{n} x_i e_i \right) = \lim_{n \to \infty} \sum_{i=1}^{n} x_i f(e_i) = \sum_{i=1}^{\infty} x_i f(e_i).$$

It follows that $\sum_{i=1}^{\infty} |x_i f(e_i)|$ exists, i.e., $y = (f(e_1), f(e_2), \ldots \in \ell_1$. Therefore, $\|y\|_1 \leqslant \|f\|_*$. Further, we have

$$|f(x)| = |\sum_{i=1}^{\infty} x_i f(e_i)| \leqslant \sup_i |x_i| \sum_{i=1}^{\infty} |f(e_i)| = \|x\|_\infty \|y\|_1.$$

This yields $\|f\|_* \leqslant \|y\|_1$. Hence, $\|y\|_1 = \|f\|_*$.

Therefore, the mapping $\varphi : c_0^* \to \ell_1$ defined by

$$\varphi(f) = (f(e_1), f(e_2), \ldots)$$

is an isometry of c_0^* onto ℓ_1.

Because the vectors $\{e_i\}$ do not form a basis for c rather $\{e_i\}$ along with the vector $e = (1, 1, \ldots)$ form a basis for c, a subtlety arises when we see that ℓ_1 is also isomorphic to c^*. Note that it can easily be shown that the mapping $\Phi : c^* \to \ell_1$ defined by

$$\Phi(f) = \left(f(e) - \sum_{i=1}^{\infty} f(e_i), f(e_1), f(e_2), \ldots \right)$$

is an isometry of c^* onto ℓ_1.

Observation

- In the case of $c^* = \ell_1$, the convergence of $\sum\limits_{i=1}^{\infty} f(e_i)$ is assured because $f \in c^*$ implies $f \in c_0^*$; hence $\varphi(f) = (f(e_1), f(e_2), \ldots) \in \ell_1$.
- By similar reasoning as above we can show that $\ell_1^* = \ell_\infty$.

Note that the basis vectors $e_i \in c_0 \subset c, i = 1, 2, \ldots$. If $x = (x_1, x_2, \ldots) \in \ell_1$, then the function defined by $e_i(x) = x_i$ can be thought of as an element of ℓ_1^* which is identified under the natural isomorphism with an element of c_0 (or c). Thus, a necessary condition for a bounded sequence in ℓ_1 to converge in the weak* topology induced on ℓ_1 by either c_0 or c is that it converge coordinatewise. In the case of $\ell_1 = c*$, however, coordinatewise convergence is also a sufficient condition for weak* convergence of bounded sequences in ℓ_1. This is because $\{e_i\}$ is a basis for c_0.

In the case of $\ell_1^* = \ell_\infty$, a bounded sequence in ℓ_∞ converges in the weak* topology induced by ℓ_1 if and only if it converges coordinatewise to an element of ℓ_∞. Similarly, weak convergence of a bounded sequence in ℓ_1 implies coordinatewise convergence.

We also notice that a bounded sequence $\{x^n\}$ in c_0 converges weakly if and only if it converges coordinatewise to $x \in c_0$, while a bounded sequence $\{x^n\}$ in c converges weakly to $x \in c$ if and only if it converges coordinatewise to x and $\lim_{n \to \infty} \lim_{i \to \infty} x_i^n = \lim_{\to \infty} x_i$.

Lemma 1.1 (Schur's Lemma) *Let* $\{x^n\} = \{(x_1^n, x_2^n, \ldots)\}$ *be a bounded sequence in* ℓ_1 *which converges to* 0. *Then,* $\lim_{n \to \infty} \|x^n\|_1 = 0$.

Notice that Schur's Lemma was instrumental in resolving an early fundamental question in functional analysis, namely whether every infinite-dimensional Banach space must necessarily contain an infinite-dimensional reflexive subspace. The answer is negative because Schur's Lemma implies that every reflexive subspace of ℓ_1 must be finite dimensional.

1.4.9 Convergence of Sequences in X^* w.r.t. Different Topologies

Let $(X, \| \cdot \|)$ be a normed space, $(X^*, \| \cdot \|_*)$ its dual space and $\{f_n\}$ a sequence in X^*. Then,
(i) $\{f_n\}$ converges to f in the norm topology on X^* (denoted by $f_n \to f$) if

$$\|f_n - f\|_* \to 0.$$

(ii) $\{f_n\}$ converges to f in the weak topology on X^* (denoted by $f_n \rightharpoonup f$) if

$$(f_n - f, g) \to 0 \text{ for all } g \in X^{**}.$$

(iii) $\{f_n\}$ converges to f in the weak* topology on X^* (denoted by $f_n \to f$ weak*ly or $f_n \rightharpoonup^* f$) if

$$(x, f_n - f) \to 0 \text{ for all } x \in X.$$

Notice that the weak topology is weaker than the norm topology, and every weakly closed set is also norm closed. The following result shows that for convex sets, the converse is also true.

Proposition 1.12 *Let C be a convex subset of a normed space X. Then, C is weakly closed iff C is closed.*

We now state a celebrated theorem in the Hilbert space that is called the Riesz representation theorem. This theorem, indeed, demonstrates that any bounded linear functional on a Hilbert space $\mathcal{H}$ can be as an inner product with a unique element in $\mathcal{H}$.

Theorem 1.23 (The Riesz representation theorem) *Let $(\mathcal{H}, \langle \cdot, \cdot \rangle)$ be a Hilbert and $f \in \mathcal{H}^*$. Then, for each $x \in \mathcal{H}$, there exists a unique element $y \in \mathcal{H}$ such that*

$$f(x) = \langle x, y \rangle.$$

Moreover, $\|f\|_* = \|y\|$.

Remark 1.8 In a Hilbert space $\mathcal{H}$ and its dual $\mathcal{H}^*$, there is one-one correspondence between $f \in \mathcal{H}^*$ and $y \in \mathcal{H}$. It follows, therefore, that $\mathcal{H}^* = \mathcal{H}$.

1.5 Compactness

Let (X, d) be a metric space and let $C \subset X$. A family of sets $\{U_i\}_{i \in I}$, I being an index set, is called a *cover* (or *covering*) of C if $C \subset \bigcup_{i \in I} U_i$. A cover of a metric space X by open sets of X is called an *open cover* of X. A subfamily of a cover which itself form a cover is called a *subcover*.

Definition 1.28 A metric space X is said to be *compact* if every open cover of X has a finite subcover. Equivalently, a subset C of X is compact if every sequence in C contains a convergent subsequence with a limit in C.

Definition 1.29 A subset C of a metric space X is said to be *totally bounded* (or *precompact*) if for each $\varepsilon > 0$, there exists a finite number of elements $x_1, x_2, \ldots, x_n$ in X such that $C \subset \bigcup_{i=1}^{n} B(x_i, \varepsilon)$. The set $\{x_1, x_2, \ldots, x_n\}$ is called a finite ε-net.

Observation

- $X = (0, 1)$ with usual metric is totally bounded, but not compact.
- $X = \mathbb{R}$ with usual metric is not totally bounded and hence not compact.
- Every subset of a totally bounded set is totally bounded.
- Every totally bounded set is bounded. However, a bounded set need not be totally bounded.

Proposition 1.13 *A metric space is totally bounded if and only if every infinite sequence in it has a Cauchy subsequence.*

Proposition 1.14 *A totally bounded metric space is separable.*

Proof Let X be a totally bounded metric space. Then, for any $n \in \mathbb{N}$, there is a finite $\frac{1}{n}$-net A_n in X. Then, the countable union of finite $\frac{1}{n}$-nets $\bigcup_{n=1}^{\infty} A_n$ is countable. Clearly it is dense in X. Hence X is separable.

Proposition 1.15 *For a metric space X, the following statements are equivalent:*

  (i)   *X is compact,*
 (ii)   *every sequence in X has a convergent subsequence,*
(iii)   *X is totally bounded and complete.*

A topological space in which every sequence has a convergent subsequence is called *sequentially compact*. The above proposition shows that sequential compactness is equivalent to compactness in the case of a metric space.

Proposition 1.16 *Let C be a subset of a complete metric space. Then, C is compact if and only if C is closed and totally bounded.*

Theorem 1.24 *Continuous image of a compact metric space is compact.*

Notice that this result is true for all topological spaces.

A subset C of a topological space is said to be *relatively compact* if its closure is compact, i.e., $\overline{C}$ is compact.

Theorem 1.25 (The Heine–Borel theorem) *A subset C of $\mathbb{R}$ is compact if and only if it is closed and bounded.*

Corollary 1.5 *A subset C of $\mathbb{R}^n$ is compact if and only if it is closed and bounded.*

Observation

- $C = [0, 1] \subset \mathbb{R}$ is compact, but $(0, 1) \subset \mathbb{R}$ is not compact.
- $\mathbb{R}^n$, $n \geq 1$ is not compact. However, every closed and bounded subset of $\mathbb{R}^n$ is compact.
- $\mathbb{R}^n$ is closed.
- The subset $C = \{\{x_n\} \in \ell_2 : |x_n| \leqslant \frac{1}{n}, n \in \mathbb{N}\}$ of ℓ_2 is compact.
- $C[0, 1]$ and ℓ_2 are not compact.
- The closed unit ball $\mathbb{B}_X = \{x \in X : \|x\| \leqslant 1\}$ in a finite dimensional normed space $(X, \|\cdot\|)$ is compact in the topology induced by the norm $\|\cdot\|$.
- The closed unit ball $\mathbb{B}_X = \{x \in X : \|x\| \leqslant 1\}$ in infinite-dimensional normed space $X = \ell_2$ is not compact in the topology induced by $\|\cdot\|_2$ norm defined by $\|x\|_2 = \left(\sum_{i=1}^{\infty} |x_i|^2 \right)^{1/2}$.

Theorem 1.26 *Let T be a continuous mapping from a compact metric space (X, d) into another metric space (Y, ρ). Then, T is uniformly continuous.*

Theorem 1.27 (Mazur's theorem) *The closed convex hull $\overline{co}(C)$ of a compact set C of a Banach space is compact.*

Proposition 1.17 *A normed space X is finite dimensional if and only if every closed and bounded subset of X is compact.*

Observation

In a normed space X it is easy to observe that

- every compact subset of X is closed, but the converse need not be true.
- every compact subset of X is bounded, but the converse need not be true.
- every compact subset of X is complete, but the converse may not be true.
- every compact subset of X is separable.

Definition 1.30 A subset of a topological space is called relatively compact if its closure is compact. A topological space is said to be locally compact if each point of the space has a compact neighborhood.

Theorem 1.28 (The Eberlein–Smulian theorem) *Let C be a weakly closed subset of a Banach space X. Then, C is weakly compact iff C is weakly sequentially compact, i.e., each sequence $\{x_n\}$ in C has a subsequence that converges weakly to a point in C.*

Theorem 1.29 (Kakutani's theorem) *Let X be a Banach space. Then, X is reflexive iff the closed unit ball $\mathbb{B}_X = \{x \in X : \|x\| \leqslant 1\}$ is weakly compact.*

Using Kakutani's theorem and Proposition 1.12, we obtain the following:

Theorem 1.30 *Let X be a Banach space. Then, X is reflexive iff every closed convex bounded subset of X is weakly compact.*

Theorem 1.31 *Let X be a reflexive Banach space and C a subset of X. Then, C is weakly compact (compactness in weak topology) iff C is bounded (boundedness in strong topology).*

Observation

1. If $\{x_n\}$ is a sequence in a Banach space X with $x_n \rightharpoonup x \in X$ and $\{\alpha_n\}$ a sequence of scalars such that $\alpha_n \to \alpha$, then $\{\alpha_n x_n\}$ converges weakly to αx.
2. If C is a nonempty subset of a Banach space X and $\{x_n\}$ a sequence in C such that $x_n \rightharpoonup x \in X$, then $x \in \overline{co}(C)$.
3. A convex subset C of a normed space X is weakly closed if and only if C is closed.
4. A weakly compact subset C of a Banach space X is bounded.
5. Any closed convex subset of a weakly compact set is itself weakly compact.

We now summarize several properties that characterize reflexivity.

Theorem 1.32 *Let X be a Banach space. Then, the following statements are equivalent:*

(a) *X is reflexive.*
(b) *$\mathbb{B}_X$ is weakly compact.*
(c) *Every bounded sequence in X in strong topology has a weakly convergent subsequence.*
(d) *X^* is reflexive.*
(e) *For any $f \in X^*$, there exists $x \in \mathbb{B}_X$ such that $f(x) = \| f \|_*$*
(e) *If $\{C_n\}$ is any descending sequence of nonempty closed convex subset of X, then $\bigcap\limits_{n=1}^{\infty} C_n \neq \varnothing$.*
(f) *$\sigma(X^*, X) = \sigma(X^*, X^{**})$, i.e., on X^* weak topology and weak* topology coincide.*

The following theorem is the fundamental result concerning the weak* topology.

Theorem 1.33 (Banach–Alaoglu's theorem) *Let X be a normed space and X^* be its dual. Then, the unit ball $\mathbb{B}_{X^*}$ in X^* is compact in the weak* topology.*

1.6 Sequence of Functions

Definition 1.31 (*Uniformly bounded sequence*) A sequence $\{f_n\}_{n \in \mathbb{N}}$ of continuous functions on an interval $I = [a, b]$ is called uniformly bounded if there is a number M such that

$$|f_n(x)| \leqslant M$$

for every function f_n belonging to the sequence, and every $x \in [a, b]$.

Definition 1.32 (*Equicontinuous sequence*) A sequence $\{f_n\}_{n \in \mathbb{N}}$ is called equicontinuous if, for every $\varepsilon > 0$, there exists $\delta > 0$ such that

$$|f_n(x) - f_n(y)| < \varepsilon$$

whenever $|x - y| < \delta$ for all functions f_n in the sequence.

The Ascoli–Arzelà theorem stated below is a fundamental result of mathematical analysis giving necessary and sufficient conditions to decide whether every sequence of a given family of real-valued continuous functions defined on a closed and bounded interval has a uniformly convergent subsequence.

It may be remarked that the main condition is the equicontinuity of the family of functions. The theorem is the basis of many proofs in mathematical analysis, including that of the Peano existence theorem in the theory of ordinary differential equations, Montel's theorem in complex analysis, and the Peter–Weyl theorem in harmonic analysis.

Theorem 1.34 (The *Ascoli–Arzelà* theorem) *If a sequence of real-valued continuous functions* $\{f_n\}_{n \in \mathbb{N}}$ *defined on a closed and bounded interval* $[a, b]$ *of the real line is uniformly bounded and equicontinuous, then there exists a subsequence* $\{f_{n_k}\}_{k=1}^{\infty}$ *that converges uniformly.*

Notice that the converse of Theorem 1.34 is also true, in the sense that if every subsequence of $\{f_n\}$ itself has a uniformly convergent subsequence, then $\{f_n\}$ is uniformly bounded and equicontinuous.

1.6.1 Lipschitz and Hölder Continuous Functions

Definition 1.33 (*Lipschitz continuous function*) A sequence $\{f_n\}$ of real-valued functions on $[a, b]$ is said to be Lipschitz continuous with the same Lipschitz constant K if:

$$|f_n(x) - f_n(y)| \leqslant K|x - y| \text{ for all } x, y \in [a, b] \text{ and all } f_n.$$

Definition 1.34 (*Hölder condition function*) A set $\mathcal{F}$ of functions f on $[a, b]$ is said to satisfies a Hölder condition of order α, $0 < \alpha \leqslant 1$, with a fixed constant M, if

$$|f(x) - f(y)| \leqslant M\,|x - y|^{\alpha}, \qquad x, y \in [a, b].$$

Observation

- If $\{f_n\}$ is a uniformly bounded sequence of real-valued functions on $[a, b]$ such that each f is Lipschitz continuous with the same Lipschitz constant K, i.e.,

$$|f_n(x) - f_n(y)| \leqslant K|x - y| \text{ for all } x, y \in [a, b] \text{ and all } f_n,$$

then there is a subsequence that converges uniformly on [a, b].
- A set $\mathcal{F}$ of functions f on $[a, b]$ that is uniformly bounded and satisfies a Hölder condition of order α, $0 < \alpha \leqslant 1$, with a fixed constant M, i.e.,

$$|f(x) - f(y)| \leqslant M\,|x - y|^{\alpha}, \qquad x, y \in [a, b]$$

is relatively compact in $C([a, b])$. In particular, the unit ball of the Hölder space $C^{0,\alpha}([a, b])$ is compact in $C([a, b])$.

1.6.2   Various Forms of Continuity of Mappings

In what follows, X and Y denote real Banach spaces and F is an operator (not necessarily linear) with Domain in X and range in Y. The domain and the range of F are denoted by $\mathcal{D}(F)$ and $\mathcal{R}(F)$, respectively. We say that F is an operator on X if $X = Y$, otherwise F is an operator from X into Y. The first and second dual spaces of X are denoted by X^* and X^{**}, respectively.

Definition 1.35 (*Convergent sequence*) A sequence $\{x_n\}$ in a Banach space X is said to be convergent to x_0 if

$$\lim_{n \to \infty} \|x_n - x_0\| = 0.$$

If $\{x_n\}$ converges to x_0, then this fact is denoted by $x_n \to x_0$.
$\{x_n\}$ converges weakly to x_0 if $f(x_n)$ converges to $f(x_0)$ in $\mathbb{R}$, i.e.,

$$\lim_{n \to \infty} |f(x_n) - f(x_0)| = 0,$$

for every linear functional $f \in X^*$.
   If $\{x_n\}$ converges weakly to x_0, then this fact is denoted by $x_n \rightharpoonup x_0$.

Definition 1.36 Let $F : \mathcal{D}(F) \subseteq X \to Y$ be a (possibly) nonlinear operator and let $x_0 \in \mathcal{D}(F)$. Then,
(a) F is said to be continuous at x_0 if for any sequence $\{x_n\}$ in $\mathcal{D}(F)$ which converges to x_0, the sequence $\{Fx_n\}$ converges to Fx_0 in Y. In other words, F is said to be continuous at x_0 if

$$x_n \to x_0 \implies F x_n \to F x_0,$$

i.e., for given $\varepsilon > 0$, $\exists$ a $\delta = \delta(\varepsilon, x_0) > 0$ such that $\|F x_n - F x_0\| < \varepsilon$ whenever $\|x - x_0\| < \delta$ for all $x \in \mathscr{D}(F)$.

(b) F is said to be weakly continuous at x_0 if for any sequence $\{x_n\}$ in $\mathscr{D}(F)$ which converges weakly to x_0 the sequence $\{F x_n\}$ converges weakly to $F x_0$ in Y. In other words, F is said to be weakly continuous at x_0 if

$$x_n \rightharpoonup x_0 \implies F x_n \rightharpoonup F x_0.$$

(c) F is said to be uniformly continuous on $\mathscr{D}(F)$ if for given $\varepsilon > 0$, there exists $\delta = \delta(\varepsilon) > 0$ such that

$$\|Fx - Fy\| < \varepsilon \quad \text{whenever} \quad \|x - y\| < \delta \quad \text{for all} \quad x, y \in \mathscr{D}(F).$$

Example 1.32 Let $X = [0, 1]$ and $Y = \mathbb{R}$, and let X and Y have absolute value norms. Then, the mapping $F : \mathscr{D} \subset X \to Y$ defined by $Fx = \frac{1}{x}$, where $\mathscr{D} = (0, 1]$, is continuous but not uniformly continuous.

Definition 1.37 Let $F : \mathscr{D}(F) \subseteq X \to Y$ be a (possibly) nonlinear operator. Then, F is said to be bounded if it maps bounded subsets of $\mathscr{D}(F)$ into bounded subsets of Y.

It is well known that for linear operators boundedness is equivalent to continuity. But there do exist nonlinear operators which are bounded but not continuous and vice versa.

Example 1.33 Let $F : \mathbb{R}^2 \to \mathbb{R}$ be defined as

$$F(x_1, x_2) = \begin{cases} \frac{x_1 x_2}{x_1^2 + x_2^2}, & (x_1, x_2) \neq (0, 0) \\ 0, & (x_1, x_2) = (0, 0) \end{cases}$$

Then, F is bounded but not continuous. To see this, we notice that

$$|F(x_1, x_2)| = \begin{cases} \frac{|x_1||x_2|}{|x_1|^2 + |x_2|^2}, & (x_1, x_2) \neq (0, 0) \\ 0, & (x_1, x_2) = (0, 0). \end{cases}$$

For $(x_1, x_2) \neq (0, 0)$, we have

$$|F(x_1, x_2)| = \begin{cases} \frac{|x_2|^2}{|x_1|^2 + |x_2|^2}, & \text{if } |x_1| \leqslant |x_2| \\ \frac{|x_1|^2}{|x_1|^2 + |x_2|^2}, & \text{if } |x_2| \leqslant |x_1| \end{cases}$$

Thus, for all $(x_1, x_2) \in \mathbb{R}^2$ we have

$$|F(x_1, x_2)| \leqslant 1$$

which shows that F is bounded.

To check the continuity, we consider the path of approach to origin along $x_2 = mx_1$. Then, we see that

$$\lim_{(x_1, x_2) \to (0,0)} F(x_1, x_2) = \lim_{x_1 \to 0} \frac{mx_1^2}{(1 + m^2)x_1^2}$$

$$= \frac{m}{1 + m^2}$$

which depends upon the choice of m. Hence F is discontinuous at $(0, 0)$.

Example 1.34 Let X be an infinite dimensional Banach space. Because the ball $\|x\| \leqslant 1$ is not compact in X, there exists a sequence $\{x_n\}$ with $\|x_n\| = 1$ such that $\|x_m - x_n\| \geqslant \varepsilon, 0 < \varepsilon < 1$. Now define $F : X \to \mathbb{R}$ as follows

$$F(x) = \begin{cases} n, & x = x_n, n = 1, 2, \ldots \\ n - \frac{n}{\varepsilon}\|x - x_n\|, & \|x - x_n\| \leqslant \frac{1}{2}\varepsilon, n = 1, 2, \ldots \\ 0 & \text{otherwise.} \end{cases}$$

Then, it is easy to observe that F is continuous but is unbounded in the ball $\|x\| \leqslant 1 + \frac{1}{2}\varepsilon$.

Theorem 1.35 *Let $A : X \to Y$ be a continuous linear operator. Then, F is weakly continuous.*

Proof Let $x_n \rightharpoonup x_0$. For a functional $g \in Y^*$, define a functional $f_g \in X^*$ as $f_g(x) = g(Ax), x \in X$. For $x, y \in X$ and scalars a, b, we have

$$f_g(ax + by) = g(A(ax + by)) = g(aAx + bAy)$$
$$= ag(Ax) + bg(Ay) = af_g(x) + bf_g(y)$$

which shows that f_g is linear. It is continuous for

$$|f_g(x)| = |g(Ax)| \leqslant \|g\|\|Ax\| \leqslant \|g\|\|A\|\|x\|.$$

Because $x_n \rightharpoonup x_0, f_g(x_n) \to f_g(x_0)$ in $\mathbb{R}$. This in turn implies that

$$|g(Ax_n) - g(Ax_0)| = |f_g(x_n) - f_g(x_0)| \to 0 \text{ as } n \to \infty,$$

i.e., $g(Ax_n) \to g(Ax_0)$ in $\mathbb{R}$. Because $g \in Y^*$ was arbitrary, it follows that

$$x_n \rightharpoonup x_0 \implies Ax_n \rightharpoonup Ax_0.$$

This shows that A is weakly continuous and hence the theorem. ∎

However, there exists continuous nonlinear operators which are not weakly continuous as we see in the following example.

Example 1.35 Let $F : L_2[0, 2\pi] \to L_2[0, 2\pi]$ be defined as

$$[Fx](s) = \int_0^{2\pi} x^2(t)dt, \quad s \in [0, 2\pi].$$

Continuity of F follows from the following inequality for a sequence $\{x_n\}$ which converges to x_0 in $L_2[0, 2\pi]$ (Fig. 1.8).

$$\|Fx_n - Fx_0\|^2 = \int_0^{2\pi} \left(\int_0^{2\pi} (x_n^2(t) - x_0^2(t))dt \right)^2 ds$$

$$\leqslant \int_0^{2\pi} \left(\int_0^{2\pi} (x_n(t) + x_0(t))^2 dt \right) \left(\int_0^{2\pi} (x_n(t) - (x_0(t))^2 dt \right) ds$$

$$= \|x_n + x_0\|^2 \left(\int_0^{2\pi} (x_n(t) - (x_0(t))^2 dt \right) ds$$

$$\leqslant \left(\sup_n \|x_n + x_0\|^2 \right) \left(\int_0^{2\pi} (x_n(t) - (x_0(t))^2 dt \right) ds$$

$$= M\|x_n - x_0\|^2, \quad \text{where } M = \sup_n \|x_n + x_0\|^2.$$

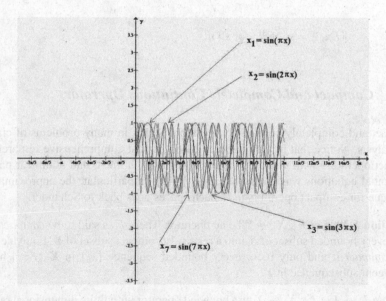

Fig. 1.8 Graphs of $\sin n\pi t$ for $n = 1, 2, 3, 7$

Because $x_n \to x_0$, it follows from the above inequality that $Fx_n \to Fx_0$, as $n \to \infty$. This shows that F is continuous. But F is not weakly continuous. For, if we consider $x_n = \sin n\pi t$, then it is easy to observe that x_n has increasing number of 0's in $[0, 2\pi]$ as n goes to infinity, but x_n is not zero function for any n. We also notice that $\{x_n\}$ does not converge to 0 in the L_∞ or L_2 norms. This dissimilarity is one of the reasons why this type of convergence is considered to be "weak".

Note that the integral

$$\int_0^{2\pi} x_n(t) y(t) dt \;\to\; 0,$$

for any square integrable function y on $[0, 2\pi]$ as $n \to \infty$, i.e., $\langle x_n, y \rangle \to \langle 0, y \rangle = 0$ as $n \to \infty$. It follows that $x_n \rightharpoonup 0$ but

$$\|Fx_n\| = \left(\int_0^{2\pi} \left(\int_0^{2\pi} \sin^2 n\pi t \, dt \right)^2 ds \right)^{1/2}$$

$$= \frac{1}{2} \left(\int_0^{2\pi} \left(\int_0^{2\pi} (1 - \cos 2n\pi t) dt \right)^2 ds \right)^{1/2}$$

$$= \frac{1}{2} \left(\int_0^{2\pi} \left(\left[t - \frac{\sin 2n\pi t}{2n\pi} \right]_0^{2\pi} \right)^2 ds \right)^{1/2}$$

$$= \pi \left(\int_0^{2\pi} 1 \, ds \right)^{1/2} = \sqrt{2}\pi^{3/2}$$

i.e., $\|Fx_n\| = \sqrt{2}\pi^{3/2}$ for all n.

1.6.3 Compact and Completely Continuous Operators

Compact and completely continuous operators occurs in many problems of classical analysis. Notice that in the nonlinear case, the first comprehensive research on compact operators with numerous applications to both linear and nonlinear partial differential equations was due to Schauder [542]. In particular, the approximation technique for compact operators in Banach spaces goes back to Schauder.

Definition 1.38 Let $F : X \to Y$ be an operator. Then, F is said to be *compact* if it maps every bounded subset of X into a relatively compact subset of Y. Equivalently, F is *compact* if and only if for every bounded sequence $\{x_n\}$ in X, $\{Fx_n\}$ has a convergent subsequence in Y.

Example 1.36 Let $F : X \to Y$ be a bounded operator with finite dimensional range. Then, F is compact.

Suppose dim $\mathcal{R}(F) = k$, then there exists a basis of $\mathcal{R}(F)$, say, $B = \{y_1, y_2, \ldots, y_k\}$ containing k independent vectors $y_1, y_2, \ldots, y_k$ such that $L(B) = \mathcal{R}(F)$. It follows that every vector in the range of F can be written as $Fx = \sum_{j=1}^{k} \alpha_j y_j$. If $\{x_n\}$ is a bounded sequence in X and $Fx_n = \sum_{j=1}^{k} \alpha_j^n y_j$, the corresponding $\{\alpha_j^n\}$ are bounded because F is bounded. As $\{\alpha_j^n\}$ has a convergent subsequence for every j, one can extract a convergent subsequence from $\{Fx_n\}$.

We have the following important characterization of compact operators.

Theorem 1.36 *Let $F : X \to Y$ be an operator and M a bounded subset of X. F is compact on M iff for each $\varepsilon > 0$ there exists a bounded operator P (dependent on ε) with finite dimensional range such that $\|Fx - Px\| < \varepsilon$ for all $x \in M$.*

Proof To prove the necessary condition, let F be a compact operator from X into Y and M a bounded subset of X. Then, $F(M)$ is relatively compact and hence totally bounded. That is, there exists a finite set of vectors $y_1, y_2, \ldots, y_n \in Y$ such that for any $x \in M$ there is a y_i which satisfied the inequality $\|Fx - y_i\| < \varepsilon$. Consider the linear span $S = [y_1, \ldots, y_n]$. Define P as

$$P(x) = \frac{1}{\sum_{j=1}^{n} \mu_j} \sum_{j=1}^{n} \mu_j y_j$$

where
$$\mu_j = \begin{cases} \varepsilon - \|Fx - y_j\|, & \text{for } \|Fx - y_j\| < \varepsilon \\ 0, & \text{for } \|Fx - y_j\| \geqslant \varepsilon. \end{cases}$$

Then, there exists an $i \in \{1, 2, \ldots, n\}$ such that $\mu_i = \varepsilon - \|Fx - y_i\| > 0$ and $\mu_j = 0$ for all $j \neq i$,

$$\|Fx - Px\| = \left\| Fx - \frac{1}{\sum_{j=1}^{n} \mu_j} \sum_{j=1}^{n} \mu_j y_j \right\| = \|Fx - y_i\| < \varepsilon.$$

Thus, the operator P satisfies the required properties with S as its range.

To prove the sufficient criterion assume that there exists an operator P satisfying the required condition for a given $\varepsilon > 0$. Suppose, if possible, F is not compact. This implies that there exists an $\varepsilon_1 > 0$ and a sequence $\{x_n\}$ in M such that

$$\|Fx_n - Fx_m\| > \varepsilon_1. \tag{1.5}$$

For our purpose we take $\varepsilon = \varepsilon_1/6$. Because P is compact there exists a subsequence $\{x_{n_k}\}$ and an integer N such that

$$\|Px_{n_k} - Px_{n_l}\| < \frac{\varepsilon_1}{6} \quad \text{for } k, l \geqslant N.$$

So we get

$$\|Fx_{n_k} - Fx_{n_l}\| \leqslant \|Fx_{n_k} - Px_{n_k}\| + \|Px_{n_k} - Px_{n_l}\| + \|Px_{n_l} - Fx_{n_l}\| < \frac{\varepsilon_1}{2} \text{ for } k, l \geqslant N,$$

contradicting (1.5).   ∎

Theorem 1.37 *Let* $F_n : X \to Y$ *be a sequence of compact operators such that* $\lim_{n \to \infty} \|F_n x - Fx\| = 0$ *uniformly for* x *in every bounded subset of* X. *Then,* F *is also compact.*

Proof Let M be any bounded subset of X. Then, there exists a number k such that $\|F_k x - Fx\| < \frac{\varepsilon}{2}$ for $x \in M$. Because $F_k(M)$ is relatively compact and hence totally bounded it follows that it has a finite $\frac{\varepsilon}{2}$-net. In view of the inequality $\|F_k x - Fx\| < \frac{\varepsilon}{2}$ for $x \in M$, this $\frac{\varepsilon}{2}$-net will be an ε-net for $F(M)$. Because Y is complete, $F(M)$ is relatively compact.   ∎

Definition 1.39 An operator $F : X \to Y$ is said to be completely continuous if for any sequence $\{Fx_n\}$ converging weakly to x, the sequence $\{Fx_n\}$ converges to Fx. That is, $x_n \rightharpoonup x \Rightarrow Fx_n \to Fx$.

Theorem 1.38 *Let F be an operator from a reflexive Banach space X into a Banach space Y. If F is completely continuous then F is continuous and compact.*

Proof Let $F : X \to Y$ be completely continuous. We need to show the compactness of F. Let M be a bounded subset of X and let $\{x_n\}$ be a sequence in M. Reflexivity of X implies that $\{x_n\}$ has a weakly convergent subsequence $\{x_{n_k}\}$. Assume that $x_{n_k} \rightharpoonup x_0$. Because F is completely continuous, $Fx_{n_k} \to Fx_0$. This proves that $F(M)$ is relatively compact.   ∎

Notice that the converse of Theorem 1.38 need not to be true, as we see in the following example.

Example 1.37 The operator F on $L_2[0, 2\pi]$ as defined by

$$[Fx](s) = \int_0^{2\pi} x^2(t)dt, \quad s \in [0, 2\pi]$$

is continuous and compact, but it is not completely continuous.
It is trivially true that F is continuous (see Example 1.34 above).
For the compactness of F, consider the ball $\overline{\mathbb{B}}_r = \{x \in L_2[0, 2\pi] : \|x\| \leqslant r\}$. For $x \in \overline{\mathbb{B}}_r$, we have

$$|Fx(s)| = \left| \int_0^{2\pi} x^2(t)dt \right| \leqslant \left| \int_0^{2\pi} r^2 dt \right| = 2\pi r^2$$

and

$$|Fx(s_1) - Fx(s_2)| = \left| \int_0^{2\pi} x^2(t)dt - \int_0^{2\pi} x^2(t)dt \right| = 0.$$

This shows that functions Fx belonging to $F(\overline{\mathbb{B}}_r) = \{y \in L_2[0, 2\pi] : \|y\| \leqslant 2\pi r^2\}$ are uniformly bounded and equicontinuous and hence by the Ascoli–Arzela theorem $F(\overline{\mathbb{B}}_r)$ contains a uniformly convergent subsequence. This in turn implies that $F(\overline{\mathbb{B}}_r)$ has a convergent subsequence in $L_2[0, 2\pi]$. That is, F is a compact operator on $L_2[0, 2\pi]$.

Furthermore, we have already proved that F is not weakly continuous and hence not completely continuous.

Observation

- For continuous linear operators, complete continuity and compactness are equivalent, provided the domain space is reflexive.

Theorem 1.39 *Let $A : X \to Y$ be a linear operator, where X is a reflexive Banach space. Then, A is completely continuous iff it is continuous and compact.*

Proof It is enough to prove the sufficient criterion. Let $x_n \rightharpoonup x_0$ in X. Because A is compact $\{Ax_n\}$ has a subsequence $\{Ax_{n_k}\}$ which converges to $y_0 \in Y$. As A is a continuous linear operator, it is also weakly continuous and hence $Ax_{n_k} \rightharpoonup Ax_0$. By uniqueness of limit, $y_0 = Ax_0$ and hence $Ax_{n_k} \to Ax_0$. Now our result follows by using a theorem on convergence of sequences which states that: $\{y_n\}$ converges to y_0 iff every subsequence of $\{y_n\}$ has in turn a subsequence which converges to y_0. ∎

For more on completely continuous linear operators, refer Rudin [534], Reed and Simon [507] and Yosida [614].

We now give an example of a class of linear integral operators which are compact. But first we have the following definitions.

Degenerate integral operator– An integral operator $K : C([0, 1]) \to C([0, 1])$ defined by

$$Kf(x) = \int_0^1 k(x, y)f(y)dy$$

is said to be degenerate if $k(x, y)$ is a finite sum of separated terms of the form

$$k(x, y) = \sum_{i=1}^n \varphi_i(x)\psi_i(y),$$

where $\varphi_i, \psi_i : [0, 1] \to \mathbb{R}$ are continuous functions. We may assume without loss of generality that $\{\varphi_1, \ldots, \varphi_n\}$ and $\{\psi_1, \ldots, \psi_n\}$ are linearly independent.

The Hilbert–Smith kernel– Given a domain (an open and connected set) Ω in n-dimensional Euclidean space $\mathbb{R}^n$, a Hilbert–Schmidt kernel is a function $k : \Omega \times \Omega \to \mathbb{C}$ with

$$\int_\Omega \int_\Omega |k(x, y)|^2 dxdy < \infty,$$

that is, the $L_2(\Omega \times \Omega, \mathbb{C})$ norm of k is finite, and the associated Hilbert–Schmidt integral operator is the operator $K : L_2(\Omega, \mathbb{C}) \to L_2(\Omega, \mathbb{C})$ given by

$$(Ku)(x) = \int_\Omega k(x, y)u(y)dy.$$

Then, K is a Hilbert–Schmidt operator with the Hilbert–Schmidt norm

$$\|K\|_{HS} = \|k\|_{L_2}.$$

Note that the Hilbert-Schmidt integral operators are both continuous (and hence bounded) and compact (as with all the Hilbert–Schmidt operators).

Example 1.38 Let A be a linear integral operator of the type

$$[Ax](s) = \int_\Omega k(s, t)x(t)\, dt, \qquad\qquad (1.6)$$

where Ω is a compact subset of $\mathbb{R}$ and K a real-valued function defined on $\Omega \times \Omega$.
Case 1. Let $k(s, t)$ be a degenerate kernel; that is,

$$k(s, t) = \sum_{j=1}^n \phi_j(s)\psi_j(t),$$

where $\phi_j(s), \psi_j(t) \in L_2(\Omega)$. Then, A is a bounded operator from $L_2(\Omega)$ to itself. For each $x \in L_2(\Omega)$, we have

$$[Ax](s) = \int_\Omega \Big[\sum_{j=1}^n \phi_j(s)\psi_j(t)x(t) \Big] dt$$

$$= \sum_{j=1}^n c_j\phi_j(s), \quad \text{where } c_j = \int_\Omega \psi_j(t)x(t)\, dt.$$

This implies that the range of A is finite dimensional and hence it is completely continuous.
Case 2. Let $k(s, t)$ be Hilbert–Schmidt; that is,

$$M = \int_\Omega \int_\Omega k^2(s, t)\, dtds < \infty.$$

A is a bounded operator on $L_2(\Omega)$ with $\|A\| \leqslant M^{1/2}$.

It is easy to show that $k(s, t)$ can be approximated in $L_2(\Omega \times \Omega)$ norm by separable kernels $k_n(s, t) = \sum_{i,j=1}^n k_{ij}\phi_i(s)\phi_j(t)$, where $\{\phi_j(s)\}_{j=1}^\infty$ is an arbitrary orthogonal set in $L_2(\Omega)$ (refer Stakgold [575]). By Case 1, A_n generated by the kernel $k_n(s, t)$, is completely continuous.

Because $A_n \to A$ (in operator norm), it follows by Theorem 1.37 that A is completely continuous. Thus, we have a class of completely continuous integral operators generated by the Hilbert–Schmidt kernels.

Remark 1.9 One can similarly show that if $\int_\Omega \int_\Omega |k(s, t)|^p \, ds dt < \infty$, then the operator generated by the kernel $k(s, t)$ is a completely continuous operator from $L_p(\Omega)$ to $L_q(\Omega)$ $(1/p + 1/q = 1)$.

We now give an example of continuous linear integral operator which is not completely continuous.

Example 1.39 Consider the convolution operator defined by the kernel $k(t)$ as

$$[Ax](s) = \int_{-\infty}^{\infty} k(s - t)x(t) \, dt.$$

If $k(t) \in L_1(-\infty, \infty)$, then A is a continuous linear operator from $L_2(-\infty, \infty)$ to itself and

$$\|Ax\| \leqslant \|k\|_{L_1} \|x\|.$$

(for reference, see Okikiolu [435]).

We will show that A is not completely continuous. For this, it suffices to show that there exists $x_n \in L_2(-\infty, \infty)$ which converges to 0 weakly while (Ax_n, x_n) does not converge to 0. Choose a function $x(s) \in L_2(-\infty, \infty)$ with support in $|s| \leqslant a$ such that

$$\int_{-\infty}^{\infty} \int_{-\infty}^{\infty} x(s)k(s - t)x(t) \, dt ds \neq 0. \tag{1.7}$$

This is possible because space of continuous functions with compact support is dense in $L_2(-\infty, \infty)$. Let us now define the sequence $x_n(s)$ as $x_n(s) = x(s - n)$. Then, we see that

$$\|x_n\| = \|x\|.$$

Let φ be a function $\in L_2(-\infty, \infty)$ with support in $|s| \leqslant b$. Then,

$$\int_{-\infty}^{\infty} \varphi(s)x_n(s) ds = \int_{-\infty}^{\infty} \varphi(s)x(s - n) ds = 0, if n > a + b.$$

Because the space of continuous functions with compact support is dense in $L_2(-\infty, \infty)$, it follows that $x_n \rightharpoonup 0$. On the other hand, we have

$$Ax(s) = \int_{-\infty}^{\infty} k(s - t)x(t) dt$$

and hence

$$Ax_n(s) = \int_{-\infty}^{\infty} k(s-t)x(t-n)dt = \int_{-\infty}^{\infty} k(s-t-n)x(t)dt.$$

So we get

$$(Ax_n, x_n) = \int_{-\infty}^{\infty} \int_{-\infty}^{\infty} x(s-n)k(s-t-n)x(t)dtds$$

$$= \int_{-\infty}^{\infty} \int_{-\infty}^{\infty} x(s)k(s-t)x(t)\,dtds \neq 0 \ \ (\text{by } (1.7)).$$

This proves our assertion.

Definition 1.40 Let $F : \mathscr{D}(F) \subseteq X \to Y$ be a (possibly) nonlinear operator and $x \in \mathscr{D}(F)$. Then,
(a) F is called demicontinuous at x if for any sequence $\{x_n\}$ converging to x, the sequence $\{Fx_n\}$ converges weakly to Fx. That is, $x_n \to x \Rightarrow Fx_n \rightharpoonup Fx$.
(b) F is called hemicontinuous at x if for any sequence $\{x_n\}$ converging to x along a line, the sequence $\{Fx_n\}$ converges weakly to Fx. That is, $F(x_n) = F(x + t_n y) \rightharpoonup Fx$ as $t_n \to 0$ for all $y \in X$.
(c) F is called locally bounded at x if the sequence $\{Fx_n\}_{n\in\mathbb{N}}$ is bounded in Y whenever $x_n \in \mathscr{D}(F)$ and $x_n \to x$.
(d) F is called locally hemibounded at x if the sequence $\{F(x+t_n y)\}_{n\in\mathbb{N}}$ is bounded in Y for every $y \in X$ and $t_n \to 0$ with $x + t_n y \in \mathscr{D}(F)$.

Remark 1.10 It is clear from the definition that F is locally bounded at $x \in \mathscr{D}(F)$ if and only if there is an open neighborhood U of x such that $F(U \cap \mathscr{D}(F))$ is bounded in Y.

Notice that every continuous operator is demicontinuous but the converse need not be true as we see in the following example.

Example 1.40 Let $X \subset L_2[0, 1]$ be defined as

$$X = \{x(t) \in L_2[0, 1] : x \text{ is absolutely continuous with } x'(t) \in L_2[0, 1] \text{ and}$$
$$x(0) = x(1) = 0\}.$$

Then, we notice that X is a dense subspace of $L_2[0, 1]$. Define an operator $A : X \to L_2[0, 1]$ as

$$[Ax](t) = x'(t).$$

It is well known that A is not a continuous operator from X to $L_2[0, 1]$. However, A is demicontinuous. Indeed, we have

$$(Ax_n, y) = \int_0^1 x_n'(t)y(t)dt = \int_0^1 x_n(t)y'(t)dt \ \text{ for all } \ y \in X.$$

Because $x_n \to 0$ in X, it follows that $\displaystyle\int_0^1 x_n(t) y'(t) dt \to 0$ and hence $(Ax_n, y) \to 0$ for all $y \in X$. Furthermore, X is dense in $L_2[0, 1]$ it follows that $(Ax_n, y) \to 0$ for all $y \in L_2[0, 1]$. That is, $Ax_n \to 0$ for all $y \in L_2[0, 1]$. This proves the demicontinuity of A.

In general we have the following relations

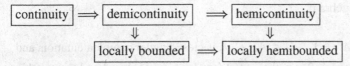

Observation

- In finite dimensional spaces demicontinuity coincides with continuity whereas hemicontinuity is just the continuity along the lines.

  Notice that every demicontinuous operator is hemicontinuous but the converse is not true as we see in the following example.

Example 1.41 Let $F : \mathbb{R}^2 \to \mathbb{R}$ be defined as

$$F(x_1, x_2) = \begin{cases} \frac{x_1^2 x_2}{x_1^4 + x_2^2}, & (x_1, x_2) \neq (0, 0) \\ 0, & (x_1, x_2) = (0, 0). \end{cases}$$

It may easily be observed that F is hemicontinuous at $(0, 0)$ but is not demicontinuous there.

Example 1.42 Every linear operator $A : X \to Y$ is hemicontinuous.

Observation

- The concepts of demicontinuity and hemicontinuity for nonlinear operators were first introduced by Browder[13] in [104] while studying the properties of monotone nonlinear operators. In Chap. 4 we discuss their interrelationships in more detail.
- From the application point of view we note that one need to impose less restrictive growth condition and regularity condition on the set of nonlinear functions generating the corresponding nonlinear operator to obtain weaker forms of continuity (refer Chap. 8) and hence the need of such a discussion in this section.

---

[13] Felix Earl Browder (July 31, 1927 - December 10, 2016) was an American mathematician known for his work in nonlinear functional analysis. He received the National Medal of Science in 1999 for his pioneering work in nonlinear functional analysis and its applications to partial differential equations, and for leadership in the scientific community. He also served as president of the American Mathematical Society from 1999 to 2000.

1.7   Measure, Measurable Function and Measure Space

In this section, we briefly review the concepts of abstract measure, measurable function and measure space. Observe that measurable sets and measurable functions play a fundamental role in integration theory.

Definition 1.41 (σ-*algebra*) Let X be a nonempty set. Then, a family $\mathcal{A}$ of subsets of X is called a σ-algebra if

(i) $X \in \mathcal{A}$,
(ii) if $A \in \mathcal{A}$, then $A^c = X - A \in \mathcal{A}$, i.e., $\mathcal{A}$ is closed under complementation, and
(iii) if $A_n \in \mathcal{A}$ for $n = 1, 2, \ldots$, then $\bigcup_{n=1}^{\infty} A_n \in \mathcal{A}$, i.e., $\mathcal{A}$ is closed under countable union and hence finite union.

Definition 1.42 (*Measurable space*) Let $\mathcal{A}$ be a σ-algebra of subsets of a nonempty set X. Then, the pair $(X, \mathcal{A})$ is called a measurable space and the elements of $\mathcal{A}$ measurable sets.

Let $(X, \mathcal{A})$ be a measurable space. Then, we observe that

(a) $\varnothing = X^c \in \mathcal{A}$, because $X \in \mathcal{A}$,
(b) if $A_n \in \mathcal{A}$ for $n = 1, 2, \ldots$, then since

$$\bigcap_{n=1}^{\infty} A_n = \bigcap_{n=1}^{\infty} (A_n^c)^c = \left(\bigcup_{n=1}^{\infty} A_n^c \right)^c,$$

it follows that $\mathcal{A}$ is closed under countable intersection and hence finite intersection,
(c) if $A_1, A_2 \in \mathcal{A}$, then since

$$A_1 - A_2 = A_1 \cap A_2^c,$$

it follows that $A_1 - A_2 \in \mathcal{A}$.

Remark 1.11 If X is any set and E a family of subsets of X, then there exists a smallest σ-algebra $\mathcal{A}^*$ in X such that $E \subset \mathcal{A}^*$.

Definition 1.43 (*Borel sets*) Let $(X, \mathcal{T})$ be a topological space. Then, there exists a smallest σ-algebra $\mathbf{B}$ in X which contains all open sets, i.e., $\mathcal{T} \subset \mathbf{B}$. The members of $\mathbf{B}$ are called Borel sets.

Observation

• All closed sets, all countable unions of closed sets and all countable intersections of open sets are Borel sets. The last two Borel sets are called respectively F_σ and G_δ sets.

Definition 1.44 (*Measurable function*) Let $(X, \mathcal{A})$ be a measurable space and $(Y, \mathcal{V})$ any topological space. Then, a mapping $f : X \to Y$ is a measurable function if $f^{-1}(V)$ is a measurable set in X for every open set V in Y.

Definition 1.45 (*Simple function*) Let $(X, \mathcal{A})$ be a measurable space and let $s : X \to [0, \infty]$ be a function. Then, s is said to be a simple function if its range is a finite set. If $c_1, c_2, \ldots, c_n$ are the distinct values of a simple function s, if

$$E_i = \{x \in X : s(x) = c_i\} \quad (i = 1, 2, \ldots, n),$$

then we have

$$s = \sum_{i=1}^{n} c_i \chi_{E_i},$$

where χ_{E_i} is the characteristic function of the set E_i.

From the above definition, it is evident that s is measurable if and only if each E_i vis measurable.

Definition 1.46 (*Measure space*) Let X be a nonempty set and $\mathcal{A}$ a σ-algebra of X. A measure μ is a nonnegative extended real-valued function, i.e., $\mu : \mathcal{A} \to [0, \infty]$ that is countably additive. That is, if $\{A_i\}$ is a disjoint countable collection of members of $\mathcal{A}$, then

$$\mu\left(\bigcup_{i=1}^{\infty} A_i\right) = \sum_{i=1}^{\infty} \mu(A_i).$$

The triplet $(X, \mathcal{A}, \mu)$ is called a measure space.

- A signed measure is a real-valued countably additive function defined on a σ-algebra.
- A complex measure is a complex-valued countably additive function defined on a σ-algebra.
- A measure space $(X, \mathcal{A}, \mu)$ is said to be finite if $\mu(X) < \infty$.

Let $(\Omega, \mathcal{A}, \mu)$ be a finite measure space and X a Banach space. A function $u : \Omega \to X$ is called *strongly measurable* if there exists a sequence $\{u_n\}$ of simple functions such that $\|u_n(x) - u(x)\|_X \to 0$ for almost all x as $n \to \infty$.

1.7.1 Integration in Normed Spaces

Definition 1.47 (*Bochner integral*) Let $(\Omega, \mathcal{A}, \mu)$ be a finite measure space and X a Banach space. Then, we define the Bochner integral of a simple function $u : \Omega \to X$ by

$$\int_E u \, d\mu = \sum_{i=1}^{n} c_i \mu(E \cap E_i)$$

for any $E \in \mathcal{A}$.

The Bochner integral of a strongly measurable function $u : \Omega \to X$ is the strong limit (if it exists) of the Bochner integral of an approximating sequence $\{u_n\}$ of simple functions. That is,

$$\int_E u \, d\mu == \lim_{n \to \infty} \int_E u_n \, d\mu.$$

Remark 1.12   1. The Bochner integral is independent of approximating sequence. 2. If u is strongly measurable, u is Bochner integrable iff $|u(\cdot)|_X$ is integrable.

1.7.2   Positive Semidefinite Operator

Let $\mathcal{H}$ be a Hilbert space. A closed densely defined operator $A : \mathscr{D}(A) \subset \mathcal{H} \to \mathcal{H}$ is said to be normal if $A^*A \subset AA^*$, or what is equivalent, if $AA^* \subset A^*A$, where A^* is the adjoint of the operator A in the Hilbert space $\mathcal{H}$.

A linear operator A defined in a Hilbert space $\mathcal{H}$ is said to be symmetric if

$$(x, Ay) = (Ax, y), \quad \text{for all } x, y \in \mathscr{D}(A).$$

A linear densely defined operator A is said to be self-adjoint if $A^* = A$.

Definition 1.48   A linear closed and densely defined operator A is said to be positive semidefinite if

$$(Ax, x) \geqslant 0 \text{ for all } x \in \mathscr{D}(A).$$

If $(Ax, x) > 0$, for all $x \in \mathscr{D}(A)$, $x \neq 0$, we say that A is positive definite.

If there exists a constant $\varepsilon > 0$ such that $(Ax, x) \geqslant \varepsilon \|x\|^2$ for all $x \in \mathscr{D}(A)$, then A is said to be strongly positive.

1.8   Nonlinear Superposition Operators

Let Ω be an arbitrary set. Let $f = f(s, u)$ be a function defined on $\Omega \times \mathbb{R}$ (or $\Omega \times \mathbb{C}$), and taking values in $\mathbb{R}$ (respectively $\mathbb{C}$). Given a function $x = x(s)$ on Ω, by applying f we get another function $y = y(s)$ on Ω, defined by $y(s) = f(s, x(s))$. In this way, the function f generates an operator

$$Fx(s) = f(s, x(s)) \tag{1.8}$$

which is usually called superposition operator (also outer superposition operator, composition operator, substitution operator or the Nemytskiĭ operator).

Definition 1.49 Assume that $f(s, x) = f : \Omega \times \mathbb{R} \to \mathbb{R}$ is a given function. For an arbitrary function $x : \Omega \to \mathbb{R}$ denote by Fx the function defined on Ω by the formula $Fx(s) = f(s, x(s))$. The operator F defined in such a way is said to be the superposition operator generated by the function f.

Another operator which is closely related to the operator (1.8) is the integral functional

$$\Phi x = \int_{\Omega} f(s, x(s)) \, ds, \tag{1.9}$$

which is of fundamental importance, for example, in variational problems of nonlinear analysis. In this section, we shall be concerned with the operator (1.9) only marginally and refer to the vast literature on variational methods in Chap. 7.

The superposition operator (1.8) has some remarkable properties. One "algebraic" property which is called the local determination of F is described in the following:

Lemma 1.2 (Jürgen and Zabreĭko [297]) *The superposition operator F has the following three (equivalent) properties:*

(a) For $D \subseteq \Omega$,

$$FP_D - P_D F = P_{\Omega \setminus D} F\theta, \tag{1.10}$$

where θ is the almost everywhere zero function.
(b) For $D \subseteq \Omega$,

$$P_D F P_D x = P_D Fx, \quad P_{\Omega \setminus D} F P_D x = P_{\Omega \setminus D} F\theta.$$

(c) If two functions x_1 and x_2 coincide on $D \subseteq \Omega$, then the functions Fx_1 and Fx_2 also coincide on D.

We suppose that the reader is familiar with the construction and the basic properties of the (Lebesgue) integral. In what follows, we shall denote by L the set of all (Lebesgue) integrable functions over Ω, equipped with the norm

$$\|x\| = \int_{\Omega} |x(s)| d\mu(s). \tag{1.11}$$

If the measure μ under consideration is fixed, we shall write simply ds instead of $d\mu(s)$.

The Carathéodory conditions– Let Ω be a measurable subset of $\mathbb{R}^n$ and $f(s, x)$ be a function of two variables s and x, where $-\infty < x < \infty, s \in \Omega$. Then, the function f is said to satisfy the Carathéodory conditions if

(i) $f(s, x)$ is continuous with respect to x for almost all $s \in \Omega$, and
(ii) $f(s, x)$ is measurable with respect of s for all values of x.

Carathéodory function– A function $f = f(s, u)$ is called Carathéodory function if it satisfies the Caratheodory conditions; i.e., if $f(s, \cdot)$ is continuous on $\mathbb{R}$ for almost all $s \in \Omega$, and $f(\cdot, u)$ is measurable on Ω for all $u \in \mathbb{R}$.

The characteristic properties of the nonlinear superposition operators associated with a Carathéodory function has been extensively studied by many researcher (see, e.g., Jügen and Zabreĭko [297], Jürgen and Väth [298], Väth [602] and the references cited therein). We state these properties established by Väth [602].

In what follows, let U, V be pseudometric spaces, and T be a measure space.

Proposition 1.18 *Let $f : D \subseteq T \times U \to V$ be a Carathéodory function and let F be its associated superposition operator. Let F map with a set $\mathcal{B}$ of measurable functions into measurable functions. If f is a Carathéodory function, then F is sequentially continuous on $\mathcal{B}$ with respect to convergence in finite measure.*

The question whether a superposition operator maps measurable functions into measurable functions is a delicate problem, even in the scalar case: Product measurability of f is neither necessary nor sufficient, see, e.g., [297]. In the most important case $D = T \times M$ the Carathéodory condition is sufficient:

Proposition 1.19 *If $f : T \times M \to V$ with $M \subseteq U$ satisfies the Carathéodory condition, then F maps measurable functions with values in M into measurable functions.*

The theory of superposition operator received a new impetus after the fundamental paper of Krasnoslskii [342] who showed a necessary and sufficient condition for the superposition operator to be continuous from the space L_p into L_q.

Theorem 1.40 *Let f satisfy the Carathéodory conditions. The superposition operator F generated by the function f maps continuously the space $L_p(\Omega)$ into $L_q(\Omega)$ $(p, q \geqslant 1)$ if and only if*

$$|f(s, x)| \leqslant a(s) + b|x|^{\frac{p}{q}},$$

for all $s \in \Omega$ and $x \in \mathbb{R}$, where $a \in L_q(\Omega)$ and $b \geqslant 0$.

The fundamental property of the superposition operator defined on the space L_1 is contained in the following theorem.

Theorem 1.41 *Assume that $f : \Omega \times \mathbb{R} \to \mathbb{R}$ satisfies the Carathéeodory conditions. Then, the superposition operator F generated by f transforms the space L_1 into itself if and only if $|f(s, x)| \leqslant a(s) + b|x|$ for all $s \in \Omega$ and $x \in \mathbb{R}$, where $a(s)$ is a from the space L_1 and b is a nonnegative constant. Moreover, the superposition F is continuous on the space L_1.*

Remark 1.13 It should be noted that the superposition F takes the values in $L_\infty(\Omega)$ if and only if the generating function f is independent on x.

1.8.1   The Nemytskiĭ Operator

In this section, we give definitions of the Nemytskiĭ operator and its continuity properties. This operator is, at present, widely used in applications.

Definition 1.50 Let $(\Omega, \mathcal{A}, \mu)$ be a complete σ-finite measure space and $f : \Omega \times \mathbb{R}^n \to \mathbb{R}$ a Carathéodory function (i.e. for all $x \in \mathbb{R}^n$, the function $s \mapsto (s, x)$ is $\mathcal{A}$-measurable and for μ-almost all $s \in \Omega$, the function $x \mapsto (s, x)$ is continuous). It is well known that such functions are jointly measurable, hence superpositionally too (i.e. $x : \Omega \to \mathbb{R}^n$ is $\mathcal{A}$-measurable, then show is $s \mapsto f(s, x(s))$). Thus, we can define the operator

$$u \mapsto N_f(x)(\cdot) = f(\cdot, x(\cdot)),$$

which sends $\mathcal{A}$-measurable functions to $\mathcal{A}$-measurable functions; that is,

$$N_f x(s) = f(s, x(s)).$$

This operator is known as the Nemytskiĭ operator.

We have the following theorem which is due to Nemytskiĭ [421].

Theorem 1.42 *Let Ω be a set of finite measure. Then, the Nemytskiĭ operator N_f transforms every sequence $x_1(s), x_2(s), \ldots$ which converge in measure into a sequence $N_f x_1(s), N_f x_2(s), \ldots$ which also converges in measure.*

Proof Let $x_k(s)$ converges in measure to $x_0(s)$. Let $\varepsilon > 0$ be given. Then, define

$$\Omega_k = \left\{ s \in \Omega : |x_0(s) - x| < \frac{1}{k} \Rightarrow |f(s, x_0(s)) - f(s, x)| < \varepsilon \right\}.$$

It is clear that $\Omega_1 \subset \Omega_2 \subset \cdots$. Because $f(s, x)$ is continuous with respect to x for almost all $s \in \Omega$ and $\Omega = \bigcup_{k=1}^{\infty} \Omega_k$, it implies that

$$\lim_{k \to \infty} \mu(\Omega_k) = \mu(\Omega). \tag{1.12}$$

Let $\eta > 0$ be given, then in view of (1.12) we can choose a number k_0 such that

$$\mu(\Omega_{k_0}) > \mu(\Omega) - \frac{\eta}{2} \text{ or equivalently } \mu(\Omega - \Omega_{k_0}) < \frac{\eta}{2}.$$

Define $F_k = \left\{ x \in \Omega : |x_0(s) - x_k(s)| < \frac{1}{k_0} \right\}$. Because $\{x_k(s)\}$ converges to $x_0(s)$ in measure, we have $\lim_{k \to \infty} \mu(\Omega - F_k) = 0$. Hence there exists an integer m such that $\mu(\Omega - F_k) < \frac{\eta}{2}$ for all $k \geqslant m$.

Let $D_k = \{s \in \Omega : |N_f x_k(s) - N_f x_0(s)| < \varepsilon\}$. Then, it is clear that $(\Omega_{k_0} \cap F_k) \subset D_k$. This in turn implies that $(\Omega - D_k) \subset \Omega - (\Omega_{k_0} \cap F_k) = (\Omega - \Omega_{k_0}) \cup (\Omega - F_k)$.

This gives

$$\mu(\Omega - D_k) \leqslant \mu(\Omega - \Omega_{k_0}) + \mu(\Omega - F_k)$$
$$< \frac{\eta}{2} + \frac{\eta}{2} = \eta \text{ for } k \geqslant m.$$

The above inequality implies that $\{N_f x_k(s)\}$ converges in measure to $N_f x_0(s)$. This completes the proof. ∎

We now have the following continuity property of the Nemytskiĭ operator N_f, which is due to Krasnoselskii [342]. Following Krasnoselskii, we state and prove the next theorem.

Theorem 1.43 *Suppose that the operator N_f maps L_p into L_q ($\frac{1}{p} + \frac{1}{q} = 1$). Then, N_f is continuous and bounded.*

Proof To prove the continuity of N_f we assume that $\mu(\Omega) < \infty$. Without loss in generality we assure that $N_f(0) = 0$. It is then sufficient to show that N_f is continuous at 0. If this is not so, there exists a sequence of function $x_n \in L_p$ converging to 0 and a positive number α such that

$$\int_\Omega |N_f x_n(s)|^q ds > \alpha, \quad n = 1, 2, \ldots \tag{1.13}$$

Because $x_n \to 0$ in L_p, without loss in generality, we can assume that

$$\sum_{n=1}^\infty \int_\Omega |N_f x_n(s)|^p ds < \infty. \tag{1.14}$$

We now construct a sequence of numbers ε_k, functions $x_{n_k}(s)$ and sets $\Omega_k \subset \Omega$ ($k = 1, 2, \ldots$) satisfying the following:

(a) $\varepsilon_{k+1} < \frac{\varepsilon_k}{2}$
(b) $\mu(\Omega_k) \leqslant \varepsilon_k$
(c) $\int_{\Omega_k} |N_f x_{n_k}(s)|^q ds > \frac{2}{3}\alpha$
(d) for any set $D \subset \Omega$, $\mu(D) \leqslant 2\varepsilon_{k+1}$ implies that

$$\int_D |N_f x_{n_k}|^q ds < \frac{\alpha}{3}.$$

Construction of the sequence $\varepsilon_k, x_{n_k}(s)$ and Ω_k ($k = 1, 2, \ldots$) is done by an inductive process. Suppose that $\varepsilon_1 = \mu(\Omega)$, $x_{n_1}(s) = x_1(s)$, $\Omega_1 = \Omega$. If ε_k, $x_{n_k}(s)$ and Ω_k have been constructed, then for ε_{k+1} we select a number which satisfies condition (d). This is possible because

$$\int_{\Omega} |N_f x_{n_k}|^q ds < \infty$$

and hence absolutely convergent. Because x_{n_k} has already been chosen ε_{k+1} will satisfy the condition (a). By Theorem 1.40, since $\{N_f x_{n_k}(s)\}$ converges to 0 in measure, there exists a number n_{k+1} and a set $F_{k+1} \subset \Omega$ such that for $s \in F_{k+1}$.

$$|N_f x_{n_{k+1}}(s)| < \left(\frac{\alpha}{3\mu(\Omega)}\right)^{1/q} \tag{1.15}$$

with $\qquad\qquad \mu(\Omega - F_{k+1}) < \varepsilon_{k+1}. \tag{1.16}$

Let $\Omega_{k+1} = \Omega - F_{k+1}$. Then, (1.16) implies that (b) is fulfilled. The condition (c) now follows from (1.13) and (1.16) and the following:

$$\int_{\Omega_{k+1}} |N_f x_{n_{k+1}}(s)|^q ds = \int_{\Omega} |N_f x_{n_{k+1}}(s)|^q ds - \int_{F_{k+1}} |N_f x_{n_{k+1}}(s)|^q ds > \frac{2}{3}\alpha$$

We now consider the sets

$$D_k = \Omega_k - \left(\bigcup_{j=k+1}^{\infty} \Omega_j\right), \quad k = 1, 2, \ldots.$$

By virtue of conditions (a) and (b)

$$\mu\left(\bigcup_{j=k+1}^{\infty} \Omega_j\right) \leq \sum_{j=k+1}^{\infty} \varepsilon_j < 2\varepsilon_{k+1}, \quad k = 1, 2, \ldots \tag{1.17}$$

Define a function $\psi(s)$ by the equation

$$\psi(s) = \begin{cases} x_{n_k}(s), & if \ s \in D_k \ (k = 1, 2, \ldots) \\ 0, & if \ s \in \bigcup_{j=1}^{\infty} D_j \end{cases} \tag{1.18}$$

From conditions (c), (d) and (1.17), it follows that

$$\int_{D_k} |N_f \psi(s)|^q = \int_{D_k} |x_{n_k}(s)|^q ds$$

$$\geq \int_{\Omega_k} |N_f x_{n_k}(s)|^q ds - \int_{\Omega_k - D_k} |N_f x_{n_k}(s)|^q ds > \frac{\alpha}{3}.$$

By (1.14), $\psi \in L_p$, and so by the assumption on N_f, $N_f \psi \in L_q$. On the other hand, $N_F \psi L_q$ since

$$\int_{\Omega} |N_f \psi(s)|^q ds \geqslant \sum_{k=1}^{\infty} \int_{D_k} |N_f \psi(s)|^q \, ds = \infty.$$

This proves the continuity of N_f.

To prove the continuity of N_f under the assumption that $\mu(\Omega| = \infty$, we proceed as follows.

It is enough to show that N_f is continuous at 0. This implies that there exists a sequence $\{x_n(s)\} \subset L_p$ such that

$$\int_{\Omega} N_f x_n(s)|^q ds > \alpha \quad (n = 1, 2, \ldots) \tag{1.19}$$

where α is some positive number. We assume that

$$\sum_{n=1}^{\infty} \int_{\Omega} |x_n(s)|^p ds < \infty, \tag{1.20}$$

We construct a sequence of functions $\{x_{n_k}(s)\}$ and subsets $\{D_k\}$ of Ω such that

(a) $\mu(D_k) < \infty$, $D_i \cap D_j = \varnothing$, $i \neq j$

(b) $\int_{D_k} |N_f x_{n_k}|^q \, ds > \alpha/2 \quad (k = 1, 2, \ldots).$

This is done in a similar inductive way as before.

Now define a function $\psi(s)$ as in (1.18). By virtue of (1.20), $\psi \in L_q$ and hence $N_f \psi \in L_q$. But by our construction (b), $N_f \psi \notin L_q$, a contradiction.

We now proceed to prove the boundedness of N_f. Since N_f is continuous at 0, there exists $\gamma > 0$ such that

$$\int_{\Omega} |N_f x(s)|^q ds \leqslant 1 \text{ whenever } \int_{\Omega} |x(s)|^p ds \leqslant \gamma. \tag{1.21}$$

Suppose that $x(s) \in L_p$. Then, there exists an integer n such that

$$n\gamma^p \leqslant \|x\|^p < (n+1)\gamma^p. \tag{1.22}$$

We divide Ω into subsets $\Omega_1, \Omega_2, \ldots, \Omega_{n+1}$ such that

$$\int_{\Omega_j} |x(s)|^p ds \leqslant \gamma^p \quad (j = 1, 2, \ldots, n+1).$$

Then, by (1.21),

$$\int_{\Omega} |N_f x(s)|^q ds \leqslant \sum_{j=1}^{n+1} \int_{\Omega_j} |N_f x(s)|^q ds \leqslant n+1,$$

and hence, $\|N_f x\| = \left(\int_\Omega |N_f x(s)|^q ds \right)^{1/q} \leqslant \left(\left(\frac{\|x\|}{\gamma} \right)^p + 1 \right)^{1/q}$

This proves the boundedness of N_f. ∎

As an immediate consequence of this result, we have the following corollary.

Corollary 1.6 *Let the function $f(s, x)$ satisfy the condition:*

$$|f(s, x)| \leqslant a(s) + b|x|^{p/q}, \tag{1.23}$$

where $b \geqslant 0$ and $a(s) \in L_q(\Omega)$ $(\frac{1}{p} + \frac{1}{q} = 1)$. Then, the corresponding Nemytskiĭ operator N_f is continuous and bounded from $L_p(\Omega)$ to $L_q(\Omega)$.

Remark 1.14 Thus, (1.23) is one of the sufficient conditions for the continuity and boundedness of the Nemytskiĭ operator from $L_p(\Omega)$ to $L_q(\Omega)$. This condition was first studied by Vainberg in [595]. Subsequently, Nemytskiĭ and Vainberg have proposed other forms of sufficient conditions for the continuity of the operator N_f. For more details, see Kransoselskiĭ [344].

However, if one considers the space C of continuous functions defined on a closed and bounded subset Ω of $\mathbb{R}^n$, we get the following theorem which is a direct consequence of the fact that continuous functions on a compact set are uniformly continuous.

Theorem 1.44 *Let the function $f(s, x)$ be continuous as a map from $\Omega \times \mathbb{R}$ to $\mathbb{R}$. Then, the Nemytskiĭ operator N_f acts on C and is continuous and bounded.*

Remark 1.15 The Nemytskiĭ theorem for the Nemytskiĭ operator (refer to Theorems 1.43–1.45) coincides with known Carathéodory lemma for the superposition operator.

Observation

- The concept of the Carathéodory operator was established in [317], and it was shown that the theory of the Carathéodory differential equations can be built up in a similar manner to the classical one, if the right-hand side of differential equations contains a Carathéodory operator instead of functions which fulfil the Carathéodory conditions. In connection with this, the problem was posed by Kartàk (see [317]) whether every Carathéodory operator can be expressed by means of a function fulfilling the Carathéodory conditions. This problem was solved in the affirmative for linear Carathéodory operators in [317]. This problem for general Carathéodory operators poses the following question: Is every Carathéodory operator equivalent to some Nemyckiĭ operator? This problem was solved in the affirmative by Vrkoč for general Carathéodory operators in [603].

1.8.2   The Hammerstein Operator

In this section, we give definitions of the Hammerstein operator and its continuity
properties. This operator is, at present, widely used in applications.

In what follows, if one considers the space C of continuous functions defined on
a closed and bounded subset Ω of $\mathbb{R}$, we get the following theorem.

Theorem 1.45 *Suppose the function $f(s, x)$ is continuous as a map from $\Omega \times \mathbb{R}$ to
$\mathbb{R}$. Then, the Nemytskiĭ operator N_f acts on C and is continuous and bounded.*

Remark 1.16 It may be remarked that N_f is compact only when the range of N_f is
a singleton (refer Krasanoselskii [344]).

Definition 1.51 Let Ω be a compact subset of $\mathbb{R}$, $k(s, t)$ a kernel defined on $\Omega \times \Omega$,
and let $f(s, x)$ be as before. Then, the Hammerstein operator F is defined as

$$[Fx](s) = \int_\Omega k(s, t) f(t, x(t)) dt.$$

If K is the linear integral operator defined by the kernel $k(s, t)$:

$$[Kx](s) = \int_\Omega k(s, t) x(t) \, dt,$$

then F can be written in the form $F = K N_f$.

The following theorem gives a sufficient condition for the Hammerstein operator F
to be continuous and compact on $L_p(\Omega)$.

Theorem 1.46 *Let the kernel $k(s, t)$ be such that*

$$\int_\Omega \int_\Omega |k(s, t)|^q \, ds dt < \infty$$

*and $f(s, x)$ be a function as defined before satisfying condition (1.23). Then, the
Hammerstein operator F is continuous and compact on L_p.*

Proof As observed above, $F = K N_f$ where

$$[Kx](s) = \int_\Omega k(s, t) x(t) dt \text{ and } [N_f x](s) = f(s, x(s)).$$

By Corollary 1.6, N_f is a continuous and bounded operator from $L_p(\Omega)$ to $L_q(\Omega)$.
Also because $k(s, t)$ is Hilbert–Schmidt, it follows that K is a continuous and compact
operator from L_q to L_p. Because F is a composition of K and N_f, it follows that F
is continuous and compact on L_p.   ∎

1.8.3 The Urysohn Operator

In this section, we give definitions of the Urysohn operator and its continuity properties. This operator is, at present, widely used in applications.

Definition 1.52 Let $K(s, t, x)$, $(s, t \in \Omega - \infty < x < \infty$, Ω a measurable subset of $\mathbb{R}^n$) given function of three variables. Then, the operator U defined as

$$[Ux](s) = \int_\Omega K(s, t, x(t)) dt \qquad (1.24)$$

is called the Urysohn operator.

Again using the fact that continuous functions on a compact set are uniformly continuous, we get the following theorems which give sufficient conditions for the Urysohn operator U to be continuous and compact on the space C and L_p, respectively.

Theorem 1.47 (Ladyzhenskii [354]) *Suppose that the function*

$$K(s, t, x) \ (s, t \in \Omega, \ -\infty < x < \infty)$$

satisfies the following conditions.
(a) $K(s, t, x)$ is continuous with respect to x for almost all $s, t \in \Omega$ and measurable with respect to t for all $s \in \Omega, -\infty < x < \infty$,
(b) for every positive number α

$$\int_\Omega \sup_{|x| \leqslant \alpha} |K(s, t, x)| dt < \infty$$

$$\lim_{\|h\| \to 0} \int_\Omega \sup_{|x| \leqslant \alpha} |K(s + h, t, x) - K(s, t, x)| dt = 0.$$

Then, the Urysohn operator U acts on C and is continuous and compact.

Theorem 1.48 (Krasnoselskii and Ladyzenskii [347]) *Let the function $K(s, t, x)$ $(s, t \in \Omega, \ -\infty < x < \infty, \ \Omega$ a bounded closed subset of $\mathbb{R}^n$) be continuous with respect to x and satisfy the inequality*

$$|K(s, t, x)| \leqslant R(s, t)(a + b|x|^{p/q})$$

for all $s, t \in \Omega, -\infty < x < \infty$ with

$$\int_\Omega \int_\Omega |R(s, t)|^p ds dt < \infty, \ a, b > 0.$$

Then, the Urysohn operator U is a compact and continuous operator on L_p.

1.9   The Sobolev Spaces

Let $\mathbb{R}^n$ denote the n-dimensional real Euclidean space and $\mathbb{N}_0 = \mathbb{N} \cup \{0\}$. For any points $x = (x_1, x_2, \ldots, x_n)$ and $y = (y_1, \ldots, y_n) \in \mathbb{R}^n$, define

$$|x| = (x_1^2 + \cdots + x_n^2)^{1/2}, \quad (x, y) = x_1 y_1 + \cdots + x_n y_n,$$

where $|\cdot|$ and $(\cdot, \cdot)$ denote, respectively, the standard 2-norm and inner product in $\mathbb{R}^n$.

For any $x \in \mathbb{R}^n$ and $\alpha = (\alpha_1, \ldots, \alpha_n)$, denote

$$x^\alpha = x_1^{\alpha_1} \cdots x_n^{\alpha_n}, \quad |\alpha| = \alpha_1 + \cdots + \alpha_n,$$

$$\text{and} \qquad D^\alpha = D_n^{\alpha_1} \cdots D_n^{\alpha_n} = \frac{\partial^{\alpha_1}}{\partial x_1^{\alpha_1}} \cdots \frac{\partial^{\alpha_n}}{\partial x_n^{\alpha_n}}.$$

We consider $u : \Omega \to \mathbb{R}$ and denote by

$$D^\alpha u = \frac{\partial^{|\alpha|} u}{\partial x_1^{\alpha_1} \partial x_2^{\alpha_2} \cdots \partial x_n^{\alpha_n}}$$

its partial derivatives of order $|\alpha|$. Here, $\alpha = (\alpha_1, \ldots, \alpha_n)^T \in \mathbb{N}_0^n$ is a multi-index of modulus $|\alpha| = \sum_{i=1}^n \alpha_i$, and for $\alpha = 0$, we set $D^0 u = u$.

Support – The support of a real or complex function f on $\mathbb{R}^n$, denoted by $\mathrm{supp}(f)$, is the closure of the set of all points $x \in \mathbb{R}^n$ at which $f(x) \neq 0$; that is,

$$\mathrm{supp}(f) = \overline{\{x \in \mathbb{R}^n : f(x) \neq 0\}}.$$

Remark 1.17 If f is a continuous function with compact support, suppose I^n is any n-cell which contains the support of f, and define

$$\int_{\mathbb{R}^n} f = \int_{I^n} f.$$

Then, the integral so defined is evidently independent of the choice of I^n, provided only that I^n contains the support of f.

For an open subset $\Omega \subset \mathbb{R}^n$, we define $C^k(\Omega)$, $k \in \mathbb{N}_0$, the linear space of continuous functions on Ω whose partial derivatives $D^\alpha u$, $|\alpha| \leqslant k$, exist and are continuous. $C^k(\Omega)$ is a Banach space with respect to the norm

$$\|u\|_{C^k(\Omega)} := \max_{0 \leqslant |\alpha| \leqslant k} \sup_{x \in \Omega} |D^\alpha u(x)|.$$

Definition 1.53 Let α be positive number $0 < \alpha < 1$. Then, the function $u(x)$ is said to satisfy Hölder's condition with exponent α in Ω or is said to be Hölder continuous if for $x \neq y$.

$$H_\alpha(u) = \sup \frac{|u(x) - u(y)|}{|x - y|^\alpha} < \infty.$$

Then, the space $C^{k,\alpha}(\Omega)$ is defined as

$$C^{k,\alpha}(\Omega) = \left\{ u \mid u \in C^k(\Omega), H_\alpha(D^\beta u) < \infty, |\beta| = k \right\}.$$

We note that $C^{k,\alpha}(\Omega)$ is a Banach space with respect to the norm

$$\|u\|_{C^{k,\alpha}(\Omega)} := \|u\|_{C^k(\Omega)} + \max_{|\beta|=k} \sup_{x,y \in \overline{\Omega}} \frac{|D^\beta u(x) - D^\beta u(y)|}{|x - y|^\alpha}.$$

Moreover, $C_0^k(\Omega)$ and $C_0^{k,\alpha}(\Omega)$ are the subspaces of functions with compact support in Ω.

Finally, $C^\infty(\Omega)$ stands for the set of functions with continuous partial derivatives of any order and

$$C_0^\infty(\Omega) = \bigcap_{k=1}^\infty C_0^k(\Omega).$$

Weak or generalized derivative– Let $u \in L_1(\Omega)$ and $\alpha \in \mathbb{N}_0^n$. The function u is said to have a weak or generalized derivative $D^\alpha u$, if there exists a function $v \in L_1(\Omega)$ such that for every $\varphi \in C_0^\infty(\Omega)$ (set of infinitely differentiable functions with compact support in Ω),

$$\int_\Omega u D^\alpha \varphi \, dx = (-1)^{|\alpha|} \int_\Omega v \varphi \, dx.$$

We then set $D^\alpha u := v$ and call v the αth weak or generalized derivative of u. More precisely, we have the following definition.

Definition 1.54 A function $u \in L_p(\Omega)$ is said to possess generalized derivatives $D^\alpha u$ of order up to k if there exists a sequence $\{u_n\} \subset C^\infty(\Omega)$ such that $\{D^\alpha u_n\}$ is Cauchy in $L_p(\Omega)$ converging to $D^\alpha u$ for $|\alpha| \leqslant k$ and $u_n \to u$ in $L_p(\Omega)$.

The notion "weak derivative" suggests that it is a generalization of the classical concept of differentiability and that there are functions which are weakly differentiable, but not differentiable in the classical sense. To this end, let us consider the following example.

Example 1.43 Let $n = 1$ and $\Omega := (-1, 1)$. The function $u(x) := |x|, x \in \Omega$, is not differentiable in the classical sense. However, one can easily see that it admits a weak derivative $D^1 u$ given by

$$D^1 u = \begin{cases} -1, & \text{when } x < 0 \\ 1, & \text{when } x > 0. \end{cases}$$

Indeed, for $\varphi \in C_0^\infty(\Omega)$, we obtain by partial integration

$$\int_{-1}^1 u(x) D^1 \varphi(x) dx = \int_{-1}^0 u(x) D^1 \varphi(x) dx + \int_0^1 u(x) D^1 \varphi(x) dx$$

$$= -\int_{-1}^0 D^1 u(x) \varphi(x) dx + (u\varphi)|_{-1}^0 - \int_0^1 D^1 u(x) \varphi(x) dx + (u\varphi)|_0^1$$

$$= -\int_{-1}^1 D^1 u(x) \varphi(x) dx + [u(0)]\varphi(0),$$

where $[u(0)] := u(0+) - u(0-)$ is the jump of u in $x = 0$. But u is continuous at $x = 0$, and hence, $[u(0)] = 0$ which gives

$$\int_{-1}^1 u(x) D^1 \varphi(x) dx = (-1)^1 \int_{-1}^1 v(x) \varphi(x) dx, \quad \text{where } D^1 u = v.$$

This allows to conclude.

It is easy to show that if u has a weak derivative $D^\alpha u$, and if $D^\alpha u$ has a weak derivative $D^\beta(D^\alpha u)$, then u has a weak derivative $D^{\beta+\alpha} u$ and $D^{\beta+\alpha} u = D^\beta(D^\alpha u)$. One can also derive other standard properties.

Definition 1.55 The Sobolev space $W^{k,p}(\Omega)$ for $p \in [1, \infty)$ is the set of all functions $u \in L_p(\Omega)$ which have generalized derivatives up to order k such that $D^\alpha u \in L_p(\Omega)$ for $|\alpha| \leqslant k$. We set $W^{0,p}(\Omega) = L_p(\Omega)$.
In other words, the Sobolev space $W^{k,p}(\Omega)$ is the closure of $C^\infty(\Omega)$ with respect to the norm

$$\|u\|_{k,p} = \Big[\sum_{|\alpha| \leqslant k} \int_\Omega |D^\alpha u|^p dx \Big]^{1/p}. \tag{1.25}$$

Definition 1.56 (The space H^k) In the special case where $p = 2$, we define the Hilbert–Sobolev space $H^k(\Omega) := W^{k,2}(\Omega)$. The space $H^k(\Omega)$ is endowed with the inner product

$$\langle u, v \rangle_{H^k} := \sum_{|\alpha| \leqslant k} \int_\Omega D^\alpha u \, \overline{D^\alpha v} \, dx.$$

Similarly, we define

$$H_0^k(\Omega) := W_0^{k,2}(\Omega).$$

Observation

- The space $W^{k,p}$ together with the norm $\| \cdot \|_{k,p} = \Big[\sum_{|\alpha| \leqslant k} \int_\Omega |D^\alpha u|^p dx \Big]^{1/p}$ is a real Banach space, provided we identify any two functions that differ on a set of n-dimensional Lebesgue measure zero.
- The space $W^{k,p}$ is separable for $p \in [1 \leqslant p < \infty)$, but for $p \in (1, \infty)$, it is uniformly convex and thus reflexive.
- For $p = 2$, $W^{k,2}(\Omega)$ is a Hilbert space with inner product

$$\langle u, v \rangle_{k,2} = \sum_{|\alpha| \leqslant k} \int_{\Omega} D^{\alpha} u \, \overline{D^{\alpha} v} \, dx.$$

- $W^{k,p}$ can be considered as the equivalence classes of Cauchy sequences of elements from C^{∞}. If $u \in W^{k,p}$, then there exists a sequence $\{u_n\} \subset C^{\infty}$ such that $D^{\alpha} u_n$ is Cauchy in L_p, $|\alpha| \leqslant k$ and $u_n \to u$ in L_p.

The spaces $W^{k,p}(\Omega)$ can be modified to incorporate boundary conditions. The closure of $C_0^{\infty}(\Omega)$ in $W^{k,p}(\Omega)$ is denoted by $W_0^{k,p}(\Omega)$ and it contains functions whose generalized derivatives up to order k vanish on the boundary $\partial\Omega$ of Ω.

For bounded domain Ω, a basic inequality of Poincare implies the existence of an absolute constant $c(\Omega)$ such that for $u \in W_0^{1,p}(\Omega)$ $(1 < p < \infty)$,

$$\|u\|_{L_p} \leqslant c(\Omega) \|\nabla u\|_{L_p}.$$

So, one can define in $W_0^{k,p}(\Omega)$ a 'short norm', given by

$$\|u\|_{k,p} = \left[\sum_{|\alpha|=k} \int_{\Omega} |D^{\alpha_u}|^p dx \right]^{1/p}.$$

which is equivalent to the norm given by (1.25).

For bounded domain Ω, $W_0^{k,p}(\Omega) \subset W_0^{k,l}(\Omega)$ for $k > 1$. Thus, each $u \in W_0^{k,p}(\Omega)$ could be associated with an element in $W_0^{1,p}(\Omega)$. This mapping is called an imbedding of $W_0^{k,p}(\Omega)$ into $W_0^{1,p}(\Omega)$. The following lemma, called Rellich's lemma, says that this mapping is actually compact.

Lemma 1.3 *Let Ω be bounded. Then, the imbedding of $W_0^{k,p}(\Omega)$ into $W_0^{k,l}(\Omega)$, $k > l$ is compact.*

In general, we have the following imbedding theorem due to Sobolev.

Theorem 1.49 (The Sobolev imbedding theorem) *Let Ω be a bounded open subset of $\mathbb{R}^n$ with smooth boundary. Then,*

(i) *if $0 \leqslant l < k$, then $W^{k,p}(\Omega) \subset W^{l,r}(\Omega)$ for $\frac{1}{r} \geq \frac{1}{p} - \frac{1}{n}(k-l)$ with continuous imbedding mapping. The imbedding is compact if $\frac{1}{r} > \frac{1}{p} - \frac{1}{n}(k-l)$ and $r < \infty$,*

(ii) *if $0 \leqslant l < k$ and $\frac{1}{p} - \frac{1}{n}(k-l) < 0$, then $W^{k,p}(\Omega) \subset C^{l,\alpha}(\Omega)$ for any α with $0 < \alpha < 1$ such that $\frac{1}{p} - \frac{1}{n}(k-l-\alpha) < 0$. The imbedding mapping is always compact in this case.*

1.9.1 Elliptic Operators

We now define elliptic operators. Consider a linear differential operator of the form $A = A(x, D) = \Sigma_{|\alpha| \leqslant k} a_{\alpha}(x) D^{\alpha}$, where $a_{\alpha}(x)$ are real-valued functions defined on

$\Omega \subset \mathbb{R}^n$. We assume that not all a_α vanish on Ω for $|\alpha| = k$, k is the order of A and

$$A' = A'(x, D) = \sum_{|\alpha|=k} a_\alpha(x) D^\alpha.$$

Definition 1.57 A differential operator A is said to be

(i) elliptic at $x_0 \in \Omega$ if for any real $\xi \neq 0$, $A'(x_0, \xi) \neq 0$.
(ii) uniformly elliptic if there exists a constant $c > 0$ such that

$$c^{-1}|\xi|^k \leqslant |A'(x, \xi)| \leqslant c|\xi|^k \text{ for } x \in \Omega, \text{ and real } \xi \neq 0.$$

Example 1.44 The Laplace operator $\Delta = D_1^2 + \cdots + D_n^2$ is uniformly elliptic and so is the biharmonic operator $\Delta = (D_1^2 + D_2^2)^2$.

Definition 1.58 Let $k = 2l$. Then, A is said to be strongly elliptic at $x_0 \in \Omega$ if $(-1)^l \operatorname{Re} A'(x_0, \xi) > 0$ for all real $\xi \neq 0$.

Notice that $-\Delta$, Δ^2 are strongly elliptic.

For a more detailed treatment on the Sobolev spaces, refer to Agman [7].

1.9.2 Semilinear Equations of Evolution

Definition 1.59 (*Strongly continuous semigroups*) Let X be a Banach space. A one-parameter family $S(t)(t \geqslant 0)$ of bounded linear operators on X is said to be a strongly continuous semigroup (C_0-semigroup, for short) if

(a) $S(0) = I$ (identity operator on X);
(b) $S(t + s) = S(t)S(s)$ for every $t, s \geqslant 0$;
(c) $\lim_{t \to 0} S(t)x = x$ for every $x \in X$ (strong continuity).

Notice that as a direct application of the uniform boundedness theorem, there exist $\omega \geqslant 0$ and $M \geqslant 1$ such that

$$\|S(t)\|_{L(X)} \leqslant Me^{\omega t}, \quad \forall t \geqslant 0.$$

This in turn entails the continuity of the map $t \mapsto S(t)x$ from $[0, 1)$ to X, for every fixed $x \in X$(cf. Pazy [468]).

The linear operator A of domain $\mathscr{D}(A) = \left\{ x \in X : \lim_{t \to 0} \frac{S(t)x-x}{t} \text{ exists} \right\}$ defined by

$$Ax = \lim_{t \to 0} \frac{S(t)x - x}{t}, \quad \forall x \in \mathscr{D}(A)$$

is the infinitesimal generator of the semigroup $S(t)$.

We now recall some basic facts on A (see, e.g., Pazy [468]).

Proposition 1.20 *A is a closed linear operator with dense domain. For every fixed $x \in \mathscr{D}(A)$, the map $t \mapsto S(t)x$ belongs to $C^1([0, \infty), \mathscr{D}(A))$ and*

$$\frac{d}{dt} S(t)x = AS(t)x = S(t)Ax.$$

1.9.3   Principle of Abstract Minimization

Problems of minimizing a finite function over some subset of $\mathbb{R}^n$ correspond one-to-one with problems of minimizing over all of $\mathbb{R}^n$ a function $f : \mathbb{R}^n \to \overline{\mathbb{R}} = \mathbb{R} \cup \{+\infty\}$, under the identifications:

$$\mathscr{D}(f) = \mathrm{dom}\, f = \text{set of feasible solutions,}$$
$$\mathrm{argmin}\, f = \text{set of optimal solutions,}$$
$$\inf f = \text{optimal value.}$$

We use the convention that argmin $f = \varnothing$ when $f \equiv \infty$. This ensures that a problem is not regarded as having an optimal solution if it does not even have a feasible solution. A lack of feasible solutions is signalled by the optimal value being ∞. The notation argmin f refers to points $\bar{x}$ giving a global minimum of f. A local minimum occurs at $\bar{x}$ if $f(\bar{x}) < \infty$ and $f(x) > f(\bar{x})$ for all $x \in V$, where

$$V \in \mathfrak{N}_{\bar{x}} := \text{the collection of all neighborhoods of } \bar{x}.$$

Then, $\bar{x}$ is a locally optimal solution to the problem of minimizing f (Fig. 1.9).

Fig. 1.9 Local and global optimality in classical case

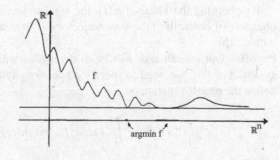

1.10 Optimal Control Theory

Optimal control is a method of automatic control in which the operating conditions of the controlled object are established and maintained such that the extremum value (minimum or maximum) of some criterion that characterizes the quality of the object's operation is achieved. The criterion of quality usually called the target function, objective function, or performance index may be a directly measurable physical quantity, such as temperature, current, voltage or pressure, or it may be efficiency, throughput or some other parameters. In this context, first we have some basic terminologies of optimal control theory.

Dynamical system – A dynamical system is a system in which a function describes the time dependence of a point in a geometrical space. Examples include the mathematical models that describe the swinging of a clock pendulum, the flow of water in a pipe and the number of fish each springtime in a lake.

Trajectory – If the given dynamical system can be solved, given an initial point it is possible to determine all its future positions, a collection of points known as a trajectory or orbit.

Optimal control theory deals with optimization problems involving a controlled dynamical system. A controlled dynamical system is a dynamical system in which the trajectory can be altered continuously in time by choosing a control parameter $\alpha(t)$ continuously in time. A deterministic controlled dynamical system is usually governed by an ordinary differential equation in the following form:

For a given set U called the control set and for $t \in \mathbb{R}^+ = [0, \infty)$, the governing ODE is

$$\left.\begin{aligned} \dot{x}(t) &= f(t, x(t), \alpha(t)), \quad t > 0 \\ x(0) &= \quad x_0 \in \mathbb{R}^n \end{aligned}\right\} \tag{1.26}$$

where $\alpha : \mathbb{R}^n \to U$ is a function called control, $f : \mathbb{R}^+ \times \mathbb{R}^n \times U \to \mathbb{R}^n$ and the unknown is the curve $x : \mathbb{R}^+ \to \mathbb{R}^n$, which we interpret as the dynamical evolution of the state of some "system".

By choosing the value of $\alpha(t)$, the state trajectory $x(t)$ can be controlled. The objective of controlling the state trajectory is to minimize a certain cost associated with (1.26).

Payoffs– Our overall task will be to determine what is the "best" control for our system. For this, we need to specify a specific payoff (or reward) criterion. Let us define the payoff functional

$$P[\alpha(\cdot)] := \int_0^T r(t, x(t), \alpha(t))dt + g(x(T)), \tag{P}$$

where $x(\cdot)$ solves ODE (1.26) for the control $\alpha(\cdot)$. Here, $r : \mathbb{R} \times \mathbb{R}^n \times U$ and $g : \mathbb{R}^n \to \mathbb{R}$ are given, and we call r the running payoff and g the terminal payoff. The terminal time $T > 0$ is given as well.

Optimal control– Our aim is to find a control $\alpha^*(\cdot)$, which maximizes the payoff. In other words, we want

$$P[\alpha^*(\cdot)] \geqslant P[\alpha(\cdot)]$$

for all controls $\alpha(\cdot) \in U$. Such a control $\alpha^*(\cdot)$ is called optimal.

To illustrate the theory, consider the following important problem studied by Bushaw and Lasalle in the field of aerospace engineering. Let $x(t)$ represent the deviation of a rocket trajectory from some desired flight path. Then, provided the deviation is small, it is governed to a good approximation by

$$\left.\begin{aligned} \dot{x}(t) &= A(t)x(t) + B(t)\alpha(t), \quad t > 0 \\ x(0) &= x_0, \quad x(T) = 0, \quad \alpha(t) \in [-1, 1] \end{aligned}\right\} \tag{1.27}$$

where A and $\dot{B}$ are appropriate matrix valued functions. The vector x_0 is the initial deviation, $\alpha(t)$ is the rocket thrust at time t and T is the final time. The problem is that of finding a rocket thrust history to minimize

$$P[\alpha(\cdot)] := \int_0^T c(t, x(t), \alpha(t)) \, dt \tag{1.28}$$

where $x(\cdot)$ solves ODE (1.27) for the control $\alpha(\cdot)$ and c is a certain cost function. Here, the aim is to reduce the deviation to zero over the time interval $[0, T]$ and at the same time to make the value of $P[\alpha(\cdot)]$ as small as possible. The cost function c is chosen so that $J(\alpha)$ has the interpretation of, say, fuel consumption or (via a change of independent variable) the time taken to reduce the deviation to zero.

Subsequently, Pontryagin and his collaborator studied the following optimal control problem : minimize

$$\left.\begin{aligned} \int_0^T c(t, x(t), \alpha(t)) &+ g(x(T)) \\ \text{subject to} \qquad \dot{x}(t) &= f(t, x(t), \alpha(t)) \\ x(0) \in C_0, \quad x(T) &\in C_1, \quad u(t) \in U \end{aligned}\right\} \tag{1.29}$$

where c, f, g are appropriate functions, and $C_0 \in \mathbb{R}^n$, $C_1 \in \mathbb{R}^n$. Minimization is carried out over control functions $\alpha(\cdot)$ and the corresponding solutions to the differential equations which satisfy the constraints. The importance of the formulation of the optimal control problem cannot be overemphasized. It incorporates a large number of dynamic optimization problems of practical interest. In particular, it also subsumes the classical problem of calculus of variations. Indeed, put

$$f(t, x, u) = u, \quad g = 0.$$

Then, the above problem reduces to: minimize

$$\int_0^T c(t, x(t), \dot{x}(t))\, dt$$

subject to $\qquad\qquad x(0) \in C_0,\ x(T) \in C_1.$

Let us consider another example of a moon lander problem. This model asks us to bring a moon lander to a soft landing on the lunar surface, using the least amount of fuel.

First of all, we introduce the notation

$$h(t) = \text{height at time } t$$

$$v(t) = \text{velocity} = \dot{h}(t)$$

$$m(t) = \text{mass of spacecraft (changing as fuel is burned)}$$

$$\alpha(t) = \text{thrust at time } t.$$

We now assume that $0 \leqslant \alpha(t) \leqslant 1$, and by Newton's law, we have

$$m\ddot{h} = -gm + \alpha,$$

the right-hand side being the difference of the gravitational force and the thrust of the rocket. This system is modelled by the ODE

$$\left.\begin{aligned} \dot{v} &= -g + \frac{\alpha(t)}{m(t)} \\ \dot{h}(T) &= v(t) \\ \dot{m}(t) &= -k\alpha(t) \end{aligned}\right\} \qquad (1.30)$$

Summarizing these equations for $t \in \mathbb{R}^+$, we obtain

$$\dot{x}(t) = f(t, x(t), \alpha(t))$$

where $x(t) = (v(t), h(t), m(t))$. We want to minimize the amount of fuel used up, that is, to maximize the amount remaining once we have landed. Let us take a payoff functional

$$P[\alpha(\cdot)] = m(\tau),$$

where τ denotes the first time that $h(\tau) = v(\tau) = 0$. We have also the extra constraints $h(t) \geqslant 0,\ m(t) \geqslant 0$. Because the final time is not given in advance, this is usually called a variable endpoint problem (Fig. 1.10).

Fig. 1.10 A moon lander
landing on the moon

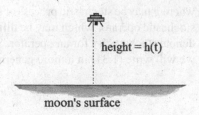

height = h(t)

moon's surface

1.11 Nonlinear Stochastic Operator Equations

We begin this section reviewing some aspects of random operator. The notion of
random variable plays a crucial role in the formation of nonlinear stochastic equations. A random variable or stochastic variable is a variable whose possible values
are numerical outcomes of a random phenomenon. As a function, a random variable
is required to be measurable.

Probability space– A probability space is a triple (Ω, Σ, P), where Ω is a set, Σ a σ-
algebra of subsets of Ω, $P : \Sigma \to [0, 1]$ a mapping such that $P(\varnothing) = 0$, $P(\Omega) = 1$,
and

$$P(\bigcup_{i=1}^{\infty} A_i) = \sum_{i=1}^{\infty} P(A_i),$$

provided $A_i \cap A_j = \varnothing$ for all $i \neq j$.

A typical point in Ω is denoted by ω and is called a sample point. A set $A \in \Sigma$
is called an event. We call P a probability measure on Ω, and $P(A) \in [0, 1]$ is
probability of the event A.

Random variable– A random variable X is a mapping $X : \Omega \to \mathbb{R}$ such that for all
$t \in \mathbb{R}$

$$\{\omega | X(\omega) \leqslant t\} \in \Sigma.$$

We usually employ capital letters to denote random variables. Often, the dependence
of X on ω is not explicitly displayed in the notation.

Expected value– Let X be a random variable, defined on some probability space
(Ω, Σ, P). The expected value of X is

$$E[X] := \int_{\Omega} X \, dP.$$

Stochastic process– A stochastic process is a collection of random variables
$X(t)(0 \leqslant t < \infty)$, each defined on the same probability measure space (Ω, Σ, P).

A general dynamical system can be viewed as nonlinear and stochastic and is
represented by the operator equation

$$\mathscr{F} u = g \tag{1.31}$$

where g may be stochastic process or simply a function, and $\mathscr{F}$ represents a nonlinear stochastic operator which may be differential or algebraic operator. Script letter will denote stochasticity for an operator. Since $\mathscr{F}u$ may have linear and nonlinear parts, we will write (1.31) in a more general form as

$$\mathscr{L}u + \mathscr{N}u = g. \tag{1.32}$$

Here, $\mathscr{L}$ denotes a linear stochastic operator and $\mathscr{N}$ is a nonlinear stochastic term. If both operators are deterministic, (1.32) is written as

$$Lu + Nu = g \tag{1.33}$$

where g again may be deterministic or stochastic. In (1.32), the linear operator $\mathscr{L}$ may be decomposed into deterministic and stochastic (linear) operators; thus,

$$\mathscr{L} = L + \mathscr{R}. \tag{1.34}$$

It may be convenient to use L as the average of $\mathscr{L}$; i.e., $L = \langle \mathscr{L} \rangle$. Then, $\mathscr{R} = \mathscr{L} - L$ is a zero-mean random operator. Suppose t is our independent variable and $\mathscr{L}$ is an nth-order differential (stochastic) operator given by

$$\mathscr{L} = \sum_{i=0}^{n} a_i(t, \omega) \frac{d^i}{dt^i} \tag{1.35}$$

where $a_i(t, \omega)$ may be stochastic processes defined on a suitable probability space, then

$$L = \sum_{i=0}^{n} \langle a_i(t, \omega) \rangle \frac{d^i}{dt^i} \tag{1.36}$$

and

$$\mathscr{R} = \sum_{i=0}^{n-1} \alpha_i(t, \omega) \frac{d^i}{dt^i} \tag{1.37}$$

where the fluctuation component x_i of each coefficient a_i is given by

$$\alpha_i(t, \omega) = a_i - \langle a_i \rangle. \tag{1.38}$$

Because it is necessary that L be invertible and the choice of $L = \langle \mathscr{L} \rangle$ may make the decomposition of the Green function difficult, we can choose a simple L more easily invertible such as L equal to the highest-order derivative only. In this case, we have

$$\mathscr{L} = L + R + \mathscr{R} \tag{1.39}$$

where $\mathscr{R}$ is again a random operator and the deterministic operator is written as $L + R$, where R is simply the remainder operator or remaining part of the operator where L is specified as the highest-ordered term and $\mathscr{R}$ is the term containing random processes.

Observe that $\mathscr{N}u$, the nonlinear part, may be deterministic written as Nu or it may also have a stochastic part which we identify as $\mathscr{M}u$; that is,

$$\mathscr{N}u = Nu + \mathscr{M}u.$$

Assume that Nu is a nonlinear function $f(u)$. Later, we will consider expressions where nonlinear term is a function of u and one or more derivatives of u such as $f(u, u')$ and various composite and product functions such as $u^3 u'$, uu'', $u^2 u'''$, uu^{iv} or $f(u, u', \ldots, u^{(m)})$. The operator $\mathscr{F}$ may involve derivatives with respect to one or more independent variables such as x, y, z, t or mixed derivatives. We will defer these cases. We will assume the same probability space for each process; it is easy enough to make them different, but it obscures notational convenience. We shall observe the following:

Case I. If the independent variable is t, then L may be $\frac{d^2}{dt^2}$, for example, or $\frac{d^2}{dt^2} + \alpha(t)\frac{d}{dt} + \beta(t)$.

Case II. If the independent variables are x, y, z, t, then we may have $L = L_x + L_y + L_z + L_t$, where, for example, $L_x = \frac{d^2}{dx^2}$, $L_y = \frac{d}{dy}$. until treat multidimensional equations we consider a single independent variable t. Thus, finally we have the operator equation $\mathscr{F}u = g$ as

$$Lu + Ru + \mathscr{R}u + Nu + \mathscr{M}u = g \qquad (1.40)$$

Note that there are five terms on the LHS of (1.40), among them any one, two, three, or four may vanish; so we include a very wide range of possibilities in the single equation (1.40) whose solution we will consider in stochastic system.

1.11.1 Stochastic Control Theory

Now assume $f : \mathbb{R}^n \times A \to \mathbb{R}^n$, and turn attention to the controlled stochastic differential equation:

$$\left.\begin{array}{l} \dot{\mathbf{X}}(s) = f(\mathbf{X}(s), \mathbf{A}(s)) + \boldsymbol{\xi}(s) \quad (t \leqslant s \leqslant T) \\ \mathbf{X}(t) = x_0. \end{array}\right\} \qquad \text{(SDE)}$$

Definition 1.60 (i) A control $\mathbf{A}(\cdot)$ is a mapping of $[t, T]$ into A, such that for each time $t \leqslant s \leqslant T$, $\mathbf{A}(s)$ depends only on s and observations of $\mathbf{X}(\tau)$ for $\tau \in [t, s]$

(ii) The corresponding payoff functional is

$$P_{x,t}[\mathbf{A}(\cdot)] = E\Big(\int_t^T r(\mathbf{X}(s), \mathbf{A}(s))ds + g(\mathbf{X}(T)) \Big), \tag{P}$$

the expected value over all sample paths for the solution of (SDE). As usual, we are given the running payoff r and terminal payoff g.

Our goal is to find an optimal control $\mathbf{A}^*(\cdot)$, such that

$$P_{x,t}[\mathbf{A}^*(\cdot)] = \max_{\mathbf{A}(\cdot)} P_{x,t}[\mathbf{A}(\cdot)].$$

Dynamic programming – To apply dynamic programming method, we firstly define the value function

$$v(x,t) := \sup \mathbf{A}(\cdot) P_{x,t}[\mathbf{A}(\cdot)].$$

The overall plan to find an optimal control $\mathbf{A}^*(\cdot)$ will be

(i) to find a Hamilton–Jacobi–Bellman type of PDE that v satisfies, and then
(ii) to utilize a solution of this PDE in designing $\mathbf{A}^*$.

1.12 Variational Method in the Hilbert Space

In this section, we review certain topics in the elementary theory of the Hilbert spaces which lead directly to abstract variational or weak formulation of boundary value problems. But, first we introduce the notion of projection operator and some related basic results.

Projection – Let K be a nonempty closed subset of $\mathbb{R}^n$ and $x \in \mathbb{R}^n$. A point $\bar{x} \in K$ is said to be the projection of x on K or best approximation of x on K, denoted by $\bar{x} = P_K(x)$, if

$$\|x - \bar{x}\| = \min_{y \in K} \|x - y\|.$$

If $x \in K$, then the projection is unique and $\bar{x} = x$. We note that the projection of x on K may not always exist (e.g. if K is open) and when it exists it may not be unique (e.g. if $K = \{x \in \mathbb{R}^2 : \|x\| \geqslant 1\}$ and x is the origin). However, under closedness and convexity assumptions, the following assertion holds.

Proposition 1.21 *Let K be a nonempty closed convex subset of $\mathbb{R}^n$ and x a point in $\mathbb{R}^n$ with $x \notin K$. Then, there exists a unique point $\bar{x} \in K$ such that*

$$\|x - \bar{x}\| = \min_{y \in K} \|x - y\|. \tag{1.41}$$

Also, the unique point $\bar{x}$ satisfies the following inequality:

$$\langle x - \bar{x}, y - \bar{x} \rangle \geqslant 0, \quad \text{for all } y \in K. \tag{1.42}$$

The inequality (1.42) shows that $x - \bar{x}$ and $y - \bar{x}$ subtend a nonacute angle between them. The projection $P_K(x)$ of x on K can be interpreted as the result of applying to x the operator $P_K : \mathbb{R}^n \to K$, which is called the projection onto K.

Corollary 1.7 *Let K be a nonempty closed convex subset of $\mathbb{R}^n$. Then, for all $x, y \in \mathbb{R}^n$,*

$$\| P_K(x) - P_K(y) \| \leqslant \| x - y \|, \tag{1.43}$$

that is, the projection operator P_K is nonexpansive. In particular, P_K is continuous on K.

The geometric interpretation of the nonexpansivity of P_K is given in Fig. 1.11b. We observe that if strict inequality holds in (1.41), then the projection operator P_K reduces the distance. However, if the equality holds in (1.41), then the distance is conserved.

In what follows, let $\mathcal{H}$ denote a Hilbert space with norm $\| \cdot \|$, scalar product $(\cdot, \cdot)$ and dual space $\mathcal{H}^*$. A subset K of $\mathcal{H}$ is called closed if $\{x_n\} \subset K$ and $\lim_{n \to \infty} x_n = x$ imply $x \in K$. The subset K is convex if $x, y \in K$ and $0 \leqslant t \leqslant 1$ imply $tx + (1 - t)y \in K$. The following minimization principle is fundamental.

Theorem 1.50 *Let K be a closed, convex, nonempty subset of the Hilbert space $\mathcal{H}$, and let $f \in \mathcal{H}^*$. Define $\varphi(x) = \frac{1}{2}\|x\|^2 - f(x)$, $x \in \mathcal{H}$. Then, there exists a unique*

$$x \in K : \varphi(x) \leqslant \varphi(y), \quad \text{for all } y \in K. \tag{1.44}$$

Lemma 1.4 *For each closed convex nonempty subset K of $\mathcal{H}$, there is a projection operator $P_K : \mathcal{H} \to K$ for which $P_K(x)$ is that point of K closest to $x \in \mathcal{H}$; it is characterized by*

$$P_K(x) \in K : (P_K(x) - x, y - P_K(x)) \geqslant 0, \quad y \in K. \tag{1.45}$$

It follows from this characterization that the function P_K satisfies

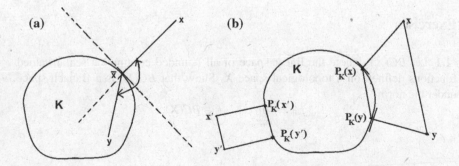

Fig. 1.11 a Projection of a point x onto K and **b** nonexpansiveness of projection operator

$$\|P_K(x) - P_K(y)\|^2 \leqslant (P_K(x) - P_K(y), x - y), \quad x, y \in \mathcal{H}. \tag{1.46}$$

For, if $\lambda = \langle P_K(x) - P_K(y), P_K(y) - y \rangle$ and $\mu = \langle P_K(y) - P_K(x), P_K(x) - x \rangle$. By Lemma 1.4, $\lambda \geqslant 0$ and $\mu \geqslant 0$. Write $\mu = \langle P_K(x) - P_K(y), x - P_K(x) \rangle$. Then,

$$\begin{aligned}\lambda + \mu &= \langle 2(P_K(x) - P_K(y)), x - y + P_K(y) - P_K(x) \rangle \\ &= 2\{\langle P_K(x) - P_K(y), x - y \rangle - \|P_K(x) - P_K(y)\|^2\}.\end{aligned}$$

Because $\lambda + \mu \geqslant 0$, this yields (1.46).

But the Cauchy–Schwarz inequality implies

$$(P_K(x) - P_K(y), x - y) \leqslant \|P_K(x) - P_K(y)\| \|x - y\|.$$

Combining the above inequality with (1.46) yields

$$\|P_K(x) - P_K(y)\|^2 \leqslant \|P_K(x) - P_K(y)\| \|x - y\|.$$

From this, we see that P_K is a contraction; i.e.,

$$\|P_K(x) - P_K(y)\| \leqslant \|x - y\|, \quad x, y \in \mathcal{H} \tag{1.47}$$

and that, in view of (1.46), P_K satisfies the angle condition

$$(P_K(x) - P_K(y), x - y) \geqslant 0, \quad x, y \in \mathcal{H}. \tag{1.48}$$

That is, the operator P_K is monotone (cf., Sect. 4.1).

Proposition 1.22 (Zarantonello, [618]) *For every element x in $\mathcal{H}$, $P_K(x)$ is characterized by the following properties:*
(1) $\langle P_K(x) - x, y \rangle \geqslant 0$ *for all y in K,*
(2) $\langle P_K(x) - x, P_K(x) \rangle = 0$.

Exercises

1.1 Let $BC(X)$ denote the linear space of all bounded continuous scalar-valued functions defined on a topological space X. Show that $BC(X)$ is a Banach space under the norm

$$\|f\| = \sup_{x \in X} |x|, \quad f \in BC(X).$$

1.2 Prove that the linear space $C[a, b]$ of real-valued continuous functions defined on $[a, b]$ is a Banach space under the norm

$$\|f\| = \sup_{x \in X} |x|, \quad f \in C[a, b].$$

1.3 Let X be a nonzero normed linear space. Prove that X is a Banach space if and only if $S = \{x \in X : \|x\| = 1\}$ is complete.

1.4 Show that $\varphi(x) = \left((|x_1|)^{1/2} + (|x_2|)^{1/2}\right)^2$ does not define a norm on a vector space of all ordered pairs $x = (x_1, x_2)$ of real numbers.

1.5 Show that $\sum_{n=1}^{\infty} 2^{-n}|x_n|$ defines a norm on ℓ_∞ and on ℓ_1.

1.6 Show that $\{e_j\}$, where $e_j = (0, 0, \cdots, 0, 1, 0, \cdots)$ with 1 in the jth place, is not a Hamel basis for the space ℓ_p, $1 \leqslant p < \infty$, but it is a Schauder basis for these spaces.

1.7 Let X be a Banach space, and let $\{x_n\}$ be a sequence of elements of X such that $\sum_{n=1}^{\infty} \|x_n\| < \infty$. Then, show that the infinite series $\sum_{n=1}^{\infty} x_n$ is convergent and defines an element of X.

1.8 Show that an infinite-dimensional Banach space cannot have a countable basis.

1.9 Show that a Banach space having a Schauder basis is separable.

1.10 Let $(X, \|\cdot\|)$ be a normed linear space. Show that if $x_n \rightharpoonup x$, then $\overline{\lim_n} \|x_n\| \geq \|x\|$, where $\overline{\lim_n} \|x_n\| = \lim_k \sup_{n>k} \|x_n\|$.

1.11 Let $(X, \mathcal{A}, \mu)$ be a measure space, then show that $\mathcal{H} := L_2(X, \mathcal{A}, \mu)$ with inner product

$$(f, g) = \int_X f \cdot \bar{g} \, d\mu$$

is a Hilbert space. Show also that every Hilbert space is "equivalent" to a Hilbert space of this form.

1.12 Show that the spaces c_0, ℓ_p such that $1 < p < \infty$ all lack Schur's property.

1.13 Let $(X, \mathcal{A}, \mu)$ be a measure space and $\mu(X) < \infty$. Prove that for $p \geq q \geq 1$,

$$L_p(X, \mu) \subset L_q(X, \mu).$$

Chapter 2
Geometry of Banach Spaces and Duality Mapping

The enchanting charms of this sublime science, mathematics reveal only to those who have the courage to go deeply into it.

Carl Friedrich Gauss

The study of convex sets is a branch of geometry, analysis and linear algebra that has numerous connections with other areas of mathematics and serves to unify many apparently diverse mathematical phenomena.

Victor Klee (1950)

In this chapter, we are mainly concern with geometrical structures such as convexity and smoothness of Banach spaces. Indeed, various kind of convexity and smoothness of Banach spaces play an important role in the existence and approximation of fixed points of nonlinear mappings. The necessary concepts of the geometry of normed spaces—strict convexity and uniform convexity—are also discussed. This chapter also deals with useful properties of duality mappings that interplay with these geometrical structures of Banach spaces. In Sect. 2.1, we deal with strict convexity while Sect. 2.2 mainly concern with uniform convexity. In Sect. 2.3, we discuss modulus of convexity. In Sect. 2.4, we mainly concern with smoothness of Banach spaces. Section 2.5 mainly deals with the concept of duality mapping from a Banach space X to its dual X^*. A discussion on these is important for a better understanding of the properties of duality mapping.

© Springer Nature Singapore Pte Ltd. 2018
H. K. Pathak, *An Introduction to Nonlinear Analysis and Fixed Point Theory*,
https://doi.org/10.1007/978-981-10-8866-7_2

2.1 Strict Convexity

Throughout this section, X denotes a real Banach space endowed with norm $\| \cdot \|$ and X^* its dual with norm $\| \cdot \|_*$. For $x \in X$, $j \in X^*$, the duality pair (x, j) is the value of j at x. Alternatively, we will also use $j(x)$ for the same.

Note that the basic property of a norm of a Banach space X is that it is always convex, i.e.

$$\|\lambda x + (1 - \lambda)\|y\| \leqslant \lambda\|x\| + (1 - \lambda)\|y\| \text{ for all } x, y \in X \text{ and } \lambda \in [0, 1].$$

However, we notice that a number of Banach spaces do not have equality when $x \neq y$, i.e.

$$\|\lambda x + (1 - \lambda)\|y\| < \lambda\|x\| + (1 - \lambda)\|y\| \text{ for all } x, y \in X \text{ and } \lambda \in [0, 1].$$

This suggests strict convexity of norm.

Definition 2.1 Let C be a subset of X, then $e \in C$ is said to be extreme point of C if e is not an interior point of any segment with endpoints in C. That is,

$$e \neq \lambda x + (1 - \lambda)y \text{ where } x, y \in C \text{ and } 0 < \lambda < 1.$$

Definition 2.2 A Banach space X is said to be *strictly convex* if every point of the unit sphere $S_X = \{x \in X : \|x\| = 1\}$ is an extreme point.

Alternatively, X is strictly convex iff $x \neq y$, $\|x\| = \|y\| = 1$ implies that $\|\lambda x + (1 - \lambda)y\| < 1$, $0 < \lambda < 1$.

This means that the mid-point $\frac{x+y}{2}$ of two distinct points x and y in the unit sphere S_X of X does not lie on S_X. In other words, if $x, y \in S_X$ with $\|x\| = \|y\| = \left\|\frac{x+y}{2}\right\|$, then $x = y$.

Example 2.1 Consider $\mathbb{R}^2$ with norm $\| \cdot \|_2$ defined as $\|x\|_2 = \sqrt{x_1^2 + x_2^2}$, $x = (x_1, x_2) \in \mathbb{R}^2$. The space $\mathbb{R}^2$, equipped with this norm, is strictly convex. This is very easily seen by looking at the unit sphere given in Fig. 1.4.

However, one can easily see that $\mathbb{R}^2$ is not strictly convex with respect to the 1-norm and ∞-norm defined as

$$\|x\|_1 = |x_1| + |x_2|,$$

$$\|x\|_\infty = \max(|x_1|, |x_2|).$$

Example 2.2 (*i*) Consider $X = \mathbb{R}^n$, $n \geq 2$ with 2- norm $\| \cdot \|_2$ defined by

$$\|x\|_2 = \Big(\sum_{i=1}^n x_i^2 \Big)^{1/2}, \quad x = (x_1, x_2, \dots, x_n) \in \mathbb{R}^n.$$

Then, X is strictly convex.

(*ii*) Consider $X = \mathbb{R}^n$, $n \geq 2$ with 1-norm $\| \cdot \|_1$ defined by

$$\|x\|_1 = \sum_{i=1}^n |x_i|, \quad x = (x_1, x_2, \dots, x_n) \in \mathbb{R}^n.$$

Then, X is not strictly convex. To see it, let $x = (1, 0, 0, \dots, 0)$ and $y = (0, 1, 0, \dots, 0)$. It is easy to see that $\|x\|_1 = 1 = \|y\|_1 = \left\| \frac{x+y}{2} \right\|_1$, but $x \neq y$.

(*iii*) Consider $X = \mathbb{R}^n$, $n \geq 2$ with ∞-norm $\| \cdot \|_\infty$ defined by

$$\|x\|_\infty = \max_{1 \leq i \leq n} |x_i|, \quad x = (x_1, x_2, \dots, x_n) \in \mathbb{R}^n.$$

Then, X is not strictly convex. To see it, let $x = (1, 1, 0, \dots, 0)$ and $y = (0, 1, 0, \dots, 0)$. It is easy to see that $\|x\|_\infty = 1 = \|y\|_\infty = \left\| \frac{x+y}{2} \right\|_\infty$, but $x \neq y$.

Example 2.3 The spaces ℓ_1 and ℓ_∞ are not strictly convex.
(i) ℓ_1 is not strictly convex; for, let $x = (1, 0, 0, \dots)$ and $y = (0, 1, 0, 0, \dots)$. Then,

$$\|x\|_1 = \|y\|_1 = 1 \quad \text{and also} \quad \left\| \frac{x+y}{2} \right\|_1 = \left\| \Big(\frac{1}{2}, \frac{1}{2}, 0, 0, \dots \Big) \right\|_1 = 1.$$

(ii) ℓ_∞ is not strictly convex; for, let $x = (1, 0, 0, \dots)$ and $y = (1, 1, 1, \dots)$. Then,

$$\|x\|_\infty = \|y\|_\infty = 1 \quad \text{and} \quad \left\| \frac{x+y}{2} \right\|_\infty = \left\| \Big(1, \frac{1}{2}, \frac{1}{2}, \dots \Big) \right\|_\infty = 1.$$

Example 2.4 The space $C[0, 1]$ is not strictly convex. For, if $x = t$, $y = t^2$, then we have

$$\|x\| = \|y\| = 1 \quad \text{and} \quad \left\| \frac{x+y}{2} \right\|_1 = 1.$$

Proposition 2.1 *The following assertions are equivalent:*

(*i*)   *X is strictly convex.*
(*ii*)   *The boundary of the unit ball $\mathbb{B}_X$ contains no line segments.*
(*iii*)   *If $x \neq y$ and $\|x\| = \|y\| = 1$, then $\|x + y\| < 2$.*
(*iv*)   *If for $x, y, z \in X$ we have $\|x - y\| = \|x - z\| + \|z - y\|$, then there exists $\lambda \in [0, 1]$ so that $z = \lambda x + (1 - \lambda)y$.*

(v)   *Any $j \in X^*$ assumes its supremum at most in one point of the unit ball $\mathbb{B}_X$.*

Proof The implications $(i) \Rightarrow (ii)$, $(i) \Rightarrow (iii)$ and $(iii) \Rightarrow (ii)$ are straightforward, so we omit the details.

$(ii) \Rightarrow (i)$. Let $x, y \in X, x \neq y, \|x\| = \|y\| = 1$ and $t_0 \in (0, 1)$ such that $t_0 x + (1 - t_0)y = 1$; we shall prove that the segment line $[x, y]$ is on the unit ball $\mathbb{B}$, which is impossible.

Let us take $t_0 < t < 1$. Then from the equality

$$1 = t_0 x + (1 - t_0)y = \frac{t_0}{t}[tx + (1 - t)y] + \left(1 - \frac{t_0}{t}\right)y$$

we obtain

$$1 = \|t_0 x + (1 - t_0)y\| \leqslant \frac{t_0}{t}\|tx + (1 - t)y\| + \left(1 - \frac{t_0}{t}\right).$$

On simplification, this yields $\|tx + (1 - t)y\| \geqslant 1$; hence, $\|tx + (1 - t)y\| = 1$.

The case, when $0 < t < t_0$, can be proved analogously.

$(i) \Rightarrow (iv)$. Let $x, y, z \in X$ be that $\|x - y\| = \|x - z\| + \|z - y\|$; without loss of generality, we can suppose that $\|x - z\| \neq 0$, $\|z - y\| \neq 0$ and $\|x - z\| \leqslant \|z - y\|$. Then, we have

$$\left\| \frac{1}{2} \frac{x - z}{\|x - z\|} + \frac{1}{2} \frac{z - y}{\|z - y\|} \right\|$$
$$\geqslant \left\| \frac{1}{2} \frac{x - z}{\|x - z\|} + \frac{1}{2} \frac{z - y}{\|x - z\|} \right\| - \left\| \frac{1}{2} \frac{z - y}{\|x - z\|} - \frac{1}{2} \frac{z - y}{\|z - y\|} \right\|$$
$$= \frac{1}{2} \frac{\|x - y\|}{\|x - z\|} - \frac{1}{2} \frac{\|z - y\| - \|x - z\|}{\|x - z\|}$$
$$= \frac{1}{2} \frac{\|x - z\| + \|z - y\| - \|z - y\| + \|x - z\|}{\|x - z\|} = 1.$$

Hence, $\left\| \frac{x-z}{\|x-z\|} + \frac{z-y}{\|z-y\|} \right\| = 2$. Because X is strictly convex, we must have $\frac{x-z}{\|x-z\|} = \frac{z-y}{\|z-y\|}$ and this after simplification yields

$$z = \frac{\|z - y\|}{\|x - z\| + \|z - y\|}x + \frac{\|x - z\|}{\|x - z\| + \|z - y\|}y$$

or

$$z = \frac{\|z - y\|}{\|x - z\| + \|z - y\|}x + \left(1 - \frac{\|z - y\|}{\|x - z\| + \|z - y\|}\right)y$$

Thus, there exists $\lambda \in [0, 1]$ such that $z = \lambda x + (1 - \lambda)y$, where $\lambda = \frac{\|z-y\|}{\|x-z\|+\|z-y\|}$.

$(iv) \Rightarrow (iii)$. Let $x, y \in X$ be such that $x \neq y$ and $\|x\| = \|y\| = \left\|\frac{x+y}{2}\right\| = 1$; then $\|x + y\| = \|x\| + \|y\|$. Comparing this equality with $\|x - y\| = \|x - z\| + \|z - y\|$, we see that there exists $\lambda \in (0, 1)$ so that $z = 0 = \lambda x - (1 - \lambda)y$, i.e. $x = \frac{1-\lambda}{\lambda}y$. Hence, $\lambda = \frac{1}{2}$ so that $x = y$, which is a contradiction.

$(i) \Rightarrow (v)$. Suppose that for $j \in X^*$, there are two vectors $x \neq y$, $\|x\| = \|y\| = 1$ with $(x, j) = (y, j) = \|j\|_*$; then for $\lambda \in (0, 1)$, we have

$$\|j\|_* = \lambda(x, j) + (1 - \lambda)(y, j) = (\lambda x + (1 - \lambda)y, j)$$
$$\leqslant \|\lambda x + (1 - \lambda)y\| \|j\|_*.$$

Because X is strictly convex, this inequality yields $1 \leqslant \|\lambda x + (1 - \lambda)y\| < 1$, which is absurd.

$(v) \Rightarrow (iii)$. Suppose $x, y \in X$, so that $x \neq y$, $\|x\| = \|y\| = 1$ and $\|x + y\| = 2$. By the Hahn–Banach theorem, there exists an $j \in X^*$ such that $\|j\| = 1$ and $\left(\frac{x+y}{2}, j\right) = \left\|\frac{x+y}{2}\right\| = 1$ hence $(x, j) + (y, j) = 2$. Further, $(x, j) \leqslant \|x\|\|j\|_* = 1$ and $(y, j) \leqslant \|y\|\|j\|_* = 1$, it follows that

$$(x, j) = (y, j) = \|j\|_* = 1$$

which contradicts (iii).

Proposition 2.2 *A convex set C in a strictly convex space X has at most one point of minimum norm.*

Proof Suppose $\rho = \inf_{y \in C} \|y\|$. Then, we need to show that C meets the sphere $S(0, \rho) = \{x : \|x\| = \rho\}$ in at most one point. Suppose, if possible, there exist two distinct points x_1 and x_2 in C such that

$$\rho = \|x_1\| = \|x_2\|.$$

By convexity of C, $\frac{x_1+x_2}{2} \in C$. As this point is distinct from both minimum norm points x_1 and x_2 in C, it follows that $\rho \leqslant \left\|\frac{x_1+x_2}{2}\right\|$. Moreover,

$$\left\|\frac{x_1 + x_2}{2}\right\| \leqslant \frac{1}{2}\Big[\|x_1\| + \|x_2\|\Big] = \rho.$$

Therefore, $\left\|\frac{x_1+x_2}{2}\right\| = \rho$.

Notice that the point $\frac{x_1+x_2}{2}$ is an interior point of the sphere $S(0, \rho)$. By strict convexity of X, $\frac{x_1+x_2}{2}$ cannot be an extreme point. It follows that $x_1 = x_2$. ∎

Proposition 2.3 *Let X be a Banach space. Then, the following statements are equivalent:*

(a) *X is strictly convex.*
(b) *For each nonzero $j \in X^*$, there exists at most one x in X with $\|x\| = 1$ such that $(x, j) = j(x) = \|j\|_*$.*

Proof $(a) \implies (b)$. Let us consider $C = \{y \in X : (y, j) = \|j\|_*\}$. Then, we see that C is a convex subset of X; for, if $y_1, y_2 \in C$ and $0 < \lambda < 1$, by linearity of j

$$\begin{aligned}
(\lambda y_1 + (1 - \lambda)y_2, j) &= j(\lambda y_1 + (1 - \lambda)y_2) \\
&= \lambda j(y_1) + (1 - \lambda)j(y_1) \\
&= \lambda\|j\|_* + (1 - \lambda)\|j\|_* \\
&= \|j\|_*.
\end{aligned}$$

This shows that $y_1, y_2 \in C$ and $0 < \lambda < 1 \implies \lambda y_1 + (1 - \lambda)y_2$. Further, since $\|j\|_* = (y, j) \leqslant \|j\|_*\|y\|$, it follows that $\|y\| \geqslant 1$ for all $y \in C$. By Proposition 2.2, there is at most one $x \in C$ such that $\inf_{y \in C} \|y\| = \|x\|$, and $\|x\| = 1$.

$(b) \implies (a)$. Suppose $x, y \in S_X$ with $x \neq y$ such that $\left\|\frac{x+y}{2}\right\| = 1$. By Corollary 1.3, there exists a functional $j \in S_{X^*}$ such that

$$\|j\|_* = 1 \text{ and } \left(\frac{x+y}{2}, j\right) = \left\|\frac{x+y}{2}\right\|.$$

Because $(x, j) \leqslant \|x\|\|j\|_* = 1$ and $(y, j) \leqslant \|y\|\|j\|_* = 1$, we must have $(x, j) = (y, j)$. This implies, by hypothesis, that $x = y$. From this, we conclude that $(b) \implies (a)$ is proved. ∎

Proposition 2.4 *Let X be a strictly convex Banach space. If $\|x + y\| = \|x\| + \|y\|$ for $0 \neq x \in X$ and $y \in X$, then there exists $\lambda > 0$ such that $y = \lambda x$.*

Proof Let $x, y \in X \setminus \{0\}$ be such that $\|x + y\| = \|x\| + \|y\|$. From Corollary 1.3, there exists $j \in X^*$ such that

$$(x + y, j) = \|x + y\|, \text{ and } \|j\|_* = 1.$$

Because $\langle x, j \rangle \leqslant \|j\|_*\|x\| = \|x\|$ and $\langle y, j \rangle \leqslant \|j\|_*\|y\| = \|y\|$, we must have $\langle x, j \rangle = \|x\|$ and $\langle y, j \rangle = \|y\|$. Otherwise, we have

$$\|x + y\| = (x + y, j) = \langle x, j \rangle + \langle y, j \rangle < \|x\| + \|y\| = \|x + y\|,$$

a contradiction. This means that $\langle \frac{x}{\|x\|}, j \rangle = \langle \frac{y}{\|y\|}, j \rangle = 1$. Because X is strictly convexity of X, it follows from Proposition 2.3 that $\frac{x}{\|x\|} = \frac{y}{\|y\|}$. Therefore, $y = \lambda x$ where $\lambda = \frac{\|y\|}{\|x\|}$. ∎

Proposition 2.5 *Let X be a Banach space. Then, the following statements are equivalent:*

(a)   *X is strictly convex.*
(b)   *For every $1 < p < \infty$,*

$$\|\lambda x + (1 - \lambda)y\|^p < \lambda\|x\|^p + (1 - \lambda)\|y\|^p \text{ for all } x, y \in X, x \neq y \text{ and } \lambda \in (0, 1).$$

Proof $(a) \implies (b)$. Suppose X is strictly convex. Assume that $x, y \in X$ with $x \neq y$. By strict convexity of X,

$$\|\lambda x + (1 - \lambda) y\|^p < (\lambda \|x\| + (1 - \lambda)\|y\|)^p \quad \text{for all } \lambda \in (0, 1). \qquad (2.1)$$

If $\|x\| = \|y\|$, then the above inequality gives

$$\|\lambda x + (1 - \lambda) y\|^p < \|x\|^p = \lambda \|x\|^p + (1 - \lambda)\|y\|^p.$$

We now assume that $\|x\| \neq \|y\|$ and consider the function $t \mapsto t^p$, for $1 < p < \infty$. Then, we see that this is a convex function and $(\frac{a+b}{2})^p < \frac{a^p + b^p}{2}$ for all $a, b \geqslant 0$ and $a \neq b$.

Thus, from (2.1) with $t = \frac{1}{2}$, we obtain

$$\left\| \frac{x+y}{2} \right\|^p \leqslant \left(\frac{\|x\| + \|y\|}{2} \right)^p < \frac{1}{2} (\|x\|^p + \|y\|^p). \qquad (2.2)$$

We now discuss the problem in the following two cases:
Case 1. If $0 < \lambda \leqslant \frac{1}{2}$, then we have from (2.1) by using (2.2)

$$
\begin{aligned}
\|\lambda x + (1 - \lambda) y\|^p &= \left\| 2\lambda \frac{x+y}{2} + (1 - 2\lambda) y \right\|^p \\
&\leqslant \left(2\lambda \left\| \frac{x+y}{2} \right\| + (1 - 2\lambda)\|y\| \right)^p \\
&< 2\lambda \left\| \frac{x+y}{2} \right\|^p + (1 - 2\lambda)\|y\|^p \\
&< \lambda(\|x\|^p + \|y\|^p) + (1 - 2\lambda)\|y\|^p \\
&= \lambda \|x\|^p + (1 - \lambda)\|y\|^p.
\end{aligned}
$$

Case 2. If $\frac{1}{2} < \lambda < 1$, take $\mu = 2\lambda - 1$; then, we have from (2.1)

$$
\begin{aligned}
\|\lambda x + (1 - \lambda) y\|^p &= \left\| \mu x + (1 - \mu) \frac{x+y}{2} \right\|^p \\
&\leqslant \left(\mu \|x\| + (1 - \mu) \left\| \frac{x+y}{2} \right\| \right)^p \\
&< \mu \|x\|^p + (1 - \mu) \left\| \frac{x+y}{2} \right\|^p \\
&< \mu \|x\|^p + \frac{1}{2}(1 - \mu)(\|x\|^p + \|y\|^p) \quad \text{(by(2.2))} \\
&= \lambda \|x\|^p + (1 - \lambda)\|y\|^p.
\end{aligned}
$$

$(b) \implies (a)$. It is obvious.   ∎

2.2 Uniform Convexity

In 1936, Clarkson [146] introduced the notion of uniform convexity of norm in a Banach space. Stated in geometric terms, a norm is uniformly convex if whenever the mid-point of a variable chord in the unit sphere of the space approaches the boundary of the sphere, and the length of the chord approaches zero.

Definition 2.3 A Banach space X is said to be *uniformly convex* if for given $\varepsilon > 0, 0 < \varepsilon \leqslant 2$, there exists a $\delta = \delta(\varepsilon) > 0$ such that

$$\left\| \frac{x + y}{2} \right\| \leqslant 1 - \delta \text{ whenever } \|x\| \leqslant 1, \|y\| \leqslant 1, \text{ and } \|x - y\| \geqslant \varepsilon.$$

In other words, a Banach space X is uniformly convex if x and y are in the unit ball $\mathbb{B}_X = \{x \in X : \|x\| \leqslant 1\}$ with $\|x - y\| \geqslant \varepsilon > 0$, the mid-point of x and y lies inside the unit ball $\mathbb{B}_X$ at a distance of at least δ from the unit sphere $\mathbb{S}_X$.

Equivalently, X is uniformly convex if whenever $x_n, y_n \in X$ with

$$\|x_n\| = \|y_n\| = 1 \text{ and } \|x_n + y_n\| \to 2, \text{ then } \|x_n - y_n\| \to 0.$$

In the sequel, as a direct consequence of the definition of uniform convexity, we shall exploit the following property: minimizing sequences in closed convex subsets are convergent. That is, if $C \subset X$ is nonvoid, closed and convex and $x_n \in C$ is such that $\lim_{n \to \infty} \| x_n \| = \inf_{y \in C} \| y \|$, then there exists a unique $x \in C$ such that

$$\| x \| = \inf_{y \in C} \| y \| \quad \text{and} \quad \lim_{n \to \infty} x_n = x.$$

Example 2.5 Every Hilbert space $\mathcal{H}$ is a uniformly convex space.

This follows from the parallelogram law:

$$\|x - y\|^2 + \|x + y\|^2 = 2[\|x\|^2 + \|y\|^2].$$

Let us take $x, y \in B_{\mathcal{H}}$ with $x \neq y$ and $\|x - y\| \geq \varepsilon$. Then

$$\|x + y\|^2 \leqslant 4 - \varepsilon^2.$$

It follows from the above inequality that

$$\left\| \frac{x + y}{2} \right\| \leqslant 1 - \delta(\varepsilon),$$

where $\delta(\varepsilon) = 1 - \sqrt{1 - \varepsilon^2/4}$. Thus, we conclude that $\mathcal{H}$ is uniformly convex.

Alternatively, taking x_n for x and y_n for y and putting $\|x_n\| = \|y_n\| = 1$, we get

$$\|x_n - y_n\|^2 = -[\|x_n + y_n\|^2 - 4].$$

Thus, if $\|x_n + y_n\| \to 2$ the R.H.S. goes to zero as $n \to \infty$. Hence, the result follows.

Example 2.6 The spaces ℓ_1 and ℓ_∞ are not uniformly convex. To see that ℓ_1 is not uniformly convex, let us take $x = (1, 0, 0, 0, \ldots), y = (0, 0, -1, 0, \ldots) \in \ell_1$ and $\varepsilon = 1$. Then,

$$\|x\|_1 = \|y\|_1 = 1 \quad \text{and} \quad \|x - y\|_1 = 2 > 1 = \varepsilon.$$

But $\left\|\frac{x+y}{2}\right\|_1 = 1$ and there is no $\delta > 0$ such that $\|\frac{x+y}{2}\|_1 \leqslant 1 - \delta$. Thus, ℓ_1 is not uniformly convex.

The space ℓ_∞ is also not uniformly convex. To see this, let us take $x = (1, 1, 0, 0, \ldots), y = (-1, 1, 0, 0, \ldots) \in \ell_\infty$ and $\varepsilon = 1$. Then

$$\|x\|_\infty = \|y\|_\infty = 1 \quad \text{and} \quad \|x - y\|_\infty = 2 > 1 = \varepsilon.$$

But $\left\|\frac{x+y}{2}\right\|_\infty = 1$ and there is no $\delta > 0$ such that $\left\|\frac{x+y}{2}\right\|_\infty \leqslant 1 - \delta$. Thus, ℓ_∞ is not uniformly convex.

Example 2.7 $L_p, \ell_p, 1 < p < \infty$ are uniformly convex spaces. For the proof of this fact, refer Day [159].

For the class of uniformly convex Banach spaces, we have the following important results:

Theorem 2.1 *Every uniformly convex space X is strictly convex.*

Proof Let X be a uniformly convex Banach space. It is immediate from the Definition 2.3 that every uniformly convex space is strictly convex. ∎

Observation

- The Banach spaces ℓ_p, ℓ_p^n, n being nonnegative integer, and $L_p[a, b]$ with $1 < p < \infty$ are uniformly convex.
- The Banach spaces $\ell_1, c, c_0, \ell_\infty, L_1[a, b], C[a, b]$ and $L_\infty[a, b]$ are not strictly convex.

Theorem 2.2 (Milman's theorem) *Every uniformly convex Banach space X is reflexive.*

This theorem, originally due to Milman [399], has been proved by many authors in different ways (see, for instance, Pettis [487], Kakutani [302] and many others). However, we give the proof based on James' characterization of reflexivity. James has shown that a Banach space is reflexive iff for every $f \in X^*$ attain its norm on the unit ball. One may observe that this is equivalent to every $f \in S_{X^*}$ attaining its norm on S_X. The proof of Milman's theorem is prefixed by the following lemma.

Lemma 2.1 *Let X be a uniformly convex Banach space, $S_{X^*} := \{f \in X^* : \|f\|_* = 1\}$ the unit sphere in X^*, and let $f \in S_{X^*}$. Let $\{x_n\}$ be a sequence in S_X such that $j(x_n) \to 1$. Then, $\{x_n\}$ is a Cauchy sequence.*

Proof Suppose that $\{x_n\}$ is not a Cauchy sequence, then there exists $\varepsilon > 0$ and two subsequences $\{x_{n_i}\}$ and $\{x_{n_j}\}$ of $\{x_n\}$ such that $\|x_{n_i} - x_{n_j}\| \geq \varepsilon$. The uniform convexity of X assures that $\exists \delta = \delta(\varepsilon) > 0$ such that $\left\| \frac{x_{n_i}+x_{n_j}}{2} \right\| < 1 - \delta$. We now observe that

$$\left| f\left(\frac{x_{n_i} + x_{n_j}}{2} \right) \right| \leqslant \|f\|_* \left\| \frac{x_{n_i} + x_{n_j}}{2} \right\| < \|f\|_*(1 - \delta) = 1 - \delta$$

and $f(x_n) \to 1$, yield a contradiction. Thus, the sequence $\{x_n\}$ which we call a norming sequence for $f \in X^*$ is a Cauchy sequence in X.

Proof of Milman's theorem Let $f \in X^*$ and $\{x_n\}$ a norming sequence in S_X for f, so that $f(x_n) \to 1$. By Lemma 2.1, $\{x_n\}$ is a Cauchy sequence. Because X is a Banach space and S_X is closed, $\{x_n\}$ converges to some $x \in S_X$. By continuity of f,

$$f(x) = f(\lim_{n \to \infty} x_n) = \lim_{n \to \infty} f(x_n) = 1$$

so that f attain its norm on S_X. Thus, by James' characterization of reflexivity for Banach spaces, X is reflexive.   ∎

Combining the above two results, we have the following theorem.

Theorem 2.3 *Every uniformly convex space X is strictly convex and reflexive.*

But the converse of the above theorem is false. In fact, Day in [158, 159] has proved that there exists a large class of spaces which are reflexive and strictly convex but are not isomorphic to any uniformly convex space. We very briefly discuss the example given by him.

Example 2.8 Let $X_n, n = 1, 2, \ldots$ be a sequence of Banach spaces and let $1 < p < \infty$. Define a space X as follows:

$$X = \left\{ x = (x_1, x_2, \ldots), x_n \in X_n \text{ and } \|x\| = \left(\sum_{j=1}^n \|x_n\|_n^p \right)^{1/p} < \infty \right\}$$

X is a Banach space. It can be shown that X is strictly convex and reflexive iff all X_n are strictly convex and reflexive. However, it is possible to make a simple choice of the spaces X_n so that X cannot be given a uniformly convex norm defining the same topology.

Observation

- The spaces ℓ_1 and L_1 are neither reflexive nor strictly convex.
- A reflexive Banach space is not necessarily uniformly convex. For example, consider a finite dimensional Banach space in which the surface of the unit ball has a "flat" part. We note that such a Banach space is reflexive because of finite dimension. But the flat portion in the surface of the ball makes it nonuniformly convex.
- It is well known that a Banach space X is reflexive if and only if every bounded sequence of elements of X contains a weakly convergent sequence.

Definition 2.4 A Banach space X is said to be locally uniformly convex if for given $\varepsilon > 0$ and $x \in S_X$, there exists a $\delta = \delta(x, \varepsilon) > 0$ such that

$$\|x - y\| \geq \varepsilon \Rightarrow \left\| \frac{x+y}{2} \right\| \leqslant 1 - \delta \text{ for all } y \in S_X.$$

Definition 2.5 X is said to be weakly uniformly convex at $x^* \in S_{X^*}$, if whenever $x_n, y_n \in X$ with

$$\|x_n\| = \|y_n\| = 1 \text{ and } \langle x^*, x_n + y_n \rangle \to 2, \text{ then } \|x_n - y_n\| \to 0.$$

X is weakly uniformly convex if it is uniformly convex at any $x^* \in S_{X^*}$.

Remark 2.1 It is clear that the uniform convexity implies both local and the weak uniform convexity.

Observation

- Every uniformly convex Banach space is locally uniformly convex.
- Every locally uniformly convex Banach space is strictly convex.

We now discuss the relationship between uniform convexity of the norm and differentiability of the norm.

Definition 2.6 Let X be a normed space with norm $\| \cdot \|$ and $\mathbb{B}_X = \{x \in X : \|x\| \leqslant 1\}$, the closed unit ball in X. Then

(i) the norm in X is said to be weakly differentiable at $x_0 \in X$ if and only if

$$\lim_{t \to 0} \left(\frac{\|x_0 + tx\| - \|x_0\|}{t} \right) \text{ exists for every } x \in X.$$

(ii) If the convergence to the limit is uniform in the closed unit ball $\mathbb{B}_X$, the norm is said to be strongly differentiable at x_0.

(iii) If the norm in X is everywhere differentiable and convergence to the limit is uniform with respect to x_0 and x, when $\|x_0\| = 1$ and $\|x\| \leqslant 1$, then the norm is called uniformly strongly differentiable.

In 1940, Smulian [573] proved the following two theorems giving the necessary and sufficient condition for the strong differentiability of the norm in a Banach space X and its dual X^*.

Theorem 2.4 *For the strong differentiability of the norm in X^* at the point f_0, $\|f_0\| = 1$, it is necessary and sufficient that the following condition be satisfied:*

$$\lim_{n \to \infty} f_0(x_n) = \|f_0\|, \|x_n\| = 1 \implies \{x_n\} \text{ is convergent.}$$

Theorem 2.5 *For the strong differentiability of the norm in X at the point x_0, $\|x_0\| = 1$, it is necessary and sufficient that the following condition be satisfied:*

$$\lim_{n \to \infty} f_n(x_0) = \|x_0\|, f_n \in X^*, \|f_n\| = 1 \implies \{f_n\} \text{ is convergent.}$$

Smulian [573] has further shown that the norm in X is uniformly strongly differentiable if and only if X^* is uniformly convex.

As an easy consequence of the above fact, we have the following theorem for uniformly convex Banach spaces.

Theorem 2.6 *If X is a uniformly convex Banach space and*

$$x_n \rightharpoonup x_0 \text{ and } \|x_n\| \to \|x_0\|, \text{ then } x_n \to x_0.$$

In 1955, Lovaglia [374] introduced a weaker type of convexity called local uniform convexity. Geometrically, this differs from uniform convexity in that it is required that one endpoint of the variable chord remain fixed.

Definition 2.7 A space X is called locally uniformly convex if and only if given $\varepsilon > 0$ and an element x with $\|x\| = 1$, there exists $\delta(\varepsilon, x) > 0$ such that

$$\left\| \frac{x + y}{2} \right\| \leqslant 1 - \delta(\varepsilon, x) \text{ whenever } \|x - y\| \geqslant \varepsilon \text{ and } \|y\| = 1.$$

It is clear that uniform convexity implies local uniform convexity and local uniform convexity implies strict convexity. There exist spaces which are locally uniformly convex but are not isomorphic to any uniformly convex space.

Example 2.9 Let X be as defined in Example 1.9 (with $p = 2$) where x_n is the n-dimensional space of vectors x_n,

$$x_n = (x_n^1, x_n^2, \ldots, x_n^n), \text{ with } \|x_n\|_n = \left(\sum_{j=1}^{n} |x_n^j|^n \right)^{1/n}.$$

Then, each X_n is uniformly convex and is therefore locally uniformly convex. Lovaglia [374] has shown that the space X so constructed is locally uniformly convex and is not isomorphic to any uniformly convex space.

The following theorem, which is due to Lovaglia, gives the relationship between local uniform convexity and differentiability of the norm is a Banach space X.

Theorem 2.7 *If X^* is locally uniformly convex, then norm is X is strongly differentiable.*

2.3 Modulus of Convexity

Definition 2.8 Let X be a Banach space. Then, a function $\delta_X : [0, 2] \to [0, 1]$ is said to be the modulus of convexity of X if

$$\delta_X(\varepsilon) = \inf \left\{ 1 - \left\| \frac{x+y}{2} \right\| : \|x\| \leqslant 1, \|y\| \leqslant 1, \|x - y\| \geqslant \varepsilon \right\}.$$

Example 2.10 Let $\mathcal{H}$ be a Hilbert space. As we see in Example 2.5 that

$$\delta_{\mathcal{H}}(\varepsilon) = 1 - \sqrt{1 - \frac{\varepsilon^2}{4}}, \ \varepsilon \in (0, 2].$$

We now give the modulus of convexity for $\ell_p (2 \leqslant p < \infty)$ spaces. The following result gives an analogue of the parallelogram law in $\ell_p (2 \leqslant p < \infty)$ spaces. But first we recall the useful Lemmas A.1.1 and A.1.4 of Appendix A.

Proposition 2.6 *In $\ell_p (2 \leqslant p < \infty)$ spaces, we have the following inequality:*

$$\|x + y\|^p + \|x - y\|^p \leqslant 2^{p-1}(\|x\|^p + \|y\|^p) \ \text{for all } x, y \in \ell_p. \tag{2.3}$$

Proof For $a, b \in \mathbb{R}$ and $p \in [2, \infty)$, applying Lemmas A.1.1 and A.1.4 of Appendix A, we get

$$\begin{aligned}
|a + b|^p + |a - b|^p &= \left[(|a + b|^p + |a - b|^p)^{2/p} \right]^{p/2} \\
&\leqslant \left[|a + b|^2 + |a - b|^2 \right]^{p/2} \ \text{(since } 2/p \leqslant 1) \\
&\leqslant \left[2|a|^2 + 2|b|^2 \right]^{p/2} = 2^{p/2} \left[|a|^2 + |b|^2 \right]^{p/2} \\
&\leqslant 2^{p/2} 2^{(p-1)/2} \left(|a|^p + |b|^p \right) = 2^{p-1} \left(|a|^p + |b|^p \right).
\end{aligned}$$

Hence for $x = \{x_n\}_{n=1}^{\infty}, y = \{y_n\}_{n=1}^{\infty} \in \ell_p$, the above inequality yields

$$\sum_{n=1}^{\infty} |x_n + y_n|^p + \sum_{n=1}^{\infty} |x_n - y_n|^p \leqslant 2^{p-1} \left(\sum_{n=1}^{\infty} |x_n|^p + \sum_{n=1}^{\infty} |y_n|^p \right).$$

Thus, for any points $x, y \in \ell_p (2 \leqslant p < \infty)$, the following analogue of the parallelogram is satisfied:

$$\|x + y\|^p + \|x - y\|^p \leqslant 2^{p-1}(\|x\|^p + \|y\|^p) \text{ for all } x, y \in \ell_p.$$

Example 2.11 For the ℓ_p $(2 \leqslant p < \infty)$ space, the modulus of convexity $\delta_{\ell_p}(\varepsilon)$ is given by

$$\delta_{\ell_p}(\varepsilon) = 1 - \left(1 - \left(\frac{\varepsilon}{2}\right)^p\right)^{1/p}, \quad \varepsilon \in (0, 2).$$

To see this, let $\mathbb{B}$ denote the closed unit ball in ℓ_p, $\varepsilon \in (0, 2)$ and $x, y \in \mathbb{B}$ such that $\|x - y\| \geq \varepsilon$. Then from (2.3), we have

$$\|x + y\|^p \leqslant 2^{p-1}\left(|x|^p + |y|^p\right) - \|x - y\|^p \leqslant 2^{p-1}(1+1) - \varepsilon^p = 2^p - \varepsilon^p$$

which implies that

$$\left\|\frac{x+y}{2}\right\| \leqslant \left(1 - \left(\frac{\varepsilon}{2}\right)^p\right)^{1/p} = 1 - \left[1 - \left(1 - \left(\frac{\varepsilon}{2}\right)^p\right)^{1/p}\right] \leqslant 1 - \delta_{\ell_p}(\varepsilon),$$

where $\delta_{\ell_p}(\varepsilon) \geqslant 1 - \left(\frac{\varepsilon}{2}\right)^p\right)^{1/p}$.

Observation

- $\delta_X(0) = 0$ and $\delta_X(\varepsilon) \geqslant 0$ for all $\varepsilon \geq 0$.
- $\delta_{\ell_p}(\varepsilon) \geqslant 0$ for all $\varepsilon > 0$ $(1 < p < \infty)$.
- $\delta_{\mathcal{H}}(\varepsilon) = 1 - \sqrt{1 - (\varepsilon/2)^2}$, $\mathcal{H}$ being Hilbert space.
- $\delta_{\ell_p}(\varepsilon) = 1 - \left(1 - \left(\frac{\varepsilon}{2}\right)^p\right)^{1/p}$ $(2 \leqslant p < \infty)$.
- The modulus of convexity $\delta_{\ell_p}(\varepsilon)$ for ℓ_p $(1 < p \leqslant 2)$ satisfies the following implicit relation

$$\left|1 - \delta_{\ell_p}(\varepsilon) + \frac{\varepsilon}{2}\right|^p + \left|1 - \delta_{\ell_p}(\varepsilon) - \frac{\varepsilon}{2}\right|^p = 2.$$

- For any Banach space X and Hilbert space $\mathcal{H}$, Hilbert space is the most convex Banach space, i.e.

$$\delta_X(\varepsilon) \leqslant \delta_{\mathcal{H}}(\varepsilon) \text{ for all } \varepsilon > 0.$$

Theorem 2.8 *A Banach space X is strictly convex if and only if $\delta_X(2) = 1$.*

Proof Suppose X is a strictly convex Banach space with modulus of convexity δ_X. Assume that $x, y \in X$ such that $x \neq -y$, $\|x\| = \|y\| = 1$ and $\|x - y\| = 2$. By strict convexity of we have

$$1 = \left\|\frac{x-y}{2}\right\| = \left\|\frac{x + (-y)}{2}\right\| < 1,$$

a contradiction. Hence, $x = -y$. Therefore, the modulus of convexity of X gives $\delta_X(2) = 1$.

Conversely, suppose $\delta_X(2) = 1$. Let $x, y \in X$ such that $\|x\| = \|y\| = \left\|\frac{x+y}{2}\right\| = 1$. Then

$$\left\|\frac{x-y}{2}\right\| = \left\|\frac{x+(-y)}{2}\right\| \leqslant 1 - \delta_X(\|x-y\|) = 1 - \delta_X(\|x-(-y)\|) = 1 - \delta_X(2) = 0.$$

This implies that $x = y$. Thus, $\|x\| = \|y\|$ and $\|x + y\| = 2 = \|x\| + \|y\|$ imply that $x = y$. Therefore, X is strictly convex.

Theorem 2.9 *A Banach space X is uniformly convex if and only if $\delta_X(\varepsilon) > 0$ for all $\varepsilon \in (0, 2]$.*

Proof Suppose X is a uniformly convex Banach space. Then for $\varepsilon > 0$, there exists $\delta_X(\varepsilon) > 0$ such that

$$0 < \delta_X(\varepsilon) \leqslant 1 - \left\|\frac{x+y}{2}\right\|$$

for all $x, y \in X$ with $\|x\| \leqslant 1$, $\|y\| \leqslant 1$ and $\|x - y\| \geqslant \varepsilon$. Therefore, $\delta_X(\varepsilon) > 0$.

Conversely, suppose X is a Banach space with modulus of convexity δ_X such that $\delta_X(\varepsilon) > 0$ for all $\varepsilon \in (0, 2]$. Let $x, y \in X$ such that $\|x\| = 1$, $\|y\| = 1$ with $\|x - y\| \geqslant \varepsilon$ for fixed $\varepsilon \in (0, 2]$. By the modulus of convexity $\delta_X(\varepsilon)$, we have

$$0 < \delta_X(\varepsilon) \leqslant 1 - \left\|\frac{x+y}{2}\right\|,$$

which implies that

$$\left\|\frac{x+y}{2}\right\| \leqslant 1 - \delta(\varepsilon),$$

where $\delta(\varepsilon) = \delta_X(\varepsilon)$ is independent of x and y. Therefore, X is uniformly convex.

Observation

- The modulus of convexity is a real-valued function defined from $[0, 2]$ to $[0, 1]$ which is continuous on $[0, 2)$.
- A Banach space X is strictly convex if and only if $\delta_X(2) = 1$.
- A Banach space X is uniformly convex if and only if $\delta_X(\varepsilon) > 0$ for $\varepsilon \in (0, 2]$.

2.4 Smoothness of Banach Spaces

Let C be a nonempty closed convex subset of a normed space X such that the belongs to int C. A linear functional $j \in X^*$ is said to be tangent to C at a point $z \in \partial C$ if $j(z) = \sup_{x \in C} j(x)$.

If $H = \{x \in X : j(x) = 0\}$ is the hyperplane, then the set $H + z = \{x + z : x \in H\}$ is called tangent hyperplane to C at z.

Definition 2.9 A Banach space X is said to be smooth if for each $x \in S_X = \{x \in X : \|x\| = 1\}$, there exists a unique functional $j_x \in X^*$ such that $(x, j_x) = \|x\|$ and $\|j_x\| = 1$.

Geometrical meaning of smoothness of a Banach space By smoothness of a Banach space X, we shall mean that at each point $x \in S_X$, there is exactly one supporting hyperplane $\{j_x = 1\}$. This means that the hyperplane $\{j_x = 1\}$ is tangent at x to the unit ball $\mathbb{B}_X$, and this unit ball is contained in the half space $\{j_x \leqslant 1\}$.

2.4.1 Modulus of Smoothness

We now introduce the modulus of smoothness of a Banach space.

Definition 2.10 Let X be a Banach space. Then, a function $\rho_X : \mathbb{R}^+ \to \mathbb{R}^+$ is said to be modulus of smoothness of X if

$$\rho_X(t) = \sup \left\{ \frac{\|x + y\| + \|x - y\|}{2} - 1 : \|x\| = 1, \|y\| = t \right\}$$

$$= \sup \left\{ \frac{\|x + ty\| + \|x - ty\|}{2} - 1 : \|x\| = \|y\| = 1 \right\}, \quad t \geqslant 0.$$

The following result gives us an important relation between the modulus of X (respectively, X^*) and that of modulus of smoothness of X^* (respectively, X).

Theorem 2.10 *Let X be a Banach space. Then, the following equalities hold:*
(a) $\rho_{X^*}(t) = \sup\{\frac{t\varepsilon}{2} - \delta_X(\varepsilon) : 0 \leqslant \varepsilon \leqslant 2\}$ for all $t > 0$.
(b) $\rho_X(t) = \sup\{\frac{t\varepsilon}{2} - \delta_{X^*}(\varepsilon) : 0 \leqslant \varepsilon \leqslant 2\}$ for all $t > 0$.

Proof (a) By the definition of modulus of smoothness of X^*, we have

$$\rho_{X^*}(t) = \sup \left\{ \frac{\|x^* + ty^*\| + \|x^* - ty^*\|}{2} - 1 : \|x^*\| = \|y^*\| = 1 \right\}, \quad t \geqslant 0.$$

Therefore, for $t > 0$ we have

$$2\rho_{X^*}(t) = \sup \left\{ \|x^* + ty^*\| + \|x^* - ty^*\| - 2 : x^*, y^* \in S_{X^*} \right\}$$

$$= \sup \left\{ \|x\|\|x^* + ty^*\| + \|y\|\|x^* - ty^*\| - 2 : x, y \in S_X, x^*, y^* \in S_{X^*} \right\}$$

$$= \sup \left\{ (x, x^* + ty^*) + (y, x^* - ty^*) - 2 : x, y \in S_X, x^*, y^* \in S_{X^*} \right\}$$

$$= \sup \left\{ (x, x^*) + t(x, y^*) + (y, x^*) - t(y, y^*) - 2 : x, y \in S_X, x^*, y^* \in S_{X^*} \right\}$$

$$= \sup \left\{ (x, x^*) + (y, x^*) + t[(x, y^*) - (y, y^*)] - 2 : x, y \in S_X, x^*, y^* \in S_{X^*} \right\}$$

$$= \sup \left\{ (x + y, x^*) + t(x - y, y^*) - 2 : x, y \in S_X, x^*, y^* \in S_{X^*} \right\}$$

$$= \sup \left\{ \|x + y\| + t\|x - y\| - 2 : x, y \in S_X \right\}, \text{ as } \|x^*\| = 1 = \|y^*\|$$

$$= \sup \left\{ \|x + y\| + t\varepsilon - 2 : x, y \in S_X, \|x - y\| \leqslant \varepsilon, 0 \leqslant \varepsilon \leqslant 2 \right\}$$

$$= \sup \left\{ t\varepsilon - 2\delta_X(\varepsilon) : 0 \leqslant \varepsilon \leqslant 2 \right\}.$$

Part(b) can be proved using parallel argument as used in Part (a).

Observation

- $\rho_X(0) = 0$ and $\rho_X(t) \geqslant 0$ for all $t \geqslant 0$
- ρ_X is an increasing continuous convex function.
- $\frac{\rho_X(t)}{t}$ is an increasing function and $\rho_X(t) \leqslant t$ for all $t > 0$.

Definition 2.11 A Banach space X is said to be uniformly smooth if

$$\rho'_X(0) = \lim_{t \to 0} \frac{\rho_X(t)}{t} = 0.$$

Example 2.12 The ℓ_p spaces$(1 < p \leqslant 2)$ are uniformly smooth. Indeed, we see that

$$\lim_{t \to 0} \frac{\rho_{\ell_p}(t)}{t} = \lim_{t \to 0} \frac{(1 + t^p)^{1/p} - 1}{t} = 0.$$

Observation

- $\ell_p, L_p \ (1 < p < \infty)$, Wm Sobolev space are uniformly convex and uniformly smooth Banach spaces, where m is a positive integer.
- The spaces $c_0, \ell_1, L_1, \ell_\infty, L_\infty$ are not smooth.

Having discussed the geometry of Banach spaces, we now define duality mapping. This notion of duality mapping was first introduced by Beurling and Livingston [57] and extensively studied by Asplund [24], Browder [89, 90], Kachurovskii [300] in connection with monotone operators.

2.5 Duality Mappings

Throughout this section, unless stated otherwise, X and X^* denote a Banach space and its dual, respectively.

Let $T : X \to 2^{X^*}$ a multivalued mapping. Then domain $\mathscr{D}(T)$, range $\mathscr{R}(T)$, inverse T^{-1} and graph $G(T)$ of T are defined as:

$$\mathscr{D}(T) = \{x \in X; Tx \neq \varnothing\},$$

$$\mathscr{R}(T) = \cup_{x \in \mathscr{D}(T)} Tx,$$

$$T^{-1}(y) = \{x \in X : y \in Tx\},$$

$$G(T) = \{(x, y) \in X \times X^* : y \in Tx, x \in \mathscr{D}(T)\}.$$

The mapping T is said to be

(i)   monotone if $(x - y, j_x - j_y) \geqslant 0$ for all $x, y \in \mathscr{D}(T)$ and $j_x \in Jx, j_y \in Jy$.
(ii)   strictly monotone if $(x - y, j_x - j_y) > 0$ for all $x, y \in \mathscr{D}(T)$ with $x \neq y$ and
       $j_x \in Jx, j_y \in Jy$.
(iii) α- monotone if there exists a continuous strictly increasing function
       $\alpha : [0, \infty) \to [0, \infty)$ with $\alpha(0) = 0$ and $\alpha(t) \to \infty$ as $t \to \infty$ such that

$$(x - y, j_x - j_y) \geqslant \alpha(\|x - y\|)\|x - y\|$$

       for all $x, y \in \mathscr{D}(T)$ and $j_x \in Jx, j_y \in Jy$.
(iv)   strongly monotone if for some constant $k > 0$,

$$(x - y, j_x - j_y) \geqslant k\|x - y\|^2$$

       for all $x, y \in \mathscr{D}(T)$ and $j_x \in Jx, j_y \in Jy$.
(v)   injective if $Tx \cap Ty = \varnothing$ for $x \neq y$.

The mapping $T : \mathscr{D}(T) \subset X \to 2^{X^*}$ is said to be coercive on a subset C of $\mathscr{D}(T)$ if
there exists a function $c : (0, \infty) \to [-\infty, \infty]$ with $c(t) \to \infty$ as $t \to \infty$ such that
$(x, Tx) \geqslant c(\|x\|)\|x\|$ for all $x \in C$.

In other words, T is coercive on C if $\lim\limits_{\|x\| \to \infty} \frac{(x, Tx)}{\|x\|} = \infty, x \in C$.

Definition 2.12 A multivalued mapping $J : X \to 2^{X^*}$ is called (normalized) duality
mapping if it satisfies the following property:

$$Jx = \{j \in X^* : (x, j) = \|x\|\|j\|_*, \ \|x\| = \|j\|_*\}.$$

Proposition 2.7 *Let X be a Banach space and let $J : X \to 2^{X^*}$ be the normalized
duality mapping. Then, we have the following:*

(a) $J(0) = 0$.
(b) J *is homogeneous, i.e.* $J(\lambda x) = \lambda J(x)$ *for all* $x \in X$ *and* $\lambda \in \mathbb{R}$.
(c) $\|x\|^2 - \|y\|^2 \geqslant 2(x - y, j)$ *for all* $x, y \in X$ *and* $j_y \in Jy$.
(d) $\|x + y\|^2 \geqslant \|x\|^2 + 2(y, j_x)$ *for all* $x, y \in X$, *where* $j_x \in Jx$.
(e) $\|x + y\|^2 \leqslant \|y\|^2 + 2(x, j_{x+y})$ *for all* $x, y \in X$, *where* $j_{x+y} \in J(x + y)$.
(f) $(x - y, j_x - j_y) \geqslant 0$ *for all* $x, y \in X$, $j_x \in Jx$ *and* $j_y \in Jy$.
(g)   *If X is strictly convex, then J is one-one.*

Proof (a) It is obvious.
(b) If $\lambda = 0$, then it is obvious that $J(0x) = 0J(x)$. So, we assume that $\lambda \neq 0$ and
that $j \in J(\lambda x)$. First, we show that $J(\lambda x) \subset \lambda J(x)$. By our assumption, we have

$$(\lambda x, j) = \|x\|\|j\|_* \ \text{and} \ \|\lambda x\| = \|j\|_*,$$

and it follows that $(\lambda x, j) = \|j\|_*^2$. Hence

$$(x, \lambda^{-1}j) = \lambda^{-1}(\lambda x, \lambda^{-1}j) = \lambda^{-2}(\lambda x, j) = \lambda^{-2}\|\lambda x\|\|j\|_* = \|\lambda^{-1}j\|_*^2 = \|x\|^2,$$

showing thereby that $\lambda^{-1}j \in Jx$, i.e., $j \in \lambda Jx$. Thus, we have $J(\lambda x) \subset \lambda Jx$. Similarly, we can show that $\lambda Jx \subset J(\lambda x)$. Therefore, $J(\lambda x) = \lambda Jx$.

(c) Suppose that $x, y \in X$ and $j \in Jy$, then

$$\|x\|^2 - \|y\|^2 - 2(x - y, j_y) = \|x\|^2 - \|y\|^2 - 2(x, j_y) + 2(y, j_y)$$
$$= \|x\|^2 + \|y\|^2 - 2(x, j_y) \geqslant \|x\|^2 + \|y\|^2 - 2\|x\|\|y\|$$
$$= (\|x\| - \|y\|)^2 \geqslant 0.$$

This implies that

$$\|x\|^2 - \|y\|^2 \geqslant 2(x - y, j_y). \tag{2.4}$$

We can similarly show that

$$\|y\|^2 - \|x\|^2 \geqslant 2(y - x, j_x) \text{ for all } x, y \in X \text{ and } j_x \in Jx \tag{2.5}$$

(d) Replacing y by $x + y$ in (2.5), we obtain the required inequality.
(e) Replacing x by $x + y$ in (2.5), we obtain the required result.
(f) Suppose that $j_x \in Jx$ and $j_y \in Jy$ for $x, y \in X$, then

$$(x - y, j_x - j_y) = (x, j_x) - (x, j_y) - (y, j_x) + (y, j_y)$$
$$\geqslant \|x\|x^2 + \|y\|^2 - \|x\|\|j_y\|_* - \|y\|\|j_x\|_*$$
$$\geqslant \|x\|x^2 + \|y\|^2 - 2\|x\|\|y\| = (\|x\| - \|y\|)^2 \geqslant 0.$$

(g) Suppose that $j \in Jx \cap Jy$ for $x, y \in X$. Now

$$j \in Jx \cap Jy \Rightarrow j \in Jx \text{ and } j \in Jy$$
$$\Rightarrow \|j\|_* = \|x\| = \|y\|, \text{ and so } \|x\|^2 = \|y\|^2 = (x, j) = (y, j).$$

It follows that

$$\|x\|^2 = ((x + y)/2, j) \leqslant \|(x + y)/2\|\|x\|$$

which yields

$$\|x\| = \|y\| = \|(x + y)/2\| \leqslant \|x\|.$$

Hence, $\|x\| = \|y\| = \|(x + y)/2\|$. Because X is strictly convex, we must have $x = y$. Therefore, J is one-one.

Proposition 2.8 *Let X be a Banach space and $J : X \to 2^{X^*}$ be a normalized duality mapping, then for $x, y \in X$, the following statements are equivalent:*
(i) $\|x\| \leqslant \|x + ty\|$ for all $t > 0$.
(ii) There exists $j \in Jx$ such that $(y, j) \geqslant 0$.

Proof $(i) \Rightarrow (ii)$. For $t > 0$, suppose that $j_t \in J(x + ty)$. Then, $(x + ty, j_t) = \|x + ty\| \|j_t\|_*$. Define $f_t = \frac{j_t}{\|j_t\|_*}$, so that $\|f_t\|_* = 1$. Because $f_t \in \|j_t\|^{-1} J(x + ty)$, it follows that

$$\|x\| \leqslant \|x + ty\| = \|j_t\|_*^{-1}(x + ty, j_t)$$
$$= (x + ty, f_t) = (x, f_t) + t(y, f_t)$$
$$\leqslant \|x\| + t(y, f_t).$$

By the Banach–Alaoglu theorem which states that the unit ball in X^* is weak*ly compact, the net $\{f_t\}$ has a limit point $f \in X^*$ such that

$$\|f\|_* \leqslant 1, \ (x, f) \geqslant \|x\| \text{ and } (y, f) \geqslant 0.$$

Thus, we have

$$\|x\| \leqslant (x, f) \leqslant \|x\| \|f\|_* \leqslant \|x\|,$$

which gives that

$$(x, f) = \|x\| \text{ and } \|f\|_* = 1.$$

Now setting $j = f\|x\|$, we obtain $\|j\|_* = \|x\|$. Therefore, $j \in Jx$ and $(y, j) \geqslant 0$.
$(ii) \Rightarrow (i)$. Suppose $x, y \in X$, with $x \neq 0$, there exists $j \in Jx$ such that $(y, j) \geqslant 0$. Then for $t > 0$, we have

$$\|x\|^2 = (x, j) \leqslant (x, j) + (ty, j) = (x + ty, j) \leqslant \|x + ty\| \|x\|.$$

Thus, we obtain
$$\|x\| \leqslant \|x + ty\| \text{ for all } t > 0. \qquad \blacksquare$$

Observation

- $\mathscr{D}(J) = X$.
- J is bounded.
- If X is strictly convex and $x, y \in X$, then $x \neq y \implies Jx \cap Jy = \varnothing$.
- If X is a Hilbert space, J reduces to the canonical isomorphism between X and X^*.

Example 2.13 In a Hilbert space $\mathcal{H}$, the normalized duality mapping is the identity. To see this, let $x \in \mathcal{H}$ with $x \neq 0$. Note that $\mathcal{H} = \mathcal{H}^*$ and

$$\langle x, x \rangle = \|x\| \cdot \|x\| \Rightarrow x \in Jx.$$

We now show that Jx is singleton. If not, then there exists $y(\neq x) \in Jx$ such that $\langle x, y \rangle = \|x\| \cdot \|y\|$ and $\|x\| = \|y\|$. Furthermore,

$$\|x - y\|^2 = \|x\|^2 + \|y\|^2 - 2\langle x, y \rangle = 0.$$

This implies that $x = y$. Therefore, $Jx = \{x\}$.

Definition 2.13 A continuous strictly increasing function $\mu : \mathbb{R}^+ \to \mathbb{R}^+$ is said to be gauge function if $\mu(0) = 0$ and $\lim\limits_{t \to \infty} \mu(t) = \infty$.

Remark 2.2 For the gauge function μ, the function $\Phi : \mathbb{R}^+ \to \mathbb{R}^+$ defined by

$$\Phi(t) = \int_0^t \mu(s)ds$$

is a continuous convex strictly increasing function on $\mathbb{R}^+$. It follows, therefore, that Φ has a continuous inverse function Φ^{-1}.

Definition 2.14 Let X be a Banach space and μ a gauge function. Then, the mapping $J_\mu : X \to 2^{X^*}$ defined by

$$J_\mu(x) = \{j \in X^* : \langle x, j \rangle = \|x\| \cdot \|j\|_*, \ \|j\|_* = \mu(\|x\|)\}, \quad x \in X$$

is called the duality mapping with gauge function μ.

In particular, when $\mu(t) = t$, the duality mapping $J_\mu = J$ is called the normalized duality mapping.

In case, $\mu(t) = t^{p-1}$, $p > 1$, the duality mapping $J_\mu = J_p$ is called generalized duality mapping and is given by

$$J_p x = \{j \in X^* : \langle x, j \rangle = \|x\| \cdot \|j\|_*, \|j\|_* = \|x\|^{p-1}\}, \quad x \in X.$$

Note that for $p = 2$, $J_p = J_2 = J$ is the normalized duality mapping.

Definition 2.15 (*Sign function*) For a complex number z, the sign function is defined by

$$sgn\ z = \begin{cases} \frac{z}{|z|} & \text{if } z \neq 0 \\ 0 & \text{if } z = 0. \end{cases}$$

Deduction Using the definition of sign function, we can easily deduce the following:

$$(i) \quad |sgn\ z| = \begin{cases} 1 & \text{if } z \neq 0 \\ 0 & \text{if } z = 0. \end{cases}$$

$$(ii) \quad z\ sgn\ \bar{z} = \begin{cases} \frac{z\bar{z}}{|z|} = |z| & \text{if } z \neq 0 \\ 0 & \text{if } z = 0. \end{cases}$$

Example 2.14 In the ℓ_2 space, the duality mapping is given by

$$J(x) = (|x_1|\text{sgn}\ x_1, |x_2|\text{sgn}\ x_2, \ldots, |x_i|\text{sgn}\ x_i, \ldots), \quad x = \{x_i\} \in \ell_2.$$

Theorem 2.11 *Let X be any Banach space with dual X^*. Then, Jx is a nonempty closed convex subset of X^* for every $x \in X$. If X^* is strictly convex, then J reduces to a single-valued mapping and is bounded.*

Proof To prove that $Jx \neq \varnothing$, it suffices to find for each $x \in X$ with $\|x\| = 1$ an element $j \in X^*$ with $\|j\|_* = 1$ such that $(x, j) = 1$. The existence of such j follows immediately from the Hahn–Banach theorem. Indeed, if x is a nonzero element in X then by the Hahn–Banach theorem, there exists $f \in X^*$ such that $(x, f) = \|x\|$ and $\|f\|_* = 1$. Set $j = \|x\|f$. Then $(x, j) = (x, \|x\|f) = \|x\|(x, f) = \|x\|^2$ and $\|j\|_* = \|x\|$. It follows that Jx is nonempty for each $x \neq 0$.
We now show that J is convex. To this end, let $j_1, j_2 \in Jx$. Then

$$(x, j_1) = \|x\|^2 = (x, j_2), \ \|j_1\|_* = \|x\|, \ \|j_2\|_* = \|x\|$$

and hence

$$\begin{aligned}
\|x\|^2 &= t(x, j_1) + (1 - t)(x, j_2) = tj_1(x) + (1 - t)j_2(x) \\
&= (tj_1 + (1 - t)j_2)(x) = (x, tj_1 + (1 - t)j_2) \\
&\leqslant \|x\| \|tj_1 + (1 - t)j_2\|_* \leqslant \|x\|[t\|j_1\|_* + (1 - t)\|j_2\|_*] \\
&= \|x\|[t\|x\| + (1 - t)\|x\|] = \|x\|^2.
\end{aligned}$$

The above inequality implies that $\|tj_1 + (1 - t)j_2\|_* = \|x\|$ which together with the fact that $(x, tj_1 + (1 - t)j_2) = \|x\|^2$ implies $tj_1 + (1 - t)j_2 \in Jx$. This proves the convexity of Jx.

Similarly, one can show that Jx is a closed subset of X^*. Suppose now X^* is strictly convex, then by the dual analogue of Proposition 2.3. there exists exactly one $j \in X^*$ such that $(x, j) = \|x\|^2, \|x\| = \|j\|_*$. This implies that J is single valued. Boundedness of J follows from the equality $\|Jx\| = \|x\|$. ∎

Example 2.15 In the $L_2[0, 1]$ space, the duality mapping is given by

$$Jx = \begin{cases} \frac{|x| \, \mathrm{sgn}\, x}{\|x\|}, & x \neq 0 \\ 0, & x = 0. \end{cases}$$

Example 2.16 A duality mapping in $L_p[0, 1](1 < p < \infty)$ space is given by

$$Jx = \begin{cases} \frac{|x|^{p-1} \, \mathrm{sgn}\, x}{\|x\|^{p-2}}, & x \neq 0 \\ 0, & x = 0. \end{cases}$$

Notice that $L_q[0, 1]$, the dual space of $L_p[0, 1]$, is strictly convex, where $\frac{1}{p} + \frac{1}{q} = 1$, and so the duality mapping J is single valued.
  If $x = 0$, then $(Jx, x) = 0 = \|x\| \|Jx\|, \|x\| = 0 = \|Jx\|$. So, we take $x \neq 0$. In this case,

$$\|Jx\| = \left(\int_0^1 \frac{|x|^{q(p-1)}|\mathrm{sgn}\, x|^q}{\|x\|^{q(p-2)}} dt \right)^{\frac{1}{q}} = \frac{1}{\|x\|^{p-2}} \left(\left(\int_0^1 |x|^p dt \right)^{\frac{1}{p}} \right)^{\frac{p}{q}}$$

$$= \frac{1}{\|x\|^{p-2}} (\|x\|)^{\frac{p}{q}} = \frac{1}{\|x\|^{p-2}} \|x\|^{p-1} = \|x\|$$

and

$$(Jx, x) = \frac{1}{\|x\|^{p-2}} \int_0^1 |x|^{p-1}|\mathrm{sgn}\, x||x|\, dt = \frac{1}{\|x\|^{p-2}} \int_0^1 |x|^p\, dt$$

$$= \frac{\|x\|^p}{\|x\|^{p-2}} = \|x\|^2 = \|x\|\|Jx\|.$$

This shows that J is, indeed, a duality mapping on $L_p[0, 1]$.

In the sequel, we discuss the continuity properties of J.

Theorem 2.12 *If X is a reflexive Banach space with strictly convex dual X^*, then duality mapping J is demicontinuous. Further, if X^* is locally uniformly convex, then J is continuous.*

Proof It suffices to prove the demicontinuity of J on the unit sphere $S_X = \{x \in X : \|x\| = 1\}$. Suppose $x_n \to x_0$ in X with $\|x_n\| = 1$, $n \in \mathbb{N}$. Because $\|Jx_n\| = \|x_n\| = 1$ and X is a reflexive Banach space, $\{Jx_n\}$ has a weakly convergent subsequence $\{Jx_{n_k}\}$ and let $Jx_{n_k} \rightharpoonup y \in X^*$. Then, strong convergence of $\{x_{n_k}\}$ and weak convergence of $\{Jx_{n_k}\}$ gives

$$(y, x_0) = \lim_{k \to \infty} (Jx_{n_k}, x_{n_k}) = \|x_{n_k}\|^2 = 1.$$

Also since $\|y\| \geqslant (y, x_0) = \lim_{k \to \infty} \|Jx_{n_k}\| \geqslant \|y\|$, we have

$$(y, x_0) = \|y\|\|x_0\|, \quad \|y\| = \|x_0\|.$$

This implies that $y = Jx_0$ and $Jx_{n_k} \rightharpoonup Jx_0$. Because this is true for every weakly convergent subsequence of $\{Jx_n\}$, it follows that $Jx_n \rightharpoonup Jx_0$ and hence the demicontinuity of J.

Now suppose that X^* is locally uniformly convex, then by Theorem 2.7, norm in X is strongly differentiable. We will show the continuity of J on unit sphere: $\|x\| = 1$.

Let $x_n \to x_0$ in X with $\|x_n\| = 1$. By what we have just proved, $Jx_n \rightharpoonup Jx_0$. We now show that this convergence is strong. Because $Jx_n \rightharpoonup Jx_0$, it implies that $(Jx_n, x_0) \to (Jx_0, x_0) = \|x_0\|$. By Theorem 2.5, it follows that $Jx_n \to Jx_0$. ∎

Remark 2.3 Recently, Troyanski has proved in [591] that every reflexive Banach space X has an equivalent norm in which both X and X^* are locally uniformly convex. Hence, continuity of J is guaranteed on a reflexive Banach space X.

Remark 2.4 Duality mapping from a Banach space X to its dual X^* need not be weakly continuous. In fact, we have the following negative result by Browder [102] which implies that there exists no weakly continuous duality mapping on L_p spaces of periodic functions on $[0, 2\pi]$.

Theorem 2.13 *Let $1 < p < \infty$, $p \neq 2$. Then the Banach space $L_p[0, 2\pi]$ of periodic functions on $[0, 2\pi]$ does not have a weakly continuous duality mapping on L_p.*

Remark 2.5 Duality mapping J satisfies many other nice properties. These will be discussed in detail in Chap. 4 on monotone operators.

Exercises

2.1 Let X be a Banach space. Show that X is strictly convex if and only if the function $f(x) = \|x\|^2$ is strictly convex.

2.2 Let X be a strictly convex Banach space and let $x, y \in X$ with $x \neq y$. If $\|x - z\| = \|x - w\|$, $\|z - y\| = \|w - y\|$ and $\|x - y\| = \|x - z\| + \|z - y\|$, show that $z = w$.

2.3 Let X be a uniformly convex Banach space and let $x, y, z \in X$ be such that $\|z - x\| = t\|x - y\|$, $\|z - y\| = (1 - t)\|x - y\|$, for some $t \in [0, 1]$. Then $z = (1 - t)x + ty$.

2.4 Let X be a Banach space. Define a function $\gamma : (0, 2] \to \mathbb{R}$ by

$$\gamma(t) = \inf\{(x - y, x^* - y^*) : x, y \in \mathbb{S}_X, \|x - y\| \geq t, x^* \in J(x), y^* \in J(y)\},$$

for all $t \in (0, 2]$. Show that X is uniformly convex if and only if $\gamma(t) > 0$ for all $t \in (0, 2]$.

2.5 Let the two norms $\| \cdot \|_1$ and $\| \cdot \|_2$ be defined on ℓ_1, where $\|x\|_1 = \sum\limits_{n \in \mathbb{N}} |x_n|$ and $\|x\|_2 = \left(\sum\limits_{n \in \mathbb{N}} |x_n|^2 \right)^{1/2}$, $x = \{x_n\}_{n \in \mathbb{N}} \in \ell_1$. Consider the norm

$$\|x\| = (\|x\|_1^2 + \|x\|_2^2)^{1/2}, \quad x = \{x_n\}_{n \in \mathbb{N}} \in \ell_1.$$

Show that the norm $\| \cdot \|$ is equivalent to the ℓ_1-norm and that it is strictly convex.

2.6 Let X be a uniformly convex Banach space and let δ be the modulus of convexity of X. Let $0 < \varepsilon < r \leqslant R$. Show that $\delta(\frac{\varepsilon}{R}) > 0$ and

$$\|\lambda x + (1 - \lambda)y\| \leqslant r\left\{1 - 2\lambda(1 - \lambda)\delta\left(\frac{\varepsilon}{R}\right)\right\}$$

for all $x, y \in X$ with $\|x\| \leqslant r$, $\|y\| \leqslant r$ and $\|x - y\| \geqslant \varepsilon$ and $\lambda \in [0, 1]$.

2.7 Show that a reflexive locally uniformly convex Banach space is weakly uniformly convex.

2.8 Let X be a Banach space, J the duality mapping of X, and let $x \in X$. Show that $J(x)$ is a bounded closed convex subset of X^*.

2.9 Let X be a Banach space and let $J : X \to 2^{X^*}$ be the normalized duality mapping. Show the following:
(a) If X is reflexive with strict convex dual X^*, then J is demicontinuous.
(b) If X is uniformly convex, then for $x, y \in \overline{B}_r(0)$, $j_x \in Jx$, $j_y \in Jy$

$$(x - y, j_x - j_y) \geqslant \alpha_r(\|x - y\|)\|x - y\|,$$

where $\alpha_r : \mathbb{R}^n \to \mathbb{R}^n$ is a function satisfies the conditions:

$$\alpha_r(0) = 0, \alpha_r(t) > 0 \text{ for all } t > 0 \text{ and } t \leqslant s \Rightarrow \alpha_r(t) \leqslant \alpha_r(s).$$

Chapter 3
Differential Calculus in Banach Spaces

*Mathematics is not a deductive science-that's a cliche. When
you try to prove a theorem, you don't just list the hypotheses,
and then start to reason. What you do is trial-and-error,
experimentation, and guesswork.*

Paul Halmos (1985)
*It is worth noting that the notation facilitates discovery. This, in
a most wonderful way, reduces the mind's labors.*

Gottfried Wilhelm Leibniz, (1646–1716)
Differentiation means linearlization.

The differential calculus is one of the fundamental techniques of nonlinear functional
analysis. Very often we will use this notion. In this chapter we develop the calculus
in real Banach spaces. Section 3.1 deals with definitions on Gâteaux and Fréchat
derivative with illustrative examples. We also give a variant of mean value theorem.
Properties of the derivative are discussed in Sect. 3.2, while in Sect. 3.3 we discuss
partial derivatives. Section 3.4 deals with higher derivative. Subsequently, we give
Taylor's theorem, inverse function and implicit function theorems. In Sect. 3.5 we
discuss an important concept of nonlinear analysis-subdifferential of convex functionals. We show that the subdifferential of norm functional is precisely the duality
mapping J. In Sect. 3.6, we discuss about differentiability of norms of Banach spaces.
Finally, we conclude the chapter in Sect. 3.7 in which we discuss about asymptotic
behaviour of generalized nonexpansive sequences.

3.1 Gâteaux and Fréchet Derivatives

In the following, X and Y are real (or complex) Banach spaces, $U \subseteq X$, U open and
$F : U \to Y$ any operator not necessarily linear with $\mathscr{D}(F) = U$.

© Springer Nature Singapore Pte Ltd. 2018
H. K. Pathak, *An Introduction to Nonlinear Analysis and Fixed Point Theory*,
https://doi.org/10.1007/978-981-10-8866-7_3

3.1.1 Gâteaux Derivative

Definition 3.1 Let X and Y be two (real or complex) Banach spaces. Then an operator $F : X \to Y$ is said to be Gâteaux differentiable at $x \in X$ if there exists a continuous linear operator $A : X \to Y$, i.e. $A \in \mathcal{L}(X, Y)$ such that

$$\lim_{t \to 0} \frac{F(x + th) - F(x)}{t} = Ah \text{ for every } h \in X.$$

That is,

$$\lim_{t \to 0} \left\| \frac{1}{t}[F(x + th) - F(x) - tAh] \right\| = 0 \text{ for every } h \in X. \tag{3.1}$$

A is called the Gâteaux derivative of F at x, and its value at h is denoted by $dF(x, h)$. We shall alternatively write $dF(x)h$ or $F'(x)h$ for the same.

Equivalently, one can define the Gâteaux derivative of F at $x \in U \subseteq X$, U being open in X, as follows: writing $\varphi(t) = F(x + th)$ for fixed $x \in X$ and $h \in X$, we say that F has Gâteaux derivative $dF(x)$ at $x \in X$ iff

$$\left. \frac{d}{dt} \varphi(t) \right|_{t=0} = dF(x, h).$$

We say that f is Gâteaux differentiable, if it is Gâteaux differentiable at every $x \in X$.

Uniqueness of Gâteaux Derivative

As in one-dimensional case, Gâteaux derivative $F'(x)$ is unique. Suppose, if possible, there exists another continuous linear operator B for which (3.1) is true, then for $t > 0$ we have

$$\|Bh - F'(x)h\| \leqslant t^{-1} \|F(x + th) - F(x) - tBh\|$$
$$+ t^{-1} \|F(x + th) - F(x) - tF'(x)h\|.$$

In view of (3.1), the RHS of the above inequality tends to zero as $t \to 0$ and hence

$$\|Bh - F'(x)h\| = 0 \text{ for every } h \in X.$$

This implies that $B = F'(x)$.

Observation

- Suppose $F : X \to Y$ is an operator, where $X = \mathbb{R}^n$, $Y = \mathbb{R}$ and $\mathbf{e}_1 = (1, 0, 0, \ldots, 0)$, $\mathbf{e}_2 = (0, 1, 0, \ldots, 0)$, $\ldots$, $\mathbf{e}_n = (0, 0, 0, \ldots, 1)$ are Hamel basis of $\mathbb{R}^n$, then $dF(x)(\mathbf{e}_i)$, i.e. the Gâteaux derivative of F at x in the direction $\mathbf{e}_i$, is the ith partial derivative of F at x. Note that $x = (x_1, x_2, x_3, \ldots, x_n) = x_1\mathbf{e}_1 + x_2\mathbf{e}_2 + x_3\mathbf{e}_3 + \cdots + x_n\mathbf{e}_n$, so that

$$dF(x)(\mathbf{e}_i) = \lim_{t \to 0} \frac{F(x + t\mathbf{e}_i) - F(x)}{t}$$

$$= \lim_{t \to 0} \frac{F(x_1, x_2, \ldots, x_i + t, \ldots, x_n) - F(x_1, x_2, \ldots, x_i, \ldots, x_n)}{t}$$

$$= \frac{\partial F(x)}{\partial x_i}.$$

- The constant mapping $F : X \to Y$ defined by $F(x) = a \ \forall x \in X$ has the derivative $F'(x) = 0$, where 0 denotes null operator.
- The existence of Gâteaux derivative of an operator $F : X \to Y$ does not necessarily imply continuity of the operator F. To observe this, let us consider the following examples.

Example 3.1 Let $F : \mathbb{R}^2 \to \mathbb{R}$ be defined by

$$F(x) = \begin{cases} \frac{2x_1^3}{x_2}, & \text{if } x = (x_1, x_2) \neq (0, 0) \\ 0, & \text{if } x = (x_1, x_2) = (0, 0). \end{cases}$$

Then

$$dF(\mathbf{0})h = \lim_{t \to 0} \frac{F(0 + th) - F(0)}{t} = \lim_{t \to 0} \frac{F(th)}{t}$$

$$= \lim_{t \to 0} \frac{2t^3 h_1^3}{t h_2} = 0 = \hat{0}h \text{ for every } h \in \mathbb{R}^2.$$

Thus, $dF(\mathbf{0})$ exists and being null operator it is a continuous linear operator but F is discontinuous at $\mathbf{0} = (0, 0)$.

Example 3.2 Let $F : \mathbb{R}^2 \to \mathbb{R}$ be defined by

$$F(x_1, x_2) = \begin{cases} \frac{2x_2 \exp(-x_1^{-2})}{x_2^2 + \exp(-2x_1^{-2})}, & x_1 \neq 0 \\ 0, & x_1 = 0. \end{cases}$$

Then it is easy to see that F is Gâteaux differentiable at $(0, 0)$ but is not continuous at the origin.

Definition 3.2 For a functional $f : X \to \mathbb{R}$, the mapping $x \mapsto f'(x)$ is called the gradient of f and is denoted by ∇f. So ∇f is a mapping from $X \to X^*$.

Example 3.3 For operators F on finite dimensional spaces, we can give concrete representation of the Gâteaux derivative $F'(x)$.

Let $F : \mathbb{R}^n \to \mathbb{R}^m$ be given by $F = (f_1, f_2, \ldots, f_m)$ and $A = (a_{ij})$ be a $m \times n$ matrix. We choose h as the *jth* coordinate vector $e_j = (0, \ldots, 1, \ldots, 0)$. Then

$$\lim_{t \to 0} \left\| \frac{1}{t}[F(x + th) - F(x) - tAh] \right\| = 0$$

implies that

$$\lim_{t \to 0} \left| \frac{1}{t} \left[f_i(x + te_j) - f_i(x) - ta_{ij} \right] \right| = 0; \ i = 1, 2, \ldots, m; \ j = 1, 2, \ldots, n.$$

This shows that f_i has partial derivatives at x and

$$\partial_j f_i(x) = \frac{\partial f_i(x)}{\partial x_j} = a_{ij}; \ i = 1, 2, \ldots, m; \ j = 1, 2, \ldots, n.$$

Hence, $F'(x)$ has the matrix representation

$$F'(x) = \begin{bmatrix} \partial_1 f_1(x) & \cdots & \partial_n f_1(x) \\ \vdots & & \vdots \\ \partial_1 f_m(x) & \cdots & \partial_n f_m(x) \end{bmatrix}.$$

This matrix is called *Jacobian matrix* of F at $\mathbf{x} \in \mathbb{R}^n$.

In the case when f is a functional, i.e. $f : \mathbb{R}^n \to \mathbb{R}, f'(x)$, denoted by ∇f is a row vector $(\partial_1 f, \partial_2 f, \ldots, \partial_n f)$ and is called *gradient of* f at $\mathbf{x} \in \mathbb{R}^n$.

Remark 3.1 Let $A : X \to Y$ be a continuous linear operator. Then

$$A(x + th) - A(x) - tAh = A(x) + A(th) - A(x) - A(th) = 0 \ \text{ for every } \ h \in X.$$

It follows that

$$\lim_{t \to 0} \left\| \frac{1}{t}[A(x + th) - A(x) - tAh] \right\| = 0 \ \text{ for every } \ h \in X.$$

Hence, $A'(x) = A$ for all $x \in X$.

Example 3.4 Let $F : L_p[0, 1] \to L_p[0, 1]$ be the Hammerstein operator given by

$$[Fx](s) = \int_0^1 k(s, t) f(t, x(t)) dt.$$

First of all, assume that the kernel k and the function f are such that the derivative can be taken under the integral sign. Then

$$\left[\frac{d}{d\tau} F(x + \tau h) \right](s) = \frac{d}{d\tau} \left[\int_0^1 k(s, t) f(t, (x + \tau h)(t)) dt \right]$$

$$= \int_0^1 k(s, t) \frac{d}{d\tau} f(t, (x + \tau h)(t)) dt$$

$$= \int_0^1 k(s,t) f_x(t, (x+\tau h)(t)) h(t) dt.$$

Hence,

$$[F'(x)h](s) = \frac{d}{d\tau} F(x+\tau h)\Big|_{\tau=0} (s)$$

$$= \int_0^1 k(s,t) f_x(t, x(t)) h(t) dt. \tag{3.2}$$

This shows that the Gâteaux derivative $F'(x)$ of the Hammerstein operator F is the linear integral operator with kernel $k(s,t) f_x(t, x(t))$ and is given by (3.2).
We have the following mean value theorem for functional f on X.

Theorem 3.1 *Let the functional $f : X \to \mathbb{R}$ has Gâteaux derivative $df(x, h)$ at every point $x \in X$. Then for points $x, x + h \in X$, there exists a constant τ, $0 < \tau < 1$ such that*

$$f(x+h) - f(x) = df(x+\tau h, h).$$

Proof Writing $\varphi(t) = f(x+th))$, then we see that

$$\varphi'(t) = \lim_{s \to 0} \left[\frac{\varphi(t+s) - \varphi(t)}{s} \right]$$

$$= \lim_{s \to 0} \left[\frac{f(x+th+sh) - f(x+th)}{s} \right]$$

$$= df(x+th, h).$$

Now using the mean value theorem for scalar functions, we get

$$\varphi(1) - \varphi(0) = \varphi'(\tau), \quad 0 < \tau < 1.$$

This gives

$$f(x+h) - f(x) = df(x+\tau h, h).$$

Remark 3.2 If dim $X > 1$, then in general there is no equality in the mean value theorem. To see this, we consider the following examples.

Example 3.5 (1) Let $F : \mathbb{R} \to \mathbb{C}$ be defined by $F(x) = e^{ix}$ for all $x \in \mathbb{R}$. Take $h = 2\pi$, then for all $x \in \mathbb{R}$,

$$F(x+h) - F(x) = 0, \quad \text{but} \quad F'(x+th) = ie^{ix+2t\pi i} \neq 0.$$

(2) Let $F : \mathbb{R}^2 \to \mathbb{R}^2$ be defined as $F(x_1, x_2) = (x_1^3, x_2^2)$. Then at any point $z = (z_1, z_2)$,

$$F'(z) = \begin{bmatrix} 3z_1^2 & 0 \\ 0 & 2z_2 \end{bmatrix}.$$

If we take $x = (0, 0)$ and $y = (1, 1)$, then it is clear that there is no z in the line joining x and y for which $Fy - Fx = F'(z)(y - x)$.

However, we have the following variant of the mean value theorem for operator $F : X \to Y$.

Theorem 3.2 *Let $F : X \to Y$ has Gâteaux derivative $F'(x)$ at every point $x \in X$. Then for points $x, x + h \in X$ and $e \in Y^*$, there exists a constant τ, $0 < \tau < 1$ such that*

$$(e, \ F(x + h) - F(x)) = (e, \ dF(x + \tau h, h)).$$

Further, F satisfies the Lipschitz condition

$$\|F(x + h) - F(x)\| \leqslant \|dF(x + \tau h)\| \|h\|.$$

Proof For $e \in Y^*$, define a functional f as

$$f(x) = (e, F(x)).$$

Then

$$\frac{f(x + th) - f(x)}{t} = \left(e, \ \frac{F(x + th) - F(x)}{t} \right).$$

Taking the limit as $t \to 0$ and using the continuity of inner product, we get $df(x, h) = (e, dF(x, h))$. Because mean value theorem is valid for functional f on x, we get

$$f(x + h) - f(x) = df(x + \tau h, h), \ \ 0 < \tau < 1.$$

This gives

$$(e, \ F(x + h) - F(x)) = (e, \ dF(x + \tau h, h)).$$

For the second part, since e is arbitrary, by Hahn–Banach's theorem we choose e of unit norm such that

$$\|(e, F(x + h) - F(x))\| = \|F(x + h) - F(x)\|.$$

Using the result of the first part, we get

$$\begin{aligned} \|F(x + h) - F(x)\| &= |(e, F(x + h) - F(x))| \\ &= |(e, dF(x + \tau h, h))| \\ &\leqslant \|dF(x + \tau h, h)\| \\ &\leqslant \|dF(x + \tau h)\| \|h\|. \end{aligned}$$

This completes the proof. ∎

A stronger differentiability notion what we call Fréchet differentiability is given in the next definition.

3.1.2 Fréchet Derivative

Definition 3.3 Let X and Y be (real or complex) Banach spaces. An operator $F : X \to Y$ is said to be Fréchet differentiable at $x \in X$ if there exists a continuous linear operator $A : X \to Y$, i.e. $A \in \mathcal{L}(X, Y)$ such that

$$F(x + h) - F(x) = Ah + w(x, h), \quad \text{and} \quad \lim_{\|h\| \to 0} \frac{\|w(x, h)\|}{\|h\|} = 0.$$

A is called the Fréchet derivative of F at x and is denoted by $F'(x)$. Its value at h will alternatively be written as $dF(x, h)$ or $F'(x)h$.

Observation

- From the context, it will be clear whether $F'(x)$ is Fréchet or Gâteaux derivative.
- As in the case of Gâteaux derivative, Fréchat derivative is unique.

Example 3.6 Let $f : \mathbb{R} \to \mathbb{R}$ be differentiable, x a continuous scaler function on $[0, 1]$, and let $F : C[0, 1] \to C[0, 1]$ be defined by

$$[Fx](t) = f(x(t)), \quad t \in [0, 1].$$

Then $F'(x) : C[0, 1] \to C[0, 1]$ is given by

$$[F'(x)u](t) = f'(x(t)) \cdot u(t), \quad t \in [0, 1],$$

i.e. the Fréchet derivative of F at x is the multiplication operator, which multiplies each continuous function u by $f' \circ x$.

Example 3.7 Let $f : \mathbb{R}^3 \to \mathbb{R}$ have continuous second-order partial derivatives with respect to three coordinate variables. Consider the function $F : C[a, b] \to \mathbb{R}$ defined by

$$F(x) = \int_a^b f(x(t), \dot{x}(t), t) dt.$$

Then the Fréchet derivative of F at x, i.e. $F'(x)$ and its value at h is given by

$$F'(x)h = \int_a^b \left[\frac{\partial f}{\partial x} - \frac{d}{dt} \left(\frac{\partial f}{\partial \dot{x}} \right) \right] h \, dt + \left[\frac{\partial f}{\partial \dot{x}} h \right]_a^b, \quad \text{where } \dot{x} \equiv \frac{dx}{dt}.$$

To verify this, we notice that

$$F(x+h) - F(x) = \int_a^b [f(x(t)+h(t), \dot{x}(t)+\dot{h}(t), t) - f(x(t), \dot{x}(t), t)]dt$$

$$= \int_a^b \left[\frac{\partial f}{\partial x}(x(t), \dot{x}(t), t)h(t) + \frac{\partial f}{\partial \dot{x}}(x(t), \dot{x}(t), t)\dot{h}(t)\right]dt + \omega(h, h)$$

where $\omega(h, h) = o(\|h\|_{C[a,b]})$, that is,

$$\frac{|\omega(h, h)|}{\|h\|_{C[a,b]}} \to 0 \quad \text{as} \quad \|h\|_{C[a,b]} \to 0.$$

Hence,

$$F'(x)h = \int_a^b \left[\frac{\partial f}{\partial x}(x(t), \dot{x}(t), t)h(t) + \frac{\partial f}{\partial \dot{x}}(x(t), \dot{x}(t), t)\dot{h}(t)\right]dt$$

$$= \int_a^b \left[\frac{\partial f}{\partial x} - \frac{d}{dt}\left(\frac{\partial f}{\partial \dot{x}}\right)\right]h dt + \left[\frac{\partial f}{\partial \dot{x}}h\right]_a^b, \quad \text{after integration by parts.}$$

Example 3.8 Let X, Y be two isomorphic Banach spaces and $GL(X, Y)$ the space of all continuous linear operators T from X onto Y, which are continuously invertible. Then the mapping $F : GL(X, Y) \to GL(Y, X)$ defined by

$$F(T) = T^{-1} \quad \text{for all } T \in GL(X, Y)$$

is differentiable at every point T_0 and

$$F'(T_0)H = -T_0^{-1}HT_0^{-1}.$$

If $\|H\| < \|T_0^{-1}\|^{-1}$, then

$$\|F(T_0 + H) - F(T_0) - (-T_0^{-1}HT_0^{-1})\|$$

$$= \|(T_0+H)^{-1} - T_0^{-1} + T_0^{-1}HT_0^{-1}\| = \|(I + T_o^{-1}H)^{-1}T_0^{-1} - T_0^{-1} + T_0^{-1}HT_0^{-1}\|$$

$$= \|\sum_{j=0}^{\infty}(-T_0^{-1})^j T_0^{-1} - T_0^{-1} + T_0^{-1}HT_0^{-1}\| = \|\sum_{j=2}^{\infty}(-T_0^{-1})^j T_0^{-1}\|$$

$$\leqslant \|\sum_{j=2}^{\infty}(-T_0^{-1})^j\|\|T_0^{-1}\| \leqslant \|H\|^2 \frac{\|T_0^{-1}\|^3}{1 - \|T_0^{-1}H\|}.$$

Letting $\|H\| \to 0$ in the above inequality, we obtain

$$\lim_{\|H\|\to 0} \frac{\|F(T_0 + H) - F(T_0) - (-T_0^{-1}HT_0^{-1})\|}{\|H\|} = 0.$$

Thus, the derivative of F at T_0 is given by

$$F'(T_0)H = -T_0^{-1}HT_0^{-1}.$$

Example 3.9 Let X, Y, Z be Banach spaces and the Banach space $X \times Y$ endowed with the norm $\|(u, v)\| = \max(\|u\|, \|v\|)$. Let $F : X \times Y \to Z$ be a continuous bilinear map, i.e.

$$\|F\| = \sup_{\|x\| \leqslant 1, \|y\| \leqslant 1} \|F(x, y)\| < \infty.$$

Let $(x_0, y_0) \in X \times Y$. Then

$$\|F(u, v)\| \leqslant \|F\| \cdot \|u\| \cdot \|v\| \leqslant \|F\| \cdot \|\langle u, v \rangle\|^2$$

yields

$$\lim_{\|\langle u, v \rangle\| \to 0} \frac{\|F(u, v)\|}{\|\langle u, v \rangle\|} = 0.$$

Therefore, F is Fréchet differentiable and $F'(x_0, y_0)$ is the linear map $F'(x_0, y_0) :$ $X \times Y \to Z$ given by

$$(u, v) \mapsto F(x_0, v) + F(u, y_0).$$

Remark 3.3 It is clear that every Fréchet differentiable operator is Gâteaux differentiable. But the converse is not true. To effect this, consider the following example.

Example 3.10 Let $f : \mathbb{R}^2 \to \mathbb{R}$ be defined as

$$f(x) = \begin{cases} \frac{x_1^3 x_2}{x_1^4 + x_2^2}, & x \neq (0, 0) \\ 0, & x = (0, 0). \end{cases}$$

f is Gâteaux differentiable at 0 with Gâteaux derivative 0. But it is not Fréchet differentiable at 0. To see this, let us take $h = (h_1, h_2) \neq (0, 0)$. Then, we have

$$\lim_{t \to 0} \frac{f(th) - f(0)}{t} = \lim_{t \to 0} \frac{th_1^3 h_2}{(t^2 h_1^4 + h_2^2)} = 0 = \hat{0}(h) \text{ for every } h \in \mathbb{R}^2$$

which shows that f is Gâteaux differentiable at 0 with Gâteaux derivative 0. However, we see that

$$\frac{\|f(h) - f(0) - \hat{0}(h)\|}{\|h\|} = \frac{|h_1^3 h_2|}{(h_1^4 + h_2^2)} \frac{1}{\sqrt{h_1^2 + h_2^2}} = \frac{1}{2\sqrt{1 + h_1^2}} \text{ for } h_2 = h_1^2.$$

Letting $h \to 0$, i.e. $h_1 \to 0$, we get

$$\lim_{\|h\|\to 0}\frac{\|w(0,h)\|}{\|h\|} = \lim_{\|h\|\to 0}\frac{\|f(h) - f(0) - \hat{0}(h)\|}{\|h\|} = \lim_{h_1\to 0}\frac{1}{2\sqrt{1 + h_1^2}} = \frac{1}{2}.$$

The following theorem gives a relationship between Gâteaux and Fréchet derivatives.

Theorem 3.3 *If the Gâteaux derivative $F'(x)$ exists in some neighbourhood of the point x and is continuous at x, then F is also Fréchet differentiable at x and is equal to $F'(x)$.*

Proof We write

$$w(x,h) = (Fx + h) - F(x) - F'(x)h.$$

Then

$$(e, w(s,h) = (e, F(x + h) - F(x)) - (e, F'(x)h), \ e \in Y^*.$$

By mean value theorem, we get

$$(e, w(x,h)) = (e, F'(x + \tau h)h - F'(x)h), \ 0 < \tau < 1.$$

By Hahn–Banach's theorem, e can be so chosen that

$$\|w(x,h)\| = |(e, w(x,h))| \ \text{and} \ \|e\| = 1.$$

Hence, we get

$$\|w(x,h)\| \leqslant \|F'(x + \tau h) - F'(x)\|\|h\|.$$

This implies that

$$\frac{\|w(x,h)\|}{\|h\|} \leqslant \|F'(x + \tau h) - F'(x)\|.$$

Because $F'(x)$ is continuous, the RHS of the above inequality goes to zero as $h \to 0$. This proves our theorem. ∎

3.2   Properties of the Derivative

The existence of Fréchet derivative at x implies the continuity of F at x. But this result is not true for Gâteaux derivatives. However, we have the strong hemicontinuity of Gâteaux differentiable operators. In this section too, we assure that $\mathscr{D}(F) = X, X$ and Y are real Banach spaces.

Theorem 3.4 *Let $F : X \to Y$ be Fréchet differential at $x \in X$. Then F is continuous at x. More precisely, there is a $\delta > 0$ and a $c \geqslant 0$ (depending on x) such that*

$$\|F(x+h) - F(x)\| \leqslant c\|h\| \quad \text{whenever} \quad \|h\| \leqslant \delta. \tag{3.3}$$

Proof By the definition of Fréchet derivative at x, it follows that there exists a $\delta > 0$ such that

$$\|F(x+h) - F(x) - F'(x)h\| \leqslant \|h\|,$$

whenever $\|h\| \leqslant \delta$.

Let $\omega_F(x, h) = F(x+h) - F(x) - F'(x)h$. Then we see that

$$\begin{aligned}
\|F(x+h) - F(x)\| &= \|F(x+h) - F(x) - F'(x)h + F'(x)h\| \\
&\leqslant \|F(x+h) - F(x) - F'(x)h\| + \|F'(x)h\| \\
&\leqslant \|\omega_F(x, h)\| + \|F'(x)\|\|h\|
\end{aligned}$$

Because $\lim\limits_{\|h\| \to 0} \frac{\omega_F(x,h)}{\|h\|} = 0$, there exists a $\delta > 0$ such that

$$\|h\| \leqslant \delta \Rightarrow \|\omega_F(x, h)\| \leqslant \|h\|.$$

Thus,

$$\|F(x+h) - F(x)\| \leqslant (1 + \|F'(x)\|)\|h\|.$$

Hence with $c = 1 + \|F'(x)\|$, result holds. Thus, we see that for a given $\varepsilon > 0$ there exists a $\delta = (\frac{\varepsilon}{c}) > 0$ such that

$$\|h\| < \delta \Rightarrow \|F(x+h) - F(x)\| < \varepsilon$$

which shows that F is continuous at x. ∎

Observation

- Let $F : X \to Y$ be an operator and $x \in X$. If F is Gâteaux differentiable at x, then F need not be continuous at x. To effect this, let $F : \mathbb{R}^2 \to \mathbb{R}$ be defined by

$$F(x_1, x_2) = \begin{cases} \frac{x_1}{x_2}(x_1^2 + x_2^2), & x_2 \neq 0 \\ 0, & x_2 = 0. \end{cases}$$

For $h = (h_1, h_2) \neq (0, 0)$ and $\hat{0} \in (\mathbb{R}^2)^*$, we have

$$\frac{1}{t}[F(0 + th) - F(0) - t\hat{0}(h)] = \frac{1}{t} \cdot \frac{th_1}{th_2} \cdot t^2(h_1^2 + h_2^2) = t(h_1^2 + h_2^2).$$

Hence,

$$\lim_{t \to 0} \|\frac{1}{t}[F(0 + th) - F(0) - t\hat{0}(h)]\| = \lim_{t \to 0} t(h_1^2 + h_2^2) = 0$$

which implies that $\hat{0}$ is the Gâteaux derivative of F at 0, but F is not continuous at 0.

Definition 3.4 Let $F : X \to Y$ be an operator and $x \in X$. F is called strongly hemi-continuous at x if for every sequence x_n converging to x along a line, $F(x_n)$ converges to Fx.

Theorem 3.5 *Let $F : X \to Y$ be Gâteaux differentiable at $x \in X$. Then F is strongly hemicontinuous at x.*

Proof Let F be Gâteaux differentiable at x and $h \in X$ be any fixed element. Then $\phi(t) = F(x + th)$ is differentiable at 0 with $\phi'(0) = F(x)h$. Because $\phi(t)$ is a function of real variable, it follows that it is also continuous at 0. That is, $\phi(t) \to \phi(0)$ as $t \to 0$, which in turn implies that $F(x + th) \to F(x)$ as $t \to 0$. This proves the result. ∎

Example 3.11 Let $f : \mathbb{R}^2 \to \mathbb{R}$ be defined as

$$f(x_1, x_2) = \begin{cases} \frac{2x_2 \exp(-x_1^{-2})}{x_2^2 + \exp(-2x_1^{-2})}, & x_1 \neq 0 \\ 0, & x_1 = 0. \end{cases}$$

f is Gâteaux differentiable at 0 but is not continuous there.

Many of the properties of the ordinary derivative carry over to the Gâteaux and Fréchet derivatives. For example:

(i) $(cF)'(x) = cF'(x)$ for scalars c, and
(ii) $(F_1 + F_2)'(x) = F_1'(x) + F_2'(x)$.

We have the following chain rule for the derivatives.

Theorem 3.6 (Chain rule for derivatives) *Suppose X, Y, Z be real Banach spaces and $G : X \to Y$ is Gâteaux differentiable at x and $F : Y \to Z$ is Fréchet differentiable at $G(x)$. Then $H = FoG$ is Gâteaux differentiable at x and*

$$H'(x) = (FoG)'x = F'(G(x))G'(x).$$

If in addition, $G(x)$ is Fréchet differentiable then $H'(x)$ is the Fréchet derivative.

Proof For $t \neq 0$ we have

$$\frac{1}{|t|} \|H(x + th) - Hx - tF'(Gx)G'(x)h\|$$

$$\leqslant \frac{1}{|t|} \|F(G(x + th)) - F(G(x)) - tF'(Gx)(G(x + th) - G(x))\|$$

$$+ \|F'(Gx)(G(x + th) - Gx - tG'(x)h\|. \tag{3.4}$$

Because G is Gâteaux differentiable, second term of RHS of (3.4) tends to zero as $t \to 0$. For any value of t such that $G(x + th) \neq G(x)$, the first term may be multiplied and divided by $\|G(x + th) - G(x)\|$. Now, since F is Fréchet differentiable and $\|G(x + th) - Gx\| \to 0$ (by hemicontinuity), the first term of RHS of (3.4) tends to zero. Hence, the LHS of (3.4) tends to zero.

This implies that H is Gâteaux differrentiable. Similarly one can show that H is Fréchet differentiable, if G is Fréchet differentiable. ∎

Corollary 3.1 *If $F : X \to Y$ is a function, which is Gâteaux differentiable at every point of the interval*

$$[x, x + h] := \{u \in X : u = \lambda x + (1 - \lambda)(x + h), \lambda \in [0, 1]\},$$

then

$$F(x + h) - F(x) = \int_0^1 F'(x + th)h \, dt.$$

We now discuss the relationship between the compactness of an operator and compactness of its derivative.

Theorem 3.7 *Suppose $F : X \to Y$ is compact and continuous and has Fréchet derivative $F'(x)$ at x. Then $F'(x)$ is compact and continuous.*

Proof Assume not. Then there exists $\varepsilon > 0$ and a sequence $\{x_n\}$ with $\|x_n\| \leqslant 1$ such that

$$\|F'(x)x_n - F'(x)x_m\| > 3\varepsilon, m \neq n.$$

But

$$F(x + h) - F(x) = F'(x)h + w(x, h)$$

and

$$\|w(x, h)\| \leqslant \varepsilon\|h\| \text{ if } \|h\| < \delta.$$

for some δ. Hence,

$$\|F(x + \delta x_m) - F(x + \delta x_n)\| \geqslant \delta\|F'(x)(x_m - x_n)\| - \|w(x, \delta x_m)\| - \|w(x, \delta x_n)\|$$
$$> 3\delta\varepsilon - \delta\varepsilon - \delta\varepsilon = \delta\varepsilon,$$

which is impossible because F is compact. ∎

The converse of the above theorem is true, provided F' is compact as an operator from X to $\mathcal{L}(X, Y)$. We state this as a theorem, for details see Vainberg [597].

Theorem 3.8 *Suppose*
(a) *$F'(x)$ is compact for each $x \in X$;*
(b) *F' is compact as an operator from X to $\mathcal{L}(X, Y)$.*
Then F is a compact and continuous operator from X to Y.

One can go a step further and ask: Is F completely continuous under the assumptions of Theorem 3.8? Answer is yes, in view of the following theorem.

Theorem 3.9 *Let $F : X \to Y$ and $F' : X \to \mathcal{L}(X, Y)$ be compact operators. Then F is completely continuous.*

Proof Let $x_n \rightharpoonup x_0$ in X. We first show that $Fx_n \rightharpoonup Fx_0$ in Y.

If possible, let us assume that Fx_n does not converge weakly to Fx_0. So there exists an $\varepsilon > 0$ and a subsequence $\{x_{n_k}\}$ and $e_0 \in Y^*$ such that

$$|(e_0, Fx_{n_k} - Fx_0)| > \varepsilon. \tag{3.5}$$

In view of mean value theorem, there exists t_{n_k}, $0 < t_{n_k} < 1$ such that

$$(e_0, Fx_{n_k} - Fx_0) = (e_0, F'(x_0 + t_{n_k}(x_{n_k} - x_0))(x_{n_k} - x_0)). \tag{3.6}$$

Because $F' : X \to \mathcal{L}(X, Y)$ is compact, there exists a subsequence t_{n_k} of t_{n_k} (which we again denote by t_{n_k} such that

$$\lim_{k \to \infty} \|F'x_0 + t_{n_k}(x_{n_k} - x_0)) - A\| = 0 \tag{3.7}$$

for some $A \in \mathcal{L}(X, Y)$.

We have

$$(e_0, F'(x_0 + t_{n_k}(x_{n_k} - x_0))(x_{n_k} - x_0)) = (e, A(x_{n_k} - x_0)). \tag{3.8}$$

Because $x_{n_k} \rightharpoonup x_0$ and A is continuous and hence weakly continuous, the firs term of RHS of (3.8) goes to zero as $k \to \infty$. The second term goes to zero in view of (3.7). So from (3.6) it implies that

$$\lim_{k \to \infty} (e_0, Fx_{n_k} - Fx_0) = 0.$$

This contradicts (3.5), and hence the weak convergence of Fx_n to Fx_0.

Now compactness of F gives that $\{Fx_n\}$ has a convergent subsequence $\{Fx_{n_k}\}$. Weak convergence of $\{Fx_n\}$ to Fx_0 implies that $\{Fx_{n_k}\}$ actually converges to Fx_0. Thus, every subsequence of $\{Fx_n\}$ has in turn a subsequence which converges to Fx_0 which implies that $\{Fx_n\}$ converges to Fx_0. This proves the main result. ∎

Observation

- Every Fréchet differentiable function is Gâteux differentiable.
- If the operator $F : X \to Y$ is Fréchet differentiable at $x \in X$, then it is continuous at x. However, if F is Gâteux differentiable at $x \in X$, then it is not necessary that F is continuous at x.
- If the operator $F : X \to Y$ is Gâteux differentiable at $x \in X$, then $F(x + ty) \to F(x)$ as $t \to 0$. That is, if $x_n \to x$ along a line, then $F(x_n) \to F(x)$.

Lemma 3.1 *Let X, Y be Banach spaces, let $f : B_X(0, r) \to Y$ be Fréchet differentiable and $\|f'(x)\|_{\mathcal{L}(X,Y)} \leqslant \alpha$ for every $x \in B_X(0, r)$ and some $\alpha \geqslant 0$. Then f is Lipschitz continuous with Lipschitz constant less than or equal to α.*

Proof Suppose $x_1, x_2 \in B_X(0, r)$. By the Hahn–Banach theorem, there is $j \in Y^*$ of unit norm such that

$$\|f(x_1) - f(x_2)\|_Y = |j(f(x_1) - f(x_2))|.$$

For $t \in [0, 1]$ set $\varphi(t) = jf(tx_1 + (1 - t)x_2)$. Now applying the Lagrange mean value theorem to φ, there exists $\tau \in (0, 1)$ such that

$$|jf(x_1) - jf(x_2)| = |\varphi(1) - \varphi(0)| \leqslant |\varphi'(\tau)| = |jf'(\tau x_1 + (1 - \tau)x_2)(x_1 - x_2)|.$$

Hence, we obtain

$$\|f(x_1) - f(x_2)\|_Y \leqslant \|f'(\tau x_1 + (1 - \tau)x_2)(x_1 - x_2)\|_Y \leqslant \alpha\|x_1 - x_2\|_X.$$

This proves our claim.

3.3 Partial Derivatives

Suppose we have given two Banach spaces X and Y, then one can readily observe that the vector space $X \times Y$ is a Banach space with any of the (equivalent) Euclidean norms

$$\|(x, y)\|_p = \left(\|x\|_X^p + \|y\|_Y^p\right)^{1/p}, \quad \|(x, y)\|_\infty = \max\left\{\|\|_X, \|y\|_Y\right\} \quad (p \geqslant 1).$$

For what follows, we always use the ∞-norm, so that

$$B_{X \times Y}((x_0, y_0), r) = B_X(x_0, r) \times B_Y(y_0, r).$$

Suppose X, Y, Z are Banach spaces, $A \in \mathcal{L}(X, Z)$ and $B \in \mathcal{L}(Y, Z)$, then the operator $T : X \times Y \to Z$ defined by

$$T(x, y) = Ax + By$$

belongs to $\mathcal{L}(X \times Y, Z)$. Conversely, any $T \in \mathcal{L}(X \times Y, Z)$ has the above representation with

$$Ax = T(x, 0) \quad \text{and} \quad By = T(0, y).$$

It is then immediate to observe that $\mathcal{L}(X, Z) \times \mathcal{L}(Y, Z)$ and $\mathcal{L}(X \times Y, Z)$ are isomorphic Banach spaces.

Let U be an open set in $X \times Y$, $u_0 = (x_0, y_0) \in U$, and let $F : U \to Z$ be Fréchet differentiable at $u_0 = (x_0, y_0) \in U$. Then one can easily check that the partial derivatives $D_x F(u_0)$ and $D_y F(u_0)$ exist (i.e. the Fréchet derivatives of $F(\cdot, y_0) : X \to Z$ in x_0 and of $F(x_0, \cdot) : Y \to Z$ in y_0, respectively).

The following result is not very hard to prove. So, we omit the details.

Theorem 3.10 *Let $F : U \to Z$ be a continuous map from an open set $U \subset X \times Y$ into Z. Then F is continuously differentiable in $x_0, y_0) \in U$ if and only if F is partially differentiable and partial derivatives are continuous mappings, that is,*

$$(x_0, y_0) \mapsto D_x F(x_0, y_0), \quad U \mapsto \mathcal{L}(X, Z)$$

and

$$(x_0, y_0) \mapsto D_y F(x_0, y_0), \quad U \mapsto \mathcal{L}(Y, Z).$$

The (total) derivative of F in u_0 is given by

$$F'(u_0)(x, y) = D_x F(u_0)(x) + D_y F(u_0)(y).$$

3.4 Higher Derivatives

In the following, X and Y are real Banach spaces.

Suppose the operator $F : X \to Y$ is Fréchet differentiable, then we define an operator $F_1 : X \to \mathcal{L}(X, Y)$ by $F_1 x = dF(x)$.

Definition 3.5 We say that $F : X \to Y$ is twice Fréchet differentiable if the map $F_1 : X \to \mathcal{L}(X, Y)$, defined above, is Fréchet differentiable. The second derivative of F is denoted by $d^2 F$ and is a map from X to $\mathcal{L}(X, \mathcal{L}(X, Y)) = \mathcal{L}(X^2, Y)$. We shall alternatively denote $F''(x)$ for the same. Similarly, we can define the second Gâteaux derivative of F.

Definition 3.6 $d^2 F(x)$ is said to be symmetric if $D^2 F((x, x_1), x_2) = d^2 F(x, X_2)$, $x_1)$ for all $x_1, x_2 \in X$. In this case, we write the value of $F''(x)$ at (x_1, x_2) as $F''(x)x_1 x_2$.

Example 3.12 Consider a functional $f : \mathbb{R}^n \to \mathbb{R}$. We now intend to give the concrete representation of the second Gâteaux derivative $f''(x)$.

If $f''(x)$ exists, then by definition

$$\lim_{t \to 0} \left(\frac{1}{|t|} \right) \|[f'(x + th) - f'(x) - tf''(x)h]\| = 0.$$

Because $f''(x) \in \mathcal{L}(\mathbb{R}^n, \mathcal{L}(\mathbb{R}^n, \mathbb{R}))$, we shall evaluate $f''(x)$ by finding $(f''(x)h)k$ where $h, k \in \mathbb{R}^n$. Choosing h as the ith coordinate vector $e_i = (0, \ldots, 1, \ldots, 0)$ and

k as the jth coordinate vector e_j, we get

$$\lim_{t \to 0}(1/|t|)|f'(x + te_i)e_j - f'(x)e_j - tf''(x)e_ie_j| = 0, \quad i, j = 1, 2, \ldots, n.$$

This gives

$$
\begin{aligned}
f''(x)e_ie_j &= \lim_{t \to 0}(1/|t|)[\partial_j f(x + te_i) - \partial_j f(x)] \\
&= \partial_i \partial_j f(x), \quad i, j = 1, 2, \ldots, n.
\end{aligned}
$$

So if $h = \sum_{i=1}^{n} h_i e_i, k = \sum_{j=1}^{n} k_j e_j$, we get

$$
\begin{aligned}
f''(x)hk &= \sum_{i=1}^{n} \sum_{j=1}^{n} h_i k_j f''(x)e_i e_j \\
&= \sum_{i=1}^{n}, \sum_{j=1}^{n} h_i k_j \partial_i \partial_j f(x) \\
&= (H_f(x)h, k),
\end{aligned}
$$

where $H_f(x)$ is the Hessian matrix

$$
H_f(x) = \begin{bmatrix}
\partial_1 \partial_1 f(x) & \cdots & \partial_i \partial_1 f(x) & \cdots & \partial_n \partial_1 f(x) \\
\vdots & & \vdots & & \vdots \\
\partial_1 \partial_j f(x) & \cdots & \partial_i \partial_j f(x) & \cdots & \partial_n \partial_j f(x) \\
\vdots & & \vdots & & \vdots \\
\partial_1 \partial_n f(x) & \cdots & \partial_i \partial_n f(x) & \cdots & \partial_n \partial_n f(x)
\end{bmatrix}.
$$

If now $F : \mathbb{R}^n \to \mathbb{R}^m$ is an operator with components $f_1, f_2, \ldots, f_m$ then the concrete representation of $[F''(x)hk] \in \mathbb{R}^m$ is given by

$$[F''(x)hk]^T = ((H_1(x)h, k), (H_2(x)h, k), \ldots, (H_m(x)h, k)),$$

where $H_1(x), H_1(x), \ldots, H_m(x)$ are the Hessian matrices of $f_1, f_2, \ldots, f_m$ at x.

From the above example, it is clear that $d^2 F(x)$ or $F''(x)$ is symmetric iff each of the Hessian matrices are symmetric for the operator $F : \mathbb{R}^n \to \mathbb{R}^m$. We have, however, a more basic result regarding the symmetry of $F''(x)$. We state this result without proof.

Theorem 3.11 *If $F : X \to Y$ has a second Fréchet derivative $d^2 F(x)$, then it is symmetric.*

In this manner, $dF(x), d^2 F(x), \ldots, d^n F(x)$ are inductively defined and $d^n F(x) \in \mathcal{L}(X^n, Y)$ is symmetric. As in the case of functions of one variable $C^n(U), U$

an open subset of X will denote the space of n-times continuously differentiable functions. We are now ready to state and prove Taylor's theorem.

3.4.1 Taylor's Theorem

In the following theorem, the integral under consideration is the Bochner integral of Banach space-valued functions.

Theorem 3.12 (Taylor's Theorem) *Suppose $F \in C^n(U)$ where U is an open subset of X containing the line segment from x_0 to x. Then*

$$F(x_0 + x) = F(x_0) + dF(x_0)x + \cdots + \frac{1}{n-1!}d^{n-1}F(x_0)x\ldots x +$$

$$\frac{1}{n-1!}\int_0^1 (1-t)^{n-1}[d^n F(x_0 + tx)x\ldots x]dt$$

$$= \sum_{k=0}^n \frac{1}{k!}[d^k F(x_0)x\ldots x] + q(x),$$

where $q(x)$ is such that $\|q(x)\| = 0(\|x\|^n)$.

Proof Let $e \in Y^*$. Set $\phi(t) = (e, F(x_0 + tx))$. Then, using Taylor's theorem for a function of real variable, we get

$$\phi(t) = \phi(0) + t\phi'(0) + \cdots + \frac{t^{n-1}}{n-1!}\phi^{n-1}(0) + \frac{1}{n-1!}\int_0^t (t-s)^{n-1}\phi^n(s)ds.$$

Now, using the fact that

$$\phi^k(t) = (e, d^k F(x_0 + tx)x\ldots x),$$

we get

$$(e, F(x_0 + tx)) = (e, F(x_0)) + (e, dF(x_0 + tx)x) + \cdots + \frac{1}{n-1!}(e, d^{n-1}F(x_0 + tx)x\ldots x)$$

$$+ \frac{1}{n-1!}\left(e, \int_0^t (t-x)^{n-1}[d^n F(x_0 + sx)x\cdots x]ds\right).$$

Because the above inequality holds true for arbitrary $e \in Y^*$, we obtain the first part of the result. For the second part, we first note that

$$\int_0^1 (1-t)^{n-1}dt = \frac{1}{n}.$$

Next, we define

$$q(x) = \frac{1}{n-1!} \int_0^1 (1-t)^{n-1}[(d^n F(x_0 + tx) - d^n F(x_0))x \ldots x]dt.$$

Because $d^n F(x)$ is continuous for every $x \in U$, $t \to \|d^n F(x_0 + tx) - d^n F(x_0)\|$ is continuous on $[0, 1]$ and hence bounded. So we get

$$\|q(x)\| \leqslant \frac{1}{n!} \sup_{t \in [0,1]} \|d^n F(x_0 + tx) - d^n F(x_0)\| \|x\|^n.$$

This proves the theorem.   ∎

3.4.2 Inverse Function Theorem and Implicit Function Theorem

Next we state and prove two major results of differential calculus. These are the implicit function theorem and the inverse function theorem. The implicit function theorem deals with the following situation. Let $F(x, y)$ and suppose that

$$F(x_0, y_0) = c.$$

Can we find a function $x \mapsto y = \varphi(x)$, which at least locally satisfies

$$F(x, \varphi(x)) = c?$$

We want φ to be differentiable provided F is differentiable. Moreover, in the neighbourhood, where

$$F(x, \varphi(x)) = c$$

is valid, $\varphi(x)$ should be the unique solution. To better understand this problem, consider the following simple example.

Example 3.13 Let $F : \mathbb{R}^2 \to \mathbb{R}$ be defined by

$$F(x, y) = x^2 + y^2 - 4.$$

Let us consider the 0-level set of F, namely the set of those $x, y \in \mathbb{R}$ that satisfy $F(x, y) = 0$, i.e. $x^2 + y^2 - 4 = 0$. This is of course the circle with center at the origin and of radius 2. We now look for a function $\varphi(x)$, such that

$$F(x, \varphi(x)) = 0$$

for all x in the domain of φ. Evidently $\varphi(x) = \pm\sqrt{4 - x^2}$ and so φ need not be unique unless we restrict its domain. Also we see that near $x_0 = \pm 2$, φ could be either square root, so it is not uniquely determined. Note that at $x_0 = \pm 2$, φ is not differentiable and

$$\frac{\partial F}{\partial y} = 0.$$

Thus, in order to produce a unique differentiable function φ, such that

$$F(x, \varphi(x)) = 0,$$

we need to look locally and impose some condition like

$$\frac{\partial F}{\partial y} \neq 0.$$

The proof of the implicit function theorem uses the Banach fixed point theorem, which we state here in the form needed and postpone the proof of the general version until Sect. 5.2.

Proposition 3.1 (Banach fixed point theorem) *Suppose X is a Banach space, C is a closed subset of X and $T : C \to C$ satisfies*

$$\|T(x_1) - T(x_2)\|_X \leqslant \lambda\|x_1 - x_2\|_X \quad \forall x_1, x_2 \in C, \ \text{for some } \lambda \in [0, 1),$$

then there exists unique $x \in C$, such that $x = T(x)$. Moreover, if we have a parametrized family $\{T(v)_{v \in V}$ (with V being an open subset of a Banach space Y) satisfying the above contraction condition with $\lambda \in [0, 1)$ independent of v, then the unique solution $x = x(v)$ of $x = T(v)x$ depends continuously on v.

Theorem 3.13 (Implicit function theorem) *Suppose X, Y, Z are Banach spaces, $U \subset X \times Y$ is an open set, $u_0 = (x_0, y_0) \in U$, and $F : U \to Z$ is a function. Assume that*

(a)   *F is a continuous differentiable function and $F(u_0) = 0$;*
(b)   *$D_y F(u)$ exists for every $u = (x, y) \in U$;*
(c)   *$D_y F$ is continuous at u_0 and $D_y F(u_0) \in \mathcal{L}(X, Y)$ is invertible, i.e. $D_y F(u_0)$ is an isomorphism.*

Then there exist neighbourhoods U_1 of x_0 and U_2 of y_0, such that $U_1 \times U_2 \subseteq U$ and a unique continuously differentiable function $\varphi : U_1 \to U_2$, such that

$$F(x, \varphi(x)) = 0 \ \forall x \in U_1$$

and

$$D\varphi(x) = D_y F(x, \varphi(x))^{-1} D_x F(x, \varphi(x)) \ \forall x \in U_1.$$

Proof Let $T_0 = D_y f(x_0, y_0) \in \mathcal{L}(Y, Z)$. By (c) T_0 is an isomorphism. Then we see that the equation $F(x, y) = 0$ can be equivalently rewritten as

$$y = y - T_0^{-1} F(x, y). \tag{3.9}$$

The advantage of passing to (3.9) is that we can apply Proposition 3.1. Namely for every x, we look for a fixed point of $y \mapsto y - T_0^{-1} F(x, y)$ and to do this we employ Proposition 3.1. Let us set

$$G(x, y) = y - T_0^{-1} F(x, y)$$

Because $T_0^{-1} \circ T_0 = I_Y$, we have

$$G(x, y_1) - G(x, y_2) = T_0^{-1}[T_0(y_1 - y_2) - F(x, y_1) - F(x, y_2)].$$

Because f is C^1 at (x_0, y_0) and T_0 is an isomorphism, we can find $r_1 > 0$ and $\delta > 0$, such that if $\|x - x_0\|_X \leqslant r_1$, $\|y_1 - y_0\|_Y \leqslant \delta$, $\|y_2 - y_0\|_Y \leqslant \delta$, then

$$\|G(x, y_1) - G(x, y_2)\|_Y \leqslant \frac{1}{2}\|y_1 - y_2\|_Y. \tag{3.10}$$

Also because of the continuity of $G(\cdot, y_0)$, we can find $r_2 > 0$, such that if $\|x - x_0\|_X \leqslant r_2$, then

$$\|G(x, y_0) - G(x_0, y_0)\|_Y < \frac{\delta}{2}. \tag{3.11}$$

Therefore, from (3.10) and (3.11), if $r = \min\{r_1, r_2\}$ and $\|x - x_0\|_X \leqslant r$, $\|y_1 - y_0\|_Y \leqslant \delta$, we have

$$\begin{aligned}
\|G(x, y) - y_0\|_Y &= \|G(x, y) - G(x_0, y_0)\|_Y \\
&\leqslant \|G(x, y) - G(x, y_0)\|_Y + \|G(x, y_0) - G(x_0, y_0)\|_Y \\
&\leqslant \frac{1}{2}\|y - y_0\|_Y + \frac{\delta}{2} \\
&\leqslant \delta.
\end{aligned} \tag{3.12}$$

Thus, we see that $G(x, \cdot)$ maps $\overline{B}_\delta(y_0) = \{y \in Y : \|y - y_0\|_Y \leqslant \delta\}$ onto itself as well as $B_\delta(y_0) = \{y \in Y : \|y - y_0\|_Y < \delta\}$ onto itself (see (3.11) and (3.12)), for all $x \in \overline{B}_r(x_0) = \{x \in X : \|x - x_0\|_X \leqslant r\}$. We can apply Proposition 3.1 to obtain the parametric family $\{y \mapsto G(x, y)\}_{x \in \overline{B}_r(x_0)}$. So for every $x \in \overline{B}_r(x_0)$, we can find unique $y = y(x) \in \overline{B}_\delta(y_0)$, such that

$$G(x, y) = y,$$

hence $F(x, y) = 0$ and the function $\varphi(x) = y(x)$ is continuous. Let $U_1 = B_r(x_0)$ and $U_2 = B_\delta(y_0)$. By choosing $r > 0$ and $\delta > 0$ small enough, it is evident that we can have that $U_1 \times U_2 \subset U$. We claim that the function $\varphi : B_r(x_0) \to Y$ is continuously differentiable. To this end, let $(x_1, y_1) \in U_1 \times U_2, y_1 = \varphi(x_1)$ (recall that $F(x, \cdot)$ maps U_2 into itself). Exploiting the differentiability of F at (x_1, y_1), we have

$$F(x, y) = A(x - x_1) + B(y - y_1) + \omega(x, y) \quad \forall \ (x, y) \in U,$$

with $A = D_x F(x_1, y_1), \ B = D_y F(x_1, y_1)$ and

$$\lim_{(x,y) \to (x_1, y_1)} \frac{\|\omega(x, y)\|_Z}{\|(x - x_1, y - y_1)\|_{X \times Y}} = 0.$$

We now see that

$$F(x, \varphi(x)) = 0 \quad \forall x \in U_1.$$

Hence,

$$\varphi(x) = -B^{-1}A(x - x1) + y_1 - B^{-1}\omega(x, \varphi(x)). \tag{3.13}$$

Also we can find $\delta_1, \delta_2 > 0$, such that if $\|x - x_1\|_X \leq \delta_1, \ \|y - y_1\|_Y \leq \delta_2$, then

$$\|\omega(x, y)\|_Z \leq \frac{1}{2\|B^{-1}\|_{\mathcal{L}}}(\|x - x_1\|_X + \|y - y_1\|_Y),$$

whence

$$\|u(x, \varphi(x))\|_Z \leq \frac{1}{2\|B^{-1}\|_{\mathcal{L}}}(\|x - x_1\|_X + \|\varphi(x) - \varphi(x_1)\|_Y). \tag{3.14}$$

From (3.13) and (3.14), it follows that

$$\|\varphi(x) - \varphi(x_1)\|_Y \leq \|B^{-1}A\|_{\mathcal{L}}\|x - x_1\|_X + \frac{1}{2}\|x - x_1\|_X + \frac{1}{2}\|\varphi(x) - \varphi(x_1)\|_Y.$$

Simplifying the above inequality, we obtain

$$\|\varphi(x) - \varphi(x_1)\|_Y \leq \alpha\|x - x_1\|_X, \tag{3.15}$$

where $\alpha = 2\|B^{-1}A\|_{\mathcal{L}} + 1$. Define

$$u(x) = -B^{-1}\omega(x, \varphi(x)).$$

From (3.13), we have

$$\varphi(x) - \varphi(x_1) = B^{-1}A(x - x_1) + u(x) \tag{3.16}$$

and since

$$\|u(x)\|_Y \leqslant \|B^{-1}\|_{\mathcal{L}} \|\omega(x, \varphi(x))\|_Z$$

and φ is continuous, we have

$$\lim_{x \to x_1} \frac{\|u(x)\|_Y}{\|x - x_1\|_X} = 0. \tag{3.17}$$

From (3.16) and (3.17), it follows that φ is Fréchet-differentiable at $x_1 \in U_1$ and

$$\varphi'(x_1) = -B^{-1}A = -(D_y F(x_1, y_1))^{-1} D_x F(x_1, y_1),$$

which means that φ is continuously differentiable.

Using the implicit function theorem (see Theorem 3.13), we can prove the inverse function theorem.

Theorem 3.14 (Inverse Function Theorem) *Suppose X, Y are Banach spaces, $U \subseteq Y$ is an open set, $y_0 \in U$ and $F : U \to X$ such that*
(i) F is a continuously differentiable function, and
(ii) $F'(y_0) \in \mathcal{L}(Y, X)$ is an isomorphism.
Then there exists a neighbourhood U_0 of y_0, $U_0 \subseteq U$ and V_0 a neighbourhood of $x_0 = F(y_0)$, such that $F : U_0 \to V_0$ is a diffeomorphism and

$$(F^{-1})'(x_0) = [F'(y_0)]^{-1}.$$

(The derivative of the inverse is the inverse of the derivative.)

Proof Define

$$G(x, y) = F(y) - x.$$

Then $D_y G(x_0, y_0) = F'(y_0)$, which by hypothesis is an isomorphism. So by virtue of Theorem 3.13, we can find a neighbourhood V_0 of x_0 and a continuously differentiable map $\varphi : V_0 \to Y$, such that $\varphi(V_0) \subseteq U_0$ for a neighbourhood U_0 of y_0,

$$G(x, \varphi(x)) = 0 \ \forall x \in V_0,$$

i.e. $F\varphi(x) = x$ for all $x \in V_0$ and $\varphi(x_0) = y_0$.

We now consider F restricted to $\varphi(V_0)$. Since $F\varphi(x) = x$, we see that φ is injective on V_0, hence a bijection from V_0 onto $\varphi(V_0)$. In addition, $\varphi(V_0) = F^{-1}(V_0)$ is open because F is continuous. So we set

$$U_0 = \varphi(V_0)$$

and we have that $F : U_0 \to V_0$ is a bijection.
Finally since

$$\varphi'(x_0) = -(D_y G(x_0, \varphi(x_0)))^{-1} D_x G(x_0, \varphi(x_0)),$$

we have

$$F'(x_0) \circ \varphi'(x_0) = I_X,$$

hence

$$\varphi'(x_0) = (F^{-1})'(x_0).$$

It follows that

$$(F^{-1})'(x_0) = [F'(y_0)]^{-1}.$$

3.5 Subdifferential of Convex Functions

We consider function $f : X \to (-\infty, \infty]$, where X is a normed linear space. We define the domain $\mathscr{D}(f)$, range $\mathscr{R}(f)$ and epigraph $epi(f)$ as follows:

$$\mathscr{D}(f) = \{x \in X : f(x) < \infty\}$$
$$\mathscr{R}(f) = \{f(x) : x \in \mathscr{D}(f)\}$$
$$epi(f) = \{(x, \rho) \in \mathscr{D}(f) \times \mathbb{R} : f(x) \leqslant \rho\}.$$

Definition 3.7 Let X be a normed linear space and $f : X \to (-\infty, \infty]$ a function. The f is said to be a convex function if

$$f[\lambda x + (1 - \lambda)y] \leqslant \lambda f(x) + (1 - \lambda)f(y), \quad 0 < \lambda < 1 \qquad (3.18)$$

for all $x, y \in X$. f is strictly convex if we have strict inequality in (3.18).

We have the following characterization of convexity of f, through the convexity of epigraph $epi(f)$.

Proposition 3.2 *Let X be a linear space and $f ; X \to (-\infty, \infty]$ a function. Then f is convex iff $epi(f)$ is a convex subset of $X \times \mathbb{R}$.*

Proof Suppose f is convex and (x, ρ) and $(y, \mu) \in epi(f))$. Then we see that

$$\lambda(x, \rho) + (1 - \lambda)(y, \mu) = (\lambda x + (1 - \lambda)y, \lambda\rho + (1 - \lambda)\mu)$$

and

$$f(\lambda x + (1 - \lambda)y) \leqslant \lambda f(x) + (1 - \lambda)f(y)$$
$$\leqslant \lambda\rho + (1 - \lambda)\mu \text{ for all } \lambda \in (0, 1).$$

This implies that $\lambda(x, \rho) + (1 - \lambda)(y, \mu) \in epi(f)$.
Conversely, suppose that $epi(f)$ is convex. Then for $x, y \in \mathscr{D}(f)$ and $(x, \rho), (y, \mu) \in epi(f)$, we have

$$\lambda(x, \rho) + (1 - \lambda)(y, \mu) = (\lambda x + (1 - \lambda)y, \lambda f(x) + (1 - \lambda)f(y)) \in epi(f) \text{ for all } \lambda \in (0, 1).$$

Thus, by the definition of $epi(f)$, it follows that

$$f(\lambda x + (1 - \lambda)y) \leqslant \lambda f(x) + (1 - \lambda)f(y).$$

This shows that f is convex and hence the proposition. ∎

Proposition 3.3 *Let C be a nonempty closed convex subset of a Banach space X and $f; C \to (-\infty, \infty]$ a convex function. Then f is lower semicontinuous in the norm topology iff f is lower semicontinuous in the weak topology.*

Proof By setting $D_\rho = \{x \in C : f(x) \leqslant \rho\}, \rho \in \mathbb{R}$, we see that D_ρ is convex. Indeed, for all $x, y \in D_\rho, 0 < \lambda < 1$,

$$f(\lambda x + (1 - \lambda)y) \leqslant \lambda f(x) + (1 - \lambda)f(y)$$
$$\leqslant \lambda\rho + (1 - \lambda)\rho = \rho.$$

It follows from Proposition 1.2 that D_ρ is closed if and only if D_ρ is weakly closed, i.e. D_ρ is closed in weak topology. Thus, by Proposition 3.2, it follows that f is l.s.c in the norm topology iff f is l.s.c. in the weak topology. ∎

Definition 3.8 A convex function is said to be proper if its epigraph is nonempty and contains no vertical lines. Thus, f is proper iff $\mathscr{D}(f)$ is nonempty and $f|_{\mathscr{D}(f)}$ is finite. So if f_0 is a proper convex function on a convex set C, then one can extend it to f in the entire space X by defining

$$f(x) = \begin{cases} f_0(x), & x \in C, \\ \infty, & x \notin C. \end{cases}$$

Here, we notice that $\mathscr{D}(f) = C$.

Example 3.14 One of the most important convex functions is the norm functional

$$f(x) = \|x\|, \quad x \in X.$$

Example 3.15 To each subset C of X, we associate the indicator function I_C of C defined as

$$I_C(x) = \begin{cases} 0, & x \in C, \\ \infty, & x \notin C. \end{cases}$$

Observation

- I_C is convex iff C is convex.
- $epi(I_C)$ is half cylinder with cross section C.

- I_C is proper iff C is nonempty.
- $\mathscr{D}(I_C) = C$.

Example 3.16 The support function $\delta^*(.|C)$ of a convex subset C of X is defined as

$$\delta^*(x^*|C) = \sup_{x \in C}(x^*, x).$$

$\delta^*(.|C)$ is a convex function on X^*.

Example 3.17 Let C be a convex subset of a Hilbert space X. Define the distance function $D(x, C) = \inf\{\|x - y\|, y \in C\}$. Then $D(x, C)$ is a convex function.

Example 3.18 For a convex subset C of X, define the gauze functional $\chi(x|C) = \inf\{\lambda \geqslant 0 : x \in \lambda C\}$. Then $\chi(x|C)$ is a convex function.

Example 3.19 Let A be a self adjoint continuous operator on a Hilbert space X. Consider the quadratic functional $\varphi(x) = \frac{1}{2}(Ax, x)$. Then φ is differentiable and $\nabla\varphi = A$. It can be shown that φ is convex iff A is positive semidefinite.

Remark 3.4 For Gâteaux differentiable functions φ, we will derive an important relation between the monotonicity and convexity in Chap. 4.

Remark 3.5 Indicator function I_C plays a very important role in convex analysis while the quadratic functional of Example 3.19 is important in variational analysis.

Let X be a topological space and $f : X \to (-\infty, \infty]$ a proper function. Then f is said to be lower semicontinuous (l.s.c.) at $x_0 \in X$ if

$$f(x_0) \leqslant \liminf_{x \to x_0} f(x) = \sup_{V \in U_{x_0}} \inf_{x \in V} f(x),$$

where U_{x_0} is a base of neighbourhoods of the point $x_0 \in X$. f is said to be lower semicontinuous on X if it is lower semicontinuous at each point of X, i.e. for each $x \in X$

$$x_n \to x \Rightarrow f(x) \leqslant \liminf_{n \to \infty} f(x_n).$$

Proposition 3.4 *Let X be a topological space and $f ; X \to (-\infty, \infty]$ a function. Then the following statements are equivalent: Let X be a linear space and $f ; X \to (-\infty, \infty]$ a function. Then*

(1) *f is lower semicontinuous.*
(2) *For each $\rho \in \mathbb{R}$, the level set $\{x \in X : f(x) \leqslant \rho\}$ is closed.*
(3) *The epigraph of the function f is closed.*

Proof (1) $\Rightarrow$ (2). Let $\rho \in \mathbb{R}$ and let $x_0 \in X$ with $f(x_0) > \rho$. Because f is lower semicontinuous at x_0, we must have

$$f(x_0) \leqslant \sup_{V \in U_{x_0}} \inf_{x \in V} f(x).$$

This implies that there exists $V_0 \in U_{x_0}$ such that $\inf_{x \in V_0} f(x) > \rho$, i.e. $f(x) > \rho$ for all $x \in V_0$. It follows that $V_0 \subset \{x \in X : f(x) > \rho\}$ which shows that $\{x \in X : f(x) > \rho\}$ is open and hence $\{x \in X : f(x) \leqslant \rho\}$ is closed.

(2) $\Rightarrow$ (1). Let $x_0 \in \mathscr{D}(f)$, $\varepsilon > 0$, and let U_{x_0} denote a base for neighbourhood of the point x_0. Set $V_\varepsilon = \{x \in X : f(x) > f(x_0) - \varepsilon\}$. Because each level set of f is closed, the set $\{x \in X : f(x) \leqslant f(x_0) - \varepsilon\}$ must be closed. It follows that $V_\varepsilon \in V_{x_0}$. Furthermore, $f(x) > f(x_0) - \varepsilon$ for all $x \in V_\varepsilon$. Therefore, $f(x) \geq f(x_0) - \varepsilon$ and hence $\sup_{V \in U_{x_0}} \inf_{x \in V} f(x) \geq f(x_0) - \varepsilon$. As ε is arbitrarily chosen positive number, so letting $\varepsilon \to 0$ we obtain $\liminf_{x \to x_0} f(x) \geqslant f(x_0)$ and so we conclude that (1) holds.

(1) $\Leftrightarrow$ (3). To see this, let us define $\varphi : X \times R \to (-\infty, \infty]$ by $\varphi(x, \rho) = f(x) - \rho$. Then

$$f \text{ is } l.s.c. \text{ on } X \Leftrightarrow f(x_0) \leqslant \liminf_{x \to x_0} f(x)$$
$$\Leftrightarrow f(x_0) - \rho \leqslant \liminf_{x \to x_0} f(x) - \rho = \liminf_{x \to x_0} (f(x) - \rho)$$
$$\Leftrightarrow \varphi(x_0, \rho) \leqslant \liminf_{x \to x_0} \varphi(x, \rho)$$
$$\Leftrightarrow \varphi \text{ is } l.s.c. \text{ on } X \times \mathbb{R}.$$

Because $epi(f)$ is a level set of φ, we conclude that the conclusion holds.

We now discuss some regularity properties of convex functions. By virtue of Proposition 3.3, the lower semicontinuity of $f : X \to (-\infty, \infty]$ can alternatively be defined as follows:

f is lower semicontinuous if the level set $\{x \in X : f(x) \leqslant \rho\}$ is a closed subset of X for all $\rho \in \mathbb{R}$.

We now ready to discuss some regularity properties of convex functions. It may easily be observed that we immediately have the following theorem regarding the lower semicontinuity of f.

Theorem 3.15 *The convex function f is lower semicontinuous on its domain if $epi(f)$ is closed.*

We give below an important continuity result of convex functions.

Theorem 3.16 *The convex function f is continuous in int $\mathscr{D}(f)$ iff f is bounded on some open subset U of $\mathscr{D}(f)$.*

Proof If f is continuous in int $\mathscr{D}(f)$, it trivially follows that f is bounded on some open subset of $\mathscr{D}(f)$.

Let us now assume that f is bounded on an open subset U of X. If $x_0 \in U$ is arbitrary, we will show that f is continuous there. Without loss in generality, we may assure that $x_0 = 0$ and $f(x_0) = 0$ and U an open ball $\{x : \|x\| < r\}$.

Let U_ε denote that ball $\{x : \|x\| < r\varepsilon\}$. For $0 < \varepsilon < 1$ and $x \in U_\varepsilon$, $\frac{x}{\varepsilon} \in U$ and so convexity of f gives

$$f(x) = f\left(\varepsilon \frac{x}{\varepsilon} + (1 - \varepsilon)0\right) \leqslant \varepsilon f\left(\frac{x}{\varepsilon}\right) + (1 - \varepsilon)f(0)$$

$$\leqslant \varepsilon k + (1 - \varepsilon)f(0) = \varepsilon k, \quad (k \text{ is the bound for } f \text{ on } U).$$

Also

$$0 = f\left(\frac{x}{1 + \varepsilon} + \frac{\varepsilon}{1 + \varepsilon}\left(-\frac{x}{\varepsilon}\right)\right) \leqslant \frac{f(x)}{1 + \varepsilon} + \frac{\varepsilon}{1 + \varepsilon}f\left(-\frac{x}{1 + \varepsilon}\right) \leqslant \frac{f(x)}{1 + \varepsilon} + \frac{\varepsilon k}{1 + \varepsilon},$$

which implies that

$$f(x) \geqslant -\varepsilon k.$$

So $-\varepsilon k \leqslant f(x) \leqslant \varepsilon k$ for $x \in U_\varepsilon$ and hence the continuity of f at 0 and consequently at x_0.

For any arbitrary point $x_0 \in int \, \mathcal{D}(f)$, there exists $\rho < 1$ such that $\rho x_0 \in int \, \mathcal{D}(f)$. Again, by convexity of f, for all $\left(1 - \frac{1}{\rho}\right)x + x_0 \in \left(1 - \frac{1}{\rho}\right)U + x_0$ we get

$$f\left(\left(1 - \frac{1}{\rho}\right)x + x_0\right) = f\left(\left(1 - \frac{1}{\rho}\right)x + \frac{1}{\rho}\rho x_0\right)$$

$$\leqslant \left(1 - \frac{1}{\rho}\right)f(x) + \frac{1}{\rho}f(\rho x_0)$$

$$\leqslant \left(1 - \frac{1}{\rho}\right)k + \frac{1}{\rho}f(\rho x_0).$$

This implies that f is bounded on $\left(1 - \frac{1}{\rho}\right)U + x_0$ containing x_0 and now result follows by what we have just proved. ∎

We immediately get the following two corollaries:

Corollary 3.2 *The convex function f is continuous in $int \, \mathcal{D}(f)$ iff $epi(f)$ has nonempty interior.*

Corollary 3.3 *Every real-valued convex function on X is continuous.*

It is interesting to note that convex functions defined on X are differentiable except on a countable subset of its domain of continuity. More precisely we have the following result due to Asplund [23].

Theorem 3.17 *Every continuous convex function on a Banach space X is Fréchet differentiable on a dense G_δ subset of its domain of continuity.*

For more on the regularity properties of the convex functions, refer to Fenchel [229], Moreau [407] and Rockfeller [523, 524]. For Gâteaux differentiable convex functions, we have the following inequality.

Theorem 3.18 *If the convex function $f(x)$ has the Gâteaux derivative $df(x_0)$ at some point $x_0 \in int \mathscr{D}(f)$, then*

$$f(x) \geqslant f(x_0) + (df(x_0), x - x_0) \ \text{for all} \ x \in X.$$

Proof Because f is convex, we have

$$f(x_0 + t(x - x_0)) - f(x_0) \leqslant tf(x) + (1 - t)f(x_0) - f(x_0)$$
$$= t[f(x) - f(x_0)], \ 0 < t < 1.$$

This implies that

$$f(x) - f(x_0) \geqslant \lim_{t \to 0+} \frac{[f(x_0 + t(x - x_0)) - f(x_0)]}{t}$$
$$= (df(x_0), x - x_0).$$

Thus, we conclude that $f(x) \geqslant f(x_0) + (df(x_0), x - x_0)$ for all $x \in X$. ∎

The above inequality leads us to the following definition.

Definition 3.9 An element $x_0^* \in X^*$ is said to be a subgradient of f at x_0 if $f(x) \geqslant f(x_0) + (x_0^*, x - x_0)$ for all $x \in X$. The set of all subgradients at x_0 is denoted by $\partial f(x_0)$.

Example 3.20 For the function $f(x) = |x|$, $x \in \mathbb{R}$, we have

$$\partial f(x) = \begin{cases} \{-1\}, & \text{if} \ x < 0, \\ [-1, 1], & \text{if} \ x = 0, \\ \{1\}, & \text{if} \ x > 0. \end{cases}$$

In this example, we see that at points where f only has one subgradient, it coincides with the Gâteaux derivative. This property holds true in general.

Definition 3.10 The multivalued $\partial f : X \to 2^{X^*}$, where $\partial f(x)$ is the set of all subgradients of f at x is called subdifferential of f at x. If $\partial f(x) \neq \varnothing$, then f is said to be subdifferentiable at x. We denote by $\mathscr{D}(\partial f)$ the set $\{x \in X : \partial f(x) \neq \varnothing\}$.

Observation

- The notion of subdifferentiability allows us to consider optimization problems for subdifferentiable functions:

$$\inf_{v \in X} f(v).$$

Obviously, a necessary optimality condition for $u \in X$ to be a minimizer of f is

$$0 \in \partial f(u).$$

Another important example is that of a constrained optimization problem for a Gâteaux differentiable function

$$f : \inf_{v \in C} f(v),$$

where $C \subset X$ is supposed to be a closed convex set. Then, we can restate the constrained as an unconstrained problem by means of the indicator function I_C of C:

$$\inf_{v \in X} \left(f(v) + I_C(v) \right)$$

and get the necessary optimality condition

$$0 \in f'(u) + \partial I_C(u).$$

- For Gâteaux differentiable convex functions, subdifferential reduces to the gradient mapping ∇f.
- The concept of subdifferentiability is of recent origin. For a more explicit treatment on it, refer Moreau [408, 409].

Theorem 3.19 *Let f be a real-valued convex function on X which is every where Gâteaux differentiable. Then ∂f reduces to single-valued mapping $\nabla f : X \to X^*$.*

Proof It suffices to show that $\partial f(x_0) \subset \{\nabla f(x_0)\}$, that is, if $x_0^* \in \partial f(x_0)$ then

$$(x_0^*, h) = (\nabla f(x_0), h) \text{ for all } h \in X.$$

By definition

$$\frac{(x_0^*, th)}{t} \leqslant \frac{f(x_0 + th) - f(x_0)}{t}, t > 0.$$

Taking limit as $t \to 0$, we get

$$(x_0^*, h) = \frac{(x_0^*, th)}{t} \leqslant \lim_{t \to 0+} \left[\frac{f(x_0 + h) - f(x_0)}{t} \right]$$

$$= (\nabla f(x_0), h).$$

Now, taking $-t$ for t we get

$$(x_0^*, h) \geqslant (\nabla f(x_0), h)$$

and hence the result.   ∎

From the definition of subdifferential, we immediately get the following theorem.

Theorem 3.20 $\partial f(x)$ *is a weak* closed convex subset of* X^* *for each* $x \in X$ *and* $\mathscr{D}(\partial f) \subset \mathscr{D}(f)$. *Further, subdifferential is a homogeneous operator, that is* $\partial f(\lambda x) = \lambda \partial f(x)$.

In view of the above theorem, it is interesting to determine how big is $\mathscr{D}(\partial f)$ with respect to $\mathscr{D}(f)$. In this respect, we have the following theorem which is due to Moreau [407]. We only state this.

Theorem 3.21 $\mathscr{D}(\partial f) = \{x \in \mathscr{D}(f) : f \text{ is continuous at } x\}$.

We now define the concept of conjugate functional. It is helpful in investigating the subdifferentiability of lower semicontinuous functions though we will not be needing this in subsequent chapters.

Definition 3.11 Let f be a function on X. Then conjugate functional f^* of f is defined by

$$f^*(x^*) = \sup\{(x^*, x) - f(x) : x \in X\}.$$

One can show that if f is a proper lower semicontinuous function so is f^* and $f^{**} = f$, that is,

$$f(x) = \sup\{(x^*, x) - f^*(x^*) : x^* \in X^*\}.$$

From the definition of subgradient it follows that if $x_0^* \in \partial f(x_0)$, then

$$f(x) \geqslant f(x_0) + (x_0^*, x - x_0) \text{ for all } x \in X.$$

That is

$$(x_0^*, x_0) \geqslant f(x_0) + (x_0^*, x) - f(x) \text{ for all } x \in X,$$

which gives

$$(x_0^*, x_0) \geqslant f(x_0) + f^*(x_0^*).$$

Also from the definition of conjugate functional, it follows that

$$f^*(x_0^*) \geqslant (x_0^*, x_0) - f(x_0),$$

which gives

$$(x_0^*, x_0) \leqslant f(x_0) + f^*(x_0^*).$$

Combining these two inequalities, we obtain

$$(x_0^*, x_0) = f(x_0) + f^*(x_0^*).$$

Conversely if $(x_0^*, x_0) = f(x_0) + f^*(x_0^*)$, it follows that

$$((x_0^*, x_0) \geqslant f(x_0) + (x_0^*, x) - f(x) \text{ for all } x \in X.$$

This gives that $x_0^* \in \partial f(x_0)$. Similarly one can show $x_0 \in \partial f^*(x_0^*)$. Thus, we have proved that

$$x_0^* \in \partial f(x_0) \text{ and } x_0 \in \partial f^*(x_0^*) \text{ iff } (x_0^*, x_0) = f(x_0) + f^*(x_0^*).$$

This notion of conjugacy is due to Fenchel [230]. For more detail treatment on it, refer Moreau [410] and Rockafellar [525].

We now state without proof a theorem on the subdifferentiability of a convex function which is due to Brøndsted and Rockafellar [85].

Theorem 3.22 *Let f be a lower semicontinuous function on a Banach space X, then $\mathcal{D}(\partial f) = \mathcal{D}(f)$.*

Remark 3.6 Monotone properties of the subdifferential of convex functions will be discussed in detail in Chap. 4 on monotone operators, Φ-accretive operators and their generalizations.

Example 3.21 Let X be a real Hilbert space. Define a functional $f(x)$ as $f(x) = \|x\|$ $\forall x \in X$. Then we observe the following:

- At $x \neq 0, f$ is Gâteaux differentiable with $\nabla f(x) = \frac{x}{\|x\|}$.
- At $x = 0, f$ is not differentiable but it is subdifferentiable. In fact, the set $\partial f(0)$ consists of all vectors x^* such that $\|x\| \geq (x^*, x)$ for all $x \in X$. This implies that $\partial f(0) = \{x^* \in X : \|x^*\| \leq 1\}$.

Example 3.22 Let $f(x) = \begin{cases} -[1 - \|x\|^2]^{1/2}, & \|x\| \leq 1 \\ \infty, & \text{otherwise.} \end{cases}$

By definition, we find that $\mathcal{D}(f) = \{x : \|x\| \leq 1\}$.
Now we observe that f is subdifferentiable at x, $\|x\| < 1$. But $\partial f(x) = \varnothing$ for $\|x\| \geq 1$. Thus,

$$\mathcal{D}(\partial f) \neq \mathcal{D}(f).$$

Example 3.23 Let us consider the indicator function I_C of a convex subset C of a Hilbert space X. $x_0^* \in \partial I_C(x_0)$ iff

$$I_C(x) \geq I_C(x_0) + (x_0^*, x - x_0) \text{ for all } x \in X.$$

This implies that $x \in C$ and $(x_0^*, x - x_0) \leq 0$ for every $x \in C$, which implies that $x_0^* \perp C$ at x. So $\partial I_C(x_0)$ is the normal cone to C at x.

We finally give the following theorem for the subdifferentiability of the norm functional $j(x) = \frac{\|x\|^2}{2}$ on a Banach space X.

Theorem 3.23 *The subdifferential of the norm functional j is precisely the duality mapping J.*

Proof By definition

$$\partial j(x_0) = \{x_0^* : j(x) \geqslant j(x_0) + (x_0^*, x - x_0) \text{ for all } x \in X\}.$$

Let $\varepsilon > 0$ be arbitrary, then writing $x = (1 \pm \varepsilon)x_0$, for $x_0^* \in \partial j(x_0)$ we have

$$\frac{1}{2}(1 \pm \varepsilon)^2\|x_0\|^2 \geqslant \frac{1}{2}\|x_0\|^2 \pm \varepsilon(x_0^*, x).$$

That is

$$\frac{1}{2}(1 + \varepsilon^2 \pm 2\varepsilon)\|x_0\|^2 \geqslant \frac{1}{2}\|x_0\|^2 \pm \varepsilon(x_0^*, x_0)$$

or

$$\pm\varepsilon(x_0^*, x_0) \leqslant \pm\varepsilon\|x_0\|^2 + \frac{1}{2}\varepsilon^2\|x_0\|^2.$$

This gives

$$\|x_0\|^2 - \frac{\varepsilon}{2}\|x_0\|^2 \leqslant (x_0^*, x_0) \leqslant \|x_0\|^2 + \frac{\varepsilon}{2}\|x_0\|^2.$$

Taking the limit with $\varepsilon \to 0$, we get

$$(x_0^*, x_0) = \|x_0\|^2$$

and this also gives

$$\|x_0^*\| \geqslant \|x_0\|. \qquad (3.19)$$

For $\|x_0\| = 1$, using the definition of $\partial j(x_0)$ we get that for

$$x_0^* \in \partial j(x_0), \quad (x_0^*, x) - \|x_0\|^2 \leqslant \frac{1}{2}[\|x\|^2 - \|x_0\|^2].$$

This gives $(x_0^*, x) \leqslant 1$ for all x with $\|x\| = 1$ and hence $\|x_0^*\| \leqslant 1$ for $\|x_0\| = 1$. Now, since ∂j is a homogeneous operator, we have

$$\partial j(x_0) = \|x_0\|\partial j\left(\frac{x_0}{\|x_0\|}\right)$$

and hence we get the inequality

$$\|x_0^*\| \leqslant \|x_0\| \text{ for arbitrary } x_0^* \in \partial j(x_0). \qquad (3.20)$$

Combining (3.19) and (3.20), we get that for $x_0^* \in \partial j(x_0)$,

$$\|x_0^*\| = \|x_0\|^2.$$

We have already proved that

$$(x_0^*, x_0) = \|x_0\|^2.$$

So we have

$$\partial j(x_0) = \left\{ x_0^* : (x_0^*, x_0) = \|x_0\|^2 \text{ and } \|x_0^*\| = \|x_0\| \right\} = J(x_0).$$

This completes the proof.   ∎

Observation

Let X be a Banach space and $f : X \to (-\infty, \infty]$ a proper convex function. Then we observe the following:

- If f is continuous, then $\partial f(x) \neq \varnothing$ for every $x \in X$.
- For every $x \in X$, $\partial f(x)$ is always a closed convex set in X^*.
- $\partial f(x)$ is a homogeneous operator, i.e. $\partial(\lambda f(x)) = \lambda \partial f(x)$ for every $x \in X$ and $\lambda \in \mathbb{R}$.
- If f is lower semicontinuous on X, then $\mathscr{D}(\partial f) = \mathscr{D}(f)$.
- f has a maximum value at $x_0 \in \mathscr{D}(\partial f)$ if and only if $0 \in \partial f(x_0)$.

3.6 Differentiability of Norms of Banach Spaces

Definition 3.12 Let X be a Banach space endowed with norm $\| \cdot \|$ and let $S_X = \{x \in X : \|x\| = 1\}$ be the unit sphere of X. Then the norm $\| \cdot \|$ is Gâteaux differentiable at point $x \in S_X$ if for $y \in S_X$

$$\frac{d}{dt}\|x + ty\|\Big|_{t=0} = \lim_{t \to 0} \frac{\|x + ty\| - \|x\|}{t} \tag{3.21}$$

exists and it is customary to denote this fact by $\langle y, \nabla\|x\| \rangle$.

(a) $\nabla\|x\|$ is called the gradient of $\varphi(x) = \|x\|$ at $x \in S_X$.

(b) The norm of X is said to be Gâteaux differentiable if it is Gâteaux differentiable at each point of S_X.

(c) The norm of X is said to be uniformly Gâteaux differentiable if for each $y \in S_X$, the limit (3.21) is approached uniformly for $x \in S_X$.

Definition 3.13 A Banach space X is said to be smooth if the limit (3.21) exists for all $x, y \in S_X$. It is also said to be uniformly smooth if the limit (3.21) is attained uniformly for $x, y \in S_X$.

Definition 3.14 Let $S_X = \{x \in X : \| x \| = 1\}$. Then the norm of X is called Fréchet differentiable if for each $x \in S_X$, the limit

$$\lim_{t \to 0} \frac{\| x + ty \| - \| x \|}{t}$$

exists uniformly for each $y \in S_X$.

Example 3.24 Let $\mathcal{H}$ be a Hilbert space. Then the norm of $\mathcal{H}$ is Gâteaux differentiable with $\nabla \|x\| = \frac{x}{\|x\|}, x \neq 0$. Indeed, for each $x \in \mathcal{H}$ with $x \neq 0$, we have

$$\lim_{t \to 0} \frac{\|x + ty\| - \|x\|}{t} = \lim_{t \to 0} \frac{\|x + ty\|^2 - \|x\|^2}{t(\|x + ty\| + \|x\|)}$$

$$= \lim_{t \to 0} \frac{2t\langle y, x \rangle + t^2 \|y\|^2}{t(\|x + ty\| + \|x\|)} = \left\langle y, \frac{x}{\|x\|} \right\rangle.$$

Therefore, the norm of $\mathcal{H}$ is Gâteaux differentiable with $\nabla \|x\| = \frac{x}{\|x\|}$.

Remark 3.7 It may be remarked that at $x \neq 0$, $\varphi(x) = \|x\|$ is Gâteaux differentiable with $\nabla \|x\| = \frac{x}{\|x\|}$. However, we see that at $x = 0$, $\varphi(x) = \|x\|$ is not differentiable, but it is subdifferentiable. Indeed, for the Hilbert space $\mathcal{H}$ we have $\mathcal{H}^* = \mathcal{H}$ and so

$$\partial \varphi(0) = \partial \|0\| = \{j \in \mathcal{H} : \langle x, j \rangle \leqslant \|x\| \ \forall x \in \mathcal{H}\}$$
$$= \{j \in \mathcal{H} : \|j\|_* \leqslant 1\}.$$

Notice that a proper convex continuous function φ is Gâteaux differentiable if and only if it has a unique subgradient. Using this fact, we establish a relation between smoothness and Gâteaux differentiability of the norm.

Theorem 3.24 *A Banach space X endowed with norm $\| \cdot \|$ is smooth if and only if the norm is Gâteaux differentiable on $X \setminus \{0\}$.*

Proof We have
the norm $\| \cdot \|$ is Gâteaux differentiable at $x \in \mathbb{S}_X$
$\Leftrightarrow \partial \|x\| = \{j \in X^* : \langle x, j \rangle = \|x\| \|j\|_* = \|x\|^2 = 1\}$ is singleton
$\Leftrightarrow \exists$ a unique $j \in X^*$ such that $\langle x, j \rangle = \|x\|$ and $\|j\|_* = 1$
$\Leftrightarrow X$ is smooth. ∎

Theorem 3.25 *Let X be a Banach space. Then we have the following:*

(a) *If X^* is strictly convex, then X is smooth.*
(b) *If X^* is smooth, then X is strictly convex.*

Proof We prove these two results contrapositively as follows:
(a) Suppose, if possible, X not smooth. Then there exist $x_0 \in S_X$ and $j_1, j_2 \in S_{X^*}$ with $j_1 \neq j_2$ such that $\langle x_0, j_1 \rangle = \langle x_0, j_2 \rangle = 1$. It follows that x_0 determines a continuous linear functional on X^* that takes the maximum value on S_{X^*} at two distinct points j_1 and j_2. This shows that X^* is not strictly convex.
(b) Suppose, if possible, X not strictly convex. Then there exist $j \in S_{X^*}$ and $x_1, x_2 \in S_X$ with $x_1 \neq x_2$ such that $\langle x_1, j \rangle = \langle x_2, j \rangle = 1$. It follows that two supporting hyperplanes pass through a single element $j \in \mathbb{S}_{X^*}$ such that

$$\langle x_1, j \rangle = \langle x_2, j \rangle = 1, j \in X^*.$$

Therefore, X^* is not smooth.   ∎

Suppose that X is a reflexive Banach space, then the dual spaces X and X^* can be equivalently renormed as strictly convex spaces such that the duality is preserved. We can make use of these facts to prove the following result.

Theorem 3.26 *Let X be a reflexive Banach space. Then we have the following:*
(a) X is smooth if and only if X^* is strictly convex.
(b) X is strictly convex if and only if X^* is smooth.

The following result establishes a relation between smoothness of a Banach space and a property of the duality mapping with gauge function μ.

Theorem 3.27 *Let X be a Banach space. Then X is smooth if and only if J_μ with gauge function μ is single-valued and*

$$\frac{d}{dt}\Phi(\|x + ty\|)\Big|_{t=0} = \langle y, J_\mu(x)\rangle \ \ \forall x, y \in X. \tag{3.22}$$

Corollary 3.4 *Let X be a Banach space and $J : X \to 2^{X^*}$ a duality mapping. Then the following statements are equivalent :*
(a) X is smooth.
(b) J is single-valued.
(c) The norm of X is Gâteaux differentiable with $\nabla\|x\| = \frac{Jx}{\|x\|}$.

The following result shows that uniform smoothness has a close relation with differentiability of norm.

Theorem 3.28 *Every uniformly smooth Banach space is smooth.*

Proof Suppose X is a uniformly smooth Banach space. Then we intend to so that X is smooth. Suppose, on the contrary, that X is not smooth. Then there exist $x \in X\setminus\{0\}$, and $i, j \in X^*$ such that $i \neq j$, $\|i\|_* = \|j\|_* = 1$ and $(x, i) = (x, j) = \|x\|$. Suppose $y \in X$ is such that $\|y\| = 1$ and $(y, i - j) > 0$. Then, for each $t > 0$, we have

$$\begin{aligned}
0 &< t(y, i - j) = t(y, i) - t(y, j) \\
&= \frac{(x + ty, i) + (x - ty, j)}{2} - 1 \\
&\leqslant \frac{\|x + ty\| + \|x - ty\|}{2} - 1.
\end{aligned}$$

It follows, therefore, that

$$0 < (y, i - j) \leqslant \frac{\rho_X(t)}{t} \ \text{ for each } \ t > 0.$$

Hence, X is not uniformly smooth.

We now establish the result concerning Fréchet differentiability of the norm of Banach spaces.

Theorem 3.29 *Let X be a Banach space with a Fréchet differentiable norm. Then the duality mapping $J : X \to X^*$ is norm-to-norm continuous.*

Proof Let $\{x_n\}$ be a sequence in $\mathbb{S}_X$ such that $x_n \to x$. Then it suffices to show that $Jx_n \to Jx$. Because X has a Fréchet differentiable norm,

$$\lim_{t \to 0} \frac{\|x + ty\| - \|x\|}{t} = (y, Jx) \text{ uniformly in } y \in \mathbb{S}_X.$$

That is, for any given $\varepsilon > 0$, there exists $\delta > 0$ such that

$$\left| \frac{\|x + ty\| - \|x\|}{t} - (y, Jx) \right| < \frac{\varepsilon}{6} \text{ for all } y \in \mathbb{B}_X \text{ and all } t \text{ with } 0 < |t| \leqslant \delta.$$

This implies that

$$\|x + ty\| - \|x\| < t\left((y, Jx) + \frac{\varepsilon}{6}\right) \text{ and } \|x - ty\| - \|x\| < -t\left((y, Jx) - \frac{\varepsilon}{6}\right).$$

Thus, we have

$$\|x + ty\| - 1 < t\left((y, Jx) + \frac{\varepsilon}{6}\right) \text{ and } \|x - ty\| - 1 < -t\left((y, Jx) - \frac{\varepsilon}{6}\right).$$

Note that

$$0 \leqslant 1 - (x, Jx_n) = (x_n, Jx_n) - (x, Jx_n) = (x_n - x, Jx_n)$$
$$= \|x_n - x\| \|Jx_n\|_* = \|x_n - x\| \to 0,$$

i.e. $(x, Jx_n) \to 1$ as $n \to \infty$. Then there exists $n_0 \in \mathbb{N}$ such that

$$|(x, Jx_n) - 1| \leqslant t\varepsilon/6 \text{ for all } n \geqslant n_0.$$

This gives

$$1 \leqslant (x, Jx_n) + t\varepsilon/6 \text{ for all } n \geqslant n_0.$$

For all $n \geqslant n_0$, we have

$$1 - t\varepsilon/6 \leqslant (x, Jx_n) = (x, Jx + Jx_n) - 1 \text{ (as } (x, Jx) = \|x\|^2 = 1)$$
$$= (x + ty, Jx) + (x - ty, Jx_n) - t(y, Jx - Jx_n) - 1$$
$$\leqslant \|x + ty\| \|Jx\|_* + \|x - ty\| \|Jx_n\|_* - t(y, Jx - Jx_n) - 1$$
$$\leqslant 1 + t\left((y, Jx) + \frac{\varepsilon}{6}\right) + 1 - t\left((y, Jx) - \frac{\varepsilon}{6}\right) - t(y, Jx - Jx_n) - 1$$
$$= t\frac{\varepsilon}{3} - t(y, Jx - Jx_n) + 1.$$

The above inequality yields

$$(y, Jx - Jx_n) \leqslant \frac{\varepsilon}{2} \text{ for all } n \geqslant n_0 \text{ and } y \in \mathbb{S}_X.$$

Similarly, we can show that

$$(y, Jx_n - Jx) \leqslant \frac{\varepsilon}{2} \text{ for all } n \geqslant n_0 \text{ and } y \in \mathbb{S}_X.$$

Thus,

$$|(y, Jx_n - Jx)| \leqslant \frac{\varepsilon}{2} \text{ for all } n \geqslant n_0 \text{ and } y \in \mathbb{S}_X.$$

It follows that

$$\|Jx_n - Jx\| \leqslant \frac{\varepsilon}{2} < \varepsilon \text{ for all } n \geqslant n_0.$$

Hence, we conclude that $x_n \to x$ in X implies that $Jx_n \to Jx$ in X^*. The following result easily follows.

Theorem 3.30 *Let X be a Banach space. Then the following two statements are equivalent:*
(a) X has uniformly Fréchet differentiable norm.
(b) X^* is uniformly smooth.

Next, we state the result without proof showing the duality between uniform convexity and uniform smoothness.

Theorem 3.31 *Let X be a Banach space. Then*
(i) X is uniformly smooth if and only if X^* is uniformly convex.
(ii) X is uniformly convex if and only if X^* is uniformly smooth.

3.7 Asymptotic Behaviour of Generalized Nonexpansive Sequences

In this section, we consider a generalized nonexpansive sequence and we use the mean point to obtain the weak convergence of $\{\frac{x_n}{n}\}$, in the case when E is reflexive and strictly convex. In addition, we obtain the strong convergence of $\{\frac{x_n}{n}\}$, in the case when E^* has a Fréchet differentiable norm.

Let E be a real Banach space; the norms of both E and its dual E^* will be denoted by $\|.\|$. The duality pairing between E and E^* will be denoted by $(\cdot, \cdot)$. The duality mapping J from E into the family of nonempty closed convex subsets of E^* is denoted by

$$J(x) = \left\{ x^* \in E^* : (x, x^*) = \| x \|^2 = \left\| x^* \right\|^2 \right\}.$$

It may be observed that for $x, y \in E$ and $j \in J(x)$,

$$(x - y, j) = \|x\|^2 - (y, j) \geq \|x\|^2 - \frac{1}{2} \left(\|y\|^2 + \|j\|^2\right) = \frac{1}{2} \left(\|x\|^2 - \|y\|^2\right).$$

Observation

- If E is reflexive and strictly convex and K is a nonempty closed convex subset of E, then the nearest point projection mapping P_K of E onto K is well defined, i.e. K is a Chebyshev set (see [39, 249]).

We denote weak convergence and strong convergence in E, respectively, by $\rightharpoonup$ and $\longrightarrow$ and let $\{x_n\}_{n \geq 0}$ be a generalized nonexpansive sequence in E. Let $K = \bigcap_{n=1}^{\infty} \overline{co} \{\{x_i - x_{i-1}\}_{i \geq n}\}$. Djafari Rouhani [189] considered nonexpansive sequences and obtained an interesting result on the weak convergence of $\{\frac{x_n}{n}\}$ under the assumption that E is reflexive and strictly convex. Recently, Jung and Park [295] dropped the strict convexity requirement in the result of Djafari Rouhani, that is, instead of the weak limit of $\{\frac{x_n}{n}\}$, they dealt with the mean point of $\{\frac{x_n}{n}\}$ concerning a Banach limit under the assumption that E is reflexive.

Definition 3.15 A sequence $\{x_n\}_{n \geq 0} \subset E$ is said to be a generalized nonexpansive sequence if it satisfies

$$\| x_{i+1} - x_{j+1} \|^2 \leq \| x_i - x_j \|^2 + \varepsilon^2(i, j) \tag{3.23}$$

for all $i, j \geq 0$, where $\varepsilon(i, j) \leq 1$ for all $i, j \geq 0$ and for any given $\varepsilon > 0$, there exists a $j_0 \geq 0$ such that $\limsup_{n \to \infty} \varepsilon(n, j) \leq \varepsilon$ for any $j \geq j_0$, i.e. $\lim_{j \to \infty} \limsup_{n \to \infty} \varepsilon(n, j) = 0$.

Let μ be a mean on the integers $\mathbb{N}$, i.e. a linear functional μ defined on ℓ_∞ such that
(a) $\mu(a) \geq 0$ if $a_n \geq 0 \; \forall \, n$,
(b) $\mu(a) = \mu(\sigma a)$ where σ denotes the right shift

$$\sigma a = \sigma(a_1, a_2, a_3, \ldots) = (a_2, a_3, a_4, \ldots),$$

(c) $\mu(a) = 1$ if $a = (1, 1, 1, \ldots)$.
Then we know that μ is a mean on $\mathbb{N}$ if and only if

$$\inf \{a_n : n \in \mathbb{N}\} \leq \mu(a) \leq \sup \{a_n : n \in \mathbb{N}\}$$

for every $a = (a_1, a_2, \ldots) \in \ell_\infty$. For convenience we use $\mu_n(a_n)$ instead of $\mu(a)$. A mean μ on $\mathbb{N}$ is called a Banach limit (see, [534]) if

$$\mu_n(a_n) = \mu_n(a_{n+1})$$

for every $a = (a_1, a_2, \ldots) \in \ell_\infty$. The Hahn–Banach theorem guarantees the existence of a Banach limit [584]. We know that if μ is a Banach limit, then

$$\liminf_{n \to \infty} a_n \leqslant \mu_n(a_n) \leqslant \limsup_{n \to \infty} a_n$$

for every $a = (a_1, a_2, \ldots) \in \ell_\infty$. Let E be a reflexive Banach space and let $\{x_n\}$ be a bounded sequence in E. We now show for a Banach limit μ , there exists a point x_0 in E such that

$$\mu_n(x_n, x^*) = (x_0, x^*) \qquad \forall\, x^* \in E^*.$$

In fact, the function $\mu_n(x_n, x^*)$ is linear in x^*. Also since

$$\left| \mu_n\left(x_n, x^*\right) \right| \leqslant \left(\sup_n \|x_n\| \right) \cdot \|x^*\|,$$

it follows that the function $\mu_n(x_n, x^*)$ is also bounded in x^*. Thus, there is a $x_0^{**} \in E^{**}$ such that $\mu_n(x_n, x^*) = (x_0^{**}, x^*)$ for every $x^* \in E^*$. Because E is reflexive, we can find a point x_0 in E such that $\mu_n(x_n, x^*) = (x_0, x^*)$ for every $x^* \in E^*$. This point x_0 is called a mean point of $\{x_n\}$ concerning μ. Furthermore [295], we also know that this mean point $x_0 \in \bigcap_{n \geq 1}^{\infty} \overline{co}\,\{x_n\}$.

Lemma 3.2 *Let* $\{a_n\}_{n \geq 0}$ *be a sequence of nonnegative real numbers with* $a_0 = 0$, *the series of nonnegative terms* $\sum_{i,j} \varepsilon(i,j)$ *be convergent, and satisfying* $a_{n+p} \leqslant a_n + a_p + n \cdot \varepsilon(n+p, n)$, $\forall\, n \geq 0, p \geq 1$. *Then the sequence* $\{\frac{a_n}{n}\}$ *converges as* $n \to \infty$ *and* $\lim\limits_{n \to \infty} \frac{a_n}{n} = \inf\limits_{n \geq 1} \frac{a_n}{n}$.

Proof Let $p \geq 1$ be fixed. Then by the division algorithm, for all $n \geq p$, there exists $k \geq 1$ such that $n = kp + i$; $0 \leqslant i < p$.

Because the series of nonnegative terms $\sum_{i,j} \varepsilon(i,j)$ converges, there exists $\eta > 0$ such that $\sum_{i,j} \varepsilon(i,j) \leqslant \eta$. Now, for any $p \geq 1$ (for notational purposes $\sum_2^1 = 0$), we have

$$a_{kp} \leqslant k \cdot a_p + \sum_{j=2}^{k} \varepsilon(jp, (j-1)p)$$

$$\leqslant k \cdot a_p + \sum_{l,m} \varepsilon(l, m) \leqslant k \cdot a_p + \eta.$$

Thus, we have

$$\frac{a_{kp}}{kp+i} \leqslant \frac{k \cdot a_p + \eta}{kp+i} \leqslant \frac{a_p}{p} + \frac{\eta}{k} \qquad \forall\, p \geq 1.$$

Hence, we have

$$\frac{a_n}{n} = \frac{a_{kp+i}}{kp+i} \leqslant \frac{a_{kp} + a_i + \varepsilon(kp+i, kp)}{kp+i} \leqslant \frac{a_p}{p} + \frac{2 \cdot \eta}{k} + \frac{a_i}{k}$$

$$\leqslant \frac{a_p}{p} + \frac{2 \cdot \eta}{k} + \frac{\max\limits_{0 \leqslant i < p} a_i}{k}.$$

Now letting $n \to \infty$, we have $k \to \infty$ and so for all $p \geq 1$, we have $\limsup\limits_{n \to \infty} \frac{a_n}{n} \leqslant \frac{a_p}{p}$.
Therefore,

$$\limsup_{n \to \infty} \frac{a_n}{n} \leqslant \inf_{p \geqslant 1} \frac{a_p}{p} \leqslant \liminf_{n \to \infty} \frac{a_n}{n}.$$

Hence,

$$\lim_{n \to \infty} \frac{a_n}{n} = \inf_{n \geqslant n_0} \frac{a_n}{n}.$$

The following well-known lemma will be useful later (cf. [189]).

Lemma 3.3 E^* *has a Fréchet differentiable norm if and only if E is reflexive and strictly convex, and has the following property: if $x_n \rightharpoonup x$ and $\| x_n \| \to \| x \|$ for a sequence $\{x_n\}$ in E, then $\{x_n\}$ converges strongly to x.*

Let D be a subset of E. Then we denote the closure of D by $\overline{D}$ and the closed convex hull of D by $\overline{co}D$, respectively. For a point x in E, we denote its distance from D by $d(x, D) = \inf_{y \in D} \| x - y \|$.

We now deal with a generalized nonexpansive sequence $\{x_n\}$ in E and study the mean point of $\{\frac{x_n}{n}\}$ concerning a Banach limit. We begin with the following lemmas which will play crucial roles in the proof of our main result. We shall also use the following basic inequality (see Lemma A.1.1)

$$(a + b)^q \leqslant a^q + b^q \tag{3.24}$$

for $0 < q \leqslant 1$ and $a, \ b \geq 0$.

Lemma 3.4 *Let E be a Banach space and let $\{x_n\}$ be a generalized nonexpansive sequence in E. Then $\lim\limits_{n \to \infty} \| \frac{x_n}{n} \|$ exists and*

$$\lim_{n \to \infty} \left\| \frac{x_n}{n} \right\| = \inf_{n \geqslant 1} \| x_n - x_0 \|.$$

Proof Let $a_n = \| x_n - x_0 \| \ \forall n \geq 1$. Now applying (3.24) to the generalized nonexpansive sequence $\{x_n\}$ successively, we obtain for all $p \geq 1$ that

$$a_{n+p} = \| x_{n+p} - x_0 \| \leqslant \| x_{n+p} - x_n \| + \| x_n - x_0 \|$$

$$\leqslant \| x_p - x_0 \| + \sum_{j=1}^{n} [\varepsilon(j+p,j) - \varepsilon(j-1+p,j-1)] + \| x_n - x_0 \|$$

$$\leqslant a_n + a_p + \varepsilon(n+p,n).$$

Hence, the result follows from Lemma 3.2. ∎

Lemma 3.5 *Let $\{a_n\}_{n \geqslant 1}$ be a sequence of positive real numbers (i.e. $a_n > 0$ for each n) and $b_n = \sum_{i=1}^{n} a_n$. Assume that $b_n \uparrow \sum_{i=1}^{\infty} a_i = \infty$. If $\{x_n\}$ is a sequence of real numbers such that $x_n \to x$, then we have*

$$\lim_{n \to \infty} \frac{1}{b_n} \sum_{i=1}^{n} a_i x_i = x.$$

Proof Let $\varepsilon > 0$. Choose some k such that $|x_n - x| < \frac{\varepsilon}{2}$ for each $n \geqslant k$. Put $M = \max\{|x_i - x| : i \doteq 1, \ldots, k\}$, and then select $l > k$ such that $\frac{Mb_k}{b_n} < \frac{\varepsilon}{2}$ for all $n \geq l$. Now notice that if $n \geqslant l$, then

$$\left| \frac{1}{b_n} \sum_{i=1}^{n} a_i x_i - x \right| = \left| \frac{1}{b_n} \sum_{i=1}^{n} a_i x_i - \frac{1}{b_n} \sum_{i=1}^{n} a_i x \right|$$

$$\leqslant \frac{1}{b_n} \sum_{i=1}^{k} a_i |x_i - x| + \frac{1}{b_n} \sum_{i=k+1}^{k} a_i |x_i - x|$$

$$\leqslant \frac{Mb_k}{b_n} + \frac{\varepsilon}{2} < \frac{\varepsilon}{2} + \frac{\varepsilon}{2} = \varepsilon$$

and the conclusion follows.

Lemma 3.6 *Let E be a reflexive Banach space and let $\{x_n\}$ be a generalized nonexpansive sequence in E. Let*

$$K = \bigcap_{n=1}^{\infty} \overline{co} \left\{ \{x_i - x_{i-1}\}_{i \geqslant n} \right\}.$$

Then $\lim_{n \to \infty} \| \frac{x_n}{n} \| = d(0, K) = \inf_{n \geqslant 1} \| \frac{x_n - x_0}{n} \|$.

Proof Let $k \geqslant 1$ be fixed and $j_n \in J(x_n - x_{k-1})$ for $n \geq k$. Now the generalized sequence $\{x_n\}$ yields for $n \geq k$ that

$$(x_k - x_{k-1}, j_n) \geqslant \frac{1}{2} \| x_n - x_{k-1} \|^2 - \frac{1}{2} \| x_n - x_k \|^2$$

$$\geqslant \frac{1}{2} \| x_n - x_{k-1} \|^2 - \frac{1}{2} \| x_{n-1} - x_{k-1} \|^2 - \frac{1}{2} (\varepsilon(n,k) - \varepsilon(n-1,k-1))^2.$$

Hence, we obtain

$$\frac{2}{n^2}\left(x_k - x_{k-1}, \sum_{i=k}^{n} j_i\right) \geq \left\|\frac{x_n - x_{k-1}}{n}\right\|^2 - \frac{1}{n^2}\sum_{i=k}^{n}(\varepsilon(i,k) - \varepsilon(i-1, k-1))^2$$

$$\geq \left\|\frac{x_n - x_{k-1}}{n}\right\|^2 - \frac{1}{n^2}\cdot 2\sum_{i,j}\varepsilon^2(i,j) \quad \forall k \geq 1. \qquad (3.25)$$

Let $S_n = \frac{2}{n^2}\sum_{i=k}^{n} j_i$ for $n \geqslant k$. Then we have

$$\| S_n \| \leqslant \frac{2}{n^2}\sum_{i=k}^{n} \| x_i - x_{k-1} \| = \frac{2}{n^2}\sum_{i=k}^{n} i\left\|\frac{x_i - x_{k-1}}{i}\right\|.$$

Because $\{\frac{x_n}{n}\}$ is bounded by Lemma 3.4, it then follows that $\{S_n\}$ is bounded. Hence, by weak* compactness of the closed unit ball of E^* the sequence $\{S_n\}$ has weak* cluster point $j \in E^*$ (obviously independent of $k \geq 1$). Since $\sum_{i,j}\varepsilon(i,j)$ is bounded $\forall k \geq 1$, we obtain from Lemma 3.4 and (3.25)

$$(x_k - x_{k-1}, j) \geq \lim_{n\to\infty}\left\|\frac{x_n}{n}\right\|^2 \quad \forall k \geq 1.$$

Hence, for any $n \geq 1$, we have

$$\left(\frac{x_n - x_0}{n}, j\right) \geq \lim_{n\to\infty}\left\|\frac{x_n}{n}\right\|^2. \qquad (3.26)$$

From Lemma 3.5, replacing a_i by i, b_n by $\sum_{i=1}^{n} i = \frac{n(n+1)}{2}$ and note that $\lim_{n\to\infty} b_n = \infty$, we obtain

$$\limsup_{n\to\infty}\frac{2}{n^2}\sum_{i=k}^{n} i\left\|\frac{x_i - x_{k-1}}{i}\right\| = \limsup_{n\to\infty}\left[\frac{n(n+1)}{n^2}\cdot\frac{2}{n(n+1)}\sum_{i=k}^{n} i\left\|\frac{x_i - x_{k-1}}{i}\right\|\right]$$

$$= \lim_{n\to\infty}\left(1 + \frac{1}{n}\right)\cdot\limsup_{n\to\infty}\frac{2}{n(n+1)}\sum_{i=k}^{n} i\left\|\frac{x_i - x_{k-1}}{i}\right\|$$

$$= \lim_{n\to\infty}\left\|\frac{x_n - x_{k-1}}{n}\right\|.$$

Now using above inequality, we also have

$$\| j \| \leq \liminf_{n \to \infty} \| S_n \| \leq \liminf_{n \to \infty} \frac{2}{n^2} \sum_{i=k}^{n} i \left\| \frac{x_i - x_{k-1}}{i} \right\|$$

$$\leq \limsup_{n \to \infty} \frac{2}{n^2} \sum_{i=k}^{n} i \left\| \frac{x_i - x_{k-1}}{i} \right\| = \lim_{n \to \infty} \left\| \frac{x_n - x_{k-1}}{n} \right\| = \lim_{n \to \infty} \left\| \frac{x_n}{n} \right\|$$

and so it follows that

$$(x_k - x_{k-1}, j) \geq \lim_{n \to \infty} \left\| \frac{x_n}{n} \right\|^2 \geq \| j \|^2 \; \forall \, k \geq 1.$$

Hence, for any $z \in \overline{co} \left\{ \{x_{i+1} - x_i\}_{i \geq 0} \right\}$

$$\frac{1}{2} \lim_{n \to \infty} \left\| \frac{x_n}{n} \right\|^2 + \frac{1}{2} \| z \|^2 \geq \frac{1}{2} \| j \|^2 + \frac{1}{2} \| z \|^2$$

$$\geq (z, j) \geq \lim_{n \to \infty} \left\| \frac{x_n}{n} \right\|^2 \geq \| j \|^2. \qquad (3.27)$$

Because $K \subset \overline{co} \left\{ \{x_{i+1} - x_i\}_{i \geq 0} \right\}$, it follows from (3.27) that

$$\| j \| \leq \lim_{n \to \infty} \left\| \frac{x_n}{n} \right\| \leq \inf_{z \in K} \| z \| = d(0, K).$$

Because $\{ \frac{x_n}{n} \}$ is bounded, it follows that $\{ \frac{x_n - x_0}{n} \}$ and E is reflexive, therefore, by Eberlein–Smulian's theorem the sequence $\{ \frac{x_n - x_0}{n} \}$ contains a weakly convergent subsequence $\{ \frac{x_{n_i} - x_0}{n_i} \}$. Suppose $\frac{x_{n_i} - x_0}{n_i} \rightharpoonup q$ for some $q \in K$. Then we have

$$\| q \| \leq \liminf_{i \to \infty} \left\| \frac{x_{n_i} - x_0}{n_i} \right\| = \lim_{n \to \infty} \left\| \frac{x_n}{n} \right\|.$$

Hence,

$$\lim_{n \to \infty} \left\| \frac{x_n}{n} \right\| = d(0, K).$$

This completes the proof.   ∎

In 2004, Pathak et al. [467] used Lemmas 3.4 and 3.6 to prove the following result.

Theorem 3.32 *Let E be a reflexive Banach space and let $\{x_n\}$ be a generalized nonexpansive sequence in E. Let*

$$K = \bigcap_{n=1}^{\infty} \overline{co} \left\{ \{x_i - x_{i-1}\}_{i \geq n} \right\}$$

and $d = d(0, K)$. *Then,* $d = d\left(0, \overline{co}\left\{\frac{x_n - x_0}{n}\right\}\right)$, *and there exists a point* z_0 *with* $\| z_0 \| = d$ *such that* $z_0 \in \overline{co}\{\frac{x_n - x_0}{n}\}$.

Proof In view of Lemma 3.3, we may assume that $\{\frac{x_n - x_0}{n}\}_{n \geq 1}$ is bounded. So, it follows from reflexiveness of E that for a Banach limit μ, there exists $z_0 \in \overline{co}\{\frac{x_n - x_0}{n}\}$ such that

$$\mu_n\left(\frac{x_n - x_0}{n}, x^*\right) = (z_0, x^*) \ \forall \, x^* \in E^*. \tag{3.28}$$

Now, for $j_0 \in J(z_0)$ we have

$$\| z_0 \|^2 = (z_0, j_0) = \mu_n\left(\frac{x_n - x_0}{n}, j_0\right)$$

$$\leqslant \mu_n\left(\left\|\frac{x_n - x_0}{n}\right\|\right) \cdot \| j_0 \| = d \cdot \| j_0 \| = d \cdot \| z_0 \|,$$

and hence $\| z_0 \| \leqslant d$. By the proof of Lemma 3.6 and (3.26), there exists a functional $j \in E^*$ with $\| z_0 \| \leqslant d$ such that

$$\left(\frac{x_n - x_0}{n}, j\right) \geq d^2 \qquad \forall \, n \geq 1. \tag{3.29}$$

As a result, we have $(z_0, j) \geq d^2$. Because $\| j \| \leqslant d$, we obtain

$$d^2 \geq \| z_0 \| \cdot \| j \| \geq (z_0, j) \geq d^2$$

and hence $\| z_0 \| = \| j \| = d$. From (3.29), it follows that $(z_0, j) \geq d^2$ for every $z \in \overline{co}\{\frac{x_n - x_0}{n}\}$ and so

$$\| z \| \cdot d = \| z \| \cdot \| j \| \geq (z, j) \geq d^2.$$

Hence, $\| z \| \geq d$ for every $z \in \overline{co}\{\frac{x_n - x_0}{n}\}$. As a result, we obtain

$$d = d\left(0, \overline{co}\left\{\frac{x_n - x_0}{n}\right\}\right)$$

Now suppose that there is another point w_0 satisfying (3.28). Then for $j \in J(z_0 - w_0)$, we have

$$\| z_0 - w_0 \|^2 = (z_0 - w_0, j) = \mu_n\left(\frac{x_n - x_0}{n} - \frac{x_n - x_0}{n}, j\right) = 0,$$

and hence $z_0 = w_0$. This completes the proof. ∎

Corollary 3.5 *Suppose* E, $\{x_n\}$, K *and* d *are as in Theorem 3.32. Then we have the following:*

(i)   If E is strictly convex, then the weak $\lim\limits_{n\to\infty} \frac{x_n}{n}$ exists and coincides with $P_K 0$ with $\| P_K 0 \| = d$.

(ii) E^* has a Fréchet differentiable norm, strong $\lim\limits_{n\to\infty} \frac{x_n}{n}$ exists and coincides with $P_K 0$.

Proof (i) Because a reflexive Banach space E is strictly convex, the set

$$\left\{ z \in \overline{co} \left\{ \frac{x_n - x_0}{n} \right\} : \| z \| = d \right\}$$

consists of exactly one point and $d(0, K) = \| P_K 0 \|$. It may be observed that this point equals z_0 in Theorem 3.32. Let $\{\frac{x_{n_i}}{n_i}\}$ be a subsequence of $\{\frac{x_n}{n}\}$ such that $\{\frac{x_{n_i}}{n_i}\}$ weakly to $p \in K$. Then since

$$\|p\| \leq \liminf_{i\to\infty} \left\| \frac{x_{n_i}}{n_i} \right\| = \lim_{n\to\infty} \left\| \frac{x_n}{n} \right\| = \| P_K 0 \|$$

we have $p = z_0 = P_K 0$. It follows that $\{\frac{x_n}{n}\}$ converges weakly to $P_K 0$. This completes the proof.

(ii) This is an immediate consequence of (i) and Lemma 3.3.   ∎

Remark 3.8 (1) Let $\{x_n\}_{n\geq 0}$ be a nonexpansive sequence in E (i.e. $\| x_{i+1} - x_{j+1} \| \leq \| x_i - x_j \|$ for all $i, j \geq 0$). Then $\lim\limits_{n\to\infty} \left\| \frac{x_n}{n} \right\|$ ([188], Theorem 3.1) and $\{x_n\}_{n\geq 0}$ also satisfies (3.23). Thus, Theorem 3.32 is a partial generalization of Theorem 3.3 in [295].

(2) Because our studies are equivalent to the study of the asymptotic behaviour of the sequence $\left\{\frac{T^n x}{n}\right\}_{n\geq 1}$ in E, T is a nonexpansive mapping from an arbitrary set K of E into itself and $x \in K$, Theorem 3.32 is a partial improvement of Theorem 5 in [515].

(3) Let $\{x_n\}_{n\geq 0}$ be an almost nonexpansive sequence in E (i.e. $\| x_{i+1} - x_{j+1} \| \leq \| x_i - x_j \| + \varepsilon(i, j)$, where $\{\varepsilon(i, j)\}_{i,j\geq 0}$ is bounded and $\lim\limits_{i,j\to\infty} \varepsilon(i, j) = 0$. Then $\lim\limits_{n\to\infty} \left\| \frac{x_n}{n} \right\|$ exists and $\{x_n\}_{n\geq 0}$ also satisfies (3.23). For the proof of this fact, see Propositions 3.3 and 3.4 in [190]. Thus, all the conditions of Theorem 3.32 and Corollary 3.5 are satisfied for almost nonexpansive sequences $\{x_n\}_{n\geq 0}$ in E.

(4) We need the condition $d = d\left(0, \overline{co}\left\{\frac{x_n - x_0}{n}\right\}\right)$ in Corollary 3.5, however, it is of interest in view of using the mean point.

(5) Our result may also be applied to the asymptotic behaviour of curves in E, and thus to the asymptotic behaviour of unbounded trajectories for the quasi-autonomous dissipative system

$$\frac{du}{dt} + Au \ni f$$

where A is an accretive (possibly multivalued) operator in $E \times E$; see [188] for the case E is a Hilbert space.

A Simple Application

We now illustrate potential application of our methods to obtain solution of certain inequality.

Theorem 3.33 *Let $f_1, f_2 \in L_2\left([0,1]\right)$ satisfy the following inequality for all $g_1, g_2 \in L_2\left([0,1]\right)$ such that*

$$\left|\int_0^1 g_1(t)\,dt\right|^2 - \left|\int_0^1 g_2(t)\,dt\right|^2 \leqslant \left|\int_0^1 (f_1(t)g_1(t) + f_2(t)g_2(t))\,dt\right|^2$$

$$\leqslant \int_0^1 (|g_1(t)| + |g_2(t)|)^2\,dt.$$

Then $f_1 = \lambda$ and $f_2 = \mu$ for some $\lambda, \mu \in [-1, 1]$.

Proof Let $E = L_2\left([0,1]; \mathbb{R}^2_\infty\right)$, where $\mathbb{R}^2_\infty$ denotes $\mathbb{R}^2$ endowed with the sup norm. Then E is reflexive and $E^* = L_2\left([0,1]; \mathbb{R}^2_1\right)$. We now consider the generalized nonexpansive sequence $\{x_n\}_{n \geq 0} \subset E$ defined by $x_{2k} = \left(2k, \frac{1}{2^k}\right)$ and $x_{2k+1} = \left(2k, \frac{1}{2^{k+1}}\right) \forall\, k \geq 0$. Finally, we apply Theorem 3.32. ∎

Exercises

3.1 Let $f : \mathbb{R}^2 \to \mathbb{R}$ be defined by

$$f(x_1, x_2) = \begin{cases} \frac{x_1^3 x_2}{x_1^4 + x_2}, & \text{if } (x_1, x_2) \neq (0, 0) \\ 0, & \text{if } (x_1, x_2) = (0, 0). \end{cases}$$

Show that f has Gâteaux differentiable at the origin.

3.2 Let a functional $f : \ell_1 \to \mathbb{R}$ be defined by

$$f(\{x_n\}) = \sum_{n=1}^\infty |x_n|.$$

Show that f is not Fréchet differentiable at any point of ℓ_1 but is Gâteaux differentiable at those points of the unit sphere for which $\{x_n\}$ has all nonzero coordinates.

3.3 Let a functional $F : C[0,1] \to \mathbb{R}$ be defined by

$$F(y) = \int_0^1 (ty(t) + +y^2(t))dt.$$

Find the value of y which gives the minimum value of F.

3.4 Let $K(s, t)$ be a real-valued function in the square reason $0 \leqslant s \leqslant t, 0 \leqslant t \leqslant 1$, and the norm functional $F(f) = \sup_{0 \leqslant t \leqslant 1} |f(t)|$ on $C[0,1]$ be defined by the following relation:

$$F(f) = \int_0^1 f^2(t)dt - \lambda \int_0^1 \int_0^1 K(s,t)f(s)f(t)dsdt - 2 \int_0^1 f(t)g(t)dt$$

where g is a fixed of $C[0, 1]$. Show that F is Fréchet differentiable.

3.5 Let $f : \mathbb{R} \to \mathbb{R}$ be a convex function. Show that for any reals $t_1 < t_2 < t_3$

$$\frac{f(t_2) - f(t_1)}{t_2 - t_1} \leqslant \frac{f(t_3) - f(t_1)}{t_3 - t_1} \leqslant \frac{f(t_3) - f(t_2)}{t_3 - t_2}.$$

3.6 Show that the functions defined on $\mathbb{R}$ into itself have subdifferentiable as indicated below:
(*i*) $F(x) = x^2$, $\partial F(x) = 2x, x \in \mathbb{R}$,
(*ii*) $F(x) = e^x$, $\partial F(x) = e^x, x \in \mathbb{R}$,
(*iii*) $F(x) = |x|$, $\partial F(x) = \begin{cases} -1, & x < 0, \\ [-1, 1], & x = 0 \\ 1, & x > 0. \end{cases}$

3.7 If $f : X \to \mathbb{R}$ is continuous and convex, then show that its subdifferential map is norm-to-weak* upper semicontinuous.

3.8 Let $\varphi : \mathbb{R} \to \mathbb{R}$ be convex, even and lower semicontinuous and $f(x) = \varphi(\|x\|)$, $x \in \mathbb{R}$. Show that f is convex, lower semicontinuous and $f^*(x^*) = \varphi^*(\|x^*\|)$, $x^* \in X^*$. In particular if $p > 1$, $\frac{1}{p} + \frac{1}{q} = 1$ and $f(x) = \frac{1}{p}\|x\|^p$, then

$$f^*(x^*) = \frac{1}{q}\|x^*\|^q.$$

3.9 Let X be a Banach space with a uniform Gâteaux differentiable norm, J the duality mapping of X, and let M be any bounded subset of X. Show that J is norm-to-weak* uniformly continuous on M.

Chapter 4
Monotone Operators, Strongly ϕ-Accretive Operators and Their Variants

> *Pure mathematicians just love to try unsolved problems—they love a challenge.*
>
> Andrew Wiles

> *Mathematics is an organ of knowledge and an infinite refinement of language. It grows from the usual language and world of intuition as does a plant from the soil, and its roots are the numbers and simple geometrical intuitions. We do not know which kind of content mathematics (as the only adequate language) requires; we cannot imagine into what depths and distances this spiritual eye (mathematics) will lead us.*
>
> Erich Kahler (1941)

> *Pure mathematics is, in its way, the poetry of logical ideas.*
>
> Albert Einstein

In this chapter we introduce the reader to the theory of monotone operators, ϕ-accretive operators and their generalizations. The concept of monotone operator was first introduced by Minty in his paper of 1962 [400], wherein he gave a surjectivity theorem for such operators. Since then this theory is widely developed and has found useful applications in the investigation of the solvability of nonlinear operator equations and in particular of partial differential equations and integral equations. Our purpose is to give a systematic treatment (with historical development) of various topics in the theory of monotone operators needed for such an investigation.

Sections 4.1 and 4.2 deal with monotone and maximal monotone operators and their properties. Section 4.3 gives surjectivity results for such operators. These are in turn applied to get existence and uniqueness results for operator equations. We mainly discuss the work of Browder and Minty. There is a brief discussion regarding the range of the sum of maximal monotone operators encompassing the results of Browder and Rockfeller.

© Springer Nature Singapore Pte Ltd. 2018
H. K. Pathak, *An Introduction to Nonlinear Analysis and Fixed Point Theory*,
https://doi.org/10.1007/978-981-10-8866-7_4

In Sect. 4.4, we give some results regarding the approximate solvability of operator equations involving monotone operators. These are on the lines of Petryshyn and Bruck, Jr. These results will help the reader to develop computer algorithms for the approximate solution of operator equations.

Section 4.5 deals with monotone properties of the subdifferential of convex functionals. As a corollary, we deduce the maximal monotonicity of the duality mapping J. In Sect. 4.6, we introduce various generalization of monotonicity concepts, pseudomonotonicity, generalized pseudomonotonicity, etc., introduced by Brezis, Browder, Hess and Petryshyn.

We conclude the Chapter in Sect. 4.7 by introducing the concept of ϕ-accretive operator and its generalizations. Some surjectivity results are also presented in the setting of Banach spaces.

4.1   Monotone Operators

To begin with, firstly, we notice that the concept of monotonicity is relatively old, for as early as 1935 Golomb [254, pp. 66–72] used monotonicity conditions for operators of a Hilbert space when he examined a nonlinear Hammerstein integral equation. Further, this notion was used by Vainberg [596] and Zarantonello [617] in 1960 to solve many problems of nonlinear equations and functional equations, respectively. Secondly, we observe that there exist a number of different monotonicity notions, but we restrict ourselves to the notions of monotonicity in the sense of Kachurovskiĭ [299, 300]. So, we start our discussion with the introduction of the notion of monotone operators in Hilbert spaces.

Let $\mathcal{H}$ be a real Hilbert space with inner product and norm $\langle \cdot, \cdot \rangle$ and $\| \cdot \|$, respectively. A map $F : \mathcal{D}(F) \subset \mathcal{H} \to \mathcal{H}$ (not necessarily continuous) is said to be

- monotone if $\langle Fx - Fy, x - y \rangle \geqslant 0$ for all $x, y \in \mathcal{D}(F)$;
- strictly monotone if $\langle Fx - Fy, x - y \rangle \geqslant 0$ for all $x, y \in \mathcal{D}(F)$ with, $x \neq y$;
- strongly monotone if $\langle Fx - Fy, x - y \rangle \geqslant c\|x - y\|^2$ for some $c > 0$ and for all $x, y \in \mathcal{D}(F)$, and
- F is dissipative if $-F$ is monotone.

Evidently, every strongly monotone map is injective. If F and G are monotone and $c > 0$, then $F + G$ is monotone and $F + cI$ is strongly monotone.

We now discuss the notion of generalized monotonicity that appears in the works of Browder [104], Brezis [79], Minty [400, 401, 403], Webb [606] and Skrypnik [563]. In 1960, R. I. Kachurovskiĭ [299] introduced the concept of monotonicity for mappings which map a Banach space into its dual space as follows:

Let X be a real Banach space with dual X^*. For $x \in X$, $y \in X^*$, (y, x) denotes the evaluation of y at x.

Definition 4.1 Let $T : \mathscr{D}(T) \subset X \to X^*$ be any operator (possibly nonlinear). Then,

(i) T is said to be monotone if

$$(Tx - Ty, x - y) \geq 0 \ \text{ for all } \ x, y \in \mathscr{D}(T). \tag{4.1}$$

(ii) T is called strictly monotone if the above inequality is strict, i.e.,

$$(Tx - Ty, x - y) > 0 \ \text{ for all } \ x, y \in \mathscr{D}(T) \text{ with } x \neq y. \tag{4.2}$$

(iii) T is called strongly monotone if there exists a constant $c > 0$ such that

$$(Tx - Ty, x - y) \geqslant c\|x - y\|^2 \ \text{ for all } \ x, y \in \mathscr{D}(T). \tag{4.3}$$

(iv) T is dissipative if $-T$ is monotone.

Remark 4.1 If the Banach space X under consideration is complex, then for the LHS of the inequality in (4.1), we consider only the real part.

Suppose $\mathcal{H}$ is a Hilbert space and T an operator on $\mathcal{H}$, then for $x, y \in \mathcal{H}$, (y, x) will denote the inner product in $\mathcal{H}$ and again (4.1) will define the monotonicity of T in $\mathcal{H}$. In tune with this argument, without loss in generality, we assume that the Banach space X or Hilbert space $\mathcal{H}$ is real.

Example 4.1 Let $f : \mathbb{R} \to \mathbb{R}$ be a monotone increasing function. Then, it follows that f is a monotone operator in the sense of Definition 4.1.

Example 4.2 Let $\mathcal{H}$ be a Hilbert space. Let $A : \mathcal{H} \to \mathcal{H}$ be a compact and self-adjoint linear operator. Then, by virtue of the spectral theorem for compact self-adjoint operators, A is monotone if all the eigenvalues of A are nonnegative.

Example 4.3 $A : L_2[0, 1] \to L_2[0, 1]$ be the differential operator $-\frac{d^2}{dt^2}$ with domain $\mathscr{D}(A)$ defined as

$$\mathscr{D}(A) = \left\{x \in L_2[0, 1] : x'(t), x''(t) \in L_2[0, 1] : x(0) = 0 = x(1)\right\}.$$

Then, A is a demicontinuous monotone operator.

Example 4.4 Consider the integral operator of convolution type (discussed in Example 1.38) defined by the kernel $k(t)$:

$$[Kx](t) = \int_{-\infty}^{\infty} k(t - \tau)x(\tau)d\tau = k * x.$$

If $k \in L_1(-\infty, \infty)$, then K is a continuous noncompact linear operator from $L_2(-\infty, \infty)$ to itself. Let $\hat{k}(iw)$ denote the Fourier transform of $k(t)$, then by Parseval's equality, we have

$$(Kx, x) = (k * x, x) = Re(\hat{k}(iw)\hat{x}(iw), \hat{x}(iw))$$

$$= Re \int_{-\infty}^{\infty} \hat{k}(iw)|\hat{x}(iw)|^2 dw. \tag{4.4}$$

From (4.4), it follows that K is monotone iff $\hat{k}(iw) \geqslant 0$ for almost all $w \in \mathbb{R}$.

Example 4.5 Let $\mathcal{H}$ be a Hilbert space and $T : \mathcal{H} \to \mathcal{H}$ a nonexpansive operator:

$$\|Tx - Ty\| \leqslant \|x - y\| \text{ for all } x, y \in \mathcal{H}.$$

Then, $I - T$ is monotone. Indeed, using Cauchy–Schwarz inequality, we have

$$\langle (I - T)x - (I - T)y, x - y \rangle = \langle x - y, x - y \rangle - \langle Tx - Ty, x - y \rangle$$
$$\geqslant \|x - y\|^2 - \|Tx - Ty\|\|x - y\|$$
$$\geqslant \|x - y\|^2 - \|x - y\|^2 = 0 \text{ for all } x, y \in \mathcal{H}.$$

Example 4.6 For a closed convex subset K of a Hilbert space $\mathcal{H}$, let $P_K x$ denote the point of minimum distance of K from x:

$$P_K x = \{z \in K : \|z - x\| = \inf_{y \in K} \|y - x\|\}.$$

Then, P_K is a monotone operator on $\mathcal{H}$. This can be seen as follows. By definition, we have

$$\|P_K x_1 - x_1\|^2 \leqslant \|P_K x_2 - x_1\|^2 \text{ and } \|P_K x_2 - x_2\|^2 \leqslant \|P_K x_1 - x_2\|^2 \text{ for all } x_1, x_2 \in \mathcal{H}.$$

Expanding the norm in terms of inner product, we get

$$\langle P_K x_1 - P_K x_2, x_1 \rangle \geqslant \frac{1}{2}\Big[\|P_K x_1\|^2 - \|P_K x_2\|^2\Big]$$

and

$$\langle P_K x_2 - P_K x_1, x_2 \rangle \geqslant \frac{1}{2}\Big[\|P_K x_2\|^2 - \|P_K x_1\|^2\Big].$$

Combining the above two inequalities, we get

$$\langle P_K x_1 - P_K x_2, x_1 - x_2 \rangle \geqslant \langle P_K x_1 - P_K x_2, x_1 \rangle + \langle P_K x_1 - P_K x_2, -x_2 \rangle$$
$$\geqslant \frac{1}{2}\Big[\|P_K x_1\|^2 - \|P_K x_2\|^2\Big] + \frac{1}{2}\Big[\|P_K x_2\|^2 - \|P_K x_1\|^2\Big]$$
$$\geqslant 0,$$

which is the required inequality needed for the monotonicity.

Example 4.7 (i) Let $f(s, x)$ be a function defined on $[0, 1] \times \mathbb{R}$ to $\mathbb{R}$, which satisfies Carathéodory condition. Let N_f be the Nemytskiǐ operator defined by f:

$$[N_f x](s) = f(s, x(s))$$

is monotone.

(ii) Let $J = [0.T] \subset \mathbb{R}$, $f : J \times \mathbb{R}^n \to \mathbb{R}^n$ measurable in $t \in J$ and continuous in $x \in \mathbb{R}^n$, $f(\cdot, 0) \in X = L_2(J)$ and

$$(f(t, x) - f(t, y) \cdot (x - y) \geqslant 0 \text{ for } t \in J \text{ and } x, y \in \mathbb{R},$$

where the dot indicates the inner product of $\mathbb{R}^n$ while $(u, v) = \int_J u(t) \cdot v(t)\, dt$ for $u, v \in X$. Let

$$\mathscr{D}(F) = \{u \in X : f(\cdot, u(\cdot)) \in X\} \text{ and } (Fu)(t) = f(t, u(t)).$$

Then, the superposition operator $F : \mathscr{D}(F) \to X$ is monotone.

In Sect. 1.9, we have given sufficient conditions for N_f to map $L_2[0, 1]$ into itself. If, in addition, f is monotone increasing with respect to the second variable, then the Nemytskiǐ operator N_f is monotone.

Example 4.8 Consider the duality mapping $J : X \to X^*$. We assume that X^* is strictly convex so that J is single valued. This mapping J is monotone:

$$\begin{aligned}
(Jx_1 - Jx_2, x_1 - x_2) &= (Jx_1, x_1) + (Jx_2, x_2) - (Jx_1, x_2) - (Jx_2, x_1) \\
&\geqslant \|x_1\|\|Jx_1\| + \|Jx_2\|\|x_2\| - \|x_2\|\|Jx_1\| - \|x_1\|\|Jx_2\| \\
&= (\|x_1\| - \|x_2\|)(\|Jx_1\| - \|Jx_2\|) \\
&= (\|x_1\| - \|x_2\|)^2 \geqslant 0.
\end{aligned}$$

Let us now assume that X is strictly convex. Then following Browder [89], we get the identity

$$(Jx_1 - Jx_2, x_1 - x_2) = (\|x_1\| - \|x_2\|)^2 + (\|Jx_1\|\|x_2\| - (Jx_1, x_2)) + (\|x_1\|\|Jx_2\| - (Jx_2, x_1)).$$

Because each term in the identity is nonnegative, it follows that if J is not strictly monotone, then there exists $\|x_1\| = \|x_2\|$ such that

$$\frac{(Jx_1, x_2)}{x_2} = \|Jx_1\| = \frac{(Jx_1, x_1)}{x_1}$$

which is equivalent to the statement

$$\left(Jx_1, \frac{x_2}{\|x_2\|}\right) = \left(Jx_2, \frac{x_1}{\|x_1\|}\right) = \|Jx_1\| = \|x_1\| = \|x_2\|.$$

Thus, there exists an element $x_1^* = Jx_1$ which attains its maximum at two different points of the unit sphere, contradicting the strict convexity of X. Thus, if X is strictly convex, then J is strictly monotone.

Example 4.9 Suppose $\Omega \subset \mathbb{R}^n$ is a bounded domain and $p \geqslant 2$. Consider the Sobolev space $H^{1,p}(\Omega) = \{f : D_i(f) \in L_p(\Omega), 0 \leqslant i \leqslant n\}$ with the norm $\|f\|_{1,p} = \left(\sum_0^n \|D_i f\|_p^p\right)^{1/p}$ (we shall denote $D_0 f \equiv f$ and $D_i f$ the derivative of f w.r.to the ith variable).

Suppose $H_0^{1,p}(\Omega)$ denotes the closure of $C_0^m(\Omega)$ in the norm $\|\cdot\|_{1,p}$. Assume that $H_0^{1,p}(\Omega)$ is endowed with the equivalent norm $\|f\|_{1,p}^0 = \left(\sum_{i=1}^n \|D_i f\|_p^p\right)^{1/p}$. Let us consider the pseudo-Laplacian operator $T : H_0^{1,p}(\Omega) \to H_0^{1,p}(\Omega)^*$ defined by

$$T(f) = -\sum_{i=1}^n D_i \left(|D_i f|^{p-2} D_i f\right), \quad f \in H_0^{1,p}(\Omega).$$

We now show that the operator T is well defined and bounded. To this end, for $f, g \in H_0^{1,p}(\Omega)$, we have

$$|\langle Tf, g \rangle| = \left| \sum_{i=1}^n \int_\Omega |D_i f|^{p-2} D_i f \cdot D_i g \, dx \right|$$

$$\leqslant \sum_{i=1}^n \left\| |D_i f|^{p-1} \right\|_q \|D_i g\|_p$$

$$\leqslant \left(\sum_{i=1}^n \left\| |D_i f|^{p-1} \right\|_q^q \right)^{1/q} \cdot \left(\sum_{i=1}^n \left\| |D_i g|^p \right\| \right)^{1/p}$$

$$= \left(\sum_{i=1}^n \|D_i f\|_p^p \right)^{1/q} \cdot \left(\sum_{i=1}^n \|D_i g\|_p^p \right)^{1/p}$$

$$= (\|f\|_{1,p}^0)^{p-1} \cdot \|g\|_{1,p}^0.$$

It follows that T is well defined and it is bounded. Moreover,

$$\langle Tf - Tg, f - g \rangle \geqslant (\|f\|_{1,p}^0)^p \cdot (\|g\|_{1,p}^0)^p - (\|f\|_{1,p}^0)^{p-1} \cdot \|g\|_{1,p}^0 - (\|g\|_{1,p}^0)^{p-1} \cdot \|f\|_{1,p}^0$$

$$\geqslant \left[(\|f\|_{1,p}^0)^{p-1} - (\|g\|_{1,p}^0)^{p-1} \right] \left(\|f\|_{1,p}^0 - \|g\|_{1,p}^0 \right)$$

$$\geqslant 0.$$

This entails that the operator T is strictly monotone.

We now give some sufficient conditions for the monotonicity of an operator. The following theorem, due to Minty [400], shows that monotonicity is essentially a local property.

Theorem 4.1 *Let C be a convex subset of a Banach space and $T : X \to X^*$ with $\mathscr{D}(T) = C$ be any nonlinear operator. Then, T is monotone iff to each $x \in C$ there exists a ball $B(x)$ such that T is monotone on $C \cap B(x)$.*

Proof Suppose x_1, x_2 are two distinct points of C, then the line segment $tx_1 + (1-t)x_2$ is compact and is contained in C. It follows that there exists a finite subcovering of neighbourhoods $(\{B(x) : x \in C\})$ of the hypothesis. We now choose $\varepsilon > 0$ to be smaller than their smallest radius and such that $\|x_1 - x_2\|/\varepsilon$ is an integer n.

Suppose $t_m = \frac{m\varepsilon}{\|x_1 - x_2\|}$, $m = 0, 1, \ldots, n$, then $y_m = t_m x_1 + (1 - t_m)x_2$ lie on the line segment joining $y_0 = x_2$ and $y_n = x_1$ and $y_m - y_{m-1} = \triangle y$ is independent of m. For each m, y_m and y_{m-1} lie in one of the balls $B(x)$ and consequently

$$(Ty_m - Ty_{m-1}, \triangle y) \geqslant 0.$$

Summing over m, we get

$$(Ty_n - Ty_0, \triangle y) \geqslant 0.$$

But $n \triangle y = x_1 - x_2$. So multiplying throughout by n, we get

$$(Tx_1 - Tx_2, x_1 - x_2) \geqslant 0.$$

This completes the proof. ∎

Theorem 4.2 *Let X be any real Banach space and $T : X \to X^*$ be a nonlinear operator. If the Gâteaux derivative $T'(x)$ exists for every $x \in X$ and is positive semidefinite, then T is monotone.*

Proof By Theorem 4.1, we have

$$(Tx_1 - Tx_2, x_1 - x_2) = (T'(x_2 + \tau(x_1 - x_2))(x_1 - x_2), x_1 - x_2)$$
$$= (T'(x)(x_1 - x_2), x_1 - x_2),$$

where $x = x_2 + \tau(x_1 - x_2)$, $0 < \tau < 1$.

Because $T'(x)$ is positive semidefinite, our result follows. ∎

Definition 4.2 Let X be a Banach space and $T : \mathscr{D}(T) \subseteq X \to X^*$ an operator. Then, T is said to be

(i) *accretive* if for each $x, y \in \mathscr{D}(T)$ and $t > 0$, the following inequality holds:

$$\|x - y\| \leqslant \|x - y + t(Tx - Ty)\|. \tag{4.5}$$

(ii) *dissipative* if $I - T$ is accretive, I being identity operator.
(iii) *expansive* if $\|Tx - Ty\| \geqslant \|x - y\|$ for all $x, y \in \mathscr{D}(T)$.

Observation

- This definition of accretivity is due to Kato [320].
- The notion of monotone operators could be considered as a generalization of the concept of accretive operators in view of the following theorem on a Hilbert space $\mathcal{H}$.

Theorem 4.3 *Let $\mathcal{H}$ be a Hilbert space. Then, $T : \mathcal{H} \rightarrow \mathcal{H}$ is monotone iff $(I + \lambda T)$ is accretive for every $\lambda > 0$.*

Proof For $x, y \in \mathcal{D}(T)$, we have

$$
\|(x - y) + \lambda(Tx - Ty)\|^2 = \|x - y\|^2 + 2\lambda(Tx - Ty, x - y) \\
+ \lambda^2 \|Tx - Ty\|^2. \tag{4.6}
$$

If T is monotone, from (4.6), it trivially follows that $(I + \lambda T)$ is accretive. Conversely, if $(I + \lambda T)$ is accretive, then we have

$$
2\lambda(Tx - Ty, x - y) + \lambda^2 \|Tx - Ty\|^2 \geqslant 0
$$

for every $\lambda > 0$. This gives

$$
(Tx - Ty, x - y) \geqslant -\lambda \|Tx - Ty\|^2 \quad \text{for } \lambda > 0.
$$

Taking the limit $\lambda \rightarrow 0$, we get the required inequality needed for monotonicity. ∎

We now discuss the continuity and boundedness property of monotone operators.

Definition 4.3 $T : \mathcal{D}(T) \subseteq X \rightarrow X^*$ is said to be locally bounded at $x \in \mathcal{D}(T)$ if $x_n \rightarrow x$ implies that $\{Tx_n\}$ is bounded in X^*. T is called hemibounded at x if x_n converging to x along a line implies that $\{Tx_n\}$ is bounded in X^*.

Recall the definitions of demicontinuity and hemicontinuity defined in Sect. 1.6.3. In the context of these definitions of $T : X \rightarrow X^*$, convergence of Tx_n to Tx refers to the convergence in the weak* topology of X^*. So $T : X \rightarrow X^*$ is said to be demicontinuous if $x_n \rightarrow x$ in X implies that Tx_n converges to Tx in the weak* topology of X^*. Similarly, we can redefine hemicontinuity of T.

Theorem 4.4 *Let X be Banach space and $T : X \rightarrow X^*$ be monotone with $\mathcal{D}(T) = X$. Then, T is hemibounded at $x_0 \in X$.*

Proof Let $x_n = x_0 + t_n x, 0 \leqslant t_n \leqslant 1$. Then, $x_n \rightarrow x_0$ along line. By monotonicity of T, we get

$$
(T(x_0 + x) - Tx_n, x_0 + x - x_n) = (T(x_0 + x) - Tx_n, x(1 - t_n)) \\
= (1 - t_n)(T(x_0 + x) - Tx_n, x) \geqslant 0.
$$

This implies that $(Tx_n, x) \leqslant (T(x_0 + x), x)$. By the principle of uniform boundedness, it suffices to show that $|(Tx_n, z)| \leqslant M_z$ for every $z \in X$. We have

$$(Tx_n, z) \leqslant t_n(Tx_n, x) - t_n(T(x_0 + z), x) + (T(x_0 + z), z)$$

$$\leqslant t_n(T(x_0 + x), x) - t_n(T(x_0 + z), x) + (T(x_0 + z), z),$$

which gives

$$(Tx_n, z) \leqslant t_n(Tx_{n}, x) - t_n(Tx_0 + z), x) + (T(x_0 + z), z)$$

$$t_n(T(x_0 + x), x) - t_n(T(x_0 + z), x) + (T(x_0 + z), x),$$

which gives

$$(Tx_n, z) \leqslant |(T(x_0 + x), x)| + |(T(x_0 + z), x)| + |(T(x_0 + z), z)|.$$

Taking $-z$ for z, we get the required boundedness of $\{Tx_n, z)\}$ and hence the theorem. ∎

Observation

- Theorem 4.4 is also true if $\mathscr{D}(T)$ is just dense in X.
- If $\mathscr{D}(T)$ is not dense in X in Theorem 4.4, then one can show that T is hemibounded at an interior point of $\mathscr{D}(T)$.
- Theorem 4.4 is due to Kato [321]. For reference, also see Rockfeller [526] and Showalter [554]. Kato, in fact, proved that if $\mathscr{D}(T)$ is open, then hemiboundedness is equivalent to local boundedness.

In view of the above observation and Theorem 4.4, we get the following important result of monotone operators which we separately state.

Theorem 4.5 *Let X be a Banach space and $T : \mathscr{D}(T) \subseteq X \to X^*$ be monotone. Then, T is locally bounded at $x \in int \, \mathscr{D}(T)$.*

Let us now consider a demicontinuous monotone operator $T : X \to X^*$ with dense domain. Then, by what we have just proved it follows that T is hemicontinuous and locally bounded. It turns out that the converse is also true. We now state and prove this result, which is due to Kato [321].

Theorem 4.6 *Let X be a Banach space and $T : \mathscr{D}(T) \subseteq X \to X^*$ be monotone with $\mathscr{D}(T)$ dense in X. Then, T is demicontinuous at $x \in int \, \mathscr{D}(T)$ iff it is hemicontinuous and locally bounded.*

Proof We need only to show that hemicontinuity and local boundedness of T imply demicontinuity. Let $x_n \to x, x_n \in \mathscr{D}(T)$. Put $t_n = \|x_n - x\|^{1/2}$ and define

$$z_n = x + t_n y, \quad y \in \mathscr{D}(T).$$

Then, since $t_n \to 0$, hemicontinuity of T implies that $Tz_n \to Tx$ in the weak* topology of X^*. By monotonicity of T, we have

$$(Tx_n - Tz_n, x_n - x - t_n y) \geqslant 0. \tag{4.7}$$

Because T is locally bounded, $\{Tx_n\}$ as well as $\{Tz_n\}$ are bounded.

$$t_n^{-1}(Tx_n - Tz_n, x_n - x) \to 0,$$

since

$$\|t_n^{-1}(x_n - x)\| = t_n^{-1}\|x_n - x\| = t_n^{-1}t_n^2 = t_n \to 0 \text{ as } n \to \infty.$$

Also, $(Tz_n, y) \to (Tx, y)$. Hence, dividing by t_n and letting $n \to \infty$ in (4.7), we get

$$\liminf_{n\to\infty}(Tx_n - Tx, y) \geqslant 0 \quad \text{for every } y \in \mathscr{D}(T). \tag{4.8}$$

Because $\mathscr{D}(T)$ is dense in X, it follows that (4.8) is true for all $y \in X$. Replacing y by $-y$, we have

$$\lim_{n\to\infty}(Tx_n - Tx, y) = 0 \text{ for all } y \in X.$$

This implies the demicontinuity of T.   ∎

The above theorem shows that the notion of monotone hemicontinuous bounded operator is stronger than the notion of demicontinuous operator.

Observation

- For monotone operators on finite dimensional spaces with dense domain, continuity is equivalent to hemicontinuity (refer Kato [321]).
- For infinite dimensional case, in view of Theorem 4.5, the assumption of local boundedness is redundant at $x \in \text{int } \mathscr{D}(T)$.

In view of above observation, we have the following theorem.

Theorem 4.7 *Let X be a Banach space and $T : \mathscr{D}(T) \subseteq X \to X^*$ be monotone. Then, T is demicontinuous at $x \in \text{int } \mathscr{D}(T)$ iff it is hemicontinuous there.*

Remark 4.2 One can extend the definition of monotonicity for operators on locally convex spaces without much difficulty. For continuity properties of monotone operators on locally convex spaces, refer Kravvaritis [348].

We now define J-monotonicity for mappings from a normed space X into itself. This definition was first introduced by Browder and de Figueredo [116].
J-monotone mapping – Let X be a normed space and $T : \mathscr{D}(T) \subseteq X \to X$ a mapping. Then, T is said to be J-monotone if

$$(Tx - Ty, J(x - y)) \geqslant 0 \text{ for all } x, y \in \mathscr{D}(T).$$

In view of a result of Kato [321], we have the following boundedness and continuity result for J-monotone operators. Let $T : \mathcal{D}(T) \subseteq X \to X$ be J-monotone. Then,

(i) T is locally bounded at $x \in \text{int } \mathcal{D}(T)$ iff it is locally hemibounded at x;
(ii) T is demicontinuous at $x \in \text{int } \mathcal{D}(T)$ iff it is hemicontinuous at x.

In what follows X and Y are real Banach spaces and X^*, Y^* their duals. Consider the operator $f : X \to Y, \mathcal{D}(f) = X$.

Definition 4.4 An operator $T : \mathcal{D}(T) \subseteq X \to Y^*$ is said to be f-monotone if

$$(Tx - Ty, J(x - y)) \geqslant 0 \text{ for all } x, y \in \mathcal{D}(T).$$

If $X = Y$ and $f = I$, then an I-monotone operator is called monotone. Note that for a linear operator $T : X \to X^*$, the monotonicity is equivalent to the nonnegativity, i.e. $(Tx, x) \geqslant 0$, for all $x \in \mathcal{D}(T)$.

Theorem 4.8 *Let $T : \mathcal{D}(T) \subseteq X \to Y^*$ be J-monotone, where $f : X \to Y$ is positively homogeneous, surjective and uniformly continuous on the unit ball of X. Then,*

(i) T is locally bounded at $x \in \text{int } \mathcal{D}(T)$ iff it is locally hemibounded at x;
(ii) T is demicontinuous at $x \in \text{int } \mathcal{D}(T)$ iff it is hemicontinuous at x.

4.2 Maximal Monotone Operator and Its Properties

In this section, we shall be mainly concerned with this multivalued mapping $T : X \to 2^{X^*}$ where X is a real Banach space.

Recall that a graph $G(T)$ of F is a subset of $X \times X^*$ given by

$$G(T) = \left\{ [x, y] \in X \times X^* : y \in Tx, x \in \mathcal{D}(T) \right\}.$$

We say that $T \subseteq T_1$ iff $G(T) \subseteq G(T_1)$.

Definition 4.5 A multivalued operator $T : X \to 2^{X^*}$ is called monotone if

$$(y_1 - y_2, x_1 - x_2) \geqslant 0 \text{ for all } x_1, x_2 \in \mathcal{D}(T) \text{ and } y_1 \in Tx_1, y_2 \in Tx_2.$$

Similarly, one can define strict monotonicity and strong monotonicity for T.

Definition 4.6 Let G be any nonempty subset of $X \times X^*$. G is called monotone if

$$(y_1 - y_2, x_1 - x_2) \geqslant 0 \text{ for all } [x_1, y_1], [x_2, y_2] \in G.$$

Definition 4.7 A monotone operator $T : \mathscr{D}(T) \subset X \to 2^{X^*}$ is called maximal monotone if it has no proper monotone extensions. That is, if for $[x, y] \in X \times X^*$, we have

$$(y - v, x - u) \geqslant 0 \text{ for all } u \in \mathscr{D}(T) \text{ and } v \in Tu, \text{ then } y \in Tx.$$

Similarly, we say that $G \subset X \times X^*$ is maximal monotone if for every monotone G_1 such that $G \subset G_1$, we have $G = G_1$.

Proposition 4.1 *Let X be a real reflexive Banach space. Let $T : X \to 2^{X^*}$ be a multivalued operator. Then,*

(i) T is monotone (maximal) iff $G(T)$ is monotone (maximal);
(ii) T is monotone (maximal) iff T^{-1} is monotone (maximal).

We first give the characterization of maximality for single-valued monotone operators—linear as well as nonlinear, which is due to Browder [63].

Theorem 4.9 *Let $T : X \to X^*$ be a single-valued hemicontinuous monotone operator. Suppose that $\mathscr{D}(T)$ is dense in X, with respect to weak topology. Then, T is maximal monotone iff T is not the restriction of any properly larger monotone single-valued mapping with domain in X and range in X^*.*

Proof One way is obvious. So now assume that T is maximal in the family of single-valued monotone operators from X to X^*. Let $[x_0, y_0] \in X \times X^*$ such that for all $x \in \mathscr{D}(T)$, we have $(Tx - y_0, x - x_0) \geqslant 0$. It suffices to show that $y_0 = Tx_0$.

We claim that $x_0 \in \mathscr{D}(T)$. If not, then T could be extended to T_1 by setting $T_1 x_0 = y_0$. Suppose now $Tx_0 - y_0 \neq 0$. Because $\mathscr{D}(T)$ is dense in X with respect to the weak topology of X, there exists an $x_1 \in \mathscr{D}(T)$ such that $(Tx_0 - y_0, x_1 - x_0) < 0$. Set $x_t = (1 - t)x_0 + tx_1 \in \mathscr{D}(T), 0 < t < 1$. Then, $0 \leqslant (Tx_t - y_0, x_t - x_0) = t(Tx_t - y_0, x_1 - x_0)$ which implies that $(Tx_t - y_0, x_1 - x_0) \geqslant 0$. As $t \to 0, x_t \to x_0$ and hence $Tx_t \to Tx_0$ in the weak* topology and so we get $(Tx_0 - y_0, x_1 - x_0) \geqslant 0$, contradicting the choice of x_1. Hence, $y_0 = Tx_0$.   ∎

From this theorem, we immediately get the following two important corollaries.

Corollary 4.1 *Let T be a hemicontinuous monotone single-valued operator with $\mathscr{D}(T) = X$, Then, T is maximal monotone.*

Corollary 4.2 *Let $T : X \to X^*$ be monotone and suppose that the range of T is all of X^* and T^{-1} is single valued and hemicontinuous. Then, T is maximal monotone.*

Theorem 4.10 *Let $L : X \to X^*$ be a single-valued monotone linear mapping from a dense subspace $\mathscr{D}(L)$ of X into X^*. Then, L is maximal monotone iff L is maximal among all single-valued monotone linear mappings L from X to X^*.*

Proof Suppose $[x_0, y_0] \in X \times X^*$ be such that for all $x \in \mathscr{D}(L)$, we have $(Lx - y_0, x - s_0) \geqslant 0$. Then, $x_0 \in \mathscr{D}(L)$. If not, then the operator $L_1 : x + \lambda x_0 \to Lx + \lambda y_0, \lambda \in \mathbb{R}$, is a proper linear monotone extension of L. We then have

$$(L(x_0 + tx) - y_0, (x_0 + tx) - x_0) \geqslant 0 \text{ for } x \in \mathscr{D}(L) \text{ and } t > 0.$$

This gives $(Lx_0 - y_0, x) \geqslant -t(Lx, x)$ for all $t > 0$. Taking limit as $t \to 0$ and using the fact that $\mathscr{D}(L)$ is dense in X, we get $(Lx_0 - y_0, x) = 0$ for all $x \in X$. This implies that $Lx_0 = y_0$, hence the result. ∎

We now give a sufficient criterion for the maximal monotonicity of a multivalued monotone operator. This result is due to Browder [90]. In the next section, we shall characterize maximal monotonicity of multivalued operators.

Theorem 4.11 *Let $T : X \to 2^{X^*}$ be a multivalued hemicontinuous monotone operator with $\mathscr{D}(T) = X$. If $T(x)$ is a closed convex subset of X^* for each $x \in X$, then T is maximal monotone.*

Proof Suppose that there exists an extension T_1 of T such that T is monotone and $y_0 \in T_1(x_0)$. We must show that $y_0 \in T(x_0)$. Suppose this is not so. Because $T(x_0)$ is a closed convex subset of X^*, there exists z in X such that

$$(y_0, z) > (y, z) \text{ for all } y \in T(x_0). \tag{4.9}$$

Let $x_t = x_0 + tz, t > 0$. Then, in view of monotonicity of T_1, we have

$$t(y_t - y_0, z) \geqslant 0 \text{ for all } y_t \in T x_t.$$

This gives

$$(y_t - y_0, z) \geqslant 0 \text{ for all } y_t \in T x_t. \tag{4.10}$$

Because $x_t \to x_0$ along a line, hemicontinuity of T gives that $y_t \to y_1$ in weak* topology of X^* for some $y_1 \in T x_0$. So in view of (4.10), we get

$$(y_1 - y_0, z) = \lim_{t \to 0}(y_t - y_0, z) \geqslant 0,$$

which is a contradiction to (4.9). ∎

Now we begin with the convexity property of the domain and range of a maximal monotone operator. In the following, for any subset D of X, co$D$, int$D$ and $\overline{D}$ will denote convex hull, interior and (strong) closure of D, respectively.

Definition 4.8 $D \subset X$ is called almost convex if int $co(D) \subset D$.

Definition 4.9 $D \subset X$ is called virtually convex if for each relatively compact subset K of co D and each $\varepsilon > 0$, there exists a continuous mapping $P : K \to D$ such that $\|P(x) - x\| < \varepsilon$ for every $x \in K$.

It follows from the definition that if D is almost convex, then cl D is convex. We have the following relationship between almost convexity and virtual convexity; see Rockfeller [527].

Proposition 4.2 *If X is a finite dimensional space, then $D \subset X$ is virtually convex iff D is almost convex.*

We now state and prove the following important convexity result for maximal monotone operators on finite dimensional spaces. This is due to Minty [401].

Theorem 4.12 *If T is a maximal monotone operator on finite dimensional Hilbert space X, then $\mathscr{D}(T)$ is almost convex.*

We first state the following lemma which was proved by Minty in [402].

Lemma 4.1 *Let $x_1, x_2, \ldots, x_n$ and $y_1, y_2, \ldots, y_n$ in X be such that*

$$(y_i - y_j, x_i - x_j) \geqslant 0, \quad 1 \leqslant i, j \leqslant n \tag{4.11}$$

and let $x \in X$ be any point. Then, there exists a point $y \in X$ such that $(y_i - y, x_i - x) \geqslant 0$ for all $i = 1, 2, \ldots, n$.

Proof of the Theorem It suffices to show that if $0 \in$ int co $\mathscr{D}(T)$, then $0 \in \mathscr{D}(T)$. Consider an open ball S such that $0 \in X \subset$ co $\mathscr{D}(T)$. There exists a finite set of vectors of S which generate X and since each vector in co $\mathscr{D}(T)$ is a finite linear combination, with positive coefficients, of vectors of $\mathscr{D}(T)$, we can find a set $\{x_1, \ldots, x_n\}$ which generates X. Consider the family C_α of subsets of X defined as

$$C_\alpha = \{y : (y_\alpha - y, x_\alpha) \geqslant 0 \text{ for all } x_\alpha \in \mathscr{D}(T) \text{ and } y_\alpha \in Tx_\alpha\}$$

We intend to show that $\bigcap_\alpha C_\alpha \neq \emptyset$, since gan.maths@gmail.com then the maximality of T would imply that $y \in T(0)$, that is, $0 \in \mathscr{D}(T)$.

It is clear that $\{C_\alpha\}$ is a family of closed sets and since T is monotone, the previous lemma implies that this family has finite intersection property. Hence, in order to show that $\bigcap_\alpha C_\alpha \neq \emptyset$, it is enough to show that there exists a finite subfamily which is compact. So consider

$$C = \bigcap_{i=1}^{n} \{y : (y_i - y, x_i) \geqslant 0, \quad y_i \in Tx_i\}$$

where $x_1, x_2, \ldots, x_n$ are as defined before. The set

$$C^* = \bigcap_{i=1}^{n} \{y : (y, x_1) \leqslant 0\}$$

consists of zero vector, for if $y \neq 0 \in C^*$ then

$$y = \sum_{i=1}^{n} c_i x_i, \quad c_i \geqslant 0 \text{ and } 0 < \|y\|^2 = \sum_{i=1}^{n} (y, c_i x_i) = \sum_{i=1}^{n} c_i (y, x_i).$$

So that at least one of the (y, z_i) is positive. Hence, by the resolution theorem for polyhedra of convex sets (refer Goldman [253], Theorem 1), C is the sum of bounded polyhedra set and the vector 0. So C is bounded: being closed, it follows that C is compact. This completes the proof. ■

Remark 4.3 Since T is maximal monotone iff T^{-1} is maximal monotone, it also follows that the range of a maximal monotone operator is almost convex on a finite dimensional Hilbert space X.

In 1970, Rockfeller [527] generalized the above result to certain infinite dimensional Banach spaces.

Definition 4.10 X is said to be smoothly reflexive if X has an equivalent norm which is everywhere Fréchet differentiable except at the origin and whose dual norm in X^* is everywhere Fréchet differentiable except at the origin.

It is known fact that every smoothly reflexive space is reflexive. In 1971, Troyanskii [591] proved that every reflexive Banach space is smoothly reflexive. We state the Rockfeller's result.

Theorem 4.13 *If X is a reflexive Banach space and $T : X \to X^*$ is maximal monotone, then $\mathscr{D}(T)$ and $\mathscr{R}(T)$ are virtually convex.*

Remark 4.4 In view of Proposition 4.2, the result of Theorem 4.10 follows as a simple corollary of Theorem 4.13. Also, in view of our earlier observation, it follows that cl $\mathscr{D}(T)$ and cl $\mathscr{R}(T)$ are convex sets for maximal monotone operator $T : X \to X^*$.

We now discuss boundedness and continuity properties of a maximal monotone operator. The following theorem generalizes Theorems 4.5 and 4.7 to multivalued case. This theorem is due to Rockfeller [527]. We state this theorem along with some important corollaries.

Theorem 4.14 *Let X be a Banach space and $T : X \to 2^{X^*}$ a maximal monotone multivalued operator. Suppose int co $\mathscr{D}(T) \neq \emptyset$ or that X is reflexive and there exists a point of $\mathscr{D}(T)$ at which T is locally bounded. Then, int $\mathscr{D}(T)$ is a nonempty convex set whose closure is cl $\mathscr{D}(T)$. Further T is locally bounded at each point of $\mathscr{D}(T)$, whereas T is not locally bounded at any boundary point of $\mathscr{D}(T)$.*

Corollary 4.3 *Suppose the hypothesis of Theorem 4.14 are satisfied and $D \subset \mathscr{D}(T)$ where T is single valued. Then, $D \subset$ int $\mathscr{D}(T)$ and T is demicontinuous on D.*

Corollary 4.4 *Suppose the hypothesis of Theorem 4.14 are satisfied. Then, $\mathscr{D}(T)$ is virtually convex and, in particular, cl $\mathscr{D}(T)$ is convex. If in addition $\mathscr{D}(T)$ is dense in X, then $\mathscr{D}(T)$ is all of X.*

We now discuss the topological properties of $G(T)$ where T is a maximal monotone multivalued operator with $\mathscr{D}(T) = X$. The following theorem is due to Kenderov [317], and it generalizes earlier results of Browder [89].

Theorem 4.15 *Let T be a maximal monotone operator from a Banach space X to 2^{X^*} with $\mathscr{D}(T) = X$. Then, $G(T)$ is a closed subset of $X \times X^*$ with respect to strong topology of X and weak* topology of X^*.*

Proof Let $[x_n,\, y_n] \in G(T)$ converges to $[x_0,\, y_0]$ in the appropriate topology. By Theorem 4.14, T is locally bounded and hence there exists $c > 0$ such that $\|y_n\| \leqslant c$. For $[x, y] \in X \times X^*$, we have

$$|(y_n - y, x_n - x) - (y_0 - y, x_0 - x)| \leqslant |(y_n - y, x_n - x_0)| + |(y_n - y_0, x - x_0)|$$
$$\leqslant \|x_n - x_0\|(c + \|y\|) + |(y_n - y_0, x - x_0)|$$
$$\to 0 \text{ as } n \to \infty.$$

Therefore, $(y_0 - y, x_0 - x) = \lim_{n\to\infty}(y_n - y, x_n - x) \geqslant 0$, for $[x, y] \in G(T)$, and so by maximal monotonicity of T, we get $[x_0, y_0] \in G(T)$. This proves the theorem. ∎

Corollary 4.5 *Let $T : X \to 2^{X^*}$ abe a maximal monotone operator with $\mathscr{D}(T) = X$. Then, the set Tx, $x \in X$ is a weak* compact and convex subset of X^*.*

Proof We have by the above theorem, Tx as a weak* closed subset of X^* and is bounded by Theorem 4.14, and hence, it is weak* compact. To prove that it is convex, we observe that for $y_i \in Tx$, $i = 1, 2$ we have

$$(ty_1 + (1 - t)y_2 - v, x - u) = t(y_1 - v, x - u) + (1 - t)y_2 - v, x - u)$$
$$\geqslant 0 \text{ for all } [u, v] \in G(T).$$

So, by the maximal monotonicity, it implies that $ty_1 + (1 - t)y_2 \in Tx$. ∎

Definition 4.11 T is called upper (lower) semicontinuous if for any closed (open) subset C of X, $T^{-1}(C)$ is closed (open.) T is, called continuous if T is both upper and lower semicontinuous.

Theorem 4.16 *Let $T : X \to 2^{X^*}$ be a maximal monotone operator with $\mathscr{D}(T) = X$. Then, T is upper semicontinuous.*

Proof Let $x_0 \in X$ and V be a weak* neighbourhood of Tx_0 in X^*. Since T is locally bounded at x_0, there exists a neighbourhood U of x_0 such that T is bounded on U. Result now follows from Theorem 4.15 which implies that $G(T)$ is closed in $X \times X^*$ with respect to strong topology of X and weak* topology of X^*. ∎

Definition 4.12 A property P is said to be satisfied almost every where on X if it is satisfied on X except on a set of first category. That is, P is true on X except on a countable subset of nowhere dense set.

We have the following interesting result due to Kenderov [324] which states that a multivalued maximal monotone operator is almost single valued.

Theorem 4.17 *Let X be a Banach space with strictly convex dual X^*. Assume that $T : X \to 2^{X^*}$ is maximal monotone with $\mathscr{D}(T) = X$. Then, the set $\{x \in X : Tx$ has more than one element$\}$ is of first category.*

4.3 Surjectivity Theorems

In this section, we shall be interested in the solvability of the operator equation $y \in Fx$, where $F : X \to 2^{X^*}$ is a multivalued monotone operator and $y \in X^*$. As a special case, we will obtain solvability of the single-valued operator equation.

To begin with, we introduce Kirzbraun theorem which is of interest by itself. This will then be used to obtain a surjectivity result for monotone operators. The classical Kirzbraun theorem in finite dimensional spaces is the following.

Theorem 4.18 *Let B_i and B_i' be closed balls in a finite dimensional space $\mathbb{R}^n$:*

$$B_i = \{x : \|x - x_i\| \leqslant \rho_i\},$$
$$B_i' = \{x : \|x - x_i'\| \leqslant \rho_i\}, \ 1 \leqslant i \leqslant m.$$

If balls are such that

$$\|x_i' - x_j'\| \leqslant \|x_i - x_j\|, \ 1 \leqslant i, j \leqslant m \tag{4.12}$$

then $\bigcap_{i=1}^m B_i \neq \emptyset$ implies that $\bigcap_{i=1}^m B_i' \neq \emptyset$.

Theorem 4.19 *Let $x_i, x_i'(i = 1, 2, \ldots, m)$ be points in $\mathbb{R}^n$ satisfying (4.12). Let p be a given point, then there exists a point p' such that*

$$\|p' - x_i'\| \leqslant \|p - x_i\|, \ 1 \leqslant i \leqslant m. \tag{4.13}$$

Proof Without loss in generality, we assume that $p \neq x_i$, $i = 1, \ldots, m$. Consider the function

$$f(x) = \max_{1 \leqslant i \leqslant m} \frac{\|x - x_i\|}{\|p - x_i\|}.$$

This is a continuous function on $\mathbb{R}^n$ and $f(x) \to \infty$ as $\|x\| \to \infty$, and hence, it attains its minimum at a point $x = p'$:

$$f(p') = \min f(x) = \lambda.$$

We will show that p' satisfies (4.13). If $\lambda = 0$, then we are done; therefore, assume that $\lambda > 0$.

Let
$$\frac{\|p' - x_i'\|}{\|p - x_i\|} = \lambda, \text{ if } 1 \leqslant i \leqslant k$$

$$(1 \leqslant k \leqslant m) \tag{4.14}$$

$$< \lambda, \text{ if } k + 1 \leqslant i \leqslant m.$$

We claim that $p' \in co(x_1', \ldots, x_k')$, otherwise by displacing p, we can change the ratio $\frac{\|p' - x_i\|}{\|p - x_i\|}$ so that it becomes less than λ, contradicting the definition of λ.

In view of (4.14), we are done if we can show that $\lambda \geqslant 1$. If possible let $\lambda < 1$. Putting $R_1 = (x_1 - p)$, $R_i' = (x_i' - p')$ in (4.14), we get

$$(R_i', R_i') < (R_i, R_i), \quad 1 \leqslant i \leqslant k. \tag{4.15}$$

Also, writing (4.12) in equivalent form, we get

$$(R_i - R_j, R_i - R_j) \geqslant (R_i' - R_j', R_i' - R_j'), \quad 1 \leqslant i, j \leqslant k. \tag{4.16}$$

Subtracting (4.16) from (4.15), we get

$$(R_i', R_j') > (R_i, R_j), \quad 1 \leqslant i, j \leqslant k. \tag{4.17}$$

But since $p_i' \in co(x_1', x_2', \ldots, x_k')$ we can write

$$p' = \sum_{i=1}^{k} c_i x_i, \quad \sum_{i=1}^{k} c_i = 1, \quad c_i \geqslant 0.$$

This gives

$$0 = \sum_{i=1}^{k} c_i x_i' - p' = \sum_{i=1}^{k} c_i x_i' - \sum_{i=1}^{k} c_i p',$$

that is

$$\sum_{i=1}^{k} c_i R_i' = 0. \tag{4.18}$$

Multiplying (4.17) by $c_i c_j$ and summing over for i, j, we obtain the inequality

$$\left\| \sum_{i=1}^{k} c_i R_i' \right\|^2 > \left\| \sum_{i=1}^{k} c_i R_i \right\|^2$$

which is a contradiction because LHS vanishes in view of (4.18) and RHS is always nonnegative.   ∎

The above result was first obtained by Kirzbraun in 1934 [336]. It was rediscovered by Valentine [600] in 1943. We have presented here the proof due to Schoenberg [543]. There is an infinite dimensional extension of the above theorem to Hilbert spaces.

Theorem 4.20 *Let $\{x_\alpha\}, \{y_\alpha\}, \alpha \in I$, an index set, be arbitrary subsets of a Hilbert space $\mathcal{H}$ such that*

$$\|x_\alpha - x_\beta\| \leqslant \|y_\alpha - y_\beta\|, \ \alpha, \beta \in I.$$

Then, for every $y \in \mathcal{H}$, there exists and $x \in \mathcal{H}$ such that

$$\|x - x_2\| \leqslant \|y - y_\alpha\|, \ \alpha \in I.$$

Proof Let $S_\alpha = \{x : \|x - x_\alpha\| \leqslant \|y - y_\alpha\|, \alpha \in I\}$. We need to show that $\cap_{\alpha \in I} S_\alpha \neq \emptyset$. $\{S_\alpha\}_{\alpha \in I}$ is a family of weakly compact subsets of $\mathcal{H}$, and so by Smulian's theorem, it suffices to show that every finite subfamily $\{S_n\}$ has nonempty intersection. This follows from the previous theorem by considering $\{S_n\}$ in a finite dimensional space generated by $\{x_n\}$ and $\{y_n\}$.

Recall that a multivalued mapping $T : \mathcal{H} \rightarrow 2^{\mathcal{H}}$ is called accretive if

$$\|x_\alpha - x_\beta\| \leqslant \|y_\alpha - y_\beta\| \text{ for all } x_\alpha, x_\beta \in \mathcal{H} \text{ and } y_\alpha \in Tx_\alpha, \ y_\beta \in Tx_\beta.$$

This completes the proof.   ∎

The above theorem implies that an accretive mapping T can always be extended to T_1 such that $T \subset T_1$. Consequently, we get the following surjectivity result.

Theorem 4.21 *Suppose $T : \mathcal{H} \rightarrow 2^{\mathcal{H}}$ is a maximal accretive mapping. Then, $\mathcal{R}(T) = \mathcal{H}$ and T^{-1} is nonexpansive.*

If $T : \mathcal{H} \rightarrow 2^{\mathcal{H}}$ is maximal monotone, then in view of Theorem 4.3, $(I + \lambda T)$ is maximal accretive for every $\lambda > 0$, and so we immediately get the following surjectivity theorem due to Minty [400].

Theorem 4.22 *Suppose $T : \mathcal{H} \rightarrow 2^{\mathcal{H}}$ is a maximal monotone operator. Then, $\mathcal{R}(I + \lambda T) = \mathcal{H}$ and $(I + \lambda T)^{-1}$ is nonexpansive and maximal monotone.*

The converse of the above theorem is also true. In fact, we have the following characterization of maximal monotonicity which is of fundamental importance for the study of such operators; refer Brezis [76].

Theorem 4.23 *Let $T : \mathcal{H} \rightarrow 2^{\mathcal{H}}$ be an operator. Then, the following are equivalent:*

(a) T is maximal monotone;
(b) T is monotone and $(I + T)(\mathcal{H}) = \mathcal{H}$;

(c) *for every* $\lambda > 0, (I + \lambda T)^{-1}$ *is a nonexpansive mapping defined on the whole space* $\mathcal{H}$.

Because every single-valued hemicontinuous monotone mapping defined on the whole space is maximal monotone, we immediately get the following important theorem as a simple corollary.

Theorem 4.24 *Let* $T : \mathcal{H} \rightarrow \mathcal{H}$ *be a single-valued hemicontinuous monotone mapping with* $\mathcal{D}(T) = \mathcal{H}$. *Then,* $(I + T)(\mathcal{H}) = \mathcal{H}$ *with* $(I + T)^{-1}$ *continuous.*

In terms of strongly monotone operators, Theorem 4.24 is reformulated as follows.

Let $T : \mathcal{H} \rightarrow \mathcal{H}$ be a hemicontinuous strongly monotone operator with $\mathcal{D}(T) = \mathcal{H}$. Then, T is $1 - 1$ and onto with T^{-1} continuous.

Observation

- The reformulated Theorem 4.24 was first proved by Browder in [92] and is a natural generalization of Lax–Milgram lemma for linear operators.

For the sake of completeness, we state this lemma as a corollary.

Corollary 4.6 (Lax–Milgram lemma) *Let A be a strongly positive definite continuous linear operator from* $\mathcal{H}$ *into itself, then A is* $1 - 1$ *and onto with* A^{-1} *continuous.*

If we merely assume monotonicity on T, we can still get the surjectivity of T provided we impose some infinity condition on T. In this direction, we first state and prove the existence result by Minty [400]. To prove this result we need the following lemma.

Lemma 4.2 *Let X be a Banach space and* B_0 *an open ball containing* 0. *Let* $T : B_0 \rightarrow X^*$ *be hemicontinuous at* $x_0 \in B_0$. *If*

$$(Tx - y_0, x - x_0) \geqslant 0 \text{ for all } x \in B_0,$$

then $Tx_0 = y_0$.

Proof For every $x \in X, x_0 + tx \in B_0$, if t is sufficiently small. So for $t > 0$, we have

$$(T(x_0 + tx) - y_0, x) \geqslant 0.$$

Let now $t \rightarrow 0^+$, then by hemicontinuity of T, $T(x_0 + tx) \rightarrow Tx_0$, and so we have

$$(Tx_0 - y_0, x) \geqslant 0 \text{ for all } x \in X.$$

Taking $-x$ for x, we obtain the reverse inequality and hence

$$(Tx_0 - y_0, x) = 0 \text{ for all } x \in X.$$

This implies that $Tx_0 = y_0$.   ∎

Theorem 4.25 *Let X be a reflexive Banach space and B_0 an open ball containing 0 and B_1 a closed ball contained in B_0. Let $T : B_0 \to X^*$ be a hemicontinuous singlevalued monotone operator for which $(Tx, x) > 0$ for $x \in \partial B_1$. Then, $Tx = 0$ has a solution in B_1.*

Proof By Lemma 4.2 it suffices to show that

$$\{x_0 \in B_0 : (Tx, x - x_0) \geqslant 0 \text{ for all } x \in B_0\} \neq \emptyset.$$

We first show that

$$\bigcap_{x \in B_0} \{x_0 \in X : (Tx, x - x_0) \geqslant 0\} \neq \emptyset.$$

This follows from Lemma 4.1 and an argument similar to that used in the Theorem 4.12. We now claim that $x_0 \in B_0$. Suppose not, then choose $z_0 \in \partial B_1$ such that $x_0 = \lambda z_0, \lambda > 1$ and hence $(Tz_0, (1 - \lambda)z_0) \geqslant 0$. This gives $(Tz_0, z_0) \leqslant 0$, which is a contradiction to our assumption on T. Hence the theorem. ∎

Definition 4.13 $T : X \to X^*$ is called coercive if $\lim\limits_{\|x\| \to \infty} \frac{(Tx,x)}{\|x\|} = \infty$.

For a function f defined on $\mathbb{R}$, this corresponds to the conditions:

$$\lim_{x \to \infty} f(x) = +\infty, \quad \lim_{x \to -\infty} f(x) = -\infty \tag{4.19}$$

It is a known fact that if f is monotone increasing and continuous function on $\mathbb{R}$ satisfying (4.19), then f maps $\mathbb{R}$ onto $\mathbb{R}$. We have a generalization of this to operators defined on reflexive Banach Spaces.

Theorem 4.26 *Let X be a reflexive Banach space and $T : X \to X^*$ be a hemicontinuous, monotone, coercive mapping with $\mathscr{D}(T) = X$. Then, $\mathscr{R}(T) = X^*$.*

Proof Let $y \in X^*$ and define $Gx = Tx - y$. Then, G is hemicontinuous and monotone and

$$(Gx, x) = (Tx, x) - (y, x)$$
$$\geqslant (Tx, x) - \|x\| \, \|y\|. \tag{4.20}$$

Since T is coercive, there exists a constant $\rho \geqslant 0$ such that

$$(Tx, x) > \|x\| \, \|y\| \text{ for } \|x\| \geqslant \rho.$$

So (4.20) gives

$$(Gx, x) > \|x\| \, \|y\| - \|x\| \, \|y\| = 0 \text{ for } \|x\| \geqslant \rho.$$

So $(Gx, x) > 0$ on $\partial B\rho$. Hence, by the previous theorem, there exists x_0 such that $Gx_0 = 0$. This proves the theorem. ∎

We now state and prove two theorems (due to Rockfeller), which generalize Theorems 4.25 and 4.26, respectively. They are obtained as easy consequences of Theorem 4.14.

Theorem 4.27 *Let X be a reflexive Banach space and let $T : X \to 2^*$ be a maximal multivalued monotone operator. Suppose there exists a subset $C \subset X$ such that $0 \in$ int co $T(C)$. Then, there exists an $x \in X$ such that $0 \in T(x)$.*

Proof Since $T(C) \subset \mathscr{R}(T) = \mathscr{R}(T^{-1})$, this theorem follows immediately from Theorem 4.14 if applied to T^{-1}.

Theorem 4.28 *Let X be a reflexive Banach space and let $T : X \to 2^{X^*}$ be a maximal monotone multivalued operator. In order that $\mathscr{R}(T)$ be all of X^*, it is necessary and sufficient that the sequence $\{y_n\} \subset X^*$ has no convergent subsequence whenever*

$$y_n \in T(x_n) \ (n = 1, 2, \ldots) \ and \ \lim_{n \to \infty} \|x_n\| = \infty.$$

Proof The stated condition says that T^{-1} is locally bounded at every point of X^*. By Theorem 4.14 applied to T^{-1}, this is equivalent to $\mathscr{D}(T^{-1})$ being an open, convex subset of X^* with no boundary points and the only such nonempty subset is X^* itself, i.e. $\mathscr{R}(T) = X^*$. This completes the proof. ∎

Remark 4.5 The necessary and sufficient condition in the above theorem is satisfied if the following condition (c) is satisfied:

(c) If $y_n \in T(x_n)$ and $\lim_{n \to \infty} \|x_n\| = \infty$ then $\lim_{n \to \infty} \|y_n\| = \infty$.

One of the sufficient conditions for (c) to be satisfied is that T be coercive.

In applied problem, one encounters operators F of the type $T = L + T_0$ when L is densely defined linear operator and F_0 is a monotone nonlinear operator defined on all of X. We state the following perturbation theorem, which is due to Browder [92].

Theorem 4.29 *Let X be a reflexive Banach space, T_1 a maximal monotone operator from X to 2^{X^*} with dense domain containing 0. Assume that T_2 is a monotone single-valued operator from X to X^* defined on the whole space such that it is hemicontinuous, bounded and coercive. Then, $\mathscr{R}(T) = X^*$ where $T = T_1 + T_2$.*

As a direct consequence of this theorem, we obtain the following result regarding the maximal monotonicity of the sum, which is of interest by itself. We present Browder's proof.

Remark 4.6 In the remaining part of this section, we can assume without loss in generality that the duality mapping $J : X \to X^*$ is singlevalued. This follows from our earlier mentioned result in Sect. 2.3 which states that in a reflexive Banach space, one can redefine the norm with respect to which X is locally uniformly convex and hence the duality mapping J is singlevalued.

Theorem 4.30 *Let X be a strictly convex reflexive Banach space. Let $T_1 : X \to 2^{X^*}$ be as defined in Theorem 4.29 and $T_2 : X \to X^*$ be hemicontinuous, bounded and monotone with $\mathcal{D}(T_2) = X$. Then, $T = T_1 + T_2$ is maximal monotone.*

Lemma 4.3 *Let T, T_1, T_2 (single valued) be monotone operators from X to 2^{X^*} with $\mathcal{D}(T_2) = X, \mathcal{D}(T) = \mathcal{D}(T_1)$ and $T = T_1 + T_2$. If T is maximal monotone then T_1 is also maximal monotone.*

Proof Suppose $[x_0, y_0] \in X \times X^*$ be such that $(y_1 - y_0, x - x_0) \geqslant 0$ for all $x \in \mathcal{D}(T_1)$ and $y_1 \in T_1(x)$. Let $z_0 = T_2(x_0)$ and $y_2 = T_2(x)$. Then, $y = y_1 + y_2 \in T(x)$ and for such a choice of y we have

$$(y - z_0 - y_0, x - x_0) = (y_1 + y_2 - z_0 - y_0, x - x_0)$$
$$= (y_1 - y_0, x - x_0) + (y_2 - z_0, x - x_0) \geqslant 0,$$

by monotonicity of T and our earlier assumption. So maximal monotonicity of F gives that $x_0 \in \mathcal{D}(T)$ and $z_0 + y \in T(x_0)$. Since $z_0 = T_2(x_0)$, we have $y_0 \in T_1(x_0)$, thereby providing the maximal monotonicity of T_1.

Proof of the Theorem We first observe that the duality mapping $J : X \to X^*$ is a singlevalued, strictly monotone, continuous, bounded and coercive mapping with domain equal to X. So T_1 and $T_2 + J$ satisfy all conditions of Theorem 4.29, and hence, $\mathcal{R}(T_1 + T_2 + J) = X^*$. We now claim that if for a sequence $\{x_n\}$ in X we have

$$(Jx_n \ Jx, x_n - x) \to 0,$$

then $\{x_n\}$ converges weakly to x. To prove this, we first observe that the following identity

$$(Jx_n - Jx, x_n - x) = (\|x_n\| - \|x\|)^2 + (\|Jx_n\|\|x\| - (Jx_n, x)) + (\|x_n\|\|Jx\| - (Jx, x_n)).$$

Since each term in the above identity is nonnegative, $(Jx_n - Jx, x_n - x) \to 0$ implies that $\|x_n\| \to \|x\|$. This in turn implies that $(Jx, x_n) \to (Jx, x)$ in view of the following inequality

$$(Jx, x_n) - (Jx, x) \leqslant \|Jx\|\|x_n\| - \|x\|^2 = \|x_n\|\|x\| - \|x\|^2 = \|x\|[\|x_n\| - \|x\|].$$

Again, since $\{x_n\}$ is a bounded sequence in a reflexive Banach space, it has a subsequence (which we do not differentiate) converging weakly to some element $y \in X$. We are through if we can show that $x = y$.

Since $\{x_n\}$ converges weakly to y, we have $(Jx, x_n) \to (Js, y)$. But $(Jx, x_n) \to (Jx, x)$, and hence, $(Jx, x) = (Jx, y)$. Strict convexity of X gives $x = y$.

We now show that $(T_1 + T_2 + J)^{-1}$ is demicontinuous. For this consider $y_n \to y$, where $y_n = (T_1 + T_2 + J)x_n$ and $y = (T_1 + T_2 + J)x$. In view of the duality mapping $J, (T_1 + T_2 + J)$ is coercive and hence $(T_1 + T_2 + J)^{-1}$ is bounded. This gives that $\{x_n\}$ is a bounded sequence. So we have

$$((T_1 + T_2 + J)x_n - (T_1 + T_2 + J)x, \ x_n - x)$$
$$= ((T_1 + T_2)x_n - (T_1 + T_2)x, \ x_n - x) + (Jx_n - Jx - x).$$

Since both sums on RHS are nonnegative, it follows that $(Jx_n - Jx, X_n - x) \to 0$. So by our earlier observation regarding the duality mapping $x_n \to x$ in X. Also, $(T_1 + T_2 + J)^{-1}$ is single valued because $(T_1 + T_2 + J)$ is strictly monotone, and therefore, by Corollary 4.1, it follows that $(T_1 + T_2 + J)$ is a maximal monotone. So by Lemma 4.3, we conclude that $T = T_1 + T_2$ is maximal monotone. ∎

We now deduce a surjectivity result from Theorem 4.29.

Theorem 4.31 *Let T be a maximal monotone operator from a reflexive Banach space X to 2^{X^*}. Further, let T be coercive with $0 \in \mathcal{D}(T)$. Then, $\mathcal{R}(T) = X^*$.*

Proof We apply Theorem 4.29 to T and εJ, $\varepsilon > 0$ and get that $\mathcal{R}(T + \varepsilon J) = X^*$. So for a given $w_0 \in X^*$, there exists $x_\varepsilon X$ such that $(T + \varepsilon J)x_\varepsilon \to w_0$. Moreover, this x_ε is unique as $(T + \varepsilon J)$ is strictly monotone. Taking inner product with x_ε, we get

$$(w_0, x_\varepsilon) = (y_\varepsilon, x_\varepsilon) + \varepsilon \|x_\varepsilon\|^2, \ y_\varepsilon \in Fx_\varepsilon.$$

Coercivity of F implies that

$$(w_0, x_\varepsilon) \geqslant c(\|x_\varepsilon\|)\|x_\varepsilon\| \text{ with } c(r) \to \infty \text{ as } r \to \infty.$$

This gives

$$\|w_0\|\|x_\varepsilon\| \geqslant c(\|x_\varepsilon\|)\|x_\varepsilon\|,$$

which in turn implies that $\|x_\varepsilon\| \leqslant M$ for all ε. So $\{x_\varepsilon : \|x_\varepsilon\| \leqslant M\}$ is a bounded set in a reflexive Banach space, and hence, there exists a subsequence of it which we again denote by $\{x_\varepsilon\}$ such that $x_\varepsilon \to x_0$ in X *as* $\varepsilon \to 0$. Also,

$$y_\varepsilon = w_0 - \varepsilon Jx_\varepsilon \to w \text{ as } \varepsilon \to 0.$$

So for each $x \in D(T)$, we have

$$(y - y_\varepsilon, x - x_\varepsilon) \geqslant 0 \text{ for all } y \in Tx.$$

Taking limit we get
$$(y - w_0, x - x_0) \geqslant 0.$$

Maximal monotonicity of T implies that $x_0 \in \mathcal{D}(T)$ and $w_0 \in Tx_0$. This completes the proof. ∎

We now state and prove two theorems which generalize Theorems 4.22 and 4.23 to operators on Banach spaces.

Theorem 4.32 *Let X be a reflexive Banach space and $T : X \to 2^{X^*}$ maximal monotone. Then, for any $\lambda > 0$, $\mathscr{R}(J + \lambda T) = X^*$ and $(J + \lambda T)^{-1}$ is single-valued demicontinuous maximal monotone operator from X^* to X.*

Proof If $0 \in \mathscr{D}(T)$, then the first part of the result follows from Theorem 4.29. If $0 \in \mathscr{D}(T)$, let $a \in \mathscr{D}(T)$. Now define operators T_1 and T_2 as

$$T_1 x = \lambda T(x + a) \quad \text{and} \quad T_2 x = J(x + a).$$

Now $0 \in \mathscr{D}(T_1)$ and T_1, T_2 satisfy the hypothesis of Theorem 4.29 and hence $\mathscr{R}(T_1 + T_2) = X^*$, and this in turn implies that $\mathscr{R}(\lambda T + J) = X^*$.

Also by Theorem 4.30, $(\lambda T + J)$ is maximal monotone and so is $(\lambda T + J)^{-1}$. Since $(\lambda T + J)^{-1}$ is single valued with $\mathscr{D}(\lambda T + J)^{-1} = X^*$, it follows by Corollary 4.4 that it is a demicontinuous maximal monotone operator. ∎

Theorem 4.33 *Let X be a reflexive Banach space and $T : X \to 2^{X^*}$ abe monotone. In order that T is maximal monotone, it is necessary and sufficient that $\mathscr{R}(J + T) = X^*$.*

Proof By Zorn's lemma, there exists a maximal monotone operator T' such that $T \subset T'$. By the previous theorem for every $u^* \in X^*$, there exists $x \in X$ and $x^* \in X^*$, such that $x^* \in T'(x)$ and $x^* + Jx = u^*$. So $T = T'$ iff every $u^* \in X^*$ can be written as $x^* + Jx$ for some $x \in X$, $x^* \in Tx$, that is iff $\mathscr{R}(T + J) = X^*$. ∎

We terminate this section by stating a surjectivity result for J-monotone operators from a Banach space X into itself. This result is due to Browder and de Figueredo [116].

Theorem 4.34 *Let X be a reflexive Banach space and $T : X \to X$ a J-monotone operator. Further, let there exists $\rho > 0$ such that $(Jx, Tx) > 0$ for $\|x\| \geqslant \rho$. Then, $\mathscr{R}(T) = X$.*

4.4 Solutions of Operator Equations Using Constructive Method

In the previous section, we discussed a number of general existence theorems for operator equations involving monotone operators. These results do not give a constructive procedure for the generation of the solutions which are proved to exist. In this section, we intend to dwell on the constructive solutions. We shall mainly be concerned with approximation schemes developed by Petryshyn more than decade ago and some recent results of Bruck, Jr.

In the following, unless otherwise stated, X is a real reflexive Banach space

Definition 4.14 A Banach space X is said to posses a property $(\pi)_c$ if there exists a sequence $\{X_n\}$ of finite dimensional subspaces of X, a sequence of linear projections $\{P_n\}$ defined on X and a constant $c > 0$ such that

$$P_n X = X_n, \quad X_n \subset X_{n+1} \quad n = 1, 2, \ldots, \quad \overline{\bigcup_{n=1}^{\infty} X_n} = X,$$

$$\|P_n\| \leqslant c, n = 1, 2, 3, \ldots, \quad P_n P_j = P_j, \quad n \geqslant j.$$

Example 4.10 Every separable Hilbert space has the property $(\pi)_1$.

Example 4.11 Let X be a Banach space with Schauder basis $\{\phi_j\}_{j=1}^{\infty}$. Then,

$$x = \sum_{j=1}^{\infty} \alpha_j \phi_j, \ \alpha_j = (\phi_j^*, x)$$

where $\phi_j^*, \in X^*$ and $(\phi_j^*, \phi_i) = \delta_{ij}$. Define X_n as

$$X_n = [\phi_1, \phi_2, \ldots, \phi_n] \text{ and } P_n : X \to X_n \text{ as}$$

$$P_n x = \sum_{j=1}^{n} \alpha_j \phi_j^* P_n^2 = P \text{ and } \|P_n\| \leqslant 1.$$

Thus, X has the property $(\pi)_1$.

We shall be interested in the constructive results regarding the existence and uniqueness of the 'exact' equation

$$Tx = y, \quad y \in X \tag{4.21}$$

as a strong or weak limit of solutions $x_n \in X_n$ of the 'approximate' equations

$$T_n x_n = P_n y. \tag{4.22}$$

Here, F is an operator from X into itself and $F_n = P_n F P_n$.

Definition 4.15 Equation (4.21) is said to be projectionally and strongly solvable or PS-solvable if there exists an integer $N > 0$ such that for each $n \geqslant N$ and given $y \in X$, (4.22) has a unique solution $x_n \in X_n$ such that $x_n \to x$ in X and x is the unique solution of (4.21).

We now state and prove an approximate solvability result obtained by Petryshyn [477].

Theorem 4.35 *Suppose that there exist a continuous monotonically increasing real function $\alpha(r)$ defined for $r \geqslant 0$ with $\alpha(0) = 0$ and $\lim_{r \to \infty} \alpha(r) = \infty$ and an integer $N > 0$ such that the following conditions are satisfied:*

(a) T_n is continuous in X_n for each $n \geqslant N$ and $T_n x \to GTx$ for each x in X;
(b) for each $n \geqslant N$ and all x, $ly \in X_n$, $\|T_n x - T_n y\| \geqslant \alpha(\|x - y\|)$,

(c) if $\{x_{n_k}\}$ is any arbitrary subsequence of $\{x_n\}$ and $\{x_{n_k}\}$ is a sequence in X with $x_{n_k} \in X_{n_k}$, $x_{n_k} \to x$ and $T_{n_k}x_{n_k} \to y$, then $Tx = y$.

Then, the Eq. (4.21) is PS-solvable.

Proof First of all we show the existence of the approximate solution $x_n \in X_n$ of (4.22). We note that $\mathscr{R}(T_n)$ is closed. For this, consider $\{y_m\} \subset \mathscr{R}(T_n)$ such that $y_m \to y$. Then, $y_m = T_n x_m, x_m \in X_n$ and so (b) gives

$$\|y_l - y_m\| = \|T_n x_l - T_n x_m\| \geqslant \alpha(\|x_l - x_m\|), n \geqslant N.$$

Since $\{y_m\}$ is Cauchy, $\alpha(\|x_l, x_m\|) \to 0$ as $l, m \to \infty$.

Let $\beta(r)$ be the inverse of $\alpha(r)$, then

$$\|x_l - x_m\| \leqslant \beta(\|y_l - y_m\|) \to 0 \text{ as } l, m \to \infty.$$

Since X_n is complete, there exists $x \in X_n$ such that $x_m \to x$ and hence $T_n x_m \to T_n x = y \in \mathscr{R}(T_n)$ and so $\mathscr{R}(T_n)$ is closed.

On the other hand, T_n is clearly one to one and continuous for each $n \geqslant N$ and hence by Brouwer's theorem on invariance of domain, T_n is an open mapping and $\mathscr{R}(T_n)$ is an open set in X_n.

Because X_n is connected and $\mathscr{R}(T_n)$ is a nonempty set in X_n which is both open and closed, it must be the full space X_n. In view of (b), we also get the uniqueness of the solution.

Thus, for each $n \geqslant N$ and each given $y \in X$, there exists a unique solution $x_n \in X_n$ such that $T_n x_n = P_n y$. For such a sequence $\{x_n\}$, we have, in view of (b),

$$\|P_n y\| = \|T_n x_n\| \geqslant \|T_n x_n - T_n(0)\| - \|T_n(0)\|$$
$$\geqslant \alpha(\|x_n\|) - c\|T(0)\|.$$

Hence,

$$\alpha(\|x_n\|) \leqslant \|P_n y\| + c\|T(0)\| \leqslant c(\|y\| + \|T(0)\|),$$

which gives

$$\|x_n\| \leqslant \beta(c\|y\| + c\|T(0)\|) = M.$$

Since X is reflexive and $\{x_n\}$ is a bounded sequence in X, there will exist a subsequence $\{x_{n_k}\}$ of $\{x_n\}$ which converges weakly to $x \in X$. Further, in view of (4.22), $T_{n_k} x_{n_k} \to y$ as $k \to \infty$ and so (c) implies that $Fx = y$ and this solution is unique. Since the solution is unique, we conclude that the selection of subsequence was unnecessary and the entire sequence $\{x_n\}$ converges weakly to x.

We now show that $x_n \to X$. To effect this, consider $\alpha(\|x_n - x\|)$. We have

$$\alpha(\|x_n - x\|) \leqslant \|T_n x_n - T_n x\|$$
$$= \|(P_n y - y) + (Tx - T_n x)\|$$
$$\leqslant \|P_n y - y\| + \|T_n x - Tx\|,$$

which gives

$$\|x_n - x\| \leqslant \beta(\|P_n y - y\| + \|T_n x - Tx\|).$$

Since $P_n y \to y$ and $T_n x \to Tx$, we get the required result.   ∎

From the above result, we now derive an approximate solvability result involving monotone operators on a Hilbert space.

Lemma 4.4 *Let T be a demicontinuous monotone operator on a Hilbert space X. Then, it satisfies condition (c) of the above theorem. Further, if it is continuous, it satisfies condition (a) also.*

Proof Let n be a fixed integer and $x \in X_n$ so that $P_n x = x$. Then, using monotonicity of T, we get

$$(Tx_m - T(P_n x), x_m - P_n x) \geqslant 0. \tag{4.23}$$

Let now $x_m \rightharpoonup x_0$, this gives

$$(T(P_n x), x_m - P_n x) \to (T(P_n x), x_0 - P_n x). \tag{4.24}$$

Since $x_m \in X_m$, $P_m x_m = x_m$ and for $m \geqslant n$, $P_m P_n = P_n$; we get

$$(Tx_m, x_m - P_n x) = (Tx_m, P_m x_m - P_m P_n x)$$
$$= (P_m Tx_m, , x_m - P_n x)$$
$$= (T_m x_m, x_m - P_n x).$$

Now as $x_m \rightharpoonup x_0$ and $T_m x_m \to y$, the above equality gives

$$(T_m x_m, x_m - P_n x) \to (y, x_0 - P_n x),$$

which in turn implies that

$$(Tx_m, x_m - P_n x) \to (y, x_0 - P_n x). \tag{4.25}$$

Combining (4.24) and (4.25), we get

$$(Tx_m - T(P_n x), x_m - P_n x) \to (y - P_n x, x_0 - P_n x) \text{ as } m \to \infty.$$

In view of (4.23), this gives us

$$(y - Tx, x_0 - x) \geqslant 0 \text{ for each } x \in X_n.$$

Since c is arbitrary, we get

$$(y - Tx, x_0 - x) \geqslant 0 \text{ for all } x \in \bigcup_{n=1}^{\infty} X_n.$$

Again, since $\bigcup_{n=1}^{\infty} X_n$ is dense in X and T is demicontinuous, it follows that

$$(y - Tx, x_0 - x) \geqslant 0 \text{ for all } x \in X.$$

Also, F is a demicontinuous monotone operator defined on the whole space, and hence, it is maximal monotone and so we get $Tx_0 = y$.

For the second part of the lemma we first note that $P_n Fx \to Fx$ for all $x \in X$. Choosing $X = X_j$ and noting that $P_j x_j = x_j$ and $P_n P_j = P_j, n \geqslant j$, we get $P_n T P_n x \to Tx$ for all $x \in X_j$. This gives $T_n x \to Tx$ for all $x \in \bigcup_{j=1}^{\infty} X_j$. Now continuity of T and the fact that $\bigcup_{j=1}^{\infty} X_j$ is dense in X proves the second part of the lemma.

From this lemma, we immediately obtain the following approximate solvability result for monotone operators.

Theorem 4.36 *Let T be a continuous and strongly monotone operator from a Hilbert space X into itself, then the Eq. (4.21) is PS-solvable.*

There is a direct generalization of Lemma 4.4 to J-monotone operators defined on a Banach space X which is such that the duality mapping $J : X \to X^*$ is weakly continuous.

Lemma 4.5 *Let X be a Banach space with strictly convex dual X^* and satisfying property $(\pi)_1$. Let $T : X \to X^*$ be J-monotone. If $\{x_n\}$ is a sequence in X with $x_n \in X_n$ such that $x_n \rightharpoonup x_0$ and $P_n T x_n \to y$. Then, $T x_0 = y$ provided that T is demicontinuous and J is continuous or T is continuous.*

This lemma gives us the following theorem, due to Petryshyn [474], for the approximate solvability of operator equations involving J-monotone operators.

Theorem 4.37 *Suppose that X satisfies the conditions of Lemma 4.5. Let T be a continuous mapping of X into itself such that*

$$(J(x - y), Tx - Ty) \geqslant \alpha(\|x - y\|) \|J(x - y)\|, \text{ for } x, y \in X.$$

Then, the Eq. (4.21) is PS-solvable.

There is a direct generalization of the approximate solvability result for operators from a space X into its dual X^*. For this, we need to define first an approximation scheme.

Let X be a Banach space with dual X^* and $\{x_n\} \subset X$ and $\{X_n^*\} \subset X^*$ be sequence of finite dimensional spaces. Let T be an operator from X to X^*, $P_n : X \to X_n$ and $Q_n : X^* \to X_n^*$ be sequence of linear projections. Corresponding to the 'exact' equation

$$Tx = y, \ y \in X^*, \tag{4.26}$$

We define 'approximate' equations

$$T_n x_n = Q_n y \tag{4.27}$$

where

$$T_n = Q_n F P_n.$$

Definition 4.16 A quadruple $\{\{x_n\}, \{x_n^*\}, \{P_n\}, \{Q_n\}\}$ is called an approximation scheme for the Eq. (4.26) if P_n and Q_n are continuous and for each $x \in X, x^* \in X^*$, $P_n x \to x$ and $Q_n x^* \to x^*$.

Definition 4.17 Equation (4.26) is said to be PS-solvable if there exists an integer $N > 0$ such that for $n \geqslant N$, $y \in X^*$, (4.27) has a unique solution x_n and $x_n \to x$ in X and x is the unique solution of (4.26).

We have a theorem similar to Theorem 4.35. We state this without proof (refer Petryshyn [475] for a proof).

Theorem 4.38 *Let X be Banach space with dual X^* and let*

$$\Gamma = \{\{X_n\}, \{X_n^*\}, \{P_n\}\{Q_n\}\}$$

be an approximation scheme for Eq. (4.26). Suppose there exist an $N \geqslant 0$ and continuous function $\alpha(r) : \mathbb{R}^+ \to \mathbb{R}^+$ with $\alpha(0) = 0$ and $\lim_{r \to \infty} \alpha(r) = \infty$ such that T_n is continuous from X_n to X_n and for $n \geqslant N$, $Q_n x \to Qx$ for each $x \in X$ and $\|T_n x - T_n y\| \geqslant \alpha(\|x - y\|)$ for all $x, y \in X_n$.

Then, the Eq. (4.26) is PS-solvable iff the operator T satisfies the condition: If X_m is an arbitrary subsequence of X_n and $\{x_m\}$ is a bounded sequence in X with $x_m \in X_m$ and $T_m x_m \to y$, then there exists a subsequence $\{x_{m_k}\}$ of $\{x_m\}$ and an element $x_0 \in X$ such that

$$x_{m_k} \to x_0 \text{ and } T_{m_k} x_{m_k} \to T x_0 (= y) \text{ as } k \to \infty.$$

As before, as a corollary, we can obtain a theorem similar to Theorem 4.38 for monotone operators defined from X to X^*.

These constructive schemes suffer from one handicap—lack of effective control for the error at each stage of the approximation. We now discuss some recent results of Bruck Jr. which arrest this handicap.

In what follows, F is a multivalued monotone operator on a Hilbert space X with domain $\mathscr{D}(T)$.

Definition 4.18 A single-valued section T_0 of T is defined as

$$T_0 x = T x, \quad x \in \mathscr{D}(T) \text{ and } \{T_0 x\} \text{ is a singleton.}$$

The following theorem, due to Bruck Jr. [122], gives the iterative solution of the equation $y \in x + Tx$ with explicit error estimate.

Theorem 4.39 *Suppose T is a multivalued monotone operator with $\mathscr{D}(T)$ an open subset of a Hilbert space $\mathcal{H}$ and $y \in \mathscr{R}(I + T)$. Then, there exists a neighbourhood $N \subset \mathscr{D}(T)$ of $\bar{x} = (I + T)^{-1} y$ and a real number $\sigma_1 > 0$ such that for any $\sigma \geqslant \sigma_1$, any initial guess $x_1 \in N$ and a single-valued section T_0 of T, the sequence $\{x_n\}$ generated from x_1 by*

$$x_{n+1} = x_n - (n + \sigma)^{-1}(x_n + T_0 x_n - y) \tag{4.28}$$

remains in $\mathscr{D}(T)$ and converges to $\bar{x}$ with estimate $\|x_n - \bar{x}\| = O(n^{-1/2})$. The sequence $\{x_n + T_0 x_n\}$ converges $(C, 1)$ to y.

Proof First, we note that $(I + T)^{-1}$ is single valued and so $\bar{x} = (I + T)^{-1} y$ is uniquely defined. Choose $\bar{u} \in T\bar{x}$ such that $y = \bar{x} + \bar{u}$.

Now, as T is monotone and $\mathscr{D}(T)$ is open, it is locally bounded at $\bar{x}$ (refer Theorem 4.9). So there exists a neighbourhood $N = B(\bar{x}, d)$ such that $N \subset \mathscr{D}(T)$ and $T(N)$ is bounded. Put $\sigma_1 = [\text{diam } T(N)/d]^2$. Then,

$$\sigma_1 > 0 \text{ and diam } T(N) \geqslant d\sigma^{1/2} \text{ if } \sigma \geqslant \sigma_1. \tag{4.29}$$

Put $t_n = (n + \sigma)^{-1}, d_n = (n + \sigma - 1)^{-1/2}$. Let x_1 be an initial guess, then define x_n by (4.28). We will show that $\{x_n\}$ is well defined and satisfies.

$$\|x_n - \bar{x}\| \leqslant d_n d\sigma^{1/2}. \tag{4.30}$$

This is proved by induction. For $n = 1$, x_1 is well defined and

$$\|x_1 - \bar{x}\| < d_1 d\sigma^{1/2} = d$$

(follows from the definition of d_n). Next, we assume that (4.29) is true for a particular value of n, and then,

$$\|x_n - \bar{x}\| \leqslant d_n d\sigma^{1/2} \leqslant d_1 d\sigma^{1/2} = d$$

and so $x_n \in N \subset \mathscr{D}(T) = \mathscr{D}(T_0)$. This implies that x_{n+1} is well defined and also since $y = \bar{x} + \bar{u}$, (4.28) gives

$$x_{n+1} - \bar{x} = (1 - t_n)(x_n - \bar{x}) - t_n(T_0 x_n - \bar{u}).$$

Therefore,

$$\|x_{n+1} - \bar{x}\|^2 = (1 - t_n)^2 \|x_n - \bar{x}\|^2 - 2t_n(1 - t_n)(T_0 x_n - \bar{u}, x_n - \bar{x}) + t_n^2 \|T_0 x_n - \bar{u}\|^2$$

$$\leqslant (1 - t_n)^2 \|x_n - \bar{x}\|^2 + t_n^2 \|T_0 x_n - \bar{u}\|^2 \tag{4.31}$$

using the fact that $T_0 x_n \in T x_n, u \in T x, T$ is monotone and $t_n(1 - t_n) > 0$. But we also have $\|T_0 x_n - \bar{u}\| \leqslant$ diam $T(N)$, since $T_0 x_n$ and $\bar{u} \in T(N)$, and hence, by combining (4.29) and (4.31) and the induction hypothesis $\|x_n - \bar{x}\| \leqslant d_n d\sigma^{1/2}$, we get

$$\|x_{n+1} - \bar{x}\|^2 \leqslant [(1 - t_n)^2 d_n^2 + t_n^2] d^2 \sigma$$

$$= d_{n+1}^2 d^2 \sigma,$$

which completes the induction process. Since $d_n = 0(n^{-1/2})$, we have the error estimate.

For the second part of the theorem, we first observe that we have the identity

$$x_{n+1} = \frac{\sigma}{n + \sigma} x_1 - \frac{1}{n + \sigma} \sum_{t=1}^{n} (T_0 x_i - y).$$

This gives

$$n^{-1} \sum_{t=1}^{n} T_0 x_i = y - x_{n+1} + \sigma n^{-1} (x_1 - x_{n+1})$$

and this in turn implies that

$(C, 1) \lim\limits_{n \to \infty} T_0 x_n = y - \bar{x}.$

Since $x_n \to \bar{x}$ strongly, also notice that $(C, 1) \lim\limits_{n \to \infty} x_n = \bar{x}$, so finally, we have

$(C, 1) \lim\limits_{n \to \infty} (x_n + T_0 x_n) = y.$

This completes the proof.   ∎

As a corollary, we immediately get the following result for continuous single-valued monotone operators.

Corollary 4.7 *Suppose T is a continuous single-valued monotone operator with open domain and $y \in \mathcal{R}(I + T)$. Then, there exists a neighbourhood $N \subset \mathcal{D}(T)$ of $(I + T)^{-1}y$ such that for any initial guess $x_1 \in N$ the sequence x_n generated as*

$$x_{n+1} = \frac{n}{n + 1} x_n - \frac{1}{n + 1} (T x_n - y)$$

remains in $\mathcal{D}(T)$ and converges to $\bar{x}$ with estimate $\|x_n - \bar{x}\| = O(n^{-1/2})$.

In his subsequent paper [123], Bruck Jr. has generalized the above result. He proves that if T is a maximal strongly monotone operator and x is a given point of a Hilbert space $\mathcal{H}$, then under appropriate conditions on the nonnegative sequences $\{\lambda_n\}$, $\{\theta_n\}$ the iteration

$$x_{n+1} = x_n - \lambda_n(T x_n + \theta_n(x_n - x_0))$$

converges strongly to $T^{-1}(0)$, provided that $\{x_n\}$ remains bounded and certain boundedness conditions are satisfied.

If $\theta_n = 0$, the above iteration scheme reduces to the previous one, where λ_n is of the form $\lambda_n = (cn - \sigma)^{-1}$, c is a constant of strong monotonicity. But these results are local in the sense that the convergence is guaranteed only when the initial value is chosen close enough to the solution.

We now state a global version of this result, which is due to Nevanlinna [422].

Let T be a maximal monotone operator on a Hilbert space $\mathcal{H}$. Nevanlinna has shown that if T is defined on the whole space and is either continuous or grows only linearly at infinity, then we can find sequences $\{\lambda_n\}$, $\{\theta_n\}$ such that if $\{x_n\}$ is defined recursively as

$$x_{n+1} \in x_n - \lambda_n(T x_n + \theta_n x_n), \tag{4.32}$$

then x_n converges strongly to a solution of $T x \ni 0$, if there exists any, otherwise it tends to infinity. Following Nevanlinna, we now precisely define how sequence $\{\lambda_n\}$, $\{\theta_n\}$ are to be selected.

Definition 4.19 Two sequences $\{\lambda_n\}$, $\{\theta_n\}$ of positive real numbers are acceptable paired if $\{\theta_n\}$ is nonincreasing, $\lim_{n\to\infty} \theta_n = 0$, and there exists an increasing sequence $\{N(i)\}$ of integers such that

$$\liminf \theta_{N(i)} \sum_{j=N(i)}^{N(i+1)-1} \lambda_j > 0, \quad \limsup \theta_{N(i)} \sum_{j=N(i)}^{N(i+1)-1} \lambda_j < \infty$$

and

$$\lim(\theta_{N(i)} - \theta_{N(i+1)}) \sum_{j=N(i)}^{N(i+1)-1} \lambda_j = 0.$$

An example of acceptably paired sequences is

$$\lambda_n = n^{-1}, \quad \theta_n = (\log\log n)^{-1}, \quad N(i) = i^i.$$

We state Nevanlinna's result, without proof.

Theorem 4.40 *Let T be a maximal monotone operator in a Hilbert space $\mathcal{H}$ with $\mathcal{D}(T) = \mathcal{H}$. Let it satisfies a growth condition of the type*

$$\|y\| \leqslant c(1 + \|x\|) \text{ for all } x, y \in Tx. \tag{4.33}$$

Assume that $\{\lambda_n\} \in l_2$ and that $\{\lambda_n\}, \{\theta_n\}$ are acceptable paired and $\{x_n\}$ satisfies (4.32) with any initial value $x_0 \in \mathcal{H}$. If $T^{-1}(0) \neq \emptyset$, then x_n converges strongly to x where x is the element in T^{-1} with minimum norm and if $T^{-1}(0) = \emptyset$, then $\|x_n\| \to \infty$ as $n \to \infty$ in such a way that $\frac{\theta_n x_n}{n}$ converges to $-a_0$ where a_0 is the element in $\overline{\mathcal{R}(T)}$ with minimum norm.

If T is strongly monotone, that is, $T = bI + T_0$, where T_0 is maximal monotone, we get the following simple result.

Theorem 4.41 *Let T_0 be a maximal monotone operator with $\mathcal{D}(T_0) = H$ and assume that T_0 satisfies the growth condition (4.33). Assume that $b > 0$ and $\{\lambda_n\} \in l_2 - l_1$. Then, $\{x_n\}$ defined by*

$$x_{n+1} \in x_n - \lambda_n(T_0 x_n + bx_n),$$

converges to the unique solution of $bx + T_0 x = 0$.

Remark 4.7 This theorem generalizes the result of Zarantonello [617], who proved the existence of solution of the equation $bx + T_0 x = 0$ with Lipschitz continuity on T_0 in addition to monotonicity.

Observation

- Browder in [93] obtained an error estimate for an iteration procedure for J-monotone operators defined on a Banach space with uniformly convex dual X^*. The duality mapping in such spaces is uniformly continuous from bounded subsets of X to X^*. In particular, let $R > 0$ be given, then there exists a nonnegative real valued continuous function w on $B(0, 2R)$ with $w(0) = 0$ such that

$$\|Ju - Jv\| \leqslant w(\|u - v\|) \text{ for each } u, v \text{ in } B(0, 2R).$$

Theorem 4.42 *Let X be a Banach space with a uniformly convex dual X^*, T be a J-monotone, bounded, continuous mapping from a ball $B(0, R)$ to X such that*

(i) there exists R_1 with $0 < R_1 < R$ such that for all u in X with $R_1 \leqslant \|u\| \leqslant R$, $(Ju, Tu) \geqslant 0$ and

(ii) there exists $\delta > 0$ and a continuous function $q : \mathbb{R}^+ \to \mathbb{R}^+$ with $q(0) = 0$ such that if u and v are pair of elements with $\|Tu\| \leqslant \delta$, for $v_0 \in B(0, R_1)$, define a sequence $\{V_{rn}\} = \{V_k\}$ as

$$V_k = (1 - (n + k)^{-1})V_{k_{-1}} - n^{-1}T(V_{k_{-1}}), 1 \leqslant k \leqslant rn,$$

where n and r are so chosen that

$$n^{-1}(M + R) \leqslant R - R_1,$$

$$(M + R)(r + 1)w(n^{-1}(M + R)) < (R - n^{-1}(M + R))^2.$$

Then, $\{V_{rn}\}$ converges to the unique solution u_0 of the equation $T(u_0) = 0$ as $r \to \infty$ and $n \to \infty$ and $(r + 1)w(n^{-1}(M + R)) \to 0$. Further, we have the error estimate

$$\|V_{rn} - u_0\| \leqslant [k_1(r + 1)w(r^{-1}(M + R)) + k_2 n^{-1}]^{1/2} + q(k_3 r^{-1}(\log r + 1))$$

for suitable constants k_1, k_2 and k_3.

4.5   Subdifferential and Monotonicity

In this section, we shall be concerned with the monotone property of the derivative of a convex functional and its subdifferential. As a corollary of this property, we derive the maximal monotonicity of the duality mapping $J : X \to 2^{X^*}$.

In the following, f will denote a convex function $f : X \to (-\infty, \infty]$ and ∇f its gradient, and X is a Banach space.

Theorem 4.43 *Let f be a proper convex function defined on X. If f is differentiable, then ∇f is monotone.*

Proof Let $x_1, x_2 \in X$. Let ϕ be the convex function defined by

$$\phi(t) = f(x_1 + t(x_2 - x_1)).$$

Then,

$$\phi'(t) = (\nabla f(x_1 + t(x_2 - x_1)), x_2 - x_1).$$

This gives

$$(\nabla f(x_2) - \nabla f(x_1), x_2 - x_1) = \phi'(1) - \phi'(0) \geqslant 0,$$

thereby proving the monotonicity of ∇f.

Theorem 4.44 *Let f be a proper differentiable function defined on X. If ∇f is monotone, then f is convex.*

Proof We have

$$[\lambda f(x) + (1 - \lambda)f(y)] - f(\lambda x + (1 - \lambda)y)$$
$$= \lambda[f(x) - f(\lambda x + (1 - \lambda)y)] + (1 - \lambda)[f(y) - f(\lambda x + (1 - \lambda)y)]$$
$$= \lambda(\nabla f(\lambda x + (1 - \lambda)y + t_1(1 - \lambda)(y - x)),$$
$$(1 - \lambda)(y - x)) + (1 - \lambda)(\nabla f(\lambda x + (1 - \lambda)y + t_2(1 - \lambda)(x - y)), \lambda(x - y)),$$

for some $0 < t_1, t_2 < 1$, in view of Lagrange theorem for functionals.

Rearranging the above equality, we get

$$[\lambda f(x) + (1 - \lambda)f(y)] - [f(\lambda x + (1 - \lambda)y)]$$

$$= \frac{\lambda}{t_1 + t_2}[(\nabla f(z_1) - \nabla f(z_2), z_1 - z_2)],$$

where

$$z_1 = \lambda x + (1 - \lambda)y + t_1(1 - \lambda)(y - x), z_2 = \lambda x + (1 - \lambda)y + t_2(1 - \lambda)(y - x).$$

Using monotonicity of ∇f, we get our result.   ■

Combining the above two theorems, we get the following.

Theorem 4.45 *A differentiable function f on X with $\mathcal{D}(f) = X$ is convex iff ∇f is monotone.*

There is a generalization of the above theorem to subdifferentials. In this direction, we have the following result due to Rockfeller [529].

Theorem 4.46 *Let X be a Banach space and f be a lower semicontinuous proper convex function on X. Then, ∂f is a maximal monotone operator on X.*

Proof Let $y \in \partial f(x_1)$ and $y_2 \in \partial f(x_2)$. Then,

$$f(x) - f(x_1) \geqslant (y_1, x - x_1) \text{ and } f(x) - f(x_2) \geqslant (y_2, x - x_2)$$

for all $x \in X$.

If we take $x = x_2$ in the first inequality and $x = x_1$ in the second inequality and add, we get the monotonicity of ∂f. Now suppose that $x_0 \in X$ and $y_0 \in X^*$ have the property that $(y - y_0, x - x_0) \geqslant 0$ whenever $y \in \partial f(x)$. We must show that $y_0 \in \partial(x_0)$. Replacing f by $h(x) = f(x + x_0) - (y_0, x)$, if necessary, we may assume that $x_0 = 0$ and $y_0 = 0$. So it suffices to show that if $0 \in \partial f(0)$, then there exist some $\overline{x}$ and $\overline{y}$ such that $\overline{y} \in \partial f(\overline{x})$ and $(\overline{y}, \overline{x}) < 0$.

Because $0 \in \partial f(0)$, it follows by definition that $f(0)$ is not the minimum of f on X. Thus, there exists some x_0 with $f(0) \geqslant f(x_0)$. Let $Q(\lambda) = f(\lambda x_0)$ for all $\lambda \in \mathbb{R}$. Then, Q is a lower semicontinuous proper convex function on a real line and $Q(0) > Q(1)$. Hence, by the theory of one-dimensional convex functions (refer Rockfeller [529]), it follows that there exists some $\lambda_0, 0 < \lambda \leqslant 1$ such that $Q(\lambda_0) < \infty, Q'(\lambda_0) < 0$, where Q' is the left hand derivative of Q. In terms of the derivative df of f, it implies that

$$f(\lambda_0 x_0) < \infty, \ -(df(\lambda_0 x_0), -x_0) < 0.$$

Putting $x = \lambda_0 x_0$, we get

$$f(x) < \infty, \ (df(x), -x_0) > 0.$$

Let $\varepsilon > 0$, then by Theorem 2 of Rockfeller [529] and the above inequality, there exists some x^* with $x^* \in \partial_\varepsilon f(x)$ and $(x^*, -x) > 0$. Here, ∂_ε is the approximate subdifferential defined as

$$\partial_\varepsilon f(x) = \{x^* \in X^* : f(y) \geqslant [f(x) - \varepsilon] + (x^*, y - x) \text{ for all } y \in X\}.$$

Since $\partial_\varepsilon f$ approximates ∂f (refer Brøsted and Rockfeller [85]), it follows that there exists $\overline{x}$ and $\overline{y}$ such that

$$\|\overline{x} - x\| \leqslant \varepsilon^{1/2}, \ \|\overline{y} - x^*\| \leqslant \varepsilon^{1/2}, \ \overline{y} \in \partial f(\overline{x}).$$

Since $(x^*, x) < 0$, we also have $(\overline{y}, \overline{x}) < 0$ when ε is sufficiently small. This proves the theorem. ∎

Since the duality mapping J is the subdifferential of the norm functional $j(x) = \frac{\|x\|^2}{2}$ with $\mathscr{D}(j) = X$, we get the following monotonicity result for J.

Theorem 4.47 *The duality mapping $J : X \to 2^{X^*}$ is a maximal monotone operator.*

We can now combine the above theorem with the results of Sect. 2.1 and get the following theorem which completely describes the properties of the duality mapping J.

Theorem 4.48 *The duality mapping $J : X \to 2^{X^*}$ is a maximal monotone operator with $\mathscr{D}(J) = X$. For each $x \in X$, Jx is closed convex set. If X is reflexive with X and X^* strictly convex, then J is a single valued, bounded, continuous, strictly monotone and coercive mapping.*

We now examine the converse of Theorem 4.46. Does every monotone operator arise from a convex function? Answer is no. For if A is positive semidefinite operator on a Hilbert space X and $f(x)$ is a quadratic functional (Ax, x), then $f(x)$ is convex iff A is self-adjoint.

Definition 4.20 $T : X \to 2^{X^*}$ is said to be cyclically monotone if for every cyclic sequence $x_0, x_1, \ldots, x_n \in \mathscr{D}(T)$ and for every $y_1 \in Tx_i, i = 1, 2, \ldots, n$, we have

$$\sum_{i=1}^{n} (y_1, x_1 - x_{i-1}) \geqslant 0.$$

It is clear that every cyclically monotone operator is monotone. In the one-dimensional cases, the converse is also true. But in general, every monotone one-dimensional operator need not be cyclically monotone.

Example 4.12 Let f be a proper convex function on a Hilbert space $\mathcal{H}$. Then, ∂f is a cyclically monotone operator. For this, consider $x_0, x_1, \ldots, x_n = x_0 \in \mathcal{H}$ and $y_1, 2, \ldots, n$. Then, the definition of ∂f implies that

$$f(x_i) - f(x_{i-1}) \leqslant (y_i, x_i - x_{i-1}), \quad i = 1, 2, \ldots, n.$$

By adding, we get

$$\sum_{i=1}^{n}(y_1, x_1 - x_{i-1}) \geqslant 0,$$

which proves the result.

Observation

- Every monotone mapping $T : \mathbb{R} \to 2^{\mathbb{R}}$ is cyclically monotone.
- Any maximal monotone map $T : \mathbb{R} \to 2^{\mathbb{R}}$ is the subdifferential of a proper convex l.s.c. function.

We now show that if T is a cyclically monotone operator, then T can be embedded into a subdifferential of some proper convex function on X. We state and prove the theorem due to Rockfeller [529].

Theorem 4.49 *Let $T : X \to 2^{X^*}$ be a cyclically monotone operator. Then, there exists a proper convex function f on X such that $T \subset \partial f$.*

Proof Consider $x_0 \in \mathscr{D}(T)$ and $y_0 \in Tx_0$. For each $x \in X$, let

$$f(x) = \sup\{(y_n, x - x_n) + \cdots + (y_0, x_1 - x_0)\}$$

where $y_i \in T(x_i), i = 1, 2, \ldots, n$ and the supremum is taken over all possible finite sets of such pairs (y_i, x_i). We will show that f is a proper convex function with $F \subset \partial f$. We first note that f is a supremum of a nonempty collection of linear functions one for each choice of $(y_1, x_1), (y_2, x_2), \ldots, (y_n, x_n)$. Hence, $f(x) > -\infty$ for all x and also convexity condition is satisfied. Furthermore, $f(x_0) = 0$ because T is cyclically monotone. Hence, f is a proper convex function.

Now choose any $\overline{x}$ and $\overline{y}$ with $\overline{y} \in T(\overline{x})$. We will show that $\overline{y} \in \partial f(\overline{x})$. It is enough to show that for each $\alpha < f(\overline{x})$, we have

$$f(x) \geqslant \alpha + (\overline{y}, x - \overline{x}) \text{ for all } x. \tag{4.34}$$

Given $\alpha < f(\overline{x})$, we can choose pairs (y_i, x_i) with $y_i \in T(x_i), i = 1, 2, \ldots, k$ and

$$\alpha < (y_k, \overline{x} - x_k) + \cdots + (y_0, x_1 - x_0). \tag{4.35}$$

Let $x_{k+1} = \overline{x}, y_{k+1} = \overline{y}$. Then,

$$f(x) \geqslant (y_{k+1}, x - x_{k+1}) + (y_k, x_{k+1} - x_k) + \cdots + (y_0, x_1 - x_0).$$

Using (4.35) and the above inequality, we get (4.34) and hence the theorem. ∎

We have the following corollary from the above theorem.

Corollary 4.8 *If T is a maximal cyclically monotone operator, then $T = \partial g$ for some lower semicontinuous proper convex function g.*

Proof From the above theorem, it is clear that there exists some proper convex function f such that $T = \partial f$. Define the function g as

$$g(x) = \liminf_{y \to x} f(y) \text{ for all } x.$$

g is lower semicontinuous proper convex function with $\partial g(x) \geqslant \partial f(x)$ for all x. Since $T = \partial f$ is maximal, it follows that $T = \partial f = \partial g$. ∎

As a matter of fact, we have a stronger theorem due to Rockfeller, which completely characterizes the subdifferential of a lower semicontinuous function. We state this without proof.

Theorem 4.50 *Let X be a Banach space, and let T be a multivalued operator from X to 2^{X^*}. In order that there exists a lower semicontinuous proper convex function f on X such that $\partial f = T$, it is necessary and sufficient that T be a maximal cyclically monotone operator. Moreover, the function f is unique up to an arbitrary constant.*

We finally give an easy application of Theorem 4.38 to obtain the solvability of the operator equation $x + Tx = y$.

Theorem 4.51 *Let $T : X \to X^*$ be the Gâteaux derivative of some lower semicontinuous convex function f with $\mathscr{D}(T) = X$. Then, the operator equation $x + Tx = y$ has a unique solution x, and this solution depends continuously on y.*

Proof By the previous theorem, $T = \nabla f$ is maximal monotone. Now the result follows by a surjectivity theorem of Sect. 4.3. ∎

The following proposition gives some basic properties of duality mapping:

Proposition 4.3 *Let X be a real Banach space. For $1 < p < \infty$, the duality mapping $J_p : X \to 2^{X^*}$ has the following properties:*

(1) *$J_p(x) \neq \phi$ for all $x \in X$ and $\mathscr{D}(J_p)(:$ the domain of $J_p) = X$,*

(2) *$J_p(x) = \| x \|^{p-2} \cdot J_2 x$ for all $x \in X (x \neq 0)$,*

(3) *$J_p(\alpha x) = \alpha^{p-1} \cdot J_2 x$ for all $\alpha \in [0, \infty)$,*

(4) *$J_p(-x) = -J_p(x)$,*

(5) *$\| x \|^p - \| y \|^p \geq p\langle x - y, j \rangle$ for all $x, y \in X$ and $j \in J_p y$;*

(6) *if X is smooth, then J_p is norm-to-weak* continuous;*

(7) *if X is uniformly smooth, then J_p is uniformly norm-to-norm continuous on each bounded subset of X,*

(8) J_p is bounded; i.e., for every bounded subset $A \subset X$, $J_p(A)$ is a bounded subset in X^*,

(9) J_p can be equivalently defined as the subdifferential of the functional $\psi(x) = p^{-1} \cdot \| x \|^p$ (Asplund [24]); i.e.,

$$J_p(x) = \partial\psi(x) = \{f \in X^* : \psi(y) - \psi(x) \geqslant \langle y - x, f\rangle \forall \, y \in X\},$$

(10) X is a uniformly smooth Banach space (equivalently, X^* is a uniformly convex Banach space) if and only if J_p is single valued and uniformly continuous on any bounded subset of X (Xu and Roach [611], Browder [94]).

Proposition 4.4 *Let X be a real Banach space and $J_p : E \to 2^{X^*}$, $1 < p < \infty$, be a duality mapping. Then, for any given $x, y \in X$, we have*

$$\|x + y\|^p \leqslant \|x\|^p + p\langle y, j_p\rangle$$

for all $j_p \in J_p(x + y)$.

Proof From Proposition 4.3(9), it follows that $J_p(x) = \partial\psi(x)$ (subdifferential of the functional $\psi(x)$), where $\psi(x)) = p^{-1} \cdot \| x \|^p$. Also, the definition of subdifferential of ψ yields

$$\psi(x) - \psi(x + y) \geqslant \langle x - (x + y), j_p\rangle$$

for all $j_p \in J_p(x + y)$. Now substituting $\psi(x)$ by $p^{-1} \cdot \| x \|^p$, we have

$$\| x + y \|^p \; \leqslant \; \| x \|^p + p\langle y, j_p\rangle$$

for all $j_p \in J_p(x + y)$. This completes the proof. ∎

Remark 4.8 If X is a uniformly smooth Banach space, it follows from Proposition 4.3(10) that $J_p, 1 < p < \infty$, is a single-valued mapping. We now define functions $\Psi, \psi : X \times X \to \mathbb{R}$ by

$$\Psi(x, y) = \|x\|^p - p\langle x - y, J_p(y)\rangle - \|y\|^p$$
$$\Psi(x, y) = \psi(x, y) + \| x - y \| \tag{4.36}$$

for all $x, y \in X$.

It is obvious from the definition of Ψ and Proposition 4.3(5) that

$$\Psi(x, y) \geqslant 0 \tag{4.37}$$

for all $x, y \in X$.

Also, we see that

$$\Psi(x, y) = \| x \|^p - p\langle x, J_p(y)\rangle + (p-1) \| y \|^p$$
$$\geq \| x \|^p - p \| x \| \| J_p(y) \| + (p-1) \| y \|^p$$
$$= \| x \|^p - p \| x \| \| y \|^{p-1} + (p-1) \| y \|^p . \qquad (4.38)$$

In particular, for $p = 2$, we have $\Psi(x, y) \geq (\|x\| - \|y\|)^2$. Further, we can show the following two propositions.

Proposition 4.5 *Let X be a smooth Banach space, and let $\{y_n\}, \{z_n\}$ be two sequences of X. If $\Psi(y_n, z_n) \to 0$ and if either $\{z_n\}$ or $\{y_n\}$ is bounded, then $\{y_n\}$ or $\{z_n\}$ is also bounded and $y_n - z_n \to 0$.*

Proof It follows from $\Psi(y_n, z_n) \to 0$ that $\psi(y_n, z_n) \to 0$ and $|\|y_n\| - \|z_n\|| \leq \|y_n - z_n\| \to 0$ because of (4.36) and (4.37). Therefore, if $\{z_n\}$ is bounded, then $\{y_n\}$ (and also if $\{y_n\}$ is bounded, then $\{z_n\}$) is also bounded and $y_n - z_n \to 0$. ∎

Proposition 4.6 *Let X be a reflexive, strictly convex and smooth Banach space, let C be a nonempty closed convex subset of X, and let $x \in X$. Then, there exists a unique element $x_0 \in C$ such that*

$$\Psi(x_0, x) = \inf\{\Psi(z, x) : z \in C\}. \qquad (4.39)$$

Proof Since X is reflexive and $\|z_n\| \to \infty$ implies $\Psi(z_n, x) \to \infty$, there exists $x_0 \in C$ such that $\Psi(x_o, x) = \inf\{\Psi(z, x) : z \in C\}$. Since X is strictly convex, $\|.\|^p$ is a strictly convex function, that is, $\|\lambda x_1 + (1-\lambda)x_2\|^p < \lambda\|x_1\|^p + (1-\lambda)\|x_2\|^p$ for all $x_1, x_2 \in X$ with $x_1 \neq x_2$, $1 \leq p < \infty$ and $\lambda \in (0, 1)$. Then, the function $\Psi(., y)$ is also strictly convex. Therefore, $x_0 \in C$ is unique. ∎

For each nonempty closed convex subset C of a reflexive, convex and smooth Banach space X, we define the mapping R_C of X onto C by $R_C x = x_0$, where x_0 is defined by (4.39). For the case $p = 2$, it is easy to see that the mapping is coincident with the metric projection in the setting of Hilbert space,. In our discussion, instead of the metric projection, we make use of the mapping R_C. Finally, we shall prove two results concerning Proposition 4.6 and the mapping R_C. The first one is the usual analogue of a characterization of the metric projection in a Hilbert space.

Proposition 4.7 *Let X be a smooth Banach space, let C be a convex subset of E, let $x \in E$ and let $x_0 \in C$. Then,*

$$\Psi(x_0, x) = \inf\{\Psi(z, x) : z \in C\} \qquad (4.40)$$

if and only if

$$\langle z - x_0, J_p(x_0) - J_p(x)\rangle \geq 0 \ \forall z \in C. \qquad (4.41)$$

Proof First, we shall show that (4.40) $\Rightarrow$ (4.41). Let $z \in C$, and let $\lambda \in (0, 1)$. It follows from $\Psi(x_0, x) \leqslant \Psi((1 - \lambda)x_0 + \lambda z, x)$ that

$$
\begin{aligned}
0 \leqslant\ & \|(1 - \lambda)x_0 + \lambda x\|^p - p\langle(1 - \lambda)x_0 + \lambda z - x, J_p(x)\rangle \\
& - \|x\|^p - \|x_0\|^p + p\langle x_0 - x, J_p(x)\rangle + \|x\|^p \\
=\ & \|(1 - \lambda)x_0 + \lambda z\|^p - \|x_0\|^p - p\lambda\langle z - x_0, J_p(x)\rangle \\
\leqslant\ & p\lambda\langle z - x_0, J_p((1 - \lambda)x_0 + \lambda z)\rangle - p\lambda\langle z - x_0, J_p(x)\rangle \\
=\ & p\lambda\langle z - x_0, J_p((1 - \lambda)x_0 + \lambda z) - J_p(x)\rangle,
\end{aligned}
$$

which implies

$$
\langle z - x_0, J_p((1 - \lambda)x_0 + \lambda z) - J_p(x)\rangle \geq 0.
$$

Tending $\lambda \downarrow 0$, since J_p is norm-to-weak* continuous, we obtain

$$
\langle z - x_0, J_p(x_0) - J_p(x)\rangle \geq 0
$$

which shows (4.41).

Next, we shall show that (4.41) $\Rightarrow$ (4.40). For any $z \in C$, we have

$$
\begin{aligned}
\Psi(z, x) - \Psi(x_0, x) =\ & \|z\|^p - p\langle z - x, J_p(x)\rangle - \|x\|^p \\
& - \|x_0\|^p + p\langle x_0 - x, J_p(x)\rangle + \|x\|^p \\
=\ & \|z\|^p - \|x_0\|^p - p\langle z - x_0, J_p(x)\rangle \\
\geqslant\ & p\langle z - x_0, J_p(x_0)\rangle - p\langle z - x_0, J_p(x)\rangle \\
=\ & p\langle z - x_0, J_p(x_0) - J_p(x)\rangle \geq 0
\end{aligned}
$$

which proves (4.40). ∎

Proposition 4.8 *Let X be a reflexive, strictly convex and smooth Banach space, let C be a nonempty closed convex subset of X, let $x \in X$, let $R_C x \in C$, and let $\|y - x\| = \|y - R_C x\| + \|R_C x - x\|$ for all $y \in L[x, R_C x] \cap C$. Then,*

$$
\Psi(y, R_C x) + \Psi(R_C x, x) \leqslant \Psi(y, x) \tag{4.42}
$$

for all $y \in L[x, R_C x] \cap C$.

Proof It follows from Proposition 4.7 that

$$
\begin{aligned}
& \Psi(y, x) - \Psi(y, R_C x) - \Psi(R_C x, x) \\
&= \|y\|^p - p\langle y - x, J_p(x)\rangle - \|x\|^p + \|y - x\| - \|y\|^p \\
&\quad + p\langle y - R_C x, J_p(R_C x)\rangle + \|R_C x\|^p - \|y - R_C x\| - \|R_C x\|^p \\
&\quad + p\langle R_C x - x, J_p(x)\rangle + \|x\|^p - \|R_C x - x\| \\
&= -p\langle y - x, J_p(x)\rangle + p\langle y - R_C x, J_p(R_C x)\rangle + p\langle R_C x - x, J_p(x)\rangle
\end{aligned}
$$

$$= p\langle y - R_C x, J_p(R_C x) - J_p(x)\rangle \geqslant 0$$

for all $y \in L[x, R_C x] \cap C$. This completes the proof.   ∎

4.6 Pseudomonotone Operator and Its Generalizations

In this section, we consider some useful generalizations of monotone operators—
pseudomonotone operators, generalized pseudomonotone operators and mappings
of type S^+ and (M). The original definition of pseudomonotonicity for a mapping
$T : X \to 2^{X^*}$ as given by Brezis [77] involves the following two conditions.

(i) T is finitely continuous(continuous from finite dimensional subspace of X into
 2^{X^*} and for any closed bounded filter $\{u_\alpha\}$ of elements of X such that $\{u_\alpha\}$
 converges to u in X while $\lim_\alpha \sup(Tu_\alpha, u_\alpha - u) \leqslant 0$, the relation $\lim_\alpha \inf$
 $(Tu_\alpha, u_\alpha - u) \geqslant (Tu, u - v)$ holds for each $v \in X$.
(ii) The function $g_v(u) = (Tu, u - v)$ is bounded from below on X, uniformly for
 bounded v in X.

For the case, when X is a reflexive Banach space, we can dispense with filters and
consider only ordinary sequences. For our purpose, we shall consider the definition
of pseudomonotonicity and generalized pseudomonotonicity given by Browder and
Hess [118].

Definition 4.21 Let $T : X \to 2^{X^*}$ be a multivalued operator. Then, T is said to be
pseudomonotone if the following conditions hold:

(a) The set Tx is a nonempty, bounded, closed and convex subset of X for all x in
 X.
(b) F is upper semicontinuous from each finite dimensional subspace Y to X to the
 weak topology on X^*.
(c) If $\{x_n\}$ is a sequence in X converging weakly to x and if $y_n \in Tx_n$ is such that
 $\lim\sup_{n\to\infty} (y_n, x_n - x) \leqslant 0$, then to each element $v \in X$, there exists $w(v) \in Tx$
 such that $\lim\inf_{n\to\infty} (y_n, x_n - v) \geqslant (w(v), x - v)$.

Definition 4.22 $T : X \to 2^{X^*}$ is called generalized pseudomonotone if the following is satisfied:
 For any sequence $\{x_n\}$ in X and a corresponding sequence $\{y_n\}$ in X^* with $y_n \in$
Tx_n, converging weakly to x, $\{y_n\}$ converging weakly to y such that $\lim\sup_{n\to\infty} (y_n, x_n -$
$x) \leqslant 0$, the element y belongs to Tx and (y_n, x_n) converges to (y, x) and $n \to \infty$.

Definition 4.23 $T : X \to 2^{X^*}$ is said to be of type S^+ if each sequence $\{x_n\}$ in
X converging weakly to x in X, $y_n \in Tx_n$ for which $\lim\sup_{n\to\infty} (y_n, x_n - x) \leqslant 0$, x_n
converges to x and $y_n \to y \in Tx$.

We immediately get the following proposition from the Definition 4.22.

Proposition 4.9 *Let* $T : X \to 2^{X^*}$ *be a multivalued operator. Then, T is generalized pseudomonotone iff* T^{-1} *is generalized pseudomonotone from* X^* *to* 2^X.

We have the following two theorems due to Browder and Hess [118], relating generalized pseudomonotonicity with pseudomonotonicity and maximal monotonicity.

Theorem 4.52 *A maximal monotone mapping* $T : X \to 2^{X^*}$ *is generalized pseudomonotone.*

Proof Recall that the mapping $T : X \to 2^{X^*}$ is monotone if for any $x, u \in \mathscr{D}(T)$ and $y \in Tx$ and $w \in Tu$, we have

$$(y - w, x - u) \geqslant 0. \tag{4.43}$$

Let $\{x_n\}$ be a sequence in X converging weakly to $x \in X$, $\{y_n\}$ a sequence in X^* with $y_n \in Tx_n$ and with $y_n \to y$ in X^*. Suppose that $\limsup_{n \to \infty} (y_n, x_n - x) \leqslant 0$. That is, $\limsup_{n \to \infty} (y_n, x_n) \leqslant (y, x)$. Let $[u, w]$ be an arbitrary element of the graph $G(T)$. By the monotonicity of T, $(y_n - w, x_n - u) \geqslant 0$ for each n. Further,

$$(y_n, x_n) = (y_n - w, x_n - u) + (y_n, u) + (w, x_n) - (w, u),$$

where

$$(y_n, u) + (w, x_n) - (w, u) \to (y, u) + (w, x) - (w, u).$$

Hence, $(y, x) \geqslant \limsup_{n \to \infty} (y_n, x_n) \geqslant (y, u) + (w, x) - (w, u)$, which implies that $(y - w, x - u) \geqslant 0$. Since the last relation holds for $[u, w] \in G(T)$ and F is maximal monotone, $y \in Tx$. Consequently, $(y_n - y, x_n - x) \leqslant 0$ for all n. That is,

$$\liminf_{n \to \infty} (y_n, x_n) \geqslant \lim_{n \to \infty} [(y_n, x) + (y, x_n - x)] = (y, x).$$

This implies that $(y_n, x_n) \to (y, x)$, which proves that T is generalized pseudomonotone.

Theorem 4.53 *Let X be a reflexive Banach space and* $T : X \to 2^{X^*}$ *a pseudomonotone mapping. Then, T is generalized pseudomonotone.*

Proof Let $[x_{n,}, y_n]$ be a sequence in $G(T)$ converging weakly to $[x, y]$ in $X \times X^*$ while $\limsup_{n \to \infty} (y_n, x_n - x) \leqslant 0$. Since T is pseudomonotone, for each $v \in X$, there exists $w(v) \in Tx$ such that

$$\liminf_{n \to \infty} (y_n, x_n - v) \geqslant (w(v), x - v).$$

By passing over to an infinite subsequence, we may assume that $(y_n, x_n) \to p$ for some real number p. Then,

$$\limsup_{n \to \infty} (y_{n,}, x_n - x) = p - (y, x) \leqslant 0, \quad \text{that is } p \leqslant (y, x).$$

Furthermore,

$$p - (y, v) \geqslant \liminf_{n \to \infty} (y_n, x_n - v) \geqslant (w(v), x - v).$$

That is,

$$(y, x - v) \geqslant (w(v), x - v), \quad \text{for all } v \in X. \tag{4.44}$$

We assert that $y \in Tx$. By condition(a) in the definition of pseudomonotonicity, Tx is a closed, convex subset of X^*. If $y \in Tx$, there would exist an element $u \in X$ such that

$$(y, u) < \inf_{z \in T(x)} (z, u).$$

Choosing $v = x - u$ in (4.44), we obtain a contradiction. Finally, we note that $\liminf_{n \to \infty} (y_n, x_n - x) \geqslant (y, x - x) = 0$, that is $\liminf_{n \to \infty} (y_n, x_n) \geqslant (y, x)$. Since we already know that $\limsup_{n \to \infty} (y_n, x_n) \leqslant (y, x)$, it follows that $(y_{n,}, x_n) \to (y, x)$. ∎

We have the converse of the above theorem which is as follows.

Theorem 4.54 *Let X be a reflexive Banach space and $T : X \to 2^{X^*}$ a bounded generalized pseudomonotone mapping. Assume that for each $x \in X$, Tx is a nonempty closed convex subset of X^*. Then, T is pseudomonotone.*

We now show that in the definition of pseudomonotonicity conditions (a) and (c) together with a weaker condition (b'), defined as follows, imply the condition (b) in a reflexive Banach space.

(b') T is locally bounded on each finite dimensional subspace X_n of X.

Let $\{x_n\}$ be a sequence in a finite dimensional subspace y of X converging to $x \in Y$. Let $y_n \in Tx_n$, and suppose that for a given weak neighbourhood V of Tx, y_n lies outside V for each n. By the local boundedness of T on Y, $\{y_n\}$ is bounded and since X is reflexive, it converges (by passing over to subsequences of $\{y_n\}$ which we again denote by $\{y_n\}$) to some element $y \in X^*$. Then,

$$\lim_{n \to \infty} (y_n, x_n - x) = (y, x - x) = 0$$

and hence, $(y, x - v) = \lim_{n \to \infty} (y_n, x_n - v) \geqslant (w(v), x - v)$ for each element $v \in X$ with $w(v) \in Tu$. Using the separation argument for convex sets, it follows that all y_n lie outside the neighbourhood V of Fx in the weak topology of X^*.

In view of the local behaviour of a maximal monotone operator in the interior of $D(T)$ (refer Sect. 4.3), we have the following theorem.

Theorem 4.55 *Let* X *be a reflexive Banach space and* $T : X \to 2^{X^*}$ *a maximal monotone operator with* $\mathscr{D}(T) = X$. *Then,* T *is pseudomonotone.*

Proof (a) In view of theorems of Sect. 4.3 on maximal monotone operators, it follows that Tx is a closed and convex subset of X^*. Also, since T is locally bounded at each $x \in \text{int } \mathscr{D}(T) = X$, it implies that Fx is bounded.
(b) This follows from the Theorem 4.14.
(c) Let now $\{x_n\}$ be a sequence in X with $x_n \to x$, and let $y_n \in Tx_n$ such that $\limsup_{n \to \infty} (y_n, x_n - x) \leqslant 0$. Let y denote an arbitrary element of Tx. Then, monotonicity of T gives

$$(y, x_n - x) \leqslant (y_n, x_n - x),$$

where the LHS of the above inequality tends to zero. Hence,

$$(y_n, x_n - x) \to 0 \text{ as } n \to \infty.$$

Let now $[u, w] \in G(T)$ be arbitrary. Since

$$(y_n, x_n - u) = (y_n, x_n - x) + (y_n, x - u),$$

it follow that

$$\liminf_{n \to \infty}(y_n, x_n - u) = \liminf_{n \to \infty}(y_n, x - u).$$

But$(w, x_n - u) \leqslant (y_n, x_n - u)$ with the LHS converging to $(y, x - u)$. Consequently,

$$(w, x - u) \leqslant \liminf_{n \to \infty} (y_n, x - u). \tag{4.45}$$

For a given $v \in X$ and $t \geqslant 0$ set $u_i - x + t(v - x)$. Let $w_i \in Tu_i$, and then, using $[u_i, w_i]$ for $[u, w]$ in (4.45), we get

$$(w_i, x - v) \leqslant \liminf_{n \to \infty}(y_n, x - v).$$

By the local boundedness of F at x, we can assume the existence of sequence $t_k \to 0_+, u_{t_k} \to x$ and $w_{t_k} \rightharpoonup w(v)$. Using the maximal monotonicity of T, we get $w(v) \in Tx$ and further

$$(w(v), x - v) \leqslant \liminf_{n \to \infty} (y_n, x - v)$$
$$\leqslant \liminf_{n \to \infty} (y_n, x_n - v),$$

which proves the theorem. ∎

In the following two theorems, we investigate the sum of two generalized pseudomonotone and pseudomonotone mappings. We state the theorems without proof. But first, we have the following definition.

Definition 4.24 Let $T : X \to 2^{X^*}$ be a multivalued mapping, then T is said to be quasi-bounded if for each $M > 0$ there exists $K(M) > 0$ such that whenever $[u, w] \in G(T)$ and $(w, u) \leqslant M\|u\|$, $\|u\| \leqslant M$, then $\|w\| \leqslant K(M)$.

Theorem 4.56 *Let X be a reflexive Banach space, T_1 and T_2 be generalized pseudomonotone mappings from X into 2^{X^*}. Suppose that T_1 is quasi-bounded and that there exists a continuous function $h : \mathbb{R}^+ \to \mathbb{R}^+$ such that*

$$(w, u) \geqslant -h(\|u\|)\|u\| \text{ for all } [u, w] \in G(T_2).$$

Then, $T_1 + T_2$ is generalized pseudomonotone.

The class of pseudomonotone operators is invariant under addition of operators without any further restriction.

Theorem 4.57 *Let X be a reflexive Banach space, and T_1 and T_2 be pseudomonotone mappings from X into 2^{X^*}. Then, $T_1 + T_2$ is pseudomonotone.*

In the following theorem, we investigate the compact perturbation of monotone operators. But first, we have the following definition.

Definition 4.25 A multivalued operator $T : X \to 2^{X^*}$ is said to be compact if for every bounded sequence $\{x_n\}$ in $\mathscr{D}(T)$ every $y_n \in Tx_n$ has a subsequence which converges in X^*.

Theorem 4.58 *Let X be a reflexive Banach space $T : X \to 2^{X^*}$ a multivalued mapping such that $T = T_1 + T_2$, where T_1 satisfies the condition*

$$(y_1 - y_2, x_1 - x_2) \geqslant \varphi(\|x_1 - x_2\|) \text{ for all } x_1, x_2 \in \mathscr{D}(T_1)$$

and $y_1 \in Tx_1, y_2 \in Tx_2$. where $\varphi : \mathbb{R}^+ \to \mathbb{R}^+$ a continuous increasing function with $\varphi(0) = 0$. If T_2 is compact, then T is of type S^+.

Proof Let $x \rightharpoonup x$ and $w_n \in Tx_n$ and $\limsup\limits_{n\to\infty}(w_n, x_n - x) \leqslant 0$. Then, we have

$$
\begin{aligned}
\varphi(\|x_n - x\|) &\leqslant (y_n - y, x_n - x)\\
&= (w_n - w, x_n - x) - (z_n - z, x_n - z)\\
&= (w_n, x_n - x) - (w, x_n - x) - (z_n, x_n - x) + (z, x_n - x), \quad (4.46)
\end{aligned}
$$

where $w_n = y_n + z_n, w = y + z$; $w_n \in Tx_n, y_n \in T_1x_n, w \in Tx, y \in T_1x,$ and $z \in T_2x$.

Because $x_n \rightharpoonup x$ and T_2 is compact, there exists a subsequence of $\{z_n\}$, which is in turn denoted by $\{z_n\}$ such that $z_n \to z_0$. So we get (from (4.46)) that

$$\limsup\limits_{n\to\infty} \varphi(\|x_n\|) \leqslant \limsup\limits_{n\to\infty}(w_n, x_n - x) \leqslant 0,$$

thereby implying that $x_n \to x$. Since T is demicontinuous, it follows that $w_n \rightharpoonup w \in Tx$. Also, we have $(w_{n_,}, x_n) \to (w, x)$. This proves that T is of type S^+.   ∎

We now state without proof a surjectivity theorem for pseudomonotone operators.

Theorem 4.59 *Let X be a reflexive Banach space and T a pseudomonotone mapping from X into 2^{X^*}. Suppose that T is coercive then $\mathcal{R}(T) = X^*$.*

Because the proof of this theorem is quite involved, we skip it. Interested readers may refer Browder and Hess [118] for its proof.

We now introduce the concept of operators of type (M) for single-valued mappings, which is originally due to Brezis [77].

Definition 4.26 A mapping $T : X \to X^*$ is said to be of type (M) if the following conditions hold:

(a) If a sequence $\{x_n\}$ in X converges weakly to x in X and $\{Tx_n\}$ converges weakly to y in X^* and $\limsup\limits_{n \to \infty}(Tx_{n_,}, x_n) \leqslant (y, x)$, then $Tx = y$.
(b) T is continuous from finite dimensional subspaces of X into X^* endowed with weak* topology.

Now we shall present the notion of pseudomonotone introduced by Karamardian [314].

Definition 4.27 A multivalued operator $T : X \to 2^{X^*}$ with domain $\mathcal{D}(T)$ and range $\mathcal{R}(T)$ is said to be pseudomonotone if $\langle x_1 - x_2, y_2 \rangle \geq 0$ implies $\langle x_1 - x_2, y_1 \rangle \geq 0$ for each $x_i \in \mathcal{D}(T)$ and $y_i \in Tx_i, i = 1, 2$.

It is obvious that each monotone operator is pseudomonotone, but the converse is not true.

In 2012, Pathak and Cho [453] introduced the concept of occasionally pseudomonotone operator as follows:

Definition 4.28 A multivalued operator $T : X \to 2^{X^*}$ is said to be occasionally pseudomonotone if, for any $x_i \in \mathcal{D}(T)$, there exist $y_i \in Tx_i, i = 1, 2$ such that $\langle x_1 - x_2, y_2 \rangle \geq 0$ implies $\langle x_1 - x_2, y_1 \rangle \geq 0$.

It is clear that every monotone operator is pseudomonotone and every pseudomonotone operator is occasionally pseudomonotone, but the converse implications need not be true. To this end, we observe the following examples.

Example 4.13 Let $X = \mathbb{R}^3$ and $T : X \to 2^{X^*}$ be a multivalued operator defined by

$$Tx = \{y = A_r x : r \in \mathbb{R}\} \ \forall \, x \in X$$

where

$$A_r = \begin{pmatrix} 0 & 0 & -1 \\ 0 & -r & 0 \\ 1 & 0 & 0 \end{pmatrix}.$$

Then, for any $x_1 = (x_1^{(1)}, x_2^{(1)}, x_3^{(1)})^T$, $x_2 = (x_1^{(2)}, x_2^{(2)}, x_3^{(2)})^T$ in $\mathbb{R}^3$, $y_1 = A_r x_1$ and $y_2 = A_r x_2$, we have

$$\langle x_1 - x_2, \ y_1 - y_2 \rangle = -r(x_2^{(1)} - x_2^{(2)})^2.$$

Thus, if $r \leqslant 0$, then T is monotone. However, if $r > 0$, then T is neither monotone nor pseudomonotone. Indeed, for $x_1 = (0, 1, 0)$, $y_1 = A_r x_1 = (0, -r, 0)$, $x_2 = (0, 0, 0)$, we have $\langle x_1 - x_2, y_2 \rangle = 0 \geq 0$ but $\langle x_1 - x_2, y_1 \rangle = -r < 0$.

Further, we see that T is occasionally pseudomonotone. To effect this, consider $x_1 = (x_1^{(1)}, x_2^{(1)}, x_3^{(1)})^T$ and $x_2 = (x_1^{(2)}, x_2^{(2)}, x_3^{(2)})^T$ in $\mathbb{R}^3$, $y_i = A_0 x_i$, $i = 1, 2$, and then, we have

$$\langle x_1 - x_2, \ y_2 \rangle = 0 \geq 0 \text{ implies } \langle x_1 - x_2, y_1 \rangle = 0 \geq 0.$$

Example 4.14 The rotation operator on $\mathbb{R}^2$ given by

$$A = \begin{pmatrix} 0 & 1 \\ -1 & 0 \end{pmatrix}$$

is monotone, and hence, it is pseudomonotone. Thus, it follows that A is also occasionally pseudomonotone.

Maximality of a pseudomonotone and occasionally pseudomonotone operators are defined as similar to maximality of monotone operator.

Notice that the class of mappings of type (M) includes in it the class of hemi-continuous monotone mappings and the class of pseudomonotone mappings(refer Brezis [77]).

We now state and prove the following important proposition regarding linear mappings from X to X^*, due to de Figueiredo and Gupta [163].

Proposition 4.10 *Let A be a linear mapping from X to X^*. Then, A is bounded iff A satisfies condition (a) of the above definition.*

Proof If A is bounded, then (a) is trivially satisfied since every continuous linear operator is also weakly continuous. Suppose now A satisfies (a), and let $\{x_n\}$ be any sequence in X such that $x_n \to x$ and $Ax_n \to y$. This gives $(Ax_n, x_n) \to (y, x)$, and so by (a), we have $Ax = y$. Thus, the graph of A is closed, and hence, by closed graph theorem, A is bounded. ∎

For nonlinear mappings, we first observe that compact operator need not be of type (M) as we see in the following example.

Example 4.15 Define $T : \ell_2 \to \ell_2$ as $T(x) = (1 - \|x\|, 0, 0, \ldots)$. Then, T is compact. Let $\{x_n\}$ be the sequence defined as

$$x_n^j = \frac{1}{2}\delta_n^j, \ \delta_n^j = 1, \text{ if } n = j \text{ and } \delta_n^j = 0, \text{ if } n \neq j.$$

Since $Tx_n = (\frac{1}{2}, 0, \ldots)$ for all n, we see that $x_n \rightharpoonup 0$, $Tx_n \to (\frac{1}{2}, 0, \ldots) = g$ and $\limsup_{n\to\infty}(Tx_{n,}, x_n) = 0 = (g, 0)$. Since $T(0) \neq g$, we see that T is not of type (M).

However, compact perturbation of strongly monotone operator (or in particular identity operator on a Hilbert space) is of type (M). This sum of two operators of type (M) need not be of type (M). However, we have the following result.

Proposition 4.11 *Let $T_1 : X \to X^*$ be a mapping of type (M) and $T_2 : X \to X^*$ be a bounded linear monotone mapping or a completely continuous mapping. Then, the mapping $T_1 + T_2$ is also of type (M).*

Proof Let $\{x_n\} \subset X$ be such that $x_n \rightharpoonup x$, $(T_1 + T_2)(x_n) \rightharpoonup y$ and

$$\limsup_{n\to\infty}((T_1 + T_2)(x_n), x_n) \leqslant (y, x).$$

If T_2 is a bounded linear monotone operator, then the functional $f : X \to \mathbb{R}$ defined by $f(x) = (T_2x, x)$ is weakly lower semicontinuous.

We have $(T_2(x_n), x_n - x) \geqslant (T_2x, x_n - x)$.

Hence,

$$\liminf_{n\to\infty}(T_2(x_n), x_n - x) \geqslant \liminf_{n\to\infty}(T_2(x), x_n - x) = 0.$$

Since

$$\liminf_{n\to\infty}(T_2(x_n), x_n - x) = \liminf_{n\to\infty}(T_2(x_n), x_n) - (T_2(x), x),$$

it follows that $\liminf_{n\to\infty} f(x_n) \geqslant f(x)$. That is, f is weakly lower semicontinuous.

Similarly, if T_2 is completely continuous, the mapping f defined as before is weakly lower semicontinuous.

Now

$$\limsup_{n\to\infty}(T_1x_n, x_n) = \limsup_{n\to\infty}(T_1 + T_2)x_{n,}, x_n) - (T_2(x_n), x_n)$$

$$= \limsup_{n\to\infty}((T_1 + T_2)(x_n), x_n) - \liminf_{n\to\infty}(T_2(x_n), x_n)$$

$$\leqslant (y - T_2(x), x).$$

Since T is of type (M), it follows that $T_1x = y - T_2x$ and consequently $(T_1 + T_2)x = y$. This proves the proposition. ∎

We now state and prove surjectivity theorem due to de Figueiredo and Gupta [163]. This is similar to Theorem 4.59.

Theorem 4.60 *Let X be a Banach space and $T : X \to X^*$ be a mapping of type* *(M). If T is coercive, then the range of T is all of X^*.*

Proof Since for any $w \in X^*$, the mapping T_w defined by

$$T_w(x) = T(x) - w$$

is a bounded coercive mapping of type (M) it suffices to show that $0 \in \mathcal{R}(T)$. Let $\Lambda = \{D| \text{ is a finite subset of } X \text{ such that } 0 \in D\}$ and co D denote the convex hull of D.

Since the mapping T is coercive and finitely continuous, it follows by Proposition 7.3 of Browder [94] that there exists a constant $R > 0$ and an element $x_D \in$ co D, for each $D \in \Lambda$, such that $\|x_D\| \leqslant R$ and $(Tx_D, x - x_D) \geqslant 0$. Suppose now,

$$M = \inf_{D \in \Lambda} \{(Tx_D, x_D)\}.$$

Clearly, $M > -\infty$, since the subset $\{x_D|D \in \Lambda\}$ is a bounded subset of X and the mapping T is bounded.

For $D_0 \in \Lambda$, set $V_{D_0} = \bigcup\{x_D|D \in \Lambda, D \supset D_0\}$. We observe the following:

(i) V_{D_0} is contained in the ball of radius R in X for each D_0 in Λ.
(ii) The family $\{\overline{V}_{D_0}|D_0 \in \Lambda\}$(where $\overline{V}_{D_0}$ denotes the weak closure of V_{D_0} in X) is a family of weakly closed subsets of X, since the Banach space X is reflexive. But the family$\{\overline{V}_{D_0}|D_0 \in \Lambda\}$ has finite intersection property, and hence,

$$\bigcap\{\overline{V}_{D_0}|D_0 \in \Lambda\} \neq \emptyset.$$

Let now, $x_0 \in \bigcap\{\overline{V}_{D_0}|D_0 \in \Lambda\}$. We assert that $Tx_0 = 0$. Suppose on the contrary $Tx_0 \neq 0$, and let $x \in X$ such that $(Tx_0, x) < M$. Let $D_1 \in \Lambda$ be such that $x \in D$, and $x_0 \in D_1$. Since $x_0 \in \overline{V}_{D_1}$ it follows by Proposition 7.2 of Browder [94] that there exists an infinite sequence $\{D_i\}_{i=2}^{\infty}, D_i \in \Lambda, D_i \supset D_1$ for each i such that $x_{D_i} \rightharpoonup x_0$. We may assume that there exists an element $y_0 \in X^*$ such that $Tx_{D_i} \rightharpoonup y_0$. It then follows from the relation $(Tx_D, x_{D_i}) \leqslant (y_0, u)$ for every $u \in$ co D_1.

Taking $u = x_0$ and using the condition (a) of the Definition 4.27, we get $Tx_0 = y_0$. Again, taking $u = x \in D_1 \subset \text{co}D_1$, this relation gives $M \leqslant \limsup_{i \to \infty} (Tx_{D_i}, x_{D_i}) \leqslant (y_0, x) = (Tx_0, x) < M$, a contradiction. Thus, $Tx_0 = 0$ and hence the proof of the theorem is complete. ∎

For a separate treatment on the theory of monotone operators, refer Kachurovskiĭ [300], Brezis [76] or Pascali and Sburlan [445].

4.7   Strongly ϕ-Accretive Operator and Surjectivity Theorems

Let X and Y be Banach spaces with Y^* the dual of Y, and let $\phi : X \rightarrow Y^*$ be a mapping satisfying:

 (i) $\phi(X)$ is dense in Y^*
(ii) for each $x \in X$ and each $\alpha \geq 0$, $\|\phi(x)\| \leqslant \|x\|$, $\|\phi(\alpha x)\| = \alpha \|\phi(x)\|$.

The following notion of strongly ϕ-accretive mappings is due to Browder [94].

Definition 4.29 A mapping $P : X \rightarrow Y$ is said to be strongly ϕ-accretive if there exists a constant $c > 0$ such that, for all $u, v \in X$,

$$\langle Pu - Pv, \phi(u - v) \rangle \geq c \|u - v\|^2. \tag{4.47}$$

We note that ϕ-accretive mappings were introduced in an effort to unify the theories for monotone mappings (when $Y = X^*$) and for accretive mappings (when $Y = X$). These mappings have been studied by Browder [94, 107, 108, 114], Kirk [328] and Ray [502].

The following result of Browder [114] is of fundamental importance..

Theorem 4.61 *Let X and Y be Banach spaces and $P : X \rightarrow Y$ a strongly ϕ-accretive mapping. If Y^* is uniformly convex and P is locally Lipschitzian, then $P(X) = Y$.*

A mapping $P : X \rightarrow Y$ is said to be locally strongly ϕ-accretive (cf. [328]) if for each $y \in Y$ and $r > 0$, there exists a constant $c > 0$ such that: if $\|Px - y\| \leqslant r$, then, for all $u \in X$ sufficiently near to x, we have

$$\langle Pu - Px, \phi(u - x) \rangle \geq c \|u - x\|^2. \tag{4.48}$$

Ray [502] extended Browder's theorem [114] by applying a theorem of Ekeland [221] and showed that a localized class of strongly ϕ-accretive mappings must be surjective under appropriate geometric assumptions on Y and continuity assumptions on P. Indeed, he proved the following.

Theorem 4.62 *Let X and Y be Banach spaces and $P : X \rightarrow Y$ a locally Lipschitzian and locally strongly ϕ-accretive mapping. If Y^* is strictly convex and J is continuous, and if $P(X)$ is closed in Y, then $P(X) = Y$.*

Park and Park [443] proved the following surjectivity theorem.

Theorem 4.63 ([443], Theorem 2). *Let X and Y be Banach spaces and $P : X \rightarrow Y$ a locally Lipschitzian and locally strongly ϕ-accretive mapping. If the duality mapping J of Y is strongly upper semicontinuous and $P(X)$ is closed, then $P(X) = Y$.*

Note that if P is strongly ϕ-accretive, then $P(X)$ is closed in Y. Therefore, as a consequence of Theorem 4.63, we have the following:

Corollary 4.9 ([443], Theorem 1). *Let X and Y be Banach spaces and $P : X \to Y$ a locally Lipschitzian and strongly ϕ-accretive mapping. If the duality mapping J of Y is strongly upper semicontinuous and $P(X)$ is closed, then $P(X) = Y$.*

Exercises

4.1 Let F be a nonexpansive mapping from a Hilbert space $\mathcal{H}$ into itself. Show that the operator $T = I - F$ is monotone.

4.2 Let $\mathcal{H}$ be a Hilbert space and $T : \mathscr{D}(T) \subseteq \mathcal{H} \to \mathcal{H}$ be such that

$$\|x - y\| \leqslant \|Tx - Ty\|$$

for all $x, y \in \mathscr{D}(T)$. Show that T is monotone if and only if $I + \lambda T$ is accretive for every $\lambda > 0$.

4.3 Let $\mathcal{H}$ be a Hilbert space. Show that a mapping $T : \mathcal{H} \to 2^{\mathcal{H}}$ is monotone if and only if

$$\|x - y + t(u - v)\| \geq \|x - y\| \ \forall u \in Tx, v \in Ty \text{ and } t \geq 0.$$

4.4 Let X be Banach space and $T : \mathscr{D}(T) \subseteq X \to X^*$ be monotone with $\mathscr{D}(T)$ dense in X. Then, prove that T is hemibounded at $x_0 \in X$.

4.5 Let X be a real reflexive Banach and $T : X \to 2^{X^*}$ a maximal monotone and coercive mapping, then show that T is surjective.

4.6 Let X be a real reflexive Banach and $T : X \to 2^{X^*}$ a maximal monotone mapping, then show that T is surjective if and only if T^{-1} is locally bounded.

4.7 Let X be a real reflexive Banach and $T : X \to 2^{X^*}$ a maximal monotone mapping, then show that T is maximal monotone if and only if $T + J$ is surjective.

4.8 Let X be a real reflexive Banach and $T : X \to 2^{X^*}$ a maximal monotone mapping, then show that $\overline{\mathscr{D}(T)}$ and $\overline{\mathscr{R}(T)}$ are convex.

4.9 Let X be a Banach space. Prove that a mapping $T : X \to 2^{X^*}$ is cyclically maximal monotone if and only if T is the subdifferential of a proper convex l.s.c. function.

4.10 Let $\mathcal{H}$ be a Hilbert space and $A : \mathcal{H} \to \mathcal{H}$ a linear maximal monotone operator. Show that A is cyclically maximal monotone if and only if A is self-adjoint.

4.11 Let X be a reflexive Banach space and A_1, A_2 two maximal monotone mappings. If Int $\mathscr{D}(A_1) \cap \mathscr{D}(A_2) \neq \emptyset$, then show that $A_1 + A_2$ is maximal monotone.

4.12 Let X be a Banach space, $A_1 : X \to 2^{X^*}$ a monotone mapping and $A_2 : X \to X^*$ a monotone operator with $\mathscr{D}(A_2) = X$. If the sum $A = A_1 + A_2$ is maximal monotone, then prove that A_1 is also maximal monotone.

4.13 Let $\mathcal{H}$ be a Hilbert space, $A : \mathcal{H} \to 2^{\mathcal{H}}$ be maximal monotone and $B : \mathcal{H} \to \mathcal{H}$ be monotone with $\mathscr{D}(B)$ closed and such that

$$\|Bx - By\| \leqslant \alpha \|x - y\|, \quad \forall x, y \in \mathscr{D}(B),$$

for some $\alpha \in (0, 1)$, then prove that $A + B$ is maximal monotone.

Chapter 5
Fixed Point Theorems

> The goal of this work, which was submitted to the University of
> Lwow as a dissertation in 1920, is to define a class of abstract
> function spaces, and to prove some theorems about them (e.g. a
> fixed point theorem).
>
> Stefan Banach (1922)
>
> Everything should be made as simple as possible, but not
> simpler.
>
> Albert Einstein

In Sect. 5.1 we discuss the Banach's contraction mapping theorem and some consequences of this theorem. We also deal with contractive mappings considered by
Edelstein [212] and certain generalizations of contraction mapping theorem, mainly
the ones obtained by Boyd and Wongs [75], Kannan [308, 309], Reich [509] and
Husain and Sehgal [283] and others. This section ends with a recent fixed point theorem due to Caristi [128, 129]. Caristi's theorem finds many applications in the field
of nonlinear functional analysis.

Section 5.2 deals with nonexpansive mappings. We present fixed point theorems
concerning nonexpansive mappings obtained by Browder [96], Göhde [251] and
Kirk [326, 327]. We also discuss approximations of fixed points of nonexpansive
mappings and generalized nonexpansive mappings. In this direction, we concentrate
on the work of Dotson, Jr. [198], Bose and Mukherjee [69] and Reich [512]. We also
introduce the concepts of asymptotic center and asymptotic radius, which are due
to Edelstein. In Sect. 5.3, we discuss Brouwer's fixed point theorem and Schauder's
fixed point theorems. We also deal with several important consequences of Schauder's
fixed point theorem.

Section 5.4 deals with fixed point theorems for multifunctions. We give the fixed
point theorems obtained by Himmelberg [273], Sehgal and Morrison [549], Naddler,
Jr. [416] and others. We also present fixed point theorems for H^+-type multivalued

© Springer Nature Singapore Pte Ltd. 2018
H. K. Pathak, *An Introduction to Nonlinear Analysis and Fixed Point Theory*,
https://doi.org/10.1007/978-981-10-8866-7_5

contraction mapping and H^+-type weak contraction mapping obtained by Pathak and Shahzad [465, 466] and is motivated by their applications to differential and integral equations. We conclude this section with a discussion on the multivalued version of Caristi's fixed point theorem.

In Sect. 5.5, we discuss common fixed point theorems for family of commuting mappings. We mainly concentrate on the works of Markov [387], Kakutani [301], de Marr [165] and Browder [106]. Section 5.6 deals with the behaviour of fixed points sets of different types of mappings.

In Sect. 5.7, we discuss some fixed point theorems in ordered Banach spaces. We are mainly concerned on the works of Krasnoselski [345], Amannn [17], Leggett and Williams [363], Williams and Leggett [607], Gatica and Smith [241], Harjani and Sadarangani [267], Dhage [172, 177] and Agarwal and Regan [6].

In Sect. 5.8, we focus our discussion on fixed point theorems in Banach algebra. We dwell on the main results of Krasnoselski [346], Burton [126], Dhage [169, 171, 173], Pathak and Deepmala [456]. Our treatment is brief and is motivated by their applications to differential and integral equations. Finally, we conclude this chapter in Sect. 5.9 by presenting some well-known lattice-theoretic fixed point theorems.

5.1 Fixed Point Theorems

Fixed point- Let $T : X \rightarrow X$ be a mapping of a set X into itself. An element $x \in X$ is said to be a fixed point or invariant point of the mapping T if $Tx = x$.

The knowledge of the existence of fixed points has relevant applications in many branches of analysis, topology, economic, biological science, and many other applied sciences. Let us show for instance the following simple but indicative example.

Example 5.1 Suppose we are given a system of n equations in n unknowns of the form

$$g_j(x_1, x_2, \ldots, x_j, \ldots, x_n) = 0, \quad j = 1, 2, \ldots, n \tag{5.1}$$

where $g_j : \mathbb{R}^n \rightarrow \mathbb{R}$ are continuous real-valued functions of the real variables x_j. Let $h_j(x) = g_j(x) + x_j$, $j = 1, 2, \ldots, n$. Define $h : \mathbb{R}^n \rightarrow \mathbb{R}^n$ by

$$h(x) = (h_1(x), h_2(x), \ldots, h_n(x)) \text{ for all } x \in \mathbb{R}^n.$$

Assume now that h has a fixed point $\bar{x} \in \mathbb{R}^n$. Then it is easily seen that $\bar{x}$ is the solution of the system of Eq. (5.1).

Fixed point theorem- By a fixed point theorem, we shall understand a statement which asserts that under certain conditions (on the mapping T and on the space X) a mapping T of X into itself admits one or more fixed points.

Fixed point space- A space X is called a fixed point space provided every continuous function $T : X \rightarrow X$ has a fixed point.

Example 5.2 (i) Any bounded and closed interval $J = [a, b] \subset \mathbb{R}$ is a fixed point space. Indeed, for any given continuous function $T : J \rightarrow J$, we have $a - T(a) \leqslant 0$

and $b - T(b) \geqslant 0$; the intermediate value theorem ensures that the equation $x - T(x) = 0$ has a solution in J, and therefore T has a fixed point.

(ii) Let $X = \mathbb{R}$ and $T : X \to X$ be a mapping defined by $Tx = x + a$ for some fixed element $a \neq 0$. Then T has no fixed point. Thus, the real line $\mathbb{R}$ is not a fixed point space.

(iii) Let $X = \mathbb{R}$ and $T : X \to X$ be a mapping defined by $Tx = \frac{1}{3}x$. Then $x = 0$ is the only fixed point of T.

(vi) Let $X = \mathbb{R}$ and $T : X \to X$ be a mapping defined by $Tx = x^2$. Then T has 0 and 1 as fixed points.

(v) Let $X = \mathbb{R}$ and $T : X \to X$ be a mapping defined by $Tx = x$. Then T has infinitely many fixed points. Indeed, every point of X is a fixed point of T.

(vi) Let $X = C[0, 1]$, the space of complex-valued continuous functions on the closed interval $[0, 1]$. Let T be defined as

$$(Tx)(t) = x(0) + \int_0^t x(s) \, ds.$$

Any function $x(t) = ke^t$, $t \in [0, 1]$, k a real constant, is a fixed point of T.

Observation

- The property of being a fixed point space is topologically invariant. To see this, let us consider the graph of any continuous function $T : [a, b] \to \mathbb{R}$, for example, the graph of

$$T(x) = \begin{cases} x \sin(1/x) & \text{if } 0 < x \leqslant 1, \\ 0 & \text{if } x = 0, \end{cases}$$

being a homeomorphic to $[a, b]$, is a fixed point space.

- If X is not a fixed point space, it may still be true that every map having some well-defined general property will have a fixed point. For example, the Banach contraction principle asserts that every complete metric space is a fixed point space for contraction maps.

- We have seen that $\mathbb{R}$ is not a fixed point space. However, $\mathbb{R}$ can be made a fixed point space relative to the class of compact maps. For let $T : \mathbb{R} \to \mathbb{R}$ be compact; then $T(\mathbb{R})$ is contained in some finite interval $[a, b]$; in particular, T maps $[a, b]$ into itself, so has a fixed point.

Fixed point theory is broadly divided into three major areas: (i) topological fixed point theory, (ii) metric fixed point theory and (iii) lattice-theoretic fixed point theory.

I. Topological Fixed Point Theory

Historically, the first theorem of this type involves a space X which is a topologically simple subset of $\mathbb{R}^n$ and a mapping of X into itself which is continuous. This is Brouwer's fixed point theorem which asserts the existence of a fixed point whenever X is the unit ball in $\mathbb{R}^n$ and T is continuous. In this theorem X can be replaced by any homeomorph thereof. Such theorems, where the spaces are subsets of $\mathbb{R}^n$, are not of much use in nonlinear functional analysis where one is generally concerned with infinite dimensional subsets of some function spaces. This was first investigated by Birkhoff and Kellogg [59] in 1922. Subsequently, Schauder [542] extended Brouwer's theorem to the case where X is a compact and convex subset of a normed linear space. This theorem was extended to locally convex topological vector space by Tychonoff [593].

II. Metric Fixed Point Theory

The celebrated contraction principle due to S. Banach, which appeared in the literature in 1922, is one of the most important and useful result in the metric fixed point theory that too with an elegant proof. It is, indeed, one of the most widely used fixed point theorems in a large variety of problems in analysis, metric theory, fractal theory, integration theory and many others. An important generalization of Banach contraction theorem, obtained by Boyd and Wong [75], is worth recognizable. Caristi's fixed point theorem [128] is also one of the most important results in metric fixed point theory because of its applications in different areas of science, economics, engineering and social sciences.

III. Lattice-Theoretic Fixed Point Theory

In 1955, Taraski [588] formulated and proved an elementary fixed point theorem which holds in arbitrary complete lattices. It is one of the important results in lattice-theoretic fixed point theory because of its wide range of applications in theories of simply ordered sets, real functions, Boolean algebra as well as general set theory and topology.

Observe that Banach [32] obtained a fixed point theorem for contraction mappings. Edelstein [212] considered contractive mapping and proved a fixed point theorem for such mapping. In 1970s, Kannan [308, 309], Husain and Sehgal [283] and Caristi [128] have considered several generalizations of contraction mapping.

Definition 5.1 Let (X, d) and (Y, ρ) be metric spaces and $T : X \to Y$ be a mapping. Then T is said to be *Lipschitz continuous* if there is $L > 0$ such that

$$\rho(Tx, Ty) \leqslant Ld(x, y) \text{ for all } x, y \in X.$$

(a) If $L = 1$, T is said to be *nonexpansive*.
(b) If $L < 1$, T is said to be a *contraction*.
(c) T is said to be contractive if for all x, y in X and $x \neq y$, we have

$$\rho(Tx, Ty) < d(x, y).$$

Notice that a contractive mapping can have at most one fixed point. Note also that

$$\boxed{\text{contraction}} \Rightarrow \boxed{\text{contractive}} \Rightarrow \boxed{\text{nonexpansive}} \Rightarrow \boxed{\text{Lipschitz}}$$

but the converse implications are not true in general. Moreover, all such mappings are continuous.

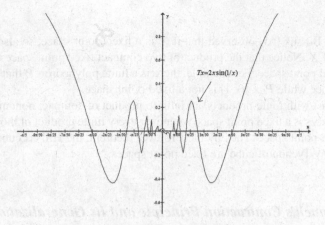

Fig. 5.1 T is continuous, but not Lipschitz continuous

We now furnish an example of a non-Lipschitzian mapping that is continuous.

Example 5.3 Let $T : [-\frac{\pi}{10}, \frac{\pi}{10}] \to [-\frac{\pi}{10}, \frac{\pi}{10}]$ be a mapping defined by

$$T(x) = \begin{cases} 2x \sin(1/x) & \text{if } x \neq 0 \\ 0 & \text{if } x = 0 \end{cases}$$

Then we see that T is continuous, but not Lipschitz continuous (see Fig. 5.1).

Example 5.4 Let $T : C[0, 1] \to C[0, 1]$ be defined as

$$(Tx)(t) = x(0) + \lambda \int_0^t x(s)ds, \lambda \in \mathbb{R}.$$

T is a contraction map if $|\lambda| < 1$.

Definition 5.2 Let X be a Banach space and let $A \subset X$. A retraction of X onto A is a continuous map $\Re : X \to A$ such that $\Re|_A$ is the identity map of A. If such a map $\Re$ exists, we say that A is a retract of X.

Example 5.5 (i) Let X be a Banach space and let $B_r = \{x : \|x\| \leqslant r\}$. The mapping $\Re_r$ defined as

$$\Re_r(x) = \begin{cases} x & \text{if } x \in B_r \\ \frac{rx}{\|x\|} & \text{if } x \notin B_r \end{cases}$$

is a retraction mapping.

(ii) If X is a Banach space, then the mapping $\Re_r$ is Lipschitz with Lipschitz constant 2.

Observation

1. In 1931, Borsuk [65] observed that if X is a fixed point space, so also is every retract of X. Notice that the product of two compact fixed point spaces need not be a fixed point space. For example, there is a finite polyhedron P that is a fixed point space while $P \times [0, 1]$ is not a fixed point space.
2. In contrast with finite products, an infinite product of compact nonempty fixed point spaces is a fixed point space whenever every finite product of those spaces is a fixed point space. Thus by the Brouwer theorem, the Hilbert cube I^∞ and in fact any Tychonoff cube are fixed point spaces.

5.1.1   Banach's Contraction Principle and Its Generalizations

The following constructive fixed point theorem was proved by S. Banach [32] and is known as Banach's contraction principle.

Theorem 5.1 (Banach's contraction principle) *Let (X, d) be a complete metric space and $T : X \to X$ is a contraction mapping with Lipschitz constant $\alpha \in (0, 1)$. Then*

  (i) T has a unique fixed point u in X.
 (ii) For an arbitrary point x_0 in X, the sequence $\{x_n\}$ generated by the Picard iteration process as defined by $x_{n+1} = Tx_n$, $n \in \mathbb{N} \cup \{0\}$ converges to u.
(iii) $d(x_n, u) \leqslant \frac{\alpha^n}{1-\alpha} d(x_0, x_1)$ for all $n \in \mathbb{N} \cup \{0\}$.

Proof (i) Choose any $x_0 \in X$ and define the iterate sequence $x_{n+1} = Tx_n$ for all $n \in \mathbb{N} \cup \{0\}$. We have

$$d(x_{r+1}, x_{s+1}) = d(Tx_r, Tx_s) \leqslant \alpha d(x_r, x_s), \quad r \neq s, \ r, s \in \mathbb{N} \cup \{0\}$$

and hence by induction on n,

$$d(x_{n+1}, x_n) \leqslant \alpha^n d(x_1, x_0).$$

If $n \in \mathbb{N} \cup \{0\}$ and $m \geq 1$,

$$
\begin{aligned}
d(x_{n+m}, x_n) &\leqslant d(x_{n+m}, x_{n+m-1}) + \cdots + d(x_{n+1}, x_n) \\
&\leqslant (\alpha^{n+m-1} + \cdots + \alpha^n) d(x_1, x_0) \\
&\leqslant (\alpha^n + \alpha^{n+1} + \alpha^{n+2} + \cdots) d(x_1, x_0) \\
&\leqslant \frac{\alpha^n}{1-\alpha} d(x_1, x_0).
\end{aligned}
\tag{5.2}
$$

The RHS of (5.2) tends to zero as $n \to \infty$. Hence, the sequence $\{x_n\}$ is Cauchy and since X is complete, $\{x_n\}$ converges to an element u of X.

As $d(x_{n+1}, Tu) = d(Tx_n, fu) \leqslant \alpha d(x_n, u)$ and $d(x_n, u) \to 0$ as $n \to \infty$, we have

$$Tu = \lim_{n \to \infty} x_{n+1} = u,$$

i.e. u is a fixed point of T.

(ii) Fur uniqueness, suppose v is another fixed point of T, $v(\neq u) \in X$ and $v = Tv$, then $d(u, v) > 0$ and

$$d(u, v) = d(Tu, Tv) \leqslant \alpha d(u, v) < d(u, v),$$

a contradiction. Hence $d(u, v) = 0$, that is $u = v$.

(iii) By the triangle inequality, we have

$$d(x_n, u) \leqslant d(x_n, x_p) + d(x_p, u) \leqslant \frac{\alpha^n}{1 - \alpha} d(x_1, x_0) + d(x_p, u)$$

for $n < p$ by (5.2). Letting $p \to \infty$, we obtain

$$d(x_n, u) \leqslant \frac{\alpha^n}{1 - \alpha} d(x_1, x_0)$$

which provides a control on the convergence rate of $\{x_n\}$ to the fixed point u. ∎

Example 5.6 (i) Let $X = \mathbb{R}$ and $T : \mathbb{R} \to \mathbb{R}$ a mapping defined by $Tx = \frac{2}{3}x + 1$, $x \in \mathbb{R}$. Then T is a contraction and $\mathfrak{F}(T) = \{3\}$.
(ii) Let $X = \mathbb{C}$ and $T : \mathbb{C} \to \mathbb{C}$ a mapping defined by $T(z) = az$, for $|a| < 1$ is a contraction mapping and $\mathfrak{F}(T) = \{0\}$.

Observation

1. Brouwer's fixed point theorem is existential by its nature.
2. The elegant Banach's fixed point theorem solves:
   ($1°$) the problem on the existence of a unique solution to an equation,
   ($2°$) gives a practical method to obtain approximate solutions and
   ($3°$) gives an estimate of such solutions.
3. There exists a mapping that is not a contraction, but it has a unique fixed point. To this end, consider $X = [0, 1]$ and $T : X \to X$ a mapping defined by $Tx = 1 - x$, $x \in X$. Then T has a unique fixed point $\frac{1}{2}$, but it is not a contraction.

Notice that the applications of the Banach's fixed point theorem and its generalizations are very important in diverse disciplines of mathematics, statistics, engineering and economics. Some examples of applications of Banach contraction principle are given below:

Example 5.7 Let $X = \mathbb{R}$ be the Banach space of real numbers endowed with the norm $\| \cdot \|$ given by $\|x\| = |x|$ and $[a, b] \subset \mathbb{R}$; $f : [a, b] \to [a, b]$, a differentiable function such that

$$|f'(x)| < c < 1.$$

Suppose we wish to find solution of the equation $f(x) = x$.

For any $x, y \in [a, b]$, we have by Lagrange's mean value theorem that

$$f(x) - f(y) = (x - y)f'(z), \quad y < z < x.$$

It follows that

$$\begin{aligned}|f(x) - f(y)| &= |x - y||f'(z)| \\ &\leqslant c|x - y|, \quad 0 < c < 1.\end{aligned}$$

Thus, f is a contraction on $[a, b]$ into itself. Since $[a, b]$ is a closed subset of $\mathbb{R}$, it follows that $[a, b]$ is complete and so by Theorem 5.1, there exists a unique fixed point $\bar{x} \in [a, b]$, i.e. $f(\bar{x}) = \bar{x}$.

Example 5.8 Let T be an operator on a Banach space X that satisfies the condition $\|T\| < 1$. Such an operator is a contraction since

$$d(Tx, Ty) = \|Tx - Ty\| \leqslant \|T\|\|x - y\| = \|T\|d(x, y).$$

We now intend to obtain the solution the equation

$$x = u + Tx \tag{†}$$

where $u \in X$ is given and x is unknown. Set $\phi(x) = u + Tx, x \in X$. Then we see that $\bar{x}$ is a solution of (†) if and only if $\bar{x}$ is a fixed point of ϕ. Since

$$\|\phi(x) - \phi(y)\| \leqslant \|T\|\|x - y\|$$

so ϕ is a contraction. Then Theorem 5.1 yields a unique solution $\bar{x}$ of (†). Furthermore, it gives an iterative procedure for obtaining the solution. If the initial choice $x_0 = 0$ and x_n defined by

$$x_{n+1} = u + Tx_n, \quad n \geq 0,$$

then by the prior error estimates in the m-th approximation, we have

$$\|x_m - \bar{x}\| \leqslant \frac{\|T\|^m \|u\|}{1 - \|T\|}.$$

Remark 5.1 The completeness of X plays here a crucial role. Indeed, contractions on incomplete metric spaces may fail to have fixed points. To see it, consider the following example.

Example 5.9 Let $X = (0, 1]$ with the usual distance. Define $T : X \to X$ as $T(x) = \frac{x}{2}$ for all $x \in X$.

Theorem 5.1 gives a sufficient condition for T in order to have a unique fixed point.

Example 5.10 Consider the mapping $T : [0, 4] \to [0, 4]$ defined by

$$T(x) = \begin{cases} \frac{1}{2} + 4x, & x \in [0, \frac{1}{2}] \\ \frac{3}{2} + x, & x \in (\frac{1}{2}, \frac{3}{2}) \\ 2, & x \in (\frac{3}{2}, 4]. \end{cases}$$

Notice that T is not even continuous, but it has a unique fixed point ($x = 2$).

The next corollary takes into account the above situation and provides existence and uniqueness of a fixed point under more general conditions. Indeed, the following result shows that when some power of T, say T^k, is a contraction mapping of X for some positive integer k, then T has a unique fixed point.

Corollary 5.1 *Let T be a continuous mapping of a complete metric space X into itself such that T^k is a contraction mapping of X for some positive integer k. Then T has a unique fixed point.*

Proof By Theorem 5.1, T^k has a unique fixed point u in X and $u = \lim_{n \to \infty} (T^k)^n x_0$, x_0 being arbitrary point in X. Also $\lim_{n \to \infty} (T^k)^n (T x_0) = u$. Hence

$$u = \lim_{n \to \infty} (T^k)^n (T x_0) = \lim_{n \to \infty} T (T^k)^n x_0 = T(\lim_{n \to \infty} (T^k)^n x_0) = Tu.$$

Since each fixed point of T is also a fixed point of T^k, the uniqueness of the fixed point of T follows from the uniqueness of the fixed point of T^k.   ∎

Remark 5.2 The continuity condition of T in Corollary 5.1 is not necessary. The following example illustrate the contention.

$$X = \mathbb{R}, \quad T(x) = \begin{cases} 1, & x \text{ is rational} \\ 0, & x \text{ is irrational,} \end{cases}$$

T is not continuous and hence not a contraction mapping.

$$T^2(x) = \begin{cases} T(1) = 1, & x \text{ is rational} \\ T(0) = 1, & x \text{ is irrational,} \end{cases}$$

T^2 is a contraction and hence continuous. Notice that T^2 and T both have the same fixed point 1.

Theorem 5.2 *Let T be a mapping of a complete metric space X into itself such that T^k is a contraction mapping of X for some positive integer k. Then T has a unique fixed point in X.*

Proof Let $u \in X$ such that $u = T^k u$, i.e. u is the unique fixed point of T^k. Applying T to this ($u = T^k u$), we have

$$Tu = T^{k+1}(u) = T^k(Tu).$$

This implies that Tu is the fixed point of T^k. By uniqueness we have $u = Tu$, i.e. u is a fixed point of T also. Further to prove the uniqueness of the fixed point of T, let v be another fixed point of T. Then

$$v = Tv = T(Tv) = \cdots = T^k(v),$$

i.e. v is also a fixed point of T^k. Therefore, $u = v$. Thus, u is a unique fixed point of T. ■

Now we give an example of a mapping T which is not a contraction but T^k is a contraction for some k.

Example 5.11 Let $T : C[a, b] \to C[a, b], (\infty < a < b < \infty)$ with uniform norm, be defined as

$$[T(f)](t) = \int_a^t f(x)dx.$$

Then, using integral formula of A-2, it can be shown that

$$[T^k(f)](t) = \frac{1}{(k-1)!} \int_a^t (t - x)^{k-1} f(x)dx$$

For sufficiently large values of k, the mapping T^k is a contraction, whereas T is not a contraction if $(b - a) > 1$.

Observation

- From Banach's contraction (BC), it follows that the mapping T is continuous. Further, we use the continuity of the mapping T to prove Banach's fixed point theorem. Thus, it is natural to consider the following question:
  Do there exist some contractive conditions which do not force the mapping T to be continuous?

In 1968, Kannan [308] answered this question affirmatively and proved a fixed point theorem for the following contractive condition, which is called Kannan's contraction (KC):

Theorem K. Let (X, d) be a complete metric space and $T : X \to X$ be a mapping such that there exists a number $r \in [0, \frac{1}{2})$ such that

$$d(Tx, Ty) \leqslant r[d(Tx, x) + d(Ty, y)] \tag{KC}$$

for all $x, y \in X$. Then, T has a unique fixed point in X.

We now illustrate the fact that there exists a mapping T which is not continuous, but the mapping T is Kannan's contraction. To this end, let us consider the following example: Let $X = \mathbb{R}$ be a usual metric space and $T : X \to X$ be a mapping defined by

$$T(x) = \begin{cases} 0 & \text{if } x \in (-\infty, 2], \\ \frac{1}{2} & \text{if } x \in (2, +\infty). \end{cases}$$

Clearly, T is not continuous at $x = 2 \in \mathbb{R}$. Further, it is easy to see that T satisfies Kannan's contraction (KC) with $k = \frac{1}{5}$.

In 1972, Chatterjea [140] proved a fixed point theorem for the following contractive condition, which is called Chatterjea's contraction (CHC):

Theorem C. Let (X, d) be a complete metric space and $T : X \to X$ be a mapping such that there exists a number $r \in [0, \frac{1}{2})$ such that

$$d(Tx, Ty) \leqslant r[d(Tx, y) + d(Ty, x)] \tag{CHC}$$

for all $x, y \in X$. Then, T has a unique fixed point in X.

In 1977, Rhoades [519] observed that Banach's contraction (BC), Kannan's contraction (KC) and Chatterjea's contraction (CHC) are independent.

In what follows, let (X, d) denote a complete metric space. In the last five decades, one may observe that many authors have improved, extended and generalized Banach's fixed point theorem in metric spaces as follows:

(1) In 1969, Meir and Keeler [393] introduced the following contractive condition: For any $\varepsilon > 0$, there exists $\delta > 0$ such that

$$\varepsilon \leqslant d(x, y) < \varepsilon + \delta \implies d(Tx, Ty) < \varepsilon \tag{MK}$$

Note that if T satisfies Meir–Keeler's contraction (MK), then T is contractive, i.e.

$$d(Tx, Ty) < d(x, y) \text{ for all } x, y \in X \text{ with } x \neq y.$$

(2) In 1971, Reich [508] introduced the following contractive condition: there exist nonnegative numbers $a, b, c \in [0, 1)$ such that $a + b + c < 1$ and

$$d(Tx, Ty) \leqslant ad(x, y) + bd(x, Tx) + cd(y, Ty) \tag{RC}$$

for all $x, y \in X$.

(3) In 1971, Ćirić [142] introduced the following contractive condition: there exist nonnegative numbers $a, b, c, e \in [0, 1)$ such that $a + b + c + 2e < 1$ and

$$d(Tx, Ty) \leqslant ad(x, y) + bd(x, Tx) + cd(y, Ty) + e[d(x, Ty) + d(y, Tx)]$$
$$\text{(CRC1)}$$

for all $x, y \in X$.

(4) In 1972, Zamfirescu [616] introduced the following contractive condition:

$$d(Tx, Ty) \leqslant \max\{d(x, y), \frac{1}{2}[d(x, Tx) + d(y, Ty)], \frac{1}{2}[d(x, Ty) + d(y, Tx)]\}$$
$$\text{(ZC)}$$

for all $x, y \in X$.

(5) In 1973, Hardy and Rogers [266] introduced the following contractive condition: There exist nonnegative numbers $a_1, a_2, a_3, a_4, a_5 \in [0, 1)$ such that $a_1 + a_2 + a_3 + a_4 + a_5 < 1$ and

$$d(Tx, Ty) \leqslant a_1 d(x, Tx) + a_2 d(y, Ty) + a_3 d(x, Ty) + a_4 d(y, Tx) + a_5 d(x, y)$$
$$\text{(HRC)}$$

for all $x, y \in X$.

(6) In 2004, Berinde [54] introduced the following contractive condition: There exist $r \in [0, 1)$ and $L \geqslant 0$ such that, for all $x, y \in X$,

$$d(Tx, Ty) \leqslant rd(x, y) + Ld(y, Tx)]. \tag{VBC}$$

5.1.2   A Converse to the Banach Contraction Mapping Principle

Let (X, d) be a complete metric space and let $T : X \to X$ be a contraction mapping. From the Banach fixed point theorem, we infer the following properties:

(i) each iteration T^n ($n = 1, 2 \cdots$) of T has a unique fixed point, say $u \in X$,
(ii) the sequence of iterates $\{T^n x\}$ converges to u for all $x \in X$. Moreover, one can easily show that
(iii) there exists an open neighbourhood U of u with the property that given any open set V containing u there exists a positive integer n_0 such that

$$T^n(U) \subset V \quad \forall n \geqslant n_0.$$

We are interested to find a metric d on X such that (X, d) is a complete metric space and T is a contraction on X. Clearly, in the light of Theorem 5.1, a necessary condition is that each iterate T^n has a unique fixed point. Consider now the situation when we know only that T has property (i). Surprisingly enough, as noted by Polish mathematician C. Bessaga in 1959, the condition turns out to be sufficient as well.

Theorem 5.3 (Bessaga [56]) *Let X be an arbitrary set, and let $T : X \to X$ be a map such that T^n has a unique fixed point $u \in X$ for every $n \geqslant 1$. Then for every*

$\lambda \in (0, 1)$, *there is a metric* $d = d_\lambda$ *on* X *that makes* X *a complete metric space, and* T *is a contraction on* X *with Lipschitz constant equal to* λ.

Proof Firstly, we choose $\lambda \in (0, 1)$. Let Z be the subset of X consisting of all elements z such that $T^n(z) = u$ for some $n \in \mathbb{N}$. Secondly, we define the following equivalence relation on $X \setminus Z$: we say that

$$x \sim y \text{ if and only if } T^n(x) = T^m(y) \text{ for some } n, m \in \mathbb{N}.$$

For arbitrary $x \in X$, a corresponding equivalence class is denoted by $[x]$, i.e.

$$[x] = \{y \in X : y \sim x\}.$$

In each equivalence class, we choose a fixed element $\hat{x}$ called representative element.

Notice that if $T^n(x) = T^m(y)$ and $T^{n'}(x) = T^{m'}(y)$, then $T^{n+m'}(x) = T^{m+n'}(x)$. But since $x \notin Z$, this yields $n + m' = m + n'$, that is, $n - m = n' - m'$. At this point, by means of the axiom of choice, we select an element from each equivalence class. We now proceed defining the distance of u from a generic $x \in X$ by setting $d(u, u) = 0$, $d(x, u) = \lambda^{-n}$ if $x \in Z$ with $x \neq u$, where $n = \min\{m \in \mathbb{N} : T^m(x) = u\}$, and $d(x, u) = \lambda^{n-m}$ if $x \notin Z$, where $n, m \in \mathbb{N}$ are such that $T^n(\hat{x}) = T^m(x)$, $\hat{x}$ being the selected representative of the equivalence class $[x]$. From the above discussion, it is clear that d is well defined.

Finally, for any $x, y \in X$, we set

$$d(x, y) = \begin{cases} d(x, u) + d(y, u) & \text{if } x \neq y, \\ 0 & \text{if } x = y, \end{cases}$$

We now observe the following:

1. It is straightforward to verify that d is a metric on X.
2. d is complete. To see this, we observe that the only Cauchy sequences which do not converge to u are ultimately constant.
3. We are left to show that T is a contraction with Lipschitz constant equal to λ. To this end, let $x \in X$, $x \neq u$.
   (*i*) If $x \in Z$, we have

   $$d(T(x), T(u)) = d(f(x), u) \leqslant \lambda^{-n} = \lambda\lambda^{-(n+1)} = \lambda d(x, u).$$

   (*ii*) If $x \notin Z$, we have

   $$d(T(x), T(u)) = d(f(x), u) = \lambda^{n-m} = \lambda\lambda^{n-(m+1)} = \lambda d(x, u)$$

   since $x \sim T(x)$.

Finally, the thesis follows directly from the definition of the distance. ∎

5.1.3 Fixed Point Theorems for (ε, k)-Uniformly Locally Contractive Mappings

We now focus on another class of mappings related to contraction mappings which are defined on some special classes of metric spaces.

Definition 5.3 Let (X, d) be a metric space and $\varepsilon > 0$. A finite sequence $x_0, x_1, \ldots, x_n$ of points of X is called on ε-chain joining x_0 and x_n if

$$d(x_{i-1}, x_i) < \varepsilon \quad (i = 1, 2, \ldots, n).$$

The metric space (X, d) is said to be ε-chainable if for each pair (x, y) of its points, there exists an ε-chain joining x and y.

Definition 5.4 A mapping $T : X \to X$ is called (ε, k)-uniformally locally contractive if there exists $\varepsilon > 0$ and k with $0 \leqslant k < 1$ such that

$$d(x, y) < \varepsilon \Rightarrow d(Tx, Ty) \leqslant kd(x, y) \quad \text{for each } x, y \in X.$$

The following theorem gives a generalization of contraction mapping principle to a class of mappings on ε-chainable spaces.

Theorem 5.4 (Edelstein [212]) *Let (X, d) be a complete ε-chainable metric space and $T : X \to X$ be an (ε, k)-uniformly locally contractive mapping. Then T has a unique fixed point u in X and $u = \lim_{n \to \infty} T^n x_0$ where x_0 is an arbitrary element of X.*

Proof Since (X, d) is ε-chainable, we define

$$d_\varepsilon(x, y) = \inf \sum_{i-1}^{n} d(x_{i_{-1}}, x_i) \text{ for } x, y \in X.$$

where the infimum is taken over all ε-chains $x_0, \ldots, x_n$ joining $x_0 = x$ and $x_n = y$.
   Then the metric d_ε on X satisfies

(i) $d(x, y) \leqslant d_\varepsilon(x, y)$
(ii) $d(x, y) = d_\varepsilon(x, y)$ for $d(x, y) < \varepsilon$.

From the above, it follows that a sequence is Cauchy with respect to d_ε if any only if it is Cauchy with respect to d and is convergent with respect to d_ε if any only if it is convergent with respect to d. So (X, d_ε) is complete whenever (X, d) is complete. With given $x, y \in X$ and any ε-chain $x_0, x_1, \ldots, x_n$ with $x_0 = x, x_n = y$, we have

$$d(x_{i-1}, x_i) < \varepsilon \quad (i = 1, 2, \ldots, n)$$

and hence
$$d(Tx_{i-1}, Tx_i) \leqslant kd(x_{i-1}, x_i) < \varepsilon \quad (i = 1, 2, \ldots, n).$$

So $T x_0, T x_1, \ldots, T x_n$ is an ε-chain joining fx and fy and

$$d_\varepsilon(Tx, Ty) \leqslant \sum_{i=1}^{n} d(Tx_{i-1}, Tx_1) \leqslant k \sum_{i=1}^{n} d(x_{i-1}, x_i).$$

Furthermore, $x_0, x_1, \ldots, x_n$ being an arbitrary ε-chain, we have

$$d_\varepsilon(Tx, Ty) \leqslant k \, d_\varepsilon(x, y).$$

Hence, f has a unique fixed point $u \in X$ given by

$$\lim_{n \to \infty} d_\varepsilon(T^n x_0, u) = 0 \text{ for any } x_0 \in X. \qquad \text{(by Theorem 5.1)} \qquad (5.3)$$

But by (i) we have $\lim_{n \to \infty} d(T^n x_0, u) = 0$. This completes the proof. ∎

The following example illustrates the application of the above theorem.

Example 5.12 Let C be a connected compact subset of a domain D in the complex plane. Let T be a complex analytic function in D which maps C in to itself and satisfies $|T'(x)| < 1$ for every z in C. Then there is a unique point z in C with $T(z) = z$.

Since T' is continuous and C is compact, there is a constant k with $0 < k < 1$ such that $|T'(z)| < k$ for all $z \in C$. For each point $w \in C$, there exists $r_w > 0$ such that $T(z)$ is analytic in the disc $B_{2r_{w_i}}(w)$ of center w and radius $2r_w$ and satisfies $|T'(z)| < k$ in it. By compactness of C, we can choose $w_1, w_2, \ldots, w_n \in C$ such that $\{B_{2r_{w_i}}(w_i)\}, i = 1, 2, 3, \ldots, n$ covers C.

Let $\varepsilon = \min\{r_{w_i}, i = 1, 2, \ldots, n\}$. We show that T is an ε-contractive mapping. For any two points z_1, z_2 in C, with $|z_1 - z_2| < \varepsilon$, are in some $B_{2r_{w_i}}(w_1)$ and thus

$$|T(z_2) - T(z_1)| = \left| \int_{z_1}^{z_2} T'(w) dw \right| \leqslant k |z_2 - z_1|.$$

Hence, $T(z) = z$ has a unique solution.

5.1.4 Fixed Point Theorems for Contractive Mappings

Definition 5.5 A mapping T of a metric space X into itself is said to be contractive if

$$d(Tx, Ty) < d(x, y), \quad x, y \in X (x \neq y)$$

and is said to be ε-contractive if

$$0 < d(x, y) < \varepsilon \Rightarrow d(Tx, Ty) < d(x, y).$$

Theorem 5.5   (Edelstein [213]) *Let T be an ε-contractive mapping of a metric space X into itself, and let x_0 be a point of X such that the sequence $\{T^n x_0\}$ has a subsequence convergent to a point u of X. Then u is a periodic point of T, i.e. there exists a positive integer k such that $T^k u = u$.*

Proof Let $\{T^{n_i} x_0\}$ be a convergent subsequence converging to u, i.e. $\lim_{i \to \infty} T^{n_i} x_0 = u$ where $\{n_i\}$ is a strictly increasing sequence and let $x_1 = T^{n_i} x_0$. Given $\varepsilon > 0$ there exists N such that

$$d(x_1, u) < \frac{\varepsilon}{4} \text{ for } i \geqslant N.$$

Choose any $i \geqslant N$ and let $k = n_{i_{+1}} - n_i$. Then

$$d(x_{i+1}, T^k u) = d(T^k x_i, T^k u) \leqslant d(x_i, u) < \frac{\varepsilon}{4}$$

and

$$d(T^k u, u) \leqslant d(T^k u, x_{i_{+1}}) + d(x_{i_{+1}}, u) < \frac{\varepsilon}{2}.$$

Suppose that $v = T^k u \neq u$. Then T being ε-contractive,

$$d(Tu, Tv) < d(u, v) \text{ or } \frac{d(Tu, Tv)}{d(u, v)} < 1.$$

The function $(x, y) \to \frac{d(Tx, Ty)}{d(x, y)}$ is continuous at (u, v). So there exist $\delta, \alpha > 0$ with $0 < \alpha < 1$ such that

$$d(x, u) < \delta, d(y, v) < \delta \Rightarrow d(Tx, Ty) < \alpha d(x, y).$$

As $\lim r \to \infty \, T^k x_r = T^k u = v$, there exists $N' \geqslant N$ such that $d(x_r, u) < \delta, d(T^k x_r, v) < \delta$ for $r \geqslant N'$ and so

$$d(Tx_r, T(T^k x_r)) < \alpha d(x_r, T^k x_r). \tag{5.4}$$

Also $d(x_r, T^k x_r) \leqslant d(x_r, u) + d(u, T^k u) + d(T^k u, T^k x_r)$

$$< \frac{\varepsilon}{4} + \frac{\varepsilon}{2} + \frac{\varepsilon}{4} = \varepsilon \text{ for } r \geq N' > N. \tag{5.5}$$

From (5.4) and (5.5), we have

$$d(Tx_r, T(T^k x_r)) < \alpha d(x_r, T^k x_r) < \alpha \varepsilon < \varepsilon \text{ for } r \geqslant N'.$$

Since T is ε-contractive, we have

$$d(T^q x_r, T^q(T^k x_r)) < \alpha d(x_r, T^k x_r) \text{ for } r \geqslant N', q > 0. \tag{5.6}$$

Setting $q = n_{r+1} - n_r$ in (5.6) we have

$$d(x_{r+1}, T^k x_{r+1}) < \varepsilon d(x_r, T^k x_r) \text{ for any } r \geq N'.$$

Hence $\quad d(x_s, T^k x_s) < \alpha^{s-r} d(x_r, T^k x_r) < \alpha^{s-r} \varepsilon.$

Finally, by triangular inequality we have

$$d(u, v) \leq d(u, x_s) + d(x_s, T^k x_s) + d(T^k x_s, v) \to 0 \text{ as } s \to \infty.$$

This contradicts the assumption that $d(u, v) > 0$. Thus

$$u = v = T^k u.$$

This completes the proof. ∎

Remark 5.3 A contractive mapping is clearly continuous, and if such a mapping has a fixed point, then this fixed point is obviously unique. However, we notice that a contractive mapping of a complete metric space into itself need not have a fixed point. It can be seen from the following examples.

Example 5.13 (*i*) Let $X = [1, \infty)$ with the usual metric $d(x, y) = |x - y|$. Let $T : X \to X$ be given by $Tx = x + 1/x$. Then X is a complete metric space but T is not a contraction mapping. Indeed, one may observe that

$$|Tx - Ty| = \frac{xy - 1}{xy}|x - y| < |x - y| \text{ for all } x, y \in X$$

and so T is contractive but has no fixed point. Note that T is not a contraction because although the ratio $\frac{xy-1}{xy}$ is less than 1 for all $x, y \in X$, it approaches 1 as x, y become large.

(*ii*) Let $X = \mathbb{R}$ with the usual metric $d(x, y) = |x - y|$. Let $T : X \to X$ be given

Fig. 5.2 T is contractive, but has no fixed point

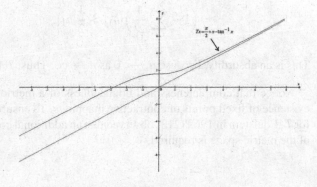

by $Tx = \frac{\pi}{2} + x - \tan^{-1} x$. Then X is a complete metric space and T is a contractive mapping, i.e.

$$|Tx - Ty| < |x - y| \text{ for all } x, y \in X \text{ with } x \neq y,$$

but T has no fixed point (Fig. 5.2).

(iii) Let $X = c_0$, the space of all real sequences $x = \{x_n\}$ with $\lim_{n \to \infty} x_n = 0$ and $d(x, y) = \|x - y\| = \sup_{n \in \mathbb{N}} |x_n - y_n|, x = \{x_n\}, y = \{y_n\} \in c_0$. Let $\mathbb{B}_X = \{x \in c_0 : \|x\| \leqslant 1\}$. For each $x \in \mathbb{B}_X$, define $T : \mathbb{B}_X \to \mathbb{B}_X$ by

$$T(x) = ((1 + \|x\|)/2, (1 - 1/2^2)x_1, (1 - 1/2^3)x_2, \ldots, (1 - 1/2^n)x_{n-1}, \ldots).$$

For any two distinct points $x, y \in \mathbb{B}_X$, we have

$$\|Tx - Ty\| = \sup \left\{ \frac{\|x\| - \|y\|}{2}, \left(1 - \frac{1}{2^n}\right)|x_{n-1} - y_{n-1}| : n = 2, 3, \ldots \right\}$$
$$\leqslant \sup \left\{ \frac{\|x - y\|}{2}, \left(1 - \frac{1}{2^n}\right)|x_{n-1} - y_{n-1}| : n = 2, 3, \ldots \right\}$$
$$< \|x - y\|.$$

We now show that T has no fixed point in $\mathbb{B}_X$. Suppose, on the contrary, there exists a point $u \in \mathbb{B}_X$ such that $Tu = u$. Then $u_1 = (1 + \|u\|)/2 > 0$, and for $n \geqslant 2$, we have

$$|u_n| = \left(1 - \frac{1}{2^n}\right)|u_{n-1}|.$$

This yields for $n \geqslant 2$,

$$|u_n| = \left(1 - \frac{1}{2^n}\right)|u_{n-1}| = \left(1 - \frac{1}{2^n}\right)\left(1 - \frac{1}{2^{n-1}}\right)|u_{n-2}|$$
$$= \prod_{i=0}^{n-2}\left(1 - \frac{1}{2^{n-i}}\right)|u_1| \geqslant \left(1 - \sum_{i=0}^{n-2}\frac{1}{2^{n-i}}\right)|u_1|$$
$$= \left(1 - \sum_{j=2}^{n}\frac{1}{2^j}\right)|u_1| > \frac{1}{2}|u_1|.$$

This is an absurdity, because $u_n \to 0$ as $n \to \infty$. Thus, T has no fixed point in $\mathbb{B}_X$.

Notice that completeness and boundedness of a metric space do not ensure the existence of fixed points of contractive mappings. To ensure existence of fixed point for T, Edelstein in 1962 [213] observed that an additional condition, i.e. compactness of the metric space is required.

Theorem 5.6 (Edelstein [213]) *Let T be a contractive mapping of a metric space X into itself, and let x_0 be a point of X such that the sequence $\{T^n x_0\}$ has a convergent subsequence which converges to a point $u \in X$. Then u is a unique fixed point of T.*

Corollary 5.2 *If T is a contractive mapping of a metric space X into a compact subset of X, then T has a unique fixed point u in X. Moreover, for any $x \in X$, and $\lim\limits_{n \to \infty} T^n x = u$.*

The following theorem guarantees the existence of a fixed point for a contractive mapping.

Theorem 5.7 *Let T be a self-mapping on a compact metric space (X, d). Suppose T is contractive, then T has a unique fixed point u in X and $\{T^n x\}$ converges to u for all x in X.*

Proof Define a mapping $\varphi : X \to \mathbb{R}^+$ by

$$\varphi(x) = d(x, Tx) \ \forall x \in X.$$

Then we notice that φ is continuous. Indeed, by contractiveness of T, we have

$$
\begin{aligned}
|\varphi(x) - \varphi(y)| &= |d(x, Tx) - d(y, Ty)| \\
&\leqslant |d(x, Tx) - d(y, Tx)| + |d(Tx, y) - d(Ty, y)| \\
&\leqslant d(x, y) + d(Tx, Ty) < 2d(x, y).
\end{aligned}
$$

Let $\varepsilon > 0$ be given. Then there exists a $\delta = \frac{\varepsilon}{2} > 0$ such that

$$d(x, y) < \delta \Rightarrow |\varphi(x) - \varphi(y)| < 2\delta = \varepsilon.$$

Therefore, φ is continuous. Clearly, φ is bounded below. Because X is compact and $\varphi : X \to \mathbb{R}^+$ is continuous, there exists a minimizer $u \in X$, i.e. there exists $u \in X$ such that

$$\varphi(u) \leqslant \varphi(x) \ \forall x \in X.$$

We show that u is a fixed point of T. Suppose contrary that u is not a fixed point of T. Then $Tu \neq u$. By contractiveness of T, we have

$$\varphi(Tu) = d(Tu, T^2 u) < d(u, Tu) = \varphi(u)$$

which contradicts that u is a minimizer of φ. Hence, u is a fixed point of T. The uniqueness of u follows on the lines of the proof of Theorem 5.1.

For the second part, let x be an arbitrary point of X. If $T^n x \neq u$, then

$$d(T^{n+1} x, u) = d(T(T^n x), Tu) < d(T^n x, u) \ \forall n \in \mathbb{N}.$$

This shows that $\{d(T^{n+1}x, u)\}$ is a strictly decreasing sequence of nonnegative real numbers. It follows that the sequence $\{d(T^{n+1}x, u)\}$ converges to its infimum. Suppose that $\lim_{n\to\infty} d(T^{n+1}x, u) = \gamma$, for some $\gamma \geq 0$.

Because $\{T^n x\}$ is a sequence of points of a compact metric space, there exists a subsequence $\{T^{n_i} x\}$ which converges to some point, say, $v \in X$.

$$d(v, u) = \lim_{i\to\infty} d(T^{n_i} x, u) = \lim_{i\to\infty} d(T^{n_i+1}x, u) = \gamma.$$

If $\gamma \neq 0$, then $v \neq u$ and so we have

$$\gamma = d(v, u) > d(Tv, Tu) = d(Tv, u)$$
$$= \lim_{i\to\infty} d(T(T^{n_i}x), u) = \lim_{i\to\infty} d(T^{n_i+1}x, u) = \gamma,$$

a contradiction. Thus, $\gamma = 0$ and therefore, $\lim_{n\to\infty} T^n x = u$.

The following example shows that in general, even in a Hilbert space for a contractive mapping T we cannot have that $T^n x \to u$ for every $x \in \mathbb{B}_X$ and $Tu = u$.

Example 5.14 Let $X = \ell_2 = \{x_1, x_2, \ldots, x_i, \ldots\}$, x_i is real for each $i \in \mathbb{N}$ and $\sum_{i=1}^{\infty} |x_i|^2 < \infty.\}$. Let $\mathbb{B}_X = \{x \in X : \|x\|_2 = (\sum_{i=1}^{\infty} |x_i|^2)^{1/2} \leqslant 1\}$. For each $x = (x_1, x_2, \ldots, x_i, \cdots) \in \mathbb{B}_X$, define a mapping $T : \mathbb{B} \to \mathbb{B}$ by

$$T(x) = (0, a_1 x_1, a_2 x_2, \ldots, a_i x_i, \ldots),$$

where $a_1 = 1, a_i = (1 - 1/i^2)$, $i = 2, 3, \ldots$. Now we can easily show that T is contractive with fixed point $(0, 0, \ldots, 0, \ldots)$. Consider $x = (1, 0, \ldots, 0, \ldots) \in \mathbb{B}_X$, then

$$T^n x = \left(0, 0, \ldots, \prod_{i=1}^{n} a_i, 0, \ldots\right) \text{ for all } n \in \mathbb{N}.$$

Thus, we find that

$$\|T^n x\| = \frac{n+1}{2n} \to \frac{1}{2} \text{ as } n \to \infty,$$

which shows that $T^n x \nrightarrow 0$.

Observation

- In Theorem 5.7, we can relax the compactness of X by requiring that $\overline{T(X)}$ be compact (just applying the theorem on the restriction of T on $\overline{T(X)}$).

Arguing like in Corollary 5.1, it is also immediate to prove the following:

Corollary 5.3 *Let X be a compact metric space and let $T : X \to X$. If T^n is a contractive, for some $n \geq 1$, then T has a unique fixed point $x \in X$.*

Observation

- The problem of defining a family of functions $\mathscr{F} = \{\lambda(x, y)\}$ satisfying $0 \leqslant \lambda(x, y) < 1$, $\sup \lambda(x, y) = 1$ and such that Banach's theorem holds when the constant λ is replaced with any $\lambda(x, y) \in \mathscr{F}$, was suggested to Rakotch[1] [497] by Professor H. Hanani.
- It is quite interesting to observe that the condition $d(Tx, Ty) \leqslant \lambda(x, y) d(x, y)$ with $\lambda(x, y) < 1$ does not guarantee the existence of a fixed point. Indeed, let $X = [1, \infty)$ be the set of real numbers with usual metric $d(x, y) = |x - y|$ and let $T : X \to X$ be defined by $Tx = x + \frac{1}{2x}$. Then

$$d(Tx, Ty) = \lambda(x, y) d(x, y); \quad \lambda(x, y) = 1 - \frac{1}{2xy} < 1,$$

but there is no $x \in X$ such that $Tx = x$.

In 1962, E. Rakotch introduced and studied a class of mappings satisfying the condition $d(Tx, Ty) \leqslant \lambda(x, y) d(x, y)$, where $\lambda(x, y) = \lambda(d(x, y))$, i.e. λ depends on the distance between x and y only. Indeed, Rakotch proved the following:

Theorem 5.8 (Rakotch [497]) *Let (X, d) be a complete metric space and $T : X \to X$ be a self-mapping of X satisfying*

$$d(Tx, Ty) \leqslant \lambda(x, y) d(x, y) \text{ for all } x, y \in X, \tag{5.7}$$

where (i) $\lambda(x, y) = \lambda(d(x, y))$, (ii) $0 \leqslant \lambda(t) < 1$ for every $t > 0$ and (iii) $\lambda(t)$ is a monotonically decreasing function of t, then there exists a unique fixed point of T.

Sehgel [547] obtained an interesting generalization of Theorem 5.6 which we discuss now.

Theorem 5.9 *Let T be a continuous mapping from a metric space X into itself such that for all x, y in X with $x \neq y$, we have*

$$d(Tx, Ty) < \max \{d(x, Tx), d(y, Ty), d(x, y)\}. \tag{5.8}$$

Suppose that, for some z in X, the sequence $\{T^n z\}$ has a cluster point u. Then the sequence $\{T^n z\}$ converges to u and u is the unique fixed point of T.

Proof If $T^n z = T^{n+1} z$ for some nonnegative integer n, then it is obvious that $\lim_{n \to \infty} T^n z = u$ and by (5.8), u is the unique fixed point of T. Hence, we may assume that for all nonnegative integers $n, d(T^n z, T^{n+1} z) > 0$. Let $U(y) = d(y, Ty)$. Then U is a continuous function and $U(T^n z) < U(T^{n-1} z) < U(z)$ for all positive integers n by condition (5.8). Let $r = \lim_{n \to \infty} U(T^n z)$. Let $\{n_i\}$ be a sequence of positive integers such that $T^{n_i} z$ converges to u. Then we have

---

[1] Rakotch's thesis for the degree of Master of Science, submitted to the Department of Mathematics of the Technion, Israel Institute of Technology, Haifa.

$U(u) = U(Tu) = r$. By (5.8) we have $u = Tu$ and $r = 0$. Now we prove that the sequence $\{T^n z\}$ converges to u. Given $\varepsilon > 0$, there exists a positive integer k such that

$$\max \{U(T^k z), d(T^k z, u)\} < \varepsilon.$$

By condition (5.8), we have for all positive integers $n \geqslant k$

$$d(T^n z, u) = d(T^n z, T^n u) < \max \{U(T^{n-1} z, d(T^{n-1} z, u)\}$$
$$< \max \{U(T^{n-1} z, d(T^{n-2} z, u)\} < \max \{U(T^k z), d(T^k z, u)\}$$
$$< \varepsilon.$$

Hence, $\lim_{n \to \infty} T^n z = u$, and this concludes the proof. ∎

Remark 5.4 For more on contractive mappings and its generalization, see Rhoades [519].

The following example shows that a contractive mapping T satisfies condition (5.8) but the converse is not true in general.

Example 5.15 Let $T : [0, 5] \to [0, 5]$ be defined as follows:

$$Tx = \begin{cases} \frac{x}{2} & \text{if } x \in [0, 4] \\ -2x + 10 & \text{if } x \in [4, 5]. \end{cases}$$

Then $T(4) - T(5) = 2$ showing that T is not contractive. But one can check that T satisfies (5.8).

We now state an easy consequence of Banach's contraction theorem as follows.

Theorem 5.10 *Let X be a Banach space, x_0 a point in X, and let $B_{x_0, \rho} = \{x \in X : \|x - x_0\| \leqslant \rho\}$ the closed ball in X with center at x_0 and radius ρ. Let $F : B_{x_0, \rho} \to X$ be given by*

$$Fx = x + \varphi(x) \text{ for all } x \in B_{x_0, \rho},$$

where φ is a Lipschitz continuous function with constant $\alpha < 1$. Then $Fx = y$ has a unique solution for every $y \in B_{F(x_0), (1-\alpha)\rho}$ and F^{-1} satisfies Lipschitz conditions with constant $1/(1 - \alpha)$.

Recently there have been numerous generalization of the Banach contraction theorem by weakening its hypothesis while retaining the convergence property of the successive iterates to the unique fixed point of the mapping.

The following theorem was obtained by Boyd and Wong [75].

Theorem 5.11 *Let X be a complete metric space, and let $T : X \to X$ be a mapping. Suppose there exists a function $\varphi : \mathbb{R}^+ \to \mathbb{R}^+$ upper semicontinuous from the right such that*

$$d(Tx, Ty) \leqslant \varphi(d(x, y)) \text{ for all } x, y \in X. \tag{5.9}$$

If $\varphi(t) < t$ for each $t > 0$, then

(i) *T has a unique fixed point u in X,*
(ii) *$\{T^n x\}$ converges to u for every x in X.*

Proof For any x in X, let $x_n = T^n x, n = 1, 2, \ldots$ and $d_n = d(x_n, x_{n+1}) = d(T^n x, T^{n+1} x)$. We may presume that $d_n > 0$ for $n \geqslant 0$. Then for $n > 1$,

$$d_n = d(Tx_{n-1}, Tx_n) \leqslant \varphi(d_{n-1}) < d_{n-1}.$$

Thus, the sequence $\{d_n\}$ is monotonically decreasing and bounded below so it is convergent. Let $\lim\limits_{n \to \infty} d_n = d$. We show that $d = 0$. If $d > 0$, then $d_{n+1} \leqslant \varphi(d_n)$. By the upper semicontinuity from the right of the function φ, we obtain

$$d = \lim_{n \to \infty} d_{n+1} \leqslant \limsup_{n \to \infty} \varphi(d_n) \leqslant \varphi(d)$$

which contradicts the property of φ. Thus, $d = 0$ and $d_n \to 0$ as $n \to \infty$.

We now claim that the sequence $\{x_n\}$ is Cauchy. Suppose, on the contrary, that $\{x_n\}$ is not a Cauchy sequence. Then there exists $\varepsilon > 0$ and for each positive integer k there exist integers $n(k)$ and $m(k)$ with $n(k) > m(k) \geq k$ such that $d(x_{m(k)}, x_{n(k)}) \geqslant \varepsilon$. Without loss of generality, we can assume that $n(k)$ is the smallest integer greater than $m(k)$ satisfying the above inequality. Let $r_k = d(x_{m(k)}, x_{n(k)})$. Then

$$\varepsilon \leqslant r_k \leqslant d(x_{m(k)}, x_{n(k)-1}) + d(x_{n(k)-1}, x_{n(k)}) \leqslant \varepsilon + d_{n(k)-1}.$$

This implies that $\lim\limits_{k \to \infty} r_k = \varepsilon$. Also we have

$$\varepsilon \leqslant r_k \leqslant d(x_{m(k)}, x_{m(k)+1}) + d(x_{m(k)+1}, x_{n(k))+1}) + d(x_{n(k)+1,}, x_{n(k)})$$
$$\leqslant d_{m(k)} + \varphi(r_k) + d_{n(k)} \to \varphi(\varepsilon) \text{ as } k \to \infty.$$

Thus $\varepsilon \leqslant \varphi(\varepsilon)$ which is a contradiction. Thus, $\{T^n x\}$ is a Cauchy sequence for any $x \in X$.

Since $\{T^n x\}$ is a Cauchy sequence and X is complete, $\lim_{n \to \infty} T^n x = u \in X$. Since T is continuous, $T(u) = T(\lim_{n \to \infty} T^n x) = \lim_{n \to \infty} T^{(n+1)} x = u$. Uniqueness of u follows from condition (5.9). ∎

Remark 5.5 The condition $\varphi(t) < t$ for all $t > 0$ cannot be dispensed with in Theorem 5.11. Even if condition $\varphi(t) < t$ does not hold for at least one value of $t > 0$, then T may have no fixed point or else more than one fixed point.

Example 5.16 Let $X = (-\infty, -1] \cup [1, \infty)$ be a metric space with the absolute value metric. Define $T : X \to X$ by

$$Tx = \begin{cases} \frac{1}{2}(x+1) & \text{if } x \geq 1 \\ \frac{1}{2}(x-1) & \text{if } x \leqslant -1. \end{cases}$$

and $\qquad\qquad Sx = -Tx$ for all $x \in X.$

Then T and S satisfy condition (5.9) with

$$\varphi(t) = \begin{cases} \frac{1}{2}t & \text{if } t < 2 \\ \frac{1}{2}t + 1 & \text{if } t \geq 2. \end{cases}$$

Notice that the function φ satisfies all the conditions of Theorem 5.11 except that $\varphi(2) = 2$. It may be observed that T has two fixed points -1 and 1, while S has no fixed points.

The following is a fixed point theorem concerning Kannan–Reich-type mapping proved by Hardy and Rogers [266].

Theorem 5.12 *Let T be a mapping from a complete metric space X into itself satisfying the following:*

$$d(Tx, Ty) \leqslant a[d(x, Tx) + d(y, Ty)] + b[d(y, Tx) + d(x, Ty)] + cd(x, y),$$
$$\tag{5.10}$$

for any x, y in X where a, b and c are nonnegative numbers such that $2a + 2b + c < 1$. Then T has a unique fixed point u in X. In fact for any $x \in X$, the sequence $\{T^n x\}$ converges to u.

The following is an example of discontinuous mapping which satisfies Kannan–Reich-type condition (5.10).

Example 5.17 Let $T : [-2, 1] \to [-2, 1]$ be defined as

$$Tx = \begin{cases} \frac{x}{2} & \text{if } x \in [0, 1) \\ -\frac{1}{8} & \text{if } x = 1 \\ -\frac{x}{2} & \text{if } x \in [-2, 0). \end{cases}$$

Remark 5.6 Take $c \neq 0$, $a = b = 0$ in Theorem 5.11 above. Then we get Banach contraction mapping theorem as a special case.

Corollary 5.4 (Kannan, Theorem 1 [308, 309]) *Let T be a mapping of a complete metric space X into itself. Suppose that there exists a number $r \in [0, \frac{1}{2})$ such that*

$$d(Tx, Ty) \leqslant r[d(x, Tx) + d(y, Ty)], \text{ for all } x, y \in X. \tag{5.11}$$

Then T has a unique fixed point in X.

Corollary 5.5 (Chatterjea, Theorem 1 [140]) *Let T be a mapping of a complete metric space X into itself. Suppose that there exists a number $r \in [0, \frac{1}{2})$ such that*

$$d(Tx, Ty) \leqslant r[d(x, Ty) + d(y, Tx)], \text{ for all } x, y \in X. \tag{5.12}$$

Then T has a unique fixed point in X.

Notice that Banach contraction mappings are continuous. We now furnish an example to show that Kannan's mappings are not necessarily continuous.

Example 5.18 Let $X = [0, 4]$ be endowed with the usual metric $d(x, y) = |x - y|$. Define $T : X \to X$ by

$$T(x) = \begin{cases} \frac{x}{3}, & \text{if } x \leqslant 3 \\ \frac{x}{4}, & \text{if } 3 < x \leqslant 4. \end{cases}$$

For $x, y \in [0, 3]$ we have

$$d(T(x), T(y)) = \frac{1}{3}|x - y| = \frac{1}{3}|x - T(x) + T(x) - T(y) + T(y) - y|$$
$$\leqslant \frac{1}{3}(d(x, T(x)) + d(T(x), T(y))) + d(y, T(y))$$

The above inequality reduces to

$$d(T(x), T(y)) \leqslant \frac{1}{2}(d(x, T(x)) + d(y, T(y))).$$

Similarly, we can obtain the same inequality for $x, y \in (3, 4]$.
    For $x \in [0, 3]$ and $y \in (3, 4]$, we have

$$d(T(x), T(y)) = \left| \frac{x}{3} - \frac{y}{4} \right| \leqslant 1 < \frac{9}{8} \leqslant \frac{1}{2} d(y, T(y)).$$

It can easily be observed that the similar inequality hold for $x \in (3, 4]$ and $y \in [0, 3]$. Thus, T satisfies (5.10), but T is discontinuous.

We state without proof the following extension of Theorem 5.1 due to Čirič appeared in 1974 in [144].

Theorem 5.13 *Let X be a complete metric space, and let $T : X \to X$ be such that*

$$d(Tx, Ty) \leqslant \max\{d(x, y), d(x, Tx), d(y, Ty), d(x, Ty), d(y, Tx)\}$$

for some $\lambda < 1$ and every $x, y \in X$. Then T has a unique fixed point $u \in X$. Moreover, $d(T^n x_0, u) = O(\lambda^n)$ for any $x_0 \in X$.

Example 5.19 Let $X = [0, 1]$ be endowed with usual metric d. Consider the map $T : X \to X$ defined by

$$T(x) = \begin{cases} \frac{1}{2} + 2x, & \text{if } x \in [0, \frac{1}{4}] \\ \frac{1}{2}, & \text{if } x \in (\frac{1}{4}, 1]. \end{cases}$$

Here we notice that T is not continuous. However, it is easy to check that it is continuous at the fixed point $(x = 1/2)$. Also notice that the mapping T fulfills the hypotheses of Theorem 5.11 with $\lambda = 2/3$.

In 1975, Hussain and Sehgal [283] proved a fixed point theorem which generalizes the Kannan–Reich and Čirič type of generalized contraction mapping theorems. An extension of Hussain and Sehgal's result was obtained by Singh and Meade [562] in 1977. Both results deal with common fixed point theorems of a pair of mappings. We state below a theorem of Hussain and Sehgal for a fixed point of a mapping.

Theorem 5.14 *Let (X, d) be a complete metric space and T be a mapping of X into itself. Let $\phi : (\mathbb{R}^+)^5 \to \mathbb{R}^+$ be continuous and nondecreasing in each coordinate variable and let T satisfies the following condition:*

$$d(Tx, Ty) \leqslant \phi\{d(x, Tx), d(y, Ty), d(x, Ty), d(y, Tx), d(x, y)\} \text{ for all } x, y \in X.$$

If $\phi(t, t, a_1 t, a_2 t) < t$, for $t > 0$ and $a_1 \in \{0, 1, 2\}$ with $a_1 + a_2 = 2$, then there exists a unique u such that $fu = u$.

Remark 5.7 Singh and Meade [562] proved the above theorem under the assumption that ϕ is upper semicontinuous and $a_1 + a_2 = 3$.

The following theorem is due to Guseman Jr. [259]. It was first proved by Sehgal [548]. This class of mappings have at each point an iterate that is contraction and yet none of the iterates of F is a contraction.

Definition 5.6 Let (X, d) and (Y, d) be metric spaces and $T : X \to Y$ be a mapping from X in to Y. The mapping T is called a local power contraction mapping if there exists a constant $0 < k < 1$ and for each $x \in X$, there exists a positive integer $n = n(x)$ such that $d(T^n x, T^n y) \leqslant kd(x, y)$ for all y in X.

Theorem 5.15 *Let T be a mapping from a complete metric space X into itself. Suppose there exists $K \subset X$ such that the following conditions are satisfied.*

 (i) $T(K) \subset K$,
 (ii) T is a local power contraction mapping on K,
 (iii) For some $z \in K$, $\{\overline{T^n z : n \geqslant 1}\} \subset K$.

Then T has a unique fixed point u and $T^n x \to u$ as $n \to \infty$ for each $x \in K$.

Proof First we prove that for each $x \in K$, the number $r(x) = \sup_n \{d(x, T^n x)\} < \infty$.

For each $x \in K$, let $m(x) = \sup\{d(x, T^n x) : 1 \leqslant n \leqslant n(x)\}$. Now if n is an arbitrary positive integer, there is an integer $s \geqslant 0$ such that $sn(x) < n \leqslant (s + 1)n(x)$, and this gives

$$d(T^n x, x) \leqslant d(T^{n(x)}(T^{n-n(x)} x), T^{n(x)} x) + d(T^{n(x)} x, x)$$

$$\leqslant kd(T^{n-n(x)} x, x) + m(x)$$

$$\leqslant m(x) + km(x) + \cdots + k^s m(x) < \frac{m(x)}{1 - k}.$$

Thus, the above inequality yields

$$\sup_{n}\{d(x, T^n x)\} \leqslant \frac{m(x)}{1-k} < \infty.$$

This proves that $r(x) < \infty$.

Let $z_1 = T^{n(z)}z$ and $z_{i+1} = T^{n(z_i)}z_i$. Then we have

$$d(z_{i+1}, z_i) \leqslant k^i d(T^{n(z_i)}, z) \leqslant k^i r(z)$$

and

$$d(z_j, z_i) \leqslant \sum_{k=1}^{j-1} d(z_{k+1}, z_k) \leqslant \frac{k^i}{1-k} r(z) \text{ for } j > i.$$

This shows that $\{z_i\}$ is Cauchy. By completeness of X and conditions (iii), $z_i \to u \in K$ as $i \to \infty$. Obviously $T^{n(u)}z_i \to T^{n(u)}u$ as $i \to \infty$. But

$$d(T^{n(u)}z_i, z_i) = d(T^{n(z_{i-1})}(T^{n(u)}z_{i-1}), T^{n(z_{i-1})}z_{i-1})$$
$$\leqslant kd(T^{n(u)}z_{i-1}, z_{i-1})$$
$$\leqslant k^i d(T^{n(u)}z, z) \text{ (by repeatedly using the above inequality)}$$
$$\to 0 \text{ as } i \to \infty.$$

Thus, $d(T^{n(u)}u, u) = \lim_{i \to \infty} d(T^{n(u)}z_i, z_i) = 0$.

Hence, u is the unique fixed point of $T^{n(u)}$ in K. Since

$$T(u) = T(T^{n(u)}u) = T^{n(u)}(Tu), \text{ we have } Tu = u.$$

Now let $t(x) = \sup\{d(T^m x, u) : 1 \leqslant m \leqslant n(u) - 1\}$. If $n = an(u) + s$ with integers a and s such that $a > 0, 0 \leqslant s < n(u)$, then

$$d(T^n x, u) = d(T^{an(u)+s}x, T^{n(u)}u) \leqslant k^a d(T^s, u) \leqslant k^a t(x).$$

Above inequality implies that $T^n x \to u$ as $n \to \infty$. ∎

The following example illustrates that there exists a continuous mapping on a metric space satisfying the condition of the above theorem where no iterate of the mapping is a contraction.

Example 5.20 Let $X = [0, 1]$ with the metric $d(x, y) = |x - y|$. We note that X is of the form

$$X = \left\{ \bigcup_{n=1}^{\infty} \left[\frac{1}{2^n}, \frac{1}{2^{n-1}} \right] \right\} \bigcup \{0\}.$$

$$Tx = \begin{cases} \frac{n+2}{n+3}\left(x - \frac{1}{2^{n-1}}\right) + \frac{1}{2^n}, & x \in \left[\frac{3n+5}{2^{n+1}(n+2)}, \frac{1}{2^{n-1}}\right] \\ \frac{1}{2^{n+1}}, & x \in \left[\frac{1}{2^n}, \frac{3n+5}{2^{n+1}(n+2)}\right], \end{cases}$$

and $T(0) = 0$.

From the definition, it follows that T is a nondecreasing continuous function on $[0, 1]$ with values in $[0, 1]$ and 0 is the unique fixed point of T. We show that T is a local power contraction but not a power contraction.

For $x \in \left[\frac{1}{2^n}, \frac{1}{2^{n-1}}\right] = I_n$ and for any y in X,

$$|Tx - Ty| \leqslant \frac{n+3}{n+4}|x - y|$$

and hence

$$|T^{n+3}x - T^{n+3}y| \leqslant \frac{1}{2}|x - y|.$$

Also $|T^n(0) - T^n y| \leqslant \frac{1}{2}|0 - y|.$

Let $0 \leqslant k \leqslant 1$ and N be a given positive integer. Let $n > \left(\frac{Nk}{1-k}\right) - 2$. By the uniform continuity of the iterates of T, there is a $\delta > 0$ such that

$$|T^i x - T^i y| < \frac{n+N+3}{2^{n+N+1}(n+N+2)}$$

for $|x - y| < \delta$ and $1 \leqslant i \leqslant N$. Taking $x = \frac{1}{2^{n-1}}$ and y any member of I_n such that $0 < |x - y| < \delta$, it can be shown that $T^i x$ and $T^i y$ are both member of

$$\left[\frac{3(n+i)+5}{2^{n+i+1}(n+i+2)}, \frac{1}{2^{n+i+1}}\right], \quad i = 1, 2, \ldots, N.$$

Then we have

$$|Tx - Ty| = \frac{n+2}{n+3}|x - y|$$

$$|T^2 x - T^2 y| = \frac{n+2}{n+4}|x - y|$$

$$\vdots$$

$$|T^N x - T^N y| = \frac{n+2}{n+2+N}|x - y| > k|x - y|.$$

5.1.5 Caristi–Kirk's Fixed Point Theorem and Generalizations

In 1976, Caristi [128, 129] proved the following fixed point theorem.

Theorem 5.16 (Caristi [128, 129]) *Let* (X, d) *be a complete metric space, and let* $T : X \to X$ *an arbitrary map of X into itself (with T not necessarily continuous). Assume there exists a lower semicontinuous function* $\varphi : X \to [0, \infty)$ *such that*

$$d(x, T(x)) \leqslant \varphi(x) - \varphi(T(x)), \quad \text{for all } x \in X. \tag{5.13}$$

Then T has (at least) a fixed point in X.

Proof For $x, y \in X$, define the relation $x \leqslant y$ iff $d(x, y) \leqslant \varphi(x) - \varphi(y)$. Then $(X, \leq)$ is a partially ordered set. Let $x_0 \in X$ be an arbitrary element of X. Using Zorn's lemma, we obtain a maximal totally ordered subset E of X containing x_0. Assume $E = \{x_\alpha\}_{\alpha \in I}$, where I is totally ordered and

$$x_\alpha \leqslant x_\beta \Leftrightarrow \alpha \leqslant \beta, \quad (\alpha, \beta \in I).$$

Now $\{\varphi(x_\alpha)\}_{\alpha \in I}$ is a decreasing net in $\mathbb{R}^+$, so there exists $r \geqslant 0$ such that $\varphi(x_\alpha) \to r$ as $\alpha \uparrow$. Let $\varepsilon > 0$. Then there exists an $\alpha_0 \in I$ such that

$$\alpha \geqslant \alpha_0 \Rightarrow r \leqslant \varphi(x_\alpha) \leqslant r + \varepsilon.$$

Let $\beta \geqslant \alpha \geqslant \alpha_0$. Then

$$d(x_\alpha, x_\beta) \leqslant \varphi(x_\alpha) - \varphi(x_\beta) \leqslant \varepsilon.$$

This implies that $\{x_\alpha\}_{\alpha \in I}$ is a Cauchy net in X. Since X is complete, there exists $x \in X$ such that $x_\alpha \to x$ as $\alpha \uparrow$. But, φ is lower semicontinuous, and hence $\varphi(x) \leqslant r$. If $\beta \geqslant \alpha$, then

$$d(x_\alpha, x_\beta) \leqslant \varphi(x_\alpha)) - \varphi(x_\beta).$$

Letting $\beta \uparrow$, we have $d(x_\alpha, x) \leqslant \varphi(x_\alpha) - r \leqslant \varphi(x_\alpha) - \varphi(x)$; giving us $x_\alpha \leqslant x, \alpha \in I$. In particular, $x_0 \leq x$. Now E is maximal and thus $x \in E$. But we have

$$d(x, Tx) \leqslant \varphi(x) - \varphi(Tx),$$

implying that $x_\alpha \leqslant x \leqslant Tx$, $\alpha \in I$. Again by maximality $Tx \in E$. Therefore, $Tx \leqslant x$ and thus $x = Tx$. ∎

If we assume the continuity of T, we obtain a slightly stronger result, even relaxing the continuity hypothesis on φ.

Theorem 5.17 *Let (X, d) be a complete metric space, and let $T : X \to X$ be a continuous map of X into itself (with T not necessarily continuous). Assume there exists a function $\varphi : X \to [0, \infty)$ such that*

$$d(x, T(x)) \leqslant \varphi(x) - \varphi(T(x)), \quad for\ all\ \ x \in X.$$

Then f has a fixed point in X. Moreover, for any $x_0 \in X$ the sequence $\{T^n(x_0)\}$ converges to a fixed point of T.

Proof Choose x_0 in X, and let $x_n = T^n(x_0)$. Then

$$d(x_{n+1,}\, x_n) \leqslant \varphi(x_n) - \varphi(x_{n+1}). \tag{5.14}$$

This implies that $\{\varphi(x_n)\}$ is a decreasing sequence. Summing (5.14) from 0 to N, we have

$$\sum_{n=0}^{N} d(x_n, x_{n+1}) \leqslant \varphi(x_0) - \varphi(x_{N+1}) \leqslant \varphi(x_0).$$

Hence,

$$\sum_{n=0}^{N} d(x_n, x_{n+1}) \leqslant \varphi(x_0).$$

This implies that $\{x_n\}$ is a Cauchy sequence in X. By completeness of X, there exists $u \in X$ such that $x_n \to u$. By continuity of T, it follows that $T(u) = u$.

Remark 5.8 Theorem 5.1 is a particular case, obtained for $\varphi(x) = d(x, T(x))/(1 - k)$, $k \in (0, 1)$.

Observation

1. Caristi's theorem is equivalent to a theorem of Ekeland [221] which turned out to be an abstraction of a lemma of Bishop and Phelps [60]. For applications, one may refer to Kirk and Caristi [333], Kirk [328] and Downing and Kirk [201, 202].
2. Caristi's proof involves transfinite induction. The proof given above is by Kirk [329] and implicit in a paper by Brønstedt [83].

The following theorem due to Downing and Kirk [202] is a generalization of the Caristi's fixed point theorem.

Theorem 5.18 (Downing and Kirk [202]) *Let X and Y be complete metric spaces and $f : X \to X$ an arbitrary mapping. Suppose there exists a closed mapping $g : X \to Y$, a lower semicontinuous mapping $\varphi : g(X) \to \mathbb{R}^+$, and a constant $c > 0$ such that for each $x \in X$,*

$$d(x, f(x)) \leqslant \varphi(g(x)) - \varphi(g(f(x)))$$
and $\qquad cd(g(x), g(f(x))) \leqslant \varphi(g(x)) - \varphi(g(f(x))). \tag{5.15}$

Then there exists $u \in X$ such that $f(u) = u$.

Proof For $x, y \in X$, define $x \leqslant y$ provided that

$$d(x, y) \leqslant \varphi(g(x)) - \varphi(g(y))$$

and $\qquad cd(g(x), g(y)) \leqslant \varphi(g(x)) - \varphi(g(y)).$

Let $\{x_\alpha\}_{\alpha \in I}$ be any chain, that is, $(I, \leqslant)$ be a totally ordered set with $x_\alpha \leqslant x_\beta$ iff $\alpha \leqslant \beta$. Then $\{\varphi(g(x_\alpha))\}_{\alpha \in I}$ is a decreasing net in $\mathbb{R}^+$. Hence, there exists $r \geqslant 0$ such that $\varphi(g(x_\alpha)) \downarrow r$. Let $\varepsilon > 0$. Then there exists $\alpha_0 \in I$ such that

$$r \leqslant \varphi(g(x_\alpha)) \leqslant r + \varepsilon \text{ for } \alpha \geqslant \alpha_0.$$

Hence, for $\beta \geqslant \alpha \geqslant \alpha_0$,

$$d(x_\alpha, x_\beta) \leqslant \varphi(g(x_\alpha)) - \varphi(g(x_\beta)) \leqslant \varepsilon,$$

and $\qquad cd(g(x_\alpha), g(x_\beta)) \leqslant \varphi(g(x_\alpha)) - \varphi(g(x_\beta)) \leqslant \varepsilon.$

Above inequalities imply that $\{g(x_\alpha)\}$ is a Cauchy net in Y and $\{x_\alpha\}$ is a Cauchy net in X. By completeness of X and Y, there exists $u \in X$ and $v \in Y$ such that $x_\alpha \to u$ and $g(x_\alpha) \to v$. Since g is a closed mapping, we have $g(u) = v$. By lower semicontinuity of φ, we have $\varphi(g(u)) \leqslant r$. Furthermore, if $\alpha, \beta \in I$ with $\alpha \leqslant \beta$, then

$$d(x_\alpha, x_\beta) \leqslant \varphi(g(x_\alpha)) - \varphi(g(x_\beta)) \leqslant \varphi(g(x_\alpha)) - r$$

and $\qquad cd(g(x_\alpha), g(x_\beta)) \leqslant \varphi(g(x_\alpha)) - r.$

By taking limits with respect to β, we have

$$d(x_\alpha, u) \leqslant \varphi(g(x_\alpha)) - r \leqslant \varphi(g(x_\alpha)) - \varphi(g(u))$$

and $\qquad cd(g(x_\alpha), g(u)) \leqslant \varphi(g(x_\alpha)) - \varphi(g(u)).$

This shows that $x_\alpha \leqslant u, \alpha \in I$, and thus every totally ordered set (a chain) in $(X, \leqslant)$ has an upper bound. By Zorn's lemma, there exists a maximal element $x \in X$. By (5.15), we have $x \leqslant f(x)$, and hence $x = f(x)$. $\blacksquare$

Remark 5.9 The above theorem reduces to the Theorem 5.16 if

$$X = Y, f = T, g = I, \text{ the identity mapping and } c = 1.$$

5.1.6 *Fixed Points via Generalized Contractions*

In this section, we present some global and local fixed point results for a self-mapping on a topologically complete/complete metric spaces by relaxing the requirement of compactness of the space as required in Theorem 2.2 of Tarafdar and Yuan [587]. Our results on metric space structures are motivated with relaxing the dependency of the function $\lambda(x, y)$ as well as the contraction principle on two arbitrary points of the space. In fact, our contraction conditions are so designed that it depends only on one arbitrary point of the space.

Let (X, d) be a metric space and let T be a mapping of X into itself. An orbit of T at a point x of X is the set

$$\mathcal{O}(x, T) := \{x, Tx, \ldots, T^n x, \ldots\}.$$

Recall that a mapping $T : X \to X$ is said to be a Banach contraction mapping if it satisfies the following inequality

$$d(T(x), T(y)) \leqslant \lambda \, d(x, y), \tag{5.16}$$

for all x, y in X, where $0 < \lambda < 1$.

Observation

- It is well known that a Banach contraction mapping T on a complete metric space X has a unique fixed point. Following the Banach contraction principle, many researchers introduced various concepts of locally contraction mappings, or of weakly contraction mappings, replacing the constant coefficient λ by a function $\lambda(x, y)$ satisfying $0 \leqslant \lambda(x, y) < 1$, sup $\lambda(x, y) = 1$ and such that the Banach theorem still holds. Significant initial results in this area were obtained by Edelstein [212] and Rakoch [498], and then by Boyd and Wong [75], Meir and Keeler [393], Ćirić [143] and many other authors.
- Pathak and Shahzad [464] provide one more affirmative answer to an open question of Rhoades [520, p. 242], whether or not there exists a contractive definition which is strong enough to generate a fixed point but which does not force the map to be continuous at the fixed point. It may be observed that in most of the fixed point theorems in the literature either explicitly assume continuity or, as shown by Rhoades [520] and Hicks and Rhoades [271], the contractive definitions themselves imply continuity at the fixed point. However, an affirmative answer was given by Pathak et al. in [455, Example 2.1].

Let X be a topological space and $Y \subset X$ be equipped with relativised topology.

Definition 5.7 A mapping $T : Y \subset X \to X$ is said to be a weak topological contraction, if Y is T-invariant and T is continuous and closed such that for each nonempty and closed subset A of Y with $T(A) = A$, A is a singleton set. Further, if $\delta(T^n(Y)) \to 0$ as $n \to \infty$ then the mapping T is said to be a strong topological contraction.

Remark 5.10 If X is a bounded metric space (i.e. $\delta(X)$, the diameter of X is finite) and T is a Banach contraction, then clearly T is a weak topological contraction.

We now introduce a concept that is more general, refined, viable and productive than the celebrated Banach contraction principle in the following definition.

Definition 5.8 Let (X, d) be a metric space. A mapping $T : Y \subset X \to X$ is said to be a metric p-contraction (or simply p-contraction) mapping if Y is T-invariant and it satisfies the following inequality

$$d(T(x), T^2(x)) \leqslant p(x)d(x, T(x)) \tag{5.17}$$

for all x in Y, where $p : Y \to [0, 1]$ is a function such that $p(x) < 1$ for all $x \in Y$ and $\sup_{x \in Y} p(Tx) = \lambda < 1$. Further, if $\bigcap_{n=0}^{\infty} T^n(Y)$ is a singleton set, where $T^n(Y) := T(T^{n-1}(Y))$ for each $n \in \mathbb{N}$ and $T^0(Y) := Y$, then T is said to be strongly p-contraction.

If $p(x) = \alpha < 1$ for all x in Y, then the p-contraction mapping is said to be a fundamental contraction. A fundamental contraction is also known as a Banach operator. Note that when $Y = X$ and $y = T(x)$, the Banach contraction principle satisfies the fundamental contraction. If $p(x) \leqslant 1$ for all $x \in Y$ and $\sup_{x \in Y} p(Tx) = 1$, then the p-contraction mapping is said to be fundamentally p-nonexpansive.

Especially when $p(x) = 1$ for all x in Y, then the fundamental p-nonexpansive mapping is said to be a fundamentally nonexpansive. Notice that if $\sup_{x \in Y} p(x) < 1$ (or $\leqslant 1$), then $\sup_{x \in Y} p(Tx) < 1$ (or $\leqslant 1$) since $T(Y) \subseteq Y$.

We now convince the reader that the concept of p-contraction is more general than the Banach contraction principle. To see this, we observe the following very simple example.

Example 5.21 Let $X = [0, 0.6]$ equipped with metric topology $\mathfrak{I}_d$, d being usual metric on X. Let $T : X \to X$ be a mapping defined by $T(x) = x^2 \ \forall x \in X$. Let $p : X \to [0, 1]$ be a function defined by $p(x) = x(1 + x) \ \forall \ x \in X$. Then, for all x in X we have

$$d(T(x), T^2(x)) = \left| x^2 - x^4 \right| = x(1 + x) \cdot x(1 - x) = p(x) \cdot x(1 - x) = p(x) \cdot d(x, T(x)).$$

Note that $\sup_{x \in X} p(x) = 0.96 < 1$. Also, $A = \{0\}$ is the only closed subset of X such that $T(A) = A$ and A is singleton. Moreover, $\delta(T^n(X)) \to 0$ as $n \to \infty$. Thus, T is a strong p-contraction. However, one can easily check that T is not a Banach contraction. To see this, let us consider $x = 0.5$ and $y = 0.6$ in (5.16).

The following example shows that a p-contraction mapping need not be continuous.

Example 5.22 Let X be a circumference:

$$X = \{(x, y) : x = \cos t, \ y = \sin t, \ 0 \leqslant t \leqslant 2\pi\}$$

equipped with usual metric d. Let $T : X \to X$ be a mapping defined by $T(t) = \frac{1}{2}t$, if $t \in [0, \frac{\pi}{2}) \bigcup (\pi, 2\pi]$, $T(t) = t - \frac{\pi}{2}$, if $t \in [\frac{\pi}{2}, \pi]$. Let $p : X \to [0, 1]$ be a function defined by $p(t) = \frac{1}{2}$, if $t \in [0, \frac{\pi}{2}) \bigcup (\pi, 2\pi]$, $p(t) = \frac{|t - \frac{\pi}{2}|}{\pi}$, if $t \in [\frac{\pi}{2}, \pi]$. Note that T and p both are discontinuous at $t = \frac{\pi}{2}$. Also, $\sup_{t \in [0, 2\pi]} p(t) = \frac{1}{2}$. Further, we can easily check that T is a p-contraction on X.

Remark 5.11 It is interesting to note that a *p*-contraction, in general, does not satisfy Caristi's condition (5.13) but every fundamental contraction does. However, if $p(Tx) \leqslant p(x)$ for all $x \in X$, then a *p*-contraction $T : X \to X$ satisfies Caristi's condition.

Let X be a nonempty set. We shall denote by 2^X the family of all nonempty subsets of X. Suppose $f : X \to 2^X$ is a set-valued mapping, then a point $x_0 \in X$ is said to be an *end point* of f if x_0 is fixed point of f and $f(x_0) = \{x_0\}$. Let $\mathbb{R}^+$ and $\mathbb{N}$ denote $[0, \infty)$ and the set of all natural numbers, respectively, throughout this section.

Tarafdar and Yuan [587] introduced the following definition of *set-valued topological contraction*.

Definition 5.9 Let X be a topological space. Then $f : X \to 2^X$ is said to be a (set-valued) *topological contraction* if f is upper semicontinuous with closed values such that for each nonempty closed subset A of X with $f(A) = A$, A is a singleton set; i.e. A is an endpoint of f.

In [587], the following theorem was proved.

Theorem 5.19 *Let X be a compact Hausdorff topological space and $f : X \to 2^X$ be (set-valued) topological contraction with closed values. Then f has a unique endpoint $x_0 \in X$ such that $\{x_0\} = \bigcap_{n=0}^{\infty} f^n(X)$, where $f^n(x) := f(f^{n-1}(X))$ for each $n \in \mathbb{N}$ and $f^0(X) :\equiv X$.*

Definition 5.10 (*Ćirić* [143]) A metric space (X, d) is said to be T-orbitally complete if T is a self-mapping of X and if any Cauchy subsequence $\{T^{n_i}x\}$ in orbit $\mathcal{O}(x, T)$, $x \in X$, converges in X.

Definition 5.11 (*Ćirić* [143]) An operator $T : X \to X$ on X is said to be orbitally continuous if $T^{n_i}x \to u$ then $T(T^{n_i}x) \to Tu$ as $i \to \infty$.

Definition 5.12 (*Ćirić* [143]) An operator $T : X \to X$ on X is said to be weakly orbitally continuous if $T^{n_i}x \to u$ then $d(T^{n_i}x, T(T^{n_i}x)) \to d(u, Tu)$ as $i \to \infty$.

It is obvious that a complete metric space is orbitally complete with respect to any self-mapping of X, and that a continuous mapping is always orbitally continuous and an orbitally continuous mapping is always weakly orbitally continuous, but the converse implications need not be true. To see this, one may observe the following simple example that a discontinuous mapping $T : X \to X$ on a noncomplete metric space X can be orbitally continuous and that X can be T-orbitally complete.

Example 5.23 Let X be the set of all negative reals and all nonnegative rationals with the discrete metric. Define $T : X \to X$ as follows:

$$Tx = \begin{cases} \frac{x}{3}, & \text{if } x \text{ is nonzero rational} \\ 1, & \text{if } x = 0 \\ \frac{x}{4}, & \text{if } \quad x \text{ is irrational.} \end{cases}$$

In 2008, Pathak and Shahzad [464] proved the following generalized contraction principle.

Theorem 5.20 (Pathak and Shahzad [464]) *Let $(X, \mathscr{T}_d)$ be a topologically complete space endowed with metric topology $\mathscr{T}_d$, and let $T : X \to X$ be a strong topological contraction mapping. Then T has precisely one fixed point w in X such that $\{w\} = \bigcap_{n=0}^{\infty} T^n(X)$, where $T^n(X) := T(T^{n-1}(X))$ for each $n \in \mathbb{N}$ and $T^0(X) := X$.*

Proof Let $\mathscr{F}$ be the family of all nonempty closed subsets F of X such that $T(F) \subset F$. Since T is a strong contraction, it follows that $\delta(T^n(F)) \to 0$ as $n \to \infty$. Clearly, $\mathscr{F}$ is nonempty as $X \in \mathscr{F}$. Partially order $\mathscr{F}$ partially by set inclusion relation. Let $\mathscr{C}$ be a chain in $\mathscr{F}$. Then $M := \bigcap_{F \in \mathscr{C}} F$ is a lower bound of $\mathscr{C}$ as $T(M) \subset M$, M is a closed set in X and $M \neq \varnothing$ by the Cantor's intersection theorem. By Zorn's lemma, there is a minimal element $U \in \mathscr{F}$. Now we claim that $T(U) = U$. If not, then $S := T(U)$ is a proper subset of U. As T is closed mapping so that S is also closed. Hence, $T(S) \subset T(U) = S$, which contradicts the minimality of U. Therefore, $T(U) = U$. Since T is a strong topological contraction mapping, we must have $U = \{w\}$ for some $w \in X$ and $T(w) = w$. Thus, T has precisely one fixed point.

For each $n \in \mathbb{N}$, let $A_n := T^n(X)$ and $A_0 := X$. Then the sequence $\{A_n := T^n(X)\}_{n=0}^{\infty}$ is closed and decreasing by virtue of $T(X) \subset X$ and the fact that T is a closed mapping. Therefore, $\hat{S} = \bigcap_{n=0}^{\infty} A_n \neq \varnothing$. Now we claim that $T(\hat{S}) = \hat{S}$; i.e. $T(\bigcap_{n=0}^{\infty} T^n(X)) = \bigcap_{n=0}^{\infty} T^n(X)$. Further, it is obvious that $T(\hat{S}) \subset \hat{S}$. We shall now prove the converse. Take any $x \in \hat{S}$ and $B_n := T^{-1}(x) \bigcap A_n$ for each $n \in \mathbb{N}$, where $B_0 := X$ and $T^{-1}(x) := \{y \in X : x = T(y)\}$. Then $B_n \neq \varnothing$. Indeed, since $x \in \bigcap_{n=0}^{\infty} A_n \subset T^{n+1}(X)$, there exists $z \in A_n$ such that $x = T(z)$, so that $z \in T^{-1}(x) \bigcap A_n$. Since T is continuous and a closed mapping, the graph of T is closed [337, Theorem 7.3.8]. Hence, $T^{-1}(x)$ is closed in X and B_n is closed due to closedness of X. Therefore, the decreasing sequence $\{B_n\}_{n=0}^{\infty}$ has nonempty intersection; i.e. $\bigcap_{n=0}^{\infty} B_n \neq \varnothing$. Let $z \in \bigcap_{n=0}^{\infty} B_n$; i.e. $z \in T^{-1}(x) \bigcap (\bigcap_{n=0}^{\infty} A_n)$, so that $z \in T^{-1}(x)$. Then it follows that $x = T(z) \in T(\bigcap_{n=0}^{\infty} B_n)$. Therefore, $\hat{S} \subset T(\hat{S})$ and we have $T(\hat{S}) = \hat{S}$; i.e.

$$T\left(\bigcap_{n=0}^{\infty} T^n(X)\right) = \bigcap_{n=0}^{\infty} T^n(X).$$

Since T is a strong topological contraction, we must have $S = \{w\}$ which is the unique fixed point of T. ∎

Remark 5.12 The first part of the proof in Theorem 5.20 shows the existence of fixed points for T by using Zorn's lemma, and the second part of the proof in Theorem 5.20 also shows that we can prove the existence of fixed points for T without Zorn's lemma.

Theorem 5.21 (Pathak and Shahzad [464]) *Let (X, d) be a metric space, and let $T : X \to X$ be a strongly p-contraction mapping. Suppose T is orbitally complete, then T has a unique fixed point.*

Proof Let x be an arbitrary point of X. Construct an iterative sequence $\{x_n\}_{n=0}^{\infty}$ defined by $x_0 = x$, $x_n = T(x_{n-1})$ for $n \in \mathbb{N}$. Now using (5.17), we observe that $d(x_1, x_2) \leqslant p(x_0)d(x_0, x_1), d(x_2, x_3) \leqslant p(x_1)d(x_1, x_2)$, i.e. $d(x_2, x_3) \leqslant p(x_0)p(x_1)$ $d(x_0, x_1)$. Continuing this process, we obtain

$$d(x_n, x_{n+1}) \leqslant \prod_{i=1}^{n} p_i \, d(x_0, x_1) \tag{5.18}$$

where $p_i = p(x_{i-1}) = p(T^{i-1}(x_0))$, $i \in \mathbb{N}$. Since $\max\{p(x_0), \sup_{x \in X} p(Tx)\} \leqslant \lambda <$ 1, it follows from (5.18) that

$$d(x_n, x_{n+1}) \leqslant \lambda^n \, d(x_0, x_1) \ \forall \, n \in \mathbb{N}. \tag{5.19}$$

For $m > n (m, n \in \mathbb{N})$, we have

$$\begin{aligned} d(x_m, x_n) &\leqslant d(x_n, x_{n+1}) + d(x_{n+1}, x_{n+2}) + \cdots + d(x_{m-1}, x_m) \\ &\leqslant [\lambda^n + \lambda^{n+1} + \cdots + \lambda^{m-n}] \, d(x_0, x_1) \\ &\leqslant [\lambda^n + \lambda^{n+1} + \cdots] \, d(x_0, x_1) \\ &< \frac{\lambda^n}{1 - \lambda} \, d(x_0, x_1). \end{aligned}$$

It follows that the sequence $\{x_n\}_{n=0}^{\infty}$ is a Cauchy sequence in X. Since X is T-orbitally complete, it follows that there exists a Cauchy subsequence $\{T^{n_i}(x)\}$ of $\{x_n\}$ in orbit $O(x, Tx)$, $x \in X$, which converges to a point z in X. We now show that z is, indeed, fixed point of T. To effect this, suppose that $z \neq T(z)$. Since T is a p-contraction, it follows that

$$d(T(x_{n_i}), T^2(x_{n_i})) \leqslant p(x_{n_i})d(x_{n_i}, T(x_{n_i})) \leqslant \sup_{x \in X} p(Tx)d(x_{n_i}, T(x_{n_i})).$$

Now taking limit superior, we get

$$\begin{aligned} d(z, T(z)) &= \limsup_{i \to \infty} d(T(x_{n_i}), T^2(x_{n_i})) \leqslant \sup_{x \in X} p(Tx) \limsup_{i \to \infty} d(x_{n_i}, T(x_{n_i})) \\ &= \sup_{x \in X} p(Tx)d(z, T(z)) < d(z, T(z)), \end{aligned}$$

a contradiction. Thus, z is a fixed point of T. Note that so far we have not used the strongly p-contractiveness of T in its full generality as in Theorem 5.20 instead we have used only p-contractiveness of T. Now we shall use the strongly p-contraction of T in its full strength to prove the existence of uniqueness of fixed point of T. To this end, we see that $\bigcap_{n=0}^{\infty} T^n(X)$ is a singleton set, say $\{w\}$, for some w in X, then since $\lim_{i \to \infty} T(T^{n_i}x) = z$ for any arbitrary x in X, it follows that $z = w$. Thus, T has precisely one fixed point. ∎

If T is a p-contraction with $\sup_{x \in Y} p(Tx) = \lambda < 1$ replaced by $\sup_{x \in Y} p(x) = \lambda < 1$, then we have

Corollary 5.6 *Let (X, d) be a metric space, and let $T : X \to X$ be a strongly p-contraction mapping with $\sup_{x \in Y} p(x) = \lambda < 1$. Suppose T is orbitally complete. Then T has a unique fixed point.*

Remark 5.13 We observe that Corollary 5.6 (without uniqueness) follows from the following Theorem A on setting $\phi(x) = (1 - \sup_{x \in X} p(x))^{-1} d(x, Tx)$.

Theorem A (see Bollenbacher and Hicks [63]). Suppose that $T : X \to X$ and $\phi : X \to [0, \infty)$ where X is a metric space. If there exists an $x \in X$ such that X is T-orbitally complete at x and

$$d(y, Ty) \leqslant \phi(y) - \phi(Ty) \text{ for all } y \in O(x, T)$$

then the following hold:
(a) $\lim T^n x = \hat{x}$ exists;
(b) $T\hat{x} = \hat{x}$ if and only if $f(x) = d(x, Tx)$ is T-orbitally lower semicontinuous at x (i.e. if $\{x_n\} \subset \mathcal{O}(x, T)$ and $\lim x_n = \hat{x}$ imply that $f(\hat{x}) \leqslant \liminf f(x_n)$).

The following result was obtained in [271]

Corollary B. Suppose that X is a complete metric space and that $0 < k < 1$. If $T : X \to X$ and there exists an $x \in X$ such that

$$d(Ty, T^2 y) \leqslant k\, d(y, Ty) \text{ for all } y \in \mathcal{O}(x, T),$$

then the following hold:
(a) $\lim_{n \to \infty} T^n x = \hat{x}$ exists;
(b) $T\hat{x} = \hat{x}$ if and only if $f(x) = d(x, Tx)$ is T-orbitally lower semicontinuous at x (i.e. if $\{x_n\} \subset \mathcal{O}(x, T)$ and $\lim x_n = \hat{x}$ imply that $f(\hat{x}) \leqslant \liminf_{n \to \infty} f(x_n)$).

We now illustrate that the set of fixed point of T is, indeed, nonempty in a complete metric space.

Example 5.24 Let $X = [1, 4]$ equipped with usual metric d. Let $T : X \to X$ be a mapping defined by $T(x) = \frac{1}{2}\left(x + \frac{1}{x}\right) \ \forall x \in X$. Let $p : X \to [0, 1]$ be a function defined by $p(x) = \frac{1}{4}\left(x - \frac{1}{x}\right) \ \forall x \in X$. Then for all x in X, we have

$$d(T(x), T^2(x)) = \left| \frac{1}{2}\left(x + \frac{1}{x}\right) - \left(\frac{1}{4}\left(x + \frac{1}{x}\right) + \frac{1}{x + \frac{1}{x}}\right) \right|$$

$$= \left| \frac{1}{4}\left(x + \frac{1}{x}\right) - \frac{1}{x + \frac{1}{x}} \right| = \frac{1}{4}\left| \frac{(x + \frac{1}{x})^2 - 4}{x + \frac{1}{x}} \right|$$

$$\leqslant \frac{1}{8}\left|\left(x + \frac{1}{x}\right)^2 - 4\right| = \frac{1}{8}\left(x - \frac{1}{x}\right)^2$$

$$= \frac{1}{4}\left(x - \frac{1}{x}\right) \cdot \frac{1}{2}\left(x - \frac{1}{x}\right) = p(x) \cdot d(x, T(x)).$$

We observe that the iterative sequence $\{x_n\}$ initialized at any point x_0 has a Cauchy subsequence $\{x_{n_i}\}$ in orbit $0(x, Tx)$, $x_0 \in X$, converges to a point 1 in X. Clearly, $\sup_{x \in X} p(x) = \frac{15}{16} < 1$. Further, we see that $\bigcap_{n=0}^{\infty} T^n(X) = \{1\}$. Thus all the conditions of Theorem 5.20 are satisfied and 1 is the unique fixed point of T.

Remark 5.14 We cannot relax the condition $\sup_{x \in X} p(Tx) = \lambda < 1$ in Theorem 5.20. To see this, we observe the following simple example.

Example 5.25 Let $X = [0, \infty)$ equipped with usual metric d. Let $T : X \to X$ be a mapping defined by $T(x) = 2$ for $x \in [0, 1]$ and $T(x) = x + \frac{1}{x}$ for $x \in (1, \infty)$. Let $p : X \to [0, 1]$ be a function defined by $p(x) = \frac{1}{2}$ if $x \in [0, 1]$ and $p(x) = \frac{x^2}{x^2+1}$ if $x \in (1, \infty)$. For $x \in [0, 1]$ we have

$$d(T(x), T^2(x)) = \frac{1}{2} \leqslant \frac{1}{2} \cdot |x - 2| = p(x) \cdot d(x, T(x))$$

where $p(x) = \frac{1}{2}$, and for x in $(1, \infty)$ we have

$$d(T(x), T^2(x)) = \left| \left(x + \frac{1}{x} \right) - \left(\left(x + \frac{1}{x} \right) + \frac{1}{x + \frac{1}{x}} \right) \right|$$

$$= \frac{1}{x + \frac{1}{x}} = \frac{x^2}{x^2 + 1} \cdot \frac{1}{x} = p(x) \cdot d(x, T(x))$$

where $p(x) = \frac{x^2}{x^2+1}$. Thus, $\sup_{x \in X} p(x) = 1$. Hence, the rate of contractivity p does not satisfy the requirement of p-contraction. As a consequence, one may see that there exists no x in $(1, \infty)$ such that $x + \frac{1}{x} = x$.

Note that even if $p : X \to [0, 1]$ is a continuous mapping in a complete metric space X, we cannot relax the condition $\sup_{x \in X} p(Tx) = \lambda < 1$ in Theorem 5.20. To see this, we observe the following simple example.

Example 5.26 Let $X = \mathbb{R}$ be equipped with discrete/usual metric d. Define $T : X \to X$ by $T(x) = x + 1$ and $p(x) = 1$ $\forall x \in X$, then we have

$$d(T(x), T^2(x)) = d(x, T(x)) \ \forall x \in X.$$

Thus, all the conditions of Theorem 5.20 except $\sup_{x \in X} p(x) = \lambda < 1$ are satisfied. Clearly T has no fixed point in the complete metric space $\mathbb{R}$.

We also note that a fundamentally p-contraction mapping need not be continuous, even at a point z to which all sequences $\{T^n x\}$, $x \in X$, converge. To see this, we observe the following simple example.

Example 5.27 Let $X = [-1, 1]$ equipped with usual metric d. Let $T : X \to X$ be a mapping defined by $T(x) = \frac{2x}{3}$ for $x \geq 0$; $T(x) = 1$ *for* $x < 0$. Let $p : X \to [0, 1]$ be a function defined by $p(x) = \frac{2}{3} \forall x \in X$. Clearly, the mapping T is a fundamentally

p-contraction which has a unique fixed point $x = 0$ and is T-orbitally continuous. Note that $x = 0$ is the point of discontinuity of T.

The p-contraction principle has a very useful local version that involves an open ball B in a metric space X and a nonself-p-contraction mapping $T : B \to X$ of B into X which does not displace the center of the ball too.

Corollary 5.7 *Let (X, d) be a metric space, x_0 a point in X, $B = B(x_0, r) = \{x \mid d(x, x_0) < r\}$ and let $T : B \to X$ be a p-contraction mapping. Let $q : \{x_0\} \times \overline{B} \to [0, 1]$ be a function such that $\sup_{x \in \overline{B}} q(x_0, x) = \beta < 1$ and $d(T(x), T(x_0)) \leqslant q(x_0, x) d(x_0, x) \ \forall x \in \overline{B}$. Suppose $\overline{B}$ is T-orbitally complete and T is orbitally continuous. If $d(T(x_0), x_0) < (1 - \beta)r$, then T has a fixed point in $\overline{B}$. Further, if $\bigcap_{n=0}^{\infty} T^n(\overline{B})$ is a singleton set, where $T^n(\overline{B}) := T(T^{n-1}(\overline{B}))$ for each $n \in \mathbb{N}$ and $T^0(\overline{B}) := \overline{B}$, then T has a unique fixed point.*

Proof Choose $\varepsilon < r$ so that $d(T(x_0), x_0) \leqslant (1 - \beta)\varepsilon < (1 - \beta)r$. We note that T maps the closed ball $\overline{B} = \{x \mid d(x, x_0) \leqslant \varepsilon\}$ into itself; for if $x \in \overline{B}$, then

$$
\begin{aligned}
d(T(x), x_0) &\leqslant d(T(x), T(x_0)) + d(T(x_0), x_0) \\
&\leqslant q(x_0, x) d(x_0, x) + (1 - \beta)\varepsilon \\
&\leqslant \sup_{x \in \overline{B}} q(x_0, x) \cdot d(x_0, x) + (1 - \beta)\varepsilon \\
&\leqslant \beta \varepsilon + (1 - \beta)\varepsilon = \varepsilon.
\end{aligned}
$$

Therefore, T is a p-contraction mapping from $\overline{B}$ into itself. Since $\overline{B}$ is T-orbitally complete and T is orbitally continuous, it follows from p-contraction that there exists a z in $\overline{B}$ such that $z = Tz$. Now using the same arguments as in Theorem 5.20, the uniqueness of z follows. ∎

Now we shall show how Theorem 5.20 can be applied for investigation of solvability of an integral equation of the form

$$
y(x) = \lambda \int_a^x K[x, t, y(t)]dt + f(x) \tag{5.20}
$$

by successive approximation, where λ is an arbitrary parameter, $f(x)$ is a given continuous function on $[a, b]$, $K(x, t, s)$ is continuous for $x \in [a, b]$, $y(x)$ is unknown continuous function on $[a, b]$ and satisfies the following generalized Lipschtzian condition

$(i) \qquad |K(x, t, s_2) - K(x, t, s_1)| \leqslant L \cdot |s_2 - s_1|^{1+\delta}, \delta \geq 0,$

where L is a constant. Note that when $K[x, t, y(t)] = K_1(x, t) \cdot y(x)$, the above equation becomes an integral equation of Volterra type.

Define an operator $T : C[a, b] \to C[a, b]$ of a space of continuous functions on $[a, b]$ into itself as follows:

$$T[y(x)] = \lambda \int_a^x K[x, t, y(t)]dt + f(x) \qquad (5.21)$$

which satisfies the following condition:

(ii) $\bigcap_{n=0}^{\infty} T^n(C[a, b])$ is a singleton set, where $T^n(C[a, b]) := T(T^{n-1}(C[a, b]))$ for each $n \in \mathbb{N}$ and $T^0(C[a, b]) := C[a, b]$.

For $y \in C[a, b]$ we obtain

$$d(T(y), T^2(y)) = \max_{x \in [a,b]} |T[y(x)] - T[T(y(x))]|$$

$$\leqslant |\lambda| \cdot L \cdot (x - a). \max_{x \in [a,b]} |y(x) - T(y(x))|^{\delta} \max_{x \in [a,b]} |y(x) - T(y(x))|$$

$$= |\lambda| \cdot L \cdot (x - a). \max_{x \in [a,b]} |y(x) - T(y(x))|^{\delta} \cdot d(y, T(y)).$$

Hence, for $|\lambda| < \frac{1}{L \cdot (b-a) \cdot \sup_{y \in C[a,b]} \max_{x \in [a,b]} |y(x) - T(y(x))|^{\delta}}$, the mapping T is a p-contraction and by Theorem 5.20 the equation $T[y(x)] = y(x)$; i.e. the Eq. (5.20) has a solution, say, w in $C[a, b]$. Now using (ii), the uniqueness of w follows.

5.1.7   p-Contraction Mappings on Compacta and Localized Meir–Keeler-Type Theorem

Recall that a mapping $T : (X, d) \to (X, d)$ is d-continuous if $F : X \to R$ defined by $F(x) = d(x, Tx)$ is continuous on X. For p-contraction mappings on compact metric space, we prove the following result:

Theorem 5.22 (Pathak and Shahzad [464]) *Let T be a d-continuous p-contraction mapping with $\sup_{x \in X} p(x) < 1$ on a compact metric space (X, d). Then T has a fixed point z in X. Further, if $\bigcap_{n=0}^{\infty} T^n(X)$ is a singleton set, where $T^n(X) := T(T^{n-1}(X))$ for each $n \in \mathbb{N}$ and $T^0(X) := X$, then T has a unique fixed point.*

Proof Define a function $F : X \to \mathbb{R}^+$ by

$$F(x) = d(x, T(x)).$$

Since T is d-continuous, it follows that F is continuous. So, as X is compact, $\exists$ a point $z \in X$ such that

$$F(z) = d(z, T(z)) = \min_{x \in X}\{d(x, T(x))\}.$$

Assume that $z \neq T(z)$, then we have

$$FT(z) = d(T(z), T^2(z)) = p(z) \, d(z, T(z)) \leqslant \sup_{x \in X} p(x) \, d(z, T(z)) < d(z, T(z)) = F(z),$$

a contradiction. Hence, z is a fixed point of T. The uniqueness of z follows as in Theorem 5.20. ∎

To justify the above result, one may consider Example 5.20 again; then we see that $F(x) = d(x, T(x))$ attains its minimum at $x = 1$ in X. Clearly, T has a unique fixed point 1 in X.

We shall denote

A. $M(x, y) = \max\{d(x, y), d(x, Tx), d(y, Ty), \frac{1}{2}[d(x, Ty) + d(y, Tx)]\}$; and

A'. $m(x, T(x)) = \max\{d(x, T(x)), \frac{1}{2}d(x, T^2(x))\}$.

Let $\phi : \mathbb{R}^+ \to \mathbb{R}^+$ be a nondecreasing function such that $0 < \phi(t) < t$ for all $t \in \mathbb{R}^+ - \{0\}$.

In [290], Jachymski proved the following:

Theorem C. Let T be a self-mapping of a metric space (X, d). If T satisfies conditions: (B) for any $\varepsilon > 0$ there exists a $\delta > 0$ such that, for any $x, y \in X$,

$$\varepsilon < M(x, y) < \varepsilon + \delta \Rightarrow d(Tx, Ty) \leqslant \varepsilon; \text{ and}$$

(C) $\quad d(Tx, Ty) < M(x, y)$, for any $x, y \in X$ with $M(x, y) > 0$.

Then, for any $x \in X$, the sequence $\{T_n x\}$ is Cauchy. Hence, if (X, d) is complete and T is continuous, then T has a unique fixed point z and $T^n x \to z$, for any $x \in X$.

In 2008, Pathak and Shahzad [464] give some new localized Meir–Keeler-type conditions ensuring convergence of the successive approximations. In fact, they dropped the continuity requirements of the mapping T in Theorem C by induction of a certain function, which is not necessarily continuous, and modifying conditions (B) and (C) appropriately. They also obtain some fixed point theorems based on an asymptotic regularity condition which generalize previously known results.

Theorem 5.23 *Let (X, d) be a metric space and T a self-mapping of X onto itself such that for any $x \in X$*

(i) $d(T(x), T^2(x)) \leqslant \phi(m(x, T(x)))$,

(ii) *if* $d(T(x), T^2(x)) < \phi(m(x, T(x)))$ *then for any $\varepsilon > 0$ with $\varepsilon \leqslant m(x, T(x)) - \phi(m(x, T(x)))$, there exists a $\delta > 0$ with $\delta < \varepsilon$ such that*

$$m(x, T(x)) < \varepsilon + \delta \text{ implies } d(T(x), T^2(x)) \leqslant \varepsilon.$$

Suppose X is T-orbitally complete and T is weakly orbitally continuous. Then T has a fixed point, say z in X. Further, if $\bigcap_{n=0}^{\infty} T^n(X)$ is a singleton set, where $T^n(X) := T(T^{n-1}(X))$ for each $n \in \mathbb{N}$ and $T^0(X) := X$, then T has a unique fixed point. Moreover, T is continuous at z if and only if for any sequence $\{x_n\}$ in X converging to z, $\lim_{n \to \infty} m(x_n, T(x_n)) = 0$.

Proof Observe that (ii) implies

$$d(T(x), T^2(x)) \leqslant m(x, T(x)) - \phi(m(x, T(x))) \tag{5.22}$$

for $x, y \in X$ with $m(x, T(x)) > 0$. Moreover, if $m(x, T(x)) = 0$, then $T(x) = T^2(x)$ for each $x \in X$; that is, $T(x)$ is a fixed point of T for each x in X. Let x_0 be an arbitrary point in X. Define $x_{n+1} := Tx_n$ and $d_n := d(x_n, x_{n+1})$ for $n \in \mathbb{N}$. We shall now show that $d_n \to 0$. We shall restrict to the case when $d_n > 0$. Then it follows from (iii) that $d_{n+1} < m(x_n, x_{n+1}) - \phi(m(x_n, x_{n+1}))$; i.e. $d_{n+1} < m(x_n, x_{n+1})$. Assume that $d_{n+1} \geqslant d_n$; then

$$d_{n+1} < \max\{d(x_n, x_{n+1}), \frac{1}{2}d(x_n, x_{n+2})\}$$

$$\leqslant \max\{d(x_n, x_{n+1}), \frac{1}{2}[d(x_n, x_{n+1}) + d(x_{n+1}, x_{n+2})]\}$$

$$\leqslant \max\{d(x_n, x_{n+1}), d(x_{n+1}, x_{n+2})\} = \max\{d_n, d_{n+1}\} = d_{n+1},$$

a contradiction. Thus, $d_{n+1} < d_n$, so $\{d_n\}$ converges to some d in $\mathbb{R}^+$. We now show that $d = 0$. Suppose the contrary, that $d > 0$. Then there exists j in $\mathbb{N}$ such that

$$d < d_n < d + \delta \text{ for all } n \geqslant j.$$

It follows from (5.22) that $d_{n+1} \leqslant d$, for all $n \geqslant j$, which contradicts the above inequality. Thus, we get $d = 0$. Now fix an $\varepsilon > 0$ with $\varepsilon < m(x, T(x)) - \phi(m(x, T(x)))$ for any $x \in X$. By our hypothesis $\delta < \varepsilon$. Since $\lim_{n \to \infty} d_n = 0$, there exists j in $\mathbb{N}$ such that $d_n < \frac{1}{2}\delta$, for all $n \geqslant j$. We now apply induction to show that, for any $n \in \mathbb{N}$,

$$d(x_j, x_{j+n}) < \varepsilon + \frac{1}{2}\delta. \tag{5.23}$$

Clearly, (5.23) holds for $n = 1$. Assume that (B) holds for some n. We shall claim that it holds for $n + 1$ as well. Using the triangle inequality, (5.23) and noting the fact that $d_n < \frac{1}{2}\delta$, for all $n \geqslant j$ we have $d(x_j, x_{j+n+1}) \leqslant d(x_j, x_{j+n}) + d_{j+n} < \varepsilon + \frac{1}{2}\delta + \frac{1}{2}\delta = \varepsilon + \delta$. This proves our claim. Observe that (5.23) implies that $\{x_n\}$ is a Cauchy sequence. Since X is T-orbitally complete, there exists $z \in X$ such that $\lim_{n \to \infty} x_n = z$. Also, $\lim_{n \to \infty} T(x_n) = \lim_{n \to \infty} x_{n+1} = z$. Now we shall show that $z = Tz$. Suppose, on the contrary, that $z \neq Tz$. Then using (i) and weak orbital continuity of T, for sufficiently large values of n we have

$$d(x_{n+1}, T(x_{n+1})) = d(T(x_n), T^2(x_n))$$

$$\leqslant \phi\left(\max\left\{d(x_n, T(x_n)), \frac{1}{2}d(x_n, T^2(x_n))\right\}\right)$$

$$\leqslant \phi\left(\max\left\{d(x_n, T(x_n)), \frac{1}{2}[d(x_n, T(x_n)) + d(x_{n+1}, T(x_{n+1}))]\right\}\right).$$

On letting $n \to \infty$ this yields $d(z, Tz) < \phi(d(z, Tz)) < d(z, Tz)$, a contradiction. Hence, $z = Tz$. The uniqueness of z follows from the same arguments as were applied in Theorem 5.20. For the second part of the theorem, let T be continuous at the fixed point z and $\{x_n\}$ a sequence in X converging to z. Then $T(x_n) \to Tz = z$, $T^2(x_n) = T(x_{n+1}) \to Tz = z$, $d(x_n, T(x_n)) \to 0$, and $d(x_n, T^2(x_n)) \to 0$. Hence, we have

$$\lim_{n \to \infty} m(x_n, T(x_n)) = 0. \qquad (5.24)$$

For the converse part, assume that $\{x_n\}$ is a sequence in X converging to z such that (5.24) holds. Then $d(x_n, T(x_n)) \to 0$ and $d(T(x_n), z) \to 0$ as $n \to \infty$. This implies that $T(x_n) \to z = Tz$; i.e. T is continuous at z. ∎

We now see in the following example that our Meir–Keeler-type (ε, δ) contraction is localized in the sense that our choices of ε, δ are localized; i.e. $\varepsilon = \varepsilon(x)$ and $\delta = \delta(x, \varepsilon)$.

Example 5.28 Let $X = [0, 4]$, equipped with usual metric d. Define $T : X \to X$ by

$$Tx = \frac{1}{4} \text{ if } x \leqslant 1, \quad Tx = 0 \text{ if } x > 1.$$

$$d(T(x), T^2(x)) = 0 \text{ and } 0 \leqslant m(x, T(x)) \leqslant \frac{3}{4} \text{ when } x \leqslant 1,$$

$$d(T(x), T^2(x)) = \frac{1}{4} \text{ and } 1 < m(x, T(x)) \leqslant 4 \text{ when } x > 1.$$

Hence, T satisfies the contractive condition (i) with $\phi(t) = t/4$ for $t > 1$ and $\phi(t) = t/3$ for $t \leqslant 1$. Also, T satisfies the contractive condition (ii). To this end, we observe that:

(1) When $x \leqslant 1$, then $\varepsilon \leqslant m(x, Tx) - \phi(m(x, Tx)) = m(x, Tx) - \frac{1}{3}m(x, Tx) = \frac{2}{3}m(x, Tx)$ so we may choose δ in the interval $\frac{1}{3}m(x, Tx) < \delta < \frac{2}{3}m(x, Tx)$ for $\varepsilon = \frac{2}{3}m(x, Tx)$.

(2) When $x > 1$, then $\varepsilon \leqslant m(x, Tx) - \phi(m(x, Tx)) = m(x, Tx) - \frac{1}{4}m(x, Tx) = \frac{3}{4}m(x, Tx)$ so we may choose δ in the interval $\frac{1}{4}m(x, Tx) < \delta < \frac{3}{4}m(x, Tx)$ for $\varepsilon = \frac{3}{4}m(x, Tx)$. However, $\phi(t)$ is neither upper semicontinuous at $t = 1$ nor nondecreasing in $\mathbb{R}^+ - \{0\}$. Moreover, $\delta(\varepsilon)$ is not lower semicontinuous at $\varepsilon = 1$. It may also be observed that $\lim_{n \to \infty} m(x_n, T(x_n)) = 0$ for any sequence $\{x_n\}$ converging to $\frac{1}{4}$ and that T is continuous at the fixed point $\frac{1}{4}$.

Remark 5.15 (1) It follows from the second part of Theorem 5.23 that T is discontinuous at the fixed point z if and only if for any sequence $\{x_n\}$ in X converging to z, $\lim_{n \to \infty} m(x_n, T(x_n)) \neq 0$.

(2) In contrast to the Meir–Keeler-type (ε, δ) condition, our choices of ε, δ are localized; i.e. $\varepsilon = \varepsilon(x)$ and $\delta = \delta(x, \varepsilon)$.

(3) The condition $\delta < \varepsilon$ in our Theorem 5.23 cannot be dispensed with. To see this, we illustrate this fact in Example 5.29 given below.

It is well known that a Meir–Keeler-type (ε, δ) condition is not sufficient to ensure the existence of a fixed point of a contractive-type mapping. To show it, we consider the following example.

Example 5.29 Let $X = [0, 2]$ and d be the usual metric on X. Define $T : X \to X$ by

$$Tx = 1 + x^2 \text{ if } x < 1, \quad Tx = 0 \text{ if } x \geqslant 1.$$

Then T satisfies the contractive condition

$$\varepsilon < m(x, y) < \varepsilon + \delta \text{ implies } d(Tx, Ty) < \varepsilon$$

with $\delta(\varepsilon) = 1$ for $\varepsilon \geqslant 1$ and $\delta(\varepsilon) = 1 - \varepsilon$ when $\varepsilon < 1$ but T does not have a fixed point. Observe that the above contractive condition is slightly stronger than the contractive condition (ii) of Theorem 5.23, but weaker than the joint conditions (i) and (ii) of the same theorem. To see this, we observe that:

For $x < 1$, we have $m(x, T(x)) = \frac{1}{2}|x - 1|$ and $d(T(x), T^2(x)) = \frac{1}{4}|x - 1|$. Define $\phi(t) = t/2$; then for any given $\varepsilon > 0$ with $\varepsilon \leqslant m(x, T(x)) - \phi(m(x, T(x)))$, we see that the contraction condition

$$\varepsilon < m(x, T(x)) < \varepsilon + \delta \text{ implies } d(T(x), T^2(x)) \leqslant \varepsilon$$

is satisfied for some $\delta > \varepsilon$.

Corollary 5.8 *Let T be a self-mapping of a metric space (X, d) such that:*

(i) $d(T(x), T^2(x)) \leqslant \phi(d(x, T(x)))$ *for all $x \in X$,*
(ii) *if $d(T(x), T^2(x)) < \phi(d(x, T(x)))$ then for any $\varepsilon > 0$ with $\varepsilon \leqslant d(x, T(x)) - \phi(d(x, T(x)))$, there exists a $\delta > 0$ with $\delta < \varepsilon$ such that*

$$d(x, T(x)) < \varepsilon + \delta \text{ implies } d(T(x), T^2(x)) \leqslant \varepsilon.$$

Suppose X is T-orbitally complete and T is weakly orbitally continuous. Then T has a fixed point, say z in X. Further, if $\bigcap_{n=0}^{\infty} T^n(X)$ is a singleton set, where $T^n(X) := T(T^{n-1}(X))$ for each $n \in \mathbb{N}$ and $T^0(X) := X$, then T has a unique fixed point. Moreover, if T is continuous then for any sequence $\{x_n\}$ in X converging to z, $\lim_{n \to \infty} d(x_n, T(x_n)) = 0$.

Proof The first part can be immediately extracted from Theorem 5.23, simply by replacing $m(x, T(x))$ with $d(x, T(x))$ in (∗) and the rest of the proof. For the second part, we observe that T is continuous and so it follows that $\lim_{n \to \infty} d(x_n, T(x_n)) = 0$. This completes the proof. ∎

Remark 5.16 Our Corollary 5.8 unifies Boyd and Wong's theorem [75] and Theorem 1.2 of Matkowski [390]. It may be observed that in both cases ϕ has, among others, the following properties:

$$\phi(t) < t \text{ and } \lim_{n \to \infty} \phi^n(t) = 0, \text{ for any } t > 0.$$

Matkowski and Mis [391] observe that these conditions are not sufficient for the existence of a fixed point. However, the existence of a fixed point can be guaranteed if some additional conditions are assumed for ϕ; e.g. ϕ is right upper semicontinuous [75] or nondecreasing [390]. Furthermore, our condition (ϕ) seems weaker than the corresponding condition of Jachymski [290] because our choices of ε, δ are localized; i.e. $\varepsilon = \varepsilon(x)$ and $\delta = \delta(x, \varepsilon)$.

5.2 Nonexpansive Mappings

Nonexpansive maps appear for the first time in the work of Kolmogoroff [339], where they were used in the axiomatic treatment of measure theory. Pontrjagin and Schnirelmann [490] used the notion in dimension theory and established the following result: if X is a compact metric space and dim $X \geq r$, then there exists a nonexpansive map $\varphi : X \to \mathbb{R}^r$ such that dim $\varphi(X) = r$. In 1965, the first fixed point theorems of a general nature for nonlinear nonexpansive mappings in noncompact settings were proved by Browder [96] and Göhde [251] independently. They proved that a nonexpansive self-mapping of a bounded, closed and convex subset K of a uniformly convex Banach space X has a fixed point. Later on, Kirk [326, 327] proved the same result under slightly weaker assumptions that the space X is reflexive and K is a bounded, closed and convex subset of X which has "normal structure". Other recent works include papers of Browder [101, 103], dealing with the relationship of nonexpansive mappings to the theory of monotone operators in Hilbert space and, in more general setting to the theory of $J-$monotone operators and accretive operators. Theorems for approximating fixed points using interactive techniques for general nonexpansive mappings are given by Browder and Petryshyn [120, 121]. Senter and Dotson, Jr. [551]. Weak convergence of successive approximations is dealt by Opial [436]. The generalizations of nonexpansive mappings have been dealt with by Kannan [310], Wong [608], Goebel, Kirk and Shimmi [248].

In this section, we discuss the results of Browder and Kirk. Also we shall briefly deal with the works of Senter and Dotson, Jr. [551], Dotson, Jr. [197] and Goebel, Kirk and Shimmi [248]. We also discuss the iteration schemes for nonexpansive mappings due to Reich [512].

Definition 5.13 Let X be a Banach space, K a closed, bounded and convex subset of X. Let $T : K \to K$ be a mapping. Then T is said to be nonexpansive if for each pair of elements x and y of K we have

$$\|Tx - Ty\| \leqslant \|x - y\|.$$

Example 5.30 If $\mathcal{H}$ is a Hilbert space, then the retraction mapping R_r is nonexpansive.

The following example (Sadovski [536]) shows that nonexpansive mapping may fail to have fixed points in general Banach space. This implies that we have to impose some restriction on the mapping or on the space to obtain fixed points.

Example 5.31 Let B be the unit ball of the sequence space c_0 (the Banach space of all sequences converging to zero with norm $\|x\| = \sup_i |x_i|$). Then B is a closed, convex and bounded set in c_0. Put

$$Tx = T(x_1, x_2, \ldots) = (1, x_1, x_2, \ldots).$$

Then $\|Tx - Ty\| = \|x - y\|$. Notice that the equation $Tx = x$ is satisfied only if $x = (1, 1, 1, \ldots)$ which is not in c_0.

Example 5.32 Consider the translation mapping $T : X \to X$ defined by $Tx = x + a$ where $a \neq 0$ is a given element in the Banach space X. This is a nonexpansive and fixed point free mapping. This shows that a nonexpansive mapping of a Banach space into itself need not have a fixed point. Also by considering the identity mapping, we conclude that the fixed point of a nonexpansive mapping need not be unique.

5.2.1 Fixed Point Theorems for Nonexpansive Mappings

Theorem 5.24 *Let X be a Banach space and let K be a closed, convex and bounded subset of X. If $T : K \to K$ is nonexpansive and $(I - T)(K)$ is a closed subset of X, then T has a fixed point in K.*

Proof We assume without loss of generality that $0 \in K$ and that K is contained in a ball of radius r centerd at 0, so that $\|Tx\| \leqslant r$ for all $x \in K$. Let $t_n < 1$ be a sequence of positive numbers tending to 1. Consider the mapping $T_n x = t_n T x$. Let $x_n \in K$ be the fixed point of the contraction mapping T_n. We have

$$\|Tx_n - x_n\| = \|Tx_n - t_n T x_n\| = (1 - t_n)\|Tx_n\| \leqslant (1 - t_n)r.$$

Thus, $\|Tx_n - x_n\| \to 0$ as $t_n \to 1$. Hence, $0 \in (I - T)(K)$ and $u = Tu$ for some $u \in K$. This completes the proof. ∎

In 1965, Browder [106] (for further reference, see also Browder- Petryshyn [121])) proved the following result:

Theorem 5.25 *Let $\mathcal{H}$ be a Hilbert space and K be a closed, bounded and convex subset of $\mathcal{H}$. If $F : K \to K$ is a nonexpansive mapping then F has a fixed point.*

Proof It can be shown that $I - F$ is a restriction of a monotone operator $I - FR$ where R is the retraction of $\mathcal{H}$ onto K. Further we claim that for any continuous monotone operator T on $\mathcal{H}$, $T(K)$ is closed. Combining these two results, we conclude that $(I - F)(K)$ is closed. Then the assertion of the theorem follows from Theorem 5.24 above (refer to Smart [571]). ∎

In 1965, Browder [106] and Göhde [251] (independently) generalized the above result to uniformly convex Banach spaces. Indeed, they proved the following result:

Theorem 5.26 *Let X be a uniformly convex Banach space, F is a nonexpansive mapping of the bounded, closed and convex subset K of X into K. Then F has a fixed point.*

Proof Let K be a closed, bounded and convex subset of a uniformly convex Banach space X, $F : K \to K$ a nonexpansive mapping, and let $\mathcal{F}$ be a family of nonempty, closed and convex subsets of K which are invariant under F, i.e. $F(K_\alpha) \subset K_\alpha$, for $K_\alpha \in \mathcal{F}$. $\mathcal{F}$ is nonempty since K is an element of $\mathcal{F}$.

Let us order the set $\mathcal{F}$ by defining $K_1 \leqslant K_2$ if $K_1 \subseteq K_2$, $\mathcal{F}$ becomes a partially ordered set which is inductive since the intersection of the elements K_α of a linearly ordered subfamily of $\mathcal{F}$ is also a closed and convex subset of K. This subset is invariant under F and is nonempty, since all the sets K_α are weakly compact subsets of the reflexive Banach space X. Hence, $\mathcal{F}$ has a minimal element K_0. Also K_0 is the convex closure of $F(K_0)$ since if K_1 is the convex closure, K_1 is a closed and convex subset of K_0 which is nonempty and invariant under F and by the minimality of K_0 in $\mathcal{F}$, we have $K_0 = K_1$.

We will be through if we can show that K_0 has exactly one element. Since K_0 is nonempty, it suffices to show that K_0 does not contain two distinct points in it. Suppose it does and let r_0 be the diameter of K_0. Choose two points x_1 and x_2 in K_0 such that $\|x_1 - x_2\| \geqslant \frac{r_0}{2}$. Let $x \in K_0$ be the midpoint of the segment joining x_1 to x_2 and then $x - y$ is the midpoint of the segment joining $x_1 - y$ to $x_2 - y$, $y \in K_0$ and $\|x_1 - y\|$ and $\|x_2 - y\|$ are both $< r_0$. By the uniform convexity of the space X, there exists a constant $s > 0$ such that $\|x - y\| \leqslant (1 - s)r_0 < r_0$. Let $r_1 = (1 - s)r_0 < r_0$ and let $K_2 = \cap_{y \in K_0}\{u : u \in K_0, \|u - y\| \leqslant r_1\}$. K_2 is a nonempty, closed and convex subset of K_0 since it is the intersection of closed and convex sets, and x lies in K_2. K_2 is a proper subset of K_0 since r_1 is less than the diameter of K_0. Finally, K_2 is invariant under F. Indeed, suppose $u \in K_2$, $y \in K_0$. For any $\varepsilon > 0$, we can find a convex linear combination $Fz_j, z_j \in K_0$, such that

$$\left\| y - \sum_{j=1}^{k} \lambda_j F z_j \right\| < \varepsilon \ \left(0 \leqslant \lambda_j \leqslant 1, \ \sum_{j=1}^{k} \lambda_j = 1\right).$$

Thus

$$\|Fu - y\| \leqslant \left\| Fu - \sum_{j=1}^{k} \lambda_j F z_j \right\| + \varepsilon \leqslant \sum_{j=1}^{k} \lambda_j \|Fu - F z_j\| + \varepsilon,$$

while $\|Fu - Fz_j\| \leqslant \|u - z_j\| \leqslant r_1$, since F is nonexpansive and u lies in K_2. But K_0 is a minimal element of $\mathcal{F}$. So we have reached a contradiction to the assumption that K_0 has at least two points. Hence, $K_0 = \{x_0\}$ and $Fx_0 = x_0$. ∎

Remark 5.17 If K is compact or F is compact, Theorem 5.24 is a consequence of Schauder fixed point theorem (see the next section), while if F is weakly continuous, it is a special case of the Tychonoff fixed point theorem, since in a reflexive Banach space, every bounded, closed and convex set K is weakly compact.

We now state and prove a theorem due to Kirk [326] who used a characterization of reflexivity due to Smulian [573] and a concept (normal structure) of Brodski and Milman [82] to prove the fixed point theorem for mappings which do not increase distances.

Smulian proved that a necessary and sufficient condition that a Banach space X be reflexive is that: (P) Every bounded descending sequence (transfinite) of nonempty, closed and convex subsets of X has a nonempty intersection.

Definition 5.14 Let X be a Banach space. For $K \subset X$, let $\delta(K)$ denote the diameter of K. Then

(i) A point $x \in K$ is a diametral point of K if sup $\{\|x - y\| : y \in K\} = \delta(K)$.
(ii) A convex set $K \subset X$ is said to have normal structure if for each bounded convex subset H of K which contains more than one point, there is some point $x \in H$ which is not a diametral point of H, i.e. there exists $x \in H$ such that

$$\sup\{\|x - y\| : y \in H\} < \delta(H).$$

Theorem 5.27 *Every compact convex set K in a Banach space has normal structure.*

Proof Assume that K does not have normal structure. Without loss of generality, we assume that diam $K > 0$. For any $x_1 \in K$, there exists $x_2 \in K$ such that diam $K = \|x_1 - x_2\|$. Then $\frac{1}{2}(x_1 + x_2) \in K$ as K is convex. Next we find x_2 in K such that

$$\text{diam } K = \left\| x_3 - \frac{x_1 + x_2}{2} \right\|$$

and proceeding in this manner we obtain a sequence $\{x_n\}$ such that

$$\text{diam } K = \left\| x_{n+1} - \frac{x_1 + x_2 + \cdots + x_n}{n} \right\|, \quad n \geqslant 2.$$

Then

$$\text{diam } K \leqslant \frac{1}{n} \sum_{i=1}^{n} \|x_{n+1} - x_i\| \leqslant \text{diam } K.$$

This implies that diam $K = \|x_{n+1} - x_i\|, 1 \leqslant i \leqslant n$.

Thus, the sequence $\{x_n\}$ has no convergent subsequences, and this contradicts the compactness of K.   ∎

Theorem 5.28 *Every closed, convex and bounded subset K of a uniformly convex Banach space X has normal structure.*

Proof Without loss of generality, we assume that K is contained in the unit ball $\{x : \|x\| \leqslant 1\}$. Let K_1 be a closed and convex subset of K. Let x_1 be any element in K_1 and $\varepsilon = \frac{1}{2}$. Take $x_2 \in K_1$ such that $\|x_2 - x_1\| \geqslant \frac{\text{diam } K_1}{2}$. Then for any $x \in K_1$, we have

$$\left\| x - \frac{x_1 + x_2}{2} \right\| = \left\| \frac{x - x_1}{2} + \frac{x - x_2}{2} \right\|$$

$$= \sqrt{\left\{ 2\left[\left\| \frac{x - x_1}{2} \right\|^2 + \left\| \frac{x - x_2}{2} \right\|^2 \right] - \left\| \frac{x - x_1}{2} + \frac{x - x_2}{2} \right\|^2 \right\}}$$

$$\leqslant \sqrt{\left\{ 2\left[\frac{(\operatorname{diam} K_1)^2}{2} \right] - \frac{\|x_2 - x_1\|^2}{4} \right\}}$$

$$= \operatorname{diam} K_1 \sqrt{\left(1 - \left(\frac{\frac{\operatorname{diam} K_1}{2} / \operatorname{diam} K_1}{2} \right)^2 \right)}$$

$$= \operatorname{diam} K_1 \left[1 - \left(1 - \sqrt{\left(1 - \left(\frac{\frac{\operatorname{diam} K_1}{2} / \operatorname{diam} K_1}{2} \right)^2 \right)} \right) \right]$$

$$\leqslant \operatorname{diam} K_1 \left[1 - \delta_X \left(\frac{\operatorname{diam} K_1}{2} \Big/ \operatorname{diam} K_1 \right) \right]$$

$$= \operatorname{diam} K_1 \left[1 - \delta_X \left(\frac{1}{2} \right) \right],$$

where $\delta_X(\varepsilon)$ is the modulus of convexity of X. This proves the assertion of the theorem since $\delta_X(\varepsilon) > 0$. ∎

In 1965, Kirk further generalized the result of Browder (see Theorem 5.25 above) to all reflexive spaces with normal structure: those spaces such that all non-trivial closed, bounded, convex sets C have a smaller radius than diameter of the space.

Theorem 5.29 *Let K be a nonempty, bounded, closed and convex subset of a reflexive Banach space X, and suppose that K has a normal structure. If F is a nonexpansive mapping of K into itself, then F has a fixed point.*

The following two lemmas are crucial for the proof of Theorem 5.29. Let K denote a nonempty, bounded, closed and convex subset of the Banach space X. Then define $r_x(K), r(K)$ and K_0 as follows:

$$r_x(K) = \sup\{\|x - y\| : y \in K\}$$
$$r(K) = \inf\{r_x(K) : x \in K\} \text{ and}$$
$$K_0 = \{x \in K : r_x(K) = r(K)\}.$$

Lemma 5.1 *If X is reflexive, then K_0 is nonempty and convex.*

Proof Let $K(x, n) = \left\{ y \in K : \|x - y\| \leqslant r(K) + \frac{1}{n} \right\}$. It is easily seen that the sets $C_n = \cap_{x \in K} K(x, n)$ form a decreasing sequence of nonempty, closed and convex sets, and hence $K_0 = \cap_{n=1}^{\infty} C_n$ is closed and convex, and by (P) is nonempty.

Lemma 5.2 *Let K be a closed and convex subset of X which contains more than one point. If K has normal structure, then $\delta(K_0) < \delta(K)$.*

Proof By normal structure, K contains at least one nondiametral point x. Hence, $r_x(K) < \delta(K)$. If z and w are any two points of K_0, then $\|z - w\| \leqslant r_x(K) = r(K)$. Hence,

$$\delta(K_0) = \sup\{\|z - w\| : w, z \in K_0\} \leqslant r(K) \leqslant r_x(K) < \delta(K).$$

Proof of Theorem 5.29 Let $\mathcal{F}$ denote the collection of all nonempty, closed and convex subsets of K each of which is mapped into itself by the mapping F. By property (P) and Zorn's lemma $\mathcal{F}$ has a minimal element which is denoted by K.

The proof is completed by showing that K consists of a single point. Suppose it is not true.

Let $x \in K_0$. Then $\|Fx - Fy\| \leqslant \|x - y\| \leqslant r(K)$ for all $y \in K$, and hence $F(K)$ is contained in the spherical ball $\overline{U}$ centerd at Fx with radius $r(K)$. Since $F(K \cap \overline{U}) \subset K \cap \overline{U}$, the minimality of K implies that $K \subset \overline{U}$. Hence, $Fx \in K_0$ and K_0 is mapped into itself by F. By Lemma 5.1, $K_0 \in \mathcal{F}$. If $\delta(K) > 0$, then by Lemma 5.2, K_0 is properly contained in K. Since this contradicts the minimality of K, $\delta(K) = 0$ and K consists of a single point. ∎

Corollary 5.9 *If the condition that K be bounded is replaced by the requirement that the sequence $\{F^n p\}$ be bounded for some $p \in K$, then F has a fixed point.*

Remark 5.18 Theorem 5.26 follows from the above theorem since a uniformly convex Banach space is reflexive and K, being a closed, convex and bounded subset of a uniformly convex Banach space, has a normal structure.

Theorem 5.29 can be restated as:

Let K be a nonempty, weakly compact, convex subset of a Banach space X and suppose K has normal structure. Then every nonexpansive mapping $F : K \to K$ has a fixed point.

Remark 5.19 The following example due to Kirk shows that in Theorem 5.3.6, reflexivity of X cannot be dispensed with. To observe this, let us consider $C[0, 1]$, the space of continuous function on $[0, 1]$. $C[0, 1]$ is not reflexive. Let

$$K = \{f \in C[0, 1] : f(0) = 0, \ f(1) = 1, \ 0 \leqslant f(x) \leqslant 1\}.$$

Notice that K is bounded, closed and convex. Define

$$F(f(x)) = xf(x).$$

Clearly, F maps K into itself and F is nonexpansive. But F has no fixed point.

Concerning Kirk's theorem, there arises a natural question what further generalizations are possible? After five decades, it remains an open question as to whether or not every reflexive Banach space $(X, \| \cdot \|)$ has the fixed point property for nonexpansive maps.

Very recently, in 2009, Benavides [48] proved the following intriguing result:

Theorem 5.30 *Given a reflexive Banach space$(X, \| \cdot \|)$, there exists an equivalent norm $\| \cdot \|_\sim$ on X such that $(X, \| \cdot \|_\sim)$ has the fixed point property for nonexpansive mappings.*

Notice that Theorem 5.30 improves a theorem of van Dulst [208] for separable reflexive Banach spaces.

In contrast to this result, the non-reflexive Banach space $(\ell_1, \| \cdot \|_1)$, the space of all absolutely summable sequences, with the absolute sum norm $\| \cdot \|_1$, fails the fixed point property for nonexpansive mappings. To see this, let

$$K := \left\{ (t_n)_{n \in \mathbb{N}} : \text{ each } t_n \geqslant 0 \text{ and } \sum_{n=1}^{\infty} t_n = 1 \right\}.$$

This is a bounded, closed and convex subset of ℓ_1. Let $T : K \to K$ be the right shift map on K; i.e.,

$$T(t_1, t_2, t_3, \cdots) := (0, t_1, t_2, t_3, \cdots).$$

Clearly, T is $\| \cdot \|_1$-nonexpansive (being an isometry) and fixed point free on K.

In 2008, in another significant development, Lin [370] provided the first example of a non-reflexive Banach space $(X, \| \cdot \|)$ with the fixed point property for non-expansive mappings. Lin verified this fact for $(\ell_1, \| \cdot \|_1)$ with the equivalent norm $\| \cdot \|_\sim$ given by

$$\| \cdot \|_\sim = \sup_{k \in \mathbb{N}} \frac{8^k}{1 + 8^k} \sum_{n=k}^{\infty} |x_n|, \text{ for all } x = (x_n)_{n \in \mathbb{N}} \in \ell_1.$$

Notice that the Banach space $(c_0, \| \cdot \|_\infty)$ of real-valued sequences that converge to zero, with the absolute supremum norm $\| \cdot \|_\infty$ is another nonreflexive Banach space of great importance in Banach space theory. It also fails the fixed point property for nonexpansive mappings. To see this, let us consider

$$K := \{ (t_n)_{n \in \mathbb{N}} : \text{each } t_n \geqslant 0; 1 = t_1 \geqslant t_2 \geqslant \cdots \geqslant t_n \geqslant t_{n+1} \to 0, \text{ as } n \to \infty \}.$$

Let $T : K \to K$ be the natural right shift map given by

$$T(t_1, t_2, t_3, \cdots) := (1, t_1, t_2, t_3, \cdots).$$

Then T is a $\| \cdot \|_1$-nonexpansive (isometric, actually) map with no fixed points in the closed, bounded and convex set K.
We now discuss the theorems proved by Dotson, Jr. [197] concerning the existence of fixed points of nonexpansive mappings on a certain class of nonconvex sets.

Definition 5.15 Let K be a subset of a Banach space X, and let $\mathcal{F} = \{f_\alpha\}_{\alpha \in K}$ be a family of functions from $[0, 1]$ into K, having the property that for each $\alpha \in K$ we have $f_\alpha(1) = \alpha$. Such a family $\mathcal{F}$ is said to be contractive provided there exists a function $\varphi : (0, 1) \to (0, 1)$ such that for all α and β in K and for all t in $(0, 1)$, we have

$$\|f_\alpha(t) - f_\beta(t)\| \leqslant \varphi(t)\|\alpha - \beta\|,$$

such a family is said to be jointly continuous if $t \to t_0$ in $[0, 1]$ and $\alpha \to \alpha_0$ in K then $f_\alpha(t) \to f_{\alpha_0}(t_0)$ in K.

Theorem 5.31 *Let K be a compact subset of a Banach space X, and suppose there exists a contractive, jointly continuous family of functions associated with K as described in Definition 5.15. Then any nonexpansive mapping T of K into itself has a fixed point in K.*

Proof For each $n = 1, 2, 3, \ldots$, let $k_n = \frac{n}{n+1}$, and define a self-mapping $T_n : K \to K$ by

$$T_n(x) = f_{Tx}(k_n) \quad \text{for all } x \in K.$$

Since $T(K) \subset K$ and $k_n < 1$, each T_n is a well-defined map from K into itself. Moreover, for each n and for all $x, y \in K$ we have

$$\|T_n x - T_n y\| = \|f_{Tx}(k_n) - f_{Ty}(k_n)\| \leqslant \varphi(k_n)\|x - y\|.$$

So, for each n, T_n is a contraction mapping on K. As a compact subset of the Banach space X, K is a complete metric space. Therefore, each T_n has a unique fixed point $x_n \in K$. Since K is compact, there exists a subsequence $\{x_{n_i}\}$ of $\{x_n\}$ such that $x_{n_i} \to u \in K$. Since $T_{n_i} x_{n_i} = x_{n_i}$, we have $T_{n_i} x_{n_i} \to u$. By the continuity of T, we have $Tx_{n_i} \to Tu$. Also by the joint continuity of the family, we have

$$T_{n_i} x_{n_i} = f_{Tx_{n_i}}(k_{n_i}) \to f_{Tu}(1) = Tu.$$

This gives $u = Tu$.   ∎

Corollary 5.10 *Let K be star-shaped (with p as star center) compact subset of a Banach space X and let T be a nonexpansive mapping of K into itself. Then T has a fixed point in K.*

Proof A set K is said to be star-shaped or star convex with p as star center if all line segments joining p to other points of K lie in K i.e. if

$$x \in K, \ \alpha p + (1 - \alpha)x \in K, \ 0 \leqslant \alpha \leqslant 1.$$

We are through if we can find a family described in Definition 5.15. Define $f_\alpha(t) = (1 - t)p + t\alpha$ so that

$$T_n x = f_{Tx}(k_n) = (1 - k_n)p + \overset{\cdot}{d}_n Tx \text{ where } k_n = \frac{n}{n+1}.$$

It can be checked that $\|f_\alpha(t) - f_\beta(t)\| \leqslant t\|\alpha - \beta\|$.

So we can take $\phi(t) = t$ for $0 < t < 1$. Also it is obvious that $f_\alpha(t)$ is jointly continuous in t and α. Now the result follows from the above theorem. ∎

Remark 5.20 Theorem 5.26 remains valid (Bose and Mukherjee [68]) if K is a compact subset of a complete metric space X and if there is an associated family as described in Definition 5.15.

Definition 5.16 A family $\mathcal{F} = \{f_\alpha\}_{\alpha \in K}$ of functions from $[0, 1]$ into a set K will be called jointly weakly continuous provided that if

$$t \to t_0 \text{ in } [0, 1] \text{ and } \alpha \to \alpha_0 \text{ in } K \Rightarrow f_\alpha(t) \rightharpoonup f_{\alpha_0}(t_0) \text{ in } K.$$

Now we state a theorem due to Dotson Jr. without proof.

Theorem 5.32 *Suppose K is a weakly compact subset of Banach space X, and suppose there exists a contractive, jointly weakly continuous family $\mathcal{F}$ of functions associated with K. Then any nonexpansive weakly continuous self-mapping F of K has a fixed point in K.*

We now briefly give some results concerning the approximation of fixed points of some class of mappings.

Definition 5.17 Let X be a Banach space and let C be a convex subset of X. A self-mapping T of C is said to be quasi-nonexpansive provided T has a fixed point in C and if $p \in C$ is a fixed point of T, then

$$\|Tx - p\| \leqslant \|x - p\| \text{ for all } x \in C.$$

The class of nonexpansive mappings is strictly contained in this class as the following example illustrates.

Example 5.33 Let $f : \mathbb{R} \to \mathbb{R}$ be defined as

$$f(x) = \begin{cases} 0 & \text{if } x = 0 \\ x \, \sin \frac{1}{x} & \text{if } x \neq 0. \end{cases}$$

Obviously $x = 0$ is the only fixed point of f. Moreover, f is quasi-nonexpansive as

$$|f(y) - 0| = \left| y \, \sin \frac{1}{y} \right| \leqslant |y| = |y - 0|$$

But f is not nonexpansive. To see it, let $x = \frac{2}{\pi}$ and $y = \frac{2}{3\pi}$. Then

$$|f(x) - f(y)| = \frac{8}{3\pi} > \frac{4}{3\pi} = |x - y|.$$

Now we prove a theorem concerning the quasi-nonexpansive mapping due to Petryshyn and Williamson [485, 486].

Theorem 5.33 *Let C be a closed subset of a Banach space X and $T : C \to X$ be a quasi-nonexpansive mapping. Suppose there exists a point x_0 in C such that $x_n = T^n x_0 \in C$. Then the sequence $\{x_n\}$ converges to a fixed point of T in C if and only if $\lim\limits_{n \to \infty} d(x_n, \mathfrak{F}(T)) = 0$ where $\mathfrak{F}(T)$ denotes the set of fixed points of T.*

Proof The necessity of the condition is obvious. Suppose $\lim\limits_{n \to \infty} d(x_n, \mathfrak{F}(T)) = 0$. We prove first that the sequence $\{x_n\}$ is Cauchy. Let $\varepsilon > 0$ and N be such that

$$d(x_n, \mathfrak{F}(T)) \leqslant \frac{\varepsilon}{2} \text{ for all } n \geqslant N.$$

For $m, n \geqslant N$ and any point $p \in \mathfrak{F}(T)$ we have

$$\|x_n - x_m\| \leqslant \|x_n - p\| + \|p - x_m\| = \|T^n(x_0) - p\| + \|T^m(x_0) - p\|$$
$$\leqslant 2\|T^N(x_0) - p\| \text{ by quasi-nonexpansivity of } T$$
$$\leqslant 2\|x_N - p\|.$$

This implies that $\|x_n - x_m\| \leqslant 2d(x_N, \mathfrak{F}(T))$. This shows that $\{x_n\}$ is a Cauchy sequence. Let $\{x_n\}$ converges to u in C. Then $d(u, \mathfrak{F}(T)) = 0$. Since $\mathfrak{F}(T)$ is closed due to quasi-nonexpansiveness of T, we have $u \in \mathfrak{F}(T)$, i.e. $u = Tu$. ∎

In a uniformly convex Banach space, Senter and Dotson, Jr. have given conditions under which certain types of iterates (Mann type [384]) of a quasi-nonexpansive mapping converge to a fixed point of the mapping.

Definition 5.18 A mapping $T : C \to C$ with nonempty fixed point set $\mathfrak{F}(T)$ in C is said to satisfy

Condition I. If there is a nondecreasing function $f : [0, \infty] \to [0, \infty]$ with $f(0) = 0$, $f(r) > 0$ for $r \in (0, \infty)$ such that $\|x - Tx\| < f(d(x, \mathfrak{F}(T))$ for all $x \in C$, where $D(x, \mathfrak{F}(T)) = \inf\{\|x - z\| : z \in \mathfrak{F}(T)\}$. Similarly, the above mapping T is said to satisfy

Condition II. If there is a real number $\alpha > 0$ such that $\|x - Tx\| \geqslant \alpha D(x, \mathfrak{F}(T))$ holds for all $x \in C$. It can be easily shown that a mapping which satisfies condition II also satisfies condition I.

We now state a key theorem of Senter and Dotson, Jr. Let $\mathbb{Z}^+$ denote the set of positive integers. Let $x_1 \in C$ and let $M(x_1, t_n, T)$ be a sequence $\{x_n\}$ defined by $x_{n+1} = (1 - t_n)x_n + t_n T(x_n)$ where $t_n \in [\beta, \gamma]$, $0 < \beta < \gamma < 1$ and $n \in \mathbb{Z}^+$.

Theorem 5.34 *Suppose X is a uniformly convex Banach space, C is a closed and convex subset of X and T is a quasi-nonexpansive mapping of C into itself. If F satisfies condition I, then for arbitrary $x_1 \in C$, $M(x_1, t_n, T)$ converges to a member of $\mathfrak{F}(T)$.*

Corollary 5.11 *Suppose X is a uniformly convex Banach space, C is a closed, convex and bounded subset of X and T is a nonexpansive mapping of C into itself. If T satisfies condition I, then for arbitrary $x_1 \in C$, $M(x_1, t_n, T)$ converges to a member of $\mathfrak{F}(T)$.*

Proof Notice that $\mathfrak{F}(T)$ is nonempty by Theorem 5.26. Let p be a fixed point of T. Then

$$\|Tx - p\| = \|Tx - Tp\| \leqslant \|x - p\|, \quad x \in C.$$

Hence, T is quasi-nonexpansive. The result follows from the theorem.

A notion more general than the class of nonexpansive mappings, what we call generalized nonexpansive mappings, has been considered by Bose and Mukherjee [68] and Shimi [552] and many others. Approximation of fixed points of such mappings has been studies by Bose and Mukherjee [68] and Shimi [552]. We state below the theorem of Goebel, Kirk and Shimi and that of Bose and Mukherjee without proof.

Theorem 5.35 *Let X be a uniformly convex Banach space, C a nonempty, bounded, closed and convex subset of X and $T : C \to C$ a continuous mapping such that*

$$\|Tx - Ty\| \leqslant a\|x - y\| + b[\|x - Tx\| + \|y - Ty\|] + c[\|x - Ty\| + \|y - Tx\|]$$

for all $x, y \in C$ where $a + 2b + 2c \leqslant 1, a, c \geqslant 0$ and $b > 0$. Then the sequence $\{x_n\}$, defined by $x_{n+1} = (1 - t_n)x_n + t_n Tx_n, n = 1, 2, \ldots$ where $x_1 \in C$ and $t_n \in [\beta, \gamma], 0 < \beta < \gamma < 1$ converges to a fixed point of T.

Remark 5.21 Browder [101] has generalized Theorem 5.26 to mappings which he calls semicontractive. On the other hand, Kirk [207, 209] concentrated his efforts on weakening the condition of uniform convexity on X. Nussbaum [428] has obtained more general results for locally almost nonexpansive mappings.

Now we state the theorems of Reich [512] concerning the iteration scheme for nonexpansive mappings in uniformly convex Banach space X with a Fréchet differentiable norm.

Definition 5.19 A sequence $\{x_n\}$ in a Banach space X is weakly almost convergent to $y \in X$ if $\frac{1}{p} \sum_{k=0}^{p-1} x_{n+k} \rightharpoonup y$ uniformly in $n \in \mathbb{N}$.

Definition 5.20 A multivalued operator $T : X \to 2^X$ is said to be m-accretive if $\mathscr{R}(I + T) = X$ and $\|x_1 - x_2\| \leqslant \|x_1 - x_2 + r(y_1 - y_2)\|$ for all $y_i \in Tx_i, i = 1, 2,$ and $r > 0$.

Theorem 5.36 *Let K be a closed and convex subset of a uniformly convex Banach space X with a Fréchet differentiable norm. If $T : K \to K$ is a nonexpansive mapping with a fixed point, then $\{T^n x\}$ is weakly almost convergent to a fixed point of T for each $x \in K$.*

Theorem 5.37 *Let X be a uniformly convex Banach space with a Fréchet differentiable norm, $T_r (r > 0)$ the resolvent of an m−accretive operator $T : X \to 2^X$ with $0 \in \mathscr{R}(T)$, and $\{r_n\}$ a positive sequence. Suppose that either*

(a) $\{r_n\}$is bounded away from zero, or
(b) the modulus of convexity of X satisfy $\delta(\varepsilon) \geqslant M\varepsilon^p$ for some $M > 0$ and $p \geqslant 2$,
     and $\sum\limits_{n=1}^{\infty} r_n^p = \infty$.

If $x_j \in X$ and $x_{n+1} = T_{r_n} x_n$ for $n \geqslant 1$, then $\{x_n\}$ converges weakly to a zero of T.

Theorem 5.38 *Let K be a closed and convex subset of a uniformly convex Banach space X with a Fréchet differentiable norm, $T : K \to K$ a nonexpansive mapping with a fixed point, and $\{c_n\}$ a real sequence such that*

(i) $0 \leqslant c_n \leqslant 1$,
(ii) $\sum\limits_{n=1}^{\infty} c_n(1 - c_n) = \infty$.

If $x_1 \in K$ and $x_{n+1} = c_n T x_n + (1 - c_n)x_n$ for $n \geqslant 1$, then $\{x_n\}$ converges weakly to a fixed point of F.

The original notion of asymptotic center was given by Edelstein [214] in 1972. He proved some of its properties and used it to prove a fixed point theorem for a class of mappings which includes nonexpansive mapping. Subsequently, Lim [366, 367], Yanagi [612] and others dealt with it extensively. Caristi [128, 129], Lim [366, 367], Downing and Kirk [202], and Yanagi [612] considered inward mappings and proved fixed point theorem for such mappings (both single-valued and multivalued). Here we present a fixed point theorem obtained by Lim [367] after giving some preliminaries.

Definition 5.21 Let K be a subset of a Banach space X and let $\{x_n\}$ be a bounded sequence in X. We define

$$AR(K, \{x_n\}) = \inf\{\lim_{n \to \infty}\ \sup\ \|y - x_n\| : y \in K\} \text{ and}$$

$$A(K, \{x_n\}) = \{z \in K : \lim_{n \to \infty}\ \sup \|z - x_n\| = AR(K, \{x_n\})\}.$$

The set $A(K, \{x_n\})$ is called the asymptotic center of $\{x_n\}$ relative to K. The number $AR(K, \{x_n\})$ is called the asymptotic radius of $\{x_n\}$ relative to K. Let $r(y) = \lim_{n \to \infty} \sup\|y - x_n\|$. Then

$$r = AR(K, \{x_n\}) = \inf\{r(y) : y \in K\} \text{ and } A(K, \{x_n\}) = \{z \in K : r(z) = r\}.$$

We list below some important properties of asymptotic radius and asymptotic centers. The notion of asymptotic center is a generalization of the notion of Chebyshev center. For Chebyshev center and Chebyshev radius, one may refer to Garkavi [239].

Let K be a nonempty subset of a Banach space X.

(a) If $r(x) \leqslant d, r(y) \leqslant d$ and $\|x - y\| \geqslant \varepsilon$, then $r\left(\frac{x+y}{2}\right) \leqslant d\left[1 - \delta\left(\frac{\varepsilon}{d}\right)\right]$.

(b) (i) $r(x)$ is convex and nonexpansive
   (ii)$|r(x) - r(y)| \leqslant \|x - y\| \leqslant r(x) + r(y)$.

(c) If K is convex, then $A(K, \{x_n\})$ is a convex set.

(d) If K is closed, then $A(K, \{x_n\})$ is closed.

(e) If K is weakly compact, then $A(K, \{z_n\})$ is a nonempty set.

(f) If X is uniformly convex and K is bounded, closed and convex, then $A(K, \{x_n\})$ consists of exactly one point, i.e. the asymptotic center is unique.

(g) $A(K, \{X_n\}) \subset \partial K \cup A(X, \{z_n\})$.

(h) There exists a subsequence $\{x_{n_i}\}$ of $\{x_n\}$ such that

$$AR(K, \{x_{n_{i_j}}\}) = AR(K, \{x_{n_i}\})$$

and $$A(K, \{x_{n_i}) \subset A(K, \{x_{n_{i_j}}\})$$

for any subsequence $\{x_{n_{i_j}}\}$ of $\{x_{n_i}\}$.

(i) If K is a weakly compact subset of a Banach space or a closed and convex subset of a reflexive Banach space, then the asymptotic center $A(K, \{x_n\})$ is a nonempty, closed, convex and bounded. (j) If X is uniformly convex, then asymptotic center is unique.

Next we define inward mapping. For a historical account of fixed point theorems for inward mappings, one may refer to Downing and Kirk [202]. Halpern [264] first considered this type of mapping in his Ph.D. thesis.

Definition 5.22 Let K be a nonempty and convex subset of X. If $x \in K$, we define the inward set of x relative of K, denoted $I_K(x)$ as follows:

$$I_K(x) = \{(1 - \alpha)x + \alpha y : y \in K, \alpha \geqslant 1\}.$$

A mapping $T : K \to X$ is said to be inward if $Tx \in I_K(x)$ for every $x \in K$. T is said to be weakly inward if $Tx \in \overline{I_K(x)}$ for every $x \in K$.

Proposition 5.1 (Lim [367]) *Let K be a closed and convex subset of a uniformly convex Banach space X and $\{x_n\}$ a bounded sequence in K. If x is the asymptotic center of $\{x_n\}$ with respect to K, then it is also the asymptotic center with respect to $\overline{I_K(x)}$.*

Proof Let u be the asymptotic center of $\{x_n\}$ with respect to $\overline{I_K(x)}$ and assume that $u \neq x$. Since $K \subseteq \overline{I_K(x)}$, we have $u \in \overline{I_K(x)} - K$ and $r(u) < r(x)$ by the uniqueness of the asypmtotic center (property (j)). By continuity of $r(\cdot)$ (property (b)), there exists $z \in \overline{I_K(x)} - K$ such that $r(z) < r(x)$. Hence, $z = (1 - \alpha)x + \alpha y$ for some $y \in K$ and $\alpha > 1$. Since $r(\cdot)$ is a convex function

$$r(y) = r\left[\frac{1}{\alpha}z + \left(1 - \frac{1}{\alpha}\right)x\right] \leqslant \frac{1}{\alpha}r(z) + \left(1 - \frac{1}{\alpha}\right)r(x)$$

$$\leqslant \frac{1}{\alpha}r(x) + \left(1 - \frac{1}{\alpha}\right)r(x) = r(x),$$

contradicting the definition of x. Hence, $u = x$.   ∎

We now state a result due to Caristi.

Proposition 5.2 *Let K be a nonempty, closed and convex subset of a Banach space X and let $T : K \to X$ be a contraction and weakly inward mapping. Then T has a unique fixed point.*

Theorem 5.39 *Let K be a closed, convex and bounded subset of a uniformly convex Banach space X and let $T : K \to X$ be a nonexpansive weakly inward mapping. Then T has a unique fixed point.*

Proof Let $x_0 \in K$. Define T_n by

$$T_n(x) = (1 - \alpha_n)x_0 + \alpha_n Tx \quad \text{where} \quad 0 < \alpha_n < 1.$$

and $\lim_{n \to \infty} \alpha_n = 1$. Each T_n is clearly a contraction with Lipschitz constant $\alpha_n < 1$ and so T_n has a unique fixed point x_n, in view of Proposition 5.2.

Moreover $\quad \|x_n - Tx_n\| = \left\|x_n - \frac{x_n}{\alpha_n} - \left(\frac{1}{\alpha_n} - 1\right)x_0\right\|$

$$= \left(\frac{1}{\alpha_n} - 1\right)\|x_n - x_0\| \to 0 \text{ as } n \to \infty.$$

Since K is bounded. Let x be the asymptotic center of $\{x_n\}$ with respect to K. Then

$$r(Tx) = \limsup_n \|Tx - x_n\|$$

$$\leqslant \limsup_n \|Tx - Tx_n\|$$

$$\leqslant \limsup_n \|x - x_n\| = r(x) \text{ as } T \text{ is nonexpansive.}$$

Since $Tx \in \overline{I_K(x)}$ and by Proposition 5.1. x is the asymptotic center of $\{x_n\}$ with respect to $\overline{I_K(x)}$, we conclude that $Tx = x$ by the uniqueness of the asymptotic center. This completes the proof.   ∎

5.2.2 Fixed Point Theorems for Pseudocontractive Mappings

We now discuss fixed point theorems for pseudocontractive mappings. This class of mappings includes the class of nonexpansive mappings.

Definition 5.23 Let K be a nonempty subset of a normed linear space X. A mapping $F : K \to X$ is said to be pseudocontractive if for each $k > 0$ and for all $x, y \in K$,

$$\|x - y\| \leqslant \|(1 + k)(x - y) - k(Fx - Fy)\|.$$

In a real Hilbert space with the corresponding norm, this is equivalent to

$$\|Fx - Fy\|^2 \leqslant \|x - y\|^2 + \|(I - F)(x) - (I - F)(y)\|^2 \text{ for all } x, y \in K$$

and this in turn is equivalent to

$$(Fx - Fy, x - y) \leqslant \|x - y\|^2 \text{ for all } x, y \in X.$$

Notice that fixed point theorems for pseudocontractive mappings play an important role in the theory of nonlinear mappings because of their connection with the accretive operators (refer Kirk and Schoneberg [335]). Browder [103] and Kato [320], independently of each other, characterized pseudocontractive mappings as those mapping F for which the mapping $T = I - F$ is accretive, i.e.

$$\text{Re}\langle j, Tx - Ty \rangle \geqslant 0 \text{ for some } j \in J(x - y)$$

where $J : X \to 2^{X^*}$ is the normalized duality mapping.

Before discussing the theorems proved by Reinermann and Schoneberg [517, 518] and Kirk and Ray [334], we give the following Lemma, due to Crandall and Pazy [151].

Lemma 5.3 *Let $\mathcal{H}$ be a real Hilbert space, let $\{x_n\}$ be a bounded sequence in $\mathcal{H}$ and $\{r_n\} \subset \mathbb{R}^+$ be strictly decreasing sequence such that*

$$(r_n x_n - r_m x_m, x_n - x_m) \leqslant 0 \text{ for all } m, n.$$

Then there exists $x \in \mathcal{H}$ such that $x_n \to x$.

Lemma 5.4 *Let $\mathcal{H}$ be a real Hilbert space and K a nonempty subset of $\mathcal{H}$. Let $F : K \to \mathcal{H}$ be continuous and pseudocontractive. Further let $\{x_n\}$ be a bounded sequence in K and $\{k_n\}, 0 < k_n < 1$ for all n, be strictly increasing with $k_n \to 1$ and $k_n F(X_n) = x_n$ for all n. Then F has a fixed point.*

Proof Define $r_n = \frac{1}{k_n} - 1$. Then

$$(r_n x_n - r_m x_m, x_n - x_m) = \left(\left(\frac{1}{k_n} - 1 \right) x_n - \left(\frac{1}{k_m} - 1 \right) x_m, x_n - x_m \right)$$

$$= (Fx_n - Fx_m, x_n - x_m) - \|x_n - x_m\|^2 \leqslant 0.$$

Then by Lemma 5.3, $x_n \to x$. Since $Fx_n - x_n = (1 - k_n)Fx_n$ by taking limit as $n \to \infty$, we have $Fx - x = 0$. This entails $Fx = x$. ∎

Theorem 5.40 *Let $\mathcal{H}$ be a real Hilbert space and let K be a closed star-shaped subset of $\mathcal{H}$. Suppose $F : K \to \mathcal{H}$ is a nonexpansive such that $F(\partial K) \subset K$ and there exists $x_0 \in K$ with bounded iterative sequence $\{F^n(x_0)\}$. Then F has a fixed point.*

Proof We may assume 0 to be a star center of K. Let $\{k_n\}$, be a sequence with $0 < k_n < 1$ such that $k_n \to 1$. Then $k_n F$ is a contraction satisfying $k_n F(\partial K) \subset K$. By Assad and Kirk [25], there exists x_n such that $k_n F x_n = x_n$ for each n. Next we show that $\{x_n\}$ is bounded. Let $S = \{F^n x_0 : n = 1, 2, \ldots\}$ and B be a closed ball about the origin with $x \subset B$. We prove that $x_n \in B$ for each n. Let $\varepsilon > 0$, $x \in (K \cap H) - B$ and $F(x) = (1 + \varepsilon)x$. By straightforward computation, we have for every $y \in S$,

$$\|Fx - Fy\|^2 \geqslant \operatorname{dist}(x, S)^2 + \varepsilon^2 \|x\|^2.$$

Then choosing $y \in S$ with $\operatorname{dist}(x, S)^2 \geqslant \|x - y\|^2 - \frac{\varepsilon^2}{2}\|x\|^2$, we find

$$\|Fx - Fy\|^2 \geqslant \|x - y\|^2 + \frac{\varepsilon^2}{2}\|x\|^2.$$

This contradicts the nonexpansiveness of F. Hence, $x_n \in B$ for every n. Now the result follows by Lemma 5.4.   ∎

Theorem 5.41 (Kirk and Ray [334]) *Let K be a closed and convex subset of a uniformly convex Banach space X. Let $F : K \to K$ be a Lipschitzian pseudocontractive mapping. Suppose for some $a \in K$, the set*

$$G(a, Fa) = \{z \in K : \|z - a\| \geqslant \|z - Fa\|\}$$

is bounded. Then F has a fixed point in K.

Proof Let k be the Lipschitz constant and chose $\alpha \in (0, 1)$ such that $\alpha k < 1$. For each $y \in K$ let $F_y : K \to K$ be defined as $F_y(x) = (1 - \alpha)y + \alpha F(x)$. Since F_y is a contraction, it has a fixed point $T_\alpha(y)$ for each $y \in K$, that is

$$T_\alpha(y) = (1 - \alpha)y + \alpha F(T_\alpha(y)), \ y \in K.$$

If $r > 0$ then $\|u - v\| \leqslant \|(1 + r)(u - v) - r(Fu - Fv)\|$. Also if r is chosen so small that $\alpha(1 + r) > r$, then it can be shown that

$$\|T_\alpha(u) - T_\alpha(v)\| \leqslant \|u - v\|.$$

Thus, $T_\alpha : K \to K$ is nonexpansive on K.

The existence of a fixed point of T_α gives the fixed point of F. This is accomplished by showing that $G(a, T_\alpha(a))$ is bounded for some $a \in K$, then T_α has a fixed point in K (by an earlier result of Kirk and Ray [334]).   ∎

5.2.3 Fixed Point Theorems for Asymptotically Nonexpansive Mappings

We now discuss some fixed point theorems for asymptotically nonexpansive mappings due to Goebel and Kirk [246].

Definition 5.24 Let K be a nonempty subset of a Banach space X. A mapping $T : K \to K$ is said to be asymptotically nonexpansive if

$$\|T^n x - T^n y\| \leqslant k_n \|x - y\| \text{ for all } x, y \in K \text{ and } \lim_{n \to \infty} k_n = 1.$$

Concerning Lin's theorem about ℓ_1 there arises a natural question whether there is a c_0- analogue of ℓ_1. It remains an open question as to whether or not there exists an equivalent norm $\| \cdot \|_\sim$ on $(c_0, \| \cdot \|_\infty)$ such that $(c_0, \| \cdot \|_\sim)$ has the fixed point property for nonexpansive mappings. However, if we weaken the nonexpansive condition to "asymptotically nonexpansive", then the answer is "no".

In 2000, Dowling, Lennard and Turett [200] showed that for every equivalent renorming $\| \cdot \|_\sim$ of $(c_0, \| \cdot \|_\infty)$, there exists a closed, bounded, convex set K and an asymptotically nonexpansive mapping $T : K \to K$ such that T has no fixed point.

In contrast to this, note that in 1972, Goebel and Kirk [246] proved the following:

Theorem 5.42 *Let K be a nonempty, closed, convex and bounded subset of a uniformly convex Banach space and let T be a asymptotically nonexpansive mapping of K into itself. Then T has a fixed point in K.*

Proof For each $x \in K$ and $r > 0$, let $B_r(x)$ denote the spherical ball centerd at x with radius r. Let y be a fixed element of K and suppose R_y consist of those numbers r for which there exists an integer k such that

$$K \cap \left[\bigcup_{i=k}^{\infty} B_r(T^i y) \right] \neq \varnothing.$$

If d is the diameter of K, then $d \in R_y$. Hence, $R_y \neq \varnothing$. Let $r_0 = \text{glb } R_y$ and for each $\varepsilon > 0$ we define

$$C_\varepsilon = \bigcup_{k=1}^{\infty} \left[\bigcap_{i=k}^{\infty} B_{r_0 + \varepsilon}(T^i y) \right].$$

Thus, for each $\varepsilon > 0$, the sets $C_\varepsilon \cap K$ are nonempty and convex. By reflexivity of X, we have

$$C = \bigcap_{\varepsilon > 0} \left[\overline{C_\varepsilon} \cap K \right] \neq \varnothing.$$

We note that for $x \in C$ and $\eta > 0$, there exists an integer N such that $\|x - T^i y\| \leqslant r_0 + \eta$ for $i \geqslant N$. Let $x \in C$ and suppose $\{T^n x\}$ does not converge to x, that is,

$Tx \neq x$. Then there exists $\varepsilon > 0$ and a subsequence $\{T^{n_i}x\}$ of $\{T^n x\}$ such that $\|T^{n_i}x - x\| \geqslant \varepsilon, i = 1, 2, \dots$. For $m > n$, we have $\|T^n x - T^m x\| \leqslant k_n \|x - T^{m-n}x\|$, since T is asymptotically nonexpansive. Assume $r_0 > 0$ and choose $\alpha > 0$ so that

$$\left[1 - \delta\left(\frac{\varepsilon}{r_0 + \alpha}\right)\right](r_0 + \alpha) < r_0.$$

Select n such that

$$\|x - T^n x\| \geqslant \varepsilon \text{ and } k_n\left(r_0 + \frac{\alpha}{2}\right) \leqslant r_0 + \alpha.$$

If $N \geqslant n$ is sufficiently large, then $m > N$ implies

$$\|x - T^{m-n}y\| \leqslant r_0 + \frac{\alpha}{2}.$$

Also we have

$$\|T^n x - T^m y\| \leqslant k_n \|x - T^{m-n}y\| \leqslant r_0 + \alpha, \|x - T^m y\| \leqslant r_0 + \alpha.$$

By uniform convexity of X, if $m > N$, we have

$$\left\|\frac{1}{2}(x + T^n x) - T^m y\right\| \leqslant \left[1 - \delta\left(\frac{\varepsilon}{r_0 + \alpha}\right)\right](r_0 + \alpha) < r_0$$

and this contradicts the definition of r_0. Thus, we arrive at the conclusion that $r_0 = 0$ and $Tx = x$. But $r_0 = 0$ implies that $\{T^n y\}$ is a Cauchy sequence giving $T^n y \to x = Tx$ as $n \to \infty$. Therefore, C is a singleton set and this point is the fixed point of T.   ∎

Definition 5.25 A mapping $T : K \to K$ is said to be asymptotically regular if for any $x \in K$, $\|T^n x - T^{n+1}x\| \to 0$ as $n \to \infty$.

Bose [71] proved the following theorem as an extension of Opial's convergence theorem (for nonexpansive mappings) to the class of asymptotically nonexpansive mappings.

Theorem 5.43 *Let X be a uniformly convex Banach space having weakly continuous duality mapping and K a nonempty, bounded, closed and convex subset of X. Suppose $T : K \to K$ is asymptotically nonexpansive and asymptotically regular. Then for any $x \in K$, the sequence $\{T^n x\}$ converges weakly to a fixed point of T.*

Before giving the proof of this theorem, we deal with a few preliminaries.

Definition 5.26 Let X be a Banach space. X satisfies Opial's condition if for each x_0 in X and each sequence $\{x_n\}$ weakly converging to x_0, the inequality

$$\liminf_{n \to \infty} \|x_n - x\| > \liminf_{n \to \infty} \|x_n - x_0\|$$

holds for all $x \neq x_0$.

An equivalent definition is obtained by replacing the above inequality by

$$\limsup_{n \to \infty} \|x_n - x\| > \limsup_{n \to \infty} \|x_n - x_0\| \text{ for all } x \neq x_0.$$

Weak Opial's condition A Banach space X is said to satisfy weak Opial's condition if the following holds: If a sequence $\{x_n\}$ is weakly convergent to x_0 in X, then for every x in X,

$$\liminf_{n \to \infty} \|x_n - x\| \geqslant \liminf_{n \to \infty} \|x_n - x_0\|;$$

equivalently

$$\limsup_{n \to \infty} \|x_n - x\| \geqslant \limsup_{n \to \infty} \|x_n - x_0\|.$$

Observation

- Every Hilbert space and $\ell_p (1 \leqslant p < \infty)$ spaces satisfy Opial's condition.
- Every Banach spaces with weakly continuous duality mappings satisfy weak Opial's condition.
- Opial [436] has, in fact, proved that on a uniformly convex Banach space having weakly continuous duality mapping (or a Hilbert space), the above inequality is strict for $x \neq x_0$.

Lemma 5.5 *Let K be a nonempty, bounded, closed and convex subset of a uniformly convex Banach space having weakly continuous duality mapping. If a sequence $\{x_n\} \subset K$ converges weakly to a point x_0, then x_0 is the asymptotic center of $\{x_n\}$ in K.*

Bose [72] has also proved that in a uniformly convex Banach space if $T : K \to K$ is asymptotically nonexpansive, then the asymptotic center of $\{T^n x\}$ in K for any $x \in K$ is a fixed point of T.

Lemma 5.6 *Let K and X be as in Lemma 5.5 and let $T : K \to K$ be an asymptotically nonexpansive mapping. Suppose x_0 is the asymptotic center of the sequence $\{T^n x\}$ for some $x \in K$. If the weak limit u_0 of a subsequence $\{T^{n_i} x\}$ is a fixed point of T, then it must coincide with x_0 which is a fixed point of T.*

Proof Let r and r' be the asymptotic radii, respectively, of $\{T^n x\}$ and $\{T^{n_i} x\}$. We have $r' < r$. By Lemma 5.5, u_0 is the asymptotic center of $\{T^{n_i} x\}$ in K. Hence, given any $\varepsilon > 0$ we can choose an integer i_0 such that

$$\|u_0 - T^{n_{i_0}} x\| \leqslant r' + \frac{\varepsilon}{2}.$$

Since u_0 is a fixed point of T and T is asymptotically nonexpansive, we can choose an integer N such that

$$\|u_0 - T^{n_{i_0}+j} x_0\| \leq k_j \left(r' + \frac{\varepsilon}{2}\right) \leqslant r' + \varepsilon \leqslant r + \varepsilon \text{ for all } j > N.$$

Thus, $\lim\sup_{n\to\infty} \|u_0 - T^n x\| = r$ and by uniqueness of the asymptotic center we have $u_0 = x_0$.

We now present the proof of the main theorem (Theorem 5.43).

Proof of Theorem 5.43 First of all we show that asymptotic regularity of T implies that every weak cluster point of $\{T^n x\}$ is a fixed point of T. Then by Lemma 5.6 all weak cluster points of $\{T^n x\}$ coincide with the asymptotic center x_0 of $\{T^n x\}$ in K. But we know that x_0 is a fixed point of T. Hence, $\{T^n x\}$ converges weakly to a fixed point of T.

Suppose $\{T^{n_i} x\}$ converges weakly to u_0. Then by Lemma 5.5, u_0 is the asymptotic center of $\{T^{n_i} x\}$ in K. Let the asymptotic radius be r. We have $(I - T)T^{n_i} x \to 0$ as $i \to \infty$ by asymptotic regularity of T. For any integer k, the sequence $\{T^{n_i+k} x\}$ converges weakly to u_0 and thus all such sequences have the same asymptotic center u_0 in K. We now show that all these sequences have the same asymptotic radius r. By asymptotic regularity of T, we have

$$\|u_0 - T^{n_i+1} x\| - \|u_0 - T^{n_i} x\| \leqslant \|(u_0 - T^{n_i+1} x) - (u_0 - T^{n_i} x)\|$$

$$= \|T^{n_i} x - T^{n_i+1} x\| \to 0 \text{ as } i \to \infty.$$

Hence, $\qquad \lim\sup_{i\to\infty} \|u_0 - T^{n_i+1} x\| = \lim\sup_{i\to\infty} \|u_0 - T^{n_i} x\| = r$.

Next we show that u_0 is a fixed point of T by proving that $T^j u_0 \to u_0$ as $j \to \infty$. Then by continuity of T, u_0 is a fixed point of F. Assume that $\{T^j u_0\}$ does not converge to u_0. Then there is $\alpha > 0$ and a sequence $\{j_m\}$ of integers so that $\|u_0 - T^{j_m} u_0\| \geqslant \alpha$ for all m. Since X in uniformly convex, one can choose $\varepsilon > 0$ such that

$$(r + \varepsilon)\left[1 - \delta\left(\frac{\alpha}{r + \varepsilon}\right)\right] < r$$

where δ is the modulus of convexity. Since all the sequences

$$\{T^{n_i+k} x\}, k = 0, 1, 2, \ldots$$

have the same asymptotic center u_0 and the same asymptotic radius r, there exist integers $N(k)$ such that

$$\|u_0 - T^{n_i+k} x\| \leqslant r + \frac{\varepsilon}{2} \text{ for all } i \geqslant N(k). \tag{5.25}$$

For any m

$$\|T^{j_m} u_0 - T^{n_i+j_m} x\| \leqslant k_{j_m} \|u_0 - T^{n_i} x\| \leqslant k_{j_m}\left(r + \frac{\varepsilon}{2}\right)$$

for $i \geqslant N(0)$. Let M be an integer such that

$$k_{j_m}\left(r+\frac{\varepsilon}{2}\right)\leqslant r+\varepsilon$$

for all $m\geqslant M$. Then we have

$$\|T^{j_m}u_0-T^{n_i+j_m}x\|\leqslant r+\varepsilon \text{ for all } i\geqslant N(0) \text{ and all } m>M, \tag{5.26}$$

and

$$\|u_0-T^{n_i+j_m}x\|\leqslant r+\varepsilon \text{ for } i\geqslant N(j_m). \tag{5.27}$$

Since $\|u_0-T^{j_m}u_0\|\geqslant\alpha$, (5.26) and (5.27) and uniform convexity of X imply that

$$\left\|\frac{u_0+T^{j_m}u_0}{2}-T^{n_i+j_m}x\right\|\leqslant(r+\varepsilon)\left[1-\delta\left(\frac{\alpha}{r+\varepsilon}\right)\right]<r, \text{ for all } i\leqslant\max\{N(0),N(j_M)\},$$

a contradiction to the fact that the sequence $\{T^{n_i+j_m}x\}_{i=1}^{\infty}$ has asymptotic radius r in K. This completes the proof. ∎

Next we consider the theorem proved by Passty [447] which extends Theorem 5.44. The proof of the Theorem 5.44 depends on Opial's Lemma [436] which does not carry over to $L_p(p\neq2)$. So new techniques are required to deal with mapping on such spaces. These techniques were provided by Baillon [30] and simplified by Bruck, Jr. [124] when the space has Fréchet differentiable norm. Passty [447] extended the Definition 5.24 to sequences of mapping which are not necessarily powers of a given mapping. We state his general theorem without proof and obtain the result in which we are interested, as a corollary.

Definition 5.27 The sequence $\{T_n\}$ of self-mappings of K is said to be asymptotically nonexpansive if

$$\|T_nx-T_ny\|\leqslant k_n\|x-y\| \text{ for all } x,y\in K \text{ with } \lim_{n\to\infty}k_n=1.$$

We denote the set of fixed points of the mapping T by $\mathfrak{F}(T)$.

Theorem 5.44 *Let X be a uniformly convex Banach space with a Fréchet differentiable norm and K a closed and convex subset of X. Let C be a subset of K and $S=\{T_n\}_{n=1}^{\infty}$, an asymptotically nonexpansive sequence of self-mappings of K such that*

(a) $C\subset\bigcap_{n=1}^{\infty}\mathfrak{F}(T_n).$

  Assume also that there exists x_0 in K for which
(b) $T_{n_i}x_0\rightharpoonup z$ *implies that $z\in C$, and*
(c) $T_nT_mx_0-T_nx_0\to0$ *as $n\to\infty$ for all (fixed) m. Then either*
   (i) $C=\varnothing$ *and $\|T_nx_0\|\to+\infty$ or*
   (ii) $C\neq\varnothing$ *and T_nx_0 converges weakly to an element of C.*

Note that the hypothesis (c) may be interpreted as asymptotic regularity of X at x_0.

Corollary 5.12 *Let K be a closed, convex and bounded subset of X and let $T : K \to K$ be weakly continuous and asymptotically nonexpansive. If T is asymptotically regular at $x_0 \in K$, then $\{T^n x_0\}$ converges weakly to a fixed point of T.*

Proof In order to apply the above theorem, take $C = \mathfrak{F}(T)$ and $S = \{T^n\}$. It is clear that conditions (a) and (c) hold. It suffices to prove that condition (b) holds. Note that $T^{n_j} x_0 \rightharpoonup z$ and weak continuity implies that $T^{n_j+1} x_0 \rightharpoonup Tz$. But asymptotic regularity implies that $\{T^{n_j+1} x_0\}$ and $\{T^{n_j} x_0\}$ must have the same limit, namely z. Hence, $Tz = z \in C \neq \varnothing$ by Goebel and Kirk [246]. So $\{T^n x_0\}$ must converge weakly to an element of C.

Finally we show by an example due to Goebel and Kirk [246] that the class of asymptotically nonexpansive mappings is wider than the class of nonexpansive mappings. To see this, let B denote the unit ball in the Hilbert space ℓ_2 and let T be defined as follows:

$$T(x_1, x_2, x_3, \cdots) \to (0, x_1^2, a_2 x_2, a_3 x_3, \cdots)$$

where (a_i) is a sequence of numbers such that $0 < a_i < 1$ and $\prod_{i=2}^{\infty} a_i = \frac{1}{2}$. Then it is easy to see that T is Lipschitzian and $\|Tx - Ty\| \leqslant 2\|x - y\|, x, y \in B$. Moreover, $\|T^i x - T^i y\| \leqslant 2 \prod_{i=2}^{\infty} a_i \|x - y\|$ for $i = 2, 3, \cdots$. Thus

$$\lim_{i \to \infty} k_i = \lim_{i \to \infty} 2 \prod_{j=2}^{i} a_j = 1.$$

Clearly the transformation T is not nonexpansive.

5.3   Fixed Point Theorems of Brouwer and Schauder

Brouwer's fixed point theorem is basic for many fixed point theorems. It states that a continuous map, which maps a convex, bounded and closed set in $\mathbb{R}^n$ into itself, has a fixed point. Though Brouwer[2] obtained his result in 1910, Poincare proved a slightly different version of it in 1886 which was subsequently rediscovered by Bhol in 1904.

There are several proofs of Brouwer's theorem-topological, analytical and degree theoretic. We give the proof of this theorem only in the next chapter by using degree theoretic arguments. In this section, we dwell on some of its important consequences. However Schauder's theorem is discussed in detail.

Unlike contraction mapping theorem, Brouwer's theorem does not give any computational scheme for obtaining a fixed point. However in 1967 Scarf [546] gave

---

[2]Brouwer's theorem has a long history. Ideas leading to the proof of Brouwer's theorem were discovered by Henri Poincaré as early as 1886. Brouwer himself proved the theorem for n = 3 in 1909. In 1910, Hadamard gave the first proof for arbitrary n, and Brouwer gave another proof in 1912. However in 1904, a result which is equivalent to Brouwer's theorem was published by P. Bohl.

some sort of algorithm for computing a fixed point of a mapping with some additional conditions. Since then many algorithms have been devised. We do not discuss these algorithms in this book, and interested readers may refer to the book edited by Karamardian [315].

Definition 5.28 A topological space X is said to possess the fixed point property if every continuous function of X into itself has a fixed point.

It is easy to prove that the unit closed interval $[0, 1]$, a finite closed interval $[a, b]$ and the closed unit disk in the plane has the fixed point property. In all other important cases, it is rather difficult to establish the existence of the fixed point property.

Theorem 5.45 *Let X and Y be topological spaces. If X is homeomorphic to Y and X has the fixed point property, then Y has the fixed point property.*

Definition 5.29 (*Partition of the unity*) Suppose $V_1, \ldots, V_n$ are open subsets of a locally compact Hausdorff space X, $K \subset X$ is compact, and

$$K \subset V_1 \cup \cdots \cup V_n.$$

Then for every $j = 1, \ldots, n$ there exists $\varphi_j \in C(X), 0 \leqslant \varphi_j \leqslant 1$, supported on V_j such that
$$\varphi_1(x) + \cdots + \varphi_n(x) = 1, \quad \forall x \in K.$$

The collection $\varphi_1, \ldots, \varphi_n$ is said to be a partition of the unity for K subordinate to the open cover $\{V_1, \ldots, V_n\}$.

The existence of a partition of the unity is a straightforward consequence of the Urysohn lemma (see, for instance, Munkre [414], p. 225). We are often interested to find partitions of the unity for a compact set $K \subset X$ whose members are continuous functions defined on K. Clearly, in this case X need not be locally compact.

Theorem 5.46 *Any nonempty, closed and convex subset K of $\mathbb{R}^n$ or of a Hilbert space $\mathcal{H}$ or of a Banach space X is a retract of any larger subset.*

For a proof of this theorem, refer to Bourbaki [73] or Dugundji [207].

Theorem 5.47 *If Y has the fixed point property and X is a retract of Y, then X has the fixed point property.*

Proof Let R be a retraction mapping of Y onto X. Let F be a continuous mapping of X into itself. Then FR is a continuous mapping of Y into X. Since FR maps Y into itself, there is a fixed point u such that $FRu = u$. But $u \in X$ and hence $Ru = u$ and this implies that $Fu = u$. ∎

5.3.1   Brouwer's Fixed Point Theorem

Let $\mathbb{B}^n = \{x \in \mathbb{R}^n : \|x\| \leqslant 1\}$ and $\mathbb{S}^{n-1} = \{x \in \mathbb{R}^n : \|x\| = 1\}$ denote the closed unit ball and (n-1)-sphere of $\mathbb{R}^n$, respectively.

Lemma 5.7 *The set $\mathbb{S}^{n-1}$ is not a retract of $\mathbb{B}^n$.*

Proof The lemma can be easily proved by means of algebraic topology tools. Indeed, we may observe that a retraction r induces a homomorphism $r_* : H_{n-1}(\mathbb{B}^n) \to H_{n-1}(\mathbb{S}^{n-1})$, where H_{n-1} denotes the $(n-1)$-dimensional homology group (see, e.g. Massy [392]). The natural injection $j : \mathbb{S}^{n-1} \to \mathbb{B}^n$ induces in turn a homomorphism $j_* : H_{n-1}(\mathbb{S}^{n-1}) \to H_{n-1}(\mathbb{B}^n)$, and the composition $r \circ j$ is the identity map on $\mathbb{S}_{n-1}$. Hence, $(r \circ j)_* = r_* \circ j_*$ is the identity map on $H_{n-1}(\mathbb{S}^{n-1})$. But since $H_{n-1}(\mathbb{B}^n) = 0$, j_* is the null map. On the other hand, $H_{n-1}(\mathbb{B}^n) = \mathbb{Z}$ if $n \neq 1$, and $H_0(\mathbb{S}^0) = \mathbb{Z} \oplus \mathbb{Z}$. This leads to a contradiction.   ■

In the following result, we illustrate the analytic proof which is less evident, and make use of differential forms. However, it provides a weaker result, namely, it shows that there exist no retraction of class C^2 from $\mathbb{B}^n$ to $\mathbb{S}^{n-1}$. But, one may note this will be enough for our scopes.

Proof First of all, we associate to a C^2 function $h : \mathbb{B}^n \to \mathbb{B}^n$ the differential (exterior) form given by

$$\omega_h = h_1 dh_2 \wedge \cdots \wedge dh_n.$$

Then the Stokes theorem (cf. Rudin [533], Chap.10) entails

$$\mathscr{D}_h := \int_{\mathbb{S}^{n-1}} S^{n-1}\omega_h = \int_{B^n} d\omega_h = \int_{B^n} dh_1 \wedge \cdots \wedge dh_n = \int_{B^n} det[J_h(x)]dx$$

where $J_h(x)$ denotes the $(n \times n)$-Jacobian matrix of h at x. Assume now that there is a retraction r of class C^2 from $\mathbb{B}^n$ to $\mathbb{S}^{n-1}$. From the above formula, we see that $\mathscr{D}_r$ is determined only by the values of r on $\mathbb{S}^{n-1}$. But $r|_{\mathbb{S}^{n-1}} = i|_{\mathbb{S}^{n-1}}$, where $i : \mathbb{B}^n \to \mathbb{B}^n$ is the identity map. Thus, $\mathscr{D}_r = \mathscr{D}_i = vol(\mathbb{B}^n)$. On the other hand $\|r\| \equiv 1$, and this implies that the vector $J_r(x)r(x)$ is null for every $x \in \mathbb{B}^n$. So 0 is an eigenvalue of $J_r(x)$ for every $x \in \mathbb{B}^n$, and therefore $det[J_r] \equiv 0$ which implies $\mathscr{D}_r = 0$.   ■

Theorem 5.48 (Brouwer's fixed point theorem) *The closed unit ball $\mathbb{B}^n$ of $\mathbb{R}^n$ has the fixed point property.*

Alternatively, we can state as follows:

Let $f : \mathbb{B}^n \to \mathbb{B}^n$ be a continuous function. Then f has a fixed point $\bar{x} \in \mathbb{B}^n$.

Proof We shall rely on the analytic proof. So, let $f : \mathbb{B}^n \to \mathbb{B}^n$ be of class C^2. Suppose, if possible, f had no fixed point, then

$$r(x) = t(x)f(x) + (1 - t(x))x$$

where

$$t(x) = \frac{\|x\|^2 - \langle x, f(x)\rangle - \sqrt{(\|x\|^2 - \langle x, f(x)\rangle)^2 + (1 - \|x\|^2)\|x - f(x)\|^2}}{\|x - f(x)\|^2}$$

is a retraction of class C^2 from $\mathbb{B}^n$ to $\mathbb{S}^{n-1}$, against the conclusion of Lemma 5.7. Graphically speaking, $r(x)$ is the intersection with $\mathbb{S}^{n-1}$ of the line obtained extending the segment connecting $f(x)$ to x. Hence, such an f has a fixed point. Finally, let $f : \mathbb{B}^n \to \mathbb{B}^n$ be continuous. Appealing to the Stone–Weierstrass theorem, we find a sequence $f_j : \mathbb{B}^n \to \mathbb{B}^n$ of functions of class C^2 converging uniformly to f on $\mathbb{B}^n$. Denote $\bar{x}_j$ the fixed point of f_j. Then there is $\bar{x} \in \mathbb{B}^n$ such that, up to a subsequence, $x_j \to \bar{x}$. Therefore,

$$\|f(\bar{x}) - \bar{x}\| \leqslant \|f(\bar{x}) - f(\bar{x}_j)\| + \|f(\bar{x}_j) - f_j(\bar{x}_j)\| + \|\bar{x}_j - \bar{x}\| \to 0 \text{ as } j \to \infty,$$

which yields $f(\bar{x}) = \bar{x}$.

Remark 5.22 An alternative approach to prove Brouwer's fixed point theorem makes use of the concept of topological degree as depicted in the next chapter.

We now prove the following version of Brouwer's FPT.

Theorem 5.49 *Every nonempty, compact and convex subset of K of $\mathbb{R}^n$ (a finite dimensional normed linear space) has the fixed point property.*

Proof For r sufficiently large, the ball B_r of radius r contains K. Then by Theorem 5.46, K is a retract of B_r. Since B_r is homeomorphic to $\mathbb{B}^n$, Theorem 5.45 asserts that B_r has the fixed point property. Then Theorem 5.47 proves that K has the fixed point property. ∎

The following theorem establishes a connection between Brouwer's FPT and retraction. For its proof refer to Istratescu [288].

Theorem 5.50 *The Brouwer's FPT is equivalent to the following assertion: there exists no retraction of $\mathbb{B}^n$ on to the boundary $\partial \mathbb{B}^n$ which is continuously differentiable.*

Theorem 5.51 *Let $F : \mathbb{R}^n \to \mathbb{R}^n$ be a continuous mapping and suppose that for some $r > 0$ and all $\eta > 0$,*

$$F(x) + \eta x \neq 0 \text{ for any } x \text{ with } \|x\| = r.$$

Then there exists a point x_0, $\|x_0\| \leqslant r$ such that $F(x_0) = 0$.

Proof Suppose there is no such point x_0 with $\|x_0\| \leqslant r$ such that $F(x_0) = 0$. Then define the mapping $G : B_r \to B_r$ by

$$G(x) = -\frac{r F(x)}{\|F(x)\|}.$$

Since $\|Gx\| = r$ and F is continuous on B_r, G is a continuous mapping of B_r into itself. Hence, by Brouwer's FPT there exists a point y such that $G(y) = y$ and $\|y\| = r$), that is,

$$F(y) + \frac{\|F(y)\|}{r} y = 0.$$

But this contradicts the property of F. Hence, there exists a point x_0 with $\|x_0\| \leqslant r$ such that $F(x_0) = 0$. ∎

Definition 5.30 The Hilbert cube $\mathcal{H}_0$ is the subset of ℓ_2 consisting of all points $x = (x_1, x_2, \ldots)$ such that $\|x_k\| \leqslant 1/k$ for all k.

Theorem 5.52 *Every compact convex subset K of a Banach space X is homeomorphic under a linear mapping to a compact convex subset of $\mathcal{H}_0$.*

Proof Without loss of generality, we assume that K is a subset of the unit ball in X. Let $\{x_n\}$ be a dense sequence spanning the linear span of K. Let us choose f_k in X^* such that

$$f_k(x_k) = \frac{\|x_k\|}{k}, \quad \|f_k\| = \frac{1}{k}, \quad k = 1, 2, \ldots.$$

Then the mapping $F : X \to \ell_2$ defined by $Fx = (f_1(x), f_2(x), \ldots)$ maps K into $\mathcal{H}_0$. F is a bounded linear mapping from X into ℓ_2 and is one-to-one on span of K. Because if $x \neq y$ in span K, we have

$$|f_k(x) - f_k(y)| \geqslant |f_k(x_k)| - |f_k(x - y - x_k)|$$
$$\geqslant \frac{\|x_k\|}{k} - \frac{\|(x - y) - x_k\|}{k} > 0,$$

if x_k is sufficiently close to $x - y$. Hence, F is a homeomorphism of K onto $F(K)$, as it is one-to-one and continuous on the compact set K. So $F(K)$ is compact and convex as linear homeomorphism preserves these properties. ∎

Theorem 5.53 *The Hilbert cube $\mathcal{H}_0$ has the fixed point property.*

Proof Let $P_n : \mathcal{H}_0 \to \mathcal{H}_0$ be the mapping defined by

$$P_n(x_1, x_2, \ldots) = (x_1, x_2, \ldots, x_n, 0, 0, \ldots).$$

For n sufficiently larger, we have

$$\|P_n x - x\| \leqslant \left(\sum_{k=n+1}^{\infty} \frac{1}{k^2} \right)^{1/2} < \varepsilon \text{ for all } x \in \mathcal{H}_0.$$

$\mathcal{H}_0$ is compact since $P_n \mathcal{H}_0$ is compact. The set $C_n = P_n \mathcal{H}_0$ is homeomorphic to the closed unit ball in $\mathbb{R}^n$ and hence can be considered as a compact convex subset of

$\mathbb{R}^n$. Since the mapping $P_n F : C_n \to C_n$, where F is any continuous mapping of $\mathcal{H}_0$ into itself, is continuous, $P_n F$ has a fixed point $y_n \in C_n \subset \mathcal{H}_0$ by Brouwer's FPT, such that

$$\|y_n - F y_n\| \leqslant \left(\sum_{k=n+1}^{\infty} \frac{1}{k^2} \right)^{1/2}.$$

Since $\mathcal{H}_0$ is compact, $\{y_n\}$ has a convergent subsequence. The limit of this sequence is a fixed point of F (Refer Smart [571]). ∎

Theorem 5.54 *Any convex and closed subset K of the Hilbert cube $\mathcal{H}_0$ has the fixed point property.*

Proof K is a retract of $\mathcal{H}_0$ by Theorem 5.46. By Theorems 5.47 and 5.52, K has the fixed point property. ∎

5.3.2 The Schauder Fixed Point Theorem

Theorem 5.55 (Schauder's fixed point theorem) *Let K be a nonempty, closed and convex subset of a normed linear space X. Let T be a continuous mapping of K into a compact subset of K. Then T has a fixed point in K.*

Proof Let A be a compact subset of K and $T : K \to A$ a continuous map, so that $T(K) \subset A$. A is contained in a closed, convex and bounded subset B of X. We have $T(B \cap K) \subset T(K) \subset A \subset B$. Thus, $T(B \cap K)$ is contained in a compact subset of $B \cap K$, and so there is no loss of generality if K is assumed to be bounded. If A_0 is countable dense subset of the compact metric space A, then the set of all rational linear combinations of elements of A_0 is a countable dense subset of the closed linear subspace E_0 spanned by A_0 and $A \subset E_0$. Then $T(K \cap E_0) \subset T(K) \subset A$, a compact subset of E_0 and $K \cap E_0$ is closed and convex. Hence, without loss of generality, we may assume that K is bounded, closed and convex subset of a separable normed linear space X with a strictly convex norm. Given a positive integer n, there exists a $\frac{1}{n}$-net $T x_1, T x_2, \ldots, T x_n$ say in $T(K)$, such that

$$\min_{1 \leqslant k \leqslant n} \|T x - T x_k\| < \frac{1}{n}, \quad x \in K. \tag{5.28}$$

Let E_n denote the convex hull of $T x_1, \ldots, T x_n$. Then $K_n \triangle K \cap E_n$ is a closed bounded subset of E_n and therefore compact. Since the norm is strictly convex, the metric projection P_n of X onto the convex compact subset K_n exists. Then $T_n \triangle P_n T$ is a continuous mapping of the nonempty, convex and compact subset K_n into itself. By the Brouwer's fixed point theorem, T_n has a fixed point $u_n \in K_n$, i.e.

$$T_n u_n = u_n. \tag{5.29}$$

By (5.28), since $Tx_k \in K_n (k = 1, 2, \ldots, n)$ we have

$$\|Tx - T_n x\| < \frac{1}{n} \text{ as } T_n x = P_n Tx. \tag{5.30}$$

The sequence $\{T u_n\}$ of $T(K)$ has a subsequence $\{T u_{n_k}\}$ converging to a point $u \in K$. By (5.29) and (5.30), we have

$$\|u_{n_k} - u\| = \|T_{n_k} u_{n_k} - u\| \leqslant \|T_{n_k} u_{n_k} - T_{n_k}\| + \|T u_{n_k} - u\|$$

$$\leqslant \frac{1}{n} + \|T u_{n_k} - u\|.$$

Taking limit $k \to \infty$, we have $\lim\limits_{k \to \infty} u_{n_k} = u$. By the continuity of T, we have

$$\lim_{k \to \infty} T u_{n_k} = Tu.$$

Since $\{T u_{n_k}\}$ converges to u as $k \to \infty$, $Tu = u$. ∎

As a consequence of Schauder's theorem, we get the following result.

Theorem 5.56 *Let K be a nonempty, compact and convex subset of a normed linear space X and let T be a continuous mapping of K into itself. Then T has a fixed point in K.*

Proof K is homeomorphic to a compact convex subset C of $\mathcal{H}_0$ (by Theorem 5.52). By Theorem 5.54, C has the fixed point property. So, in view of Theorem 5.45, K has the fixed point property. Thus, T has a fixed point in K. ∎

Remark 5.23 (*a*) Theorem 5.54 with additional hypothesis that K be complete follows from Theorem 5.55. Because, if K is complete convex subset and $T(K)$ is contained in compact subset A of K, then the closed and convex hull of A is a compact convex subset K_0 of K, and $T K_0 \subset K_0$.

(*b*) Schauder's fixed point theorem was generalized to locally convex topological vector space by Tychonoff [593], and this generalization is known as Schauder–Tychonoff theorem.

Theorem 5.57 *Let T be a compact and continuous mapping of a normed linear space X into itself and let $T(X)$ be bounded. Then T has a fixed point.*

Proof Let K be the closed and convex hull of $T(X)$. Then K is bounded and $T(K)$ is contained in a compact subset of K. By Theorem 5.55 T has a fixed point. ∎

Theorem 5.58 *Let X be a reflexive Banach space, K a closed and convex subset of X and T a weakly continuous mapping of K into a bounded subset of K. Then T has a fixed point K.*

Proof Since K is closed and convex, it is also weakly closed. As X is reflexive, each bounded weakly closed subset of X is weakly compact. Hence, the weakly closed and convex hull of $T(K)$ is weakly compact. The result follows by Theorem 5.55. ∎

Theorem 5.59 *Let X be a normed linear space. Suppose that K is a nonempty, closed, and convex subset of X, and T a continuous mapping of K into itself such that $T(K)$ is relatively compact in X. Then T has a fixed point.*

Proof Let B be the convex hull of $T(K)$ in X, and let S be the closure of B in X. Then $S \subset K$. Since $T(K)$ is relatively compact, it follows that S is compact. S is also nonempty and convex. As T is continuous on K, we have

$$T(S) = T(\overline{B}) = T(\overline{B} \cap K) \subset \overline{T}(K) = \overline{B} = S.$$

By application of Theorem 5.56 to the restriction of T to X, we get the desired result. ∎

Let X be a Banach space and T a mapping of X into itself. It is useful to know condition under which $I + T$ maps X onto itself. If T is a linear mapping, a sufficient condition for above to hold is that $\|T\| < 1$. For, if y is any element of X, the series $\sum_{n=0}^{\infty} (-1)^n T^n y$ is convergent in X and has a sum say $x \in X$. Then x satisfies the equation $x + Tx = y$. But for nonlinear mapping T, we have to depend on Schauder's theorem to prove the following.

Theorem 5.60 *Let T be a weakly continuous mapping transforming bounded sets into weakly relatively compact set, and let it satisfies $\lim\limits_{\|x\| \to \infty} \sup \frac{\|Tx\|}{\|x\|} < 1$. Then $I + T$ maps X onto itself.*

Proof Let y be any point in X. It suffices to show that the mapping P defined by $P(x) = y - Tx$ has a fixed point in X. Let $B_r = \{x : \|x\| \leqslant r\}$ with r suitably chosen Whatever be the value of r, P restricted to B_r is continuous and $P(B_r)$ is relatively compact due to our hypothesis on T. It remains to show that $P(B_r)$ is contained in B_r. Let $M(r)$ denote the supremum of $\|Tx\|$ for $\|x\| \leqslant r$. Then $\lim\limits_{\|x\| \to \infty} \sup \frac{\|Tx\|}{\|x\|} < 1$ and the fact that T is locally bounded shows that $\lim\limits_{r \to \infty} \sup \frac{M(r)}{r} < 1$. On the other hand, if $\|x\| \leqslant r$, then $\|Px\| \leqslant \|y\| + M(r)$. So, if r be chosen large enough, $P(x)$ will belong to B, whenever x has that property. Our result follows by an application of Theorem 5.54 in X (with respect to the weak topology of X). ∎

Remark 5.24 The above result is simplified if we assume that X is a reflexive Banach space. Because, the weak continuity of T itself ensures that T transforms bounded sets into weakly relatively compact sets (each bounded set being weakly relatively compact as a consequence of the reflexivity of the space).

Theorem 5.61 (Krasnoselski fixed point theorem) *Let K be a nonempty, complete and convex subset of a normed linear space X. Let T be a continuous mapping of K into a compact subset of X. Let $S : K \to X$ be a contradiction mapping with*

Lipschitz constant α and let $Tx + Sy \in K$ for all $x, y \in K$ for all $x, y \in K$. Then there is a point $u \in K$ such that $Tu + Su = u$.

Proof Define $F : K \times K \rightharpoonup K$ by $F(x, y) = Tx + Sy$. Then we have

$$\|F(x, y) - F(x, y')\| = \|Sy - Sy'\| \leqslant \alpha \|y - y'\|; \quad x, y, y' \in K. \tag{5.31}$$

Also

$$\|F(x, y) - F(x'y)\| = \|Tx - Tx'\|; \quad x, x', y \in K.$$

Hence, for each fixed x, the mapping $y \to F(x, y)$ is a contraction mapping of the complete metric space K into itself. Therefore, it has a unique fixed point in K which is denoted by Ax. That is, $Ax = F(x, Ax)$, $(x \in K)$. Then we have

$$\begin{aligned}
\|Ax - Ax'\| &= \|F(x, Ax) - F(x', Ax)\| \\
&= \|F(x, Ax) - (F(x'Ax) + F(x', Ax) - F(x', Ax')\| \\
&\leqslant \|Tx - Tx'\| + \alpha \|Ax - Ax'\|.
\end{aligned}$$

Therefore, $\|Ax - Ax'\| \leqslant \frac{1}{1-\alpha} \|Tx - Tx'\|$.

This shows that the mapping A is continuous and that $A(K) \subset K$ is precompact since TK is compact. Since K is complete, $\overline{A(K)} \subset K$ is compact. By Schauder's theorem, A has a fixed point u in K, i.e. $Au = u$. But this implies that

$$u = Au = F(u, Au) = F(u, u) = Tu + Su.$$

This completes the proof.   ∎

Under the conditions of Schauder's theorem, we have no method for approxiamting a fixed point of mapping. However in a special situation, this can be done as we see in the following theorem due to Krasnoselski [343].

Theorem 5.62 *Let K be a bounded, closed and convex subset of a uniformly convex Banach space X. Let T be a nonexpansive mapping of K into a compact subset of K. Let x_0 be an arbitrary point of K. Then the sequence defined by*

$$x_{n+1} = \frac{1}{2}(x_n + Tx_n) \ (n = 0, 1, 2, \ldots)$$

converges to a fixed point of T in K.

Proof Let $\mathfrak{F}(T)$ denote the set of fixed points of T in K. By Schauder's theorem the set $\mathfrak{F}(T)$ is nonempty. First we show that

$$\|x_{n+1} - y\| \leqslant \|x_n - y\|, \quad (y \in \mathfrak{F}(T), \ n = 0, 1, 2, \ldots) \tag{5.32}$$

Since $y = Ty$, we have

$$\|x_{n+1} - y\| = \frac{1}{2}(x_n + Tx_n) - \frac{1}{2}(y + Ty)\| = \left\|\frac{1}{2}(x_n - y) + \frac{1}{2}(Tx_n - Ty)\right\|$$

$$\leqslant \frac{1}{2}\|x_n - y\| + \frac{1}{2}\|Tx_n - Ty\| = \frac{1}{2}\|x_n - y\| + \frac{1}{2}\|x_n - y\| = \|x_n - y\|.$$

Suppose there exists an $\varepsilon > 0$ and a positive integer N such that

$$\|x_n - Tx_n\| \geqslant \varepsilon \ \forall \ n \geqslant N. \tag{5.33}$$

Then $\qquad \|x_n - y - (Tx_n - Ty)\| \geqslant \varepsilon \ \forall \ n \geqslant N, y \in \mathfrak{F}(T).$

Also $\qquad \|Tx_n - Ty\| \leqslant \|x_n - y\| \leqslant \|x_0 - y\|.$

Since the space is uniformly convex, there exists a constant $\delta, 0 < \delta < 1$, such that

$$\|x_{n+1} - y\| = \left\|\frac{1}{2}(x_n - y) + \frac{1}{2}(Tx_n - Ty)\right\|$$

$$\leqslant \delta \max \{\|x_n - y\|, \|Tx_n - Ty\|\}$$

$$< \|x_n - y\| \ \forall \ n \geqslant N.$$

Therefore $\lim_{n\to\infty} x_n = y$ where $Ty = y$.

If there does not exist an $\varepsilon > 0$ for which (5.33) holds, there exists a subsequence such that $\lim_{k\to\infty} (x_{n_k} - Tx_{n_k}) = 0$, and such that $\{Tx_{n_k}\}$ converges. But this implies that $\lim_{k\to\infty} x_{n_k} = u = \lim_{k\to\infty} Tx_{n_k}$ and hence $Tu = u$. So

$$\|x_{n+1} - u\| \leqslant \|x_n - u\|.$$

As $\lim_{k\to\infty} \|x_{n_k} - u\| = 0$, we have $\lim_{n\to\infty} \|x_n - u\| = 0$. ∎

The following theorem, due to Altman [10, 11], is proved by Schauder's theorem. A proof of this theorem using the concept of degree theory can be found in Berger and Berger [53].

Theorem 5.63 *Let X be a normed linear space. Let T be a continuous mapping of $B_r = \{x : \|x\| \leqslant r\}$ into a compact subset of X such that*

$$\|Tx - x\|^2 \geqslant \|Tx\|^2 - \|x\|^2, \text{ for all } x \text{ such that } \|x\| = r. \tag{5.34}$$

Then T has a fixed point in B_r.

Proof Suppose T has no fixed point in B_r, then we have

$$\|Tx - x\| + \|x\| \geqslant \|Tx\|, \quad (\|x\| = r). \tag{5.35}$$

The above inequality (5.35) follows because

$$(\|Tx - x\| + \|x\|)^2 - \|Tx\|^2 = \|Tx - x\|^2 + \|x\|^2 - \|Tx\|^2 + 2\|x\|\|Tx - x\|$$
$$\geqslant 2\|x\|\|Tx - x\| > 0.$$

Let P be the mapping defined by

$$Px = \begin{cases} x, & x \in B_r \\ \dfrac{rx}{\|x\|}, & x \notin B_r. \end{cases}$$

Then P is a continuous projection of X onto B_r. Let $\tilde{T} \triangle PT$. Then $\tilde{T}$ maps B_r continuously into a compact subset of B_r i.e. $PTu = u$. If $Tu \in B_r$, then $PTu = Tu$, and $Tu = u$. If $Tu \notin B_r$ (T has no fixed point in B_r), then

$$\|Tu\| > r \text{ and } u = PTu = \frac{r}{\|Tu\|}Tu.$$

It follows that $\|u\| = r$ and we have

$$\|Tu - u\| + \|u\| = \left\| \frac{\|Tu\|}{r}u - u \right\| + \|u\|$$
$$= \left[\frac{\|Tu\|}{r} - 1 + 1 \right] \|u\| = \|Tu\|.$$

This contradicts (5.34), and hence the proof. ∎

Theorem 5.64 (Rothe [530]) *Let X be a normed linear space and T be a continuous mapping of B_r into a compact subset of X such that*

$$T(\partial B_r) \subset B_r \text{ i.e. } \|Tx\| \leqslant \|x\| \ \forall x \in \partial B_r.$$

Then T has a fixed point.

Proof We show that Rothe's condition: $\|Tx\| \leqslant \|x\|$ for all $x \in \partial B_r$ implies Altman's condition on the boundary:

$$\|Tx - x\|^2 \geqslant \|Tx\|^2 - \|x\|^2.$$

For $x \in \partial B_r$, we have

$$\|Tx - x\|^2 \geqslant [\|x\| - \|Tx\|]^2 = \|x\|^2 - 2\|x\| \|Tx\| + \|Tx\|^2$$
$$\geqslant \|x\|^2 - 2\|x\|^2 + \|Tx\|^2 = \|Tx\|^2 - \|x\|^2.$$

Hence, the proof follows from Altman's theorem (Theorem 5.63). ∎

This theorem is also true if we replace B_r by any closed and convex subset of a normed linear space X.

Theorem 5.65 (Potter [491]) *Let X be a normed linear space. Let K be any closed and convex subset of X and T be a continuous mapping of K into a compact subset of X such that $T(\partial K) \subset K$. Then T has a fixed point.*

Proof We note that the result is trivial if int $K = \varnothing$. Assume without loss of generality, that $0 \in$ int K. Define radial retraction R of X onto K by

$$Rx = \frac{x}{\max(1, \rho(x))},$$

where $\rho(x) = \inf\{\alpha : x \in \alpha K\}$ is the Minkowski functional. Then R is a continuous retraction of X onto K and if $Rx \in$ int K then $Rx = x$ and if $x \notin K$, then $Rx \in \partial K$. Consider the mapping $RT = \tilde{T}$. Then $\tilde{T}$ maps K continuously into a compact subset of K. Hence, by Schauder's theorem, $\tilde{T}$ has a fixed point $u \in K$ i.e. $\tilde{T}u = u$. If $u \in \partial K$, then $Tu \in K$ and $u = RTu = Tu$. If $u \in$ int K, then RTu is in int K. That is,

$$u = RTu = Tu.$$

This completes the proof.   ∎

5.3.3 The Schauder–Tychonoff Fixed Point Theorem

We first extend Brouwer's fixed point theorem to a more general situation.

Lemma 5.8 *Let K be a nonvoid, compact and convex subset of a finite dimensional real Banach space X. Then every continuous function $f : K \to K$ has a fixed point $\bar{x} \in K$.*

Proof Without loss of generality, we may assume that X is homeomorphic to $\mathbb{R}^n$ for some $n \in \mathbb{N}$. Also, we can assume $K \subset \mathbb{B}^n$. For every $x \in \mathbb{B}^n$, let $p(x) \in K$ be the unique point of minimum norm of the set $x - K$. Notice that $p(x) = x$ for every $x \in K$. Moreover, p is continuous on $\mathbb{B}^n$. Indeed, given $x_n, x \in \mathbb{B}^n$, with $x_n \to x$, we have

$$\|x - p(x)\| \leqslant \|x - p(x_n)\| \leqslant \|x - x_n\| + \inf_{k \in K} \|x_n - k\| \longrightarrow \|x - p(x)\|$$

as $n \to \infty$. Thus, $x - p(x_n)$ is a minimizing sequence as $x_n \to x$ in $x - K$, and this implies the convergence $p(x_n) \to p(x)$. Define now $g(x) = f(p(x))$. Then g maps continuously $\mathbb{B}^n$ onto K. From Theorem 5.48, there is $\bar{x} \in K$ such that $g(\bar{x}) = \bar{x} = f(\bar{x})$.   ∎

Remark 5.25 If there is a compact and convex set $K \subset \mathbb{R}^n$ such that $h(K) \subset K$, then h has a fixed point $\bar{x} \in K$.

Theorem 5.66 (Schauder–Tychonoff fixed point theorem) *Let X be a locally convex space, $K \subset X$ nonvoid and convex, $K_0 \subset K$, K_0 compact. Given a continuous map $f : K \to K_0$, there exists $\bar{x} \in K_0$ such that $f(\bar{x}) = \bar{x}$.*

Proof Let $\mathcal{B}$ denote the local base for the topology of X generated by the separating family of seminorms $\mathscr{P}$ on X. Given $U \in \mathcal{B}$, from the compactness of K_0, there exist $x_1, \ldots, x_n \in K_0$ such that

$$K_0 \subset \bigcup_{j=1}^{n} (x_j + U).$$

Let $\varphi_1, \ldots, \varphi_n \in C(K_0)$ be a partition of the unity for K_0 subordinate to the open cover $\{xj + U\}$, and define

$$f_U(x) = \sum_{j=1}^{n} \varphi_j(f(x))x_j, \quad \forall x \in K$$

then

$$f_U(K) \subset K_U := co(\{x_1, \ldots, x_n\}) \subset K$$

and Lemma 5.8 yields the existence of $x_U \in K_U$ such that $f_U(x_U) = x_U$. Then

$$x_U - f(x_U) = f_U(x_U) - f(x_U) = \sum_{j=1}^{n} \varphi_j(f(x_U))(x_j - f(x_U)) \in U \qquad (5.36)$$

for $\varphi_j(f(x_U)) = 0$ whenever $x_j - f(x_U) \notin U$. Appealing again to the compactness of K_0, there exists

$$\bar{x} \in \bigcap_{W \in \mathcal{B}} \overline{\{f(x_U) : U \in \mathcal{B}, U \subset W\}} \subset K_0. \qquad (5.37)$$

Select now $p \in \mathscr{P}$ and $\varepsilon > 0$, and let

$$V = \{x \in X : p(x) < \varepsilon\} \in \mathcal{B}.$$

Since f is continuous on K, there is $W \in \mathcal{B}$, $W \subset V$, such that

$$f(x) - f(\bar{x}) \in V$$

whenever $x - \bar{x} \in 2W, x \in K$. Moreover, by (5.37), there exists $U \in B, U \subset W$, such that

$$\bar{x} - f(x_U) \in W \subset V. \qquad (5.38)$$

Combining (5.36) and (5.38) we get

$$x_U - \bar{x} = x_U - f(x_U) + f(x_U) - \bar{x} \in U + W \subset W + W = 2W$$

which yields

$$f(x_U) - f(\bar{x}) \in V. \tag{5.39}$$

Hence, (5.38) and (5.39) entail

$$p(\bar{x} - f(\bar{x})) \leqslant p(\bar{x} - f(x_U)) + p(f(x_U) - f(\bar{x})) < 2\varepsilon.$$

Being p and ε arbitrary, we conclude that $p(\bar{x} - f(\bar{x})) = 0$ for every $p \in \mathscr{P}$, which implies the equality $f(\bar{x}) = \bar{x}$.

Remark 5.26 In general, it is not possible to extend Theorem 5.65 to noncompact settings. This fact was already envisaged in our previous discussion about nonexpansive maps. Let us recall another famous example in Hilbert spaces.

Example 5.34 (*Kakutani*) Consider the Hilbert space ℓ_2. For a fixed $\varepsilon \in (0, 1]$, let $f_\varepsilon : \overline{B}_{\ell_2}(0, 1) \to \overline{B}_{\ell_2}(0, 1)$ be given by

$$f_\varepsilon(x) = (\varepsilon(1 - \|x\|), x_0, x_1, \ldots), \quad \forall x = (x_0, x_1, x_2, \ldots) \in \ell_2.$$

Then it is clear that f_ε has no fixed points in $\overline{B}_{\ell_2}(0, 1)$, but it is Lipschitz continuous with Lipschitz constant slightly greater than 1. Indeed, we observe that

$$\|f_{\ell_2}(x) - f_\varepsilon(y)\| \leqslant \sqrt{1 + \varepsilon^2} \, \|x - y\| \, \forall x, y \in \overline{B}_{\ell_2}(0, 1).$$

We now recall the well-known definition of lower and upper semicontinuous real functions.

Definition 5.31 Let X be a topological space. A function $f : X \to (-\infty, \infty]$ is said to be lower semicontinuous if $f^{-1}((\alpha, \infty])$ is open for every $\alpha \in \mathbb{R}$. Similarly, a function $g : X \to [-\infty, \infty)$ is said to be upper semicontinuous if $-g$ is lower semicontinuous.

Observation

• As a direct consequence of the definition (refer to Definition 5.31 above), we observe that the supremum of any collection of lower semicontinuous functions is lower semicontinuous. Moreover, if f is lower semicontinuous and X is compact, then f attains its minimum on X. Indeed, if it is not so, denoting $m = \inf_{x \in X} f(x) \in [-\infty, \infty)$, the sets $f^{-1}((\alpha, \infty])$ with $\alpha > m$ form an open cover of Y that admits no finite subcovers.

The next result is the famous Ky Fan inequality.

Theorem 5.67 (Ky Fan [225]) *Let* $K \subset X$ *be a nonvoid, compact and convex. Let* $\Phi : K \times K \to \mathbb{R}$ *be map such that*

(a) $\Phi(\cdot, y)$ *is lower semicontinuous* $\forall y \in K$;
(b) $\Phi(x, \cdot)$ *is concave* $\forall x \in K$.

Then there exists $x_0 \in K$ such that $\sup_{y \in K} \Phi(x_0, y) \leqslant \sup_{y \in K} \Phi(y, y)$.

Proof Let us fix $\varepsilon > 0$. In correspondence with every $x \in K$, there are $y_x \in K$ and an open neighbourhood U_x of x such that

$$\Phi(z, y_x) > \sup_{y \in K} \Phi(x, y) - \varepsilon, \forall z \in U_x \cap K.$$

Since K compact, there exist some $x_1, \ldots, x_n \in K$ there holds

$$K \subset U_{x_1} \cup \cdots \cup U_{x_n}.$$

Let $\varphi_1, \ldots, \varphi_n \in C(K)$ be a partition of the unity for K subordinate to the open cover $\{U_{x_j}\}$. We now define

$$f(x) = \sum_{j=1}^{n} \varphi_j(x) y_{x_j}, \forall x \in K.$$

Then it is clear that the map f is continuous, and

$$f(co(\{y_{x_1}, \ldots, y_{x_n}\})) \subset co(\{y_{x_1}, \ldots, y_{x_n}\}).$$

Hence, by Lemma 5.8 f admits a fixed point $\bar{x} \in K$. Therefore,

$$\sup_{y \in K} \Phi(y, y) \geqslant \Phi(\bar{x}, \bar{x}) \geqslant \sum_{j=1}^{n} \varphi_j(\bar{x}) \Phi(\bar{x}, y_{x_j})$$

$$\geqslant \sum_{j=1}^{n} \varphi_j(\bar{x}) (\sup_{y \in K} \Phi(x_j, y) - \varepsilon)$$

$$\geqslant \inf_{x \in K} \sup_{y \in K} \Phi(x, y) - \varepsilon$$

$$= \sup_{y \in K} \Phi(x_0, y) - \varepsilon$$

for some $x_0 \in K$. Letting $\varepsilon \to 0$ we obtain

$$\sup_{y \in K} \Phi(y, y) \geqslant \sup_{y \in K} \Phi(x_0, y).$$

This yields the desired inequality.   ∎

5.3.4 Extension of Continuous Mappings

We begin this section with the statement of a theorem on the extension of continuous mappings.

Theorem 5.68 *Let X be a Banach space, $A \subset X$ a closed subset and $T : A \to Y$ a continuous map from A into the Banach space Y. Then there exists a continuous extension $\tilde{T}$ of T with $\tilde{T} : X \to Y$, $\tilde{T}|_A = T$ and*

$$\tilde{T}(X) \subset coT(A).$$

Corollary 5.13 *Let C be a closed and convex subset of X. Then there exists a mapping $R : X \to C$ such that $R|_C = id$.*

Proof Take $T = id|_C$. Now apply the extension Theorem 5.68.

Corollary 5.14 (Brouwer's fixed point theorem) *Let C be a compact convex set in $\mathbb{R}^n$, $f : C \to C$ a continuous mapping. Then f has a fixed point.*

Proof Choose $r > 0$ such that $\overline{B}(0, r) \supset C$. Let $\tilde{f} : \mathbb{R}^n \to \mathbb{R}^n$ be an extension of f with $\tilde{f}(\mathbb{R}^n) \subset coC$. Then

$$\tilde{f}(\overline{B}(0, r)) \subset coC \subset C \subset \overline{B}(0, r).$$

Thus, $\tilde{f}$ has a fixed point $\tilde{x} \in C$. Therefore, $f(\tilde{x}) = \tilde{x} \in C$.

Corollary 5.15 (Schauder's fixed point theorem) *Let K be a compact convex subset of X, and $T : K \to K$ a continuous mapping. Then T has a fixed point.*

5.4 Fixed Point Theorems for Multifunctions

In this section, we discuss fixed point theorems concerning multivalued mappings or multifunctions. The study of fixed point problems of such mappings was initiated by Kakutani in 1941 in finite dimensional spaces. It was extended to infinite dimensional Banach spaces by Bohnenblust and Karlin in 1950 and to locally convex spaces by Fan [225, 226] in 1952. Fan's result also generalizes Schauder–Tychonoff's theorem.

Fixed point theorems for multifunctions provide natural setting for many problems in control theory (refer Dauer [156]) involving differential equations. Also they have been effectively used in tackling problems in economics and game theory, we shall state relevant results in this direction.

The developments of geometric fixed point theory for multifunctions were initiated by Nadler, Jr. [416] and subsequently pursued by Markin [385, 386], Assad and Kirk [25]. Browder [98], Goebel [245], Lami-Dozo [356], Reich [514] and others. Since then fixed point theorems for multifunctions have been extensively studied, we shall cover these developments in detail.

We recall the definition of upper semicontinuous function for a multivalued mapping.

Definition 5.32 A multifunction $T : X \to 2^Y$ is said to be upper semicontinuous (u.s.c.) if and only if the set $\{x \in X | T(x) \cap B \neq \varnothing\}$ is closed for each closed subset B of Y.

If Y is a compact Hausdorff space, and if $T(x)$ is closed for each x, then T is u.s.c. if and only if T has a closed graph. If $A \subseteq X$, then $T(A) = \bigcup_{x \in A} T(x)$ and if $B \subseteq Y, T^{-1}(B) = \{x \in X : T(x) \cap B \neq \varnothing\}$. It is obvious that if $B = \bigcup_i B_i \subseteq Y$, then $T^{-1}(B) = \bigcup_i T^{-1}(B_i)$. Also the multifunction T is upper semicontinuous if and only if for each closed set $A \subseteq Y, T^{-1}(A)$ is a closed subset of X. It is point closed (convex, compact) if and only if for each $x \in X, T(x)$ is closed (convex, compact) subset of Y.

Definition 5.33 A point $x_0 \in X$ is said to be a fixed point of the multifunction $T : X \to 2^X$ if $x_0 \in T(x_0)$.

We state below, without proof, a theorem due to Fan [225].

Theorem 5.69 *Let X be a normed linear space, K a nonempty, compact and convex subset of X, and T a mapping that assigns to each $x \in K$ a nonempty, closed and convex subset $T(x)$ of K. Suppose T is upper semicontinuous, then there exists a fixed point of T in K.*

Kakutani in 1941 [301] proved the above theorem for $X = \mathbb{R}^n$. Recently Himmelberg [273] generalized Fan's result, and Sehgal and Morrison [549] have further generalized Himmelberg's work.

Theorem 5.70 (Himmelberg [273]) *Let K_1 be a nonempty and convex subset of a normed linear space X. Let $T : K_1 \to K_1$ be an u.s.c. multifunction such that $T(x)$ is closed and convex for each $x \in K_1$ and $T(K_1)$ is contained in some compact subset C of K_1. Then T has a fixed point in K_1.*

For the proof of Theorem 5.70, we need the following result.

Theorem 5.71 *Let K be a nonempty and compact subset of a normed linear space X, and $T : K \to K$ be an u.s.c. multifunction such that $T(x)$ is closed for all $x \in K$ and convex for all x in some dense almost convex subset A of K. Then T has a fixed point.*

Proof of Theorem 5.70 Without loss of generality, we may assume that X is complete (conditions on K_1 and T remain unchanged).

Let $A = co\, C$ and $K = \bar{A}$. Then K is compact, $A \subset K_1$ and $T(A) \subset C \subset A$. Let $H = T \bigcap (A \times A)$. Evidently, H is a relatively closed subset of $A \times A$ and has the same values on A and T. Consider the relation $\bar{H} \subset X \times X$ with closure relative to $K \times K$.

Note that $\bar{H}$ is a multifunction from K onto K, i.e. $\bar{H}^{-1}(K) = K$, because $\bar{H}^{-1}(K)$ is closed and contains A. Moreover, $\bar{H}(K) \subset C \subset A$ and $H = \bar{H} \cap (A \times A)$. So $\bar{H}(x) = H(x) = F(x)$ for all $x \in A$. Thus, by Theorem 5.71, $\bar{H}$ has a fixed point, say x, in K. But $x \in \bar{H}(x) \subset C \subset A$. So $x \in T(x)$.

Remark 5.27 Theorem 5.69 is also true if the space under consideration is a locally convex linear topological space. The Himmelberg's proof is in that setting.

As a consequence of the above theorem, we get the following min-max theorem.

Theorem 5.72 *Let K_1 and K_2 be compact subsets of the normed linear spaces X_1 and X_2, respectively. Let A_1 and A_2 be dense almost convex subsets of K_1 and K_2, respectively. Let f be a continuous real-valued function on $K_1 \times K_2$. If for any $x_0 \in A_1 y_0 \in A_2$, the sets $\{x \in K_1 : f(x, y_0) = \max_{\xi \in K_1} f(\xi, y_0)\}$ and $\{y \in K_2 : f(x_0, y) = \min_{\eta \in K_2} f(x_0, \eta)\}$ are convex. Then*

$$\max_{x \in K_1} \min_{y \in K_2} f(x, y) = \min_{y \in K_2} \max_{x \in K_1} f(x, y)$$

The following theorem of Sehgal and Morrison [549] is a generalization of the theorem of Himmelberg.

Theorem 5.73 *Let S be a nonempty and convex subset of a normed linear space X and K a compact subset of S. Let Y be a regular separated topological space and $f : X \rightarrow Y$ be a point closed, u.s.c. and $g : K \rightarrow Y$ be a point compact, u.s.c. multifunctions. If for $x \in S$*

(i) $f(x) \cap g(K) \neq \varnothing$,
(ii) $g^{-1}(f(x))$ is convex,

then there is an $x \in K$ such that $f(x) \cap g(x) \neq \varnothing$.

Remark 5.28 Theorem 5.70 comes out as a particular case of Theorem 5.73 by taking $X = Y$ and g as the identity mapping of K onto itself.

Plunket [488], Ward [605] and others have shown that the spaces which have the fixed point property for continuous compact multivalued mappings constitute a fairly small subclass of those spaces which have fixed point property for continuous single-valued mappings.

5.4.1 Fixed Point Theorems for Multivalued Contraction Mappings

Now we discuss some fixed point theorems for multivalued contraction mappings, due to Nadler, Jr. [416]. These theorems do not place severe restrictions on the images of points and, in general, the space is required to be a complete metric space. We also briefly touch upon the works of Assad and Kirk [25] and Markin [381] on multivalued contraction, and of Smithson [572] on contractive multifunctions.

Let (X, d) be a metric space. Let $CB(X)$ denote the set of nonempty, closed and bounded subsets of X and $\mathcal{K}(X)$ denote the set of nonempty and compact subsets of X. Denote

$$d(a, B) = \inf\{d(a, b) : b \in B \subset X\}, \quad a \in X,$$

$$\rho(A, B) = \sup_{a \in A} d(a, B),$$

$$H(A, B) = \max\{\rho(A, B), \rho(B, A)\},$$

$$H^+(A, B) = \frac{1}{2}\{\rho(A, B) + \rho(B, A)\}$$

for all $A, B \in CB(X)$. It is well known that H is a metric on $CB(X)$ and is called the Hausdorff–Pompeu metric induced by d. In Proposition 5.3 below, we show that H^+ is also a metric on $CB(X)$.

Alternatively, we may define Hausdorff–Pompeu metric H on $CB(X)$ as follows.

Definition 5.34 The Hausdorff–Pompeu metric H on $CB(X)$ induced by d is given by

$$H(A, B) = \inf\{\varepsilon | A \subset N(\varepsilon, B) \text{ and } B \subset N(\varepsilon, A)\} \text{ for } A, B \in CB(X)$$

where $N(\varepsilon, C) = \{x \in X | d(x, c) < \varepsilon \text{ for some } c \in C\}$, $\varepsilon > 0$ and $C \in CB(X)$.

Proposition 5.3 H^+ *is a metric on* $CB(X)$.

Proof Let $A, B \in CB(X)$ such that $H^+(A, B) = 0$. Then this is equivalent to $\rho(A, B) = 0$ and $\rho(B, A) = 0$; i.e. $\inf_{y \in B} d(x, y) = 0$ for any $x \in A$ and $\inf_{x \in A} d(y, x) = 0$ for any $y \in B$. Therefore, these are equivalent to $x \in \overline{B} = B$ for any $x \in A$, and $y \in \overline{A} = A$ for any $y \in B$, $\overline{B}$ being closure of B. It follows that $A \subset B$ and $B \subset A$. Hence, $A = B$.

The symmetry of the function H^+ follows directly from the definition.

To show the triangle inequality, let $A, B, C \in CB(X)$. Then for any $(x, y, z) \in A \times B \times C$, we have

$$d(x, z) \leqslant d(x, y) + d(y, z),$$

whence

$$\inf_{z \in C} d(x, z) \leqslant d(x, y) + \inf_{z \in C} d(y, z) \leqslant d(x, y) + \rho(B, C).$$

Since the above inequality holds for any $y \in B$, we get

$$\inf_{z \in C} d(x, z) \leqslant \inf_{y \in B} d(x, y) + \rho(B, C) \leqslant \rho(A, B) + \rho(B, C).$$

Hence,

$$\rho(A, C) \leqslant \rho(A, B) + \rho(B, C). \qquad (5.40)$$

Interchanging the roles of A and C, we get

$$\rho(C, A) \leqslant \rho(C, B) + \rho(B, A). \qquad (5.41)$$

Adding (5.40) and (5.41), and then dividing by 2, we get

$$H^+(A, C) \leqslant H^+(A, B) + H^+(B, C). \qquad (5.42)$$

Notice that the two metrics H and H^+ are equivalent [352] since

$$\frac{1}{2} H(A, B) \leqslant H^+(A, B) \leqslant H(A, B).$$

In the light of this equivalence and referring to Kuratowski [352], we conclude that $(CB(X), H^+)$ is complete whenever (X, d) is complete. Indeed, it is a simple consequence of the completeness of the Hausdorff–Pompeu metric H. Moreover, $\mathcal{K}(X)$ is a closed subspace of $(CB(X), H^+)$.

Notice also that $H^+ : CB(X) \times CB(X) \to \mathbb{R}$ is a continuous function. To see this, we observe that the inequality

$$H^+(A, B) \leqslant H^+(A, C) + H^+(C, B)$$

holds for any $A, B, C \in CB(X)$. Now pick any $(A_0, B_0) \in CB(X) \times CB(X)$. Then for a given $\varepsilon > 0$, we can choose a positive number $\delta = \frac{\varepsilon}{2}$ such that

$$\left| H^+(A, B) - H^+(A_0, B_0) \right| \leqslant H^+(A, A_0) + H^+(B_0, B) < \delta + \delta = 2\delta = \varepsilon$$

whenever $H^+(A, A_0) < \delta$, $H^+(B_0, B) < \delta$. This shows that H^+ is continuous at (A_0, B_0).

Observation

- The metric H depends on the metric d of X.
- By definitions of ρ and H it is easy to see that

$$d(a, B) \leqslant \rho(A, B) \leqslant H(A, B)$$

for all $a \in A$ and $A, B \in CB(X)$.
- One can easily see that for a given $a \in A$ and an $\varepsilon > 0$, there exists $b \in B$ such that

$$d(a, b) \leqslant d(a, B) + \varepsilon.$$

- Two equivalent metrics on X may not generate equivalent Hausdorff–Pompeu metrics for $CB(X)$ (Kelley) [323].

In a classical approach, one can easily prove Propositions 5.4 and 5.5 stated below.

Proposition 5.4 *If $a, b \in X$ and $A, B \in CB(X)$, then the relations:*

(1) $d(a, b) = H^+(\{a\}, \{b\})$,
(2) $A \subset \overline{S}(B; r_1), B \subset \overline{S}(A; r_2) \Rightarrow H^+(A, B) \leqslant r$ *where* $r = (r_1 + r_2)/2$, *and*
(3) $H^+(A, B) < r \Rightarrow \exists r_1, r_2 > 0$ *such that* $(r_1 + r_2)/2 = r$ *and* $A \subset S(B; r_1)$, $B \subset S(A; r_2)$ *hold.*

Proof The relation (1) follows immediately from the definition of the function H^+. To proof relation (2), from the inclusions $A \subset \overline{S}(B; r_1), B \subset \overline{S}(A; r_2)$, it follows that

$$\forall x \in A, \exists y_x \in B \quad \text{such that} \quad d(x, y_x) \leqslant r_1$$

and

$$\forall y \in B, \exists x_y \in A \quad \text{such that} \quad d(x_y, y) \leqslant r_2.$$

From here it follows that

$$\inf_{y \in B} d(x, y) \leqslant r_1 \text{ for every } x \in A, \text{ and } \inf_{x \in A} d(x, y) \leqslant r_2 \text{ for every } y \in B.$$

Hence,

$$\sup_{x \in A} \left(\inf_{y \in B} d(x, y) \right) \leqslant r_1 \text{ and } \sup_{y \in B} \left(\inf_{x \in A} d(x, y) \right) \leqslant r_2.$$

Therefore $H^+(A, B) \leqslant r$ where $r = \frac{r_1 + r_2}{2}$.

To proof relation (3), let $H^+(A, B) = k < r$. Then there exist $k_1, k_2 > 0$ such that $k = \frac{k_1 + k_2}{2}$ and

$$\sup_{x \in A} (\inf_{y \in B} d(x, y)) = k_1, \ \sup_{y \in B} (\inf_{x \in A} d(x, y)) = k_2.$$

As $0 < k < r$, it follows that there exist $r_1, r_2 > 0$ such that $k_1 < r_1, k_2 < r_2$ and $r = \frac{r_1 + r_2}{2}$. Then from the above inequalities, it follows that

$$\inf_{y \in B} d(x, y) \leqslant k_1 < r_1 \text{ for every } x \in A \text{ and } \inf_{x \in A} d(x, y)) \leqslant k_2 < r_2 \text{ for every } y \in B.$$

Then, for any $x \in A$ there exists $y_x \in B$ such that

$$d(x, y_x) < \inf_{y \in B} d(x, y) + r_1 - k_1 \leqslant r_1.$$

and, for any $y \in B$ there exists $x_y \in A$ such that

$$d(x_y, y) < \inf_{x \in A} d(x, y) + r_2 - k_2 \leqslant r_2.$$

Hence, for any $x \in A$ and $y \in B$ it follows that

$$x \in \bigcup_{y \in B} S(y; r_1) \text{ and } y \in \bigcup_{x \in A} S(x; r_2),$$

that is,

$$A \subset S(B; r_1) \text{ and } B \subset S(A; r_2).$$

Remark 5.29 From the relations (2) and (3), it follows immediately that the relations

(2′) $A \subset S(B; r_1)$, $B \subset S(A; r_2) \Rightarrow H^+(A, B) \leqslant r$ where $r = (r_1 + r_2)/2$, and
(3′) $H^+(A, B) < r \Rightarrow \exists r_1, r_2 > 0$ such that $(r_1 + r_2)/2 = r$ and $A \subset \overline{S}(B; r_1)$, $B \subset \overline{S}(A; r_2)$ hold.

Proposition 5.5 *If $A, B \in CB(X)$, then the equalities*

(4) $H^+(A, B) = \inf\{r > 0 : A \subset S(B; r_1), B \subset S(A; r_2), r = (r_1 + r_2)/2\}$,
(4′) $H^+(A, B) = \inf\{r > 0 : A \subset \overline{S}(B; r_1), A \subset \overline{S}(B; r_2), r = (r_1 + r_2)/2\}$ *hold.*

Proof From the relation (2′), it follows that

$$H^+(A, B) \leqslant \inf\left\{r > 0 : A \subset S(B; r_1), A \subset S(B; r_2), r = \frac{r_1 + r_2}{2}\right\}.$$

To prove the opposite inequality, let $H^+(A, B) = k$, and let $t > 0$. Then $H^+(A, B) < k + t$. From (3) it follows that $\exists t_1, t_2 > 0$ with $\frac{t_1 + t_2}{2} = t$ such that $A \subset S(B; k + t_1)$ and $B \subset S(A; k + t_2)$. Hence,

$$\{r > 0 : A \subset S(B; r_1), B \subset S(A; r_2)\} \supset \{k + t : t > 0, A \subset S(B; k + t_1), B \subset S(A; k + t_2)\}.$$

From this inclusion relation, it follows that

$$\inf\{r > 0 : A \subset S(B; r_1), B \subset S(A; r_2)\} \leqslant \inf\{k + t : t > 0\} = k = H^+(A, B).$$

In conclusion, we have

$$H^+(A, B) = \inf\left\{r > 0 : A \subset S(B; r_1), B \subset S(A; r_2), r = \frac{r_1 + r_2}{2}\right\}.$$

Theorem 5.74 *If the metric space (X, d) is complete, then so is $(CB(X), H^+)$ and also $\mathcal{K}(X)$ is a closed subspace of $(CB(X), H^+)$.*

Proof Let (X, d) be a complete metric space and let $\{A_n\}_{n \in \mathbb{N}}$ be a Cauchy sequence in $CB(X)$. We claim that the sequence $\{A_n\}_{n \in \mathbb{N}}$ is convergent to the set $B = Ls\, A_n = \{x \in X : \forall \varepsilon > 0, \forall m \in \mathbb{N}, \exists n \in \mathbb{N}, n \geq m \text{ such that } S(x; \varepsilon) \cap A_n \neq \varnothing\}$.

Since the sequence $\{A_n\}_{n \in \mathbb{N}}$ is Cauchy, for any $\varepsilon > 0$ there exists $m(\varepsilon) \in \mathbb{N}$ such that

$$H^+(A_n, A_{m(\varepsilon)}) < \varepsilon \text{ for any } n \in \mathbb{N}, n \geq m(\varepsilon).$$

Hence, by relation (4), it follows that $\exists\, \varepsilon_1, \varepsilon_2 > 0$ with $\frac{\varepsilon_1+\varepsilon_2}{2} = \varepsilon$ and $m(\varepsilon_1), m(\varepsilon_2) \in \mathbb{N}$ such that $\min\{m(\varepsilon_1), m(\varepsilon_2)\} \geq m(\varepsilon)$, $A_n \subset S(A_{m(\varepsilon_1)}; \varepsilon_1)$ for any $n \in \mathbb{N}$, $n \geq m(\varepsilon_1)$ and $A_{m(\varepsilon_2)} \subset S(A_n; \varepsilon_2)$ for any $n \in \mathbb{N}$, $n \geq m(\varepsilon_2)$.

From the properties of upper topological limit Ls it follows that $B \subset \overline{\bigcup_{k \geq n} A_k}$ for any $n \in \mathbb{N}$. Therefore $B \subset \overline{S}(A_{m(\varepsilon_1)}; \varepsilon_1)$, whence the relation

(i) $\qquad\qquad\qquad B \subset \overline{S}(A_{m(\varepsilon_1)}; 4\varepsilon_1)$ holds.

On the other hand, taking $\overline{\varepsilon}_k = \frac{\varepsilon_1}{2^k}$, $k \in \mathbb{N}$, it follows that there exists $n_k = m(\overline{\varepsilon}_k) \in \mathbb{N}$ such that

$$H^+(A_n, A_{n_k}) < \overline{\varepsilon}_k, \; \forall n \geq n_k.$$

Next, we choose n_k such that the sequence $\{n_k\}_{k \in \mathbb{N}}$ to be strictly increasing. Let $p \in A_{n_0} = A_{m(\varepsilon_1)}$ arbitrarily, and let there be the sequence $\{p_{n_k}\}_{k \in \mathbb{N}}$ such that $p_{n_0} = p$ and $p_{n_k} \in A_{n_k}$ with the property that $d(p_{n_k}, p_{n_{k-1}}) < \frac{\varepsilon_1}{2^{k-2}}$. It follows that the sequence $\{p_{n_k}\}_{k \in \mathbb{N}}$ is a Cauchy sequence in the complete metric space (X, d). Hence, it is convergent to a point $l \in X$.

Since $d(p_{n_k}, p_{n_0}) < 4\varepsilon_1$, it follows that $d(l, p) \leq 4\varepsilon_1$. Therefore, $\inf_{y \in B} d(p, y) \leq 4\varepsilon_1$; that is, $p \in \overline{S}(B; 4\varepsilon_1)$, which implies that

(ii) $\qquad\qquad\qquad A_{n_0} \subset \overline{S}(B; 4\varepsilon_1).$

Keeping in view the relations (i) and (ii), (3) yields $H^+(A_{n_0}, B) \leq 4\varepsilon_1$. Taking into account the fact that H^+ is a metric on $CB(X)$, we get

$$H^+(A_n, B) \leq H^+(A_n, A_{n_0}) + H^+(A_{n_0}, B) < 5\varepsilon_1,$$

for any $n \geq m(\varepsilon_1) = n_0$. Thus, the sequence $\{A_n\}_{n \in \mathbb{N}}$ converges to $B = Ls\, A_n$; that is, $(CB(X), H^+)$ is a complete metric space. This proves the first assertion of our theorem.

To prove the second assertion, we just require to show that $C(X)$ is a complete subspace of $(CB(X), H^+)$. Let $\{A_n\}_{n \in \mathbb{N}}$ be a Cauchy sequence in $C(X)$. Then, $\{A_n\}_{n \in \mathbb{N}}$ is a Cauchy sequence in $CB(X)$. Let $A \in CB(X)$ be such that $A = \lim_{n \to \infty} A_n$. Then for any $\varepsilon > 0$, there exists $m(\varepsilon) \in \mathbb{N}$ such that

$$H^+(A_n, A) < \frac{\varepsilon}{2} \; \forall n \geq m(\varepsilon), n \in \mathbb{N}.$$

Hence, by relation (4), it follows that $\exists\, \varepsilon_1, \varepsilon_2 > 0$ with $\frac{\varepsilon_1+\varepsilon_2}{2} = \varepsilon$ and $m(\varepsilon_1), m(\varepsilon_2) \in \mathbb{N}$ such that $\min\{m(\varepsilon_1), m(\varepsilon_2)\} \geq m(\varepsilon)$, $A_n \subset S(A; \frac{\varepsilon_1}{2})$ for any $n \in \mathbb{N}$, $n \geq m(\varepsilon_1)$ and $A \subset S(A_n; \frac{\varepsilon_2}{2})$ for any $n \in \mathbb{N}$, $n \geq m(\varepsilon_2)$.

Suppose $n_0 \geq m(\varepsilon_2)$ is a fixed natural number. Then $A \subset S(A_{n_0}; \frac{\varepsilon_2}{2})$. Since A_{n_0} is compact in X, it follows that it is totally bounded. Hence, there exist $x_i^{\varepsilon_2}, i \in \overline{1, p}$ such that $A_{n_0} \subset \bigcup_{i=1}^{p} S(x_i^{\varepsilon_2}; \frac{\varepsilon_2}{2})$, whence $A \subset \bigcup_{i=1}^{p} S(x_i^{\varepsilon_2}; \varepsilon_2)$. Therefore $A \in \mathcal{K}(X)$.

Definition 5.35 Let (X, d_1) and (Y, d_2) be two metric spaces. Let $F : X \to CB(Y)$. T is said to be a multivalued contraction mapping if and only if

$$H(Tx, Ty) \leqslant k\, d_1(x, y), \quad x, y \in X, \tag{5.43}$$

where $0 \leqslant k < 1$ is a fixed real number.

The following is a simple consequence of the definition of the Hausdorff–Pompeu metric H. Let $A, B \in CB(X)$ and $a \in A$. If $\varepsilon > 0$, then there exists $b = b(a) \in B$ such that

$$d(a, b) \leqslant H(A, B) + \varepsilon, \tag{5.44}$$

that can easily be transformed to

$$hd(a, b) \leqslant H(A, B) \text{ where } h \in (0, 1). \tag{5.45}$$

Indeed, if $H(A, B) = 0$, then $a \in B$ and so (5.45) holds for $b = a$. If $H(A, B) > 0$, then

$$\varepsilon = (h^{-1} - 1)H(A, B) > 0. \tag{5.46}$$

As observed above, for any $\varepsilon > 0$ there exists $b \in B$ such that

$$d(a, b) \leqslant d(a, B) + \varepsilon \leqslant H(A, B) + \varepsilon. \tag{5.47}$$

By inserting the value of ε from (5.46) in (5.47), we obtain (5.45).

Notice also that if $A, B \in \mathcal{K}(X)$ and $a \in A$, then there exists $b \in B$ such that

$$d(a, b) \leqslant H(A, B).$$

Example 5.35 Let $X = [0, 1]$ and $\psi : X \to X$ such that

$$\psi(x) = \begin{cases} \frac{1}{2}x + \frac{1}{2}, & 0 \leqslant x \leqslant \frac{1}{2} \\ -\frac{1}{2}x + 1, & \frac{1}{2} \leqslant x \leqslant 1., \end{cases}$$

Define $T : X \to 2^X$ by $T(x) = \{0\} \cup \{\psi(x)\}$ for each $x \in X$. Then one can easily check that T is multivalued contraction mapping and the set of fixed points of T is $\{0, \frac{2}{3}\}$.

Theorem 5.75 (Nadler [416]) *Let (X, d) be a complete metric space. If $T : X \to CB(X)$ is a multivalued contraction mapping, then T has a fixed point.*

Proof Construct a sequence $\{x_n\}$ in X in the following way. Set $h = \sqrt{k}$. Choose $x_0 \in X$. Denote by x_1 any fixed element in Tx_0. Since $Tx_0, Tx_1 \in CB(X)$ and $x_1 \in Tx_0$, there is a point $x_2 \in Tx_1$ such that

$$hd(x_1, x_2) \leqslant H(Tx_0, Tx_1).$$

In general, if x_n is chosen, then we choose $x_{n+1} \in Tx_n$ such that

$$hd(x_n, x_{n+1}) \leqslant H(Tx_{n-1}, Tx_n).$$

Thus, from (5.40) we have

$$hd(x_n, x_{n+1}) \leqslant k\, d(x_{n-1}, x_n) = h^2 d(x_{n-1}, x_n).$$

Hence, we obtain

$$d(x_n, x_{n+1}) \leqslant hd(x_{n-1}, x_n).$$

Repeating the above argument n-times we get

$$d(x_n, x_{n+1}) \leqslant h^n d(x_0, x_1).$$

For $m > n \geq 1$, we have

$$\begin{aligned}
d(x_n, x_m) &\leqslant d(x_n, x_{n+1}) + d(x_{n+1}, x_{n+2}) + \cdots + d(x_{m-1}, x_m) \\
&\leqslant (h^n + h^{n+1} + \cdots + h^m)d(x_0, x_1) \\
&\leqslant (h^n + h^{n+1} + \cdots)d(x_0, x_1) \\
&= \frac{h^n}{1-h}d(x_0, x_1).
\end{aligned}$$

As $h \in (0, 1)$, for a given $\varepsilon > 0$, we can choose $N \in \mathbb{N}$ so large that $\frac{h^n}{1-h}d(x_0, x_1) < \varepsilon$. Thus, we have

$$d(x_n, x_m) < \varepsilon \ \text{ for all } \ m, n \geq N.$$

From this inequality, we conclude that $\{x_n\}$ is a Cauchy sequence. Since X is complete, the sequence $\{x_n\}$ converges to some point $u \in X$.

From (5.45) and by triangle inequality, we have

$$\begin{aligned}
d(u, Tu) &\leqslant d(u, x_{n+1}) + d(x_{n+1}, Tu) \\
&\leqslant d(u, x_{n+1}) + H(Tx_n, Tu) \\
&\leqslant d(u, x_{n+1}) + kd(x_n, u) \to 0 \ \text{ as } \ n \to \infty.
\end{aligned}$$

Hence, $d(u, Tu) = 0$. Since Tu is closed, it follows that $u \in Tu$.

Pathak and Shahzad [466] introduce the notion of H^+-contraction for multifunctions.

Definition 5.36 Let (X, d) be a metric space. A multivalued map $T : X \to CB(X)$ is called H^+-contraction if
(1°) there exists k in (0, 1) such that

$$H^+(Tx, Ty) \leqslant kd(x, y) \text{ for every } x, y \in X,$$

(2°) for every x in X, y in $T(x)$ and $\varepsilon > 0$, there exists z in $T(y)$ such that

$$d(y, z) \leqslant H^+(T(y), T(x)) + \varepsilon.$$

In [465], Pathak and Shahzad introduced the notion of H^+-type multivalued weak contractive mapping.

Definition 5.37 Let (X, d) be a metric space. A mapping $T : X \to CB(X)$ is called an H^+-type multivalued weak contractive mapping if the condition (2°) holds and there exists $0 < k < 1$ such that
(3°)$H^+(Tx, Ty) \leqslant k \max\{d(x, y), d(x, Tx), d(y, Ty), [d(x, Ty) + d(y, Tx)]/2\}$,
for all x, y in X.

In [466], Pathak and Shahzad proved the following result.

Theorem 5.76 *Every H^+-type multivalued contraction mapping $T : X \to CB(X)$ with Lipschitz constant $0 < k < 1$ has a fixed point.*

Proof Let $x_0 \in X$ be arbitrary. Fix an element x_1 in Tx_0. From (2°) it follows that we can choose $x_2 \in Tx_1$ such that

$$d(x_1, x_2) \leqslant H^+(Tx_0, Tx_1) + \varepsilon \tag{5.48}$$

In general, if x_n be chosen, then we choose $x_{n+1} \in Tx_n$ such that

$$d(x_n, x_{n+1}) \leqslant H^+(Tx_{n-1}, Tx_n) + \varepsilon. \tag{5.49}$$

Set $\varepsilon = (\frac{1}{\sqrt{k}} - 1)H^+(Tx_{n-1}, Tx_n)$. Then from (2°), it follows that

$$d(x_n, x_{n+1}) \leqslant H^+(Tx_{n-1}, Tx_n) + \left(\frac{1}{\sqrt{k}} - 1\right) H^+(Tx_{n-1}, Tx_n) = \frac{1}{\sqrt{k}}H^+(Tx_{n-1}, Tx_n).$$

Thus, we have
$$\sqrt{k}\, d(x_n, x_{n+1}) \leqslant H^+(Tx_{n-1}, Tx_n). \tag{5.50}$$

Now, from (1°) we have

$$\sqrt{k}\, d(x_n, x_{n+1}) \leqslant k\, d(x_{n-1}, x_n) = (\sqrt{k})^2\, d(x_{n-1}, x_n).$$

Hence, for all $n \in \mathbb{N}$ we have

$$d(x_n, x_{n+1}) \leqslant \sqrt{k}\, d(x_{n-1}, x_n).$$

Repeating the same argument n-times, we get

$$d(x_n, x_{n+1}) \leqslant k^{\frac{n}{2}} \, d(x_0, x_1).$$

This implies that $\{x_n\}$ is a Cauchy sequence. Since X is complete, there exists $u \in X$ such that $\lim_{n \to \infty} x_n = u$.

Since

$$\frac{1}{2} \Big\{ \rho(Tx_n, Tu) + \rho(Tu, Tx_n) \Big\} = H^+(Tx_n, Tu) \leqslant k \, d(x_n, u),$$

it follows that

$$\liminf_{n \to \infty} \Big\{ \rho(Tx_n, Tu) + \rho(Tu, Tx_n) \Big\} = 0.$$

Since

$$\liminf_{n \to \infty} \rho(Tx_n, Tu) + \liminf_{n \to \infty} \rho(Tu, Tx_n) \leqslant \liminf_{n \to \infty} \{ \rho(Tx_n, Tu) + \rho(Tu, Tx_n) \},$$

we have

$$\liminf_{n \to \infty} \rho(Tx_n, Tu) + \liminf_{n \to \infty} \rho(Tu, Tx_n) = 0.$$

This implies that

$$\liminf_{n \to \infty} \rho(Tx_n, Tu) = 0.$$

Since $\lim_{n \to \infty} d(x_{n+1}, u) = 0$ exists, and

$$d(u, Tu) \leqslant \rho(Tx_n, Tu) + d(x_{n+1}, u),$$

it follows that

$$\begin{aligned}
d(u, Tu) &\leqslant \liminf_{n \to \infty} [\rho(Tx_n, Tu) + d(x_{n+1}, u)] \\
&= \liminf_{n \to \infty} \rho(Tx_n, Tu) + \lim_{n \to \infty} d(x_{n+1}, u) = 0.
\end{aligned}$$

This implies that $d(u, Tu) = 0$, and since Tu is closed it must be the case that $u \in Tu$.

Remark 5.30 As $max\{a, b\} \geq \frac{1}{2}\{a + b\} \ \forall \, a, b \geq 0$, it follows that multivalued contraction (5.43) always implies multivalued H^+-contraction but the converse implication need not be true. To see this, we observe the following.

Example 5.36 Let $X = \big\{ 0, \frac{1}{4}, 1 \big\}$ and $d : X \times X \to \mathbb{R}$ be a standard metric. Let $T : X \to CB(X)$ be such that

$$T(x) = \begin{cases} \{0\}, & \text{for } x = 0, \\ \{0, \frac{1}{4}\}, & \text{for } x = \frac{1}{4}, \\ \{0, 1\}, & \text{for } x = 1, \end{cases}$$

It is routine to check that multivalued H^+-contraction (1°) is satisfied for all $x, y \in X$ and for any $k \in [\frac{2}{3}, 1)$. Further, we see that for every x in X, y in $T(x)$ and $\varepsilon > 0$, there exists z in $T(y)$ such that $d(y, z) \leqslant H^+(T(y), T(x)) + \varepsilon$. Indeed,
(i) if $x = 0$, $y \in T(0) = \{0\}$, $\varepsilon > 0$, there exists $z \in T(y) = \{0\}$ such that

$$0 = d(y, z) \leqslant H^+(T(y), T(x)) + \varepsilon,$$

(iia) if $x = \frac{1}{4}$, $y \in T(x) = T(\frac{1}{4}) = \{0, \frac{1}{4}\}$, say $y = 0$, $\varepsilon > 0$, there exists $z \in T(y) = \{0\}$ such that

$$0 = d(y, z) < \frac{1}{8} + \varepsilon = H^+(T(y), T(x)) + \varepsilon,$$

(iib) if $x = \frac{1}{4}$, $y \in T(x) = T(\frac{1}{4}) = \{0, \frac{1}{4}\}$, say $y = \frac{1}{4}$, $\varepsilon > 0$, there exists $z(= \frac{1}{4}) \in T(y) = \{0, \frac{1}{4}\}$ such that

$$0 = d(y, z) < 0 + \varepsilon = H^+(T(y), T(x)) + \varepsilon,$$

(iiia) if $x = 1$, $y \in T(x) = T(1) = \{0, 1\}$, say $y = 0$, $\varepsilon > 0$, there exists $z \in T(y) = \{0\}$ such that

$$0 = d(y, z) < \frac{1}{2} + \varepsilon = H^+(T(y), T(x)) + \varepsilon,$$

(iiib) if $x = 1$, $y \in T(x) = T(1) = \{0, 1\}$, say $y = 1$, $\varepsilon > 0$, there exists $z(= 1) \in T(y) = \{0, 1\}$ such that

$$0 = d(y, z) < 0 + \varepsilon = H^+(T(y), T(x)) + \varepsilon.$$

Thus, the condition (2°) is also satisfied. Clearly, $0, \frac{1}{4}, 1$ are fixed points of T. However, we observe that the map T does not satisfy the assumptions of Theorem 5.75. Indeed, for $x = 0$ and $y = 1$, we have

$$H(T(0), T(1)) = H(\{0\}, \{0, 1\}) = 1 > k\,d(0, 1),$$

for all $k \in (0, 1)$.

Example 5.37 Let $X = \left[0, \frac{2\sqrt{2}}{3}\right] \cup \{1\}$ and $d : X \times X \to \mathbb{R}$ be a standard metric. Let $T : X \to CB(X)$ be such that

$$T(x) = \begin{cases} \left[\frac{11x}{50(x+1)}, \frac{11}{50}\right], & \text{for } x \in \left[0, \frac{2\sqrt{2}}{3}\right] \\ \{\frac{11}{50}\}, & \text{for } x = 1. \end{cases}$$

Set $k = 0.99$. We discuss the following cases:
Case 1. When $x, y \in \left[0, \frac{2\sqrt{2}}{3}\right]$, $y > x$, we note that

$$H^+(Tx, Ty) = \frac{11}{100} \cdot \frac{y-x}{1+x+y+xy} \leqslant \frac{11}{100} \cdot \frac{y-x}{1+y-x} < 0.99\frac{y-x}{1+y-x} \leqslant 0.99\,d(x, y).$$

Case 2. When $x \in \left[0, \frac{2\sqrt{2}}{3}\right]$ and $y = 1$, we note that $H^+(Tx, Ty) = \frac{11}{100}\left|1 - \frac{x}{1+x}\right| \leqslant$ $0.99(1 - x)$ is true if $\frac{11}{100} \cdot \frac{1}{1+x} \leqslant 0.99(1 - x)$ i.e. if $\frac{1}{9} \leqslant 1 - x^2$ i.e. if $0 \leqslant x \leqslant \frac{2\sqrt{2}}{3}$. To check the condition $(2°)$, we consider the following cases:

Case (i). For any $x \in \left[0, \frac{2\sqrt{2}}{3}\right]$, $y \in Tx = \left[\frac{11x}{50(x+1)}, \frac{11}{50}\right]$ and $\varepsilon > 0$, there exists $z(= y) \in Ty = \left[\frac{11y}{50(y+1)}, \frac{11}{50}\right]$ such that $0 = d(y, z) \leqslant \frac{11}{100} \cdot \frac{y-x}{1+x+y+xy} + \varepsilon = H^+(T(y), T(x)) + \varepsilon$. Note that $\frac{11y}{50(y+1)} \leqslant \frac{11y}{50} \leqslant y \leqslant \frac{11}{50}$ i.e. $y \in Ty$.

Case (ii). For $x = 1$, $y \in Tx = \left\{\frac{11}{50}\right\}$, i.e. $y = \frac{11}{50}$ and $\varepsilon > 0$, there exists $z(= \frac{792}{6100})$ $\in Ty = \left[\frac{121}{3050}, \frac{11}{50}\right]$ such that

$$d(y, z) = \frac{11}{122} < \frac{11}{122} + \varepsilon = H^+(T(y), T(x)) + \varepsilon.$$

This proves the condition $(2°)$. Thus, all the requirements of Theorem 5.76 are satisfied and $0 \in T0$ is the unique fixed point of T. However, we note that when $y = 1$ and $x \to \frac{2\sqrt{2}}{3}$ from the left, then

$$H(Tx, Ty) = \frac{11x}{50(1 + x)} > 1 - x.$$

Thus, T does not satisfy the assumptions of Theorem 5.75.

Proposition 5.6 *Suppose X and $CB(X)$ are as in the preceding theorem, and let $T_i : X \to CB(X)$, $i = 1, 2$, be two H^+-type multivalued contraction mappings with Lipschitz constant $k < 1$. Then if $Fix(T_1)$ and $Fix(T_2)$ denote the respective fixed point sets of T_1 and T_2,*

$$H^+(Fix(T_1), Fix(T_2)) \leqslant \frac{1}{1 - \sqrt{k}} \sup_{x \in X} H^+(T_1x, T_2x).$$

Proof Let $\varepsilon > 0$ be given. Select $x_0 \in Fix(T_1)$, and then select $x_1 \in T_2x_0$. From $(2°)$ it follows that we can choose $x_2 \in T_2x_1$ such that

$$d(x_1, x_2) \leqslant H^+(T_2x_0, T_2x_1) + \varepsilon.$$

Now define $\{x_n\}$ inductively so that $x_{n+1} \in T_2(x_n)$ and

$$d(x_n, x_{n+1}) \leqslant H^+(T_2x_{n-1}, T_2x_n) + \varepsilon. \tag{5.51}$$

Set $\varepsilon = (\frac{1}{\sqrt{k}} - 1)H^+(T_2x_{n-1}, T_2x_n)$. Then from (5.51), it follows that

$$d(x_n, x_{n+1}) \leqslant H^+(T_2x_{n-1}, T_2x_n) + \left(\frac{1}{\sqrt{k}} - 1\right) H^+(T_2x_{n-1}, T_2x_n) = \frac{1}{\sqrt{k}}H^+(T_2x_{n-1}, T_2x_n).$$

Thus, we have

$$\sqrt{k}\, d(x_n, x_{n+1}) \leqslant H^+(T_2 x_{n-1}, T_2 x_n). \tag{5.52}$$

Now applying (1°) for T_2, we have

$$\sqrt{k}\, d(x_n, x_{n+1}) \leqslant k\, d(x_{n-1}, x_n) = (\sqrt{k})^2\, d(x_{n-1}, x_n).$$

Hence, for all $n \in \mathbb{N}$, we have

$$d(x_n, x_{n+1}) \leqslant \sqrt{k}\, d(x_{n-1}, x_n).$$

Repeating the same argument n-times, we get

$$d(x_n, x_{n+1}) \leqslant k^{\frac{n}{2}}\, d(x_0, x_1).$$

This implies that $\{x_n\}$ is a Cauchy sequence with limit, say z. Since T_2 is continuous, we have

$$\lim_{n \to \infty} H(T_2 x_n, T_2 z) = 0.$$

Also, since $x_{n+1} \in T_2(x_n)$ it must be the case that $z \in T_2 z$; that is, $z \in \text{Fix}(T_2)$. Furthermore, using (5.52) we have

$$d(x_0, z) \leqslant \sum_{n=0}^{\infty} d(x_{n+1}, x_n) \leqslant (1 + \sqrt{k} + (\sqrt{k})^2 + \cdots) d(x_1, x_0) \leqslant \frac{1}{1 - \sqrt{k}} (H^+(T_2 x_0, T_1 x_0) + \varepsilon).$$

Reversing the roles of T_1 and T_2 and repeating the argument as above lead to the conclusion that for each $y_0 \in \text{Fix}(T_2)$, there exist $y_1 \in T_1 y_0$ and $w \in \text{Fix}(T_1)$ such that

$$d(y_0, w) \leqslant \frac{1}{1 - \sqrt{k}} (H^+(T_1 y_0, T_2 y_0) + \varepsilon).$$

Since $\varepsilon > 0$ is arbitrary, the conclusion follows.

Theorem 5.77 *Suppose X and $CB(X)$ are as in the preceding theorem, and let $T_i : X \to CB(X), i = 1, 2, \ldots$ be a sequence of H^+-type multivalued contraction mappings with Lipschitz constant $k < 1$. If $\lim_{n \to \infty} H^+(T_n x, T_0 x) = 0$ uniformly for $x \in X$, then*

$$\lim_{n \to \infty} H^+(Fix(T_n), Fix(T_0)) = 0.$$

Proof Let $\varepsilon > 0$ be given. Since $\lim_{n \to \infty} H^+(T_n x, T_0 x) = 0$ uniformly for $x \in X$, it is possible to choose $N \in \mathbb{N}$ so that for $n \geq N$, $\sup_{x \in X} H^+(T_n x, T_0 x) < (1 - \sqrt{k})\varepsilon$. By Proposition 5.6, $H^+(\text{Fix}(T_n), \text{Fix}(T_0)) < \varepsilon$ for all $n \geq N$. Hence, the conclusion follows.

Theorem 5.78 (Pathak and Shahzad [465]) *Let (X, d) be a complete metric space. Let $T : X \to CB(X)$ be an H^+-type k-multivalued weak contractive mapping with $0 < k < 1$. Then, T has a fixed point.*

Proof Notice first that for each $A, B \in CB(X)$, $a \in A$ and $\alpha > 0$ with $H^+(A, B) < \alpha$,, there exists $b \in B$ such that $\max\{d(a, b), d(a, Ta), d(b, Tb), \frac{1}{2}[d(a, Tb) + d(b, Ta)]\} < \alpha$. Now, let $L > 0$ be such that $k < L < 1$. Then

$$H^+(Tx, Ty) < L \max\{d(x, y), d(x, Tx), d(y, Ty), [d(x, Ty) + d(y, Tx)]/2\}, \tag{5.53}$$

for any $x, y \in X, x \neq y$.

Now we choose a sequence $\{x_n\}$ recursively in X in the following way. Let $x_0 \in X$ be arbitrary. Fix an element x_1 in Tx_0. From (2°) it follows that we can choose $x_2 \in Tx_1$ such that

$$d(x_1, x_2) \leqslant H^+(Tx_0, Tx_1) + \varepsilon \tag{5.54}$$

In general, if x_n be chosen, then we choose $x_{n+1} \in Tx_n$ such that

$$d(x_n, x_{n+1}) \leqslant H^+(Tx_{n-1}, Tx_n) + \varepsilon. \tag{5.55}$$

Set $\varepsilon = (\frac{1}{\sqrt{L}} - 1)H^+(Tx_{n-1}, Tx_n)$. Then from (5.55), it follows that

$$d(x_n, x_{n+1}) \leqslant H^+(Tx_{n-1}, Tx_n) + \left(\frac{1}{\sqrt{L}} - 1\right) H^+(Tx_{n-1}, Tx_n) = \frac{1}{\sqrt{L}} H^+(Tx_{n-1}, Tx_n).$$

Thus, we have

$$\sqrt{L}\, d(x_n, x_{n+1}) \leqslant H^+(Tx_{n-1}, Tx_n) \tag{5.56}$$

for each $n \in \mathbb{N}$.

Thus, from (5.53) we have

$$
\begin{aligned}
\sqrt{L}\, d(x_n, x_{n+1}) &< L \max\{d(x_{n-1}, x_n), d(x_{n-1}, Tx_{n-1}), d(x_n, Tx_n), \\
&\qquad [d(x_{n-1}, Tx_n) + d(x_n, Tx_{n-1})]/2\} \\
&\leqslant (\sqrt{L})^2 \max\{d(x_{n-1}, x_n), d(x_{n-1}, x_n), d(x_n, x_{n+1}), d(x_{n-1}, x_{n+1})/2\} \\
&\leqslant (\sqrt{L})^2 \max\{d(x_n, x_{n-1}), d(x_n, x_{n+1}), [d(x_{n-1}, x_n) + d(x_n, x_{n+1})]/2\} \\
&= (\sqrt{L})^2 \max\{d(x_n, x_{n-1}), d(x_n, x_{n+1})\}.
\end{aligned}
$$

It follows that

$$d(x_n, x_{n+1}) < \sqrt{L} \max\{d(x_n, x_{n-1}), d(x_n, x_{n+1})\} \tag{5.57}$$

for each $n \in \mathbb{N}$. Note that if $x_n = x_{n+1}$ for some $n \in \mathbb{N}$ then, $x_n = x_{n+1} \in Tx_n$, that is, x_n is a fixed point of T and we are finished. So, we may assume that $d(x_{n+1}, x_n) > 0$ for each $n \in \mathbb{N}$. Suppose that $d(x_{n-1}, x_n) < d(x_n, x_{n+1})$ for some $n \in \mathbb{N}$, then inequality (5.57) gives

$$d(x_n, x_{n+1}) < \sqrt{L}\, d(x_n, x_{n+1}),$$

a contradiction. So we must have $d(x_{n-1}, x_n) \geq d(x_n, x_{n+1})$ for each $n \in \mathbb{N}$. Hence, for all $n \in \mathbb{N}$, (5.57) yields

$$d(x_n, x_{n+1}) < c\, d(x_{n-1}, x_n), \qquad (5.58)$$

where $c = \sqrt{L}$. Repeating the same argument n-times as in (5.58), we obtain

$$d(x_n, x_{n+1}) < c^n\, d(x_0, x_1). \qquad (5.59)$$

It is obvious that $\{x_n\}$ is bounded. Indeed, for any $n \in \mathbb{N}$, we have

$$d(x_0, x_n) \leq \sum_{i=0}^{n-1} d(x_i, x_{i+1}) < (1 + c + c^2 + \cdots c^n) d(x_0, x_1)$$

$$< (1 + c + c^2 + \cdots) d(x_0, x_1) = \frac{1}{1-c} d(x_0, x_1) < \infty.$$

Further, by virtue of (5.59), one may observe that $\{x_n\}$ is a Cauchy sequence. Since X is complete, there exists $u \in X$ such that $\lim_{n\to\infty} x_n = u$. Assume that $u \neq Tu$, i.e. $d(u, Tu) > 0$. Now using (5.53) we have

$$\frac{1}{2}\Big\{\rho(Tx_n, Tu) + \rho(Tu, Tx_n)\Big\}$$
$$= H^+(Tx_n, Tu)$$
$$< L \max\{d(x_n, u), d(x_n, Tx_n), d(u, Tu), [d(x_n, Tu) + d(u, Tx_n)]/2\}$$
$$\leq L \max\{d(x_n, u), d(x_n, x_{n+1}), d(u, Tu), [d(x_n, Tu) + d(u, x_{n+1})]/2\},$$

it follows that

$$\frac{1}{2} \liminf_{n\to\infty} \Big\{\rho(Tx_n, Tu) + \rho(Tu, Tx_n)\Big\} \leq L\, d(u, Tu)).$$

Since $\lim_{n\to\infty} d(x_{n+1}, u) = 0$ exists, and

$$d(u, Tu) = \frac{1}{2}[d(u, Tu) + d(Tu, u)] \leq \frac{1}{2}[\rho(Tx_n, Tu) + \rho(Tu, Tx_n)] + d(x_{n+1}, u),$$

it follows that

$$d(u, Tu) \leq \frac{1}{2} \liminf_{n\to\infty} [\rho(Tx_n, Tu) + \rho(Tu, Tx_n)] + \liminf_{n\to\infty} d(x_{n+1}, u)$$
$$\leq L\, d(u, Tu)) + \lim_{n\to\infty} d(x_{n+1}, u) = L\, d(u, Tu)) < d(u, Tu)),$$

a contradiction. This implies that $d(u, Tu) = 0$, and since Tu is closed it must be the case that $u \in Tu$.

Notice that every multivalued contraction mapping with respect to Pompeu–Hausdorff metric H is an H^+-type multivalued weak contractive mapping, but the converse implication need not be true. To effect this, we observe the following example:

Example 5.38 Let $X = [-2, 2]$ and $d : X \times X \to \mathbb{R}$ be a standard metric. Let $T :$ $X \to CB(X)$ be defined by $Tx = \{\frac{x}{4}\}$, if $x \in [-1, 2]$ and $Tx = \{2\}$, otherwise. It is clear that if $x, y \in [-1, 2]$ or $x, y \in [-2, -1)$, then

$$H^+(Tx, Ty) \leqslant \frac{1}{4} d(x, y).$$

If $x \in [-1, 2]$ and $y \in [-2, -1)$, then we have

$$H^+(Tx, Ty) = \frac{1}{2}\left[\left|2 - \frac{x}{4}\right| + \left|2 - \frac{x}{4}\right|\right] = |2 - \frac{x}{4}| \leqslant 2 + \frac{1}{4} = \frac{3}{4} \cdot 3 \leqslant \frac{3}{4} \cdot \max\{d(y, Ty), d(x, Tx)\}.$$

It follows that

$$H^+(Tx, Ty) \leqslant k \max\{d(x, y), d(x, Tx), d(y, Ty), [d(x, Ty) + d(y, Tx)]/2\}$$

for all $x, y \in X$ and $k \in [\frac{3}{4}, 1)$. To check the condition (2°), we consider the following cases:

Case 1. If $x \in [-2, -1)$, then for any $y \in Tx = \{2\}$ there exists $z \in Ty = \{\frac{1}{2}\}$ such that for any $\varepsilon > 0$

$$d(y, z) = \frac{3}{2} \leqslant \frac{3}{2} + \varepsilon = H^+(Ty, Tx) + \varepsilon.$$

Case 2. If $x \in [-1, 2]$, then for any $y \in Tx = \{\frac{x}{4}\}$ there exists $z \in Ty = \{\frac{x}{16}\}$ such that for any $\varepsilon > 0$

$$d(y, z) = \frac{3|x|}{16} \leqslant \frac{3|x|}{16} + \varepsilon = H^+(Ty, Tx) + \varepsilon.$$

Thus, all the conditions of Theorem 5.78 are satisfied. Moreover, $0 \in T0 = \{0\}$ is a fixed point of T.

Notice that the map T does not satisfy the assumptions of Theorems 5.75 and 5.76. Indeed, for $x = -1$ and $y \to -1$ from the left, we have

$$H(T(-1), T(y)) = H^+(T(-1), T(y)) = 2 + \frac{1}{4} > k\, d(-1, y),$$

for all $k \in (0, 1)$.

We also notice that since

$$[d(x, Ty) + d(y, Tx)]/2 \leqslant \max\{d(x, Ty), d(y, Tx)\}$$

for all $x, y \in X$, it follows that every weak contractive mapping is quasi-contraction.

Using the technique of the proof of Theorem 5.78, one can easily prove the following result.

Theorem 5.79 *Let (X, d) be a complete metric space. Let $T : X \to CB(X)$ be a H^+-type k-multivalued quasi-contraction mapping with $0 < k < \frac{1}{2}$. Then, T has a fixed point.*

Next we state a theorem concerning multivalued contraction mapping in a complete metric space which is metrically convex. Due to the last assumption, significant weakening can be made regarding the domain and range of the mappings.

Definition 5.38 Let (X, d) be a complete metric space. X is said to be metrically convex if for each $x, y \in X$ with $x \neq y$ there exists $z \in X, x \neq z \neq y$ such that

$$d(x, z) + d(z, y) = d(x, y).$$

Menger and Blumenthal [395] have shown that in such a space every two points are the endpoints of at least one metric segment.

Remark 5.31 If K is a closed subset of a complete and metrically convex space X and if $X \in K, y \notin K$, then there exists a point z in the boundary of ∂K such that

$$d(x, z) + d(z, y) = d(x, y).$$

The following theorem was proved by Assad and Kirk [25].

Theorem 5.80 *Let (X, d) be a complete and metrically convex space, K a nonempty and closed subset of X and F a multivalued contraction mapping from X into $CB(X)$. If $F(x) \subset K$ for each $x \in \partial K$, then there exists a fixed point of F in K.*

For application purpose, we need the contraction mapping theorem in a convex setting. Some new fixed point theorems in Banach spaces are obtained by the application of Theorem 5.80. For example, if H is a closed and convex subset of a Banach space X and F is a contraction mapping of K into H where K is a nonempty and closed subset of H, then F has a fixed point if F maps the boundary of K relative to H back into K. Such hypothesis are not new in analysis. For mappings which are compact and continuous, H is often taken as a positive cone in X and K the intersection of H with the closed unit ball.

We now discuss contractive multifunctions and state a theorem, proved by Smithson [572], which extends Edelstein's FPT for contractive single-valued mappings to multifunctions.

Definition 5.39 An orbit $\mathcal{O}(x)$ of a multifunction $F : X \to CB(X)$ at the point x is a sequence $\{x_n : x_n \in F(x_{n-1})\}$ where $x_0 = x$.

An orbit $\mathcal{O}(x)$ of a multifunction F is called regular iff

$$d(x_{n+1}, x_{n+2}) \leqslant d(x_n, x_{n+1}) \text{ and } d(x_{n+1}, x_{n+2}) \leqslant H(F(x_n), F(x_{n+1})).$$

Definition 5.40 A multifunction $F : X \to CB(X)$ is said to be contractive iff for each $x_1, x_2 \in X$ with $x_1 \neq x_2$, $H(F(x_1), F(x_2)) < d(x_1, x_2)$.

An immediate consequence of the definition is the following : If $y_1 \in F(x_1)$, then there is an element $y_2 \in F(x_2)$ such that $d(y_1, y_2) < d(x_1, x_2)$.

Remark 5.32 Let F be a point compact, contractive multifunction. Define an orbit $\mathcal{O}(x)$ by choosing $x_n \in F(x_{n_{-1}})$ such that

$$d(x_{n-1}, x_n) = d(x_{n-1}, F(x_{n-1})) = \inf\{d(x_{n-1}, y) : y \in F(x_{n-1})\}.$$

Since F is contractive, the orbit $\mathcal{O}(x)$ is regular.

Theorem 5.81 *Let F be a point closed, contractive multifunction. If there is a regular orbit $\mathcal{O}(x)$ for F which contains a subsequence $\{x_{n_i}\}$ converging to y_0 such that $x_{n_{i+1}} \to y_1$, then $y_1 = y_0$, i.e. F has a fixed point.*

Corollary 5.16 *If F is point closed, contractive multifunction on the compact metric space X into itself, then F has a fixed point.*

Remark 5.33 Reich [509] and Bose and Mukherjee [66] have extended the work of Nadler, Jr. and obtained fixed point theorems for generalized multivalued contraction mappings. We shall discuss these in Sect. 5.5 (common fixed point theorems) as corollaries of common fixed point theorems concerning such mappings.

Definition 5.41 A mapping $F : X \to CB(X)$ is said to be nonexpansive if

$$H(Fx, Fy) \leqslant \|x - y\| \text{ for all } x, y \in X.$$

In the following, K is a nonempty, convex and weakly compact subset of a Banach space X, and $\mathcal{K}(X)$ denotes the family of nonempty and compact subsets of X. We will say that a mapping $F : X \to 2^X$ is demiclosed if

$$x_n \rightharpoonup x \text{ and } y_n \in Fx_n \to y \Rightarrow y \in Fx.$$

Proposition 5.7 (Lami Dozo [356]) *Let $F : K \to \mathcal{K}(X)$ be nonexpansive and let X satisfy Opial's condition. Then $I - F$ is demiclosed.*

Proof Since the domain of $I - F$ is weakly compact, it is enough to prove that the graph of $I - F$ is sequentially closed. Let $(x_n, y_n) \in G(I - F)$ where $G(I - F)$ denotes the graph of $I - F$ such that $x_n \rightharpoonup x$, $y_n \to y$. Then $x \in K$ and we have to prove that $y \in (I - F)(x)$. Since $y_n \in x_n - Fx_n$, $y_n = x_n - z_n$ for some $z_n \in Fx_n$. As F is nonexpansive, there exists $z_n' \in Fx$ such that

$$\|z_n - z_{n'}\| \leqslant \|x_n - x\|. \tag{5.60}$$

In (5.60) taking limits, we have

$$\liminf_{n \to \infty} \|x_n - x\| \geqslant \liminf_{n \to \infty} \|z_n - z_n'\|$$
$$\geqslant \liminf_{n \to \infty} \|x_n - y_n - x_n'\|. \tag{5.61}$$

But Fx is compact and $y_n \to y$. Hence, there exists a subsequence of $\{z_n'\}$, again denoted by $\{z_n'\}$, converging to $z \in Fx$. So from (5.61), we get

$$\liminf_{n \to \infty} \|x_n - x\| \geqslant \liminf_{n \to \infty} \|x_n - y - z\|.$$

By Opial's condition , we have $y + z = x$. Thus, $y = x - z \in x - Fx$. ∎

Theorem 5.82 (Lami Dozo [366]) *Let X be a Banach space which satisfies Opial's condition. If K is a nonempty, convex and weakly compact subset of X and $F : K \to \mathcal{K}(K)$ is a nonexpansive mapping, then F has a fixed point in K.*

Proof Let $x_0 \in K$ be a fixed element and let $\{k_n\}$; $0 < k_n < 1$ and $k_n \to 1$. Define mappings

$$F_n x = k_n Fx + (1 - k_n)x_0. \tag{5.62}$$

Then $F_n : K \to C(K)$ and each F_n is a contraction. By Theorem 5.74, there exists $x_n \in K$ such that $x_n \in F_n x_n$. Since K is weakly compact, there exists a subsequence of $\{x_n\}$, again denoted by $\{x_n\}$, converging weakly to $x \in K$. From (5.62), we have

$$x_n = k_n z_n + (1 - k_n)x_0 \text{ where } z_n \in Fx_n.$$

So $\|x_n - z_n\| = (1 - k_n)\|x_0 - z_n\|$. Hence, $y_n = x_n - z_n \in (I - F)x_n$ and $y_n \to 0$. This means that $x_n, y_n) \in G(I - F)$ and $x_n \rightharpoonup x$, $y_n \to 0$. So by demiclosedness of $(I - F), 0 \in (I - F)x$, that is $x \in Fx$. ∎

In the following, we discuss some FPT of multivalued nonexpansive mappings on nonconvex domain, more precisely on star-shaped domains. First we present the results of Itoh and Takahasi [289] and then the result of Yanagi [612].

Theorem 5.83 *Let K be a weakly compact star-shaped subset of a Banach space X which satisfies Opial's condition. Let F be a nonexpansive mapping from K into $\mathcal{K}(X)$ and for each $x \in \partial K$, let $Fx \subset K$. Then F has a fixed point in K.*

Proof Let x_0 be the star-center of K. Choose a sequence $\{k_n\}$ such that $0 < k_n < 1$ and $k_n \to 0$ as $n \to \infty$. Define mappings F_n from K into $C(X)$ by

$$F_n x = k_n x_0 + (1 - k_n) F x, \; x \in K.$$

Then each F_n is a $(1 - k_n)$-contradiction and $F_n x \subset K$ for each $x \in \partial K$. By Theorem 5.75, there exists $x_n \in K$ such that $x_n \in F_n x_n$. This implies that there is a $y_n \in F x_n$ such that $x_n = k_n x_0 + (1 - k_n) y_n$. By weak compactness of K, we can assume without loss in generality, that $\{x_n\}$ converges weakly to some element $x \in K$.

Since $\|x_n - y\| \to 0$ as $n \to \infty$ and $I - F$ is demiclosed, we have $0 \in (I - F)x$, i.e. $x \in F x$.   ∎

We get the following corollary for single-valued mappings.

Corollary 5.17 *Let K be a weakly compact star-shaped subset of a Banach space X which satisfies Opial's condition. Let F be a nonexpansive mapping of K into X such that $F(\partial K) \subset K$. Then F has a fixed point in K.*

In the following theorem, we can drop the Opial's condition on X provided the set K is assumed to be compact.

Theorem 5.84 *Let K be a compact star-shaped subset of a Banach space X and let $F : K \to \mathcal{K}(X)$ be a nonexpansive mapping such that $F(\partial K) \subset K$. Then F has a fixed point in K.*

Corollary 5.18 *If K is as in the above theorem and $F : K \to X$ nonexpansive such that $F(\partial K) \subset K$, then F has a fixed point.*

Theorem 5.85 (Yanagi [612]) *Let K be a nonempty, weakly compact and star-shaped subset of a uniformly convex Banach space X and let $F : K \to \mathcal{K}(X)$ be nonexpansive. If for each $x \in \partial K$, $F x \subset K$, and $\eta x + (1 - \eta) F x \subset K$ for some $\eta \in (0, 1)$ for $F x \subset int (K)$, then there exists a fixed point of F in K.*

Proof Define $F_n = k_n x_0 + (1 - k_n) F x$ as in Theorem 5.78, where x_0 is the star-center of K. By Theorem 5.75 each F_n has a fixed point x_n, i.e. $x_n \in F x_n$ for each n. Also there exists $y_n \in F x_n$ such that

$$x_n = k_n x_0 + (1 - k_n) y_n. \tag{5.63}$$

Since $\{x_n\}$ is bounded, by property of asymptotic center and radius, we have

$$AR(K, \{x_{n_i}\}) = AR(K, \{x_n\}) \; and$$

$$A(K, \{x_n\}) \subset A(K, \{x_{n_i}\}) (\text{refer Definition } 5.20(h)).$$

Let $z \in A(K, \{x_n\})$. Since $F z$ is compact, there exists $z_n \in F z$ such that

$$\|z_n - y_n\| \leqslant H(F z, F x_n) \leqslant \|z - x_n\|. \tag{5.64}$$

Also we can extract a subsequence $\{z_{n_i}\}$ of $\{z_n\}$ such that

$$z_{n_i} \to z_0 \in Fz. \tag{5.65}$$

Since $A(K, \{x_n\}) \subset A(K, \{x_{n_i}\})$, we have $z \in A(K, \{x_{n_i}\})$.

$$\text{Now} \quad \|x_{n_i} - y_{n_i}\| = \frac{k_{n_i}}{1 - k_{n_i}} \|x_0 - x_n\| \to 0, \text{ as } n \to \infty. \tag{5.66}$$

Thus

$$
\begin{aligned}
\limsup_{i \to \infty} \|z_0 - x_{n_i}\| &\leqslant \limsup_{i \to \infty} \|z_0 - z_{n_i}\| + \limsup_{i \to \infty} \|z_{n_i} - y_{n_i}\| \\
&\quad + \limsup_{i \to \infty} \|y_{n_i} - x_{n_i}\| \\
&= \limsup_{i \to \infty} \|z_{n_i} - y_{n_i}\| \text{ by } (5.65) \text{ and } (5.66) \\
&\leqslant \limsup_{i \to \infty} \|z - x_{n_i}\| \text{ by } (5.43) \\
&= \inf\{\limsup_{i \to \infty} \|y - x_{n_i}\| : y \in K\}. \tag{5.67}
\end{aligned}
$$

If $z \in \partial K$, then

$$w = \eta z + 1(1 - \eta)z_0 \in K$$

for some $\eta \in (0, 1)$ by hypothesis and so by uniform convexity of X, we have for some $\delta \in (0, 1)$,

$$
\begin{aligned}
\limsup_{i \to \infty} \|w - x_{n_i}\| &= \|\eta z + (1 - \eta)z_0 - x_{n_i}\| \\
&\leqslant \eta \|z - x_{n_i}\| + (1 - \eta)\|z_0 - x_{n_i}\|.
\end{aligned}
$$

We have

$$
\begin{aligned}
\limsup_{i \to \infty} \|w - x_{n_i}\| &\leqslant \eta \,[\inf \,\{\limsup_{i \to \infty} \|y - x_{n_i}\| : y \in K\}] \\
&\quad + (1 - \eta)[\inf \,\{\limsup_{i \to \infty} \|y - x_{n_i}\| : y \in K\} \\
&= \inf \,\{\limsup_{i \to \infty} \|y - x_{n_i}\| : y \in K\}.
\end{aligned}
$$

This gives $z = z_0 \in Fz$. Again if $z \in A(X, \{x_{n_i}\})$, we have

Fig. 5.3 The inward set
$I_K(x)$ of x relative to
$K, x \in K$

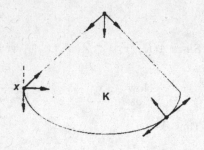

$$AR(X, \{x_{n_i}\}) \leqslant \limsup_{i \to \infty} \|z_0 - x_{n_i}\|$$

$$\leqslant \limsup_{i \to \infty} \|z_0 - z_{n_i}\| + \limsup_{i \to \infty} \|z_{n_i} - y_{n_i}\|$$

$$+ \limsup_{i \to \infty} \|y_{n_i} - x_{n_i}\|$$

$$= \limsup_{i \to \infty} \|z_{n_i} - y_{n_i}\|$$

$$\leqslant \limsup_{i} \|z - x_{n_i}\| \text{ (by 5.64)}$$

$$= AR(X, \{x_{n_i}\}).$$

Thus, $AR(X, \{x_{n_i}\}) \leqslant \limsup\limits_{i \to \infty} \|z_0 - z_{n_i}\| \leqslant AR(X, \{x_{n_i}\})$, and hence $z_0 \in A(X, \{x_{n_i}\})$.

Uniform convexity of X gives us $z = z_0 \in Fz$.

Definition 5.42 Let K be a nonempty subset of a Banach space X. We recall that the inward set of x relative to $K, x \in K$ is the set

$$I_k(x) = \{(1 - k)x + ky : y \in K, k \geqslant 0\}.$$

A mapping $F : K \to CB(X)$ is said to be inward if $Fx \subseteq I_K(x)$ for all $x \in K$ and weakly inward if $Fx \subseteq \overline{I}_K(x)$ for all $x \in K$ (Fig. 5.3).

Theorem 5.86 (Downing and Kirk [202]) *Let K be a nonempty, closed and convex of a Banach space X. Suppose F is an upper semicontinuous mapping of K into the family of nonempty and closed subsets of X, satisfying the following conditions (for a fixed $k \in (0, 1)$) :*
(a) For each $x \in K$, there exists $\delta = \delta(x) > 0$ such that

$$y \in B_\delta(x) \cap K \Rightarrow d(y, Fy) \leqslant d(y, Fx) + k\|x - y\|$$

(b) $F_1(x) \cap \overline{I}_K(x) \neq \varnothing$ for each $x \in K$ where given $x \in K$ and

$$\alpha \geqslant 1, F_\alpha(x) = \{z \in Fx : \|x - z\| \leqslant \alpha d(x, Fx)\} \text{ or}$$

(b') *corresponding to each $x \in K$, there exist constants*

$$\alpha = \alpha(x) > 1, \beta = \beta(x) \in (0, 1) \text{ such that } (1 - \beta)x + \beta F_\alpha(x) \subset K.$$

Then there exists $x_0 \in K$ such that $x_0 \in Fx_0$.

Corollary 5.19 *Let K and X be as defined before and $F : K \to CB(X)$ be a multivalued contraction mapping which satisfies either condition (b) or (b'). Then F has a fixed point.*

Proof In this case, F is automatically upper semicontinuous. Further, we have

$$d(y, Fy) \leqslant d(y, Fx) + H(Fx, Fy)$$
$$\leqslant d(y, Fx) + k\|x - y\| \text{ for all } x, y \in K.$$

So condition (a) is satisfied and hence the result.

Corollary 5.20 *Let K and X be as defined before and $F : K \to CB(X)$ be a multivalued contraction mapping for which $Fx \in \overline{I}_K(x), x \in K$. Then F has a fixed point.*

Proof Under the stated assumptions, (b) is automatically satisfied and hence the result.

Corollary 5.21 *Let K and X be as defined before. Suppose the mapping $F : K \to X$ is continuous, weakly inward and satisfies the following condition for a fixed $k \in (0, 1)$: for each $x \in K$ there exists $\delta = \delta(x) > 0$ such that*

$$y \in B_\delta(x) \cap K \Rightarrow \|y - Fy\| \leqslant \|y - Fx\| + k\|x - y\|.$$

Then F has a fixed point in K.

The following theorem was proved by Downing and Kirk using Corollaries 5.19–5.21 of Theorem 5.86 in conjunction with an elegant approach of Goebel [245]. Before discussing the theorem, we state the following lemmas needed for its proof.

Definition 5.43 A sequence $\{x_n\}$ is said to be regular if all its subsequences $\{x_{n_i}\}$ have the same asymptotic radius and a sequence $\{x_n\}$ is said to be almost convergent if all its subsequences $\{x_{n_i}\}$ have the same asymptotic center.

Lemma 5.9 (Goebel [245]) *In a uniformly convex space, each regular sequence is almost convergent.*

Lemma 5.10 (Goebel [245]) *Any bounded sequence in a Banach space contains a regular subsequence.*

Lemma 5.11 *Let X be uniformly convex, K a bounded, closed and convex subset of X. Let $\{x_i\}$ be a sequence in K with asymptotic center $y \in K$ and asymptotic radius r. For $\alpha \in (0, 1)$, let $z_i = (1 - \alpha)y + \alpha x_i, i = 1, 2, \ldots$. Then $A(K, \{z_i\}) = y$ and $AR(K, \{z_i\}) = \alpha r$.*

Theorem 5.87 *Let X be a uniformly convex Banach space, K a nonempty, bounded, closed and convex subset of X, and F a nonexpansive set-valued mapping from K into $\mathcal{K}(x)$. If $Fx \subset I_K(x)$ for all $x \in K$, then F has a fixed point in K.*

Proof Let $x_0 \in K$ and $\{k_n\} \subset (0, 1)$ with $k_n \to 0$. For each n, define F_n by $F_n x = k_n x_0 + (1 - k_n) Fx, x \in K$. Then F_n is a setvalued contraction with Lipschitz constant $1 - k_n$. Since $I_K(x)$ is convex for each $x \in K$,

$$F_n x \subset I_K(x), \quad x \in K. \tag{5.68}$$

It can be shown that $F(K) = \cup_{x \in K} Fx$ is a bounded set. ∎

By Corollary 5.20, each F_n has a fixed point x_n in K. Since $x_n \in F_n x, d(x_n, Fx_n) \leqslant H(F_n x_n, Fx_n)$ and by uniform convergence of $\{F_n\}$, we have

$$d(x_n, Fx_n) \to 0 \text{ as } n \to \infty. \tag{5.69}$$

By Lemmas 5.9 and 5.10 (passing to a subsequence if necessary), we may assume that $\{x_n\}$ is regular and almost convergent. Let $z = A(K, \{x_n\})$ and $r = AR(K, \{x_n\})$. For each n choose $y_n \in Fx_n$ such that

$$\|x_n - y_n\| = d(x_n, Fx_n). \tag{5.70}$$

Since F is compact-valued, select again $z_n \in Fz$ such that

$$\|z_n - y_n\| \leqslant H(Fz, Fx_n) \leqslant \|z - x_n\|. \tag{5.71}$$

Let $\{x_{n_i}\}$ converges to $z_0 \in Fz$ as $i \to \infty$. Since $Fz \subset I_K(z)$ there exists $\alpha \in (0, 1)$ such that $(1\ \alpha)z + \alpha z_0 \in K$. Also $z = A(K, \{x_{n_i}\})$ and $r = AR(K, \{x_{n_i}\})$, since $\{x_n\}$ is regular and almost convergent. If $w_i = (1 - \alpha)z + \alpha x_{n_i}, i = 1, 2, \ldots,$ then by Lemma 5.11, we have

$$z = A(K, \{w_i\}) \quad \text{and} \quad \alpha r = AR(K, \{w_i\}).$$

Let $w = (1 - \alpha)z\ \alpha z_0, u_i = (1 - \alpha)z + \alpha z_{n_i}$ and $v_i = (1 - \alpha)z + \alpha y_{n_i}$. Then for each i, we have

$$\|w_i - w\| \leqslant \|w - u_i\| + \|u_i - v_i\| + \|v_i - w_i\|$$
$$= \alpha \|z_0 - z_{n_i}\| + \alpha \|z_{n_i} - y_{n_i}\| + \alpha \|y_{n_i} - x_{n_i}\|.$$

By (5.69)–(5.71), we obtain

$$\limsup_{i \to \infty} \|w_i - w\| \leqslant \alpha \limsup_{i \to \infty} \|z_{n_i} - y_{n-i}\|$$

$$\leqslant \alpha \limsup_{i \to \infty} \|z - x_{n_i}\|$$

$$= \alpha r.$$

Since $AR(K, \{w_i\}) = \alpha r, z = w = (1 - \alpha)z + \alpha z_0$ by uniqueness of the asymptotic center. Hence, $z = z_0 \in Fz$. This completes the proof.

The following theorem of Lim [366] generalizes Theorem 5.87.

Theorem 5.88 *Let K be a closed, convex and bounded subset of a uniformly convex Banach space X and let $F : K \to \mathcal{K}(X)$ be a nonexpansive mapping such that*

$$Fx \subset \overline{I}_K(x) \text{ for every } x \in K.$$

Then there exists a fixed point of F in K.

Definition 5.44 Let K be a convex subset of X. The mapping $F : K \to CB(X)$ is said to be semiconvex on K if for any $x, y \in K, z = kx + (1 - k)y$, where $0 \leqslant k \leqslant 1$, and any $x_1 \in Fx, y_1 \in Fy$, there exists $z_1 \in Fz$ such that $\|z_1\| \leqslant \max \{\|x_1\|, \|y_1, \|\}$.

Definition 5.45 The mapping $F : K \to CB(X)(K \in CB(X))$ is said to be a generalized contraction if for each $x \in K$, there is a number $\alpha(x) < 1$ such that

$$H(Fx, Fy) \leqslant \alpha(x)\|x - y\| \text{ for each } y \in K.$$

Theorem 5.89 (Yanagi [612]) *Let K be a nonempty, weakly compact and convex subset of a Banach space X and let $F : K \to \mathcal{K}(X)$ be a nonexpansive and weakly inward mapping. If $I - F$ is demiclosed or semiconvex on K, then F has a fixed point.*

Proof Let $x_0 \in K$. Choose a sequence $\{k_m\}$ such that $0 < k_n < 1$ and $k_n \to 0$. Define the mapping $F_n x = k_n x_0 + (1 - k_n)Fx$ for all $x \in K$. Then by Theorem 5.80 for each n, F_n has a fixed point x_n. Hence, there exists $y_n \in Fx_n$ such that

$$x_n = k_n x_0 + (1 - k_n)y_n.$$

If $I - F$ is demiclosed on K, then by weak compactness of K there exists a subsequence $\{x_{n_i}\}$ of $\{x_n\}$ which converges in K weakly to an element z in K. Further, we have

$$\|x_{n_i} - y_{n_i}\| = \frac{k_{n_i}}{1 - k_{n_i}}\|x_0 - x_{n_i}\| \to 0 \text{ as } i \to \infty.$$

Thus, $0 \in (I - F)(x)$, that is, $z \in Fz$. Suppose $I - F$ is semiconvex on K, then we have

$$d(x_n, Fx_n) \leqslant \|x_n - y_n\| \leqslant \frac{k_n}{1 - k_n}\|x_0 - x_n\| \to 0 \text{ as } n \to \infty,$$

and so $\inf\{d(x, Fx) : x \in K\} = 0$. Define $H_r = \{x \in K : d(x, Fx) \leqslant r\}$ where $r > 0$. It can be seen due to semiconvexity that H_r are weakly closed and convex subsets for every $r > 0$ (See, Ko [338]). The family $\{H_r : r > 0\}$ has the finite intersection property. Hence, by weak compactness of K, we have $\bigcap\{H_r : r > 0\} \neq \varnothing$. It follows, therefore, that any point in $\bigcap\{H_r : r > 0\}$ is a fixed point of F.   ∎

This theorem has the following interesting corollaries:

Corollary 5.22 *Let K be a nonempty, weakly compact and convex subset of a Banach space X which satisfied Opial's condition (or weak Opial's condition). If $F : K \to \mathcal{K}(X)$ is nonexpansive (or a generalized contraction) mapping which is also weakly inward, then F has a fixed point.*

Proof Since X satisfies Opial's condition and F is nonexpansive, by Proposition 5.7, it follows that $I - F$ is demiclosed. Hence, the result follows from the above theorems. Similarly, if X satisfies weak Opial's condition and F is a generalized contraction, then $I - F$ is demiclosed.   ∎

Corollary 5.23 *Let K be a nonempty, compact and convex subset of a Banach space X and let $F : K \to \mathcal{K}(X)$ be nonexpansive and weakly inward. Then F has a fixed point.*

Definition 5.46 Let K be a nonempty subset of a Banach space X. Then a mapping $F : K \subset X \to 2^X$ is called a pseudocontraction (Browder [94]) if, for each $r > 0$, each $x, y \in K$ and each $u \in Fx, v \in Fy$, we have

$$\|x - y\| \leqslant \|(1 + r)(x - y) - r(u - v)\|.$$

If F is pseudocontraction, then $T = I - F$ is accretive.

Observation

- For single-valued mappings, it is known that a nonexpansive mapping is a pseudo-contraction. Following example of Downing and Ray [203] shows that the same is not true in the setvalued case: Let $F : \mathbb{R}^2 \to C(\mathbb{R}^2)$ be defined by

$$F(x, y) = \{(a, 0) : x - 1 \leqslant a \leqslant x + 1\}$$

  F is a setvalued nonexpansive mapping which is not a pseudocontraction since $I - F$ is not accretive. This indicates that there may not be any firm connection between fixed point theory of nonexpansive mappings and pseudocontractions in the setvalued case as it exists in the single-valued case. However, the following theorem of Downing and Ray [203] is of interest.

Theorem 5.90 *Let K be a closed and convex subset of Banach space X having the fixed point property for single-valued nonexpansive mappings. Suppose $F : K \to CB(K)$ is a Lipschitzian pseudocontraction. Then F has a fixed point in K.*

Finally we conclude this section with a few results obtained by Aubin and Siegel [27] recently. Let X be a complete metric space and F a setvalued mapping from X into the family of nonempty subsets of X. We discuss the problem of existence of fixed points $x \in Fx$ and the existence of stationery points $\{x^*\} = Fx^*$. This has relevance in control theory. We merely state the theorems of our interest after giving required definitions.

Definition 5.47 A function $\phi : X \to \mathbb{R}^+$ is called a weak entropy of F if for all $x \in X$, there exists $y \in Fx$ such that

$$d(x, y) \leqslant \phi(x) - \phi(y). \tag{5.72}$$

A mapping $F : X \to 2^X$ with $\mathscr{R}(F) \neq \varnothing$ is said to be weakly dissipative if there exists a weak entropy of F.

Definition 5.48 A sequence $\{x_n\}$ of elements in X is called a trajectory (or orbit) starting at x if $x_0 = x$ and $x_{n+1} \in Fx_n$, $n \geqslant 0$. The set of all such trajectories is denoted by $\mathscr{F}(F, x)$.

Theorem 5.91 *A weakly dissipative closed mapping F from X into family of nonempty subsets of X has a fixed point. Furthermore, for any $x_0 \in X$, there exists a trajectory starting at x_0 and converging to a fixed point.*

Proof Let ϕ be a weak entropy of F. Given any $x \in X$ we can use (5.72) to construct a sequence $\{x_i\} \in \mathscr{F}(F, x)$ such that

$$d(x_p, x_q) \leqslant \sum_{n=p}^{q-1} d(x_n, x_{n+1}) \leqslant \phi(x_p) - \phi(x_q), \quad q \geqslant p.$$

The sequence $\{x_i\}$ is a Cauchy sequence since $\{\phi(x_i)\}$ is nonincreasing and bounded from below. Let it converge to u. Since $x_{n+1} \in Fx_n$, (u, u) is in the closure of the graph of F. So by assumption on F, $u \in Fu$. ∎

Theorem 5.92 *Let $F : X \to \mathcal{K}(X)$ be Lipschitz with Lipschitz constant $0 < k < 1$. Then it is weakly dissipative and hence it has a fixed point when X is complete.*

Definition 5.49 Given a mapping F from X into the family of nonempty subsets of X, a function $\phi : X \to \mathbb{R}^+$ is called an entropy of F if

$$d(x, y) \leqslant \phi(x) - \phi(y) \quad \text{for all } x \in X \text{ and all } y \in Fx.$$

A mapping F is called dissipative if it has an entropy.

The following theorem gives the stationary point ($Fx = \{x\}$) for dissipative mappings. This generalizes Theorem 5.16 to multifunctions.

Theorem 5.93 *Let F be a mapping of X (a complete metric space) into the family of nonempty subsets of X. Let F be dissipative with lower semicontinuous entropy $\varnothing$. Then F has a stationary point.*

5.4.2 Coincidence and Fixed Points for Multivalued Mappings

In this section, we consider some problems on coincidence point and fixed point theorems for multivalued mappings. Applying the characterizations of $\mathcal{P}$-functions, some existence theorems are presented for coincidence point and fixed point distinct from Nadler's fixed point theorem, Berinde–Berinde's fixed point theorem, Mizoguchi–Takahashi's fixed point theorem and Du's fixed point theorem for nonlinear multivalued contractive mappings in complete metric spaces.

Definition 5.50 A function $\varphi_{\mathcal{MT}} : (0, \infty) \rightarrow [0, 1)$ is said to be an $\mathcal{MT}$-function if it satisfies Mizoguchi–Takahashi's condition; that is,

$$\limsup_{r \rightarrow t^+} \varphi_{\mathcal{MT}}(r) < 1 \text{ for each } t \in [0, \infty).$$

Definition 5.51 A function $\varphi : (0, \infty) \rightarrow [0, \frac{1}{2})$ is said to be a $\mathcal{P}$-function if it satisfies the following condition:

$$\limsup_{r \rightarrow t^+} \varphi(r) < \frac{1}{2} \text{ for each } t \in [0, \infty).$$

Let X be a metric space and $T : X \rightarrow CB(X)$ a multifunction. Throughout this section, $\mathcal{F}(T)$ denotes the set of fixed points of T.

Theorem 5.94 (Nadler [416]) *Let (X, d) be a complete metric space and $T : X \rightarrow CB(X)$ be a k-contraction, i.e. assume that there exists a nonnegative number $k < 1$ such that*

$$H(Tx, Ty) \leqslant k d(x, y) \text{ for all } x, y \in X.$$

Then $\mathcal{F}(T) \neq \varnothing$.

In 2007, M. Berinde and V. Berinde [55] proved the following fixed point theorem.

Theorem 5.95 (Berinde and Berinde [55]) *Let (X, d) be a complete metric space, $T : X \rightarrow CB(X)$ be a multivalued map, $\varphi_{\mathcal{MT}} : (0, \infty) \rightarrow [0, 1)$ be an $\mathcal{MT}$-function and $L \geq 0$. Assume that*

$$H(Tx, Ty) \leqslant \varphi_{\mathcal{MT}}(d(x, y)) d(x, y) + L d(y, Tx) \text{ for all } x, y \in X.$$

Then $\mathcal{F}(T) \neq \varnothing$.

Notice that if we let $L = 0$ in Theorem 5.95, then we can obtain Mizoguchi–Takahashi's fixed point theorem [406] which is a partial answer of Problem 9 in Reich [516]. Indeed, S. Reich established the following:

Theorem 5.96 (Reich [510]) *Let* (X, d) *be a complete metric space,* $T : X \to \mathcal{K}(X)$ *be a multivalued map,* $\varphi_{\mathcal{MJ}} : (0, \infty) \to [0, 1)$ *be an* $\mathcal{MJ}$-*function. Assume that*

$$H(Tx, Ty) \leqslant \varphi_{\mathcal{MJ}}(d(x, y)) \, d(x, y) \; \text{ for all } x, y \in X, x \neq y.$$

Then $\mathfrak{F}(T) \neq \varnothing$.

Reich [511] posed the question whether Theorem 5.96 above is also true for the map $T : X \to CB(X)$. Mizoguchi and Takahashi [406] in 1989 responded to this conjecture and proved the following theorem which additionally is more general than the Nadler's theorem.

Theorem 5.97 (Mizoguchi and Takahashi [406]) *Let* (X, d) *be a complete metric space,* $T : X \to CB(X)$ *be a multivalued map,* $\varphi_{\mathcal{MJ}} : (0, \infty) \to [0, 1)$ *be an* $\mathcal{MJ}$-*function. Assume that*

$$H(Tx, Ty) \leqslant \varphi_{\mathcal{MJ}}(d(x, y)) \, d(x, y) \; \text{ for all } x, y \in X.$$

Then $\mathfrak{F}(T) \neq \varnothing$.

Recently, W.-S., Du [206] proved an interesting result as follows:

Theorem 5.98 (Du [206]) *Let* (X, d) *be a complete metric space,* $T : X \to CB(X)$ *be a multivalued map,* $g : X \to X$ *be a continuous self-map and* $\varphi_{\mathcal{MJ}} : (0, \infty) \to [0, 1)$ *be an* $\mathcal{MJ}$-*function. Assume that*
(a) Tx is g-invariant (i.e. $g(Tx) \subseteq Tx$) for each $x \in X$;
(b) there exists $h : X \to [0, \infty)$ such that

$$H(Tx, Ty)) \leqslant \varphi_{\mathcal{MJ}}(d(x, y)) \, d(x, y) + h(gy)d(gy, Tx) \text{ for all } x, y \in X.$$

Then $\mathcal{COP}(g, T) \cap \mathfrak{F}(T) \neq \varnothing$.

It may be observed that Mizoguchi–Takahashi's fixed point theorem is a generalization of Nadler's fixed point theorem, but its primitive proof in [406] is difficult. Another proof in [154] is not yet simple. Recently, Suzuki [581] gave a very simple proof of Theorem 3.6. Several characterizations of $\mathcal{P}$-functions are first given in this work. Applying the characterizations of $\mathcal{P}$-functions, we establish some existence theorems for coincidence point and fixed point in complete metric spaces. From these results, we can obtain new results which are different from Berinde–Berinde's fixed point theorem, Mizoguchi–Takahashi's fixed point theorem and Du's for nonlinear multivalued contractive maps.

Inspiring from the characterization of $\mathcal{MJ}$-functions studied by W.-S. Du [206], Pathak, Agarwal and Cho [454] establish some characterizations of $\mathcal{P}$-function. The proof of the following Lemma 5.5.4 is essentially same as that of Lemma 2.1 of Du [206]. However, for the sake of completeness we give the proof sketch briefly.

Lemma 5.12 *Let $\varphi : [0, \infty) \to [0, \frac{1}{2})$ be a function. Then the following statements are equivalent.*

(a) *φ is a $\mathcal{P}$-function.*

(b) *For each $t \in [0, \infty)$, there exist $r_t^{(1)} \in [0, \frac{1}{2})$ and $\varepsilon_t^{(1)} > 0$ such that $\varphi(s) \leqslant r_t^{(1)}$ for all $s \in (t, t + \varepsilon_t^{(1)})$.*

(c) *For each $t \in [0, \infty)$, there exist $r_t^{(2)} \in [0, \frac{1}{2})$ and $\varepsilon_t^{(2)} > 0$ such that $\varphi(s) \leqslant r_t^{(2)}$ for all $s \in [t, t + \varepsilon_t^{(2)}]$.*

(d) *For each $t \in [0, \infty)$, there exist $r_t^{(3)} \in [0, \frac{1}{2})$ and $\varepsilon_t^{(3)} > 0$ such that $\varphi(s) \leqslant r_t^{(3)}$ for all $s \in (t, t + \varepsilon_t^{(3)}]$.*

(e) *For each $t \in [0, \infty)$, there exist $r_t^{(4)} \in [0, \frac{1}{2})$ and $\varepsilon_t^{(4)} > 0$ such that $\varphi(s) \leqslant r_t^{(4)}$ for all $s \in [t, t + \varepsilon_t^{(4)})$.*

(f) *For any nonincreasing sequence $\{x_n\}_{n \in \mathbb{N}} \in [0, \infty)$, we have $0 \leqslant \sup_{n \in \mathbb{N}} \varphi(x_n) < \frac{1}{2}$.*

(g) *φ is a function of contractive factor [205]; that is, for any strictly decreasing sequence $\{x_n\}_{n \in \mathbb{N}} \in [0, \infty)$, we have $0 \leqslant \sup_{n \in \mathbb{N}} \varphi(x_n) < \frac{1}{2}$.*

Proof (i) $(a) \Leftrightarrow (b)$.

We first show $(a) \Rightarrow (b)$. Suppose that φ is a $\mathcal{P}$-function. Then for each $t \in [0, \infty)$, there exists $\varepsilon_t > 0$ such that

$$\sup_{t < s < t + \varepsilon_t} \varphi(s) < \frac{1}{2}.$$

By the denseness of $\mathbb{R}$, there also exists $r_t \in [0, \frac{1}{2})$ such that

$$\sup_{t < s < t + \varepsilon_t} \varphi(s) \leqslant r_t < \frac{1}{2}$$

which says that $\varphi(s) \leqslant r_t$ for all $s \in (t, t + \varepsilon_t)$. The converse part (i.e. $(b) \Rightarrow (a)$) is obvious.

$(b) \Rightarrow (c)$.

Clearly, $(c) \Rightarrow (b)$ is true for $r_t^{(1)} := r_t^{(2)}$ and $\varepsilon_t^{(1)} := \varepsilon_t^{(2)}$. Conversely, assume (b) holds. Let $t \in [0, \infty)$ be given. Then, by our hypothesis, there exist $r_t^{(1)} \in [0, \frac{1}{2})$ and $\varepsilon_t^{(1)} > 0$ such that $\varphi(s) \leqslant r_t^{(1)}$ for all $s \in (t, t + \varepsilon_t^{(1)})$. Put $\varepsilon_t^{(2)} = \varepsilon_t^{(1)}$ and

$$r_t^{(2)} = \max\{r_t^{(1)}, \varphi(t), \varphi(t + r_t^{(1)})\}.$$

Then $r_t^{(2)} \in [0, \frac{1}{2})$ and $\varphi(s) \leqslant r_t^{(2)}$ for all $s \in [t, t + \varepsilon_t^{(2)}]$. So we prove $(b) \Rightarrow (c)$.

(iii) The implications $(c) \Rightarrow (d) \Rightarrow (b)$ and $(c) \Rightarrow (e) \Rightarrow (b)$.

(iv) Let us prove $(e) \Rightarrow (f)$. Suppose that (e) holds. Let $\{x_n\}_{n \in \mathbb{N}}$ be a nonincreasing sequence in $[0, \infty)$. Then $t_0 := \lim_{n \in \mathbb{N}} x_n = \inf_{n \in \mathbb{N}} x_n$ exists. By our hypothesis, there exist $r_{t_0} \in [0, \frac{1}{2})$ and $\varepsilon_{t_0} > 0$ such that $\varphi(s) \leqslant r_{t_0}$ for all $s \in [t_0, t_0 + \varepsilon_{t_0})$. On the other hand, there exists $N \in \mathbb{N}$, such that

$$t_0 \leqslant x_n < t_0 + \varepsilon_{t_0}.$$

for all $n \in \mathbb{N}$ with $n \geq N$. Hence, $\varphi(x_n) \leqslant r_{t_0}$ for all n $n \geq N$. Let

$$\eta := \max\{\varphi(x_1), \varphi(x_2), \ldots, \varphi(x_{N-1}), r_{t_0}\} < \frac{1}{2}.$$

Then $\varphi(x_n) \leqslant \eta$ for all $n \in \mathbb{N}$. Hence, $0 \leqslant \sup_{n \in \mathbb{N}} \varphi(x_n) \leqslant \eta < \frac{1}{2}$ and (f) holds.

(v) The implication $(f) \Rightarrow (g)$ is obvious.

(vi) Finally, we prove $(g) \Rightarrow (e)$.

Assume that φ is a function of contractive factor. On the contrary, suppose that there exists $\hat{t} \in [0, \infty)$ such that for each $r \in [0, \frac{1}{2})$ and each $\varepsilon > 0$ there is $s \in [\hat{t}, \hat{t} + \varepsilon)$ with the property $\varphi(s) > r$. So, for $r_1 := \varphi(\hat{t}) \in [0, \frac{1}{2})$ and for $\varepsilon_1 := 1 > 0$ it must exists $s_1 \in [\hat{t}, \hat{t} + \varepsilon_1)$ with the property $\varphi(s_1) > r_1$. The last inequality also implies that $s_1 \neq \hat{t}$ and thus $\hat{t} < s_1$. Choose $\varepsilon_2 > 0$ satisfying $\hat{t} + \varepsilon_2 \leqslant s_1$, and set

$$r_2 := \max\left\{\varphi(s_1), \frac{1}{2} - \frac{1}{4}\right\}.$$

Then, for r_2 and for ε_2 as indicated, we can find $s_2 \in [\hat{t}, \hat{t} + \varepsilon_2)$ with $\varphi(s_2) > r_2$. This also entails that $\hat{t} < s_2 < s_1$. Continuing this process, we can construct a strictly decreasing sequence $\{s_n\} \subset [\hat{t}, \infty) \subset [0, \infty)$ such that

$$\varphi(s_n) > r_n := \max\left\{\varphi(s_{n-1}, \frac{1}{2} - \frac{1}{n}\right\} \geq \frac{1}{2} - \frac{1}{n}$$

for all $n \in \mathbb{N}$. This yields $\sup_{n \in \mathbb{N}} \geq \frac{1}{2}$ which contradicts that φ is a function of contractive factor. Therefore, we show that $(g) \Rightarrow (e)$ is true. By $(i) - (vi)$, we complete the proof.

Notice that if we define $\varphi_{\mathcal{MJ}}(t) = 2\varphi(t)$ for all $t \in [0, \infty)$, then $\varphi_{\mathcal{MJ}}(t)$ is essentially an $\mathcal{MJ}$-function.

In 2015, Pathak, Agarwal and Cho [454] proved the following result.

Theorem 5.99 *Let (X, d) be a complete metric space, $T : X \to CB(X)$ be a multivalued map, $f, g : X \to X$ be continuous self-maps and $\varphi : (0, \infty) \to [0, \frac{1}{2})$ be a $\mathcal{P}$-function. Assume that*

(a_1) *for each $x \in X$, $\{fy = gy : y \in Tx\} \subseteq Tx$.*

(b_1) *there exists $\hat{h}, \hat{k} : X \to [0, \infty)$ such that*

$$H(Tx, Ty) \leqslant \varphi(d(x, y)) \, [d(x, Ty) + d(y, Tx)]$$
$$+ \hat{h}(fy)(d(fy, Tx) + \hat{k}(gy)(d(gy, Tx) \text{ for all } x, y \in X.$$

Then $\mathcal{COP}(f, g, T) \cap \mathfrak{F}(T) \neq \varnothing$.

Proof By (a_1), we note that for each $x \in X$, we have $d(fy, Tx) = 0$ and $d(gy, Tx) = 0$ for all $y \in Tx$. Then for each $x \in X$, we get from (b_1) and (i) that

$$H(Tx, Ty) \leqslant \varphi(d(x, y)) \, d(x, Ty) \text{ for all } y \in Tx. \tag{5.73}$$

Further, for each $y \in Tx$, $d(y, Ty) \leqslant H(Tx, Ty)$. Therefore, form (5.73), for each $x \in X$ we have

$$d(y, Ty) \leqslant \varphi(d(x, y)) \, d(x, Ty) \text{ for all } y \in Tx.$$

Let $x_0 \in X$ be arbitrary and fixed. Since $Tx_0 \neq \emptyset$, there exists $x_1 \in X$ such that $x_1 \in Tx_0$. If $d(x_0, x_1) = 0$, then $x_0 = x_1 \in \mathfrak{F}(T)$. Hence, $x_0 \in \mathfrak{F}(T)$ and we are done. Otherwise, if $x_0 \neq x_1$, let $\kappa : [0, \infty) \to [0, \frac{1}{2})$ be defined by $\kappa(t) = \frac{1/2 + \varphi(t)}{2}$. Clearly, $0 \leqslant \varphi(t) < \kappa(t) < 1/2$ for all $t \in [0, \infty)$, it follows that

$$\begin{aligned}
d(x_1, Tx_1) &< \kappa(d(x_0, x_1)) \, [d(x_0, Tx_1) + d(x_1, Tx_0)] \\
&= \kappa(d(x_0, x_1)) \, d(x_0, Tx_1) \\
&\leqslant \kappa(d(x_0, x_1)) \, [d(x_0, x_1) + d(x_1, Tx_1)].
\end{aligned}$$

This implies that

$$d(x_1, Tx_1) < \psi(d(x_0, x_1)) \, d(x_0, x_1). \tag{5.74}$$

where $\psi(d(x_0, x_1)) = \frac{\kappa(d(x_0, x_1))}{1 - \kappa(d(x_0, x_1))} < 1$. We claim that there exists $x_2 \in Tx_1$ such that

$$d(x_1, x_2) < \psi(d(x_0, x_1)) \, d(x_0, x_1). \tag{5.75}$$

If this claim is false; i.e.

$$d(x_1, x_2) \geqslant \psi(d(x_0, x_1)) \, d(x_0, x_1),$$

then we have

$$d(x_1, x_2) \geqslant \inf_{y \in Tx_1} d(x_1, y) \geq \psi(d(x_0, x_1)) \, d(x_0, x_1),$$

i.e.

$$d(x_1, Tx_1) \geqslant \psi(d(x_0, x_1)) \, d(x_0, x_1),$$

a contradiction to (5.74).

If $d(x_1, x_2) = 0$, then $x_1 = x_2 \in \mathfrak{F}(T)$ and hence $x_1 \in \mathfrak{F}(T)$. Otherwise there exists $x_3 \in Tx_2$ such that

$$d(x_2, x_3) < \psi(d(x_1, x_2)) \, d(x_1, x_2). \tag{5.76}$$

Let $d_n = d(x_{n-1}, x_n)$, $n \in \mathbb{N}$. Proceeding with the same argument as above and assuming that $x_{n-1} \neq x_n$, for otherwise x_{n-1} is a fixed point of T, we obtain a sequence $\{x_n\}_{n \in \mathbb{N}}$ in X such that

$$d_{n+1} < \psi(d_n) \, d_n. \tag{5.77}$$

where $x_n \in Tx_{n-1}$ for each $n \in \mathbb{N}$.

Since φ is a $\mathcal{P}$-function, by Lemma 5.12, $0 \leqslant \sup_{n \in \mathbb{N}} \varphi(d_n) \leqslant \sup_{n \in \mathbb{N}} \kappa(d_n) < 1/2$. Then it follows that

$$0 < \sup_{n \in \mathbb{N}} \psi(d_n) = \frac{\sup_{n \in \mathbb{N}} \kappa(d_n)}{1 - \sup_{n \in \mathbb{N}} \kappa(d_n)} < 1.$$

Let $\alpha := \sup_{n \in \mathbb{N}} \psi(d_n)$. So $\alpha \in (0, 1)$. Since $\psi(t) < 1$ for all $t \in [0, \infty)$, by (2.6), the sequence $\{d_n\}_{n=1}^{\infty}$ is a strictly decreasing sequence of positive real numbers bounded below by 0, and thus convergent. By (5.77), we get

$$d_{n+1} < \psi(d_n)d_n \leqslant \alpha \, d_n \text{ for each } n \in \mathbb{N}. \tag{5.78}$$

Further, we claim that $\{d_n\}_{n=1}^{\infty}$; i.e. $\{d(x_n, x_{n+1})\}_{n=1}^{\infty}$ converges to 0. Suppose that $\lim_{n \to \infty} d(x_n, x_{n+1}) = \gamma > 0$. Then for $0 < \varepsilon < \gamma$, there exists a natural number n_0 such that

$$0 < \delta = \gamma - \varepsilon < d(x_n, x_{n+1}) \; \forall n \geqslant n_0,$$

which is a contradiction, since $\theta(d(x_n, x_{n+1})) \to 0$. From (5.78), we get

$$d(x_n, x_{n+1}) \leqslant \alpha^n \, d(x_0, x_1). \tag{5.79}$$

For any $m > n > n_0$, by (5.79), we have

$$d(x_m, x_n) \leqslant \sum_{j=n}^{m-1} d(x_j, x_{j+1}) \leqslant \sum_{j=n}^{m-1} \alpha^j d(x_0, x_1) = \left(\sum_{j=n}^{m-1} \alpha^j \right) d(x_0, x_1)$$

$$< \frac{\alpha^n}{1 - \alpha} d(x_0, Tx_0) \to 0, \text{ as } n \to \infty.$$

Therefore, $\lim_{n \to \infty} \sup\{d(x_m, x_n) : m > n\} = 0$. Hence, $\{x_n\}$ is a Cauchy sequence in X. Since X is complete, there exists $\xi \in X$ such that $x_n \to \xi$ as $n \to \infty$. Since, $x_{n+1} \in Tx_n$, we have from (a) that

$$fx_{n+1} = gx_{n+1} \in Tx_n, \text{ for each } n \in \mathbb{N}. \tag{5.80}$$

Since f and g are continuous and $\lim_{n \to \infty} x_n = \xi$, we have

$$\lim_{n \to \infty} fx_{n+1} = \lim_{n \to \infty} gx_{n+1} = f\xi = g\xi. \tag{5.81}$$

From (ii) and (b_1), we have

$$d(x_{n+1}, T\xi) \leqslant H(Tx_n, T\xi)$$
$$\leqslant \varphi(d(x_n, \xi)) \{d(x_n, T\xi) + d(\xi, Tx_n)\}$$
$$+ \hat{h}(f\xi)(d(f\xi, fx_{n+1}) + \hat{k}(g\xi)d(g\xi, gx_{n+1})$$
$$\leqslant \varphi(d(x_n, \xi)) \{d(x_n, T\xi) + d(\xi, x_{n+1})\}$$
$$+ \hat{h}(f\xi)(d(f\xi, fx_{n+1}) + \hat{k}(g\xi)d(g\xi, gx_{n+1})$$
$$< \frac{1}{2} \{d(x_n, T\xi) + d(\xi, x_{n+1})\} + \hat{h}(f\xi)(d(f\xi, fx_{n+1}) + \hat{k}(g\xi)d(g\xi, gx_{n+1}).$$

$$(5.82)$$

We claim that $\xi \in T\xi$; that is, $d(\xi, T\xi) = 0$. If not, then using the continuity of the function $x \mapsto d(x, T\xi)$, we obtain from (5.82) that

$$\lim_{n \to \infty} d(x_{n+1}, T\xi)$$
$$\leqslant \lim_{n \to \infty} \left[\frac{1}{2} \{d(x_n, T\xi) + d(\xi, x_{n+1})\} + \hat{h}(f\xi)(d(f\xi, fx_{n+1}) + \hat{k}(g\xi)d(g\xi, gx_{n+1})\right].$$

Therefore, $d(\xi, T\xi) \leqslant \frac{1}{2}d(\xi, T\xi)$, a contradiction. Hence, $d(\xi, T\xi) = 0$, and since $T\xi$ is closed it must be the case that $\xi \in T\xi$. By (a), $f\xi = g\xi \in T\xi$. Therefore, $\xi \in \mathcal{COP}(f, g, T) \cap \mathfrak{F}(T)$ and the proof is complete.

The single-valued version of Theorem 5.99 may be stated as follows:

Theorem 5.100 *Let (X, d) be a complete metric space, $T : X \to X$ be a single-valued map, $f, g : X \to X$ be continuous self-maps and $\varphi : (0, \infty) \to [0, \frac{1}{2})$ be a $\mathcal{P}$-function. Assume that*
(a_1') *for each $x \in X$, $Tx(= y)$ is a coincidence point of f and g i.e. $fy = gy$ for $y = Tx$.*
(b_1') *there exists $\hat{h}, \hat{k} : X \to [0, \infty)$ such that*

$$d(Tx, Ty) \leqslant \varphi(d(x, y)) [d(x, Ty) + d(y, Tx)]$$
$$+ \hat{h}(fy)(d(fy, Tx) + \hat{k}(gy)(d(gy, Tx) \text{ for all } x, y \in X.$$

Then $\mathcal{COP}(f, g) \cap \mathfrak{F}(T) \neq \varnothing$.

The following example shows the generality of Theorem 5.99 over Nadler's fixed point theorem.

Example 5.39 Let $X = \{0, \frac{3}{4}, 1\}$ and $d : X \times X \to \mathbb{R}$ be a standard metric. Let $T : X \to CB(X)$ be a multivalued mapping and $f, g : X \to X$ be two mappings defined by

$$Tx = \begin{cases} \{0\}, & \text{if } x = 0 \\ \{0, \frac{3}{4}\}, & \text{if } x = \frac{3}{4} \\ \{0, 1\}, & \text{if } x = 1, \end{cases}$$

and $f, g = I_X$ (: the identity mapping on X). Define a function $\varphi : [0, \infty) \to [0, \frac{1}{2})$ by $\varphi(t) := \frac{1}{3}$ for all $t \in [0, \infty)$ and two mappings $\hat{h}, \hat{k} : X \to [0, \infty)$ by $\hat{h}(x) := 1$ and $\hat{k}(x) := 0$ for all $x \in X$. Then we observe that the following hold:

(a_1) for each $x \in X$, $\{fy = gy : y \in Tx\} \subseteq Tx$;
(b_1) $\mathcal{COP}(f, g, T) \bigcap \mathfrak{F}(T) = \{0, \frac{3}{4}, 1\}$;
(c) f and g are continuous.

Observe that $\lim \sup_{s \to t^+} \varphi(s) = \frac{1}{3} < \frac{1}{2}$ for all $t \in [0, \infty)$ and so φ is a $\mathcal{P}$-function. Moreover, it is routine to check that all the conditions of Theorem 5.99 are satisfied for all $x, y \in X$. Moreover, $\mathcal{COP}(f, g) \cap \mathfrak{F}(T) \neq \varnothing$.

Notice that the mapping T does not satisfy the assumptions of Theorem 5.94. Indeed, by taking $0 \leqslant k < 1$, $x = 0$ and $y = 1$, we have

$$H(T0, T1) = H(\{0\}, \{0, 1\}) = 1 > kd(0, 1),$$

a contradiction.

By applying the same arguments as in Theorem 5.99, we can obtain generalized version of Kannan-type fixed point theorem for multivalued mappings as follows:

Theorem 5.101 *Let (X, d) be a complete metric space, $T : X \to CB(X)$ be a multivalued map, $f, g : X \to X$ be continuous self-maps and $\varphi : (0, \infty) \to [0, \frac{1}{2})$ be a $\mathcal{P}$-function. Assume that*
$(a°)$ *for each $x \in X$, $\{fy = gy : y \in Tx\} \subseteq Tx$.*
$(b°)$ *there exists $\hat{h}, \hat{k} : X \to [0, \infty)$ such that*

$$H(Tx, Ty) \leqslant \varphi(d(x, y)) \, [d(x, Tx) + d(y, Ty)]$$
$$+ \hat{h}(fy)(d(fy, Tx) + \hat{k}(gy)(d(gy, Tx) \text{ for all } x, y \in X.$$

Then $\mathcal{COP}(f, g, T) \cap \mathfrak{F}(T) \neq \varnothing$.

A slight variant of Example A of Du [205] given below illustrates the generality of Theorem 5.101 over Mizoguchi–Takahashi's fixed point theorem and Du's fixed point theorem.

Example 5.40 Let ℓ_∞ be the Banach space consisting of all bounded real sequences with supremum norm $\| \cdot \|_\infty$ and let $\{e_n\}$ be the canonical basis of ℓ_∞. Let $\{\tau_n\}$ be a sequence of positive real numbers satisfying $\tau_2 = \tau_{10^7}$ and $2\tau_{n+2} < \tau_n$ for $n \in \mathbb{N} \setminus \{2\}$ (e.g. let $\tau_2 = \tau_{10^7}$, $2\tau_{n+2} < \tau_n$ and $\tau_n = \frac{1}{2^n}$ for $n \in \mathbb{N} \setminus \{2\}$). Thus, the sequence $\{\tau_n\}$ convergent. Put $v_n = \tau_n e_n$ for all $n \in \mathbb{N}$ and let $X = \{v_n\}_{n \in \mathbb{N}}$ be a bounded and complete subset of ℓ_∞. Then $(X, \| \cdot \|_\infty)$ be a complete metric space and $\|v_n - v_m\|_\infty = \tau_n$ if $m > n$.

Let $T : X \to CB(X)$ and $f, g : X \to X$ be defined by

$$Tv_n := \begin{cases} \{v_1, v_2, v_{10^7}\}, & \text{if } n \in \{1, 2, 10^7\}, \\ X \setminus \{v_1, v_2, \ldots, v_n, v_{n+1}\}, & \text{if } n \geq 3 \text{ and } n \neq 10^7 \end{cases}$$

and

$$fv_n = gv_n := \begin{cases} v_2, & \text{if } n \in \{1, 2, 10^7\}, \\ v_{n+1}, & \text{if } n \geq 3 \text{ and } n \neq 10^7 \end{cases}$$

respectively. Then we observe that the following hold:

(a) for each $x \in X$, $\{fy = gy : y \in Tx\} \subseteq Tx$.

(b_1) $\mathcal{COP}(f, g, T) \cap \mathfrak{F}(T) = \{v_1, v_2, v_{10^7}\}$.

(c) f and g are continuous.

Since

1. $\|fv_1 - fv_2\|_\infty = 0 < \tau_1 = \|v_1 - v_2\|_\infty$

2. $\|fv_1 - fv_{10^7}\|_\infty = 0 < \tau_1 = \|v_1 - v_{10^7}\|_\infty$

3. $\|fv_1 - fv_m\|_\infty = \tau_m < \tau_1 = \|v_1 - v_m\|_\infty$ for any $3 \leqslant m < 10^7 - 1$

4. $\|fv_1 - fv_m\|_\infty = 0 < \tau_1 = \|v_1 - v_m\|_\infty$ for $m = 10^7 - 1$

5. $\|fv_1 - fv_m\|_\infty = \tau_{10^7} < \tau_1 = \|v_1 - v_m\|_\infty$ for any $m > 10^7$

6. $\|fv_2 - fv_m\|_\infty = \tau_{m+1} < \tau_m = \|v_2 - v_m\|_\infty$ for any $3 \leqslant m < 10^7 - 1$

7. $\|fv_2 - fv_m\|_\infty = 0 < \tau_m = \|v_2 - v_m\|_\infty$ for $m = 10^7 - 1$

8. $\|fv_2 - fv_m\|_\infty = 0 = \|v_2 - v_m\|_\infty$ for $m = 10^7$

9. $\|fv_2 - fv_m\|_\infty = \tau_2 = \|v_2 - v_m\|_\infty$ for any $m > 10^7$

10. $\|fv_{10^7} - fv_m\|_\infty = \tau_{m+1} < \tau_m = \|v_{10^7} - v_m\|_\infty$ for any $3 \leqslant m < 10^7 - 1$

11. $\|fv_{10^7} - fv_m\|_\infty = 0 < \tau_{10^7-1} = \|v_{10^7} - v_m\|_\infty$ for $m = 10^7 - 1$

12. $\|fv_{10^7} - fv_m\|_\infty = \tau_{10^7} = \|v_{10^7} - v_m\|_\infty$ for any $m > 10^7$

13. $\|fv_n - fv_m\|_\infty = \tau_{n+1} < \tau_n = \|v_n - v_m\|_\infty$ for any $3 \leqslant n < 10^7 - 1, n > 10^7$ and $m > n$.

Since f is nonexpansive and $f = g$ on X, it follows that f and g both are continuous on X.

Define $\varphi : [0, \infty) \to [0, \frac{1}{2})$ by

$$\varphi(t) := \begin{cases} \frac{\tau_{n+2}}{\tau_n}, & \text{if } t = \tau_n \text{ for some } n \in \mathbb{N} \\ 0, & \text{otherwise.} \end{cases}$$

Also define $\hat{h}, \hat{k} : X \to [0, \infty)$ by

$$\hat{h}(v_n) = \hat{k}(v_n) := \begin{cases} 0, & \text{if } n \in \{1, 2, 10^7\} \\ \frac{\tau_1}{2\tau_n}, & \text{if } n \geq 3 \text{ and } n \neq 10^7. \end{cases}$$

Notice that $\lim \sup_{s \to t^+} \varphi(s) = 0 < \frac{1}{2}$ for all $t \in [0, \infty)$, therefore φ is a $\mathscr{P}$-function. We claim that

$$H_\infty(Tx, Ty) \leqslant \varphi(\|x - y\|_\infty)\{\|x - Tx\|_\infty + \|y - Ty\|_\infty\}$$
$$+ \hat{h}(fy)\|fy - Tx\|_\infty + \hat{k}(gy)\|gy - Tx\|_\infty \quad \text{for all } x, y \in X \ (*)$$

where H_∞ is the Hausdorff metric induced by $\|\cdot\|_\infty$. In order to verify that T satisfies $(*)$, we consider the following thirteen possible cases:

Case 1. $\varphi(\|v_1 - v_2\|_\infty)\{\|v_1 - Tv_1\|_\infty + \|v_2 - Tv_2\|_\infty)\} + \hat{h}(fv_2)\|fv_2 - Tv_1\|_\infty + \hat{k}(gv_2)\|gv_2 - Tv_1\|_\infty = 0 = H_\infty(Tv_1, Tv_2)$.

Case 2. $\varphi(\|v_1 - v_{10^7}\|_\infty)\{\|v_1 - Tv_1\|_\infty + \|v_{10^7} - Tv_{10^7}\|_\infty\} + \hat{h}(fv_{10^7})\|fv_{10^7} - Tv_1\|_\infty) + \hat{k}(gv_{10^7})\|gv_{10^7} - Tv_1\|_\infty) = 0 = H_\infty(Tv_1, Tv_{10^7})$.

Case 3. For any m with $3 \leqslant m < 10^7 - 1$, we have
$\varphi(\|v_1 - v_m\|_\infty)\{\|v_1 - Tv_1\|_\infty + \|v_m - Tv_m\|_\infty\} + \hat{h}(fv_m)\|fv_m - Tv_1\|_\infty + \hat{k}(gv_m)\|gv_m - Tv_1\|_\infty = \frac{\tau_3 \tau_m}{\tau_1} + 2(\frac{\tau_1}{2\tau_{m+1}})\tau_{m+1} = \frac{\tau_3 \tau_m}{\tau_1} + \tau_1 > \tau_1 = H_\infty(Tv_1, Tv_m)$.

Case 4. For $m = 10^7 - 1$, we have
$\varphi(\|v_1 - v_m\|_\infty)\{\|v_1 - Tv_1\|_\infty + \|v_m - Tv_m\|_\infty\} + \hat{h}(fv_m)\|fv_m - Tv_1\|_\infty + \hat{k}(gv_m)\|gv_m - Tv_1\|_\infty = \frac{\tau_3 \tau_{10^7-1}}{\tau_1} + 2(\frac{\tau_1}{2\tau_{10^7}})\tau_{10^7} = \frac{\tau_3 \tau_{10^7-1}}{\tau_1} + \tau_1 > \tau_1 = H_\infty$
(Tv_1, Tv_m).

Case 5. For any $m > 10^7$, we have
$\varphi(\|v_1 - v_m\|_\infty)\{\|v_1 - Tv_1\|_\infty + \|v_m - Tv_m\|_\infty\} + \hat{h}(fv_m)\|fv_m - Tv_1\|_\infty + \hat{k}(gv_m)\|gv_m - Tv_1\|_\infty = \frac{\tau_3 \tau_m}{\tau_1} + 2(\frac{\tau_1}{2\tau_{m+1}})\tau_{10^7} > \frac{\tau_3 \tau_m}{\tau_1} + \tau_1 > \tau_1 = H_\infty(Tv_1, Tv_m)$.

Case 6. For any m with $3 \leqslant m < 10^7 - 1$, we have
$\varphi(\|v_2 - v_m\|_\infty)\{\|v_2 - Tv_2\|_\infty + \|v_m - Tv_m\|_\infty\} + \hat{h}(fv_m)\|fv_m - Tv_2\|_\infty + \hat{k}(gv_m)\|gv_m - Tv_2\|_\infty = \frac{\tau_{m+2}\tau_m}{\tau_m} + 2(\frac{\tau_1}{2\tau_{m+1}})\tau_{m+1} = \tau_{m+2} + \tau_1 > \tau_1 = H_\infty$
(Tv_2, Tv_m).

Case 7. For $m = 10^7 - 1$, we have
$\varphi(\|v_2 - v_m\|_\infty)\{\|v_2 - Tv_2\|_\infty + \|v_m - Tv_m\|_\infty\} + \hat{h}(fv_m)\|fv_m - Tv_2\|_\infty + \hat{k}(gv_m)\|gv_m - Tv_2\|_\infty = \frac{\tau_{m+2}\tau_m}{\tau_m} + 2(\frac{\tau_1}{2\tau_{10^7}})\tau_{10^7} = \tau_{10^7+1} + \tau_1 > \tau_1 = H_\infty$
(Tv_2, Tv_m).

Case 8. For $m = 10^7$, we have
$\varphi(\|v_2 - v_m\|_\infty)\{\|v_2 - Tv_2\|_\infty + \|v_m - Tv_m\|_\infty\} + \hat{h}(fv_m)\|fv_m - Tv_2\|_\infty + \hat{k}(gv_m)\|gv_m - Tv_2\|_\infty = \frac{\tau_{m+2}\tau_m}{\tau_m} + 2(\frac{\tau_1}{2\tau_{10^7+1}})\tau_{10^7}) > \tau_{10^7+2} + \tau_1 > \tau_1 = H_\infty$
(Tv_2, Tv_m).

Case 9. For any $m > 10^7$, we have

$\varphi(\|v_2 - v_m\|_\infty)\{\|v_2 - Tv_2\|_\infty + \|v_m - Tv_m\|_\infty\} + \hat{h}(fv_m)\|fv_m - Tv_2\|_\infty + \hat{k}(gv_m)\|gv_m - Tv_2\|_\infty = \frac{\tau_{10^7} + 2\tau_m}{\tau_{10^7}} + 2(\frac{\tau_1}{2\tau_{m+1}})\tau_{10^7} > \frac{\tau_{10^7} + 2\tau_m}{\tau_{10^7}} + \tau_1 > \tau_1 = H_\infty(Tv_1, Tv_m)$.

Case 10. For any m with $3 \leqslant m < 10^7 - 1$, we have

$\varphi(\|v_{10^7} - v_m\|_\infty)\{\|v_{10^7} - Tv_{10^7}\|_\infty + \|v_m - Tv_m\|_\infty)\} + \hat{h}(fv_m)\|fv_m - Tv_{10^7}\|_\infty + \hat{k}(gv_m)\|gv_m - Tv_{10^7}\|_\infty = \frac{\tau_{10^7} + 2\tau_m}{\tau_{10^7}} + 2(\frac{\tau_1}{2\tau_{m+1}})\tau_{m+1} = \frac{\tau_{10^7} + 2\tau_m}{\tau_{10^7}} + \tau_1 > \tau_1 = H_\infty(Tv_{10^7}, Tv_m)$.

Case 11. For $m = 10^7 - 1$, we have

$\varphi(\|v_{10^7} - v_m\|_\infty)\{\|v_{10^7} - Tv_{10^7}\|_\infty + \|v_m - Tv_m\|_\infty)\} + \hat{h}(fv_m)\|fv_m - Tv_{10^7}\|_\infty + \hat{k}(gv_m)\|gv_m - Tv_{10^7}\|_\infty = \frac{\tau_{10^7+1}\tau_m}{\tau_{10^7-1}} + 2(\frac{\tau_1}{2\tau_{10^7}})\tau_{10^7} = \tau_{10^7+1} + \tau_1 > \tau_1 = H_\infty(Tv_{10^7}, Tv_m)$.

Case 12. For any $m > 10^7$, we have

$\varphi(\|v_2 - v_m\|_\infty)\{\|v_2 - Tv_2\|_\infty + \|v_m - Tv_m\|_\infty\} + \hat{h}(fv_m)\|fv_m - Tv_2\|_\infty + \hat{k}(gv_m)\|gv_m - Tv_2\|_\infty = \frac{\tau_{10^7} + 2\tau_m}{\tau_{10^7}} + 2(\frac{\tau_1}{2\tau_{m+1}})\tau_{10^7} > \frac{\tau_{10^7} + 2\tau_m}{\tau_{10^7}} + \tau_1 > \tau_1 = H_\infty(Tv_2, Tv_m)$.

Case 13. For any n with $3 \leqslant n < 10^7 - 1$, $n > 10^7$ and $m > n$, we have

$\varphi(\|v_n - v_m\|_\infty)\{\|v_n - Tv_n\|_\infty + \|v_m - Tv_m\|_\infty)\} + \hat{h}(fv_m)\|fv_m - Tv_n\|_\infty + \hat{k}(gv_m)\|gv_m - Tv_n\|_\infty = \frac{\tau_{n+2}}{\tau_n}\{\tau_n + \tau_m\} = \tau_{n+2} + \frac{\tau_{n+2}\tau_m}{\tau_n} > \tau_{n+2} = H_\infty(Tv_n, Tv_m)$.

Hence, by Cases 1–13 we prove that T satisfies $(*)$. Therefore, all assumptions of Theorem 5.101 are satisfied. Further, Theorem 5.101 claims that $\mathcal{COP}(f, g, T) \cap \mathfrak{F}(T) \neq \varnothing$. Notice that if we consider a convergent sequence $\{\tau_n\}_{n=1}^\infty$ defined by $\tau_2 = \tau_{10^7}$, $4\tau_{n+2} < \tau_n$, $\tau_n = \frac{1}{5^n}$ for all $n \in \mathbb{N} \setminus \{2\}$ then we observe by defining $\varphi_{\mathcal{MJ}}(t) = 2\varphi(t)$ for all $t \in [0, \infty)$ that

(i) $H_\infty(Tv_1, Tv_m) = \tau_1 > 2\tau_3 + (\frac{\tau_1}{2\tau_{m+1}})\tau_{m+1} = 2\tau_3 + \frac{\tau_1}{2}$

$> \varphi_{\mathcal{MJ}}(\|v_1 - v_m\|_\infty)\|v_1 - v_m\|_\infty + \hat{k}(gv_m)\|gv_m - Tv_1\|_\infty$

for any m with $3 \leqslant m < 10^7 - 1$, so Du's fixed point theorem is not applicable here.

(ii) $H_\infty(Tv_1, Tv_m) = \tau_1 > 2\tau_3 = \varphi_{\mathcal{MJ}}(\|v_1 - v_m\|_\infty)\|v_1 - v_m\|_\infty$

for any m with $3 \leqslant m < 10^7 - 1$, so Mizoguchi–Takahashi's fixed point theorem is not applicable here.

Theorem 5.102 *Let (X, d) be a complete metric space, $T : X \to CB(X)$ be a multivalued map, $g : X \to X$ be a continuous self-map and $\varphi : (0, \infty) \to [0, \frac{1}{2})$ be a $\mathcal{P}$-function. Assume that*

(a) Tx is g-invariant (i.e. $g(Tx) \subseteq Tx$) for each $x \in X$;

(b_1'') there exists $\hat{k} : X \to [0, \infty)$ such that

$H(Tx, Ty) \leqslant \varphi(d(x, y)) \, [d(x, Tx) + d(y, Ty)] + \hat{k}(gy)(d(gy, Tx)$ *for all $x, y \in X$.*

Then $\mathcal{COP}(g, T) \cap \mathfrak{F}(T) \neq \varnothing$.

Proof Define $\hat{h}(t) = 0$ for all $t \in [0, \infty)$. So (b°) implies(b''_1) and hence the conclusion follows from Theorem 5.101.

As a direct consequence of Theorem 5.101, we obtain the following result.

Theorem 5.103 *Let (X, d) be a complete metric space, $T : X \to CB(X)$ be a multivalued map, $g : X \to X$ be continuous self-maps and $\varphi : (0, \infty) \to [0, \frac{1}{2})$ be a $\mathcal{P}$-function. Assume that*
(a) Tx is g-invariant (i.e. $g(Tx) \subseteq Tx$) for each $x \in X$;
(b$_2$) there exists $L \geq 0$ and a function $k : X \to [0, L]$ such that

$$H(Tx, Ty) \leqslant \varphi(d(x, y)) \ [d(x, Tx) + d(y, Ty)] + k(gy)(d(gy, Tx) \text{ for all } x,$$
$y \in X$.
Then $\mathcal{COP}(g, T) \cap \mathfrak{F}(T) \neq \varnothing$.

Example 5.41 Let ℓ_∞, $\| \cdot \|_\infty$, $\mathcal{H}_\infty$, $\{e_n\}$, $\{\tau_n\}$, X, T, f, g, θ and φ be the same as in Example 5.40. Let $\{\tau_n\}$ be a sequence of positive real numbers satisfying $\tau_1 = \tau_2 = \tau_3$, $\tau_{n+1} < \tau_n$ for all $n \geq 3$ and $L' = L + \tau \leqslant \tau_n$ for all $n \in \mathbb{N}$ and for some $\tau > 0$ (e.g. let $\tau_1 = L' + \frac{1}{2} = \tau_2 = \tau_3$, $\tau_n = L' + \frac{1}{n+2}$ for $n \geq 3$). Thus, the sequence $\{\tau_n\}$ is convergent. Put $v_n = \tau_n e_n$ for all $n \in \mathbb{N}$ and let $X = \{v_n\}_{n \in \mathbb{N}}$ be a bounded and complete subset of ℓ_∞. Then $(X, \| \cdot \|_\infty)$ be a complete metric space and $\|v_n - v_m\|_\infty = \tau_n$ if $m > n$.

Let $T : X \to CB(X)$ and $g : X \to X$ be defined by

$$Tv_n := \begin{cases} \{v_1, v_2, v_3\}, & \text{if } n \in \{1, 2, 3\} \\ X \setminus \{v_1, v_2, \ldots, v_n, v_{n+1}\}, & \text{if } n \geq 4 \end{cases}$$

and

$$gv_n := \begin{cases} v_2, & \text{if } n \in \{1, 2, 3\} \\ v_{n+1}, & \text{if } n \geq 4 \end{cases}$$

respectively. Then we observe that the following hold.
(a) Tx is g-invariant (i.e. $g(Tx) \subseteq Tx$) for each $x \in X$;
(b$_1$) $\mathcal{COP}(g, T) \cap \mathfrak{F}(t) = \{v_1, v_2, v_3\}$.
(c) g is continuous.
Since

1. $\|g\dot{v}_1 - gv_2\|_\infty = 0 < \tau_1 = \|v_1 - v_2\|_\infty$

2. $\|gv_1 - gv_3\|_\infty = 0 < \tau_1 = \|v_1 - v_3\|_\infty$

3. $\|gv_1 - gv_m\|_\infty = \tau_2 = \tau_1 = \|v_1 - v_m\|_\infty$ for any $m \geq 4$

4. $\|gv_2 - gv_3\|_\infty = 0 < \tau_2 = \|v_2 - v_3\|_\infty$

5. $\|gv_2 - gv_m\|_\infty = \tau_2 = \|v_2 - v_m\|_\infty$ for any $m \geq 4$

6. $\|gv_3 - gv_m\|_\infty = \tau_2 = \tau_3 = \|v_3 - v_m\|_\infty$ for any $m \geq 4$

7. $\|gv_n - gv_m\|_\infty = \tau_{n+1} < \tau_n = \|v_n - v_m\|_\infty$ for any $n \geq 4$ and $m > n$

Since g is nonexpansive, it follows that g is continuous on X.
Define $\varphi : [0, \infty) \to [0, \frac{1}{2})$ by

$$\varphi(t) := \begin{cases} \frac{\tau_n}{L'}, & \text{if } t = \tau_n \text{ for } n \in \{1, 2, 3\} \\ \frac{\tau_{n+2}}{L'}, & \text{if } t = \tau_n \text{ for } n \geq 4 \\ 0, & \text{otherwise.} \end{cases}$$

Define $k : X \to [0, L]$ by

$$k(v_n) := \begin{cases} \frac{L}{\sqrt{2}}, & \text{if } n \in \{1, 2, 3\} \\ L(1 - \frac{\tau_n}{\tau_1}), & \text{if } n \geq 4. \end{cases}$$

Notice that $\lim\sup\limits_{s \to t^+} \varphi(s) = 0 < \frac{1}{2}$ for all $t \in [0, \infty)$, therefore φ is a $\mathcal{P}$-function.
In order to verify that T satisfies

$$H_\infty(Tx, Ty) \leqslant \varphi(\|x - y\|_\infty)\{\|x - Tx\|_\infty + \|y - Ty\|_\infty\}$$
$$+ k(gy)\|gy - Tx\|_\infty \quad \text{for all } x, y \in X \qquad (**)$$

we need to consider the following thirteen cases:

Case 1. $\varphi(\|v_1 - v_2\|_\infty)\{\|v_1 - Tv_1\|_\infty + \|v_2 - Tv_2\|_\infty)\} + k(gv_2)\|gv_2 - Tv_1\|_\infty$
$\quad = 0 = H_\infty(Tv_1, Tv_2).$

Case 2. $\varphi(\|v_1 - v_3\|_\infty)\{\|v_1 - Tv_1\|_\infty + \|v_3 - Tv_3\|_\infty\} + k(gv_3)\|gv_3 - Tv_1\|_\infty$
$\quad = 0 = H_\infty(Tv_1, Tv_3).$

Case 3. For any $m \geq 4$, we have
$\varphi(\|v_1 - v_m\|_\infty)\{\|v_1 - Tv_1\|_\infty + \|v_m - Tv_m\|_\infty\} + k(gv_m)\|gv_m - Tv_1\|_\infty$
$\quad = \frac{\tau_1 \tau_m}{L'} + L(\tau_1 - \tau_{m+1}) \geq \tau_1 + L(\tau_1 - \tau_{m+1}) > \tau_1 = H_\infty(Tv_1, Tv_m).$

Case 4. $\varphi(\|v_2 - v_3\|_\infty)\{\|v_2 - Tv_2\|_\infty + \|v_3 - Tv_3\|_\infty\} + k(gv_3)\|gv_3 - Tv_2\|_\infty$
$\quad = 0 = H_\infty(Tv_2, Tv_3).$

Case 5. For any $m \geq 4$, we have
$\varphi(\|v_2 - v_m\|_\infty)\{\|v_2 - Tv_2\|_\infty + \|v_m - Tv_m\|_\infty\} + k(gv_m)\|gv_m - Tv_2\|_\infty$
$\quad = \frac{\tau_2 \tau_m}{L'} + L(\tau_1 - \tau_{m+1}) \geq \tau_2 + L(\tau_1 - \tau_{m+1}) > \tau_2 = H_\infty(Tv_2, Tv_m).$

Case 6. For any $m \geq 4$, we have
$\varphi(\|v_3 - v_m\|_\infty)\{\|v_3 - Tv_3\|_\infty + \|v_m - Tv_m\|_\infty\} + k(gv_m)\|gv_m - Tv_3\|_\infty$
$\quad = \frac{\tau_3 \tau_m}{L'} + L(\tau_1 - \tau_{m+1}) \geq \tau_3 + L(\tau_1 - \tau_{m+1}) > \tau_3 = H_\infty(Tv_3, Tv_m).$

Case 7. For any $n \geq 4$ and $m > n$, we have
$$\varphi(\|v_n - v_m\|_\infty)\{\|v_n - Tv_n\|_\infty + \|v_m - Tv_m\|_\infty\} + k(gv_m)\|gv_m - Tv_n\|_\infty$$
$$= \frac{\tau_{n+2}(\tau_n + \tau_m)}{L'} \geq \tau_{n+2} + \frac{\tau_{n+2}\tau_m}{L'} > \tau_{n+2} = H_\infty(Tv_n, Tv_m).$$
Hence, we conclude that T satisfies $(**)$. Applying Theorem 5.99, we have

$$\mathcal{COP}(f, g, T) \cap \mathfrak{F}(T) \neq \varnothing.$$

The following result easily follows from Theorem 5.103.

Theorem 5.104 *Let* (X, d) *be a complete metric space,* $T : X \to CB(X)$ *be a multivalued map,* $g : X \to X$ *be continuous self-maps and* $\varphi : (0, \infty) \to [0, \frac{1}{2})$ *be a* $\mathcal{P}$-*function. Assume that*
(a) Tx *is* g-*invariant (i.e.* $g(Tx) \subseteq Tx)$ *for each* $x \in X$;
(b_3) there exists $L \geq 0$ *such that*
$$H(Tx, Ty) \leqslant \varphi(d(x, y)) [d(x, Tx) + d(y, Ty)] + L(d(gy, Tx) \text{ for all}$$
$x, y \in X$.
Then $\mathcal{COP}(g, T) \cap \mathfrak{F}(T) \neq \varnothing$.

Proof Define $k : X \to [0, L]$ by $k(x) = L$ for all $x \in X$. Then, we see that (b_3) implies (b_2) and hence the conclusion follows from Theorem 5.103.

Remark 5.34 Since (b_2) implies $k(x) \leqslant L$ for all $x \in X$, it follows that (b_2) implies (b_3). As a consequence, we find that (b_2) and (b_3) are indeed equivalent.

The following intersection theorem is also immediate from Theorem 5.101.

Theorem 5.105 *Let* (X, d) *be a complete metric space,* $T : X \to CB(X)$ *be a multivalued map,* $g : X \to X$ *be a continuous self-map and* $\varphi : (0, \infty) \to [0, \frac{1}{2})$ *be a* $\mathcal{P}$-*function. Assume that*
(a) Tx *is* g-*invariant (i.e.* $g(Tx) \subseteq Tx)$ *for each* $x \in X$;
(b_4) there exists $L \geq 0$ *and a function* $\bar{k} : X \to [L, \infty)$ *such that*
$$H(Tx, Ty) \leqslant \varphi(d(x, y)) [d(x, Tx) + d(y, Ty)] + \bar{k}(gy)(d(gy, Tx) \text{ for all}$$
$x, y \in X$.
Then $\mathcal{COP}(g, T) \cap \mathfrak{F}(T) \neq \varnothing$.

By applying Theorem 5.99, we can obtain generalized version of primitive Chatterjea's fixed point theorem [140] for multivalued mappings as follows:

Theorem 5.106 *Let* (X, d) *be a complete metric space,* $T : X \to CB(X)$ *be a multivalued map and* $\varphi : (0, \infty) \to [0, \frac{1}{2})$ *be a* $\mathcal{P}$-*function and* $\tilde{k} : X \to [0, \infty)$ *be a function. Assume that*
$$H(Tx, Ty) \leqslant \varphi(d(x, y)) [d(x, Ty) + d(y, Tx)] + \tilde{k}(y)(d(y, Tx) \text{ for all } x, y \in X.$$
Then $\mathfrak{F}(T) \neq \varnothing$.

By applying Theorem 5.101, we can obtain generalized version of primitive Kannan's fixed point theorem [308] for multivalued mappings as follows:

Theorem 5.107 *Let* (X, d) *be a complete metric space,* $T : X \to CB(X)$ *be a multivalued map and* $\varphi : (0, \infty) \to [0, \frac{1}{2})$ *be a* $\mathcal{P}$-*function and* $\tilde{k} : X \to [0, \infty)$ *be a function. Assume that*
$$H(Tx, Ty) \leqslant \varphi(d(x, y)) [d(x, Tx) + d(y, Ty)] + \tilde{k}(y)(d(y, Tx) \text{ for all } x, y \in X.$$
Then $\mathfrak{F}(T) \neq \varnothing$.

5.4.3   Coincidence and Common Fixed Points of Nonlinear Hybrid Mappings

The concept of commutativity of single-valued mappings [296] was extended in [306] to the setting of a single-valued mapping and a multivalued mapping on a metric space. This concept of commutativity has been further generalized by different authors, viz. weakly commuting [307], compatible [561] and weakly compatible [449]. It is interesting to note that in all the results obtained so far, concerning common fixed points of hybrid mappings, the (single-valued and multivalued) mappings under consideration satisfy either the commutativity condition or one of its generalizations (see, for instance, [42, 420, 449, 459, 460]).

In this section, we present some existence results of fixed points of hybrid contractions which do not satisfy any of the commutativity conditions or its above-mentioned generalizations in the field of hybrid fixed point theory.

For a metric space (X, d), let $(CB(X), H)$ and $(CL(X), H)$ denote, respectively, the hyperspace of nonempty, closed, bounded and nonempty, closed subsets of X, where H is the Hausdorff metric induced by d. For $f : X \to X$ and $T : X \to CL(X)$ we shall use the following notations:

$$L(x, y) = \max\{d(fx, fy), d(fx, Tx), d(fy, Ty), \frac{1}{2}(d(fx, Ty) + d(fy, Tx))\},$$

$$N(x, y) = [\max\{d^2(fx, fy), d(fx, Tx) \cdot d(fy, Ty), d(fx, Ty) \cdot d(fy, Tx),$$
$$\frac{1}{2}d(fx, Tx) \cdot d(fy, Tx), \frac{1}{2}d(fx, Ty) \cdot d(fy, Ty)\}]^{\frac{1}{2}}.$$

We recall some definitions.

Definition 5.52 Mappings f and T are said to be *commuting* at a point $x \in X$ if $fTx \subseteq Tfx$. The mappings f and T are said to be *commuting* on X if $fTx \subseteq Tfx$ for all $x \in X$.

Definition 5.53 Mappings f and T are said to be *weakly commuting at a point* $x \in X$ if
$$H(fTx, Tfx) \leqslant d(fx, Tx).$$

The mappings f and T are said to be *weakly commuting on X* if
$$H(fTx, Tfx) \leqslant d(fx, Tx)$$

for all $x \in X$.

Definition 5.54 The mappings f and T are said to be *compatible* if $fTx \in CB(X)$ for all $x \in X$ and $\lim_{n \to +\infty} H(Tfx_n, fTx_n) = 0$, whenever $\{x_n\}$ is a sequence in X, such that $Tx_n \to M \in CB(X)$ and $fx_n \to t \in M$, as $n \to +\infty$.

Definition 5.55 The mappings f and T are said to be *f-weak compatible* if $fTx \in CB(X)$ for all $x \in X$ and the following limits exist and satisfy the inequalities:
(i) $\lim_{n \to \infty} H(Tfx_n, fTx_n) \leqslant \lim_{n \to \infty} H(Tfx_n, Tx_n)$, (ii) $\lim_{n \to \infty} d(fTx_n, fx_n) \leqslant \lim_{n \to \infty} H(Tfx_n, Tx_n)$,·
whenever $\{x_n\}$ is a sequence in X, such that $Tx_n \to M \in CB(X)$ and $fx_n \to t \in M$ as $n \to \infty$.

Let $C(T, f)$ denote the set of all coincidence points of the mappings f and T, that is $C(T, f) = \{u : fu \in Tu\}$.

Definition 5.56 The mappings f and T are said to be *coincidentally commuting* if they commute at their coincidence points.

Definition 5.57 Mappings f and T are said to be *coincidentally idempotent* if $ffu = fu$ for every $u \in C(T, f)$, that is, if f is idempotent at the coincidence points of f and T.

Definition 5.58 Mappings f and T are said to be *occasionally coincidentally idempotent* (or, in brief, oci) if $ffu = fu$ for some $u \in C(T, f)$.

It should be remarked that *coincidentally idempotent* pairs of mappings are *occasionally coincidentally idempotent*, but the converse is not necessarily true as shown in Example 5.46 of this section.
We recall the following lemma:

Lemma 5.13 ([449]) *Let $T : Y \to CB(X)$ and $f : Y \to X$ be f-weak compatible. If $\{fw\} = Tw$ for some $w \in Y$ and $H(Tx, Ty) \leqslant h \ (a \cdot L(x, y) + (1 - a) \cdot N(x, y))$ for all x, y in Y, where $0 < h < 1, 0 \leqslant a \leqslant 1$, then $fTw = Tfw$.*

We remark that the above-mentioned lemma has been used in [449, 459, 460], to prove the existence of fixed points of hybrid mappings.
In 2013, Pathak and Rodríguez-López [463] proved a fixed point result for hybrid mappings under a general integral-type contractivity condition for occasionally coincidentally idempotent mappings.

Theorem 5.108 *Let Y be an arbitrary nonempty set, (X, d) be a metric space, $f : Y \to X$ and $T : Y \to CB(X)$ be such that*

$$T(Y) \subseteq f(Y), \tag{5.83}$$

that is, $\cup_{y \in Y} T(y) \subseteq f(Y)$,

$$\exists q \in (0, 1) \text{ such that } \int_0^{H(Tx, Ty)} \psi(t) \, dt \leqslant q \int_0^{L(x,y)} \psi(t) \, dt, \ \forall x, y \in Y, \tag{5.84}$$

$$f(Y) \text{ is complete,} \tag{5.85}$$

$\psi : \mathbb{R}^+ \to \mathbb{R}^+$ *is a Lebesgue measurable mapping which is nonnegative, summable on each compact interval, and such that*

$$\psi(x) > 0, \ \forall x > 0, \tag{5.86}$$

which trivially implies that

$$\int_0^\varepsilon \psi(t)\, dt > 0 \ \text{for each } \varepsilon > 0 \tag{5.87}$$

and

$$\int_0^\varepsilon \psi(t)\, dt < \int_0^{\tilde{\varepsilon}} \psi(t)\, dt \ \text{for each } 0 < \varepsilon < \tilde{\varepsilon}. \tag{5.88}$$

Suppose also that

$$\int_0^{\mu\varepsilon} \psi(t)\, dt \leqslant \gamma(\mu) \int_0^\varepsilon \psi(t)\, dt, \quad \text{for each } \mu > 1 \text{ and } \varepsilon > 0, \tag{5.89}$$

where $\gamma : (1, +\infty) \longrightarrow \mathbb{R}^+$ *is such that*

$$0 < \gamma(q^{-1/2}) \cdot q < 1 \tag{5.90}$$

and

$$\gamma(q^{-1/2}) \cdot q \cdot \gamma\left(\frac{1}{\gamma(q^{-1/2}) \cdot q}\right) \leqslant 1. \tag{5.91}$$

Then T and f have a coincidence point. Further, if f and T are occasionally coincidentally idempotent, then f and T have a common fixed point.

Proof In view of (5.83) and Nadler's Remark in [416], given the point $x_0 \in Y$, we can construct two sequences $\{x_n\}$ in Y and $\{y_n\}$ in X such that, for each $n \in \mathbb{N}$,

$$y_n = fx_n \in Tx_{n-1} \ \text{and} \ d(y_n, y_{n+1}) \leqslant q^{-1/2} \cdot H(Tx_{n-1}, Tx_n).$$

Indeed, since $Tx_0 \subseteq f(Y)$, there exists $x_1 \in Y$ such that $fx_1 = y_1 \in Tx_0$. Besides, given $y_1 \in Tx_0$, by Nadler's Remark in [416] and using that $q^{-1/2} > 1$, we can choose $y_2 \in Tx_1 \subseteq f(Y)$ such that $d(y_1, y_2) \leqslant q^{-1/2} \cdot H(Tx_0, Tx_1)$ and $y_2 = fx_2$ for a certain $x_2 \in Y$. The continuation of this process allows to construct the two above-mentioned sequences $\{x_n\}$ and $\{y_n\}$ inductively.

We claim that $\{y_n\}$ is a Cauchy sequence. Using the inequality in (5.84) and also property (5.89), which is trivially valid for $\varepsilon = 0$, it follows, for $n \geq 2$, that

$$\int_0^{d(fx_{n-1}, fx_n)} \psi(t)\,dt \leqslant \int_0^{q^{-1/2}H(Tx_{n-2}, Tx_{n-1})} \psi(t)\,dt$$

$$\leqslant \gamma(q^{-1/2}) \int_0^{H(Tx_{n-2}, Tx_{n-1})} \psi(t)\,dt$$

$$\leqslant \gamma(q^{-1/2}) \cdot q \int_0^{L(x_{n-2}, x_{n-1})} \psi(t)\,dt,$$

where

$$L(x_{n-2}, x_{n-1})$$
$$= \max\{d(fx_{n-2}, fx_{n-1}), d(fx_{n-2}, Tx_{n-2}), d(fx_{n-1}, Tx_{n-1}),$$
$$\frac{1}{2}(d(fx_{n-2}, Tx_{n-1}) + d(fx_{n-1}, Tx_{n-2}))\}$$
$$\leqslant \max\{d(fx_{n-2}, fx_{n-1}), d(fx_{n-2}, fx_{n-1}), d(fx_{n-1}, fx_n), \frac{1}{2}d(fx_{n-2}, fx_n)\}$$
$$\leqslant \max\{d(fx_{n-2}, fx_{n-1}), d(fx_{n-1}, fx_n), \frac{1}{2}(d(fx_{n-2}, fx_{n-1}) + d(fx_{n-1}, fx_n))\}$$
$$= \max\{d(fx_{n-2}, fx_{n-1}), d(fx_{n-1}, fx_n)\}.$$

Suppose that

$$d(fx_{n-1}, fx_n) > \lambda \cdot d(fx_{n-2}, fx_{n-1}), \text{ for some } n \in \mathbb{N} \text{ with } n \geq 2,$$

where $\lambda = \gamma(q^{-1/2}) \cdot q \in (0, 1)$, hence $d(fx_{n-1}, fx_n) > 0$ and

$$0 < \max\{d(fx_{n-2}, fx_{n-1}), d(fx_{n-1}, fx_n)\} < \frac{1}{\lambda}d(fx_{n-1}, fx_n),$$

so that

$$\int_0^{d(fx_{n-1}, fx_n)} \psi(t)\,dt \leqslant \gamma(q^{-1/2}) \cdot q \int_0^{\max\{d(fx_{n-2}, fx_{n-1}), d(fx_{n-1}, fx_n)\}} \psi(t)\,dt$$

$$< \gamma(q^{-1/2}) \cdot q \int_0^{\frac{1}{\lambda}d(fx_{n-1}, fx_n)} \psi(t)\,dt$$

$$\leqslant \gamma(q^{-1/2}) \cdot q \cdot \gamma\left(\frac{1}{\lambda}\right) \int_0^{d(fx_{n-1}, fx_n)} \psi(t)\,dt$$

$$\leqslant \int_0^{d(fx_{n-1}, fx_n)} \psi(t)\,dt,$$

where we have also used (5.88) (a consequence of (5.86)), (5.89)–(5.91). The previous inequalities imply that

$$\int_0^{d(fx_{n-1}, fx_n)} \psi(t)\, dt < \int_0^{d(fx_{n-1}, fx_n)} \psi(t)\, dt,$$

which is a contradiction. In consequence,

$$d(fx_{n-1}, fx_n) \leqslant \lambda \cdot d(fx_{n-2}, fx_{n-1}), \text{ for every } n \in \mathbb{N}, n \geq 2,$$

where $\lambda = \gamma(q^{-1/2}) \cdot q \in (0, 1)$, by hypothesis, and hence $\{fx_n\}$ is a Cauchy sequence in $f(Y)$. This is clear from the following inequality, valid for $n, m \in \mathbb{N}$, $n > m$,

$$\begin{aligned} d(fx_n, fx_m) &\leqslant \sum_{j=m+1}^{n} d(fx_j, fx_{j-1}) \\ &\leqslant \sum_{j=m+1}^{n} \lambda^{j-1} d(fx_1, fx_0) = \frac{\lambda^m - \lambda^n}{1 - \lambda} d(fx_1, fx_0) \\ &\leqslant \frac{\lambda^m}{1 - \lambda} d(fx_1, fx_0), \end{aligned}$$

which tends to zero as $m \to +\infty$.

Since $f(Y)$ is complete, then the sequence $\{fx_n\}$ has a limit in $f(Y)$, say u. Let $w \in f^{-1}(u)$ and prove that $fw \in Tw$.

Suppose that $fw \notin Tw$, then, by (5.84), we have

$$\int_0^{d(fx_{n+1}, Tw)} \psi(t)\, dt \leqslant \int_0^{H(Tx_n, Tw)} \psi(t)\, dt \leqslant q \int_0^{L(x_n, w)} \psi(t)\, dt,$$

where

$$\begin{aligned} L(x_n, w) &= \max\{d(fx_n, fw), d(fx_n, Tx_n), d(fw, Tw), \frac{1}{2}(d(fx_n, Tw) + d(fw, Tx_n))\} \\ &= d(fw, Tw), \text{ for } n \text{ large.} \end{aligned}$$

Here, we have used that $d(fx_n, fw) = d(fx_n, u) \to 0$, as $n \to +\infty$, $d(fx_n, Tx_n) \leqslant d(fx_n, fx_{n+1}) \to 0$, as $n \to +\infty$, $d(fw, Tw) > 0$ due to $fw \notin Tw$ and Tw closed, and

$$\frac{1}{2}(d(fx_n, Tw) + d(fw, Tx_n)) \leqslant \frac{1}{2}(2d(fx_n, fw) + d(fw, Tw) + d(fx_n, Tx_n))$$

$$\to \frac{1}{2}d(fw, Tw), \text{ as } n \to +\infty.$$

Hence, for n large enough, we have

$$\int_0^{d(fx_{n+1},Tw)} \psi(t)\,dt \leqslant q \int_0^{d(fw,Tw)} \psi(t)\,dt.$$

Making n tend to $+\infty$ in the previous inequality, we have

$$\int_0^{d(fw,Tw)} \psi(t)\,dt \leqslant q \int_0^{d(fw,Tw)} \psi(t)\,dt$$

and, therefore, since $q < 1$ and $d(fw, Tw) > 0$, we get

$$\int_0^{d(fw,Tw)} \psi(t)\,dt < \int_0^{d(fw,Tw)} \psi(t)\,dt,$$

which is a contradiction. Hence, $fw \in Tw$, that is, w is a coincidence point for T and f.

Although this fact is not relevant to the proof, we note that $H(Tx_{n-1}, Tx_n) \to 0$, since $\lim_{n\to\infty} d(y_{n-1}, y_n) = 0$. Indeed,

$$\int_0^{H(Tx_{n-1},Tx_n)} \psi(t)\,dt \leqslant q \int_0^{L(x_{n-1},x_n)} \psi(t)\,dt,$$

where

$$\begin{aligned}
L(x_{n-1}, x_n) &\leqslant \max\{d(fx_{n-1}, fx_n), d(fx_n, fx_{n+1})\} \\
&\leqslant \max\{d(fx_{n-1}, fx_n), \lambda d(fx_{n-1}, fx_n)\} \\
&= d(fx_{n-1}, fx_n),
\end{aligned}$$

therefore

$$\int_0^{H(Tx_{n-1},Tx_n)} \psi(t)\,dt \leqslant q \int_0^{d(fx_{n-1},fx_n)} \psi(t)\,dt.$$

Then $\lim_{n\to+\infty} \int_0^{H(Tx_{n-1},Tx_n)} \psi(t)\,dt = 0$ and, by the properties of ψ, we get $H(Tx_{n-1}, Tx_n) \to 0$ as $n \to +\infty$. From the definition of $\{y_n\}$, we deduce that $d(fx_n, Tx_n) \leqslant H(Tx_{n-1}, Tx_n)$, for every n and, therefore, $\lim_{n\to\infty} d(fx_n, Tx_n) = 0$, so that $\{x_n\}$ is asymptotically T-regular with respect to f. However, this property can be deduced directly from the fact that

$$0 \leqslant d(fx_n, Tx_n) \leqslant d(fx_n, fx_{n+1}) \to 0, \text{ as } n \to +\infty.$$

Now, if f and T are occasionally coincidentally idempotent, then $ffw = fw$ for some $w \in C(T, f)$. Then, we have

$$\int_0^{H(Tfw,Tw)} \psi(t)\,dt \leqslant q \int_0^{L(fw,w)} \psi(t)\,dt, \tag{5.92}$$

where

$$L(fw, w) = \max\{d(ffw, fw), d(ffw, Tfw), d(fw, Tw),$$
$$\frac{1}{2}(d(ffw, Tw) + d(fw, Tfw))\}$$

$$= \max\{d(fw, fw), d(fw, Tfw), d(fw, Tw), \frac{1}{2}(d(fw, Tw) + d(fw, Tfw))\}$$

$$= d(fw, Tfw) \leqslant H(Tw, Tfw).$$

If $Tfw \neq Tw$, then from inequality (5.92) and using (5.87) (which is guaranteed by (5.86)), we have that

$$\int_0^{H(Tfw, Tw)} \psi(t)\, dt \leqslant q \int_0^{H(Tfw, Tw)} \psi(t)\, dt < \int_0^{H(Tfw, Tw)} \psi(t)\, dt,$$

which is a contradiction. Hence, $Tfw = Tw$. Thus, we have $fw = ffw$ and $fw \in Tw = Tfw$, i.e. fw is a common fixed point of f and T. ∎

Let Φ denote the family of maps ϕ from the set $\mathbb{R}^+$ of nonnegative real numbers to itself such that

$$\phi(t) \leqslant qt \text{ for all } t \geq 0 \text{ and for some } q \in (0, 1). \tag{5.93}$$

Corollary 5.24 *Let Y be an arbitrary nonempty set, (X, d) be a metric space, $f : Y \to X$ and $T : Y \to CB(X)$ be such that $T(Y) \subseteq f(Y)$,*

$$\int_0^{H(Tx, Ty)} \psi(t)\, dt \leqslant \phi\Big(\int_0^{L(x, y)} \psi(t)\, dt \Big) \tag{5.94}$$

for all x, y in Y, where $\phi \in \Phi$ (satisfying (5.93) for a certain $q \in (0, 1)$),

$$f(Y) \text{ is complete,}$$

$\psi : \mathbb{R}^+ \to \mathbb{R}^+$ is a Lebesgue measurable mapping which is nonnegative, summable on each compact interval, and such that (5.86) holds. Suppose also that (5.89)–(5.91) hold for a certain $\gamma : (1, +\infty) \longrightarrow \mathbb{R}^+$ and q determined by (5.93). Then T and f have a coincidence point. Further, if f and T are occasionally coincidentally idempotent, then f and T have a common fixed point.

Proof It is a consequence of Theorem 5.108 since (5.93) and (5.94) imply that

$$\int_0^{H(Tx, Ty)} \psi(t)\, dt \leqslant \phi\Big(\int_0^{L(x, y)} \psi(t)\, dt \Big) \leqslant q \int_0^{L(x, y)} \psi(t)\, dt,$$

for all x, y in Y and $q \in (0, 1)$. ∎

Remark 5.35 Condition

$$\int_0^{\mu\varepsilon} \psi(t)\,dt \leqslant \mu \int_0^{\varepsilon} \psi(t)\,dt, \qquad \text{for each } \mu > 1, \text{ and } \varepsilon > 0, \qquad (5.95)$$

implies the validity of hypothesis (5.89) in Theorem 5.108 for the particular case of γ the identity mapping. Moreover, for $0 < q < 1$, hypotheses (5.90) and (5.91) are trivially satisfied for this choice of γ. Indeed, using that $0 < q < 1$, we get

$$0 < \gamma(q^{-1/2}) \cdot q = q^{-1/2} \cdot q = q^{1/2} < 1,$$

and

$$\gamma(q^{-1/2}) \cdot q \cdot \gamma \left(\frac{1}{\gamma(q^{-1/2}) \cdot q} \right) = q^{-1/2} q \frac{1}{q^{-1/2} q} = 1.$$

Remark 5.36 Assuming (5.90), condition (5.91) is trivially valid if $\lambda \cdot \gamma(\frac{1}{\lambda}) \leqslant 1$, for every $\lambda \in (0, 1)$, or, equivalently, $\gamma(\frac{1}{\lambda}) \leqslant \frac{1}{\lambda}$, for every $\lambda \in (0, 1)$, that is, $\gamma(z) \leqslant z$, for every $z > 1$. Note that this last condition is trivially valid for γ the identity mapping. Moreover, if $\gamma(z) \leqslant z$, for every $z > 1$, then $\gamma(z) < z^2$, for every $z > 1$ and, therefore, if $q \in (0, 1)$, then $\gamma(q^{-1/2}) < q^{-1}$, obtaining (5.90) if $\gamma(q^{-1/2}) > 0$.

Remark 5.37 According to Remark 5.36, for $q \in (0, 1)$ fixed and ψ satisfying (5.86), an admissible function γ can be obtained by taking

$$\gamma(z) \geq \sup_{\varepsilon > 0} \frac{\int_0^{z\varepsilon} \psi(t)\,dt}{\int_0^{\varepsilon} \psi(t)\,dt}, \ z > 1,$$

provided that $\gamma(q^{-1/2}) > 0$ and $\gamma(z) \leqslant z$, for every $z > 1$.

Example 5.42 Taking ψ as the constant function $\psi(t) = K > 0, t > 0$, in the statement of Theorem 5.108, condition (5.89) is reduced to

$$K\mu\varepsilon \leqslant \gamma(\mu)K\varepsilon, \qquad \text{for each } \mu > 1 \text{ and } \varepsilon > 0,$$

so that we must choose γ as a nonnegative function satisfying that $\gamma(z) = z$ for $z > 1$ (obviously, $\gamma(q^{-1/2}) > 0$ since $q \in (0, 1)$), in order to guarantee conditions (5.89)–(5.91).

Example 5.43 A simple calculation provides that, for function $\psi(t) = t, t > 0$, condition (5.89) is written as $\gamma(z) \geq z^2$ for $z > 1$ and, therefore, in this case, condition (5.90) is never fulfilled. If we take $\psi(t) = Kt^m, t > 0$, for $K > 0$ and $m > 0$ fixed, then (5.89) implies that $\gamma(z) \geq z^{m+1} > z$, for $z > 1$.

Example 5.44 Now, we choose $\psi(t) = Kt^m, t > 0$, where $K > 0$ and $-1 < m < 0$ are fixed. Note that the case $m = 0$ has already been studied in Example 5.42. In this case $-1 < m < 0$, condition (5.89) is reduced to

$$K\frac{(\mu\varepsilon)^{m+1}}{m+1} \leqslant \gamma(\mu)K\frac{\varepsilon^{m+1}}{m+1}, \qquad \text{for } \mu > 1 \text{ and } \varepsilon > 0,$$

which is equivalent to $\gamma(z) \geq z^{m+1}$, for $z > 1$. Note that this inequality implies, for $0 < q < 1$, that $\gamma(q^{-1/2}) > 0$. If we add the hypothesis $\gamma(z) \leqslant z$, for $z > 1$, then we guarantee the validity of conditions (5.90) and (5.91) due to Remark 5.36. Hence, we can take any nonnegative function γ satisfying that

$$z^{m+1} \leqslant \gamma(z) \leqslant z, \qquad \text{for } z > 1.$$

Of course, $\gamma(z) = z$ and $\gamma(z) = z^{m+1}$ are valid choices.

Example 5.45 Take $\psi(t) = e^{-t}, t > 0$. Condition (5.89) is equivalent to

$$1 - e^{-\mu\varepsilon} \leqslant \gamma(\mu)(1 - e^{-\varepsilon}), \qquad \text{for } \mu > 1 \text{ and } \varepsilon > 0,$$

that is,

$$\gamma(\mu) \geq \frac{1 - e^{-\mu\varepsilon}}{1 - e^{-\varepsilon}}, \qquad \text{for } \mu > 1 \text{ and } \varepsilon > 0.$$

Now, for each $z > 1$ fixed, we calculate $\sup\limits_{\varepsilon>0} \dfrac{1 - e^{-z\varepsilon}}{1 - e^{-\varepsilon}}$, which is obviously positive and we check that its value is equal to z.

It is easy to prove that, for $z > 1$ fixed, the function $\varepsilon \in (0, +\infty) \longrightarrow \mathscr{R}_z(\varepsilon) = \dfrac{1 - e^{-z\varepsilon}}{1 - e^{-\varepsilon}}$ is decreasing on $(0, +\infty)$. Indeed, the sign of its derivative coincides with the sign of the function $\nu(\varepsilon) = ze^{-z\varepsilon}(1 - e^{-\varepsilon}) - (1 - e^{-z\varepsilon})e^{-\varepsilon}$ and also with the sign of $\tau(\varepsilon) = ze^{-z\varepsilon}(e^{\varepsilon} - 1) + e^{-z\varepsilon} - 1$, for $\varepsilon \in (0, +\infty)$. Now, function τ is strictly negative on $(0, +\infty)$, since $\tau(0) = \tau(0^+) = 0$ and $\tau'(\varepsilon) = z(1 - z)e^{-z\varepsilon}(e^{\varepsilon} - 1) < 0$, for $\varepsilon > 0$.

Moreover, $\lim_{\varepsilon \to 0^+} \mathscr{R}_z(\varepsilon) = z$, for each $z > 1$, in consequence, $\sup\limits_{\varepsilon>0} \mathscr{R}_z(\varepsilon) = z$, for every $z > 1$. Therefore, if $\gamma(z) \geq z$, for every $z > 1$, then (5.89) follows. Note also that, if $q \in (0, 1)$, then $\gamma(q^{-1/2}) > 0$. Finally, for $q \in (0, 1)$, if we take $\gamma : (1, +\infty) \longrightarrow \mathbb{R}^+$ such that $\gamma(z) = z$, for $z > 1$, we deduce the validity of (5.89)–(5.91).

The following example shows that Theorem 5.108 is a proper generalization of the fixed point results in [420, 449, 459, 460].

Example 5.46 Let $X = \mathbb{R}^+$ be endowed with the Euclidean metric, $f : X \to X$ and $T : X \to CB(X)$ be defined by

$$fx = 4(x^2 + x) \quad \text{and} \quad Tx = [0, x^2 + 7] \; \forall x \in X.$$

Let $\phi : \mathbb{R}^+ \to \mathbb{R}^+$ be defined by $\phi(t) = \frac{1}{4}t$, for all $t \in \mathbb{R}^+$. Then mappings f and T are not commuting and also do not satisfy any of its generalizations, viz. weakly commuting, compatibility, weak compatibility. Also the mappings f and T are non-coincidentally commuting. Note that $f1 \in T1$, but $ff1 \neq f1$ and so f and T are not coincidentally idempotent, but $f0 \in T0$ and $ff0 = f0$ thus f and T are occasionally coincidentally idempotent. For all x and y in X, we have

$$
\int_0^{H(Tx,Ty)} \psi(t)\,dt = \int_0^{|x^2-y^2|} \psi(t)\,dt = \int_0^{(\frac{x+y}{4})\cdot\frac{1}{(x+y+1)}\cdot(4\,|x-y|.(x+y+1))} \psi(t)\,dt
$$

$$
= \int_0^{(\frac{x+y}{4})\cdot\frac{1}{(x+y+1)}\cdot(4\,|x^2-y^2+x-y|)} \psi(t)\,dt \leqslant \int_0^{\frac{1}{4}d(fx,fy)} \psi(t)\,dt
$$

$$
\leqslant \frac{1}{4}\int_0^{d(fx,fy)} \psi(t)\,dt \leqslant \frac{1}{4}\int_0^{L(x,y)} \psi(t)\,dt = \phi\left(\int_0^{L(x,y)} \psi(t)\,dt\right).
$$

Note that these inequalities are valid if

$$
\int_0^{\frac{1}{4}d(fx,fy)} \psi(t)\,dt \leqslant \frac{1}{4}\int_0^{d(fx,fy)} \psi(t)\,dt,
$$

which is satisfied taking, for instance, the constant function $\psi \equiv 1$. On the other hand, γ is chosen as the identity map and it satisfies (5.90) and (5.91).

Note that 0 is a common fixed point of f and T. We remark that the results of [420, 449, 459] and [460] cannot be applied to these mappings f and T.

Theorem 5.109 *In Theorem 5.108, we can assume instead of condition (5.84) one of the inequalities*

$$
\int_0^{H(Tx,Ty)} \psi(t)\,dt \leqslant q\left(a\int_0^{L(x,y)} \psi(t)\,dt + b\int_0^{N(x,y)} \psi(t)\,dt\right), \quad \forall x,y \in Y, \quad (5.96)
$$

or

$$
\int_0^{H(Tx,Ty)} \psi(t)\,dt \leqslant q\int_0^{aL(x,y)+bN(x,y)} \psi(t)\,dt, \quad \forall x,y \in Y, \quad (5.97)
$$

where $a,\,b \geq 0$, $a+b \leqslant 1$ and $q \in (0,1)$.

Similarly, in Corollary 5.24, we can consider one of the contractivity conditions

$$
\int_0^{H(Tx,Ty)} \psi(t)\,dt \leqslant \phi\left(a\int_0^{L(x,y)} \psi(t)\,dt + b\int_0^{N(x,y)} \psi(t)\,dt\right), \quad \forall x,y \in Y, \quad (5.98)
$$

or

$$
\int_0^{H(Tx,Ty)} \psi(t)\,dt \leqslant \phi\left(\int_0^{aL(x,y)+bN(x,y)} \psi(t)\,dt\right), \quad \forall x,y \in Y, \quad (5.99)
$$

where $a,\,b \geq 0$, $a+b \leqslant 1$ and $\phi \in \Phi$ (satisfying (5.93) for a certain $q \in (0,1)$) and the conclusion follows.

Proof It follows from the inequality

$$N(x, y) \leqslant L(x, y), \text{ for every } x, y$$

and the nonnegative character of a, b and ψ. Indeed, $d^2(fx, fy) \leqslant [L(x, y)]^2$,

$$d(fx, Tx) \cdot d(fy, Ty) \leqslant [L(x, y)]^2,$$

$$d(fx, Ty) \cdot d(fy, Tx) \leqslant \frac{1}{4}(d(fx, Ty) + d(fy, Tx))^2 \leqslant [L(x, y)]^2,$$

$$\frac{1}{2}d(fx, Tx) \cdot d(fy, Tx) \leqslant [\max\{d(fx, Tx), \frac{1}{2}(d(fx, Ty) + d(fy, Tx))\}]^2 \leqslant [L(x, y)]^2,$$

$$\frac{1}{2}d(fx, Ty) \cdot d(fy, Ty) \leqslant [\max\{d(fy, Ty), \frac{1}{2}(d(fx, Ty) + d(fy, Tx))\}]^2 \leqslant [L(x, y)]^2,$$

hence, for instance,

$$a \int_0^{L(x,y)} \psi(t)\, dt + b \int_0^{N(x,y)} \psi(t)\, dt \leqslant (a+b) \int_0^{L(x,y)} \psi(t)\, dt \leqslant \int_0^{L(x,y)} \psi(t)\, dt.$$

Note that, in cases (5.98) and (5.99), it is not necessary to assume the nondecreasing character of the function ϕ since, using that $\phi \in \Phi$, we deduce (5.96) and (5.97), respectively. ∎

Of course, function $\phi \equiv 0$ is admissible in the results of this section. Note that, taking $a = 1$ and $b = 0$ in the inequalities of Theorem 5.109, we obtain the corresponding contractivity conditions of Theorem 5.108 and Corollary 5.24. On the other hand, taking $a = 0$ and $b = 1$ in Theorem 5.109, we have the following results, which are also Corollaries of Theorem 5.108.

Corollary 5.25 *Let Y be an arbitrary nonempty set, (X, d) be a metric space, $f : Y \to X$ and $T : Y \to CB(X)$ be such that conditions (5.83), (5.85) hold and*

$$\int_0^{H(Tx,Ty)} \psi(t)\, dt \leqslant q \int_0^{N(x,y)} \psi(t)\, dt, \quad \forall\, x, y \in Y, \qquad (5.100)$$

where $0 < q < 1$ and $\psi : \mathbb{R}^+ \to \mathbb{R}^+$ is a Lebesgue measurable mapping which is nonnegative, summable on each compact interval, and such that (5.86) holds. Assume also that (5.89)–(5.91) are fulfilled for a certain $\gamma : (1, +\infty) \longrightarrow \mathbb{R}^+$. Then f and T have a coincidence point. Further, if f and T are occasionally coincidentally idempotent, then f and T have a common fixed point.

Corollary 5.26 *Let Y be an arbitrary nonempty set, (X, d) be a metric space, f : $Y \to X$ and $T : Y \to CB(X)$ be such that conditions (5.83), (5.85) hold and*

$$\int_0^{H(Tx,Ty)} \psi(t)\, dt \leqslant \phi\Big(\int_0^{N(x,y)} \psi(t)\, dt\Big), \quad \forall\, x, y \in Y, \tag{5.101}$$

where $\phi \in \Phi$ (satisfying (5.93) for $q \in (0, 1)$) and $\psi : \mathbb{R}^+ \to \mathbb{R}^+$ is a Lebesgue measurable mapping which is nonnegative, summable on each compact interval, and such that (5.86) holds. Assume also that (5.89)–(5.91) are fulfilled for a certain $\gamma : (1, +\infty) \longrightarrow \mathbb{R}^+$. Then f and T have a coincidence point. Further, if f and T are occasionally coincidentally idempotent, then f and T have a common fixed point.

Let $\eta : [0, \infty) \to [0, 1)$ be a function having the following property (see, for instance, [42, 279]):

(𝒫) For $t \geq 0$, there exist $\delta(t) > 0$, $s(t) < 1$, such that $0 \leqslant r - t < \delta(t)$ implies $\eta(r) \leqslant s(t)$.

This property obviously holds if η is continuous since η attains its maximum (less than 1) on each compact $[t, t + \delta(t)]$.

Definition 5.59 A sequence $\{x_n\}$ is said to be asymptotically T-regular with respect to f if $\lim_{n\to\infty} d(fx_n, Tx_n) = 0$.

The following theorem is related to the main results of Hu [279, Theorem 2], Jungck [296], Kaneko [306, 307], Nadler [416, Theorem 5] and Beg and Azam [42, Theorem 5.4 and Corollary 5.5].

Theorem 5.110 *Let Y be an arbitrary nonempty set, (X, d) be a metric space, $f : Y \to X$ and $T : Y \to CL(X)$ be such that condition (5.83) holds and*

$$\int_0^{H(Tx,Ty)} \psi(t)\, dt < \eta(d(fx, fy)) \int_0^{d(fx,fy)} \psi(t)\, dt, \tag{5.102}$$

for all $x, y \in Y$, where $\eta : [0, \infty) \to [0, 1)$ satisfies $(\mathscr{P})$ and $\psi \geq 0$ is nonincreasing. Suppose also that Tx is a compact set, for every $x \in Y$. If $f(Y)$ is complete, then
(i) there exists an asymptotically T-regular sequence $\{x_n\}$ with respect to f in Y.
(ii) f and T have a coincidence point.
Further, if f and T are occasionally coincidentally idempotent, then f and T have a common fixed point.

Proof For some x_0 in Y, let $y_0 = fx_0$ and choose x_1 in Y such that $y_1 = fx_1 \in Tx_0$. Then, by (5.102), we have

$$\int_0^{H(Tx_0,Tx_1)} \psi(t)\, dt < \eta(d(fx_0, fx_1)) \int_0^{d(fx_0,fx_1)} \psi(t)\, dt.$$

Using (5.83), we can choose $x_2 \in Y$ such that $y_2 = fx_2 \in Tx_1$ and satisfying that

$$d(y_1, y_2) = d(fx_1, y_2) = d(fx_1, Tx_1) \leqslant H(Tx_0, Tx_1),$$

hence

$$\int_0^{d(y_1, y_2)} \psi(t)\, dt = \int_0^{d(fx_1, fx_2)} \psi(t)\, dt$$

$$< \eta(d(fx_0, fx_1)) \int_0^{d(fx_0, fx_1)} \psi(t)\, dt$$

$$\leqslant \int_0^{d(fx_0, fx_1)} \psi(t)\, dt.$$

Note that, in the previous inequalities, we have used that $d(fx_0, fx_1) > 0$. If $d(fx_0, fx_1) = 0$, then $fx_0 = fx_1 \in Tx_0$ and $\{x_n\}$ is asymptotically T-regular with respect to f.

By induction, we construct a sequence $\{x_n\}$ in Y and $\{y_n\}$ in $f(Y)$, such that, for every n,

$$d(fx_{n-1}, y_n) = d(fx_{n-1}, Tx_{n-1}) = \min_{y \in Tx_{n-1}} d(fx_{n-1}, y) \leqslant H(Tx_{n-2}, Tx_{n-1}),$$

and $y_n = fx_n \in Tx_{n-1}$.

Also, we have

$$\int_0^{d(y_{n+1}, y_{n+2})} \psi(t)\, dt = \int_0^{d(fx_{n+1}, fx_{n+2})} \psi(t)\, dt$$

$$= \int_0^{d(fx_{n+1}, Tx_{n+1})} \psi(t)\, dt \leqslant \int_0^{H(Tx_n, Tx_{n+1})} \psi(t)\, dt$$

$$\leqslant \eta(d(fx_n, fx_{n+1})) \int_0^{d(fx_n, fx_{n+1})} \psi(t)\, dt$$

$$< \int_0^{d(fx_n, fx_{n+1})} \psi(t)\, dt = \int_0^{d(y_n, y_{n+1})} \psi(t)\, dt.$$

It follows that the sequence $\{d(y_n, y_{n+1})\}$ is decreasing and converges to its greatest lower bound, say t. Clearly $t \geq 0$. If $t > 0$, then by property $(\mathscr{P})$ of η, there will exist $\delta(t) > 0$, and $s(t) < 1$ such that

$$0 \leqslant r - t < \delta(t) \text{ implies } \eta(r) \leqslant s(t).$$

For this $\delta(t) > 0$ there exists $N \in \mathbb{N}$ such that $0 \leqslant d(y_n, y_{n+1}) - t < \delta(t)$, whenever $n \geq N$. Hence, $\eta(d(y_n, y_{n+1})) \leqslant s(t)$, whenever $n \geq N$. Let $K = \max\{\eta(d(y_0, y_1)), \eta(d(y_1, y_2)), \ldots, \eta(d(y_{N-1}, y_N)), s(t)\}$. Then for $n = 1, 2, 3, \ldots$, we have

$$\int_0^{d(y_n, y_{n+1})} \psi(t)\, dt < \eta(d(y_{n-1}, y_n)) \int_0^{d(y_{n-1}, y_n)} \psi(t)\, dt$$

$$\leqslant K \int_0^{d(y_{n-1}, y_n)} \psi(t)\, dt$$

$$\leqslant K^n \int_0^{d(y_0, y_1)} \psi(t)\, dt \to 0, \quad \text{as } n \to \infty,$$

which contradicts the assumption that $t > 0$. Thus, $\lim_{n \to \infty} d(y_n, y_{n+1}) = 0$; i.e. $d(fx_n, Tx_n) \to 0$, as $n \to +\infty$. Hence, the sequence $\{x_n\}$ is asymptotically T-regular with respect to f.

We claim that $\{fx_n\}$ is a Cauchy sequence. Let $n, m \in \mathbb{N}$ with $n < m$, then, by the nonincreasing character of ψ, we get

$$\int_0^{d(y_n, y_m)} \psi(t)\, dt \leqslant \int_0^{d(y_n, y_{n+1}) + d(y_{n+1}, y_{n+2}) + \cdots + d(y_{m-1}, y_m)} \psi(t)\, dt$$

$$= \int_0^{d(y_n, y_{n+1})} \psi(t)\, dt + \int_{d(y_n, y_{n+1})}^{d(y_n, y_{n+1}) + d(y_{n+1}, y_{n+2})} \psi(t)\, dt$$

$$+ \cdots + \int_{d(y_n, y_{n+1}) + d(y_{n+1}, y_{n+2}) + \cdots + d(y_{m-2}, y_{m-1})}^{d(y_n, y_{n+1}) + d(y_{n+1}, y_{n+2}) + \cdots + d(y_{m-1}, y_m)} \psi(t)\, dt$$

$$\leqslant \int_0^{d(y_n, y_{n+1})} \psi(t)\, dt + \int_0^{d(y_{n+1}, y_{n+2})} \psi(t)\, dt$$

$$+ \cdots + \int_0^{d(y_{m-1}, y_m)} \psi(t)\, dt$$

$$= \sum_{i=n}^{m-1} \int_0^{d(y_i, y_{i+1})} \psi(t)\, dt.$$

Now, we recall that

$$\int_0^{d(y_{n+1}, y_{n+2})} \psi(t)\, dt \leqslant \eta(d(y_n, y_{n+1})) \int_0^{d(y_n, y_{n+1})} \psi(t)\, dt,$$

for every n, which implies that

$$\int_0^{d(y_{n+2},y_{n+3})} \psi(t)\,dt \leqslant \eta(d(y_{n+1},y_{n+2}))\int_0^{d(y_{n+1},y_{n+2})} \psi(t)\,dt$$

$$\leqslant \eta(d(y_{n+1},y_{n+2}))\eta(d(y_n,y_{n+1}))\int_0^{d(y_n,y_{n+1})} \psi(t)\,dt.$$

Following this procedure, we prove that

$$\int_0^{d(y_j,y_{j+1})} \psi(t)\,dt \leqslant \prod_{i=n}^{j-1} \eta(d(y_i,y_{i+1}))\int_0^{d(y_n,y_{n+1})} \psi(t)\,dt, \text{ for every } j = n+1,\ldots,m-1.$$

Therefore,

$$\int_0^{d(y_n,y_m)} \psi(t)\,dt \leqslant \sum_{i=n}^{m-1} \int_0^{d(y_i,y_{i+1})} \psi(t)\,dt$$

$$= \int_0^{d(y_n,y_{n+1})} \psi(t)\,dt + \sum_{i=n+1}^{m-1} \int_0^{d(y_i,y_{i+1})} \psi(t)\,dt$$

$$\leqslant \left[1 + \sum_{i=n+1}^{m-1} \prod_{l=n}^{i-1} \eta(d(y_l,y_{l+1}))\right]\int_0^{d(y_n,y_{n+1})} \psi(t)\,dt.$$

We check that the right-hand side in the last inequality tends to 0 as $n, m \to +\infty$. Since $\int_0^{d(y_n,y_{n+1})} \psi(t)\,dt \to 0$, as $n \to +\infty$, it suffices to show that

$$\sum_{i=n+1}^{m-1} \prod_{l=n}^{i-1} \eta(d(y_l,y_{l+1})) \text{ is bounded (uniformly on } n, m).$$

Indeed, we check that $\sum_{i=n+1}^{m-1}\prod_{l=n}^{i-1}\eta(z_l)$ is bounded, for any sequence $\{z_l\}$ with nonnegative terms and tending to 0 as $l \to +\infty$, using the property $(\mathscr{P})$ of the function η. Given $t = 0$, by $(\mathscr{P})$, there exist $\delta(0) > 0$, $s_0 < 1$, such that $0 \leqslant r < \delta(0)$ implies $\eta(r) \leqslant s_0$. Since $\{z_l\} \to 0$, given $\delta(0) > 0$, there exists $l_0 \in \mathbb{N}$ such that, for every $l \geq l_0$, we have $0 \leqslant z_l < \delta(0)$. This implies that $\eta(z_l) \leqslant s_0$, for every $l \geq l_0$.

In consequence, for $n \geq l_0$, we get

$$0 \leqslant \sum_{i=n+1}^{m-1} \prod_{l=n}^{i-1} \eta(z_l) \leqslant \sum_{i=n+1}^{m-1} \prod_{l=n}^{i-1} s_0 = \sum_{i=n+1}^{m-1} (s_0)^{i-n} = \frac{s_0 - (s_0)^{m-n}}{1 - s_0} < \frac{s_0}{1 - s_0},$$

and this expression is bounded independently of m, n.

Hence, $\{fx_n\}$ is a Cauchy sequence in $f(Y)$. Since $f(Y)$ is complete, $\{fx_n\}$ converges to some p in $f(Y)$. Let $z \in f^{-1}(p)$. Then $fz = p$. Next, we have

$$\int_0^{d(fz,Tz)} \psi(t)\,dt \leqslant \int_0^{d(fx_{n+1},fz)+d(fx_{n+1},Tz)} \psi(t)\,dt$$

$$= \int_0^{d(fx_{n+1},Tz)} \psi(t)\,dt + \int_{d(fx_{n+1},Tz)}^{d(fx_{n+1},Tz)+d(fz,fx_{n+1})} \psi(t)\,dt$$

$$\leqslant \int_0^{d(fx_{n+1},Tz)} \psi(t)\,dt + \int_0^{d(fz,fx_{n+1})} \psi(t)\,dt$$

$$\leqslant \int_0^{H(Tx_n,Tz)} \psi(t)\,dt + \int_0^{d(fz,fx_{n+1})} \psi(t)\,dt$$

$$\leqslant \eta(d(fx_n,fz)) \int_0^{d(fx_n,fz)} \psi(t)\,dt + \int_0^{d(fz,fx_{n+1})} \psi(t)\,dt.$$

Letting $n \to \infty$, we get $\int_0^{d(fz,Tz)} \psi(t)\,dt \leqslant 0$. Thus, we have $d(fz, Tz) = 0$. Hence, $fz \in Tz$.

Now, if f and T are occasionally coincidentally idempotent, then $ffw = fw$ for some $w \in C(T, f)$. Then we have

$$\int_0^{H(Tfw,Tw)} \psi(t)\,dt \leqslant \eta(d(ffw, fw)) \int_0^{d(ffw,fw)} \psi(t)\,dt = 0.$$

Thus, $Tfw = Tw$. It follows that $ffw = fw \in Tw = Tfw$. Hence, fw is a common fixed point of T and f. ∎

Now we state some fixed point theorems for Kannan type multivalued mappings which extend and generalize the corresponding results of Shiau, Tan and Wong [553] and Beg and Azam [41, 42]. A proper blend of the proof of Theorem 5.108 and those of [460, Theorem 6, Theorem 7, Theorem 8 respectively] and [459, Theorems 3.1, 3.2, 3.3] will complete the proof.

Theorem 5.111 *Let Y be an arbitrary nonempty set, (X, d) be a metric space, $f : Y \to X$ and $T : Y \to CB(X)$ be such that (5.83) holds and*

$$\int_0^{H^r(Tx,Ty)} \psi(t)\,dt$$

$$\leqslant \alpha_1(d(fx, Tx)) \int_0^{d^r(fx,Tx)} \psi(t)\,dt + \alpha_2(d(fy, Ty)) \int_0^{d^r(fy,Ty)} \psi(t)\,dt,$$

$$(5.103)$$

for all $x, y \in Y$, where $\alpha_i : \mathbb{R}^+ \to [0, 1)(i = 1, 2)$ are bounded on bounded sets, r is some fixed positive real number and $\psi : \mathbb{R}_+ \to \mathbb{R}^+$ is a Lebesgue measurable mapping which is summable on each compact interval, and $\int_0^\varepsilon \psi(t)\,dt > 0$, for each $\varepsilon > 0$. Suppose that there exists an asymptotically T-regular sequence $\{x_n\}$ with respect to f in Y. If $T(Y)$ is complete or

$$\exists k \in \mathbb{N} \text{ such that } fx_{n+k} \in Tx_n, \text{ for every } n \in \mathbb{N}, \text{ and } f(Y) \text{ is complete,} \quad (5.104)$$

then f and T have a coincidence point. Further, if f and T are occasionally coincidentally idempotent, then f and T have a common fixed point.

Proof By hypotheses,

$$\int_0^{H^r(Tx_n,Tx_m)} \psi(t)\,dt$$

$$\leqslant \alpha_1(d(fx_n,Tx_n)) \int_0^{d^r(fx_n,Tx_n)} \psi(t)\,dt + \alpha_2(d(fx_m,Tx_m)) \int_0^{d^r(fx_m,Tx_m)} \psi(t)\,dt.$$

Since $\{x_n\}$ is asymptotically T-regular with respect to f in Y, then $\{\alpha_1(d(fx_n,Tx_n))\}_n$ and $\{\alpha_2(d(fx_m,Tx_m))\}_m$ are bounded sequences and

$$\int_0^{d^r(fx_n,Tx_n)} \psi(t)\,dt \to 0, \quad \int_0^{d^r(fx_m,Tx_m)} \psi(t)\,dt \to 0, \quad \text{as } n, m \to +\infty.$$

This provides the property $H(Tx_n, Tx_m) \to 0$ as $n, m \to +\infty$, so that $\{Tx_n\}$ is a Cauchy sequence in $(CB(X), H)$.

If $T(Y)$ is complete, there exists $K^* \in T(Y) \subseteq f(Y)$ such that $H(Tx_n, K^*) \to 0$ as $n \to +\infty$. Let $u \in Y$ be such that $f(u) \in K^*$. Then

$$\int_0^{d^r(fu,Tu)} \psi(t)\,dt$$

$$\leqslant \int_0^{H^r(K^*,Tu)} \psi(t)\,dt$$

$$\leqslant \int_0^{(H(K^*,Tx_n)+H(Tx_n,Tu))^r} \psi(t)\,dt$$

$$= \int_0^{H^r(Tx_n,Tu)} \psi(t)\,dt + \int_{H^r(Tx_n,Tu)}^{H^r(Tx_n,Tu)+\text{ terms containing } H(K^*,Tx_n)} \psi(t)\,dt$$

$$\leqslant \alpha_1(d(fx_n,Tx_n)) \int_0^{d^r(fx_n,Tx_n)} \psi(t)\,dt + \alpha_2(d(fu,Tu)) \int_0^{d^r(fu,Tu)} \psi(t)\,dt$$

$$+ \int_{H^r(Tx_n,Tu)}^{H^r(Tx_n,Tu)+\text{ terms containing } H(K^*,Tx_n)} \psi(t)\,dt,$$

where the number of terms containing $H(K^*, Tx_n)$ is a finite number depending on r, and therefore fixed. Calculating the limit as $n \to +\infty$, and taking into account that the length of the intervals in the last integral tends to zero, we get

$$\int_0^{d^r(fu,Tu)} \psi(t)\,dt(1 - \alpha_2(d(fu,Tu))) \leqslant \lim_{n\to+\infty} \alpha_1(d(fx_n,Tx_n)) \int_0^{d^r(fx_n,Tx_n)} \psi(t)\,dt$$

$$= 0.$$

Therefore,

$$\int_0^{d^r(fu,Tu)} \psi(t)\,dt \leqslant 0$$

and, by the properties of ψ, we get $d^r(fu, Tu) = 0$, which implies that $fu \in Tu$ and u is a coincidence point.

Now, suppose that $f(Y)$ is complete. Note that Tx_n is closed and bounded for every $n \in \mathbb{N}$. Take $k > 1$ fixed. By the results in [416], we can affirm that for every $y_1 \in Tx_n$, there exists $y_2 \in Tx_m$ such that $d(y_1, y_2) \leqslant kH(Tx_n, Tx_m)$.

Given $n, m \in \mathbb{N}$, we choose $y_1 \in Tx_n$ and, for this $y_1 \in Tx_n$ fixed, we choose $y_2 \in Tx_m$ such that $d(y_1, y_2) \leqslant kH(Tx_n, Tx_m)$. Then

$$
\begin{aligned}
d(fx_n, fx_m) &\leqslant d(fx_n, y_1) + d(y_1, y_2) + d(y_2, fx_m) \\
&\leqslant d(fx_n, Tx_n) + kH(Tx_n, Tx_m) + d(Tx_m, fx_m).
\end{aligned}
$$

By the hypothesis on $\{x_n\}$ and the Cauchy character of $\{Tx_n\}$, we deduce that $\{fx_n\}$ is a Cauchy sequence. Since $f(Y)$ is complete, there exists $f(u) \in f(Y)$ such that $\{f(x_n)\} \to f(u)$. By hypotheses, $d(fx_{n+k}, Tu) \leqslant H(Tx_n, Tu)$, for every n, hence

$$
\begin{aligned}
\int_0^{d^r(fx_{n+k},Tu)} \psi(t)\,dt &\leqslant \int_0^{H^r(Tx_n,Tu)} \psi(t)\,dt \\
&\leqslant \alpha_1(d(fx_n, Tx_n)) \int_0^{d^r(fx_n,Tx_n)} \psi(t)\,dt \\
&\quad + \alpha_2(d(fu, Tu)) \int_0^{d^r(fu,Tu)} \psi(t)\,dt,
\end{aligned}
$$

and taking the limit as $n \to +\infty$, we get

$$\int_0^{d^r(fu,Tu)} \psi(t)\,dt \leqslant \alpha_2(d(fu, Tu)) \int_0^{d^r(fu,Tu)} \psi(t)\,dt.$$

In this case,

$$(1 - \alpha_2(d(fu, Tu))) \int_0^{d^r(fu,Tu)} \psi(t)\,dt \leqslant 0$$

and $d(fu, Tu) = 0$, which implies that $fu \in Tu$. Now, if f and T are coincidentally idempotent, then $ffw = fw$ for some $w \in C(T, f)$. Hence,

$$\int_0^{H^r(Tfw,Tw)} \psi(t)\,dt$$

$$\leqslant \alpha_1(d(ffw,Tfw)) \int_0^{d^r(ffw,Tfw)} \psi(t)\,dt + \alpha_2(d(fw,Tw)) \int_0^{d^r(fw,Tw)} \psi(t)\,dt$$

$$= \alpha_1(d(fw,Tfw)) \int_0^{d^r(fw,Tfw)} \psi(t)\,dt.$$

Since $ffw = fw \in Tw$, we get

$$\int_0^{d^r(fw,Tfw)} \psi(t)\,dt \leqslant \int_0^{H^r(Tw,Tfw)} \psi(t)\,dt \leqslant \alpha_1(d(fw,Tfw)) \int_0^{d^r(fw,Tfw)} \psi(t)\,dt.$$

Therefore

$$\int_0^{d^r(fw,Tfw)} \psi(t)\,dt\,(1 - \alpha_1(d(fw,Tfw))) \leqslant 0,$$

obtaining $d(fw,Tfw) = 0$ and $fw \in Tfw$. Since $0 \leqslant \int_0^{H^r(Tfw,Tw)} \psi(t)\,dt \leqslant 0$, we deduce that $H(Tfw,Tw) = 0$ and $Tfw = Tw$. In consequence, $ffw = fw \in Tw = Tfw$ and fw is a common fixed point of T and f. ∎

Remark 5.38 In the statement of Theorem 5.111, condition (5.104) can be replaced by the more general one

$$f(Y) \text{ is complete.}$$

To complete the proof with this more general hypothesis, take into account that, for $y \in Y$, $T(y)$ is a closed set in X and $T(Y) \subseteq f(Y)$. Using that $f(Y)$ is complete, we deduce that $(CL(f(Y)), H)$ is complete. Hence, $\{Tx_n\}$ is a sequence in $CL(f(Y))$ and it is a Cauchy sequence in $(CL(f(Y)), H)$. Therefore, there exists $K^* \in CL(f(Y))$ such that $H(Tx_n, K^*) \to 0$ as $n \to +\infty$. Note also that K^* is a closed set in the complete space $f(Y)$, then K^* is complete and, therefore, a closed set, then $K^* \in CL(X)$. Once we have proved that $H(Tx_n, K^*) \to 0$ as $n \to +\infty$ in $(CL(f(Y)), H)$, the proof follows analogously.

Theorem 5.112 *In addition to the hypotheses of Theorem 5.111, suppose that Tx_n is compact, for all $n \in \mathbb{N}$. If $f(z)$ is a cluster point of $\{fx_n\}$, then z is a coincidence point of f and T.*

Proof Let $y_n \in Tx_n$ be such that $d(fx_n, y_n) = d(fx_n, Tx_n) \to 0$, this is possible since Tx_n is compact. It is obvious that a cluster point of $\{fx_n\}$ is a cluster point of $\{y_n\}$. Let $f(z)$ be a cluster point of $\{fx_n\}$ and $\{y_n\}$, then we check that $fz \in Tu$, where u is obtained in the proof of Theorem 5.111. Note that, for every $y \in Tu$,

$$d(fz, y) \leqslant d(fz, fx_n) + d(fx_n, y_n) + d(y_n, y) = d(fz, fx_n) + d(fx_n, Tx_n) + d(y_n, y),$$

hence

$$d(fz, Tu) = \inf_{y \in Tu} d(fz, y) \leqslant d(fz, fx_n) + d(fx_n, Tx_n) + \inf_{y \in Tu} d(y_n, y)$$
$$= d(fz, fx_n) + d(fx_n, Tx_n) + d(y_n, Tu)$$
$$\leqslant d(fz, fx_n) + d(fx_n, Tx_n) + H(Tx_n, Tu).$$

In consequence,

$$\int_0^{d^r(fz,Tu)} \psi(t)\, dt \leqslant \int_0^{(d(fz,fx_n)+d(fx_n,Tx_n)+H(Tx_n,Tu))^r} \psi(t)\, dt.$$

Using this, there exists a subsequence fx_{n_k} converging to fz, the properties of $\{x_n\}$, and the inequality

$$\int_0^{H^r(Tx_{n_k},Tu)} \psi(t)\, dt$$
$$\leqslant \alpha_1(d(fx_{n_k}, Tx_{n_k})) \int_0^{d^r(fx_{n_k},Tx_{n_k})} \psi(t)\, dt + \alpha_2(d(fu, Tu)) \int_0^{d^r(fu,Tu)} \psi(t)\, dt$$
$$\xrightarrow{k \to +\infty} 0,$$

then, taking the limit when $n_k \to +\infty$, we get $\int_0^{d^r(fz,Tu)} \psi(t)\, dt \leqslant 0$, and $fz \in Tu$. To prove that $fz \in Tz$, using that $fu \in Tu$, we get

$$\int_0^{d^r(fz,Tz)} \psi(t)\, dt \leqslant \int_0^{H^r(Tu,Tz)} \psi(t)\, dt$$
$$\leqslant \alpha_1(d(fu, Tu)) \int_0^{d^r(fu,Tu)} \psi(t)\, dt + \alpha_2(d(fz, Tz)) \int_0^{d^r(fz,Tz)} \psi(t)\, dt$$
$$= \alpha_2(d(fz, Tz)) \int_0^{d^r(fz,Tz)} \psi(t)\, dt.$$

This implies that

$$(1 - \alpha_2(d(fz, Tz))) \int_0^{d^r(fz,Tz)} \psi(t)\, dt \leqslant 0$$

and, by the properties of α_2 and ψ, we deduce that $d(fz, Tz) = 0$, which proves that z is a coincidence point of f and T. ∎

The following result extends [460, Theorem 3.3].

Theorem 5.113 *Let Y be an arbitrary nonempty set, (X, d) be a metric space, $f : Y \to X$ and $T : Y \to CB(X)$ be such that (5.83) and (5.103) hold, where $\alpha_i : \mathbb{R}^+ \to [0, 1)(i = 1, 2)$ are bounded on bounded sets, and such that*

$$\alpha_1(d(fx, Tx)) + \alpha_2(d(fy, Ty)) \leqslant 1, \text{ for every } x, y,$$

r is some fixed positive real number and $\psi : \mathbb{R}^+ \to \mathbb{R}^+$ *is a Lebesgue measurable mapping which is summable on each compact interval, and* $\psi(x) > 0$, *for each* $x > 0$. *Suppose that*

$$\inf\{d(fz_n, Tz_n) : n \in \mathbb{N}\} = 0, \text{ for every sequence } \{z_n\} \text{ in } Y$$

with $fz_n \in Tz_{n-1}, \forall n.$ (5.105)

If $T(Y)$ *is complete or* $f(Y)$ *is complete, then* f *and* T *have a coincidence point. Further, if* f *and* T *are occasionally coincidentally idempotent, then* f *and* T *have a common fixed point.*

Proof Using Theorem 5.111, it suffices to prove that there exists an asymptotically T-regular sequence $\{x_n\}$ with respect to f in Y. Let $x_0 \in Y$ and take $\{x_n\}$ in Y such that $fx_n \in Tx_{n-1}$, for every $n \in \mathbb{N}$. Then

$$\int_0^{d^r(fx_n, Tx_n)} \psi(t)\,dt \leqslant \int_0^{H^r(Tx_{n-1}, Tx_n)} \psi(t)\,dt$$

$$\leqslant \alpha_1(d(fx_{n-1}, Tx_{n-1})) \int_0^{d^r(fx_{n-1}, Tx_{n-1})} \psi(t)\,dt$$

$$+ \alpha_2(d(fx_n, Tx_n)) \int_0^{d^r(fx_n, Tx_n)} \psi(t)\,dt.$$

Hence,

$$(1 - \alpha_2(d(fx_n, Tx_n))) \int_0^{d^r(fx_n, Tx_n)} \psi(t)\,dt$$

$$\leqslant \alpha_1(d(fx_{n-1}, Tx_{n-1})) \int_0^{d^r(fx_{n-1}, Tx_{n-1})} \psi(t)\,dt,$$

or also, using the hypothesis on α_1 and α_2,

$$\int_0^{d^r(fx_n, Tx_n)} \psi(t)\,dt \leqslant \frac{\alpha_1(d(fx_{n-1}, Tx_{n-1}))}{(1 - \alpha_2(d(fx_n, Tx_n)))} \int_0^{d^r(fx_{n-1}, Tx_{n-1})} \psi(t)\,dt$$

$$\leqslant \int_0^{d^r(fx_{n-1}, Tx_{n-1})} \psi(t)\,dt.$$

The properties of ψ imply that $d^r(fx_n, Tx_n) \leqslant d^r(fx_{n-1}, Tx_{n-1})$, for every $n \in \mathbb{N}$, and $\{d(fx_n, Tx_n)\}_{n \in \mathbb{N}}$ is nonincreasing and bounded below, therefore it is convergent to the infimum, that is

$$d(fx_n, Tx_n) \to \inf\{d(fx_n, Tx_n) : n \in \mathbb{N}\} = 0,$$

and $\{x_n\}$ is asymptotically T-regular with respect to f in Y.   ∎

Remark 5.39 Note that condition (5.105) in Theorem 5.113 cannot be replaced by

$$\inf\{d(fx, Tx) : x \in Y\} = 0,$$

since the infimum taking the sequence $\{z_n\}$ could be positive (we calculate the infimum in a smaller set).

Remark 5.40 In Theorem 5.113, condition (5.105) can be replaced by the following:

$$\inf\{H(Tz_{n-1}, Tz_n) : n \in \mathbb{N}\} = 0, \text{ for every sequence } \{z_n\} \text{ in } Y$$
$$\text{with } fz_n \in Tz_{n-1}, \forall n. \tag{5.106}$$

Indeed, since
$$d(fz_n, Tz_n) \leqslant H(Tz_{n-1}, Tz_n), \forall n,$$

then

$$0 \leqslant \inf\{d(fz_n, Tz_n) : n \in \mathbb{N}\} \leqslant \inf\{H(Tz_{n-1}, Tz_n) : n \in \mathbb{N}\} = 0,$$

and $d(fz_n, Tz_n) \to 0$.

Remark 5.41 In Theorem 5.113, if we are able to obtain a sequence $\{x_n\}$ with an infinite number of terms which are different, then we can relax condition (5.105) to the following:

$$\inf\{d(fx, Tx) : x \in B\} = 0, \text{ for every infinite set } B \text{ of } Y. \tag{5.107}$$

5.4.4 Common Fixed Points of Asymptotically I-Contractive Mappings

In this section, we extend and generalize a famous result by Browder [96] Göhde [251] and Kirk [326] recently extended by Luc in [376] and a recent result of Penot [472] by using the notion of asymptotically I-compact subset of a Banach space. However, it may be remarked that here no compactness assumption is involved. Instead we use asymptotic contractiveness concepts; a comparison of this concept with other notions of asymptotic conditions (e.g. uniform asymptotic introduced in [471] and asymptotic contractiveness for single map introduced in [376]) will be made later on. Note that the meaning of the word "asymptotic" is subtle. Indeed, the word "asymptotic" is not related to the iterations of the map, but refers to the behaviour of the map at infinity.

Recall that a subset C of a linear space X is star-shaped with respect to q (or, briefly, star-shaped) if there exists a $q \in C$ such that

$$kx + (1 - k)q \in C$$

for any $k \in [0, 1]$ and $x \in C$. Of course, if C is convex, then it is star-shaped with respect to any $q \in C$. Here q is called the star center of C.

Definition 5.60 Let C be a subset of a linear space X and let $T, I : C \to X$. Then C is said to be (I, T)-star-shaped with respect to q (or, briefly, (I, T)-star-shaped) if there exists a $q \in C$ such that

$$kT(x) + (1 - k)q \in I(C)$$

for any $k \in [0, 1]$ and $x \in C$.

If I and T both are identity maps on C, then the definition of (I, T)-star-shaped reduces to the ordinary definition of star-shaped.

Definition 5.61 Let X be a Banach space and let C be a subset of X. Let $T, I : C \to X$. Let C be an (I, T)-star-shaped subset of X. We say that T is asymptotically I-contractive on C if, for some $q \in C$,

$$\lim_{n \to \infty} \sup_{x \in C, \, \|x\| > n, \, \|I(x)\| > n} \frac{\|T(x) - T(q)\|}{\|I(x) - I(q)\|} < 1. \qquad (5.108)$$

Note that this condition is independent of the choice of $q \in C$. To see this, let us consider $q' \in C$ such that $q' \neq q$, then

$$\lim_{n \to \infty} \sup_{x \in C, \, \|x\| > n, \, \|I(x)\| > n} \frac{\|T(x) - T(q')\|}{\|I(x) - I(q')\|}$$

$$\leqslant \lim_{n \to \infty} \sup_{x \in C, \, \|x\| > n, \, \|I(x)\| > n} \left[\frac{\|T(x) - T(q)\| + \|T(q')\|}{\|I(x) - I(q)\|} \cdot \frac{\|I(x) - I(q)\|}{\|I(x) - I(q')\|} \right]$$

$$\doteq \lim_{n \to \infty} \sup_{x \in C, \, \|x\| > n, \, \|I(x)\| > n} \left[\left\{ \frac{\|T(x) - T(q)\|}{\|I(x) - I(q)\|} + \frac{\|T(q')\|}{\|I(x) - I(q)\|} \right\} \cdot \frac{\|I(x) - I(q)\|}{\|I(x) - I(q')\|} \right]$$

$$< 1.$$

If I is the identity map on C, then T is said to be asymptotically contractive on C if, for some $q \in C$,

$$\limsup_{x \in C, \, \|x\| \to \infty} \frac{\|T(x) - T(q)\|}{\|x - q\|} < 1. \qquad (5.109)$$

It may be observed that the notion of asymptotically I-contractive map enables us to extend to unbounded sets the result of [96, 246, 296] valid for I-nonexpansive self-mappings on closed star-shaped bounded subsets of uniformly convex Banach spaces. Note that every convex subset of a Banach space is star-shaped but the converse need not be true. For example, one may observe that $C = \{(x, 0) : x \in [0, \infty)\} \bigcup \{(0, y) : y \in [0, \infty)\}$ is a star-shaped subset of $\mathbb{R}^2$ with respect to $(0, 0)$, but it is neither bounded nor convex. Define $T, I : C \to X$ by $T(x, 0) = (\frac{x}{2}, 0)$, if $x \in [0, 1]$, $T(x, 0) = (0, 0)$ if $x > 1$ and $T(0, y) \doteq (0, 0)$ if $y \geq 0$; $I(x, 0) = (\frac{x}{2}, 0)$, if $x \in [0, 1]$, $I(x, 0) = (0, 0)$ if $x > 1$ and $I(0, y) = (\frac{y}{2}, 0)$, if $y \in [0, 1]$, $I(0, y) = (0, 0)$ if $y > 1$. Clearly, C is (I, T)-star-shaped with respect to $q = (0, 0)$. Observe that $I(C)$ is bounded, closed and convex.

Recall that a mapping T is I-nonexpansive in C, if $\|T(x) - T(y)\| \leq \|Ix - Iy\|$ $\forall x, y \in C$.

Definition 5.62 Let X be a Banach space and let C be a subset of X. Let $T, I : C \to X$. Let C be an (I, T)-star-shaped subset of X with respect to some $q \in C$(or, briefly, (I, T)-star-shaped). Then T is said to be radially asymptotically I-contractive with respect to some $q \in C$ in the sense that for any unit vector u in the asymptotic cone (or horizon cone)

$$C_\infty := \limsup_{t \to \infty} t^{-1} C := \{v \in X : \exists (t_n) \to \infty, (v_n) \to v, t_n v_n \in C \ \forall n \in \mathbb{N}\}$$

of C one has

$$\limsup_{t \to \infty, \, q+tu \in C} \frac{\|T(q + tu) - T(q)\|}{\|I(q + tu) - I(q)\|} < 1.$$

If I is the identity map on C, then the above inequality reduces to

$$\limsup_{t \to \infty, \, q+tu \in C} \frac{1}{t} \|T(q + tu) - T(q)\| < 1.$$

In such case, T is said to be radially asymptotically contractive with respect to some $q \in C$.

Recall that two mappings $T : C \to X$ and $I : C \to X$ are said to be weakly compatible in C if $TI(v) = IT(v)$ whenever $T(v) = I(v)$ for some v in C. We now prove the following variant of the main result of Jungck [296].

Proposition 5.8 *Let C be a subset of a Banach space $(X, \| \cdot \|)$ and let $T, I : C \to X$ be two nonself-maps satisfying the inequality*

$$\|Tx - Ty\| \leq \lambda \|Ix - Iy\| \ \forall x, y \in C, \tag{5.110}$$

where $0 < \lambda < 1$. If $T(C) \subset I(C)$ and $I(C)$ is closed, then T and I have a coincidence point v in C. Further, if T and I are weakly compatible in C, then T and I have a unique common fixed point.

Proof Let $x_0 \in C$ be arbitrary. Since $Tx_0 \in I(C)$, there is some $x_1 \in C$ such that $Ix_1 = Tx_0$. Then choose $x_2 \in C$ such that $Ix_2 = Tx_1$. In general, after having chosen

$x_n \in C$ we choose $x_{n+1} \in C$ such that $Ix_{n+1} = Tx_n$. We now show that $\{Ix_n\}$ is a Cauchy sequence. From (5.110) we have

$$\|Ix_n - Ix_{n+1}\| = \|Tx_{n-1} - Tx_n\| \leqslant \lambda \|Ix_{n-1} - Ix_n\|.$$

Repeating the above argument n-times, we get

$$\|Ix_n - Ix_{n+1}\| \leqslant \lambda \|Ix_{n-1} - Ix_n\| \leqslant \cdots \leqslant \lambda^n \|Ix_0 - Ix_1\|.$$

It then follows that, for any $m > n$,

$$\|Ix_n - Ix_m\| \leqslant \frac{\lambda^n}{1 - \lambda} \|Ix_0 - Ix_1\| \to 0 \text{ as } m > n \to \infty.$$

Thus, $\{Ix_n\}$ is a Cauchy sequence. Since $I(C)$ is closed in X and so complete, there is some $u \in I(C)$ such that

$$\lim_{n \to \infty} Ix_n = \lim_{n \to \infty} Tx_{n-1} = u.$$

Since $u \in I(C)$, there exists a $v \in C$ such that $Iv = u$. From (5.110) we get

$$\|Tx_n - Tv\| \leqslant \lambda \|Ix_n - Iv\|.$$

Taking the limit as $n \to \infty$ we obtain

$$\|u - Tv\| \leqslant \lambda \|u - Iv\|.$$

Hence, $Tv = u$; i.e. v is a coincidence point of T and I. Since T and I are weakly compatible, they commute at v. Using (5.110),

$$\|TIv - Tv\| \leqslant \lambda \|I^2 v - Iv\| = \lambda \|ITv - Tv\| = \lambda \|TIv - Tv\|,$$

which implies that $TIv = Tv = Iv$. It then follows that $Iv = TIv = ITv$, and Iv is a common fixed point of T and I. The uniqueness of the common fixed point Iv follows from (5.110).   ∎

Following essentially the same reasoning as in Propositions 10.1 and 10.2 in Goebel and Kirk [247], we can easily prove the following Propositions 5.9 and 5.10, respectively.

Proposition 5.9 *Suppose C is a subset of a uniformly convex Banach space X and suppose $T : C \to X$ and $I : C \to X$ are two nonself-maps such that mapping T is I-nonexpansive in C and $I(C)$ is bounded and convex. Then for $\{u_n\}, \{v_n\}, \{z_n\}$ in C and $Iz_n = \frac{1}{2}(Iu_n + Iv_n)$,*

$$\lim_{n \to \infty} \|Iu_n - Tu_n\| = 0, \quad \lim_{n \to \infty} \|Iv_n - Tv_n\| = 0 \implies \lim_{n \to \infty} \|Iz_n - Tz_n\| = 0.$$

Proposition 5.10 *Suppose C is a subset of a uniformly convex Banach space X and suppose $T : C \to X$ and $I : C \to X$ are two nonself-maps such that T is I-nonexpansive in C, $I(C)$ is bounded, closed and convex and satisfy $\inf\{\|Ix - Tx\| : x \in C\} = 0$. Then T and I have a coincidence point in C.*

The following proposition is an easy consequence of Propositions 5.9 and 5.10.

Proposition 5.11 *Let X be a uniformly convex Banach space, C a subset of X, $T : C \to X$ and $I : C \to X$ be two nonself-maps such that mapping T is I-nonexpansive in C, $I(C)$ is bounded, closed and convex subset of X. Then the mapping $f = I - T$ is demiclosed on X.*

In [472], Penot prove the following result.

Proposition 5.12 *Let X be a uniformly convex Banach space and let C be a closed and convex subset of X. Let $T : C \to X$ be a nonexpansive map which is asymptotically contractive on C and such that $T(C) \subset C$. Then T has a fixed point.*

In 2007, Pathak, Rhoades and Khan [462] extended and generalized the above result of Penot [472] for a pair of maps in the following.

Theorem 5.114 *Let X be a uniformly convex Banach space, C a subset of X. Let $T, I : C \to X$ and assume that C an (I, T)-star-shaped subset of X. Let T be an I-nonexpansive map which is asymptotically I-contractive in C and such that $kT(C) + (1 - k)C \subset I(C)$ for any $k \in [0, 1]$, $I(C)$ is bounded, closed and convex and I is continuous. Then T and I have a coincidence point $\bar{x}$ in C. Further, if $I^2\bar{x} = I\bar{x}$, and T and I are weakly compatible in C, then T and I have a unique common fixed point.*

Proof Let (t_n) be a sequence in $(0, 1)$ with limit 0 and let C be an (I, T)-star-shaped subset of X with respect to some $q \in C$. For $n \in \mathbb{N}$, define $T_n : C \to X$ by

$$T_n(x) := (1 - t_n)T(x) + t_n q \tag{5.111}$$

so that, by the (I, T)-star-shaped property of C, $T_n(x) \in I(C)$ for each $x \in C$. Since T is I-nonexpansive, it follows that

$$\|T_n(x) - T_n(y)\| = (1 - t_n)\|T(x) - T(y)\|$$
$$\leqslant (1 - t_n)\|I(x) - I(y)\|$$

i.e. T_n is I-contractive with rate $(1 - t_n)$. Proposition 5.8 ensures that T_n and I have a coincidence point $x_n \in C$.

We shall show that the sequence (x_n) is bounded. If this is not the case, taking a subsequence if necessary, we may assume that $(\|x_n\|) \to \infty$. Let $\alpha \in (0, 1)$ and $\rho > 0$ be such that $\|T(x) - T(q)\| \leqslant \alpha\|I(x) - I(q)\|$ for $x \in C$ satisfying $\|x\| \geq \rho$. Then, for sufficiently large n, we have

$$\|x_n\| = \|(1 - t_n)T(x_n) + t_n q\|$$
$$\leqslant (1 - t_n)(\alpha\|I(x_n) - I(q)\| + \|T(q)\|) + t_n\|q\|.$$

Noting that $I(C)$ is bounded, dividing both sides by $\|x_n\|$ and taking limits, we get $1 \leqslant \alpha$, a contradiction. Thus, (x_n) and hence $(T(x_n))$ is bounded, and

$$\|I(x_n) - T(x_n)\| = t_n\|q - T(x_n)\| \to 0, \quad \text{as } n \to \infty.$$

Since X is reflexive, taking a subsequence if necessary, we may assume that (x_n) has a weak limit, say, $\bar{x}$. Since $I - T$ is demiclosed (i.e. its graph is sequentially closed in the product of the weak topology with the norm topology), we get that $I(\bar{x}) - T(\bar{x}) = 0$; i.e. $\bar{x}$ is a coincidence point of I and T. Further, if $I^2\bar{x} = I\bar{x}$, and T and I weakly commute in C, we have

$$I\bar{x} = I^2\bar{x} = IT\bar{x} = TI\bar{x}$$

showing that $I\bar{x}$ is a common fixed point of T and I.   ∎

Observation

- One can add that the set of common fixed points is closed, convex and bounded. The first two properties are proved in the usual way; the boundedness property follows immediately from (5.108).
- The preceding result can also be deduced from the classical result of [96, 251, 326] by applying it to the restriction of T to a sufficiently large ball in X. This direct way can be deduced from the preceding proof. It also follows from the observation that T is asymptotically I-contractive on C iff there exists some $c \in (0, 1)$ and $r > 0$ such that

$$\|T(x)\| \leqslant c\|I(x)\| \qquad \forall x \in C \backslash r\mathbb{B}_X,$$

  whenever $\|x\| \to \infty$ implies $\|I(x)\| \to \infty$, where $\mathbb{B}_X$ is the closed unit ball of X, so that $T(C \cap r\mathbb{B}_X) \subset I(C \cap r\mathbb{B}_X)$.
- If I is the identity map on C, then T is asymptotically contractive on C iff there exists some $c \in (0, 1)$ and an $r > 0$ such that

$$\|T(x)\| \leqslant c\|x\| \qquad \forall x \in C \backslash r\mathbb{B}_X,$$

  where $\mathbb{B}_X$ is the closed unit ball of X, so that $T(C \cap r\mathbb{B}_X) \subset C \cap r\mathbb{B}_X$.

Definition 5.63 A subset C of a uniformly convex Banach space X is said to be *asymptotically I-compact* if, for any sequence (x_n) of C such that $\|x\| \to \infty$ implies $\|I(x)\| \to \infty$ and that $(r_n) := (\|I(x_n)\|) \to \infty$, the sequence $(r_n^{-1}I(x_n))$ has a convergent subsequence. Locally compact convex sets and epigraphs of hypercoercive functions $T : C \to X$; i.e. $epi\,T = \{(y, t) \in C \times \mathbb{R} : T(y) \leqslant t\}$ with respect to $I : C \to X$ are asymptotically I-compact in the sense that $T(x)/\|I(x)\| \to \infty$ as

$\|I(x)\| \to \infty$. If I is the identity map on C, then C is called asymptotically compact (see [473]).

We now compare the preceding result with Theorem 5.1 of [375]. There, C is assumed to be *asymptotically compact* in the sense of [1, 377, 471, 531] (see also [161, 470, 615]); i.e. for any sequence (x_n) of C such that $(r_n) := (\|x_n\|) \to \infty$, the sequence $(r_n^{-1} x_n)$ has a convergent subsequence. Obviously this assumption is satisfied in finite dimensions; but it is a rather restrictive assumption in infinite dimensional spaces. However, locally compact convex sets and epigraphs of hypercoercive functions (i.e. functions T such that $T(x)/\|x\| \to \infty$ as $\|x\| \to \infty$) are asymptotically compact.

On the other hand, the asymptotic condition imposed on T in [376] is milder than the one considered here. Our asymptotic condition is obviously satisfied if T is asymptotically I-contractive on C. In fact, if T is I-nonexpansive, T is radially asymptotically I-contractive if and only if it is directionally asymptotically I-contractive in the sense that for any unit vector $u \in C_\infty$ one has

$$\limsup_{t \to \infty, v \to u, q+tv \in C} \frac{\|T(q+tu) - T(q)\|}{\|I(q+tu) - I(q)\|} < 1,$$

whenever $\|x\| \to \infty$ implies $\|I(x)\| \to \infty$ and one has the following relationships with our assumption.

Lemma 5.14 *Any asymptotic I-contraction $T : C \to X$ is a directional asymptotic I-contraction. If C is asymptotically I-compact, the converse holds.*

Proof The first part of the assertion is immediate. To prove the second part, assume that T is not an asymptotic I-contraction; i.e. for any $q \in C$ there exists a sequence (x_n) in C such that $(\|x_n\|) \to \infty$ and $\lim_n t_n^{-1} \|T(x_n) - T(q)\| \geq 1$ for $t_n := \|I(x_n) - I(q)\|$. Since C is asymptotically I-compact, the sequence $(u_n) := (t_n^{-1}(I(x_n) - I(q)))$ has a convergent subsequence with limit $u \in C_\infty$. Since $\lim_n t_n^{-1} \|T(x_n) - T(q)\| \geq 1$, T is not an asymptotic I-contraction.

For the case when I is the identity map on C, we recover Lemma 3 of Penot [472] in the following.

Corollary 5.27 *Any asymptotic contraction $T : C \to X$ is a directional asymptotic contraction. If C is asymptotically compact, the converse holds.*

It follows from Lemma 5.14 above that Theorem 5.1 of [376] is a direct consequence of Theorem 5.114. We also observe that Corollary 3 of [472] (stated below) is an immediate consequence of a corollary in [326].

Corollary 5.28 *([376]) Let X be a uniformly convex Banach space and let C be a closed and convex subset of X. Let $T : C \to X$ be a nonexpansive map which is radially asymptotically contractive on C and such that $T(C) \subset C$. Then T has a fixed point.*

382   5   Fixed Point Theorems

It may be remarked that the assumption of uniform convexity in Corollary 5.27 above is not needed. It is sufficient to know that bounded, closed and convex sets have the fixed point property for nonexpansive maps (see, for instance, Kirk [326]). We now present a criterion in order that T be asymptotially I-contractive. It relies on the following notion introduced in [471]. Here X is any normed linear space, and $\mathbb{B}_X$ denotes its closed unit ball.

Definition 5.64 A cone K of X is a firm (outer) asymptotic cone of a subset C of X if for any $\varepsilon > 0$ there exists some $r > 0$ such that for any $x \in C \setminus r\mathbb{B}_X$ one has $d(x, K) < \varepsilon \|x\|$.

We now introduce, in more general form, a variant of concepts due to Krasnoselski [345].

Definition 5.65 Given a firm asymptotic cone K of a subset C of X, a positively homogeneous mapping $T_\infty : K \to X$ is said to be a firm (outer) asymptotic derivative of $T : C \to X$ with respect to $I : C \to X$ if for any $\varepsilon > 0$, there exists a $\rho > 0$ such that, for any $x \in C \setminus \rho\mathbb{B}_X$, there exists a $v \in K$ satisfying $\|x - v\| < \varepsilon \|I(x)\|$,

$$\|T(x) - T_\infty(v)\| < \varepsilon \|I(x)\|.$$

If I is the identity map on C, $T_\infty : K \to X$ is called a firm (outer) asymptotic derivative of $T : C \to X$.

Note that this condition is satisfied when $T : C \to X$ has a firm (or strong) asymptotic derivative (or F-derivative at infinity) with respect to $I : C \to X$ in the sense that there exists a continuous linear mapping $T_\infty : X \to X$ such that

$$\lim_{r \to \infty} \sup_{x \in C \setminus rB_X} \frac{1}{\|I(x)\|} \|T(x) - T_\infty(x)\| = 0.$$

The following criterion was established in ([345], Sect. 3.2.2).

Lemma 5.15 *Suppose $T : X \to X$ is Gâteaux differentiable on $X \setminus r\mathbb{B}_X$ for some $r > 0$ and there exists a continuous linear mapping $A : X \to X$ such that $\|T'(x) - A\| \to 0$ as $\|x\| \to \infty$. Then T has a firm (or strong) asymptotic derivative $T_\infty = A$.*

A weaker condition than the above is that of asymptotable.

Asymptotable map - A map $T : C \to X$ is said to be asymptotable if there exists a positively homogeneous map $T_\infty : C_\infty \to X$ such that, for any $u \in C_\infty$, one has $t^{-1}T(tv) \to T_\infty(u)$ as $(t, v) \to (\infty, u)$ with $tv \in C$ (see [472]).

For asymptotable maps, the following criterion was established in [472].

Lemma 5.16 *([472]) If $T : C \to X$ is asymptotable and if C is asymptotically compact, then T_∞ is a firm asymptotic semiderivative of T.*

We now state and prove the announced criterion for asymptotic I-contractiveness of the mapping T.

Proposition 5.13 *Let K be a firm asymptotic cone of a subset C of X. Suppose T : $C \to X$ has a firm asymptotic semiderivative $T_\infty : K \to X$ with respect to $I : C \to X$, which is asymptotically I-contractive on K. Then T is asymptotic I-contractive on C.*

Proof From the observation following Definition 5.61, we can take $q = 0$ in that definition applied to T_∞ and K, so that there exists some $c \in (0, 1)$ such that

$$\|T_\infty(v)\| \leqslant c \|I(v)\|$$

for $v \in K$ with sufficiently large norm. Since K is a cone and T_∞ is positively homogeneous, this relation is satisfied for any $v \in K$. Let $c' \in (c, 1)$ and let $\varepsilon > 0$ be such that $c + 3\varepsilon < c'$. Then, taking $\rho > 0$ associated with the ε in Definition 5.64, for any $x \in C \backslash \rho \mathbb{B}_X$ we can pick a $v \in K$ satisfying $\|x - v\| < \varepsilon \|x\|$, $\|T(x) - T_\infty(v)\| < \varepsilon \|I(x)\|$, so that we get

$$\|T(x) - T(q)\| = \|T(x) - T_\infty(v) + T_\infty(v) - T(q)\|$$

$$\leqslant \|T(x) - T_\infty(v)\| + \|T_\infty(v)\| + \|T(q)\|$$

$$\leqslant \varepsilon \|I(x)\| + \|T_\infty(v)\| + \|T(q)\|$$

$$\leqslant \varepsilon \|I(x)\| + c \|I(v)\| + \|T(q)\|$$

$$\leqslant 2\varepsilon \|I(x)\| + c \|I(x)\| + \|T(q)\|$$

$$\leqslant (c + 2\varepsilon) \|I(x) - I(q)\| + \|T(q)\| + (c + 2\varepsilon) \|I(q)\|$$

$$\leqslant (c + 3\varepsilon) \|I(x) - I(q)\|$$

$$\leqslant c' \|I(x) - I(q)\|$$

provided

$$\varepsilon \|I(x) - I(q)\| \geq \|T(q)\| + (c + 2\varepsilon) \|I(q)\|,$$

which occurs when $\|I(x)\| \geq \varepsilon^{-1}(\|T(q)\| + (c + 3\varepsilon) \|I(q)\|)$.   ∎

Finally, combining Propositions 5.11 and 5.13 yields.

Theorem 5.115 *Let X be a uniformly convex Banach space, C a nonempty subset of X and $T, I : C \to X$. Let C be an (I, T)-star-shaped subset of X and let K be a firm asymptotic cone of C. Suppose that $I(C)$ is bounded, closed and convex and I is continuous. Let T be an I-nonexpansive map which has a firm asymptotic semiderivative $T_\infty : K \to X$ which is asymptotically I-contractive on K. Then T and I have a coincidence point v in C. Further, if $I^2 v = Iv$, and T and I are weakly compatible in C, then T and I have a unique common fixed point.*

If I is the identity map on C, we obtain the following result.

Corollary 5.29 ([472]) *Let X be a uniformly convex Banach space and let C be a closed and convex subset of X. Let K be a firm asymptotic cone of C. Let $T : C \to X$ be a nonexpansive map which has a firm asymptotic semiderivative $T_\infty : K \to X$ which is asymptotically contractive on K. Then T has a fixed point.*

Observation

- The result in Theorem 5.115 does not involve any compactness assumptions. However, such compactness assumption can be used as criteria ensuring its hypothesis, according to Lemmas 5.14 and 5.15. These criteria clearly shows the links with the results by Luc [376].
- The result of Theorem 5.115 can be extended to real-world nonconvex situations or to more general spaces as in [94, 326, 331, 332, 469, 619]. We also refer the interested reader to [161, 377, 470, 615] for the use of asymptotic compactness in various fields.

5.5 Common Fixed Point Theorems

There appears in literature several generalizations of the famous Banach Contraction Principle. One such generalization was given by Presic ([494, 496]) as follows.

Theorem 5.116 (Presic [496]) *Let (X, d) be a metric space, k a positive integer, $T : X^k \longrightarrow X$ be a mapping satisfying the following condition:*

$$d(T(x_1, x_2, \ldots, x_k), T(x_2, x_3, \ldots, x_{k+1}))$$
$$\leqslant q_1.d(x_1, x_2) + q_2.d(x_2, x_3) + \cdots + q_k.d(x_k, x_{k+1}) \qquad (5.112)$$

where $x_1, x_2, \ldots, x_{k+1}$ are arbitrary elements in X and $q_1, q_2, \ldots, q_k$ are nonnegative constants such that $q_1 + q_2 + \cdots + q_k < 1$. Then there exists some $x \in X$ such that $x = T(x, x, \ldots, x)$. Moreover if $x_1, x_2, \ldots, x_k$ are arbitrary points in X and for $n \in N$ $x_{n+k} = T(x_n, x_{n+1}, \ldots, x_{n+k-1})$, then the sequence $\{x_n\}$ is convergent and $\lim x_n = T(\lim x_n, \lim x_n, \ldots, \lim x_n)$.

Note that for $k = 1$, the above theorem reduces to the well-known Banach Contraction Principle. Čirič and Presic [145] generalizing the above theorem proved the following:

Theorem 5.117 (Čirič and Presic [145]) *Let (X, d) be a metric space, k a positive integer, $T : X^k \longrightarrow X$ be a mapping satisfying the following condition:*

$$d(T(x_1, x_2, \ldots, x_k), T(x_2, x_3, \ldots, x_{k+1})) \leqslant \lambda.max\{d(x_1, x_2), d(x_2, x_3), \ldots, d(x_k, x_{k+1})\}$$
$$(5.113)$$

where $x_1, x_2, \ldots, x_{k+1}$ are arbitrary elements in X and $\lambda \in (0, 1)$. Then there exists some $x \in X$ such that $x = T(x, x, \ldots, x)$. Moreover if $x_1, x_2, \ldots, x_k$ are arbitrary *points in X and for $n \in N$ $x_{n+k} = T(x_n, x_{n+1}, \ldots, x_{n+k-1})$, then the sequence $< x_n >$ is convergent and $\lim x_n = T(\lim x_n, \lim x_n, \ldots, \lim x_n)$. If in addition T satisfies $d(T(u, u, \ldots, u), T(v, v, \ldots, v)) < d(T(u, v))$, for all $u, v \in X$. Then x is the unique point satisfying $x = T(x, x, \cdots, x)$.*

Definition 5.66 (*George et al.* [240]) Let (X, d) be a metric space, k a positive integer, $T : X^k \to X$ and $f : X \to X$ be mappings.

(*a*) An element $x \in X$ is said to be a coincidence point of f and T if and only if $f(x) = T(x, x, \ldots, x)$. If $x = f(x) = T(x, x, \ldots, x)$ then we say that x is a *common fixed point* of f and T. If $w = f(x) = T(x, x, \ldots, x)$ then w is called a point of coincidence of f and T.
(*b*) Mappings f and T are said to be *commuting* if and only if $f(T(x, x, \ldots, x)) = T(fx, fx, \ldots, fx)$ for all $x \in X$.
(*c*) Mappings f and T are said to be *weakly compatible* if and only if they commute at their coincidence points.

Remark 5.42 For $k = 1$, the above definitions reduce to the usual definition of commuting and weakly compatible mappings in a metric space.

The set of coincidence points of f and T is denoted by $C(f, T)$.

Lemma 5.17 (*Pacurar* [439]) *Let X be a nonempty set, k a positive integer and $f : X^k \to X$, $g : X \to X$ two weakly compatible mappings. If f and g have a unique point of coincidence $y = f(x, x, \ldots, x) = g(x)$, then y is the unique common fixed point of f and g.*

5.5.1 Common Fixed Point Theorems in b-Metric Spaces

In this section, we give a gentle proof of a generalized Čirič–Presic-type fixed point theorem in b-metric space. To begin with, let us consider a function $\phi : \mathbb{R}^k \to \mathbb{R}$, such that

1. ϕ is an increasing function, i.e.

$$x_1 < y_1, x_2 < y_2, \ldots, x_k < y_k \implies \phi(x_1, x_2, \ldots, x_k) < \phi(y_1, y_2, \ldots, y_k).$$

2. $\phi(t, t, t, \ldots, t) \leqslant t$, for all $t \in X$.
3. ϕ is continuous in all variables.

Theorem 5.118 *Let (X, d) be a b-metric space with $s \geq 1$. For any positive integer k, let $f : X^k \to X$ and $g : X \to X$ be mappings satisfying the following conditions:*

$$f(X^k) \subseteq g(X) \tag{5.114}$$

$$d(f(x_1, x_2, \ldots, x_k), f(x_2, x_3, \ldots, x_{k+1}))$$

$$\leqslant \lambda\phi(d(gx_1, gx_2), d(gx_2, gx_3), d(gx_3, gx_4), \ldots, d(gx_k, gx_{k+1})) \qquad (5.115)$$

where $x_1, x_2, \ldots, x_{k+1}$ are arbitrary elements in X, $\lambda \in (0, \frac{1}{s^k})$

$$g(X) \text{ is complete} \qquad (5.116)$$

and

$$d(f(u, u, \ldots, u), f(v, v, \ldots, v)) < d(gu, gv), \qquad (5.117)$$

for all $u, v \in X$. Then f and g have a coincidence point, i.e. $C(f, g) \neq \varnothing$. In addition, if f and g are weakly compatible then f and g have a unique common fixed point. Moreover, for any $x_1 \in X$, the sequence $\{y_n\}$ defined by $y_n = g(x_n) = f(x_n, x_{n+1}, cdots, x_{n+k-1})$ converges to the common fixed point of f and g.

Proof For arbitrary $x_1, x_2, \ldots, x_k$ in X let

$$R = \max\left(\frac{d(gx_1, gx_2)}{\theta}, \frac{d(gx_2, gx_3)}{\theta^2}, \cdots, \frac{d(gx_k, f(x_1, x_2, \ldots, x_k))}{\theta^k}\right) \qquad (5.118)$$

exists in X, where $\theta = \lambda^{\frac{1}{k}}$. By (5.113) we define sequence $\{y_n\}$ in $g(X)$ as $y_n = gx_n$ for $n = 1, 2, \ldots, k$ and $y_{n+k} = g(x_{n+k}) = f(x_n, x_{n+1}, \ldots, x_{n+k-1}), n = 1, 2, \ldots$. Let $\alpha_n = d(y_n, y_{n+1})$. By the method of mathematical induction, we will prove that

$$\alpha_n \leqslant R\theta^n \ \forall \ n \in \mathbb{N}. \qquad (5.119)$$

Clearly by the definition of R, (5.119) is true for $n = 1, 2, \ldots, k$. Let the k inequalities $\alpha_n \leqslant R\theta^n, \alpha_{n+1} \leqslant R\theta^{n+1}, \cdots, \alpha_{n+k-1} \leqslant R\theta^{n+k-1}$ be the induction hypothesis. Then we have

$$
\begin{aligned}
\alpha_{n+k} &= d(y_{n+k}, y_{n+k+1}) \\
&= d(f(x_n, x_{n+1}, \ldots, x_{n+k-1}), f(x_{n+1}, x_{n+2}, \ldots, x_{n+k})) \\
&\leqslant \lambda\phi(d(gx_n, gx_{n+1}), d(gx_{n+1}, gx_{n+2}), c \ldots, (gx_{n+k-1}, gx_{n+k})) \\
&\quad d(gx_n, f(x_n, x_n, \ldots, x_n)), d(gx_{n+k}, f(x_{n+k}, x_{n+k}, \ldots, x_{n+k}) \\
&= \lambda\phi(\alpha_n, \alpha_{n+1}, \ldots, \alpha_{n+k-1}) \\
&\leqslant \lambda\phi(R\theta^n, R\theta^{n+1}, \ldots, R\theta^{n+k-1}) \\
&\leqslant \lambda\phi(R\theta^n, R\theta^n, \ldots, R\theta^n) \\
&\leqslant \lambda R\theta^n \\
&= R\theta^{n+k}
\end{aligned}
$$

Thus, inductive proof of (5.119) is complete. Now for $n, p \in \mathbb{N}$, we have

$$d(y_n, y_{n+p}) \leqslant sd(y_n, y_{n+1}) + s^2 d(y_{n+1}, y_{n+2}) + \cdots + s^{p-1} d(y_{n+p-1}, y_{n+p}),$$
$$\leqslant s R \theta^n + s^2 R \theta^{n+1} + \cdots + s^{p-1} R \theta^{n+p-1}$$
$$\leqslant s R \theta^n (1 + s\theta + s^2 \theta^2 + \cdots)$$
$$= \frac{s R \theta^n}{1 - s\theta}.$$

Hence, sequence $\{y_n\}$ is a Cauchy sequence in $g(X)$ and since $g(X)$ is complete, there exists $v, u \in X$ such that $\lim_{n \to \infty} y_n = v = g(u)$.

$$d(gu, f(u, u, \ldots, u)) \leqslant s[d(gu, y_{n+k}) + d(y_{n+k}, f(u, u, \ldots, u)]$$
$$= s[d(gu, y_{n+k})) + d(f(x_n, x_{n+1}, \ldots, x_{n+k-1}), f(u, u, \ldots, u)]$$
$$= sd(gu, y_{n+k}) + sd(f(x_n, x_{n+1}, \ldots, x_{n+k-1}), f(u, u, \ldots, u))$$
$$\leqslant sd(gu, y_{n+k}) + s^2 d(f(u, u, \ldots, u), f(u, u, \ldots, x_n))$$
$$+ s^3 d(f(u, u, \ldots, x_n), f(u, u, \ldots, x_n, x_{n+1}))$$
$$+ \cdots + s^{k-1} d(f(u, x_n, \ldots, x_{n+k-2}), f(x_n, x_{n+1}, \ldots, x_{n+k-1}))$$
$$\leqslant sd(gu, y_{n+k}) + s^2 \lambda \phi \{ d(gu, gu), d(gu, gu), \ldots, d(gu, gx_n) \}$$
$$+ s^3 \lambda \phi \{ d(gu, gu), d(gu, gu), \ldots, d(gu, gx_n), d(gx_n, gx_{n+1}) \} + \cdots$$
$$+ s^{k-1} \lambda \phi \{ d(gu, gx_n), d(gx_n, gx_{n+1}), \ldots, d(gx_{n+k-2}, gx_{n+k-1}) \}.$$
$$= sd(gu, y_{n+k}) + s^2 \lambda \phi (0, 0, \ldots, d(gu, gx_n))$$
$$+ s^3 \lambda \phi (0, 0, \ldots, d(gu, gx_n), d(gx_n, gx_{n+1})) + \cdots$$
$$+ s^{k-1} \lambda \phi (d(gu, gx_n), d(gx_n, gx_{n+1}), \ldots, d(gx_{n+k-2}, gx_{n+k-1})).$$

Taking the limit when n tends to infinity, we obtain $d(gu, f(u, u, \ldots, u)) \leqslant 0$. Thus, $gu = f(u, u, u, \ldots, u)$, i.e. $C(g, f) \neq \varnothing$. Thus, there exists $v, u \in X$ such that $\lim_{n \to \infty} y_n = v = g(u) = f(u, u, u, \ldots, u)$. Since g and f are weakly compatible, $g(f(u, u, \ldots, u) = f(gu, gu, gu, \ldots, gu)$. By (5.117) we have

$$d(ggu, gu) = d(gf(u, u, \ldots, u), f(u, u, \ldots, u))$$
$$= d(f(gu, gu, gu, \ldots, gu), f(u, u, \ldots, u))$$
$$< d(ggu, gu)$$

implies $d(ggu, gu) = 0$ and so $ggu = gu$. Hence, we have $gu = ggu = g(f(u, u, \ldots, u)) = f(gu, gu, gu, \cdots, gu)$, i.e. gu is a common fixed point of g and f, and $\lim_{n \to \infty} y_n = g(\lim_{n \to \infty} y_n) = f(\lim_{n \to \infty} y_n, \lim_{n \to \infty} y_n, \ldots, \lim_{n \to \infty} y_n)$. Now suppose x, y be two fixed points of g and f. Then

$$d(x, y) = d(f(x, x, x, \ldots, x), f(y, y, y, \ldots, y))$$
$$< d(gx, gy)$$
$$= d(x, y)$$

This implies $x = y$. Hence, the common fixed point is unique.

Remark 5.43 Taking $s = 1$, $g = I$ and $\phi(x_1, x_2, \ldots, x_k) = \max\{x_1, x_2, \ldots, x_k\}$ in Theorem 5.118, we obtain Theorem 5.117, i.e. the result of Čirič and Presic [145].

Remark 5.44 For $\lambda \in (0, \frac{1}{s^{k+1}})$, we can drop the condition (5.117) of Theorem 5.118. In fact we have the following :

Theorem 5.119 *Let (X, d) be a b-metric space with $s \geq 2$. For any positive integer k, let $f : X^k \to X$ and $g : X \to X$ be mappings satisfying conditions (5.113), (5.114) and (5.116) with $\lambda \in (0, \frac{1}{s^{k+1}})$. Then all conclusions of Theorem 5.117 hold.*

Proof As proved in Theorem 5.118, there exists $v, u \in X$ such that $\lim_{n \to \infty} y_n = v = g(u) = f(u, u, \ldots, u)$, i.e. $C(g, f) \neq \varnothing$. Since g and f are weakly compatible, $g(f(u, u, \ldots, u)) = f(gu, gu, gu, \cdots, gu)$. By (5.115) we have

$$
\begin{aligned}
d(ggu, gu) &= d(gf(u, u, \ldots, u), f(u, u, \ldots, u)) \\
&= d(f(gu, gu, gu, \ldots, gu), f(u, u, \ldots, u)) \\
&\leqslant sd(f(gu, gu, gu, \ldots, gu), f(gu, gu, \ldots, gu, u)) \\
&\quad + s^2 d(f(gu, gu, \ldots, gu, u), f(gu, gu, \ldots, u, u)) \\
&\quad + \cdots + s^{k-1} d(f(gu, gu, \ldots, u, u), f(u, u, \ldots, u)) \\
&\quad + s^{k-1} d(f(gu, u, \ldots, u, u), f(u, u, \ldots, u)) \\
&\leqslant s\lambda\phi(d(ggu, ggu), \ldots, d(ggu, ggu), d(ggu, gu)) \\
&\quad + s^2 \lambda\phi(d(ggu, ggu), \ldots, d(ggu, gu), d(gu, gu)) \\
&\quad + \cdots + s^{k-1} \lambda\phi(d(ggu, gu), \ldots, d(gu, gu), d(gu, gu)) \\
&= s\lambda\phi(0, 0, 0, \ldots, d(ggu, gu)) + s^2 \lambda\phi(0, 0, \ldots, 0, d(ggu, gu), 0) \\
&\quad + \cdots + s^{k-1} \lambda\phi(d(ggu, gu), 0, 0, \ldots, 0) \\
&= s\lambda[1 + s + s^2 + s^3 + \cdots + s^{k-2} + s^{k-2}]d(ggu, gu) \\
&\leqslant s\lambda[1 + s + s^2 + s^3 + \cdots + s^{k-2} + s^{k-1}]d(ggu, gu) \\
&= s\lambda \frac{s^k - 1}{s - 1} d(ggu, gu).
\end{aligned}
$$

$s\lambda\frac{s^k-1}{s-1} < 1$ implies $d(ggu, gu) = 0$ and so $ggu = gu$. Hence, we have $gu = ggu = g(f(u, u, \ldots, u)) = f(gu, gu, gu, \cdots, gu)$, i.e. gu is a common fixed point of g and f, and $\lim_{n \to \infty} y_n = g(\lim_{n \to \infty} y_n) = f(\lim_{n \to \infty} y_n, \lim_{n \to \infty} y_n, \ldots, \lim_{n \to \infty} y_n)$. Now suppose x, y be two fixed points of g and f. Then

$$
\begin{aligned}
d(x, y) &= d(f(x, x, x, \ldots, x), f(y, y, y, \ldots, y)) \\
&\leqslant sd(f(x, x, \ldots, x), f(x, x, \ldots, x, y)) + s^2 d(f(x, x, \ldots, x, y), \\
&\quad f(x, x, x, \ldots, x, y, y)) + \cdots + s^{k-1} d(f(x, x, y, \ldots, y), f(y, y, \ldots, y)) \\
&\quad + s^{k-1} d(f(x, y, y, \ldots, y), f(y, y, \ldots, y))
\end{aligned}
$$

$$\leqslant s\lambda\phi\{d(fx, fx), d(fx, fx), \ldots, d(fx, fy)\} + s^2\lambda\phi\{d(fx, fx),$$
$$d(fx, fx), \ldots, d(fx, fy), d(fy, fy)\}$$
$$+ \cdots + s^{k-1}\lambda\phi\{d(fx, fy), d(fy, fy), \ldots, d(fy, fy)\}.$$

$$= s\lambda\phi(0, 0, \ldots, d(fx, fy)) + s^2\lambda\phi(0, 0, \ldots, d(fx, fy), 0) + \cdots$$
$$+ s^{k-1}\lambda\phi(d(fx, fy), 0, 0, 0, \ldots, 0)).$$
$$= \lambda[s + s^2 + s^3 + \cdots + s^{k-1} + s^{k-1}]d(fx, fy)$$
$$= s\lambda[1 + s + s^2 + s^3 + \cdots + s^{k-2} + s^{k-2}]d(fx, fy)$$
$$\leqslant s\lambda[1 + s + s^2 + s^3 + \cdots + s^{k-2} + s^{k-1}]d(fx, fy)$$
$$= s\lambda\frac{s^k - 1}{s - 1}d(fx, fy).$$
$$= s\lambda\frac{s^k - 1}{s - 1}d(x, y).$$

This implies $x = y$. Hence, the common fixed point is unique.

Example 5.47 Let $X = \mathbb{R}$ and $d : X \times X \to X$ such that $d(x, y) = |x - y|^3$. Then, d is a b-metric on X with $s = 4$. Let $f : X^2 \to X$ and $g : X \to X$ be defined as follows:

$$f(x, y) = \frac{x^2 + y^2}{13} + \frac{18}{13} \quad \forall (x, y) \in \mathbb{R}^2,$$

and

$$gx = x^2 - 2 \quad \forall x \in \mathbb{R}.$$

We will prove that f and g satisfy condition (5.115).

$$d(f(x, y), f(y, z)) = |f(x, y) - f(y, z)|^3$$
$$= \left|\frac{x^2 - z^2}{13}\right|^3 = \left|\frac{x^2 - y^2 + y^2 - z^2}{13}\right|^3$$
$$\leqslant 4\left(\left|\frac{x^2 - y^2}{13}\right|^3 + \left|\frac{y^2 - z^2}{13}\right|^3\right)$$
$$= \frac{4}{13^3}[|x^2 - y^2|^3 + |y^2 - z^2|^3]$$
$$= \frac{8}{13^3}\frac{1}{2}[|x^2 - y^2|^3 + |y^2 - z^2|^3]$$
$$\leqslant \frac{8}{13^3}Max\{|x^2 - y^2|^3, |y^2 - z^2|^3\}$$
$$= \frac{8}{13^3}Max\{d(gx, gy), d(gy, gz)\}$$

Thus, f and g satisfy condition (5.115). with $\lambda = \frac{8}{13^3} \in (0, \frac{1}{4^3})$. Clearly $C(f, g) = 2$, f and g commute at 2. Finally, 2 is the unique common fixed point of f and g. But f and g do not satisfy condition (5.117) as at $x = -1$ and $y = 1$, $d(f(x, x), f(y, y)) = d(f(-1, -1),$ $f(1, 1)) = d(\frac{2}{13} + \frac{18}{13}, \frac{2}{13} + \frac{18}{13}) = 0 = d(-1, -1) = d(g(-1),$ $g(1)) = d(gx, gy)$.

5.5.2 Common Fixed Points of a Family of Mappings

In this section, we discuss common or simultaneous fixed points of a pair of mappings, both single-valued and multivalued and common fixed points of a family of mappings.
Common fixed point - Let K be an arbitrary set and let $\mathscr{F}$ be a family of mappings $F : K \to K$. A point $u \in K$ is called a common fixed point for the family if $F(u) = u$ for all $F \in \mathscr{F}$.

The first result for families of mappings was proved by Markov in 1936. Kakutani gave a direct proof of Markov's theorem in 1938 and also proved a fixed point theorem for groups of affine equicontinuous mappings. Ryll-Nardzewski obtained an important extension of the results of Markov and Kakutani in 1966. Day [160] also proved a more general theorem. For further work in this direction one can refer to Greenleaf [256], Huff [280] and Mitchell [405]. For a commuting family of nonexpansive mappings, we shall refer the work of de Marr [165], Browder [106], Belluce and Kirk [46] and Kuhfittig [350].

We begin with the following interesting problem posed by Isbel [285] in 1957: if $\mathscr{F}$ is a family of commuting continuous self-mappings of $[0, 1]$ then do there exist common fixed point for $\mathscr{F}$? Boyce [74] and Huneke [281] independently answered this question in the negative. The result is stated below:

There exist two commuting continuous self-mappings of $[0, 1]$ without a common fixed point.

Definition 5.67 Let K be a convex subset of a normed linear space X and let F be a self-mapping of K. F is said to be an affine mapping if

$$F(\alpha x + (1 - \alpha)y) = \alpha F x + (1 - \alpha) F y$$

for all $x, y \in K$ and $0 < \alpha < 1$.

First we state and prove Markov–Kakutani's theorem. Markov [387] in 1936, proved the following result by using Tychonoff's theorem. We give the proof due to Kakutani [302].

Theorem 5.120 *Let K be a compact convex subset of a normed linear space X. Let $\mathscr{F}$ be a commuting family of continuous affine mappings which map K into itself. Then there exists a point u in K such that $Fu = u$ for each $F \in \mathscr{F}$.*

Proof Let n be a positive integer, and let $F_n = n^{-1}(I + F + \cdots + F^{n-1})$ where $F \in \mathscr{F}$. Let $\mathscr{K}$ be the family of all sets $F_n(K)$ for $n \geqslant 1$ and $F \in \mathscr{F}$. Then each set in $\mathscr{K}$ is convex and compact as $F \in \mathscr{F}$ is affine and continuous, and $F_n(K) \subset K$.

Because F_n and G_m belonging to $\mathscr{F}$ commute, it follows that

$$F_n G_m(K) \subset F_n(K) \cap G_m(K), F, G \in \mathscr{F}.$$

Thus, any finite subfamily of K has a nonvoid intersection. Hence, compactness of $K \in \mathscr{K}$ implies that there is a $u \in \cap\{K : K \in \mathscr{K}\}$.

If $F \in \mathscr{F}$ and $Fu \neq u$, there is a neighbourhood V of the origin of X such that $Fu - u \in V$. If n is an arbitrary positive integer, then there exists a $v \in K$ such that $u = F_n v$. Hence, $Fu - u = n^{-1}(F^n - 1)v \in V$. As $F^n v \in K$, it follows that $n^{-1}(K - K)$ is not a subset of V for any positive integer n. But $K - K = \phi(K \times K)$, where $\phi(x, y) = x - y$. This shows that $K - K$ is compact. We arrive at a contradiction since a compact subset of a normed linear space is bounded. So $Fu = u$ for $F \in \mathscr{F}$. ∎

Definition 5.68 A family of linear mapping on a normed linear space X is said to be equicontinuous on a subset K of X if for every neighbourhood V of the origin in X, there is a neighbourhood U of the origin in X such that if $x, y \in K$ and $x - y \in U$ then $F(x - y) \in V$ for all $F \in \mathscr{F}$. The same definition holds good for arbitrary families of continuous mappings.

The following theorem was proved by Kakutani [302].

Theorem 5.121 *Let K be a compact convex subset of a normed linear space X, and let $\mathscr{F}$ be a group of affine mapping which is equicontinuous on K and such that $F(K) \subset K$ for all $F \in \mathscr{F}$. Then there exists a point $u \in K$ such that $Fu = u$ for all $F \in \mathscr{F}$.*

Proof By Zorn's lemma, K contains a minimal nonempty compact convex subset K_1 such that $\mathscr{F}(K_1) \subseteq K_1$. If K_1 contains just one point, the assertion of the theorem is proved. If this is not the case, the compact set $K_1 - K_1$ contains some point other than the origin and, consequently, there exists a spherical ball or neighbourhood V_1 centerd at the origin such that $\overline{V}_1$ does not contain $K_1 - K_1$. By the equicontinuity of $\mathscr{F}$ on the set K_1, there exists a spherical ball or neighbourhood V_1 centerd at the origin such that $\overline{V}_1$ does not contain $K_1 - K_1$. By the equicontinuity of $\mathscr{F}$ on the set K_1, there exists neighbourhood U_1 centerd at the origin such that $x_1 - x_2 \in U_1$ whenever $x_1, x_2 \in K_1$. Then $\mathscr{F}(x_1 - x_2) \subset V_1$. Define $U_2 = co(\mathscr{F}U_1) = co\{Fx : F \in \mathscr{F}, x \in U_1\}$. Since $\mathscr{F}$ is a group $\mathscr{F}U_2 = U_2$ and from continuity of $F \in \mathscr{F}$ we have $\mathscr{F}\overline{U}_2 = \overline{U}_2$. Set $r = \inf\{a : a > 0, aU_2 \supseteq K_1 - K_1\}$, and $U = rU_2$. It follows that for each t, $0 < t < 1$, the set $K_1 - K_1$ is not contained in $(1 - t)\overline{U}$, while $K_1 - K_1 \supseteq (1 + t)U$. The family of open sets $\{\frac{1}{2}U + x\}$, $x \in K_1$ is a open covering of K_1. From compactness of K_1, we select a finite subcovering of K_1 as $\{\frac{1}{2}U + x_i\}$, $i = 1, 2, \ldots, n$ and let $p = \frac{x_1 + x_2 + \cdots + x_n}{n}$. If x is any point in K_1, then $x_i - x \in \frac{1}{2}U$ for some i, $1 \leqslant i \leqslant n$. But $x_i - x \in (1 + t)U$ for $1 \leqslant i \leqslant n$ and $t > 0$. Thus, we have

$$p \in \frac{1}{n}\left[\frac{1}{2}U + (n-1)(1+t)U\right] + x.$$

Taking $t = \frac{1}{4(n-1)}$, we have $p \in \left(1 - \frac{1}{4n}\right)U + x$. We define

$$K_2 = K_1 \bigcap \left[\bigcap_{x \in K_1} \left\{\left(1 - \frac{1}{4n}\right)\overline{U} + x\right\}\right].$$

The set K_2 is nonempty, and since $\left(1 - \frac{1}{4n}\right)\overline{U}$ does not contain $K_1 - K_1$, $K_2 \neq K_1$. The closed set K_2 is clearly convex. Moreover as $F(\alpha\overline{U} + x) \subset \alpha\overline{U} + Fx$ for $F \in \mathscr{F}$, $x \in K_1$. Also it follows that $F(K_1) = K_1$ for $F \in \mathscr{F}$, since $\mathscr{F}$ is a group and $F(K_1) \subset K_1$. This implies that $\mathscr{F}K_2 \subset K_2$, contradicting the minimality of K_1, as K_2 is a proper subset of K_1.   ∎

We state an interesting extension of Kakutani's fixed point theorem for families of mappings, due to Ryll-Nardzewski [532].

Theorem 5.122 *Let X be a Banach space and let K be a nonempty, convex and weakly compact subset of X. Suppose S is a semigroup of mappings from K into itself and S is noncontracting (or distal), i.e. for $x \neq y$, 0 is not in the norm closure of the set $\{Fx - Fy : F \in S\}$. Then there is a common fixed point for S.*

Now we discuss common fixed points of families of nonlinear mappings which are nonexpansive. The first result in this direction was obtained by de Marr in 1963.

Theorem 5.123 (de Marr [165]) *Let K be a compact convex subset of a Banach space X and let $\mathscr{F}$ be a family of commuting nonexpansive mappings of K into itself. Then there is a common fixed point for the family.*

For the proof, we need the following two lemmas stated without proof.

Lemma 5.18 *Let K be a compact subset of a Banach space X. Let $d = \operatorname{diam} K$ and suppose $d > 0$. Then there exists $x \in \operatorname{co} K$ such that $\sup\{\|x - z\| : z \in K\} < d$.*

Lemma 5.19 *Let K be a convex subset of a Banach space X. Let $F : X \to X$ be a nonexpansive mapping. Suppose K is invariant under the mapping F and there exists a compact subset C of K such that $C = \{Fx : x \in C\}$ and is not a singleton set. Then there exists a closed and convex set K_1 such that $K_1 \cap K$ is invariant under F and $C \cap K_1'$ is nonempty (K_1' is the complement of K_1).*

Proof of Theorem 5.123. By Zorn's lemma, K contains a minimal nonempty, compact and convex subset K_1 such that $\mathscr{F}(K_1) \subset K_1$. Theorem is proved if K_1 consists of a single point. Now we show that if K_1 contains at least two distinct elements, it leads to a contradiction. Using Zorn's lemma again, we find a nonempty compact subset K_2 of K_1 which is invariant under the mapping $F \in \mathscr{F}$. We prove $K_2 = \{Fx : x \in K_2\}$ for each F in $\mathscr{F}$. Suppose on the contrary there exists a mapping $G \in \mathscr{F}$ such that $G(K_2) \neq K_2$. $G(K_2)$ is compact since G is continuous. Let x be in $G(K_2)$. Then

$x = Gy$ for some $y \in K_2$. By commutativity of $\mathscr{F}$, we have $LGy = GLy \in G(K_2)$ and thus $L(GK_2) \subset G(K_2)$ for all $L \in \mathscr{F}$. This clearly contradicts the minimality of K_2 and thus $K_2 = \{Fx, x \in K_2\}$ for each F in $\mathscr{F}$.

Suppose K_2 has at least two points, otherwise the theorem is proved. Then by Lemma 2 there exists a subset K_2 such that $K_2' \cap K_2 \neq \varnothing$. Then $K_3 \cap K_2$ is nonempty, compact and convex and not equal to K_1 and invariant under $\mathscr{F}$ and this leads to a contradiction. Thus, K_1 reduces to a point, and this point is the common fixed point of $\mathscr{F}$.

The following two theorems were proved by Belluce and Kirk [46].

Theorem 5.124 *Let K be a closed, bounded and convex subset of a Banach space X. If $\mathscr{F}$ is a family of commuting nonexpansive mappings from K into itself and C is a compact subset of K with the property that for all $x \in K$, $C \cup \overline{F^n x} \neq \varnothing$, for some $F \in \mathscr{F}$, then $\mathscr{F}$ has a common fixed point in K.*

Notice that Theorem 5.124 generalizes Theorem 5.123.

Theorem 5.125 *Let K be a weakly compact and convex subset of a strictly convex Banach space X. Suppose $\mathscr{F}$ is a family of commuting nonexpansive mappings of K into K with the property that for each $F \in \mathscr{F}$, the F-closure of K is nonempty. Then $\mathscr{F}$ has a common fixed point in K.*

The next theorem proved by Browder [96] is the nonlinear extension of Markov–Kakutani's theorem and an extension of Theorem 5.123.

Theorem 5.126 *Let K be a closed, bounded and convex subset of a uniformly convex Banach space and let $\mathscr{F}$ be a commuting family of nonexpansive mappings of K into itself. Then the family $\mathscr{F}$ has a common fixed point.*

We now consider a family of mappings which satisfy generalized Kannan–Reich condition. We give the proof of the existence of common fixed point of a pair of such setvalued mappings. This result is due to Bose and Mukerjee [66] and contains Wong's [608] result for single-valued mapppings.

Let (X, d) be a bounded metric space and let H denote the Hausdorff metric on the space $CL(X)$, the space of nonempty and closed subset of X. Let $F_i, i = 1, 2$, be mappings from X into $CL(X)$ satisfying the following condition:

$$
\begin{aligned}
H(F_1(x), F_2(y)) \leqslant {} & a_1 d(x, F_1(x)) + a_2 d(y, F_2(y)) + a_3 d(y, F_1(x)) \\
& + a_4 d(x, F_2(y)) + a_5 d(x, y)
\end{aligned}
\tag{5.120}
$$

for any $x, y \in X$ where a_1, a_2, a_3, a_4 and a_5 are nonnegative numbers and $a_1 + a_2 + a_3 + a_4 + a_5 < 1$ and $a_1 = a_2$ or $a_3 = a_4$.

Theorem 5.127 *Let (X, d) be a complete bounded metric space and let $F_i : X \to CL(X)$, $i = 1, 2$ be multivalued mappings satisfying condition (5.120). Then F_1 and F_2 have a common fixed point.*

Proof Let $x_0 \in X$. Consider the sequence $\{x_n\}$ where

$$x_{2n+1} \in F_1(x_{2n+2}) \text{ and } x_{2n+2} \in F_2(x_{2n+1}).$$

Assume $H(F_1(x_0), F_2(x_1)) \neq 0$. For if it is zero, then F_1 and F_2 have a common fixed point. Then by the definition if a number $h > H(F_1(x_0), F_2(x_1))$ there exists $x_2 \in F_2(x_1)$ such that $d(x_1, x_2) \leqslant h$. Let

$$h = p^{-1} H(F_1(x_0), F_2(x_1))$$

where $p = (a_1 + a_2 + a_3 + a_4 + a_5)^{1/2}$. Then

$$
\begin{aligned}
d(x_1, x_2) &\leqslant p^{-1} H(F_1(x_0), F_2(x_1)) \\
&\leqslant p^{-1}[a_1 d(x_0, F_1(x_0)) + a_2 d(x_1, F_2(x_1)) + a_3 d(x_1, F_1(x_0)) \\
&\quad + a_4 d(x_0, F_2(x_1)) + a_5 d(x_0, x_1)] \\
&\leqslant p^{-1}[a_1 d(x_0, x_1) + a_2 d(x_1, x_2) + a_3 d(x_0, x_1) + a_4 d(x_1, x_2) + a_5 d(x_0, x_1)].
\end{aligned}
$$

That is $d(x_1, x_2) \leqslant \frac{a_1 + a_4 + a_5}{p - a_2 - a_4} d(x_0, x_1)$.

Proceeding in a similar manner, there exists $x_2 \in F_1(x_2)$ such that

$$d(x_2, x_3) \leqslant \frac{a_2 + a_3 + a_5}{p - a_1 - a_3} d(x_1, x_2).$$

We have $0 < r, s < 1$ if $a_3 = a_4$ and $0 < rs < 1$ when $a_1 = a_2$ or $a_3 = a_4$ where

$$r = \frac{a_1 + a_4 + a_5}{p - a_2 - a_4} \text{ and } \frac{a_2 + a_3 + a_5}{p - a_1 - a_3}.$$

Further

$$d(x_{2n+1}, x_{2n+2}) \leqslant (rs)^n r d(x_0, x_1) \text{ and}$$

$$d(x_{2n}, x_{2n+1}) \leqslant (rs)^n d(x_0, x_1).$$

It is easily seen that the sequence $\{x_n\}$ is Cauchy and hence converges to some point $u \in X$. Consider $D(u, F_2(u))$. We have

$$
\begin{aligned}
D(u, F_2(u)) &\leqslant d(u, x_{n+1}) + d(x_{n+1}, F_2(u)) \\
&\leqslant d(u, X_{n+1}) + H(F_1(x_n), F_2(u)) \text{ (here } n \text{ is taken to be even)} \\
&\leqslant d(u, x_{n+1}) + a_1 d(x_n, x_{n+1}) + a_2 d(u, F_2(u) + a_3 d(u, x_{n+1}) \\
&\quad + a_4 d(x_n, u) + a_4 d(u, F_2(u)) + a_5 d(x_n, u).
\end{aligned}
$$

Taking limit $n \to \infty$, we have

$$D(u, F_2(u)) \leqslant (a_2 + a_4)d(u, F_2(u)), \text{ i.e. } d(u, F_2(u)) = 0.$$

Since $F_2(u)$ is closed we have $u \in F_2(u)$. Similarly it can be shown that $u \in F_1(u)$. ∎

Remark 5.45 In the above theorem instead of a bounded complete metric space, we can take any complete metric space with the following modification: $F_1 : (X, d) \rightarrow CB(X)$ where $CB(X)$ denotes the space of bounded and closed subsets of X endowed with the Hausdorff metric.

Corollary 5.30 *Let (X, d) be a complete metric space and let $F_i : X \rightarrow X, i = 1, 2$ be mappings satisfying the following condition:*

$$d(F_1(x), F_2(y)) \leqslant a_1 d(x, F_1(x)) + a_2 d(y, F_2(y)) + a_3 d(y, F_1(x))$$
$$+ a_4 d(x, F_2(y)) + a_5 d(x, y)$$

for any x, y in X where a_1, a_2, a_3, a_4 and a_5 are nonnegative numbers and $a_1 + a_2 + a_3 + a_4 + a_5 < 1$ and $a_1 = a_2$ or $a_3 = a_4$. Then F_1 and F_2 have a common fixed point.

Corollary 5.31 *Let (X, d) be a complete bounded metric space and let $F : X \rightarrow CL(X)$ be multivalued mapping satisfying the following condition:*

$$H(F(x), F(y)) \leqslant a_1 d(x, F(x)) + a_2 d(y, F(y)) + a_3 d(y, F(x))$$

$$+ a_4 d(x, F(y)) + a_5 d(x, y),$$

for any $a, y \in X$ where a_1, a_2, a_3, a_4 and a_5 are nonnegative numbers and $a_1 + a_2 + a_3 + a_4 + a_5 < 1, a_1 = a_2$ and $a_3 = a_4$. Then F has a fixed point.

The following theorem proved by Bose and Mukerjee [67] is a generalization of a theorem of Iseki [286]. Here F is a single-valued mapping.

Theorem 5.128 *Let $\{F_n\}$ be a sequence of self-mappings of X such that*

$$d(F_i x, F_j y) \leqslant a_1 d(F_i x, x) + a_2 d(F_j y, y) + a_3 d(F_i x, y)$$
$$+ a_4 d(F_i y, x) + a_5 d(x, y) \quad (j > i)$$

for all $x, y \in X$ where a_1, a_2, a_3, a_4 and a_5 are nonnegative numbers and $\sum_{k=1}^{5} a_k < 1$ and $a_3 = a_4$. Then the sequence $\{F_n\}$ has a unique common fixed point.

The following common fixed point theorem was proved by Husain and Sehgal [283], and later on it was improved upon by Singh and Meade [562] in a slightly different form.

Theorem 5.129 *Let F and G be self-mappings of a metric space X. Suppose there exists a* $\phi : (\mathbb{R}^+)^5 \to \mathbb{R}^+$ *which is continuous and nondecreasing in each coordinate variables and satisfies the relation* $\phi(t, t, a_1 t, a_2 t, t) < t$ *for* $t > 0, a_i \in \{0, 1, 2\}$ *with* $a_1 + a_2 = 2$. *Let F and G satisfy*

$$d(Fx, Fy) \leqslant \phi[d(x, Fx), d(y, Gy), d(x, Gy), d(y, Fx), d(x, y)]$$

for all $x, y \in X$. *Then there exists a unique* $u \in X$ *such that* $Fu = u = Gu$.

We now present an iteration scheme which converges strongly in one case and weakly in another case to a common fixed point of a finite family of nonexpansive mappings. These results were obtained by Kuhfittig [350] in 1981. We end this section with another iteration scheme converging weakly in a more general setting than Kuhfittig's proved by Bose and Sahani [70].

Let K be a convex subset of a Banach space X. Suppose $\{F_i\}_{i=1}^k$ be a family of nonexpansive mappings of K into itself. Define the following mappings:

$$U_0 = I, \text{ the identity mapping.}$$

For $o < \alpha < 1$, let

$$U_r = (1 - \alpha)I + \alpha F_r U_{r-1}, r = 1, 2, \ldots, k.$$

Theorem 5.130 *Let K be a compact convex subset of a strictly convex Banach space X and* $\{F_i\}_{i=1}^k$ *be a finite family of nonexpansive self-mappings of K with a nonempty set of common fixed points. Then for any* $x \in K$, *the sequence* $\{U_k^n x\}_{n=1}^\infty$ *converges strongly to a common fixed point of* $\{F_i\}_{i=1}^k$.

Theorem 5.131 *If X is uniformly convex Banach space satisfying Opial's condition (in particular if X is a Hilbert space). Let K be a closed and convex subset of X and* $\{F_i\}_{i=1}^k$ *be a family of nonexpansive self-mappings of K with a nonempty set of common fixed points, then for any* $x \in K$, *the sequence* $\{U_k^n x\}_{n=1}^\infty$ *converges weakly to a common fixed point.*

Remark 5.46 The proof of Theorem 5.130 is based on an iteration scheme for a nonexpansive mapping by Edelstein [216] and the proof of Theorem 5.131 is based on an approximation scheme for a nonexpansive mapping given by Opial [436].

Remark 5.47 If the family of mappings $\{F_i\}_{i=1}^k$ is commutative, then the set of common fixed point of $\{F_i\}_{i=1}^k$ is nonempty by de Marr's Theorem (Theorem 5.123). Hence, in Theorem 5.130 if we take the family to be commutative, we can drop the condition that the set of common fixed points by nonempty.

Remark 5.48 Suppose in Theorem 5.131, we assume further that K is bounded and $\{F_i\}_{i=1}^k$ is commutative. Then by Browder's theorem (Theorem 5.126), the set of common fixed points of $\{F_i\}_{i=1}^k$ is nonempty since X is strictly convex and reflexive.

Thus, we can drop the assumption that the set of common fixed points of F_i is nonempty with the addition of aforementioned conditions (i.e. K is bounded and F_i commutative.)

Theorem 5.132 *Let K be a closed and convex subset of a uniformly convex Banach space X with a Fréchet differentiable norm, $\{F_i\}_{i=1}^{k}$ a family of nonexpansive self-mappings of K with a nonempty set of fixed points, and $\{c_n\}$ a real sequence such that*

(i) $0 \leqslant c_n \leqslant 1$,

(ii) $\displaystyle\sum_{n=1}^{\infty} c_n(1 - c_n) = \infty$.

If $x_1 \in K$ and $x_{n+1} = (1 - c_n)x_n + c_n F_k U_{k-1} x_n$ for $n \geqslant 1$, then $\{x_n\}$ converges weakly to a common fixed of $\{F_i\}_{i=1}^{k}$.

Proof One can easily prove that U_j and $F_j U_{j-1}$, $j = 1, 2, \ldots, k$ are nonexpansive and map K into itself. Also it is easy to show that the families $\{U_j\}_{j=1}^{k}$ and $\{F_j\}_{j=1}^{k}$ have the same set of common fixed points. Since $F_k U_{k-1}$ is a nonexpansive self-mapping of K, the sequence $\{x_n\}$ defined in the theorem converges weakly to a fixed point v of $F_k U_{k-1}$ by a theorem of Reich (Theorem 5.38). We shall next show that v is a common fixed point of F_k and $U_{k-1}(k \geqslant 2)$. To this end we first show that $F_{k-1} U_{k-2} v = v(k \geqslant 2)$. Suppose this is not so. Let

$$z = U_{k-1}v = (1 - \alpha)v + \alpha F_{k-1} U_{k-2}v.$$

Then $z \neq v$. By hypothesis, there exists a point w such that

$$F_j w = w, \ j = 1, 2, \ldots, k.$$

Since $\{F_j\}$ and $\{U_j\}$ have the same common fixed points, it follows that

$$F_{k-1} U_{k-2} w = w.$$

By nonexpansiveness we have

$$\|F_{k-1} U_{k-2}v - w\| \leqslant \|v - w\| \tag{5.121}$$

and $\qquad \|F_k z - w\| \leqslant \|z - w\|$.

Again $\qquad\qquad F_k z = F_k U_{k-1} v = v$.

Since uniformly convex Banach space is strictly convex, it follows that

$$\|v - w\| \leqslant \|z - w\| = \|(1 - \alpha)v + \alpha F_{k-1}U_{k-2}v - w\|$$
$$= \|(1 - \alpha)(v - w) + \alpha(F_{k-1}U_{k-2}v - w)\|$$
$$< \ \max \ \{\|v - w\|, \|F_{k-1}U_{k-2}v - w\|\}$$

which contradicts (5.121). Hence, $F_{k-1}U_{k-2}v = v$. As $U_{k-1} = (1 - \alpha)I + \alpha F_{k-1}$ U_{k-2}, we have

$$U_{k-1}v = (1 - \alpha)v + \alpha v = v \ \text{ and } \ v = F_k U_{k-1}v = F_k v.$$

Thus, v is a common fixed point of F_k and U_{k-1}. Since $F_{k-1}U_{k-2}v = v$, we repeat the above argument to show that $F_{k-2}U_{k-2}v = v$ and consequently v must be a common fixed point of F_{k-1} and U_{k-2}. Continuing in this manner, we can prove that $F_1 U_0 v = v$ and that v is a common fixed point of F_2 and that v is a common fixed point of F_2 and U_1. Hence, v is a common fixed point of $\{F_i\}_{i=1}^{k}$.   ∎

Remark 5.49 Remark 5.48 is applicable to the above theorem.

As an application of the above theorem, we have the following: suppose we have a system of equations of the form

$$x - T_i x = f_i, i = 1, 2, \ldots, k. \tag{5.122}$$

where each T_i is a nonexpansive self-mapping of X and each f_i is a given element of X. Consider the family of mappings defined by

$$F_i x = f_i + T_i x, i = 1, 2, \ldots, k.$$

Then each F_i is a nonexpansive self-mapping of X. Also x is a solution of (5.122) if and only if x is a common fixed point of $\{F_i\}_{i=1}^{k}$.

Observation

- Since the above theorem remains valid when $K = X$, the iteration scheme of the theorem can be applied to obtain an approximate solution of the above system of equations in a specified sense.

5.6 Sequences of Contractions, Generalized Contractions and Fixed Points

We first pose the following question: Does the convergence of a sequence of mapping $\{F_i\}$ in a metric space to a mapping F imply the convergence of the sequence of their fixed points to a fixed point of F? In this context, we consider two types of convergence of mappings-

(i) pointwise convergence,
(ii) uniform convergence.

The first theorem regarding the continuity of fixed points of contraction mappings was proved by Bonsal [64]. Subsequently, Nadler, Jr. [417] obtained results concerning sequence of contraction mappings and also gave an application suggested by Dorroh. The behaviour of the fixed points of set-valued mappings has been considered by Nadler, Jr. [416] and Markin [385, 386]. Both established conditions implying the strong convergence of the fixed points of a sequence of set-valued contractions. A similar result concerning the weak convergence of the fixed points of setvalued nonexpansive mappings in a Banach space was obtained by Markin [386] who used it to obtain a stability result for generalized differential equations.

Theorem 5.133 (Bonsal [64]) *Let (X, d) be a complete metric space, and let T and $T_n (n = 1, 2, \ldots)$ be contraction mappings of X into itself with the same Lipschitz constat $k < 1$, and with fixed points u and u_n, respectively. Suppose that $\lim_{n \to \infty} T_n x = Tx$ for every $x \in X$. Then $\lim_{n \to \infty} u_n = u$.*

Proof From Theorem 5.1, we have

$$d(u_r, T_r^n x_0) \leqslant \frac{k^n}{1 - k} d(T_r x_0, x_0), x_0 \in X.$$

Set $n = 0$ and $x_0 = u$, then

$$d(u_r, u) \leqslant \frac{1}{1 - k} d(T_r u, u) = \frac{1}{1 - k} d(T_r u, Tu).$$

Since $d(T_r u, Tu) \to 0$ as $r \to \infty$, we have $\lim_{r \to \infty} d(u_r, u) = 0.$ ∎

Observation

- In the above theorem, all contraction mappins are assumed to have same Lipschitz constant. This is rather a strong condition. The following two theorems of Nadler, Jr. avoid such strong condition.

Theorem 5.134 *Let (X, d) be a metric space, let $T_i : X \to X$ be a mapping with at least one fixed point x_i, for each $i = 1, 2, \ldots$ and let $T_0 : X \to X$ be a contraction mapping with fixed point x_0. If the sequence $\{T_i\}$ converges uniformly to T_0, then the sequence $\{x_i\}$ converges to x_0.*

Theorem 5.135 *Let (X, d) be a locally compact metric space, let $T_i : X \to X$ be a contraction mapping with fixed point x_i for each $i = 1, 2, \ldots$. Let $T_0 : X \to X$ be a contraction mapping with fixed point x_0. If the sequence $\{T_i\}$ is equicontinuous and converges pointwise to T_0, then the sequence $\{x_i\}$ converges to x_0.*

Proof Let $\varepsilon > 0$ and assume that ε is sufficiently small such that $B_\varepsilon(x_0) = \{x \in X \,|\, d(x_0, x) \leqslant \varepsilon\}$ is a compact subset of X. Since $\{T_i\}$ is an equicontinuous sequence of functions converging pointwise to T_0 and since $B_\varepsilon(x_0)$ is compact, the sequence $\{T_i\}$ converges uniformly on $B_\varepsilon(x_0)$ to T_0. Choose N such that for $i \geqslant N$, $d(T_i x, T_0 x) < (1 - \alpha_0)\varepsilon$ for all $x \in B_\varepsilon(x_0)$ where $\alpha_0 < 1$ is a Lipschitz constant for T_0. Then

$$d(T_i x, x_0) \leqslant d(T_i x, T_0 x) + d(T_0 x, T_0 x_0)$$
$$< (1 - \alpha_0)\varepsilon + \alpha_0 d(x, x_0)$$
$$\leqslant (1 - \alpha_0)\varepsilon + \alpha_0 \varepsilon = \varepsilon \;\; \forall\, x \in B_\varepsilon(x_0) \text{ and } i \geqslant N.$$

This proves that T_i maps $B_\varepsilon(x_0)$ into itself. Letting T_i to be the restriction of T_i to $B_\varepsilon(x_0)$ for each $i \geqslant N$, we have each T_i to be a contraction mapping of $B_\varepsilon(x)$ into itself. Since $B_\varepsilon(x_0)$ is a complete metric space, T_i has a fixed point for each $i \geqslant N$. From the definition of T_i and the fact that T_i has only one fixed point, it follows that x_i is the fixed point of T_i, i.e. $x_i \in B_\varepsilon(x_0)$ for $i \geqslant N$. This proves that the sequence $\{x_n\}$ converges to x_0 by Theorem 5.134. ∎

The following characterization of finite dimensional space was given by Nadler, Jr. [417].

Theorem 5.136 *A separable or a reflexive Banach space is finite dimensional if and only if, whenever for a pointwise convergent sequence of contraction mappings, the sequence of fixed point converges to the fixed point of the pointwise limit mapping.*

We now give a theorem concerning multivalued contractions by Nadler, Jr. [416].

Theorem 5.137 *Let (X, d) be a complete metric space, let $T_i : X \to \mathcal{K}(X)$, the space of all compact subsets of X, be a multivalued contraction mapping with fixed point x_i for each $i = 1, 2, \ldots$ and let $T_0 : X \to \mathcal{K}(X)$ be a multivalued contraction mapping. Suppose any of the following holds:*

(i) *each of the mapping $f_1, f_2, \ldots$ has the same Lipschitz constant $\eta < 1$ and the sequence $\{T_i\}$ converges pointwise to T_0;*
(ii) *the sequence $\{T_i\}$ converges uniformly to T_0;*
(iii) *the space (X, d) is locally compact and the sequence $\{T_i\}$ converges pointwise to T_0.*

Then there is a subsequence $\{x_j\}$ of $\{x_i\}$ such that $\{x_j\}$ converges to a fixed point of T_0.

To prove Theorem 5.137, we need the following lemma stated without proof.

Lemma 5.20 *Let (X, d) be a metric space, let $T_i : X \to CB(X)$ be a multivalued contraction mapping with fixed point x_i for each $i = 1, 2, \ldots$, and let $T_0 : X \to CB(X)$ be a multivalued contraction mapping. If the sequence $\{T_i\}$ converges pointwise to T_0 and if $\{x_{i_j}\}$ is a convergent subsequence of $\{x_i\}$, then $\{x_{i_j}\}$ converges to a fixed point of T_0.*

Proof of Theorem 5.137 For each $i = 0, 1, 2, \ldots$ we define $\hat{T}_i : \mathcal{K}(X) \to \mathcal{K}(X)$ by

$$\hat{T}_i(A) = \bigcup_{a \in A} T_i(a) \text{ for all } A \in \mathcal{K}(X).$$

Then it is easy to show that $\hat{T}_i$ is a contraction mapping (Nadler, Jr. [416]) and therefore has a unique fixed point $A_i \in \mathcal{K}(X)$ for $i = 0, 1, 2, \cdots$. We observe the following:

(1) If the sequence $\{T_i\}$ converges pointwise to T_0 as assumed in (i) and (iii), then $\{T_i\}$ converges uniformly on compact subsets of X to T_0 (Rudin, [533], p. 156) and hence, the sequence $\{\hat{T}_i\}$ converges pointwise on $\mathcal{K}(X)$ to $\hat{T}_i$.

(2) If the sequence $\{T_i\}$ converges uniformly to T_0 as in (ii), then the sequence $\{\hat{T}_i\}$ converges uniformly on compact subsets of X to T_0(Rudin, [533], p. 156) and hence, the sequence $\{\hat{T}\}$ converges pointwise on $\mathcal{K}(X)$ to $\hat{T}_0$. Then by Theorems 5.133 and 5.134, the sequence $\{A_i\}$ converges to A_0. Hence, $S = \bigcup\{A_i | i = 0, 1, 2, \cdots 2, \ldots\}$ is a compact subset of X.

By the iteration procedure of Banach (Lusternik and Sobolov [378], p. 40–42), the sequence $\{\hat{T}_i^n(x_i)\}_{n=1}^{\infty}$ converges to A_i and therefore it follows that $x_i \in A_i$ for each $i = 1, 2, \ldots$, since $x_i, \in \hat{T}_i^n(x_i)$ for all $n = 1, 2, \ldots$. Thus, $\{x_i\}$ is a sequence in the compact set X. Hence, $\{x_i\}$ has a convergent subsequence $\{x_{i_j}\}$ which, by Lemma 5.20, converges to a fixed point of T_0.

Let A be a closed bounded subset of a Hilbert space $\mathcal{H}$, d the metric of $\mathcal{H}$, and H the Hausdorff metric on the closed subsets of A generated by d. It is assumed that the family of setvalued mappings $T_k, k = 0, 1, \ldots$ satisfies the following conditions:

(i) $T_k(x)$ is a nonempty, closed and convex subset of A for each $x \in A$;
(ii) each T_k is a setvalued contraction, i.e. there is a $\eta \in (0, 1)$ such that

$$H(T_k(x), T_k(y)) \leqslant \eta d(x, y) \text{ for } x, y \in A \text{ and } k = 0, 1, 2, \ldots;$$

(iii) $\lim_{k \to \infty} H(T_k(x), T_0(x)) = 0$ uniformly for all $x \in A$.

Theorem 5.138 (Markin [385]) *If the conditions (i) to (iii) as above are satisfied, then the fixed point sets of the sequence $\{T_k\}, k = 1, 2, 3, \ldots$ converge to the fixed point set of T_0 in the Hausdorff metric.*

Next theorem, also proved by Markin [386], deals with the weak convergence of the fixed points of setvalued nonexpansive mappings in a Banach space.

Theorem 5.139 *Let X be a strictly convex Banach space with a weakly continuous duality map, and let B be a weakly compact convex subset of X. Suppose that $\{T_i\}$ is a sequence of nonexpansive mappings from X into $K_c(X)$ (the space of nonempty, compact and convex subsets of X) converging to T_0 in the Hausdorff metric H and leaving N invariant. If x_i is a fixed point of T_i in B for $i = 1, 2, \ldots$, and $x_i \rightharpoonup x_0$, then x_0 is a fixed point of T_0.*

Proof For any nonexpansive map T of X into $K_c(X)$, the map $I - T_i, i = 0, 1, 2, \ldots,$ are J-monotone. Since x_i is a fixed point of T_i, we have $0 \in (I - T_i)(x_i), i = 1, 2, \ldots$ By the J—monotonicity property, for any $v \in X$, there is a $v_i \in (I - T_i)(v)$ such that

$$(J(v - x_i), v_i - 0) \geqslant 0, i = 1, 2, \ldots \tag{5.123}$$

The pointwise convergence of the sequence $\{T_i\}$ implies that

$$\lim_{i \to \infty} H((I - T_i)(v), (I - T_0)(v)) = 0 \text{ for any } v \in X.$$

Since the sets $(I - T_i)(v))$ are compact for $i = 0, 1, 2, \ldots$ the set $\bigcup_{i=0}^{\infty}(I - T_i)(v)$ is also compact for $v \in X$ (Castiang [132]) and the sequence $\{v_i\}$ can be assumed convergent to a point $v_0 \in (I - T_0)(v)$. Taking the limit in (5.123) we have

$$(J(v - x_0), v_0 - 0) \geqslant 0. \tag{5.124}$$

Since the mapping $I - T_0$ is J-monotone, the inequality (5.124) implies that $0 \in (I - T_0)(x_0)$. This proves that x_0 is a fixed point of T_0.

Definition 5.69 Let (X, d) be a complete metric space. A mapping $T : X \to X$ is said to be of generalized Kannan–Reich type if

$$d(Tx, Ty) \leqslant a_1 d(Tx, x) + a_2 d(Ty, y) + a_3 d(Tx, y)$$
$$+ a_4 d(Ty, x) + a_5 d(x, y) \text{ for all } x, y \in X$$

where a_i's are nonnegative numbers such that $\sum_{n=1}^{5} a_n < 1$ (here $a_1 = a_2$ and $a_3 = a_4$ by symmetry).

The following two theorems were proved by Bose and Mukherjee [67].

Theorem 5.140 *Let $\{T_n\}$ be a sequence of self-mappings of X having at least one fixed point x_n each and let $\{T_n\}$ converge uniformly to T_0, a mapping of generalized Kannan–Reich type. Let x_0 be the unique fixed point of T_0. Then $x_n \to x_0$.*

Theorem 5.141 *Let $\{T_n\}$ be a sequence of mappings of generalized Kannan–Reich type and let $\{T_n\}$ converges pointwise to T, a generalized Kannan–Reich-type mapping. Let x_n and x_0 be fixed points of T_n and T, respectively. Then $x_n \to x_0$.*

The following theorem of Bose and Mukherjee [67] is a generalization of a theorem of Wong [608] to multivalued mappings.

Theorem 5.142 *Let $\{T_n\}$ be a sequence of multivalued mappings of X into $CB(X)$ satisfying the following condition:*

$$H(T_n x, T_n y) \leqslant a_n^1 d(T_n x, x) + a_n^2 d(T_n y, y) + a_n^3 d(T_n x, y)$$
$$+ a_n^4 d(T_n y, x) + a_n^5 d(x, y)$$

for all $x, y \in X$ where $a_n^j \geqslant 0$, $j = 1, 2, 3, 4, 5$ and $\sum_{j=1}^{5} a_n^j < 1$.

Let $a_n^j \to a^j$ as $n \to \infty$ and $\sum_{j=1}^{5} a^j < 1$. Let $\{T_n\}$ converge to T_0 pointwise and let x_n be the fixed point of T_n. If x_0 is any cluster point of the sequence $\{x_n\}$, then $x_0 \in T_0 x_0$.

Proof Let $x_{n_i} \to x_0$. For simplicity of notation, we write i in place of n_i. Then

$$
\begin{aligned}
d(x_0, T_0 x_0) &\leqslant d(x_0, x_i) + d(x_i, T_0 x_0) \\
&\leqslant d(x_0, x_i) + H(T_i x_i, T_0 x_0) \text{ as } x_i \in T_i x_i, \\
&\leqslant d(x_0, x_i) + H(T_i x_i, T_i x_0) + H(T_i x_0, T_0 x_0) \\
&\leqslant d(x_0, x_i) + a_i^1 d(T_i x_i, x_i) + a_i^2 d(T_i x_0, x_0) + a_i^3 d(T_i x_i, x_0) \\
&\quad + a_i^4 d(T_i x_0, x_i) + a_i^5 d(x_0, x_i) + H(T_i x_0, T_0 x_0).
\end{aligned}
$$

After some simplification, we have

$$(1 - a_i^2 - a_i^4) d(x_0, T_0 x_0) \leqslant (1 + a_i^3 + a_i^4 + a_i^5) d(x_0, x_i)$$
$$+ (1 + a_i^2 + a_i^4) H(T_i x_0, T_0 x_0).$$

Taking limit as $i \to \infty$, we have

$$(1 - a^2 - a^4) d(x_0, T_0 x_0) = 0,$$

i.e. $d(x_0, T_0 x_0) = 0$ as $1 - a^2 - a^4 \neq 0$.

Since $T_0 x_0$ is closed, we have $x_0 \in T_0 x_0$.

Remark 5.50 In this case, $a_n^1 = a_n^2$ and $a_n^3 = a_n^4$ due to symmetry. Hence, each T_n has a fixed point by Corollary 5.31.

We now state a theorem concerning sequences of mappings of Husain–Sehgal type (refer Theorem 5.8).

Theorem 5.143 *Let T and a sequence $\{T_n\}$ be self-mappings of a complete metric space X such that $T_n \to T$ uniformly. Suppose for each $n \geqslant 1$, T_n has a fixed point x_n and T is a mapping of Husain–Sehgal type. If u is the fixed point of T and sup $d(x_n, u) < \infty$, then $x_n \to u$.*

5.7   Fixed Point Theorems in Ordered Banach Spaces

In this section, we discuss methods developed for the investigation of various questions concerning positive solutions of operator equations.

We mainly deal with fixed point theorems developed by Krasnoselski [345]. Towards the end, we follow them up with fixed point theorems due to Leggett and Williams [363], Amann [17] and Gatica and Smith [241]. Our choice of theorems is motivated by their applications to differential and integral equations arising in applied problems.

Definition 5.70 Let X be a real Banach space. A set $K \subset X$ is called cone if the following conditions hold good:

  (i)   the set K is closed;
 (ii)   if $u, v \in K$, then $\alpha u + \beta v \in K$ for all $\alpha, \beta \geqslant 0$;
(iii) $x, -x \in K \Rightarrow x = 0$

From (ii) it is obvious that cone K is a convex set.
Solid cone - Cone K is called solid cone if it contains its interior points.
Reproducing cone - Cone K is called reproducing if $x \in X$ has a representation $x = u - v, u, v \in K$.
Notice that every solid cone is reproducing.

Definition 5.71 The linear space X is called partially ordered if, for certain pair of points $x, y \in X$, the relation $x \leqslant y$ is defined and it satisfies the following properties:

  (i) $x \leqslant y \Rightarrow tx \leqslant ty$ for $t \geqslant 0$ and $tx \geqslant ty$ for $t < 0$;
 (ii) $x \leqslant y$ and $y \leqslant x \Rightarrow x = y$;
(iii) $x_1 \leqslant y_1$ and $x_2 \leqslant y_2 \Rightarrow x_1 + x_2 \leqslant y_1 + y_2$;
 (iv) $x \leqslant y$ and $y \leqslant z \Rightarrow x \leqslant z$.

In a Banach space X with a cone K, a partial order relation is introduced in the following manner:

$$x \leqslant y \text{ if } y - x \in K.$$

It can be easily seen that all the above properties are satisfied. The partial order relation defined by K also satisfies the following property.
(v) One can pass to the limit in the inequalities. That is,

If $\|x_n - x\| \to 0$, $\|y_n - y\| \to 0$ and $x_n \leqslant y_n$ $(n = 1, 2, \ldots)$, then $x \leqslant y$.

We note that $K = \{x \in X : x \geqslant 0\}$. The elements of

$$\dot{K} = K - \{0\} = \{x \in X : x > 0\}.$$

are called positive. If K has nonempty interior, we write $x \ll y$ iff $y - x \in \text{int } K$. We write $x \not\leqslant y$ iff $y - x \notin K$.

Ordered Banach space - A real Banach space X ordered by a cone K is called an ordered Banach space and is denoted by (X, K).

Order interval - For every pair $x, y \in X$, the set $\langle x, y \rangle = \{z \in X; x \leqslant z \leqslant y\}$ is called the order interval or conical segment. It is nonempty if and only if $x \leqslant y$.

Order bounded - A subset $D \subset X$ is called order bounded if D is contained in some order interval.

Order convex - D is said to be order convex whenever $x, y \in D$ implies $\langle x, y \rangle \subset D$.

Order convex hull - For every subset $D \subset X$ we define $[D]$, the order convex hull of D, to be the smallest order convex subset of X containing D, i.e. $[D] = \cup \{\langle x, y \rangle : x, y \in D\}$.

Definition 5.72 Let $F : X \to X$ be a nonlinear operator. Then the operator

1. F is called called isotone or increasing if $x \leqslant y \Rightarrow Fx \leqslant Fy$.
2. F is called strictly increasing if $x < y \Rightarrow Fx < Fy$.
3. F is called strongly increasing if int $K \neq \varnothing$ and if $x < y \Rightarrow F(y) - F(x) \in$ int K.

Definition 5.73 A cone K is called normal if a $\delta > 0$ exists such that $\|e_1 + e_2\| \geqslant \delta$, whenever $e_1, e_2 \in K$ and $\|e_1\| = \|e_2\| = 1$. The cone of nonnegative functions in the space C(or L_p) is a normal cone.

The following theorem (refer Krasnoselski [345]) gives some important characterizations of normal cones. A Banach space X ordered by a cone K is called an ordered Banach space and is denoted by (X, K).

Theorem 5.144 *Let (X, K) be an ordered Banach space. Then the following statements are equivalent:*

 (i) K is normal;
 (ii) every order interval is bounded;
(iii) there exists an equivalent increasing norm on X;
(iv) the order convex hull of every bounded set is bounded.

Definition 5.74 The space X is called regularly partially ordered if each bounded monotonic sequence in it has a limit. A cone which generates a regular partial ordering is called regular. Every regular cone is normal.

Example 5.48 Let X be a compact Hausdorff space. Let $C(X)$ denote the Banach space of all continuous real-valued functions on X with the sup norm. $C(X)$ is an ordered Banach space with the natural ordering. Its (positive) cone, denoted by $C_+(X)$, is normal and had nonempty interior. The sup norm is monotone, and f is an interior point of $C_+(X)$ if and only if $f(x) > 0$ for every $x \in X$.

Example 5.49 Let Ω be a Lebesgue measurable subset of $\mathbb{R}^n$ of positive measure. Each of the Banach spaces $L_p(\Omega)$ (real valued), $1 \leqslant p \leqslant \infty$ is an ordered Banach space with respect to the natural ordering and its cone is

$$L_{p+}(\Omega) = \{f \in L_p(\Omega) : f(x) \geqslant 0 \text{ for almost all } x \in \Omega\}.$$

Notice that each $L_{p+}(\Omega)$ is a normal cone since L_p norm is monotone. We also notice that each of the cones $L_{p+}(\Omega)$, $1 \leqslant p < \infty$, is generating but has empty interior whereas $L_{\infty+}(\Omega)$ has nonempty interior.

Definition 5.75 A nonlinear operator $F : X \to X$ is said to be Fréchet differentiable with respect to the cone K at x_0 if there exists $F'(x_0) \in \mathcal{L}(X, X)$ such that

$$F(x_0 + h) - F(x_0) = F'(x_0)h + w(x_0, h), \quad h \in K$$

where

$$\lim_{h \in K, \|h\| \to 0} \frac{\|w(x_0, h)\|}{\|h\|} = 0.$$

$F'(x_0)$ is called the Fréchet derivative of F at x_0 with respect to the cone K (or along K).

Definition 5.76 The function $y(t)$ with values in X is differentiable at infinity if the ratio $(1/t)y(t)$ converges in norm to some element $y'(\infty) \in X$ as $t \to \infty$.

(i) The operator F is said to be differentiable at infinity in the direction $h \in K$ if the function $y(t) = F(th)$ is differentiable at infinity. The operator F is said to be differentiable at infinity if F is differentiable in all directions $h \in X (h \neq 0)$ were the derivatives $y'(\infty)$ of the function $y(t) = F(th)$ are representable in the form $y'(\infty) = F'(\infty)h(h \in X)$ for some $F'(\infty) \in L(X, X)$.

(ii) F is said to be differentiable at infinity in the cone if the preceding statement holds for all $h \in K(h \neq 0)$, and $F'(\infty)$ is called the derivative at infinity with respect to the cone.

(iii) $F'(\infty)$ is called asymptotic derivative of the operator F, and F is asymptotically linear if

$$\lim_{R \to \infty} \sup_{\|x\| \geqslant R} \frac{\|Fx - F'(\infty)x\|}{\|x\|} = 0. \tag{5.125}$$

(iv) $F'(\infty)$ is called asymptotic derivative with respect to the cone K, and F is said to be asymptotically linear with respect to the cone K if (5.125) hold for all $x \in K$.

Definition 5.77 A cone K is said to allow the plastering K_1 if a cone K_1 can be found such that every nonzero element $x_0 \in K$ is an interior point of the cone K_1 and, furthermore, it lies in the cone K_1 together with the spherical ball of radius $b\|x_0\|$, where b does not depend on the element x_0.

Definition 5.78 Let X be a Banach space with K as a cone.

1. The linear operator $F : X \to X$ is called positive if it maps the cone K into itself. The linear positive operator is isotone.
2. The linear operator F is called strictly positive if $F(\dot{K}) \subset \dot{K}$.
3. If K has nonempty interior, then F is called strongly positive if $F(\dot{K}) \subset \text{int} K$.

5.7.1 Existence of a Characteristic Vector

Let us assume that the linear operator T acts in the space X with the cone K and leaves this cone invariant, i.e. T is a positive operator. We want to investigate the existence of characteristic vectors of the operator T belonging to K. These characteristic vectors are called positive and corresponding characteristic values are nonnegative. We remark that we do not call positive the characteristic values of the operator T if the corresponding characteristic vectors are not in K, even if they are positive.

The following theorem is an illustration of the use of fixed point theorems.

Theorem 5.145 (Krasnoselski [345]) *Let the linear positive operator T be continuous which maps bounded subsets into compact sets. Let the relation*

$$T^p u \geqslant \alpha u \quad (\alpha > 0) \tag{5.126}$$

be satisfied for some nonzero element u such that $-u \notin K$, $u = v - w$ $(v, w \in K)$ and p is some natural number. Then the operator T has at least one characteristic vector x_0 in K:

$$T x_0 = \eta_0 x_0 \tag{5.127}$$

where the positive characteristic number η_0 satisfies the inequality

$$\eta_0 \geqslant p\sqrt{\alpha}. \tag{5.128}$$

Proof The existence of a single characteristic vector $x_0 (\|x_0\| = 1)$ for the operator T in the cone K is shown. The operators T_n $(n = 1, 2, 3, \ldots)$ are defined as follows:

$$T_n x = \frac{T\left(x + \frac{v}{n}\right)}{\left\| T\left(x + \frac{v}{n}\right)\right\|} \tag{5.129}$$

$x \in B = K \cap \mathbb{B}_1$ where $\mathbb{B}_1 = \{x : \|x\| \leqslant 1\}$.

Each T_n transforms B into itself and is continuous and maps bounded sets into compact sets together with the operator T. Since $T\left(x + \frac{v}{n}\right) \geqslant \frac{1}{n} T v \neq 0$, we have $\left\| T\left(x + \frac{v}{n}\right)\right\| > 0$ for $x \in \mathbb{B}_1$. By Schauder's theorem, every operator T_n has a fixed point $x_n \in B$. From Eq. (5.129), we have $\|x_n\| = 1$. Writing

$$\lambda_n = \left\| T\left(x + \frac{v}{n}\right)\right\|$$

we get the equation

$$T\left(x_n + \frac{v}{n}\right) = \lambda_n x_n, \quad n = 1, 2, 3, \ldots. \tag{5.130}$$

The relations

$$x_n \geqslant \frac{1}{\lambda_n} T x_n, \quad x_n \geqslant \frac{1}{n\lambda_n} T v, \quad n = 1, 2, \ldots \tag{5.131}$$

follow from (5.130). It follows from Eqs. (5.126) and (5.130) that

$$x_n \geqslant \frac{1}{\lambda_n^{p-1}} T^{p-1} x_n = \frac{1}{\lambda_n^p} T^p \left(x_n + \frac{v}{n} \right) \geqslant \frac{1}{n\lambda_n^p} T^p v \geqslant \frac{1}{n\lambda_n^p} T^p u \geqslant \frac{\alpha}{n\lambda_n^p} u$$

$$(n = 1, 2, 3, \ldots)$$

So, there exists a maximal $\beta_n > 0$ such that $x_n \geqslant \beta_n u$. Furthermore, the relations

$$x_n \geqslant \frac{1}{\lambda_n^p} T^p \left(x_n + \frac{v}{n} \right) \geqslant \frac{1}{\lambda_n^p} \left(\beta_n + \frac{1}{n} \right) T^p u \geqslant \frac{\alpha}{\lambda_n^p} \left(\beta_n + \frac{1}{n} \right) u. \tag{5.132}$$

follow from (5.126) and (5.132). Moreover, the maximality of numbers β_n gives

$$\frac{\alpha}{\lambda_n^p} \left(\beta_n + \frac{1}{n} \right) \leqslant \beta_n \quad (n = 1, 2, 3, \ldots),$$

which is equivalent to the following inequality

$$\lambda_n^p \geqslant \alpha + \frac{\alpha}{n\beta_n} \quad (n = 1, 2, 3, \ldots). \tag{5.133}$$

Because the operator T is compact, a subsequence of indices $n_i (i = 1, 2, \ldots)$ can be chosen such that the sequence $T \left[x_{n_i} + \left(\frac{v}{n_i} \right) \right]$ strongly converges to some element y^*. By (5.133), $\{\lambda_{n_i}\}$ converges to some number λ_0 which satisfies inequality (5.128). Then the element $\{x_{n_i}\}$ converges in norm to the element $x_0 = \frac{1}{\lambda_0} y^*$. To obtain (5.127), it is sufficient to take limit in the equation

$$T \left(x_{n_i} + \frac{v}{n_i} \right) = \lambda_{n_i} x_{n_i} \quad (i = 1, 2, \ldots).$$

The above theorem imposes two restrictions on T: condition (5.126) and T are continuous and T maps bounded sets into compact sets. We state three theorems due to Krasnoselski, without proofs, where these restrictions on T have been replaced by assumption involving some properties of the space X and the cone K.

Definition 5.79 A linear positive operator T is uniformly positive operator if $\|Tx\| \geqslant b\|x\| (x \in K)$ where $b > 0$.

Theorem 5.146 *Let a continuous linear operator T be uniformly positive and transforms bounded sets into compact sets. Then T has at least one positive characteristic vector.*

Theorem 5.147 *Let the space X be weakly complete and the unit sphere in X be weakly compact. Let the cone K allow plastering. Then every linear operator T, leaving the cone K invariant, has at least one characteristic vector.*

Theorem 5.148 *Let the space X be weakly complete and the unit sphere in the space X be weakly compact. Let the cone K be normal and the linear operator T satisfy the condition*

$$a(x)u_0 \leqslant Tx \leqslant \rho a(x)u_0 \quad (x \in K)$$

where

$$a(x) \geqslant 0, \; \rho \geqslant 1. \tag{5.134}$$

Then T has at least one positive characteristic vector.

We now study conditions under which the equation

$$Tx = x \tag{5.135}$$

with a positive nonlinear operator T has at least one nonzero solution x^* in the cone K. In majority of cases, $T(0) = 0$. Thus, it is a question of a second solution in the cone K.

Theorem 5.149 *Let an operator T, isotone on the segment $\langle x_0, u_0 \rangle$ into itself, that is*

$$Tx_0 \geqslant x_0 \text{ and } Tu_0 \leqslant u_0. \tag{5.136}$$

Then for the existence on $\langle x_0, u_0 \rangle$ of at least one fixed point x^* for the operator T, it suffices that any one of the following three conditions be satisfied:

 (i) *The cone K is regular, and the operator T is continuous;*
 (ii) *the cone K is normal, and the operator T is continuous and transforms bounded sets into compact sets;*
 (iii) *the cone K is normal, the space X is weakly complete, the unit sphere in X is weakly compact, and the operator T is weakly continuous.*

Proof When condition (i) is satisfied, let us construct the iterative sequence $\{x_n\}$ defined by

$$x_n = Tx_{n-1} \quad (n = 1, 2, \ldots). \tag{5.137}$$

It follows from condition (5.136) and from isotone property of the operator T that the sequence given by (5.137) is monotone and bounded:

$$x_0 \leqslant x_1 \leqslant x_2 \leqslant \cdots \leqslant x_n \leqslant \cdots$$

$$x_n \leqslant u_0 \quad (n = 0, 1, 2, \ldots).$$

Since the cone K is regular, the sequence (5.137) converges in norm to some element $x^* \in \langle x_0, u_0 \rangle$. Taking limit in (5.137) as $n \to \infty$, we obtain the equality $Tx^* = x^*$.

If condition (ii) or (iii) is satisfied, then statement of the theorem follows from Schauder's or Tychonoff's fixed point theorem, respectively. Note that with condition (ii) or condition (iii) holding, the fixed point x^* of the operator T also can be obtained as the limit of the sequence (5.137). In case (ii), convergence follows from its compactness and in case (iii), the sequence (5.137) converges weakly to x^*.

Remark 5.51 Tychonoff's fixed point theorem quoted above is the following: If a weakly continuous operator T maps a weakly compact and weakly complete set C into itself, then T has a fixed point in C.

The conditions of Theorem 5.149 do not guarantee the uniqueness of the fixed point for the operator T on a conical segment $\langle x_0, u_0 \rangle$. Following example illustrates this.

Example 5.50 Let X be a one-dimensional space of real numbers, and K be a set of nonnegative numbers. Let T be defined as

$$Tx = x + \frac{\sin x}{2} \quad (x \geqslant 0). \tag{5.138}$$

This operator (a function) is isotone since the derivative is positive. Set $x_0 = \frac{\pi}{2}$, $u_0 = \frac{7\pi}{2}$. Then inequalities (5.136) are satisfied. At the same time, the operator T has three fixed points π, 2π and 3π on $\left[\frac{\pi}{2}, \frac{7\pi}{2}\right]$ (Fig. 5.4).

Definition 5.80 Let Y be a nonempty subset of some ordered space X. A fixed point x of a mapping $T : Y \rightarrow X$ is called a minimal (maximal) if every fixed point y of T in Y satisfies $x \leqslant y$ $(y \leqslant x)$. There is at most one minimal (maximal) fixed point.

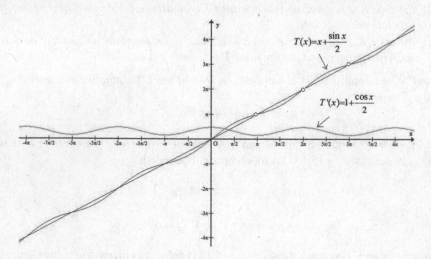

Fig. 5.4 Nonuniqueness of the fixed point for the operator T

Theorem 5.150 (Amann [17]) *Let (X, K) be an ordered Banach space and let D be an order convex subset of X. Suppose that $F : D \to X$ is increasing (isotone) mapping which is continuous and maps every order interval in D into a relatively compact set. If there exists $\overline{y}, \hat{y} \in D$ with $\overline{y} \leqslant \hat{y}$ such that $\overline{y} \leqslant F(\overline{y})$ and $F(\hat{y}) \leqslant \hat{y}$, then F has a minimal fixed point $\overline{x}$ in $(\overline{y} + K) \cap D$. Moreover, $\overline{x} \leqslant \hat{y}$ and $\overline{x} = \lim_{k \to \infty} F^k(\overline{y})$, that is, the minimal fixed point can be computed by the iteration sequence $x_0 = \overline{y}, x_{k+1} = F(x_k)$. Furthermore, the sequence $\{x_k\}$ is increasing.*

Proof Since F is increasing, the mapping F maps the order interval $\langle \overline{y}, \hat{y} \rangle$ into itself. Hence, the sequence $\{x_k\}$ is well defined, and it is relatively compact since it is contained in $F \langle \overline{y}, \hat{y} \rangle$. Hence, it has at least one limit point. It can be shown by induction that the sequence $\{x_k\}$ is increasing. This implies that it has precisely one limit point, say, $\overline{x}$ and the whole sequence converges to $\overline{x}$. Since F is continuous, $\overline{x}$ is a fixed point of F. If x is any fixed point in D such that $x \geqslant \overline{y}$, then by replacing $\hat{y}$ by x in the above discussion, it follows that $\overline{x} \leqslant x$. Thus $\overline{x}$ is the minimal fixed point of F in $(\overline{y} + K) \cap D$. It is of interest to note that the existence of minimal fixed point is not claimed in D.

Corollary 5.32 *Let (X, K) be an ordered Banach space and let $\langle \overline{y}, \hat{y} \rangle$ be a nonempty order interval in X. Suppose that $F : (\langle \overline{y}, \hat{y} \rangle) \to X$ is an increasing (isotone) continuous mapping with $F(\langle \overline{y}, \hat{y} \rangle)$ relatively compact such that $\overline{y} \leqslant F(\overline{y})$ and $F(\hat{y}) \leqslant \hat{y}$. Then F has a minimal fixed point $\overline{x}$ and a maximal fixed point $\hat{x}$.*

   Moreover, $\overline{x} = \lim_{k \to \infty} F^k(\overline{y})$ and $\overline{X}) = \lim_{k \to \infty} F^k(\hat{y})$, and the sequences $\{F^k(\overline{y})\}$ and $\{F^k(\hat{y})\}$ are increasing and decreasing, respectively.

Proof The assertion about $\overline{x}$ follows from the above theorem. The assertion about $\hat{x}$ follows by applying the preceding theorem to the mapping $F : \langle \overline{z}, \hat{z} \rangle \to X$ where $G(x) = -F(-x), \overline{z} = -\hat{y}$ and $\hat{z} = -\hat{y}$.

Corollary 5.33 *Let (X, K) be an ordered Banach space and let $F : K \to K$ be increasing and continuous and maps any order interval into a relatively compact set. Then F has a minimal fixed point $\overline{x}$ if and only if there exists $\hat{y} \in K$ such that $\hat{F}(y) \leqslant \hat{y}$. In this case $\overline{x} \leqslant \hat{y}$, and the sequence $\{F^k(0)\}$ converges increasingly towards $\overline{x}$.*

Proof If F has a minimal fixed point $\overline{x}$, then $F(\overline{x}) \leqslant \overline{x}$. Since $F(0) \geqslant 0$, the converse assertion is true by the above theorem.

Remark 5.52 The above results are true if K is normal and F is increasing, continuous and if F maps bounded sets into compact sets.

   Every order interval is bounded by Theorem 5.144. Thus F is continuous and maps every order interval into relatively compact set.

   Using the asymptotic behaviour of a mapping, the existence of fixed points of the mapping can be derived as shown in the following theorems due to Amann [19] stated without proof.

Definition 5.81 Let Y be a nonempty set and X be an ordered set.

1. A mapping $\hat{F} : Y \to X$ is said to be a majorant of the mapping $F : Y \to X$ if $F(y) \leqslant \hat{F}(y)$ for all $y \in Y$.
2. Minorants are defined by just reversing the inequality.

For, a linear operator $F : X \to X$ belonging to $L(X)$, let

$$r(F) = \lim_{k \to \infty} \|F^k\|^{1/k}.$$

It is called the spectral radius of F.

Theorem 5.151 *Let (X, K) be an ordered Banach space with K having nonempty interior. Let $F : K \to K$ be an increasing (isotone) mapping which is continuous on order intervals and maps every order interval into relatively compact set and satisfies $F'(0) = 0$. Suppose that F has strongly positive Fréchet derivative $F'(0)$ at 0 such that $r(F'(0)) > 1$. Moreover, suppose that F has a continuous, asymptotically linear majorant $\hat{F}$ which maps bounded sets into compact sets such that $r(\hat{F}'(\infty)) < 1$. Then F has at least one fixed point in the interior of the cone.*

To ensure the existence of fixed point for nonisotone operators F which is positive, we apply different fixed point principles and, in particular, Schauder's and Tychonoff's principles, we assume that the operator F maps bounded sets into compact sets and is continuous or (if the space X is weakly complete and the unit sphere weakly compact) weakly continuous. The proof reduces to finding a convex set $C \subset X$ which remains invariant under the operator F. The conical segment $\langle x_0, u_0 \rangle$, the intersections $K_0 \cap \langle x_0, u_0 \rangle$ of the conical segment $\langle x_0, u_0 \rangle$, with some cone $K_0 \subset K$, the intersection $K(r)$ of the cone K with the sphere $\|x\| \leqslant R$ and so on, can appear as the set C. We discuss only two theorems of this type due to Krasnoselski [345].

Theorem 5.152 *Let the operator F, positive on the cone K, have a strong asymptotic derivative $F'(\infty)$ with respect to a cone where the spectrum of the linear operator $F'(\infty)$ lies in the circle $|\eta| \leqslant \rho_0$, where $\rho < 1$. Then there exists at least one fixed point for the operator F in the cone K if any one of the following conditions be satisfied:*

(i) the operator F is continuous and maps bounded sets into compact sets;
(ii) the space X is weakly complete, the unit sphere is weakly compact and the operator F is weakly continuous.

Proof A norm $\|x\|_0$ can be introduced in X equivalent to the originally given one such that $\|F'(\infty)\|_0 = 1 - \eta$ where $\eta > 0$. Also it can be shown that in the new norm, the operator F satisfies the inequalities

$$\|Fx\|_0 < \|x\|_0 \quad (x \in K, \|x\|_0 \geqslant R)$$

where R is some positive number.

(Notice that for a given $\varepsilon_0 > 0$, there exists $\rho_0 > 0$ such that

$$\|x\|_0 = \frac{\|Fx\|}{\rho_0 + \varepsilon_0} + \frac{\|F^2 x\|}{(\rho_0 + \varepsilon_0)^2} + \cdots + \frac{\|F^{n_0 - 1} x\|}{(\rho_0 + \varepsilon_0)^{n_0 - 1}}, \quad x \in X$$

defines a norm.)

The above inequality implies that the operator F maps the intersection C of the cone K with the sphere $\|x\|_0 \leqslant R_0$ into itself, where

$$R_0 = \max \left\{ R, \sup_{x \in K, \|x\|_0 \leqslant R} \|Fx\|_0 \right\}.$$

This set C is bounded, convex and closed. The existence of a fixed point x_0 of F in C follows from Schauder's principle for case (i) and from Tychonoff's principle for case (ii).

Theorem 5.153 *Let the positive operator $F(F(0) = 0)$ have a Fréchet derivative $F'(0)$ and a asymptotic derivative $F'(\infty)$ with respect to a cone. Let the spectrum of the operator $F'(\infty)$ lie in the circle $|\eta| \leqslant \rho$, where $\rho < 1$. Let the operator $F'(0)$ have in K a characteristic vector h_0 such that whenever $F'(0)h_0 = \eta_0 h_0$, we have $\eta_0 > 1$. Then there exists a nonzero fixed point of F in K if the operator F is continuous and maps bounded sets into compact sets (or if the operator F is weakly continuous, the space X is weakly complete, the unit sphere in X is weakly compact and the cone K allows plastering).*

Proof Define the operators F_n by the equality

$$F_n x = \begin{cases} Fx & \text{if } x \in K, \|x\| \geqslant \frac{1}{n}, \\ Fx + \left(\frac{1}{n} - \|x\|\right) h_0 & \text{if } x \in K, \|x\| \leqslant \frac{1}{n}. \end{cases}$$

Each F_n is continuous and maps bounded sets into compact sets. They satisfy all the conditions of Theorem 5.152 and hence, have fixed points in the cone K. We shall show that the norms of these fixed points are greater than $\frac{1}{n}$ if $\frac{1}{n}$ is sufficient small. Then the fixed points of the operator F_n will be nonzero fixed points of the operator F, and this will prove the theorem. We assume that every operator F_n has a fixed point whose norm does not exceed $1/n$. We can then construct a sequence $\{x_n\} \subset K$ such that

$$0 < \|x_n\| \leqslant \frac{1}{n} \quad (n = 1, 2, \ldots) \tag{5.139}$$

and

$$Fx_n + \left(\frac{1}{n} - \|x_n\|\right) h_0 = x_n \quad (n = 1, 2, \ldots) \tag{5.140}$$

Without loss of generality, we can assume that the sequence

$$F'(0)\left(\frac{x_n}{\|x_n\|}\right) \quad (n = 1, 2, \ldots)$$

converges in norm to some element $z \in K$. Rewriting Eq. (5.139) in the form

$$\left(\frac{1}{n\|x_n\|} - 1\right) h_0 = \frac{x_n}{\|x_n\|} - \frac{Fx_n - F'(0)x_n}{\|x_n\|} - F'(0)\left(\frac{x_n}{\|x_n\|}\right), \tag{5.141}$$

it follows that

$$\|h_0\| \lim_{n \to \infty} \left(\frac{1}{n\|x_n\|} - 1\right) \leqslant 1 + \|z\|.$$

Thus, without loss of generality, we can assume that the sequence $\{\frac{1}{n\|x_n\|} - 1\}_{n=1}^{\infty}$ converges to some number α which by virtue of (5.139) is nonnegative. Then, it follows from (5.141) that the sequence $\{\frac{x_n}{\|x_n\|}\}_{n=1}^{\infty}$ converges in norm to some vector $u_0 \in K$ where $\|u_0\| = 1$. Taking limit in (5.141), we have

$$u_0 = F'(0)u_0 + \alpha h_0. \tag{5.142}$$

Notice that the number α is positive since $F'(0)$ does not have characteristic vectors in K with a characteristic value equal to unity. Hence, a number $t_0 > 0$ can be found such that $u_0 \geqslant t_0 h_0$ and whenever $u \geqslant t h_0, t \leqslant t_0$. At the same time, the inequality $u_0 \geqslant F'(0)(t_0 h_0) + \alpha h_0 = (\eta_0 t_0 + \alpha) h_0$ follows from (5.142), that is, $u_0 \geqslant t h_0$, where $t = \eta_0 t_0 + \alpha > t_0$. This completes the proof as we have arrived at contradiction.

In the following result, we assume that the positive operator F has a Fréchet derivative $F'(0)$ with respect to a cone and a asymptotic derivative $F'(\infty)$ with respect to a cone. Further, we assume that F is either continuous and maps bounded sets into compact sets or weakly continuous. For the latter case, as usual, the space X is weakly complete and weakly compact, and the cone K is plastered.

Theorem 5.154 *Let the operator $F'(0)$ not have positive eigenvalues, exceeding or equal to 1. Let the operator $F'(\infty)$ have an eigenvector*

$$g \in K; \quad F'(\infty)g = \lambda_\infty g,$$

where $\lambda_\infty > 1$ and $F'(\infty)$ does not have positive eigenvalues equal to 1. Then the operator F has at least one nonzero fixed point in K.

Let $F : \langle \overline{y}, \hat{y} \rangle \to \mathbb{R}$ be a continuous function on some nontrivial interval $\langle \overline{y}, \hat{y} \rangle \subset \mathbb{R}$ such that $(F(\overline{y}) - \hat{y})(\hat{y} - F(\hat{y})) > 0$. Then by intermediate value theorem, F has a fixed point in $\langle \overline{y}, \hat{y} \rangle$. This face has been generalized by the theorem on the compression of a cone and the theorem on the expansion of a cone, respectively. One-dimensional version of both the theorems reduces to the above-mentioned result. Compression and expansion of a cone are to be defined later.

The following theorems are obtained by Amann [17] as straightforward generalization of one-dimensional case to order intervals with certain additional conditions.

Theorem 5.155 *Let (X, K) be an ordered Banach space with K having nonempty interior. Suppose that there exist found points in X.*

$$\overline{y}_1 < \hat{y}_1 < \overline{y}_2 < \hat{y}_2$$

and a continuous, strongly increasing mapping $F : \langle \overline{y}_1, \hat{y}_2 \rangle \to X$ with $F(\langle \overline{y}_1, \hat{y}_2 \rangle)$ relatively compact such that

$$\overline{y}_1 \leqslant F(\overline{y}_1), \quad F(\hat{y}_1) < \hat{y}_1, \overline{y}_2 < F(\overline{y}_2) \quad \text{and} \quad F(\hat{y}_2) \leqslant \hat{y}_2.$$

Then F has at least three distinct fixed points x, x_1, s_2 such that

$$\overline{y}_1 \leqslant x_1 \ll \hat{y}_1, \quad \overline{y}_2 \ll x_2 \leqslant \hat{y}_2, \quad \text{and} \quad \overline{y}_2 \nleqslant x \nleqslant \hat{y}_1.$$

The following corollary follows from Corollary 5.32 to Theorem 5.150.

Corollary 5.34 *If the hypotheses of the above theorem be satisfied, then F has at least three distinct fixed points x_1, x_2, x_3 such that x_1, x_2, x_3 such that $x_1 \ll x_2 \ll x_3$.*

Theorem 5.156 *Let (X, K) be an ordered Banach space with K having nonempty interior. Suppose $\overline{y} < F(\overline{y})$ and let $F : \langle \overline{y}, \hat{y} \rangle \to X$ be a strongly increasing continuous mapping with $F(\langle \overline{y}, \hat{y} \rangle)$ relatively compact such that $\overline{y} < F(\overline{y})$ and $F(\hat{y}) < \hat{y}$. Moreover, assume that the minimal fixed point $\overline{x}$ and the maximal fixed point $\hat{x}$ are distinct, and that F has strongly positive Fréchet derivatives at $\overline{x}$ and $\hat{x}$. Then F has at least three distinct fixed points provided*

$$r(F'(\overline{x})) \neq 1 \text{ and } r(F'(\hat{x})) \neq 1.$$

Definition 5.82 *(Krasnoselski [345])* A mapping $F : K \to K$ of the cone K of an ordered Banach space (X, K) is said to be a compression of the cone if $F(0) = 0$ and if there exist numbers $R > r > 0$ such that

$$Fx \nleqslant x, \text{ if } x \in K, \quad \|x\| \leqslant r \text{ and } x \neq 0 \tag{5.143}$$

and, for all $\varepsilon > 0$,

$$(1 + \varepsilon)x \nleqslant Fx \text{ if } x \in K \text{ and } \|x\| \geqslant R. \tag{5.144}$$

Krasnoselski proved that if F is compression of the cone K and is positive, continuous on K and maps bounded sets into compact sets, then F has at least one nonzero fixed point x in K with $r \leqslant \|x\| \leqslant R$. This result remains valid when (5.143) and (5.144) are replaced by the following weaker conditions:

$$Fx \nleqslant x \text{ if } x \in K \text{ and } \|x\| = r \tag{5.145}$$

and for all $\varepsilon > 0$,

$$(1 + \varepsilon)x \nleqslant Fx \text{ if } x \in K \text{ and } \|x\| = R. \tag{5.146}$$

Leggett and Williams [363] have improved the above result by replacing (5.145) by a weaker condition:

$$Fx \nleqslant x \text{ if } x \in K(u) \text{ and } \|x\| = r, \tag{5.147}$$

where u is a fixed element of $K - \{0\}$ and

$$K(u) = \{x \in K : \alpha x \geqslant u \text{ for some positive member } \alpha\}.$$

The theorem proved by them has wider applications and is even easier to apply in situations where compression of the cone theorem is applicable.

For $0 < R \leqslant \infty$, suppose K_R denotes the subset of a cone K where

$$K_R = \{x \in K : \|x\| \geq R\}, \ 0 < R < \infty \text{ and } K_\infty = K.$$

Theorem 5.157 (Leggett and William [363]) *Let $F : K_R \to K$ be a continuous operator with $F(0) = 0$ and let F maps bounded sets into relatively compact sets, and suppose the following conditions are satisfied:*

$$\text{For all } \varepsilon > 0, \quad (1 + \varepsilon)x \nleqslant Fx \text{ for } x \in K, \|x\| = R \tag{5.148}$$

there exists a null sequence $\{u_n\}$ in K_R such that if $x_n \geq u_n, \ n = 1, 2, 3, \dots$ and $x_n \to 0$, there exists a subsequence $\{x_{n_k}\}$ such that

$$Fx_{n_k} \nleqslant x_{n_k}, \ k = 1, 2, 3, \dots \tag{5.149}$$

Then F has a nonzero fixed point in K_R.

Theorem 5.158 (Leggett and Williams [363]) *Let $F : K_R \to K$ be a continuous operator with $F(0) = 0$ and let F maps bounded sets into relatively compact sets. Suppose there exists a number $r, 0 < r < R$, and a vector $u \in K - \{0\}$ such that*

$$Fx \nleqslant x \text{ if } x \in K(u) \text{ and } \|x\| = r. \tag{5.150}$$

Assume further that for each $\varepsilon > 0$,

$$(1 + \varepsilon)x \nleqslant Fx \text{ if } x \in K \text{ and } \|x\| = R. \tag{5.151}$$

Then F has a fixed point x in K with $r \leqslant \|x\| \leqslant R$.

Proof Suppose $B : K_R \to K$ is defined as follows:

$$Bx = \begin{cases} Fx & \text{if } r \leqslant \|x\| \leqslant R \\ \|x\| r^{-1} F(r\|x\|^{-1}x) & \text{if } 0 < \|x\| < r \\ 0 & \text{if } x = 0. \end{cases}$$

It is easy to show that B is completely continuous on K_R. Because $Bx = Fx$ for $\|x\| = R$, condition (5.148) of Theorem 5.157 is satisfied. We now show that condition (5.149) of Theorem 5.158 is also satisfied. Suppose $Bx = x$ and $x \neq 0$. If $r \leqslant \|x\| \leqslant R$, then $x = Bx = Fx$, and if $0 < \|x\| < r$, then $x = Bx = \|x\| r^{-1} F(r\|x\|^{-1}x)$ and this implies that $F(y) = y$, where $y = r\|x\|^{-1}x$.

Assume that $u \in K_R$ and define a sequence $\{u_n\}$ in K_R by $u_n = ujn, n = 1, 2, \ldots$ and suppose x_n is a null sequence with $x_n \geqslant u_n$. For large n, $\|x_n\| < r$ and thus $Bx_n = \|x_n\| r^{-1} F(r\|x_n\|^{-1}x_n)$. Now $r\|x_n\|^{-1}x_n \in K(u)$ as $x_n \in K(u)$, and $r\|x_n\|^{-1}\|x_n\| = r$, so that $F(r\|x_n\|^{-1}x_n) \nleqslant (r\|x_n\|^{-1}x_n)$. Therefore, $Bx_n = \|x_n = \|x_n\| r^{-1} F(r\|x_n\|^{-1}x_n) \nleqslant \|x_n\| r^{-1} r\|x_n\|^{-1}x_n = x_n$. This shows that the condition (5.149) is satisfied. The result follows by Theorem 5.157.

We now present some multiple fixed point theorems obtained by Williams and Leggett [607] which they applied to problems in chemical reactor theory.

Let $C(\Omega)$ denote the real Banach space (with usual sup norm) of real-valued continuous functions on the compact set $\Omega \subset \mathbf{R}^n$, and let $L_\infty(\Omega)$ be the Banach space (with the ess-sup norm) of almost every where b bounded, real-valued Legesgue measurable functions on Ω. Let, for each positive number c,

$$\langle 0, c \rangle = \{x \in C(\Omega) : 0 \leqslant x(t) \leqslant c, \ t \in \Omega\} \text{ and}$$

$$\langle 0, c \rangle_\infty = \{x \in L_\infty(\Omega) : 0 \leqslant x(t) \leqslant c, \text{ a.e. in } \Omega\}.$$

Let F be a continuous mapping from $\langle 0, c \rangle_\infty$ into $C^+(\Omega)$, the set of nonnegative functions in $C(\Omega)$ such that it maps $\langle 0, c \rangle_\infty$ into a relatively compact subset of $C^+(\Omega)$.

Theorem 5.159 *Let F be a continuous mapping of $\langle 0, c \rangle_\infty$ into $\langle 0, c \rangle$, with $F(\langle 0, c \rangle_\infty)$ relatively compact. Assume that there exist numbers a and b and a nonempty and closed set $\Omega_1 \subset \Omega$ such that $0 < a < b \leqslant c$;*

$$|Fx(t) - Fx(s)| \leqslant b - a \text{ if } x \in \langle 0, c \rangle_\infty \text{ and } t, x \in \Omega_1, \tag{5.152}$$

$$Fx(t) > a \text{ (respectively, } Fx(t) \geqslant a), \ t \in \Omega_1 \tag{5.153}$$

whenever $x \in \langle 0, c \rangle_\infty$ and $a \leqslant x(s) \leqslant b$ a.e.in Ω_1. Then there exists $x \in \langle 0, c \rangle$ such that $Fx = x$ and $x(t) > a$ (respectively, $x(t) \geqslant a$) for each $t \in \Omega_1$.

Proof Suppose the mapping $B : \langle 0, c \rangle \to \langle 0, c \rangle_\infty$ is defined as follows:

$$Bx(t) = \begin{cases} a & \text{if } t \in \Omega_1 \text{ and } x(t) < a, \\ x(t) & \text{otherwise.} \end{cases}$$

Then we see that B is continuous, and the mapping $T = FB$ is continuous from $\langle 0, c \rangle$ into $\langle 0, c \rangle$ with $T(\langle 0, c \rangle)$ relatively compact. Because $\langle 0, c \rangle$ is closed, bounded and convex, by Schauder's fixed point theorem, there exists a fixed point x of T Suppose there exists $t_0 \in \Omega$, such that $x(t_0) \leqslant a$ (respectively $x(t_0) < a$). Then for each $s \in \Omega_1$, we have

$$x(s) - x(t_0) \leqslant |x(s) - x(t_0)| \leqslant |F(Bx)(s) - F(Bx)(t_0)| \leqslant |b - a|.$$

This implies that $x(s) \leqslant x(t_0) + b - a \leqslant b$. Hence, $a \leqslant Bx(s) \leqslant b$ for each $s \in \Omega$. From (5.153), it follows that $x(t_0) = F(Bx)0(t_0) > a$ (respectively, $x(t_0) \geqslant a$), a contradiction. Hence, $x(t) > a$ (respectively, $x(t) \geqslant a$) for each $t \in \Omega_1$, $Bx = x$, which entails that $Fx = x$.

Theorem 5.160 *Suppose in addition to the hypothesis of Theorem 5.159 (with $Fx(t) > a$ in (5.153)), there exists $d, 0 < d < a$, such that*

$$F : \langle 0, d \rangle_\infty \to \{x \in \langle 0, d \rangle : \|x\| < d\}. \tag{5.154}$$

Then F has at least three fixed points $\langle 0, c \rangle$.

Proof From condition (5.154) and by Schauder's fixed point theorem, there exists a fixed point x_1 of F in $\langle 0, d \rangle$. By Theorem 5.159, F also has a fixed point $x_2 \in \langle 0, c \rangle$ such that $x_2(t) > a$ for each $t \in \Omega_1$. Let $B : \langle 0, c \rangle_\infty$ be defined by

$$Bx(t) = \begin{cases} b & \text{if } t \in \Omega_1 \text{ and } x(t) > b, \\ x(t) & \text{otherwise.} \end{cases}$$

Then $T \triangleq FB$ is a continuous self-mapping of $\langle 0, c \rangle$ with $T(\langle 0, c \rangle)$ relatively compact in $\langle 0, c \rangle$ such that it maps $\langle 0, d \rangle$ into its interior in $\langle 0, c \rangle$. Let $Q = \{x \in \langle 0, c \rangle : x(t) \geqslant a \text{ for each } t \in \Omega_1\}$. This is a closed, convex and bounded subset of $C(\Omega)$. Moreover if $x \in Q$, then $a \leqslant Bx(t) \leqslant b, t \in Q_1$. This together with (5.153) implies that $a < F(Bx)(t)$ for each $t \in \Omega_1$. Hence, T maps Q into its interior in $\langle 0, c \rangle$. Then by Theorem 2 of Amann ([18], p. 360), T has a fixed point $x_3 \in \langle 0, c \rangle - Q \cup \langle 0, d \rangle$. Since $x_3 \notin Q$, there exists $t_0 \in \Omega_1$, such that $x_3(t_0) < a$. So for each $t \in \Omega_1$.

$$x_3(t) - x_3(t_0) \leqslant |x_3(t) - x_3(t_0)| = |F(Bx_3)(t) - F(Bx_3)(t_0)| \leqslant ba,$$

implying that $x_3(t) \leqslant x_3(t_0) + b - a < b$. Hence, $Bx_3 = x_3$, and this implies that $Fx_3 = x_3$. It is obvious that x_1, x_2, x_3 are distinct.

Theorem 5.161 *Suppose $F : \langle 0, b \rangle_\infty \to C^+(\Omega)$ is continuous with $F(\langle 0, b \rangle_\infty)$ relatively compact and that there exist numbers a and d and a nonempty and closed subset Ω_1, of Ω such that $0 < d < a < b$;*

$$|Fx(t) - Fx(s)| \leqslant b - a \text{ if } x \in \langle 0, \ b \rangle_\infty \text{ and } t, s \in \Omega_1, Fx(t) > a, \ t \in \Omega_1;$$

if $x \in \langle 0, b \rangle_\infty$ and $a \leqslant x(s)$ a.e. $s \in \Omega_1$; $F : \langle 0, d \rangle_\infty \to \{x \in \langle 0, d \rangle : \|x\| < d\}$ and Fx assumes its maximum on Ω_1 if $x \in \langle 0, b \rangle_\infty$. Then F has at least two fixed points in $\langle 0, b \rangle$.

We now intend to discuss few fixed point theorems proved by Gatica and Smith [241] for continuous mappings which maps bounded subsets into precompact subsets, defined in cones of real Banach spaces which are of interest in the study of boundary value problems of ordinary differential equations and periodic solutions of delay-integral equations.

Lemma 5.21 *Let C be a compact subset of K with $0 \in C$. Then 0 does not belong to the closed and convex hull of C.*

The following theorem is due to Gustafson and Schmitt [260].

Theorem 5.162 *Let $0 < r < R$ be real numbers,*

$$D = \{x \in K : r \leqslant \|x\| \leqslant R\}.$$

Let $F : D \to K$ be a continuous mapping which maps bounded subsets of D into precompact subsets of K such that

(i) $x \in D, \|x\| = R, Fx = \eta x \Rightarrow \eta \leqslant 1$;
(ii) $x \in D, \|x\| = r, Fx = \eta x \Rightarrow \eta \geqslant 1$;
(iii) $\inf_{\|x\|=r} \|Fx\| > 0$.

Then F has a fixed point in D.

Let us observe the following theorem in which the boundary conditions (i) and (ii) of Theorem 5.162 are reversed.

Theorem 5.163 *Let $0 < r < R$ be real numbers, $D = \{x \in K : r \leqslant \|x\| \leqslant R\}$, and let $F : D \to K$ be a continuous mapping which maps bounded subsets of D into precompact subsets of K such that*

(i) $x \in D, \ \|x\| = R, \ Fx = \eta x \Rightarrow \eta \geqslant 1$;
(ii) $x \in D, \ \|x\| = r, \ Fx = \eta x \Rightarrow \eta \leqslant 1$;
(iii) $\inf_{\|x\|=R} \|Fx\| > 0$.

Then F has a fixed point in D.

Proof Suppose first that $0 < r < 1$ and $R = 1/r$. Define a mapping $B : D \to K$ by $Bx = \|x\|^2 F \left(\frac{x}{\|x\|^2} \right), x \in D$. Then B is a continuous mapping which maps bounded subsets of D into precompact subsets of K satisfying the conditions of the previous theorem. Hence, B has a fixed point in D, which provides a fixed point for F.

Now suppose $0 < r < R$ be arbitrary and let $D_1 = \{x \in K : \frac{1}{2} \leqslant \|x\| \leqslant 2\}$. Define $F_1 : D_1 \to K$ by

$$F_1 x = \frac{3\|x\|}{2\|x\|(R - r) + 4r - R} F \left(\frac{2\|x\|(R - r) + 4r - R}{3\|x\|} \right), \quad x \in D_1.$$

It can be seen that F_1 satisfies all conditions so that the result of the first part of the proof can be applied. Thus, F_1 has a fixed point in D_1 which in turn provides a fixed point of F in D.

Remark 5.53 Turner [592] proved the above two theorems under the additional assumption that the cone K is reproducing.

Theorem 5.164 *Let $F : K \to K$ be a continuous mapping which maps bounded subsets into precompact subsets and suppose that F satisfies the following conditions:*
(i) there exists $r > 0$ and a continuous mapping $B : \{x \in K : \|x\| = r\} \to K$ which maps bounded subsets into precompact subsets such that $\inf_{\|x\|=r} \|Bx\| > 0$ and the equation
$$y = Fy + \eta By, \quad 0 < \eta < \infty,$$

has no solutions of norm r in K.
(ii) there exists $R > r$ such that the equation

$$z = \eta F z, \quad 0 < \eta < 1,$$

has no solution of norm R. Then F has fixed point u in K with $r \leqslant \|u\| \leqslant R$.

Theorem 5.165 *Let $F : K \to K$ be a continuous operator which maps bounded subsets into precompact subsets and which satisfies the following:*

(i) there exist $r > 0$ and a continuous operator, which maps bounded subsets into precompact subsets, $B : \{x \in K : \|x\| = r\} \to K$ such that $\|Bx\| \leqslant r$, $x \in K$, $\|x\| = r$, and $x \in K$, $x = \eta F x + (1 - \eta) Bx$, $0 < \eta < 1 \Rightarrow \|x\| \neq r$.
(ii) there exists $R > r$ and a mapping, which maps bounded subsets into precompact subsets, $C : \{x \in K : \|x\| = R\} \to K$ such that $\inf_{\|x\|=R} \|Cx\| > 0$, and $z \in K$, $z = FRz + \eta Cz$, $0 < \eta < \infty \Rightarrow \|z\| \neq R$.

Then F has a fixed point $u \in K$ such that $r \leqslant \|u\| \leqslant R$.

The following theorem is a generalization of Theorem 5.153.

Theorem 5.166 *Let $F : K \to K$ be a continuous mapping which maps bounded subsets into precompact subsets, such that $F(0) = 0$ and F is Fréchet differentiable at $x = 0$ in the direction of the cone. Assume further that F satisfies the conditions:*

(i) $F'(0)$, the Fréchet derivative of F at 0, has an eigenvector $k \in K$ corresponding to an eigenvalue $\eta_0 > 1$, and 1 is not an eigenvalue of $F'(0)$ with a corresponding eigenvector in K.

(ii) There exists $R > 0$ such that if $x \in K$, $\|x\| = R$, and $Fx = \mu x$ then $\mu \leqslant 1$.

Then F has a nonzero fixed point $u \in K$ with $\|u\| \leqslant R$.

The following result is the multivalued analogue of Krasnoselski's fixed point theorem due to Agarwal and Regan [6] in a cone.

Let $E = (E, \| \cdot \|)$ be a Banach space and $C \subseteq E$. For $\rho > 0$ let

$$\Omega_\rho = \{x \in E : \|x\| < \rho\} \text{ and } \partial\Omega_\rho = \{x \in E : \|x\| = \rho\}.$$

Theorem 5.167 (Agarwal and Regan [6]) Let $E = (E, \| \cdot \|)$ be a Banach space and $C \subseteq E$ a cone and let $\| \cdot \|)$ be increasing with respect to C. Also r, R are constants with $0 < r < R$. Suppose $A : \overline{\Omega_R} \cap C \to K(C)$ (here $K(C)$ denotes the family of nonempty, convex, compact subset of C) is an upper semicontinuous, compact map and assumes one of the following conditions

(A) $\|y\| \leqslant \|x\|$ for all $y \in A(x)$ and $x \in \partial\Omega_R \cap C$ and $\|y\| > \|x\|$ for all $y \in A(x)$ and

$$x \in \partial\Omega_r \cap C$$

or

(B) $\|y\| > \|x\|$ for all $y \in A(x)$ and $x \in \partial\Omega_R \cap C$ and $\|y\| \leqslant \|x\|$ for all $y \in A(x)$ and

$$x \in \partial\Omega_r \cap C$$

hold. Then A has a fixed point in $C \cap (\overline{\Omega_R} \backslash \Omega_r)$.

Remark 5.54 Fixed point theorems for condensing mapping and K set contractions in ordered Banach spaces will be discussed in the next chapter.

5.7.2 Fixed Point Theorems in Partially Ordered Banach Spaces

Unless otherwise mentioned, throughout this section that follows, let E denote a partially ordered real normed linear space with an order relation $\preceq$ and the norm $\| \cdot \|$. It is known that E is regular if $\{x_n\}_{n\in\mathbb{N}}$ is a nondecreasing (resp. nonincreasing) sequence in E such that $x_n \to x^*$ as $n \to \infty$, then $x_n \preceq x^*$ (resp. $x_n \succeq x^*$) for all $n \in \mathbb{N}$. Clearly, the partially ordered Banach space $C(J, \mathbb{R})$ is regular and the conditions guaranteeing the regularity of any partially ordered normed linear space E may be found in Nieto and Lopez [423] and Heikkili and Lakshmikantham [269] and the references therein.

We need the following definitions in the sequel.

Definition 5.83 A mapping $T : E \to E$ is called isotone or nondecreasing if it preserves the order relation $\preceq$, that is, if $x \preceq y$ implies $Tx \preceq Ty$ for all $x, y \in E$.

Definition 5.84 An operator T on a normed linear space E into itself is called compact if $T(E)$ is a relatively compact subset of E. T is called totally bounded if for any bounded subset S of E, $T(S)$ is a relatively compact subset of E. If T is continuous and totally bounded, then it is called completely continuous on E.

Definition 5.85 (*Dhage* [177]) A mapping $T : E \to E$ is called partially continuous at a point $a \in E$ if for $\varepsilon > 0$ there exists a $\delta > 0$ such that $\|Tx - Ta\| < \varepsilon$ whenever x is comparable to a and $\|x - a\| < \delta$. T is called partially continuous on E if it is partially continuous at every point of it. It is clear that if T is partially continuous on E, then it is continuous on every chain C contained in E. T is called partially bounded if $T(C)$ is bounded for every chain C in E. T is called uniformly partially bounded if all chains $T(C)$ in E are bounded by a unique constant.

Definition 5.86 (*Dhage* [177, 178]) An operator T on a partially normed linear space E into itself is called partially compact if $T(C)$ is a relatively compact subset of E for all totally ordered sets or chains C in E. T is called partially totally bounded if for any totally ordered and bounded subset C of E, $T(C)$ is a relatively compact subset of E. If T is partially continuous and partially totally bounded, then it is called partially completely continuous on E.

Definition 5.87 (*Dhage* [178]) The order relation $\preceq$ and the metric d on a nonempty set E are said to be compatible if $\{x_n\}_{n \in \mathbb{N}}$ is a monotone, that is, monotone nondecreasing or monotone nonincreasing sequence in E and if a subsequence $\{x_{n_k}\}_{k \in \mathbb{N}}$ of $\{x_n\}_{n \in \mathbb{N}}$ converges to x^* implies that the whole sequence $\{x_n\}_{n \in \mathbb{N}}$ converges to x^*. Similarly, given a partially ordered normed linear space $(E, \preceq, \|\cdot\|)$, the order relation $\preceq$ and the norm $\|\cdot\|$ are said to be compatible if $\preceq$ and the metric d defined through the norm $\|\cdot\|$ are compatible.

Clearly, the set $\mathbb{R}$ with usual order relation $\leqslant$ and the norm defined by the absolute value function has this property. Similarly, the space $C(J, \mathbb{R})$ with usual order relation $\leqslant$ and the supremum norm $\|\cdot\|$ are compatible.

Definition 5.88 A mapping $T : E \to E$ is called $\mathcal{D}$-set-Lipschitz if there exists a continuous nondecreasing function $\varphi : \mathbb{R}_+ \to \mathbb{R}_+$ such that

$$\|Tx - Ty\| = \varphi(\|x - y\|)$$

for all $x, y \in E$, where $\varphi(0) = 0$. The function φ is sometimes called a $\mathcal{D}$-function of T on E.

Definition 5.89 (*Dhage* [177]) Let $(E, \preceq, \|\cdot\|)$ be a partially ordered normed linear space. A mapping $T : E \to E$ is called partially nonlinear $\mathcal{D}$-Lipschitz if there exists a $\mathcal{D}$-function $\varphi : \mathbb{R}_+ \to \mathbb{R}_+$ such that

$$\|Tx - Ty\| = \varphi(\|x - y\|)$$

for all comparable elements $x, y \in E$. If $\varphi(r) = kr, k > 0$, then T is called a partially Lipschitz with a Lipschitz constant k. If $k < 1$, T is called a partially contraction with contraction constant k. Finally, T is called nonlinear $\mathcal{D}$-contraction if it is a nonlinear $\mathcal{D}$-Lipschitz with $\varphi(r) < r$ for $r > 0$.

Remark 5.55 If $\phi, \psi \; \mathbb{R}_+ \to \mathbb{R}_+$ are two $\mathcal{D}$-functions, then i) $\phi + \psi$, ii) $\lambda \phi, \lambda > 0$ and iii) $\phi \circ \psi$ are also $\mathcal{D}$-functions on $\mathbb{R}_+$ and commonly used $\mathcal{D}$-functions are $\phi(r) = kr$, $\phi(r) = \frac{Lr}{K+r}, L > 0, K > 0$, $\phi(r) = r - \log(1 + r)$ and $\phi(r) = \log(1 + r)$ etc.

The following applicable hybrid fixed point theorem is proved in Dhage [177].

Theorem 5.168 (Dhage [177]) *Let $(E, \preceq, \| \cdot \|)$ be a regular partially ordered complete normed linear space such that the order relation $\preceq$ and the norm $\| \cdot \|$ in E are compatible. Let $\mathcal{P}, \mathcal{Q} : E \to E$ be two nondecreasing operators such that*

(a) *$\mathcal{P}$ is partially bounded and partially nonlinear $\mathcal{D}$-contraction,*
(b) *$\mathcal{Q}$ is partially continuous ad partially compact, and*
(c) *there exists an element $x_0 \in E$ such that $x_0 \preceq \mathcal{P}x_0 + \mathcal{Q}x_0$.*

Then the operator equation $\mathcal{P}x + \mathcal{Q}x = x$ has a solution x^* in E and the sequence $\{x_n\}_{n=0}^{\infty}$ of successive iterations defined by $x_{n+1} = \mathcal{P}x_n + \mathcal{Q}x_n, n = 0, 1, \ldots$, converges monotonically to x^*.

We now introduce the following definition which plays a crucial role to deal with our main result.

Definition 5.90 A mapping $p : J \times \mathbb{R} \to \mathbb{R}$ is said to be exponentially continuous if for each $\lambda > 0$ and $t \in J$, $\frac{p(t,x)}{e^{\lambda t}}$ is continuous for all $x \in \mathbb{R}$.

5.7.3 Weakly Contractive Mapping Theorems in Partially Ordered Sets

In 1997, Alber and Guerre-Delabriere in [8] define weakly contractive maps. In this paper, they confine their theorems to Hilbert spaces, but acknowledge that their results are true, at least for uniformly smooth and uniformly convex Banach spaces. In [521] Rhoades extends some results appearing in [8] to arbitrary Banach spaces. If X is an arbitrary Banach space, then a self-map T of X satisfies the Banach contraction principle if there exists a constant k satisfying $0 \leqslant k < 1$ such that, for $x, y \in X$

$$\|Tx - Ty\| \leqslant k\|x - y\|. \tag{5.155}$$

As noted in the introduction of [8], inequality (5.155) can be written in the form

$$\|Tx - Ty\| \leqslant \|x - y\| - q\|x - y\|. \tag{5.156}$$

where $k = 1 - q$ with $q \in (0, 1]$. Therefore, the following definition of weakly contractive maps due to Alber and Guerre-Delabriere [8] seems to be natural.

Weakly contractive mapping- A self-map T of X is weakly contractive if, for every $x, y \in X$,

$$\|Tx - Ty\| \leqslant \|x - y\| - \psi(\|x - y\|) \tag{5.157}$$

where $\psi : [0, \infty) \to [0, \infty)$ is continuous and nondecreasing function such that they are positive on $(0, \infty)$, $\psi(0) = 0$ and $\lim_{t \to \infty} \psi(t) = \infty$. (Examples of such functions are $\psi(t) = at, a \in (0, 1)$, $\psi(t) = \frac{t^2}{t+1}$, $\psi(t) = \ln(t + 1)$.)

Now, let (X, d) be a metric space and $f : X \to X$. f is said to be weakly contractive if for $x, y \in X$

$$d(fx, fy) \leqslant d(x, y) - \psi(d(x, y))$$

where $\psi : [0, \infty) \to [0, \infty)$ satisfies the above-mentioned conditions.

Observe that weakly contractiveness implies continuity of the map f. Existence of fixed point in partially ordered sets has been considered recently in [423] and the references therein.

Notice that (5.155) can also be expressed as

$$\|Tx - Ty\| \leqslant (1 + q)\|x - y\| - (1 - q)\|x - y\|. \tag{5.158}$$

where $k = 2q$ with $q \in (0, \frac{1}{2})$. The extension of (5.158) in the context of Banach spaces to weakly $(\varphi - \psi)$-contractive maps is a natural one. A self-map f of X is said to be weakly $(\varphi - \psi)$-contractive of type (I) if, for every $x, y \in X$,

$$d(fx, fy) \leqslant \varphi(d(x, y)) - \psi(d(x, y)) \tag{5.159}$$

where $\varphi : [0, \infty) \to [0, \infty)$ is a continuous and nondecreasing function and $\psi : [0, \infty) \to [0, \infty)$ is a continuous and nonincreasing function satisfying the following conditions:

$(C1)$ $\varphi(0) - \psi(0) = 0,$
$(C2)$ φ and ψ both are positive on $(0, \infty)$, and
$(C3)$ $\varphi(t) - \psi(t) < t$ for all $t > 0$.

Obviously, weakly $(\varphi - \psi)$-contractive mappings of type (I) are continuous.

A self-map f of X is said to be weakly $(\varphi - \psi)$-contractive of type (II) if, for every $x, y \in X$, (5.159) holds, where $\varphi, \psi : [0, \infty) \to [0, \infty)$ are continuous and nondecreasing functions satisfying the following conditions:

$(D1)$ $\varphi(0) - \psi(0) = 0,$
$(D2)$ φ and ψ both are positive on $(0, \infty)$, and

(D3) $\varphi(t) - \psi(qt) \leqslant (1-q)t$ and $\varphi(qt) - \psi(0) \leqslant qt$ for all $t > 0$ and for some $q \in (0, 1)$.

Remark 5.56 Especially, when $\varphi(t) = t$ for all $t \geq 0$, our definition of weakly $(\varphi - \psi)$-contractive of type (I)/type (II) mappings reduces to the definition of weakly contractive mapping.

Definition 5.91 If $(X, \leqslant)$ is a partially ordered set and $f : X \to X$, we say that f is monotone nondecreasing (nonincreasing) if

$$x, y \in X, x \leqslant y \Rightarrow f(x) \leqslant f(y) \ (f(x) \geq f(y)).$$

This definition coincides with the notion of a nondecreasing (nonincreasing) function in the case where $X = \mathbb{R}$ and $\leqslant (\geq)$ represents the usual total order in $\mathbb{R}$. In [521], the following result was proved:

Theorem 5.169 *Let (X, d) be a complete metric space, and $f : X \to X$ is a weakly contractive map. Then f has a unique fixed point in X.*

In 2009, Harjani and Sadarangani [267] proved the following result.

Theorem 5.170 *Let $(X, \leqslant)$ be a partially ordered set, and suppose that there exists a metric d in X such that (X, d) is a complete metric space. Let $f : X \to X$ be a continuous and nondecreasing mapping such that*

$$d(fx, fy) \leqslant d(x, y) - \psi(d(x, y)) \ \text{for} \ x \geq y \tag{5.160}$$

where $\psi : [0, \infty) \to [0, \infty)$ is a continuous and nondecreasing functions such that it is positive on $(0, \infty)$, $\psi(0) = 0$ and $\lim_{x \to \infty} \psi(x) = \infty$. If there exists $x_0 \in X$ with $x_0 \leqslant f(x_0)$, then f has a fixed point.

It may be remarked that Theorem 5.170 does not guarantee the uniqueness of the fixed point. This motivates us to improve the criteria on the space as well as on the mapping to ensure the existence and uniqueness of fixed point of maps under consideration.

In 2010, Rhoades et al. [522] extended and improved the above cited results, i.e. Theorems 5.169 and 5.170 to weakly $(\varphi - \psi)$-contractive maps of type(I)/type(II) in the context of ordered metric spaces under certain restriction on the domain of maps which are extensions of those in [267, 521]. In the sequel, they applied their main result to obtain solution of first-order periodic problem and studied the possibility of optimally controlling the solution of ordinary differential equation via dynamic programming.

We now present several fixed point theorems for weakly $(\varphi - \psi)$-contractive mappings in a complete metric space endowed with a partial order. First, we shall prove a fixed point result for weakly $(\varphi - \psi)$-contractive mapping of type (I).

Theorem 5.171 *Let $(X, \leqslant)$ be a partially ordered set and suppose that there exists a metric d in X such that (X, d) is a complete metric space. Let $f : X \to X$ be a continuous and nondecreasing mapping such that*

$$d(fx, fy) \leqslant \varphi(d(x, y)) - \psi(d(x, y)) \ \text{for} \ x \geq y \qquad (5.161)$$

where $\varphi : [0, \infty) \to [0, \infty)$ is a continuous and nondecreasing function and $\psi : [0, \infty) \to [0, \infty)$ is a continuous and nonincreasing function satisfying the conditions $(C1) - (C3)$. If there exists $x_0 \in X$ with $x_0 \leqslant f(x_0)$, then f has a fixed point.

Proof If $x_0 = f(x_0)$, then the proof is finished. So we suppose that $x_0 < f(x_0)$. Since $x_0 < f(x_0)$ and f is nondecreasing function, we obtain by induction that

$$x_0 < f(x_0) \leqslant f^2(x_0) \leqslant f^3(x_0) \leqslant f^4(x_0) \cdots \leqslant f^n(x_0) \leqslant f^{n+1}(x_0) \leqslant \cdots$$

Put $x_{n+1} = f(x_n)$ for each $n \in \mathbb{N}$. Then for each $n \in \mathbb{N}$, from (5.161) and, as x_n and x_{n+1} are comparable, we have

$$d(x_{n+1}, x_n) = d(fx_n, fx_{n-1}) \leqslant \varphi(d(x_n, x_{n-1})) - \psi(d(x_n, x_{n-1})).$$

If there exists an $n_0 \in \mathbb{N}$ such that $d(x_{n_0}, x_{n_0-1}) = 0$ then $x_{n_0} = f(x_{n_0-1}) = x_{n_0-1}$ and x_{n_0-1} is a fixed point of T and the proof is finished.

On the other hand, suppose that $d(x_{n+1}, x_n) \neq 0$ for all $n \in \mathbb{N}$. Then taking into account (5.161) and our assumptions about φ and ψ, we have

$$d(x_{n+1}, x_n) = d(fx_n, fx_{n-1}) \leqslant \varphi(d(x_n, x_{n-1})) - \psi(d(x_n, x_{n-1})) < d(x_n, x_{n-1}).$$

Denoting $d(x_{n+1}, x_n)$ by ρ_n we have

$$\rho_n \leqslant \varphi(\rho_{n-1}) - \psi(\rho_{n-1}) < \rho_{n-1}. \qquad (5.162)$$

Hence, $\{\rho_n\}$ is a nonnegative nonincreasing sequence and hence possesses a limit, say, ρ^* such that $\rho^* \geq 0$. We claim that $\rho^* = 0$.

Now, from (5.162), if $\rho^* > 0$, taking limit when $n \to \infty$, we get

$$\rho^* \leqslant \varphi(\rho^*) - \psi(\rho^*) \leqslant \rho^*.$$

Thus, we have

$$\rho^* = \varphi(\rho^*) - \psi(\rho^*).$$

By (C3), for $\rho^* > 0$ we obtain

$$\rho^* = \varphi(\rho^*) - \psi(\rho^*) < \rho^*,$$

a contradiction. Therefore, $\rho^* = 0$.

Now, we show that $\{x_n\}$ is a Cauchy sequence. Fix $\varepsilon > 0$. Since $\rho_n = d(x_{n+1}, x_n) \to 0$, there exists $n_0 \in \mathbb{N}$ such that

$$d(x_{n_0+1}, x_{n_0}) \leqslant \min \left\{ \frac{\varepsilon}{2}, \ \varepsilon - \varphi(\varepsilon) + \psi(\varepsilon) \right\}.$$

We claim that $f\left(\overline{B(x_{n_0}, \varepsilon)} \cap \{y \in X : y \ge x_{n_0}\}\right) \subset \overline{B(x_{n_0}, \varepsilon)}$.

Let $z \in \overline{B(x_{n_0}, \varepsilon)} \cap \{y \in X : y \ge x_{n_0}\}$. Then there arises two cases:

Case 1. When $0 < d(z, x_{n_0}) \le \frac{\varepsilon}{2}$. In this case, as z and x_{n_0} are comparable, we have

$$
\begin{aligned}
d(f(z), x_{n_0}) &\le d(f(z), f(x_{n_0})) + d(f(x_{n_0}), x_{n_0}) = \\
&= d(f(z), f(x_{n_0})) + d(x_{n_0+1}, x_{n_0}) \\
&\le \varphi(d(z, x_{n_0})) - \psi(d(z, x_{n_0})) + d(x_{n_0+1}, x_{n_0}) \\
&< d(z, x_{n_0}) + d(x_{n_0+1}, x_{n_0}) \le \frac{\varepsilon}{2} + \frac{\varepsilon}{2} = \varepsilon.
\end{aligned}
$$

Case 2. When $\frac{\varepsilon}{2} < d(z, x_{n_0}) \le \varepsilon$. In this case, as z and x_{n_0} are comparable and φ is a nondecreasing function and ψ is a nonincreasing function, $\varphi((d(z, x_{n_0}))) \le \varphi(\varepsilon)$ and $\psi(d(z, x_{n_0}) \ge \psi(\varepsilon)$, we have

$$
\begin{aligned}
d(f(z), x_{n_0}) &\le d(f(z), f(x_{n_0})) + d(f(x_{n_0}), x_{n_0}) = \\
&= d(f(z), f(x_{n_0})) + d(x_{n_0+1}, x_{n_0}) \\
&\le \varphi(d(z, x_{n_0})) - \psi(d(z, x_{n_0})) + d(x_{n_0+1}, x_{n_0}) \\
&\le \varphi(\varepsilon) - \psi(\varepsilon) + d(x_{n_0+1}, x_{n_0}) \\
&\le \varphi(\varepsilon) - \psi(\varepsilon) + \varepsilon - \varphi(\varepsilon) + \psi(\varepsilon) = \varepsilon.
\end{aligned}
$$

This proves our claim.

Since $x_{n_0+1} \in \overline{B(x_{n_0}, \varepsilon)} \cap \{y \in X : y \ge x_{n_0}\}$, our claim gives us that $x_{n_0+2} = f(x_{n_0+1}) \in \overline{B(x_{n_0}, \varepsilon)} \cap \{y \in X : y \ge x_{n_0}\}$. Repeating this process, it follows that $x_n \in B(x_{n_0}, \varepsilon)$ for all $n \ge n_0$. Since ε is arbitrary, $\{x_n\}$ is a Cauchy sequence.

As X is complete, there exists $x^* \in X$ such that $\lim_{n\to\infty} x_n = x^*$. Again, since $\rho_n \to 0$ and f is continuous, it follows that x^* is a fixed point of f. This completes the proof. ∎

Theorem 5.172 *Let $(X, \le)$ be a partially ordered set and suppose that there exists a metric d in X such that (X, d) is a complete metric space. Let $f : X \to X$ be a continuous and nondecreasing mapping such that*

$$
d(fx, fy) \le \varphi(d(x, y)) - \psi(d(x, y)) \text{ for } x \ge y \tag{5.163}
$$

where $\varphi, \psi : [0, \infty) \to [0, \infty)$ are continuous and nondecreasing functions satisfying the conditions (D1)-(D3). If there exists $x_0 \in X$ with $x_0 \le f(x_0)$, then f has a fixed point.

Proof Following the proof of Theorem 5.171, we only have to check that $\{x_n\}$ is a Cauchy sequence. Fix $\varepsilon > 0$. Since $\rho_n = d(x_{n+1}, x_n) \to 0$, there exists $n_0 \in \mathbb{N}$ such that

$$
d(x_{n_0+1}, x_{n_0}) \le \min\left\{q\,\varepsilon, \ (1-q)\varepsilon\right\}.
$$

We claim that $f\left(\overline{B(x_{n_0}, \varepsilon)} \cap \{y \in X : y \geq x_{n_0}\}\right) \subset \overline{B(x_{n_0}, \varepsilon)}$.

Let $z \in \overline{B(x_{n_0}, \varepsilon)} \cap \{y \in X : y \geq x_{n_0}\}$. Then there arises two cases:
Case 1. When $0 < d(z, x_{n_0}) \leq q\,\varepsilon$. In this case, as z and x_{n_0} are comparable, we have

$$
\begin{aligned}
d(f(z), x_{n_0}) &\leq d(f(z), f(x_{n_0})) + d(f(x_{n_0}), x_{n_0}) = d(f(z), f(x_{n_0})) + d(x_{n_0+1}, x_{n_0}) \\
&\leq \varphi(d(z, x_{n_0})) - \psi(d(z, x_{n_0})) + d(x_{n_0+1}, x_{n_0}) \\
&\leq \varphi(q\,\varepsilon) - \psi(0) + d(x_{n_0+1}, x_{n_0}) \\
&\leq q\,\varepsilon + (1 - q)\,\varepsilon + \varepsilon.
\end{aligned}
$$

Case 2. When $q\,\varepsilon < d(z, x_{n_0}) \leq \varepsilon$. In this case, as z and x_{n_0} are comparable and φ is a nondecreasing function and ψ is a nonincreasing function, $\varphi((d(z, x_{n_0}) \leq \varphi(\varepsilon)$ and $\psi(d(z, x_{n_0})) \geq \psi(\varepsilon)$, we have

$$
\begin{aligned}
d(f(z), x_{n_0}) &\leq d(f(z), f(x_{n_0})) + d(f(x_{n_0}), x_{n_0}) = d(f(z), f(x_{n_0})) + d(x_{n_0+1}, x_{n_0}) \\
&\leq \varphi(d(z, x_{n_0})) - \psi(d(z, x_{n_0})) + d(x_{n_0+1}, x_{n_0}) \\
&\leq \varphi(\varepsilon) - \psi(q\varepsilon) + + d(x_{n_0+1}, x_{n_0}) \\
&\leq (1 - q)\,\varepsilon + q\varepsilon = \varepsilon.
\end{aligned}
$$

This proves our claim.   ∎

In what follows we prove that Theorem 5.171 is still valid for f not necessarily continuous, assuming the following hypothesis in X (which appears in Theorem 1 of [423] and Theorem 3 of [267]) if $\{x_n\}$ is a nondecreasing sequence in X such that $x_n \to x$ then

$$x_n \leq x \quad \text{for all } n \in \mathbb{N} \tag{5.164}$$

Theorem 5.173 *Let $(X, \leq)$ be a partially ordered set and suppose that there exists a metric d in X such that (X, d) is a complete metric space. Assume that X satisfies (3.4). Let $f : X \to X$ be a nondecreasing mapping such that*

$$d(fx, fy) \leq \varphi(d(x, y)) - \psi(d(x, y)) \text{ for } x \geq y$$

where $\varphi : [0, \infty) \to [0, \infty)$ is a continuous and nondecreasing function and $\psi : [0, \infty) \to [0, \infty)$ is a continuous and nonincreasing function satisfying the conditions $(C1) - (C3)$. If there exists $x_0 \in X$ with $x_0 \leq f(x_0)$, then f has a fixed point.

Proof Following the proof of Theorem 5.171, we only have to check that $f(z) = z$. In fact,

$$
\begin{aligned}
d(f(z), z) &\leq d(f(z), f(x_n)) + d(f(x_n), z) \leq \\
&\leq \varphi(d(z, x_n)) - \psi(d(z, x_n)) + d(x_{n+1}, z)
\end{aligned}
$$

and taking limit as $n \to \infty$, $d(f(z), z) \leq 0$ and this proves that $d(f(z), z) = 0$ and consequently $f(z) = z$.

We now present an example where it can be appreciated that hypotheses in Theorems 5.171 and 5.173 do not guarantee uniqueness of the fixed point. This example appears in [423]. Let $X = \{(1, 0), (0, 1)\} \subset \mathbb{R}^2$ and consider the usual order

$$(x, y) \leqslant (z, t) \Leftrightarrow x \leqslant z \text{ and } y \leqslant t.$$

Thus, $(X, \leqslant)$ is a partially ordered set, whose different elements are not comparable. Besides, (X, d) is a complete metric space considering d the Euclidean distance. Put $\varphi(t) = \frac{2}{3}(t + 1)$, $\psi(t) = \frac{2}{3}$ for all $t \in [0, \infty)$. The identity map $f(x, y) = (x, y)$ is trivially continuous and nondecreasing since elements in X are only comparable to themselves. Observe that conditions (5.161) and $(C1) - (C3)$ of Theorem 5.171 are satisfied, i.e. f is a weakly $(\varphi - \psi)$ -contractive mapping of type (I).

On the other hand, if we consider $\varphi(t) = \frac{2}{3}t$, $\psi(t) = \frac{1}{3}t$ for all $t \in [0, \infty)$. Then conditions (5.163) and (D1)-(D3) of Theorem 5.172 are satisfied, i.e. f is a weakly $(\varphi - \psi)$-contractive mapping of type (II). To see this, let us take $q = \frac{1}{2}$. Moreover, $(1, 0) \leqslant f(1, 0) = (1, 0)$ and f has two fixed points in X.

Now there arises a natural question whether there is any sufficient condition for the uniqueness of the fixed point in Theorems 5.171 and 5.173. The answer is affirmative. The conditions are:

(SC1) for $x, y \in X$ there exists a lower bound or an upper bound.
(SC2) X is such that if $\{x_n\}$ is a sequence in X whose consecutive terms are comparable, then there exists a subsequence $\{x_{n_i}\}$ of $\{x_n\}$ such that every term is comparable to the limit x.
(SC3) f maps comparable elements to comparable elements, that is, for $x, y \in X, x \leqslant y \Rightarrow f(x) \leqslant f(y)$ or $f(x) \geqslant f(y)$.

In [423] it is proved that condition $(SC1)$ is equivalent to: for $x, y \in X$ there exists a $z \in X$ which is comparable to x and y.

It may be remarked that corresponding results for Theorems 5.171 and 5.173 pertaining to uniqueness of the fixed point under conditions $(SC1) - (SC3)$ can be obtained by applying similar arguments as in [267], so we omit the details.

Now we state the following theorems without proof which ensure the uniqueness of fixed points in Theorems 5.171 and 5.173 respectively. An appropriate blend of the proofs of Theorem 5.171 and [267, Theorem 4] works.

Theorem 5.174 *Let $(X, \leqslant)$ be a partially ordered set, $\mathscr{C}$ a chain in X and suppose that there exists a metric d in $\mathscr{C}$ such that $(\mathscr{C}, d)$ is a complete metric space. Let $f : \mathscr{C} \to \mathscr{C}$ be a nondecreasing mapping such that*

$$d(fx, fy) \leqslant \varphi(d(x, y)) - \psi(d(x, y)) \text{ for all } x, y \in \mathscr{C} \tag{5.165}$$

where $\varphi : [0, \infty) \to [0, \infty)$ is a continuous and nondecreasing function and $\psi : [0, \infty) \to [0, \infty)$ is a continuous and nonincreasing function satisfying the conditions $(C1) - (C3)$. If there exists $x_0 \in \mathscr{C}$ with $x_0 \leqslant f(x_0)$, then f has a unique fixed point in $\mathscr{C}$.

Theorem 5.175 *Let $(X, \leqslant)$ be a partially ordered set, $\mathscr{C}$ a chain in X and suppose that there exists a metric d in $\mathscr{C}$ such that $(\mathscr{C}, d)$ is a complete metric space. Let $f : \mathscr{C} \to \mathscr{C}$ be a nondecreasing mapping such that*

$$d(fx, fy) \leqslant \varphi(d(x, y)) - \psi(d(x, y)) \text{ for all } x, y \in \mathscr{C} \qquad (5.166)$$

where $\varphi, \psi : [0, \infty) \to [0, \infty)$ are continuous and nondecreasing functions satisfying the conditions $(D1) - (D3)$. If there exists $x_0 \in \mathscr{C}$ with $x_0 \leqslant f(x_0)$, then f has a unique fixed point in $\mathscr{C}$.

Example 5.51 Let $X = \mathbb{R}$ and let $\leqslant$ denote usual order relation in $\mathbb{R}$. Then $(X, \leqslant)$ be a partially ordered set. Put

$$\mathscr{C} = \{0\} \cup \{\pm 2^{-n} : n \in \mathbb{N}\}$$

and let d be the usual metric on $\mathscr{C}$. Define mapping $f : \mathscr{C} \to \mathscr{C}$ by

$$fx = \frac{1}{2}x \text{ for all } x \in \mathscr{C}.$$

It is obvious that $\mathscr{C}$ is a chain in X and it is complete. The map f is continuous and nondecreasing. Put $\varphi(t) = \frac{2}{3}t + \frac{1}{1+t}$, $\psi(t) = \frac{1}{1+t}$ for all $t \in [0, \infty)$. Then conditions $(C1) - (C3)$ are satisfied. Notice that (5.163) obviously holds. Moreover, $-\frac{1}{2} \leqslant f(-\frac{1}{2}) = -\frac{1}{4}$ and f has a unique fixed point 0 in $\mathscr{C}$. Besides, we notice that the iterative sequence $\{x_n\}$ given by $x_0 = -\frac{1}{2}, x_1 = f(-\frac{1}{2}) = -\frac{1}{4}, x_2 = f(-\frac{1}{4}) = -\frac{1}{8}, \cdots$ is nondecreasing and converges to 0. Furthermore, we observe that each $x_n \leqslant 0$.

5.8   Fixed Point Theorems in Banach Algebra

It is well known that the important fixed point theorem due to Krasnoselski which combines the metric fixed point theorem of Banach with the topological fixed point theorem of Schauder in a Banach space has many applications to nonlinear integral equations. Many results have been obtained to improve and weaken the hypotheses of Krasnoselski's fixed point theorem due to several authors and the references therein. The study of the nonlinear integral equations in Banach algebras was initiated by Dhage via fixed point theorems.

In this section, we present the concept of $\mathcal{P}$-Lipschitzian maps which is appreciably weaker than $\mathcal{D}$-Lipschitzian maps and give the proof of some fixed point theorems of Dhage under some weaker conditions.

We first recall the following. In 1969, Boyd and Wong [75] introduced, without nomenclature, the concept of nonlinear contraction.

Definition 5.92   A mapping T on a Banach space X with norm $\|.\|$ is said to be nonlinear contraction if it satisfies

$$\|Tx - Ty\| \leqslant \phi(\|x - y\|), \quad \text{for all } x, y \in X, \tag{5.167}$$

where ϕ is a real continuous function such that $\phi(r) < r, r > 0$.

In 2003, Dhage [171] introduced the following:

Definition 5.93   A mapping T on a Banach space X is called $\mathcal{D}$-Lipschitzian if there exists a continuous and nondecreasing function $\phi : \mathbb{R}^+ \longrightarrow \mathbb{R}^+$ such that

$$\|Tx - Ty\| \leqslant \phi(\|x - y\|), \quad \text{for all } x, y \in X, \tag{5.168}$$

where $\phi(0) = 0$.

It is shown in Dhage [171] that every Lipschitzian mapping is $\mathcal{D}$-Lipschitzian map, but the converse may not be true.

Again, let X be a Banach space and $T : X \to X$ an operator. Then

1. T is called a compact operator if $T(X)$ is a compact subset of X.
2. T is called totally bounded if for any bounded subset S of X, $T(S)$ is a totally bounded set of X.
3. T is called completely continuous if it is continuous and totally bounded.

Note that every compact operator is totally bounded, but the converse may not be true, however, two notions are equivalent on a bounded subset of X.

The famous Krasnoselski [346] fixed point theorem states that:

Theorem 5.176   (Krasnoselski [346]) *Let S be a closed, convex and bounded subset of a Banach algebra X and let $A, B : S \longrightarrow X$ be two operators such that*

(a) A is a contraction;
(b) B is completely continuous; and
(c) $Ax + By \in S$, $\forall x, y \in S$.

Then the operator equation

$$Ax + Bx = x \tag{5.169}$$

has a solution in S.

It has been mentioned in Burton [126, Theorem 2] that hypothesis (c) of Theorem 5.176 is very strong and can be replaced with a mild one. Indeed, he proved the following modification of Krasnoselski's fixed-point theorem.

Theorem 5.177   (Burton [126]) *Let S be a closed, convex and bounded subset of a Banach algebra X and let $A : X \longrightarrow X$ and $B : S \longrightarrow X$ be two operators such that*

(a) A is a contraction;

(b) B is completely continuous; and
(c) [x = Ax + By, for all y ∈ S] ⇒ x ∈ S.

Then the operator equation (5.169) has a solution.

In 2003, Dhage [171, Theorem 2.3] proved the following fixed point theorem.

Theorem 5.178 (Dhage [171]) *Let S be a closed, convex and bounded subset of a Banach algebra X and let $A : X \longrightarrow X$ and $B : S \longrightarrow X$ be two operators such that*

(a) A is $\mathcal{D}$-Lipschitzian;
(b) $(I/A)^{-1}$ exists on $B(S)$, I being the identity operator on X;
(c) B is completely continuous; and
(d) $x = AxBy \Rightarrow x \in S, \ \forall \ y \in S.$

Then the operator equation

$$AxBx = x, \tag{5.170}$$

has a solution, whenever $M\phi(r) < r, r > 0$, where $M = \|B(S)\|$.

In 2005, Dhage [173, Theorem 2.1] improved Theorem 5.178 in the following way:

Theorem 5.179 (Dhage [173]) *Let S be a closed, convex and bounded subset of a Banach algebra X and let $A : X \longrightarrow X$ and $B : S \longrightarrow X$ be two operators such that*

(a) A is $\mathcal{D}$-Lipschitzian;
(b) B is completely continuous; and
(c) $x = AxBy \Rightarrow x \in S, \ \forall \ y \in S.$

Then the operator equation (5.170) has a solution, whenever $M\phi(r) < r, r > 0$, where $M = B(S)$.

Notice that the proof of Theorems 5.178 and 5.179 does not realize the continuity of the function ϕ involved in the definition of $\mathcal{D}$-Lipschitzian maps. Therefore, it is of interest to prove the improved version of these results under some weaker conditions. Further, it may be remarked that Dhage ([169, 171, 173] and some references therein) cites the following form of Boyd–Wong's fixed point theorem.

Theorem 5.180 (Dhage [169]) *Let $T : X \longrightarrow X$ be a nonlinear contraction on a Banach space X. Then T has a unique fixed point.*

But the Boyd–Wong fixed point theorems in its original form are as follows:

Theorem 5.181 (Boyd–Wong [75]) *Let (X, ρ) be a complete metric space, and let $T : X \longrightarrow X$ satisfies*

$$\rho(T(x), T(y)) \leqslant \psi(\rho(x, y)), \ \text{for all } x, y \in X, \tag{5.171}$$

where $\psi : \overline{P} \longrightarrow \mathbb{R}^+$ is upper semicontinuous from right on $\overline{P}$, and satisfies $\psi(t) < t$ for $t \in \overline{P} - \{0\}$, where $P = \{\rho(x, y) : x, y \in X\}$ and $\overline{P}$ denotes the closure of P. Then T has a fixed point $x_0 \in X$ and $T^n(x) \longrightarrow x_0$, for each $x \in X$.

Theorem 5.182 (Boyd–Wong [75]) *Suppose that (X, ρ) is a completely metrically convex metric space and that $T : X \longrightarrow X$ satisfies*

$$\rho(T(x), T(y)) \leqslant \psi(\rho(x, y)), \quad \text{for all } x, y \in X, \tag{5.172}$$

where $\psi : \overline{P} \longrightarrow \mathbb{R}^+$ satisfies $\psi(t) < t$ for $t \in \overline{P} - \{0\}$, where $P = \{\rho(x, y) : x, y \in X\}$ and $\overline{P}$ denotes the closure of P. Then T has a fixed point $x_0 \in X$ and $T^n(x) \longrightarrow x_0$, for each $x \in X$.

As observed by Boyd–Wong that for a metrically convex space, the condition of semi continuity may be dropped. Since every Banach space is metrically convex, it is just sufficient to consider only the second version of Boyd–Wong's theorem, i.e. Theorem 5.182 in the setting of Banach algebra, i.e. there is no need to take ψ to be continuous as Dhage considered in his papers (see, for instance, [169, 171, 173]).

5.8.1 $\mathcal{P}$-Lipschitzian Maps

We now introduce a new concept of $\mathcal{P}$-Lipschitzian mapping which is weaker than the concept of $\mathcal{D}$- Lipschitzian mapping.

Definition 5.94 A mapping T on a Banach space X is called $\mathcal{P}$-Lipschitzian if there exists a nondecreasing function $\phi : \mathbb{R}^+ \longrightarrow \mathbb{R}^+$ such that

$$\|Tx - Ty\| \leqslant \phi(\|x - y\|), \quad \text{for all } x, y \in X. \tag{5.173}$$

Sometimes we call the function ϕ a $\mathcal{P}$ function of T on X. Notice that every $\mathcal{D}$-Lipschitzian mapping is a $\mathcal{P}$-Lipschitzian mapping, but the converse need not be true. To see this, consider the following:

Example 5.52 Let $X = \mathbb{R}$, $f : X \longrightarrow X$ be defined by

$$f(x) = \begin{cases} \sin x, & \text{for } x \geq 0, \\ \frac{1}{1+|x|}, & \text{for } x < 0, \end{cases}$$

and let $\phi : \mathbb{R}^+ \longrightarrow \mathbb{R}^+$ be defined by

$$\phi(t) = \begin{cases} e^t, & \text{for } t > 0; \\ 2, & \text{for } t = 0. \end{cases}$$

Now we consider the following two cases:

Case 1: When $x \geq 0$,

$$|fx - fy| = |\sin x - \sin y| \leqslant |x - y| \leqslant e^{|x-y|} \leqslant \phi(|x - y|).$$

Case 2: When $x < 0$,

$$|fx - fy| = \left| \frac{1}{1+|x|} - \frac{1}{1+|y|} \right| \leqslant |x - y| \leqslant e^{|x-y|} \leqslant \phi(|x - y|).$$

Thus, we conclude that $\| fx - fy \| \leqslant \phi(\|x - y\|,) \quad \forall\, x, y \in X$.

We also observe that

1. ϕ is not continuous at $t = 0$,
2. ϕ is nondecreasing,
3. $\phi(0) \neq 0$.

Thus, f is a $\mathcal{P}$-Lipschitzian mapping but not $\mathcal{D}$-Lipschitzian. Hence, every $\mathcal{D}$-Lipschitzian mapping is $\mathcal{P}$-Lipschitzian map, but the converse need not be true.

Remark 5.57 Note that from Definition 5.94 and Example 5.52, it is clear that the reverse implications in the following diagram need not be true.

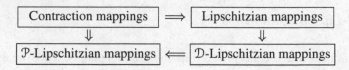

In 2012, by relaxing the hypothesis in the main theorem of Dhage [172], Pathak and Deepmala [456] proved the following fixed point theorem involving three operators on a Banach algebra.

Theorem 5.183 *Let S be a closed, convex and bounded subset of a Banach algebra X and let $A, C : X \longrightarrow X$ and $B : S \longrightarrow X$ be three operators such that*

(a) A and C are $\mathcal{P}$-Lipschitzian with $\mathcal{P}$ function ϕ_A and ϕ_C;
(b) B is completely continuous; and
(c) if $x = AxBy + Cx$ then $x \in S, \ \forall\, y \in S$.

Then the operator equation $AxBx + Cx = x$ has a solution, whenever $M\phi_A(r) + \phi_C(r) < r, \ r > 0$, where $M = \|B(S)\|$.

Proof Let $y \in S$ and define a mapping $A_y : X \longrightarrow X$ by

$$A_y(x) = AxBy + Cx, \quad \text{for all } x \in X.$$

Then we have

$$\|A_y x_1 - A_y x_2\| \leqslant \|Ax_1 - Ax_2\| \|By\| + \|Cx_1 - Cx_2\|$$
$$\leqslant M\phi_A(\|x_1 - x_2\|) + \phi_C(\|x_1 - x_2\|), \ x_1, x_2 \in X.$$

This shows that A_y is a nonlinear contraction on X, since $M\phi_A(r) + \phi_C(r) < r, \ r > 0$. Hence, by Theorem 5.176, there is a unique point $x^* \in X$ such that

$$A_y(x^*) = Ax^*By + Cx^* = x^*.$$

Therefore by (c), we have that $x^* \in S$. Define a mapping $N:S \longrightarrow S$ by $Ny = z$, where $z \in X$ is the unique solution of the equation

$$z = AzBy + Cz, \ y \in S.$$

We show that N is continuous. Let $\{y_n\}$ be a sequence in S converging to a point y. Since, S is closed, $y \in S$. Now

$$
\begin{aligned}
\|Ny_n - Ny\| &= \|AN(y_n)By_n - AN(y)By\| + \|C(Ny_n) - C(Ny)\| \\
&\leqslant \|AN(y_n)By_n - AN(y)By_n\| + \|AN(y)By_n - AN(y)By\| \\
&\quad + \|C(Ny_n) - C(Ny)\| \\
&\leqslant \|AN(y_n) - AN(y)\| \, \|By_n\| + \|AN(y)\| \, \|By_n - By\| \\
&\quad + \|C(Ny_n) - C(Ny)\| \\
&\leqslant M\phi_A(\|Ny_n - Ny\|) + \|ANy\| \, \|By_n - By\| \\
&\quad + \phi_C(\|Ny_n - Ny\|).
\end{aligned}
$$

Since, $M\phi_A(r) + \phi_C(r) < r, \ r > 0$, there exists $k \in (0, 1)$ such that $M\phi_A(r) + \phi_C(r) = kr$ and

$$\|Ny_n - Ny\| \leqslant k(\|Ny_n - Ny\|) + \|ANy\| \, \|By_n - By\|.$$

Taking the limit superior as $n \longrightarrow \infty$ on both sides, we obtain

$$\limsup_{n \to \infty} \|Ny_n - Ny\| \leqslant k \limsup_{n \to \infty}(\|Ny_n - Ny\|) + \|ANy\| \, (\limsup_{n \to \infty} \|By_n - By\|).$$

This shows that $\lim_n \|Ny_n - Ny\| = 0$ and consequently N is continuous on S. Next we show that N is a compact operator on S. Now for any $z \in S$ we have

$$\|Az\| \leqslant \|Aa\| + \|Az - Aa\| \leqslant \|Aa\| + \alpha\|z - a\| \leqslant c,$$

where $c = \|Aa\| + \text{diam}(S)$ for some fixed $a \in S$.

Let $\varepsilon > 0$ be given. Since, B is completely continuous, $B(S)$ is totally bounded. Hence, there is a set $Y = \{y_1, y_2, \ldots, y_n\}$ in S such that

$$B(S) \subset \bigcup_{i=1}^{n} B_\delta(w_i),$$

where $w_i = B(y_i), \delta = \left(\frac{1-(\alpha M+\beta)}{c}\right) \varepsilon$ and $B_\delta(w_i)$ is an open ball in X centerd at w_i of radius δ. Therefore, for any $y \in S$ we have a $y_k \in Y$ such that

$$\|By - By_k\| < \left(\frac{1 - (\alpha M + \beta)}{c}\right)\varepsilon.$$

Also, we have

$$
\begin{aligned}
\|Ny - Ny_k\| &\leqslant \|AzBy - Az_k By_k\| + \|Cz - Cz_k\| \\
&\leqslant \|AzBy - Az_k By\| + \|Az_k By - Az_k By_k\| + \|Cz - Cz_k\| \\
&\leqslant \|Az - Az_k\|\,\|By\| + \|Az_k\|\,\|By - By_k\| + \|Cz - Cz_k\| \\
&\leqslant (\alpha M + \beta)\|z - z_k\| + \|Az\|\,\|By - By_k\| \\
&\leqslant \tfrac{c}{1-(\alpha M+\beta)}\|By - By_k\| \\
&< \varepsilon.
\end{aligned}
$$

This is true for every $y \in S$ and hence

$$N(S) \subset \bigcup_{i=1}^{n} B_\varepsilon(z_i),$$

where $z_i = N(y_i)$. As a result, $N(S)$ is totally bounded. Since N is continuous, it is a compact operator on S. Now an application of Schauder's fixed point theorem yields that N has a fixed point in S. Then by the definition of N

$$x = Nx = A(Nx)Bx + C(Nx) = AxBx + Cx,$$

and so, the operator equation $x = AxBx + Cx$ has a solution in S. ∎

Similarly, we prove the following fixed point theorems involving two operators on a Banach algebra relaxing the hypothesis of Dhage [171, Theorem 2.3] and Dhage [173, Theorem 2.1], respectively.

Theorem 5.184 *Let S be a closed, convex and bounded subset of a Banach algebra X and let $A : X \longrightarrow X$ and $B : S \longrightarrow X$ be two operators such that*

(a) A is $\mathcal{P}$-Lipschitzian;
(b) $(I/A)^{-1}$ exists on $B(S)$, I being the identity operator on X;
(c) B is completely continuous; and
(d) $x = AxBy \Rightarrow x \in S$, $\forall\, y \in S$.

Then the operator equation $AxBy = x$ has a solution, whenever $M\phi(r) < r, r > 0$, where $M = \|B(S)\|$.

Proof Let $y \in S$ and define a mapping $A_y : X \longrightarrow X$ by

$$A_y(x) = AxBy, \quad \text{for all } x \in X.$$

Then we have

$$\|A_y x_1 - A_y x_2\| \leqslant \|Ax_1 - Ax_2\| \|By\|$$
$$\leqslant M\phi(\|x_1 - x_2\|), \; x_1, x_2 \in X.$$

Hence, by Theorem 5.176, there is a unique point $x^* \in X$ such that $A_y(x^*) = x^*$. Remaining proof of the theorem is similar to the proof of Theorem 2.3 [171] and Theorem 5.177. So we omit the details. ∎

Theorem 5.185 *Let S be a closed, convex and bounded subset of a Banach algebra X and let $A : X \longrightarrow X$ and $B : S \longrightarrow X$ be two operators such that*

(a) A is $\mathcal{P}$-Lipschitzian;
(b) B is completely continuous; and
(c) $x = AxBy \Rightarrow x \in S, \; \forall \, y \in S$.

Then the operator equation $AxBy = x$ has a solution, whenever $M\phi(r) < r, r > 0$, where $M = \|B(S)\|$.

Proof Let $y \in S$ and define a mapping $A_y : X \longrightarrow X$ by

$$A_y(x) = AxBy, \quad \text{for all } x \in X.$$

Following Theorem 5.184, we can show that there is a unique point $x^* \in X$ such that $A_y(x^*) = x^*$. The rest of the proof follows on the lines of the proof furnished in Theorem 5.177 (See also Theorem 2.1 of Dhage [173]).

Remark 5.58 Since every Lipschitzian and $\mathcal{D}$-Lipschitzian mappings are $\mathcal{P}$-Lipschitzian, we obtain the fixed point theorems studied in [171–173] as a particular case of Theorems 5.177, 5.184 and 5.185, which are useful to obtain the solutions of some nonlinear differential and integral equations.

The following sufficient condition guarantees the hypothesis (c) of Theorem 5.177.

Proposition 5.14 *Let S be a closed, convex and bounded subset of a Banach Algebra X such that $S = \{y \in X : \|y\| \leqslant r\}$ for some real number $r > 0$. Let $A, C : X \longrightarrow X$, $B : S \longrightarrow S$ be two operators satisfying hypothesis $(a) - (b)$ of Theorem 5.177. Further, if*

$$\|x\| \leqslant \left\| \left(\frac{I - C}{A} \right) x \right\|, \tag{5.174}$$

for all $x \in X$, then $x \in S$.

Proof The proof follows on the lines of the proof of Proposition 2.1 of [171].

5.9 Lattice-Theoretic Fixed Point Theorems

In 1927, Knaster and Tarski proved a set-theoretical fixed point theorem by which every function, on and to the family of all subsets of a set, which is increasing under set-theoretical inclusion, has at least one fixed point; see [313], where some applications of this result in set theory (a generalization of the Cantor–Bernstein's theorem) and topology are also mentioned. A generalization of this result is the lattice-theoretical fixed point theorem stated below as Theorem 5.180. The theorem in its present form and its various applications and extensions were found by the author in 1939 and discussed by him in a few public lectures in 1939–1942. (See, e.g. a reference in the American Mathematical Monthly 49(1942), 402.) An essential part of Theorem 5.180 was included in ([58], p. 54); however, the author was informed by Professor Garrett Birkhoff that a proper historical reference to this result was omitted by mistake.

Let $\mathfrak{A} = \langle A, \leq \rangle$ be a complete lattice. We shall consider functions on A to A and, more generally, on a subset B of A to another subset C of A. Such a function f is called increasing if, for any elements $x, y \in B$, $x \leq y$ implies $f(x) \leq f(y)$. By a fixed point of a function f we understand, of course, an element x of the domain of f such that $f(x) = x$.

We now state below the lattice-theoretical fixed point theorem of Tarski without proof.

Theorem 5.186 (Tarski [588]) *Let*

(i) $\mathfrak{A} = \langle A, \leq \rangle$ be a complete lattice,
(ii) f be an increasing function on A to A
(iii) P be the set of all fixed points of f.

Then the set P is not empty, and the system $(P, \leq)$ is a complete lattice; in particular, we have

$$\bigcup P = \bigcup E_x[f(x) \geq x] \in P$$

and

$$\bigcap P = \bigcap E_x[f(x) \geq x] \in P.$$

By Theorem 5.10, the existence of a fixed point for every increasing function is a necessary condition for the completeness of a lattice. The question naturally arises whether this condition is also sufficient. It has been shown that the answer to this question is affirmative (Refer to A.C. Davis [157]).

A set F of functions is called commutative if

(i) all the functions of F have a common domain, say B, and the ranges of all functions of F are subsets of B;
(ii) for any $f, g \in F$ we have $fg = gf$, that is,

$$f(g(x)) = g(f(x)) \text{ for every } x \in B.$$

Using this notion, Theorem 5.186 can be improved in the following way:

Theorem 5.187 (Generalized lattice-theoretical fixed point theorem) *Let*

(i) $\mathfrak{A} = \langle A, \leq \rangle$ be a complete lattice,

(ii) F be any commutative set of increasing functions on A to A
(iii) P be the set of all common fixed points of all the functions $f \in F$.

Then the set P is not empty and the system $(P, \leq)$ is a complete lattice; in particular, we have

$$\bigcup P = \bigcup E_x[f(x) \geq x \text{ for all } f \in F] \in P$$

and

$$\bigcap P = \bigcap E_x[f(x) \geq x \text{ for all } f \in F] \in P.$$

5.9.1 Reflexivity and Perturbed Fixed Point Property for Cascading Nonexpansive Maps in Banach Lattices

For the first time, James [291] investigated connections between spaces containing good copies of ℓ_1 or c_0 and the fixed point property. He proved that neither ℓ_1 nor c_0 is distortable. In his proof, he provided a tool which appeared to be useful in considering the question of whether ℓ_1 or c_0 could be renormed to have the fixed point property. The proof shows that both spaces ℓ_1 and c_0 admit fixed point-free isometries on bounded, closed and convex sets and all renormings of ℓ_1 or c_0 contain almost isometric copies of ℓ_1 or c_0, then perturbations of the isometries would hopefully produce nonexpansive self-maps of bounded, closed and convex subsets without fixed points in the renormed spaces.

James's distortion theorems state that Banach spaces which contain isomorphic copies of ℓ_1 (respectively, c_0) contain almost isometric copies of ℓ_1 (respectively, c_0) (See also [199], [200]).

James's distortion theorems A Banach space X contains an isomorphic copy of ℓ_1 if and only if, for every null sequence (ε_n) in $(0, 1)$, there exists a sequence $(x_n)_{n \in \mathbb{N}}$ in X such that

$$(1 - \varepsilon_k) \sum_{n=k}^{\infty} |t_n| \leqslant \left\| \sum_{n=k}^{\infty} t_n x_n \right\| \leqslant \sum_{n=k}^{\infty} |t_n|$$

holds for all $(t_n) \in \ell_1$ and for all $k \in \mathbb{N}$.

A Banach space X contains an isomorphic copy of c_0 if and only if, for every null sequence (ε_n) in $(0, 1)$, there exists a sequence $(x_n)_{n \in \mathbb{N}}$ in X such that

$$(1 - \varepsilon_k) \sup_{n \geqslant k} |t_n| \leqslant \left\| \sum_{n=k}^{\infty} t_n x_n \right\| \leqslant \sup_{n \geqslant k} |t_n|$$

holds for all $(t_n) \in c_0$ and for all $k \in \mathbb{N}$.

We now introduce the notion of cascading nonexpansive mapping. This new class of mappings strictly includes nonexpansive mappings. Cascading nonexpansive mappings are analogous to asymptotically nonexpansive mappings, but examples show that neither of these two classes of mappings contain the other.

Let C be a bounded, closed and convex subset of a Banach space $(X, \|\cdot\|)$. Let $T : C \to C$ be a mapping. Let $C_0 := C$ and

$$C_1 := \overline{co}(T(C)) \subseteq C.$$

Clearly C_1 is a bounded, closed and convex set in C. Let $x \in C_1$. Then

$$Tx \in T(C_1) \subseteq T(C) \subseteq \overline{co}(T(C)) = C_1.$$

This means that T maps C_1 into C_1. Inductively, for all $n \in \mathbb{N}$ we define

$$C_n := \overline{co}(T(C_{n-1})) \subseteq C_{n-1}.$$

It follows that T maps C_n into C_n.

Definition 5.95 Let $(X, \|\cdot\|)$ be a Banach space and C be a bounded, closed and convex subset of X. Let $T : C \to C$ be a mapping and $(C_n)_{n \in \mathbb{N}}$ be defined as above. We say that T is cascading nonexpansive if there exists a sequence $(k_n)_{n \in \mathbb{N}}$ in $[1, \infty)$ such that $k_n \to 1$, and for all $n \in \mathbb{N}$, for all $x, y \in C_n$,

$$\|Tx - Ty\| \leqslant k_n \|x - y\|.$$

Note that every cascading nonexpansive mapping is norm-to-norm continuous; and every nonexpansive map is cascading nonexpansive. Notice also that cascading nonexpansive mappings arise naturally in Banach spaces $(X, \|\cdot\|)$ that contain an isomorphic copy of ℓ_1 or c_0. We now furnish few examples of such spaces.

Example 5.53 (i) Banach spaces isomorphic to a nonreflexive Banach lattice, and nonreflexive Banach spaces with an unconditional basis. (See, for example, Lindenstrauss and Tzafriri [372] 1.c.5 and [371] 1.c.12.)
(ii) Banach spaces isomorphic to a nonreflexive symmetrically normed ideal of operators on an infinite-dimensional Hilbert space. (See Peter Dodds and Lennard [192].)

Using Strong James' Distortion Theorem for ℓ_1 and c_0, we prove the following:

Theorem 5.188 *Let $(X, \|\cdot\|)$ be a Banach space that contains an isomorphic copy of ℓ_1 or c_0. Then there exists a bounded, closed and convex set $C \subseteq X$ and an affine cascading nonexpansive mapping $T : C \to C$ such that T is fixed point free.*

Proof We need to consider the following two cases:

Case 1. Suppose $(X, \| \cdot \|)$ contains an isomorphic copy of ℓ_1. By the Strong Jamess' Distortion Theorem for ℓ_1 ([4], [5]), there exists a normalized sequence $(x_j)_{j \in \mathbb{N}}$ in X and a null sequence $(\varepsilon_n)_{n \in \mathbb{N}}$ in $(0, 1)$ such that for all $n \in \mathbb{N}$, for all $t = (t_j)_{j \in \mathbb{N}} \in c_{00}$,

$$(1 - \varepsilon_n) \sum_{j=n}^{\infty} |t_j| \leqslant \left\| \sum_{j=n}^{\infty} t_j x_j \right\| \leqslant \sum_{j=n}^{\infty} |t_j|.$$

Define the bounded, closed and convex subset C of X by

$$C := \overline{co}\{x_j : j \in \mathbb{N}\} = \left\{ \sum_{j=1}^{\infty} t_j x_j : \text{each } t_j \geqslant 0 \text{ and } \sum_{j=1}^{\infty} t_j = 1 \right\}.$$

Further, we define $T : C \to C$ by

$$T\left(\sum_{j=1}^{\infty} t_j x_j \right) = \sum_{j=1}^{\infty} t_j x_{j+1}.$$

Then it is easy to see that the mapping T is affine and fixed point free. Inductively, we see that for all $n \in \mathbb{N}$, $C_n := \overline{co}(T(C_{n-1})) = T(C_{n-1})$ is given by

$$C_n = \left\{ \sum_{j=n+1}^{\infty} t_j x_j : \text{each } t_j \geqslant 0 \text{ and } \sum_{j=n+1}^{\infty} t_j = 1 \right\}.$$

Fix $n \in \mathbb{N}_0$ and fix $x, y \in C_n$. Hence, $x = \sum_{j=n+1}^{\infty} t_j x_j$ and $y = \sum_{j=n+1}^{\infty} s_j x_j$, where each $t_j, s_j \geqslant 0$ and $\sum_{j=n+1}^{\infty} t_j = \sum_{j=n+1}^{\infty} s_j = 1$. Take $a_j = t_j - s_j$ for all $j \geqslant n + 1$. Then

$$\|x - y\| = \left\| \sum_{j=n+1}^{\infty} a_j x_j \right\| \geqslant (1 - \varepsilon_{n+1}) \sum_{j=n+1}^{\infty} |a_j|,$$

and because each $\|x_k\| = 1$, it follows that

$$\|Tx - Ty\| = \left\| \sum_{j=n+1}^{\infty} a_j x_{j+1} \right\| \leqslant \sum_{j=n+1}^{\infty} |a_j|$$

$$\leqslant \frac{1}{(1 - \varepsilon_{n+1})} \|x - y\|.$$

Consequently, T is cascading nonexpansive on C.

Case 2. Suppose $(X, \| \cdot \|)$ contains an isomorphic copy of c_0. By a strengthening of the Strong James' Distortion Theorem for c_0 ([5], Theorem 8), there exists a normalized sequence $(x_j)_{j \in \mathbb{N}}$ in X and a null sequence $(\varepsilon_n)_{n \in \mathbb{N}}$ in $(0, 1)$ such that for all $n \in \mathbb{N}$, for all $t = (t_j)_{j \in \mathbb{N}} \in c_{00}$,

$$\max_{j \geqslant n} |t_j| \leqslant \left\| \sum_{j=n}^{\infty} t_j x_j \right\| \leqslant (1 + \varepsilon_n) \max_{j \geqslant n} |t_j|$$

Define $y_k := x_1 + x_2 + \cdots + x_k$, for all $k \in \mathbb{N}$. Next we define the bounded, closed and convex subset C of X by

$$C := \overline{co}\{y_k : k \in \mathbb{N}\} = \left\{ \sum_{j=1}^{\infty} t_j x_j : 1 = t_1 \geqslant t_2 \cdots \geqslant t_k \to 0 \text{ as } k \to \infty \right\}.$$

Also, we define the function $T : C \to C$ by

$$T \left(\sum_{j=1}^{\infty} t_j x_j \right) = x_1 + \sum_{j=1}^{\infty} t_j x_{j+1}.$$

Clearly the mapping T is affine and fixed point free. Inductively, it follows that for all $n \in \mathbb{N}$, $C_n := \overline{co}(T(C_{n-1})) = T(C_{n-1})$ is given by

$$C_n = \left\{ \sum_{j=1}^{\infty} t_j x_j : 1 = t_1 = t_2 \cdots = t_{n+1} \geqslant t_{n+2} \geqslant \cdots \geqslant t_k \to 0 \text{ as } k \to \infty \right\}.$$

Fix $n \in \mathbb{N}_0$ and fix $x, y \in C_n$. So, $x = \sum_{j=1}^{\infty} t_j x_j$ and $y = \sum_{j=1}^{\infty} s_j x_j$, where $1 = t_1 = t_2 \cdots = t_{n+1} \geqslant t_{n+2} \geqslant \cdots \geqslant t_k \to 0$ as $k \to \infty$ and $1 = s_1 = s_2 \cdots = s_{n+1} \geqslant s_{n+2} \geqslant \cdots \geqslant s_k \to 0$ as $k \to \infty$. Take $a_j = t_j - s_j$ for all $j \in \mathbb{N}$. Then we see that

$$\|x - y\| = \left\| \sum_{j=n+2}^{\infty} a_j x_j \right\| \geqslant \max_{j \geqslant n+2} |a_j|,$$

and

$$\|Tx - Ty\| = \left\| \sum_{j=1}^{\infty} a_j x_{j+1} \right\| = \left\| \sum_{j=n+2}^{\infty} a_j x_{j+1} \right\|$$

$$= \left\| \sum_{k=n+3}^{\infty} a_{k-1} x_k \right\|$$

$$\leqslant (1 + \varepsilon_{n+3}) \max_{k \geqslant n+3} |a_{k-1}| = (1 + \varepsilon_{n+3}) \max_{k \geqslant n+2} |a_{k-1}|$$

$$\leqslant (1 + \varepsilon_{n+3}) \|x - y\|.$$

From the above inequality we conclude that T is cascading nonexpansive on C.

Theorem 5.189 *Let $(X, \| \cdot \|)$ be a reflexive Banach space. Then there exists an equivalent norm $\| \cdot \|_{\sim}$ on X such that for every bounded, closed and convex subset C of X, for all $\| \cdot \|_{\sim}$-cascading nonexpansive mappings $T : C \to C$, T has a fixed point in C.*

Proof Following Benavides [48], we find that there exists an equivalent norm $\| \cdot \|_{\sim}$ on X such that for every bounded, closed and convex subset K of X, for all $\| \cdot \|_{\sim}$-nonexpansive mappings $F : K \to K$, F has a fixed point in K. Fix an arbitrary bounded, closed and convex subset C of X. Let $T : C \to C$ be a $\| \cdot \|_{\sim}$-cascading nonexpansive mapping. As above, let $C_0 := C$ and $C_n := \overline{co}(T(C_{n-1}))$, for all $n \in \mathbb{N}$. By hypothesis there exists a sequence $(k_n)_{n \in \mathbb{N}}$ in $[1, \infty)$ such that $k_n \to 1$, and for all $n \in \mathbb{N}$, for all $x, y \in C_n$,

$$\|Tx - Ty\|_{\sim} \leqslant k_n \|x - y\|_{\sim}.$$

Because X is reflexive, C is weakly compact. By Zorn's Lemma there exists a nonempty, bounded, closed and convex set $D \subseteq C$ such that D is a minimal invariant set for T, that is, $T(D) \subseteq D$, and if E is a nonempty , bounded, closed and convex subset of D with $T(E) \subseteq E$, then we must have $E = D$. It follows that $\overline{co}(T(D)) = D$. Let $D_0 := D$ and $D_n := \overline{co}(T(D_{n-1}))$, for all $n \in \mathbb{N}$. Inductively, we see that for all $n \in \mathbb{N}$, $D = D_n \subseteq C_n$. Therefore, by our hypotheses on T, for all $x, y \in D$, for all $n \in \mathbb{N}$,

$$\|Tx - Ty\|_{\sim} \leqslant k_n \|x - y\|_{\sim}.$$

Letting $n \to \infty$ and noting the fact that $\lim_{n \to \infty} k_n = 1$, for all $x, y \in D$, we obtain

$$\|Tx - Ty\|_{\sim} \leqslant \|x - y\|_{\sim}.$$

Hence, T is $\| \cdot \|_{\sim}$-nonexpansive on D. Finally, it follows by Benavides [48] that T has a fixed point in $D \subseteq C$.

Combining Theorems 5.188 and 5.189, we get the following "fixed point property" characterization of reflexivity in Banach lattices, Banach spaces with an unconditional basis, and symmetrically normed ideals of operators on an infinite-dimensional Hilbert space.

In 2014, Lennard and Nezir [364] proved the following result that is stated below without proof.

Theorem 5.190 Let $(X, \| \cdot \|)$ be a Banach lattice, or a Banach space with an unconditional basis, or a symmetrically normed ideal of operators on an infinite-dimensional Hilbert space. Then the following are equivalent.
(1) X is reflexive.
(2) There exists an equivalent norm $\| \cdot \|_\sim$ on X such that for all bounded, closed and convex sets $C \subseteq X$ and for all $\| \cdot \|_\sim$-cascading nonexpansive mappings $T : C \to C$, T has a fixed point in C.

Exercises

5.1 Define a contraction mapping and prove that every contraction mapping T defined on a Banach space X into itself has a fixed point $\bar{x} \in X$.

5.2 Let $X = [0, 1]$ be a metric space with the usual metric and $T : X \to X$ be a mapping defined by $T(x) = \frac{1}{13}(x^3 + x^2 + 1)$ for all $x \in X$. Prove that T is a contraction mapping with Lipschitz constant $\alpha = \frac{5}{13}$.
[Hint: For all $x, y \in X$, we have

$$
\begin{aligned}
|Tx - Ty| &= \frac{1}{13} \left| (x^3 - y^3) + (x^2 - y^2) \right| \\
&\leq \frac{1}{13} \left[\left| (x - y)(x^2 + xyy^2) \right| + |x - y| \, |x + y| \right] \\
&\leq \frac{1}{13} \left[3 \, |x - y| + 2 \, |x - y| \right] = \frac{5}{13} |x - y|,
\end{aligned}
$$

which shows that T is a contraction mapping with Lipschitz constant $\frac{5}{13}$.]

5.3 Let $X = \{x \mid x \in \mathbb{Q}, x \geqslant 1\}$ be a metric space with the usual metric and $T : X \to X$ be defined by $Tx = \frac{x}{2} + x^{-1}$ for all $x \in X$. Show that T is a contraction mapping with Lipschitz constant $\alpha = \frac{1}{2}$.

5.4 Let $X = [1, \infty)$ be a metric space with the usual metric and $T : X \to X$ be defined by $Tx = \frac{99}{100}(x + \frac{1}{x})$ for all $x \in X$. Show that T is a contraction mapping with Lipschitz constant $\alpha = \frac{99}{100}$.

5.5 Let $X = \mathbb{R}$ be a metric space with the usual metric and $T : X \to X$ be a mapping defined by $T(x) = \cos x$ for all $x \in X$. Show that the mapping T is not contraction but contractive.
[Hint: For all $x, y \in \mathbb{R}$, we have

$$|\cos x - \cos y| = \left|2 \sin \left(\frac{x-y}{2}\right) \sin \left(\frac{x+y}{2}\right)\right|$$

$$= 2 \left|\sin \left(\frac{x-y}{2}\right)\right| \left|\sin \left(\frac{x+y}{2}\right)\right|$$

$$\leqslant 2 \left|\sin \left(\frac{x-y}{2}\right)\right|$$

$$< 2 \left|\frac{x-y}{2}\right| \quad \text{(since } |\sin \theta| < |\theta| \text{ for all } \theta \neq 0)$$

$$= |x - y|,$$

which shows that T is not contraction but contractive.]

5.6 Let $X = \mathbb{R}$ be a metric space with the usual metric and $T : X \to X$ be a mapping defined by $T(x) = \sin x$ for all $x \in X$. Show that the mapping T is contractive.

5.7 If T is a contraction mapping, show that for any positive integer n, T^n is a contraction mapping. However, if T^n is a contraction for $n > 1$ then T need not be a contraction.

5.8 Let X be a complete metric space and let $T : X \to X$ be a Kannan mapping. Show that T has a unique fixed point in X. Show also that a Kannan mapping need not be continuous.

5.9 Show that if T is a function from a nonempty and compact metric space X to itself such that
$$d(Tx, Ty) < d(x, y) \ \forall x, y \in X, \ x \neq y,$$
then T has a unique fixed point.

5.10 Let (X, d) be an arbitrary metric space and $T : X \to X$ a mapping which satisfies
$$d(Tx, Ty) < d(x, y) \ (x, y \in X)$$
whenever $x \neq y$. Assume that for some $x \in X$, the iterated sequence $\{x_n\}$ defined by $x_n = T^n x$ has a subsequence which converges to $u \in X$. Show that u is a fixed point of T.

5.11 Let (X, d) be a complete metric space and $T : X \to X$ a continuous mapping. Assume also that there exist an integer n and a positive number $k \in (0, \frac{1}{2})$ such that
$$d(Tx, Ty) \leqslant k \left(d(x, T^n z) + d(y, T^n z)\right) \ \forall x, y, z \in X.$$
Prove that T has a fixed point.

5.12 Let (X, d) be an arbitrary metric space and $T : X \to X$ a continuous mapping. Assume that for some $x \in X$, the orbit $\{T^n x\}$ contains a convergent subsequence $\{T^{n_i} x\}$. Show that if $d(T^{n_i} x, T^{n_i+1} x) \to 0$, then T has a fixed point.

5.13 Prove that every continuous map of a closed, bounded and convex set in $\mathbb{R}^n$ into itself has a fixed point.

5.14 Prove that the unit ball $K^\infty = \left\{ x = \{x_i\} \in \ell_2 \,\big|\, \|x\|^2 = \sum_{i=1}^\infty x_i^2 \leqslant 1 \right\}$ in the Hilbert space ℓ_2 is not a fixed point space.
   [Hint: Show that $T : K^\infty \to K^\infty$ given by $x = (x_1, x_2, \ldots) \mapsto (\sqrt{1 - \|x\|^2},$ $x_1, x_2, \ldots)$ is a continuous map without fixed points.]

5.15 Prove that the function $f : \mathbb{R} \to \mathbb{R}$ defined by $f(x) = \frac{1}{3}(x^3 + 1)$ has three fixed points α, β, γ, where $-2 < \alpha < -1, 0 < \beta < 1, 1 < \gamma < 2$.

5.16 Let X be a complete metric space and $T : X \to X$ a contraction mapping satisfying

$$d(Tx, Ty) \leqslant cd(x, y) \ \forall \, x, y \in X \text{ and } 0 < c < 1.$$

Prove that if $\bar{x}$ is the unique fixed point of T, then

$$d(\bar{x}, x) \leqslant \frac{1}{1-c} d(Tx, x) \ \forall x \in X.$$

5.17 Let X be a Banach space, $m \geq 1$, and let T be a continuous linear operator on X such that $\|T^m\| < 1$. Fix $u \in X$ and define

$$\Phi(v) = u + Tv, v \in X.$$

(*a*) Show that Φ^m is a contraction.
(*b*) Show that the equation $v = u + Tv$ has a unique solution $v \in X$.

5.18 Let U be an open bounded subset in a Banach space X with $0 \in U$ and $F, G :$ $\overline{U} \to X$ be two contractive maps such that $F|_{\partial U} = G|_{\partial U}$. Show that

$$\mathfrak{F}(F) \neq \varnothing \Longleftrightarrow \mathfrak{F}(G) \neq \varnothing.$$

5.19 Let T be a contraction mapping of a complete metric space X into itself and S be another mapping of X into itself such that for all $x \in X$, and a suitable $\eta > 0$

$$d(Tx, Sx) \leqslant \eta.$$

Using induction, show that for any $x \in X$,

$$d(T^m x, S^m x) \leqslant \eta \frac{1 - c^m}{1 - c} \ (m = 1, 2, \ldots).$$

5.20 Let (X, d) be a complete metric space and let $T : X \to K(X)$, where $K(X)$ is the class of all nonempty and compact subsets of X. Assume that there exists a map $\varphi : (0, \infty) \to [0, 1)$ such that

$$\forall_{t\in(0,\infty)}\left\{\limsup_{r\to t^+}\varphi(r)<1\right\}$$

and

$$\forall_{x,y\in X,x\neq y}\left\{H(Tx,Ty)\leqslant\varphi(d(x,y))\,d(x,y)\right\}.$$

Then T has a fixed point.

5.21 Let (X,d) be a complete metric space and let $T:X\to CB(X)$. Assume that there exists a map $\varphi:(0,\infty)\to[0,1)$ such that

$$\forall_{t\in[0,\infty)}\left\{\limsup_{r\to t^+}\varphi(r)<1\right\}$$

and

$$\forall_{x,y\in X,x\neq y}\left\{H(Tx,Ty)\leqslant\varphi(d(x,y))\,d(x,y)\right\}.$$

Then T has a fixed point.

5.22 Let (X,d) be a complete metric space, $T:X\to CB(X)$ a multivalued map and $L>0$, where $CB(X)$ is the class of all nonempty, closed and bounded subsets of X. Assume that

$$H(Tx,Ty)\leqslant\varphi(d(x,y))d(x,y)+Ld(y,Tx)\quad\forall x,y\in X,$$

where φ is a function from $[0,\infty)$ into $[0,1)$ satisfying $\limsup_{r\to t^+}\varphi(r)<1$ for all $t\in[0,\infty)$. Then there exists $z\in X$ such that $z\in Tz$.

5.23 Let (X,d) be a complete metric space and let $T:X\to CL(X)$, where $CL(X)$ is the class of all nonempty and closed subsets of X. Assume that the following conditions hold:
(i) the map $f:X\to\mathbb{R}$ defined by $f(x)=d(x,Tx)$, $x\in X$, is lower semicontinuous;
(ii) there exist $b\in(0,1)$, $I_b^x=\left\{y\in Tx:bd(x,y)\leqslant d(x,Tx)\right\}$, and $\varphi:[0,\infty)\to[0,b)$ such that

$$\forall_{t\in[0,\infty)}\left\{\limsup_{r\to t^+}\varphi(r)<b\right\}$$

and

$$\forall_{x\in X}\exists_{y\in I_b^x}\left\{d(y,Ty)\leqslant\varphi(d(x,y))\,d(x,y)\right\}.$$

Then T has a fixed point.

5.24 Let X be an ordered Banach space, and T an increasing mapping of X into itself. Let the ordering of X be such that every bounded and nonempty subset of X has a supremum and infimum. Further, suppose that there exist $u_-, u_+ \in X$ such that

$$u_- \leqslant Tu_- \leqslant Tu_+ \leqslant u_+.$$

Show that T has maximul and minimul fixed points u_{min}, u_{max} satisfying the relation

$$u_- \leqslant u_{min} \leqslant u_{max} \leqslant u_+.$$

Chapter 6
Degree Theory, k-Set Contractions and Condensing Operators

A mathematical theory is not to be considered complete until you have made it so clear that you can explain it to the first man whom you meet on the street.

David Hilbert

Mathematics is an organ of knowledge and an infinite refinement of language. It grows from the usual language and world of intuition as does a plant from the soil, and its roots are the numbers and simple geometrical intuitions. We do not know which kind of content mathematics (as the only adequate language) requires; we cannot imagine into what depths and distances this spiritual eye (mathematics) will lead us.

Erich Kähler (1941)

It is by logic that we prove, but by intuition that we discover.

Henri Poincaré

The notion of "degree" of a map was first defined by Brouwer, who showed that the degree is homotopy invariant, and used it to prove the Brouwer fixed point theorem. Note that topological degree theory is a generalization of the winding number of a curve in the complex plane. It is closely connected to fixed point theory, and can be used to estimate the number of solutions of an equation. For a given equation, if one solution of an equation is easily found, then degree theory can often be used to prove existence of a second, nontrivial, solution.

Notice that there are different types of degree for different types of maps, say, for example, for maps between Banach spaces, there is the Brouwer degree (Kronecker characteristics, Poincare index) in $\mathbb{R}^n$, and for compact mappings in normed spaces, there is the Leray–Schauder degree.

In topology, the degree of a continuous mapping between two compact oriented manifolds of the same dimension is a number that represents the number of times that the domain manifold wraps around the range manifold under the mapping. The degree is always an integer-valued characteristic assigned to a map, but may be positive or negative depending on the orientations. The map should be reasonable in the sense that it should satisfy the standard additivity, homotopy and normalization properties.

© Springer Nature Singapore Pte Ltd. 2018
H. K. Pathak, *An Introduction to Nonlinear Analysis and Fixed Point Theory*,
https://doi.org/10.1007/978-981-10-8866-7_6

This means that the degree is an algebraic "count" of solutions to an operator equation that is not affected by small perturbations or even larger deformations.

Observation

- In the finite dimensional case, the degree known under the names "Brouwer degree" is well defined for continuous maps, while in the infinite dimensional spaces it is defined for special classes of maps such as
  (i) Leray–Schauder degree for compact fields,
  (ii) Caccioppoli–Smale degree for smooth maps,
  (iii) Nussbaum–Sadovskii degree for the condensing fields,
  (iv) Browder–Petryshin–Skrypnik degree for monotone operators etc.
- The notion of degree allows to define an important topological characteristic of a singular point of a vector field (i.e. a point where the field either vanishes or is not defined). Namely, a degree of a vector field on a small sphere centered at an isolated singular point is called an index of the point.

Notice that fixed point theorems and the local topological degree are closely related. The topological methods give qualitative information only, that is, they do not give any quantitative information. The topological degree theory has an important advantage over the fixed point theory in the sense that it gives information about the number of distinct solutions, continuous families of solutions and stability of solutions. But for our purpose, the discussion will be restricted to the existence of fixed points. First we deal with the local topological degree in n-dimensional Euclidean space $\mathbb{R}^n$. One encounters new situations when one tries to extend the definition of the degree of a continuous mapping F from finite dimensional setting to infinite dimensional setting. A large class of mappings for which the degree of a mapping can be defined in infinite dimensional setting is the class of compact operators. This is known as Leray–Schauder [365] degree theory. In Sect. 6.2, we discuss Leray–Schauder degree for a completely continuous vector field. Section 6.3 deals with the Skrypnik degree for a pseudomonotone mapping while in Sect. 6.4 we discuss the Browder–Petryshyn degree for A-proper mappings.

A generalization of Leray–Schauder degree has been given by Browder and Nussbaum [119]. But we shall discuss in Sect. 6.5, the generalized degree theory as developed by Sadovskii [536], who extended the concept of degree to the class of limit-compact operators. In Sect. 6.6, we discuss various notions of measure of noncompactness along with some basic properties possesses by these notions and present a generalization of Darbo's fixed point theorem. Subsequently, we shall use this theory in Sect. 6.7 to discuss k-set contractions and condensing mappings. In this connection, for more recent results, especially for degree theory of more general mappings of monotone type, refer to Browder [115].

6.1 The Topological Degree

The concept of a local degree in n-dimensional Euclidean space, that is, of the degree with respect to a neighbourhood of an isolated solution of the system $fx = y$, goes back to Kronecker [349]. A detailed discussion of Kronecker's index and some of its application was given by Hadamard [263]. Brouwer [86] extended this local degree to a degree in the large and used it as basis of many of his famous results. This "global" degree is by nature a concept of combinatorial topology. In this connection, one can refer Alexandroff and Hopf [9] and Cronin [152].

Definition 6.1 If f_1 and f_2 are continuous maps of the space X into the space Y, we say that f_1 is homotopic to f_2 if there is a continuous map $H : [0, 1] \times X \to Y$ such that

$$H(0, x) = f_1(x) \text{ and } H(1, x) = f_2(x) \text{ for each } x \in X.$$

The map H is called a homotopy between f_1 and f_2. If f_1 is homotopic to f_2, we write $f_1 \simeq f_2$.

6.1.1 Axiomatic Definition of the Brouwer Degree in $\mathbb{R}^n$

Throughout this section, let $\mathbb{R}$ denote the set of real numbers and let $\mathbb{R}^n = \{x = (x_1, x_2, \ldots, x_n) : x_i \in \mathbb{R} \text{ for } i = 1, 2, \ldots, n\}$ be endowed with standard 2-norm $|x| = \left(\sum_{i=1}^{n} x_i^2 \right)^{\frac{1}{2}}$. For subsets $A \subset \mathbb{R}^n$ we use the usual symbol $\overline{A}$, ∂A to denote the closure and the boundary of A, respectively. The open and the closed ball of center x_0 and radius $r > 0$ will be denoted by

$$B_r(x_0) = \{x \in \mathbb{R}^n : |x - x_0| < r\} = x_0 + B_r(O) \text{ and } \overline{B}_r(x_0) = \overline{B_r(x_0)}.$$

Unless otherwise stated, Ω is always an open bounded subset of $\mathbb{R}^n$.

The Brouwer[1] degree is a tool, which allows an answer to the following: Let $\Omega \subset \mathbb{R}^n$ be open and bounded, and $f : \overline{\Omega} \to \mathbb{R}^n$ a continuous function, and let $y \notin f(\partial\Omega)$, the f-image of the boundary $\partial\Omega$ of Ω. Then we raise a natural question: Does a given equation

$$f(x) = y$$

has a solution $x \in \Omega$? More precisely, for each admissible triple (f, Ω, y) we associate an integer $deg(f, \Omega, y)$ such that

$$deg(f, \Omega, y) \neq 0 \Rightarrow \text{ the existence of a solution } x \in \Omega \text{ of this equation } f(x) = y.$$

---

[1]L. E. J. Brouwer was a Dutch mathematician and philosopher, who worked in topology, set theory, measure theory and complex analysis. He was the founder of the mathematical philosophy of intuitionism.

Uniqueness of *deg.* The integer associated with $deg(f, \Omega, y)$ is uniquely defined by the following properties:

The first condition is concerned with the normalization property. If f is the identity map, and $y \in \mathbb{R}^n$, then the equation $f(x) = y$ has a solution $x \in \Omega$ if and only if $y \in \Omega$, i.e.

$$(d1) \qquad deg(I, \Omega, y) = \begin{cases} 1, & \text{for } y \in \Omega \\ 0, & \text{for } y \notin \overline{\Omega}. \end{cases}$$

The second condition is a natural formulation of the desire that d should yield information on the location of solutions. Suppose that $f(x) = y$ has finitely many solutions in $\Omega_1 \bigcup \Omega_2$, where Ω_1 and Ω_2 are two disjoint open subsets of Ω, but it has no solution in $\overline{\Omega} \setminus (\Omega_1 \bigcup \Omega_2)$. Then the number of solutions in Ω is the sum of the numbers of solutions in Ω_1 and Ω_2. This condition suggests that d should be additive in its argument Ω, i.e.

$$(d2) \qquad deg(f, \Omega, y) = d(f, \Omega_1, y) + deg(f, \Omega_2, y)$$

whenever Ω_1 and Ω_2 are two disjoint open subsets of Ω, such that $y \notin f(\overline{\Omega} \setminus (\Omega_1 \bigcup \Omega_2))$.

The third condition reflects the desire that for complicated f the number $deg(f, \Omega, y)$ can be calculated by $deg(g, \Omega, y)$, with simpler g, at least if f can be continuously deformed into g such that at no stage of the deformation we get solution on the boundary $\partial \Omega$. This leads to the following condition

$$(d3) \qquad deg(H(t, \cdot), \Omega, y(t)) \text{ is independent of } t \in [0, 1]$$

whenever $H : [0, 1] \times \overline{\Omega} \to \mathbb{R}^n$ and $y : [0, 1] \to \mathbb{R}^n$ are continuous and $y(t) \notin H(\partial \Omega)$ for all $t \in [0, 1]$.

Theorem 6.1 *There exists a unique $\mathbb{H}$-valued mapping d, which associates every admissible triple (f, Ω, y), where $\Omega \subset \mathbb{R}^n, f : \overline{\Omega} \to \mathbb{R}^n$ is continuous and $y \in \mathbb{R}^n \setminus f(\partial \Omega)$ an integer $deg(f, \Omega, y)$, with the following property:*

1. **Existence:** *If $deg(f, \Omega, y) \neq 0$, then there exist $x \in \Omega$ such that $f(x) = y$.*
2. **Normalization:** *$deg(id, \Omega, y) = 1$ if $y \in \Omega$, $deg(id, \Omega, y) = 0$, if $y \notin \Omega$.*
3. **Homotopy invariance:** *Let $H : [0, 1] \times \Omega \to \mathbb{R}^n$ be continuous and $y \notin H(t, \partial \Omega)$ for $t \in [0, 1]$, then $deg(H(t, \cdot), \Omega, y)$ is independent from t.*
4. **Additivity:** *If $\bigcup_{i=1}^{m} \Omega_i \subset \Omega, \bigcup_{i=1}^{m} \overline{\Omega}_i \subset \overline{\Omega}, \Omega_i$ is open, disjoint, $y \notin \bigcup_{i=1}^{m} f(\partial \Omega_i)$, then*

$$deg(f, \Omega, y) = \sum_{i=1}^{m} deg(f, \Omega_i, y).$$

5. *If $g : \overline{\Omega} \to \mathbb{R}^n$ is continuous and $|f - g| < dist(y, f(\partial \Omega))$, then*

$$deg(f, \Omega, y) = deg(g, \Omega, y).$$

6. *If $z \in \mathbb{R}^n, |y - z| < dist(y, f(\partial \Omega))$, then*

$$deg(f, \Omega, y) = deg(f, \Omega, z).$$

7. *If $g : \overline{\Omega} \to \mathbb{R}^n$ is continuous and $f|_{\partial\Omega} = g|_{\partial\Omega}$, then*

$$deg(f, \Omega, y) = deg(g, \Omega, y).$$

8. $deg(f, \Omega, y) = deg(f(\cdot) - y, \Omega, 0)$.
9. **Excision property:** *Let C be a closed subset of Ω, $C \neq \overline{\Omega}$ and $y \notin f(C)$, then*

$$deg(f, \Omega, y) = deg(f, \Omega \setminus C, y).$$

10. **Reduction property:** *Let $m < n$, $\Omega \subset \mathbb{R}^n$ be open and bounded $f : \overline{\Omega} \to \mathbb{R}^n$ continuous, $y \in R^m \setminus (I - f)(\partial\Omega)$. Then*

$$deg(I - f, \Omega, y) = deg(I - f|_{\overline{\Omega} \cap \mathbb{R}^m}, \Omega \cap \mathbb{R}^m, y).$$

Since Brouwer's basic paper in 1912, many efforts have been made to define the degree of a mapping by strictly analytic method without involving the concepts of combinatorial topology. In this connection, one can see the works of Nagumo [418], Heinz [270] and Schwartz [545]. However, we include here only the analytic definition of the degree of a mapping.

Let $f : \Omega \subset \mathbb{R}^n \to \mathbb{R}^n$ and $y \in \mathbb{R}^n$ be a given vector. It is of considerable interest to know in advance the number of solutions of the system $fx = y$ in some specified set $C \subset \mathbb{R}^n$. For example, we may be interested in the qualitative theory of the structure of the real solutions of the system $fx = y$. We assume that f is continuous. We start by defining an invariant which will be useful in classifying systems of the form $fx = 0$ where $f = (f_1(x_1, \ldots, x_n), f_2(x_1, \ldots, x_n), \cdots, f_n(x_1, \ldots, x_n))$ and in giving information about the structure of their real solutions. This invariant is known as the degree of the map and is defined below.

Definition 6.2 We now set our objective to characterize all mappings

$$f : \Omega \subset \mathbb{R}^n \to \mathbb{R}^n, f = (f_1(x_1, x_2, \ldots, x_n), \cdots, f_n(x_1, x_2, \ldots, x_n)),$$

f_i are continuous functions by an integer invariant called the degree of the mapping F with the property that two functions that are "near" to each other in some sense are assigned the same integer. To define this integer invariant, we proceed in the following manner:

Let $C(\overline{\Omega})$ denote the set of continuous vector-valued functions defined on $\overline{\Omega} \subset \mathbb{R}^n$ with the topology of uniform convergence, where Ω is a bounded open set in $\mathbb{R}^n$. Let $C^1(\overline{\Omega})$ denote the subset of continuously differentiable vector-valued functions on $\overline{\Omega}$. The idea is to have a measure of the number of solutions of $f(x) = y$ in Ω by an integer $deg(f, \Omega, y)$ which depends continuously on both f and y. This integer is called the degree of the mapping f relative to the point y and the set Ω. First of all we define $deg(f, \Omega, y)$ on a dense set of functions f in $C(\overline{\Omega})$ and on a dense set of points $y \in \Omega$. Then we extend the definition to all points in Ω and all functions in $C(\overline{\Omega})$ by a limiting process. It is assumed that $f(x) = y$ has no solution for $x \in \partial \Omega$. The definition proceeds by counting the algebraic number of solutions of $f(x) = y$ in Ω in the following three steps:

(i) Suppose $f \in C^1(\overline{\Omega})$ and the Jacobian determinant of F i.e.

$$det \, J_f(x) = det \left(\frac{\partial f_i}{\partial x_j} \right),$$

does not vanish at any point x in the solution set $\{x : x \in \Omega, f(x) = y\}$. Then we define

$$deg(f, \Omega, y) = \sum_{x \in f^{-1}(y)} \text{sgn det } J_f(x), \tag{6.1}$$

where $deg(f, \Omega, y) = 0$ if $f^{-1}(y) = \varnothing$.

Notice that this number is a finite integer as the set $f^{-1}(y)$ is discrete by the inverse function theorem and this discrete set has not limit point in the compact set $\overline{\Omega}$.

(ii) To define $deg(f, \Omega, y)$ for functions $f \in C^1(\overline{\Omega})$ whose solution set contains some x where the Jacobian determinant det $J_f(x) = 0$, we need the following special case of Sard's theorem, which is stated without proof (the interested reader can see also Berger and Berger [53], p. 35).

Let $\Omega \subset \mathbb{R}^n$ be an open subset. We will use $\circ(|h|)$ to describe those expressions which, roughly speaking, are of higher than first order in h as $h \to 0$. We recall that a function $F : \Omega \to \mathbb{R}^n$ is differentiable at $x_0 \in \Omega$ if there is a matrix $f'(x_0)$ such that $f(x_0 + h) = f(x_0) + f'(x_0)h + \circ(|h|)$, where $x_0 + h \in \Omega$ and $\frac{|\circ(|h|)|}{|h|}$ tends to zero as $|h| \to 0$.

We use $C^k(\Omega)$ to denote the space of k-times continuously differentiable functions. If f is differentiable at x_0, we call $J_f(x_0) = \text{det} f'(x_0)$ the Jacobian of f at x_0. If $J_f(x_0) = 0$, then x_0 is said to be a critical point of f and we use $S_f(\Omega) = \{x \in \Omega : J_f(x) = 0\}$ to denote the set of critical points of f, in Ω. If $f^{-1}(y) \cap S_f(\Omega) = \varnothing$, then y is said to be a *regular value* of f. Otherwise, y is said to be a *singular value* of f.

Lemma 6.1 (Sard's lemma) *Let $\Omega \subset \mathbb{R}^n$ be open and $f \in C^1(\Omega)$, that is, $f : \Omega \to \mathbb{R}^n$ is continuously differentiable function. Then $\mu_n(f(S_f(\Omega))) = 0$, where μ_n is the n-dimensional Lebesgue measure.*

Proof To begin the proof, we first need to know about Lebesgue measure μ_n is that $\mu_n(J) = \prod_{i=1}^n (b_i - a_i)$ for the interval $J = [a, b] \subset \mathbb{R}^n$ and that $M \subset \mathbb{R}^n$ has measure zero, i.e. $\mu_n(M) = 0$ if and only if to every $\varepsilon > 0$ there exists at most countably many intervals J_i such that $M \subset \bigcup_{i=1}^{\infty} J_i$ and $\sum_{i=1}^{\infty} \mu_n(J_i) \leqslant \varepsilon$.

Then it is easy to see that at most countable union of sets of measure zero also has measure zero.

Since Ω is open, we can write $\Omega = \overset{\infty}{\underset{i=1}{\cup}} Q_i$, where Q_i is a cube for $i = 1, 2, \ldots$. We only need to show that $\mu_n(f(S_f(Q))) = 0$ for a cube $Q \subset \Omega$. In fact, let l be the lateral length of Q. By the uniform continuity of f' on Q, for any given $\varepsilon > 0$, there exists an integer $m > 0$ such that

$$|f'(x) - f'(y)| \leqslant \varepsilon$$

for all $x, y \in Q$ with $|x - y| \leqslant \frac{\sqrt{n}l}{m}$. Therefore, we have

$$|f(x) - f(y) - f'(y)(x - y)| \leqslant \int_0^1 |f'(y + t(x - y)) - f'(y)||x - y| dt$$
$$\leqslant \varepsilon |x - y|$$

for all $x, y \in Q$ with $|x - y| \leqslant \frac{\sqrt{n}l}{m}$. Let us decompose Q into r cubes, Q^i, of diameter $\frac{\sqrt{n}l}{m}$, $i = 1, 2, \ldots, r$. Since $\frac{l}{m}$ is the lateral length of Q^i, we have $r = m^n$. Now, suppose that $Q^i \cap S_f(\Omega) \neq \varnothing$. Choosing $y \in Q^i \cap S_f(\Omega)$, we have $f(y + x) - f(y) = f'(y)x + R(y, x)$ for all $x \in Q^i - y$, where $|R(y, x + y)| \leqslant \frac{\sqrt{n}l}{m}$. Therefore, we have

$$f(Q^i) = f(y) + f'(y)(Q^i - y) + R(y, Q^i).$$

But $f'(y) = 0$, so $f'(y)(Q^i - y)$ is contained in an $(n-1)$-dimensional subspace of $\mathbb{R}^n$. Thus, $\mu_n(f'(y)(Q^i - y)) = 0$, so we have

$$\mu_n(f(Q^i)) \leqslant 2^n \varepsilon^n \left(\frac{\sqrt{n}l}{m}\right)^n.$$

Obviously, $f(S_f(Q)) \subset \cup_{i=1}^r f(Q^i)$, so we have

$$\mu_n(f(S_f(Q)) \leqslant r 2^n \varepsilon^n \left(\frac{\sqrt{n}l}{m}\right)^n = 2^n \varepsilon^n (\sqrt{n}l) n.$$

By letting $\varepsilon \to 0^+$, we obtain $\mu_n(f(S_f(Q))) = 0$. Therefore, $\mu_n(f(S_f(\Omega))) = 0$. This completes the proof.

Now let $y \in f(S_f(\Omega))$, then y can be approximated by points y_i whose solution sets contain only x at which $\det J_f(x) \neq 0$. So we define

$$deg(f, \Omega, y) = \lim_{i \to \infty} deg(f, \Omega, y_i).$$

Observation

- This definition is justified since it can be shown that it is independent of the approximating sequence $\{y_i\}$, the limit exists and is finite.

(*iii*) To extend the definition to all continuous functions $f \in C(\overline{\Omega})$, we use the fact that $C^1(\overline{\Omega})$ is dense in $C(\overline{\Omega})$. Let $\{f_i\} \subset C^1(\overline{\Omega})$ be such that $f_i \to f$ uniformly in $\overline{\Omega}$. Then we define

$$d(f, \Omega, y) = \lim_{i \to \infty} d(f_i, \Omega, y).$$

Observation

- It may be observe again that this definition is justified as it is independent of the approximating sequence $\{f_i\}$ and a finite limit exists.

The following two lemmas stated without proof justify steps (i) and (iii) in the above definition.

Lemma 6.2 *Suppose (i)* $f \in C^2(\Omega) \cap C(\overline{\Omega})$; *(ii) det* $J_f(x) \neq 0$ *for* $x \in f^{-1}(y)$ *or* $f^{-1}(z)$ *and (iii)* $f(x) \neq y$ *for* $x \in \partial\Omega$. *Then whenever* z *is sufficiently near* y, *we have*

$$deg(f, \Omega, z) = deg(f, \Omega, y).$$

Lemma 6.3 *Suppose (i)* $f, g \in C^1(\Omega) \cup C(\overline{\Omega})$, *(ii) det* $J_f(x) \neq 0$ *for* $x \in f^{-1}(y)$ *and (iii)* $f(x) \neq y$ *for* $x, \in \partial\Omega$. *Then whenever* g *is sufficiently close to* f *in* C^1 *sense (i.e. for* ε *sufficiently small,* $\sup\{|f - g| + \sum_{i=1}^n |f_{x_i} - g_{x_i}|\} < \varepsilon$) *we have*

$$deg(g, \Omega, y) = deg(f, \Omega, y).$$

Here f_{x_i} *denotes the partial derivative of* f *with respect to* x_i.

Remark 6.1 Justification of step (i) follows from Lemma 6.2 (extended to $f \in C^1(\Omega)$) and Lemma 6.3 by approximating a given function in $C^1(\Omega)$ by an appropriate $C^2(\Omega)$ function.

From Lemma 6.1 it follows that Lemma 6.3 holds without the hypothesis (ii). The justification of step (iii) follows if Lemma 6.3 can be extended to function g close to f in the sense (i.e. $\sup |f - g| < \varepsilon$). This can be seen by a homotopy argument. Let

$$H(t, x) = tf + (1 - t)g \text{ for } t \in [0, 1],$$

and suppose that $H(x, t) \neq y$ for $x \in \partial\Omega$. Then we define

$$t \sim t' \text{ if } deg(H(t, x), \Omega, y) = deg(H(t', x), \Omega, y).$$

"$\sim$" is an equivalence relation. Then the associated disjoint equivalence classes are open in the space $[0, 1]$. Thus $[0, 1]$ is a union of open sets, namely the open equivalence classes. This contradicts the connectedness of $[0, 1]$ unless there is exactly one equivalence class. Hence $0 \sim 1$ and $deg(f, \Omega, y) = deg(g, \Omega, y)$.

Let $f : \overline{\Omega} \to \mathbb{R}^n$ be a continuous mapping and $f(x) = y$ has no solutions for $x \in \partial\Omega$ where Ω is an arbitrary bounded open set in $\mathbb{R}^n$. We state below a few basic properties of the function $deg(f, \Omega, y)$ with indication of proofs of these results.

6.1.2 Invariance Properties

(a) **Boundary value dependence**: $deg(f, \Omega, y)$ is uniquely determined by the action of f on $\partial\Omega$.

(b) **Homotopy invariance**: Suppose $H(t, x) = y$ has no solution $x \in \partial\Omega$ for any $t \in [0, 1]$ then $deg(H(t, x), \Omega, y)$ is a constant independent of $t \in [0, 1]$ provided $H(t, x)$ is a continuous function of t and x.

(c) **Continuity**: $deg(f, \Omega, y)$ is a continuous function of $f \in C(\overline{\Omega})$ (with respect to uniform convergence) and $y \in \Omega$.

(d) **Poincare–Böhl**: If the vectors $f(x) - y$ and $g(x) - y$ never point in opposite directions for $x \in \partial\Omega$, then $deg(f, \Omega, y) = deg(g, \Omega, y)$ provided each is defined.

We now indicate the proofs of these properties. Notice that continuity is an immediate consequence of the analytic definition of $deg(f, \Omega, y)$. From this follows homotopy invariance property. Because $\phi(t) = deg(H(t, x), \Omega, y)$ is a continuous function of t and $\phi(t)$ is integer valued, $\phi(t)$ must be a constant. Boundary value dependence follows immediately by considering the homotopy

$$H(t, x) = tf(x) + (1 - t)\tilde{f}(x)$$

where $f = \tilde{f}$ on $\partial\Omega$. Similarly, property (d) follows by means of the homotopy

$$H(t, x) = t[f(x) - y] + (1 - t)[g(x) - y].$$

Some More Properties

(e) **Domain decomposition**: If $\{\Omega_i : 1 \leqslant i \leqslant m\}$ is a finite collection of disjoint open subsets of Ω and $f(x) \neq y$ for $x \in (\Omega \setminus \cup_{i=1}^{m}\Omega_i)$, then

$$deg(f, \Omega, y) = \sum_{i=1}^{m} deg(f, \Omega_i, y).$$

(f) **Cartesian product formula**: If $y \in \Omega \subset \mathbb{R}^n, y' \in \Omega' \subset \mathbb{R}^m$ and $f : \Omega \to \mathbb{R}^n$, $g : \Omega' \to \mathbb{R}^m$, then

$$deg((f, g), (y, y'), \Omega \times \Omega') = deg(f, \Omega, y) \times deg(g, \Omega', y')$$

provided the right-hand side is defined.

(g) If $f(x) \neq y$ in $\overline{\Omega}$, then $deg(f, \Omega, y) = 0$.

(h) **Odd mapping**: Let Ω be a symmetric domain about the origin, and $f(-x) = -f(x)$ on $\partial\Omega$, with $f : \Omega \to \mathbb{R}^n$, and $f(x) \neq 0$ on $\overline{\Omega}$, then $deg(f, \Omega, 0)$ is an odd integer.

(i) **Excision**: If M is a closed subset of $\overline{\Omega}$ on which $f(x) \neq y$, then

$$deg(f, \Omega, y) = deg(f, \Omega \setminus M, y).$$

(j) **Index theorem**: If the solutions of $f(x) = y$ are isolated on Ω and $f(x) \neq y$ on $\partial \Omega$, then

$$deg(f, \Omega, y) = \sum_{x_i \in f^{-1}(y)} d(f, B_\varepsilon(x_i), y),$$

where $B_\varepsilon(x_i) = \{x : |x - x_i| < \varepsilon\}$. Here ε is sufficiently small so that x_i is the only solution of $f(x) = y$ in $\overline{B}_\varepsilon(x_i)$ and is independent of ε.

(k) **Borsuk–Ulam theorem**: If $f : \mathbb{S}^n \to \mathbb{R}^n$ is a continuous mapping, then there is a point $x \in \mathbb{S}^n$ such that $f(x) = f(-x)$. Here $\mathbb{S}^n$ is the unit sphere in $\mathbb{R}^{n+1}$ with center at the origin.

For the proof of the above properties, the reader may refer to Berger and Berger [53].

6.1.3 Applications of the Brouwer Degree

The properties of the degree of a mapping provide interesting information concerning the solutions of the system $f(x) = y, f = (f_1, \ldots, f_n); y = (y_1, \ldots, y_n)$. Here $f_i(x)$ are continuous real-valued functions in $\overline{\Omega}$ and y_i are real numbers. Some results are given below.

Theorem 6.2 *If* $deg(f, \Omega, y) \neq 0$, *then* $f(x) = y$ *has solutions in* Ω.

Proof As $deg(f, \Omega, y) \neq 0$, $f(x) = y$ necessarily has solutions in Ω by excision property discussed above.

Theorem 6.3 (Brouwer's fixed point theorem) *Let* $D \subset \mathbb{R}^n$ *be a nonempty compact convex set and let* $f : D \to D$ *be a continuous mapping of* D *into itself. Then* f *has a fixed point. The same is true if* D *is only homeomorphic to a compact convex set.*

Proof **Step 1**: First we proof the result for a special case. Suppose that $D = \overline{B}_r(0)$, where $B_r(0) = \{x \in \mathbb{R}^n : |x| < r\}$. We may assume that $f(x) \neq x$ on ∂D and $deg(I - f, B_r(0), 0) = 0$, then we shall obtain a contradiction by showing that $d(I - f, B_r(0), 0) = 1$. Define $H : [0, 1] \times \overline{B}_r(0) \to \overline{B}_r(0)$ by

$$H(t, x) = x - tf(x) \ \forall x \in \overline{B}_r(0) \text{ and } t \in [0, 1].$$

Clearly, H is continuous and rewriting $H(x, t)$ as $H(t, x) = t(I - f)(x) + (1 - t)I(x)$ we see that $I - f$ is homotopic to I. We also observe that under the given hypothesis, the homotopy $H(t, x) = x - tf(x)$, $t \in [0, 1]$ has no zero on $\partial B_r(0)$. To see this, we observe that $\|f\| \leqslant r$ and so

$$|H(t, x)| \geq |x| - t|f(x)| \geq (1 - t)r > 0$$

in $\partial \overline{B}_r(0) \times [0, 1)$ and $f(x) \neq x$ for $|x| = r$. Thus, by homotopy invariance property, we have

$$deg(I - f, D, 0) = deg(I - tf, B_r(0), 0) = deg(I, B_r(0), 0) = 1$$

and this proves existence of an $\bar{x} \in \overline{B}_r(0)$ such that $\bar{x} - f(\bar{x}) = 0$.

Step 2: We now consider the general case. So, let D be a compact and convex set. By Theorem 5.68 we have a continuous extension $\tilde{f} : \mathbb{R}^n \to D$ of f with $f(\mathbb{R}^n) \subset \overline{cof(D)} \subset D$ since

$$\frac{1}{\sum\limits_{i=1}^{m} 2^{-i}\varphi(x)} \sum_{i=1}^{m} 2^{-i}\varphi(x)f(a_i)$$

is defined for $m = m(x)$ being sufficiently large and belongs to co $f(D)$. Now we choose a ball $\overline{B}_r(0) \supset D$, and applying first step we can find a fixed point $\tilde{x}$ in $\overline{B}_r(0)$. But $\tilde{f}(\tilde{x}) \in D$ and therefore $\tilde{x} = \tilde{f}(\tilde{x}) = f(\tilde{x})$.

Step 3: Finally assume that $D = h(D_0)$ with D_0 compact convex and h a homeomorphism. Then, by Step 2, the mapping $h^{-1}fh : D_0 \to D_0$ has a fixed point x_0 and therefore $h^{-1}fh(x_0) = x_0$ yields

$$f(h(x_0)) = h(x_0) \in D.$$

That is, $h(x_0)$ is a fixed point of f in D. ∎

Theorem 6.4 *Let $f : \mathbb{R}^n \to \mathbb{R}^n$ be a continuous mapping and $0 \in \Omega \subset \mathbb{R}^n$ with Ω an open bounded subset. If $(f(x), x) > 0$ for all $x \in \partial\Omega$, then $deg(f, \Omega, 0) = 1$.*

Proof Put $H(t, x) = tx + (1 - t)f(x)$ for all $(t, x) \in [0, 1] \times \overline{\Omega}$. Then applying the same argument as in Theorem 6.3 we see that $0 \notin H([0, 1] \times \partial\Omega)$, and so we have

$$deg(f, \Omega, 0) = deg(I, \Omega, 0) = 1.$$

This completes the proof.

Example 6.1 Consider the system of ordinary differential equations

$$x' = f(t, x(t)), \quad t \in \mathbb{R}, \quad x \in \mathbb{R}^n, \tag{6.2}$$

where $x' = \frac{dx}{dt}$ and $f : \mathbb{R} \times \mathbb{R}^n \to \mathbb{R}^n$ is T-periodic, i.e. $f(t + T, x) = f(t, x)$ for all $(t, x) \in \mathbb{R} \times \mathbb{R}^n$. So it is natural to look for T-periodic solutions. Suppose, for simplicity, that f is continuous and that there is an open ball $B_r(0) \subset \mathbb{R}^n$ such that the initial value problems

$$x'(t) = f(t, x(t)), \quad x(0) = y \in B_r(0) \tag{6.3}$$

have a unique solution $x(t, y)$ on $[0, \infty)$. Now, let $P_t y = x(t; y)$ and assume also that the function f satisfies the boundary condition

$$(f(t, y), y) = \sum_{i=1}^{n} f(t, y) y_i < 0 \text{ for } t \in [0, T] \text{ and } |y| = r,$$

where $| \cdot |$ denotes the Euclidean norm in $\mathbb{R}^n$. Then, we have

$$P_t : \overline{B}_r(0) \to \overline{B}_r(0) \quad \text{for all } t \in \mathbb{R}^+,$$

because

$$\frac{d}{dt} |x(t)|^2 = 2(x'(t), x(t)) = 2(f(t, x(t)), x(t)) < 0$$

if the solution x of (6.3) takes a value in $\partial B_r(0)$ at time t. Moreover, since (6.3) has a unique solution, it is not hard to prove that P_t is continuous. Thus, by the Brouwer fixed point theorem, P_T has a fixed point y_T in $\overline{B}_r(0)$; that is, Eq. (6.2) has a solution such that

$$x(0; y_T) = y_T = x(T; y_T).$$

Now, one may easily verify that $u : [0, \infty) \to \mathbb{R}^n$, defined by

$$u(t) = x(t - kT; y_T) \text{ on } [kT, (k+1)T],$$

is a T-periodic solution of (6.3). The map P_T is usually called the Poincare operator of $x' = f(t, x(t))$. The preceding discussion thus implies the following important theorem, which relates the Brouwer fixed point theorem to the existence of periodic solutions.

Theorem 6.5 *The vector $x(\cdot, y)$ is a T-periodic solution if and only if y is a fixed point of P_T.*

The next example shows that it is impossible to retract the whole unit ball continuously onto its boundary such that the boundary remains pointwise fixed.

Example 6.2 There is no continuous map $f : \overline{B}_1(0) \to \partial B_1(0)$ such that $f(x) = x$ for all $x \in \partial B_1(0)$.

If the assumption is false, then we see that $g = -f$ would have a fixed point $\bar{x}$, by Theorem 6.3, but this implies $|\bar{x}| = 1$ and therefore $\bar{x} = -f(\bar{x}) = -\bar{x}$, which absurd. Note that this result is in fact equivalent to Brouwer's theorem for the ball. To this end, suppose that the mapping $f : \overline{B}_1(0) \to \overline{B}_1(0)$ is continuous and has no fixed point. Let $g(x)$ be the point where the line segment from $f(x)$ to x hits $\partial B_1(0)$, i.e. $g(x) = f(x) + t(x)(x - f(x))$, and this yields

$$1 = |g(x)|^2 = (f(x) + t(x)(x - f(x)), f(x) + t(x)(x - f(x)))$$
$$= (f(x), f(x)) + + 2t(x)(f(x), x - f(x)) + t^2(x)(x - f(x), x - f(x))$$

Fig. 6.1 Nonexistence of
the retraction of the whole
unit ball continuously onto
its boundary

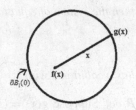

$$= |f(x)|^2 2t(x)(f(x), x - f(x)) + t^2(x)|x - f(x)|^2,$$

where $t(x)$ is the positive root of

$$t^2(x)|x - f(x)|^2 + 2t(x)(f(x), x - f(x)) + |f(x)|^2 = 1.$$

Because $t(x)$ is continuous, g would be such a retraction which does not exist by
assumption (Fig. 6.1).

Theorem 6.6 *Let $0 \in \Omega$, and let $f : \overline{\Omega} \to \mathbb{R}^n$ be a continuous map such that*

$$f(x) \neq \lambda x \ \text{for all } x \in \partial\Omega \ \text{and all } \lambda > 1. \tag{LS}$$

Then f has a fixed point x in Ω; that is, $x = f(x)$ for some $x \in \overline{\Omega}$.

Proof Consider the homotopy $H(t, x) = x - tf(x)$ for $x \in \overline{\Omega}$ and $t \in [0, 1]$. We claim
that $H(t, x) \neq 0$ for $t \in [0, 1]$ and $x \in \partial\Omega$. If not, there would exist $t_0 \in [0, 1]$ and
$x_0 \in \partial\Omega$ such that $H(t_0, x_0) = x_0 - t_0 f(x_0) = 0$.

If $t_0 = 0$ then $x_0 = 0$, contradicting the fact that $x_0 \in \partial\Omega$ and $0 \in \Omega$. If $t_0 \in (0, 1)$,
then $x_0 - t_0 f(x_0) = 0$ implies that $f(x_0) = (1/t_0)x_0$ and $1/t_0 > 1$, in contradiction to
(LS). If $t_0 = 1$, then f has a fixed point $x_0 \in \partial\Omega$ and there is nothing to prove. Thus
we may assume that $H(t, x) \neq 0$ for $x \in \partial\Omega$ and $t \in [0, 1]$. It follows from this and
homotopy invariance that

$$deg(I - f, \mathbb{B}, 0) = deg(I, \mathbb{B}, 0),$$

where $\mathbb{B}$ is the open unit ball in $\mathbb{R}^n$. But $deg(I, \mathbb{B}, 0) = 1$, and therefore f has a fixed
point in $\overline{\Omega}$.

A simple application of Brouwer's fixed point theorem is the following existence
principle for system of equations.

Proposition 6.1 *Let $B = \overline{B}_r(0) \subset \mathbb{R}^n$ be the closed ball with radius r and $g_i : B \to
\mathbb{R}$ be continuous mappings, $i = 1, 2, \dots, n$. If for all $x = (x_1, x_2, \dots, x_n) \in \mathbb{R}^n$, $|x| =
r$*

$$\sum_{i=1}^{n} g_i(x)x_i \geq 0 \tag{6.4}$$

then the system of equations

$$g_i(x) = 0, \quad i = 1, 2, \ldots, n \tag{6.5}$$

has a solution $\hat{x}$ with $|\hat{x}| \leqslant r$.

Proof Suppose $g(x) = (g_1(x), g_2(x), \ldots, g_n(x))$ and assume that $g(x) \neq 0$ for all $x \in B$. Define

$$f(x) = -\frac{rg(r)}{|g(x)|}.$$

Then we see that f is a continuous map of the compact convex set B into itself. Therefore, there exists a fixed point $\hat{x}$ of f with $|\hat{x}| = |f(\hat{x})| = r$. Furthermore,

$$\sum g_i(\hat{x})x_i = -\frac{1}{r}|g(\hat{x})| \sum f_i(\hat{x})x_i$$

$$= -\frac{1}{r}|g(\hat{x})| \sum x_i^2 < 0.$$

This contradicts (6.4). Hence the Eq. (6.5) has a zero.

Surjective maps– We now show that a certain growth condition of $f \in C(\mathbb{R}^n)$ implies $f(\mathbb{R}^n) = \mathbb{R}^n$. To this end, first consider $f_0(x) = Ax$ with a positive-definite matric A. Because $det\, A \neq 0, f_0$ is surjective. We also have

$$(f_0(x), x) \geqslant c|x|^2 \text{ for some } c > 0 \text{ and every } x \in \mathbb{R}^n,$$

where $(\cdot, \cdot)$ is the inner product in $\mathbb{R}^n$ and $|\cdot| = (\cdot, \cdot)^{1/2}$ its Euclidean norm, and therefore

$$\frac{(f(x), x)}{|x|} \to \infty \text{ as } |x| \to \infty.$$

This condition is sufficient for surjectivity in the nonlinear case too, since we can prove the following.

Theorem 6.7 *Let $f : \mathbb{R}^n \to \mathbb{R}^n$ be a continuous mapping. If $\lim\limits_{|x| \to \infty} \frac{(f(x),x)}{|x|} = +\infty$, then $f(\mathbb{R}^n) = \mathbb{R}^n$.*

Proof For a given $y \in \mathbb{R}^n$, let $H(t, x) = tx + (1 - t)f(x) - y$. At $|x| = r$ we have

$$(H(t, x), x) = t(x, x) + (1 - t)(f(x), x) - y$$

$$= t|x|^2 + (1 - t)(f(x), x) - y$$

$$\geqslant r[tr + (1 - t)\frac{(f(x), x)}{|x|} - |y|$$

$$> 0$$

for $t \in [0, 1]$ and $r > |y|$ being sufficiently large. Therefore, $deg(f, B_r(0), y) = 1$ for such an r, i.e. $f(x) = y$ has a solution in $B_r(0)$.

Theorem 6.8 (Invariance of a normal) *Suppose that $0 \in \Omega \subset \mathbb{R}^n$ and that n is odd. If $f \in C(\overline{\Omega})$ and $0 \notin f(\partial\Omega)$, then there are $y \in \partial\Omega$ and $\lambda \neq 0$ such that $f(y) = \lambda y$.*

Proof Define the homotopies $H(t, x)$ and $K(t, x)$ by

$$H(t, x) = (1 - t)f(x) + tx,$$
$$K(t, x) = (1 - t)f(x) - tx,$$

where $x \in \Omega$ and $t \in [0, 1]$. If no $y \in \partial\Omega$ and $\lambda > 0$ can be found to satisfy $f(y) = \lambda y$, then $H(t, x) \neq 0$ and $K(t, x) \neq 0$ for $x \in \partial\Omega$ and $0 < t \leqslant 1$. Since $0 \notin f(\partial\Omega)$, $H(0, x)$ and $K(0, x)$ are also nonzero for $x \in \partial\Omega$. Hence homotopy invariance applied to $H(t, x)$, and $K(t, x)$ respectively yields

$$deg(f, \Omega, 0) = deg(I, D, 0),$$
$$deg(f, \Omega, 0) = deg(-I, D, 0).$$

Now $deg(I, \Omega, 0) = 1$, and it is seen from the definition of the degree that

$$deg(-I, \Omega, 0) = (-1)^n.$$

We thus have $1 = (-1)^n$, whence n is even, contrary to our hypothesis.

Observation

- The condition that n be odd is necessary for Theorem 6.8 to be true. To this end, a counterexample may be given by the map of the unit disc $\mathbb{D}$ in $\mathbb{R}^2$ given in polar coordinates by $(r, \theta) \mapsto (r, \theta + r)$.

Recall that Ω is said to be symmetric with respect to the origin if $\Omega = -\Omega$, and a map f on Ω is said to be odd if $f(-x) = -f(x)$ on Ω.

Theorem 6.9 (Borsuk's Theorem) *Let $\Omega \subset \mathbb{R}^n$ be open bounded and symmetric with $0 \in \Omega$. If $f \in C(\overline{\Omega})$ is odd and $0 \notin f(\partial\Omega)$, then $deg(f, \Omega, 0)$ is odd.*

Proof I. Without loss of generality, we may assume that $f \in C^1(\overline{\Omega})$ with $J_f(0) \neq 0$. To see this, approximate f by a differentiable function $g_1 \in C^1(\overline{\Omega})$ and consider the odd part g_2 with $g_2(x) = \frac{1}{2}(g_1(x) - g_1(-x))$. Choose δ which is not an eigenvalue of $g_2'(0)$. Then $\tilde{f} = g_2 - \delta I$ is continuously differentiable, odd with $J_{\tilde{f}}(x) \neq 0$, and

$$|f - \tilde{f}| = \sup_{x \in \Omega} \left| \frac{1}{2}(f(x) - g_1(x)) - \frac{1}{2}(f(-x) - g_1(-x) - \delta x) \right|$$
$$\leqslant |f - g_1| + \delta \, diam \, \Omega$$

can be chosen sufficiently small, hence

$$deg(f, \Omega, 0) = deg(\tilde{f}, \Omega, 0).$$

II. Let $f \in C^1(\overline{\Omega})$ and $J_f(0) \neq 0$. To prove the theorem, it suffices to show that there is an odd $g \in C^1(\overline{\Omega})$ sufficiently close to f such that $0 \notin g(S_g)$, since then

$$deg(f, \Omega, 0) = deg(g, \Omega, 0) = sgn \, J_g(0) + 2 \sum_{0 \neq x \in g^{-1}(0)} sgn \, J_g(x),$$

where the sum is even since $g(x) = 0$ iff $g(-x) = 0$ and $J_g(\cdot)$ is even.

III. Such a map g will be defined by induction as follows. Consider $\Omega_k = \{x \in \Omega : x_i \neq 0$ for some $i \leqslant k\}$ and an odd $\varphi \in C^1(\mathbb{R})$ such that $\varphi'(0) = 0$ and $\varphi(t) = 0$ iff $t = 0$.

We now consider $\tilde{f}(x) = \frac{f(x)}{\varphi(x_1)}$ on the open bounded $\Omega_1 = \{x \in \Omega : x_l \neq 0\}$. By quotient rule, we see that

$$\tilde{f}'(x) = \frac{\varphi(x_1) f'(x) - f(x) \varphi'(x)}{\varphi(x_1)^2} = \frac{1}{\varphi(x_1)} (f'(x) - \varphi'(x) \cdot \frac{f(x)}{\varphi(x_1)})$$
$$= \frac{1}{\varphi(x_1)} (f'(x) - \varphi'(x_1) \tilde{f}(x)).$$

By Sard's lemma, we find $y^1 \notin \tilde{f}(S_{\tilde{f}}(\Omega_1))$ with $|y^1|$ as small as necessary in the sequel. Hence, 0 is a regular value for $g_1(x) = f(x) - \varphi(x_1) y^1$ on Ω_1, since $g_1'(x) = \varphi(x_1) \tilde{f}'(x)$ for $x \in \Omega_1$ such that $g_1(x) = 0$. Now, suppose that we have already an odd $g_k \in C^1(\overline{\Omega})$ close to f on $\overline{\Omega}$ such that $0 \notin g_k(S_{g_k}(\Omega_k))$ for some $k < n$. Then we define $g_{k+1}(x) = g_k(x) - \varphi(x_{k+1}) y^{k+1}$ with $|y^{k+1}|$ small and such that 0 is a regular value for g_{k+1} on $\{x \in \Omega : x_{k+1} \neq 0\}$.

Evidently, $g_{k+1} \in C^1(\overline{\Omega})$ is odd and close to f on $\overline{\Omega}$. If $x \in \Omega_{k+1}$ and $x_{k+1} = 0$, then $x \in \Omega_k$, $g_{k+1}(x) = g_k(x)$ and $g_{k+1}'(x) = g_k'(x)$, hence $J_{g_{k+1}}(x) \neq 0$, and therefore $0 \notin g_{k+1} f(S_{g_{k+1}})$. Thus, $g = g_n$ is odd, close to f on $\overline{\Omega}$, and such that $0 \notin g(S_g(\Omega \setminus \{0\}))$, since $\Omega_n = \Omega \setminus \{0\}$. By the induction step, we also have $g'(0) = g_1'(0) = f'(0)$, hence $0 \notin g(S_g(\Omega))$.

Observation

- So far we have applied homotopy invariance of the degree as it stands. However, we see that it is also useful to use the converse, namely if two maps f and g have different degree then a certain H that connects f and g cannot be a homotopy. Along these lines, we have the following theorem.

Theorem 6.10 (Hedgehog Theorem) *Let $\Omega \subset \mathbb{R}^n$ be open bounded with $0 \in \Omega$ and let $f : \partial\Omega \to \mathbb{R}^n \setminus \{0\}$ be continuous. Suppose also that the space dimension is odd. Then there exist $x \in \partial\Omega$ and $\lambda \neq 0$ such that $f(x) = \lambda x$.*

Proof Without loss of generality, we may assume that $f \in C(\overline{\Omega})$, by Proposition 6.1. Since n is odd, we have $deg(-I, \Omega, 0) = -1$. Now there arises two cases:

Case I. If $deg(f, \Omega, 0) \neq -1$, then $H(t, x) = (1 - t)f(x) - tx$ must have a zero $(t_0, x_0) \in (0, 1) \times \partial\Omega$. Therefore, $f(x_0) = t_0(1 - t_0)^{-1}x_0$.

Case II. If $deg(f, \Omega, 0) = -1$, then we apply the same argument to $H(t, x) = (1 - t)f(x) + tx$.

Remark 6.2 Since the dimension in Theorem 6.10 is odd, it does not apply to $\mathbb{C}^n$. As a matter of fact, the rotation by $\frac{\pi}{2}$ of the unit sphere $\mathbb{S}$ in $\mathbb{C} = \mathbb{R}^2$, i.e. $f(x_1, x_2) = (-x_2, x_1)$, is a simple counter example. Notice that the rotational matrix R_θ in $\mathbb{R}^2$ is given by

$$R_\theta = \begin{pmatrix} \cos\theta & -\sin\theta \\ \sin\theta & \cos\theta \end{pmatrix}.$$

Therefore,

$$R_{\frac{\pi}{2}}(x_1, x_2)^T = \begin{pmatrix} \cos\frac{\pi}{2} & -\sin\frac{\pi}{2} \\ \sin\frac{\pi}{2} & \cos\frac{\pi}{2} \end{pmatrix}\begin{pmatrix} x_1 \\ x_2 \end{pmatrix} = \begin{pmatrix} 0 & -1 \\ 1 & 0 \end{pmatrix}\begin{pmatrix} x_1 \\ x_2 \end{pmatrix} = \begin{pmatrix} -x_2 \\ x_1 \end{pmatrix}.$$

In case $\Omega = B_1(0)$ the theorem tells us that there is at least one normal such that f changes at most its orientation. In other words, there is no continuous nonvanishing tangent vector field on $\mathbb{S} = \partial B_1(0)$, i.e. an $f : S \to \mathbb{R}^n$ such that

$$f(x) \neq 0 \text{ and } (f(x), x) = 0 \text{ on } S.$$

In particular, $n = 3$ this means that hedgehog cannot be combed without leaving tufts or whorls'. However, we note that $f(x) = (x_2, -x_1, \ldots, x_{2m}, -x_{2m-1})$ is a nonvanishing tangent vector field on $S \subset \mathbb{R}^{2m}$.

6.1.4 The Leray–Schauder Degree

We begin this section with Leray–Schauder lemma without proof which facilitates the basic properties of the degree $deg(f, \overline{\Omega}, 0)$ used subsequently in the investigation of properties of the degree of mappings in Banach spaces. In the sequel, $\mathbb{R}^n$ is n-dimensional Euclidean space and Ω is a bounded open set in $\mathbb{R}^n$. We consider a continuous mapping $f(x) = (f_1(x_1, \ldots, x_n), \cdots, f_n(x_1, \ldots, x_n)$ from the closure $\overline{\Omega}$ of the domain Ω to $\mathbb{R}^n$.

Lemma 6.4 (Leray–Schauder [365]) *Let $f : \overline{\Omega} \to \mathbb{R}^n$ be a continuous mapping such that*

$$f_n(x_1, \ldots, x_n) = x_n \text{ for } (x_1, x_2, \ldots, x_n) \in \Omega. \tag{6.6}$$

Suppose that the condition

$$f(x) \neq 0 \ \text{for} \ x \in \partial\Omega = \overline{\Omega} \setminus \Omega$$

is fulfilled and that the intersection $\Omega' = \Omega \cap \{x : x_n = 0\}$ *is nonempty Then*

$$deg(f, \overline{\Omega}, 0) = deg(f', \overline{\Omega'}, 0), \tag{6.7}$$

where f' *is the mapping of* Ω' *into* $\mathbb{R}^{n-1}$ *defined by the equality*

$$f'(x_1, \ldots, x_{n-1}) = (f_1(x_1, \ldots, x_{n-1}, 0), \cdots, f_{n-1}(x_1, \ldots, x_{n-1}, 0)).$$

We now extend the definition of degree of a mapping to infinite dimensional spaces. In this direction, we notice that there exists a continuous mapping of the unit ball $\mathbb{B}_X = \{x : \|x\| \leqslant 1\}$ of an infinite dimensional space X into itself without having a fixed point. The following example illustrates the difficulty incurred in extending the notion of Brouwer's degree to infinite dimensional spaces.

Example 6.3 Let $X = \ell_2$, the infinite dimensional Hilbert space. Let F be an operator from $\mathbb{B}_X$ into itself defined as follows:

$$Fx = \left(\sqrt{1 - \|x\|^2}, x_1, x_2, \ldots\right) \ \forall x \in \mathbb{B}_X.$$

Then we see that
(*i*) F is a continuous mapping as it is the sum of two continuous mappings:

$$Fx = \left(\sqrt{1 - \|x\|^2}, 0, 0, \ldots\right) + \left(0, x_1, x_2, \ldots\right).$$

(*ii*) F maps $\mathbb{B}_X$ into itself for if $\|x\| \leqslant 1$, then

$$\|fx\|^2 = 1 - \|x\|^2 + \sum_{i=1}^{\infty} \|x_i\|^2 = 1.$$

Suppose F has a fixed point $\bar{x}$. Since $\|f\bar{x}\| = 1$ we have $\|\bar{x}\| = 1$. So

$$F\bar{x} = (0, \bar{x}_1, \bar{x}_2, \ldots) = (\bar{x}_1, \bar{x}_2, \ldots).$$

This implies that $\bar{x} = 0$. But this contradicts the fact $\|\bar{x}\| = 1$. Hence, F has no fixed point on the unit ball $\mathbb{B}_X = \{x : \|x\| \leqslant 1\}$.

Thus the direct generalization of the Brouwer fixed point theorem is false. As the proof of the Brouwer's FPT is a consequence of the elementary properties of the degree of a mapping, such a degree function cannot, in general, be defined for the operator $(I - F)$, where I is the identity operator. A large class of operators for which Brouwer's *FPT* is valid and for which the associated degree of mapping $(I - F)$ can be defined is the class of compact continuous operators. These operators were studied by Schauder in 1920. The degree of such operators $I - F$, where F is continues and compact, i.e. compact perturbation of the identity operator, is known as Leray–Schauder degree.

The following lemma is needed to define Leray–Schauder degree.

Lemma 6.5 *Let* $F : \overline{\Omega} \to \mathbb{R}^n$ *be a continuous mapping, where* Ω *is a bounded domain in* $\mathbb{R}^{n+m}$, $m > 0$. *Then* $deg(I - F, \Omega, y) = deg(I - F, \Omega \cap \mathbb{R}^n, y)$ *for all* $y \in \mathbb{R}^n$ *provided* $x - F(x) \neq y$ *for* $x \in \partial\Omega$.

Proof We prove the result for the case when $F \in C^1(\overline{\Omega})$ and the jacobian determinant in $\mathbb{R}^{n+m}$ of $x - F(x)$ does not vanish at any solution of $x - F(x) = y$ for $y \in \mathbb{R}^n$, $x \in \Omega$, where $F(x) = (f_1(x), f_2(x), \ldots, f_n(x), 0, \ldots, 0.)$ By analytic definition of the degree of a mapping we have

$$deg(I - F, \Omega, y) = \sum_{x - F(x) = y} \text{sgn det } (J_{n+m}(x))$$

$$= \sum_{x - F(x) = y} sgn \begin{pmatrix} J_n(x) & 0 \\ 0 & I_m \end{pmatrix}$$

$$= \sum_{x - F(x) = y} \text{sgn det}(J_n(x))$$

$$= deg(I - F, \Omega \cap \mathbb{R}^n, y),$$

where I_m is the $m \times m$ identity matrix.

We now give the definition of Leray–Schauder degree with the help of Brouwer's degree and its various properties.

Theorem 6.11 *Suppose* Ω *is any bounded open subset of a Banach space* X *which meets every finite dimensional subspace of* X *in a bounded open set. Let* $Tx = y$ *have no solutions on* $\partial\Omega$, *where* $T = I - F$ *and* F *is a compact and continuous operator defined on* $\overline{\Omega}$. *Then an integer-valued function* $deg(T, \Omega, y)$ *can be defined satisfying the properties discussed for the degree of a mapping in finite dimensional spaces.*

Proof We first approximate the compact mapping $F : \Omega \to X$ by a sequence $\{F_n\}$ of compact mappings with finite dimensional range defined by $F_n : \Omega \to X_n$, where X_n is a finite dimensional subspace of X such that $\lim_{n \to \infty} \| F_n x - F x \| = 0$. Then by Lemma 6.5, we compute the degrees of mappings $T_n = I - F_n$ as

$$deg(I - F_n, \Omega, y) = deg(I - F_n, \Omega \cap X_n, y) = deg(T_n, \Omega_n, y)$$

where Ω_n is a finite dimensional bounded open subset of Ω and $y \in X_n$. Finally $deg(I - F_n, \Omega, y)$ is defined as the limit of $deg(T_n, \Omega_n, y)$ as $n \to \infty$.

Observation
It can be shown that

(1°) $deg(I - F_n, \Omega_n, y)$ is independent of both the finite dimensional subspace X_n of X containing y and range of F_n with the help of Lemma 6.5.

(2°) $deg(I - F, \Omega, y)$ is well defined by proving that the limit exists and is independent of the approximating sequence $\{F_n\}$.

We discuss below how the properties of the degree are carried over to infinite dimensional case for this class of mappings.

Theorem 6.12 *If $deg(I - F, \Omega, y) \neq 0$, then there is an $x \in \Omega$ such that $x - Fx = y$.*

Proof Let $\{F_i\}$ be a sequence of mappings with finite dimensional ranges such that $\| F_i x - Fx \| \leqslant \frac{1}{i}$ for $x \in \overline{\Omega}$, and let d_i be their associated degrees. Then for sufficiently large i, $d_i \neq 0$ because

$$deg(I - F, \Omega, y) \neq 0.$$

Hence by the finite dimensional property of degree, there is a point

$$x_i \in \Omega \cap X_i \text{ satisfying } x_i - F_i x_i = y.$$

As F is compact and $\{x_i\}$ is bounded, $\{Fx_i\}$ has a convergent subsequence which we again denote by $\{Fx_i\}$. We have

$$\|x_i - Fx_i - y\| \leqslant \|fx_i - F_i x_i\| + \|x_i - F_i x_i - y\| \leqslant \frac{1}{i}.$$

As $i \to \infty$, $\{x_i\}$ converges to an element x(say) and x satisfies $x - Fx = y$. This completes the proof.

Corollary 6.1 (Leray–Schauder fixed point theorem) *If $deg(I - F, \Omega, 0) \neq 0$, then there is an $x \in \Omega$ such that $x = Fx$.*

Theorem 6.13 (Homotopy Invariance) *Let $H(t, x)$ be a compact continuous homotopy mapping of $[0, 1] \times X \to X$. If Ω is bounded open set as in Theorem 6.11 and $x - H(t, x) \neq y$ for $x \in \partial\Omega$ and $t \in [0, 1]$, then*

$$deg(x - H(t, x), \Omega, y)$$

is constant for $t \in [0, 1]$.

Proof By Theorem 1.37 $H(t, x)$ can be approximated by finite dimensional mapping uniformly, for $t \in [0, 1]$. Now the proof follows from the finite dimensional result on homotopy invariance.

Theorem 6.14 (Schauder's FPT) *Let B_r be an open ball containing the origin in the Banach space X, and F a compact mapping of $\overline{B}_r$ into itself. Then the equation $Fx = x$ has at least one solution.*

Proof The operator $tF(x)$ is a compact mapping of $[0, 1] \times \overline{B}_r \to \overline{B}_r$. By the homotopy invariance property of the degree (refer to Theorem 6.13 above), we have

$$deg(x - tF(x), B_r, 0) = deg(x, B_r, 0) = 1$$

provided $x - tF(x) \neq 0$ on ∂B_r where $t \in [0, 1]$. So $Fx = x$ has at least one solution. For if $x - Fx \neq 0$ on ∂B_r, then

$$deg(x - F(x), B_r, 0) = 1.$$

6.2 Degree Theory of Completely Continuous Vector Fields

The preceding section was devoted to the description of classical and up-to-date results concerning the so-called homotopic theory of continuous mappings or vector fields in Banach spaces. In this section, we are mainly concerned with the mapping degree theory and the vector field rotation theory.

Notice that at the same time there exist two main variants of this theory. One of them is based on the study of continuous mappings from one Banach space (or manifold) into another. Another variant is based on the study of vector fields in Banach spaces (or on Banach manifolds). As a matter of fact, both variants essentially coincide.

So we begin with a natural question—Can we extend the mapping degree theory and vector field rotation theory to continuous mappings and continuous vector fields in infinite dimensional linear spaces, at least to continuous mappings and continuous vector fields in Hilbert and Banach spaces? The answer to this question seems impossible, for example, because there exists only one homotopy class of continuous vector fields (mappings) which are defined on the closed unit ball of an infinite-dimensional Banach space X, take their values in X and do not vanish on the unit sphere of X.

In this context, Leray and Schauder offered a simple construction for the rotation of vector fields with completely continuous operators (indeed they considered mappings of type $\Phi = I - A$ with completely continuous operators A) and obtained the main properties of this rotation (degree). It turned out that

- the Leray–Schauder rotation (degree) for completely continuous vector fields was completely analogous to the classical Brouwer–Hopf theory in finite dimensional spaces and was its natural generalization.
- numerous applications were reduced just to the analysis of vector fields with completely continuous operators.

As a consequence, the middle of the twentieth century was the time of prosperity and triumph for the Leray–Schauder theory, and now this constitutes an elegant and rich branch of nonlinear analysis. So, our study on Leray–Schauder degree theory of completely continuous vector field begins with the following definitions.

Fig. 6.2 a Vector field
$\Phi(x, y) = (-x, -y)$ **b** Vector
field on a sphere $\mathbb{S}$

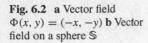

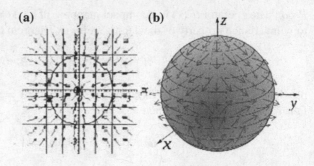

Fig. 6.3 Vector field for the
differential equation
$x' = x, y' = -y$

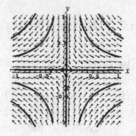

Vector field − A vector field on a set $\mathfrak{M}$ from some Banach space X into another space Y is simply a mapping of $\mathfrak{M} \subseteq X$ into Y (Fig. 6.2). Here one can visualize the vector field $\Phi : \mathfrak{M} \to Y$ as a vector with beginning at $x \in \mathfrak{M}$.

Example 6.4 Sketch the diagram of the vector field for the differential equation $x' = x, y' = -y$.

Solution The diagram of the vector field for the given differential equation is as shown in Fig. 6.2.

Completely continuous vector field − A vector field Φ on $\mathfrak{M} \subset X$ is called completely continuous if it has a representation $\Phi = I - A$, where I is the identity mapping and A is a compact and continuous (completely continuous) operator.

From the above definition, we infer that the class of completely continuous vector fields coincides with the class of compact continuous perturbations of the identity operator.

We now deal with completely continuous vector fields in X which are defined on different sets from X. For what follows, assume that $\mathfrak{M}$ is a subset of X, $C(\mathfrak{M})$ is the family of all completely continuous vector fields defined on $\mathfrak{M}$.

Singular vector field − A vector field $\Phi \in C(\mathfrak{M})$ is called nonsingular if $\Phi \neq 0$ on $\mathfrak{M}$. The set $\mathcal{N}(\Phi, \mathfrak{M})$ of points from $\mathfrak{M}$ at which Φ vanishes is called the zero set or the singularity set of the vector field Φ on the set $\mathfrak{M}$.

Suppose Ω is a bounded open set in X. Denote by $\mathcal{F}_{\mathcal{LS}}(\Omega)$ the family of completely continuous vector fields which are defined on the closure $\overline{\Omega}$ of the set Ω and are not zero on the boundary $\partial \Omega$ of the set Ω.

Homotopy– Two completely continuous vector fields $\Phi_0, \Phi_2 \in \mathcal{F}_{\mathcal{LS}}(\Omega)$ are called homotopic if there exists completely continuous operator $A(x, t)$ defined on $\overline{\Omega} \times [0, 1]$ and taking in values in X, for which the following properties hold:

(i) $x - A(x, 0) = \Phi_0 x, \quad x - A(x, 1) = \Phi_1 x \quad (x \in \overline{\Omega})$,
(ii) $x \neq A(x, t) \quad (x \in \partial\Omega, t \in [0, 1])$.

Observation

• The relation of the homotopy is an equivalence relation, and therefore any family $\mathcal{F}_{\mathcal{LS}}(\Omega)$ is a union of classes of completely continuous vector fields that are mutually homotopic.

The set of all homotopies on the closure $\overline{\Omega}$ of the set Ω is denoted by $\mathcal{H}_{\mathcal{LS}}(\Omega)$.

6.2.1 The Leray–Schauder Degree for a Completely Continuous Vector Field

The rotation of a completely continuous vector field– For the sake of simplicity, we give the definition of the rotation of a vector field on the surface of the unit sphere $\mathbb{S}$ of a Banach space X.

Suppose A is a completely continuous operator defined on $\mathbb{S}$. Consider the vector field $\Phi = I - A$ defined on $\mathbb{S}$. This means that at every point $x \in \mathbb{S}$, we are given the vector $\Phi x = x - Ax$. Because the operator A is completely continuous, this vector field is completely continuous.

Suppose a completely continuous vector field Φ on $\mathbb{S}$ does not contain the null vector. This entails that there exists an $\alpha > 0$ such that $\|\Phi x\| > \alpha$ for $x \in \mathbb{S}$. Select a finite dimensional space X_n in X which has the following property: For each $x \in \mathbb{S}$, every element Ax is at a distance less than or equal to $\frac{\alpha}{3}$ from X_n.

Let us denote by P_n the operator (possibly a linear one) which projects the compact set $A\mathbb{S}$ on X_n, i.e. P_n is such an operator that $P_n A\mathbb{S} \subset X_n$. Furthermore, assume that

$$\|P_n Ax - Ax\| \leqslant \frac{\alpha}{2} \quad (x \in \mathbb{S}).$$

Let γ be the degree of the mapping,

$$Bx = \frac{x - P_n Ax}{\|x - P_n Ax\|} \quad (x \in \mathbb{S}_n)$$

of the unit finite dimensional sphere $\mathbb{S}_n = X_n \cap \mathbb{S}$ upon itself.
We now observe the following facts:

(1°) One can prove that the number γ does not depend on the selection of the approximating space X_n nor on the choice of the projection operator P_n.

(2°) The number γ is the rotation of the completely continuous vector field Φ on $\mathbb{S}$.

(3°) In an entirely analogous manner, one defines the rotation of a field on the boundary $\partial\Omega$ of an arbitrary open set $\Omega \subset X$.

Remark 6.3 Leray and Schauder give the definition of the topological degree of the mapping $I - A$ of a closed region $\overline{\Omega}$. In this definition, it is assumed that the completely continuous operator A is defined not only on $\partial\Omega$ but on the entire set Ω. It follows that the topological degree of the mapping, as defined by Leray and Schauder, coincides with the rotation of the field $I - A$ on $\partial\Omega$.

Notice that every completely continuous operator A, given on $\partial\Omega$, can be extended over Ω with the preservation of complete continuity of the operator. This fact leads to the conclusion that the definition of the rotation of a vector field is no more general than the topological degree of a mapping.

Thus, the concepts of the "topological degree of a mapping" and of "the rotation of a completely continuous vector field" are entirely equivalent (see also Krasnosel'skiĭ [346]).

If a completely continuous operator A is defined on $\overline{\Omega}$ and has no fixed points on $\partial\Omega$, then the rotation $\gamma(I - A, \Omega)$ coincides with the Leray–Schauder degree of $I - A$ on Ω with respect to the origin, that is,

$$\gamma(I - A, \Omega) = deg(I - A, \Omega, 0).$$

Let Ω be a bounded open set in X. Denote by $\mathcal{F}_{\mathcal{BH}}(\Omega)$ the family of continuous vector fields which are defined on the closure $\overline{\Omega}$ of the set Ω and are not zero on the boundary $\partial\Omega$ of the set Ω.

For continuous vector fields, the following result is known as Poincaré–Bohl's theorem.

Theorem 6.15 *Two continuous vector fields* $\Phi_0, \Phi_1 \in \mathcal{F}_{\mathcal{BH}}(\Omega)$ *are homotopic if at each* $x \in \partial\Omega$ *the vectors* $\Phi_0 x$ *and* $\Phi_1 x$ *do not point in opposite directions (in particular, coincide). The condition about nonopposite directions can be written as inequality*

$$\|\Phi_0 x - \Phi_1 x\| < \|\Phi_0 x\| + \|\Phi_1 x\| \quad (x \in \partial\Omega) \tag{6.8}$$

or

$$(\Phi_0 x, \Phi_1 x) > -\|\Phi_0 x\|\|\Phi_1 x\| \quad (x \in \partial\Omega) \tag{6.9}$$

where $\|\cdot\|$ *is the usual or a strictly convex norm in* X, $(\cdot, \cdot)$ *is the usual scalar product in* X.

In particular, the Poincaré–Bohl conditions hold if

$$\|\Phi_1 x - \Phi_0 x\| < \|\Phi_0 x\| \quad (x \in \partial\Omega); \tag{6.10}$$

the corresponding theorem is known as the Rouché theorem.

There exists a unique integer function, $\gamma(\Phi, \Omega) = \gamma_{\mathcal{LS}}(\Phi, \Omega)$ that is defined on the union of all sets $\mathcal{F}_{\mathcal{LS}}(\Omega)$ and has the following properties:

I. If Φ_0 and Φ_1 are two homotopic fields from $\mathcal{F}_{\mathcal{LS}}(\Omega)$, then

$$\gamma(\Phi, \Omega) = \gamma(\Phi, \Omega).$$

II. If Ω_1 and Ω_2 are bounded open sets in X, $\Phi \in \mathcal{F}_{\mathcal{LS}}(\Omega_1) \cap \mathcal{F}_{\mathcal{LS}}(\Omega_2)$, and Φ does not vanish on the set $\Omega_1 \cap \Omega_2$, then

$$\gamma(\Phi, \Omega) = \gamma(\Phi, \Omega_1) + \gamma(\Phi, \Omega_2).$$

III. If $\Phi(x) = x - x_0$ and $x_0 \in \Omega$, then $\gamma(\Phi, \Omega) = 1$.

The function $\gamma(\cdot, \cdot)$ is called the rotation, and the value $\gamma(\Phi, \Omega)$ of this function is called the rotation (winding number or index) of the vector field Φ on the boundary $\partial\Omega$ of the set Ω.

Note that the rotation $\gamma(\Phi, \Omega)$ is a function of the restriction of the field Φ on the boundary $\partial\Omega$ of the set Ω, and the set Ω itself, rather than the boundaries of the set. One knows such closed and bounded sets which are boundary of different bounded open sets in X. The rotation of some vector field Φ which is defined on the union of the closures of both (or more) of these bounded open sets with common boundary can be different, depending on the set.

We also notice that the rotation $\gamma(\Phi, \Omega)$ of the vector field Φ on the boundary $\partial\Omega$ of the bounded open set Ω in essential depends on the values Φ on the boundary $\partial\Omega$. More precisely, if $\Phi_0 = I - A_0$ and $\Phi_1 = I - A$, where A_0 and A are completely continuous operators, are two completely continuous vector fields defined on the closure $\overline{\Omega}$ of the set Ω, and their restrictions on $\partial\Phi$ are the same, then,

$$\gamma(\Phi_0, \Omega) = \gamma(\Phi_1, \Omega),$$

that is,

$$deg(I - A_0, \Omega, 0) = deg(I - A_1, \Omega, 0).$$

However, we see that this common rotation depends on the set Ω whose boundary is $\partial\Omega$; if there exist two or more such bounded open sets, the rotation $\gamma(\Phi, \Omega)$ can be different depending on the choice of Ω.

The main result of the homotopic and homological theory of continuous vector fields is the proof of the existence and uniqueness theorem for the rotation of a vector field. In this context, we observe that Hopf [276–278] not only gave the proof of the existence and uniqueness for the rotation of vector fields, but obtained such fundamental theoretical facts like the homotopic classification theorem and the theorem on the connections between the rotation of a vector field and the essentiality of its zero set.

Usually, the rotation is determined for nonsingular continuous vector fields which are defined only on the boundary an of a bounded open set O but not on this set itself. Indeed, both approaches are equivalent. Due to the famous Tietze–Urysohn's theorem, every continuous mapping $F : \mathfrak{M} \to X$ defined on a closed subset $\mathfrak{M}$ of the space X can be extended to a continuous mapping $\tilde{F}$ defined on the whole space X; moreover, one can demand that the additional condition $\tilde{F}x \in \overline{co}F(\mathfrak{M})(x \in X)$ be fulfilled. Owing to this fundamental result, any nonsingular continuous vector field Φ defined on the boundary $\partial\Omega$ of a bounded open set Ω, is the restriction of continuous vector fields $\tilde{\Phi}$ defined on the closure $\overline{\Omega}$, of the set Ω, (and even on the whole space X); all these continuations $\tilde{\Phi}$ are mutually homotopic (on $\overline{\Omega}$), and therefore have the same rotation $\gamma(\tilde{\Phi}, \Omega)$ coinciding, by definition, with the rotation $\gamma(\Phi, \Omega)$.

6.2.1.1   The Main Theorems on Rotation of a Vector Field

As in the finite dimensional case, the properties I–III imply some simple and important corollaries.

Theorem 6.16 *Let Ω be a bounded open set in X. If a completely continuous vector field $\Phi \in C(\Omega)$ has no zeros in Ω then $\gamma(\Phi, \Omega) = 0$; in other words, if $\gamma(\Phi, \Omega) \neq 0$ then Φ has at least one zero in Ω.*

Theorem 6.17 *Let Φ be a completely continuous vector field which is defined on the union of the closures of two bounded open sets Ω_1 and Ω_2 and $\Phi \in \mathcal{F}_{\mathcal{LS}}(\Omega_1) \cap \mathcal{F}_{\mathcal{LS}}(\Omega_2)$. Then*

$$\gamma(\Phi, \Omega_1) = \gamma(\Phi, \Omega_2) \tag{6.11}$$

provided that Φ is nonsingular on $\Omega_1 \triangle \Omega_2$.

Index– Let Φ be a vector field from $\mathcal{F}_{\mathcal{BH}}(\Omega)$ and $\mathcal{N}(\Phi, \Omega)$ the set of zeros of Φ in Ω. Assume that N is an isolated subset of $\mathcal{N}(\Phi, \Omega)$ or, in other words, there exists a neighbourhood $O(N)$ of N which does not meet $\mathcal{N}(\Phi, \Omega) \setminus N$. Then the rotation $\gamma(\Phi, O(N))$ of Φ on the boundaries of all neighbourhoods $O(N)$ of N containing no other points of $\mathcal{N}(\Phi, \Omega)$ is the same; this rotation is called the index of N and denoted by $\text{ind}(N, \Phi)$.

Theorem 6.18 *Let Ω be a bounded domain in X, $\Phi \in \mathcal{F}_{\mathcal{LS}}(\Omega)$ a completely continuous vector field, and $\mathcal{N}(\Phi, \Omega)$ be a union of isolated subsets $N_1, N_2, \ldots, N_s$. Then*

$$\gamma(\Phi, \Omega) = \sum_{i=1}^{s} \text{ind}(N_i, \Phi). \tag{6.12}$$

Isolated zero– A zero $x_0 \in \mathcal{N}(\Phi, \Omega)$ is called isolated if there exists a neighbourhood $O(x_0)$ of X which contains no other points from $\mathcal{N}(\Phi, \Omega)$.

Index of a zero– If x_0 is an isolated zero of the vector field Φ, then the rotation $\gamma(\Phi, B(x_0, r))$ of Φ is the same for all balls $B(x_0, r)$ with center x_0 and sufficiently small radius r; this common rotation is called the index of the zero x_0 and denoted by $\text{ind}(x_0, \Phi)$ (Fig. 6.4). If all points of $\mathcal{N}(\Phi, \Omega)$ are isolated, then any one of them

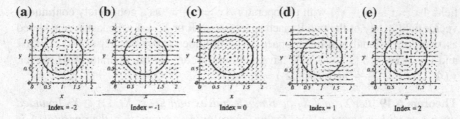

Fig. 6.4 The origin is an isolated zero of the vector fields, and distinct indexes are counted as shown in Fig. 6.4a–e

is a component of $\mathcal{N}(\Phi, \Omega)$, and $\mathcal{N}(\Phi, \Omega)$ is a finite set, say, $\{x_0, \ldots, x_s\}$. Then the following equality is true:

$$\gamma(\Phi, \Omega) = \mathrm{ind}(x_0, \Phi) + \cdots + \mathrm{ind}(x_s, \Phi). \qquad (6.13)$$

6.2.2 Invariance Principles

It may be observed that when we deal with completely continuous vector fields in infinite-dimensional Banach spaces X, there appears a crucially new problem. In fact, we usually deal with operators that are defined by some analytical formulas, but usually have no natural Banach space X in which these operators should be considered. The choice of the Banach space X in which the corresponding vector fields $\Phi = I - A$ turn out to be completely continuous (and with other useful properties) demands serious additional investigations. In any case it is difficult to expect that such a choice is unique. But if the same vector field $\Phi = I - A$ is completely continuous vector field in two different Banach spaces X_1 and X_2, there arises the essential question on relations between all geometric characteristics of Φ in X_1 and X_2.

Suppose X_1 and X_2 are two Banach spaces embedded in some linear space X, and Ω_1 and Ω_2 are bounded open sets in X_1 and X_2, correspondingly. Assume that $\Phi = I - A$ is a vector field that is defined on some set $\mathcal{M} \subset X$, $\overline{\Omega}_1, \overline{\Omega}_2 \subset \mathcal{M}$ and

$$A(\overline{\Omega}_1) \subset X_1, \quad A(\overline{\Omega}_2) \subset X_2.$$

If the operator A is completely continuous on $\overline{\Omega}_1$ as an operator acting in X_1, as well as on $\overline{\Omega}_2$ as an operator acting in X_2, we can try to find conditions implying the coincidence of the rotations $\gamma(I - A, \Omega_1)$ and $\gamma(I - A, \Omega_2)$ on the boundaries of the sets Ω_1 and Ω_2, correspondingly.

The first among these conditions is almost evident: the sets of fixed points of the operator A in the closures of the open sets Ω_1 and Ω_2, correspondingly, in the spaces X_1 and X_2 must coincide. In this case, we say that the sets Ω_1 and Ω_2 have a common core for the operator A. But this condition is not sufficient. One can consider the vector field $\Phi x = x - (x - x^n)$ with the operator $Ax = x - x^n$ as a completely continuous vector field in any of the two Banach spaces $X = \mathbb{R}$ or $X = \mathbb{C}$ (the last is considered as two-dimensional real linear space). The bounded open sets $\Omega_1 = (-1, 1) \subset \mathbb{R}$ and $\Omega_2 = x \in \mathbb{C} : \|x\| < 1 \subset \mathbb{C}$ have a common core for the operator A. However, $\gamma(\Phi, \Omega_1) = \frac{1-(-1)^n}{2}$ and $\gamma(\Phi, \Omega_2) = n$.

Theorem 6.19 *Let X_l and X_2 be Banach spaces and $\Omega_1 \subset X_l$, $\Omega_2 \subset X_2$ bounded open sets with a common core for an operator A. Assume that the operator A is completely continuous on $\overline{\Omega}_1$ as an operator in X_l and on $\overline{\Omega}_2$ as an operator in X_2, and one from the following conditions holds:*

(a) *The operator A is continuous on $\overline{\Omega}_1$ and $\overline{\Omega}_2$ as an operator from X_l to X_2 and from X_2 to X_l, correspondingly.*

(b) *The norms of the spaces X_l and X_2 are compatible, the operator A is locally bounded on $\overline{\Omega}_1$ and $\overline{\Omega}_2$ as an operator from X_l to X_2 and from X_2 to X_l, correspondingly, and compact on $X_2 \cap \overline{\Omega}_1$ and $X_1 \cap \overline{\Omega}_2$ as an operator from X_l to X_2 and from X_2 to X_l, correspondingly.*

(c) *The norms of the spaces X_l and X_2 are compatible, and there exist convex and compact sets S_1 and S_2 in $X_l \cap X_2$ for which*

$$A(S_1 \cap \Omega_1) \subset S_2, \quad A(S_2 \cap \Omega_2) \subset S_1$$

and $x_0 \in coAx_0$, S_j implies $x_0 \in S_j$ $(j = 1, 2)$.

Then the equality

$$\gamma(\Phi, \Omega_1) = \gamma(\Phi, \Omega_2) \tag{6.14}$$

holds true.

6.3 The Skrypnik Degree Theory

Skrypnik's 1973 monograph [564] developed a new topological degree theory for mappings of class α from a reflexive space X to its dual X^*, and applied it to the solvability of nonlinear elliptic PDEs of abstract and concrete nature.

Let X be a real Banach space, X^* its conjugate space and Γ a fundamental subspace of X^* such that the unit ball of the space X is Γ-weakly compact (i.e. a compact set in the Γ-weak topology). The simplest but main case is when X is a reflexive Banach space and $\Gamma = X^*$. Another important case is when $X = Z^*$, where Z is a Banach space, and $\Gamma = Z \subseteq X^{**}$.

Let us consider an operator Φ defined on a set $\mathfrak{M}$ of the space X and taking its values in the space X^*. We need some definitions for such operators.

An operator $\Phi : \mathfrak{M} \to X^*$ is said to satisfy the condition $\alpha(\mathfrak{M})$ if each sequence $x_n \in \mathfrak{M}$ which Γ-weakly converges to x_0, and for which

$$\limsup_{n \to \infty} \langle \Phi x_n, x_n - x_0 \rangle \leqslant 0$$

strongly converges to x_0. Here and below $\langle v, u \rangle$ is a "scalar product", i.e. the value of the functional $v \in X^*$ at the element $u \in X$.

An operator $\Phi : \mathfrak{M} \to X^*$ is called demicontinuous if it maps each sequence x_n strongly converging to x_0 into a sequence $\Phi x_n \, \Gamma$-weakly converging to Φx_0.

In what follows, $\mathcal{F}_S(\Omega)$ denotes the class of all demicontinuous vector fields Φ defined on the closure $\overline{\Omega}$ of a bounded open set $\Omega \subseteq X$, taking their values in Y, not vanishing on $\overline{\Omega}$, and satisfying the condition $\alpha(\partial\Omega)$.

Let U be an arbitrary finite dimensional subspace of the space X and $B = \{u_1, \ldots, u_n\}$ its arbitrary basis. Then we can consider the operator $\Phi_{(B)}$ defined on the set $\Omega_{(U)} = \Omega \cap U$, taking its values in U and defined by the formula

$$\Phi_{(B)}u = \sum_{j=1}^{n} \langle \Phi u, u_j \rangle u_j \quad (u \in U). \tag{6.15}$$

Now suppose that $\overline{\Omega}$, the closure of a bounded open set Ω, is contained in the set $\mathfrak{M}$, the vector field Φ is defined on $\overline{\Omega}$, takes it values in Γ, is demicontinuous, has the property $\alpha(\partial\Omega)$, where $\partial\Omega$ is the boundary of the set Ω and, at last, does not vanish on $\partial\Omega$. In this case, there exists a finite dimensional subspace U_* such that for any finite dimensional subspace $U \supseteq U_*$ and any basis B of the subspace U of the vector field $\Phi_{(B)}$ does not vanish on the boundary $\partial\Omega_{(U)}$ and the rotations $\gamma(\Phi_{(U)}, \Omega_{(U)})$ and $\gamma(\Phi_{(U_*)}, \Omega_{(U_*)})$ are the same. This common rotation $\gamma(\Phi_{(B)}, \Omega_{(U)})$ is called the rotation of the vector field $\Phi : \overline{\Omega} \to X^*$ on the boundary $\partial\Omega$ of the set Ω.

The rotation $\gamma_S(\Phi, \Omega)$ of vector fields from $\mathcal{F}_S(\Omega)$ possesses almost all main properties I–III of "usual" rotations which were considered above. Of course, one need to introduce the corresponding notion of homotopy.

Definition 6.3 Two demicontinuous vector fields Φ_0 and Φ_1 are called homotopic if there exists a demicontinuous family of vector fields $\Phi(\lambda, x)$ defined on the set $[0, 1] \times \overline{\Omega}$, taking values from X^* and satisfying the following conditions:

$$\Phi(\lambda, \cdot) \in \alpha(\partial\Omega) \quad (\lambda \in [0, 1])$$
$$\Phi(0, x) = \Phi_0 x, \; \Phi(1, x) = \Phi_1 x \quad (x \in \Omega)$$
$$\Phi(\lambda, x) \neq 0 \quad (\lambda \in [0, 1], \; x \in \partial\Omega)$$

Pseudomonotone operator– An operator Φ is called pseudomonotone if each sequence $x_n \in \mathfrak{M}$ which Γ-weakly converges to x_0, and for which the sequence Φx_n for which

$$\limsup_{n \to \infty} \langle \Phi x_n, x_n - x_0 \rangle \leqslant 0 \tag{6.16}$$

satisfies the condition

$$\lim_{n \to \infty} \langle \Phi x_n, x_n - x_0 \rangle = 0. \tag{6.17}$$

Observation

(i) For the class of demicontinuous vector fields with the property $\alpha_0(\partial\Omega)$ in case of a separable space X, one can give the definition of the rotation described in detail above but using only sequences of finite dimensional subspaces U_n whose linear hulls are dense in U_n.

(ii) For the class of pseudomonotone demicontinuous vector fields in case when there exists a demicontinuous operator Q with the property $\alpha_0(\partial\Omega)$ for which $\langle Qx, x \rangle > 0$ $(x \neq 0)$ in X one can define the rotation, $\gamma(\Phi, \Omega)$ of an original vector field Φ on the boundary $\partial\Omega$ of a bounded set Ω as common rotation, $\gamma(\varepsilon Q + \Phi, \Omega)$ for sufficiently small positive ε.

6.3.1   Degree of Mappings of Class α

Throughout this section, X stands for a real separable reflexive Banach space, and X^* for its dual. We denote the strong and weak convergence by $\rightarrow$ and $\rightharpoonup$, respectively; it will be clear from the context in what space the convergence takes place. For arbitrary elements $u \in X$ and $h \in X^*$, henceforth $\langle h, u \rangle$ denotes the action of the functional h on the element u.

In the sequel, Ω is a bounded open subset of the space X and $\partial\Omega$ is its boundary. We consider an operator A, in general nonlinear, defined on a subset $\mathfrak{M} \subset X$, with values in X^*.

Definition 6.4 We say that A satisfies the condition $\alpha_0(\mathfrak{M})$ if whenever $\{u_n\} \subset \mathfrak{M}$, for which $u_n \rightharpoonup u$, $Au_n \rightharpoonup 0$, and

$$\limsup_{n\to\infty} \langle Au_n, u_n - u_0 \rangle \leqslant 0 \tag{6.18}$$

it follows that the sequence u_n converges strongly to u_0.

Definition 6.5 Let $\mathfrak{M} \subset \overline{\Omega}$. We say that an operator $A : \overline{\Omega} \to X^*$ satisfies the condition $\alpha(\mathfrak{M})$, if each sequence $\{u_n\} \subset \mathfrak{M}$ converging weakly to some u_0 and for which (6.18) holds is in fact strongly convergent to u_0.

Definition 6.6 For $\mathfrak{M} \subset \overline{\Omega}$ we denote by $A_0(\Omega, \mathfrak{M})$ (respectively $A(\Omega, \mathfrak{M})$ the set of all bounded demicontinuous mappings $A : \overline{\Omega} \to X^*$ satisfying condition $\alpha_0(\mathfrak{M})$ (respectively condition $\alpha(\mathfrak{M})$). When $\mathfrak{M} = \overline{\Omega}$, we write $A_0(\Omega), A(\Omega)$ instead of $A_0(\Omega, \overline{\Omega})$, $A(\Omega, \overline{\Omega})$.

We define $Deg(A, \overline{\Omega}, 0)$—the degree of a mapping A on the set Ω with respect to the origin of the space X^*—under the conditions:

(a) $A \in A_0(\Omega, \overline{\Omega})$;
(b) $Au \neq 0$ for any element $u \in \overline{\Omega}$.

Let $\{v_i\}$, $i = 1, 2, \ldots$, be any complete system of the space X and suppose that for every n the elements $v_1, \ldots, v_n$ are linear independent. Further, let F_n denote the linear hull of the elements $v_1, \ldots, v_n$.

Let us define for every $n = 1, 2, \ldots$ the finite dimensional approximation A_n of the mapping A in the following way:

$$A_n u = \sum_{i=1}^{n} \langle Au, v_i \rangle v_i \ \text{for } u \in \overline{\Omega}_n, \ \Omega_n = \Omega \cap F_n. \tag{6.19}$$

Theorem 6.20 *Let A be an operator satisfying conditions (a), (b). Then there exists N such that for $n \geqslant N$ the following assertions hold:*

(1) the equation $A_n u = 0$ has no solutions belonging to $\partial \Omega_n$;
(2) the degree $deg(A_n, \overline{\Omega}_n, 0)$ of the mapping A_n on the set Ω_n with respect to $0 \in F_n$ is defined and independent of n.

Proof We proof the first assertion by contradiction. Assume that there is a sequence $u_k \in \partial \Omega_{n_k}$ such that $n_k \to \infty$ as $k \to \infty$ and $A_{n_k} u_k = 0$. We may consider that u_k converges weakly to some element $u_0 \in X$. In addition, $A_{n_k} u_k = 0$ implies the weak convergence of Au_k to zero. We show that u_k converges strongly to $u_0 \in \partial \Omega$ and $Au_0 = 0$. This suffices for the proof of $\cdot(1)$ by virtue of assumption (b).

Let us select any subsequence $w_k \in F_{n_k}$ such that $w_k \to u_0$. Then

$$\langle Au_k, u_k - u_0 \rangle = \langle Au_k, w_k - u_0 \rangle + \langle Au_k, u_k - w_k \rangle.$$

The second term of the right-hand side vanishes by the assumption that $A_{n_k} u_k = 0$ and $u_k - w_k \in F_{n_k}$. The first term tends to zero as $k \to \infty$ due to the boundedness of the operator A and the choice of the subsequence w_k. Thereby

$$\lim_{k \to \infty} \langle Au_k, u_k - u_0 \rangle = 0.$$

Hence, by virtue of the condition $\alpha_0(\partial \Omega)$, we obtain the strong convergence of u_k to u_0, which guarantees that $u_0 \in \partial \Omega$ and $Au_0 = 0$, a contradiction to assumption (b). Thus, for sufficiently large n, we observe that the assertion (1) ensures that the degree $deg(A_n, \overline{\Omega}_n, 0)$ of the finite dimensional mapping A_n is well defined.

We now prove the independence of $deg(A_n, \overline{\Omega}_n, 0)$ of n by using the auxiliary mapping

$$\tilde{A}_n u = \sum_{i=1}^{n-1} \langle Au, v_i \rangle v_i + \langle h_n, u \rangle v_n$$

where h_n is some element of the space X^* satisfying the conditions

$$\langle h_n, v_i \rangle = 0 \ \text{ for } i < n, \ \ \langle h_n, v_n \rangle = 1.$$

By the Leray–Schauder lemma, we infer that

$$deg(A_{n-1}, \overline{\Omega}_{n-1}, 0) = deg(\tilde{A}_n, \overline{\Omega}_n, 0). \tag{6.20}$$

Notice that the right-hand side of (6.20) being defined for large n by virtue of (1). We now prove the following equality

$$deg(\tilde{A}_n, \overline{\Omega}_n, 0) = deg(A_n, \overline{\Omega}_n, 0). \tag{6.21}$$

In order to prove (6.21), by virtue of the properties of the degree of finite dimensional mappings, it suffices to establish that

$$[tA_n + (1-t)\tilde{A}_n]u \neq 0 \quad \text{for } u \in \partial\Omega_n, \quad t \in [0, 1]. \tag{6.22}$$

We prove the assertion (6.22) by contradiction. Let us assume that there exist sequences $u_k \in \partial\Omega_{n_k}, t_k \in [0, 1]$ such that

$$[t_k A_{n_k} + (1-t_k)\tilde{A}_{n_k}]u = 0 \quad n_k \to \infty \text{ as } n \to \infty. \tag{6.23}$$

Thus, we have

$$\left.\begin{aligned} \langle Au_k, v_i \rangle = 0 \ \text{ for } \leqslant n_k - 1, \\ t_k\langle Au_k, v_{n_k} \rangle + (1-t_k)\langle h_{n_k}, u_k \rangle = 0. \end{aligned}\right\} \tag{6.24}$$

By (1) and (6.24) we obtain that $Au_k \rightharpoonup 0$ and for sufficiently large k the inequality $0 < t_k < 1$ is satisfied. We may assume that the sequence $\{u_k\}$ converges weakly to some element $\tilde{u}_0 \in X$ and we choose a sequence $\tilde{w}_k$ such that $\tilde{w}_k \in F_{n_k-1}, \tilde{w}_k \to \tilde{u}_0$. Then

$$\langle Au_k, u_k - \tilde{u}_0 \rangle = \langle Au_k, \tilde{w}_k - \tilde{u}_0 \rangle - \frac{1-t_k}{t_k}\langle h_{n_k}, u_k \rangle^2. \tag{6.25}$$

Notice that the equality (6.25) is the consequence of the following two equalities which are obtained on the basis of (6.24) and the choice of $\tilde{w}_k$:

$$\langle Au_k, \tilde{w}_k \rangle = 0, \quad \langle Au_k, u_k \rangle = -\frac{1-t_k}{t_k}\langle h_{n_k}, u_k \rangle^2.$$

The first term on the right-hand side of (6.25) tends to zero and hence, by virtue of the condition $\alpha_0(\partial\Omega)$, we deduce the strong convergence of u_k to $\tilde{u}_0 \in \partial\Omega$. The first equality in (6.24) yields $A\tilde{u}_0 = 0$, which contradicts condition (b).

This establishes assertion (6.22), and together with it the equality (6.21). It follows from (6.20) and (6.21) that the $deg(A_n, \overline{\Omega}_n, 0)$ is independent of n for sufficiently large n; this concludes the proof of the theorem.

In view of Theorem 6.16, we infer that $\lim\limits_{n\to\infty} deg(A_n, \overline{\Omega}_n, 0)$ exists and we denote it by $D\{v_i\}$.

Theorem 6.21 *Suppose that the conditions (a), (b) are satisfied. Then the limit*

$$D\{v_i\} = \lim_{n\to\infty} deg(A_n, \overline{\Omega}_n, 0) \tag{6.26}$$

does not depend on the choice of the sequence $\{v_i\}$.

Proof It suffices to show that $D\{v_i'\} = D\{v_i\}$ for any other sequence $\{v_i'\}$ having the same properties as the sequence $\{v_i\}$. Thus, we can assume that for each n, the system $v_1, v_2, \ldots, v_n, v_1', v_2', \ldots, v_n'$ is linearly independent. Otherwise, an auxiliary system $\{\tilde{v}_i\}$ can be constructed such that both the systems $v_1, v_2, \ldots, v_n, \tilde{v}_1, \tilde{v}_2, \ldots, \tilde{v}_n$ and $\tilde{v}_1, \tilde{v}_2, \ldots, \tilde{v}_n, v_1', v_2', \ldots, v_n'$ are linearly independent for any n, and the proof reduces to establishing the equality $D\{v_i\} = D\{v_i'\}$.

Let L_{2n} be the linear space spanned by the elements $v_1, v_2, \ldots, v_n, v_1', v_2', \ldots, v_n'$ and define for $n = 1, 2, \ldots$ the finite dimensional mapping

$$A_{2n,t}u = \sum_{i=1}^{n}\{\langle Au, v_i \rangle v_i + [t\langle Au, v_i' \rangle + (1-t)\langle f_i^{(n)}, u \rangle]v_i'\}, \tag{6.27}$$

where $u \in L_{2n} \cap \overline{\Omega}$, $t \in [0, 1]$ and $f_i^{(n)}$ are elements of the space X^*, satisfying the conditions

$$\langle f_i^{(n)}, v_j \rangle = 0, \ (f_i^{(n)}, v_j') = \delta_{ij}, \ i, j = 1, 2, \ldots, n, \tag{6.28}$$

δ_{ij} being the Kronecker symbol equal to zero for $i \neq j$ and to 1 for $i = j$.

Now, by virtue of the definition of $D\{v_i\}$ and Theorem 6.20, the proof of Theorem 6.21 is reduced to establishing the equality

$$deg(A_n, \overline{\Omega}_n, 0) = deg(A_n', \overline{\Omega}'_n, 0) \tag{6.29}$$

for large n. Here A_n' and $\overline{\Omega}'_n$ are determined by (6.19) in which v_i is replaced by v_i'.

To prove the equality (6.29), it suffices to verify that, for large n,

$$A_{2n,t}u \neq 0 \ \text{for} \ u \in \partial(L_{2n} \cap \overline{\Omega}), \quad t \in [0, 1]. \tag{6.30}$$

In fact, by the virtue of the Leray–Schauder lemma, under the condition (6.30) we have

$$deg(A_{2n,0}, L_{2n} \cap \overline{\Omega}, 0) = deg(A_n, \overline{\Omega}_n, 0). \tag{6.31}$$

On the other hand, due to properties of the degree of finite dimensional mappings and (6.30), we obtain

$$deg(A_{2n,0}, L_{2n} \cap \overline{\Omega}, 0) = deg(A_{2n,1}, L_{2n} \cap \overline{\Omega}, 0). \tag{6.32}$$

By (6.31), (6.32), taking into account that in the definition of $A_{2n,1}$ and L_{2n} the arguments v_i and v_i' have a symmetric role, we derive (6.29). Hence, the proof of Theorem 6.21 is reduced to the verification of (6.30).

We prove (6.30) by the contradiction, assuming that there are sequences $\bar{u}_k, \bar{t}_k$ such that

$$A_{2n_k, \bar{t}_k}\bar{u}k = 0, \ u_k \in \partial(L_{2n_k} \cap \overline{\Omega}), \bar{t}_k \in [0, 1], \ n_k \to \infty. \tag{6.33}$$

From this we have

$$\left.\begin{array}{ll} \langle A\bar{u}_k, v_i \rangle = 0, & i = 1, 2, \ldots, n_k, \\ \bar{t}_k \langle A\bar{u}_k, v_i' \rangle + (1 - \bar{t}_k)\langle f_i^{(n_k)}, \bar{u}_k \rangle = 0, & i = 1, 2, \ldots, n_k. \end{array}\right\} \tag{6.34}$$

It follows from Theorem 6.20 that $0 < \bar{t}_k < 1$ for k large. We may consider that $\bar{u}_k \rightharpoonup u_0, \bar{t}_k \to \bar{t}_0, A\bar{u}_k \rightharpoonup 0$. Select a sequence $\bar{w}_k \in F_{n_k}$ such that $\bar{w}_k \to \bar{u}_0$. Using the formula

$$\bar{u}_k = \sum_{i=1}^{n_k} (c_i^{(k)} v_i + \bar{c}_i^{(k)} v_i'), \quad \bar{c}_i^{(k)} = \langle f_i^{(n_k)}, \bar{u}_k \rangle,$$

from (6.34) we obtain

$$\langle A u_k, \bar{u}_k - \bar{u}_0 \rangle = \langle A\bar{u}_k, \bar{w}_k - \bar{u}_0 \rangle + \sum_{i=1}^{n_k} \langle A\bar{u}_k, v_i' \rangle \langle f_i^{(n_k)}, \bar{u}_k \rangle$$

$$= \langle A\bar{u}_k, \bar{w}_k - \bar{u}_0 \rangle - \frac{1 - \bar{t}_k}{\bar{t}_k} \sum_{i=1}^{n_k} \langle f_i^{(n_k)}, \bar{u}_k \rangle^2.$$

Hence, we have

$$\limsup_{k \to \infty} \langle A u_k, \bar{u}_k - \bar{u}_0 \rangle \leqslant 0,$$

and by virtue of condition $\alpha_0(\partial\Omega)$, we deduce the strong convergence of $\bar{u}_k$ to $\bar{u}_0 \in \partial\Omega$. The demicontinuity of the operator A implies $A\bar{u}_0 = 0$, which contradicts the assumption (b). This completes the proof of the theorem.

Theorems 6.19 and 6.20 justify the following definition.

Definition 6.7 For an operator $A : \Omega \to X^*$, satisfying conditions (a), (b), by its degree on the set Ω with respect to the point $0 \in X^*$, we mean the number $\lim_{n \to \infty} deg(A_n, \overline{\Omega}_n, 0)$ where A_n, Ω_n are determined in accordance with (6.19). This degree is denoted by $Deg(A, \overline{\Omega}, 0)$.

6.3.2 The Degree of a Pseudomonotone Mapping

In this section, we mainly discuss the degree of a pseudomonotone mapping in a nonseparable space. Throughout this section, X is a real reflexive Banach space. We denote by $\mathcal{F}(X)$ the set of all finite dimensional subspaces of X. Let $F \in \mathcal{F}(X)$ and let $dim\, F = \lambda$. So, let $v_1, v_2, \ldots, v_\lambda$ be a basis in F. We define the finite dimensional mapping

$$A_F(u) = \sum_{i=1}^{\lambda} \langle A u, v_i \rangle v_i \text{ for } u \in \overline{\Omega}_F, \quad \Omega_F = \Omega \cap F. \tag{6.35}$$

Theorem 6.22 *Let $A : \Omega \to X^*$ be a demicontinuous operator satisfying condition $\alpha(\partial\Omega)$, where $\partial\Omega$ is the boundary of a bounded open set $\Omega \subset X$, and $Au \neq 0$ for $u \in \partial\Omega$. Then there exists a subspace $F_0 \in \mathcal{F}(X)$ such that any subspace F belonging to $\mathcal{F}(X)$ and containing F_0 satisfies the properties:*

(1) the equation $A_F(u) = 0$ has no solution belonging to $\partial\Omega_F$;
(2) $\deg(A_F, \overline{\Omega}_F, 0) = \deg(A_{F_0}, \overline{\Omega}_{F_0}, 0)$, where $\deg$ is the degree of the finite-dimensional mapping.

Proof **Step I. Existence of** F_0. First we establish the existence of a subspace $F_0 \in \mathcal{F}(X)$ such that for $F \supset F_0$, $F \in \mathcal{F}(X)$, the set

$$Z_{F_0}^F = \{u \in \partial\Omega_F : \langle Au, u \rangle \leqslant 0, \ \langle Au, v \rangle = 0 \ \text{for} \ v \in F_0\}$$

is empty.

Suppose, on the contrary, that for any $F_0 \in \mathcal{F}(X)$ there is some $F_1 \in \mathcal{F}(X)$, $F_1 \supset F_0$, such that $Z_{F_0}^F \neq \varnothing$. Denote

$$G_{F_0} = \bigcup_{F \supset F_0} Z_{F_0}^F$$

and let $\overline{G}_{F_0}^{(w)}$ be the weak closure of G_{F_0}. Then the system of sets $\{\overline{G}_F^{(w)} : F \in \mathcal{F}(X)\}$ has the finite intersection property, and the reflexivity of the space X implies (see, Dunford and Schwartz [209]) the existence of a u_0 such that

$$u_0 \in \bigcap_{F \in \mathcal{F}(X)} \overline{G}_F^{(w)}. \tag{6.36}$$

We now prove that $u_0 \in \partial\Omega$ and $Au_0 = 0$. This will contradict the conditions of the theorem. Let w be an arbitrary element of X and take $F_0 \in \mathcal{F}(X)$ such that $u_0 \in F_0$, $w \in F_0$. By (6.36) we have $u_0 \in \overline{G}_{F_0}^{(w)}$ and, therefore, there exists a sequence $u_n \in \overline{Z}_{F_0}^{F_n}$, $F_n \supset F_0$, such that $u_n \rightharpoonup u_0$, $u_n \in \partial\Omega_{F_n}$, and

$$\langle Au_n, u_n \rangle \leqslant 0, \quad \langle Au_n, u_0 \rangle = 0, \quad \langle Au_n, w \rangle = 0. \tag{6.37}$$

It follows from (6.37), we infer that

$$\langle Au_n, u_n - u_0 \rangle \leqslant 0.$$

Whence, by condition $\alpha(\partial\Omega)$, the sequence u_n converges strongly to u_0. By the last equality in (6.37) and the demicontinuity of the operator A, we obtain $\langle Au_0, w \rangle = 0$ and, as w is arbitrary, we derive $Au_0 = 0$. This contradicts the conditions of the theorem.

In this way, we have proved the existence of a subspace $F_0 \in \mathcal{F}(X)$ such that for any $F \supset F_0$, $F \in \mathcal{F}(X)$, the set $Z_{F_0}^F$ is empty. This subspace F_0 satisfies the requirements of the theorem.

Step II. Validation of assertions (1) and (2). The validity of the assertion (1) for the selected F_0 follows directly from the equality $Z_{F_0}^F = \varnothing$. Let us prove the second assertion. Let $F \supset F_0$, $F \in \mathcal{F}(X)$; we choose a basis in F of the form $v_1, v_2, \ldots, v_{\lambda_0}, w_1, w_2, \ldots, w_\mu$, where v_i, $i = 1, 2, \ldots, \lambda_0$, is a basis in F_0. We consider on Ω_F two mappings

$$
\left.
\begin{aligned}
A_F(u) &= \sum_{i=1}^{\lambda_0} \langle Au, v_i \rangle v_i + \sum_{i=1}^{\mu} \langle Au_i, w_i \rangle w_i \rangle, \\
\tilde{A}_F(u) &= \sum_{i=1}^{\lambda_0} \langle Au, v_i \rangle v_i + \sum_{i=1}^{\mu} \langle f_{i,F}, u \rangle w_i \rangle
\end{aligned}
\right\}
\tag{6.38}
$$

where $f_{i,F}$ is an element in X^* satisfying the conditions: $\langle f_{i,F}, w_j \rangle = \delta_{ij}$ for $j = 1, 2, \ldots, \mu$, $\langle f_{i,F}, v_k \rangle = 0$ for $k = 1, 2, \ldots, \lambda_0$; δ_{ij} is the Kronecker symbol.

By the Leray–Schauder lemma, we have $deg(\tilde{A}_F, \overline{\Omega}_F, 0) = deg(A_{F_0}, \overline{\Omega}_{F_0}, 0)$ and, therefore, to prove the theorem it suffices to verify that

$$
tA_F(u) + (1-t)\tilde{A}_F(u) \neq 0 \text{ for } u \in \partial\Omega_F, \ t \in [0,1].
\tag{6.39}
$$

If (6.39) were not satisfied, then for some $u_0 \in \partial\Omega_F$, $t_0 \in [0,1]$, we would have

$$
\left.
\begin{aligned}
\langle Au_0, v_i \rangle &= 0, \ i = 1, 2, \ldots, \lambda_0, \\
t_0 \langle Au_0, w_i \rangle + (1-t_0)\langle f_{i,F}, u_0 \rangle &= 0, i = 1, 2, \ldots, \mu.
\end{aligned}
\right\}
\tag{6.40}
$$

Due to the first already proved assertion of the theorem, we have $t_0 \neq 0$. Let $u_0 = \sum_{i=0}^{\lambda_0} a_i v_i + \sum_{j=1}^{\mu} b_j w_j$ and compute $\langle Au_0, u_0 \rangle$ such that

$$
\langle Au_0, u_0 \rangle = \sum_{j=1}^{\mu} \langle Au_0, w_j \rangle = -\frac{t_0}{1-t_0} \langle f_{j,F}, u_0 \rangle^2 \leqslant 0.
\tag{6.41}
$$

The existence of u_0 satisfying (6.40), (6.41) contradicts to $Z_{F_0}^F = \varnothing$. Thus (6.39) holds, which proves the theorem.

Theorem 6.22 justifies the introduction of the following definition.

Definition 6.8 Let $A : \Omega \to X^*$ be a demicontinuous operator satisfying condition $\alpha(\partial\Omega)$, where $\partial\Omega$ is the boundary of a bounded open set $\Omega \subset X$, and $Au \neq 0$ for $u \in \partial\Omega$. Then, for a subspace $F_0 \in \mathcal{F}(X)$ such that any subspace F belonging to $\mathcal{F}(X)$ and containing F_0 satisfying the properties (1) and (2) of Theorem 6.22, the number

$$
Deg(A, \overline{\Omega}, 0) = deg(A_{F_0}, \overline{\Omega}_{F_0}, 0)
$$

is called the degree of the mapping A on the set $\overline{\Omega}$ with respect to the point $0 \in X^*$.

In what follows, X stands for a real nonseparable reflexive Banach space, and X^* for its dual. We denote the strong and weak convergence by $\rightarrow$ and $\rightharpoonup$, respectively; it will be clear from the context in what space the convergence takes place. For arbitrary elements $u \in X$ and $h \in X^*$, henceforth $\langle h, u \rangle$ denotes the action of the functional h on the element u.

Let Ω be a bounded open set in X. Assume that on the space X there is an operator $A_0 : X \rightarrow X^*$, $A_0 \in A(\partial\Omega, \Omega)$ the family $A(\partial\Omega, \Omega)$ was introduced by Definition 6.6. In the case of uniformly convex spaces X, X^*, as the operator A_0 we may take the duality map, which will be outlined at the end of this section.

Let $A : \overline{\Omega} \rightarrow X^*$ be a demicontinuous pseudomonotone operator and suppose that $0 \notin \overline{A(\partial\Omega)}$, where the bar denotes the strong closure and $\partial\Omega$ the boundary of Ω. Under these conditions, we define the degree of the mapping A. Our assumptions guarantee the validity of the inequality

$$\|Au\|_{X^*} \geqslant \delta_0 \quad \text{for } u \in \partial\Omega$$

with a positive number δ_0.

Let us introduce the mapping $\varepsilon A_0 + A : \overline{\Omega} \rightarrow X^*$. Set

$$M = \sup_{u \in \overline{\Omega}} \|A_0 u\|_{X^*}.$$

If $0 < \varepsilon < \frac{\delta_0}{M}$, then for the mapping $\varepsilon A_0 + A$, which satisfies all conditions of Theorem 3.1, the degree

$$Deg(\varepsilon A_0 + A, \overline{\Omega}, 0) \tag{6.42}$$

is defined. It is required to verify the condition $\alpha(\partial\Omega)$ for the operator $\varepsilon A_0 + A$. Let $u_n \in \partial\Omega$ be a sequence such that $u_n \rightharpoonup u_0$ and

$$\limsup_{n \to \infty} \langle \varepsilon A_0 u_n + A u_n, u_n - u_0 \rangle \leqslant 0. \tag{6.43}$$

Passing, if necessary, to a subsequence, from (6.43) we obtain one of the inequalities

$$\limsup_{n \to \infty} (A_0 u_n, u_n - u_0) \leqslant 0, \quad \limsup_{n \to \infty} (A u_n, u_n - u_0) \leqslant 0. \tag{6.44}$$

If the second inequality holds, then, by the pseudomonotonicity of the operator A, we have $\lim_{n \to \infty} (A u_n, u_n - u_0) = 0$ which, together with (6.43), implies the validity of the first inequality in (6.44). Thus, the first inequality in (6.44) is always true, and the strong convergence of u_n to u_0 is obtained from the condition $\alpha(\partial\Omega)$ for the operator A_0.

Similarly, as in the proof of Theorem 6.22, we can infer (see also the next section regarding the properties of the degree of a mapping) that, for $0 < \varepsilon < \frac{\delta}{M}$, the degree defined by (6.42) does not depend on ε, so that the limit

$$\lim_{\varepsilon \to 0} Deg(\varepsilon A_0 + A, \overline{\Omega}, 0)$$

exists. One can prove that this limit does not depend on the choice of the mapping A_0. Thus, we can introduce the following concept.

Definition 6.9 For a demicontinuous pseudomonotone operator A with $0 \notin \overline{A(\partial \Omega)}$, the number

$$Deg(A, \overline{\Omega}, 0) = \lim_{\varepsilon \to 0} Deg(\varepsilon A_0 + A, \overline{\Omega}, 0),$$

is called the degree of the mapping A on the set Ω with respect to the point $0 \in X^*$. $A_0 \in A(\Omega, \partial \Omega)$ and Deg is the degree introduced by Definition 6.8.

6.4   The Browder–Petryshyn Degree Theory

In Sect. 6.1 we studied the Leray–Schauder degree theory which is, indeed, a very natural method to solve an infinite dimensional equation $Tx = y$. This method involves an approximation technique in which we approximate the original equation by finite dimensional equations. Notice also that the well-known Galerkin method has proved to be a very efficient tool in finite dimensional approximation. In the 1960s, Browder and Petryshyn systematically studied the finite dimensional method for a large class of mappings, which they called A-proper mappings, and they developed a similar theory to the Brouwer degree.

Let X and Y be two real separable Banach spaces, X_n and Y_n be two sequences of finite dimensional spaces, dim $X_n =$ dim Y_n and $P_n : X \to X_n$ and $Q_n : Y \to Y_n$ be two sequences of (in general nonlinear) continuous operators. In this situation, an approximation scheme for the pair of Banach spaces X and Y is said to be given. Usually, the finite dimensional spaces X_n and Y_n are subspaces of the spaces X and Y, correspondingly, but in what follows it is natural to suppose that $X_n = Y_n (n = 1, 2, \ldots)$.

6.4.1   A-Proper Mappings

We begin with the following definition.

Definition 6.10 Let X and Y be real separable Banach spaces.

(1) If there is a sequence of finite dimensional subspaces $X_n \subset X$ and a sequence $\{P_n\}$ of linear projections $P_n : X \to X_n$ such that $P_n x \to x$ for all $x \in X$, then we say that X has a projection scheme $\{X_n, P_n\}$.

(2) If X and Y have projection schemes $\{X_n, P_n\}$ and $\{Y_n, Q_n\}$, respectively, and dim $X_n = \dim Y_n$ for all positive integers n, then we call $\prod = \{X_n, P_n; Y_n, Q_n\}$ an operator projection scheme.

Example 6.5 Suppose $X = C[0, 1]$. For $n \in \mathbb{N}$, partition $[0, 1]$ into n equal parts and set $0 = t_0 < t_1 < t_2 < \cdots < t_n = 1$, where $t_i = \frac{i}{n}$. Suppose X_n is the subspace of all $x \in X$ which are linear in every subinterval $[t_i, t_{i+1}]$ and $P : X \to X_n$ be the projection satisfying $P_n x(t_i) = x(t_i)$ for $i = 1, 2, \ldots, n$. Then we see that $\{X_n, P_n\}$ is a projection scheme for X.

Example 6.6 Let X be a Banach space with a Schauder basis $\{e_i : i \in \mathbb{N}\}$. For $x = \sum_{i=1}^{\infty} \alpha_i(x)e_i$, let us define

$$X_n = \text{span}\{e_1, e_2, \ldots, e_n\}, \quad P_n x = \sum_{i=1}^{n} \alpha_i(x)e_i.$$

Then one can readily see that $\{X_n, P_n\}$ is a projection scheme.

Remark 6.4 In the case of a separable Hilbert space, we may choose an orthonormal basis $\{e_i : i \in \mathbb{N}\}$, then the projection $P_n x = \sum_{i=1}^{n} \alpha_i(x, e_i)e_i$ satisfies $P_n^* = P_n$ and $\|P_n\| = 1$.

Example 6.7 Let X be a reflexive Banach space with a projection scheme $\{X_n, P_n\}$ such that

$$P_n P_m = P_{\min\{m,n\}} \quad m, n \in \mathbb{N}.$$

Then $\{P_n^* X^*, P_n^*\}$ is a projection scheme for X^*.

Example 6.8 Let X and Y be Banach spaces. Let $\{e_n\}$ be a Schauder basis of X and $\{e_n'\}$ be a Schauder basis of Y. Put $X_n = \text{span}\{e_1, e_2, \ldots, e_n\}$ and $Y_n = \text{span}\{e_1', e_2', \ldots, e_n'\}$. For $x = \sum_{i=1}^{\infty} \alpha_i e_i$ and $y = \sum_{i=1}^{\infty} \beta_i e_i'$, set

$$P_n x = \sum_{i=1}^{n} \alpha_i e_i \quad \text{and} \quad Q_n y = \sum_{i=1}^{n} \beta_i e_i'.$$

Then we see that $\prod = \{X_n, P_n; Y_n, Q_n\}$ is an operator projection scheme.

Petryshyn offered the following definition.

Definition 6.11 Let X, Y be real Banach spaces and $\prod = \{X_n, P_n; Y_n, Q_n\}$ be an operator projection scheme. Suppose that Ω is a bounded open set in X, then a vector field (mapping) $\Phi : \overline{\Omega} \subset X \to Y$ is called approximately proper or, for short, A-proper (respectively, pseudo A-proper) with respect to $\prod$ if, for any bounded $x_m \in \Omega \cap X_m$ and $Q_m \Phi x_m \to y$, there exists a subsequence $\{x_{m_k}\}$ such that $x_{m_k} \to x \in \Omega$ and $\Phi x = y$, (respectively, there exists $x \in \mathscr{D}(\Phi)$, such that $\Phi x = y$). We denote by $A_{\prod}(\Omega, Y)$ the class of all A-proper mappings $\Phi : \overline{\Omega} \to Y$.

Definition 6.12 Let X, Y be real Banach spaces and $\prod = \{X_n, P_n; Y_n, Q_n\}$ be an operator projection scheme. Suppose that Ω is a bounded open set in X. A family of mappings $H(t, x) : [0, 1] \times \Omega \to Y$ is called A-proper homotopy with respect to $\prod$ if

(1) for any bounded sequence $\{x_m\}$ in $\Omega \cap X_m$, $t_m \to t_0$ and $Q_m H(t_m, x_m) \to y$, there exists a subsequence $\{x_{m_k}\}$ of $\{x_m\}$ such that $x_{m_k} \to x \in \Omega$ and $H(t_0, x) = y$;

(2) $Q_n H(t, x) : [0, 1] \times \Omega \cap X_n \to Y_n$ is continuous for $n = 1, 2, \ldots$.

6.4.2   Browder–Petryshyn's Degree for A-Proper Mappings

In this section, we discuss mainly generalized degree for A-Proper mappings. In what follows, let X, Y be real separable Banach spaces and $\prod = \{X_n, P_n; Y_n, Q_n\}$ be an operator projection scheme. Let $\Omega \subset X$ be an open bounded subset and L be a dense subspace of X with $\bigcup_{n=1}^{\infty} X_n \subset L$. We let $\Omega_L = \Omega \cap L$.

Lemma 6.6 Let $\Phi \in A_{\prod}(\Omega \cap L, Y)$. Suppose that $p \notin \Phi(\overline{\Omega} \cap L)$. Then there exists an integer $n_0 > 0$ such that

$$Q_n p \notin Q_n \Phi(\partial(\Omega \cap X_n)) \quad \text{for all } n > n_0.$$

Proof We proof the lemma by contradiction. So, assume that the assertion of the lemma is false. Then there exists $n_k \to \infty$ and $x_{n_k} \in \partial\Omega \cap X_{n_k}$ such that $Q_{n_k} p = Q_{n_k} \Phi x_{n_k}$. Obviously, $x_{n_k} \in \partial\Omega \cap L$. Thus we have $Q_{n_k} \Phi x_{n_k} \to p$ as $k \to \infty$ and the A-properness of Φ guarantees the existence of a subsequence $\{x_{n_{k_l}}\}_{l=1}^{\infty}$ such that $x_{n_{k_l}} \to x_0 \in \partial\Omega \cap L$, and $\Phi x_0 = p$, which is a contradiction. This completes the proof.

Definition 6.13 Let $\Phi \in A_{\prod}(\Omega \cap L, Y)$. Suppose that $p \notin \Phi(\overline{\Omega} \cap L)$ and $Q_n \Phi$ is continuous. We define a generalized degree $Deg(\Phi, \Omega, p)$ by

$$Deg(\Phi, \Omega \cap L, p) = \{l \in \mathbb{Z} \cup \{\pm\infty\} : deg(Q_{n_j} \Phi, \Omega \cap X_{n_j}, Q_{n_j} p) \to l$$
$$\text{for some } n_j \to \infty\},$$

where $\mathbb{Z}$ is the set of all integers.

By Lemma 6.6, we know that there exists an integer $n_0 > 0$ such that $p \notin Q_n \Phi(\partial\Omega \cap X_n)$ and $Q_n \Phi$ is continuous, so the Brouwer degree $deg(Q_n \Phi, \Omega \cap X_n, Q_n p)$ is well defined for $n > n_0$. Thus, $Deg(\Phi, \Omega \cap L, p)$ is nonempty and the definition is well defined.

Theorem 6.23 Let $\Phi \in A_{\prod}(\Omega \cap L, Y)$ and $p \notin \Phi(\overline{\Omega} \cap L)$. Then the generalized degree has the following properties:

(1) If $Deg(\Phi, \Omega \cap L, p) \neq \{0\}$, then $\Phi x = p$ has a solution in $\Omega \cap L$;

(2) *If $\Omega_i \subset \Omega$ for $i = 1, 2$, $\Omega_1 \cap \Omega_2 = \varnothing$; and $p \notin (\overline{\Omega} \setminus \Omega_1 \cup \Omega_2) \cap L$, then*

$$Deg(\Phi, \Omega \cap L, p) \subseteq Deg(\Phi, \Omega_1 \cap L, p) + Deg(\Phi, \Omega_2 \cap L, p),$$

here we use the convention that $+\infty + (-\infty) = \mathbb{Z} \cup \{\pm\infty\}$;
(3) *If $H(t, x) : [0, 1] \times \overline{\Omega} \cap L \to Y$ is a A-proper homotopy and $p \notin H(t, x)$ for all $(t, x) \in [0, 1] \times \Omega \cap L$, then $Deg(H(t, \cdot), \Omega \cap L, p)$ does not depend on $t \in [0, 1]$;*
(4) *If $0 \in \Omega$, Ω is symmetric about 0, $\Phi : \Omega \cap L \to Y$ is an odd A-proper mapping and $0 \notin \Phi(\partial\Omega \cap L)$, then $Deg(\Phi, \Omega \cap L, 0)$ contains no even numbers.*

Proof (1) Suppose that $Deg(\Phi, \Omega \cap L, p) \neq \{0\}$, then there exists $n_k \to \infty$ such that

$$deg(Q_{n_k}\Phi, \Omega \cap X_{n_k}, Q_{n_k}p) \neq 0.$$

Thus, there exists $x_{n_k} \in \Omega \cap L$ such that $Q_{n_k}\Phi x_{n_k} = Q_{n_k}p$. Further, by the A-properness of Φ, there is a subsequence $\{x_{n_{k_j}}\}$ with $x_{n_{k_j}} \to x_0 \in \Omega \cap L$ and $Tx_0 = p$.
(2) Because $p \notin (\overline{\Omega} \setminus \Omega_1 \cup \Omega_2) \cap L$, there exists $n_0 > 0$ such that $Q_n p \notin (\overline{\Omega} \setminus \Omega_1 \cup \Omega_2) \cap X_n$ for all $n > n_0$. Therefore, we have

$$deg(Q_n\Phi, \Omega, p) = deg(Q_n\Phi, \Omega_1, p) + deg(Q_n\Phi, \Omega_2, p) \text{ for all } n > n_0.$$

Set $l = \lim\limits_{j \to \infty} deg(Q_{n_j}'\Phi, \Omega \cap X_{n_j}, Q_{n_j}p)$, then we have

$$l = \lim\limits_{j \to \infty} [deg(Q_{n_j}\Phi, \Omega_1 \cap X_{n_j}, Q_{n_j}p) + deg(Q_{n_j}\Phi, \Omega_2 \cap X_{n_j}, Q_{n_j}p)].$$

Now there arises three cases:
Case I. If $\lim\limits_{j \to \infty} deg(Q_{n_j}\Phi, \Omega_1 \cap X_{n_j}, Q_{n_j}p)$ and $\lim\limits_{j \to \infty} deg(Q_{n_j}\Phi, \Omega_2 \cap X_{n_j}, Q_{n_j}p)$ are both equal to $+\infty$ or $-\infty$, then $l = +\infty$ or $l = -\infty$ and so the conclusion holds.
Case II. If one of them equals to $+\infty$ and the other one is $-\infty$, then, by the convention, we have $l \in \mathbb{Z} \cup \{\pm\infty\}$.
Case III. If

$$\limsup_{j \to \infty} |deg(Q_{n_j}\Phi, \Omega_1 \cap X_{n_j}, Q_{n_j}p)| < +\infty$$

and

$$\limsup_{j \to \infty} |deg(Q_{n_j}\Phi, \Omega_2 \cap X_{n_j}, Q_{n_j}p)| < +\infty,$$

then the conclusion is obvious.
(3) We claim that there exists an integer $n_0 > 0$ such that

$$Q_n p \notin \bigcup_{t \in [0,1]} H(t, \partial\Omega \cap X_n) \text{ for all } n > n_0.$$

Suppose the assertion is false. Then there exist $n_j \to \infty, t_j \to t_0, x_{n_j} \in \partial\Omega \cap X_{n_j}$ such that $Q_{n_j}p = Q_{n_j}H(t_j, x_{n_j})$. Therefore, $\{x_{n_j}\}$ has a subsequence $\{x'_{n_j}\}$ which converges to $x_0 \in \partial\Omega \cap L$ and $H(t_0, x_0) = p$, which is a contradiction. Thus, the Brouwer degree $deg(Q_n H(t, \cdot), \Omega \cap X_n, Q_n p)$ does not depend on $t \in [0, 1]$ for $n > n_0$, so $Deg(H(t, \cdot), \Omega \cap L, p)$ does not depend on $t \in [0, 1]$.

(4) Because $\Omega \cap X_n$ is symmetric about 0, by Borsuk's theorem, we have

$$deg(Q_n\Phi, \Omega \cap X_n, 0) \text{ is odd for } n \text{ sufficiently large.}$$

Thus, $Deg(\Phi, \Omega \cap L, 0)$ contains no even numbers. This completes the proof.

6.5   The Sadovskii Degree of Limit-Compact Operators

In this section, we discuss the extension of degree theory to noncompact nonlinear operators. The degree theory of limit-compact operators is based on the classical theory of the degree of mappings $I - F$ (with F compact and continuous) developed by Leray and Schauder [365] as discussed earlier. A generalization of the Leray–Schauder degree has been given by Browder and Nussbaum. Here we are giving the generalized degree theory as developed by Sadovskii [536].

Now we describe a procedure by which one can construct for any operator F : $\mathfrak{M} \to X$ that maps a subset $\mathfrak{M}$ of a normed linear space X into X, a certain closed convex set $F^\infty(\mathfrak{M})$, the so-called limit range of F. We observe the following:

(1°) The limit range of F, i.e. the set $F^\infty(\mathfrak{M})$ necessarily contains all fixed points of F.

(2°) The behaviour of F on $\mathfrak{M} \cap F^\infty(\mathfrak{M})$ can be shown to be decisive in certain sense for F on the whole of $\mathfrak{M}$.

Using the concept of the limit range, we define the class of limit-compact operators. Frequently we need to consider not individual operators but families of operators, i.e. $F : \Delta \times \mathfrak{M} \to X$ where Δ be an arbitrary set. For such an operator, we construct a transfinite sequence of sets $\{K_\alpha\}$ in accordance with the formula

(a) $K_0 = \overline{co}F(\Delta \times \mathfrak{M})$

(b) $K_\alpha = \overline{co}F(\Delta \times (\mathfrak{M} \cap K_{\alpha-1}))$ if $\alpha - 1$ exists   (6.45)

(c) $K_\alpha = \bigcap_{\beta < \alpha} K_\beta$ if $\alpha - 1$ does not exist.

The sequence of sets K_α possesses some important properties that are given in the following lemma.

Lemma 6.7 (a) Each of the sets K_α is closed and convex.
(b) $F[\Delta \times (\mathfrak{M} \cap K_\alpha)] \subseteq K_{\alpha+1}$.
(c) If $\eta < \alpha$, then $K_\alpha \subseteq K_\eta$.
(d) The sets K_α are invariant under F in the sense that $F[\Delta \times (\mathfrak{M} \cap K_\alpha)] \subseteq K_\alpha$.

(e) *There exists an order number δ such that $K_\alpha = K_\delta$ for $\alpha \geqslant \delta$.*

Definition 6.14 The limit set K_α of the transfinite sequence (6.45) is called the limit range of the operator F on the set $\Delta \times \mathfrak{M}$ and is denoted by $F^\infty(\Delta \times \mathfrak{M})$. The operator $F : \Delta \times \mathfrak{M} \to X$ is said to be limit-compact if its restriction to the set $\Delta \times [\mathfrak{M} \cap F^\infty(\Delta \times \mathfrak{M})]$ is a compact operator, that is, the set $F[\Delta \times [\mathfrak{M} \cap F^\infty(\Delta \times \mathfrak{M}))]$ is relatively compact. In particular, F is limit-compact on $\Delta \times \mathfrak{M}$ if $F^\infty(\Delta \times \mathfrak{M}) = \varnothing$,

Let K be a closed convex set and U an open (bounded) set in a normed linear space X. We assume that K consists of more than one point and that $K \cap U \neq \varnothing$. By $\overline{U}_K$ and ∂U_K we denote, respectively, the closure and boundary of the set $U_K = U \cap K$ in the relative topology of K.

Let F be a compact and continuous operator from $U_K \to K$ having no fixed points in ∂U_K. Let $deg(I - F, \overline{U}_K, 0)$ denote the degree of the operator $I - F$ on $\overline{U}_K$ with respect to K. It is convenient to extend the definition of the degree $deg(I - F, \overline{U}_K, 0)$ to the case when $\overline{U}_K$ is empty or consists of single element, taking it equal to zero in the first case and to unity in the second.

Observation

- If $deg(I - F, \overline{U}_K, 0) \neq 0$, then F has at least one fixed pint in U_K.

Let D be a closed convex set in X with induced topology. Now we shall define the degree of the operator $I - F$ with a limit-compact F and state a theorem which plays a crucial role in establishing the properties of this degree. Let F be a continuous operator from $\overline{U}_D$ to D, and let F be limit-compact with respect to $\overline{U}_D$. We assume that $F(x) \neq x$ for $x \in \partial \overline{U}_D$. Our aim is to define an integral characteristic.

$$deg(I - F, \overline{U}_D, 0),$$

the so-called degree of the operator $I - F$ on $\overline{U}_D$ with respect to D.

We quote the following lemma without proof.

Lemma 6.8 *The degree $deg(I - F, \overline{U}_K, 0)$ where $K = F^\infty(U_D)$ is defined, that is, the following conditions are satisfied.*

(a) $F(\overline{U}_K) \subseteq K$
(b) the operator F is compact on $\overline{U}_K$ (i.e. the set $F(\overline{U}_K)$ is relatively compact),
(c) F does not have fixed point on ∂U_K.

Definition 6.15 Under the above assumption, the degree $deg(I - F, \overline{U}_K, 0)$ which exists by virtue of the preceding lemma is called the degree of the operator $I - F$ on $\overline{U}_R$ with respect to R and is denoted by $deg(I - F, \overline{U}_R, 0)$.

Remark 6.5 Throughout this section on degree theory, we adopt the usual notation instead of the rotation of the vector fields.

The following theorem is of great importance in the subsequent development of the degree theory.

Theorem 6.24 *Suppose the conditions of Lemma 6.8 are satisfied. Let C be a closed convex set in X satisfying the following conditions:*

(a) $K \subseteq C \subseteq D$;
(b) *the set $F(\overline{U}_X)$ is relatively compact;*
(c) $F(\overline{U}_C) \subseteq C$.

Then $deg(I - F, \overline{U}_D, 0) = deg(I - F, \overline{U}_C, 0)$.

By using the above theorem, properties of the degree of the operator $I - F$ with F limit-compact can be derived from corresponding properties of the operator $I - F$ where F is compact and continuous.

In other words, properties of the degree of the operator $I - F$ where F is limit-compact are similar to the properties of the degree of the operator $I - F$ so restricted that F is compact and continuous. However, there are some differences. Most important among them is the fact that the number of limit-compact homotopies is relatively small.

Theorem 6.25 *If $deg(I - F, \overline{U}_D, 0) \neq 0$, then F has at least one fixed point in U_D.*

Proof By definition, $deg(I - F, \overline{U}_D, 0) = deg(I - F, \overline{U}_K, 0)$ where $K = F^{\infty}(U_D)$. But $deg(I - F, \overline{U}_K, 0) \neq 0$ implies that the operator F has a fixed point on the set $U_K \subseteq U_D$.

Now we shall give a criterion for the degree to be nonvanishing, in particular, this is criterion for the existence of a fixed point. We state the following theorem by Sadovskii, without proof.

Theorem 6.26 *Let D be a closed convex set in a normed linear space X and let F : $D \to D$ be a limit-compact continuous operator. Suppose that one of the following conditions is satisfied:*

(a) $F^{\infty}(D) \neq \varnothing$,
(b) *D has a nonempty subset B such that $coF(B) \supseteq B$,*
(c) *D has a nonempty compact subset C such that $F(C) \subseteq C$,*
(d) *for some point $x_0 \in D$, the set of points of the sequence $\{F^n x_0 : n = 0, 1, 2, \ldots\}$ is relatively compact in the space X,*
(e) *the space X is reflexive, and the set D is nonvoid and bounded.*

Then

$$deg(I - F, D, 0) = 1.$$

Remark 6.6 If for the open set U, the whole space X is chosen; then

$$U_D = \overline{U}_D = D \text{ and } deg(I - F, \overline{U}_D, 0) = deg(I - F, D, 0).$$

6.6 Measures of Noncompactness

The concept of measure of noncompactness has played an important role in nonlinear functional analysis, especially in the study of metric and topological fixed point theory. It may be observed that several papers have been published on the existence and behaviour of solutions of a wide class of nonlinear differential and integral equations via measure of noncompactness.

We shall now discuss the k-set contractions, introduced by Kuratowski [351] and Darbo [155] and the condensing operators (also called 1-set contraction) introduced by Sadovskii [535]. "Contraction of sets" are operators whose application reduces a certain numerical characteristic of the sets, known as a measure of noncompactness. If a k-set contraction is applied to a set, it does not immediately become compact or even necessarily have a smaller diameter, it merely becomes more compact (if $k < 1$) than it was. Darbo utilized the Kuratowski measure of noncompactness. Sadovskii, who considered condensing map first, used a different measure of noncompactness called "ball measure of noncompactness" introduced by Gohberg, Goldenstein and Markus [250], apparently unaware of Kuratowski's measure of noncompactness.

The approach initiated by Kuratowski and Darbo was taken up again in 1967 by Ambrosetti [20, 21], Szufla [583], Furi and Vignoli [236], Nussbaum [428], Petryshyn [476] and others, who in their turn did not know any of the works of Soviet authors. Furi and Vignoli introduced the concept of condensing operator in a metric space. Himmelberg, Porter and Van Vleck [275] introduced the condensing multifunction.

In the following discussion, we shall only use Kuratowski[2] measure of noncompactness. Later on we shall discuss the relationship between Kuratowski measure of noncompactness and ball measure of noncompactness.

Let X be a metric space. Recall that $A \subset X$ is bounded if A is contained in some ball. A bounded subset A of X said to be is relatively compact, if there exists for any $\varepsilon > 0$ a finite covering of A by balls of radius ε and

$$\text{diam}(A) = \sup\{\|x - y\| : x, y \in A\}$$

is called the diameter of A.

For a nonempty subset A of X, we denote by $\overline{X}$, coA and $\overline{co}A$ the closure, convex hull and the convex closure of A, respectively. We denote the standard algebraic operations on sets by the symbols $\lambda A = \{\lambda x : x \in A\}$, $\lambda \in \mathbb{K}$ and $A + B = \{x + y : x \in A, y \in B\}$. Furthermore, the family of all bounded subsets of X is denoted by $B(X)$.

Definition 6.16 Let X be a metric space and A a bounded subset of X. Then, the function $\chi : B(X) \to \mathbb{R}^+$ defined by

---

[2]Kazimierz Kuratowski (2 February 1896–18 June 1980) was a Polish mathematician and logician. He was one of the leading representatives of the Warsaw School of Mathematics.

$$\chi(A) = \inf\{\delta > 0 : A \text{ admits a finite cover by sets of diameter} \leqslant \delta\}$$

is called the (Kuratowski) measure of noncompactness and the function $\beta : B(X) \to \mathbb{R}^+$ defined by

$$\beta(A) = \inf\{\rho > 0 : A \text{ can be covered by finitely many balls of radius } \rho\}$$

is called the ball (Sadovskii) measure of noncompactness.

Observation

- $\chi(\varnothing) = 0$ because diam $\varnothing = 0$.
- For $A \in B(X)$, we always have $\beta(A) \leqslant \chi(A) \leqslant 2\beta(A)$, but there exist $A \in B(X)$, such that the strict inequality hold.

The properties of the Kuratowski measure of noncompactness are listed in the following proposition.

Proposition 6.2 *Let X be a Banach space and $A, B, A_i \in B(X)$. Then*

(1°) $\chi(A) = 0$ *iff A is relatively compact.*
(2°) $0 \leqslant \chi(A) \leqslant \text{diam} A$.
(3°) $A \subseteq B \Rightarrow \chi(A) \leqslant \chi(B)$.
(4°) $\chi(A) = \chi(\overline{A}); \chi(\lambda A) = |\lambda| \chi(A), \lambda \in \mathbb{R}$.
(5°) *If $A_i = \overline{A}_i, A_{i+1} \subset A_i$, and $\lim \chi(A_i) = 0$, then $A_\infty = \bigcap\limits_{i=1}^{\infty} A_i \neq \varnothing$ and*
  $\chi(A_\infty) = 0$.
(6°) $\chi(A_1 \cup A_2) = \max\{\chi(A_1), \chi(A_2)\}; \chi(A_1 + A_2) \leqslant \chi(A_1) + \chi(A_2)$.
(7°) $\chi(A) = \chi(coA) = \chi(\overline{co}A)$.
(8°) $\chi(N\varepsilon(A)) \leqslant \chi(A) + 2\varepsilon$, *where $N_\varepsilon(A) = \{x \in X : D(x, A) < \varepsilon\}$.*
(9°) $\chi(x + A) = \chi(A)$ *(i.e. χ is invariant under translation).*

Proof (1°) A is relatively compact iff for every $\varepsilon > 0$, there exists a finite covering by balls of diameter ε.
(2°) A can be covered by A with diam A, so the result follows.
(3°) Every cover of B is a cover of A, it follows that $\chi(A) \leqslant \chi(B)$.
(4°) Since $A \subset \overline{A}$, it follows that $\chi(A) \leqslant \chi(\overline{A})$. But, if $S_1, S_2, \ldots, S_n$ is a cover of A, i.e. $A \subset \bigcup_{i=1}^{n} S_i$, then $\overline{A} \subset \bigcup_{i=1}^{n} \overline{S}_i$ with diam $S_i = $ diam $\overline{S}_i$, so $\chi(A) = \chi(\overline{A})$.
For the second part, just note that diam$(\lambda A) = |\lambda|$ diam A.
(5°) We have

$$A_i = \overline{A}_i, A_{i+1} \subset A_i \Rightarrow A_i \text{ is closed and } A_\infty \subset A_i, \forall i$$
$$\Rightarrow \chi(A_\infty) \subset \chi(A_i) \to 0 \text{ as } i \to \infty \text{ and } A_\infty \text{ is closed.}$$

We now show that $A_\infty \neq \varnothing$.
(6°) Let $A = A_1 \cup A_2$ and $\eta = \max\{\chi(A_1), \chi(A_2)\}$. Then it follows from $A_1 \subset A$ that $\chi(A_1) \leqslant \chi(A)$ and $\eta \leqslant \chi(A)$. Conversely, let for any given $\varepsilon > 0$ with given coverings $A_j \subset S_{jk}, j = 1, 2$ with diam$S_{jk} \leqslant \chi(A_j) + \varepsilon \leqslant \eta + \varepsilon$. All of these $S_{jk}'s$

together form a covering of A, so that $\chi(A) \leqslant \eta + \varepsilon$, i.e. $\chi(A) \leqslant \eta$. Thus, we obtain $\chi(A) = \eta$, i.e. $\chi(A_1 \cup A_2) = \max\{\chi(A_1), \chi(A_2)\}$.

For the second part, let $S_{11}, S_{12}, \ldots, S_{1m}$ be a cover of A_1, $S_{21}, S_{22}, \ldots, S_{2n}$ a cover of A_2, then all sets $\{S_{1i} + S_{2j} : i = 1, 2, \ldots, m; j = 1, 2, \ldots, n\}$ form a cover of $A_1 + A_2$ and for $i = 1, 2, \ldots, m; j = 1, 2, \ldots, n$ we have

$$diam(S_{1i} + S_{2j}) \leqslant diam\, S_{1i} + diam\, S_{2j}.$$

Thus, the result follows. The proof of the results $(7°)$–$(9°)$ follows easily, so we omit the details.

Note that the assertions $(2°)$–$(4°)$ were established by Kuratowski [351] while $(5°)$ and $(6°)$ were obtained by Darbo [155].

Assume that E is a Banach space with the norm $\| \cdot \|$, the family of all nonempty and bounded subsets of E is denoted by $\mathfrak{M}_E$ and its subfamily consisting of all relatively compact sets is denoted by $\mathfrak{N}_E$. The following definition of measure of noncompactness is introduced and studied by Banas and Goebel [35].

Definition 6.17 A mapping $\mu : \mathfrak{M}_E \to \mathbb{R}^+ = [0, \infty)$ is said to be a measure of noncompactness in E if it satisfies the following conditions:
$1°$ $\varnothing \neq \mu^{-1}(\{0\}) \subset \mathfrak{N}_E$.
$2°$ $\mu(\bar{X}) = \mu(X)$.
$3°$ $\mu(coX) = \mu(X)$.
$4°$ $X \subset Y \Rightarrow \mu(X) \leqslant \mu(Y)$
$5°$ If (X_n) is a nested sequence of closed sets from $\mathfrak{M}_E$ such that $\lim_{n\to\infty} \mu(X_n) = 0$, then the intersection set $X_\infty = \bigcap_{n=1}^{\infty} X_n$ is nonempty.

The family $\mu^{-1}(\{0\})$ described in $1°$ is called the kernel of the measure of noncompactness μ and denoted by $\ker \mu$. Furthermore, we observe that the intersection set X_∞ from axiom $5°$ is a member of the kernel $\ker \mu$. As $\mu(X_\infty) \leqslant \mu(X_n)$ for any n, we have that $\mu(X_\infty) = 0$. This yields that $X_\infty \in \ker \mu$.

A measure μ of noncompactness is said to be sublinear if

$6°$ $\mu(X_1 + X_2) \leqslant \mu(X_1) + \mu(X_2)$ for all $X_1, X_2 \in \mathfrak{M}_E$, and
$7°$ $\mu(\lambda X) \leqslant |\lambda| \mu(X)$ for all $\lambda \in \mathbb{R}$ and $X \in \mathfrak{M}_E$.

The following well-known result of Schauder plays a key role in the topological fixed point theory (cl. [4, 35, 246] and the references therein).

Theorem 6.27 (Schauder [542]) *Let Ω be a nonempty, bounded, closed and convex subset of a Banach space E. Then each continuous and compact map $T : \Omega \to \Omega$ has at least one fixed point in the set Ω.*

The generalization of Schauder's fixed point which is called the Darbo fixed point theorem is formulated below.

Theorem 6.28 (Darbo [155]) *Let Ω be a nonempty, bounded, closed and convex subset of a Banach space E and let $T : \Omega \to \Omega$ be a continuous mapping. Assume that there exists a constant $k \in [0, 1)$ such that*

$$\mu(TX)) \leqslant k \, \mu(X)$$

for any nonempty subset X of Ω, where μ is a measure of noncompactness defined in E. Then T has a fixed point in the set Ω.

Next, a generalization of fixed point theorem of Darbo was proved by Dhage [175]. But first we recall the following useful definition introduced by Dhage [175].

Definition 6.18 A mapping $T : X \to X$ is called $\mathcal{D}$-set-Lipschitz if there exists an upper semicontinuous nondecreasing function $\varphi : \mathbb{R}^+ \to \mathbb{R}^+$ such that $\mu(T(A)) \leqslant \varphi(\mu(A))$ for all $A \in B(X)$ with $T(A) \in B(X)$, where $\varphi(0) = 0$. The function φ is sometimes called a $\mathcal{D}$-function of T on X. Especially when $\varphi(r) = kr, k > 0, T$ is called a k-set-Lipschitz mapping and if $k < 1$, then T is called a k-set contraction on X. Further, if $\varphi(r) < r$ for $r > 0$, then T is called a nonlinear $\mathcal{D}$-set contraction on X.

Remark 6.7 (Dhage [176]) If $\phi, \psi : \mathbb{R}^+ \to \mathbb{R}^+$ are two $\mathcal{D}$-functions, then $i)\phi + \psi, ii)\lambda\phi, \lambda > 0$ and $iii)\phi \circ \psi$ are also $\mathcal{D}$-functions on $\mathbb{R}^+$, and commonly used $\mathcal{D}$-functions are $\psi(r) = kr, \psi(r) = \frac{Lr}{K+r}, L > 0, K > 0$, and $\psi(r) = \log(1 + r)$ etc. A few details of $\mathcal{D}$-functions appear in Dhage [176] and the references cited therein.

Lemma 6.9 (Dhage [175]) *If φ is a $\mathcal{D}$-function on $\mathbb{R}^+$ into itself with $\varphi(r) < r$ for $r > 0$, then $\lim_{n \to \infty} \varphi^n(t) = 0$ for all $t \in [0, \infty)$ and vice versa.*

Using Lemma 6.9, Dhage [175] proved the following applicable measure theoretic fixed point result.

Theorem 6.29 (Dhage [175]) *Let C be a closed, convex and bounded subset of a Banach space X and let $T : C \to C$ be a continuous and nonlinear $\mathcal{D}$-set contraction. Then T has a fixed point.*

Remark 6.8 Let us denote by $\mathfrak{F}(T)$ the set of all fixed points of the operator T which belong to C. It can be shown that the set $\mathfrak{F}(T)$ existing in Theorem 6.29 belongs to the family $ker \, \mu$. Indeed, if $\mathfrak{F}(T) \notin ker \, \mu$, then $\mu(\mathfrak{F}(T)) > 0$ and $T(\mathfrak{F}(T)) = \mathfrak{F}(T)$. Now from nonlinear set contractivity, it follows that $\mu(T(\mathfrak{F}(T))) \leqslant \phi(\mu(\mathfrak{F}(T)))$ which is a contradiction since $\phi(r) < r$ for $r > 0$. Hence, $\mathfrak{F}(T) \in ker \, \mu$. This particular property of the measures has been used in the study of attractivity of solutions for the nonlinear functional integral equation in question.

Recently, a generalization of fixed point theorem of Darbo was proved by Aghajani, Banas and Sabzali [5] as follows:

Theorem 6.30 *Let Ω be a nonempty, bounded, closed and convex subset of a Banach space E and let $T : \Omega \to \Omega$ be a continuous operator such that*

$$\psi(\mu(TX)) \leqslant \psi(\mu(X)) - \phi(\mu(TX))$$

for every nonempty subset X of Ω, where μ is an arbitrary measure of noncompactness and $\phi, \psi : \mathbb{R}^+ \to \mathbb{R}^+$ are given functions such that ψ is continuous and ϕ is lower semicontinuous on $\mathbb{R}^+$. Moreover, $\phi(0) = 0$ and $\phi(t) > 0$ for $t > 0$. Then T has at least one fixed point in Ω.

Definition 6.19 Let $\mathcal{R}$ denote the class of those functions $\beta : \mathbb{R}^+ \to [0, 1)$ which satisfies the condition

$$\beta(t_n) \to 1 \text{ implies } t_n \to 0.$$

Clearly, $\beta(t) = \exp(-t^2)$ and $\beta(t) = \frac{\ln(1+t)}{t}$ are examples of functions satisfying condition of Definition 6.19.

Obviously, the identity mapping on $\mathbb{R}^+$ into itself also satisfies the requisite conditions of Definition 6.19.

In 2015, Dhage, Dhage and Pathak [164] proved a generalization of Darbo's fixed point theorem and compliment of Theorem 6.29, formulated as follows:

Theorem 6.31 *Let Ω be a nonempty, bounded, closed and convex subset of a Banach space E, and let $T : \Omega \to \Omega$ be a continuous operator such that*

$$\psi(\mu(TX)) \leqslant \psi(\mu(X)) - \phi(\beta(\mu(TX))) \tag{6.46}$$

for every nonempty subset X of Ω and each $\beta \in \mathcal{R}$, where μ is an arbitrary measure of noncompactness and $\phi : \mathbb{R}^+ \to \mathbb{R}^+$ is nondecreasing function such that ϕ is a lower semicontinuous on $\mathbb{R}^+$ such that $\phi(0) = 0$ and $\phi(t) > 0$ for $t > 0$. Then T has at least one fixed point in Ω.

Proof Define a sequence $\{\Omega_n\}$ as $\Omega_0 = \Omega$ and $\Omega_n = \overline{co}T\Omega_{n-1}$ for $n = 1, 2, \dots$. If there exists a natural number n_0 such that $\mu(\Omega_{n_0}) = 0$, then Ω_{n_0} is compact. By Theorem 6.28, T has a fixed point in Ω. Next, we assume that $\mu(\Omega_{n_0}) > 0$ for $n = 1, 2, \dots$. Using (6.46), we get

$$\psi(\mu(\Omega_{n+1})) = \psi(\mu(\overline{co}T\Omega_n)) \leqslant \psi(\mu(\Omega_n)) - \phi(\beta(\mu(\Omega_{n+1}))). \tag{6.47}$$

Now, taking into account that $\Omega_{n+1} \subset \Omega_n$, on the basis of axiom 2° of Definition 6.17, the sequence $\{\mu(\Omega_n)\}$ is nonincreasing and nonnegative. From this we infer that $\mu(\Omega_n) \to r$ when $n \to \infty$, where $r \geq 0$ is a nonnegative real number. Since β is continuous, it follows that $\beta(\mu(\Omega_n)) \to \beta(r)$ as $n \to \infty$. Now, in view of (6.47) we obtain

$$\limsup_{n \to \infty} \mu(\Omega_{n+1}) \leqslant \limsup_{n \to \infty} \mu(\Omega_n) - \liminf_{n \to \infty} \phi(\mu(\beta(\Omega_{n+1}))).$$

This yields $r \leqslant r - \liminf_{n\to\infty} \phi(\beta(\mu(\Omega_{n+1})))$. Since ϕ is nondecreasing, we obtain $\phi(r) \leqslant \liminf_{n\to\infty} \phi(\beta(\mu(\Omega_{n+1}))) = 0$. From this, in view of the fact that $\phi(0) = 0$, we deduce that $\beta(\mu(\Omega_n)) \to 0$ as $n \to \infty$. By the definition of β we infer that $\mu(\Omega_n) \to 0$ as $n \to \infty$. Now, using axiom 6° of Definition 6.17 we derive that the set $\Omega_\infty = \bigcap_{n=1}^{\infty} \Omega_n$ is nonempty, closed, convex and $\Omega_\infty \subset \Omega$. Notice that $T(\Omega_\infty) \subset \Omega$, i.e. T maps Ω_∞ into itself and $\Omega_\infty \in \ker\mu$. Now taking into account Schauder fixed point principle (cf. Theorem 5.55), we infer that the operator T has a fixed point x in the set Ω_∞. Since $\Omega_\infty \subset \Omega$, it follows that $x \in \Omega$. This completes the proof.

Taking $\beta : \mathbb{R}^+ \to \mathbb{R}^+$ to be an identity mapping on $\mathbb{R}^+$, then we obtain the following fixed point result as a corollary with interesting consequences.

Corollary 6.2 *Let Ω be a nonempty, bounded, closed and convex subset of a Banach space E and let $T : \Omega \to \Omega$ be a continuous operator satisfying the inequality*

$$\mu(TX)) \leqslant \mu(X) - \phi(\mu(X)) \tag{6.48}$$

for every nonempty subset X of Ω, where μ is an arbitrary measure of noncompactness and $\phi : \mathbb{R}^+ \to \mathbb{R}^+$ is a lower semicontinuous on $\mathbb{R}^+$ such that $\phi(0) = 0$ and $\phi(t) > 0$ for $t > 0$. Then T has at least one fixed point in Ω.

When $\phi(r) = (1-k)r$, $0 \leqslant k < 1$, Corollary 6.2 reduces to Theorem 6.28 above due to Darbo [155]. Again, when $\mu(X) > 0$, then from condition (6.48), we obtain the following Sadovskii's fixed point theorem for condensing mappings characterized by the inequality $\mu(TX) < \mu(X)$. The mappings satisfying this contractive inequality are called condensing mappings on Banach spaces.

Theorem 6.32 (Sadovskii [536]) *Let Ω be a nonempty, bounded, closed and convex subset of a Banach space E and let $T : \Omega \to \Omega$ be a continuous and condensing mapping. Then T has at least one fixed point in Ω.*

A slight variant of Theorem 6.30 can be formulated as given below.

Theorem 6.33 *Let Ω be a nonempty, bounded, closed and convex subset of a Banach space E and let $T : \Omega \to \Omega$ be a continuous operator such that*

$$\mu(TX)) \leqslant \mu(X) - \phi(\mu(X)) \tag{6.49}$$

for every nonempty subset X of Ω, where μ is an arbitrary measure of noncompactness and $\phi : \mathbb{R}^+ \to \mathbb{R}^+$ is a nonincreasing lower semicontinuous function on $\mathbb{R}^+$ such that $\phi(0) = 0$ and $\phi(t) > 0$ for each $t > 0$. Then T has at least one fixed point in Ω.

It has been shown in Banas and Goebel [35] that μ is a sublinear measure of noncompactness in X. In what follows, denote by Φ the family of all functions $\varphi : \mathbb{R}^+ \to \mathbb{R}^+$ being nondecreasing and upper semicontinuous on $\mathbb{R}^+$ and such that
(i) $\varphi(t) < t$ for all $t > 0$;

(ii) $c = \sup_{t>0} k(t) < 1$ where $k(t) = \frac{\varphi(t)}{t}$.

Notice that each $\varphi \in \Phi$ satisfies the following condition

$$\lim_{n\to\infty} \varphi^n(t) = 0 \text{ for any } t > 0.$$

A function $\varphi : \mathbb{R}^+ \to \mathbb{R}^+$ is said to be superadditive if the following condition holds:

$$\varphi(t) + \varphi(s) \leqslant \varphi(t+s) \ \forall \ t, s \in \mathbb{R}^+.$$

Clearly, $\varphi = \ln(1+t)$ and $\varphi = t - \ln(1+t)$ for all $t \in \mathbb{R}^+$ are examples of superadditive functions.

6.7 k-Set Contractions and Condensing Maps

Kuratowski also introduced the notion of a class of maps which are called k-set contractions. Let X_1 and X_2 be metric spaces and $T : X_1 \to X_2$ a continuous map.

Definition 6.20 T is said to be a k-set contractions if for every bounded set $A \subset X_1$, $T(A)$ is bounded and
$$\chi_2(T(A)) \leqslant k\chi_1(A),$$

where χ_i denotes the measure of noncompactness in X_i. If we consider a family of operators, i.e. $T : \Lambda \times X_1 \to X_2$ where Λ is an arbitrary set, then T is said to be k-set contraction if for any bounded set $A \subset X_1$, $T(\Lambda \times A)$ is bounded and $\chi_2[T(\Lambda \times A)] \leqslant k\chi_1(A)$.

If T is a k-set contraction with $k < 1$, T is called a strict-set contraction. If for every $x \in X_1$, there exists a neighbourhood N_x of x such that $T|_{N_x}$ is a strict-set contraction, then T is called a local strict-set contraction.

If for every bounded set $A \subset X$, such that $\chi(A) > 0$, $T(A)$ is bounded and $\chi_2(T(A)) < \chi_1(A)$, we say that T is a condensing map. Similarly we define a local condensing map.

Example 6.9 Let X be a Banach space, $U \subset X$ and $K : U \to X$ Lipschitz continuous with Lipschitz constant $k < 1$. $C : U \to X$ is compact. Then $T = K + C$ is a k-set contraction. To this end, we observe the following:
Let $B \subset U$ be a bounded set. Then $T(B) \subset K(B) + C(B)$. This implies that

$$\chi(T(B)) \leqslant \chi(K(B)) + \chi(C(B)) \leqslant k\chi(B).$$

Observation

- Any strict-set contraction is a condensing map, but simple examples show that the converse is false (*Nussbaum* [429]).

- Every condensing map is a 1-set contraction.
- Every compact map is a 0-set contraction.

The following proposition holds for k-set contractions.

Proposition 6.3 *(a) If $T_i : D \to X$ is a k_i-set contraction, $i = 1, 2$, and $T_3 : T_1(D)$ $\to X$ is a k_3-set contraction, then $(T_1 + T_2) : D \to X$ is $(k_1 + k_2)$-set contraction, and $T_3 T_1 ; D \to X$ is a $k_1 k_3$-set contraction.*
(b) $T : D \to X$ is compact iff T is 0-set contraction.
(c) If $T : D \to X$ is α-Lipschitzian, i.e.

$$\|Tx - Ty\| \leqslant \alpha \|x - y\| \ \text{for all} \ x, y \in D,$$

then T is k-set contraction with $k = \alpha$.
(d) If $U : D \to K$ is compact and $H : D \to X$ is α-Lipschitzian, then $T = U + H$ is a k-set contraction with $k = \alpha$.
(e) If $T : D \to X$ is semicontractive type with constant $k \leqslant 1$ then T is k-set contraction.

Observation

- The assertions (a) to (d) follow from the definition of k-set contraction, and (e) has been established by Nussbaum [429].
- If A is a closed and bounded subset of a Banach space X and $T: A \to X$ is a condensing map, then $I - T$ is a proper map, i.e. $(I - T)^{-1}$ maps compact sets into compact sets. This, in turn, implies that $I - T$ takes closed sets to closed sets.

Using above results, Darbo [155] has proved the following fixed point theorem.

Theorem 6.34 (Darbo [155]) *Let A be a closed, bounded convex subset of a Banach space X and let $T : A \to A$ be a strict-set contraction. Then T has a fixed point.*

The following extension of Darbo's theorem was proved by Nussbaum [428, 429].

Theorem 6.35 (Nussbaum [428, 429]) *Let A be a closed, bounded, convex subset of a Banach space X such that int A is nonempty. Let $T : A \to X$ be a strict-set contraction. Assume that for some $x_0 \in A$, we have*

$$t x_0 + (1 - t)Tx \neq x \ \text{for all} \ x \in \partial A \ \text{and} \ 0 < t \leqslant 1.$$

Then T has a fixed point.

In the following, we state two theorems giving conditions under which a condensing operator turns out to be a limit-compact operator (Sadovskii [536]).

Proposition 6.4 *Let A be a closed set in a normed linear space X and Λ a compact topological space. Suppose that the continuous operator $T : \Lambda \times A \to X$ is condensing. Then T is limit-compact on $\Lambda \times A$.*

Proposition 6.5 *If a continuous operator T is condensing and carries a closed convex set A into itself $(T(\Lambda \times A) \subseteq A$ where Λ is a compact topological space), then T is limit-compact on $\Lambda \times A$.*

Now we state the fixed point theorem concerning a condensing operator. This was proved by Sadovskii, and it contains many well-known fixed point theorems, such as the Schauder's theorem, Banach's contraction theorem (under more stringent restriction on the set), Krasnosel'skiĭ's theorem, Darbo's theorem, etc.

Theorem 6.36 *Let A be a nonempty closed convex set in a normed linear space X. Let $T : A \rightarrow A$ be a continuous condensing map. Then $d(I - T, A, 0) = 1$, i.e. T has a fixed point.*

Proof By Proposition 6.5, the operator T is limit-compact. We shall show that the requirement of Theorem 6.26 is satisfied. Consider $S = \{T^n x_0 : n = 0, 1, 2, \ldots\}$, where x_0 is any point in A. Obviously, $T(S) \neq \varnothing$ and $S = T(S) \cup \{x_0\}$. Therefore, $r(T(S)) = r(S)$. It then follows from the definition of the condensing operator that the set $\overline{S}$ is compact and this shows that the condition (d) of the Theorem 6.26 is satisfied. Hence, $d(I - T, A, 0) \neq 0$, i.e. T has a fixed point. ∎

Corollary 6.3 (Schauder's theorem) *Let A be a nonempty closed convex subset of a normed linear space X. Let T be a continuous mapping of A into a compact subset of A. Then T has a fixed point in A.*

Proof In this case, T is condensing as T is a 0-set contraction. This follows from Proposition 6.3 (b). ∎

Corollary 6.4 (Banach's contraction theorem with stringent condition on the set (convexity)) *Let A be a closed and convex subset of a normed linear space X and let T be a contraction mapping from A into itself. Then T has a fixed point in A.*

Proof It follows from Proposition 6.3 (c). ∎

Corollary 6.5 (Krasnosel'skiĭ's theorem) *Let A be a nonempty closed convex subset of a normed linear space X. Let T be a continuous mapping of A into a compact subset of X and let S be a contraction mapping from A into X, and let $Tx + Sy \in A$; $x, y \in A$. Then there is a point $u \in A$ such that $Tu + Su = u$.*

Proof It follows from Proposition 6.3 (d).

Corollary 6.6 (Darbo's theorem) *Let A be a closed, bounded and convex subset of a normed linear space X. Let T be a strict-set contraction from A into itself. Then T has a fixed point in A.*

Proof This follows easily since any strict-set contraction is condensing.

We now introduce a more general notion of measure of noncompactness.

Definition 6.21 Let X be a normed linear space. Let $\mathscr{F}$ denote a subset of 2^X (generally bounded subsets) which together with every set A contains the set co A. Let $(B, \leqslant)$ be a partially ordered set. A measure of noncompactness in X is a function $\Psi : \mathscr{F} \to B$ such that $\Psi(\overline{co}A) = \Psi(A)$ for any $A \in \mathscr{F}$. A measure of noncompactness is said to be monotonic if

$$A_1 \subseteq A_2 \Rightarrow \Psi(A_1) \leqslant \Psi(A_2).$$

Example 6.10 The measure of noncompactness Ψ satisfying Proposition 6.3 with $X = C[a, b]$ is the following: (A is assumed to be bounded in our discussion).

$$\Psi(A) = \lim_{\delta \to 0} \ \sup_{x \in A} \ \sup_{0 \leqslant \tau \leqslant \delta} \|x - x_\tau\|$$

where x_τ is a shift of the function $x(t)$:

$$x_\tau(t) = \begin{cases} x(t + \tau) & \text{if } a \leqslant t \leqslant b - \tau, \\ x(b) & \text{if } b - \tau \leqslant t \leqslant b. \end{cases}$$

This also defines a measure of noncompactness in the space $L_p(a, b)(p \geqslant 1)$ where x_τ is taken as a shift of the function x or

$$x_\tau = \frac{1}{2\tau} \int_{t-\tau}^{i+\tau} x(s)ds$$

(the function $x(t)$ is assumed to be continued to the interval $[a - \tau, b + \tau]$ by zero).

Let X be a Banach space with Ψ as measure of noncompactness. A measure of noncompactness on $C([0, \tau], X)$ is defined as

$$\Psi_C(A) = \Psi(A_{[0,\tau]})$$

where $A_{[0,\tau]} = \{x(t) : x \in A, t \in [0, \tau]\}$ and norm $\|x\|_C = \max\{\|x(t)\| : t \in [0, \tau]\}$. Let $C^1([0, \tau], X)$ be the Banach space of continuously differentiable functions $x : [0, \tau] \to X$ with norm

$$\|x\|_{C^1} = \|x\|_C + \|x'\|_C.$$

Then a measure of noncompactness on $C^1[0, \tau], X]$ is $\Psi_{C^1} : \mathscr{F}_{C^1} \to C[0, \tau]$, (range of Ψ can be an arbitrary partially order set) given by

$$\Psi_{C^1}(A)(t) = \Psi(A'_t) \text{ where } A'_t = \{x'(t) : x \in A\}.$$

Example 6.11 Various examples of condensing operators associated with differential equations have been studies by Ambrosetti [20, 21] and others. We give the one discussed by Sadovskii [536].

Consider the Cauchy problem for an ordinary differential equation

$$x' = f(t, x(t))$$
$$x(0) = 0 \tag{6.50}$$

f is a real-valued function defined on $[0, \tau] \times \mathbb{R}$.

The problems (6.50) are equivalent to the integral equation

$$x(t) = \int_0^t f[s, x(s)]ds$$

which is written in operator from as $KNx = x$. Here $N : C[0, \tau] \to C[0, \tau]$ is given by

$$(Nx)(t) = f[t, x(t)]$$

and $K : C[0, \tau] \to C[0, \tau]$ is given by

$$(Kx)(t) = \int_0^t x(s)ds.$$

The operator KN is condensing under some suitable condition. If α is a measure of noncompactness in $\mathbb{R}$, then it generates a measure of noncompactness in $C[0, \tau]$ as defined in Example 6.10. Under some suitable conditions, it has been shown by Sadovskii [536] that KN is Ψ_{C^1} condensing. KN acts in both the spaces $C([0, \tau], \mathbb{R})$ and $C^1([0, \tau], \mathbb{R})$ and in both cases the set of its fixed points coincides with the set of solutions of the problem (6.50).

The following notion of measure of noncompactness was introduced by Gohberg, Goldenstein and Markus [250] and is called ball measure or Hausdorff measure of noncompactness. Let X be a normed linear space, and let A be a bounded subset of X. $\chi_X(A)$ is defined as

$$\chi_X(A) = \inf\{d > 0 | A \text{ can be covered by a finite number of balls with centers}$$
$$\text{in } X \text{ and radius } d, \text{ i.e. the set } A \text{ has a finite } d\text{-mesh in } X\}.$$

One can show that χ has most of the properties of r- the Kuratowski measure of noncompactness and in particular, Proposition 6.3 holds for χ_x. However, as has been noted by Furi and Vignoli and Nussbaum χ_x do differ in one crucial way. If $A \subset B \subset X$ and B inherits its metric from X, then the measure of noncompactness of A as a subset of B is the same as the measure of noncompactness of A as a subset of X, and this is reflected in the notation. But, in general $\chi_B(A)$ need not be equal to $\chi_X(A)$. For example, if $x = \{x \in X : \|x\| = 1\}$ in an infinite dimensional Hilbert space X, then $\chi_X(S) = \sqrt{2}$ while $\chi_X(S) = 1$. If X is a Banach space and A any subset of X, a trivial argument shows that

$$\chi_X(A) \leqslant r(A) \leqslant 2\chi_X(A).$$

To say anything more precise, it seems necessary to know more about A or the Banach space X. It is shown by Nussbaum that $r(S) = 2$. So in this case $r(S) = 2\chi_X(S)$.

As in the case of r, we have ball k-set contractions and ball-condensing mappings, corresponding to χ. If in Propositions 6.3 and 6.4, D is taken to be X, then all of its assertions remain valid for ball k-set contractions. In general, one cannot say much about the precise relation between k-set contractions or ball k-set contractions or between condensing and ball-condensing mappings. However, Nussbaum [429] in his thesis, for $T \in B(X, Y)$ obtained a number of interesting results concerning the relation between k-set contractions and ball k-set contraction.

Now we discuss some recent fixed point theorems concerning strict-set contractions, which leave the cone of an ordered Banach space invariant, proved by Cac and Gatica [127]. In this connection, refer to the works of Amann [17], Martin [388] and Potter [491].

Definition 6.22 Let α be any function which assigns to each bounded subset A of a Banach space X a nonnegative real number $\alpha(A)$. α is said to be a generalized measure of noncompactness on X if it satisfies the following:
For bounded subset A and B of X and λ real, we have

  (i) $\alpha[\overline{co}(A)] = \alpha(A)$
 (ii) $\alpha(A + B) \leqslant \alpha(A) + \alpha(B)$
(iii) $\alpha(\lambda A) = |\lambda|\alpha(A)$
 (iv) $\alpha(A \cup B) = \max\{\alpha(A), \alpha(B)\}$
  (v) There exist positive constants m, M such that $m\alpha(A) \leqslant \alpha(A) \leqslant M\alpha(A)$ for every bounded subset A of X.

Let X and Y be Banach spaces with generalized measures of noncompactness α_X and α_Y, respectively. As before we define $k-$set contraction and strict-set contraction using α_X and α_Y instead of r_1 and r_2, respectively.

The following two theorems concerning strict-set contractions were proved by Amann [19] using the degree theory for k-set contractions as developed by Nussbaum [429] (generalization of Theorems 6.16 and 6.11 of Krasnoselskii [345]).

Theorem 6.37 *Let X be an ordered Banach space with total cone $K(X = \overline{K - K})$ and let $T : K \to K$ be a strict-set contraction. Assume that*

(a) *T is asymptotically linear along K, that $T(0) = 0$ and that T is differentiable at 0 along K.*
(b) *$T'(0)$ does not have a positive eigenvector corresponding to an eigenvalue greater than or equal to one and one is not an eigenvalue with corresponding positive eigenvector of $T'(\infty)$, but there exists $\lambda_\infty > 1$ and $v \in K, v \neq 0$ such that $T'(\infty)v = \lambda_\infty v$.*

Then T has a fixed point $u > 0$.

Proof The proof of this theorem follows from Theorem 5.154.

Theorem 6.38 *The assertion of the above theorem remains valid if condition (b) is replaced by the following:*
(b′) $T'(\infty)$ does not have a positive eigenvector corresponding to an eigenvalue greater than or equal to one, and one is not an eigenvalue with corresponding positive eigenvector for $T'(0)$, but there exists $\lambda_0 > 1$ and $v \in K, v \neq 0$ such that $T'(0)y = \lambda_0 v$.

We state and prove the following proposition which is needed for the proof of the next theorem (Krasnoselskii [345], Lemma 4.5 has proved this lemma for compact mappings T).

Proposition 6.6 *Let T be a strict-set contraction from $K_{r,R} = \{x \in K : r \leqslant \|x\| \leqslant R, \ 0 \leqslant R\}$ into K satisfying the conditions: $Tx = v_0(x \in K, \|x\| = r)$ and $Tx = u_0(x \in K, \|x\| = R)$ where $\|v_0\| \leqslant r < R \leqslant \|u_0\|$. Then the mapping T has at least one fixed point in $K_{r,R}$.*

Proof Consider only the nontrivial case $\|v_0\| < r < R < \|u_0\|$. Let the mapping T be extended to a mapping $T_1 : K_{0,\|u_0\|} \to K_{0,\|u_0\|}$ as follows:

$$T_1 x = \begin{cases} u_0 & \text{if } R = \|x\| \leqslant \|u_0\| \\ Tx & \text{if } r \leqslant \|x\| \leqslant R \\ v_0 & \text{if } \|x\| \leqslant r. \end{cases}$$

T_1 so defined is a strict-set contraction. Define $A_0 = K_{0,\|u_0\|}$ and $A_{n+1} = \overline{co}(T_1(A_n))$, $n \geqslant 0$. Then $A_{n+1} \subset A_n$ for all n and $\lim_{n \to \infty} \alpha(A_n) = 0$. Hence by Proposition 6.3(c),
$C_0 = \overset{\infty}{\underset{n=1}{\cap}} A_n$ is a nonempty compact convex subset of K. Also note that $T_1(C_0) \subseteq C_0$
and that $u_0, v_0 \in C_0$. Define $C_1 = (C_0 \cap K_{r,R}) \cup \{x \in K : \|x\| = r \text{ or } \|x\| = R\}$ and $T_2 : C_1 \to K$ by

$$T_2 x = \begin{cases} u_0 & \text{if } \|x\| = R \\ Tx & \text{if } x \in C_0 \cap K_{r,R} \\ v_0 & \text{if } \|x\| = r. \end{cases}$$

T_2 is a compact continuous mapping with its range contained in C_0. By Dugundji's extension theorem [207], we obtain a mapping $T_3 : K_{r,R} \to C_0$ as an extension of T_2. Then T_3 has a fixed point $x_0 \in K_{r,R}$ (by Lemma 6.5 of Krasnoselskii [345]). But x_0 belongs to $C_0 \cap K_{r,R}$ and T_3 coincides with F in $C_0 \cap K_{r,R}$ by definition of T_3.

Definition 6.23 Let T be a mapping from K into itself. T is called an expansion of the cone K if $T(0) = 0$ and number $r, R > 0$ can be found such that for all $\varepsilon > 0$

$$Tx \ngeqslant (1 + \varepsilon)x, \quad x \in K, \|x\| \leqslant r, \ x \neq 0 \tag{6.51}$$

and

$$Tx \nleqslant x, x \in K, \quad \|x\| \geqslant R. \tag{6.52}$$

Notice that Krasnoselskii has proved that if T is a compact compression or a compact expansion of the cone K, then it has at least one nonzero fixed point in K. Potter [491] has proved a similar result for a strict-set contraction and compression of the cone. Cac and Gatica [127] has proved the same result for a strict-set contraction and expansion of the cone K and this is presented below.

Theorem 6.39 *Let T be a strict-set contraction and an expansion of the cone K. Then T has at least one nonzero fixed point in K.*

Proof Let $k \in [0, 1]$ such that $\alpha[T(A)] \leqslant k\alpha(A)$ for every bounded subset A of K. Choose μ such that $k + \mu < 1$. Let $R^* = \frac{R}{k+\mu}$, $2\delta = R^* - R$. Here $r, R > 0$ are as given in (6.51) and (6.52). Define the mapping $\overline{T} : K_{r/2,R^*} \to K$ as follows.

$$\overline{T}(x) = \begin{cases} \frac{2\|x\|-r}{r} Fx, & \frac{r}{2} \leqslant \|x\| \leqslant r, \\ Tx, & r \leqslant \|x\| \leqslant R, \\ \frac{\|x\|}{R} T\left[\frac{R}{\|x\|}x\right] + (\|x\| - R)h, & R \leqslant \|x\| \leqslant R + \delta, \\ \frac{\|x\|}{R} \frac{R^*-\|x\|}{\delta} T\left[\frac{R}{\|x\|}x\right] + \delta h, & R + \delta \leqslant \|x\| \leqslant R^*, \end{cases}$$

where $h > 0$ is chosen such that

$$\delta\|h\| > R^* + \frac{R^*}{R}\left[\sup_{y\in K, \|y\|\leqslant R} \|Ty\|\right]. \tag{6.53}$$

It can be shown that the mapping $\overline{T} : K_{r/2,R^*} \to K$ is a strict-set contraction. Moreover $\overline{T}x = 0$ whenever $\|x\| = r/2$ and $\|\overline{T}x\| = \delta\|h\| > R^*$ whenever $\|x\| = R^*$. Hence by Proposition 6.6, $\overline{T}$ has a fixed point $u \in K_{r/2,R^*}$. Next it is claimed that $r \leqslant \|u\| \leqslant R$. Then by the definition of $\overline{T}$, u is a nonzero fixed point of T. We have $\|u\| > r/2$. Suppose $\|u\| < r$. Then

$$T(u) = \left(\frac{r}{2\|u\|-r}\right)\overline{T}(u) = \left[1 + \frac{2r-2\|u\|}{2\|u\|-r}\right]u.$$

Since $\frac{2r-2\|u\|}{2\|u\|-r} > 0$, this violates (6.53). Suppose $R \leqslant \|u\| \leqslant R + \delta$. Then

$$u = \frac{\|u\|}{R} T\left(\frac{R}{\|u\|}u\right) + (\|u\| - R) \text{ and } T\left(\frac{R}{\|u\|}u\right) \leqslant \frac{R}{\|u\|}u$$

and this violates (6.52). Finally if we assume that $r + \delta \leqslant \|u\| \leqslant R^*$, then we arrive at a contradiction to (6.53).

Finally we conclude this section by discussing some results concerning nonlinear contractions obtained by Eisenfeld and Lashmikantham [219]. Consider the mapping F from a metric space X into itself satisfying the condition

$$d(Tx, Ty) \leqslant \psi(d(x, y)), \quad \text{for all} \ \ x, y \in X$$

where ψ is a mapping from $\mathbb{R}^+$ into itself. If ψ is a linear mapping $\psi x = kx$, then the $T : X \to X$ is said to be Lipschitz. Instead of choosing ψ to be a linear mapping from the cone $\mathbb{R}^+$ into itself, they have taken ψ to be a nonlinear mapping from a cone K of a Banach space into itself. This makes it possible to have greater flexibility in the choice of ψ and to have the advantage of stronger convergence properties and more accurate estimates. This comparison mapping ψ is positive (in the sense that it takes values in a cone), isotone and has a unique fixed point which is the zero element of the cone. For a regular cone (such as in L_P, $1 \leqslant p < \infty$), ψ has to be upper semicontinuous from above (i.e. from the right). In case of normal cones which are not regular (such as in $C[0, 1]$), ψ need to be completely continuous. The complete continuity condition which is also employed by Krasnoselskii [345] may be replaced by a weaker compactness type condition in terms of Kuratowski measure of noncompactness with upper semicontinuity from the right.

Semimonotone norm— The norm in an ordered Banach space X is called semi-monotone if for arbitrary

$$x, y \in K, x \leqslant y \Rightarrow \|x\| \leqslant N\|y\|$$

where the constant N does not depend on x and y.

Notice that a necessary and sufficient condition for the cone K to be normal is that the norm is semimonotone.

Regular cone— A cone is said to be regular if each decreasing sequence $\{u_n\}_0^\infty$ bounded from below (i.e. there is a v such that $u_n \geqslant v, n \geqslant 0$) is convergent. A decreasing sequence $u_0 \geq u_1 \geq u_2 \geq \cdots$ in a space with a normal cone is convergent if it has a convergent subsequence.

Definition 6.24 Following Proposition 6.3 (c), a mapping Ψ of a complete metric space (X, d) into itself is said to be quasi-compact if the sequence of measure of noncompactness $r(A_n)$ of closed sets $A_{n+1} = \Psi \overline{A}_n, n \geqslant 0$, approaches zero.

Definition 6.25 A mapping Ψ on a partially ordered set into itself is said to be upper semicontinuous from the right if whenever

$$u_0 \geqslant u_1 \geqslant \cdots \quad \text{and} \quad \Psi u_0 \geqslant \Psi u_1 \geqslant \cdots$$

are both monotonic convergent sequences and $w = \lim_{n \to \infty} u_n$ is in the domain of Ψ then $\Psi w \geq \lim_{n \to \infty} \Psi u_n$.

Proposition 6.7 *Let Ψ be a mapping from a partially ordered set into itself which is isotone and upper semicontinuous from the right. Suppose the sequence of iterates*

$u_n = \Psi^n u_0$, of an element u_0, is decreasing and convergent to a vector u which lies in the domain of Ψ. Then u is a fixed point of Ψ.

We need this for the proof of the following proposition.

Proposition 6.8 *Let f be isotone and upper semicontinuous from the right from an interval $[0, a]$ of real numbers into itself such that $f(x) = x$ if and only if $x = 0$. Let Ψ be a mapping from a complete metric space (X, d) into itself such that*

$$r(\Psi(A))) \leqslant f(r(A)) \tag{6.54}$$

for every subset A of X. Then Ψ is quasi-compact.

Proof Let $A_{n+1} = \overline{\Psi(A_n)}, n \geqslant 0$ and let $K_n = r(A_n), n \geqslant 0$. Then by condition (6.54), $K_{n+1} \leqslant f(K_n), n \geqslant 0$. Define $t_{n+1} = f(t_n), t_0 = K_0$. Since f is isotone, it follows that $k_n \leqslant t_n, n \geqslant 0$ and $\{t_n\}$ is a decreasing sequence. Let $\lim_{n \to \infty} t_n = t_\infty$. By Proposition 6.8, t_∞ is a fixed point of f and hence it is identically zero. This means that k_n converges to zero thereby proving the assertion.

Remark 6.9 If the mapping Ψ in the above proposition is a strict-set contraction, the Ψ is quasi-compact as one can take $f(x) = kx$ with $k < 1$.

Using Propositions 6.7 and 6.8, Eisenfeld and Lakshmikantham proved the following theorem which we state without proof.

Theorem 6.40 *Let Ψ be an isotone, upper semicontinuous mapping from a conical segment $\langle 0, u_0 \rangle$ into itself satisfying the following condition:*
(c) Either Ψ is quasi-compact and the cone is normal or the cone is regular (or both). Then the sequence of iterates $\{\Psi^n u_0\}$ is decreasing and convergent to fixed point w of Φ. Furthermore, $v \leqslant \Psi$ implies $v \leqslant w$. In particular, w is the maximal fixed point of Ψ in the segment $\langle 0, u_0 \rangle$.

Definition 6.26 Let X be a set and let ρ be a mapping from $X \times X$ into a cone K of a Banach space. The mapping ρ is said to be K-metric on X if the following conditions are satisfied:

(i) $\rho(x, y) = \rho(y, x)$,
(ii) $\rho(x, y) = 0$ iff $x = y$, and
(iii) $\rho(x, y) \leqslant \rho(x, z) + \rho(z, y)$.

Cauchy sequence– A sequence $\{x_n\}$ in the K–metric space (X, ρ) is said to be Cauchy if $\lim \rho(x_n, x_m) = 0$.
Convergent sequence– A sequence $\{x_n\}$ in the K-metric space (X, ρ) is said to be convergent if there is a $x \in X$ such that $\lim_{n \to \infty} \rho(x_n, x) = 0$.
Complete K-metric space– A K-metric space is complete if every Cauchy sequence is convergent.
K-convergent sequence – A convergent sequence $\{x_n\}$ is said to be K-convergent to x if there is sequence $u_n \to 0$ in K such that $\rho(x_n, x) \leqslant u_n$.

The following two theorems of Eisenfeld and Lakshmikantham [219, 220]give iterative schemes to obtain the fixed point.

Theorem 6.41 *Let (X, ρ) be a complete $K-$metric space. Let Ψ be an isotone, upper semicontinuous (from the right) mapping from the conical segment $< 0, u_0 >$ into itself such that the condition (c) of Theorem 6.40 is satisfied. Suppose also that 0 is the unique fixed point of Ψ. Let T be a mapping from X into itself such that*

$$\rho(Tx, Ty) \leqslant u_0, \quad x, y \in X$$

and $$\rho(Tx, Ty) \leqslant \Psi[\rho(x, y)] \text{ whenever } \rho(x, y) \leqslant u_0.$$

Then for any $x_0 \in X$, the sequence of iterates $x_n = T^n x_0$, K-converges to the unique fixed point u of T.

Proof First, we prove that the sequence $\{x_n\}$ is Cauchy. Let $m \geqslant n \geqslant 1$. Since $\rho(x_n, x_m) = \rho(Tx_{n-1}, Tx_{m-1}) \leqslant u_0$, we have

$$\rho(x_{n+1}, x_{m+1}) \leqslant \Psi[\rho(x_n, x_m)].$$

Similarly we have

$$\rho(x_{n+1}, x_{m+1}), \leqslant \Psi^n[\rho(Tx_0, Tx_{m-n})] \leqslant \Psi^n u_0. \tag{6.55}$$

By Theorem 6.40, $\Psi^n u_0$ decreases to the fixed point 0 of Ψ. Hence the sequence $\{x_n\}$ is Cauchy. Let $y = \lim_{n \to \infty} x_n$. By taking $m \to \infty$ in the inequality (6.55), we get $\rho(x_{n+1}, y) \leqslant \Psi^n u_0$. Thus the sequence $\{x_n\}$, K-converges to y as $\rho(x_{n+1}, y) \leqslant u_0$, $\rho(Tx_{n+1}, Ty) \leqslant \rho(x_{n+1}, y) \leqslant \Psi^{n+1} u_0 \to 0$ as $n \to \infty$. So $\rho(x_{n+2}, Ty) \to 0$, thereby implying that $y = Ty$. Suppose z is also a fixed point of T, i.e. $Tz = z$. Then

$$\rho(y, z) = \rho(Ty, Tz) \leqslant u_0.$$

So $\rho(y, z) \leqslant \Psi[\rho(y, z)]$. By Theorem 6.40, $\rho(y, z) \leqslant 0$. This implies that $\rho(y, z) = 0$, i.e. $y = z$.

Next theorem is an improvement of the above theorem.

Theorem 6.42 *Let (X, ρ) be a K-metric space and let Ψ be an isotone mapping from a segment $< 0, u_0 >$ into itself such that $\lim_{n \to \infty} \Psi^n u_0 \to 0$.*
Suppose T is a closed mapping from a subset D of X into X such that

$$\rho(Tx, Ty) \leqslant \Psi\rho(x, y), x, y \in D, \rho(x, y) \leqslant u_0.$$

Suppose further that $x_0 \in D$ and $x_n = T^n x \in D$, $n = 1, 2, \ldots$ and that $\rho(x_n, x_0) \leqslant u_0$. Then $\{x_n\}$ K-converges to a fixed point of T.

Exercises

6.1 Let $\Omega \subset \mathbb{R}$ be an open interval with $0 \in \Omega$ and let $f(x) = \alpha x^n$, $\alpha \neq 0$, then

$$deg(f, \Omega, 0) = \begin{cases} 0, & \text{if } n \text{ is even} \\ \text{sgn } \alpha, & \text{if } n \text{ is odd}. \end{cases}$$

6.2 Let $f : \mathbb{R}^2 \to \mathbb{R}^2$ be defined by
(i) $f(x, y) = x^3 - 3xy^2$,   (ii) $f(x, y)n = -y^3 + 3x^2y$.
Let $O = (0, 0)$ and $a = (1, 0)$; then show that $deg(f, S_2(O), a) = 3$.

6.3 Prove that for $m \in \mathbb{Z}$ there exists $\Omega \subset \mathbb{R}$ open and bounded and $f \in C(\overline{\Omega})$ with $0 \notin f(\partial\Omega)$ so that $deg(f, \Omega, 0) = m$.

6.4 Let $\Omega \subset \mathbb{R}^n$ be open bounded, $f \in C(\overline{\Omega})$, $y \in C(\overline{\Omega})$ and $|g(x)| < |f(x)|$ on $\partial\Omega$. Show that $deg(f + g, \Omega, 0) = deg(f, \Omega, 0)$. For analytic functions, this result is known as Rouche's theorem.
[Hint: Use homotopy invariance. To achieve this, consider $H(t, x) = t(f + g)(x) + (1 - t)f(x)$, $(t, x) \in [0, 1] \times \overline{\Omega}$.]

6.5 Let $\Omega = B_1(0) \subset \mathbb{R}^n$, $f \in C(\overline{\Omega})$ and $0 \notin f(\overline{\Omega})$. Then there exist $x, y \in \partial\Omega$ and $\lambda > 0$, $\mu < 0$ such that $f(x) = \lambda x$ and $f(y) = \mu y$, i.e. f has a positive and negative eigenvalue, each with an eigenvector in $\partial\Omega$.

6.6 Let Ω symmetric about $0 \in \Omega \subset \mathbb{R}^n$. If $f : \overline{\Omega} \to \mathbb{R}^n$ is a continuous function, $0 \notin f(\partial\Omega)$, and

$$\frac{f(x)}{|f(x)|} \neq \frac{f(-x)}{|f(-x)|} \quad \text{for all } x \in \partial\Omega,$$

then $deg(f, \Omega, 0)$ is an odd number. This result is known as antipode theorem.

6.7 Let $\Omega \subset \mathbb{R}^n$ be open, bounded, symmetric with $0 \in \Omega$. Let $f : \overline{\Omega} \to \mathbb{R}^n$ be continuous such that for all $\lambda \geq 1$ and $x \in \partial\Omega$, $f(x) \neq 0, f(-x) - \lambda f(x) \neq 0$. Then prove that $deg(f, \Omega, 0)$ is odd.

6.8 Let $\Omega \subset \mathbb{R}^n$ be open, bounded, symmetric with $0 \in \Omega$. Let $f : \overline{\Omega} \to \mathbb{R}^n$ be continuous and $m < n$. Then show that there exists an $x \in \partial\Omega$ such that $f(x) = f(-x)$. This result is known as Borsuk–Ulam's theorem.

6.9 Let A be a real $n \times n$ matrix with det $A \neq 0$ and $f \in C(\mathbb{R}^n)$ such that $|x - Af(x)| \leqslant \alpha|x| + \beta$ on $\mathbb{R}^n$ for some $x \in [0, 1)$ and $\beta \geqslant 0$. Then $f(\mathbb{R}^n) = \mathbb{R}^n$.

6.10 Let $\Omega = B_1(0) \subset \mathbb{R}^{2m+1}$, and $f : \partial\Omega \to \partial\Omega$ continuous. Then there exists an $x \in \partial\Omega$ such that either $x = f(x)$ or $x = -f(x)$.

6.11 Let $\Omega \subset \mathbb{R}^n$ be open, bounded and symmetric with respect to $0 \in \Omega$ and $\{A_1, A_2, \ldots, A_p\}$ be a covering of $\partial\Omega$ by closed sets $A_j \subset \partial\Omega$ such that $A_j \cap (-A_j) = \varnothing$ for $j = 1, 2, \ldots, p$. Then prove that $p \geq n + 1$.

6.12 Let $\mathbb{B}_X$ be the unit ball in an infinite dimensional Banach space X. Then show that the measure of noncompactness $\chi(\mathbb{B}_X) = 2$.

6.13 Let $\Omega \subset X$ be open, bounded and symmetric with respect to $0 \in \Omega$ and $T: \overline{\Omega} \to X$ compact and odd. Let $Y \subset X$ be a proper linear subspace and $(I - T)(\overline{\Omega}) \subset Y$. Then prove that there exists a fixed point $x_0 \in \partial\Omega$ of T.

6.14 Let $\Omega \subset X$ be open, bounded and symmetric with respect to $0 \in \Omega$ and $T: \overline{\Omega} \to X$ compact. Let $(I - T)(\overline{\Omega}) \subset Y \subsetneq X$. Then prove that there is an $x \in \partial\Omega$ such that

$$x - T(x) = -x - T(-x).$$

6.15 Let $T: X \to X$ be compact, and $L: X \to X$ compact and linear. If there exists a $\lambda \in \mathbb{R}$, such that for all $x \in X$, $\|x\| = r$ we have

$$\|Tx - \lambda Lx\| < \|x - \lambda Lx\|,$$

then show that $Deg(I - T, B_r(0), 0)$ is odd.

6.16 Let $X = c_0$ and $T(x) = \frac{1}{3}\sum_{i=1}^{\infty}\alpha_i^3 e_i$, for $x = \sum_{i=1}^{\infty}\alpha_i e_i \in c_0$ and $\{e_i\}_{i\in\mathbb{N}}$ the canonical basis in c_0; prove that T is A-proper but $T'(\frac{e_i}{\sqrt{i}})$ is not.

6.17 Let $X = c_0$ and $Tx = -\sum_{i=1}^{\infty}\left(1 + \frac{1}{i}\right)\alpha_i e_i$ for $x = \sum_{i=1}^{\infty}\alpha_i e_i \in c_0$ and $\{e_i\}_{i\in\mathbb{N}}$ the basis in c_0; prove that T is A-proper and $deg(T, S_1(0), 0) = \{-1, 1\}$.

Chapter 7
Variational Methods and Optimization

As long as a branch of knowledge offers an abundant of problems, it is full of vitality.

David Hilbert
Optimization is a cornerstone for the development of civilization.

Yuqi He

The purpose of this chapter is to give an introduction of variational methods and optimization theory in a rather convincing manner along with results of nonlinear analysis leading to an applied environment. So, we have chosen variational principles as the starting point of our discussion in the framework of Banach space theory that leads to optimization with the observation that by applying the techniques involved in variational methods and optimization one can deal with some real-world problems that arise in nonlinear analysis. We initiate our discussion by presenting some variational principles and their applications. The epicenter of our discussion is the so-called Ekeland variational principle (in short, EVP). Indeed, we show that EVP is equivalent to some other well-known results of nonlinear analysis, notably Takahashi's minimization theorem.

A remarkable discovery by Ekeland in 1972 laid down the foundation of the variational principle. The variational principle provides an approximate minimizer of a bounded below and lower semicontinuous function in a given neighbourhood of a point. This principle is now well as Ekeland variational principle. Notably, the localization property in EVP is very useful and explains the importance of the result. It may also be observed that is among the most important results obtained in nonlinear analysis. In fact, EVP appeared as one of the most useful tools to solve the problems from optimization, dynamical systems, optimal control theory, game theory nonlinear integral equations and some nonlinear real-world problems. In Sect. 7.1,

© Springer Nature Singapore Pte Ltd. 2018
H. K. Pathak, *An Introduction to Nonlinear Analysis and Fixed Point Theory*,
https://doi.org/10.1007/978-981-10-8866-7_7

we start our discussion with variational principle, while in Sect. 7.2 we discuss some optimization problems.

In Sect. 7.3, we present several forms of Ekeland's variational principle. We discuss equilibrium problem which is a unified model of several problems, namely minimization problem, saddle point problem, Nash equilibrium problem, fixed point problem, complementarity problem, variational inequality problems. Section 7.4 deals with mappings associated with variational inequality, while Sect. 7.5 initiated the discussion on variational calculus by presenting some fundamental problems that arise in applied sciences. In Sect. 7.6, we discuss in detail several variational methods with their applications in physical problems in applied environment. Finally, we conclude the chapter in Sect. 7.7 by presenting variational formulation for linear and nonlinear problems.

7.1   Variational Principles

Hartman and Stampacchia [268] proved the following:

Theorem 7.1 *Let K be a compact convex subset in $\mathbb{R}^n$ and $f : K \to \mathbb{R}^n$ be a continuous mapping. Then, there exists $\bar{x} \in K$ such that*

$$\langle f(\bar{x}), y - \bar{x} \rangle \geqslant 0 \text{ for all } y \in K,$$

where $\langle \cdot, \cdot \rangle$ denotes the scalar product in $\mathbb{R}^n$.

Ky Fan [227] proved the following theorem, which is called Fan's best approximation theorem:

Theorem 7.2 *Let K be a nonempty compact convex set in a normed vector space X. Then, for any continuous mapping $f : K \to X$, there exists $\bar{x} \in K$ such that*

$$\|\bar{x} - f(\bar{x})\| = \min_{y \in K} \|y - f(\bar{x})\|.$$

In particular, if $f(K) \subset K$, then $\bar{x}$ is a fixed point of f.

Theorems 7.1 and 7.2 play a very important role in nonlinear analysis, variational inequality problems, complementarity problems, optimization problems, game theory and others.

7.1.1   Equilibrium Problems

Let X be a normed vector space with the dual space X^*, K be nonempty convex subset of X and $F : K \times K \to \mathbb{R}$ be a real-valued bifunction.

In 1994, from the ideas of Theorems 7.1 and 7.2, Blum and Oettli [61] considered the equilibrium problem (shortly, (EP)) as follows:

Find $\bar{x} \in K$ such that (EP)

$$F(\bar{x}, y) > 0 \text{ for all } y \in K. \tag{EP}$$

The equilibrium problem (EP) has some special cases as follows:

(I) Minimization problem. Minimization problem (shortly, (MP)) is the following:

Find $\bar{x} \in K$ such that

$$\varphi(\bar{x}) \leqslant \varphi(y) \text{ for all } y \in K,$$

where $\varphi : K \to \mathbb{R}$ is a function. If we set $F(x, y) = \varphi(y) - \varphi(x)$ for all $x, y \in K$, then the problem (MP) is equivalent to the problem (EP).

(II) Saddle point problem. Saddle point problem (shortly, (SPP)) is the following:

Find $(\bar{x}_1, \bar{x}_2) \in K_1 \times K_2$ such that

$$L(\bar{x}_1, y_2) \leqslant L(\bar{x}_1, \bar{x}_2) \leqslant L(y_1, \bar{x}_2) \text{ for all } (y_1, y_2) \in K_1 \times K_2,$$

where $L : K_1 \times K_2 \to \mathbb{R}$ is a real-valued bifunction.

If we set $K = K_1 \times K_2$ and define a bifunction $F : K \times K \to \mathbb{R}$ by

$$F((x_1, x_2), (y_1, y_2)) = L(y_1, x_2) - L(x_1, y_2) \text{ for all } (x_1, x_2), (y_1, y_2) \in K_1 \times K_2,$$

then the problem (SPP) coincides with the problem (EP).

Let I be a finite set (a set of players) and, for each $i \in I$, K_i be a strategy set of ith player. Let $K = \sum_{i \in I} K_i$ and, for each player $i \in I$, $\varphi_i : K \to \mathbb{R} =$ loss function of the ith player depending on the strategies of all players. For any $x = (x_1, x_2, \ldots, x_n) \in K$, define $x^i = (x_1, x_2, \ldots, x_{i-1}, x_{i+1}, \ldots, x_n)$.

(III) Nash equilibrium problem. Nash equilibrium problem (shortly, (NEP)) is the following:

Find $\bar{x} \in K$ such that, for each $i \in I$,

$$\varphi_i(\bar{x}) \leqslant \varphi_i(\bar{x}^i, y_i) \text{ for all } y_i \in K_i.$$

If we define $f(x, y) = \sum_{i=1}^{n} (\varphi_i(x^i, y_i) - \varphi_i(x))$, then the problem (NEP) is same as the problem (EP).

Let K be a nonempty subset of an inner product space X and $\varphi : K \to K$ be a mapping.

(IV) Fixed point problem. Fixed point problem (shortly, (FPP)) is the following:

Find $\bar{x} \in K$ such that

$$\varphi(\bar{x}) = \bar{x}.$$

If set $F(x, y) = \langle x - \varphi(x), y - x_i \rangle$ for all $x, y \in K$, then $\bar{x}$ is a solution of the problem (FPP) if and only if $\bar{x}$ is a solution of the problem (EP).

Let X be a normed vector space with the dual space X^*, K be nonempty closed convex cone in X and

$$K^* = \{x^* \in X^* : \langle x^*, y_i \rangle \geqslant 0, \ \forall y \in K\}$$

be a polar cone of K. Let $\varphi : K \to X^*$ be a mapping.

(V) Complementarity problem. Let $K = \mathbb{R}_+^n$ and $f : K \to \mathbb{R}^n$ be a mapping. Complementarity problem (shortly, (CP)) is the following:
Find $\bar{x} \in K$ such that

$$f(\bar{x}) \in K, \ \langle f(\bar{x}), \bar{x} \rangle = 0.$$

Let K be a closed convex cone in $\mathbb{R}^n$ and $f : K \to \mathbb{R}^n$ be a mapping. The nonlinear complementarity problem (NCP) is the following:
Find $\bar{x} \in K$ such that

$$f(\bar{x}) \in K^*, \ \langle f(\bar{x}), \bar{x} \rangle = 0,$$

where $K^* = \{y \in \mathbb{R}^n : \langle y, x \rangle \geqslant 0 \text{ for all } x \in K\}$ is the dual cone of K.

Geometrically, the nonlinear complementarity problem is to find a nonnegative vector $\bar{x}$ such that its image $f(\bar{x})$ is also nonnegative and is orthogonal to $\bar{x}$.

(1) The problem (VIP) is equivalent to the problem (NCP).
(2) If set $F(x, y) = \langle \varphi(x), y - x \rangle$ for all $x, y \in K$, then the problems (VIP), (NCP) and (EP) are equivalent.

Let $\mathcal{H}$ be a real Hilbert space with the inner product $\langle \cdot, \cdot \rangle$ and the norm $\| \cdot \|$ and K be a nonempty closed convex subset of $\mathcal{H}$. Let $\varphi : K \to \mathbb{R}$ be a real-valued function, $f : \mathcal{H} \to \mathcal{H}$ be a nonlinear mapping and $F : K \times K \to \mathbb{R}$ be a bifunction.

(VI) Generalized mixed equilibrium problem. Generalized mixed equilibrium problem (shortly, (GMEP)) is the following:
Find $x \in K$ such that

$$F(x, y) + \varphi(y) - \varphi(x) + \langle fx, y - x \rangle \geqslant 0 \text{ for all } y \in K,$$

which was studied by Peng and Yao in 2008.

Special Cases of (GMEP):

(1) If $f = 0$, then the (GMEP) becomes the following mixed equilibrium problem (shortly, (MEP)):
Find $x \in K$ such that

$$F(x, y) + \varphi(y) - \varphi(x) \geqslant 0 \text{ for all } y \in K,$$

which was studied by Ceng and Yao [134] in 2008.

(2) If $\varphi = 0$, then the (GMEP) becomes the following generalized equilibrium problem (shortly, (GEP)):
Find $x \in K$ such that

$$F(x, y) + \langle fx, y - x \rangle \geqslant 0 \text{ for all } y \in K,$$

which was studied by Takahashi and Takahashi [586] in 2008.

(3) If $\varphi = 0$ and $f = 0$, then the (GMEP) becomes the following equilibrium problem (shortly, (EP)):
Find $x \in K$ such that
$$F(x, y) \geqslant 0 \text{ for all } y \in K.$$

(4) If $F(x, y) = 0$ for all $x, y \in K$, then the (GMEP) become the following generalized variational inequality problem (shortly, (GVIP)):
Find $x \in K$ such that

$$\varphi(y) - \varphi(x) + \langle fx, y - x \rangle \geqslant 0 \text{ for all } y \in K.$$

(5) If $\varphi = 0$ and $F(x, y) = 0$ for all $x, y \in K$, then the (GMEP) becomes the following variational inequality problem (shortly, (VIP)):
Find $x \in K$ such that

$$\langle fx, y - x \rangle \geqslant 0 \text{ for all } y \in K.$$

(6) If $f = 0$ and $F(x, y) = 0$ for all $x, y \in K$, then the (GMEP) becomes the following variational minimization problem (shortly, (MP)):
Find $x \in K$ such that

$$\varphi(y) - \varphi(x) \geqslant 0, \text{ i.e. } \varphi(y) \geqslant \varphi(x) \text{ for all } y \in K.$$

Note that $\operatorname{argmin}_{y \in K} f(y) = \{x \in K : f(x) \leqslant f(y)\}$ is the set of the (MP).

Here "argmin" means "the minimizer of". For example, $x^* = \operatorname{argmin}\{f(x) : x \in K\}$ means x^* being the minimizer of the function $f(x)$ on K; i.e., x^* is the solution to the problem $\min\{f(x) : x \in K\}$.

7.1.2 Variational Inequalities and Optimization Problem

We first review some definitions: Given a continuous function $f : \mathbb{R}^n \to \mathbb{R}^n$ and a closed subset K in $\mathbb{R}^n$, the well-known finite dimensional variational inequality problem [148, 231, 244, 566], denoted by $VI(K, f)$, is to find an element $\bar{x} \in K$

such that

$$\langle f(\bar{x}), x - \bar{x} \rangle \geq 0 \quad \text{for all} \quad x \in K, \tag{7.1}$$

where $\langle \cdot, \cdot \rangle$ denotes the usual linear product in $\mathbb{R}^n$.

Geometrically, the variational inequality (7.1) states that the vector $f(\bar{x})$ must be at a nonobtuse angle with all the feasible vectors emanating from $\bar{x}$. In other words, the vector $\bar{x}$ is a solution of VI(K, f) if and only if $f(\bar{x})$ forms a nonobtuse angle with every vector of the form $x - \bar{x}$, for all $x \in K$.

Proposition 7.1 *Let* $f : \mathbb{R}^n \to \mathbb{R}^n$ *be a mapping. A vector* $\bar{x} \in \mathbb{R}^n$ *is a solution of* VI(K,f) *if and only if* $f(\bar{x}) = \mathbf{0}$.

Proof Let $f(\bar{x}) = \mathbf{0}$. Then, obviously, the inequality (7.1) holds with equality. Conversely, suppose that $\bar{x}$ satisfies the inequality (7.1). Then, by taking $y = \bar{x} - f(\bar{x})$ in (7.1), we get

$$\langle f(\bar{x}), \bar{x} - f(\bar{x}) - \bar{x} \rangle = \langle f(\bar{x}), -f(\bar{x}) \rangle \geqslant 0,$$

that is $-\|f(\bar{x})\|^2 \geqslant 0$, which implies that $f(\bar{x}) = \mathbf{0}$.

Notice that both unconstrained and constrained optimization problems can be formulated as a variational inequality problem. If $F(x)$ is the gradient of a differentiable function $f : \mathbb{R}^n \to \mathbb{R}$, then the following result provides a relationship between an optimization problem and a variational inequality problem.

Proposition 7.2 *Let* K *be a nonempty convex subset of* $\mathbb{R}^n$ *and* $f : K \to \mathbb{R}$ *be a differentiable function. If* $\bar{x}$ *is a solution of the following optimization problem*

$$\min_{x \in K} f(x), \tag{OP}$$

then $\bar{x}$ *is a solution of VIP with* $F = \nabla f$.

Proof For any $x \in K$, define a function $\varphi : [0, 1] \to \mathbb{R}$ by $\varphi(t) = f(\bar{x} + t(x - \bar{x}))$, for all $t \in [0, 1]$. Since $\varphi(t)$ attains its minimum at $t = 0$, therefore, $\varphi'(0) \geqslant 0$, that is

$$\langle \nabla f(\bar{x}), x - \bar{x} \rangle \geqslant 0, \quad \text{for all} \ x \in K. \tag{VIP}$$

Hence, $\bar{x}$ is a solution of VIP with $F \equiv \nabla f$.

7.1.3 Variational Inequalities and Fixed Point Problem

Let K be a nonempty subset of $\mathbb{R}^n$ and $T : K \to K$ be a mapping. The fixed point problem (FPP) is to find $\bar{x} \in K$ such that

$$T(\bar{x}) = \bar{x}. \tag{FPP}$$

The following result provides a relationship between a variational inequality and a fixed point problem.

Proposition 7.3 *Let K be a nonempty subset of $\mathbb{R}^n$ and $T : K \to K$ be a mapping. If the mapping $f : K \to K$ is defined by*

$$f(x) = x - T(x), \tag{7.1'}$$

then variational inequality problem (VIP) coincides with fixed point problem (FPP).

Proof Let $\bar{x} \in K$ be a fixed point of the problem (FPP). Then, $f(\bar{x}) = 0$, and thus, $\bar{x}$ solves (7.1) and (7.1').

Conversely, suppose that $\bar{x}$ solves (7.1) with $f(\bar{x}) = \bar{x} - T(\bar{x})$. Then, $T(\bar{x}) \in K$, and letting $x = T(\bar{x})$ in (7.1) gives $-\|\bar{x} - T(\bar{x})\|^2 \geqslant 0$, that is $\bar{x} = T(\bar{x})$.

7.2 Optimization Problem in Banach Spaces

Let X be a nonempty set. A function $f : X \to \mathbb{R} \cup \{+\infty\}$ is said to be proper if $f(x) \neq +\infty$ for all $x \in X$. The domain of f denoted by $\mathscr{D}(f)$ is defined by

$$\mathscr{D}(f) = \{x \in X : f(x) < +\infty\}.$$

Let X be a Banach space, K a nonempty subset of X, and let $f : X \to \mathbb{R} \cup \{+\infty\}$ a proper, lower semicontinuous function, which is bounded below. Then, it is well known that the following constrained optimization problem (in short, COP)

$$\inf_{x \in K} f(x) \tag{COP}$$

has a solution and that the solution set of COP is compact.

Problem. Let X be a Banach space, $K \subseteq X$ a nonempty, noncompact set, and let $f : X \to \mathbb{R} \cup \{+\infty\}$ a proper, lower semicontinuous function, which is bounded below. Then, the following optimization problem (in short, OP)

$$\inf_{x \in K} f(x) \tag{OP}$$

or of the COP without compactness need not have a solution.

Now, there arises a natural question—Can we achieve the infimum of the OP or of the COP without compactness? The answer is affirmative, but we need some kind of coercivity assumption as well as convexity structure on K.

However, we notice that one can always obtain an approximately ε-solution; i.e., for a given $\varepsilon > 0$ and a point $x_0 \in X$, we always have

$$\inf_X f \leqslant f(x_0) \leqslant \inf_X f + \varepsilon,$$

where $\inf_K f = \inf_{x \in X} f(x)$.

To see this, let us consider the case when $X = \mathbb{R}^n$. In such case, the situation can be remedied by considering a suitable small perturbation of f. For simplicity, let us assume that $K = \mathbb{R}^n$, $m = \inf_{x \in X} f(x)$, $\varepsilon > 0$. Suppose $x_0 \in X$ be such that

$$f(x_0) \leqslant m + \varepsilon.$$

Consider the function
$$f_\varepsilon(x) = f(x) + \varepsilon \|x - x_0\|_X.$$

Then, one can readily see that $f_\varepsilon : \mathbb{R}^n \to \mathbb{R} \cup \{+\infty\}$ is proper, lower semicontinuous, and in addition, f_ε is weakly coercive, i.e.

$$f_\varepsilon(x) \to +\infty \quad \text{as} \quad \|x\|_{\mathbb{R}^n} \to +\infty.$$

By invoking the Weierstrass theorem, we infer that f_ε attains its infimum at a point $y \in \mathbb{R}^n$. Note that
$$\|y - x_0\|_{\mathbb{R}^n} \leqslant 1.$$

Indeed, if $\|y - x_0\|_{\mathbb{R}^n} > 1$, then we have

$$f_\varepsilon(y) = f(y) + \varepsilon \|y - x_0\|_{\mathbb{R}^n} > f(y) + \varepsilon \geqslant m + \varepsilon$$
$$\geqslant f(x_0) = f_\varepsilon(x_0) > f_\varepsilon(y),$$

a contradiction. Also
$$f_\varepsilon(y) \leqslant f_\varepsilon(x) \quad \forall x \in \mathbb{R}^n,$$

hence
$$f(y) + \varepsilon \|y - x_0\|_{\mathbb{R}^n} \leqslant f(x) + \varepsilon \|y - x_0\|_{\mathbb{R}^n},$$

this implies that
$$f(y) \leqslant f(x) + \varepsilon \|y - x\|_{\mathbb{R}^n}.$$

Thus, keeping in view the above argument we infer that for a given $\varepsilon > 0$ and $x_0 \in \mathbb{R}^n$ satisfying
$$f(x_0) \leqslant \inf_{x \in X} f(x) + \varepsilon,$$

that is $x_0 \in \mathbb{R}^n$ is an ε-minimizer, we can find $y \in \mathbb{R}^n$, such that $\|y - x_0\|_{\mathbb{R}^n} \leqslant 1$ and the function
$$x \mapsto f(x) + \varepsilon \|y - x\|_{\mathbb{R}^n}$$

attains its infimum at $y \in \mathbb{R}^n$.

Notice that in the aforementioned analysis, we have used the Weierstrass theorem as an analytic tool, which guarantees a minimizer for a proper, lower semicontinuous function, which is bounded below. Remember that this is an essentially finite dimensional situation, but in an infinite dimensional space this need not work. However, this work with extra conditions, such as the reflexivity of X and the weak lower semicontinuity of the function. Notice that the above argument work, in general, if we can reformulate the above principle as follows: suppose X is an infinite-dimensional Banach space with norm $\| \cdot \|_X$. Suppose $x_0 \in X$ is an ε-minimizer of $f : X \to \mathbb{R} \cup \{+\infty\}$, which is proper, lower semicontinuous function, which is bounded below, then a small Lipschitz perturbation of f attains a strict minimum at a point $y \in X$, which is relatively close to x_0 in the sense that $\|y - x_0\| \leqslant 1$. This entails that we can find a Lipschitz continuous function $h_0 : X \to \mathbb{R}$ with a small Lipschitz constant, such that $f + h_0$ attains a strict minimum at $y \in X$. Notice also that the essence of the Ekeland variational principle and its extensions lies on the fact that this principle can be formulated in the setting of any complete metric space. As a consequence, this result becomes a viable, useful and effective tool to solve various problems that arise in different areas of nonlinear analysis.

7.2.1 Convex Optimization Problems

Throughout this section, we assume that $(X, \| \cdot \|)$ is a reflexive Banach space.

Definition 7.1 A functional $f : X \to \overline{\mathbb{R}} = \mathbb{R} \cup \{+\infty\}$ is said to be coercive, if

$$f(v) \to +\infty \quad \text{for } \|v\|_X \to +\infty.$$

Theorem 7.3 (Solvability of unconstrained minimization problems) *Suppose that* $f : X \to \mathbb{R} \cup \{+\infty\}$, $f \neq +\infty$ *is a weakly semicontinuous, coercive functional. Then, the unconstrained minimization problem*

$$f(u) = \inf_{v \in X} f(v) \tag{7.2}$$

admits a solution $u \in X$.

Proof Let $\alpha := \inf_{v \in X} f(v)$ and assume that $\{v_n\}_{n \in \mathbb{N}}$ is a minimizing sequence, i.e. $f(v_n) \to \alpha$ as $n \to \infty$. Since $\alpha < +\infty$ and in view of the coercivity of f, the sequence $\{v_n\}_{n \in \mathbb{N}}$ is bounded. Consequently, in view of Theorem 1.29, there exists a subsequence v_{n_i} and $u \in X$ such that $v_{n_i} \rightharpoonup u$ as $i \to \infty$. The weak lower semicontinuity of f implies

$$f(u) \leqslant \inf_{n \in \mathbb{N}} f(v_n) = \alpha,$$

whence $f(u) = \alpha$.

Theorem 7.4 (Existence and Uniqueness) *Suppose that $f : X \to \overline{\mathbb{R}}$ is a proper convex, lower semicontinuous, coercive functional. Then, the unconstrained minimization problem (7.2) has a solution $u \in X$. If f is strictly convex, then the solution is unique.*

Proof The existence follows from Theorem 7.3. For the proof of the uniqueness, let $u_1 \neq u_2$ be two different solutions. Then, there holds

$$f\left(\frac{1}{2}(u_1 + u_2)\right) < \frac{1}{2}f(u_1) + \frac{1}{2}f(u_2) = \inf_{v \in X} f(v),$$

which is a contradiction.

7.3 Ekeland Variational Principles

In this section, we present several forms of Ekeland's variational principle. We discuss equilibrium problem which is a unified model of several problems, namely minimization problem, saddle point problem, Nash equilibrium problem, fixed point problem, complementarity problem, variational inequality problems. The essence of Ekeland's variational principles lies on the fact that it guarantees the existence of ε-solution optimization problem where neither compactness nor convexity assumption on the underlying space is needed.

Theorem 7.5 (Ekeland's variational principle-strong form) *Let (X, d) be a complete metric space and $f : X \to \mathbb{R} \cup \{+\infty\}$ a proper, lower semicontinuous and bounded below function. Let $\varepsilon > 0$ and $x_0 \in X$ be given such that*

$$f(x_0) \leqslant \inf_{x \in X} f(x) + \varepsilon.$$

Then, for a given $\lambda > 0$, there exists $x_\lambda \in X$ such that

(a) $f(x_\lambda) \leqslant f(x_0)$;
(b) $d(x_\lambda, x_0) \leqslant \lambda$;
(c) $f(x_\lambda) < f(x) + \frac{\varepsilon}{\lambda}d(x, x_\lambda)$ for all $x \in X \setminus \{x_\lambda\}$.

Proof For the sake of convenience, let us set $d_\lambda(\cdot, \cdot) = d(\cdot, \cdot)$. Then, we see that d_λ is equivalent to d and (X, d_λ) is complete.

On X, we define a relation $\preceq$ by

$$x \preceq y \text{ if and only if } f(x) \leqslant f(y) - \varepsilon d_\lambda(x, y).$$

Evidently, this order is (i) reflexive, i.e. for all $x \in X, x \preceq x$; (ii) antisymmetric, i.e. for all $x, y \in X, x \preceq y$ and $y \preceq x$ imply $x = y$; (iii) transitive, i.e. for all

$x, y, z \in X, x \preceq y$ and $y \preceq z$ imply $x \preceq z$. So we conclude that the relation $\preceq$ is a partial order.

Inductively, we define a sequence $\{S_n\}_{n \geqslant 1}$ of subsets of X as follows. Let us start with $x_1 = x_0$ (x_0 is the same as given in the statement of the theorem) and define

$$S_1 = \{x \in X : x \preceq x_1\}; \quad x_2 \in S_1 \text{ such that } f(x_2) \leqslant \inf_{S_1} + \frac{\varepsilon}{2^3},$$

$$S_2 = \{x \in X : x \preceq x_2\}; \quad x_3 \in S_2 \text{ such that } f(x_3) \leqslant \inf_{S_2} + \frac{\varepsilon}{2^4},$$

and inductively

$$S_n = \{x \in X : x \preceq x_n\}; \quad x_{n+1} \in S_n \text{ such that } f(x_{n+1}) \leqslant \inf_{S_n} + \frac{\varepsilon}{2^{n+2}}.$$

Since $x_{n+1} \preceq x_n$, we have that $S_n \supseteq S_{n+1}$ for $n \geqslant 1$. claim that S_n is closed.

Indeed, if $\{u_k\} \subset S_n$ with $u_k \to u \in X$ as $k \to \infty$. Then, $u \preceq x_n$ and so $f(u_k) \leqslant f(x_n) - \varepsilon d_\lambda(u_k, x_n)$. Now taking limit and using the lower semicontinuity of f, continuity of d and so the continuity of d_λ, we conclude that $u \in S_n$. This shows that S_n is closed.

Now, we show that the diameter of these sets S_n is diam $(S_n) \to 0$ as $n \to \infty$. If $z \in S_n$, we have that $z \preceq x_n$. Further, we observe that $z \in S_{n-1}$ and so

$$\varepsilon d_\lambda(z, x_n) \leqslant f(x_n) - f(z) \leqslant \inf_{x \in S_{n-1}} f(x) + \frac{\varepsilon}{2^{n+1}} - f(z)$$

$$\leqslant f(z) + \frac{\varepsilon}{2^{n+1}} - f(z) = \frac{\varepsilon}{2^{n+1}} \text{ for all } z \in S_n,$$

which gives

$$d_\lambda(z, x_n) \leqslant \frac{1}{2^{n+1}} \text{ for all } z \in S_n.$$

Thus, for all $z, z' \in S_n$, we have

$$d_\lambda(z, z') \leqslant d_\lambda(z, x_n) + d_\lambda(x_n, z') \leqslant \frac{1}{2^{n+1}} + \frac{1}{2^{n+1}} = \frac{1}{2^n},$$

which gives diam $S_n = \sup\{d_\lambda(z, z') : z, z' \in S_n\} \leqslant \frac{1}{2^n}$ and hence diam $(S_n) \to 0$ as $n \to \infty$. Because (X, d_λ) is complete and $\{S_n\}$ is a decreasing sequence of closed sets, by Cantor's theorem, we infer that

$$\bigcap_{n=1}^{\infty} S_n = \{x_\lambda\}.$$

We still need to show that this unique point x_0 satisfies conditions (a)–(c).

Because $x_\lambda \in S_1$, we have

$$x_\lambda \preceq x_1 = x_0 \text{ if and only if } f(x_\lambda) \leqslant f(x_0) - \varepsilon d_\lambda(x_0, x_\lambda)$$

and so $f(x_\lambda) \leqslant f(x_0)$. Hence (a) holds.

Also, we have

$$d_\lambda(x_0, x_n) = d_\lambda(x_0, x_n) \leqslant \sum_{i=1}^{n-1} d_\lambda(x_i, x_{i+1}) \leqslant \sum_{i=1}^{n-1} 2^{-i}.$$

Taking limit as $n \to \infty$, we obtain $d_\lambda(x_0, x_\lambda) \leqslant 1$ and so $d(x_0, x_\lambda) \leqslant \lambda$, i.e. (b) holds.

Finally to prove (c), suppose $x \neq x_\lambda$. Then, we see that $x \preceq x_\lambda$ cannot be true because otherwise $x \preceq x_\lambda$ implies $x \preceq x_n$ for all $n \geqslant 1$, hence

$$x \in \bigcap_{n=1}^{\infty} S_n$$

which implies that $x = x_\lambda$. So $x \npreceq x_\lambda$, which means that

$$f(x) > f(x_\lambda) - \varepsilon d_\lambda(x, x_\lambda),$$

that is

$$f(x) > f(x_\lambda) - \frac{\varepsilon}{\lambda} d(x, x_\lambda),$$

and hence (c) holds.

Observation

- Strong form of Ekeland variational principle says that for given $\varepsilon, \lambda > 0$ and x_0 an ε-approximate solution of an optimization problem, there exists a new point x_λ that is not worst than x_0 belongs to a λ-neighbourhood of x_0, and especially, x_λ satisfies inequality (c).
- If $\lambda > 0$ is large, then (b) provides little information on the whereabouts of x_λ if,
  while (c) tells us that x_λ is close to being a global minimizer of $f(\cdot) + \frac{\varepsilon}{\lambda} d(\cdot, x_\lambda)$.
  The opposite situation occurs when $\lambda > 0$ is small. Then, (b) implies that x_λ is
  close to x_0, but the inequality in (c) gives us little information.

We can deduce from Theorem 7.5, so-called, the weak formulation of Ekeland's variational principle.

Corollary 7.1 (Ekeland's variational principle-weak form) *Let (X, d) be a complete metric space and $f : X \to \mathbb{R} \cup \{+\infty\}$ a proper, lower semicontinuous, bounded from below functional. Then, for every $\varepsilon > 0$, we can find $x_\varepsilon \in X$, such that*

(a) $f(x_\varepsilon) \leqslant \inf_{x \in X} f(x) + \varepsilon;$

(b) $f(x_\varepsilon) < f(x) + \varepsilon d(x, x_\varepsilon)$ *for all* $x \in X \setminus \{x_\varepsilon\}$.

Corollary 7.2 *Let* (X, d) *be a complete metric space and* $f : X \to \mathbb{R} \cup \{+\infty\}$ *a proper, lower semicontinuous function, which is bounded below,* $\varepsilon > 0$ *and* $\bar{x} = x(\varepsilon) \in X$ *satisfies*

$$f(\bar{x}) \leqslant \inf_{x \in X} f(x) + \varepsilon.$$

Then, we can find $x_\varepsilon \in X$, *such that*

(a) $f(x_\varepsilon) \leqslant f(\bar{x})$;
(b) $d(x_\varepsilon, \bar{x}) \leqslant \sqrt{\varepsilon}$;
(c) $f(x_\varepsilon) < f(x) + \sqrt{\varepsilon} d(x, x_\varepsilon)$ *for all* $x \in X \setminus \{x_\varepsilon\}$.

If we put more structure on the space X, we can strengthen the conclusion of Theorem 7.5.

Theorem 7.6 *Let* X *be a Banach space and* $f : X \to \mathbb{R}$ *a proper, lower semicontinuous function, which is bounded below which is Gâteaux differentiable. Then, we can find* $x_\varepsilon \in X$, *such that*

$$f(x_\varepsilon) \leqslant \inf_{x \in X} f(x) + \varepsilon \quad and \quad \|f'(x_\varepsilon)\|_{X^*} \leqslant \varepsilon.$$

Proof By virtue of Corollary 7.1, we can find $x_\varepsilon \in X$, such that

$$f(x_\varepsilon) \leqslant \inf_{x \in X} f(x) + \varepsilon \quad and \quad f(x_\varepsilon) < f(x) + \varepsilon \|x - x_\varepsilon\| \quad for \ all \ x \in X.$$

Let $h \in X$ and $t > 0$ be arbitrary. Set $x = x_\varepsilon + th$, then we obtain

$$\frac{f(x_\varepsilon) - f(x_\varepsilon + th)}{t} \leqslant \varepsilon \|h\|_X.$$

Passing to the limit as $t \to 0$, we obtain

$$-(f'(x_\varepsilon), h) \leqslant \varepsilon \|h\|_X \quad \forall h \in X,$$

so $|(f'(x_\varepsilon), h)| \leqslant \varepsilon \|h\|_X$ and thus $\|f'(x_\varepsilon)\|_{X^*} \leqslant \varepsilon$. This completes the proof.

Corollary 7.3 *Suppose* X *is a Banach space and* $f : X \to \mathbb{R}$ *is a lower semicontinuous function, which is bounded below which is Gâteaux differentiable, then there exists a sequence* $\{x_n\}_{n \in \mathbb{N}} \subseteq X$, *such that*

$$f(x_n) \searrow \inf_{x \in X} f(x) \quad and \quad f'(x_n) \to 0.$$

Remark 7.1 The above corollary asserts the existence of a minimizing sequence, whose elements satisfy the first-order necessary conditions, up to any desired approximation.

Corollary 7.4 *Suppose X is a Banach space and $f : X \to \mathbb{R}$ is a lower semicontinuous function, which is bounded below which is Gâteaux differentiable, then for each minimizing sequence $\{y_n\}_{n \in \mathbb{N}}$ of f (i.e. $f(y_n) \searrow \inf_{x \in X} f(x)$), we can find another minimizing sequence $\{x_n\}_{n \in \mathbb{N}}$ of f, such that:*

(a) $f(x_n) \leqslant f(y_n)$;
(b) $\|x_n - y_n\|_X \to 0$;
(c) $\|f'(x_n)\|_{X^*} \to 0$.

Aubin and Frankowska [26] established the following form of Ekeland's variational principle which is equivalent to Theorem 7.5.

Theorem 7.7 (Ekeland's variational principle-strong form) *Let (X, d) be a complete metric space and $f : X \to \mathbb{R} \cup \{+\infty\}$ a proper, lower semicontinuous and bounded below functional. Let $x_0 \in \mathscr{D}(f)$ and $\varepsilon > 0$ be fixed. Then, there exists $x_\varepsilon \in X$ such that*

(a) $f(x_\varepsilon) - f(x_0) + \varepsilon d(x_0, x_\varepsilon) \leqslant 0$;
(b) $f(x_\varepsilon) < f(x) + \varepsilon d(x, x_\varepsilon)$ *for all* $x \in X \setminus \{x_\varepsilon\}$.

Notice that the property of Ekeland's variational principle for proper but extended real-valued lower semicontinuous and bounded below functions on a metric space characterizes completeness of the metric space.

Theorem 7.8 (Converse of Ekeland's variational principle) *A metric space (X, d) is complete if for every function $f : X \to \mathbb{R} \cup \{+\infty\}$ which is proper, lower semicontinuous and bounded below functional on X and for every given $\varepsilon > 0$, there exists $x_\varepsilon \in X$ such that*

$$f(x_\varepsilon) \leqslant \inf_{x \in X} f(x) + \varepsilon$$

and

$$f(x_\varepsilon) \leqslant f(x) + \varepsilon d(x, x_\varepsilon) \text{ for all } x \in X.$$

7.3.1   Applications to Fixed Point Theorems

The Ekeland variational principle is a very powerful tool of nonlinear analysis. Below, we show how the well-known results in fixed point theory such as Banach contraction principle and Caristi's fixed point theorem can be derived from the Ekeland variational principle. Subsequently, we show that the two results, namely Ekeland's variational inequality and Caristi's fixed point theorem, are equivalent, in the sense that the Ekeland variational principle can also be derived from Caristi's fixed point theorem. First, we state and prove Banach contraction principle followed by Caristi's fixed point theorem.

Theorem 7.9 (Banach contraction principle) *Let X be a complete metric space and $T : X \to X$ be a contraction mapping. Then, T has a unique fixed point in X.*

Proof Define the functional f by

$$f(x) := d(x, T(x)) \text{ for all } x \in X.$$

Then, it readily follows that f is bounded below and continuous on X. Choose a positive number ε such that $0 < \varepsilon < 1 - \lambda$, where λ is the Lipschitz constant. By Theorem 7.1, there exists $x_\varepsilon \in X$, depending on ε such that

$$f(x_\varepsilon) \leqslant f(x) + \varepsilon d(x, x_\varepsilon) \text{ for all } x \in X.$$

Thus, we have

$$d(x_\varepsilon, T(x_\varepsilon)) \leqslant d(x, T(x)) + \varepsilon d(x, x_\varepsilon) \text{ for all } x \in X.$$

On putting $x = T(x_\varepsilon)$ in the above inequality, we obtain

$$\begin{aligned}
d(x_\varepsilon, T(x_\varepsilon)) &\leqslant d(T(x_\varepsilon), T(T(x_\varepsilon)) + \varepsilon d(x_\varepsilon, T(x_\varepsilon)) \\
&\leqslant \lambda d(x_\varepsilon, T(x_\varepsilon) + \varepsilon d(x_\varepsilon, T(x_\varepsilon)) \\
&= (\lambda + \varepsilon) \, d(T(x_\varepsilon), T(T(x_\varepsilon)) + \varepsilon d(x_\varepsilon, T(x_\varepsilon)).
\end{aligned}$$

Suppose $x_\varepsilon \neq T(x_\varepsilon)$, then the above inequality gives $1 \leqslant \lambda + \varepsilon$, which contradicts to our assumption that $\lambda + \varepsilon < 1$. Thus, we have $x_\varepsilon = T(x_\varepsilon)$. The uniqueness of x_ε can be proved as in Theorem 5.1.

Theorem 7.10 (Caristi's fixed point theorem) *Let X be a complete metric space and $F : X \to 2^X \setminus \varnothing$ be a multivalued mapping such that*

$$d(x, y) \leqslant \varphi(x) - \varphi(y) \text{ for some } y \in F(x) \text{ and all } x \in X, \tag{7.3}$$

where $\varphi : X \to \mathbb{R} \cup \{+\infty\}$ is a proper, lower semicontinuous and bounded below functional. Then, there exists $\bar{x} \in X$ such that $\bar{x} \in F(\bar{x})$ and $\varphi(\bar{x}) < \infty$.

Proof By virtue of Corollary 7.1 with $\varepsilon = 1$, we can find $\bar{x} \in X$ such that

$$\varphi(\bar{x}) < \varphi(x) + d(x, \bar{x}) \text{ for all } x \in X \setminus \{\bar{x}\}. \tag{7.4}$$

We now claim that $\bar{x} \in F(\bar{x})$. Suppose that this is not true. Then, for all $y \in F(\bar{x})$, we have that $y \neq \bar{x}$. So, let $y \in F(\bar{x})$ be as in (7.3). Then, (7.3) and (7.4) entail

$$d(y, \bar{x}) \leqslant \varphi(\bar{x}) - \varphi(y) \text{ and } \varphi(\bar{x}) - \varphi(y) < d(y, \bar{x})$$

which cannot hold simultaneously.

Remark 7.2 We emphasize that on the multivalued function F no regularity conditions were imposed except for (7.2), which is a mild restriction. Suppose that F has compact values and

$$H(F(x), F(y)) \leqslant kd(x, y) \ \forall x, y \in X \text{ and with } k \in (0, 1).$$

Here, $H(\cdot, \cdot)$ stands for the Hausdorff–Pompeu metric on the nonempty and closed subsets of X. Then, we can apply Theorem 7.10 with

$$\varphi(x) = \frac{1}{1 - k} d(x, F(x)).$$

Indeed, let $y \in F(x)$ be such that $d(x, F(x)) = d(x, y)$. Such an element exists since $F(x)$ is compact. Then, we have

$$\begin{aligned}(1 - k)d(x, y) &= d(x, F(x)) - kd(x, y) \\ &\leqslant d(x, F(x)) - H(F(x), F(y)) \\ &\leqslant d(x, F(x)) - d(y, F(y)),\end{aligned}$$

and so

$$\varphi(y) \leqslant \varphi(x) - d(x, y).$$

Thus, the condition (7.2) is satisfied. Note that the resulting fixed point theorem is a particular case of Nadler's fixed point theorem (see Theorem 5.74). Of course, if F is single valued, we recover the well-known Banach contraction principle (see Theorem 5.1). Notice also that Banach fixed point theorem contains much more information.

Park [442] obtained some equivalent formulations of Ekeland's variational principle as follows.

Theorem 7.11 *Let (X, d) be a complete metric space and $f : X \to \mathbb{R} \cup \{+\infty\}$ a proper, lower semicontinuous and bounded below functional. Let $\varepsilon > 0$ and $x_0 \in X$ be given such that*

$$f(x_0) \leqslant \inf_{x \in X} f(x) + \varepsilon.$$

Then, the following statements are equivalent:

(a) *For a given $\lambda > 0$, there exists $\bar{x} \in \overline{S} = \overline{S}_\lambda(x_0)$ such that $f(\bar{x}) \leqslant f(x_0)$ and*

$$f(\bar{x}) < f(x) + \frac{\varepsilon}{\lambda} d(x, \bar{x}) \text{ for all } x \in X \setminus \{\bar{x}\}.$$

(b) *Let $T : \overline{S} \to 2^X$ be a set-valued map satisfying the following:*

$$\forall x \in \overline{S} \setminus T(x), \ \exists y \in X \setminus \{x\} \text{ such that } f(y) \leqslant f(x) - \frac{\varepsilon}{\lambda} d(y, x).$$

Then, T has a fixed point $\bar{x} \in \overline{S}$, i.e. $\bar{x} \in T(\bar{x})$ such that $f(\bar{x}) \leqslant f(x_0)$.
(c) Let $\varphi : \overline{S} \to X$ be a single-valued map such that

$$f(\varphi(x)) \leqslant f(x) - \frac{\varepsilon}{\lambda} d(\varphi(x), x) \text{ for all } x \in \overline{S}.$$

Then, φ has a fixed point $\bar{x} \in \overline{S}$ such that $f(\bar{x}) \leqslant f(x_0)$.
(d) Let $T : \overline{S} \to 2^X$ be a set-valued map such that for all $\bar{X} \in \overline{S}, T(x) \neq \varnothing$, and for any $y \in T(x)$, we have

$$f(y) \leqslant f(x) - \frac{\varepsilon}{\lambda} d(y, x).$$

Then, T has a stationary point $\bar{x} \in \overline{S}$, i.e. $T(\bar{x}) = \{\bar{x}\}$ such that $f(\bar{x}) \leqslant f(x_0)$.

Proof $(a) \Rightarrow (b)$: By (a), there exists a point $\bar{x} \in \overline{S}$ such that

$$f(\bar{x}) < f(x) + \frac{\varepsilon}{\lambda} d(x, \bar{x}) \text{ for all } x \in X \setminus \{\bar{x}\}. \tag{7.5}$$

Suppose, if possible, $\bar{x} \notin T(\bar{x})$, then by assumption, there exists a $y \neq \bar{x}$ such that

$$f(y) \leqslant f(\bar{x}) - \frac{\varepsilon}{\lambda} d(y, \bar{x}).$$

By putting $y = x$ in the above inequality and simplifying, we get

$$f(\bar{x}) \geqslant f(x) + \frac{\varepsilon}{\lambda} d(x, \bar{x}) \text{ for all } x \neq \bar{x},$$

a contradiction of (7.5).
$(b) \Rightarrow (c)$: Let $T : \overline{S} \Rightarrow 2^X$ be defined as follows:

$$T(x) = \{\varphi(x)\} \text{ for all } x \in \overline{S}.$$

Assume that for any $x \in \overline{S}, x \neq \varphi(x)$. Otherwise, every point of $\overline{S}$ is a fixed point of φ. This entails $x \in \overline{S} \setminus T(x)$. Further, we see that $y = \varphi(x) \in X \setminus \{x\}$ satisfies

$$f(y) \leqslant f(x) - \frac{\varepsilon}{\lambda} d(y, x).$$

Then, from (b), there exists $\bar{x} \in T(\bar{x}) = \{\varphi(\bar{x})\}$ such that $f(\bar{x}) \leqslant f(x_0)$.
$(c) \Rightarrow (d)$: By (c), one can readily see that any single-valued map φ on $\{T(x) : x \in \overline{S}\}$ has a fixed point. We now show that T has a stationary point in $\overline{S}$. Suppose, on the contrary, that T has no stationary point in $\overline{S}$. Consider a map φ such that for any $x \in \overline{S}$,

$$x \in T(x) \Rightarrow \varphi(x) \in T(x) \setminus \{x\}.$$

Then, such φ cannot have a fixed point, which is a contradiction.
$(d) \Rightarrow (a)$: Define a set-valued map $T : \overline{S} \to 2^X$ by

$$T(x) = \{y \in X : f(y) \leqslant f(x) - \frac{\varepsilon}{\lambda}\, d(y, x).$$

Then, for all $x \in \overline{S}$, $T(x) \neq \varnothing$, and by (d), T has a stationary point $\bar{x}$ which satisfies condition (a).

7.3.2   Applications to Optimization

In this section, we present the following existence result established by Takahashi [585] for a solution of an optimization problem without compactness and convexity assumptions on the underlying space.

Theorem 7.12 (Takahashi's minimization theorem) *Let (X, d) be a complete metric space and $f : X \to \mathbb{R} \cup \{+\infty\}$ a proper, lower semicontinuous and bounded below functional. Suppose that, for each $x_0 \in X$ with $\inf\limits_{x \in X} f(x) < f(x_0)$, there exists $z \in X$ such that $z \neq x_0$ and*

$$f(z) + d(x_0, z) \leqslant f(x_0).$$

Then, there exists $\bar{x} \in X$ such that $f(\bar{x}) = \inf\limits_{x \in X} f(x)$; that is, $\bar{x}$ is a solution of optimization problem.

Proof Suppose on the contrary that $\inf\limits_{x \in X} f(x) < f(y)$ for all $y \in X$ and let $x_0 \in X$ with $f(x_0) < +\infty$. Let us define inductively a sequence $\{x_n\}$ in X, starting with $x_1 = x_0$, and suppose that $x_n \in X$ is known. Then, $x_{n+1} \in S_{n+1}$ such that

$$S_{n+1} = \{x \in X : f(x) \leqslant f(x_n) - d(x_n, x)\}$$

and

$$f(x_{n+1}) \leqslant \frac{1}{2}\Big\{ \inf_{x \in S_{n+1}} f(x) + f(x_n) \Big\}. \tag{7.6}$$

We claim that $\{x_n\}$ is a Cauchy sequence. Indeed, if $m > n$, then

$$d(x_n, x_m) \leqslant \sum_{i=1}^{m-1} d(x_i, x_{i+1})$$

$$\leqslant \sum_{i=1}^{m-1} [f(x_i) - f(x_{i+1})]$$

$$= f(x_n) - f(x_m). \tag{7.7}$$

Because $\{f(x_n)\}$ is a decreasing sequence and the function f is bounded below, there exists $\varepsilon > 0$ such that

$$f(x_n) - f(x_m) < \varepsilon \quad \text{for all } m > n.$$

Therefore, (7.7) yields
$$d(x_n, x_m) < \varepsilon \quad \text{for all } m > n$$

and hence $\{x_n\}$ is a Cauchy sequence in X. Because X is a complete metric space, there exists $\bar{x} \in X$ such that $x_n \to \bar{x}$. Now letting $m \to \infty$ in (7.7), the lower semicontinuity of f and continuity of d imply that

$$d(x_n, \bar{x}) \leqslant f(x_n) - \lim_{m \to \infty} f(x_m) \leqslant f(x_n) - f(\bar{x}). \tag{7.8}$$

Further, by hypothesis, there exists a $z \in X$ such that $z \neq \bar{x}$ and

$$f(z) + d(\bar{x}, z) \leqslant f(\bar{x}). \tag{7.9}$$

Now (7.5) and (7.9) entail

$$
\begin{aligned}
f(z) &\leqslant f(\bar{x}) - d(\bar{x}, z) \\
&\leqslant f(\bar{x}) - d(\bar{x}, z) + f(x_n) - f(\bar{x}) - d(x_n, \bar{x}) \\
&= f(x_n) - [d(x_n, \bar{x}) + d(\bar{x}, z)] \\
&\leqslant f(x_n) - d(x_n, z).
\end{aligned}
$$

It follows from the above inequality that $z \in S_{n+1}$ for all $n \in \mathbb{N}$. Now using (7.6), we obtain

$$2f(x_{n+1}) - f(x_n) \leqslant \inf_{x \in S_{n+1}} f(x) \leqslant f(z).$$

Thus, we obtain

$$f(\bar{x}) \leqslant \lim_{n \to \infty} f(x_n) \leqslant f(z) \leqslant f(\bar{x}) - d(\bar{x}, z) < f(\bar{x})$$

which is a contradiction. Hence, we conclude that there exists $\bar{x} \in X$ such that $f(\bar{x}) = \inf_{x \in X} f(x)$.

Remark 7.3 Takahashi's minimization theorem 7.12 and Ekeland's variational principle Theorem 7.7 are equivalent.

Proof We first prove Theorem 7.7 by using Theorem 7.12. Let

$$X_0 = \{x \in X : f(x) \leqslant f(x_0) - \varepsilon d_\lambda(x_0, x)\}.$$

Because $x_0 \in X_0$, $X_0 \neq \varnothing$. By lower semicontinuity of f and continuity of d_λ, X_0 is closed. Further, for each $x \in X_0$,

$$\varepsilon d_\lambda(x_0, x) \leqslant f(x_0) - f(x) \leqslant f(x_0) - \inf_{x \in X} f(x) \leqslant \varepsilon$$

and so $d_\lambda(x_0, x) \leqslant 1$. Thus, $d(x_0, x) \leqslant \lambda$. We also have $f(x) \leqslant f(x_0)$. Finally, we prove conclusion (c) of Theorem 7.7. Suppose, on the contrary, that for every $x \in X_0$, there exists $y \in X$ such that $y \neq x$ and $f(y) \leqslant f(x) - \varepsilon d_\lambda(y, x)$. Then we have

$$\begin{aligned}
\varepsilon d_\lambda(x_0, y) &\leqslant \varepsilon d_\lambda(x_0, x) + \varepsilon d_\lambda(x, y) \\
&\leqslant f(x_0) - f(x) + f(x) - f(y) \\
&= f(x_0) - f(y).
\end{aligned}$$

This shows that $y \in X_0$. Then, by Theorem 7.12, there exists $\bar{x} \in X$ such that $f(\bar{x}) = \inf_{x \in X_0} f(x)$. This contradicts our hypothesis that there exists $y_0 \in X_0$ with $f(y_0) < f(\bar{x})$.

Now, we prove Theorem 7.9 by using Theorem 7.3. By Theorem 7.3, for any given $\varepsilon > 0$ and $\lambda = 1$, there exists $\bar{x} \in X$ such that

$$f(\bar{x}) < f(x) + \varepsilon d(x, \bar{x}) \quad \text{for all } x \in X \text{ with } x \neq \bar{x}. \tag{7.10}$$

We claim that $f(\bar{x}) = \inf_{x \in X} f(x)$.

Suppose, on the contrary, that there exists $u \in X$ such that $f(u) > \inf_{x \in X} f(x)$. By hypothesis, there exists $v \in X$ such that $v \neq u$ and

$$f(v) + d(u, v) \leqslant f(u)$$

contradicting (7.10). Hence, we must have $f(\bar{x}) = \inf_{x \in X} f(x)$.

Remark 7.4 By using Theorem 7.12, Takahashi [585] also derived Caristi's fixed point theorem [129], Nadler's fixed point theorem for set-valued map [416] and Fan's minimization theorem [227].

7.4 Mappings Associated with Variational Inequality

In this section, we prove sufficient conditions for the monotonicity of normal-fixed point (in short, NFP) mappings associated with variational inequality problems over a general closed convex set. Sufficient conditions for the strong monotonicity of its perturbed version is also shown. These results include some well-known results in the literature. Inspired by these results, we propose an appreciable iterative algorithm for the variational inequality problem whose normal-fixed point map is monotone.

Notice that it is more convenient to express the inequality (7.1) as

$$(x - \bar{x})^T f(\bar{x}) \geq 0 \text{ for all } x \in K. \tag{7.11}$$

It is well known that the above problem can be reformulated as nonsmooth equations such as the fixed point and normal equations. The fixed point equation is defined by

$$\pi_\alpha(x) = x - P_K(x - \alpha f(x)) = 0 \tag{7.12}$$

and the normal equation is defined by

$$\Phi_\alpha(x) = f(P_K(x - \alpha f(x)) = 0, \tag{7.13}$$

where $\alpha > 0$ is a positive scalar and $P_K(\cdot)$ denotes the projection operator on the convex set K, i.e.

$$P_K(x) = \arg\min \{\|z - x\| : z \in K\}.$$

We now introduce the normal-fixed point (or briefly, NFP) equation as follows:

$$\begin{aligned} \Psi_\alpha(x) &= f(P_K((1 + \alpha)x - \alpha P_K(x - \alpha f(x)))) \\ &+ \alpha(P_K(x - \alpha f(x)) - P_K(x)) = 0. \end{aligned} \tag{7.14}$$

where $\alpha > 0$ is a positive scalar and $P_K(\cdot)$ denotes the Euclidean projection operator on the convex set K. We call Ψ_α the normal-fixed point (or, briefly, NFP) mapping.

Notice that the following is a well-known result about the projection operator $P_K(\cdot)$.

Lemma 7.1 *Let K be a nonempty closed convex set in $\mathbb{R}^n$. For any $z \in \mathbb{R}^n$, we have*

$$(x - P_K(z))^T (P_K(z) - z) \geq 0 \ \forall x \in K.$$

Proof If $z \in K$, then it must hold $z = P_K(z)$. As a consequence,

$$(x - P_K(z))^T (P_K(z) - z) = 0 \ \forall x \in K.$$

If $z \notin K$, then the direction $P_K(z) - z$ is perpendicular to the face of the convex set K. So, it will make an acute angle θ with the direction of $x - P_K(z)$ as shown in Fig. 7.1. Thus,

$$\begin{aligned} (x - P_K(z))^T (P_K(z) - z) &= \langle x - P_K(z), \ P_K(z) - z \rangle \\ &= \|x - P_K(z)\| \ \|P_K(z) - z\| \cos \theta \\ &\geq 0 \ \forall x \in K. \end{aligned}$$

This completes the proof.

Fig. 7.1 $P_K(z)$ is that point
of K for which
$-\frac{\pi}{2} \leqslant \theta \leqslant \frac{\pi}{2}$ for every x in
K

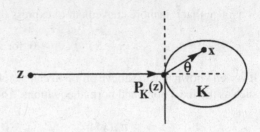

7.4.1   Normal Cone and Variational Inequalities

Let K be a nonempty subset of $\mathbb{R}^n$ and $F : K \to \mathbb{R}^n$ be a mapping. Then, one can easily see that the vector $\bar{x}$ is a solution of VIP if and only if

$$0 \in F(\bar{x}) + N_K(\bar{x}),$$

where $N_K(\bar{x})$ is the normal cone to K at $\bar{x}$ and is defined by

$$N_K(\bar{x}) = \begin{cases} \{z \in \mathbb{R}^n| \text{ for each } x \in K, \langle z, x - \bar{x} \rangle \leq 0\}, & \text{if } \bar{x} \in K, \\ \varnothing, & \text{if } \bar{x} \notin K. \end{cases} \qquad (7.15)$$

Geometrically, a vector $\bar{x}$ is a solution of VIP if and only if $-F(\bar{x}) \in N_K(\bar{x})$ as shown in the Fig. 7.2. Clearly, $\bar{x} \in K$ is a solution of VIP if and only if

$$0 \in F(\bar{x}) + N_K(\bar{x}). \qquad (7.16)$$

The inclusion (7.16) is called a generalized equation.

A very common problem arising in equilibrium analysis (for $\alpha > 0$) and optimization problems ($\alpha = 1$) is that of finding a point x such that

Fig. 7.2 The geometry of a
variational inequality

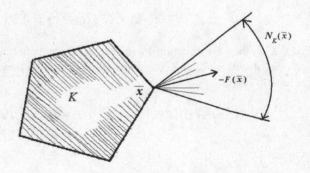

$$f(P_K((1+\alpha)x - \alpha P_K(x - \alpha f(x))))$$

$$+ \alpha(P_K(x - \alpha f(x)) - P_K(x)) = 0, \tag{7.17}$$

where α is a positive scalar.

Example 7.1 Let ψ be a C^1 function on an open subset Ω of $\mathbb{R}^n$ containing a point k_0 of K. Then, a necessary condition for k_0 to minimize ψ locally on K is that

$$0 \in d\psi(k_0) + N_K(k_0),$$

where $N_K(k_0)$ is the normal cone to K at k_0.

If k_0 satisfies (7.14) and if we put $f = \alpha \, d\psi$ and $x_0 = k_0 - d\psi(k_0)$, then we see that $P_K(x_0) = k_0$ and therefore k_0 satisfies (7.13).

In what follows, $\| \cdot \|$ denotes the 2-norm (Euclidean norm) of vector in $\mathbb{R}^n$. It is easy to observe that $\bar{x}$ solves $VI(K, f)$ if and only if $\pi_\alpha(\bar{x}) = 0$. Indeed, if $\bar{x}$ solves $VI(K, f)$, we have

$$(x - \bar{x})^T f(\bar{x}) \geq 0, \ \forall x \in K,$$

which implies that the direction $f(\bar{x})$ is perpendicular (at $\bar{x}$) to the face of the convex set K. Thus, for any given stepsize $\alpha > 0$, the projection of the point $y = \bar{x} - \alpha f(\bar{x})$ on K is $\bar{x}$; that is, $\bar{x} = P_K(y)$; i.e.,

$$\bar{x} = P_K(\bar{x} - \alpha f(\bar{x})).$$

Thus, $\pi_\alpha(\bar{x}) = 0$. So we assume that $\pi_\alpha(\bar{x}) = 0$; i.e.,

$$\bar{x} = P_K(\bar{x} - \alpha f(\bar{x})).$$

Let $y = \bar{x} - \alpha f(\bar{x}))$. Then, the above equality implies that the projection of y on the convex set K is equal to $\bar{x}$; i.e., $\bar{x} = P_K(y)$. Therefore, by Lemma 7.1, we have

$$(x - P_K(y))^T(P_K(y) - y) \geq 0 \ \forall x \in K.$$

Since $P_K(y) = \bar{x}$ and $y = \bar{x} - \alpha f(\bar{x})$, the above inequality reduces to

$$(x - \bar{x})^T(\bar{x} - (\bar{x} - \alpha f(\bar{x}))) \geq 0 \ \forall x \in K.$$

Since $\alpha > 0$, this inequality amounts to

$$(x - \bar{x})^T f(\bar{x}) \geq 0 \ \forall x \in K.$$

Thus, $\bar{x}$ is a solution of $VI(K, f)$.

Notice that if $\bar{x}$ solves $VI(K, f)$, then $\bar{x} - \frac{1}{\alpha} f(\bar{x})$ is a solution to $\Phi_\alpha(x) = 0$; conversely, if $\Phi_\alpha(\bar{u}) = 0$, then $P_K(\bar{u})$ is a solution to $VI(K, f)$. On the other hand, it turns out that if $\bar{x}$ solves $VI(K, f)$ and $\pi_\alpha(\bar{x}) = 0$, then $\bar{x} - \frac{1}{\alpha} f(\bar{x})$ is a solution to

$\Psi_\alpha(x) = 0$, and conversely, if $\pi_\alpha(\bar{u}) = 0$ and $\Psi_\alpha(\bar{u}) = 0$, then $P_K(\bar{u})$ is a solution to $VI(K, f)$.

Let $S(\pi_\alpha)$ denote the solution set of $\pi_\alpha(x)$. In view of above discussion and the definition of normal-fixed point equation, we are well in a position to unify partially the two equations, namely fixed point equation and normal equation as in the following proposition.

Proposition 7.4 *Suppose the solution set of normal-fixed point equation contains the solution set of fixed point equation, then the solution set of the fixed point equation is the solution set of normal equation.*

Proof Let $S(\Psi_\alpha) \supset S(\pi_\alpha) \neq \varnothing$, and let $\bar{u}$ be a zero solution of π_α. Then, the equation $\Psi_\alpha(\bar{u}) = 0$ reduces to $\Phi_\alpha(\bar{u}) = 0$; that is, $\bar{u}$ is also a zero solution of Φ_α.

Suppose the solution set of the normal-fixed point equation contains the solution set of normal equation. Now, there arises a natural question—Is the solution set of the normal equation also the solution set of the fixed point equation?

One may conjecture that the answer to this problem must be negative.

Definition 7.2 A function f is said to be a $P_0(P)$-function if for every pair (x, y) with $x \neq y$,

$$\max_{1 \leq i \leq n} (x - y)_i (f_i(x) - f_i(y)) \geq 0 \ (>). \tag{7.18}$$

That is, for any $x, y (x \neq y)$, there exists at least one index i such that

$$(x - y)_i (f_i(x) - f_i(y)) \geq 0 \ (>). \tag{7.18'}$$

It can easily be verified, by using the technique of Ravindran and Gowda (Appendix, p. 759 [500]) that our new class of function Ψ_α has P_0-property.

In the beginning of twenty-first century, several authors studied the P_0-property of fixed point and normal maps when K is a rectangular box in $\mathbb{R}^n$, i.e. the cartesian product of n one-dimensional intervals, for such a K, Ravindran and Gowda [500] (respectively, Gowda and Tawhid [255]) showed that $\pi_\alpha(x)$ (respectively $\Phi_\alpha(x)$) is a P_0-function if f is. Notice that the monotone maps are very important special cases of the class of P_0-function. It is worth considering the following problem:

Problem. When are the mappings Ψ_α monotone if K is a general closed convex set:

(i) Does monotonicity of f alone implies monotonicity of Ψ_α?
(ii) Does monotonicity of f and π_α implies monotonicity of Ψ_α?
(iii) Does monotonicity of f and Φ_α implies monotonicity of Ψ_α?
(iv) Does monotonicity of f, π_α and Φ_α implies monotonicity of Ψ_α?

Intuitively, we may conjecture that the normal-fixed point map is monotone if either the mapping(s) f or f and π or f and Φ_α or f, π_α and Φ_α are monotone. However, this conjecture is not true. The following exa+e shows that for a given $\alpha > 0$ the monotonicity of f in general does not imply the monotonicity of the normal-fixed point map $\Psi_\alpha(x)$.

Example 7.2 Let K be a closed convex set given by

$$K = \{x \in \mathbb{R}^3 : -\infty < x_1, x_2 < \infty, \ x_3 = 0\}$$

and

$$f(x) = \begin{pmatrix} 0 & -1 & 0 \\ 1 & 0 & 0 \\ 0 & 0 & 1 \end{pmatrix} \begin{pmatrix} x_1 \\ x_2 \\ x_3 \end{pmatrix} = \begin{pmatrix} -x_2 \\ x_1 \\ x_3 \end{pmatrix}.$$

For any $x, y \in \mathbb{R}^3$, we have that $(x - y)^T (f(x) - f(y)) = (x_3 - y_3)^2 \geq 0$. Hence, the function f is monotone on $\mathbb{R}^3$. We now show that for an arbitrary scalar $\alpha > 0$ the normal-fixed point mapping is not monotone in $\mathbb{R}^2$. Indeed, let $u = (0, 0, 0)^T$ and $y = (1, \frac{\alpha}{2}, \frac{\alpha^2}{3})^T$. It is easy to verify that $\Psi_\alpha(u) = (0, 0, 0)^T$ and $\Psi_\alpha(y) = (-\frac{\alpha}{2} - \alpha^2 + \frac{\alpha^3}{2}, 1 - \alpha^2 - \frac{\alpha^3}{2}, 0)^T$. Thus, we have

$$(u - y)^T (\Psi_\alpha(u) - \Psi_\alpha(y)) = -\alpha^2 - \frac{\alpha^4}{4} < 0,$$

which implies that $\Psi_\alpha(\cdot)$ is not monotone on $\mathbb{R}^3$.

The following example shows that for a given $\alpha > 0$ the monotonicity of mappings f and π_α, in general, does not imply the monotonicity of the normal-fixed point map Ψ_α.

Example 7.3 Let K be a closed convex set given by

$$K = \{x \in \mathbb{R}^2 : -\infty < x_1 < \infty, \ x_2 = 0\}$$

and

$$f(x) = \begin{pmatrix} 0 & -1 \\ 1 & 0 \end{pmatrix} \begin{pmatrix} x_1 \\ x_2 \end{pmatrix} = \begin{pmatrix} -x_2 \\ x_1 \end{pmatrix}.$$

For any $x, y \in \mathbb{R}^2$, we have that $(x - y)^T (f(x) - f(y)) = 0$. Hence, the function f is monotone on $\mathbb{R}^2$. We now show that for an arbitrary scalar $\alpha > 0$ the normal-fixed point mapping is not monotone in $\mathbb{R}^2$. Indeed, let $u = (0, 0)^T$ and $y = (-2\alpha^2, 1)^T$. It is easy to verify that $\Psi_\alpha(u) = (0, 0)^T$ and $\Psi_\alpha(y) = (\alpha^2, -3\alpha^2)^T$. Thus, we have

$$(u - y)^T (\Psi_\alpha(u) - \Psi_\alpha(y)) = -6\alpha^2 < 0,$$

which implies that $\Psi_\alpha(\cdot)$ is not monotone on $\mathbb{R}^2$. However, it is interesting to note that the fixed point mapping is monotone. It is easy to verify that $\pi_\alpha(u) = (0, 0)^T$ and $\pi_\alpha(y) = (-\alpha, 1)$. Thus, we have

$$(u - y)^T (\pi_\alpha(u) - \pi_\alpha(y)) = 2\alpha^3 + 1 > 0,$$

which implies that $\pi_\alpha(\cdot)$ is monotone on $\mathbb{R}^2$.

The following example shows that for a given $\alpha > 0$ the monotonicity of mappings f and Φ_α, in general, does not imply the monotonicity of the normal-fixed point map Ψ_α.

Example 7.4 Let K be a closed convex set given by

$$K = \{x \in \mathbb{R}^3 : -\infty < x_2, x_3 < \infty, \ x_1 = 0\}$$

and

$$f(x) = \begin{pmatrix} 1 & 0 & 0 \\ 0 & 0 & -1 \\ 0 & 1 & 0 \end{pmatrix} \begin{pmatrix} x_1 \\ x_2 \\ x_3 \end{pmatrix} = \begin{pmatrix} x_1 \\ -x_3 \\ x_2 \end{pmatrix}.$$

For any $x, y \in \mathbb{R}^3$, we have that $(x - y)^T(f(x) - f(y)) = (x_1 - y_1)^2 \geq 0$. Hence, the function f is monotone on $\mathbb{R}^3$. We now show that for an arbitrary scalar $\alpha > 0$ the normal-fixed point mapping is not monotone in $\mathbb{R}^3$. Indeed, let $u = (0, 0, 0)^T$ and $y = (\alpha^2, \alpha, 1)^T$. It is easy to verify that $\Psi_\alpha(u) = (0, 0, 0)^T$ and $\Psi_\alpha(y) = (0, -1 + \alpha^2 - \alpha^3, \alpha - \alpha^2 - \alpha^3)^T$. Thus, we have

$$(u - y)^T(\Psi_\alpha(u) - \Psi_\alpha(y)) = -\alpha^2 - \alpha^4 < 0,$$

which implies that $\Psi_\alpha(\cdot)$ is not monotone on $\mathbb{R}^3$. However, it is interesting to note that the normal mapping is monotone. It is easy to verify that $\Phi_\alpha(u) = (0, 0, 0)^T$ and $\Phi_\alpha(y) = (\alpha^3, -1, \alpha)$. Thus, we have

$$(u - y)^T(\Phi_\alpha(u) - \Phi_\alpha(y)) = \alpha^5 > 0,$$

which implies that $\Phi_\alpha(\cdot)$ is monotone on $\mathbb{R}^3$.

Our next example shows that monotonicity of mapping f, fixed point mapping π_α and normal mapping Φ_α does not necessarily imply the monotonocity of normal-fixed point mapping Ψ_α.

Example 7.5 Let K be a closed convex set given by

$$K = \{x \in \mathbb{R}^2 : -\infty < x_1 < \infty, \ x_2 = 0\}$$

and

$$f(x) = \begin{pmatrix} 0 & 1 \\ -1 & 0 \end{pmatrix} \begin{pmatrix} x_1 \\ x_2 \end{pmatrix} = \begin{pmatrix} x_2 \\ -x_1 \end{pmatrix}.$$

For any $x, y \in \mathbb{R}^2$, we have that $(x - y)^T (f(x) - f(y)) = 0$. Hence, the function f is monotone on $\mathbb{R}^2$. We now show that for an arbitrary scalar $\alpha > 0$ the normal-fixed point mapping is not monotone in $\mathbb{R}^2$. Indeed, let $u = (0, 0)^T$ and $y = (\frac{\alpha}{2}, 1)^T$. It is easy to verify that $\Psi_\alpha(u) = (0, 0)^T$ and $\Psi_\alpha(y) = (-\alpha^2, -\alpha^2 - \frac{\alpha}{2})^T$. Thus, we have

$$(u - y)^T (\Psi_\alpha(u) - \Psi_\alpha(y)) = -\frac{\alpha^3}{2} - \alpha^2 - \frac{\alpha}{2} < 0,$$

which implies that $\Psi_\alpha(\cdot)$ is not monotone on $\mathbb{R}^n$. However, it is interesting to note that the fixed point and normal mappings are monotone. It is easy to verify that $\pi_\alpha(u) = (0, 0)^T$ and $\pi_\alpha(y) = (\alpha, 1)$. Thus, we have

$$(u - y)^T (\pi_\alpha(u) - \pi_\alpha(y)) = \frac{\alpha^2}{2} + 1 > 0,$$

which implies that $\pi_\alpha(\cdot)$ is monotone on $\mathbb{R}^n$.

Again, we see that $\Phi_\alpha(u) = (0, 0)^T$ and $\Phi_\alpha(y) = (0, \frac{\alpha}{2})^T$. Thus, we have

$$(u - y)^T (\Phi_\alpha(u) - \Phi_\alpha(y)) = \frac{\alpha}{2} > 0,$$

which implies that $\Phi_\alpha(\cdot)$ is monotone on $\mathbb{R}^n$.

From the above examples, we conclude that a certain condition stronger than the monotonicity of f is required to guarantee the monotonicity of $\Phi_\alpha(x)$. One such condition is the so called (K, θ)-projective-coercivity condition. We call f to be (K, θ)-projective-coercive with modulus $\beta > 0$ on a set $S \subset \mathbb{R}^n$ containing a convex set K if for any $\theta > 0$ there exists a constant $\beta > 0$ such that

$$(x - y)^T ((P_K(w_{x,\theta})) - ((P_K(w_{y,\theta}))) \geq \theta \|x - y\|^2$$

and

$$(x - y)^T (f(P_K(w_{x,\theta})) - f((P_K(w_{y,\theta}))) \geq \beta[\|f(x) - f(y)\|^2 + \theta \|x - y\|]^2$$

for all $x, y \in S$ and $w_{x,\theta} = (1+\theta)x - \theta P_K(x - \theta f(x))$, $w_{y,\theta} = (1+\theta)y - \theta P_K(y - \theta f(y)) \in S$.

A special case of the (K, θ)-projective-coercive map is the θ-strongly projective-monotone map with modulus $c > 0$ on a set $S \subset \mathbb{R}^n$ if there exists a constant c such that $c > \theta > 0$,

$$(x - y)^T ((P_K(w_{x,\theta})) - ((P_K(w_{y,\theta}))) \geq \theta \|x - y\|^2$$

$$(x - y)^T (f(P_K(w_{x,\theta})) - f((P_K(w_{y,\theta}))) \geq c \|x - y\|^2 \quad \text{for all } x, y \in S.$$

Especially, when $S = K$ and $\theta = 0$, (K, θ)-projective-coercivity condition reduces to so-called cocoercivity condition on K. We recall that f is said to be cocoercive with modulus $\beta > 0$ on a set $S \subset \mathbb{R}^n$ if there exists a constant $\beta > 0$ such that

$$(x - y)^T (f(x) - f(y)) \geq \beta \|x - y\|^2 \text{ for all } x, y \in S,$$

where

$$\beta = \sup \left\{ \gamma > 0 : (x - y)^T (f(x) - f(y)) \geq \gamma \|f(x) - f(y)\|^2 \quad \text{for all } x, y \in S \right\}.$$

Evidently, such a scalar is unique and $0 < \beta < \infty$ provided that f is not a constant mapping.

To solve some variational inequality problems, Bruck [125], Zhu and Marcotte [624, 625] and many other researchers used cocoercivity condition. In the work of Gabay [238], we find that this condition is used implicitly. In [621], Zhao and Li also used the cocoercivity condition to study the strict feasibility of complementarity problems. In [625], Zhu and Marcotte studied that in an affine case the cocoercivity has a close relation to the property of positive semidefinite (psd)-plus matrices.

A special case of the cocoercive map is the strongly monotone and Lipschitzian map. We recall that a mapping f is said to be strongly monotone with modulus $c > 0$ on the set S if there is a scalar $c > 0$ such that

$$(x - y)^T (f(x) - f(y)) \geq c \|x - y\|^2 \text{ for all } x, y \in S.$$

It is evident that any cocoercive map on the set S must be monotone and Lipschitz continuous (with constant $L = \frac{1}{\beta}$), but not necessarily strongly monotone (for instance, the constant mapping) on the same set.

As a matter of fact, the aforementioned problem (P) is not completely unknown. This end, we observe the following:

(1) Gabay [238] showed, but did not explicitly state, that by using the cocoercivity condition implicitly and using properties of nonexpansive mappings, $\pi_\alpha(x)$ and $\Phi_{1/\alpha}(x)$ are monotone if the scalar α is chosen such that the map $I - \alpha f$ is nonexpansive.

(2) Gabay [238] and Sibony [555] showed that $\pi_\alpha(x)$ and $\Phi_{1/\alpha}(x)$ are monotone for the strongly monotone map f if the scalar α is chosen such that the map $I - \alpha f$ is contractive.

However, contrary to above observations, we notice that $\pi_\alpha(x)$ and $\Phi_\alpha(x)$ are still monotone (strongly monotone) even when α is chosen such that $I - \alpha f$ is not nonexpansive (contractive). To illustrate this fact, let $K = \mathbb{R}^n_+$ and $f(x) = x$. Then, we see that the function f is cocoercive with modulus $\beta = 1$. While $I - \alpha f$ is not nonexpansive for $\alpha > 2$, the map $\pi_\alpha(x)$ remains monotone. This necessitates to improve the result of Sibony [555] and Gabay [238]. In fact, we show that if f is cocoercive (strongly monotone and Lipschitz continuous, respectively), the monotonicity (strong monotonicity respectively) of the maps $\pi_\alpha(x)$ and $\Phi_\alpha(x)$ can be ensured when lies in a large interval in which the map $I - \alpha f$ may not be nonexpansive (contractive, respectively). The results derived in this section are not obtainable by the proof based on the nonexpansiveness and contractiveness of maps.

In what follows, we use the standard concept "nonexpansive" map and "contractive" map in the literature to mean a Lipschitzian map with constant $L = 1$ and $L < 1$, respectively.

The main purpose of this section is to introduce an application of the monotonicity of $\Psi_\alpha(x)$ (see, for instance, Sect. 7.4.2). This application is motivated by the globally convergent inexact Newton method for the system of monotone equations proposed by Solodov and Svaiter [565] and its modification proposed by Zhao and Li [621] (See also [565, 567, 569, 622]).

In [623], Zhao and Li observed that their modified algorithms do not require projection operations in the line-search step; however, their algorithms require the composite of f and the projection operation P_K to be nonexpansive. As a consequence, the computational cost is significantly reduced. Inspiring from the techniques of Zhao and Li [623], we propose a new modified Solodov and Svaiter method to solve the monotone equation $\Psi_\alpha(x) = 0$.

7.4.1.1   Monotonicity of $\Psi_\alpha(x)$

It is well known that if f is strongly monotone with modulus $c > 0$ and Lipschitz continuous with constant $L > 0$, then $I - \alpha f$ is contractive when $0 < \alpha < \frac{2c}{L^2}$ (see, for instance, Sibony [555] and Gabay [238]). Because P_K is nonexpansive, this in turn implies that $\pi_\alpha(x)$ and $\Phi_{1/\alpha}(x)$ are both strongly monotone for $0 < \alpha < \frac{2c}{L^2}$.

Notice also that if f is cocoercive with modulus > 0, then $I - \alpha f$ is nonexpansive for $0 < \alpha < 2$, and thus, we can easily verify that $\pi_\alpha(x)$ and $\Phi_{1/\alpha}(x)$ are monotone for $0 < \alpha < 2$ (see Gabay [238], Theorem 6.2 therein).

In 2001, Zhao and Li [623] proved an improved version of the above-mentioned results. They proved that

(i) when lies outside of the interval $(0, \frac{2c}{L^2})$, for instance, $\frac{2c}{L^2} \leq \alpha \leq \frac{4c}{L^2}$, $\pi_\alpha(x)$ and $\Phi_{1/\alpha}(x)$ are still strongly monotone although $I - \alpha f$, in this case, is not contractive, and

(ii) when lies outside of the interval $(0, 2\beta]$, for instance, $2\beta < \alpha \leq 4\beta$, $\pi_\alpha(x)$ and $\Phi_{1/\alpha}(x)$ remain monotone although $I - \alpha f$ is not nonexpansive.

This new result on monotonicity (strong monotonicity) of $\pi_\alpha(x)$ and $\Phi_{1/\alpha}(x)$ for $\alpha > 2\beta (\alpha \geq \frac{2c}{L^2})$ is not obtainable by using the nonexpansive (contractive) property of $I - \alpha f$. The reason is as follows: let f be cocoercive with modulus $\beta > 0$ on the set $S \subseteq \mathbb{R}^n$. We now verify that $I - \alpha f$ is nonexpansive on S if and only if $0 < \alpha \leq 2\beta$. It is sufficient to show that if $\alpha > 0$ is chosen such that $I - \alpha f$ is nonexpansive on S, then we must have $\alpha \leq 2\beta$. In fact, if $I - \alpha f$ is nonexpansive, then for any x, y in S we have

$$\|x - y\|^2 \geq \|(I - \alpha f)(x) - (i - \alpha f)(y)\|^2$$
$$= \|x - y\|^2 - 2\alpha(x - y)^T(f(x) - f(y)) + \alpha^2\|f(x) - f(y)\|^2,$$

which implies that

$$(x - y)^T(f(x) - f(y)) \geq \frac{\alpha}{2}\|f(x) - f(y)\|^2.$$

By the definition of β, we deduce that $\frac{\alpha}{2} \leq \beta$, the desired consequence. Similarly, let f be strongly monotone with modulus $c > 0$ and Lipschitz continuous with constant $L > 0$ on the set S, where

$$c = \sup\{\gamma > 0 : (x - y)^T(f(x) - f(y)) \geq \gamma\|f(x) - f(y)\|^2 \text{ for all } x, y \in S\}$$

and

$$L = \sup\{\gamma > 0 : \|f(x) - f(y)\| \leq \gamma\|x - y\|^2 \text{ for all } x, y \in S\}.$$

We can easily see that $0 < c < \infty$ and $L > 0$ provided that S is not a single-point set. It is also easy to show that $I - \alpha f$ is contractive if and only if $0 < \alpha < \frac{2c}{L^2}$.

Because the map $I - \alpha f$ is not contractive (nonexpansive, respectively) for $\alpha \geq \frac{2c}{L^2}$ ($\alpha > 2\beta$, respectively), our result established in this section follows directly from the proof of Sibony [555] and Gabay [238].

We now intend to improve the above-mentioned results. In fact, we prove sufficient conditions for the monotonicity of normal-fixed point maps associated with variational inequality problems over a general closed convex set. In the sequel, sufficient conditions for the strong monotonicity of its perturbed version are also shown. It may be observed that these results include some well-known results in the literature as particular cases. More precisely, Pathak [452] proposed an appreciable modified Zhao and Li's and Solodov and Svaiter's iterative algorithm for the variational inequality problem whose normal-fixed point map is monotone.

We also study the strong monotonicity of the perturbed normal-fixed point maps defined by

$$\Psi_{\alpha,\varepsilon}(x) = f(P_K((1 + \alpha)x - \alpha P_K(x - \alpha f(x))))$$
$$+ \varepsilon P_K((1 + \alpha)x - \alpha P_K(x - \alpha f(x)))$$
$$+ \alpha(P_K(x - \alpha f(x))) - P_K(x)).$$

Further, when $\alpha = 1$ and if f is a P_0-function and K is a rectangular set, then Gowda and Tawhid [255] showed that the perturbed mapping $\Psi_{\alpha,\varepsilon}(x)$ is a P_0-function. We now intend to show a sufficient condition for the strong monotonicity of $\Psi_{\alpha,\varepsilon}(x)$. The following lemma is crucial in the proof of Theorem 7.13 stated below.

Lemma 7.2 *(i) For all $x, y \in \mathbb{R}^n$,*

$$(x - y)^T (P_K(x) - P_K(y) \leq \|x - y\|]^2.$$

(ii) For any $\alpha > 0$, denote $u_{z,\alpha} = z - \alpha f(z)$ for all $z \in \mathbb{R}^n$. Then

$$\|u_{x,\alpha} - u_{w,\alpha}\|^2 \leq \|x - w\|^2 + \alpha^2 \|f(x) - f(w)\|^2.$$

(iii) For any $\alpha > 0$ and vector $b \in \mathbb{R}^n$, the following inequality holds for all $v \in \mathbb{R}^n$:

$$\alpha \|v\|^2 + v^T b \geq -\frac{\|b\|^2}{4\alpha}.$$

Proof (i) By Schwarz inequality and the nonexpansive property of projection operator, we have

$$
\begin{aligned}
(x - y)^T (P_K(x) - P_K(y) &\leq |(x - y)^T (P_K(x) - P_K(y)| \\
&\leq \|x - y\| \|P_K(x) - P_K(y)\| \\
&\leq \|x - y\|^2 \text{ for all } x, y \in \mathbb{R}^n.
\end{aligned}
$$

(ii) By monotonicity of f, we have

$$
\begin{aligned}
\|u_{x,\alpha} - u_{w,\alpha}\|^2 &= \|x - \alpha f(x) - w - \alpha f(w)\|^2 \\
&= \|x - w\|^2 - \alpha(x - w)^T (f(x) - f(w)) + \alpha^2 \|f(x) - f(w)\|^2 \\
&\leq \|x - w\|^2 + \alpha^2 \|f(x) - f(w)\|^2
\end{aligned}
$$

for all $x, w \in \mathbb{R}^n$.

This proves the result (ii).

(iii) Given $\alpha > 0$ and $b, v \in \mathbb{R}^n$, it is easy to check that the minimum value of $\alpha \|v\|^2 + v^T b$ is $-\frac{\|b\|^2}{4\alpha}$. This proves the result (iii).

We now state and prove the main result of this section.

Theorem 7.13 (Pathak [452]) *Let K be an arbitrary closed convex set in $\mathbb{R}^n$ and $K \subseteq S \subseteq \mathbb{R}^n$. Let $f : \mathbb{R}^n \to \mathbb{R}^n$ be a function.*

(i) if f is (K, α)-projective-coercive with modulus $\beta > 0$ on a set $S \subseteq \mathbb{R}^n$ containing a convex set K, then for any fixed scalar α with $\beta > \max\{\alpha + 1 + \frac{1}{4\alpha}, \frac{\alpha^2}{4}\}$, the normal-fixed point map $\Psi_\alpha(x)$ given by (3) is monotone on the set S.

(ii) if f is α-strongly projective-coercive with modulus $c > 0$ on the set $S \subseteq \mathbb{R}^n$ and f is Lipschitz continuous with constant $L > 0$, then for any two fixed scalars α, ε satisfying $c > \alpha > 0$ and $\varepsilon > \alpha(1 + \frac{\alpha L^2}{4}) + \frac{1}{4\alpha}$, the perturbed map $\Psi_{\alpha,\varepsilon}(x)$ is strongly monotone on the set S.

(iii) *if f is (K, α)-projective-coercive with modulus $\beta > 0$ on the set $S \subseteq \mathbb{R}^n$, then for any fixed scalar α with $\beta > \max\{\alpha + 1 + \frac{1}{4\alpha}, \frac{\alpha^2}{4}\}$ and $\varepsilon > 0$ the perturbed map $\Psi_{\alpha,\varepsilon}(x)$ is monotone on the set S.*

Proof Let α, ε be given two scalars such that $\alpha > \varepsilon \geq 0$. For any vector x, y in S, using the notation of (3) and Lemma 7.1, we have

$$
\begin{aligned}
(x-y)^T &(\Psi_{\alpha,\varepsilon}(x) - \Psi_{\alpha,\varepsilon}(y)) + \alpha^2 \|x - y\|^2 \\
&= (x - y)^T [f(P_K(u_{x,\alpha})) + \varepsilon P_K(u_{x,\alpha} \\
&\quad + \alpha P_K(u_{x,\alpha}) - \alpha P_K(x) - f(P_K(u_{y,\alpha})) \\
&\quad - \varepsilon P_K(u_{y,\alpha}) - \alpha P_K(u_{y,\alpha}) - \alpha P_K(x)] + \alpha^2 \|x - y\|^2 \\
&\geq \varepsilon(x - y)^T (P_K(u_{x,\alpha}) - P_K(u_{y,\alpha})) \\
&\quad + (x - y)^T (f(P_K(u_{x,\alpha})) - f(\Psi_K(u_{y,\alpha}))) \\
&\quad - \frac{1}{4} \|P_K(u_{x,\alpha}) - P_K(u_{y,\alpha})\|^2 - \alpha \|x - y\|^2.
\end{aligned}
\tag{7.19}
$$

Let f be a (K, α)-projective-coercive with modulus $\beta > 0$ on a set $S \subseteq \mathbb{R}^n$ containing a convex set K. Then, the inequality (7.19) gives

$$
\begin{aligned}
(x - y)^T &(\Psi_{\alpha,\varepsilon}(x) - \Psi_{\alpha,\varepsilon}(y)) + \alpha^2 \|x - y\|^2 \\
&\geq \varepsilon\alpha \|x - y\|^2 + \beta[\|f(x) - f(y)\|^2 \\
&\quad + \alpha \|x - y\|^2] - \frac{1}{4} \|u_{x,\alpha} - u_{y,\alpha}\|^2 - \alpha \|x - y\|^2 \\
&\geq [(\beta + \varepsilon)\alpha - \alpha] \|x - y\|^2 + \beta \|f(x) - f(y)\|^2 \\
&\quad - \frac{1}{4} [\|x - y\|^2 + \alpha^2 \|f(x) - f(y)\|^2] \\
&= \left[(\beta + \varepsilon)\alpha - \alpha - \frac{1}{4} \right] \|x - y\|^2 + \left(\beta - \frac{\alpha^2}{4} \right) \|f(x) - f(y)\|^2.
\end{aligned}
\tag{7.20}
$$

Setting $\varepsilon = 0$ in the above inequality (7.20), we have

$$
\begin{aligned}
(x - y)^T &(\Psi_{\alpha,\varepsilon}(x) - \Psi_{\alpha,\varepsilon}(y)) \\
&\geq \left[\beta\alpha - \alpha^2 - \alpha - \frac{1}{4} \right] \|x - y\|^2\| + \left(\beta - \frac{\alpha^2}{4} \right) \|f(x) - f(y)\|^2.
\end{aligned}
$$

For $\beta > \max\{\alpha + 1 + \frac{1}{4\alpha}, \frac{\alpha^2}{4}\}$, the right-hand side is nonnegative, and hence, the map Ψ_α is monotone on the set S. This proves the result (i).

Let $\beta > \max\{\alpha + 1 + \frac{1}{4\alpha}, \frac{\alpha^2}{4}\}$ and $\varepsilon > 0$. Then, the inequality (7.20) can be further written as

$$(x - y)^T (\Psi_{\alpha,\varepsilon}(x) - \Psi_{\alpha,\varepsilon}(y))$$

$$\geq \left[(\beta + \varepsilon)\alpha - \alpha^2 - \alpha - \frac{1}{4} \right] \|x - y\|^2 + \left(\beta - \frac{\alpha^2}{4} \right) \|f(x) - f(y)\|^2.$$

Clearly, the right-hand side of the above inequality is nonnegative, and hence, the perturbed map $\Psi_{\alpha,\varepsilon}$ is monotone. Thus, the result (iii) is proved.

Let f be α-strongly projective-coercive with modulus $c > 0$ on a set $S \subseteq \mathbb{R}^n$ containing a convex set K, and let f be Lipschitz continuous with constant $L > 0$. Then, the inequality (7.20) gives

$$(x - y)^T (\Psi_{\alpha,\varepsilon}(x) - \Psi_{\alpha,\varepsilon}(y)) + \alpha^2 \|x - y\|^2$$

$$\geq (\varepsilon\alpha + c - \alpha)\|x - y\|^2 - \frac{1}{4}\left[\|x - y\|^2 + \alpha^2 \|f(x) - f(y)\|^2 \right]$$

$$= \left(\varepsilon\alpha + c - \alpha - \frac{1}{4} \right) \|x - y\|^2 - \frac{\alpha^2}{4} \|f(x) - f(y)\|^2$$

which yields

$$(x - y)^T (\Psi_{\alpha,\varepsilon}(x) - \Psi_{\alpha,\varepsilon}(y))$$

$$\geq (\varepsilon\alpha + c - \alpha^2 - \alpha)\|x - y\|^2 - \frac{1}{4}\left[\|x - y\|^2 + \alpha^2 \|f(x) - f(y)\|^2 \right]$$

$$\geq \left(\varepsilon\alpha + c - \alpha - \frac{1}{4} - \frac{\alpha^2 L^2}{4} \right) \|x - y\|^2.$$

Since $c > \alpha > 0$ and $\varepsilon > \alpha(1 + \frac{\alpha L^2}{4}) + \frac{1}{4\alpha}$, it follows that $\varepsilon\alpha + c - \alpha - \frac{1}{4} - \frac{\alpha^2 L^2}{4} > 0$. Hence, the perturbed map $\Psi_{\alpha,\varepsilon}$ is strongly monotone. This proves (ii).

7.4.2 *Application to Iterative Algorithm for* VI(K, f)

Because the map Ψ_α is monotone if the function f is (K, α)-projective-coercive with modulus $\beta > 0$ and α lies in a certain interval, we can solve the (K, α)-projective-coercive variational inequality problems via solving the system of monotone equation Ψ_α. Recently, Solodov and Svaiter [565] proposed a class of inexact Newton method for monotone equations as stated below. Let $F(x)$ be a monotone mapping from $\mathbb{R}^n$ into $\mathbb{R}^n$.

Algorithm SS (see [565]). Choose any $x^0 \in \mathbb{R}^n$, $t \in (0, 1)$, and $\lambda \in (0, 1)$. Set $k := 0$.

Inexact Newton step. Choose a psd matrix G_k. Choose $\mu_k > 0$ and $\gamma_k \in [0, 1)$. Compute $d^k \in R^n$ such that

$$0 = F(x^k) + (G_k + \mu_k I)d^k + e^k,$$

where $\|e^k\| \leq \gamma_k \mu_k \|d^k\|$. Stop if $d^k = 0$. Otherwise,

Line-search step. Find $y^k = x^k + \alpha_k d^k$, where $\alpha_k = t^{m_k}$ with m_k the smallest nonnegative integer such that

$$-F(x^k + t^k d^k)^T d^k \geq \lambda(1 - \gamma_k)\mu_k \|d^k\|^2.$$

Projection step. Compute

$$x^{k+1} = x^k - \frac{F(y^k)^T(x^k - y^k)}{\|F(y^k)\|^2} F(y^k).$$

Set $k := k + 1$, and repeat.

In [565], Solodov and Svaiter pointed out that the above inexact Newton step is motivated by the idea of the proximal point algorithm. Algorithm SS has an advantage over other Newton methods in that the whole iteration sequence is globally convergent to a solution of the system of equations, provided a solution exists, under no assumption on other than continuity and monotonicity.

Setting $F(x) = \Psi_\alpha$, from Theorem 7.13 above and following the technique of Theorem 2.1 in [565], we have the following result.

Theorem 7.14 *Let f be a (K, α)-projective-coercive map with modulus $\beta > 0$. Substitute $F(x)$ in Algorithm SS by $\Psi_\alpha(x)$, where $\beta > max\{\alpha + 1 + \frac{1}{4\alpha}, \frac{\alpha^2}{4}\}$. If μ_k is chosen such that $C_2 \geq \mu_k \geq C_1 \|\Psi_\alpha(x^k)\|$, where C_1 and C_2 are two constants, then Algorithm SS converges to a solution of the variational inequality provided that a solution exists.*

While Algorithm SS can be used to solve the monotone equation Ψ_α, each line-search step needs to compute the value of $\Psi_\alpha(x^k + \beta^m d^k)$, which represents a major cost of the algorithm in calculating projection operations. Hence, in general cases, Algorithm SS has high computational cost per iteration when applied to solve $\Psi_\alpha = 0$. To reduce this major computational burden, we propose the following algorithm which, of course, needs nonexpansiveness of $f \circ P_K$ and the evaluation of the function f in line-search steps.

Algorithm 7.1 Let K be any closed convex subset of $\mathbb{R}^n$. Assume that $f \circ P_K$ is nonexpansive. Choose $x^0 \in \mathbb{R}^n$, $t \in (0, 1)$ and $\gamma \in [0, 1]$. Set $k := 0$.

Inexact Newton step. Choose a positive semidefinite matrix G_k. Choose $\mu_k > 0$ and $\gamma_k \in [0, 1)$. Compute $d^k \in \mathbb{R}^n$ such that

$$0 = \Psi_\alpha(x^k) + (G_k + \mu_k I)d^k + e^k, \tag{7.21}$$

where $\|e^k\| \leq \gamma_k \mu_k \|d^k\|$. Stop if $d^k = 0$. Otherwise,

Line-search step. Find $y^k = x^k + s_k d^k$, where $s_k = t^{m_k}$ with m_k the smallest nonnegative integer and

$$\|f(x^k + t^k d^k) - f(x^k)\| \leq \frac{(1 - \gamma)\mu_k - 2(1 + 4\alpha)t^k}{4\alpha}\|d^k\|.$$

Projection step. Compute

$$x^{k+1} = x^k - \frac{\Psi_\alpha(y^k)^T(x^k - y^k)}{\|\Psi_\alpha(y^k)\|^2}\Psi_\alpha(y^k). \tag{7.22}$$

Set $k := k + 1$. Return.

The above algorithm has the following crucial property.

Lemma 7.3 *Let $\Psi_\alpha(x)$ be given as (7.14). At kth iteration, if m_k is the smallest nonnegative integer such that (7.22) holds, then $y^k = x^k + t^{m_k}d^k$ satisfies the following equation:*

$$-\pi_\alpha(x^k + t^{m_k}d^k)^T d^k \geq \frac{1}{2}(1 - \gamma)\mu_k\|d^k\|^2.$$

Proof By the definition of $\Psi_\alpha(x)$, the nonexpansiveness of the projection operator, and (7.22), we have

$$\begin{aligned}
\|\Psi_\alpha(x^k + t^{m_k}d^k) - \Psi_\alpha(x^k)\| &= \|f(P_K((1 + \alpha)(x^k + t^{m_k}d^k) \\
&\quad - \alpha P_K((x^k + t^{m_k}d^k) - \alpha f(x^k + t^{m_k}d^k))) \\
&\quad + \alpha(P_K((x^k + t^{m_k}d^k) - \alpha f(x^k + t^{m_k}d^k)) \\
&\quad - P_K((x^k + t^{m_k}d^k)) - f(P_K((1 + \alpha)(x^k) \\
&\quad - \alpha P_K((x^k) - \alpha f(x^k))) - \alpha(P_K((x^k) \\
&\quad - \alpha f(x^k)) - P_K(x^k))\| \\
&\leq (1 + 4\alpha)t^{m_k}\|d^k\| + 2\alpha\|f(x^k + t^{m_k}d^k) - f(x^k)\| \\
&\leq \frac{1}{2}(1 - \gamma)\mu_k\|d^k\|. \tag{7.23}
\end{aligned}$$

Also,

$$\begin{aligned}
-\pi_\alpha(x^k + t^{m_k}d^k)^T d^k &= -[\pi_\alpha(x^k + t^{m_k}d^k) - \pi_\alpha(x^k)]^T d^k - \pi_\alpha(x^k)^T d^k \\
&\geq -\|\pi_\alpha(x^k + t^{m_k}d^k) - \pi_\alpha(x^k)\| \|d^k\| - \pi_\alpha(x^k)^T d^k. \tag{7.24}
\end{aligned}$$

By (7.21) and positive semidefiniteness of G_K, we have

$$-\pi_\alpha(x^k)^T d^k = (d^k)^T(G_k + \mu_k I)d^k + (e^k)^T d^k$$
$$= \mu_k(d^k)^T d^k + (d^k)^T G_k d^k + (e^k)^T d^k + (e^k)^T d^k$$
$$\geq \mu_k \|d^k\|^2 - \gamma \mu_k \|d^k\|^2$$
$$= (1 - \gamma)\mu_k \|d^k\|^2. \tag{7.25}$$

Combining (7.23)–(7.25) yields

$$-\pi_\alpha(x^k + t^{m_k}d^k)^T d^k \geq \frac{1}{2}(1 - \gamma)\mu_k\|d^k\|^2.$$

This completes the proof.

Using Lemma 7.3 and following the line of the proof of Theorem 2.1 in [565], it is not difficult to prove the following convergence result.

Theorem 7.15 *Let $f : \mathbb{R}^n \to \mathbb{R}^n$ be a continuous function such that there exists a constant $\alpha > 0$ such that $\Psi_\alpha(x)$ defined by (7.14) is monotone. Choose G_k and k such that $\|G_k\| \leq C'$ and $\mu_k = C\|\Psi_\alpha(x)\|^p$, where C', C and p are three fixed positive numbers and $p \in (0, 1]$. Then, the sequence $\{x^k\}$ generated by Algorithm 7.1 converges to a solution of variational inequality provided that solution exists.*

Observation

- Algorithm 7.1 can solve the variational inequality whose fixed point mapping $\Psi_\alpha(x)$ is monotone for some $\alpha > 0$. Since the (K, α)-projective-coercive f implies the monotonicity of the functions $\Psi_\alpha(x)$ for suitable choices of the value of α, Algorithm 7.1 can locate a solution of any solvable (K, α)-projective-coercive variational inequality problem. This algorithm is an advantage over Algorithm SS in that it does not carry out any projection operation in the line-search step; however, we need the composite of f and projection operation to be nonexpansive, and hence, the computational cost is significantly reduced.
- In this section, we have shown some sufficient conditions for the monotonicity (strong monotonicity) of the normal-fixed point (NFP) mappings associated with the variational inequalities. This algorithm can be viewed as a modification of Solodov and Svaiter's and Zhao and Li's method with significantly reduced computational cost.

7.5   Variational Calculus

The calculus of variations has its origin in the generalization of the classical notion of maxima or minima of function of a single variable or more variables. The history of calculus of variations traced back to the year 1696 when Johann Bernoulli advanced the problem of the *brachistochrone* (see, Elsgolts [222]). Notice that several physical laws can be deduced from concise mathematical principles to the effect that a certain functional in a given process attains a maximum or minimum. In mechanics, we have the principle of least action, the principle of conservation of momentum and the principle of conservation of angular momentum. Likewise, we have Fermat's principle in optics and the principle of Castigliano in the theory of elasticity.

The aim of calculus of variations is to explore methods for finding the maximum or minimum of a functional defined over a class of functions. As a matter of fact, variable quantities called functionals play an important role in many problems arising in analysis, mechanics, geometry, etc.

Functional—By a functional, we mean a quantity whose values are determined by one or several functions. Thus, one might say that a functional is a kind of function, where the independent variable is itself a function (or curve).

The following are examples of functionals:

Example 7.6 Let $y(x)$ be an arbitrarily continuously differentiable function, defined on the interval $[a, b]$. Then, the formula

$$I[y] = \int_a^b y'^2(x)\, dx$$

defines a functional on the set of all such functions $y(x)$.

Example 7.7 Let $F(x, y, z)$ be a continuous function of three variables. Then, the expression

$$I[y] = \int_a^b F[x, y(x), y'(x)]\, dx, \qquad (7.26)$$

where $y(x)$ ranges over the set of all continuously differentiable functions defined on the interval $[a, b]$, which defines a functional. Some special cases of $F(x, y, z)$ are as folows:

  (*i*) if $F(x, y, z) = \sqrt{1 + z^2}$, then $I[y]$ is the length of the curve $y = y(x)$,
 (*ii*) if $F(x, y, z) = z^2$, then $I[y]$ reduces to the case considered in Example 7.6.

We now indicate some typical examples of variational problems, by which we mean problems involving the determination of maxima and minima of functionals.

1. Find the shortest plane curve joining two points A and B, i.e. find the curve $y = y(x)$ for which the functional

$$\int_a^b \sqrt{1 + y'^2}\, dx$$

achieves its minimum. The curve in question turns out to be the straight-line segment joining A and B.

2. Let A and B be two fixed points. Then, the time it takes a particle to slide under the influence of gravity along some path joining A and B depends on the choice of the

path (curve) and hence is a functional. The curve such that the particle takes the least time to go from A to B is called the brachistochrone. The brachistochrone problem was posed by John Bernoulli in 1696 and played an important part in the development of the calculus of variations. The problem was solved by John Bernoulli, James Bernoulli, Newton and L'Hospital. The brachistochrone turns out to be a cycloid, lying in the vertical plane and passing through A and B.

3. The following variational problem, called the isoperimetric problem, was solved by Euler: among all closed curves of a given length I, find the curve enclosing the greatest area. The required curve turns out to be a circle.

All of the above problems involve functionals which can be written in the form

$$\int_a^b F(x, y, y')dx.$$

Such functionals have a "localization property" consisting of the fact that if we divide the curve $y = y(x)$ into parts and calculate the value of the functional for each part, the sum of the values of the functional for the separate parts equals the value of the functional for the whole curve. It is just these functionals which are usually considered in the calculus of variations. As an example of a "nonlocal functional", consider the expression

$$\frac{\int_a^b x\sqrt{1 + y'^2}dx}{\int_a^b \sqrt{1 + y'^2}dx}$$

which gives the abscissa of the center of mass of a curve $y = y(x), a \leqslant x \leqslant b$, made out of some homogeneous material.

Definition 7.3 A functional $I[y(x)]$ attains maximum on a curve $y = y_0(x)$, if the value of I on any curve close to $y = y_0(x)$ does not exceed $I[y_0(x)]$. This means that

$$\Delta I = I[y(x)] - I[y_0(x)] \leqslant 0.$$

Further, if $\Delta I \leqslant 0$ and $\Delta I = 0$ only on $y = y_0(x)$, we say that the functional $I[y(x)]$ attains strict maximum on $y = y_0(x)$.

If $\Delta I = I[y(x)] - I[y_0(x)] \geqslant 0$ for all curves close to $y = y_0(x)$, then we say that the functional $I[y(x)]$ attains minimum on $y = y_0(x)$ and the strict minimum is defined in the same way.

The following result easily follows from Definition 7.3, so we omit the details.

Theorem 7.16 *If a functional $I[y(x)]$ attains a maximum or minimum on $y = y_0(x)$, where the domain of definition belongs to certain class, then at $y = y_0(x)$ we have*

$$\delta I = 0.$$

7.5.1 Euler–Lagrange Equation

Let us examine the extremum of the functional

$$I[y(x)] = \int_a^b F(x, y, y') \, dx \tag{7.27}$$

subject to boundary conditions $y(a) = y_1$ and $y(b) = y_2$, where y_1 and y_2 are prescribed at the fixed points a and b.

In this section, we derive the simplest and most basic result of the calculus of variations; we solve the first problem of the calculus of variations. The problem is: we seek a function $y(x)$ from among a maximally inclusive comparison set of continuous and twice differentiable but otherwise arbitrary functions $Y(x)$ connecting given endpoints, $(a, y(a))$ and $(b, y(b))$, that make a particular definite integral,

$$I[Y(x)] = \int_a^b F(x, Y(x), Y'(x)) \, dx \tag{7.28}$$

stationary. Here, the integrand $F(x, Y, Y')$ is itself a continuous and thrice differentiable function of x, Y, and Y'. Our notation underlines the distinction between the set of comparison functions $Y(x)$ and the particular member $y(x)$ which is actual or true.

Because the integral I is not a function of one or even a countably infinite number of discrete parameters, but is a function of a function, that is a functional or expression which assigns a number to a function, a new method is required for finding the extremizing function $y(x)$. That new method is the "calculus of variations".

To initiate the method, we parameterize $Y(x)$ with ε and carefully choose ε so that $\varepsilon = 0$ reduces the comparison function $Y(x)$ to the true function $y(x)$. Thus, by construction, $I(\varepsilon)$ realizes a stationary value when $\varepsilon = 0$, that is $I'(\varepsilon) = 0$ when $\varepsilon = 0$. We will exploit this property in determining the functional form of $y(x)$.

First, construct the comparison functions $Y(x)$ out of the supposed true or extremizing function $y(x)$ and another set of arbitrary functions $\eta(x)$ scaled by the parameter ε so that

$$Y(x) = y(x) + \varepsilon \eta(x). \tag{7.29}$$

Because we limit the comparison set to continuous, twice differentiable functions and $y(x)$ is a special member of that set, the functions $\eta(x)$ are also continuous and twice differentiable. Then, we can differentiate equation (7.29) and arrive at

$$Y'(x) = y'(x) + \varepsilon \eta'(x). \tag{7.30}$$

Again, since the endpoints of all possible $Y(x)$ are the same, that is $Y(a) = y(a) = y_1$ and $Y(b) = y(b) = y_2$,

$$\eta(a) = 0 \quad \text{and} \quad \eta(b) = 0. \tag{7.31}$$

Substituting expressions (7.29) and (7.30) for $Y(x)$ and $Y'(x)$ into the integral (7.28) yields

$$I(\varepsilon) = \int_a^b F(x, y(x) + \eta(x), y'(x) + \eta'(x)) \, dx. \tag{7.32}$$

This $I(\varepsilon)$ has the desired property: $\varepsilon = 0$ is a stationary value of $I(\varepsilon)$, that is $I'(\varepsilon) = 0$ when $\varepsilon = 0$, because $\varepsilon = 0$ renders $Y(x) = y(x)$ which by construction makes the integral I stationary. Next, we differentiate $I(\varepsilon)$ with respect to ε under the integral sign of Eq. (7.32) and find that

$$I'(\varepsilon) = \int_a^b \left[\left(\frac{\partial F}{\partial Y} \right) \eta(x) + \left(\frac{\partial F}{\partial Y'} \right) \eta'(x) \right] dx \tag{7.33}$$

where we have remembered that $Y = y + \varepsilon\eta$. Integrating the second term by parts yields

$$I'(\varepsilon) = \int_a^b \eta(x) \left[\left(\frac{\partial F}{\partial Y} - \frac{d}{dx} \left(\frac{\partial F}{\partial Y'} \right) \right) \right] dx$$
$$+ \left(\frac{\partial F}{\partial Y'} \right) \eta(x) \Big|_b - \left(\frac{\partial F}{\partial Y'} \right) \eta(x) \Big|_a. \tag{7.34}$$

Since $\eta(x)$ vanishes at the endpoints, the two surface terms, $\left(\frac{\partial F}{\partial Y'} \right) \eta(x) \Big|_b$ and $\left(\frac{\partial F}{\partial Y'} \right) \eta(x) \Big|_a$, vanish and (7.34) reduces to

$$I'(\varepsilon) = \int_a^b \eta(x) \left[\left(\frac{\partial F}{\partial Y} - \frac{d}{dx} \left(\frac{\partial F}{\partial Y'} \right) \right) \right] dx. \tag{7.35}$$

Finally, recall that $I(\varepsilon)$ was constructed so that $I'(\varepsilon) = 0$ when $\varepsilon = 0$ and also that $\varepsilon = 0$ collapses $Y(x)$ into $y(x)$. Therefore, setting $\varepsilon = 0$ changes equation (7.35) into

$$\int_a^b \eta(x) \left[\left(\frac{\partial F}{\partial y} - \frac{d}{dx} \left(\frac{\partial F}{\partial y'} \right) \right) \right] dx = 0. \tag{7.36}$$

Now, the $\eta(x)$ are quite arbitrary except for continuity, smoothness and vanishing endpoint conditions; otherwise, $\eta(x)$ may have many wiggles or none at all or it may vanish over part of its range and be very large in the rest. Integral (7.36) can vanish for each and every one of these diverse possibilities as required if and only if

$$\frac{\partial F}{\partial y} - \frac{d}{dx} \left(\frac{\partial F}{\partial y'} \right) = 0. \tag{7.37}$$

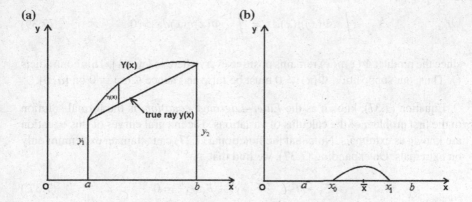

Fig. 7.3 a True ray $y(x)$, a possible ray from the comparison set $Y(x)$, and the difference function $\eta(x)$; **b** A continuous function which is positive in an interval but vanishes outside

If this were not the case and $\frac{\partial F}{\partial y} - \frac{d}{dx}\left(\frac{\partial F}{\partial y'}\right) \neq 0$ for some values of x within a subinterval of the interval $[x_1, x_2]$, then we could find a continuous, twice differentiable function $\eta(x)$ which vanished outside this subinterval but was positive wherever

$$\frac{\partial F}{\partial y} - \frac{d}{dx}\left(\frac{\partial F}{\partial y'}\right) > 0$$

and negative whenever

$$\frac{\partial F}{\partial y} - \frac{d}{dx}\left(\frac{\partial F}{\partial y'}\right) < 0,$$

thus contradicting the requirement (7.37). This argument establishes the fundamental lemma of the calculus of variations. More precisely, we have the following (Fig. 7.3a):

Lemma 7.4 (Fundamental Lemma) *Let* $\Phi(x)$ *be continuous in the closed interval* $[a, b]$. *If for every continuous function* $\eta(x)$,

$$\int_a^b \Phi(x)\eta(x)dx = 0, \tag{7.38}$$

then $\Phi(x) \equiv 0$ *on* $[a, b]$.

Proof Suppose, on the contrary, that $\Phi(x) \neq 0$ (positive, say) at a point $x = \bar{x}$ in $a \leqslant x \leqslant b$. By virtue of the continuity of Φ, it follows that $\Phi(x) \neq 0$ and maintain positive sign in a small neighborhood $x_0 \leqslant x \leqslant x_1$ of the point $\bar{x}$. Because η is an arbitrary continuous function, we might choose $\eta(x)$ such that $\eta(x)$ remains positive in $x_0 \leqslant x \leqslant x_1$, but vanishes outside this interval. It then follows from (7.38) that

$$\int_a^b \Phi(x)\eta(x)dx = \int_{x_0}^{x_1} \Phi(x)\eta(x)dx > 0 \qquad (7.39)$$

since the product $\Phi(x)\eta(x)$ remains positive everywhere in $[x_0, x_1]$. This contradicts (). Thus, our supposition $\Phi(x) \neq 0$ must be false and hence $\Phi(x) \equiv 0$ on $[a, b]$.

Equation (7.37), known as the *Euler–Lagrange equation*, is the formal solution to the first problem of the calculus of variations. The integral curves of this equation are known as extremals. Note that the functional (7.27) can attain an extremum only on extremals. On expanding (7.37), we find that

$$F_y - F_{xy'} - F_{yy'}y' - F_{y'y'}y'' = 0 \qquad (7.37')$$

which is, in general, a second-order differential equation in $y(x)$. The two arbitrary constants appearing in the solution $y(x)$ are determined from the boundary conditions $y(x_1) = y_1$ and $y(x_2) = y_2$.

7.5.2 *Variational Problems for Functionals*

In this section, we examine the extremum of the functional of the form

$$I[y(x)] = \int_a^b F(x, y_1(x), y_2(x), \ldots, y_n(x), y_1'(x), y_2'(x), \ldots, y_n'(x)) \, dx, \quad (7.40)$$

where the functional F is differentiable three times w.r. to all its arguments.

To find the necessary conditions for the extremum of the functional (7.40), we consider the following boundary conditions for $y_1(x), y_2(x), \ldots, y_n(x)$:

$$BCI : \quad y_1(a) = C_1, y_2(a) = C_2, \ldots, y_n(a) = C_n$$
$$BCII : \quad y_1(b) = C_1', y_2(b) = C_2', \ldots, y_n(a) = C_n'$$

where $C_1, C_2, \ldots, C_1', C_2', \ldots$ are constant.

Let us vary only one of the functions $y_i(x)$ $(i = 1, 2, \ldots, n)$ and keep the others fixed. Then, the functional I reduces to a functional dependent on, say, only one of the functions $y_i(x)$. Thus, the function $y_i(x)$ having a continuous derivative must satisfy Euler–Lagrange equation

$$F_{y_i} - \frac{d}{dx}F_{y_i'} = 0$$

where the boundary conditions on $y_i(x)$ at $x = a$ and $x = b$ are utilized from BC I and BC II.

Since this argument applies to any functions $y_i(x)$ ($i = 1, 2, \ldots, n$), we obtain a system of second-order differential equations:

$$F_{y_i} - \frac{d}{dx} F_{y_i'} = 0 \quad (i = 1, 2, \ldots, n). \tag{7.41}$$

These define, in general, a $2n$-parameter family of curves in the space and provide the family of extremals for the given variational problem.

7.5.3 Isoperimetric Problem

Suppose we seek to find the curve among all closed curves of a given length that encloses the greatest area. In such problems, it is necessary to a given integral

$$I = \int_a^b F(x, y, y') \, dx \tag{7.42}$$

maximum or minimum while keeping another integral

$$J = \int_a^b G(x, y, y') \, dx \tag{7.43}$$

constant, with one or more constraint, say for example $J = \alpha$(constant). Such problems are called *isoperimetric problems* and generally solved by the method of Lagrange's multiplier method. So, we take

$$I + \lambda J = \int_a^b (F + \lambda G)(x, y, y') \, dx = \int_a^b H(x, y, y') \, dx, \text{ say}$$

where $H = F + \lambda G$.

Now, the condition for $\int_a^b H(x, y, y') \, dx$ to be extremum is

$$\frac{\partial H}{\partial y} - \frac{d}{dx}\left(\frac{\partial H}{\partial y'}\right) = 0.$$

Next, we shall determine the values of constants of integration and λ by boundary conditions and the given integral J.

Example 7.8 Find the extremal of the functional $I = \int_0^\pi (y'^2 - y^2) \, dx$ under the boundary conditions $y(0) = 0$, $y(\pi) = 1$ and subject to constraint $\int_0^\pi y \, dx = 1$.

Solution. To maximize the integra $I = \int_0^\pi (y'^2 - y^2)\, dx$, we take $F = y'^2 - y^2$ and $G = y$; therefore, we write

$$H = F + \lambda G = y'^2 - y^2 + \lambda y.$$

Now, H must satisfy Euler–Lagrange equation $\frac{\partial H}{\partial y} - \frac{d}{dx}\left(\frac{\partial H}{\partial y'}\right) = 0$; therefore, we have

$$\lambda - 2y - \frac{d}{dx}(2y') = 0;$$

this entails

$$y'' + y = \frac{\lambda}{2}.$$

Solving this second-order linear differential equation and applying boundary conditions, we get

$$y = -\frac{1}{2}\cos x + C_2 \sin x + \frac{1}{2}.$$

Finally, the constraint condition $\int_0^\pi y\, dx = 1$ yields

$$y = \frac{1}{2}(1 - \cos x) + \frac{2 - \pi}{4}\sin x.$$

7.5.4 *Functionals Depending on Higher-Order Derivatives*

Let us now consider the extremum of a functional of the form

$$I[y(x)] = \int_a^b F(x, y(x), y'(x), \ldots, y^{(n)}(x))\, dx, \qquad (7.44)$$

where we assume F to be differentiable $n + 2$ times w.r.to all the arguments. The boundary conditions are taken in the form

$$y(a) = y_1, y'(a) = y_1', \ldots, y^{(n-1)}(a) = y_1^{(n-1)}, \qquad (7.45a)$$

$$y(b) = y_2, y'(b) = y_2', \ldots, y^{(n-1)}(b) = y_2^{(n-1)}. \qquad (7.45b)$$

Invoking the fundamental lemma of calculus of variations, the function $y = y(x)$ which extremizes I satisfies

$$\frac{\partial F}{\partial y} - \frac{d}{dx}F_{y'} + \frac{d^2}{dx^2}F_{y''} - \cdots + (-1)^n \frac{d^n}{dx^n}F_{y^{(n)}} = 0 \qquad (7.46)$$

which is known as the Euler–Poisson equation.

Clearly, this is a differential equation of the order $2n$, and hence, its solution involves $2n$ arbitrary constants. These are found by using the $2n$ boundary conditions (7.45a) and (7.45b).

7.5.5 Functionals Dependent on Functions of Several Independent Variables

In the preceding sections, we have considered Euler's equations in variational problems for determining extremals involves ordinary differential equations. We extend this to the problem of determining the extrema of functionals involving multiple integrals leading to one or more partial differential equations.

Let us consider the problem of finding an extremum of the functional

$$J[u(x, y)] = \iint_D F(x, y, u, u_x, u_y)\, dx\, dy \tag{7.47}$$

over a region of integration D by determining u which is continuous and has continuous derivatives up to the second order and takes on prescribed values on the boundary ∂D of D. Furthermore, we assume that F is thrice differentiable.

Invoking the fundamental lemma of calculus of variations, the function $u = u(x, y)$ which extremizes J satisfies

$$F_u - \frac{\partial}{\partial x} F_{u_x} - \frac{\partial}{\partial y} F_{u_y} = 0. \tag{7.48}$$

The extremizing function y is determined from the solution of the partial differential equation (7.48) which is known as *Euler–Ostrogradsky equation*.

Notice that if the integrand of a functional J contains derivatives of order higher than two, then by a straightforward extension of above analysis, we may derive a modified Euler–Ostrogradsky equation for determining functionals. For example, if the extremal functional is

$$J[u(x, y)] = \iint_D F(x, y, u, u_x, u_y, u_{xx}, u_{xy}, u_{yy})\, dx\, dy,$$

then we get the equation for extremals as

$$F_u - \frac{\partial}{\partial x} F_{u_x} - \frac{\partial}{\partial y} F_{u_y} + \frac{\partial^2}{\partial x^2} F_{u_{xx}} + \frac{\partial^2}{\partial x \partial y} F_{u_{xy}} + \frac{\partial^2}{\partial y^2} F_{u_{yy}} = 0.$$

7.5.6   *Variational Theory of Eigenvalue Problems*

In this section, we discuss applicability of variational theory to obtain bounds on the eigenvalues of a Stürm–Liouville problem. Assume that $P(x)$ and $Q(x)$ are two given functions, where Q is continuous and P is continuously differentiable on the interval $[a, b]$ and $P(x) > 0$ in $[a, b]$. Consider the quadratic functional

$$I[y(x)] = \int_a^b [P(x)y'^2 + Q(x)y^2]\, dx \qquad \text{(SL)}$$

subject to the boundary conditions

$$y(a) = 0, \quad y(b) = 0. \qquad \text{(BC)}$$

For the extremum of $I[y(x)]$, Euler's equation gives

$$[P(x)y'(x)]' - Q(x)y(x) = 0. \qquad \text{(E)}$$

To investigate the quadratic functional (SL) satisfying $P(x) > 0$, we consider the values of the functional (SL) for the functions $y(x)$ satisfying the constraint

$$\int_a^b w(x)(y(x))^2\, dx = 1, \qquad (7.49)$$

where $w(x)$, called weight function, is a continuous function satisfying the condition $w > 0$ in $[a, b]$. By (7.49), we can see easily that the values of $I[y]$ are bounded below. Indeed, we have

$$I[y] \geqslant \min_{x \in [a,b]} (w(x))^{-1} Q(x).$$

Suppose $y = y_1(x)$ is the function satisfying the boundary conditions (BC) and the normalizing condition (7.49) for which $I[y]$ attains the minimum value given above. Using the principle of Lagrange multipliers, it can be easily shown that $y_1(x)$ satisfies

$$[P(x)y_1'(x)]' - [Q(x) - \lambda_1 w(x)]y_1(x) = 0, \qquad (7.50)$$

where λ_1 is the Lagrange multiplier. We write the equation in the form

$$L[y_1] = \lambda_1 w(x)y_1, \qquad (7.51)$$

where the operator L is given by $L[y] = -[P(x)y'(x)]' + Q(x)y(x)$. Notice that L is a linear operator in the corresponding function space. Equation (7.51) shows that y_1 is an eigenfunction corresponding to the eigenvalue λ_1.

After having determined the function $y_1(x)$, we proceed to determine the function $y_2(x)$ which minimize the functional (SL) satisfying the boundary conditions and the constraints

$$\int_a^b wy^2\, dx = 1 \quad \text{and} \quad \int_a^b wyy_1\, dx = 0.$$

This function can again be found by the method of Lagrange multipliers to satisfy the equation $Ly_2 = Cwy_1 + \lambda_2 wy_2$. Now multiplying both sides of this equation by y_1 and integrating w.r.to x from a to b and using (7.51) and above constraints, we obtain $C = 0$. Thus, we find that

$$Ly_2 = \lambda_2 wy_2$$

showing, thereby, that $y_2(x)$ is an eigenfunction of the operator L with eigenvalue λ_2. Hence, we have an infinite sequence $y_1(x)$, $y_2(x)$, $y_3(x)$, ... of eigenfunctions of L which satisfy the given boundary conditions.

Observation

- The infinite sequence of eigenfunctions $y_1(x)$, $y_2(x)$, $y_3(x)$, ... are all mutually orthogonal w.r.to the weight function $w(x)$. All these functions satisfy the normalized condition (7.49). Moreover, it is convenient to interpret these functions as elements of Hilbert space $L_2[a, b]$ such that the operator L defined on the subspace of functions satisfying the boundary conditions (BC) is self-adjoint.
- $I[y_1] \leqslant I[y_2] \leqslant I[y_3] \leqslant \cdots$ and that the sequence $I[y_n]$ is infinitely large because if we impose an additional constraint the class of functions under consideration becomes narrower so that the minimum attained on this restricted class can only increase.

To illustrate the eigenvalue problem as discussed above, let us consider the functional

$$I[y] = \int_0^l y'^2\, dx$$

subject to the boundary conditions

$$y(0) = y(l) = 0$$

and the constraint (7.49) with $w(x) \equiv 1$. Then, (7.51) becomes $y'' + \lambda y = 0$, and the eigenvalues and the normalized eigenfunctions are given by

$$y_n(x) = \sqrt{\frac{2}{l}} \sin \frac{n\pi x}{l}, \quad \lambda_n = \left(\frac{n\pi}{l}\right)^2, \quad n = 1, 2, 3, \ldots.$$

Notice that the factor $\sqrt{\frac{2}{l}}$ is the normalization constant introduced to make the norms of the eigenfunctions equal to unity.

7.5.7 *Variational Principle of Least Action*

Suppose there is a system of n particles of masses $m_i (i = 1, 2, \ldots, n)$ located at $(x_i, y_i, z_i)(i = 1, 2, \ldots, n)$. Suppose the given system is conservative. According to Hamiltonian principle of least action for a system of n particles,

$$\int_{t_1}^{t_2} L \, dt = \int_{t_1}^{t_2} (T - U) \, dt$$

is an extremum for the system for fixed terminal times where the kinetic energy T is given by

$$T = \frac{1}{2} \sum_{i=1}^{n} m_i (\dot{x}_i^2 + \dot{y}_i^2 + \dot{z}_i^2)$$

and the potential energy U is independent of time t. The Lagrangian L is the difference between kinetic and potential energies, and the Hamiltonian H for this system is

$$H = \sum_{i=1}^{n} m_i (\dot{x}_i^2 + \dot{y}_i^2 + \dot{z}_i^2) - (T - U) = T + U,$$

which does not involve t explicitly and is the total energy of the system. Therefore, H remains constant throughout the motion. Hence, for a conservative dynamical system, we find that $T + U$ is constant along each extremal; that is, the total energy of a conservative system does not change during the motion of the system. Thus, for the given system of particles, Hamiltonian variational principle takes the form

$$\delta \int_{t_1}^{t_2} (T - U) \, dt = \delta \int_{t_1}^{t_2} [2T - (T - U)] \, dt = 0.$$

This simple form, owing to $\delta(T + U) = \delta(\text{constant}) = 0$, leads to

$$\delta \int_{t_1}^{t_2} 2T \, dt = 0, \tag{7.52}$$

which is referred to as the *variational principle of least action*.

7.6 Variational Methods

The term "variational method" refers to a large collection of optimization methods. In the classical sense, these methods involve finding the extremum of an integral depending on an unknown function and its derivatives. There are a number of quali-

tative features that are shared across variational formulation. The primary component
is an optimization problem. The problem of interest is either transformed into a variational problem or directly formulated as such based on a principle as in maximum
entropy estimation. The quantity to be optimized is typically an unknown function
which in simple cases may be reduced to a vector by taking function values at discrete points. The solution to variational problem is often in terms of fixed point
equations that capture necessary condition for optimality. Mean field equations and
Euler–Lagrange equations are prime examples of fixed point equations. However, we
notice that this classical definition and accompanying calculus of variation no longer
adequately characterize modern variational methods. Modern variational approaches
have become indispensable tools in various fields such as control theory, optimization, statistics, economics, as well as machine learning.

7.6.1 Rayleigh–Ritz Method

The essence of the Rayleigh–Ritz method lies on the fact that it tells us to forego
looking for an exact solution and to look instead for an approximate solution from
within a (well selected) finite set of functions. More explicitly, the basic idea of this
method is that the values of the functional $I[x]$ are taken not on arbitrary admissible
curves but on linear combinations of the form

$$y_n = \sum_{i=1}^{n} c_i f_i(x).$$

where c's are constants and $f_1(x), f_2(x), \ldots, f_n(x), \ldots$ constitute a suitable set of
functions. The above linear combination should be admissible in the given variational
problem. This implies certain restrictions on the above sequence. On such linear
combinations, $I[y(x)]$ reduces to a function $\psi(c_1, c_2, \ldots, c_n)$ of the coefficients
$c_1, c_2, \ldots, c_n$ that are determined from the system

$$\frac{\partial \psi}{\partial c_i} = 0 \quad (i = 1, 2, \ldots, n).$$

Passing to the limit as $n \to \infty$, if the limit exists, we get the solution of the variational
problem as

$$y = \sum_{i=1}^{\infty} c_i f_i(x).$$

If we do not take the limit, then we obtain an approximate solution of the variational
problem (see, Myskis [415]).

Notice that for the functions

$$y = \sum_{i=1}^{\infty} c_i f_i(x).$$

to be admissible, they should satisfy the boundary conditions of the variational problem apart from satisfying the requirements of continuity or smoothness. If the boundary conditions are linear and homogeneous, e.g. $y(x_1) = 0 = y(x_2)$, then the easiest way to choose $W_i(x)$ is as follows:

$$f_i(x) = (x - x_1)(x - x_2)\psi_i(x)$$

or some other functions satisfying

$$f_i(x_1) = f_i(x_2) = 0.$$

If the conditions are not homogeneous, e.g. $y(x_1) = y(x_2)$, where at least one of y_1 and y_2 are not zero, then we seek a solution of the form

$$y_n = \sum_{i=1}^{n} c_i f_i(x) + f_0(x), \tag{7.53}$$

where $f_i(x), i = 0, 1, 2, \ldots$ are known as coordinate functions. The boundary conditions are

$$f_0(x_1) = y_1, \quad f_0(x_2) = y_2 \quad f_i(x_1) = f_i(x_2) = 0 \tag{7.54}$$

for $i = 1, 2, \ldots, n$.

It is important to note that the collection of functions $f_i(x), i = 1, 2, \ldots, n$ mentioned before may be regarded as a part of an infinite sequence $\{f_n\}_{n=1}^{\infty}$ of functions f_n which are linearly independent and complete in the functions $y \in C^1[x_1, x_2]$, say, satisfying $y(x_1) = 0$, $y(x_2) = 0$. It may be readily shown the system of functions

$$f_i = x^{i-1}(x - x_1)(x - x_2), \quad i = 1, 2, \ldots \tag{7.55}$$

or

$$f_k(x) = \sin \frac{k\pi(x - x_1)}{x_2 - x_1}, \quad k = 1, 2, \ldots \tag{7.56}$$

is linearly independent and complete in $C^1[x_1, x_2]$.

Observation

1. If the Rayleigh–Ritz method is employed to determine the absolute minimum of a functional, then the approximate value of the minimum of the functional is obtained in excess, since the minimum of the functional on all admissible classes

of curves cannot exceed the minimum of the same functional on a subclass of the form

$$y_n = \sum_{i=1}^{n} c_i f_i(x).$$

2. Using the same argument, we may say that if the Rayleigh–Ritz method is used to obtain the maximum of the functional, then the approximate value of the maximum is obtained in defect.

Remark 7.5 The conditions for the convergence of the sequence mentioned above by the Rayleigh–Ritz method and the determination of the speed of convergence for some specific but frequently occurring functionals were worked out by N.M. Krylov and N. Bololyubov (see, Elsgolts [222]).

Notice that the estimates made by Krylov and Bololyubov are so complicated that they are impractical in concrete situations: for this reason, to test the accuracy of the results obtained by Rayleigh–Ritz method, we ordinarily use the following procedure (which is mathematically not rigorous but nevertheless sufficiently reliable). After calculating $y_n(x)$ and $y_{n+1}(x)$, their values are compared at several values of the interval $[x_1, x_2]$. If their values coincide with the limits of desired accuracy, then the solution of the variational problem is taken as y_n. But if these values do not agree at several chosen points, we compute y_{n+2} and compare the values y_{n+1} and y_{n+2}. This process is repeated till the values of $y_{n+k}(x)$ and y_{n+k+1} agree within the limits of desired accuracy.

Example 7.9 Find the extremum of the functional

$$I[y(x)] = \int_0^1 \left(y'^2 + y^2 \right) dx, \qquad y(0) = 0, \qquad y(1) = 1$$

using the Rayleigh–Ritz method.

Solution. It can be readily shown by solving the Euler equation that the exact solution is

$$y_{ex} = \frac{\sinh x}{\sinh 1}.$$

Using the above method, we put $f_0(x) = x$ in (7.53) and take $f_i(x)$ given by (7.55) as the coordinate functions. We first take $n = 1$, which means that we seek an approximate solution of the form

$$y = x + cx(1 - x).$$

Substituting this in the given functional of the form

$$I = \frac{11}{30}c^2 + \frac{1}{6}c + \frac{4}{3}.$$

Equating $\frac{dI}{dc}$ to zero, we find $c = -\frac{5}{22}$. Thus,

$$y_{1app.} = x - \frac{5}{22}x(1-x). \tag{7.57}$$

Next, for $n = 2$, we take the approximate solution as

$$y_{2app.} = x + c_1 x(1-x) + c_2 x^2(1-x). \tag{7.58}$$

Clearly, the functions $x(1-x)$ and $x^2(1-x)$ are linearly independent.
 Substituting (7.58) in the functional and equating to zero $\frac{\partial I}{\partial c_1}$ and $\frac{\partial I}{\partial c_2}$, we get

$$\frac{11}{15}c_1 + \frac{11}{30}c_2 + \frac{1}{6} = 0,$$
$$\frac{11}{30}c_1 + \frac{2}{7}c_2 + \frac{1}{10} = 0.$$

Solving these equations, we get $c_1 = -0.1459, c_2 = -0.1628$. Hence, (7.58) becomes

$$y_{2app.} = 0.8541\,x - 0.0169\,x^2 + 0.1628\,x^3. \tag{7.59}$$

It would be instructive to compare the exact solution with the approximate solutions (7.57) and (7.59). The following table gives the results of comparison between exact solution and first/second approximated solution:

x	0.0	0.1	0.2	0.3	0.4	0.5	0.6	0.7	0.8	0.9	1.0
y_{ex}	0.000	0.085	0.171	0.259	0.349	0.443	0.541	0.645	0.755	0.874	1.0
$y_{1app.}$	0.000	0.079	0.163	0.252	0.345	0.443	0.545	0.652	0.763	0.893	1.0
$y_{2app.}$	0.000	0.085	0.171	0.259	0.349	0.443	0.541	0.645	0.755	0.874	1.0

Thus, we find that with a two-term approximation, the exact solution practically coincides with the approximate solution.

Example 7.10 Minimize the integral

$$I[y] = \int_{-l}^{l}\left(\int_{-l}^{l}\frac{y'(s)}{x-s}ds\right)y(x)dx$$

subject to the constraint

$$J[y] = \int_{-l}^{l} y(x)dx = S = \text{constant}$$

and the boundary conditions $y(l) = y(-l) = 0$.

Solution. This is an aerodynamic problem (see [415]) where the value of the functional $I[y]$ is proportional to the drag experienced by an aeroplane with a streamlined wing of finite span $2l$ (the compressibility of the air and the frictional force being neglected). The integral $J[y]$ is the lift force, and $y(x)$ gives the dependence of the circulation of air flux round the wing of the aeroplane. Thus, the problem reduced to the determination of the distribution of circulation along the wing so that the drag of the aeroplane is minimum subject to a given lift force.

We introduce a change of the independent variable $x = -l \cos \theta$ so that the ends of the wing now correspond to $\theta = 0$ and $\theta = \pi$. We seek a solution in the form

$$y = \sum_{j=1}^{\infty} c_j \sin j\theta, \tag{7.60}$$

which satisfies the boundary conditions $y = 0$ at $\theta = 0$ and π corresponding to $y(-l) = y(l) = 0$.

Substituting (7.60) in the equation of constraint $J[y] = S$, we get $c_1 = 2S/(\pi l)$. Again, substitution of (7.60) in the expression for $I[y]$ gives

$$I[y] = \sum_{j=1}^{\infty} \sum_{k=1}^{\infty} c_j c_k k \int_0^{\pi} \left(\int_0^{\pi} \frac{\cos k\Psi}{\cos \Psi - \cos \theta} \right) \sin j\theta \sin \theta d\theta. \tag{7.61}$$

The inner integral is a singular integral for $0 < \theta < \pi$ and evaluation of its principle value gives $\pi \frac{\sin k\theta}{\sin \theta}$. Finally, substitution of this value in (7.61) gives

$$I[y] = \frac{\pi^2}{2} \sum_{k=1}^{\infty} k c_k^2.$$

Since $c_1 = \frac{2S}{\pi l}$ and the other coefficients are arbitrary, it follows that $I[y]$ attains its least value when $c_2 = c_3 = c_4 = \cdots = 0$. This leads to

$$y = \frac{2S}{\pi l} \sin \theta = \frac{2S}{\pi l} \cdot \sqrt{l^2 - x^2}, \tag{7.62}$$

which gives the required distribution of circulation. It is easy to see that this distribution is obtained for a wing whose shape in the xy-plane is an ellipse with semiaxes l and $\frac{2}{\pi l}$.

The above example shows that in the Rayleigh–Ritz method, sometimes it is possible to find the general formula for the coefficients c_i in the representation (7.53) and then pass to the limit for $n \to \infty$.

Example 7.11 Find the first eigenvalue of the problem

$$y''(x) + \lambda(1 + x^2)(1 + x^2)y = 0. \qquad y(-1) = y(1) = 0.$$

Solution. Using the extremal definition of eigenvalues, we may employ the Rayleigh–Ritz method to compute the eigenvalues of the above problem.

For the coordinate functions, choose $f_k(x) = 1 - x_{2k}$, $(k = 1, 2, \ldots)$, which are linearly independent and clearly satisfy the boundary conditions. Let us assume that

$$y(x) = c_1(1 - x^2) + c_2(1 - x^4).$$

Then, we may pose the problem of extremizing the functional

$$I[y] = \int_{-1}^{1} [y'^2 - \lambda(1+^2)y^2]dx$$

whose Euler equation coincides with the above second-order differential equation. Substitution of the above expression for y in $I[y]$ yields

$$I = c_1^2 \left(\frac{8}{3} - \frac{128}{105}\lambda\right) + c_1 c_2 \left(\frac{16}{5} - \frac{64}{45}\lambda\right) + c_2^2 \left(\frac{32}{7} - \frac{5888}{3465}\lambda\right).$$

Now, equating $\frac{\partial I}{\partial c_1}$ and $\frac{\partial I}{\partial c_2}$ to zero and using the condition for nontrivial solution yield the smallest value of λ as $\lambda_1 = 2.1775$.

7.6.2 Galerkin Method

In the preceding section, we have seen the effectiveness of the Rayleigh–Ritz method in solving variational problems. In fact, the Rayleigh–Ritz method gives rise to the following scheme of solving a boundary value problem: for a given ordinary differential equation subject to some boundary conditions, we construct a functional whose Euler's equation coincides with the given differential equation and then apply the Rayleigh–Ritz method. Consider, for example, the differential equation

$$a(x)y'' + b(x)y' + c(x)y = f(x).$$

Let us multiply the equation by an arbitrary function $\mu(x)$ such that

$$-2P = a\mu, \quad -2P' = b\mu, \quad 2Q = c\mu, \quad -R = f\mu$$

hold. Then, the first two equations give $\frac{P'}{P} = \frac{b}{a}$ and $\mu = -\frac{2}{b}P' = -\frac{2}{b}\frac{b}{a}P$, i.e. $(\log P)' = \frac{b}{a}$ and $\mu = -\frac{2}{a}P$. Integrating the first equality, we obtain

$$P = \exp\left(\int_{x_0}^{x} \frac{b}{a}dx\right).$$

The remaining two determine Q and R. With this choice of P, Q and R, it can be readily shown that the required functional is of the form

$$I[y] = \int \left[P(x)y'^2 + Q(x)y^2 + R(x)y \right] dx$$

between appropriate limits.

We now observe that there is another direct method known as Galerkin method, named after its discoverer B.G. Galerkin, that can also be used to solve the boundary value problems as described above. Note that this method can be applied to both ordinary and partial differential equations irrespective of the case whether the given ODE or PDE is linear or nonlinear. To see this, let us consider the equation

$$y'' + p(x)y' + q(x)y = f(x) \tag{7.63a}$$

with the boundary conditions

$$y(x_0) = 0, \quad y(x_1) = 0. \tag{7.63b}$$

If the ODE (7.63a) is subject to the nonhomogeneous boundary conditions $y(x_0) = y_0, y(x_1) = y_1$, then noting the fact that the equation of the straight line passing through the points (x_0, y_0) and (x_1, y_1) is given by $y - y_0 = \frac{y_1 - y_0}{x_1 - x_0} \cdot (x - x_0)$ one can readily reduced to homogeneous conditions by the change of variables $z = y - y_0 - \frac{y_1 - y_0}{x_1 - x_0} \cdot (x - x_0)$. Denote by $L := \frac{d^2}{dx^2} + p(x)\frac{d}{dx} + q(x)$, so that we can write (7.63a) in the operator form as follows:

$$Ly = f(x). \tag{7.64}$$

We now select a complete system of continuous linearly independent coordinate functions $\{\phi_n(x) : n \in \mathbb{N}\}$ which satisfy

$$\phi_n(x_0) = \phi_n(x_1) = 0 \text{ for } n = 1, 2, \ldots.$$

We now seek an approximate solution of the operator equation (7.64) of the form

$$y_n = \sum_{i=1}^{n} c_i \phi_i(x).$$

Substituting y_n in (7.63a) and choosing the coefficients c_i so that $L\left(\sum_{i=1}^{n} c_i \phi_i(x) \right) - f(x)$ is orthogonal in the interval $[x_0, x_1]$ to each of the functions $\phi_i(x), i = 1, 2, \ldots, n$, we get

$$\int_{x_0}^{x_1} \left[L\left(\sum_{i=1}^{n} c_i \phi_i(x) \right) - f(x) \right] \phi_i(x) dx = 0 \qquad i = 1, 2, \ldots, n. \qquad (7.65)$$

We now expect that y_n tends to the exact solution

$$y_{ex} = \sum_{i=1}^{\infty} c_i \phi_i(x) \quad \text{as } n \to \infty.$$

This is due to the fact that if the series obtained convergence and admits term-wise differentiation, then $Ly_{ex} - f(x)$ is orthogonal in $[x_0, x_1]$ to each function $\phi_i(x)$. But since the system $\{\phi_n(x)\}$ is complete, it follows that $Ly_{ex} - f(x) = 0$. Clearly, y_{ex} satisfies all the boundary conditions $y_{ex}(x_0) = y_{ex}(x_1) = 0$ since $\phi_i(x_0) = \phi_i(x_1) = 0$. In actual computations, one restricts to only a finite number of coordinate functions $\phi_i(x)$. In this case, however, condition of completeness is abandoned, but the functions $\phi_i(x)$ are assumed to be linearly independent and consistent with the boundary conditions

$$\phi_i(x_0) = \phi_i(x_1) = 0.$$

Observation

1. It is important to note that for Euler's equations encountered in variational problems, Galerkin's method described above coincides with Rayleigh–Ritz method.
2. It is interesting to note that Galerkin's method is also applicable to many ordinary and partial differential equations which are not Euler's equations and appear irrespective of variational problems.

To illustrate Galerkin's method, we consider two specific fluid flow problems, but first we present some basics of the fluid dynamics. The term fluid dynamics refers to the science treating the study of fluids in motion.

Definition 7.4 *Fluid* is a substance that is capable of flowing and that changes its shape at a steady rate when acted upon by a force tending to change its shape. Fluids may be divided into two kinds: (i) *liquids* which are incompressible; i.e., there volumes do not change when the pressure changes and (ii) *gases* which are compressible fluids suffering change in volume whenever the pressure changes.

Remark 7.6 Note that there is no sharp distinction between the three states of matter. Furthermore, it is more convenient to treat the fluid as having continuous structure so that at each point we can prescribe a unique velocity, a unique pressure, a unique density, etc. Notice also that for a continuous or ideal fluid we can define a fluid particle as the fluid contained within an infinitesimal volume whose size is so small that it may be regarded as a geometrical point.

There are two types of forces on the fluid system, namely *body forces* and *surface forces*. The force which is proportional to mass (or possibly the volume) of the body on which it acts is called *body force*, while the force which acts on a surface element

Fig. 7.4 Normal and
shearing stresses

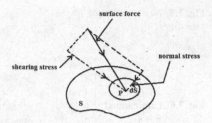

and is proportional to the surface area is called *surface force*. Suppose that the fluid element is enclosed by the surface S and P is an arbitrary point of S. Suppose dS is the surface element around P, then one may notice that the surface force on dS is, in general, not in the direction of normal at P to dS (see, for instance, Fig. 7.4). So, the force may be resolved into two components, one normal and the other tangential to the area dS.

Definition 7.5 The normal force per unit area is said to be the *normal stress* or *pressure*, while the tangential force per unit area is said to be the *tangential* or *shearing stress*.

A fluid possesses one of the two important characteristics: either viscous or non-viscous. Note that viscosity is a property of the fluid which opposes the relative motion between the two surfaces of the fluid that are moving at different velocities. More explicitly, viscosity means friction between the molecules of fluid. For example, honey has a much higher viscosity than water.

Definition 7.6 A fluid is said to be *viscous* or *real* when normal and shearing stresses exist. On the other hand, a fluid is said to be inviscid (nonviscous, frictionless) when it does not exert any shearing stress, whether at rest or in motion. This means that the pressure exerted by an inviscid fluid on any surface is always along the normal to the surface at the point. Due to shearing stress, a viscous fluid produces resistance to the body moving through it as well as between the particles of the fluid itself.

Observation

- It is quite interesting to observe that all fluids have positive viscosity and are technically said to be viscous or viscid. Zero viscosity is observed only at very low temperatures in superfluids.
- In common terms, however, a liquid is said to be viscous if its viscosity is substantially greater than that of water.
- It may be observed that air and water are treated inviscid fluid, whereas syrup and lubricant oil are treated as viscous fluid.

Suppose fluid is flowing between two parallel plates AB and CD. The plate AB is fixed, i.e. $u = 0$, while the plate CD is moving with uniform velocity u separated by distance y. A resistance force F acts between two layers such that
(i) $F \propto \frac{du}{dy}$ (velocity gradient)   (ii) $F \propto A$,

Fig. 7.5 Newton's law of
viscosity

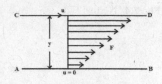

Fig. 7.6 Laminar and
turbulent flows

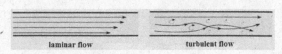

laminar flow   turbulent flow

where A is the area of cross section of upper plate. It follows that

$$F \propto A\frac{du}{dy} \text{ or } F = \mu A\frac{du}{dy}$$

where μ is called coefficient of viscosity and so

$$\frac{F}{A} = \mu\frac{du}{dy}, \text{ but } \frac{F}{A} = T = \text{shearing stress} \Rightarrow T = \mu\frac{du}{dy}.$$

If $\frac{du}{dy} = 1$, then $T = \mu$; i.e., the friction force per unit area is equal to the coefficient
of viscosity required to produce unit velocity gradient. This is called Newton's law
of viscosity (Figs. 7.5 and 7.6).

Definition 7.7 Fluids which obey Newton's law of viscosity are known as
Newtonian fluids, otherwise non-Newtonian.

Observation

- In most common cases, the viscosity of non-Newtonian fluids is dependent
  on shear force. However, we observe that some non-Newtonian fluids with
  shear-independent viscosity still exhibit normal stress differences or other non-
  Newtonian behaviour.
- Water, air and mercury are all Newtonian fluids, while paints, coaltar, ketchup,
  blood, shampoo, custard, toothpaste and molten polymers are all non-Newtonian
  fluids.
- The existence of shearing stresses and the condition of no-slip near solid walls
  constitute the main difference between the perfect (nonviscous) and real (viscous)
  fluids. Thus, for a perfect (or nonviscous) fluids, $\mu = 0$.

Definition 7.8 The flow of a fluid when each particle of the fluid traces out a definite
curve and the curves traced out by any two different fluid particles do not intersect is
said to be *laminar* (or streamline). On the other hand, a flow in which fluid particle
does not trace out a definite curve and the curves traced out by fluid particles intersect
is said to be *turbulent*.

Definition 7.9 The force on an object that resists its motion through a fluid is called *drag*.
• When the fluid is a gas like air, it is called *aerodynamic drag* or *air resistance*.
• When the fluid is a liquid like water, it is called *hydrodynamic drag*.

Definition 7.10 The Reynolds number Re is defined as

$$Re = \frac{\text{Inertia force}}{\text{Viscous force}}$$
$$= \frac{\text{Mass} \times \text{Acceleration}}{\text{Shear Stress} \times \text{Cross sectional area}}$$

$$= \frac{\text{Volume} \times \text{Density} \times \frac{\text{Velocity}}{\text{Time}}}{\text{Shear stress} \times \text{Cross sectional area}}$$
$$= \frac{\text{Cross sectional area} \times \text{Linear dimension} \times \rho \times \frac{\text{Velocity}}{\text{Time}}}{\text{Shear stress} \times \text{Cross sectional area}}$$
$$= \frac{(\text{Velocity})^2 \times \rho}{\mu(du/dy)} = \frac{V^2 \rho}{\mu(V/L)} = \frac{VL\rho}{\mu} = \frac{VL}{v}$$

where L and V denote the characteristic length and characteristic velocity, respectively, so that velocity will be proportional to V and $\frac{du}{dy}$ will be proportional to $\frac{V}{L}$.

Observation

• If for any flow problem Re is small, then we can ignore the inertia force, whereas Re is larger than we can neglect the effect of the viscous force, and consequently, the fluid may be treated as nonviscous fluid.
• When the viscous force is predominating force, Reynolds number must be the same for dynamic similarity of two flows.
• It is experimentally shown that if the value of Reynolds number exceeds a critical value the flow ceases to be laminar and the flow becomes turbulent. When $Re <$ 2000, the flow is laminar.

Problem 1. In the first problem, Djukic [191] used this technique to solve the presence of a uniform transverse magnetic field. To illustrate the problem, let us consider the steady laminar two-dimensional stagnation point flow of a power-law non-Newtonian electrically conducting fluid towards an infinite flat plate coincident with X-axis, O being the stagnation point. Further, suppose that

(i)   a uniform transverse magnetic field B acts along the Y-axis;
(ii)   the fluid is incompressible and its apparent viscosity changes with the rate of shear (power-law fluid). The external electric field is zero and the field due to polarization of charges is negligible;
(iii)   the magnetic Reynolds number is assumed to be small so that induced magnetic field can also be neglected.

Fig. 7.7 The steady laminar
two-dimensional stagnation
point flow of a power-law
non-Newtonian electrically
conducting fluid towards an
infinite flat plate

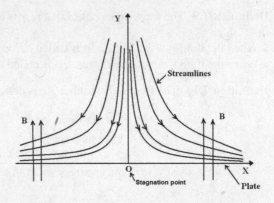

If (u, v) are the velocity components inside the boundary layer and $V(x)$ is the
x-components of velocity outside the boundary layer, the steady two-dimensional
boundary layer equation is

$$u\frac{\partial u}{\partial x} + v\frac{\partial u}{\partial y} = V\frac{dV}{dx} + v\frac{\partial}{\partial y}\left[\left|\frac{\partial u}{\partial y}\right|^{n-1} \cdot \frac{\partial u}{\partial y}\right] + \sigma B^2(V - u)/\rho, \qquad (7.66)$$

and the equation of continuity yields

$$\frac{\partial u}{\partial x} + \frac{\partial v}{\partial y} = 0. \qquad (7.67)$$

The boundary conditions are (Fig. 7.7)

$$u = v = 0 \text{ at } y = 0$$
$$u \to V(x) \text{ as } y \to \infty. \qquad (7.68)$$

Here, k is the fluid consistency index and n is the power-law index. Further, υ, ρ and
σ denote the constant $\frac{k}{\rho}$, the fluid density and the constant electrical conductivity of
the fluid, respectively. It is well known that for a plane stagnation flow,

$$V = ax,$$

where a is a constant.

We now seek similarly a solution of the system (7.66)–(7.68) by using similarity
transformations

$$\eta = \left[\frac{2a^{2-n}}{v(1+n)}\right]^{1/(1+n)} \cdot x^{(1-n)/(1+n)} \cdot y \qquad (7.69a)$$

$$u = V\frac{dF(\eta)}{d\eta}, \qquad (7.69b)$$

where $F(\eta)$ is to be determined.

Combining (7.66)–(7.69b), we obtain the following ordinary differential equation for $F(\eta)$:

$$\left(\frac{d^2 F}{d\eta^2}\right)^{n-1} \cdot \frac{d^3 F}{d\eta^3} + F \frac{d^2 F}{d\eta^2} + \frac{(1+n)}{2n}\left[1 - \left(\frac{dF}{d\eta}\right)^2 + N\left(1 - \frac{dF}{d\eta}\right)\right] = 0$$

$$\tag{7.70}$$

subject to the boundary conditions

$$F = \frac{dF}{d\eta} = 0 \ \ at \ \ \eta = 0, \tag{7.71}$$

$$\frac{dF}{d\eta} \to 1 \ \ as \ \ \eta \to \infty. \tag{7.72}$$

In (7.70), N stands for the magnetic parameter $\sigma B^2/(a\rho)$, which is constant.

To solve (7.70), we introduce the independent variable λ in place of η as

$$\lambda = \frac{dF}{d\eta}. \tag{7.73}$$

Then, (7.59) becomes

$$\phi \frac{d^2 \phi}{d\lambda^2} - \frac{1}{n+1}\left(\frac{d\phi}{d\lambda}\right)^2 - 2\lambda\phi - (1-\lambda^2)\frac{d\phi}{d\lambda} - N\left[(1-\lambda)\frac{d\phi}{d\lambda} + (n+1)\phi\right] = 0, \tag{7.74}$$

where

$$\phi = \left(\frac{2n}{n+1}\right)\left(\frac{d^2 F}{d\eta^2}\right)^{n+1} \tag{7.75}$$

is the new dependent variable. The boundary conditions for (7.74) are obtained as follows by using (7.70), (7.71) and (7.74), together with the condition of vanishing shear stress at the boundary layer edge given by $\frac{\partial u}{\partial y} \to 0$ as $y \to \infty$;

$$\frac{d\phi}{d\lambda} = -(n+1)(1+N) \ \ at \ \ \lambda = 0, \tag{7.76}$$

$$\phi = 0 \ \ at \ \ \lambda = 1.$$

Further combining (7.74) and (7.75), we have the supplementary boundary conditions for (7.74)

$$\frac{d^2 F}{d\lambda^2} = (n+1)N \ \ at \ \ \lambda = 0,$$

$$\frac{d\phi}{d\lambda} = 0 \ \ at \ \ \lambda = 1 \tag{7.77}$$

It is interesting to note that by the transformation (7.74), the infinite interval $[0, \infty]$ for η is mapped into $[0, 1]$ for λ.

To apply the Galerkin method, we seek an approximate solution of (7.74) in the form

$$\phi = (n+1)\left[b(1 - 4\lambda^3 + 3\lambda^4) + \frac{1}{2}N(1 - \lambda)^2 + \frac{1}{3}(2 - 3\lambda + \lambda^3)\right], \quad (7.78)$$

which satisfies the boundary conditions (7.76) and (7.77). Here, b is an unknown constant. Substituting (7.78) in (7.74) and interesting the result with respect to λ from $\lambda = 0$ to 1, we obtain an equation in b whose solution is

$$b = \frac{7}{48}\left\{\left[N + \frac{n+6}{4(n+2)^2} + \frac{4n(8+5N)}{7(n+2)}\right]^{1/2} - \left[N - \frac{n+6}{2(n+2)}\right]\right\} \quad (7.79)$$

Using (7.68), (7.69a), (7.75) and (7.78), the shear stress at the wall $\tau_0 = [k(\frac{\partial u}{\partial y})^n]_{y=0}$ is given in terms of shear stress coefficients $C_f (= \frac{\tau_0}{\rho V^2})$ as

$$C_f = C(n, N)R_{ex}^{-1/(n+1)}, \quad (7.80)$$

where

$$C(n, N) = \left[(n+1)\left(b + \frac{2}{3} + \frac{N}{2}\right)/n\right]^{n/(n+1)} \quad (7.81)$$

and $R_{ex} = V^{2-n}x^n/\nu$ is the local Reynolds number.

Observation

- It is important to note that the index n appearing in the constitutive relation between the shear stress and rate of strain within the framework of boundary layer approximation given by

$$\tau_{xy} = k\left|\frac{\partial u}{\partial y}\right|^{n-1} \cdot \frac{\partial u}{\partial y}$$

  is positive. Notice also that $n < 1$ for a pseudo-plastic fluid (shear-thinning), while $n > 1$ corresponds to dilatant field parameter N increases the shear stress parameter C for a fixed n.
- It is also found that for the nonmagnetic case ($N \neq 0$), the values of $C(n, N)$ versus n obtained by the Galerkin method when compared with the corresponding numerical solution by Schulman and Berkovsky [544] give very good agreement for $n < 1$ although there is some error ($<2.8\%$) for ($n > 1$). This lends credence to the fact that the Galerkin method is a useful tool in handling highly nonlinear differential equation such as (7.70).

Problem 2. In the second problem, the same method was used by Sengupta and Gupta [550] who studied the stability of the flow of an electrically conducting liquid of kinematic viscosity ν and electrical conductivity σ in a channel formed by two vertical parallel plates separated by a distance D and placed on a turn table, which is rotated with angular velocity Ω about a vertical axis taken as the z-axis. The shear flow relative to the rotating frame is induced by a uniform pressure gradient P along the channel (i.e. along the z-axis), and a uniform magnetic field H_0 acts along the y-axis taken normal to the plates, the x-axis being parallel to the plates in the horizontal direction, the origin being taken in the mid-section of the channel. Neglecting the induced magnetic field in the fluid of magnetic permeability $\mu_{e'}$, the linearized perturbation equations for the dimensionless velocity components (u', v', w') with $v' = \frac{\partial \Psi}{\partial z}$,

$$w' = \frac{\partial \Psi}{\partial Y}, \quad Y = \frac{y}{D_1}, \quad Z = \frac{z}{D_1}$$
$$\Psi = \mathrm{Re}\{ie^{ikz}h(Y)\}, \quad u' = \mathrm{Re}\{e^{ikz}g(Y)\}$$

are

$$E(D^2 - k^2)g(Y) + \left(R_0 \frac{d\overline{U}}{dY} - 1\right)kh(Y) = 0,$$

$$E(D^2 - K^2)^2 h(Y) - kg(Y) = 0,$$

where $U(y)$ is the basic velocity distribution, and k is the wave number of disturbance and

$$D \equiv \frac{d}{dY}, \quad E = \frac{v}{2\Omega D_1^2}, \quad R_0 = \frac{U_0}{2\Omega D_1}, \quad \overline{U}(Y) = \frac{U(y)}{U_0}$$

$$U_0 = \frac{PM(\cosh M - 1)}{\mu_e^2 \sigma H_0^2 \sinh M}, \quad M = \frac{\mu_e H_0 D_1}{2}\left(\frac{\sigma}{\rho v}\right)^{1/2}.$$

The above differential equation with boundary condition $h = Dh = g = 0$ at $Y = \frac{1}{2}$ was solved by the Galerkin method, and it is found that magnetic field exerts a strong stabilizing influence on the flow. This study is of some importance in the stability problem for a zonal flow.

7.6.3 Methods of Projection

Note that apart from the Rayleigh–Ritz method and Galerkin's methods such as of least squares, or collocation methods can be viewed in a more global setting as projective methods.

In what follows, let $(X, \| \cdot \|_X)$ and $(Y, \| \cdot \|_Y)$ be two normed linear spaces. Consider the equation

$$Ax = y, \tag{7.82}$$

where A maps X into Y. Let X_n be a finite dimensional (hence closed) subspace of X such that $X_n = \text{span}(\phi_1, \phi_2, \ldots, \phi_n)$. Then, a direct method for solving the variational problem associated with (7.82) is simply a construction of a numerical algorithm for finding a function

$$x_n = c_1\phi_1 + c_2\phi_2 + \cdots + c_n\phi_n$$

belonging to X_n which makes

$$\| Ax_n - y \|_Y + \| x_n - x \|_X$$

small. Solution by any one of the methods mentioned above is usually (but not always) equivalent to projecting the solution or an approximate solution onto a finite dimensional subspace of X (see, Prenter [495]).

In yet another general setting, an abstract variational problem can be stated as follows: Let $\mathcal{H}$ be a Hilbert space and $a(\cdot, \cdot) : \mathcal{H} \times \mathcal{H} \to \mathbb{R}$ be a bounded linear functional. Further, let $f : \mathcal{H} \to \mathbb{R}$ be a bounded linear functional. Then, the problem is to find an element x such that

$$a(x, y) = f(y) \qquad \text{for all} \ \ y \in \mathcal{H}. \tag{7.83}$$

It can be shown that in the above abstract setting, the variational problem has a unique solution in view of the Lax[1]–Milgram lemma (see [360]) which we now discuss.

Lax–Milgram Lemma. If the bilinear form defined above is coercive in the sense that there exists an $\alpha > 0$ such that $a(x, x) = \alpha \| x \|^2$ for every $x \in \mathcal{H}$, then the above variational problem has a unique solution.

Let us now consider an approximate problem corresponding to the variational problem given by (7.83). Suppose $\mathcal{H}^*$ is the space of bounded linear functionals on $\mathcal{H}$. Then, $f \in \mathcal{H}^*$. Suppose that $\mathcal{H}_h$ is a finite dimensional subspace of $\mathcal{H}$. Then, an approximate problem corresponding to the above variational problem consists of determination of a $u_h \in \mathcal{H}_h$ such that

$$a(u_h, v_h) = f(v_h) \quad \forall v_h \in H_h. \tag{7.84}$$

By virtue of the Lax–Milgram lemma, the above equation has a unique solution if $a(x, y)$ is coercive.

Now, it can be shown that the solution of (7.84) is equivalent to the solution of the matrix equation

$$AU = b, \tag{7.85}$$

---

[1]Peter David Lax is a Hungarian-born American mathematician working in the areas of pure and applied mathematics. He has made important contributions to integrable systems, fluid dynamics and shock waves, solitonic physics, hyperbolic conservation laws, mathematical and scientific computing, among other fields. Lax is listed as an ISI highly cited researcher.

where the transpose of the matrix A, denoted A^T, is $A^T = \left(a(w_i, w_j)\right)_{i,j}$, where W_i is basis of $\mathcal{H}_h$, with $i = 1, 2, \ldots, n(h)$; $j = 1, 2, \ldots, n(h)$, $n(h)$ being the dimension of $\mathcal{H}_h$. Further,

$$U = \left(\alpha_1, \alpha_2, \ldots, \alpha_{n(h)}\right), \qquad b = \left(f(w_1), f(w_2), \ldots, f(w_{n(h)})\right).$$

This can be shown as follows: since $\{w_i\}$ is a basis of $\mathcal{H}_h$ and $u_h, v_h \in \mathcal{H}_h$,

$$u_h = \sum_{i=1}^{n_h} \alpha_i w_i, \quad v_h = \sum_{j=1}^{n_h} \beta_j w_j$$

where $\alpha_i, \beta_j \in \mathbb{R}$. Substituting these in (7.84), we get

$$a\left(\sum_{i=1}^{n_h} \alpha_i w_i, \sum_{j=1}^{n_h} \beta_j w_j\right) = f\left(\sum_{j=1}^{n_h} \beta_j w_j\right).$$

In view of the bilinearity of $a(\cdot, \cdot)$ and linearity of f, we obtain

$$\sum_{i=1}^{n_h} \sum_{j=1}^{n_h} \alpha_i \beta_j a(w_i, w_j) = \sum_{j=1}^{n_h} \beta_i f(w_j).$$

Recalling the definitions of A^T, U and b mentioned above, we find that

$$V^T A U = V^T b, \tag{7.86}$$

where $V = (\beta_1, \beta_2, \ldots, \beta_{n(h)})$. Because (7.86) is true for all $V \in \mathbb{R}^{n(h)}$, we get the relation (7.85), which is what we set out to prove.

Now, we can solve the matrix equation (7.85) by any one of the standard methods such as Gauss's method of elimination and Cholesky's method. It may be noted that for a coercive $a(\cdot, \cdot)$, A is positive definite. Further, A is symmetric if the bilinear form is symmetric. Again, since $a(\cdot, \cdot)$ is coercive, A is invertible.

Notice that the choice of the basis functions $\{w_i\}$ is of crucial importance in facilitating numerical computations. It is convenient to choose $\{w_i\}$ so that A becomes a sparse matrix and the computing time is reasonably small.

7.6.4 Variational Inequalities

We have already seen by Lax–Milgram lemma that if $\mathcal{H}$ is a (real) Hilbert space and $a(\cdot, \cdot)$ is a bounded coercive bilinear functional, then there exists a unique $x \in \mathcal{H}$ such that

$$a(x, y) = f(y) \quad \forall \, y \in \mathcal{H} \tag{7.87}$$

where $f : \mathcal{H} \to \mathbb{R}$ is a bounded linear functional. In particular, if $a(\cdot, \cdot)$ is also symmetric, then the functional $I : \mathcal{H} \to \mathbb{R}$ defined by

$$I(y) = \frac{1}{2}a(y, y) - (f, y) \tag{7.88}$$

attains its minimum at x. Thus, (7.87) can be regarded as the Euler equation of unconstrained variational problem.

However, if we consider a constrained variational problem, i.e., we minimize I over a closed convex subset K of the Hilbert space $\mathcal{H}$, then we get an inequality instead of (7.87). This is known as a variational inequality [556]. We have the following theorem.

Theorem 7.17 *Let* $a(\cdot, \cdot)$ *be a bounded symmetric and coercive bilinear functional on a Hilbert space* $\mathcal{H}$ *and* $K \subset \mathcal{H}$ *be a closed convex subset. Let* $f : \mathcal{H} \to \mathbb{R}$ *be a bounded linear functional. Then, there exists a unique* $x \in H$ *such that*

$$a(x, y - x) \geq (f, y - x) \ \forall \, y \in K \tag{7.89}$$

To prove this, we consider

$$\langle x, y \rangle = a(x, y) \ \forall \, x, y \in \mathcal{H}.$$

Then, from the bilinearity and symmetry of $a(\cdot, \cdot)$, it follows that $\langle x, y \rangle$ is an inner product for $\mathcal{H}$. Let

$$\|\|x\|\|^2 = \langle x, x \rangle = a(x, x). \tag{7.90}$$

Because $a(\cdot, \cdot)$ is bounded and coercive, we may write

$$\alpha \, \| \, x \, \|^2 \leqslant \|\|x\|\|^2 \leqslant M \, \| \, x \, \|^2, \ \alpha > 0, M > 0$$

and hence the new norm $\|\|\cdot\|\|$ is equivalent to the original one. Thus, $\mathcal{H}$ is also a Hilbert space with respect to the new inner product. Now, by Riesz representation theorem, we can find $\bar{f} \in \mathcal{H}$ such that for any $y \in \mathcal{H}$,

$$a(\bar{f}, y) == \langle \bar{f}, y \rangle = (f, y). \tag{7.91}$$

We now consider

$$\begin{aligned}
\frac{1}{2}\|\|y - \bar{f}\|\|^2 &= \frac{1}{2}a(y - \bar{f}, y - \bar{f}) \\
&= \frac{1}{2}a(y, y) - a(y, \bar{f}) + \frac{1}{2}a(\bar{f}, \bar{f}) \\
&= \frac{1}{2}a(y, y) - (f, y) + \frac{1}{2}\|\|\bar{f}\|\|^2 \tag{7.92}
\end{aligned}$$

$$= I(y) + \frac{1}{2} \||\bar{f}\||^2.$$

Because $\||\bar{f}\||^2$ is a constant, minimizing $I(y)$ over K is equivalent to minimizing $\||y - \bar{f}\||^2$ over K. Now, one of the classical results of functional analysis is the minimization of the norm in a Hilbert space which is embodied in the following result: let $\mathcal{H}$ be a real Hilbert space and $K \subset \mathcal{H}$ be a closed convex subset. Then, there exists a unique $y \in K$ such that

$$\| x - y \| = \min_{z \in K} \| x - z \| . \tag{7.93}$$

Further, y can be characterized by

$$y \in K, \quad (x - z, z - y) \leq 0 \ \ \forall z \in K. \tag{7.94}$$

Hence, using the above characteristic in our foregoing analysis, it follows from (7.92) that

$$\langle \bar{f} - x, y - x \rangle \leq 0 \ \ \text{for every } y \in K$$

leading to $a(x, y - x) \geq (f, y - x)$, which is what we set out to prove.

The symmetry condition on $a(\cdot, \cdot)$ in the above theorem due to Stampacchia [578]: Let $\mathcal{H}$ be a Hilbert space and let $a(\cdot, \cdot)$ be a bounded coercive bilinear form on $\mathcal{H}$. Then, given $f : \mathcal{H} \to \mathbb{R}$, a bounded linear functional, there exists a unique $x \in K$ (a closed convex subset of $\mathcal{H}$) such that

$$a(x, y - x) \geq (f, y - x)$$

for every $y \in K$. Of course, in the case we will not be able to identify the problem with one of minimization. Variational inequalities have enormous applications to the study of free boundary problems (e.g. Stefan problems) in mechanics and physics. However, we will not pursue this matter any further as it is beyond the scope of this book. The interested reader will find a complete account of the theory and applications of variational inequalities in Duvaut and Lions [210].

We now state without proof some of the fundamental inequalities which are useful in applied problems:

I. Poincare's inequality: Suppose Ω is a bounded open set in $\mathbb{R}^n$ and $W^{1,p}(\Omega)$ is the Sobolev space of order 1 for $1 \leqslant p \leqslant \infty$ with seminorm $| \cdot |_{1,p,\Omega}$. Further, suppose $W_0^{1,p}(\Omega)$ is the closure of $C_0^\infty(\Omega)$ in $W^{1,p}(\Omega)$, where $C_0^\infty(\Omega)$ is the space of infinitely differentiable functions with compact support in Ω. Then, there exists a positive constant $\beta = \beta(\Omega, p)$ such that

$$|u|_{0,p,\Omega} \leqslant \beta |u|_{1,p,\Omega} \ \ \text{for every } u \in W_0^{1,p}(\Omega).$$

The above inequality is known as Poincare's inequality and is fundamental in studying weak solutions of Dirichlet boundary value problems.

II. Korn's inequality: Suppose Ω is a bounded open subset of $\mathbb{R}^3$ of class C^1 and V is the space $(H^1(\Omega))^3$, where $H^1(\Omega)$ is the Sobolev space $W^{1,2}(\Omega)$. If $v \in V$ with $v = (v_1, v_2, v_3)$ with

$$\epsilon_{ij}(v) = \frac{1}{2}\left(\frac{\partial v_i}{\partial x_j} + \frac{\partial v_j}{\partial x_i}\right), \quad 1 \leqslant i, j \leqslant 3,$$

then there exists a positive constant C depending only on Ω such that

$$\int_\Omega \epsilon_{ij}(v)\epsilon_{ij}(v) + \int v_i v_i \geqslant C\|v\|_V^2$$

for every $v \in V$ with $\|\cdot\|_V$ denoting the usual product norm on V.

The above inequality is known as Korn's inequality (see, Komkov [341]), and it plays an important role in the weak formulation of equation of linear elasticity.

III. Weyl's inequality: Consider the problem

$$\int_{-\infty}^{\infty} y^2\, dx \text{ maximum}, \quad \int_{-\infty}^{\infty} x^2 y^2\, dx \text{ and } \int_{-\infty}^{\infty} y'^2\, dx \text{ given},$$

then its Euler's equation is
$$y'' + (a + bx^2)y = 0,$$

which can be solved in terms of parabolic cylinder functions. This gives the variational basis for the following inequality

$$\int_0^\infty f^2\, dx < 2\left(\int_0^\infty x^2 f^2\, dx\right)^{1/2}\left(\int_0^\infty f'^2\, dx\right)^{1/2}$$

unless $f = Ae^{-Bx^2}$, A and B being constant.

The above inequality is known as Weyl's inequality (see, Hardy et al. [265]), and it is useful in mechanics.

IV. Garding's inequality: Suppose Lu denotes the operator

$$Lu = \sum_{0 \leqslant |r|, |s| \leqslant m} (-1)^r D^{|r|}(a_{rs}(\mathbf{x}) D^s \mathbf{u}),$$

where $\mathbf{u}$ and $\mathbf{x}$ are vectors, D is the differential operator, and $a_{rs}(\mathbf{x})$ are sufficiently smooth. Further,

(a) L is strongly elliptic with a modulus of ellipticity independent of $\mathbf{x}$ in a domain Ω;

(b)   the coefficients a_{rs} are bounded; i.e., there exists a positive constant C_1 such that $|a_{rs}| < C_1$; and

(c)   the highest order coefficients have a bounded modulus of continuity; i.e., there exists a positive constant C_2 such that

$$|a_{rs}(\mathbf{x}_1) - a_{rs}(\mathbf{x}_2)| < C_2|\mathbf{x}_1 - \mathbf{x}_2| \text{ for } |r| = m \text{ and } |s| = m$$

for all $\mathbf{x}_1, \mathbf{x}_2 \in \Omega$ and C_2 is small in a neighborhood of the origin. Then, there exist positive constants C_3 and C_4 such that

$$\|\mathbf{u}\|_m^2 \leqslant C_3 B(\mathbf{u}, \mathbf{u}) + C_4\|\mathbf{u}\|_0^2,$$

where $\|\mathbf{u}\|_0$ denotes the usual $L_2(\Omega)$ norm of $\mathbf{u}(\mathbf{x})$ and $B(\mathbf{u}, \mathbf{v})$ is the bilinear form

$$B(\mathbf{u}, \mathbf{v}) = \sum_{0 \leqslant |r|, |s| \leqslant m} D^r \mathbf{v}(a_{rs} D^s \mathbf{u}).$$

The above inequality is known as Garding's inequality (see, Komkov [341]) and finds applications in the solution of elliptic partial differential equations.

7.6.5 Matrix Inversion

In this section, we discuss a well-known variational formulation of a matrix inversion problem to find an estimation and subsequently derive finite element methods as a variational solution to Poisson differential equation.

Suppose we are given an input vectors $\{\mathbf{x}_1, \mathbf{x}_2, \ldots, \mathbf{x}_n\}$, $\mathbf{x}_i \in \mathbb{R}^d$ and corresponding scalar output values $\{y_1, y_2, \ldots, y_n\}$. We intend to find the best linear predictor in the form $y = \beta^T \mathbf{x} = \sum_{i=1}^d \beta_i x_i$, where β is the vector of parameters. More explicitly, we will assume that fitting criterion is least squares. Notice that the least squares optimal parameter setting β^* is given by $\beta^* = C^{-1}b$, where

$$C = \sum_{i=1}^n \mathbf{x}_i \mathbf{x}_i^T, \quad b = \sum_{i=1}^n y_i \mathbf{x}_i. \tag{7.95}$$

As we increase the dimension d of the input vector, evaluating $\beta^* = C^{-1}b$ can become more cumbersome. To ease this issue, we formulate a variational approach to compute $C^{-1}b$. Transformation of this problem into an optimization problem is the starting point of variational problem.

Suppose that we knew the solution to the above problem; that is, we had already evaluated β^*. We can then certainly optimize

$$I(\beta) = \frac{1}{2}(\beta^* - \beta)^T C(\beta^* - \beta) \tag{7.96}$$

with respect to β to find β^*. Here, distance measure is weighted with matrix C so that deviations of β from β^* count more in directions where input examples x vary considerably.

Notice that this is variational formulation leading to β^*, but it is important to realize that we could not yet evaluate $I(\beta)$ without first computing β^*.

7.6.6   Finite Element Method

In this section, we shall describe a method known as finite element method for solving the abstract variational problem in its approximate form (7.87), the method being economical from computational point of view. With reference to application of the above method to certain problems of elasticity, the matrix $a(w_i, w_j)$ and the vector $f(w_i)$ are often called the *stiffness matrix* and *load vector*, respectively.

Following Ciarlet (see, e.g., [556]), we now introduce the concept of a finite element.

Finite element— Let K be a polyhedron in $\mathbb{R}^n$, P_K the space of polynomials with dimension m and $\sum_K$ a set of distribution (i.e. continuous linear functionals on $D(\Omega)$ of cardinality m). The triplet $(K, P_K, \sum_K)$ is called a *finite element* if

$$\sum_K = \{L_i \in D^* | i = 1, 2, \ldots, m\}$$

is such that for given $\alpha_i \in \mathbb{R}, 1 \leqslant i \leqslant m$, the system of equations

$$L_i(p) = \alpha_i, \quad 1 \leqslant i \leqslant m$$

admits of a unique solution $p \in P_K$. The elements $L_i (i = 1, 2, \ldots, m)$ are known as degree of freedom of P_K.

If K is 2-simplex, then $(K, P_K, \sum_K)$ is called a triangular finite element, and if K is 3-simplex, then $(K, P_K, \sum_K)$ is known as tetrahedral.

Notice that many problems in physics can be reduced to solving differential equations. This includes, for example, finding the temperature distribution in material or gauging material distribution. To this end, let us consider that one of the simplest but nevertheless representative problems is the one-dimensional Poisson differential equation:

$$-u'' = f(x), \quad \forall x \in [a, b] \tag{7.97}$$

where $u''(x)$ is the second derivative of $u(x)$ with respect to scalar argument x and $f(x)$ is the "source". Assume that the solution $u(x)$ represents deformation that satisfies homogeneous boundary condition,

$$u(a) = u(b). \tag{7.98}$$

Note a number of techniques for solving this problem. The best known is perhaps finite element method that can be viewed as variational method.

Notice also that finite element method is essentially a projection method [556] concerned with the determination of a finite dimensional subspace V_h of $H^1(\Omega)$, the Hilbert–Sobolev space of order 1 and Ω, an open subset of $\mathbb{R}^n$ with a smooth boundary. This space is defined by

$$H^1(\Omega) = \left\{ v \in L_2(\Omega) : \frac{\partial v}{\partial x_i} \in L_2(\Omega), \ 1 \le i \le n \right\}.$$

where the derivatives $\frac{\partial v(x_1, x_2, ..., x_n)}{\partial x_i}$ are in the sense of distribution. This implies that

$$\left\langle v, \frac{\partial \phi}{\partial x_i} \right\rangle = -\left\langle \frac{\partial v}{\partial x_i}, \phi \right\rangle,$$

where $\phi \in D(\Omega)$, the space of infinitely differentiable functions with compact support in Ω. Further, $L_2(\Omega)$ is the space square integrable functions in Ω in the Lebesgue sense.

To find V_h of $H^1(\Omega)$, we first define a triangulation as follows:

Triangulation—Let $\Omega \subset \mathbb{R}^2$ be a given polygonal domain. Recall that a finite collection of triangles T_h is called a *triangulation* if the following conditions are satisfied:

(i) $\overline{\Omega} = \bigcup_{K \in T_h} \overline{K}$, where $\overline{K}$ denotes a triangle with boundary.

(ii) $K \cap K_1 = \varnothing$ for $K, K_1 \in T_h, \ K \ne K_1$.

(iii) $K \cap K_1 =$ a vector or a side. This means that if we consider two triangles, their boundaries may have one vertex common or one side common.

We introduce $P(K)$ as a function space defined on $K \in T_h$ such that $P(K) \subset H^1(K)$. Generally, $P(K)$ is taken as a space of polynomials of some degree. The following result then holds.

Lemma 7.5 *Let $C(\overline{\Omega})$ be the space of continuous real-valued functions on $\overline{\Omega}$ and*

$$V_h = \left\{ v_h \in C^0(\overline{\Omega}) \mid v_h|_K \in P(K), K \in T_h \right\},$$

where $v_h|_K$ is the restriction of v_h on K and $P(K) \subset H^1(K)$. Then $V_h \subset H^1(\Omega)$.

Proof Suppose $u \in V_h$ and v_i is function defined on Ω such that $v_i|_K = \frac{\partial u|_K}{\partial x_i}$. Now $v_i|_K$ is well defined since $u|_K \subset H^1(K)$. Further, from the definition of a Hilbert–Sobolev space $H^1(K)$, it follows that $v_i \in L_2(\Omega)$ as $v_i = \frac{\partial u}{\partial x_i} \in L_2(K)$. The lemma is thus established if we can show that

$$v_i = \frac{\partial u}{\partial x_i} \in D^*(\Omega)$$

where $D^*(\Omega)$ is the dual space of $D(\Omega)$. This stems from the fact that $\frac{\partial u}{\partial x_i} \in D^*(\Omega)$ implies that $u \in H^1(\Omega)$ and this in turns implies that $V_h \in H^1(\Omega)$.

For any $\phi \in D(\Omega)$, we have

$$
\begin{aligned}
(v_i, \phi) &= \int_\Omega v_i \phi dx = \sum_{K \in T_h} \int_K v_i \phi dx \\
&= \sum_{K \in T_h} \int_K \frac{\partial}{\partial x_i} u|_K \phi dx \\
&= \sum_{K \in T_h} \left[\int_{\Gamma = \partial K} u|_K \phi K_i d\Gamma - \int_K u|_K \frac{\partial \phi}{\partial x_i} dx \right]
\end{aligned}
\tag{7.99}
$$

by applying the generalized Green's formula such that Γ is the boundary of K. If η_i denotes the ith component of the outer normal at Γ, then

$$(v_i, \phi) = -\int_\Omega u \frac{\partial \phi}{\partial x_i} dx + \sum_{K \in T_h} \int_\Gamma u|_K \phi \eta_i^K d\Gamma. \tag{7.100}$$

The second integral in (7.100) vanishes since if K_1 and K_2 are two adjacent triangles, $\eta_i^{K_i} = -\eta_i^{K_2}$. Thus

$$(v_i, \phi) = -\int_\Omega u \frac{\partial \phi}{\partial x_i} dx = \left(\frac{\partial u}{\partial x_i}, \phi \right),$$

which shows that $v_i = \frac{\partial u}{\partial x_i} \in D^*(\Omega)$. This completes the proof.

Example 7.12 We consider the Neumann boundary value problem

$$-\nabla^2 u + u = f \quad \text{in } \Omega, \quad \Omega \subset \mathbb{R}^2, \frac{\partial u}{\partial n} = 0 \quad \text{on the boundary } \Gamma. \tag{7.101}$$

The variational formulation for the above problem is as follows:

$$V = H^1(\Omega), \tag{7.102}$$

$$a(u, v) = \int_\Omega \left\{ \sum_{i=1}^2 \frac{\partial u}{\partial x_i} + uv \right\} dx \tag{7.103}$$

$$= \int_\Omega (\nabla u \cdot \nabla v + uv) \, dx, \tag{7.104}$$

$$L(v) = \int_\Omega fv \, dx. \tag{7.105}$$

An internal approximation of this problem is the following:

$$V_h = \left\{ v_h \in C(\bar{\Omega}) | \forall K \in T_h, v_h|_K \in P_1(K) \right\}$$

as already pointed out in the lemma of this section. We choose a basis for V_h as follows:

$$w_i(a_{ij}) = \delta_{ij}, \quad 1 \leq j \leq n(h), \quad 1 \leq j \leq n(h),$$

where δ_{ij} is the Kronecker delta and $a_1, a_2, \ldots, a_{n(h)}$ are the vertices of the triangulation T_h. We then have for every $v_h \in V_h$,

$$v_h = \sum_{i=1}^{n(h)} v_h(a_i) w_i.$$

Then, (7.102)–(7.105) is equivalent to determining the value of u_h such that

$$a(u_h, v_h) = L(v_h) \quad \forall v_h \in V_h. \tag{7.106}$$

Since $w_i (i = 1, 2, \ldots, n(h))$ is a basis for V_h, the solution of (7.106) is taken as

$$u_h = \sum_{k=1}^{n(h)} \beta_k w_k. \tag{7.107}$$

Hence, β_k are the solutions of the following linear system:

$$\sum_{k=1}^{n(h)} a(w_k, w_l) = L(w_l) \text{ for } 1 \leqslant l \leqslant n(h) \tag{7.108}$$

with

$$a(w_k, w_l) = \sum_{K \in T_h} \int_K \left(\sum_{i=1}^{2} \frac{\partial w_k}{\partial x_i} \frac{\partial w_l}{\partial x_i} + w_k w_l \right) dx \tag{7.109}$$

$$L(w_l) = \sum_{K \in T_h} \int_K f(w_l) \, dx. \tag{7.110}$$

Hence, if we know the stiffness matrix $a(w_k, w_l)$ and the load vector $L(w_l)$, β_k can be found from (7.108). Putting these values in (7.107), the solution of (7.106) can be found.

7.6.7   Trefftz Method

If the Euler–Ostrogradsky equation arising in the problem of extremum of a functional depending on functions of two or more independent variables is linear and
homogeneous, but the domain of integration is bounded by a complex contour, the
Trefftz method (see [415]) proves very effective. We shall demonstrate this method
by considering the problem of finding a stationary value of the Dirichlet integral

$$I[u(x, y)] = \iint_D \left[\left(\frac{\partial u}{\partial x} \right)^2 + \left(\frac{\partial u}{\partial y} \right)^2 \right] dxdy \qquad (7.111)$$

satisfying the boundary condition

$$u|_\Gamma = f, \qquad (7.112)$$

where f is a given function on the boundary Γ of the domain D. This problem arises
in the solution of Laplace's equation $\nabla^2 u = 0$ in a domain D with the boundary
condition given above.

Let $u_0(x, y)$ be the unknown solution of the problem. We seek an approximate
solution of the form

$$u_{app.} = \sum_{k=1}^{n} c_k u_k(x, y), \qquad (7.113)$$

where u_k are solutions of $\nabla^2 u = 0$. If u_0 were known, we could pose the problem
of finding the coefficients c_k in such a way that $u_{app.}$ minimizes the integral

$$I\{u_{app.} - u_0\} = \iint_D \left[\left(\frac{\partial u_{app.}}{\partial x} - \frac{\partial u_0}{\partial x} \right)^2 + \left(\frac{\partial u_{app.}}{\partial y} - \frac{\partial u_0}{\partial y} \right)^2 \right] dxdy. \quad (7.114)$$

It appears natural to try to construct the approximate solution of the form (7.113)
that minimizes the functional (7.114). Equating the partial derivatives of the expression in (7.114) with respect to $c_j (j = 1, 2, \ldots, n)$, we get

$$\sum_{k=1}^{n} \iint_D \left(\frac{\partial u_j}{\partial x} \frac{\partial u_k}{\partial x} + \frac{\partial u_j}{\partial y} \frac{\partial u_k}{\partial y} \right) c_k dxdy = \iint_D \left(\frac{\partial u_j}{\partial x} \frac{\partial u_0}{\partial x} + \frac{\partial u_j}{\partial y} \frac{\partial u_0}{\partial y} \right) dxdy$$

$$(j = 1, 2, \ldots, n). \qquad (7.115)$$

But for any harmonic function v and an arbitrary function u, we have

$$\iint_D \nabla u \cdot \nabla v dxdy = \int_\Gamma u \frac{\partial v}{\partial n} dS.$$

Using (7.112) and the above relation, we get from (7.112) the equation

$$\sum_{k=1}^{n} \left(\int_{\Gamma} \frac{\partial u_j}{\partial n} u_k d\Gamma \right) c_k = \int_{\Gamma} f \frac{\partial u_j}{\partial n} d\Gamma \qquad (j = 1, 2, \ldots, n) \qquad (7.116)$$

which only involves the value of u_0 on the boundary Γ. Thus, (7.116) can be used when u_0 is unknown in the domain D. When the values of c_k determined from (7.116) are substituted in (7.113), we obtain the same expression as if we have minimized the integral (7.114) and thus determine an approximate solution of the original problem.

It should be noted that the method is also applicable when we have the boundary condition $\frac{\partial u}{\partial n} \big\|_{\Gamma} = f$ instead of (7.112). Numerical calculations reveal that the Trefftz method is highly effective.

7.6.8 Kantorovich Method

While applying the Rayleigh–Ritz method to functional $I[z(x_1, x_2, \ldots, x_n)]$ depending on the functions of several variables, the coordinate system of functions is chosen as $W_1(x_1, x_2, \ldots, x_n)$, $W_2(x_1, x_2, \ldots, x_n)$, $W_n(x_1, x_2, \ldots, x_n)$, ... and an approximate solution is sought in the form

$$z_m = \sum_{k=1}^{m} \alpha_k W_k(x_1, x_2, \ldots, x_n),$$

where $\alpha'_k s$ are constants and $W'_k s$ are independent.

In Kantorovich method [222] also, one chooses the coordinate system of functions as above, but the approximate solution is sought in the form

$$z_m = \sum_{k=1}^{m} \alpha_k(x_i) W_k(x_1, x_2, \ldots, x_n), \qquad (7.117)$$

where the coefficients $\alpha_k(x_i)$ are unknown functions of one of the independent variables. On the class of functions of the type (7.117), the functional $I[z]$ is transformed into a functional of the form $I[\alpha_1(x_i), \alpha_2(x_i), \ldots, \alpha_m(x_i)]$, which depends on $\alpha_1(x_i), \alpha_2(x_i), \ldots, \alpha_m(x_i)$. These functions are then chosen so as to extremize the functional I.

If we proceed to the limit as $m \rightarrow \infty$, then subject to certain conditions one can obtain an exact solution. But if such limit is not taken, then method will give an approximate solution which is generally more accurate than that obtained by Ritz method with the same coordinate functions and the same number of terms. The greater precision of Kantorovich method stems from the fact that the class of functions nut with constant $\alpha'_k s$. Thus, it is possible to find functions that approximate better the solution of the variational problem than the functions from the class $\alpha_k W_k(x_1, x_2, \ldots, x_n)$, where the $\alpha'_k s$ are constants.

Example 7.13 Find an approximate solution of Poisson's equation

$$\frac{\partial^2 z}{\partial x^2} + \frac{\partial^2 z}{\partial y^2} = -1 \text{ in the rectangle } D = \{(x, y) | -a \le x \le a, \ -b \le y \le b\}$$

with $z = 0$ on the boundary ∂D of D.

Solution. It may be readily shown that Poisson's equation is the Euler–Ostrogradsky equation for the functional

$$I[z] = \iint \left[\left(\frac{\partial z}{\partial x}\right)^2 + \left(\frac{\partial z}{\partial y}\right)^2 - 2z \right] dxdy.$$

We seek an approximate solution in the form

$$z_1(x, y) = (b^2 - y^2)\alpha(x),$$

which clearly satisfies $z_1 = 0$ on $y = \pm b$. Substituting this in the expression for I, we get

$$I[z_1] = \int_{-a}^{a} \left[\frac{16}{15}b^5\alpha'^2 + \frac{8}{3}b^3\alpha^2 - \frac{8}{3}b^3\alpha \right] dx$$

whose Euler equation is

$$\alpha''(x) - \frac{5\alpha}{2b^2} = -\frac{5}{4b^2}. \tag{7.118}$$

Its solution satisfying $\alpha(-a) = \alpha(a) = 0$ is

$$\alpha(x) = \frac{1}{2} \left(1 - \frac{\cosh \sqrt[2]{\frac{5}{2}} \cdot \frac{x}{b}}{\cosh \sqrt[2]{\frac{5}{2}} \cdot \frac{a}{b}} \right).$$

Thus, the approximate solution is

$$z_1(x, y) = \frac{b^2 - y^2}{2} \left(1 - \frac{\cosh \sqrt[2]{\frac{5}{2}} \cdot \frac{x}{b}}{\cosh \sqrt[2]{\frac{5}{2}} \cdot \frac{a}{b}} \right).$$

To obtain a more accurate solution, we may try a solution of the form

$$z_2(x, y) = b^2 - y^2\alpha_1(x) + b^2 - y^2\alpha_2(x).$$

7.7 Variational Formulation for Linear and Nonlinear Problems

There exist a multitude of problems in linear and nonlinear analysis which are escaped from having solution. We intend to discuss such problems in some detail by transforming them into variational formulation. Observe that during the last six decades, research on heat and mass transfer in impinging flow led to a multitude of experimental data obtained by different methods involving different hydrodynamic, thermal or material boundary conditions. However, we note that variational principles do not exist for many heat and mass transfer problems that arise in engineering applications. Coal gasification process involves the description of solid fuel conversion in fixed-bed and entrained flow gasification reactors. Notice that the effect of gasification on the dynamics and kinematics of immersed spherical and nonspherical solid particles are two such examples.

In this section, we examine the question of the existence of a variational principle in a more systematic way. In this context, we use the notions of Gâteaux derivatives in order to be able to give a general treatment of variational formulation for nonlinear differential equations.

To begin with, let us consider a vector field V and we ask the question: Can v be derived from a potential? It is well known that if $\nabla \times v = 0$, then v can be represented as the gradient of a potential, i.e. $v = \nabla \phi$.

The aforementioned elementary concept in vector calculus tells us that if we regard the Euler equation in a variational principle as the gradient of a functional, analogous to a potential, then we should not expect every differential equation to be derivable from a potential for the simple reason that every vector field cannot be derived from a potential. In order to make these concepts clearer, we should define the gradient of a functional and the derivative of a differential operator.

An equation (or a differential equation, to be more specific) is generally associated with additional conditions which specify initial, boundary and regularity conditions, as well as the functional class. The set formed by an equation and all additional conditional conditions constitutes a problem. Every problem may be expressed in the general form

$$P(u) = \phi_v, \tag{7.119}$$

where P denotes an operator which may be nonlinear. The set of elements u that satisfies the given initial or boundary conditions and the given functional class is called the domain of the operator and is denoted by $\mathscr{D}(P)$ which can be considered as a subset of a vector space U. The set of elements $v = P(v)$ constitutes the range of P denoted by $\mathscr{R}(P)$. This is supposed to be embedded in another vector space V, and ϕ_v in (7.119) is the null element of V. Suppose

$$P'_u \phi = \lim_{\varepsilon \to 0} \frac{P(u + \varepsilon \phi) - P(u)}{\varepsilon} = \left[\frac{\partial}{\partial \varepsilon} P(u + \varepsilon \phi) \right]_{\varepsilon=0} \tag{7.120}$$

exists. The limit is defined by the topology of the V-space. Then, $P'_u\phi$ is called the Gâteaux differential of the operator P in the direction ϕ, and P'_u is called the Gâteaux derivative of the operator P. The subscript u means that the differentiation of the operator is with respect to the argument u.

We now introduce a bilinear functional $\langle u, v \rangle$ that satisfies the following requirements:

(i)   It must be real-valued even if U and V are vector spaces over the complex number field.

(ii)   It must be bilinear over the real number field.

(iii)   It must be nondegenerate. This entails

if $\langle u_0, v_0 \rangle = 0$ for every $v \in V$, then $u_0 = \phi_u$;

if $\langle v_0, u \rangle = 0$ for every $u \in U$, then $v_0 = \phi_v$.

The real number $s = \langle u, v \rangle$ is called the *scalar product* of the elements $v \in V$ and $u \in U$. The V-space is called the dual or the conjugate of the U-space, and one writes $V = U^*$.

In practice, one can take the bilinear functional as follows. In many physical problems, one can take the bilinear functional $\langle v, u \rangle$ as

$$\langle v, u \rangle = \int \mathbf{v}(\mathbf{x}) \cdot \mathbf{u}(\mathbf{x}) \, d\Omega, \tag{7.121}$$

where Ω is a subset of $\mathbb{R}^n$ and $\mathbf{x}$ is a point of Ω with $\mathbf{x} = (x_1, x_2, \ldots, x_n)$. Here, $v \cdot u$ is the scalar product, i.e. a scalar-valued function formed of two vector values or two tensor-valued functions such as

$$\sum_k v_k(\mathbf{x}) u^k(\mathbf{x}) \text{ or } \sum_{h,k} v_{hk}(\mathbf{x}) u^{hk}(\mathbf{x}),$$

where the two tensors must have the same symmetry. Of course, a more general bilinear functional can be constructed as follows: if $A : U \rightarrow \dot{U}$ and $B : V \rightarrow V$ are two linear invertible operators whose domains are the entire $U-$ and $V-$spaces respectively, then the bilinear functional is

$$\langle v, u \rangle = \int B\mathbf{v}(\mathbf{x}).A\mathbf{u}(\mathbf{x}) \, d\Omega.$$

Once a bilinear functional is introduced, it is natural to introduce a topology in both the spaces such that the bilinear functional becomes continuous with respect to both the arguments. Such a topology is said to be *compatible with duality*. To illustrate this notion, one may observe the following example.

Example 7.14 Consider the operator P whose domain is that of differential functions over $[0, 1]$ which vanish at $x = 0$. Suppose the bilinear functional is the usual one given by (7.121). We may choose $\mathcal{D}(P) = U = C^1[0, 1]$ and $\mathcal{R}(P) = V = C[0, 1]$. The topologies induced by the norms

$$\|u\| = \max_{x \in [0,1]} [|u(x)| + |u'(x)|]$$

$$\|v\| = \max_{x \in [0,1]} |v(x)|$$

are compatible with duality. To show this, we first note that the two spaces are complete, and if $\|u_n - u_0\| < \varepsilon$, then

$$|\langle v, u_n \rangle - \langle v, u_0 \rangle| = |\langle v, u_n - u_0 \rangle|$$

$$= \left| \int_0^1 v(u_n - u_0)dx \right| \le (\max |v|).\varepsilon.1$$

Since v is continuous and the interval is finite, the property follows. Once the continuity is ensured, if u belongs to a dense subset D of a linear space U, the condition $\langle v_0, u \rangle = 0$ for every $u \in D$ assures that v_0 is ϕ_v.

Gradient of a functional—The gradient of a functional is defined as follows: given a functional $F(u)$ such that

$$F(u) = \int L(u)\, dV, \tag{7.122}$$

its Gâteaux differential in the direction ϕ is

$$\lim_{\varepsilon \to 0} \frac{F(u + \varepsilon \phi) - F(u)}{\varepsilon} = \int \lim_{\varepsilon \to 0} \frac{L(u + \varepsilon \phi) - L(u)}{\varepsilon}\, dV$$

$$= \int L'_u \phi\, dV.$$

Now, the Gâteaux differential $F'_u \phi$ defined by the limit on the left-hand side depends on u and ϕ. Integration by parts gives

$$F'_u \phi = \int L'_u \phi\, dV = \int \phi P(u)\, dV + \text{boundary terms} \tag{7.123}$$

and the operator $P(u)$ is called the *gradient of the functional*.

To see whether an operator $P(u)$ is the gradient a functional, we should see whether the path integral in (7.123) depends on the path of integration. Consider two paths

$$I : u \to u + \varepsilon \phi \to u + \varepsilon \phi + v\Psi,$$

$$II : u \to u + \varepsilon \Psi \to u + v\Psi + \varepsilon \phi$$

where u, ϕ and $\Psi \in X$ and ε and v are constants.

If the path integral is independent of the path chosen, then the following equation must hold:

$$\int P(u)\varepsilon\phi dV + \int P(u + \varepsilon\phi)v\Psi dV = \int v\Psi P(u)dV + \int P(u + v\Psi)\varepsilon\phi dV.$$

We can rearrange this equation as

$$\int \left[\frac{P(u + \varepsilon\phi) - P(u)}{\varepsilon}\right]\Psi dV = \int \left[\frac{P(u + v\Psi) - P(u)}{v}\right]\phi dV.$$

When $\varepsilon \to 0$ and $v \to 0$, we have from above,

$$\int \Psi P'_u \phi dV = \int \phi P'_u \Psi dV. \tag{7.124}$$

This implies that the operator P'_u is symmetric. The fact that (7.124) is the condition for the existence of a functional having the operator $P(u)$ as its gradient follows from the following theorem given by Vainberg [597].

Theorem 7.18 *Suppose that the following conditions are satisfied:*

(i) *P is an operator from a normed space X into its conjugate space X^*.*
(ii) *P has a linear Gâteaux differential $DP(x, h)$ at every point of the open ball $B_r(x_0)$.*
(iii) *The bilinear functional $(DP(x, h_1), h_2)$ is continuous in x for every point x in $B_r(x_0)$.*

Then, in order that the operator P be potential in the ball $B_r(x_0)$, it is necessary and sufficient that the bilinear functional $(DP(x, h_1), h_2)$ be symmetric for every x in $B_r(x_0)$, i.e.
$$(DP(x, h_1), h_2) = (DP(x, h_2), h_1) \tag{7.125}$$

for every h_1 and h_2 in X and every x in D.

Equation (7.125) is just the symmetry condition (7.124). Provided that an operator $P(u)$ is the gradient of a functional, i.e.

$$P(u) = \nabla F(u),$$

the functional F can be written as

$$F(u) = \int u \int_0^1 P(\lambda u)d\lambda dV. \tag{7.126}$$

The variation of $F(u)$ due to a variation δu in u is

$$\delta F = \int P(u)\delta u dV.$$

Clearly, if F is the functional arising in a variational principle, then $P(u) = 0$ is the corresponding Euler equation. Consequently, the question as to whether a variational principle exists for a given operator P depends on whether the operator has symmetric Gâteaux differential expressed by the condition (7.124).

Example 7.15 Consider the following equation (see).

$$f(u; u_{,j}, u_{,jk}) = 0. \tag{7.127}$$

Solution. From (7.120), we have

$$P'_u \phi = \left[\frac{\partial}{\partial \varepsilon} f\left(u + \varepsilon\phi, u_{,j} + \varepsilon\phi_{,jk}\right) \right]$$

$$= \frac{\partial f}{\partial u}\phi + \frac{\partial f}{\partial u_{,j}}\phi_{,j} + \frac{\partial f}{\partial u_{,jk}}\phi_{,jk}. \tag{7.128}$$

To test the symmetry requirement, we now integrate the following by parts

$$\int \Psi P'_u \phi \, dV = \int \Psi \left[\frac{\partial f}{\partial u}\phi + \frac{\partial f}{\partial u_{,j}}\phi_{,j} + \frac{\partial f}{\partial u_{,jk}}\phi_{,jk} \right] dV$$

$$= \int \phi \left\{ \left[\frac{\partial f}{\partial u} - \nabla_j \left(\frac{\partial f}{\partial u_{,j}} \right) + \nabla_k \nabla_j \left(\frac{\partial f}{\partial u_{,jk}} \right) \right] \Psi \right.$$

$$+ \left[-\frac{\partial f}{\partial u_{,j}} + 2\nabla_k \left(\frac{\partial f}{\partial u_{,jk}} \right) \right] \nabla_j \Psi$$

$$\left. + \frac{\partial f}{\partial u_{,jk}}\nabla_k\nabla_j\Psi \right\} dV + \text{boundary terms}$$

$$= \int \phi \overline{P}'_u \Psi dV + \text{boundary terms}. \tag{7.129}$$

Equation (7.129) defines the Gâteaux derivative $\overline{P}'_u$, which may be regarded as the adjoint to P_u. It should be noted that an adjoint is generally defined for a linear operator, but the notion of an adjoint is also useful for nonlinear operators.

Suppose the nonlinear operator P has Gâteaux derivative P'_u which is not symmetric. We then defines its adjoint $P^*(u, v) = \overline{P}'_v$ by (7.129). Hence, the boundary value problem in a region D with boundary Γ given by

$$P(u) = f \quad \text{in} \quad D,$$
$$B_i(u) = 0 \quad \text{on} \quad \Gamma$$

admits of a variational principle $\delta I = 0$ with

$$I(u, v) = \int_D [vP(u) - ug - vf] \, dV.$$

Here, the Euler equation are $P(u) = f$, $P^*(u, v) = g$. Now, using (7.128) and the symmetry condition (7.124), we get

$$\int \phi \overline{P}'_u \Psi dV = \int \phi P'_u \Psi dV = \int \phi \left[\frac{\partial f}{\partial u} + \frac{\partial f}{\partial u_{,j}} \nabla_j + \frac{\partial f}{\partial u_{,jk}} \nabla_j \nabla_k \right] \Psi dV.$$

Since this relation holds for arbitrary ϕ and Ψ, we must have

$$\frac{\partial}{\partial u} - \nabla_j \frac{\partial f}{\partial u_{,j}} + \nabla_k \nabla_j \frac{\partial f}{\partial u_{,jk}} = \frac{\partial f}{\partial u},$$

$$-\frac{\partial f}{\partial u_{,j}} + 2\nabla_k \left(\frac{\partial f}{\partial u_{,jk}} \right) = \frac{\partial . f}{\partial u_{,j}}$$

These two equations are equivalent to

$$\frac{\partial f}{\partial u_{,j}} - \nabla_k \left(\frac{\partial f}{\partial u_{,jk}} \right) = 0 \qquad (7.130)$$

and this is the required condition for (7.127) to be derivable from a potential.

We now recall the concept of Fréchet derivative (see [534]). Suppose X and Y are Banach spaces, Ω into Y, and $x \in \Omega$. If there exists $A \in \mathcal{L}(X, Y)$ (the Banach space of all bounded linear mapping of X into Y) such that

$$\lim_{h \to 0} \frac{\| F(x + h) - F(x) - Ah \|}{\| h \|} = 0.$$

then A is called the Fréchet derivative of F at x.

Observation

- It is well known that if a Fréchet differential exists, then so does a Gâteaux differential. Hence, if the operator has a symmetric Fréchet differential, then a variation principle exists for (7.119).
- Many natural functionals which may encounter in the real-world problem may not be bounded at all, neither from above nor from below. So we cannot look for maxima or minima. Instead, we seek saddle points by a min-max argument. The notion of Fréchet derivative plays an important role here. In this context, it may be pertinent to recall from the Lax–Milgram lemma of Sect. 7.6.3 that if $J(v) \to +\infty$ when $\| v \| \to \infty$ and is weakly lower semicontinuous, then J attains a global minimum; i.e., we can find $u \in V$ such that $J(u) = \lim_{v \in V} J(V)$.

Let us consider now the problem of finding a nontrivial stationary point of a given real C^1 functional J defined in a Banach space X and state the following:

7.7.1 Mountain Pass Lemma

Let $J : X \to \mathbb{R}$ be a C^1 functional. Further, let $u_0, u_1 \in X$, $c_0 \in \mathbb{R}$ and $R_1 > 0$ such that

(i) $\| u_1 - u_0 \| > R_1$,

(ii) $J(u_0), J(u_1) < c_0 \leq J(v)$ for all v such that $\| v - u_0 \| = R_1$. Then, J has a stationary value $c \geq c_0$ defined by

$$c = \inf_p \max_{u \in p} J(u).$$

Here, p represents any continuous path joining u_0 to u_1 in X. We take the maximum J over p and then take the infimum with respect to all possible paths. Since every such path must cross the sphere $S_{R_1}(u_0) = \left\{ v \mid \|v - u_0\| = R_1 \right\}$, where we have $J \geq c_0$, we see that $\max_p J \geq c_0$.

Now, we address a natural question—Why this lemma is called "mountain pass lemma"? To know this, we take the following observation:

- Think of J as representing the height of land at a point u. Then, u_0 is a point in a valley U bounded by a mountain range ∂U which is the boundary $\|v - u_0\| = R_1$. For any path p joining u_0 to u_1, $\max_p J$ represents how high we have to go on that path. Taking the infimum then minimizes this. But, c is the height of the lowest mountain range. When we are at that mountain pass, the earth is level, so that the Fréchet derivative J' vanishes.

Although this result is intuitively obvious, it is false even in finite dimensions. For example, in the complex plane, consider the nonnegative function

$$F(z) = |e^z - 1|^4.$$

Obviously, F achieves the minimum at 0 and 2π i. One can show that for small $r > 0$,

$$F(z) \geq c_0 > 0 \text{ for } |z| = r.$$

On the other hand, zero is the only critical value of F. Thus, F satisfies condition (ii) of MPL above, but the conclusion of MPL does not hold (Fig. 7.8).

Now, in view of above example, there arises a natural question. How to validate MPL? Notice that for the validity of MPL, we have to add an additional condition to MPL. This is known as Palais–Smale condition $(PS)_c$ (a kind of compactness condition) and is defined as follows:

- $(PS)_c$: Any sequence u_i in X for which $F'(u) \to c$ and the Fréchet derivatives $F'(ui) \to 0$ strongly in X^* (the dual space of X) has a strongly convergent subsequence $\{u_{ij}\}$ in X.

Fig. 7.8 Minima of the
complex-valued function
$F(z) = |e^z - 1|^4$ at 0 and
$2\pi i$

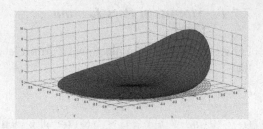

It may be noted that the Fréchet derivative $F'(u)$ of F at u represents a continuous linear functional on X, i.e. an element in X^*. It may be shown that MPL formulation above along with $(PS)_c$ gives the correct MPL.

If P is a linear operator, i.e. $P = L$, then

$$L'_u \phi = \lim_{\varepsilon \to 0} \frac{L(u + \varepsilon\phi) - L(u)}{\varepsilon} = L\phi$$

Consequently, $L'_u = L$ and the symmetry condition (7.124) becomes

$$\Psi L\phi dV = \int \phi L\Psi \, dV \qquad (7.131)$$

and the functional (7.126) then reduces to

$$F(u) = \frac{1}{2} \int uLu \, dV. \qquad (7.132)$$

7.7.2 *Variational Principles for Non-self-adjoint Equations*

Let $\mathcal{H}$ be a Hilbert space and A be a continuous linear operator such that $A : \mathcal{H} \to \mathcal{H}$ and $y \in \mathcal{H}$. We introduce a functional f_y on H by

$$f_y(x) = (Ax, y).$$

Clearly, $f_y(x)$ is a linear functional, and for $x \in \mathcal{H}$,

$$| f_y(x) | = | (Ax, y) | \leq \| Ax \| \cdot \| y \| \leq \| A \| \cdot \| y \| \cdot \| x \| .$$

Thus, f_y is bounded and so f_y is a continuous linear functional defined everywhere on H and $\| f_y \| \leq \| A \| \cdot \| y \|$. Hence, by the Riesz representation theorem (Rudin [534]),

$$f_y(x) = (x, y^*)$$

for all $x \in \mathcal{H}$ such that $y^* \in \mathcal{H}$ is uniquely determined by f_y. Clearly, if y changes, then y^* also changes. So we introduce an operator A^* such that

$$y^* = A^*y.$$

Here, A^* is defined everywhere on $\mathcal{H}$, and its range is also in $\mathcal{H}$. This operator A^* is called the adjoint operator to A. From the above equations, we find that A and A^* are connected by

$$(Ax, y) = (x, A^*y).$$

Now, we define a self-adjoint operator A^* as follows: A continuous linear operator $A : \mathcal{H} \rightarrow \mathcal{H}$ is called self-adjoint if $A^* = A$.

Given a linear differential operator L, we define the adjoint as above. When $L = L^*$, (7.131) holds and a variational principle exists with the functional given by (7.132), provided the boundary conditions are appropriate. When L is not self-adjoint, (7.131) does not hold. For such an operator, let the linear boundary value problem be

$$Lu = f \quad \text{in } V \tag{7.133a}$$

with

$$B_i u = 0 \quad \text{on } S \tag{7.133b}$$

where S is the piece-wise smooth boundary of the domain V. For a linear boundary value problem with a nonhomogeneous boundary condition given by

$$Lw = h \quad \text{in } V$$
$$B_i w = g_i \quad \text{on } S. \tag{7.134}$$

We assume that a function v can be found such that v satisfies $B_i v = g_i$ on S and v can be extended into the region V. Then, it is clear that the function u given by $u = w - v$ satisfies (7.133a) and (7.133b).

A variational principle for a non-self-adjoint problem (7.133a) subject to (7.133b) can be founded by a method closely related to the least squares method. This method minimizes the mean square residual

$$J(u) = \int_V (Lu - f)^2 \, dV \tag{7.135}$$

among function satisfying $B_i u = 0$. The variation of J is given by

$$\delta I = 2 \int_V (Lu - f)L \ \delta u \ dV$$

$$= 2 \int \delta u (L^*Lu - L^*f)dV + B(\delta u, Lu - f) \qquad (7.136)$$

Thus, the Euler equation with the essential and the natural boundary conditions is given by

$$L^*Lu = L^*f \ \text{ in } V \qquad (7.137a)$$

subject to

$$B_i u = 0, \quad B_j^* Lu = B_j^* f \ \text{ on } S. \qquad (7.137b)$$

Mikhlin [396] studied the conditions under which the solution of (7.137a) subject to (7.137b) reduced to the solution of the problem (7.133a) subject to (7.133b). If the system (7.133) admits of a solution and the inequality

$$(u, u) \leqslant K(Lu, Lu) \qquad (7.138)$$

holds, then the least square using (7.135) is equivalent to the variational method applied to (7.137a) subject to (7.137b). When the adjoint problem is unique, i.e. when the system

$$\begin{aligned} L^*v = 0 &\quad \text{in } V \\ B_i v = 0 &\quad \text{on } S \end{aligned} \qquad (7.139)$$

has the trivial solution $v = 0$, then solving (7.137a) subject to (7.137b) is equivalent to solving the problem (7.133a) subject to (7.133b). An equation similar to (7.137a) is given by

$$Ax = f, \qquad A = T^*T \qquad (7.140)$$

An equation of this type arises in a variety of problems in physics and mechanics. Laplace's equation, the biharmonic equation and the Lagrange–Germain equation describing a static deflection of a thin plate are all of this type.

Let us consider a linear operator A mapping a subset D_A of a Hilbert space $\mathcal{H}_1$ into $\mathcal{H}_1$. We assume that A is positive definite, so that $A = T^*T$. Further, the domains of T are dense in $\mathcal{H}_1$, and hence, T^* is uniquely defined. Clearly, T and T^* are the linear maps given by

$$T : D_A \subset \mathcal{H}_1 \to \mathcal{H}_2, \qquad T^* : \mathcal{H}_2 \to \mathcal{H}_1.$$

Thus, (7.140) in the space $\mathcal{H}_1$ can be rewritten as a pair of equations

$$Tx = p \quad \text{in } \mathcal{H}_2$$
$$T^*p = f \quad \text{in } \mathcal{H}_1 \tag{7.141}$$

The system (7.141) has been designated by Rall [499] as a Hamiltonian system.

Let us introduce a new Hilbert space $\mathcal{H} = \mathcal{H}_1 \bigoplus \mathcal{H}_2$, such that every element h on $\mathcal{H}$ is an ordered pair $h = (x, p)$ with x in $\mathcal{H}_1$ and p in $\mathcal{H}_2$. We define the inner product in $\mathcal{H}$ as follows:

$$\{h_1, h_2\} = (x_1, x_2)_{\mathcal{H}_1} + \langle p_1, p_2 \rangle_{\mathcal{H}_2},$$

where $(\cdot)_{\mathcal{H}_1}$ and $\langle \cdot, \cdot \rangle_{\mathcal{H}_2}$ denote, respectively, the inner product in $\mathcal{H}_1$ and $\mathcal{H}_2$. Let $\tilde{p} \in \mathcal{H}_2, \tilde{x} \in \mathcal{H}_1, x \in \mathcal{H}_1$ be arbitrary vectors and $\tilde{w} = (\tilde{x}, \tilde{p})$ a vector in $\mathcal{H}$. The corresponding value of the functional $L : \mathcal{H} \to R$ is given by

$$L(x, Tx) = \langle T\tilde{x}, \tilde{p} \rangle - \frac{1}{2}\langle \tilde{p}, \tilde{p} \rangle - (f, \tilde{x}). \tag{7.142}$$

If the gradient L_w of the above functional is uniquely defined at $w_0 = (x_0, p_0) \in \mathcal{H}$, then Eq. (7.141) are the necessary and sufficient conditions for the vanishing of L_w. Thus, the functional has a critical point (or an extremum) at w_0 if and only if these equations are satisfied.

Further, in the context of complementary variational principle, the system (7.141) has been designated by Noble (see [22]) as a Hamiltonian system. This can be easily seen if one introduced the functional

$$w(x, p) = \frac{1}{2}\langle p, p \rangle + \langle f, x \rangle, \tag{7.143}$$

which can be regarded as the Hamiltonian such that the system (7.141) can be written as

$$Tx = W_p, \qquad T^*p = W_x. \tag{7.144}$$

These are Hamilton's canonical equations in the special case when

$$T = \frac{d}{dt}, \qquad T^* = -\frac{d}{dt}.$$

Here, x and p denote the vectors of generalized displacement and momenta, respectively.

Let the Hamiltonian system (7.144) (with Hamiltonian $\mathscr{H} = W(x, p)$) be satisfied at the critical point $x = \tilde{x}, p = \tilde{p}$. But if $W(x, p)$ fails to be convex at some neighborhood of $p = \tilde{p}$ for fixed $\tilde{x}$ (or vice versa), then we do not have an obvious way of formulating complementary variational principles ([22]), or deriving by two-sided bounds on the value of the Lagrangian. A remedy for this was suggested by Komkov ([341]) who pointed out that in such cases complementary variational principles could be formulated by altering the rules of multiplication of some vector-valued functions. The appropriate algebra for such problems was shown by him to be the quaternionic algebra of Hamilton. But we do not pursue it here because this will take us too far from our discussion.

Further, standard technique introduce by Kato [319] is to write (7.140) in terms of Lagrangian coordinates q, $\dot{q}$ and t with the Lagrangian $L(q, \dot{q}, t)$ as

$$T^*Tq = f$$

and then rewrite it as a system of two equations as before:

$$Tq = p(t), \quad T^*p = f(q(t), t).$$

The Hamiltonian is similar to (7.142) and is given by

$$\mathscr{H} = \frac{1}{2}\langle p, p \rangle + V(q, t), \qquad \frac{\partial V}{\partial q} = f(q, t).$$

The Legendre transformation is defined by the mapping

$$\{q, Aq\} \rightarrow \{q, p\}$$

with

$$p = \frac{\partial L}{\partial (Aq)}, \qquad \mathscr{H} = L - \langle Aq, p \rangle.$$

This establishes the duality between a Lagrangian and a Hamiltonian formulation.

Observation

1. If a linear operator is self-adjoint, then a variational be constructed by least squares method as explained above.
2. For a nonlinear operator, a variational principle exists if the Gâteaux differential of the operator is symmetric.

Examples of Variational Calculus

Example 7.16 Find the extremals of the functional $I[y(x)] = \int x(dx^2 + dy^2)^{1/2}$.

Solution. The given functional is

$$I[y(x)] = \int x(dx^2 + dy^2)^{1/2} = \int x\sqrt{1 + y'^2}\, dx.$$

So $F(x, y, y') = x\sqrt{1 + y'^2}$, $F_y = F_{yy'} = 0$, $F_{xy'} = \frac{y'}{\sqrt{1+y'^2}}$ and $F_{y'y'} = \frac{x}{(1+y'^2)^{3/2}}$.

Hence, the Euler's equation

$$F_y - F_{xy'} - F_{yy'}y' - F_{y'y}y'' = 0$$

reduces to

$$F_{xy'} + F_{y'y'} y'' = 0$$

i.e., $\dfrac{y'}{\sqrt{1+y'^2}} + \dfrac{x}{(1+y'^2)^{3/2}} y'' = 0$, i.e. $xy'' + y'(1+y'^2) = 0$

Example 7.17 Find the extremal of the following functional

$$I[y(x)] = \int_0^1 (y'^2 + y' + 1)\, dx, \quad y(0) = 1, y(1) = 2.$$

Solution. The given functional is

$$I[y(x)] = \int_0^1 (y'^2 + y' + 1)\, dx \tag{7.145}$$

with the boundary conditions

$$y(0) = 1, \quad y(1) = 2. \tag{7.146}$$

Here, $F(x, y, y') = y'^2 + y' + 1$, $F_y = 0$, $F_{y'} = 2y' + 1$.
   Hence, Euler's equation $F_y - \dfrac{d}{dx} F_{y'} = 0$ becomes

$$0 - \frac{d}{dx}(2y' + 1) = 0 \, , \, i.e. \, y'' = 0.$$

Its solution is $y = c_1 x + c_2$.
   Applying the boundary conditions (7.146), we have

$$y(0) = c_1 \cdot 0 + c_2 = 0; \, y(1) = c_1 + c_2 = 2,$$

which gives $c_1 = c_2 = 1$. Thus, the extremals are obtained on the straight line $y = x + 1$.

Example 7.18 Prove that the extremals of the functional $I[y(x)] = \int \sqrt{x(1 + y'^2)}\, dx$ are parabolas.

Solution. Here, $F(x, y, y') = \sqrt{x(1 + y'^2)}$ which is independent of y. Therefore, Euler's equation $F_y - \dfrac{d}{dx} F_{y'} = 0$ reduces to $-\dfrac{d}{dx}\left(\dfrac{\partial F}{\partial y'}\right) = 0$ or $\dfrac{\partial F}{\partial y'} = \text{constant} = c$, say,

i.e. $\dfrac{1}{2\sqrt{x(1+y'^2)}} .x.2y' = c$

i.e. $\sqrt{x} \cdot \dfrac{y'}{\sqrt{1+y'^2}} = c$ or $xy'^2 = c^2(1 + y'^2)$

i.e. $y'^2(x - c^2) = c^2$ or $y' = \dfrac{c}{\sqrt{x-c^2}}$ or $\dfrac{dy}{dx} = \dfrac{c}{\sqrt{x-c^2}}$.

Integrating both sides w.r.to x, we obtain

$$y = c \int \frac{dx}{\sqrt{x-c^2}} + c_1$$

i.e. $$y = 2c\sqrt{x - c^2} + c_1$$

or $$(y - c_1)^2 = 4c^2(x - c^2),$$

which represents a family of parabolas. Thus, the extremals of the given functional are parabolas.

Example 7.19 Derive the fundamental equation of quantum mechanics from a variational (Schrödinger equation [411]).

Solution. Suppose m is the mass of the particle whose motion is considered in a field of potential energy V and h denotes the plank's constant. We now define an operator known as the Hamiltonian operator as follows:

$$H \equiv -k\nabla^2 + V(x, y, z), \tag{7.147}$$

where $k = \frac{h^2}{8\pi^2 m}$ and the Laplacian $\nabla^2 \equiv \frac{\partial^2}{\partial x^2} + \frac{\partial^2}{\partial y^2} + \frac{\partial^2}{\partial z^2}$. We seek a wave function Ψ (possibly complex) which extremizes the functional

$$\iiint\limits_D \Psi^*(H\Psi)\,dxdydz \tag{7.148}$$

subject to the constraints

$$\iiint\limits_D \Psi^*\Psi\,dxdydz = 1, \tag{7.149}$$

where Ψ^* is the complex conjugate of Ψ and the integration is over a fixed domain D of x, y and z.

We further assume that the admissible functions Ψ and Ψ^* either vanish at the boundaries of the domain or take on same values and derivatives at corresponding points on opposite boundaries. As a consequence,

$$\iiint\limits_D \Psi^*\nabla^2\Psi dxdydz = -\iiint\limits_D \nabla\Psi^* \cdot \nabla\Psi dxdydz.$$

Introducing the Lagrange multiplier λ, we then find the extremum of the functional

$$\iiint\limits_D K dxdydz = \iiint\limits_D \left[k(\Psi_x^*\Psi_x + \Psi_y^*\Psi_y + \Psi_z^*\Psi_z) + V\Psi^*\Psi - \lambda\Psi^*\Psi \right] dxdydz.$$

The Euler equations are

$$\frac{\partial K}{\partial \Psi} - \frac{\partial}{\partial x}\left(\frac{\partial K}{\partial \Psi_x}\right) - \frac{\partial}{\partial y}\left(\frac{\partial K}{\partial \Psi_y}\right) - \frac{\partial}{\partial z}\left(\frac{\partial K}{\partial \Psi_z}\right) = 0,$$

$$\frac{\partial K}{\partial \Psi^*} - \frac{\partial}{\partial x}\left(\frac{\partial K}{\partial \Psi_x^*}\right) - \frac{\partial}{\partial y}\left(\frac{\partial K}{\partial \Psi_y^*}\right) - \frac{\partial}{\partial z}\left(\frac{\partial K}{\partial \Psi_z^*}\right) = 0,$$

which reduces to

$$-k\nabla^2\Psi + V\Psi = \lambda\Psi \qquad (7.150)$$

This may be written as

$$H\Psi = \lambda\Psi.$$

If we multiply this by Ψ^* and integrate over the domain of x, y, z, the left side becomes the stationary integral (7.148), which is denoted by E. Hence, by (7.149), we have $\lambda = E$.

Observe that for $\lambda = E$, (7.150) reduces to the Schrödinger equation. Note that there is an interesting and important connection between Hamilton–Jacobi equation for a classical system and the Schrödinger equation for a quantum mechanical system. In fact, if we put the wave function $\Psi = e^{(i/h)S}$, where S is the action function of the classical system, then the Schrödinger equation reduces to the Hamilton–Jacobi equation provided S is much larger than Planck's constant h, a universal constant of nature. Notice that $h = 6.626196 \times 10^{-27}$ erg s. Thus, in the limit of large values of action and energy, the surfaces of constant phase for the wave function Ψ reduce to surface of constant action S for the corresponding classical system. In this case, wave mechanics reduces to classical mechanics just as wave optics reduces to geometrical optics in the limit of very small wavelengths.

Remark 7.7 It may be remarked that the Klein–Gordon equation

$$\nabla^2\Psi - \frac{1}{c^2}\frac{\partial^2\Psi}{\partial t^2} - \left(\frac{mc}{h}\right)^2\Psi = 0,$$

where c denotes the velocity of light, representing a possible wave equation for a relativistic particle (though it is not correct for an electron or proton) can be constructed in the same manner from the Lagrange functions

$$L = -\frac{h^2}{2m}\left[\nabla\Psi^* \cdot \nabla\Psi - \frac{1}{c^2}\left(\frac{\partial\Psi}{\partial t}\right)^2 + \left(\frac{mc}{h}\right)^2\Psi^*\Psi\right].$$

Exercises

7.1 Show that the functional $I[y(x)] = \displaystyle\int_{x_1}^{x_2}(y^2 + x^2y')\,dx$ assumes extreme values on the straight line $y = x$.

7.2 Show that the extremal of the functional $I[y(x)] = \displaystyle\int_0^1(x\sin y + \cos y)\,dx$, $y(0) = 0$, $y(1) = \frac{\pi}{4}$ can be found on the curve $y = \tan^{-1}x$.

7.3 Find the curve on which the functional $I[y(x)] = \int_1^2 \frac{x^3}{y^2}\, dx$ with $y(1) = 1$, $y(2) = 4$ can be extremized.

7.4 (Euler–Lagrange Equation) Find the function $y(x)$, having boundary conditions $y(0) = 0$ and $y(1) = 1$, that makes the integral

$$I = \int_{x_1}^{x_2} \left(y'^2 + yy' + y^2\right) dx$$

stationary.

7.5 Evaluate the functional $I[y(x)] = \int_0^{\log 2} \left(e^{-x} y'^2 - e^x y^2\right) dx$ and find the extremals.

[Hint: Use a substitution $x = \log u,\ y = v$.]

7.6 (Minimum Surface) A curve $y(x)$ in the xy-plane connecting the points (x_1, y_1) and (x_2, y_2) is revolved around the y-axis and generates a surface of revolution with area

$$A = \int_{x_1}^{x_2} 2\pi x \sqrt{1 + y'(x)^2}\, dx.$$

Show that the curve $y(x)$ which generates the surface of revolution with the least area can be expressed in terms of the inverse hyperbolic cosine

$$y(x) = c_1 \cosh^{-1}(x/c_1) + c_2$$

where c_1 and c_2 are constants.

7.7 Find the function $y(x)$ which connects the points $y(0) = 0$ and $y(1) = 1$ and makes the integral

$$I[y(x)] = \int_0^1 \left[y'^2 - \frac{\pi^2 y^2}{4}\right] dx$$

stationary by solving the Euler–Lagrange equation.

7.8 Find the curve with fixed boundary points (x_1, y_1) and (x_2, y_2) such that its rotation about x-axis gives rise to a surface of revolution of minimum surface area.

7.9 Show that the curve which extremizes the functional $I = \int_0^{\pi/4} \left(y'^2 - y^2 + x^2\right) dx$ under the conditions $y(0) = 0$, $y'(0) = 1$, $y\left(\frac{\pi}{4}\right) = y'\left(\frac{\pi}{4}\right) = \frac{1}{\sqrt{2}}$ is $y = \sin x$.

7.10 Find the plane curve of fixed perimeter and maximum area.

7.11 Find a function $y(x)$ for which $\int_0^1 (x^2 + y'^2)\, dx$ is a stationary, given that $\int_0^1 y^2\, dx = 2$ with boundary conditions $y(0) = 0$, $y(1) = 0$.

7.12 Find the extremals of the isoperimetric problem $I[y(x)] = \int_{x_0}^{x_1} y'^2\, dx$ given that $\int_{x_0}^{x_1} y\, dx = c$, a constant.

7.13 Show that the shortest curve joining two fixed points is a straight line.

7.14 Show that the shortest line between any two points on a cylinder is a helix.

7.15 Determine the shape of a solid of revolution moving in a flow of gas with least resistance.

7.16 (Euler's problem on buckling) If a rod made up of homogeneous and isotropic material with uniform cross section throughout its length is simply supported by a longitudinal force P acting at the other end, it is then in either stable or unstable equilibrium. This means that after a slight lateral bending, it will return to either its equilibrium position or buckle, depending on whether the magnitude of P is less than or greater than a certain critical value P_0. Determine the buckling force P_0.

7.17 Using variational theory, approximate the first eigenvalue of the problem $\nabla^2 z + \lambda z = 0$ with $z = 0$ on the boundary of the domain D, which is a circle of unit radius with center at the origin.

7.18 Using variational theory, approximate the first eigenvalue of the Mathieu equation
$$y'' + \lambda(2 + \cos x)y = 0, \quad y(0) = y(\pi) = 0.$$

7.19 Using the Rayleigh–Ritz method, find an approximate solution of the differential equation
$$y'' + x^2 y = x, \quad y(0) = y(1) = 0.$$

Determine $y_2(x)$ and y_3 and compare their values at the points $x = 0.25$, $x = 0.5$ and $x = 0.75$.

7.20 Using the Rayleigh–Ritz method, find an approximate solution of $\nabla^2 z = -1$ inside the square $S = \{(x, y) \in \mathbb{R}^2 : -l \leqslant x \leqslant l, -l \leqslant y \leqslant l\}$ which vanishes on the boundary of the square S.

7.21 Using variational methods, establish the inequality
$$\int_0^\infty (y^2 - y'^2 + y''^2)\, dx > 0$$

unless $y(x) = C \exp\left(-\frac{1}{2}x\right) \sin\left(\frac{\sqrt{3}}{2}x - \frac{\pi}{3}\right)$ when there is equality. Assume that $y, y'' \in L_2(0, \infty)$.

7.22 Let Q be the square $\{(x_1, x_2) \in \mathbb{R}^2 : 0 < x_1 < 1, 0 < x_2 < 1\}$. Using variational methods, establish the inequality

$$\int_Q f^2 \, dx \leqslant \frac{1}{2\pi^2} \int_Q |\nabla f|^2 \, dx$$

for every function $f \in H_0^1(Q)$ and the constant in the inequality is exact.

7.23 Let Q be the cube $\{(x_1, x_2, x_3) \in \mathbb{R}^3 : 0 < x_1 < 1, 0 < x_2 < 1, 0 < x_3 < 1\}$. Using variational methods, establish the inequality

$$\|f\|_{L_2}^2 \leqslant \frac{1}{3\pi^2} \|\nabla f\|_{L_2}^2$$

for every function $f \in H_0^1(Q)$.

Chapter 8
Applications of Monotone Operator Theory

> *Mathematical research should be as broad and as original as possible, with very long-range goals. We expect history to repeat itself: we expect that the most profound and useful future applications of mathematics cannot be predicted today, since they will arise from mathematics yet to be discovered.*

> Arthur M. Jaffe (1984)
> (Ordering the Universe: the Role of Mathematics)

We studied so far many nonlinear problems that arose naturally in mathematical physics, differential topology and geometry of manifolds. In mathematical physics, it arises in the problems of elasticity, Newtonian mechanics of particles, theory of gas discharge and quantum fields, etc. In differential topology, it arises in smoothness of surfaces, convex analysis, etc. In differential geometry, nonlinearity arises in study of geodesics on manifolds, surface of constant curvatures, theory of automorphic forms, etc.

In the most general setting, these problems can be described by a nonlinear boundary value problems of the type given below. Suppose Ω is an open subset of $\mathbb{R}^n$, m a positive integer, and assume that the following conditions hold:

$$\left.\begin{aligned} Au &= 0 \text{ in } \Omega \\ D^\alpha u &= 0 \text{ in } \partial\Omega \end{aligned}\right\} \tag{8.1}$$

where A is a quasi-linear differential operator of the form

$$Au = \sum_{|\alpha| \leqslant k} (-1)^{|\alpha|} D^\alpha A_\alpha(x, u, Du, D^2 u, \ldots, D^k u).$$

For each α, A_α is a function of $x \in \Omega$, the function u on Ω and all derivatives of u up to order k.

© Springer Nature Singapore Pte Ltd. 2018
H. K. Pathak, *An Introduction to Nonlinear Analysis and Fixed Point Theory*,
https://doi.org/10.1007/978-981-10-8866-7_8

There are basically two approaches to obtain the solvability of (8.1) direct and indirect approaches.

- In the direct approach, one examines the operator theoretic analogue of (8.1), directly in appropriate function spaces (mainly Sobolev spaces).
- In the indirect approach, (8.1) is converted into the equivalent integral equation of the form

$$u(s) + \int_\Omega K(s, t) f(t, u(t)) dt = v(s) \tag{8.2}$$

by means of Green's function $K(x, t)$, and then, functional analytic study is carried out in appropriate function spaces (L_p-spaces).

In this chapter, we deal with these and other aspects of nonlinear equations. In Sect. 8.1, we deal with differential equations arising in boundary value problems. In Sect. 8.2, we discuss integral equations in the most general setting. In Sect. 8.3, we give computational schemes for the solvability of nonlinear equations. Finally, we close the chapter in Sect. 8.4 by presenting strong convergence of proximal-type algorithm for occasionally pseudomonotone operators in Hilbert spaces.

8.1 Ordinary Differential Equations

We start our discussion by considering two-point boundary value problem for ordinary differential equation

$$\left.\begin{aligned} u'' &= f(x, u(x)) \text{ for a.e. } x \in [0, \pi] \\ u(0) &= u(\pi) = 0 \end{aligned}\right\} \tag{8.3}$$

where $f : [0, \pi] \times \mathbb{R} \to \mathbb{R}$ is a function satisfying Carathéodory conditions. To solve the two-point BVP for ordinary differential equation (8.3), we follow operator theoretic approach as follows:

Let $\mathcal{H} = L_2[0, \pi]$, $Eu = u''$ and $[N_f u](x) = f(x, u(x))$, where $\mathcal{D}(E)$ is given by

$$\mathcal{D}(E) = \left\{ u \in \mathcal{H} : u \text{ is absolutely continuous, } u', u'' \in \mathcal{H}; u(0) = u(\pi) = 0 \right\}.$$

Assume that $f(x, u(x))$ satisfies a growth condition

$$|f(x, u)| \leqslant a(x) + b|u|, a \in L_2[0, \pi], \ b > 0.$$

Then, we see that E is a demicontinuous operator from $\mathcal{D}(E)$ to $\mathcal{H}$ and N_f a bounded and continuous operator from $\mathcal{H}$ into itself. Moreover, (8.3) is equivalent to the operator equation

$$Eu = N_f u \tag{8.4}$$

in the space $L_2[0, \pi]$. Because the null space of E is trivial, (8.4) is equivalent to the equation

$$u - E^{-1} N_f u = 0$$

or

$$u + K N_f u = 0, \tag{8.5}$$

where $K = -E^{-1}$. Equation (8.5) is a Hammerstein equation, which will be discussed in detail in the next section. However, we shall make use of the following proposition which is to be proved later (in Sect. 8.2).

Proposition 8.1 *Let X be a reflexive Banach space with dual X^*. Let $K : X \to X^*$ be a bounded symmetric monotone operator and $N : X^* \to X$ a demicontinuous operator satisfying the condition*

$$(u_1 - u_2, Nu_1 - Nu_2) \geqslant -p\|u_1 - u_2\|^2, u_1, u_2 \in X^* \text{ for some } p \geqslant 0.$$

If $\|K\|_p < 1$, then the Hammerstein equation

$$u + KNu = v$$

has a unique solution u. This solution u varies continuously with v.

We now come back to the original problem. To this end, we first observe that the linear associated eigenvalue problem $Eu = \lambda u$ with boundary condition $u(0) = u(\pi) = 0$ has a countable system of eigenvalues $-n^2, n = 1, 2, \ldots$ with corresponding eigenfunctions $\{\phi_n\}$ forming a complete orthonormal set in $L_2[0, \pi]$. This implies that $-E^{-1}$ has eigenvalues $\frac{1}{n^2}$, $n = 1, 2, \ldots$ with eigenfunctions $\{\phi_n\}$. This gives us

$$(-E^{-1}u, u) \geqslant 0 \ \forall u \in L_2[0, \pi] \text{ and } \|-E^{-1}u\| \leqslant \|u\|.$$

Finally, we use Proposition 8.1 to get the following theorem regarding the solvability of (8.3).

Theorem 8.1 *Let $f : [0, \pi] \times \mathbb{R} \to \mathbb{R}$ be a function satisfying Carathéodory conditions such that*

$$|f(x, u)| \leqslant a(x) + b|u|, a \in L_2[0, \pi], \ b > 0.$$

Further, assume that there exists $p \geqslant 0$ such that the following condition holds

$$(f(x, u_1) - f(x, u_2))(u_1 - u_2) \geqslant -p(u_1 - u_2)^2 \ \text{for all } u_1, u_2 \in \mathbb{R}.$$

If $p < 1$, then (8.3) has a unique solution.

The problem under consideration is of particular interest when E has a nontrivial null space, the case which is often mentioned as the problem at resonance. In such a case, (8.4) is decomposed into a system of two equations, one of which, usually called the auxiliary equation, is in an infinite dimensional space and is often uniquely solvable. The other one, called the bifurcation equation, lies in a finite dimensional space. This method is called the alternative method and is essentially due to Cesari [137]. We now discuss this method, starting with a result of Kannan [312].

Example 8.1 Let $T = [0, l]$ and consider the following problem

$$\left. \begin{array}{r} -u''(x) = \lambda u(x)) \quad \text{for a.e. } x \in T \\ u(0) = u(l) = 0. \end{array} \right\} \tag{8.6}$$

We say that $\lambda \in \mathbb{R}$ is an eigenvalue of minus the scalar Laplacian operator with Dirichlet boundary condition $(-u'', W_0^{1,2}(T))$, if (8.6) has a nontrivial solution $u \in W_0^{1,2}(T)$, which is called a corresponding eigenfunction. It is well known that (8.6) has a sequence of eigenvalues $0 < \lambda_1 \leqslant \lambda_2 \leqslant \cdots \leqslant \lambda_k \to +\infty$ and the corresponding eigenfunctions $\{\phi_n\}_{n \geqslant 1} \subseteq L_2(T)$ form an orthonormal basis of $L_2(T)$. More precisely, we have

$$\lambda_n = \left(\frac{n\pi}{l}\right)^2 \quad \text{and} \quad \phi_n(x) = \sqrt{\frac{2}{l}} \sin\left(\frac{n\pi x}{l}\right), \ n \geq 1.$$

Example 8.2 Let $T = [0, l]$ and consider the following problem

$$\left. \begin{array}{r} -u''(x) = \lambda u(x)) \quad \text{for a.e. } x \in T \\ u(0) = u(l), u'(0) = u'(l). \end{array} \right\} \tag{8.7}$$

In this problem, we can say that there is an increasing sequence of eigenvalues $\lambda_0 = 0 < \lambda_1 \leqslant \lambda_2 \leqslant \cdots \leqslant \lambda_k \to +\infty$ and the corresponding sequence of eigenfunctions $\{\phi_n\}_{n \geqslant 1} \subseteq L_2(T)$ that form an orthonormal basis of $L_2(T)$. In fact, we have

$$\lambda_n = \left(\frac{2n\pi}{l}\right)^2 \quad \text{and} \quad \phi_0(x) = \frac{1}{\sqrt{l}}, \ \phi_n(x) = \sqrt{\frac{2}{l}} \cos\left(\frac{2n\pi x}{l}\right), \ n \geq 1.$$

Example 8.3 Let $T = [0, l]$ and consider the following problem

$$\left.\begin{array}{l} -u''(x) = \lambda u(x)) \quad \text{for a.e. } x \in T \\ u(0) = u'(0), \ u(l) = u'(l). \end{array}\right\} \quad (8.8)$$

In this problem, $\lambda \in \mathbb{R}$ is an eigenvalue of minus the scalar Laplacian operator with Dirichlet boundary condition $(-u'', W_0^{1,2}(T))$, if (8.8) has a nontrivial solution $u \in W_0^{1,2}(T)$, which is called a corresponding eigenfunction. It is well known that (8.8) has a sequence of eigenvalues $0 < \lambda_1 \leqslant \lambda_2 \leqslant \cdots \leqslant \lambda_k \to +\infty$ and the corresponding eigenfunctions $\{\phi_n\}_{n \geqslant 1} \subseteq L_2(T)$ form an orthonormal basis of $L_2(T)$. More precisely, we have

$$\lambda_n = \left(\frac{n\pi}{l}\right)^2 \quad \text{and} \quad \phi_n(x) = \sqrt{\frac{n^2\pi^2 + l^2}{2l}}\left[\sin\left(\frac{n\pi x}{l}\right) + \frac{n\pi}{l}\cos\left(\frac{n\pi x}{l}\right)\right], \ n \geq 1.$$

Example 8.4 Consider the eigenvalue problem

$$-u''' - \lambda u' = 0, \quad u(0) = u'(0) = u'(1) = 0$$

which has infinitely many eigenvalues $\lambda_n = (n\pi)^2, n = 1, 2, \ldots$. Each λ_n has a multiplicity of 1, and the corresponding eigenvectors $\{\phi_n(x)\} = \left\{\frac{\sqrt{2}}{n\pi}(1 - \cos(n\pi x))\right\}$ form a complete orthonormal sequence in H^k.

8.1.1 Nonlinear Differential Equations

First, we recall the following:

Let $\varphi : \mathbb{R}^+ \to \mathbb{R}^+$ be continuous, and let $T : X \to X^*$ be a mapping such that

$$(Tx - Ty, x - y) \geqslant \varphi(\|x - y\|) \quad \text{for all } x, y \in X.$$

Then, T is called monotone if $\varphi(t) = 0$, strongly monotone if $\varphi(t) = ct^2$ for some $c > 0$ and firmly monotone if $\varphi(0) = 0$ and $\varphi(t) > 0$ when $t > 0$.

Notice that the importance of mappings of monotone type stems from the fact that various classes of differential operators in divergence form give rise to equations involving operators of monotone type acting in suitable Sobolev spaces to their respective duals (see, e.g., [89, 482]). As a result, since 1960 this class of operators has been extensively studied by many authors under the basic assumption that X is reflexive (see [204, 564, 620]).

We now consider the nonlinear differential equation

$$u'' = f(x, u(x))$$

where $u(x) = (u_1(x), \ldots, u_n(x))$ and f is 2π-periodic in x satisfying Carathéodory conditions.

Let $\mathcal{H} = L_2[0, \pi]$, $Eu = u''$ and N_f a bounded and continuous operator from $\mathcal{H}$ into itself with $\mathcal{D}(E)$ given by

$$\mathcal{D}(E) = \{u \in \mathcal{H} : u \text{ is absolutely continuous}, u', u'' \in \mathcal{H}; u(0) = u(2\pi), u'(0) = u'(2\pi)\}.$$

The null space of E is nontrivial, and the linear associated eigenvalue problem

$$Eu + \lambda u = 0$$

with boundary conditions $u(0) = u(2\pi)$, $u'(0) = u'(2\pi)$ has countable eigenvalues λ_n with corresponding eigenfunctions $\{\phi_n\}$ forming a   complete orthonormal system in $L_2[0, 2\pi]$. Also, we have $\lambda_k \geqslant 0$ and $\lambda_k \to \infty$. Let $P : L_2[0, 2\pi] \to M$ be the projection operator with $M = [\phi_1, \phi_2, \ldots, \phi_m]$. Let $M_1 = (I - P)M$ and $L : M_1 \to M_1$ be the linear operator defined by

$$Lu = -\sum_{m+1}^{\infty} c_i \lambda_i^{-1} \phi_i \text{ for } u = \sum_{m+1}^{\infty} c_i \phi_i \in M_1,$$

m being such that $\lambda_{m+1} > 0$.

With these notations, we see that

$$L(I - P)Eu = (I - P)u, \quad u \in \mathcal{D}(E), \tag{8.9}$$

$$EL(I - P)N_f u = (I - P)N_f u, \quad u \in M_1, \tag{8.10}$$

$$PEu = EPu, u \in \mathcal{D}(E). \tag{8.11}$$

If $u \in \mathcal{D}(E)$ satisfies (8.4), then by applying $L(I - P)$ and using (8.9) we obtain $(I - P)u = L(I - P)N_f u$, and thus, a solution $u \in \mathcal{D}(E)$ of (8.4) is also a solution of

$$u - L(I - P)N_f u = Pu. \tag{8.12}$$

And if u is a solution of (8.12), by applying E to both sides of (8.12), we have (by using (8.10) and (8.11))

$$Eu - (I - P)N_f u = EPu = PEu \text{ or } EU - N_f u = P(Eu - N_f u).$$

Thus, a solution of (8.12) is also a solution of (8.4) iff

$$P(Eu - N_f u) = 0. \tag{8.13}$$

So we conclude that solvability of (8.4) is equivalent to the solvability of the system of equations (8.12) and (8.13).

If now u^* be any arbitrary element of M and if the equation

$$u - L(I - P)N_f u = u^* \qquad (8.14)$$

is uniquely solvable for such $u^* \in M$, then the solution u is such that $Pu = u^*$ and $PEu = EPu = Du^*$. Thus, Eq. (8.13) reduces to

$$PN_f(I - L(I - P)N_f)^{-1} u^* - Eu^* = 0. \qquad (8.15)$$

Hence, if Eq. (8.14) is uniquely solvable for each $u^* \in M$, solving (8.4) is equivalent to solving the system equations (8.14) and (8.15). Equations (8.14) and (8.15) are called the auxiliary and bifurcation equations, respectively.

We now discuss the solvability of auxiliary and bifurcation equations. The auxiliary equation

$$u - L(I - P)N_f u = u^* \qquad (A)$$

is of Hammerstein type (8.5) where $K = -L(I - P)$ and $(Ku, u) \geqslant \lambda_{m+1}\|Ku\|^2$, which gives

$$\|Ku\|^2 \leqslant \frac{1}{\lambda_{m+1}}|(Ku, u)| \leqslant \frac{1}{\lambda_{m+1}}\|Ku\|\|u\|$$

$$\Rightarrow \qquad \|Ku\| \leqslant \frac{1}{\lambda_{m+1}}\|u\|.$$

Using Proposition 8.1, we get the following proposition for the solvability of the auxiliary equation(A).

Proposition 8.2 *Suppose the Nemytskiǐ operator N_f is defined by $N_f u(t) = f(t, u(t))$ be such that $N : \mathcal{H} \to \mathcal{H}$ is a demicontinuous. Further, assume that there exist $p, m > 0$ such that for all $u, v \in \mathcal{H}$, $(N_f u - N_f v, u - v) \geqslant -p\|u - v\|^2$. If $p < \lambda_{m+1}$, then the auxiliary equation (A) has a unique 2π-periodic solution in $\mathcal{H}$ for each $u^* \in M$. Further, $[I - L(I - P)N_f]^{-1}$ is continuous and bounded.*

We now use the theory of monotone operators to obtain the solvability of the bifurcation equation

$$PN_f[I - L(I - P)N_f]^{-1} u^* - Eu^* = 0. \qquad (8.16)$$

in the finite dimensional space M.

Proposition 8.3 *Suppose the Nemytskiǐ operator N_f is defined by $N_f u(t) = f(t, u(t))$ is such that N_f is demicontinuous from $\mathcal{H}$ into itself. Further assume the following:*

(i) there exists $\lambda_m < p < \lambda_{m+1}$ such that for all $u, v \in \mathcal{H}$.

$$(N_f u - N_f v, u - v) \geqslant -p\|u - v\|^2,$$

(ii) *there exists q such that* $\lambda_m < q \leqslant p$ *and for all* $u^*, v^* \in M$,

$$(N_f u - N_f v, u^* - v^*) \leqslant -q\|u^* - v^*\|^2,$$

where u, v *are the solutions of the auxiliary equation corresponding to* u^*, v^*.

Then, the bifurcation equation (8.16) has a unique solution.

Proof Let $T : M \to M$ be the operator defined by

$$T = PN_f[I - L(I - P)N_f]^{-1} - E.$$

Let u^*, v^* be two elements of M and let $u, v \in \mathcal{H}$ such that

$$[I - L(I - P)N_f]^{-1}u^* = u, \quad [I - L(I - P)N_f]^{-2}v^* = v.$$

Notice that such u, v exist, in view of our assumptions (i) and (ii) and Proposition 8.2. Further, we see that

$$\begin{aligned}
(Tu^* - Tv^*, u^* - v^*) &= (Eu^* - Ev^*, u^* - v^*) \\
&= -(PN_f u - PN_f v, u^* - v^*) \\
&= -\lambda_m\|u^* - v^*\|^2 - (N_f u - N_f v, u^* - v^*) \\
&= -\lambda_m\|u^* - v^*\|^2 + q\|u^* - v^*\|^2 \\
&= (q - \lambda_m)\|u^* - v^*\|^2.
\end{aligned}$$

Since $q > \lambda_m$, it follows that T is a strongly monotone operator, and hence, by Theorem 4.24, we get that $Tu^* = 0$ has a unique solution. That is, the bifurcation equation (8.16) has a unique solution.

Combining these two propositions, we obtain the following theorem for the solvability of Eq. (8.4).

Theorem 8.2 *Let the Nemytskiĭ operator defined by* $(N_f u)(x) = f(x, u(x))$ *be such that* $N_f : \mathcal{H} \to \mathcal{H}$ *is demicontinuous. Further, assume the following:*

  (i) f *is* 2π-*periodic in* x,
 (ii) *there exist* p, q, m *such that* $m^2 < q \leqslant p < (m+1)^2$,
(iii) *for all* $u, v \in \mathcal{H}, (N_f u - N_f v, u - v) \geqslant -p\|u - v\|^2$,
 (iv) *for all* $u, v \in \mathcal{H}, (N_f u - N_f v, u^* - v^*) \leqslant -q\|u^* - v^*\|^2$,

where u *and* v *are the solutions of the auxiliary equation corresponding to* u^*, v^*. *Then, Eq. (8.4) has a unique* 2π *periodic solution.*

It may be of interest to consider the problem at resonance,

$$Eu + \lambda_m u = N_f u, \qquad\qquad (8.17)$$

where λ_m is any eigenvalue of the operator E. If $E' = E + \lambda_m I$, then by suitable indexing, we can assume that $\lambda_m < \lambda_{m+1}$ and the eigenvalues of E' are

$$\lambda_1 - \lambda_m \leqslant \cdots \leqslant \lambda_m - \lambda_m = 0 < \lambda_{m+1} - \lambda_m \leqslant \cdots .$$

Let $\varepsilon > 0$ be any number, $0 < \varepsilon C \lambda^{-1}(\lambda_{m+1} - \lambda_m)$, and take $q = \varepsilon$, $p = \lambda_{m+1} - \lambda_m - \varepsilon$. Then, (8.17) has a unique solution if we require for N_f to satisfy

$$(N_f u - N_f v, u - v) \geqslant -p\|u - v\|^2 \text{ for all } u, v \in \mathcal{H}$$

and

$$(N_f u - N_f v, u^* - v^*) \leqslant -q\|u^* - v^*\|^2 \text{ for all } u^*, v^* \in M.$$

Finally, we use the same conventions as above.

We finally state a result due to Kannan [312] for a periodically perturbed conservative system

$$u'' + \nabla G(u) = p(x),$$

where $p(x) \in L_2[0, 2\pi]$ and is 2π-periodic in x and G in $C^2(\mathbb{R}^n, \mathbb{R})$. This problem can be treated as a case of perturbation at resonance with $\lambda_m = 0$.

Theorem 8.3 *Let* $G \in C^2(\mathbb{R}^n, \mathbb{R})$ *be such that for all* $a \in \mathbb{R}^n$,

$$M^2 I < qI \left(\frac{\partial^2 G(a)}{\partial u_i \partial u_j} \right) \leqslant pI < (M+1)^2 I$$

when M is an integer. Then, the nonlinear differential equation

$$u'' + \nabla G(u) = p(x)$$

has a unique 2π-periodic solution.

8.1.2 Variational Boundary Value Problem

We now discuss the solvability of variational boundary value problem corresponding to a general quasi-linear partial differential equation

$$Au = \sum_{|\alpha| \leqslant k} (-1)^{|\alpha|} D^\alpha A_\alpha(x, Du, \ldots, D^k u) = 0.$$

To help build ideas, we first draw analogy with the situation for linear equations. Corresponding to the linear differential operator L given by

$$Lu = \sum_{|\alpha| \leqslant k} (-1)^{|\alpha|} D^{\alpha}(a_{\alpha\beta}(x)D^{\beta}u),$$

we define the Dirichlet form $\alpha(u, v)$ as

$$\alpha(u, v) = \sum_{|\alpha| \leqslant k, \, |\beta| \leqslant k} \int_{\Omega} a_{\alpha\beta}(x)D^{\beta}u(x)D^{\alpha}v(x)dx.$$

Consider now the Dirichlet problem

$$Lu = f \text{ in } \Omega$$

$$D^{\alpha}u = 0 \text{ on } \Omega, \; |\alpha| \leqslant k - 1. \tag{8.18}$$

If u and $a_{\alpha\beta}$ are sufficiently smooth, (8.18) is equivalent to the equation

$$\alpha(u, v) = (f, v) \text{ for all } v \in C_0^{\infty}(\Omega). \tag{8.19}$$

Hence to extend the meaning of the solution of the Dirichlet problem (8.18), it suffices to assume that (8.19) is satisfied with the condition that $u, v \in W_0^{k,2}(\Omega)$.

We now define the principal part of L as

$$L' = (-1)^k \sum_{|\alpha|=k, \, |\beta|=k} a_{\alpha\beta} D^{\alpha+\beta}(x)\xi^{\alpha+\beta}$$

The uniform strong ellipticity of L is defined as follows: there exists a positive constant c_0 such that for all real ξ and all $x \in \Omega$.

$$\sum_{|\alpha|=k, \, |\beta|=k} a_{\alpha\beta}(x)\xi^{\alpha+\beta} \geqslant c_0|\xi|^{2k}.$$

Proposition 8.4 (Garding's inequality) *Let Ω be any open set and L be a uniformly strongly elliptic operator. Assume that*

(i) $a_{\alpha\beta}(x)$ is uniformly continuous on Ω, $|\alpha| = |\beta| = k$,
(ii) $a_{\alpha\beta}(x)$ is bounded and measurable, $|\alpha| + |\beta| \leqslant 2k$.

Then, there are constants $r_0 > 0$ and $\lambda_0 > 0$ such that

$$\alpha(u, v) \geqslant r_0 c_0 \|u\|_{k,2}^2 - \lambda_0 \|u\|^2 \text{ for all } u \in C_0^k(\Omega).$$

Proposition 8.5 (Regularity theorem) *Assume that*

(i) Ω *is a bounded subset of* $\mathbb{R}^n$ *with smooth boundary*
(ii) L *is a uniformly strongly elliptic operator whose coefficients are in* $C^\infty(\Omega)$
(iii) $f \in C^\infty(\overline{\Omega})$
(iv) $u \in W^{k,2}(\Omega)$ *is a solution of (8.18).*

Then, u can be modified on a set of measure zero so that $u \in C^\infty(\overline{\Omega})$. *We now assume that coefficient* $a_{\alpha\beta}$ *are bounded. Then,* $\alpha(u, v)$ *is a bounded bilinear form in* $W^{k,2}(\Omega)$ *and hence can be written as*

$$\alpha(u, v) = (Tu, v) \text{ for all } v \text{ in } V = W_0^{k,2}(\Omega),$$

where T is a bounded linear mapping on V. If we assume the strong form of the Garding's inequality

$$\alpha(u, v) \geqslant c\|u\|_{k,2}^2, u \in W_0^{k,2}(\Omega), c > 0,$$

then T satisfies the condition

$$(Tu, u) \geqslant c\|u\|_{k,2}^2 \text{ for all } u \in V = W_0^{k,2}(\Omega).$$

Further, it follows by Lax-Milgram lemma that $Tu = f$ *has a unique solution u in* V. *So (8.18) has a unique solution* $u \in W_0^{k,2}(\Omega)$. *If f and* $a_{\alpha\beta}$ *are sufficiently smooth, it follows by regularity theorem (Proposition 8.5) that u actually belongs to* $C_0^\infty(\overline{\Omega})$ *(except on a set of measure zero). This motivates us to define a nonlinear Dirichlet form* $\alpha(u, v)$, *corresponding to the quasi-linear elliptic operator A which is given in the divergence form*

$$Au = \sum_{|\alpha|\leqslant k} (-1)^{|\alpha|} D^\alpha A_\alpha\big(x, Du(x), \ldots, D^k u(x)\big) \tag{8.20}$$

where for each α, $A - \alpha$ *is a mapping of* $\Omega \times \mathbb{R}^{S_k}$ *to* $\mathbb{R}$. *Corresponding to the representation (8.19), we define the nonlinear Dirichlet form* $\alpha(u, v)$ *as*

$$\alpha(u, v) = \sum_{|\alpha|\leqslant k} \int_\Omega A_\alpha\big(x, u(x), \ldots, D_u^k(x)\big) D^\alpha v(x)\, dx.$$

or
$$\alpha(u, v) = \sum_{|\alpha|\leqslant k} \int_\Omega A_\alpha(x, \xi_k u(x)) D^\alpha v(x)\, dx,$$

$$\tag{8.21}$$

for all u, v in $W^{k,p}(\Omega)$, *where* $\xi_k u(x) = \{D^\alpha u(x) : |x| \leqslant k\}$.

Definition 8.1 Let V be a closed subspace of $W^{k,p}(\Omega)$ with $W_0^{k,P}(\Omega) \subset V$, and let $\alpha(u, v)$ be a Dirichlet form as defined by (8.21). Let $f \in V^*$. Then, $u \in V$ is a variational solution of the Dirichlet problem $Au = f$ if

$$\alpha(u, v) = (f, v) \text{ for all } v \in V. \tag{8.22}$$

Eq. (8.22) together with the restriction that u lies in V has the force not only of requiring that u should satisfy the partial differential equation $Au = f$ but also of imposing boundary condition upon u.

In the following, $\mathbb{R}^{Sm}$ is the vector space whose elements are $\xi = \{\xi_x : |x| \leqslant k\}$. Then, A can be written as

$$Au = \sum_{|\alpha| \leqslant k} (-1)^{|\alpha|} D^\alpha A_\alpha(x, \xi_k u(x)),$$

where $\xi u(x) = \{D^\alpha u(x) : |\alpha| \leqslant k\}$.

Assumption (A)

(1) Each $A_\alpha(x, \xi_k)$ is measurable in x for fixed ξ_k in $\mathbb{R}^{S_k}$ and continuous in $\xi_k \in \mathbb{R}^{S_k}$ for almost all x in Ω.
(2) There exists a real number p with $1 < p < \infty$ such that for a given $g \in L_q(\Omega)(q = p/(p-1))$ and a constant $c > 0$, we have

$$|A_\alpha(x, \xi_k)| \leqslant c|\xi_k|^{p-1} + g(x).$$

We now define operators $F_\alpha : W^{k,p} \to L_q$ and $G_\alpha : (L_p x)^{S_k} \to L_q$ by

$$[F_\alpha u](x) = A_\alpha(x, u(x)), u \in W^{k,p}(\Omega)$$

and
$$[G_\alpha]\phi(x) = A_\alpha(x, \phi(x)), \quad \phi \in (L_p x)^{S_k},$$

respectively. Here, $(L_p x)^{S_k}$ is the product space of L_p, S_k times.

We have the following lemma.

Lemma 8.1 *Suppose that each A_α satisfies Assumption (A). Then, the operator F_α is a bounded continuous operator from W_m^p to L_q.*

Proof Let $ju = \{\xi(u)\}$, then $F_\alpha = G_\alpha o j$. G_α maps all of $(L_p x)^{S_k}$ into L_q. So by an extension of Theorem 1.44, it follows that G_α is a bounded and continuous map from $(L_p x)^{S_k}$ into L_q. As F_α is the composition of two continuous and bounded maps, it following that F_α is a bounded and continuous map from $W^{k,p}$ to L_q.

Theorem 8.4 *Let each A_α satisfies Assumption (A). Then, the Dirichlet form $\alpha(u, v)$ satisfies the inequality*
$$|\alpha(u, v)| \leqslant g_1(\|u\|_{k,p})\|v\|_{k,p},$$

where $g_1(r)$ is a function of the real variable r depending on $\|g\|$.

Proof We have

$$\alpha(u, v) = \sum_{|\alpha| \leqslant k} \int_\Omega A_\alpha(x, \xi_k u(x)) D^\alpha v(x) \, dx.$$

Assumption (A) (1) gives

$$\begin{aligned}
|\alpha(u, v)| &\leqslant \sum_{|\alpha| \leqslant k} \int_\Omega |D^\alpha v(x)| \left[g(x) + c|\xi u(x)|^{p-1} \right] dx \\
&\leqslant \sum_{|\alpha| \leqslant k} \left[\|D^\alpha v\| \right] \|g\| + c\|u\|_{k,p}^{p-1} \\
&= \|v\|_{k,p} \left[\|g\| + c\|u\|_{k,p}^{p-1} \right].
\end{aligned}$$

This proves the theorem. ∎

In view of this theorem, it follows that for a fixed a $v \in W^{k,p}$, $\alpha(u, v)$ is a bounded linear functional on V, and hence, by Riesz representation theorem, there exists $Fu \in V^*$ such that

$$\alpha(u, v) = (Fu, v) \text{ for all } v \in V.$$

Theorem 8.5 *Under the hypothesis of the above theorem, the operator F is a bounded demicontinuous operator from V to V^*.*

Proof By Theorem 8.4, we get

$$|(Fu, v)| \leqslant \|v\|_{k,p} \left[\|g\| + c\|u\|_{k,p}^{p-1} \right],$$

for all $v \in V$ and hence we get

$$\|Fu\| \leqslant c\|u\|_{k,p}^{p-1} + \|g\|.$$

Let the map $F_\alpha : W^{k,p}(\Omega) \to L_q(\Omega)$ be as defined before. Then,

$$(Fu_n - Fu, v) = \sum_{|\alpha| \leqslant k} \int_\Omega (F_\alpha u_n(x)) - F_\alpha u(x)) D^\alpha v(x) dx.$$

Applying Hölder's inequality, we get

$$|(Fu_n - Fu, v)| \leqslant \|v\|_{k,p} \Sigma |\alpha| \leqslant k \|F_\alpha u_n - F_\alpha u\|_q.$$

Suppose $u_n \to u$ in V. By Lemma 8.1, each F_α is a bounded continuous map from $W^{k,p}$ to L_q and so $\|F_\alpha u\|_q \to 0$ for each α. Therefore, it follows from the above inequality that $(Fy_n - Fu, v) \to 0$ for each $v \in V$, which proves the demicontinuity of F.

Assumption (B)

(1) For each $x \in \Omega$ and each pair ξ_k, ξ_k' in $\mathbb{R}^{S_k}$,

$$\Sigma |\alpha| \leqslant k \left[A_\alpha(x, \xi_k) - A_\alpha(x, \xi_k') \right] \left[\xi_\alpha - \xi_k' \right] \geqslant 0$$

(2) There exists a positive constant $c_0 > 0$ and a function h in $L_q(\Omega)$ such that for all ξ_k in $\mathbb{R}^{S_k}$ and all x in Ω

$$\Sigma |\alpha| \leqslant k A_\alpha(x, \xi_k) \xi_k \geqslant c_0 |\xi_k|^p - h(x), \quad 1 < p < \infty.$$

Theorem 8.6 *Let A be a quasi-linear operator of type (8.20) satisfying assumptions (A) and (B) on an open subset Ω of $\mathbb{R}^n$. Let V be a closed subspace of $W^{k,p}(\Omega)$ which contains $W_0^{k,p}(\Omega)$. Then, for each $f \in V^*$, $Au = f$ has a variational solution $u \in V$.*

Proof Let $\alpha(u, v)$ be the Dirichlet form defined by (8.21).Then, the solvability of $\alpha(u, v) = (f, v)$ is equivalent to the solvability of $Fu = f$, where F is a nonlinear operator given by $\alpha(u, v) = (Fu, v)$. We have already shown that F is a demicontinuous operator from V into V^*. Further, we have

$$(Fu_1 - Fu_2, u_1 - u_2) = \alpha(u_1, u_1 - u_2) - \alpha(u_2, u_1 - u_2)$$

$$= \int_\Omega \Sigma |\alpha| \leqslant k \left[A_\alpha(x, \xi_k u_1(x)) - A_\alpha(x, \xi_k u_2(x)) \right] \times \left[\xi_k(u_1(x)) - \xi_k(u_2(x)) \right] dx \geqslant 0,$$

in view of Assumption (B) (1). Also

$$(Fu, u) = \int_\Omega \Sigma |\alpha| \leqslant k \left[A_\alpha(x, \xi_k u(x)) \xi_k u(x) \right] dx$$

$$\geqslant c_0 \int_\Omega |\xi_k u(x)|^p dx - \|h\|_{L_q} \|u\|_{k,p}$$

$$\geqslant c_0 \|u\|_{k,p}^p - c_1 \|u\|_{k,p}, \quad c_1 = \|h\|_{L_q}.$$

Because $p > 1$, it follows that

$$\frac{(Fu, u)}{\|u\|_{k,p}} \geqslant c_0 \|u\|_{\cdot,p}^{p-1} - c_1 \to \infty \quad \text{as} \quad \|u\|_{k,p} \to \infty.$$

Thus, F is a demicontinuous, monotone, coercive mapping of V into V^*, and hence, it follows by Theorem 4.26 that $Fu = f$ has a solution $u \in V$. This proves the theorem. ∎

Remark 8.1 Assumption (B) replaces the strong ellipticity condition imposed in the theory of linear elliptic problem. In particular, Assumption (B) (1) implies ellipticity or monotonicity on differential operators A_α corresponding to all derivatives of order less than or equal to k, not just the higher-order derivatives. By applying Sobolev imbedding lemma, the ellipticity on the lower order derivatives could be eliminated as we see in the following theorem.

Assumption (C)

(i) If $\xi_k = (\zeta_k, \eta_{k-1})$ is the division of ξ_k into its kth order components ζ_k and the corresponding $(k-1)th$ order η_{k-1}, then for x in Ω and each $\eta_{k-1} \in \mathbb{R}^{S_{k-1}}$,

$$\Sigma|\alpha| = m\left[A_\alpha(x, \zeta_k, \eta_{m-1} - A_\alpha(x, \zeta_k', \eta_{k-1})\right]\left[\zeta_\alpha - \zeta_\alpha'\right]dx > 0, \quad \zeta_k \neq \zeta_k'.$$

(ii) There exist constants $c_0 > 0$ and c_2 such that for all x and ξ_k

$$\Sigma|\alpha| \leqslant kA_\alpha(x, \xi)\xi_\alpha \geqslant c_0|\xi|^p - c_2.$$

Theorem 8.7 *Let Ω be a bounded open subset of $\mathbb{R}^n$ wi smooth boundary and V a closed subspace of $W^{k,p}(\Omega)$ which contains $W_0^{k,p}(\Omega)$. Let A be a quasi-linear elliptic operator of type (8.20) satisfying assumptions (A) and (C). Then, for each $f \in V^*$, $Au = f$ has a variational solution $u \in V$.*

Remark 8.2 The result of Theorem 8.7 holds true if Assumption (C)(ii) is replaced by the following one:

Assumption (C)(ii)':

$$\sum_{|\alpha|=k} A_\alpha(x, \zeta_k, \eta_{k-1}\zeta_\alpha \geqslant c_0|\zeta|^p - c_1 \sum_{n-(n/p)\leqslant|\beta|\leqslant k-1} |\eta_\beta|^r, r < p-1, c_0 > 0.$$

Example 8.5 Consider the Dirichlet problem

$$\frac{\partial}{\partial x_1}\left(b_1(x)\frac{\partial u}{\partial x_1}\right) + \frac{\partial}{\partial x_2}\left(b_2(x)\frac{\partial u}{\partial x_2}\right) = f\left(x, u, \frac{\partial u}{\partial x_1}, \frac{\partial u}{\partial x_2}\right) \quad in\ \Omega \subseteq \mathbb{R}^2$$

$$u = 0 \text{ on } \partial\Omega. \tag{8.23}$$

In our earlier notation, this is equivalent to

$$Au = 0, \quad u \in W_0^{1,2}(\Omega)$$

where

$$Au = \sum_{|\alpha|\leqslant 1} (-1)^{|\alpha|} D^\alpha A_\alpha(x, \xi u(x)),$$

$$A_{00}(x, \xi_0, \xi_1, \xi_2) = f(x, \xi_0, \xi_1, \xi_2),$$
$$A_{01}(x, \xi_0, \xi_1, \xi_2) = b_1(x)\xi_1,$$
$$A_{10}(x, \xi_0, \xi_1, \xi_2) = b_2(x)\xi_2.$$

We assume that the function f satisfies the Carathéodory conditions and

$$|f(x, \xi_0, \xi_1, \xi_2)| \leqslant b\left[|\xi_0| + |\xi_1| + |\xi_2|\right] + g(x), \quad b > 0, \quad g \in L_2,$$

$$b_1(x) \geqslant r, b_2(x) \geqslant r, \ r > 0, \ b_1, b_2 \in L_2.$$

It is easily seen that:

(i) Assumption (A) of the above theorem is trivially satisfied in view of our assumptions on f and b_1 and b_2,

(ii) Assumption (C) is also satisfied since

$$\sum_{|\alpha|=1} \left[A_\alpha(x, \xi_0, \xi_1, \xi_2) - A_\alpha(x, \xi_0, \xi_1', \xi_2') \right] (\xi_\alpha - \xi_\alpha')$$

$$= b_1(x)(\xi_1 - \xi_1')(\xi_1 - \xi_1') + b_2(x)(\xi_2 - \xi_2')(\xi_2 - \xi_2')$$

$$= b_1(x)(\xi_1 - \xi_1')^2 + b_2(x)(\xi_2 - \xi_2')^2 > 0;$$

and

$$\sum_{|\alpha|=1} A_\alpha(x, \xi_0, \xi_1, \xi_2)\xi_\alpha = b_1(x)\xi_1{}^2 + b_2(x)\xi_2{}^2 \geqslant r(\xi_1{}^2 + \xi_2{}^2).$$

As all conditions of Theorem 8.7 are satisfied, it follows that (8.23) has a variational solution $u \in W_0^{1,2}(\Omega)$.

8.2   Integral Equations

In this section, we shall mainly be concerned with the integral equations of the following type:

(1) Nonlinear Hammerstein integral equation

$$x(s) + \int_\Omega k(s, t) f(t, x(t)) \, dt = y(s). \tag{8.24}$$

(2) Nonlinear integral equation of mixed type involving sum of Hammerstein operator

$$x(s) + \int_\Omega k_1(s, t) f_1(t, x(t)) \, dt + \int_\Omega k_2(s, t) f_2(t, x(t)) dt = y(s). \tag{8.25}$$

(3) Generalized Hammerstein integral equation

$$x(s) + \int_\Omega k_x(s, t) f(t, x(t)) \, dt = y(s), \tag{8.26}$$

where $k_x(s, t)$ is a kernel which is a function of the solution.

(4) Urysohn's integral equation

$$x(s) + \int_\Omega k(s, t, x(t)) \, dt = y(s). \qquad (8.27)$$

Our approach is operator theoretic. That is, the problem of determining the solution of the above equations is reduced to the problem of determining the solution of the appropriate nonlinear operator equations.

8.2.1 Hammerstein Operator Equation

Consider the nonlinear Hammerstein integral equation:

$$x(s) + \int_\Omega k(s, t) f(t, x(t)) dt = y(s), \qquad (8.28)$$

where k is a function defined on $\Omega \times \Omega$ with values in $\mathbb{R}$, f is a function defined on $\Omega \times \mathbb{R}$ to $\mathbb{R}$ and is such that it satisfies Carathéodory conditions (refer Sect. 1.8.1) and Ω is a measurable subset of $\mathbb{R}^n$.

We now define the linear and nonlinear operators K and N_f as follows:

$$[Kx](s) = \int_\Omega k(s, t) x(t) dt \qquad (8.29)$$

$$[N_f x](s) = f(s, x(s)). \qquad (8.30)$$

Then, one can readily see that (8.28) is equivalent to the nonlinear operator equation

$$x + K N_f x = y. \qquad (8.31)$$

The above equation is called Hammerstein operator equation. We shall obtain abstract solvability theorem for (8.31) and then derive solvability results for (8.28). By suitably redefining N_f, one can assume that $y = 0$ in (8.31), and hence, we need consider only the following operator equation

$$x + K N_f x = 0. \qquad (8.32)$$

We divide the discussion on the above equation in two parts, one for noncompact and the other K compact.

8.2.1.1   K is Noncompact

The direct application of the theory of monotone operators dates back to Dolph-Minty [195]. We state and prove their theorem as follows:

Theorem 8.8 *Let $\mathcal{H}$ be a Hilbert space and $N_f : \mathcal{H} \to \mathcal{H}$ be hemicontinuous and monotone and $K : \mathcal{H} \to \mathcal{H}$ continuous and strongly monotone. Then, (8.31) has a unique solution x for every y, and this solution x depends continuously on y and satisfies the estimate*

$$\|x\| \leqslant \frac{1}{c} \left[\|K\| \|y\| \right]$$

where c is the constant of strong monotonicity.

Proof Let x_1, x_2 be such that

$$x_1 + K N_f x_1 = y = x_2 + K N_f x_2.$$

Then,

$$\left(N_f x_1 - N_f x_2, x_1 - x_2 \right) + \left(N_f x_1 - N_f x_2, K N_f x_1 - K N_f x_2 \right) = 0.$$

Strong monotonicity of K gives $N_f x_1 = N_f x_2$, which in turn implies that $x_1 = x_2$. This proves the uniqueness.

    We now prove the existence of solution. Since $(Kx, x) \geqslant c\|x\|^2$ for some $c > 0$, it implies that K has bounded inverse K^{-1} and K^{-1} is strongly monotone. Hence, $K^{-1} + N_f$ is strongly monotone, and so by Theorem 4.24, $(K^{-1} + N_f)x = K^{-1}y$ has a solution x for every $y \in \mathcal{H}$, and this solution depends continuously on y. Also, strong monotonicity of K^{-1} gives the required estimate.   ∎

    Notice that in concrete cases K is the integral operator given by (8.29) and is therefore strongly monotone. Consequently, we can utmost make K monotone. Hence, for application purpose we need to generalize the above theorem wherein only monotonicity condition on K is imposed. We have such a theorem which is again due to Dolph-Minty [195].

Theorem 8.9 *Let $K : \mathcal{H} \to \mathcal{H}$ be linear, continuous and monotone and $N_f : \mathcal{H} \to \mathcal{H}$ be continuous, bounded and monotone. Further, let there exists $M > 0$ such that*

$$(N_f x, x) \geqslant 0 \, \text{for} \, \|x\| > M. \tag{8.33}$$

Then the operator Eq. (8.32) has a solution.

Remark 8.3 The above two theorems also hold if the underlying space is a reflexive Banach space.

Example 8.6 We now apply the above theorem to concrete Hammerstein integral equation

$$x(s) + \int_\Omega k(s, t) f(t, x(t)) dt = 0, \tag{8.34}$$

where K and f are as defined before. We consider the following assumptions:

(1) $k(s, t)$ is symmetric, positive semidefinite and Hilbert–Schmidt kernel.
(2) (*a*) $f(s, x)$ is monotone increasing with respect to x and satisfies a growth condition of the type

$$|f(s, x)| \leqslant a(s) + b|x|, \ b > 0, \ a \in L_2(\Omega).$$

(*b*) There exist constants a, b, M such that for all $s \in \Omega$

$$f(s, x) > a + bx \text{ for } x > M \text{ and}$$

$$f(s, x) < -a + bx \text{ for } x < -M.$$

As shown before, the above equation is equivalent to the operator equation

$$x + K N_f x = 0,$$

where K and N_f are as defined by (8.29) and (8.30). In view of assumptions (1) and (2(a)), it follows that K is a bounded linear monotone operator from $L_2(\Omega)$ into itself (refer Examples 1.37 and 4.7) and N_f is a continuous, bounded monotone operator from $L_2(\Omega)$ into itself (refer Example 4.2 and Theorem 2.5). Further, (2)(b) implies that there exists a constant $M > 0$ such that $(N_f x, x) \geqslant 0$ for $\|x\| > M$. As all the conditions of Theorem 8.9 are satisfied, it follows that there exists a solution x of the operator equation

$$x + K N_f x = 0.$$

This in turn implies that there exists a solution $x(s) \in L_2(\Omega)$ of (8.34).

Remark 8.4 One can get $L_p(\Omega)$ solution of (8.34) by assuming weaker conditions on K and f. We merely have to appeal to the theory of monotone operators on Banach spaces. This will be clear in our subsequent theorems.

Definition 8.2 Let X be a Banach space with dual X^* and $K : X \to X^*$ be a bounded linear monotone mapping. Then, K is said to be angle bounded with constant $c \geqslant 0$ if for all x_1, x_2 in X

$$|(K x_1, x_2) - (K x_2, x_1)| \leqslant 2c \left[(K x_1, x_1)(K x_2, x_2)\right]^{1/2}.$$

K is angle bounded iff the numerical range of the complexified operator K lies in the sector

$$\left\{ \xi \in C : |\arg \xi| \leqslant \tan^{-1} c < \frac{\pi}{2} \right\}.$$

Obviously, every symmetric bounded linear mapping is angle bounded with constant $x = 0$.

This concept of angle boundedness was first introduced by Amann [13]. For some sufficient conditions for angle boundedness of linear operators, refer Amann [14, 15].

As we shall see later, an essential tool in the proof of our main theorem is the following auxiliary result (now known as splitting lemma), which is of interest by itself.

Lemma 8.2 *Let $K : X \to X^*$ be a angle-bounded continuous monotone linear mapping. Then, there exists a Hilbert space $\mathcal{H}$, a continuous linear mapping S of X into $\mathcal{H}$ with S^* injective and a bounded skew symmetric linear mapping B of $\mathcal{H}$ such that*

(1) $K = S^*(I + B)S$
(2) $\|B\| \leqslant c$
(3) $\|S\|^2 \leqslant \|K\|$
(4) $\left[(I + B)^{-1}h, h \right]_{\mathcal{H}} \geqslant \frac{1}{[1 + \|B\|^2]} \|h\|_{\mathcal{H}}^2.$

This lemma is due to Browder and Gupta [117].

Theorem 8.10 *Let $K : X \to X^*$ be a monotone, angle-bounded continuous linear mapping with constant of angle boundedness $x \geqslant 0$. Let N_f be a hemicontinuous mapping of X^* into X such that for a given constant k*

$$\left(x_1 - x_2, N_f x_1 - N_f x_2 \right) \geqslant -k \|x_1 - x_2\|^2 \text{ for all } x_1, x_2 \in X^*.$$

Suppose $k(1 + c^2)\|K\| < 1$, then there exists a unique solution x in X^* of the Hammerstein operator equation (8.32) and the solution x satisfies the estimate $\|x\| \leqslant \|K\| \frac{(1+c^2)\|N_f(0)\|}{[1-(1+c^2)k\|K\|]}.$

Proof Consider the Hammerstein operator equation

$$x + K N_f x = 0.$$

By splitting Lemma 8.1, we have $K = S^*(I + B)S$ and so the above equation becomes

$$x + S^*(I + B)S N_f x = 0. \tag{8.35}$$

Since S^* is injective, there exists a unique $h \in \mathcal{H}$ such that $x = S^*h$ and so (8.35) becomes

$$S^*h + S^*(I + B)S N_f S^* h = 0. \tag{8.36}$$

Since S^* is injective and $(I + B)$ is invertible, the above equation is equivalent to

$$Th = 0. \tag{8.37}$$

where $T = (I + B)^{-1} + SN_f S^*$. Using the above lemma, we obtain that

$$[Th_1 - Th_2, h_1 - h_2]_{\mathcal{H}} \geq \left(\frac{1}{1+x^2} - k\|K\| \right) \|h_1 - h_2\|_{\mathcal{H}}^2$$

$$= c_1 \|h_1 - h_2\|_{\mathcal{H}}^2 \tag{8.38}$$

where

$$c_1 = \left[\frac{1}{1+c} - k\|K\| \right] > 0.$$

Since T is a hemicontinuous, strongly monotone operator on $\mathcal{H}$, it follows that there exists a unique solution h of (8.37). This in turn implies that there exists a unique solution $x = S^* h$ of the Hammerstein equation (8.32). The estimate for x follows from the strong monotonicity of T. ∎

If N_f is given to be monotone, we get the following result.

Corollary 8.1 *Let $K : X \rightarrow X^*$ be monotone continuous, angle-bounded mapping and $N_f : X^* \rightarrow X$ be hemicontinuous and monotone. Then, there exists a unique solution of (8.32). This result was first obtained by Amann [13] with an additional assumption that X^* has a dense continuous linear imbedding in a Hilbert space.*

Corollary 8.2 *Let $K : X \rightarrow X^*$ be a bounded linear mapping which is monotone and symmetric. Suppose $N_f : X^* \rightarrow X$ is a hemicontinuous mapping such that for a given k and all $x_1, x_2 \in X^*$.*

$$(x_1 - x_2, N_f x_1 - N_f x_2) \geq -k\|x_1 - x_2\|^2.$$

Let $k\|K\| < 1$. Then, (8.32) has a unique solution.

This result was first obtained by Golomb [254] for $X = L_2(\Omega)$ and Vainberg [597] for $X = L_p(\Omega)$.

As in Example 8.5, one can give an immediate application of the above theorem to nonlinear Hammerstein integral equation with less restrictive conditions on the nonlinear function $f(s, x)$.

In the following, we give an application of the above theorem to nonlinear Volterra integral equation of convolution type which occurs often in real-life problems.

Example 8.7 A nonlinear Volterra integral equation of convolution type is of the form.

$$x(t) + \int_0^t k(t - \tau) f(\tau, x(\tau) d\tau = 0. \tag{8.39}$$

Here, kernel k is a function from $[0, \infty)$ to $\mathbb{R}$ and $f(t, x)$ is defined on $[0, \infty) \times \mathbb{R}$ with values in $\mathbb{R}$.

In operator theoretic terms, (8.39) is equivalent to the Hammerstein equation

$$x + K N_f x = 0$$

where the Nemytskiǐ operator N_f is as defined before and K is the convolution operator

$$[Kx](t) = \int_0^t k(t - \tau) x(\tau) d\tau. \tag{8.40}$$

Consider the following assumptions:

(1) $K \in L_{p/2}[0, \infty) \cap L_1[0, \infty)$ and its Fourier transform $K(iw)$ is such that $\operatorname{Re} K(iw) \geqslant 0$ for all w, and there exists a constant c such that

$$|\operatorname{Im} K(iw)| \leqslant c \operatorname{Re} K(iw) \quad \text{for all } w. \tag{8.41}$$

(2) $f(t, x)$ satisfies a growth condition of the type
    (a) $|f(t, x)| = \leqslant a(t) + b|x|^{p-1}, a \in L_q[0, \infty), b > 0 \quad \left(\frac{1}{p} + \frac{1}{q} = 1, p \geqslant 2 \right)$
    and is such that

$$(b) \quad \left[\int_0^\infty (f(t, x_1)) - f(t, x_2(t))(x_1(t) - x_2(t)) dt \right] \geqslant -K \left(\int_0^\infty |x_1(t)|^q dt \right)^{2/q}$$

   for some constant K.
(3) $K(1 + c^2)\|K\| < 1$.

Because $K \in L_{p/2}[0, \infty)$, it follows by Young's inequality that K maps $L_q[0, \infty)$ to $L_p[0, \infty)$ boundedly (refer Sect. 1.8). We now show that K is monotone as an operator from $L_q[0, \infty)$ to $L_p[0, \infty)$. Since $L_2[0, \infty) \cap L_q[0, \infty)(q \leqslant 2)$ is dense in $L_q[0, \infty)$, it suffices to show that $(Kx, x) \geqslant 0$ for $x \in L_2[0, \infty) \cap L_q[0, \infty)$. Also as K is the convolution operator defined by (8.40) and the kernel $K(t) \in L_1[0, \infty)$, it follows that $Kx \in L_2[0, \infty)$ for $x \in L_2[0, \infty)$. Hence, (Kx, x) is L_2 inner product, and hence, using Parseval's equality, we have

$$(Kx, x) = \frac{1}{\pi} \int_0^\infty \operatorname{Re} K(iw)|x(iw)|^2 dw \geqslant 0.$$

We now prove that in view of inequality (8.41) of Assumption (1), K is angle bounded. For this, we use the equivalent criterion for angle boundedness : K is angle bounded iff the numerical range $N_f(K)$ of the complexified operator lies in the sector

$$= \{\tau \in C : 0 \leqslant |\arg \tau| \geqslant \tan^{-1} c < \frac{\pi}{2}\}.$$

$N_f(K)$ is defined as

$$N_f(K) = \{(Kx, x) : \|x\| = 1\}.$$

As before

$$(Kx, x) = \frac{1}{\pi} \int_0^\infty K(iw)|x(iw)|^2 dw.$$

Now by using (8.41), we get

$$0 \leqslant |\arg(Kx, x)| \leqslant \tan^{-1} c < \frac{\pi}{2}.$$

Further, assumptions 2(a), (b) imply that the Nemytskiĭ operator N_f maps $L_p[0, \infty)$ to $L_q[0, \infty)$ and is such that

$$(x_1 - x_2, N_f x_1 - N_f x_2) \geqslant -k \|x_1 - x_2\|^2$$

Applying Theorem 8.10, we claim that (8.39) has a unique solution $x(t) \in L_p[0, \infty)$.

8.2.1.2   K is Compact

In a concrete case like Example 8.5, K is the integral operator generated by some Hilbert–Schmidt kernel $K(s, t)$ and hence is compact (refer Chap. 1, Sect. 1.8).

We now study the abstract Hammerstein operator equation (8.32) with this additional assumption of compactness on operator K. As we will see, one can now weaken condition on N. A result in this direction was first obtained by Amann [16] which we shall now prove.

Theorem 8.11 *Let X be an arbitrary Banach space with dual X^* and $K : X \to X^*$ be monotone, compact and injective. Let Y be a closed subspace of X^* such that $\mathscr{R}(K) \subset Y$. Let $N_f : Y \to X$ be continuous and bounded and assume that there exists a constant $\rho > 0$ such that $\|x\| \geqslant \rho$ implies that*

$$(x, K^{-1}x) + (x, N_f x) > 0. \tag{8.42}$$

Then, the Hammerstein operator equation (8.32) has at least one solution x in Y. Moreover, every solution x satisfies $\|x\| \leqslant \rho$.

Proof Let $x \in Y$ be any solution of the Hammerstein operator equation

$$x + K N_f x = 0.$$

Then, $x \in R(K)$, and since K is injective, it satisfies the equation $K^{-1}x + N_f x = 0$. This implies that $(x, K^{-1}x)) + (x, N_f x) = 0$ and hence $\|x\| \leqslant \rho$, in view of (8.42).

Denote by σ the positive real number defined by

$$\sigma = \sup\{\|N_f x\| : \|x\| \leqslant \rho, x \in Y\}.$$

By the boundedness of N_f, this is finite.

Define a mapping $\widetilde{N}_f : Y \to X$ by

$$
\widetilde{N}_f x = \begin{cases} N_f x, & \|N_f x\| \leqslant \sigma \\ \frac{\sigma N_f x}{\|N_f x\|}, & \|N_f x\| > \sigma. \end{cases}
$$

$\widetilde{N}_f$ is continuous and is such $\|\widetilde{N}_f x\| \leqslant \sigma$ for all $x \in Y$. Hence, the mapping $K\widetilde{N}_f :$ $Y \to Y$ is continuous and compact and maps the ball $B_\delta (\delta = \|K\|\sigma)$ into itself. Hence, by Schauder's theorem, it has at least one fixed point x in B_δ. That is, there exists $x \in Y$ satisfying $x + K\widetilde{N}_f x = 0$. Or equivalently $K^{-1}x + \widetilde{N}_f x = 0$. Assume that $\|N_f x\| > \sigma$. Then,

$$
(x, K^{-1}x) + \frac{\sigma}{\|N_f x\|}(x, N_f x) = 0.
$$

or

$$
\frac{\|N_f x\|}{\sigma}(x, K^{-1}x) + (x, N_f x) = 0
$$

or

$$
(x, K^{-1}x) + (x, N_f x) = \left(1 - \frac{\|N_f x\|}{\sigma}\right)(x, K^{-1}x) \leqslant 0.
$$

This implies that $\|x\| \leqslant \sigma$, which is a contradiction. Therefore, $\widetilde{N}_f x = N_f x$, and we have $x + K N_f x = 0$. ∎

Remark 8.5 In the above theorem, one can omit the condition of injectivity on K. Amann [16] has proved that for monotone K one can suitably define new operators $\hat{K}$ and $\hat{N}$ and spaces $\hat{X}^* \hat{X}$ such that new operator $\hat{K}$ is injective and the solvability of $x + K N_f x = 0$ is equivalent to the solvability of $x + \hat{K}\hat{N}x = 0$. So we have the following theorem.

Theorem 8.12 *Let $K : X \to X^*$ be monotone and compact. Let Y be a closed subspace of X^* such that $\mathscr{R}(K) \subset Y$. Let $N_f : Y \to X$ be continuous and bounded and assume that there exists a constant ρ such that*

$$
(x, N_f(x)) > 0 \text{ for } \|x\| > \rho. \tag{8.43}
$$

Then, the Hammerstein equation (8.32) has a solution x in Y with $\|x\| \leqslant \rho$.

Equation (8.43) is called condition at infinity upon N_f. This can be considerably relaxed if K is restricted to the class of angle-bounded mappings. We have the following theorem.

Theorem 8.13 *Let $K : X \to X^*$ be angle bounded and compact and Y be a closed subspace of X^* such that $\mathscr{R}(K) \subset Y$. Let $N_f : Y \to X$ be continuous and bounded and assume that there exists a function $\varphi : \mathbb{R}^+ \to \mathbb{R}^+$ satisfying $\varphi(\rho) = 0(\rho^2)$ as $\rho \to \infty$ such that for all $x \in R(K)$*

$$(x, N_f x) > -\varphi(\|x\|).$$

Then, the Hammerstein equation (8.32) has at least one solution in Y.

We now apply the above theorem to obtain a continuous solution of the nonlinear Hammerstein integral equation.

Example 8.8 Consider again the nonlinear Hammerstein integral equation

$$x(s) + \int_\Omega k(s,t) f(t, x(t)) dt = 0.$$

Here, Ω is an arbitrary set, dt is a σ-additive, and σ-finite measure on a σ-algebra of subsets of Ω, $k : \Omega \times \Omega \to \mathbb{R}$ is a kernel which defines a bounded linear integral operator

$$Kx(s) = \int_\Omega k(s,t) x(t) dt$$

in the space $L_2 = L_2(\Omega)$ and the function x is unknown. Consider the following assumptions:

(1) $K(x,t)$ is a continuous symmetric Kernel with nonnegative eigenvalues.
(2) $f : \Omega \times \mathbb{R} \to \mathbb{R}$ is a Carathéodory function; i.e., $f(\cdot, u)$ is measurable on Ω for all $u \in \mathbb{R}$, and $f(t, \cdot)$ is continuous on $\mathbb{R}$, and there exists $\rho \geqslant 0$ such that for all $s \in \Omega$, $|x| > \rho$ implies that $xf(s,x) \geqslant 0$. We define the linear integral operator K and the nonlinear operator N_f as before. Then, the above integral equation is equivalent to
$$x + K N_f x = 0.$$

Because K is continuous and Ω is bounded, the corresponding operator K maps $L_1(\Omega)$ boundedly into its dual $L_\infty(\Omega)$. By our Assumption (1), $K : L_2(\Omega) \to L_2(\Omega)$ is monotone and symmetric (and hence angle bounded). Hence, the continuity of K and the fact that $L_2(\Omega)$ is dense in $L_1(\Omega)$ imply that K is angle bounded as a mapping from $L_1(\Omega)$ to $L_\infty(\Omega)$. Also, by continuity assumption on f, it follows that the Nemytskiĭ operator N maps $L_\infty(\Omega)$ continuously and boundedly into $L_1(\Omega)$.

Finally, for $x \in C(\Omega)$, we have

$$
\begin{aligned}
(x, N_f x) &= \int_\Omega x(s) f(x, x(s)) ds \\
&= \int_{\{s \in \Omega : |x(s)| \geqslant \rho\}} x(s) f(x, x(s)) ds + \int_{\{s \in \Omega : |x(s)| \leqslant \rho\}} x(s) f(x, x(s)) ds \\
&\geqslant \int_{\{s \in \Omega : |x(s)| \leqslant \rho\}} x(s) f(x, x(s)) ds \\
&\geqslant -\rho \max_{|x| \leqslant \rho} |f(s, x)| \\
&= -\varphi_0.
\end{aligned}
$$

Hence, with $\varphi(\rho) = \varphi_0$ for all $\rho \in \mathbb{R}^+$, all assumptions of the previous theorem are satisfied. Hence, we have a continuous solution of the Hammerstein nonlinear integral equation. For more results on Hammerstein operator equation, we refer to Petryshyn and Fitzptrik [483], Browder [95], Brezis and Browder [80]. The concept of pseudomotonicity is very effectively used in these papers.

8.2.2 Equations Involving Sum of Hammerstein Operators

Let us consider nonlinear integral equations of mixed type

$$u(s) + \int_\Omega K_1(s,t) f_1(t, u(t)) dt + \int_\Omega K_2(s,t) f_2(t, u(t)) dt = 0, \qquad (8.44)$$

where K_1, K_2 are kernel function defined on $\Omega \times \Omega$ with values in $\mathbb{R}$ and f_1 f_2 are real-valued functions on $\Omega \times \mathbb{R}$, satisfying Carathéodory conditions.

As before, we define linear operators K_1, K_2 and nonlinear operators N_1, N_2 as

$$[K_i u](s) = \int_\Omega K_i(s,t) u(t) dt$$

$$[N_i u](s) = f_i(s, u(s)), \quad i = 1, 2.$$

Then, problem of solving (8.44) is reduced to the problem of solving the operator equation

$$u + K_1 N_1 u + K_2 N_2 u = 0 \qquad (8.45)$$

in appropriate function spaces.

Equation (8.45) involves sum of the Hammerstein operators $K_1 N_1$ and $K_2 N_2$. For the sake of generality, in this section we shall consider operator equation which involves sum of n Hammerstein operators:

$$u + \sum_{j=1}^n K_j N_j u = 0. \qquad (8.46)$$

In order to apply the theory which we have developed so far, we define Banach space $\mathfrak{X} = X \times X \times \cdots \times X$ and for $U = [u_1, u_2, \ldots, u_n] \in \mathfrak{X}$, the norm $\| \cdot \|$ on $\mathfrak{X}$ as

$$\|U\| = \sqrt{\sum_{j=1}^n \|u_j\|^2}.$$

We can similarly define $\mathfrak{X}^*$. Next, we define linear mapping and nonlinear mapping K and N by

$$K(U) = [K_1 u_1, \ldots, K_n u_n] \in \mathfrak{X}^*$$

and

$$N(U) = [N_1 u, \ldots, N_n u]$$

where

$$u = \sum_{j=1}^{n} u_j, U = [u_1, u_2, \ldots, u_n].$$

It can be seen easily that (8.46) is equivalent to the operator equation

$$U + KNU = 0. \tag{8.47}$$

Equation (8.47) is now the standard Hammerstein operator equation in a new space, and so one can make use of the theory developed before to obtain the existence results. This approach was initiated by Browder [95] to tackle equations of the type (8.46). We state his result.

Theorem 8.14 *Let* $\{K_j\}_{j=1}^{n}$ *be a family of continuous monotone angle-bounded operators from* X *to* X^* *with the same constant* c *of angle boundedness and* $\|K_j\| \leqslant K_0$ *for all* i. *Let* $\{N_j\}_{j=1}^{n}$ *be the corresponding family of nonlinear hemicontinuous operators from* X^* *to* X *satisfying a condition of the type*

$$\sum_{j=1}^{n} (u_i - v_j, N_j u - N_j v) \geqslant -k \sum_{j=1}^{n} \|u_j - v_j\|^2 \tag{8.48}$$

where $k \geqslant 0$ *and*

$$u = \sum_{j=1}^{n} u_j, v = \sum_{j=1}^{n} v_i.$$

If $(1 + c^2) K_0 k < 1$, *then (8.46) has a unique solution.*

Observation

- It may be observed that condition (8.48), though a natural generalization of the monotonicity condition, is rather hard to verify. The author (Joshi [292]) has weakened this condition on the operators $N_1, N_2, \ldots, N_n$ by assuming additional hypothesis of compactness on the linear operators K_j. In the application of this theory to the case where the K_j are integral operators, this assumption of compactness is a natural one. We now state this result.

Theorem 8.15 *Let* $\{K_j\}_{j=1}^{n}$ *be a finite family of monotone compact angle-bounded operators from a Banach space* X *to its dual* X^* *with the same constant of angle boundedness* $c \geqslant 0$ *and* $\|K - j\| \leqslant K_0$ *for each* i. *Let* $\{N_i\}_{j=1}^{n}$ *be a corresponding family of hemicontinuous, bounded nonlinear operators from* X^* *to* X *which satisfy the following condition. For every n-tuple* $\{u_1, u_2, \ldots, u_n\}$

$$\sum_{j=1}^{n}(u_1, N_i u) \geqslant -k \sum_{j=1}^{n} \|u_j\|_{X^*}^2 + \sum_{j=1}^{n}(u_j, N_j 0) \qquad (8.49)$$

where

$$u = \sum_{j=1}^{n} u_i \ and \ k < (1 + c^2)^{-1} K_0^{-1}.$$

Then, Eq. (8.46) has a solution in X^*.

Gupta [258] has generalized and simplified the above result. We will state and prove his result. We need the following definition.

Definition 8.3 A bounded linear mapping $K : X \to X^*$ is said to be quasi-monotone if

$$u_k = \inf \left\{ \frac{(Ku, u)}{\|Ku\|^2} : u \in X, \ Ku \neq 0 \right\} > -\infty.$$

We note that if K is angle bounded, then K is quasi-monotone with $u_k \geqslant 0$.

Theorem 8.16 *Let $\{K_1, \ldots, K_n\}$ be a family of compact quasi-monotone linear mappings from X into X^*. Let Y be a closed subspace of X^* such that $\bigcup_{i=1}^{n} R(K_i) \subset Y$. Let $\{N_1, \ldots, N_n\}$ be a corresponding family of bounded demicontinuous nonlinear mapping from Y to X. Assume that there exists a function $\varphi : \mathbb{R}^+ \to \mathbb{R}^+$ satisfying $\lim_{\rho \to \infty} \varphi(\rho)\rho^{-2} = 0$ such that for some $\lambda < \mu = \min\{\mu_{kj} : 1 \leqslant j \leqslant n\}$ and for any n-tuple $\{u_1, u_2, \ldots, u_n\}$ in Y with $u = \sum_{j=1}^{n} u_j$ we have*

$$\sum_{j=1}^{n}(u_j, N_j u) + \lambda \sum_{j=1}^{n} \|u_j\|_{X^*}^2 \geqslant -\varphi\left(\sqrt{\left(\sum_{j=1}^{n} \|u_j\|^2\right)}\right). \qquad (8.50)$$

Then, (8.46) has at least one solution u in Y.

Proof We define $\mathfrak{X}, \mathfrak{X}^*, K, N$ as before. Then, $K : \mathfrak{X} \to \mathfrak{X}^*$ is a quasi-monotone linear mapping with $\mu = \inf\{(KU, U)/\|KU\|^2, KU \neq 0\}$ and $N : \mathfrak{X}^* \to \mathfrak{X}$ is a bounded demicontinuous mapping such that for any $u = [u_1, u_2, \ldots, u_n]$, we have

$$(U, N(U)) + \lambda \|U\|^2 = \sum_{j=1}^{n}(u_j, N_j u) + \lambda \sum_{j=1}^{n} \|u_j\|^2$$

$$\geqslant -\varphi\left(\sqrt{\left(\sum_{j=1}^{n} \|u_j\|^2\right)}\right) = -\varphi(\|U\|).$$

As observed before, the solvability of (8.46) is equivalent to the solvability of the equation

$$U + RNU = 0$$

in $\mathfrak{X}$. So it suffices to give a proof of the above theorem when $n = 1$, which we now proceed to do. Since the mapping $K_1 N_1 : Y \to Y$ is a compact continuous mapping, it suffices to show, by Leray–Schauder principle, that there is a $\rho > 0$ such $(I + t K_1 N - 1)(u) \neq 0$ for every $t \in [0, 1]$ and $u \in Y$ with $\|u\|_{\mathfrak{X}^*} = \rho$, where I denotes the identity mapping on Y. Now, let $\rho > 0$ such that

$$\mu - \lambda - \varphi(\rho)\rho^{-2} > 0,$$

which exists since $\lambda < \mu$ and $\lim_{\rho \to \infty} \varphi(\rho)\rho^{-2} = 0$.

Further, we clearly see that $u + t K_1 N_1 u \neq 0$ for $t = 0$ and $u \in Y$ with $\|u\|_Y = \rho$. Suppose now that $u + t K_1 N_1 u = 0$ for some $t > 0$ and $u \in Y$ with $\|u\|_{\mathfrak{X}^*} = \rho$. Then, we have

$$
\begin{aligned}
0 = (u, N_1 u) + t(K_1 N_1 u, N_1 u) &\geqslant (u, N - 1u) + t\mu\|K - 1N - 1u\|^2 \\
&= (u, N_1 u) + \frac{1}{t}\mu\|u\|^2 \\
&\geqslant (\mu - \lambda\varphi(\rho)\rho^{-2})\rho^2 > 0,
\end{aligned}
$$

a contradiction. Hence, we have $u + t K_1 N_1 u \neq 0$ for $t \in [0, 1]$ and $u \in Y$ with $\|u\|_{\mathfrak{X}^*} = \rho$. ∎

Example 8.9 As a concrete application of the above theorem, we consider the following nonlinear integral equation which involves sum of Hammerstein operators

$$u(s) + \int_\Omega \sum_{j=1}^n K_j(s, t) f_j(t, u(t))dt = 0. \tag{8.51}$$

Consider the following assumptions:

(1) Ω is bounded measurable subset of $\mathbb{R}^N$, and the family of kernels $\{K_j\}_{j=1}^n$ is continuous as a mapping from $\Omega \times \overline{\Omega}$ to $\mathbb{R}$. Further, we assume that there exists $\alpha > 0$ such that for each j the corresponding integral operators K_j satisfy the condition $(K_j u, u) \geqslant \alpha \|K_j u\|_{L_\infty}^2 (\Omega)$ for every $u \in L_1(\Omega)$.
(2) The family $\{f_j\}_{j=1}^n$ of nonlinear functions is continuous from $\Omega \times \Omega$ into $\mathbb{R}$ such that there exists a $\rho > 0$ satisfying $\sum_{j=1}^n f_j(t, u_j)u_j \geqslant 0$ for any n-tuple $[u_1, \ldots, u_n]$ with $u = \sum_{j=1}^n u_j$, $(\sum_{j=1}^n |u_j|) \geqslant \rho$ and for every $t \in \Omega$.
Under the assumptions (1) and (2), (8.51) has a continuous solution.

8.2.3   Generalized Hammerstein Equation

A nonlinear integral equation of generalized Hammerstein type is of

$$u(s) + \int_\Omega k(s, t, u) f(t, u(t)) dt = v(s), \tag{8.52}$$

where the kernel function $k(s, t, u)$ is defined on $\Omega \times \Omega$ and is a function of the solution u.

The integral equation of the type (8.52) was first studied by W. Petry [474–476] and later by stuart [579, 580] and Leggett [361, 362]. More recently, they have been investigated by Backwinkel-Schillings [28, 29] and Srikanth and Joshi [576].

In operator theoretic terms, one can easily observe that (8.52) is equivalent to the operator equation

$$u + K(u) N_f u = v, \tag{8.53}$$

where each $K(u)$ is the linear integral operator defined as before and N_f is the Nemytskiĭ operator.

In the following, we shall obtain existence results for operator equation of the type (8.53). Without loss in generality, we may assume that $v = 0$. The basic idea employed by Backwinkel-Schillings [29] is the following:

For a fixed $u \in X$, Eq. (8.53) is a Hammerstein equation, and hence, one can apply known theorems to get a unique solution for such a kind equation. Denoting this by Tu, we have a mapping $T : X \to X$ given implicitly by

$$Tu + K(u) N_f Tu = v.$$

Solvability of (8.53) would follow if we can show that T has a fixed point.

Theorem 8.17 *Let X be a real reflexive Banach space, and for each $u \in X$, let $K(u) : X^* \to X$ be a linear angle-bounded operator with the same constant of angle boundedness $c \geqslant 0$. Let the mapping $u \to K(u)$ be completely continuous from X to $L(X^*, X)$. Further, assume that there exists a constant $r > 0$ such that for each $u \in X$ we have $\|K(u)\| \leqslant r$. Let $N_f : X \to X^*$ be a continuous bounded mapping which satisfy the inequality*

$$(u_1 - u_2, N_f u_1, N_f u_2) \geqslant -k\|u_1 - u_2\|^2 \quad for \ u_1, u_2 \in X,$$

where $kr < (1 + c^2)$. Then, the equation $u + K(u) N_f u = 0$ has at least one solution u in X.

One can, however, examine Eq. (8.53) directly without obtaining the solution for a fixed $u \in X$. This method was adopted by Srikanth and Joshi [576] and has obtained results more general than that of Backwinkel-Schilling [29].

Theorem 8.18 *Let X be a real reflexive Banach space with dual X^*. Let $K(u) : X \in X^*$ be a family of maps (not necessarily linear) and $N_f : X^* \to X$ be any nonlinear map. Consider the following assumptions on the family $\{K(u)\}$ and the map N_f:*

(i) *If $u_n \rightharpoonup u$ in X^* and $v_n \rightharpoonup v$ in X, then*
   (a) $K(u_n)v_n \rightharpoonup K(u)v$
   (b) $\liminf\limits_{n \to \infty}(K(u_n)v_n, v_n - v) \geqslant 0.$

(ii) *N_f is a one to one generalized pseudomonotone map from X^* onto X with N_f^{-1} bounded.*

(iii) *There exist two real-valued functions c_1 and c_2 defined on $\mathbb{R}^+$ such that $c_1(r) + c_2(r) \to \infty$ as $r \to \infty$ and for each $u \in X^*$*
   (a) $(K(u)v, v) \geqslant c_1(\|u\|)\|v\|$ *for all $v \in X$*
   (b) $(N_f^{-1}v, v) \geqslant c_2(\|v\|)\|v\|$ *for all $v \in X$.*
   Then, (8.53) has a solution.

Proof Under the hypothesis of the above theorem, the solvability of the equation

$$u + K(u)N_f u = 0 \tag{8.54}$$

is equivalent to the solvability of the equation

$$K(N_f^{-1}v)v + N_f^{-1}v = 0, \tag{8.55}$$

for some v in X. Define a map $S : X \to X^*$ as follows:

$$S(v) = K(N_f^{-1}v)v + N_f^{-1}v.$$

S is well defined. We now claim that S is of type (M). Suppose $v_n \rightharpoonup v$ in X and $Sv_n \rightharpoonup g$ with $\limsup\limits_{n \to \infty} (Sv_n, v_n - v) \leqslant 0$. This gives

$$\limsup_{n \to \infty} (K(N_f^{-1}v_n)v_n + N_f^{-1}v_n, v_n - v) \leqslant 0.$$

Since $v_n \rightharpoonup v$ and N_f^{-1} is bounded, $\{N_f^{-1}v_n\}$ has a weakly convergent subsequence which we again denote by $\{N_f^{-1}v_n\}$. Let $N_f^{-1}v_n \rightharpoonup w$ in X^*. From Assumption (i)(b), it now follows that $\limsup\limits_{n \to \infty} (N_f^{-1}v_n, v_n - v) \leqslant 0$ and as N_f is generalized pseudomonotone so is N_f^{-1}, and hence, we have $N_f^{-1}v = w$. It now follows from (i)(a) and the above proved fact that

$$K(N_f^{-1}v_n)v_n + N_f^{-1}v_n \rightharpoonup K(N_f^{-1}v)v + N_f^{-1}v.$$

By the uniqueness of limit, $g = K(N_f^{-1}v)v + N_f^{-1}v$. This is $Sv = g$. Also, since N_f^{-1} is bounded generalized pseudomonotone, it is demicontinuous, and hence, X

is demicontinuous. This establishes that S is of type (M). Further, it is clear that S is bounded. Also, from Assumption (iii), it is evident that S is coercive. That is,

$$\frac{(Sv, v)}{\|v\|} \to \infty \text{ as } \|v\| \to \infty.$$

Thus, S satisfies all the requirements of Theorem 4.60, and hence, $Sv = 0$ has a solution; i.e., (8.54) has a solution. ∎

In the following theorem, we can weaken the conditions on N_f by imposing slightly stronger conditions on the family $\{K(u)\}$.

Theorem 8.19 *Let X be a real reflexive Banach space with dual X^*. Let $K(u)$: $X \to X^*$ be a family of bounded linear maps for each $u \in X^*$ and $N_f : X \to X^*$ be a nonlinear map. Consider the following assumptions on the family $\{K(u)\}$ and the map N_f:*

(i) (a) $(K(u)v, v) \geqslant 0$ *for all $u \in X^*$ and $v \in X$,*

(b) $\sup_{u \in X^*} \|K(u)\| \leqslant \alpha, \alpha > 0$,

(c) $u_n \rightharpoonup u$ *in $X^* \Rightarrow \|K(u_n) - K(u)\| \to 0$.*

(ii) *The map N_f is bounded, demicontinuous and monotone.*

(iii) *There exists a constant β such that*

$$\|N_f u\| \leqslant \beta \|u\| \text{ for } \|u\| \geqslant r, \text{ where } r > 0 \text{ and } \alpha(\beta + 1) \leqslant 4.$$

Then, the equation

$$u + K(u)N_f u = 0$$

has a solution.

As an application of the above theorem, we consider the following example.

Example 8.10 Consider the following nonlinear differential equation

$$\frac{d}{dt}\left[a(t, u(t))\frac{du(t)}{dt}\right] = f(t, u(t))$$

subject to boundary conditions : $u(0) = u(1) = 0$.

Consider the following assumptions:

(i) $a : [0, 1] \times \mathbb{R} \to \mathbb{R}$ is continuous, and there exists a constant $\varepsilon_0 \geqslant 0$ such that $a(t, y) > \varepsilon_0$ for all $t, y \in [0, 1] \times \mathbb{R}$.

(ii) $f(t, y) : [0, 1] \times \mathbb{R} \to \mathbb{R}$ is continuous, and there exists positive constants A_1, A_2 such that

$$|f(t, y)| \leqslant A_1|y| + A_2.$$

Further, for all $t \in [0, 1]$ and all y_1, y_2, we have

$$(f(t, y_1) - f(t, y_2)(y_1 - y_2) \geqslant 0.$$

Notice that the solvability of the above differential equation is equivalent to the solvability of the generalized Hammerstein equation

$$u(t) + \int_0^1 K(s, t; u) f(s, u(s)) ds = 0. \tag{8.56}$$

where the kernel $K(s, t; u)$ is given by the Green's function

$$K(s, t; u) = \frac{1}{\int_0^1 \frac{d\tau}{a(\tau, u(\tau))}} \begin{cases} \int_0^t \frac{d\tau}{a(\tau, u(\tau))} \int_s^1 \frac{d\tau}{a(\tau, u(\tau))}, 0 \leqslant t \leqslant s \leqslant 1 \\ \int_0^s \frac{d\tau}{a(\tau, u(\tau))} \int_t^1 \frac{d\tau}{a(\tau, u(\tau))}, 0 \leqslant s \leqslant t \leqslant 1 \end{cases}$$

Defining operators $K(u)$ and N_f as before, we can show that (8.56) is equivalent to the following operator equation in $L_2[0, 1]$

$$u + K(u)N_f u = 0. \tag{8.57}$$

Following Petry [479], it can be shown that family $K(u)$ satisfies Assumption (c) of the above theorem. Assumptions (ii) and (iii) of the theorem are trivially satisfied and hence (8.57) has a solution in $L_2[0, 1]$. This proves the result.

8.2.4 Urysohn Equation

Urysohn's integral equation is of the form

$$u(s) + \int_\Omega K(s, t, u(t)) dt = 0. \tag{8.58}$$

Usually, one assumes that Ω is a measurable subset of $\mathbb{R}^n$ and that $K(s, t, u)$ is a function defined on $\Omega \times \Omega \times R$ with values in R, and it satisfies the so-called Carathéodory conditions. Urysohn's equation has been discussed by Urysohn [594], Kolomy [340] and others. Attempts have been made to apply the theory of monotone operators to get existence theorem for (8.58). Our main aim is to use the theory of abstract Hammerstein operators to obtain existence theorems for (8.58) with rather simple conditions on the function K. For details, refer Joshi [293].

We define a linear operator $K : L_2(\Omega \times \Omega) \to L_2(\Omega \times \Omega)$ with range in $L_2(\Omega)$ and a nonlinear operator $N : L_2(\Omega) \to L_2(\Omega \times \Omega)$ as follows

$$[Ku](s) = \int_\Omega u(s, t) dt, \tag{8.59}$$

$$[Nu](s, t) = K(s, t, u(t)).   \tag{8.60}$$

In all our considerations in this section, Ω will be a set of finite measure in $\mathbb{R}^n$ and

$$L_2(\Omega) = \left\{ u : \int_\Omega u^2(t) dt < \infty \right\}$$

$$L_2(\Omega \times \Omega) = \left\{ v : \int_\Omega \int_\Omega v^2(s, t) ds\, dt < \infty \right\}.$$

Observe that $L_2(\Omega)$ is a closed subspace of $L_2(\Omega \times \Omega)$.

Lemma 8.3 *K is a continuous linear operator from $L_2(\Omega \times \Omega)$ to $L_2(\Omega \times \Omega)$ with range $L_2(\Omega)$.*

One of the hypotheses on the existence theorem is the compactness of the operator KN. In the following lemmas, sufficient conditions are given which ensure this.

Lemma 8.4 *Let K satisfies the Carathéodory conditions and*

$$|K(s, t, u)| \leqslant a(s, t) + b(s, t)|u|\ a, b \in L_2(\Omega \times \Omega),\ b(s, t) > 0,\ s, t \in \Omega;\ u \in \mathbb{R}.$$

Then, N is a continuous bounded operator from $L_2(\Omega)$ to $L_2(\Omega \times \Omega)$. This is similar to what we have proved in Chap. 2, Sect. 2.1 (Theorem 2.4).

We now define an operator $U : L_2(\Omega) \to L_2(\Omega)$ by

$$[Ux](s) = \int_\Omega K(s, t, x(t)) dt.   \tag{8.61}$$

Obviously, we have $U = KN$. The operator U is called Urysohn operator and was earlier defined in Chap. 1, Sect. 8.1.

We now restate Theorem 2.7 as our Lemma 8.5 given below.

Lemma 8.5 *Under the conditions of Lemma 8.4, the Urysohn operator U is a continuous and compact mapping from $L_2(\Omega)$ to $L_2(\Omega)$.*

Theorem 8.20 *Assume that $K(x, t, x)$ satisfies the Carathéodory conditions and that the operators K and N are as defined in (8.59) and (8.60) and the map KN : $L_2(\Omega) \to L_2(\Omega)$ is compact. Also, assume that $\sup_{|x| \leqslant \sigma} |K(s, t, x)|$ is in $L_1(\Omega \times \Omega)$ where $\sigma > 0$ is such that*

$$xK(s, t, x) \geqslant -c(s, t)|x|^2 \text{ for } |x| > \sigma   \tag{8.62}$$

$c \in L_{2/(2-r)}$ *for some $r \leqslant 2$, $c(s, t) \geqslant 0$ for $s, t \in \Omega$. If ρ_0 is such that*

$$\sigma a(\sigma) \rho_0^{-2} + \|c\| (\mu(\Omega))^{r/2} \rho_0^{r-2} < 1,   \tag{8.63}$$

then the Urysohn's equation (8.58) has a solution x in $L_2(\Omega)$ such that $\|x\| \leqslant \rho_0$. Here, $a(\sigma)$ denotes the L_1-norm of $\sup_{|x| \leqslant \sigma} |K(s, t, x)|$, $\|c\|$ the $L_{2/(2-r)}$ norm of c, $\|x\|$ the L_2-norm of x.

Proof The assertion will follow from Theorem 8.13. We set $X = L_2(\Omega \times \Omega)$ and $Y = L_2(\Omega)$, and then, (8.58) is equivalent to the operator equation

$$x + KNx = 0, \tag{8.64}$$

where K and N are as defined before. By Lemma 8.3, K is a bounded linear operator from X to X^* with range Y. Furthermore, we have

$$
\begin{aligned}
(Kx, x) &= \int_\Omega \int_\Omega x(s, t) ds dt \int_\Omega x(s, \tau) d\tau \\
&= \int_\Omega ds \left(\int_\Omega x(s, t) dt \right) \left(\int_\Omega x(s, \tau) d\tau \right) \\
&= \int_\Omega ds \left(\int_\Omega x(s, t) dt \right)^2 \geqslant 0,
\end{aligned}
$$

which implies that K is a monotone map from X to X^*. Also $(Kx, y) = (Ky, x)$, that is K is symmetric. Hence, K is angle bounded with the constant of angle boundedness $c = 0$. Also using (8.62), we can show that

$$(x, Nx) \geqslant \|x\|^r (\mu(\Omega))^{r/2} \|c\| - \sigma a(\sigma).$$

Since K and N satisfy all the conditions of Theorem 8.13, it follows that (8.64) has a solution x in Y with $\|x\| \leqslant \rho_0$. This in turn implies that (8.58) has a solution x in L_2 satisfying $\|x\| \leqslant \rho_0$. ∎

Remark 8.6 It may be remarked that the condition (8.63) is satisfied for all sufficiently large ρ_0 if either $r < 2$ or $r = 2$ and $\|c\|_\infty \mu(\Omega) < 1$.

Corollary 8.3 *Assume that $K(s, t, x)$ satisfies the Carathéodory conditions and*

$$|K(s, t, x)| \leqslant a(s, t) + b(s, t) |x| \text{ for } x \in R$$

$$a, b \in L_\infty, b(s, t) > 0 \text{ for } s, t \in \Omega, \|b\|_\infty \mu(\Omega) < 1.$$

Then, (8.58) has a solution x in L_2.

Notice also that as a corollary of Theorem 8.20 we can obtain existence theorem for the integral equation

$$x(s) + \int_\Omega K(s, t) f_1(t, x(t)) dt + \int_\Omega K_2(s, t) f_2(t, x(t)) dt = 0,$$

which contains a sum of Hammerstein integral operators. For details, see Joshi [293].

8.3   Solutions of Nonlinear Equations by Using Computational Schemes

In this section, we give computational schemes for the solvability of Hammerstein and generalized Hammerstein operator equations. As concrete application, we give a computational scheme for the solvability of Chandrasekhar's H-equation

$$u(s) = 1 + u(s) \int_0^1 \frac{s}{s+t} \Psi(t)u(t)dt,$$

occurring in heat radiative transfer.

We first prove the following result which gives iterative solution together with a rate of convergence for the solution of Hammerstein operator equation

$$u + KNu = w, \tag{8.65}$$

by using Bruck's Jr. iterative scheme (refer to Theorem 4.39).

Theorem 8.21 *Let X be a reflexive Banach space and $K : X \to X^*$ a symmetric monotone bounded linear operator. Let $N : X^* \to X$ be a nonlinear hemicontinuous operator such that for some constant k*

$$(u_1 - u_2, Nu_1 - Nu_2) \geqslant -k\|u_1 - u_2\|_{X^*}^2.$$

If $k\|K\| < 1$, then (8.65) has a unique solution u and the sequence $S^*v_n + w$, where v_n is defined by

$$v_{n+1} = \left(\frac{n}{n+1}\right) v_n - \left(\frac{1}{n+1}\right) SN[S^*v_n + w]$$

for any arbitrary initial choice of v_1 in $\mathcal{H}$ converges to u. Further

$$\left\| S^*v_n + w - u \right\| = 0(n^{-1/2}).$$

Here, $S : X \to \mathcal{H}$ is the linear operator splitting K through the Hilbert space $\mathcal{H}$.

Proof As proved in the previous section, the solvability of (8.65) is equivalent to the solvability of $Tv = 0$ where $T; \mathcal{H} \to \mathcal{H}$ is given by

$$Tv = v + SNS^*(v + w).$$

We have already shown that $Tv = 0$ has a unique solution. It is clear that T is of the type $1 + T_1$ where T_1 is a maximal monotone operator on $\mathcal{H}$, and hence, by Theorem 4.39, it follows that the sequence

$$v_{n+1} = \left(\frac{n}{n+1}\right) v_n - \left(\frac{1}{n+1}\right) SN(S^* v_n + w)$$

converges to the unique solution of $Tv = 0$ with error estimate

$$\|v_n - v\| = 0(n^{-(1/2)}).$$

Hence, the sequence $\{S^* v_n + w\}$ converges to the unique solution u of (8.65) with the required error estimate.

We now pass on to the generalized Hammerstein equation. Let X be a reflexive Banach space with $\{X_n, P_n\}$ as a projection scheme. We shall give constructive results regarding the solvability of the 'exact' equation

$$u + K(u)Nu = 0 \tag{8.66}$$

as a weak or strong limit of solutions $u_n \in Y_n = (P_n^* X^*)$ of the approximate equation

$$u_n + K_n(u_n)N_n u_n = 0. \tag{8.67}$$

Here, for each $u \in X^*$, $K(u) : X \to X^*$ is a linear operator and $N : X^* \to X$ is a nonlinear operator. $K_n(u) = P_n^* K(u) P_n$ maps the range X_n of P_n into Y_n and $N_n = P_n N P_n^*$ maps Y_n into X_n. ∎

Definition 8.4 Equation (8.66) is called strongly (weakly) approximately solvable if (8.67) has a solution $u_n \in Y_n$, and there exists a subsequence $\{u_{n_k}\}$ of $\{u_n\}$ such that $u_{n_k} \to u$ in X^*. It follows that $u_{n_k} \rightharpoonup u$ and u is a solution of (8.66).

In the following theorem, C denotes a closed and bounded set in X^* containing the origin with ∂C as its boundary.

Theorem 8.22 *Let X be a real reflexive Banach space with dual X^* and $\{X_n, P_n\}$ a projection scheme in X. Let for each $u \in C$, $K(u) : X \to X^*$ be a bounded linear operator and $N : X^* \to X$ be a continuous and bounded nonlinear operator. Further, we assume that the following hold:*

(a) $K(u)$ is compact and monotone for each $u \in C$,
(b) $u_n \rightharpoonup u$ in C implies that $K(u_n)v \to K(u)v$ for all $v \in X$,
(c) $(u, Nu) > 0$ for all $u \in \partial C$.
   Then, (8.66) is strongly approximately solvable. If the solution of (8.66) is unique, then the entire sequence $\{u_n\}$ converges to the unique solution.

Proof Let $T_n u = -K_n(u)N_n u$. It is clear T_n is a continuous operator from $C_n = C \cap Y_n$ into Y_n. Also, we claim that $T_n u \neq \lambda u$ whenever $\lambda > 1$ and $u \in \partial C_n$. For, if there exists $u \in \partial C_n$ with $T_n u = \lambda u$, we get

$$\lambda(u, Nu) + (K_n(u)N_n u, Nu) = 0.$$

That is

$$\lambda(u, Nu) = -(K(u)P_n Nu, P_n Nu) \leqslant 0,$$

a contradiction to our Assumption (c). Hence, it follows by Leray–Schauder principle that there exists $u_n \in C_n \subset C$ such that $u_n = T_n u_n$. That is

$$u_n + K_n(u_n)N_n u_n = 0. \tag{8.68}$$

Since $\{u_n\}$ is a bounded sequence in a reflexive Banach space, there exists a subsequence $\{u_{n_k}\}$ of it which converges weakly to $u_0 \in C$. Similarly boundedness of N implies that there exists a subsequence of $\{Nu_{n_k}\}$ (which we again denote by $\{Nu_{n_k}\}$) converging weakly to v. We claim that $P_{n_k} Nu_{n_k} \rightharpoonup v$ as $K \to \infty$. For this, consider $(w, P_{n_k} Nu_{n_k} - v)$, where $w \in X^*$ is arbitrary. We have

$$
\begin{aligned}
(w, P_{n_k} Nu_{n_k} - v) &= (w, P_{n_k} Nu_{n_k} - P_{n_k}v + P_{n_k}v - v) \\
&= (P_{n_k}^* w, Nu_{n_k} - v) + (w, P_{n_k}v - v) \\
&= (P_{n_k}^* w - w, Nu_{n_k} - v) + (w, p_{n_k}v - v) + (w, Nu_{n_k} - v) \\
&\leqslant \|P_{n_k}^* w - w\| \|Nu_{n_k} - v\| + (w, Nu_{n_k} - v) + \|w\| \|P_{n_k}v - v\| \\
&\to 0 \text{ as } k \to \infty.
\end{aligned}
$$

We now consider $K(u_0)P_{n_k} Nu_{n_k} - K(u_{n_k})P_{n_k} Nu_{n_k}$. Since $K(u_{n_k})v \to K(u_0)v$ for all $v \in X$ (in view of Assumption (b)), it follows by uniform boundedness principal that $K(u_{n_k}) \to K(u_0)$ in operator norm and hence

$$\|K(u_0)P_{n_k} Nu_{n_k} - K(u_{n_k})p_{n_k} Nu_{n_k}\| \leqslant \|K(u_0) - K(u_{n_k})\| \|P_{n_k} Nu_{n_k}\| \to 0 \text{ as } k \to \infty.$$

This in turn implies that

$$P_{n_k}^* K(u_0)P_{n_k} Nu_{n_k} - P_{n_k}^* K(u_{n_k})P_{n_k} Nu_{n_k} \to 0, \tag{8.69}$$

as $k \to \infty$. As $P_{n_k} Nu_{n_k} \rightharpoonup v$, using the compactness of $K(u_0)$ we get

$$K(u_0)P_{n_k} Nu_{n_k} \to K(u_0)v,$$

and hence, this in turn implies that

$$P_{n_k}^* K(u_0)P_{n_k} Nu_{n_k} \to K(u_0)v. \tag{8.70}$$

Now combining (8.68) and (8.69), we obtain

$$
\begin{aligned}
K_{n_k}(u_{n_k}N_{n_k}u_{n_k} &= P_{n_k}^* K(u_{n_k})P_{n_k} Nu_{n_k} \\
&\to K(u_0)v, \text{ as } k \to \infty.
\end{aligned}
$$

But (8.69) gives

$$u_{n_k} = -K_{n_k}(u_{n_k})N_{n_k}u_{n_k} \tag{8.71}$$

and hence u_{n_k} is strongly convergent. That is $u_{n_k} \to u_0$. Using the continuity assumptions on N and $K(u)$, we obtain $Nu_{N_k} \to Nu_0 = v$ and $K_{n_k}(u_{n_k})N_{n_k}u_{n_k} \to K(u_0)Nu_0$. So (8.71) gives

$$u_0 + K(u_0)Nu_0 = 0,$$

thereby proving the strong approximate solvability of (8.66). If (8.66) has unique solution, then it follows from the above proof that every subsequence of $\{u_n\}$ has in turn a subsequence which converges to the unique solution. This implies that the entire sequence $\{u_n\}$ converges to the unique solution. ■

As a corollary of the above theorem, we obtain the main existence theorem of Amann [13] along with approximate solvability (in a Banach space X with a projection scheme) for the Hammerstein operator equation

$$u + KNu = 0. \tag{8.72}$$

Corollary 8.4 *Let X be a real reflexive Banach space with projection scheme. Let $K : X \to X^*$ be a compact, monotone, linear operator and $N : X^* \to X$ a bounded, continuous nonlinear operator. Further, we assume that there exists a constant $R > 0$ such that*

$$(u, Nu) > 0 \ for \ \|u\| = R.$$

Then, (8.72) has solution u satisfying $\|u\| \leqslant R$ and is strongly approximately solvable.

One can weaken the condition of compactness on $K(u)$ if we impose stronger conditions on N, say for an example, N is a continuous and bounded operator of type (M) or (S_+).

Theorem 8.23 *Let X be a real reflexive Banach space with a projection scheme $\{X_n, P_n\}$. Let for each $u \in C$, $K(u) : X \to X^*$ be a bounded linear operator and $N : X^* \to X$ a bounded, continuous nonlinear operator. Further, we assume that the following hold:*

(a) $K(u)$ is monotone for each $u \in C$,
(b) $u_n \rightharpoonup u$ in C implies that $K(u_n)v \to K(u)v$ for all $v \in X$,
(c) $(u, Nu) > 0$ for all $u \in \partial C$.

Then, (8.66) is weakly approximately solvable if N is of type (M) and is strongly approximately solvable if N is of type (S_+). If the solution of (8.66) is unique, then the entire sequence $\{u_n\}$ converges to the unique solution.

Proof As in the previous theorem, Leray–Schauder principle gives a sequence $\{u_n\} \subset C \cap Y_n$ where u_n is a solution of the 'approximate' Eq. (8.67). Also, we get a subsequence $\{u_{n_k}\}$ of $\{u_n\}$ such that

$$u_{n_k} \rightharpoonup u_0 \text{ and } N u_{n_k} \rightharpoonup v$$

and

$$w_{n_k} = P_{n_k}^* K(u_0) P_{n_k} N u_{n_k} - K_{n_k} N_{n_k} u_{n_k} \to 0 \text{ as } k \to \infty.$$

$$u_{n_k} = -K_{n_k}(u_{n_k}) N_{n_k} u_{n_k} = w_{n_k} - P_{n_k}^* K(u_0) P_{n_k} N u_{n_k}. \tag{8.73}$$

As $P_{n_k} N u_{n_k} \rightharpoonup v$ and $K(u_0)$ is a bounded linear operator and hence weakly continuous, it follows that $P_{n_k}^* K(u_0) P_{n_k} N u_{n_k} \rightharpoonup K(u_0)v$. Since $u_{n_k} \rightharpoonup u_0$ and $w_{n_k} \to 0$ as $k \to \infty$, the above equality gives $u_0 = -K(u_0)v$. Now consider $\overline{\lim}_k (u_{n_k} - u_0, N u_{n_k})$. We have

$$\overline{\lim}_k(u_{n_k} - u_0, N u_{n_k}) = \overline{\lim}_k(w_{n_k} - u_0, N u_{n_k}) - \underline{\lim}_k(P_{n_k}^* K(u_0) P_{n_k} N u_{n_k}, N u_{n_k})$$
$$+ (K(u_0)v, v) - \underline{\lim}_k(K(u_0) P_{n_k} N u_{n_k}, N u_{n_k}). \tag{8.74}$$

Monotonicity of $K(u_0)$ gives

$$(K(u_0) P_{n_k} N u_{n_k}, P_{n_k} N u_{n_k}) = (K(u_0) P_{n_k} N u_{n_k} - K(u_0)v, P_{n_k} N u_{n_k} - v) - (K(u_0)v, v)$$
$$+ (K(u_0)v, P_{n_k} N u_{n_k}) + (K(u_0)v, P_{n_k} N u_{n_k}, v)$$
$$\geqslant -(K(u_0)v, v) + (K(u_0)v, P_{n_k} N u_{n_k} + (K(u_0) P_{n_k} N u_{n_k}, v).$$

Since $P_{n_k} N u_{n_k} \rightharpoonup v$ and $K(u_0) P_{n_k} N u_{n_k} \rightharpoonup K(u_0)v$, we get

$$\underline{\lim}_k(K(u_0) P_{n_k} N u_{n_k}, P_{n_k} N u_{n_k} \geqslant -(K(u_0)v, v) + (K(u_0)v, v) + (K(u_0)v, v)$$
$$= (K(u_0)v, v).$$

Combining this inequality with (8.74), we get

$$\underline{\lim}_k(u_{n_k} - u_0, N u_{n_k}) \leqslant (K(u_0)v, v) - (K(u_0)v, v) = 0.$$

First assume that N is of type (M). Since we have $u_{n_k} \rightharpoonup u_0$ and $N u_{n_k} \rightharpoonup v$ and $\underline{\lim}_k(u_{n_k} - u_0, N u_{n_k}) \leqslant 0$, it follows that $N u_0 = v$. As $u_0 = -K(u_0)v$, we get $u_0 + K(u_0) N u_0 = 0$. This gives weak approximate solvability of (8.66).

If N is of type (S_+), using similar arguments one can show that $u_{n_k} \to u_0$ and u_0 is a solution of (8.66). The last conclusion of the theorem follows similarly.

We now obtain an approximate solvability result for the special case $N = I$. That is, we give a constructive result regarding the existence of a solution of the "exact" equation

$$u + K(u)u = w. \tag{8.75}$$

as a strong limit of solutions $u_n \in X_n$ of the approximate equation

$$u_n + P_n K(u_n) u_n = w_n. \tag{8.76}$$

Here, $\{x_n, P_n\}$ is an approximation scheme in a Hilbert space X.   ∎

In the following, K^* denotes the conjugate of the bounded linear operator K.

Theorem 8.24 *Let X be a Hilbert space with an approximation scheme $\{x_n, P_n\}$ and A a closed convex subset of X. Assume that for each $u \in A$, $K(u)$ is a bounded linear monotone operator satisfying the following condition:*

(a) $u_k \rightharpoonup u$ in A implies that $K^*(u_k)v \to K^*(u)v$ for all $v \in X$.
Then, (8.75) is approximately strongly solvable.

Proof We first claim that $K(u) : X \to X$ is jointly weakly continuous. That is, $u_k \rightharpoonup u$ in A and $v_k \rightharpoonup v$ in X imply that $K(u_k)v_k \rightharpoonup K(u)v$. Consider $(K(u_k)v_k - K(u)v, x)$, $x \in X$. We have

$$(K(u_k)v_k - K(u)v, x) = (K(u_k)v_k - K(u)(v_k), x) + (K(u)(v)k) - K(u)(v), x)$$
$$= (v_k, K^*(u_k)x - K^*(u_k)x - K^*(u)x) + (K(u)v_k - K(u)v, x).$$

As $k \to \infty$, the first term in the RHS of the above inequality goes to zero in view of Assumption (a) and the second term goes to zero as $K(u)$ is a continuous linear operator and hence also weakly continuous. This proves our claim.

We now consider the approximate equation

$$u_n + P_n K(u_n) = w_n.$$

Let T be the operator on X_n defined by

$$Tu = u + P_n K(u)u - w_n.$$

Defined a closed bounded set C as

$$C = \{u \in A \cap X_n; (w, u) \leqslant (u, u) \leqslant (w, w)\}.$$

Then, T is a continuous operator on X_n such that

$$(Tu, u) = (u, u) + (P_n K(u)u, u) - (w_n, u)$$
$$\geqslant (u, u) - (w, u)$$
$$\geqslant 0, \text{ for } u \in C.$$

Hence, it follows by Leray–Schauder principle that $Tu = 0$ has a solution $u_n \in C$. That is

$$u_n + P_n K(u_n)u_n = w_n.$$

$\{u_n\}$ is a bounded sequence in a Hilbert space, and hence, there exists a subsequence of it which we again denote by $\{u_n\}$ such that $u_n \rightharpoonup u$. We claim that u is a solution of (8.75). As $u_n \rightharpoonup u$ and $w_n \rightharpoonup w$, it suffices to show that $P_n K(u_n)u_n \rightharpoonup K(u)u$ as $n \to \infty$. Consider $(P_n K(u_n)u_n - K(u)u, x)$, $x \in X$. We have

$$(P_n K(u)n)u_n - K(u)u, x) = (P_n K(u_n)u_n - P_n K(u)u, x) + (P_n K(u)u - K(u)u, x)$$
$$= (K(u_n)u_n - K(u)u, x) + (K(u_n)u_n - K(u)u, P_n x - x)$$
$$\quad + (P_n K(u)u - K(u)u, x)$$
$$= (K(u_n)u_n - K(u)u, x) + \|K(u_n)u_n - K(u)u\| \, \|P_n x - x\|$$
$$\quad + \|P_n K(u)u - K(u)u\| \, \|x\|$$
$$\to 0 \text{ as } n \to \infty,$$

since $K(u_n)u_n \rightharpoonup K(u)u$ and $P_n x \to x$.

This proves that (8.75) is approximately weakly solvable. We now show that $\{u_n\}$ actually converges strongly to u. For this it suffices to show that

$$\limsup_n \ (u_n, u_n - u) \geqslant 0 \text{ as } u_n \rightharpoonup u.$$

As $u_n = w_n - P_n K(u_n)u_n, \ u = w - K(u)u$, we have

$$(u_n, u_n - n) = (w_n - w + u + K(u)u - P_n K(u_n)u_n, u_n - u)$$
$$= (w_n - w, u_n - u) + (P_n K(u_n)u - P_n K(u_n)u_n, u_n)$$
$$\quad - (P_n K(u_n)u - P_n K(u)u, u_n - u)$$
$$\quad - (P_n K(u)u - K(u)u, u_n - u) + (u, u_n - u)$$
$$\leqslant \|w_n - w\| \, \|u_n - u\| - (K(u_n)u_n - K(u_n)u, u_n - u)$$
$$\quad - (K(u_n)u_n - K(u_n)u, u - P_n u)$$
$$\quad + \|P_n K(u)u - K(u)u\| \, \|u_n - u\| + (u, u_n - u)$$
$$\leqslant \|w_n - w\| \, \|u_n - u\| + \|K(u_n)u_n - K(u_n)u\| \, \|P_n u - u\|$$
$$\quad + \|P_n K(u)u - K(u)u\| \, \|u_n - u\| + (u, u_n - u)$$

in view of monotonicity of $K(u), u \in A$.

Taking lim sup of both sides, we get

$$\limsup_{n \to \infty} \ (u_n, u_n - n) \leqslant \lim_{n \to \infty} \|w_n - w\| \, \|u_n - u\| + \lim_{n \to \infty} \|K(u_n)u_n - K(u_n)u\| \times$$
$$\|P_n u - u\| + \lim_{n \to \infty} \|P_n K(u)u - K(u)u\| \, \|u_n - u\|$$
$$+ \lim_{n \to \infty} (u, u_n - u)$$
$$\to 0 \text{ as } n \to \infty,$$

since $P_n u \to u$ for all $u \in X$ and $u_n \rightharpoonup u$. This proves our theorem.   ∎

As an application of our main result, we obtain an approximate solvability result for nonlinear integral equations of the type

$$u(s) + u(s) \int_0^1 K(s, t)u(t)dt = w(s) \tag{8.77}$$

in the space $L_2(0, 1]$.

Example 8.11 Assume that $K(x, t)$ is Hilbert–Schmidt kernel. So eigenfunctions of K form a complete orthonormal set in $L_2[0, 1]$. Let $e_1, e_2, \ldots$ be eigenfunctions of K with eigenvalues $\lambda_1, \lambda_2, \ldots$. Define a sequence X_n of finite dimensional subspaces of $X = L_2[0, 1]$ and linear projections $P_n : X \to X_n$ as follows:

$$X_n = [e_1, e_2, \ldots, e_n], \ P_n u = \sum_{k=1}^n \alpha_k e_k, \ u = \sum_{k=1}^\infty \alpha_k e_k.$$

Then, the approximate equation in the finite dimensional space X_n is given by

$$u(s) + P_n \left[u(s) \int_0^1 K(s, t)u(t)dt \right] = P_n w, \tag{8.78}$$

where $u = \sum_{k=1}^n \alpha_k e_k$.
Taking inner product with e_n, we get

$$(u, e_n) + \left(P_n \left[u(s) \int_0^1 K(s, t)u(t)dt \right], e_n \right) = (P_n w, e_n),$$

which gives

$$\alpha_n + \left(\left(\sum_{k=1}^n \alpha_k e_k(s) \right) \left(\sum_{j=1}^n \lambda_j \alpha_j e_j(s) \right), e_n \right) = \beta_n, \tag{8.79}$$

where

$$w = \sum_{k=1}^\infty \beta_k e_k.$$

writing r_{jkm} for $\int_0^1 e_k(s)e_j(s)e_n(s)ds$ in (8.79), we get an equivalent system of nonlinear equations.

$$\alpha_n + \sum_{j=1}^n \sum_{k=1}^n \lambda_k \alpha_j \alpha_k r_{jkn} = \beta_n, n \geqslant 1. \tag{8.80}$$

Thus, solvability of the approximate equation (8.79) is equivalent to the solvability of (8.80). One can now use the known techniques to solve the nonlinear system given by (8.80).

We have the following theorem giving the approximate solvability of (8.77).

Theorem 8.25 *Suppose that*

(a) $K(s, t) \geqslant 0$ *a.e. on* $[0, 1] \times [0, 1]$

(b) *ess* sup $\int_0^1 K^2(s, t)dt < \infty,$

(c) $w(s) \geqslant 0$ *a.e. on* $[0, 1]$.

Then, (8.77) is approximately strongly solvable in $L_2[0, 1]$.

Proof Let $A = \{u \in L_2 : u(s) \geqslant 0 \text{ a.e. } [0, 1]\}$ and let

$$K(u)(v)(s) = v(s) \int_0^1 K(s, t)u(t)dt.$$

Then, for each $u \in A$, $K(u)$ is a bounded linear operator on L_2. For each $u \in X$ and $u_k \rightharpoonup u$ in A, we have $K^*(u_k)v \to K^*(u)v$.

Also for each $u \in A$, we have

$$(K(u)v, v) = \int_0^1 \left(\int_0^1 K(s, t)u(t)dt \right) v^2(t)dt$$

$$\geqslant 0 \text{ for all } v \in L_2.$$

That is, $K(u)$ is monotone for each $u \in A$. Thus, the family $\{K(u)\}, u \in A$ of linear operators on X satisfies all conditions of Theorem 8.24, and hence, (8.77) is approximately solvable. ∎

As a corollary of this theorem, we obtain an approximate solvability result for Chandrasekhar's H-equation

$$1 + u(s) \int_0^1 \frac{s}{s+t} \Psi(t)u(t)dt = u(s). \tag{8.81}$$

Here, the known function $\Psi(t)$ is assumed to be nonnegative, bounded and measurable. Since Eq. (8.81) is not given in the standard form, we first state a lemma which is useful in this direction. Refer Chandrasekhar [139] for a proof of this lemma.

Lemma 8.6 *Suppose that* $\int_0^1 \Psi(t)dt \leqslant \dfrac{1}{2}$ *and that* $u \in L_2$ *is a positive solution of the equation*

$$u(s) \left\{ 1 - 2 \int_0^1 \Psi(t)dt \right\}^{1/2} + u(s) \int_0^1 \frac{t}{s+t} \Psi(t)u(t)dt = 1. \qquad (8.82)$$

Then $\int_0^1 \Psi(s)u(s)ds = 1 - \left(1 - 2 \int_0^1 \Psi(s)ds \right)^{1/2}$ *and* u *is also a solution of (8.81)*

Equation (8.82) is now in the standard from (8.77) with

$$K(s,t) = \left[\frac{t}{c(s+t)} \right] \Psi(t) \text{ and } w(s) = \frac{1}{c}. \qquad (8.83)$$

Here, $c = \left(1 - 2 \int_0^1 \Psi(s)ds \right)^{-1/2}$. Notice that without loss in generality we can assume that $\int_0^1 \Psi(t)dt < \dfrac{1}{2}$. As a result, one can tackle the case $\int_0^1 \Psi(t)dt = \dfrac{1}{2}$ as limiting case of the strict inequality. Thus, K and w given by (8.83) satisfy all the requirements of Theorem 8.24, and hence, we get the following solvability result for Chandrasekhar's H-equation.

Theorem 8.26 *Let* $\int_0^1 \Psi(t)dt \leqslant \dfrac{1}{2}$, *then Chandrasekhar's* H*-equation (8.81) is approximately strongly solvable.*

8.4   Strong Convergence of a Proximal-Type Algorithm

Let $\mathcal{H}$ be a real Hilbert space with inner product $\langle \cdot, \cdot \rangle$, and let $T : \mathcal{H} \to 2^{\mathcal{H}}$ be a maximal monotone operator (or a multifunction) on $\mathcal{H}$. We consider the classical problem

$$\text{find } x \in \mathcal{H} \text{ such that } 0 \in Tx. \qquad (8.84)$$

A wide variety of problems, such as optimization and related fields, min-max problems, complementarity problem and variational inequalities, fall within this general framework. For example, if T is the subdifferential ∂f of a proper lower semicontinuous convex function $f : \mathcal{H} \to (-\infty, \infty)$, then T is a maximal monotone operator and the equation $0 \in \partial f(x)$ is reduced to $f(x) = \min \{f(z) : z \in \mathcal{H}\}$. One method of solving $0 \in Tx$ is the proximal point algorithm. Let I denote the identity operator on $\mathcal{H}$. Rockafellar's proximal point algorithm generates, for any starting point $x_0 = x \in \mathcal{H}$, a sequence $\{x_n\}$ in $\mathcal{H}$ by the rule

$$x_{n+1} = (I + r_n T)^{-1} x_n, \quad n = 0, 1, 2, \ldots, \tag{8.85}$$

where $\{r_n\}$ is a sequence of positive real numbers. Note that (8.85) is equivalent to

$$0 \in T x_{n+1} + \frac{1}{r_n}(x_{n+1} - x_n), n = 0, 1, 2, \ldots.$$

This algorithm was first introduced by Martinet [389] and generally studied by Rockafellar [525] in the framework of a Hilbert space. Later, many authors studied the convergence of (8.85) in a Hilbert space; see Brezis and Lions [81], Lions [373], Passty [446], Güler [257], Solodov and Svaiter [567] and the references therein. Rockafellar [525] proved that if $T^{-1}0 \neq \emptyset$ and $\liminf\limits_{n\to\infty} r_n > 0$, then the sequence generated by (8.85) converges weakly to an element of $T^{-1}0$. Further, Rockafellar [525] posed an open question of whether the sequence generated by (8.85) converges strongly or not. This question was solved by Güler [257], who introduced an example for which the sequence generated by (8.85) converges weakly but not strongly. On the other hand, Kamimura and Takahashi [303, 304] and Solodov and Svaiter [568] one decade ago modified the proximal point algorithm to generate a strongly convergent sequence. Solodov and Svaiter [568] introduced the following algorithm:

$$\left.\begin{array}{l} x_0 \in X, \\[2mm] 0 = v_n + \dfrac{1}{r_n}(y_n - x_n), v_n \in T x_n, \\[2mm] H_n = \{z \in X : \langle z - y_n, v_n \rangle \leqslant 0\}, \\[2mm] W_n = \{z \in X : \langle z - x_n, x_0 - x_n \rangle \leqslant 0\}, \\[2mm] x_{n+1} = P_{H_n \cap W_n} x_0, n = 0, 1, 2, \ldots. \end{array}\right\} \tag{8.86}$$

Kamimura and Takahashi [305] extended Solodov and Svaiter's result to more general Banach spaces like the spaces $L_p(1 < p < \infty)$ by further modifying the proximal point algorithm (8.86) in the following form in a smooth Banach space X:

$$\left.\begin{array}{l} x_0 \in X, \\[2mm] 0 = v_n + \dfrac{1}{r_n}((J_2)y_n - (J_2)x_n), v_n \in T x_n, \\[2mm] H_n = \{z \in X : \langle z - y_n, v_n \rangle \leqslant 0\}, \\[2mm] W_n = \{z \in X : \langle z - x_n, J_2(x_0) - J_2(x_n) \rangle \leqslant 0\}, \\[2mm] X_{n+1} = P_{H_n \cap W_n} x_0, n = 0, 1, 2, \ldots. \end{array}\right\} \tag{8.87}$$

to generate a strongly convergent sequence. They proved that if $T^{-1}0 \neq \emptyset$ and $\liminf\limits_{n\to\infty} r_n > 0$, then the sequence generated by (8.87) converges strongly to $P_{T^{-1}0} x_0$.

It is well known that the proximal point algorithm converges weakly to a zero of a maximal monotone operator, but it fails to converge strongly. Then, in [568], Solodov and Svaiter introduced the new proximal-type algorithm to generate a strongly convergent sequence and established a convergence property for the algorithm in Hilbert spaces. Further, Kamimura and Takahashi [305] extended Solodov and Svaiter's result to more general Banach spaces and obtained strong convergence of a proximal-type algorithm in Banach spaces.

In 2012, Pathak and Cho investigated strong convergence of the proximal point algorithm in Hilbert spaces, and this study extended the results of Kamimura and Takahashi.

We denote by $L[x_1, x_2]$ the ray passing through x_1, x_2. Throughout this section, unless otherwise stated, we assume that $T : X \to 2^{X^*}$ is a occasionally pseudomonotone maximal monotone operator. In this section, we study the following algorithm in a smooth Banach Space X, which is an extension of (8.87):

$$
\left.
\begin{aligned}
&x_0 \in X, \\
&0 = v_n + \frac{1}{r_n}(J_p(y_n) - J_p(x_n)), \, v_n \in Tx_n, \\
&H_n = \{z \in X : \langle z - y_n, v_n \rangle \leqslant 0\}, \\
&W_n = \{z \in X : \langle z - x_n, J_p(x_n) - J_p(x_n) \rangle \leqslant 0\}, \\
&x_{n+1} = R_{H_n \cap W_n} x_n, n = 0, 1, 2, \ldots,
\end{aligned}
\right\}
\tag{8.88}
$$

where $\{r_n\}$ is a sequence of positive real numbers.

First, we investigate the condition under which the algorithm (8.88) is well defined. Rockafellar [523] proved the following theorem.

Theorem 8.27 *Let X be a reflexive, strictly convex and smooth Banach space, and let $T : X \to 2^{X^*}$ be a monotone operator. Then, T is maximal if and only in $R(J_p + rT) = X^*$ for all $r > 0$.*

By appropriate modification of arguments in the above theorem, we can prove the following:

Theorem 8.28 *Let X be a reflexive, strictly convex and smooth Banach space, and let $T : X \to 2^{X^*}$ be an occasionally pseudomonotone operator. Then, T is maximal if and only if $R(J_p + rT) = X^* \, \forall \, r > 0$.*

Using Theorem 8.28, we can show the following result.

Proposition 8.6 *Let X be a reflexive, strictly convex and smooth Banach space. If $T^{-1}0 \neq \emptyset$, then the sequence generated by (8.88) is well defined.*

Proof From the very definition, it is obvious that both H_n and W_n are closed convex sets. Let $w \in T^{-1}0$. From Theorem 8.28, there exists $(y_0, v_0) \in E \times E^*$ such that $0 = v_0 + \frac{1}{r_0}(J_p(y_0) - J_p(x_0))$ and $v_0 \in Ty_0$. Since T is occasionally pseudomonotone and $\langle y_0 - w, 0 \rangle = 0 \geq 0, Tw \ni 0$, it implies that

$$
\langle y_0 - w, v_0 \rangle \geq 0
$$

for some $v_0 \in T y_0$. It follows that $w \in H_0$. On the other hand, it is clear that $w \in W_0 = X$. Then, $w \in H_0 \cap W_0$, and therefore, $x_1 = R_{H_0 \cap W_0} x_0$ is well defined. Suppose that $w \in H_{n-1} \cap W_{n-1}$ is well defined for some $n \geq 1$.

Again by Theorem 8.28, we obtain $(y_n, v_n) \in X \times X^*$ such that $0 = v_n + \frac{1}{r_n} (J_p(y_n) - J_p(x_n))$ and $v_n \in T y_n$. Then, the occasionally pseudomonotonicity of T and $\langle y_n - w, 0 \rangle = 0 \geq 0, T w \ni 0$ implies that $\langle y_n - w, v_n \rangle \geq 0$ for some $v_n \in T y_n$, so $w \in H_n$. It follows from Proposition 2.5 that

$$\langle w - x_n, J_p(x_0) - J_p(x_n) \rangle = \langle w - R_{H_{n-1} \cap W_{n-1}} x_0, J_p(x_0) - J_p(R_{H_{n-1} \cap W_{n-1}} x_0) \rangle \leqslant 0$$

which implies $w \in W_n$. Therefore, $w \in H_n \cap W_n$, and hence, $x_{n-1} = R_{H_n \cap W_n} x_0$ is well defined. Then, by induction, the sequence generated by (8.88) is well defined for each nonnegative integer n.   ∎

Remark 8.7 From the above proof, we obtain

$$T^{-1} 0 \subset H_n \cap W_n.$$

for each nonnegative integer n.

In 2012, Pathak and Cho [453] proved the following result.

Theorem 8.29 *Let X be a reflexive, strictly convex and uniformly smooth Banach space. If $T^{-1} 0 \neq \emptyset$, ϕ satisfies the condition (4.36) and $\{r_n\} \subset (0, \infty)$ satisfies $\liminf_{n \to \infty} r_n > 0$, then the sequence $\{x_n\}$ generated by (8.88) converges strongly to $R_{T^{-1} 0} x_0$.*

Proof It follows from the definition of W_{n+1} and Proposition 4.7 that $x_{n+1} = R_{W_{n+1}} x_0$. Further, from $x_0 \in L(x_n, R_{W_{n+1}} x_0) \cap W_{n-1}$ and Proposition 4.8, we have

$$\Psi(x_n, R_{W_{n+1}} x_0) + \Psi(R_{W_{n+1}} x_0, x_0) \leqslant \Psi(x_n, x_0)$$

and hence

$$\Psi(x_n, x_{n+1}) + \Psi(x_{n+1}, x_0) \leqslant \Psi(x_n, x_0). \tag{8.89}$$

Since the sequence $\{\Psi(x_n, x_0)\}$ is monotone decreasing and bounded below by 0, it follows that $\liminf_{n \to \infty} \Psi(x_n, x_0)$ exists and, in particular, $\{\Psi(x_n, x_0)\}$ is bounded. Then, by (4.37), $\{x_n\}$ is also bounded. This implies that there exists a subsequence $\{x_{n_i}\}$ of $\{x_n\}$ such that $x_{n_i} \rightharpoonup w$ for some $w \in X$. We shall show that $w \in T^{-1} 0$. It follows from (8.89) that $\Psi(x_n, x_{n+1}) \to 0$. On the other hand,

$$\Psi(R_{H_n}x_n, x_n) - \Psi(y_n, x_n)$$
$$= \| R_{H_n}x_n \|^p - p\langle R_{H_n}x_n - x_n, J_p(x_n)\rangle - \| x_n \|^p$$
$$\quad + \| R_{H_n}x_n - x_n \| - \| y_n \|^p + p\langle y_n - x_n, J_p(x_n)\rangle + \| x_n \|^p - \| y_n - x_n \|$$
$$= \| R_{H_n}x_n \|^p - \| y_n \|^p + p\langle y_n - R_{H_n}x_n, J_p(x_n)\rangle + \| R_{H_n}x_n - y_n \|$$
$$\geqslant p\langle R_{H_n}x_n - y_n, J_p(y_n)\rangle + p\langle y_n - R_{H_n}x_n, J_p(x_n)\rangle + \| R_{H_n}x_n - y_n\|$$
$$= p\langle y_n - R_{H_n}x_n - y_n, J_p(x_n) - J_p(y_n)\rangle + \| R_{H_n}x_n - y_n\|.$$

Since $R_{H_n}x_n \in H_n$ and $0 = v_n + \frac{1}{r_n}(J_p(y_n) - J_p(x_n))$, it follows that $\langle y_n - R_{H_n}x_n - y_n J_p(x_n) - J_p(y_n)\rangle \geq 0$ and therefore that $\Psi(R_{H_n}x_n, x_n) \geq \Psi(y_n, x_n)$. Further, since $x_{n+1} \in H_n$, we have $\Psi(x_{n+1}, x_n) \geq \Psi(R_{H_n}x_n, x_n)$ which yields

$$\Psi(x_{n+1}, x_n) \geq \Psi(R_{H_n}x_n, x_n) \geq \Psi(y_n, x_n).$$

Then, it follows from $\Psi(x_n, x_{n+1}) \to 0$ that $\Psi(y_n, x_n) \to 0$. Consequently, by Proposition 4.5, we have $y_n - x_n \to 0$, which implies $y_{n_i} \rightharpoonup w$. Moreover, since J is uniformly norm-to-norm continuous on bounded subsets and $\liminf\limits_{n\to\infty} r_n > 0$, we obtain

$$v_n = -\frac{1}{r_n}(J_p(y_n) - J_p(x_n)) \to 0.$$

It follows from $v_n \in Ty_n$, $v_n \to 0$ and $y_{n_i} \rightharpoonup w$ that

$$\lim_{i,n\to\infty} \langle z - y_{n_i}, v_n \rangle = \langle z - w, 0 \rangle = 0 \text{ for all } z \in \mathscr{D}(T).$$

Then, occasionally pseudomonotonicity of T implies that $\langle z - w, z' \rangle = 0$ for some $z' \in Tz$. Therefore, from the maximality of T, we obtain $w \in T^{-1}0$. Let $w^* \in R_{T^{-1}0}x_0$. Now, from $x_{n+1} = R_{H_n \cap W_n}x_0$ and $w^* \in T^{-1}0 \subset L(x_n, R_{W_{n+1}}x_0) \cap H_n \cap W_n$, we have

$$\Psi(x_{n+1}, x_0) \leqslant \Psi(w^*, x_0).$$

Then

$$\Psi(x_n, w^*) = \Psi(x_n, x_0) + \Psi(x_0, w^*) - p\langle x_n, x_0, J_p(w^*) - J_p(x_0)\rangle$$
$$\quad + \| x_n - w^* \| - \| x_n - x_0 \| - \| x_0 - w^* \|$$
$$\leqslant \Psi(w^*, x_0) + \Psi(x_0, w^*) - p\langle x_n - x_0, J_p(w^*) - J_p(x_0)\rangle$$
$$\quad + \| x_n - w^* \| - \| x_n - x_0 \| - \| x_0 - w^* \|$$

which yields

$$\limsup_{i\to\infty} \Psi(x_{n_i}, w^*) \leqslant \Psi(w^*, x_0) + \Psi(x_0, w^*) - p\langle w - x_0, J_p(w^*) - J_p(x_0)\rangle$$
$$\quad + \| w - w^* \| - \| w - x_0 \| - \|x_0 - w^*\|.$$

Thus, from Proposition 4.7, we have

$$\Psi(w^*, x_0) + \Psi(x_0, w^*) - p \langle w - x_0, J_p(w^*) - J_p(x_0) \rangle + \| w - w^* \|$$
$$- \| w - x_0 \| - \| x_0 - w^* \|$$
$$= p \langle w - w^*, J_p(x_0) - J_p(w^*) \rangle$$
$$\leqslant 0.$$

Then, we obtain $\limsup\limits_{i \to \infty} \Psi(x_{n_i}, w^*) \leqslant 0$ and hence $\Psi(x_{n_i}, w^*) \to 0$. It follows from Proposition 4.5 that $x_{n_i} \to w^*$. This means that the whole sequence $\{x_n\}$ generated by (8.88) converges weakly to w^* and that each weakly convergent subsequence of $\{x_n\}$ converges strongly to w^*. Therefore, $\{x_n\}$ converges strongly to $w^* \in R_{T^{-1}0}x_0$. This completes the proof. ∎

An Application

Let $f : X \to (-\infty, \infty]$ be a proper convex lower semicontinuous function. Then, the subdifferential ∂f of f is defined by

$$\partial f(x) = \{v \in X^* : f(y) - f(x) \geq \langle y - x, v \rangle, \ \forall \, y \in X\}$$

for all $z \in X$.
Using Theorem 8.29, we consider the problem of finding a minimizer of the function f.

Theorem 8.30 *Let X be reflexive, strictly convex, and uniformly smooth Banach space, and let $f : E \to (-\infty, \infty]$ be a proper convex lower semicontinuous function. Assume that $\{r_n\} \subset (0, \infty)$ satisfies $\liminf\limits_{n \to \infty} r_n > 0$ and let $\{x_n\}$ be the sequence generated by*

$$\left. \begin{aligned} &x_0 \in X, \\ &y_n = \underset{z \in X}{\mathrm{argmin}} \, \{f(z) + \frac{1}{pr_n}\|z\|^p - \frac{1}{r_n}\langle z, J_p(x_n) \rangle\}, \\ &0 = v_n + \frac{1}{r_n}(J_p(y_n) - J_p(x_n)), v_n \in Tx_n, \\ &H_n = \{z \in X : \langle z - y_n, v_n \rangle \leqslant 0\}, \\ &W_n = \{z \in X : \langle z - x_n, J_p(x_n) \rangle \leqslant 0\}, \\ &x_{n+1} = R_{H_n \cap W_n}x_0, n = 0, 1, 2, \ldots. \end{aligned} \right\} \qquad (8.90)$$

If $(\partial f)^{-1} \neq \emptyset$, then $\{x_n\}$ converges strongly to the minimizer of f.

Proof Since $f : X \to (-\infty, \infty]$ is a proper convex lower semicontinuous function, by Rockafellar [529], the subdifferential ∂f of f is a maximal monotone operator, and so it is also maximal occasionally pseudomonotone operator. We also know that

$$y_n = \operatorname*{argmin}_{z \in X} \{f(z) + \frac{1}{pr_n}\|z\|^p - \frac{1}{r_n}\langle z, J_p(x_n)\rangle\}$$

is equivalent to

$$\frac{1}{r_n}(J_p(z) - J_p(x_n)) \in \partial f(y_n) \text{ for all } z \in X.$$

This implies that

$$0 \in \partial f(y_n) + \frac{1}{r_n}(J_p(y_n) - J_p(x_n)).$$

Thus, we have $v_n \in \partial f(y_n)$ such that $0 = v_n + \frac{1}{r_n}(J_p(y_n) - J_p(x_n))$. Using Theorem 8.29, we get the conclusion. ∎

Exercises

8.1 Let X be a reflexive separable Banach space. Assume that $T : X \to X^*$ (possibly nonlinear) is bounded, pseudomonotone and coercive. Then, for arbitrary $f \in X^*$, there exists a solution $u \in X$ of equation

$$T(u) = f.$$

8.2 Let X be a reflexive separable Banach space. Assume that $T : X \to X^*$ (possibly nonlinear) is bounded, hemicontinuous, monotone and coercive. Then, for arbitrary $f \in X^*$, there exists a solution $u \in X$ of equation

$$T(u) = f.$$

Further, if T is strictly monotone then show that the solution is unique.

8.3 Let V be a closed linear subspace of the Sobolev space $W^{1,p}(\Omega)$, containing $W_0^{1,p}(\Omega)(1 < p < \infty, \Omega \subset \mathbb{R}^n$ is a bounded domain with sufficiently smooth boundary). Further, suppose the operator $T : V \to V^*$ is given by

$$\langle T(u), v\rangle = \sum_{j=1}^{n} \int_{\Omega} a_j(x, u(x), Du(x))D_j v(x)dx$$

$$+ \int_{\Omega} a_0(x, u(x), Du(x))v(x)dx, \quad u, v \in V \tag{†}$$

where the function $a_j : \Omega \times \mathbb{R}^{n+1} \to \mathbb{R}$ satisfy conditions:
(A1) Assume that the functions $a_j : \Omega \times \mathbb{R}^{n+1} \to \mathbb{R}$ satisfy the Carathodory conditions; i.e., for almost all (a.a.) fixed $x \in \Omega$, the function $\xi \mapsto a_j(x, \xi), \xi \in \mathbb{R}^{n+1}$ is

continuous and for each fixed $\xi \in \mathbb{R}^{n+1}$, $x \mapsto a_j(x, \xi)$, $x \in \Omega$ is measurable.

(A2) Assume that there exist a constant c_1 and a nonnegative function $k_1 \in L_q(\Omega)(\frac{1}{p} + \frac{1}{q} = 1)$ such that for almost all $x \in \Omega$, each $\xi \in \mathbb{R}^{n+1}$

$$|a_j(x, \xi)| \leqslant c_1 |\xi|^{p-1} + k_1(x).$$

Show that $T : V \to V^*$ is bounded and hemicontinuous.

8.4 Let V be a closed linear subspace of the Sobolev space $W^{1,p}(\Omega)$, containing $W_0^{1,p}(\Omega)(1 < p < \infty, \Omega \subset \mathbb{R}^n$ is a bounded domain with sufficiently smooth boundary). Further, suppose the operator $T : V \to V^*$ is given by (†), where the function $a_j : \Omega \times \mathbb{R}^{n+1} \to \mathbb{R}$ satisfy the following condition:

(A3) Assume that for a.a. $x \in \Omega$, all $\xi, \xi^* \in \mathbb{R}^{n+1}$

$$\sum_{j=0}^{n} [a_j(x, \xi) - a_j(x, \xi^*)](\xi_j - \xi_j^*) \geqslant 0.$$

Show that T given by (†) is monotone.

8.5 Assume that the functions a_j satisfy (A1), for a.e. $x \in \Omega$, the functions $\xi \mapsto a_j(x, \xi)$ are continuously differentiable and the matrix

$$\left(\frac{\partial a_j(x, \xi)}{\partial \xi_k} \right)_{j,k=0}^{n} \quad \text{is positive semidefinite.}$$

Show that T given by (†) is monotone.

Chapter 9
Applications of Fixed Point Theorems

*The most painful thing about mathematics is how far away you
are from being able to use it after you have learned it.*

James Newman

*I cannot abstain from playing the role of an (often unwelcome)
intermediary in this drama between mathematics and physics,
which fertilize each other in the dark, and deny and misconstrue
one another when face to face.*

Hermann Weyl (1928)

(In the preface of the first edition of Hermann Weyl's book on
group theory and quantum mechanics)

Fixed point theory is a viable, productive, conclusive and useful to solve problems
of existence and uniqueness of solution of a differential equation or an integral equation. Moreover, it encompasses various facets of analysis and a fascinating subject
endowed with sophisticated tools with an enormous number of applications in various fields of mathematics. In this chapter, we intend to give some applications of
fixed point theorems to obtain existence theorems for nonlinear differential and integral equations. Our treatment includes some standard well-known results as well as
some recent ones. We have avoided an extensive discussion on this areas instead we
concentrate on a few important problems. As usual, in most cases, the differential
equations are transformed into an equivalent operator equations involving integral
operators, and then, appropriate fixed point theorems or degree theoretic methods
are invoked to prove the existence of desired solutions by recasting the operator
equations into fixed point equations.

For earlier works, one can consult Cronin [152] and Miranda [404] for general
references, and for recent work, see Browder [94] and Martin Jr. [388]. For applications to nonlinear integral equations, refer Krasnoselskii [344] who has dealt with
this work in detail in his book. For an exhaustive discussion on positive solutions of
operator equations, see Krasnoselskii [345] and Amann [17]. For a different approach

© Springer Nature Singapore Pte Ltd. 2018
H. K. Pathak, *An Introduction to Nonlinear Analysis and Fixed Point Theory*,
https://doi.org/10.1007/978-981-10-8866-7_9

to the existence problems in differential equations, refer to Cesari [137]. Nussbaum [433] and Walther [604] deal with applications to functional differential equations, and Gustafson [262] has dealt with nodal problems in differential equations. These references have been cited to indicate the extent of the applications.

9.1　Application to Geometry of Banach Spaces

Let X be a metric space and D a nonempty subset of X. Let T be a mapping of D into X, and let $\mathfrak{F}(T)$ be the set of all fixed points of T. For a given $x_0 \in D$, the sequence of iterate $\{x_n\}$ is determined by $x_n = T(x_{n-1}) = T^n(x_0), n = 1, 2, 3, \ldots$. Let $\omega = \mathbb{N} \cup \{0\}$.

The concept of quasi-nonexpansive mapping was initiated by Tricomi in 1941 for real functions. It was further studied by Diaz and Metcalf [187] and Doston [196, 198] for mappings in Banach spaces. Recently, this concept was given by Kirk [330] in metric spaces as follows:

Definition 9.1 The mapping T is said to be quasi-nonexpansive if for each $x \in D$ and for every $p \in \mathfrak{F}(T), d(T(x), p) \leq d(x, p)$. A mapping T is conditionally quasi-nonexpansive if it is quasi-nonexpansive whenever $\mathfrak{F}(T) \neq \varnothing$.

Definition 9.2 (*Pathak* [451]) The mapping T is said to be locally quasi-nonexpansive at $p \in \mathfrak{F}(T)$ w.r.t. a sequence $\{x_n\}$ if for all $n \in \omega, d(x_{n+1}, p) \leq d(x_n, p)$.

Obviously, locally quasi-nonexpansiveness at $p \in \mathfrak{F}(T) \Rightarrow$ locally quasi-nonexpansiveness at $p \in \mathfrak{F}(T)$ w.r.t. a sequence $\{x_n\}$.

Supper drop– Let $K = K(z, r)$ be a closed ball in a Banach space X. For a sequence $\{x_n\}_{n=0}^{\infty} \not\subseteq K$ converging to x, we define $\lim\limits_{n \to \infty} \mathcal{D}_n = \mathcal{SD}(x, K)$, where $\mathcal{D}_0 = co(\{x_0\} \cup K)$, and $\mathcal{D}_{n+1} = co(\{x_n\} \cup \mathcal{D}_n)$ $\forall n \in \omega$, and $\mathcal{SD}(x, K)$ is called a super drop.

Clearly, for a constant sequence $\{x_n\} \equiv \{x\}$ converging to x, we have $\mathcal{D}_{n+1} = \mathcal{D}_n$ $\forall n \in \omega$, so that $\mathcal{D}(x, K) = co(\{x\} \cup K)$ and is called a drop. Thus, the concept of a drop is a special case of super drop. It is also clear that if $y \in \mathcal{D}(x, K)$, then $\mathcal{D}(y, K) \subset \mathcal{D}(x, K)$, and if $z = 0$, then $\|y\| \leqslant \|x\|$.

Theorem 9.1 (Daneš) *Let C be a closed subset of a Banach space X, let $z \in X - C$, and let $K = K(z, r)$ be a closed ball of radius $r < d(z, C) = R$. Let $T : C \to C$ be any map such that $T(c) \in C \cap \mathcal{D}(c, K)$ for each $c \in C$. Then, for each $x \in C$, the map T has a fixed point in $C \cap \mathcal{D}(x, K)$.*

As a consequence of Theorem 9.1, we can easily obtain the following result.

Theorem 9.2 (Supporting drops theorem) *Let C be a closed set in a Banach space X, and $z \in X - C$ a point with $d(z, C) = R > 0$. Then, for any $r < R < \rho$, there is an $x_0 \in \partial C$ with*

$$\|z - x_0\| \leqslant \rho \ \text{and} \ C \cap \mathcal{D}(x_0, K(z, r)) = \{x_0\}.$$

The following lemmas are useful to prove the next result.

Lemma 9.1 *Let T be locally quasi-nonexpansive at $p \in \mathfrak{F}(T)$ w.r.t. $\{x_n\}$, and*

$$\lim_{n \to \infty} d(x_n, \mathfrak{F}(T)) = 0.$$

Then, $\{x_n\}$ is a Cauchy sequence.

Proof Since $\lim\limits_{n \to \infty} d(x_n, \mathfrak{F}(T)) = 0$, then for any given $\varepsilon > 0$, there exists $n_1 \in \mathbb{N}$ such that for each $n \geq n_1$, $d(x_n, \mathfrak{F}(T)) < \frac{\varepsilon}{2}$. So, there exists $q \in \mathfrak{F}(T)$ such that for all $n \geq n_1$, $d(x_n, q) < \frac{\varepsilon}{2}$. Thus, for any $m, n \geq n_1$, we have that

$$d(x_m, x_n) \leq d(x_m, q) + d(x_n, q) < \frac{\varepsilon}{2} + \frac{\varepsilon}{2} = \varepsilon, \quad q \in \mathfrak{F}(T).$$

Hence, $\{x_n\}$ is a Cauchy sequence.   ∎

Theorem 9.3 (Pathak [451]) *Let $\mathfrak{F}(T)$ be a nonempty closed set. Then,*

(1) $\lim\limits_{n \to \infty} d(x_n, \mathfrak{F}(T)) = 0$ *if $\{x_n\}$ converges to a point p in $\mathfrak{F}(T)$;*
(2) *$\{x_n\}$ converges to a point in $\mathfrak{F}(T)$ if $\lim\limits_{n \to \infty} d(x_n, \mathfrak{F}(T)) = 0$, T is locally quasi-nonexpansive at $p \in \mathfrak{F}(T)$ w.r.t. $\{x_n\}$, and X is complete.*

Proof (1) Since $\mathfrak{F}(T)$ is closed, $p \in \mathfrak{F}(T)$ and the mapping $x \mapsto d(x, \mathfrak{F}(T))$ is continuous, then

$$\lim_{n \to \infty} d(x_n, \mathfrak{F}(T)) = d(\lim_{n \to \infty} x_n, \mathfrak{F}(T)) = d(p, \mathfrak{F}(T)) = 0.$$

(2) From Lemma 9.1, $\{x_n\}$ is a Cauchy sequence. Since X is complete, then $\{x_n\}$ converges to a point, say q, in X. Since $\mathfrak{F}(T)$ is closed, then $0 = \lim\limits_{n \to \infty} d(x_n, \mathfrak{F}(T)) = d(\lim\limits_{n \to \infty} x_n, \mathfrak{F}(T)) = d(q, \mathfrak{F}(T))$ implies that $q \in \mathfrak{F}(T)$.   ∎

Recall that a function $\varphi : X \to \mathbb{R}$ is called a lower semicontinuous whenever $\{x \in X : \varphi(x) \leq \alpha\}$ is closed for each $\alpha \in \mathbb{R}$.

We now give some applications of Theorem 9.3 using Caristi's fixed point theorem to geometry of Banach Spaces.

Theorem 9.4 (Pathak [451]) *Let C be a closed subset of a Banach space X, let $z \in X - C$, and let $K = K(z, r)$ be a closed ball of radius $r < d(z, C) = R$. Let x be an arbitrary element of C, let $\{x_n\}$ be a sequence in C converging to x, and let $T : C \to X$ be any continuous function defined implicitly by $T(x) \in C \cap \mathcal{S}\mathcal{D}(x, K)$ for each $x \in C$ in the sense that $T(x_n) \in C \cap \mathcal{D}_n$ for each $n \in \omega$. Then,*

(1) $\lim\limits_{n \to \infty} d(x_n, \mathfrak{F}(T)) = 0$ *if $\{x_n\}$ converges to a point p in $F(T)$;*

(2) $\{x_n\}$ *converges to a point in* $\mathfrak{F}(T)$ *if* $\lim\limits_{n\to\infty} d(x_n, \mathfrak{F}(T)) = 0$, *and* T *is locally quasi-nonexpansive at* $p \in \mathfrak{F}(T)$ *w.r.t.* $\{x_n\}$.

Proof Without loss of generality, we may assume that $z = 0$. Let $\|x\| = \eta \geq R$, and let $X = A \cap \mathcal{SD}(x, K)$. Then, it is clear that T maps X into itself. For given $y \in X$ and a sequence $\{y_n\}$ converging to y, we shall estimate $\|y - T(y)\|$ on X.

For given $y \in X$ and the corresponding sequence $\{y_n\}$, there is a sequence $\{b_n\}$ in X with $T(y_n) = tb_n + (1 - t)y_n$, $0 < t < 1$. Now $\|T(y_n)\| \leq t\|b_n\| + (1 - t)\|y_n\|$, we have

$$t(\|y_n\| - \|b_n\|) \leq \|y_n\| - \|T(y_n)\|,$$

so because $\|y_n\| - \|b_n\| \geq R - \eta$, we find that $t \leq \frac{\|y_n\| - \|T(y_n)\|}{R - \eta}$. Thus ,

$$\|y_n - T(y_n)\| \leq t\|y_n - b_n\| \leq t(\|y_n\| + \|b_n\|) \leq (\eta + r)$$

$$\leq \frac{\eta + r}{R - r}(\|y_n\| - \|T(y_n)\|).$$

Define $d(x, y) = \|x - y\| \ \forall x, y \in X$ and $\varphi(y) = \frac{\eta + r}{R - r}\|y\|$ then X is complete as a metric space and $\varphi : X \to \mathbb{R}$ is a continuous function. So, φ is a lower-semicontinuous function. Also, the above inequality takes the form

$$d(y_n, T(y_n)) \leq \varphi(y_n) - \varphi(T(y_n)).$$

Proceeding to the limit as $n \to \infty$, we obtain

$$d(y, T(y)) \leq \varphi(y) - \varphi(T(y)),$$

for each $y \in X$. Therefore, applying Caristi's fixed point theorem, we obtain that T has a fixed point $p = p(x)$ for each $x \in C$, i.e. $\mathfrak{F}(T) \neq \varnothing$. By continuity of T, $\mathfrak{F}(T)$ is closed. Hence, the conclusion follows from Theorem 9.3. ∎

Since drop is a special case of super drop, we have the following:

Corollary 9.1 *Let* C *be a closed subset of a Banach space* X, *let* $z \in X - C$, *and let* $K = K(z, r)$ *be a closed ball of radius* $r < d(z, C) = R$. *Let* x *be an arbitrary element of* C, *and let* $T : C \to X$ *be any (not necessarily continuous) function defined implicitly by* $T(x) \in C \cap \mathcal{D}(x, K)$ *for each* $x \in C$. *Then,*

(1) $\lim\limits_{n\to\infty} d(x_n, \mathfrak{F}(T)) = 0$ *if* $\{x_n\}$ *converges to a point* p *in* $\mathfrak{F}(T)$;

(2) $\{x_n\}$ *converges to a point in* $\mathfrak{F}(T)$ *if* $\lim\limits_{n\to\infty} d(x_n, \mathfrak{F}(T)) = 0$, *and* T *is locally quasi-nonexpansive at* $p \in \mathfrak{F}(T)$ *w.r.t.* $\{x_n\}$.

9.2 Application to System of Linear Equations

In this section, we present an application of celebrated Banach contraction theorem to find the solution of the following system of linear equations with n unknowns:

$$\left.\begin{array}{c} a_{11}x_1 + a_{12}x_2 + \cdots + a_{1n}x_n = b_1 \\ a_{21}x_1 + a_{22}x_2 + \cdots + a_{2n}x_n = b_2 \\ \cdots\cdots\cdots\cdots\cdots\cdots\cdots\cdots \\ a_{n1}x_1 + a_{n2}x_2 + \cdots + a_{nn}x_n = b_n. \end{array}\right\} \tag{9.1}$$

This system of linear equations (9.1) can be written as

$$\left.\begin{array}{rl} x_1 = & (1 - a_{11})x_1 - a_{12}x_2 - \cdots - a_{1n}x_n + b_1 \\ x_2 = & -a_{21}x_1 + (1 - a_{22})x_2 - a_{23}x_3 - \cdots - a_{2n}x_n + b_2 \\ x_3 = & -a_{31}x_1 - a_{32}x_2 + (1 - a_{33})x_3 - \cdots - a_{3n}x_n + b_3 \\ \cdots & \cdots\cdots\cdots\cdots\cdots\cdots\cdots\cdots\cdots\cdots\cdots\cdots\cdots \\ x_n = & -a_{n1}x_1 - a_{n2}x_2 - a_{n3}x_3 - \cdots + (1 - a_{nn})x_n + b_n. \end{array}\right\} \tag{9.2}$$

Letting $\alpha_{ij} = -a_{ij} + \delta_{ij}$, where

$$\delta_{ij} = \begin{cases} 1 & \text{for } i = j \\ 0 & \text{for } i \neq j. \end{cases}$$

Then, the system (9.2) is equivalent to the following system:

$$x_i = \sum_{j=1}^{n} \alpha_{ij}x_j + b_i, \quad i = 1, 2, 3, \ldots, n. \tag{9.3}$$

If $x = (x_1, x_2, \ldots, x_n) \in \mathbb{R}^n$, $b = (b_1, b_2, \ldots, b_n) \in \mathbb{R}^n$ and $A = (a_{ij})_{n \times n}$ a matrix, i.e.

$$A = \begin{pmatrix} a_{11} & a_{12} & \cdots & a_{1n} \\ a_{21} & a_{22} & \cdots & a_{2n} \\ \cdots & \cdots & \cdots & \cdots \\ a_{n1} & a_{n2} & \cdots & a_{nn} \end{pmatrix}, x = (x_1, x_2, \ldots, x_n)^T, b = (b_1, b_2, \ldots, b_n)^T,$$

then the system (9.3) is equivalent to

$$x = x - Ax + b. \tag{9.4}$$

Thus, we see that finding the solution of the problem (9.4) is the problem of finding fixed point of the transformation $T : \mathbb{R}^n \to \mathbb{R}^n$ defined by

$$T(x) = x - Ax + b. \tag{9.5}$$

If T is a contraction mapping, then we can use Banach Contraction Theorem 5.1 and obtain the unique solution of $T(x) = x$ by the method of successive approximation. Observe that, for $x, y \in \mathbb{R}^n$,

$$Tx - Ty = (x - Ax + b) - (y - Ty + b) = (x - y) - (Ax - Ay)$$

$$= (x - y) - A(x - y) = (I - A)(x - y).$$

Theorem 9.5 *Let $X = \mathbb{R}^n$ be a metric space with the metric $d_\infty(x, y) = \max_{1 \leqslant i \leqslant n} |x_i - y_i|$. If $\sum_{j=1}^n |\alpha_{ij}| \leqslant \alpha < 1$ for all $i = 1, 2, \ldots, n$, then the system of linear equations (9.1) in n unknown has a unique solution.*

Proof Since $X = \mathbb{R}^n$ is complete with respect to the metric d_∞, it suffices to show that the mapping T defined by (9.5) is a contraction. We have

$$d_\infty(T(x), T(y)) = \max_{1 \leqslant i \leqslant n} \left| \sum_{j=i}^n \alpha_{ij}(x_j - y_j) \right| \leqslant \max_{1 \leqslant i \leqslant n} \sum_{j=i}^n |\alpha_{ij}| |x_j - y_j|$$

$$\leqslant \max_{1 \leqslant j \leqslant n} |x_j - y_j| \left(\max_{1 \leqslant i \leqslant n} \sum_{j=i}^n |\alpha_{ij}| \right) = d_\infty(x, y) \left(\max_{1 \leqslant i \leqslant n} \sum_{j=i}^n |\alpha_{ij}| \right)$$

$$\leqslant \alpha d_\infty(x, y).$$

Thus, T is contraction mapping. Hence, by Banach's contraction theorem 5.1, the linear system (9.1) has a unique solution. ∎

We now demonstrate application of Brouwer's fixed point theorem in theory of matrices. To effect this, we need an example of continuous operator in $\mathbb{R}^n$ given by a matrix.

Example 9.1 Let $\mathbb{C}^n$ be the space of n-tuples of complex numbers. Let $A = (a_{ij})_{m \times n}$ be a matrix of order $m \times n$. Define a map $T : \mathbb{C}^n \to \mathbb{C}^n$ by

$$Tx = Ax, \quad x \in \mathbb{C}^n.$$

For any fixed i, the Cauchy–Schwarz inequality yields

$$\left| \sum_{j=1}^n a_{ij}(x_j - y_j) \right| \leqslant \left(\sum_{j=1}^n |a_{ij}|^2 \right)^{\frac{1}{2}} \|x - y\|.$$

Hence,

$$\|Ax - Ay\| = \|A(x - y)\|$$

$$\leqslant \|A\| \|x - y\| = \left(\sum_{i,j} |a_{ij}|^2 \right)^{\frac{1}{2}} \|x - y\|$$

$$\leqslant \sqrt{mn} M \|x - y\|$$

where M is the modulus of the largest element of A. Hence, for any given $\varepsilon > 0$, there exists a $\delta = \frac{\varepsilon}{M} \sqrt{mn} > 0$, and for any $x \in \mathbb{C}^n$,

$$\|x - y\| < \delta \Rightarrow \|Ax - Ay\| < \varepsilon.$$

Thus, we see that the matrix operator A is uniformly continuous, and so it is continuous.

The following result is an application of Brouwer's fixed point theorem which is important in many applied fields.

Theorem 9.6 (Perron–Frobenius) *Let* $\mathbb{A} = (a_{ij})$ *be an* $n \times n$ *matrix with strictly positive entries. Then,* $\mathbb{A}$ *has a strictly positive eigenvalue.*

Proof The matrix $\mathbb{A} = (a_{ij})$ can be viewed as a linear transformation from $\mathbb{R}^n$ to $\mathbb{R}^n$. Notice that a positive matrix transforms vectors with positive components into vectors with positive components. Geometrically, this works in the n-dimensional analogue of the first quadrant. Notice also that the eigenvalue equation maps vectors into multiples of themselves. Thus, it is reasonable, and when looking for eigenvector, we have to consider only vectors with positive components. So, if we could disregard the magnitude of the vector and concentrate on its direction then we could regard eigenvector as a "fixed" vector under the transformation, in the sense that its direction would be unchanged. We may therefore consider $\frac{x}{\|x\|}$ in place of x. This would allow us to apply the operator to the n-dimensional analogue the positive quadrant of the surface of the sphere. However, we notice that the surface of a sphere is not convex, and so we use a plane rather than a surface of the sphere.

We now introduce the set

$$K = \left\{ x \in \mathbb{R}^n : \sum_{i=1}^{n} x_i = 1, x_i \geqslant 0 \text{ for } i = 1, 2, \ldots, n \right\}.$$

Clearly, the set K is compact and convex. Define

$$Tx = \frac{\mathbb{A}x}{\|\mathbb{A}x\|_1}$$

where $\| \cdot \|_1$ is the euclidean 1-norm. Notice that if $x \in K$, then all the entries of x are nonnegative and at least one is strictly positive; hence, all the entries of $\mathbb{A}x$ are strictly positive. Then, T is a continuous function mapping K into K, and therefore, there exists $\bar{x} \in K$ such that $\mathbb{A}\bar{x} = \lambda\bar{x}$ with $\lambda = ||\mathbb{A}\bar{x}||_1 = \sum_{i=1}^{n}(\mathbb{A}\bar{x})_i$. ∎

Theorem 9.7 (Fundamental Theorem of Algebra) *Let $p(z) = a_0 + a_1 z + \cdots + a_n z^n$ be a complex polynomial of degree $n \geq 1$. Then, there exists $z_0 \in \mathbb{C}$ such that $p(z_0) = 0$.*

Proof We begin with the identification of $\mathbb{C}$ with $\mathbb{R}^2$. Assume, without loss of generality, that $a_n = 1$. Suppose $r = 2 + |a_0| + |a_1| + \cdots + |a_{n-1}|$. Now define the continuous function $f : \mathbb{C} \to \mathbb{C}$ as follows:

$$f(z) = \begin{cases} z - \frac{p(z)}{r}e^{i(1-n)\theta}, & \text{if } |z| \leq 1, \\ z - \frac{p(z)}{rz^{n-1}}, & \text{if } |z| > 1, \end{cases}$$

where $z = \rho e^{i\theta}$ with $\theta \in [0, 2\pi)$. Let us consider the compact and convex set $C = \{z : |z| \leq r\}$. In order to apply the Brouwer fixed point theorem, we need to show that $f(C) \subset C$. Indeed, if $|z| \leq 1$, we observe that

$$|f(z)| \leq |z| + \frac{|p(z)|}{r}$$
$$\leq 1 + \frac{1 + |a_0| + \cdots + |a_{n-1}|}{r} = 1 + \frac{r-1}{r} = 2 - \frac{1}{r} < 2 \leq r.$$

Further, if $1 < |z| \leq r$, we have

$$|f(z)| \leq \left| z - \frac{p(z)}{rz^{n-1}} \right| = \left| z - \frac{z}{r} - \frac{a_0 + a_1 z + \cdots + a_{n-1}z^{n-1}}{rz^{n-1}} \right|$$
$$\leq r - 1 + \frac{|a_0| + \cdots + |a_{n-1}|}{r}$$
$$\leq r - 1 + \frac{r-2}{r} = r - \frac{2}{r} < r.$$

It follows from the above inequalities that C is invariant for f. Hence, f has a fixed point $z_0 \in C$, which is clearly a root of p. This completes the proof. ∎

Remark 9.1 An alternative approach to prove Theorem 9.7 makes use of the concept of topological degree (see, e.g., Munkres [414], Chap. 9).

9.2.1 Existence of Solutions of Nonlinear Matrix Equation

In this section, we have applied Theorem 5.118 due to Pathak et al. [458] to study the existence of solutions of nonlinear matrix equation

$$X = Q + \sum_{i=1}^{m} A_i X^{\delta_i} A_i^*, \quad 0 < |\delta_i| < 1 \qquad (9.6)$$

where Q is an $n \times n$ positive semidefinite matrix and $A_i's$ are nonsingular $n \times n$ matrices, or Q is an $n \times n$ positive definite matrix and $A_i's$ are arbitrary $n \times n$ matrices and positive definite solution X is sought. Here, A_i^* denote the conjugate transpose of the matrix A_i. The existence and uniqueness of positive definite solutions and numerical methods for finding a solution of (9.6) have recently been studied by many authors (see [458] and the references therein). The Thompson metric on the open convex cone $P(N)(N \geq 2)$, the set of all $N \times N$ Hermitian positive definite matrices, is defined by

$$d(A, B) = \max \left\{ \log M(A/B), \log M(B/A) \right\} \qquad (9.7)$$

where $M(A/B) = \inf\{\lambda > 0 : A \leqslant \lambda B\} = \lambda_{\max}(B^{-1/2} A B^{-1/2})$, the maximal eigenvalue of $B^{-1/2} A B^{-1/2}$. Here, $X \leqslant Y$ means that $Y - X$ is positive semidefinite, and $X < Y$ means that $Y - X$ is positive definite.

Thompson [589] has proved that $P(N)$ is a complete metric space with respect to the Thompson metric d and $d(A, B) = \| \log(A^{-1/2} B A^{-1/2}) \|$, where $\|.\|$ stands for the spectral norm. The Thompson metric exists on any open normal convex cones of real Banach spaces [426, 589], in particular the open convex cone of positive definite operators of a Hilbert space. It is invariant under the matrix inversion and congruence transformations:

$$d(A, B) = d(A^{-1}, B^{-1}) = d(MAM^*, MBM^*) \qquad (9.8)$$

for any nonsingular matrix M. One remarkable and useful result is the nonpositive curvature property of the Thompson metric:

$$d(X^r, Y^r) \leqslant rd(X, Y), r \in [0, 1]. \qquad (9.9)$$

By the invariant properties of the metric, we then have

$$d(MX^r M^*, MY^r M^*) = |r|d(X, Y), r \in [-1, 1] \qquad (9.10)$$

for any $X, Y \in P(N)$ and nonsingular matrix M. Proceeding as in Lim [369], we prove the following:

Lemma 9.2 *For any* $A_1, A_2, \ldots, A_k \in P(N), B_1, B_2, \ldots, B_k \in P(N), d(A_1 + A_2 + \cdots + A_k, B_1 + B_2 + \cdots +, B_k) \leqslant \max\{d(A_1, B_1), d(A_2, B_2), \ldots, d(A_k, B_k)\}$.

Proof Without loss of generality, we can assume that

$$d(A_1, B_1) \leqslant d(A_2, B_2) \leqslant \cdots \leqslant d(A_k, B_k) = \log r.$$

Then,

$$B_1 \leqslant r A_1, \, B_2 \leqslant r A_2, \ldots, B_k \leqslant r A_k$$

and

$$A_1 \leqslant r B_1, \, A_2 \leqslant r B_2, \ldots, A_k \leqslant r B_k,$$

and thus,

$$B_1 + A_1 \leqslant r(A_1 + B_1), \, B_2 + A_2 \leqslant r(A_2 + B_2), \ldots, B_k + A_k \leqslant r(A_k + B_k).$$

So, we have

$$A_1 + A_2 + \cdots + A_k \leqslant r[B_1 + B_2 + \cdots + B_k]$$

and

$$B_1 + B_2 + \cdots + B_k \leqslant r[A_1 + A_2 + \cdots + A_k].$$

Hence,

$$d(A_1 + A_2 + \cdots + A_k, B_1 + B_2 + \cdots + B_k)$$
$$\leqslant \log r = d(A_k, B_k) = \max\{d(A_1, B_1), d(A_2, B_2), \ldots, d(A_k, B_k)\}.$$

This completes the proof.

For arbitrarily chosen positive definite matrices $X_{n-r}, X_{n-(r-1)}, \ldots, X_n$, consider the iterative sequence of matrices, given by

$$X_{n+1} = Q + A_1^* X_{n-r}^{\alpha_1} A_1 + A_2^* X_{n-(r-1)}^{\alpha_2} A_2 + \cdots + A_{r+1}^* X_n^{\alpha_{r+1}} A_{r+1} \qquad (9.11)$$

$\alpha_1, \alpha_2 \ldots \alpha_{r+1}$ are real numbers.

Theorem 9.8 *Suppose that* $\lambda = \max\{|\alpha_1|, |\alpha_2|, \ldots, |\alpha_{r+1}|\} \in (0, 1)$.

 (i) *Equation (9.11) has a unique equilibrium point in* $P(N)$; *that is, there exist a unique* $U \in P(N)$ *such that*

$$U = Q + A_1^* U^{\alpha_1} A_1 + A_2^* U^{\alpha_2} A_2 + \cdots + A_{r+1}^* U^{\alpha_{r+1}} A_{r+1} \qquad (9.12)$$

 (ii) *The iterative sequence* $\{X_n\}$ *defined by (9.11) converges to a unique solution of* *(9.6).*

Proof Define the mapping $f : P(N) \times P(N) \times P(N) \times \cdots \times P(N) \to P(N)$ by

$$f(X_1, X_2, X_{n-(r-2)}, \ldots, X_k) = Q + A_1^* X_1^{\alpha_1} A_1 + A_2^* X_2)^{\alpha_2} A_2 + \cdots + A_{r+1}^* X_k^{\alpha_{r+1}} A_{r+1} \qquad (9.13)$$

where $X_1, X_2, \ldots, X_k \in P(N)$.

For all $X_{n-r}, X_{n-(r-1)}, X_{n-(r-2)}, \ldots, X_{n+1} \in P(N)$, we have

$$
\begin{aligned}
&d(f(X_{n-r}, X_{n-(r-1)}, X_{n-(r-2)}, \ldots, X_n),\, f(X_{n-(r-1)}, X_{n-(r-2)}, X_{n-(r-2)}, \ldots, X_{n+1})) \\
&= d(Q + A_1^* X_{n-r}^{\alpha_1} A_1 + A_2^* X_{n-(r-1)}^{\alpha_2} A_2 + \cdots + A_{r+1}^* X_n^{\alpha_{r+1}} A_{r+1}, \\
&\quad\quad Q + A_2^* X_{n-(r-1)}^{\alpha_1} A_2 + A_3^* X_{n-(r-2)}^{\alpha_3} A_3 + \cdots + A_{r+2}^* X_{n+1}^{\alpha_{r+2}} A_{r+2} \\
&\leqslant d(A_1^* X_{n-r}^{\alpha_1} A_1 + A_2^* X_{n-(r-1)}^{\alpha_2} A_2 + \cdots + A_{r+1}^* X_n^{\alpha_{r+1}} A_{r+1}, \\
&\quad\quad A_2^* X_{n-(r-1)}^{\alpha_1} A_2 + A_3^* X_{n-(r-2)}^{\alpha_3} A_3 + \cdots + A_{r+2}^* X_{n+1}^{\alpha_{r+2}} A_{r+2} \\
&\leqslant max\{d(A_1^* X_{n-r}^{\alpha_1} A_1,\, A_2^* X_{n-(r-1)}^{\alpha_1} A_2),\, d(A_2^* X_{n-(r-1)}^{\alpha_2} A_2,\, A_3^* X_{n-(r-2)}^{\alpha_3} A_3), \\
&\quad\quad \cdots d(A_{r+1}^* X_n^{\alpha_{r+1}} A_{r+1},\, A_{r+2}^* X_{n+1}^{\alpha_{r+2}} A_{r+2}) \\
&\leqslant \max\{|\alpha_1| d(X_{n-r}, X_{n-(r-1)}),\, |\alpha_2| d(X_{n-(r-1)}, X_{n-(r-2)}), \ldots, |\alpha_{r+1}| d(X_n, X_{n+1})\} \\
&\leqslant \max\{|\alpha_1|, |\alpha_2|, \ldots, |\alpha_{r+1}|\} \max\{d(X_{n-r}, X_{n-(r-1)}),\, d(X_{n-(r-1)}, X_{n-(r-2)}), \\
&\quad\quad \ldots, d(X_n, X_{n+1})\} \\
&\leqslant \lambda \max\{d(X_{n-r}, X_{n-(r-1)}),\, d(X_{n-(r-1)}, X_{n-(r-2)}), \ldots, d(X_n, X_{n+1})\}. \quad\quad (9.14)
\end{aligned}
$$

Further, for all $X_{n-r}, X_{n-(r-1)}, X_{n-(r-2)}, \ldots, X_{n+1} \in P(N).\, X, Y \in P(N)$, we have

$$
\begin{aligned}
&d(f(X, X, \ldots, X),\, f(Y, Y, \ldots, Y)) \\
&\quad = d(Q + A_1^* X^{\alpha_1} A_1 + A_2^* X^{\alpha_2} A_2 + \cdots + A_{r+1}^* X^{\alpha_{r+1}} A_{r+1}, \\
&\quad\quad Q + A_2^* Y^{\alpha_1} A_2 + A_3^* Y^{\alpha_3} A_3 + \cdots + A_{r+2}^* Y^{\alpha_{r+2}} A_{r+2} \\
&\quad \leqslant d(A_1^* X^{\alpha_1} A_1 + A_2^* X^{\alpha_2} A_2 + \cdots + A_{r+1}^* X^{\alpha_{r+1}} A_{r+1}, \\
&\quad\quad A_2^* Y^{\alpha_1} A_2 + A_3^* Y^{\alpha_3} A_3 + \cdots + A_{r+2}^* Y^{\alpha_{r+2}} A_{r+2} \\
&\quad \leqslant \max\{d(A_1^* X^{\alpha_1} A_1,\, A_2^* Y^{\alpha_1} A_2),\, d(A_2^* X^{\alpha_2} A_2,\, A_3^* Y^{\alpha_3} A_3), \\
&\quad\quad \ldots, d(A_{r+1}^* X^{\alpha_{r+1}} A_{r+1},\, A_{r+2}^* Y^{\alpha_{r+2}} A_{r+2})\} \\
&\quad \leqslant \max\{|\alpha_1| d(X, Y),\, |\alpha_2| d(X, Y), \ldots, |\alpha_{r+1}| d(X, Y)\} \\
&\quad \leqslant \max\{|\alpha_1|, |\alpha_2|, \ldots, |\alpha_{r+1}|\} \max\{d(X, Y), d(X, Y), \ldots, d(X, Y)\} \\
&\quad \leqslant \lambda \max\{d(X, Y), d(X, Y), \ldots, d(X, Y)\} \\
&\quad < d(X, Y).
\end{aligned}
$$

Since $\lambda \in (0, 1)$, (i) and (ii) follow immediately from Theorem 5.119 with $s = 1$ and $g = I$. This completes the proof.

Numerical Experiment Illustrating the Above Convergence Algorithm

Consider the nonlinear matrix equation

$$
X = Q + A^* X^{\frac{1}{2}} A + B^* X^{\frac{1}{3}} B + C^* X^{\frac{1}{4}} C \quad\quad (9.15)
$$

where

$$A = \begin{pmatrix} 14/3 & 1/3 & 1/4 \\ 2/15 & 1/12 & 1/23 \\ 3/10 & 9/20 & 11/4 \end{pmatrix}, \qquad B = \begin{pmatrix} 2/5 & 3/2 & 4/6 \\ 10/4 & 6/13 & 7/46 \\ 5/2 & 4/7 & 6/13 \end{pmatrix}, \qquad C = \begin{pmatrix} 1/3 & 19/24 & 22/55 \\ 17/10 & 27/15 & 45/17 \\ 13/8 & 1/3 & 1/4 \end{pmatrix},$$

and $Q = \begin{pmatrix} 1 & 2 & 3 \\ 2 & 6 & 4 \\ 1 & 2 & 7 \end{pmatrix}$.

We define the iterative sequence $\{X_n\}$ by

$$X_{n+1} = Q + A^* X_{n-2}^{\frac{1}{2}} A + B^* X_{n-1}^{\frac{1}{3}} B + C^* X_n^{\frac{1}{4}} C \tag{9.16}$$

Let $R_m \, (m \geq 2)$ be the residual error at the iteration m that is

$$R_m = \| X_{m+1} - (Q + A^* X_{m+1}^{\frac{1}{2}} A + B^* X_{m+1}^{\frac{1}{3}} B + C^* X_{m+1}^{\frac{1}{4}} C \|$$

where $\| \cdot \|$ is the spectral norm. For initial values,

$$X_0 = \begin{pmatrix} 1 & 0 & 0 \\ 0 & 1 & 0 \\ 0 & 0 & 1 \end{pmatrix}, \quad X_1 = \begin{pmatrix} 1 & 1 & 0 \\ 1 & 1 & 0 \\ 1 & 0 & 1 \end{pmatrix}, \quad X_2 = \begin{pmatrix} 1 & 1 & -1 \\ -1 & 1 & 1 \\ -1 & 1 & 1 \end{pmatrix},$$

we computed the successive iterations and the error R_m using MATLAB and found that after thirty five iterations the sequence given by (9.16) converges to

$$U = X_{35} = \begin{pmatrix} 639.1810 & 54.1681 & 107.3574 \\ 54.1285 & 44.7768 & 44.1469 \\ 104.3977 & 42.1095 & 112.5509 \end{pmatrix}$$ which is clearly a solution of (9.15).

The convergence history of the algorithm (9.16) is given in Fig. 9.1.

9.3   Application to Control Theory

In 2008, Pathak and Shahzad [464] studied the possibility of optimally controlling the solution of the ordinary differential equation via dynamic programming.

Let A be a compact subset of $\mathbb{R}^m$, and let for each given $a \in A$, $F_a : \mathbb{R}^n \to \mathbb{R}^n$ be a strongly p-contraction mapping such that

$$F_a(x) = \mathbf{f}(x, a) \quad \forall x \in \mathbb{R}^n,$$

where

$$\mathbf{f} : \mathbb{R}^n \times A \to \mathbb{R}^n$$

is a given bounded, generalized Lipschitzian continuous function.

We will now study the possibility of optimally controlling the solution $\mathbf{x}(\cdot)$ of the ordinary differential equation

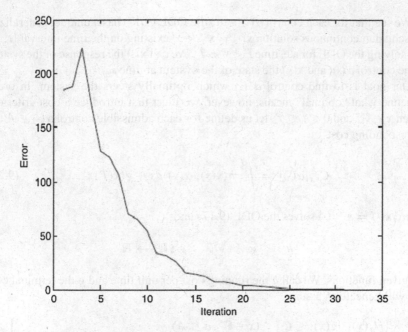

Fig. 9.1 The convergence history of the algorithm

$$\left.\begin{array}{l} \dot{x}(s) = f(x(s), \alpha(s)) \quad (t < s < T) \\ x(t) = x. \end{array}\right\} \tag{9.17}$$

Here, $\dot{} = \frac{d}{ds}$, $T > 0$ is a fixed terminal time, and $x \in \mathbb{R}^n$ is a given initial point, taken on by our solution $\mathbf{x}(\cdot)$ at the starting time $t \geq 0$. At later times $t < s < T$, $\mathbf{x}(\cdot)$ evolves according to the ODE (9.17). The function $\alpha(\cdot)$ appearing in (9.17) is a control, that is , some appropriate scheme for adjusting parameters from the set A as time evolves, thereby affecting the dynamics of the system modelled by (9.17). Let us write

$$\mathcal{A} = \{\alpha : [0, T] \to A \mid \alpha(\cdot) \text{ is measurable}\} \tag{9.18}$$

to denote the set of admissible controls. Then, since

$$|\mathbf{f}(x, a)| \leqslant C, \ |\mathbf{f}(x, a) - \mathbf{f}(y, a)| \leqslant C|x - y|^{1+\delta} \quad (x, y \in \mathbb{R}^n, a \in A) \tag{9.19}$$

for some constant $C, \delta \geq 0$, we have

$$|F_a(x) - F_a^2(x))| \leqslant p(x)|x - F_a(x)| \tag{9.20}$$

for all x in $\mathbb{R}^n$, where $p : \mathbb{R}^n \to [0, 1]$ is a function such that $\sup_{x \in \mathbb{R}^n} p(x) = \lambda < 1$.

We see that for each control $\alpha(\cdot) \in \mathcal{A}$, the ODE (9.17) has a unique, generalized Lipschitzian continuous solution $\mathbf{x}(\cdot) = \mathbf{x}^{\alpha(\cdot)}(\cdot)$, existing on the time interval $[t, T]$ and solving the ODE for a.e. time $t < s < T$. We call $\mathbf{x}(\cdot)$ the response of the system to the control $\alpha(\cdot)$, and $\mathbf{x}(s)$ the state of the system at time s.

Our goal is to find control $\alpha^*(\cdot)$ which optimally steers the system. In order to define what "optimal" means, however, we must first introduce a cost criterion. Given $x \in \mathbb{R}^n$ and $0 \leqslant t \leqslant T$, let us define for each admissible control $\alpha(\cdot) \in \mathcal{A}$ the corresponding cost

$$C_{x,t}[\alpha(\cdot)] := \int_t^T h(\mathbf{x}(s), \alpha(s))ds + g(\mathbf{x}(T)), \tag{9.21}$$

where $\mathbf{x}(\cdot) = \mathbf{x}^{\alpha(\cdot)}(\cdot)$ solves the ODE (9.17) and

$$h : \mathbb{R}^n \times A \to \mathbb{R}, \quad g : \mathbb{R}^n \to \mathbb{R}$$

are given functions. We call h the running cost per unit time and g the terminal cost and will henceforth assume

$$\left.\begin{array}{l} |H_a(x)|, |g(x)| \leqslant C \quad (x \in \mathbb{R}^n, a \in A) \\ |H_a(x) - H_a^2(x)| \leqslant p(x)|x - H_a(x)|, |g(x) - g^2(x)| \leqslant p(x)|x - g(x)| \end{array}\right\} \tag{9.22}$$

for some constant C, $p : \mathbb{R}^n \to [0, 1]$ is a function such that $\sup_{x \in \mathbb{R}^n} p(x) = \lambda < 1$ and for each given $a \in A$, $H_a : \mathbb{R}^n \to \mathbb{R}^n$ is a strongly p-contraction mapping such that

$$H_a(x) = h(x, a) \ \forall x \in \mathbb{R}^n.$$

Given now $x \in \mathbb{R}^n$ and $0 \leqslant t \leqslant T$, we would like to find if possible a control $\alpha^*(\cdot)$ which minimizes the cost functional (9.21) among all other admissible controls.

To investigate the above problem, we shall apply the method of dynamic programming. We now turn our attention to the value function $u(x, t)$ defined by

$$u(x, t) := \inf_{\alpha(\cdot) \in \mathcal{A}} C_{x,t}[\alpha(\cdot)] \quad (x \in \mathbb{R}^n, 0 \leqslant t \leqslant T). \tag{9.23}$$

The plan is this: having defined $u(x, t)$ as the least cost given, we start at the position x at time t, and we want to study u as a function of x and t. We are therefore embedding our given control problem (9.17), (9.21) into the larger class of all such problems, as x and t vary. This idea then can be used to show that u solves a certain Hamilton–Jacobi type PDE and finally to show conversely that a solution of this PDE helps us to synthesize an optimal feedback control.

Let us fix $x \in \mathbb{R}^n$, $0 \leqslant t \leqslant T$. Following the technique of Evans [224], p. 553, we can obtain the optimality conditions in the form given below:

For each $\xi > 0$ so small that $t + \xi \leqslant T$,

$$u(x,t) := \inf_{\alpha(\cdot)\in\mathcal{A}}\left\{\int_t^{t+\xi} h(\mathbf{x}(s),\alpha(s))ds + u(\mathbf{x}(t+\xi),t+\xi)\right\}, \quad (9.24)$$

where $\mathbf{x}(\cdot) = \mathbf{x}^{\alpha(\cdot)}$ solves the ODE (9.17) for the control $\alpha(\cdot)$.

9.3.1 Variational Inequalities in Problems of Stochastic Optimal Control

Belbas and Mayergoyz [43] applied fixed point methods to solve certain discrete variational and quasi-variational inequalities. Note that variational inequalities arise in optimal stochastic control [51] as well as in other problems in mathematical physics, for examples deformation of elastic bodies stretched over solid obstacles, elasto-plastic torsion [211]. We also notice that the iterative method for solutions of discrete variational inequalities is very suitable for implementation on parallel computers with single instruction, multiple-data architecture, particularly on massively parallel processors.

In this section, we intend to show the existence of iterative solution of the two-obstacle variational inequality that arises in certain stochastic optimal control problem. But, first, we introduce the following definitions:

M-matrix − A square matrix A is an M-matrix if and only if every off-diagonal entry of A is nonpositive.

Variational inequality problem − The variational inequality problem is to find a function u such that

$$\left.\begin{array}{r} \max\{Lu - f, u - \phi\} = 0 \text{ on } \Omega, \\ u = 0 \text{ on } \partial\Omega. \end{array}\right\} \quad (9.25)$$

where Ω is a nonempty q-star-shaped open bounded subset of $\mathbb{R}^N$ for some $q \in \Omega$ with smooth boundary such that $0 \in cl(\Omega)$, and L is an elliptic operator defined on Ω by

$$L = -a_{ij}(x)\frac{\partial^2}{\partial x_i \partial x_j} + b_i(x)\frac{\partial}{\partial x_i} + c(x).I_N,$$

where summation with respect to repeated indices is implied, $c(x) \geq 0$, $[a_{ij}(x)]$ is a strictly positive definite matrix, uniformly in x, for $x \in \overline{\Omega}$, f and ϕ are smooth functions defined in Ω and ϕ satisfies the condition: $\phi(x) \geq 0$ for $x \in \partial\Omega$.

The corresponding problem of stochastic optimal control can be described as follows: $L - cI$ is the generator of a diffusion process in $\mathbb{R}^N$, c is a discount factor, f is the continuous cost, and ϕ represents the cost incurred by stopping the process. The boundary condition "$u = 0$ on $\partial\Omega$" expresses the fact that stopping takes place either prior or at the time that the diffusion process exists from Ω.

A problem related to (9.25) is the two-obstacle variational inequality. Given two smooth functions ϕ and μ defined on $cl(\Omega)$ such that $\phi \leqslant \mu$ in $\Omega, \phi \leqslant 0 \leqslant \mu$ on $\partial\Omega$, the corresponding variational inequality is as follows:

$$\left.\begin{array}{r} \max\{\min[Lu - f, u - \phi), u - \mu]\} = 0 \quad \text{on} \;\; \Omega, \\ u = 0 \quad \text{on} \;\; \partial\Omega. \end{array}\right\} \tag{9.26}$$

Note that the problem (9.26) arises in stochastic game theory.

Let A be an N-square matrix corresponding to the finite difference discretizations of the operator L. We shall make the following assumptions about the matrix A:

$$A_{ii} = 1, \sum_{j:j\neq i} A_{ij} > -1, A_{ij} < 0 \;\; \text{for} \;\; i \neq j. \tag{9.27}$$

Notice that "M-matrices" arising from the finite difference discretization of continuous elliptic operators will have the property (9.27) under the appropriate conditions.

Let Q denote the set of all discretized vectors in Ω (see [601]). Let $B = I_N - A$. Then, the corresponding properties for the B-matrices are:

$$B_{ii} = 0, \sum_{j:j\neq i} B_{ij} < 1, B_{ij} > 0 \;\; \text{for} \;\; i \neq j. \tag{9.28}$$

Let $q = \max_i \sum_j B_{ij}$ and A^* be an $N \times N$ matrix such that $A_{ii}^* = 1 - q$ and $A_{ij}^* = -q$ for $i \neq j$. Then, we have $B^* = I_N - A^*$.

Now, we show the existence of iterative solutions of variational inequalities: consider the following discrete variational inequalities mentioned above:

$$\max[\min\{A(x - A^*.d(x, Tx)) - f, x - A^*.d(x, Tx) - \phi\}, \; x - A^*.d(x, Tx) - \mu] = 0, \tag{9.29}$$

where T is a mapping from $\mathbb{R}^N$ into itself implicitly defined by

$$\begin{aligned} Tx = \min[\max\{Bx + A(1 - B^*)x.d(x, Tx) + f, \\ (1 - B^*)x.d(x, Tx) + \phi\}, \;\; (1 - B^*)x.d(x, Tx) + \mu] \end{aligned} \tag{9.30}$$

for all $x \in cl(Q)$ such that the following conditions holds:

(i) for each $x \in cl(Q), Tx(= y)$ is a coincidence point of f and g, i.e. $fy = gy$ for $y = Tx$,

(ii) $\max\left(\max_i\{\max_j(1 - B_{ij}^*)x, \max_j(1 - B_{ij}^*)y\}, \max_i\{\max_j(B_{ij})x, \max_j(B_{ij})y\}\right)$
$\leqslant \varphi(d(x, y))$ for all $x, y \in cl(Q)$, where $\varphi : (0, \infty) \to [0, \frac{1}{2})$ is a $\mathcal{P}$-function, then (9.30) is equivalent to the following fixed point problem:

$$\exists x \in cl(Q) \quad \text{such that} \quad fx = gx, \;\; x = Tx \tag{9.31}$$

that is, $cl(Q) \cap \mathcal{COP}(f, g) \cap \mathfrak{F}(T) \neq \varnothing$.

In two-person game, we determine the best strategies for each player on the basis of maximin and minimax criterion of optimality. This criterion will be well stated as follows: a player lists his/her worst possible outcomes, and then, he/she choose that strategy which corresponds to the best of these worst outcomes. Here, the problem (9.29) exhibits the situation in which two players are trying to control a diffusion process; the first player is trying to maximize a cost functional, and the second player is trying to minimize a similar functional. The first player is called the maximizing player and the second one the minimizing player. Here, f represents the continuous rate of cost for both players, ϕ is the stopping cost for the maximizing player, and μ is the stopping cost for the minimizing player. This problem is fixed by inducting the constrained conditions (i) and (ii) as stated above.

Theorem 9.9 *Under the assumptions (9.27) and (9.28), a solution for (9.31) exists.*

Proof Let $(Ty)_i = \max_j (1 - B_{ij}^*) y.|y_j - (Ty)_j| + \mu_i$ for any $y \in cl(Q)$ and any $i, j = 1, 2, \ldots, N$. Now, for any $x \in cl(Q)$, since $(Tx)_i \leqslant \max_j (1 - B_{ij}^*) x.|x_j - (Tx)_j| + \mu_i$, we have

$$
\begin{aligned}
(Tx)_i - (Ty)_i &\leqslant \max_j (1 - B_{ij}^*) x.|x_j - (Tx)_j| - \max_j (1 - B_{ij}^*) y.|y_j - (Ty)_j| \\
&\leqslant \max\{\max_j (1 - B_{ij}^*) x, \max_j (1 - B_{ij}^*) y\} \\
&\quad \cdot \max_j [|x_j - (Tx)_j| + |y_j - (Ty)_j|] \\
&\leqslant \varphi(d(x, y))[d(x, Tx) + d(y, Ty)] \\
&\leqslant \varphi(d(x, y))[d(x, Tx) + d(y, Ty)].
\end{aligned}
\tag{9.32}
$$

where $\hat{h}, \hat{k} : \mathbb{R}^N \to [0, \infty)$ are some functions. If

$$
(Ty)_i = \max \left\{ B_{ij}|y_j - Ty_j| + (1 - B_{ij}^*) y.|y_j - Ty_j| + f_i, \ (1 - B_{ij}^*).|y_j - Ty_j| + \phi_i \right\},
$$

i.e. if the maximizing player succeeds to maximize a cost functional in his/her strategy which corresponds to the best of N worst outcomes from his/her list, then the game would be one sided. In this situation, we introduce the one-sided operator:

$$
T^+ x = \max \left\{ B.d(x, Tx) + A(1 - B^*)x.d(x, Tx) + f, (1 - B^*)x.d(x, Tx) + \phi \right\}.
$$

Therefore, we have

$$
(Ty)_i = (T^+ y)_i.
$$

Now, if $(Tx)_i = B_{ij}|x_j - Tx_j| + A_{ij}(1 - B_{ij}^*) x.|x_j - Tx_j| + f_i$, then since $(Ty)_i \geq B_{ij}|y_j - Ty_j| + A_{ij}(1 - B_{ij}^*) y.|y_j - Ty_j| + f_i$; by using (9.27), we have

$$(T^+x)_i - (T^+y)_i \leqslant \max_j(B_{ij})x.|x_j - (Tx)_j| - \max_j(B_{ij})y.|y_j - (Ty)_j|$$

$$+ \max_j(1 - B_{ij}^*)x.|x_j - (Tx)_j| - \max_j(1 - B_{ij}^*)y.|y_j - (Ty)_j|$$

$$\leqslant \max_i\{\max_j(B_{ij})x, \max_j(B_{ij})y\}.\max_j[|x_j - (Tx)_j| + |y_j - (Ty)_j|],$$

$$\max_i\{\max_j(1 - B_{ij}^*)x, \max_j(1 - B_{ij}^*)y\}$$

$$\cdot \max_j[|x_j - (Tx)_j| + |y_j - (Ty)_j|]$$

$$\leqslant \max\Big(\max_i\{\max_j(1 - B_{ij}^*)x, \max_j(1 - B_{ij}^*)y\},$$

$$\max_i\{\max_j(B_{ij})x, \max_j(B_{ij})y\}\Big).[d(x, Tx) + d(y, Ty)]$$

$$\leqslant \varphi(d(x, y))[d(x, Tx) + d(y, Ty)]. \tag{9.33}$$

If $(Tx)_i = (1 - B_{ij}^*)x.|x_i - Tx_i| + \phi_i$ then since $(Ty)_i \geq (1 - B_{ij}^*)y.|y_i - Ty_i| + \phi_i$, we have

$$(Tx)_i - (Ty)_i \leqslant \max_j(1 - B_{ij}^*)x.|x_j - (Tx)_j| - \max_j(1 - B_{ij}^*)y.|y_j - (Ty)_j|$$

$$\leqslant \max\{\max_j(1 - B_{ij}^*)x, \max_j(1 - B_{ij}^*)y\}.\max_j[|x_j - (Tx)_j| + |y_j - (Ty)_j|]$$

$$\leqslant \varphi(d(x, y))[d(x, Tx) + d(y, Ty)]. \tag{9.34}$$

Hence, from (9.32)–(9.34), we have

$$(Tx)_i - (Ty)_i \leqslant \varphi(d(x, y))\ [d(x, Tx) + d(y, Ty)]. \tag{9.35}$$

Since x and y are arbitrarily chosen, we have

$$(Ty)_i - (Tx)_i \leqslant \varphi(d(x, y))\ [d(x, Tx) + d(y, Ty)]. \tag{9.36}$$

Therefore, from (9.35) and (9.36), it follows that

$$|(Tx)_i - (Ty)_i| \leqslant \varphi(d(x, y))\ [d(x, Tx) + d(y, Ty)]$$

$$\leqslant \varphi(d(x, y))\ [d(x, Tx) + d(y, Ty)]$$

$$+ \tilde{h}(fy)(d(fy, Tx) + \tilde{k}(gy)(d(gy, Tx),$$

where $\hat{h}, \hat{k} : \mathbb{R}^N \to [0, \infty)$ are some functions; that is,

$$d(Tx, Ty) \leqslant \varphi(d(x, y))\ [d(x, Tx) + d(y, Ty)] + \tilde{h}(fy)(d(fy, Tx) + \tilde{k}(gy)(d(gy, Tx).$$

Hence, we see that the condition (b_1') is satisfied. Therefore, Theorem 5.5.39′ ensures the existence of a solution of (9.31). This completes the proof. ∎

9.3.2   Game Theory and Nash Equilibria

Let us consider a game with $n \geq 2$ players, under the assumption that the players do not cooperate among themselves. In this game, each player pursues a strategy, independence of the strategies of the other players. Let S_k denotes the set of all possible strategies of the kth player. Now set $S = S_1 \times S_2 \times \cdots \times S_n$. An element $\mathbf{x} \in S$ is called a strategy profile. There is a function $f : S \to \mathbb{R}^n$, denoted $f(\mathbf{x}) = (f_1(\mathbf{x}), f_2(\mathbf{x}), \ldots, f_n(\mathbf{x}))$, that takes the players choice to the payoffs for each player. If

$$\sum_{k=1}^{n} f_k(\mathbf{x}) = 0, \quad \forall \mathbf{x} \in S$$

the game is said to be of zero-sum. The aim of each player is to minimize his loss, or, equivalently, to maximize his gain.

Two-Person Zero-Sum Game

An abstract model of two-person zero-sum game consists of two players—player 1 and player 2. The game consists of series of rounds. In each round, player 1 has a choice I_i from a list $I = \{I_1, I_2, \ldots, I_m\}$ of possible m choices and player 2 has a choice II_j from a list $II = \{II_1, II_2, \ldots, II_n\}$ of possible n choices. For each k $(k = 1, 2)$, let $f_k : I \times II \to \mathbb{R}$ be the loss function of the kth player. In two-person zero-sum game, the benefit to one player is equal to the loss of the other player.

In this game, the players determine the probability with which to play each choice. Let x_i be the probability that player 1 will play choice I_i, and let $\mathbf{x} = (x_1, x_2, \ldots, x_m)$ be the associated probability vector. Similarly, let $\mathbf{y} = (y_1, y_2, \ldots, y_n)$ be the associated probability vector for player 2. We shall refer to $\mathbf{x}$ as the strategy for player 1 and $\mathbf{y}$ as the strategy for player 2. Then, the space of all possible strategies for player 1 is

$$\Delta^m = \{(x_1, x_2, \ldots, x_m) : x_1 + x_2 + \cdots + x_n = 1 \text{ and } x_i \geqslant 0 \text{ for all } i\},$$

which is called an m-simplex. Similarly, the space of all possible strategies for player 2 is Δ^n.

Now we shall incorporate simple probability to understand what the long term, or average, payoff to each player. The average payoff will be based on their strategies. Let players 1 and 2 choose $\mathbf{x}$ and $\mathbf{y}$ as their respective strategies. In a given round, suppose player 1 choose I_i and player 2 choose II_j, the payoff to player 1 will be $f_1(I_i, II_j)$. From statistical theory, over the long term, the percentage of rounds that this will happen is $p_{ij} = x_i y_j$. Summing over all possible pair of choices, the expected payoff $e_1(x, y)$ for player 1 will be

$$e_1(\mathbf{x}, \mathbf{y}) = \mathbf{x} P \mathbf{y}^T = \sum_{i=1}^{m} \sum_{j=1}^{n} x_i y_j f_1(I_i, II_j), \text{ where } P = (p_{ij}).$$

Similarly, the expected payoff for player 2 will be

$$e_2(\mathbf{x}, \mathbf{y}) = \mathbf{x}P\mathbf{y}^T = \sum_{i=1}^{m} \sum_{j=1}^{n} x_i y_j f_2(I_i, II_j).$$

Here, one may notice that e_1 and e_2 are both continuous functions of $(\mathbf{x}, \mathbf{y})$.

Definition 9.3 A Nash equilibrium is a pair of strategies such that neither player can improve their payoff if they alone change their strategy, i.e. a point $(x^*, y^*) \in I \times II$ such that

$$e_1(x, y^*) \leqslant e_1(x^*, y^*), \quad \text{for all } x \in \Delta^m,$$

and

$$e_2(x^*, y) \leqslant e_2(x^*, y^*), \quad \text{for all } y \in \Delta^n.$$

Consider the two-player zero-sum game as described above. Since the spaces of all possible strategies for player 1 and player 2 are m-simplex Δ^m and n-simplex Δ^n, respectively, it follows that the space of all possible strategies is homeomorphic to an $(n + m)$-dimensional disc. We now intend to define a continuous map that takes a pair of strategies to a new pair such that fixed point of this map is Nash equilibria. In this context, we use Brouwer's fixed point theorem to prove the fundamental theorem of Nash[1] as follows:

Theorem 9.10 *For a two-player zero-sum game, there exists a Nash equilibrium.*

Proof First of all, we observe that the space of all pairs $(\mathbf{x}, \mathbf{y})$ of choices is $S = \Delta^m \times \Delta^n$, which is homeomorphic to an $(m + n)$-dimensional disc. Suppose (I_i) denoted the strategy of playing only choice I_i by player 1, i.e. I_i is the probability vector in Δ^m that has 1 in the ith entry 0 in every other entry. Similarly, suppose (II_j) denotes the strategy of playing only II_j by player 2. For $\mathbf{x}, \mathbf{y} \in S$, let us define

$$\xi_i(\mathbf{x}, \mathbf{y}) = \max\{0, e_1((I_i), \mathbf{y}) - e_1(\mathbf{x}, \mathbf{y})\}, 1 \leqslant i \leqslant m.$$

Notice that $c_i(\mathbf{x}, \mathbf{y})$ is the amount that player 1 would increase his expected payoff by playing strategy I_i instead of $\mathbf{x}$. Similarly, the amount that player 2 would increase his expected payoff by playing strategy II_j instead of $\mathbf{y}$ is given by

$$\eta_j(\mathbf{x}, \mathbf{y}) = \max\{0, e_2(\mathbf{x}, (II_j)) - e_2(\mathbf{x}, \mathbf{y})\}, 1 \leqslant j \leqslant n.$$

---

[1] John Forbes Nash Jr. (June 13, 1928–May 23, 2015) was an American mathematician who made fundamental contributions to game theory, differential geometry and the study of partial differential equations. Nash's work has provided insight into the factors that govern chance and decision-making inside complex systems found in everyday life. John F. Nash Jr. shared the 1994 Nobel Memorial Prize (the Sveriges Riksbank Prize) in Economic Sciences with game theorists Reinhard Selten and John Harsanyi "for their pioneering analysis of equilibria in the theory of noncooperative games". In 2015, he also shared the Abel Prize with Louis Nirenberg for his work on nonlinear partial differential equations.

Notice also that ξ_i and η_j are each continuous functions of $(\mathbf{x}, \mathbf{y})$.

Define the function $g : S \to S$ by $g(\mathbf{x}, \mathbf{y}) = (\mathbf{x}', \mathbf{y})$, where

$$x_i' = \frac{x_i + \xi_i(\mathbf{x}, \mathbf{y})}{1 + \sum_{k=1}^{m} \xi_k(\mathbf{x}, \mathbf{y})}, \quad 1 \leqslant i \leqslant m, \tag{9.37}$$

and

$$y_j' = \frac{y_j + \eta_j(\mathbf{x}, \mathbf{y})}{1 + \sum_{l=1}^{n} \eta_l(\mathbf{x}, \mathbf{y})}, \quad 1 \leqslant j \leqslant n. \tag{9.38}$$

Notice that x_i' and y_j', being combination of continuous functions using sums, quotients and compositions, are continuous functions. Hence, g is a continuous function. Thus, by Brouwer's fixed point theorem, we must have a fixed point. So, let this fixed point be $(\mathbf{u}, \mathbf{v})$.

It is clear from Eqs. (9.37) and (9.38) that $(\mathbf{x}, \mathbf{y})$ is a fixed point of g, if $\xi_i(\mathbf{x}, \mathbf{y}) = 0$ and $\eta_j(\mathbf{x}, \mathbf{y}) = 0$ for all i and j. We claim that this is true for every fixed point.

Notice that $e_1(\{I_i\}, \mathbf{v}) > e_1(\mathbf{u}, \mathbf{v})$ is not true for all $i = 1, 2, \ldots, m$. For if, this is true, then

$$e_1(\mathbf{u}, \mathbf{v}) = \sum_{i=1}^{m} u_i e_1(\{I_i\}, v) > \sum_{i=1}^{m} u_i e_1(\mathbf{u}, \mathbf{v}) = e_1(\mathbf{u}, \mathbf{v}),$$

a contradiction. Therefore, for some i, $\xi_i(\mathbf{u}, \mathbf{v}) = 0$. Since $(\mathbf{u}, \mathbf{v})$ is a fixed point of g for this i, using (9.37), we obtain

$$u_i = \frac{u_i + \xi_i(\mathbf{u}, \mathbf{v})}{1 + \sum_{k=1}^{m} \xi_k(\mathbf{u}, \mathbf{v})}$$

$$\implies u_i = \frac{u_i}{1 + \sum_{k=1}^{m} \xi_k(\mathbf{u}, \mathbf{v})}$$

$$\implies 1 + \sum_{k=1}^{m} \xi_k(\mathbf{u}, \mathbf{v}) = 1 \text{ or } 1 + \sum_{i=1}^{m} \xi_i(\mathbf{u}, \mathbf{v}) = 1$$

$$\implies \sum_{i=1}^{m} \xi_i(\mathbf{u}, \mathbf{v}) = 0.$$

Therefore, $\xi_i(\mathbf{u}, \mathbf{v}) = 0$ for all $i = 1, 2, \ldots, m$. Moreover,

$$e_1(\mathbf{u}, \mathbf{v}) \geqslant e_1(\{I_i\}, \mathbf{v}) \text{ for all } i = 1, 2, \ldots, m.$$

Since $e_1(\mathbf{u}, \mathbf{v}) = \sum_{i=1}^{m} u_i e_1(\{I_i\}, \mathbf{v})$ and $e_1(\mathbf{x}, \mathbf{v}) = \sum_{i=1}^{m} x_i e_1(\{I_i\}, \mathbf{v})$, we have

$$e_1(\mathbf{u}, \mathbf{v}) \geqslant e_1(\mathbf{x}, \mathbf{v}) \text{ for all } \mathbf{x} \in \Delta^m.$$

Similarly, we have

$$e_1(\mathbf{u}, \mathbf{v}) \geqslant e_1(\mathbf{u}, \mathbf{y}) \text{ for all } \mathbf{y} \in \Delta^n.$$

Hence, we conclude that $(\mathbf{u}, \mathbf{v})$ is a Nash equilibrium.   ∎

9.4   Application to Differential Equations

Nonlinear phenomena, which are modelled with nonlinear differential equations, appear in different domains of physical sciences and engineering such as fluid dynamics, aerodynamics, nonlinear control systems, electrical engineering [325]. Many different methods have been developed for solving differential equations. However, the solution of nonlinear differential equations is still challenging.

9.4.1   Picard–Lindelof Existence Theorem for Ordinary Differential Equation

We begin with the following existence theorem for the ordinary differential equation (Picard–Lindelof theorem). We first reduce the equation to Hammerstein operator equation, and then, existence theorem is obtained as an easy application of Banach's contraction theorem.

Theorem 9.11 *Let Ω denote an open set in $\mathbb{R}^2$, $(t_0, x_0) \in \Omega$. Let $f(t, x)$ be a real-valued function defined and continuous in Ω, and let it satisfy the Lipschitz condition*

$$|f(t, x_1) - f(t, x_2)| \leqslant M|x_1 - x_2|, \quad (t, x_1), (t, x_2) \in \Omega.$$

Then, there exists a $\tau > 0$, and a unique function $\phi(t)$ continuous and differentiable in $[t_0 - \tau, t_0 + \tau]$ such that

(i) $\varphi(t_0) = x_0$,
(ii) $x(t) = \varphi(t)$ satisfies the differential equation

$$\frac{dx}{dt} = f(t, x(t)) \text{ for } t \in [t_0 - \tau, t_0 + \tau]. \tag{9.39}$$

Proof First, we prove that there exist an $\tau > 0$ and a unique function $\varphi(t)$ continuous in $[t_0 - \tau, t_0 + \tau]$ such that

$$\varphi(t) = x_0 + \int_{t_0}^{t} f(s, \varphi(s)) \, ds, \quad t_0 - \tau \leqslant t \leqslant t_0 + \tau$$

and $(t, \varphi(t)) \in \Omega$. Then, since $f(t, \varphi(t))$ is continuous, it follows that $\varphi(t)$ is differentiable in $[t_0 - \tau, t_0 + \tau]$ that satisfies (i) and (ii).

Let B denote a closed disc of center (t_0, x_0) with positive radius and contained in the open set Ω. Let m denote the least upper bound of the continuous function $|f|$ on the compact set B. Let τ and δ be so chosen that $0 < \tau < M^{-1}$, the rectangle $|t - t_0| \leqslant \tau$, $|x - x_0| \leqslant \delta$ is contained in B, and $m\tau < \delta$. Let E denote the set of all continuous functions mapping $[t_0 - \tau, t_0 + \tau]$ into $[x_0 - \delta, x_0 + \delta]$. E is a closed subset of the complete metric space $C[t_0 - \tau, t_0 + \tau]$ (with uniform norm) and is therefore complete. We define a mapping $F\varphi = \psi$ for $\varphi \in E$ by

$$\psi(t) = x_0 + \int_{t_0}^{t} f(s, \varphi(s))ds.$$

Clearly $\psi(t)$ is continuous in $[t_0 - \tau, t_0 + \tau]$. Also

$$|\psi(t) - x_0| = \left| \int_{t_0}^{t} f(s, \varphi(s))ds \right| \leqslant m|t - t_0| < \delta.$$

Thus, F maps E into itself. Also,

$$
\begin{aligned}
|F\varphi_1 - F\varphi_2| &= |\psi_2(t) - \psi(2)(t)| \\
&= \left| \int_{t_0}^{t} [f(s, \varphi_1(s)) - f(s, \varphi_2(s))]ds \right| \\
&\leqslant |t - t_0|M \sup_{s \in [t_0 - \tau, t_0 + \tau]} |\varphi_1(s) - \varphi_2(s)| \\
&\leqslant \tau M \|\varphi_1 - \varphi_2\|,
\end{aligned}
$$

that is, $\|F\varphi_1 - F\varphi_2\| \leqslant \tau M \|\varphi_1 - \varphi_2\|$
where $\tau M < 1$. This shows that F is a contraction mapping in E. By Banach's contraction theorem, there exists a $\varphi \in E$ such that $F\varphi = \varphi$, i.e.

$$\varphi(t) = x_0 + \int_{t_0}^{t} f(s, \varphi(s))ds.$$

This completes the proof. ∎

Remark 9.2 Here, $F\varphi = x_0 + KN\varphi$, where K is an integral operator with Kernel 1 and N is the Nemytskiĭ operator.

For results on the continuation properties of the solution $x(t, x_0)$, refer to Cronin [152] and Berger [52].

Theorem 9.12 *If φ_n is the solution of the differential equation (9.39) with initial condition $\varphi_n(t_0) = x_n$ and $\{x_n\}$ is a real sequence converging to x_0, then $\{\varphi_n\}$ converges to the solution φ of (9.39) with $\varphi(t_0) = x_0$ in E.*

Proof Let F_n be a mapping defined by

$$(F_n\varphi)(t) = x_n + \int_{t_0}^{t} f(s, \varphi(s))ds, \ \varphi \in E, \ n = 1, 2, 3, \ldots.$$

Then, $|(F_n\varphi)(t) - x_0| \leqslant |x_n - x_0| + m\tau < \delta$ for n sufficiently large. Hence, F_n maps E into itself for n sufficiently large. Also, the mappings F_n and F have the same Lipschitz constant $\tau M < 1$. Clearly, $\lim_{n\to\infty} F_n\psi = F\psi$ for each $\psi \in E$. Since F_n is a contraction, there exists a unique fixed point φ_n of F_n. By Theorem 5.133, we have $\lim_{n\to\infty} \varphi_n = \varphi$. ∎

9.4.2   Cauchy–Peano Existence Theorem for Ordinary Differential Equation

Next, we prove the Cauchy–Peano existence theorem for ordinary differential equation

$$\frac{dx}{dt} = f(t, x(t)), \ x(t_0) = x_0$$

with the help of Schauder's fixed point theorem. Peano proved the existence of a local solution, possibly nonunique, when f is continuous and bounded.

Theorem 9.13 *If f is continuous on the rectangle $R = \{(t, x) : |t - t_0| \leqslant a, |x - x_0| \leqslant b\}$, then there exists a solution φ(which is continuously differentiable) of the differential equation*

$$\frac{dx}{dt} = f(t, x(t)) \ on \ |t - t_0| \leqslant \alpha$$

with the initial condition $\varphi(t_0) = x_0.$

Here, $\alpha = \min\left(a, \frac{b}{m}\right)$ and $m = \max_{(t,x)\in R} |f(t, x)|.$

Proof Since $f(t, x)$ is continuous on a neighbourhood of t_0, x_0), we can choose $\tau > 0$ such that f is continuous in the rectangle $|t - t_0| \leqslant \tau$ and $|x - x_0| \leqslant m\tau$ and satisfies the inequality $|f(t, x)| \leqslant m$. Let E denote the Banach space $C[t_0 - \tau, t_0 + \tau]$. Let K be the subset of E consisting of all continuous mappings of $[t_0\tau, t_0 + \tau]$ into $[x_0 - m\tau, x_0 + m\tau]$.

Clearly, K is abounded, closed and convex subset of E. Let F be the mapping defined on K by

$$F\varphi(t) = x_0 + \int_{t_0}^{t'} f(s, \varphi(s))ds, \ |t - t_0| \leqslant \tau.$$

Then, $F(K) \subset K$. Also, $F(K)$ is an equicontinuous set since

$$|(F\varphi)(t) - (F\varphi)(t')| \leqslant \int_t^{t'} |f(s, \varphi(s))| ds \leqslant m|t - t'| \text{ for all } \varphi \in K.$$

As $F(K)$ is bounded, by Ascoli–Arzela theorem, $F(K)$ is contained in a compact set. Therefore, by Schauder's fixed point theorem (Theorem 5.55), F has a fixed point φ in K, i.e.,

$$\varphi(t) = x_0 + \int_{t_0}^{t'} f(s, \varphi(s)) ds, |t - t_0| \leqslant \tau.$$

Obviously, φ is differentiable in $\{t : |t - t_0| \leqslant \tau\}$ and provides a local solution $x(t) = \varphi(t)$ of the given differential equation. ∎

Theorem 9.14 *Let Ω be an open subset of $\mathbb{R}^2$, $(t_0, x_0) \in \Omega$, $M > 0$ a real number, and let $\{m_i\}$ be a bounded sequence of strictly positive real numbers. For each $i = 0, 1, 2, \ldots$ let $f_i(t, x)$ be a real-valued continuous function defined on Ω such that*

$$|f_i(t, x)| \leqslant M \text{ for all } (t, x) \in \Omega,$$

and

$$|f_i(t, x) - f_i(t, y)| \leqslant m_i |x - y| \text{ for all } (t, x), (t, y) \in \Omega.$$

Suppose that the sequence $\{f_i\}$ converges pointwise on Ω to f_0. Let τ be such that $0 < m_i \tau < 1$ for all $i = 0, 1, 2, \ldots$, and such that

$$W = \{(t, x) : |t - t_0| \leqslant \tau \text{ and } |x - x_0| \leqslant M\tau\} \subset \Omega.$$

Then the sequence $\{\varphi_i\}$ converges uniformly on $I = [t_0 - \tau, t_0 + \tau]$ to φ_0 where for each $i = 0, 1, 2, \ldots$, φ_i is the unique solution on I of the initial value problem

$$\frac{dx}{dt} = f_i(t, x(t)), x(t_0) = x_0.$$

Proof Let E be the set of all real-valued functions defined on I with range lying in $[x_0 - M\tau, x_0 + M\tau]$ and with Lipschitz constant less than or equal to M. Then, E, with supremum norm, is a compact metric space. For each $i = 0, 1, 2, \ldots$ and each $\varphi \in E$, define $F_i(\varphi)$ by

$$[F_i(\varphi)](t) = x_0 + \int_{t_0}^t f_i(s, \varphi(s)) ds.$$

Then, F_i is a contraction mapping from E into itself with Lipschitz constant less than or equal to $m_i \tau$. Consider

$$[F_i(\varphi)](t) - [F_0(\varphi)](t) = \int_{t_0}^t [f_i(s, \varphi(s)) - f_0(s, \varphi(s))] ds$$

for each $\varphi \in E, t \in I$, and $i = 1, 2, \ldots$. Because the sequence of integrands converges pointwise to zero and is uniformly bounded by $2M$, by Lebesgue's dominated convergence theorem, the sequence of integrals converges to zero. Hence, the sequence $\{F_i(\varphi)\}$ converges pointwise on I to $F_0(\varphi)$; i.e., $\{F_i\}$ converges pointwise on E to F_0. By Theorem 5.135, the sequence $\{\varphi_i\}$ where φ_i is the unique fixed point of F_i converges to the fixed point φ_0 of F and these fixed points are the unique solutions of the given initial value problems.   ∎

9.4.3 Existence Theorem for First-Order Periodic Problem

In this section, we study existence of solution of first-order periodic problem.

Consider the space $C(I)$, the class of real-valued continuous functions defined on $I = [0, T]$, endowed with the metric d given by

$$d(x, y) = \sup\{|x(t) - y(t)| : t \in I\} \text{ for } x, y \in C(I)$$

Clearly, $(C(I), d)$ is a complete metric space. Further, note that $C(I)$ can also be equipped with a partial order given by

$$x, y \in C(I), x \leqslant y \Leftrightarrow x(t) \leqslant y(t) \text{ for } t \in I.$$

We now prove the existence of solution for the following first-order periodic problem

$$\begin{cases} u'(t) = f(t, u(t)), \ t \in I = [0, T] \\ u(0) = u(T). \end{cases} \tag{9.40}$$

where $T > 0$ and $f : I \times \mathbb{R} \to \mathbb{R}$ is a continuous function.

Definition 9.4 A lower solution for (9.40) is a function $\alpha \in C^1(I)$ such that

$$\alpha'(t) \leqslant f(t, \alpha(t)) \text{ for } t \in I = [0, T]$$
$$\alpha(0) \leqslant \alpha(T).$$

Theorem 9.15 *Let $f : I \times \mathbb{R} \to \mathbb{R}$ be continuous, and suppose that there exist $\lambda > 0$, $0 < a < 1$ such that for all $x, y \in \mathbb{R}$ with $y \geq x$*

$$0 \leqslant f(t, y) + \lambda y - [f(t, x) + \lambda x] \leqslant \lambda \left[\ln(y - x + 1) + \frac{(y - x)^2}{2(y - x + 1)} \right] \tag{9.41}$$

If a lower solution for first-order periodic problem (9.40) exists, then there exists a unique solution of (9.40).

Proof The first-order periodic problem (9.40) can be written as

$$u'(t) + \lambda u(t) = f(t, u(t)) + \lambda u(t), \ t \in I = [0, T]$$
$$u(0) = u(T)$$

The above problem is equivalent to the integral equation

$$u(t) = \int_0^T G(t, s)[f(s, u(s)) + \lambda u(s)]ds$$

where $G(t, s)$ is the green function given by

$$G(t, s) = \begin{cases} \frac{e^{\lambda(T+s-t)}}{e^{\lambda T}-1} & \text{if } 0 \leqslant s < t \leqslant T \\ \frac{e^{\lambda(s-t)}}{e^{\lambda T}-1} & \text{if } 0 \leqslant t < s \leqslant T \end{cases}$$

Define $F : C(I) \to C(I)$ by

$$(Fu)(t) = \int_0^T G(t, s)[f(s, u(s)) + \lambda u(s)]ds.$$

Clearly, $u \in C^1(I)$ is a solution of (9.40) if $u \in C(I)$ is a fixed point of F.

By hypothesis (9.41) about f, the mapping F is nondecreasing and so, for $u \geq v$, we have

$$f(t, u) + \lambda u \geq f(t, v) + \lambda v$$

which implies, using the fact that $G(t, s) > 0$ for $(t, s) \in I \times I$, that

$$(Fu)(t) = \int_0^T G(t, s)[f(s, u(s)) + \lambda u(s)]ds$$
$$\geqslant \int_0^T G(t, s)[f(s, v(s)) + \lambda v(s)]ds = (Fv)(t)$$

for $t \in I$. For $u \geqslant v$, we have

$$d(Fu, Fv) = \sup_{t \in I} |(Fu)(t) - (Fv)(t)|$$

$$\leqslant \sup_{t \in I} \int_0^T G(t, s)[f(s, u(s)) + \lambda u(s) - f(s, v(s)) - \lambda v(s)]ds$$

$$\leqslant \sup_{t \in I} \int_0^T G(t, s) \cdot \lambda \left[\ln\left(u(s) - v(s) + 1\right) + \frac{(u(s) - v(s))^2}{2(u(s) - v(s) + 1)} \right] ds.$$

Put $\Phi(x) = \left[\ln(x + 1) + \frac{x^2}{2(x+1)} \right]$. Obviously, $\Phi : [0, \infty) \to [0, \infty)$ is nondecreasing, positive in $(0, \infty)$ $\left(\Phi'(x) = \frac{2x+1}{(x+1)^2} \right)$ and $u \geqslant v$ then

$$\left[\ln\big(u(s) - v(s) + 1\big) + \frac{\big(u(s) - v(s)\big)^2}{2\big(u(s) - v(s) + 1\big)}\right] \leqslant \left[\ln(\|u - v\| + 1) + \frac{\|u - v\|^2}{2(\|u - v\| + 1)}\right].$$

Now considering the above inequality, we obtain

$$d(Fu, Fv) \leqslant \sup_{t \in I} \int_0^T G(t, s) \cdot \lambda \left[\ln\big(u(s) - v(s) + 1\big) + \frac{\big(u(s) - v(s)\big)^2}{2\big(u(s) - v(s) + 1\big)}\right] ds$$

$$\leqslant \left[\ln(\|u - v\| + 1) + \frac{\|u - v\|^2}{2(\|u - v\| + 1)}\right] \cdot \lambda \sup_{t \in I} \int_0^T G(t, s) ds$$

$$= \left[\ln(\|u - v\| + 1) + \frac{\|u - v\|^2}{2(\|u - v\| + 1)}\right] \cdot$$

$$\lambda \sup_{t \in I} \frac{1}{e^{\lambda T} - 1} \left(\frac{1}{\lambda} e^{\lambda(T + s - t)}\Big]_0^t + \frac{1}{\lambda} e^{\lambda(s - t)}\Big]_t^T\right)$$

$$= \left[\ln(\|u - v\| + 1) + \frac{\|u - v\|^2}{2(\|u - v\| + 1)}\right] \cdot \lambda \frac{1}{\lambda(e^{\lambda T} - 1)} (e^{\lambda T} - 1)$$

$$= \left[\ln(\|u - v\| + 1) + \frac{\|u - v\|^2 + 1}{2(\|u - v\| + 1)} - \frac{1}{2(\|u - v\| + 1)}\right].$$

Thus, we have

$$d(Fu, Fv) \leqslant \left[\ln(\|u - v\| + 1) + \frac{\|u - v\|^2 + 1}{2(\|u - v\| + 1)} - \frac{1}{2(\|u - v\| + 1)}\right].$$

Put $\varphi(x) = \ln(x + 1) + \frac{x^2 + 1}{2(x + 1)}$, $\psi(x) = \frac{1}{2(x + 1)}$. Clearly, both the functions $\varphi, \psi :$ $[0, \infty) \to [0, \infty)$ are continuous, φ is nondecreasing and ψ is nonincreasing for all $x \in (0, \infty)$ and satisfying conditions (C1)–(C3).

Thus, from (9.41), for $u \geqslant v$, we obtain

$$d(Fu, Fv) \leqslant \varphi(d(u, v)) - \psi(d(u, v)).$$

Let $\alpha(t)$ be a lower solution of (9.40). Then, we will show that $\alpha \leqslant F\alpha$. Now

$$\alpha'(t) + \lambda\alpha(t) \leqslant f(t, \alpha(t)) + \lambda\alpha(t) \text{ for } t \in I$$

Multiplying by $e^{\lambda t}$, we get

$$\big(\alpha(t)e^{\lambda t}\big)' \leqslant [f(t, \alpha(t)) + \lambda\alpha(t)]e^{\lambda t} \text{ for } t \in I$$

which on integration gives

$$\alpha(t)e^{\lambda t} \leqslant \alpha(0) + \int_0^t [f(s, \alpha(s)) + \lambda\alpha(s)]e^{\lambda s}\,ds \quad \text{for } t \in I \qquad (9.42)$$

which implies that

$$\alpha(0)e^{\lambda T} \leqslant \alpha(T)e^{\lambda T} \leqslant \alpha(0) + \int_0^T [f(s, \alpha(s)) + \lambda\alpha(s)]e^{\lambda s}\,ds.$$

This gives

$$\alpha(0) \leqslant \int_0^T \frac{e^{\lambda s}}{e^{\lambda T} - 1}[f(s, \alpha(s)) + \lambda\alpha(s)]\,ds.$$

From this inequality and (9.42), we obtain

$$\alpha(t)e^{\lambda t} \leqslant \int_0^t \frac{e^{\lambda(T+s)}}{e^{\lambda T} - 1}[f(s, \alpha(s)) + \lambda\alpha(s)]\,ds + \int_t^T \frac{e^{\lambda s}}{e^{\lambda T} - 1}[f(s, \alpha(s)) + \lambda\alpha(s)]\,ds.$$

As a consequence, we have

$$\alpha(t) \leqslant \int_0^t \frac{e^{\lambda(T+s-t)}}{e^{\lambda T} - 1}[f(s, \alpha(s)) + \lambda\alpha(s)]\,ds + \int_t^T \frac{e^{\lambda(s-t)}}{e^{\lambda T} - 1}[f(s, \alpha(s)) + \lambda\alpha(s)]\,ds.$$

Hence,

$$\alpha(t) \leqslant \int_0^T G(t, s)[f(s, \alpha(s)) + \lambda\alpha(s)]\,ds = (F\alpha)(t) \quad \text{for } t \in I.$$

Finally, Theorems 5.171 and 5.172 give that F has a unique fixed point. ∎

Remark 9.3 In the proof of Theorem 9.15, the unique solution of (9.41) can be obtained as $\lim_{n\to\infty} F^n(x)$, for every $x \in C(I)$. If we choose $x(t) = \alpha(t)$, then $F^n(\alpha)$ is a monotone nondecreasing sequence uniformly convergent to the unique solution of (9.40).

Remark 9.4 The condition (9.41) of Theorem 9.15 can be replaced by

$$0 \leqslant f(t, y) + \lambda y - [f(t, x) + \lambda x] \leqslant \lambda\Phi(y - x) \quad \text{for } y \geqslant x$$

where $\Phi : [0, \infty) \to [0, \infty)$ can be written by

$$\Phi(x) = \varphi(x) - \psi(x)$$

where $\varphi : [0, \infty) \to [0, \infty)$ is a continuous and nondecreasing function and $\psi : [0, \infty) \to [0, \infty)$ is a continuous and nonincreasing function satisfying conditions (C1)–(C3). Examples of such functions are:

(i) $\varphi(x) = \ln(x+1) + \frac{x^2+1}{2(x+1)}$ and $\psi(x) = \frac{1}{2(x+1)}$ (which appears in Theorem 9.15).

(ii) $\varphi(x) = \frac{x}{x^2+1} + \arctan x$ and $\psi(x) = \frac{x}{x^2+1}$.

In 2008, Pathak and Shahzad [464] studied the possibility of optimally controlling the solution of ordinary differential equation via dynamic programming. As an application of our main result in Sect. 3, we continue our discussion to solve certain problems in control theory in an ordered set. In what follows, we use the terminology of Pathak and Shahzad [464] and Evans [224].

Let A be a compact subset of $\mathbb{R}^m$, and let for each given $a \in A$, $F_a : \mathbb{R}^n \to \mathbb{R}^n$ be a continuous and nondecreasing weakly $(\varphi - \psi)$-contractive mapping of type (I) such that

$$F_a(x) = \mathbf{f}(x, a) \ \forall x \in \mathbb{R}^n,$$

where $\mathbf{f} : \mathbb{R}^n \times A \to \mathbb{R}^n$ is a given bounded, continuous and nondecreasing weakly $(\varphi - \psi)$-contractive mapping of type (I). Consider the usual order

$$(x_1, x_2, \ldots, x_n) \leqslant (y_1, y_2, \ldots, y_n) \Leftrightarrow x_i \leqslant y_i \ i = 1, 2, \ldots, n.$$

Thus, $(\mathbb{R}^n, \leqslant)$ is a partially ordered set. Besides, $(\mathbb{R}^n, d)$ is a complete metric space considering $d(x, y) = |x - y|$ the Euclidean distance between x and y.

We will now study the possibility of optimally controlling the solution $\mathbf{x}(\cdot)$ of the ordinary differential equation

$$\begin{cases} \dot{x}(s) = f(x(s), \alpha(s)) \ \ (t < s < T) \\ x(t) = x. \end{cases} \tag{9.43}$$

Here, $\dot{} = \frac{d}{ds}$, $T > 0$ is a fixed terminal time, and $x \in \mathbb{R}^n$ is a given initial point, taken on by our solution $\mathbf{x}(\cdot)$ at the starting time $t \geq 0$. At later times $t < s < T$, $\mathbf{x}(\cdot)$ evolves according to the ODE (9.43). The function $\alpha(\cdot)$ appearing in (9.43) is a control, that is , some appropriate scheme for adjusting parameters from the set A as time evolves, thereby affecting the dynamics of the system modelled by (9.43). Let us write

$$\mathcal{A} = \{\alpha : [0, T] \to A \mid \alpha(\cdot) \text{ is measurable} \} \tag{9.44}$$

to denote the set of admissible controls. Then, since

$$|\mathbf{f}(x, a)| \leq C, \ |\mathbf{f}(x, a) - \mathbf{f}(y, a)| \leqslant \varphi(|x - y|) - \psi(|x - y|) \ \ (x, y \in \mathbb{R}^n, a \in A \text{ with } x \geqslant y) \tag{9.45}$$

for some constants $C, \delta \geqslant 0$, we have

$$|F_a(x) - F_a(y))| \leqslant \varphi(|x - y|) - \psi(|x - y|) \tag{9.46}$$

for all $x, y \in \mathbb{R}^n$ with $x \geqslant y$, where $\varphi : [0, \infty) \to [0, \infty)$ is a continuous and nondecreasing function and $\psi : [0, \infty) \to [0, \infty)$ is a continuous and nonincreasing function satisfying conditions (C1)–(C3). Suppose that there exists $x_0 \in X$ with $x_0 \leqslant f(x_0)$.

We see that for each control $\alpha(\cdot) \in \mathcal{A}$, the ODE (9.43) has a unique, generalized Lipschitzian continuous solution $\mathbf{x}(\cdot) = \mathbf{x}^{\alpha(\cdot)}(\cdot)$, existing on the time interval $[t, T]$ and solving the ODE for a.e. time $t < s < T$. We call $\mathbf{x}(\cdot)$ the response of the system to the control $\alpha(\cdot)$, and $\mathbf{x}(s)$ the state of the system at time s.

Our goal is to find control $\alpha^*(\cdot)$ which optimally steers the system. In order to define what "optimal" means, however, we must first introduce a cost criterion. Given $x \in \mathbb{R}^n$ and $0 \leqslant t \leqslant T$, let us define for each admissible control $\alpha(\cdot) \in \mathcal{A}$ the corresponding cost

$$C_{x,t}[\alpha(\cdot)] := \int_t^T h(\mathbf{x}(s), \alpha(s))ds + g(\mathbf{x}(T)), \qquad (9.47)$$

where $\mathbf{x}(\cdot) = \mathbf{x}^{\alpha(\cdot)}(\cdot)$ solves the ODE (9.43) and

$$h : \mathbb{R}^n \times A \to \mathbb{R}, \quad g : \mathbb{R}^n \to \mathbb{R}$$

are given functions. We call h the running cost per unit time and g the terminal cost and will henceforth assume

$$\begin{cases} |H_a(x)|, |g(x)| \leqslant C \\ |H_a(x) - H_a(y)| \leqslant \varphi(|x - y|) - \psi(|x - y|), \\ |g(x) - g(y)| \leqslant \varphi(|x - y|) - \psi(|x - y|) \quad (x, y \in \mathbb{R}^n, a \in A) \end{cases} \qquad (9.48)$$

for some constant C, and for each given $a \in A$, $H_a : \mathbb{R}^n \to \mathbb{R}^n$ is a bounded, continuous and nondecreasing weakly $(\varphi - \psi)$-contractive mapping of type (I) defined by

$$H_a(x) = h(x, a) \ \forall x \in \mathbb{R}^n.$$

Given now $x \in \mathbb{R}^n$ and $0 \leqslant t \leqslant T$, we would like to find if possible a control $\alpha^*(\cdot)$ which minimizes the cost functional (9.47) among all other admissible controls.

To investigate the above problem, we shall apply the method of dynamic programming. We now turn our attention to the value function $u(x, t)$ defined by

$$u(x, t) := \inf_{\alpha(\cdot) \in \mathcal{A}} C_{x,t}[\alpha(\cdot)] \qquad (x \in \mathbb{R}^n, 0 \leqslant t \leqslant T). \qquad (9.49)$$

The plan is this: having defined $u(x, t)$ as the least cost given, we start at the position x at time t, and we want to study u as a function of x and t. We are therefore embedding our given control problem (9.43), (9.47) into the larger class of all such problems, as x and t vary. This idea then can be used to show that u solves a certain

Hamilton–Jacobi type PDE and finally to show conversely that a solution of this PDE helps us to synthesize an optimal feedback control.

Let us fix $x \in \mathbb{R}^n$, $0 \leqslant t \leqslant T$. Following the technique of Evans [224], p. 553, we can obtain the optimality conditions in the form given below:

For each $\xi > 0$ so small that $t + \xi \leqslant T$,

$$u(x, t) := \inf_{\alpha(\cdot) \in \mathcal{A}} \left\{ \int_t^{t+\xi} h(\mathbf{x}(s), \alpha(s)) ds + u(\mathbf{x}(t + \xi), t + \xi) \right\}, \qquad (9.50)$$

where $\mathbf{x}(\cdot) = \mathbf{x}^{\alpha(\cdot)}$ solves the ODE (9.43) for the control $\alpha(\cdot)$.

9.4.4   Existence of Periodic Solutions for Nonlinear Equations of Evolution

We next consider the existence of periodic solutions for nonlinear equations of evolution discussed by Browder [99]. Let $\mathcal{H}$ be a Hilbert space. Consider the time dependent nonlinear equation of evolution in $\mathcal{H}$ given by

$$\frac{dx}{dt} + A(t)x = f(t, x) \qquad (9.51)$$

where $A(t)$ is a family of closed linear operators in $\mathcal{H}$ and f maps $\mathbb{R} \times \mathcal{H}$ into $\mathcal{H}$. Let $A(t)$ and $f(t, x)$ be periodic in t with common period $p > 0$. We are interested in the existence of solution $x(t)$ of Eq. (9.51) for $t \geqslant 0$ with period p, an integer. We make the following assumption.

Assumption I.

(a) For each t and each pair x and y in $\mathcal{H}$,

$$\text{Re}(f(t, x) - f(t, y), x - y) \leqslant 0. \qquad (9.52)$$

(b) For each t and each x in $\mathscr{D}(A(t))$

$$\text{Re}(A(t)x, x) \geqslant 0. \qquad (9.53)$$

We suppose that for each s in $\mathbb{R}$, the homogeneous linear equation

$$\frac{dx}{dt} + A(t)x = 0 \quad (t > s) \qquad (9.54)$$

has a solution $x(t)$ for $t > s$ for each choice of initial data $x(s) = y$ in $\mathscr{D}(A(s))$. We set $X(t, s)y = x(t)$. It follows easily from Assumption I(b) that $\|X(t, s)\| \leqslant 1$ for each $s < t$.

Definition 9.5 If $x : \mathbb{R}^+ \to \mathcal{H}$ is continuous in the strong topology, x is said to be a mild solution of Eq. (9.51) on $\mathbb{R}^+$ with initial value $x(o) = x_0$, if and only if for each $t > 0$,

$$x(t) = X(t, 0)x_0 + \int_0^t X(t, s) f(s, x(s))ds. \qquad (9.55)$$

Assumption II. There exits a mild solution x of Eq. (9.51) on $\mathbb{R}^+$ for each initial value x_0 in $\mathcal{H}$. Browder [99] has shown that Assumption II follows from Assumption I and the following two conditions:

(a) f is continuous from $\mathbb{R}^+ \times \mathcal{H}$ into $\mathcal{H}$ and maps bounded set into bounded sets.
(b) $\mathcal{D}(A(t)) = w.$ independent of t in $\mathbb{R}^+ : t \to A(t) \in \mathcal{L}(W, \mathcal{H})$ is of class C^1 and $\mathcal{D}(A^*(t)) \subset W$.

Notice that condition (b) has been weakened by Kato [322] in the following way: $[A(t) + \lambda I]^{-1}$ exists on $\mathcal{H}$ for $\lambda > 0$ and is strongly of class C^1 in t on $\mathbb{R}^+$.

Theorem 9.16 *Suppose that $A(t)$ and $f(t, x)$ satisfy Assumptions (I) and (II) and are periodic in t of period $p > 0$. Suppose further that there exists $r > 0$ such that $Re(f(t, x), x) < 0$ for $\|x\| = r$ and all t in $[0, p]$. Then, there exists and element x_0 of $\mathcal{H}$ with $\|x_0\| < r$ such that the mild solution of*

$$\frac{dx}{dt} + A(t)x = f(t, x), t > 0$$

with $x(0) = x_0$, which is periodic with period p.

Proof Let F be the mapping of B_r into $\mathcal{H}$, which assigns to each x_0 of B_r, the value at p of the mild solution $x(t)$ of Eq. (9.51) with $x(0) = x_0$, i.e.

$$F x_0 = x(p) = X(p, 0)x_0 + \int_0^p X(p, s) f(s, x(s))ds.$$

Each fixed point of F then corresponds to a periodic solution of Eq. (9.51) with period p. Since

$$\frac{1}{2}\frac{d}{dt}\left\{\|x(t)\|^2\right\} = -Re(A(t)x(t), x(t)) + Re(f(t, x(t)), x(t))$$

$$\leqslant Re(f(t, x(t)), x(t)),$$

it follows that for any value of t in $[0, p]$ for which $\|x(t)\| = r$, we have

$$\frac{d}{dt}\{\|x(t)\|^2\} < 0.$$

Hence, $\|x(p)\| \leqslant r$, and F maps B_r into itself. If x_0 and x_1 are two elements of B_r and $x_0(t)$ and $x_1(t)$, the corresponding mild solutions, then we have

$$\frac{1}{2}\frac{d}{dt}\left\{\|x_0(t) - x_1(t))\|^2\right\} = -\text{Re}\ (A(t)[x_0(t) - x_1(t)], x_0(t) - x_1(t))$$

$$+ \text{Re}\ (f(t, x_0(t)) - f(t, x_1(t)), x_0(t) - x_1(t))$$

$$\leqslant 0.$$

Hence, $\|x_0(p) - x_1(p)\| \leqslant \|x_0(0) - x_1(0)\|$,   i.e. $\|Fx_0 - Fx_1\| \leqslant \|x_0 - x_1\|$.
Hence, by Theorem 5.25, F has a fixed point in B_r and Eq. (9.51) has a periodic solution.

Let us consider the following two point boundary value problem:

$$x''(t) = f(t, x(t), x'(t)),\ \ 0 \leqslant t \leqslant T$$

$$x(0) = \alpha\ \text{and}\ x(T) = \beta.$$

The existence of a solution of this boundary value problem is discussed in Edward [218] following an argument of Bass [40]. We briefly indicate the proof.

Here, f is assumed to be continuous and bounded on $[0, T] \times \mathbb{R} \times \mathbb{R}$. Transforming the above boundary value problem into an equivalent integral equation, we get

$$x(t) = \alpha + \frac{(\beta - \alpha)t}{T} - \int_0^T G(t, s) f(s, x(s), x'(s)) ds \tag{9.56}$$

where $G(t, x) = \begin{cases} \frac{s(T-t)}{T}, & 0 \leqslant s \leqslant T \\ \frac{t(T-s)}{T}, & 0 \leqslant t \leqslant s \leqslant T \end{cases}$

We take $X = C^1[0, T]$ with norm $\|x\| = \sup\limits_{t \in [0,T]} [|x(t)| + |x'(t)|]$ and define $F : X \to X$ by

$$(Fx)(t) = \alpha + \frac{(\beta - \alpha)t}{T} \int_0^T G(t, s) f(s, x(s), x'(s)) ds.$$

It can be shown that F is continuous and $F(X)$ is relatively compact in X. Then, by Theorem 5.59, a version of Schauder's fixed point theorem, F has a fixed point in X which precisely gives the desired solution of Eq. (9.56).

For the nonnegative solution of the above two point boundary value problem where $0 \leqslant -f(t, u, v) \leqslant a + b_0(t)u$, $0 \leqslant t \leqslant T$, $u \geqslant 0$, $-\infty < v < \infty$, refer to Krasnoselskii [345].

Gatica and Smith [241] considered the following boundary value problem:

$$x'' + f(t, x) = 0;\ \ x(0) = x(L) = 0, \tag{9.57}$$

where x and f are vectors functions and f satisfies the condition:
(c) $f : [0, L] \times \mathbb{R}^{n+} \to \mathbb{R}^{n+}$ is continuous and $f(t, 0) = 0$ where

$$\mathbb{R}^{n+} = \{x \in \mathbb{R}^n : x_i \geqslant 0, 1 \leqslant i \leqslant n\}.$$

We look for a nontrivial solution of (9.54) in the cone $K = \{x : [0, L] \to \mathbb{R}^{n+} | x$ satisfies the boundary condition of (9.54) is continuous and concave on $[0, L]\}$ in the Banach space $X = C[0, L]$.

Theorem 9.17 (Gatica and Smith [241]) *Suppose the condition (c) holds and that there exist numbers r and R, $0 \leqslant r < R$ such that*

(i) $x'' + f(t, x) \geqslant 0$ *has no C^2 solutions $x \in K$, with $\|x\|_\infty = r$;*
(ii) $z'' + f(t, z) \leqslant 0$ *has no C^2 solutions $z \in K$, with $\|z\|_\infty = R$.*

Then, Eq. (9.57) has a nonzero solution $y \in K$ with $r \leqslant \|u\|_\infty \leqslant R$.

This is proved by applying Theorem 5.165 to the operator

$$(Fx)(t) = \int_0^L G(t, s) f(s, x(s)) ds$$

where $G(t, s)$ is the Green's function for the boundary value problem.

Corollary 9.2 *Suppose that the condition (c) holds and x and f are scalar-valued. Suppose that there exist nonnegative continuous functions ϕ and ψ defined on $[0, L]$ and numbers r, R with $0 < r < R$ such that*

(i) $f(t, x) \leqslant \phi(t)x, t \in [0, L], 0 \leqslant x \leqslant r$ *and* $y'' + \phi(t)y = 0$ *is disconjugate on $[0, L]$ (if every solution $y(t) \not\equiv 0$ vanishes at most once on $[0, L]$).*
(ii) $f(t, x) \geqslant \psi(t)x, t \in [0, L], x \geqslant R$ *and* $z'' + \psi(t)z = 0$ *is conjugate on $(0, L)$.*

Then, the boundary value problem (9.57) has a nonzero solution $y \in K$ with $r \leqslant \|y\|_\infty$.

Theorem 9.18 *Suppose that the condition (c) holds and x and f are scalar valued such that the following are satisfied:*

(i) $y'' + f_x(t, 0)y = 0$ *is disconjugate on $[0, L]$.*
(ii) *there exists an $R > 0$ such that* $z'' + f(t, z) \leqslant 0$ *has no C^2 solution $z \in K$ with $\|z\|_\infty = R$.*

Then, the boundary value problem (9.57) has a nonzero solution $x \in K$ with $\|x\|_\infty \leqslant R$.

Proof Define the operator $F : K \to K$ by

$$(Fx)(t) = \int_0^L G(t, s) f(x, x(s)) ds.$$

The operator F is continuous and maps bounded subsets into precompact subsets. Further, $F(0) = 0$, and F is Fréchet differentiable in the direction of the cone at 0 with

$$F'(0)h(t) = \int_0^L G(t,s)fx(s,0)h(s)ds.$$

We use Theorem 5.165 by setting $B \equiv 0$ and $Cx = h$ for all $x \in \{y \in K : \|y\| = R\}$ where $h \in K \cap C^2[0, L]$ with $\|h\| > R$. First, we show that condition (i) of Theorem 5.165 is satisfied for all small values of r. Suppose this is not true. Then, using the fact that $F(0) = 0$ and F is Fréchet differentiable at 0 with $F''(0)$ compact (Krasnoselskii [345]), it can be shown that there exist $\lambda \in [1, \infty)$ and $u \in K$ with $\|u\| = 1$ such that $F'(0)u = \lambda u$. Thus,

$$\lambda u(t) = \int_0^L G(t,s)f_x(s,0)u(s)ds.$$

Differentiating this, we get

$$u''(t) + \lambda f_x(t,0)u(t) = 0, \quad (u(0) = u(L) = 0).$$

We have

$$\lambda^{-1} f_x(t,0) \leqslant f_x(t,0).$$

But if $y'' + f_x(t,0)y = 0$ is disconjugate on $[0, L]$, then by Sturm comparison theorem $u'' + \lambda^{-1} f_x(t,0)u = 0$ must also be so. Since $u \neq 0$, we arrive at a contradiction. The condition (ii) of Theorem 5.165 can be shown to be satisfied. Hence the result.   ∎

9.4.5 Boundary Value Problems

We now discuss some boundary value problems occurring in chemical reactor theory whose multiple solutions were obtained by Williams and Leggett[2] [607] using the fixed point theorems proved by them.

----

[2]In 1982, Williams and Leggett studied uniqueness and multiplicity of solutions for the boundary value problem

$$\beta x''(t) - x'(t) + pf(x(t)) = 0, \quad 0 \leqq t \leqq 1, \quad \beta > 0, \quad p > 0,$$

$$\beta x'(0) - x(0) = 0, \quad x'(1) = 0,$$

which arises in chemical reactor theory. The reaction rate f is given by

$$f(x) = (q - x)\exp[-k/(1+x)], \quad k > 0, \quad q > 0.$$

Uniqueness is shown for (1) sufficiently small p, (2) certain regions of p and sufficiently small β, (3) large p and sufficiently large β and (4) fixed β and sufficiently large p. Regions of points (β, p, q, k) are identified where there are at least three solutions. The combination of these results gives an improved picture of the behaviour of the number of solutions as p and β vary. Read More: L.R. Williams and W. Leggett, Unique and Multiple Solutions of a Family of Differential Equations Modelling Chemical Reactions, SIAM J. Math. Anal., 13(1) (1982), 122–133.

First, consider the boundary value problem

$$x''(t) + \frac{1}{t}x'(t) + f(x(t)) = 0, 0 < t < 1 \tag{9.58}$$

$$x'(0) = x(1) = 0, \tag{9.59}$$

where f is nonnegative and continuous.

The solutions of (9.58) with given boundary conditions (9.59) are the fixed points of the operator F on $C[0, 1]$ defined by

$$(Fx)(t) = \int_0^1 G(t, s) f(x(s)) ds,$$

where G is the Green's function of (9.58)–(9.59), given by

$$G(t, s) = \begin{cases} -s(\ln s), & 0 \leqslant t \leqslant s \leqslant 1, \\ -s(\ln t), & 0 \leqslant s \leqslant t \leqslant 1, \end{cases}$$

Moreover F maps $L_\infty([0, 1])$ into $C[0, 1]$ and is continuous and maps $L_\infty([0, 1])$ into a relatively compact subset of $C[0, 1]$) (by Arzela–Ascoli theorem). Let Ω_1 be a nonempty closed subset of $[0, 1]$ which is the solution to the max–min problem

$$\min_{t \in \Omega_1} \int_{\Omega_1} G(t, s) ds = \max_{\Omega_0 \subset \Omega} \min_{t \in \Omega_0} \int_{\Omega_0} G(t, s) ds, \tag{9.60}$$

where Ω is a compact subset of $[0, 1]$. This choice of Ω_1 satisfies the requirements on F in condition (5.153) of Theorems 5.159 and 5.160. For the boundary value problem (9.47) and (9.48) with bounded nondecreasing nonlinearities like $f(x) = \beta \exp\left(-\frac{1}{x+\tau}\right) \beta > 0, \tau \geqslant 0$, this is the optimal choice of Ω_1.

In this case, solution of (9.49) is $\Omega_1 = [0, e^{-1/2}]$. For fixed x, $G(t, s)$ is a nonincreasing function of t satisfying the following inequalities:

$$\int_0^1 G(t, s) ds \leqslant \int_0^1 G(0, s) ds = \frac{1}{4}, \quad 0 \leqslant t \leqslant 1;$$

$$\int_{\omega_1} G(t, s) ds \geqslant \frac{1}{4e}, \quad t \in \Omega_1;$$

$$\int_{[0,1]-\Omega_1} G(t, s) ds \geqslant \frac{e-2}{4e}, \quad t \in \Omega_1$$

and $\quad \int_0^1 |G(t_1, s) - G(t_2, s)| ds \leqslant \int_0^1 |G(0, s) - G(1, s)| ds = \frac{1}{4e}; \quad t_1, t_2 \in \Omega_1.$

Theorem 9.19 *Suppose there exists nonnegative numbers a, b, c, d and m such that* $0 < d < a < b \leqslant c$:

$$0 \leqslant m \leqslant f(x) \leqslant 4e(b - a), 0 \leqslant x \leqslant c; \tag{9.61}$$

$$f(x) > 4ea - (e - 2)m, \quad a \leqslant x \leqslant b; \tag{9.62}$$

$$f(x) < 4d, \qquad\qquad 0 \leqslant x \leqslant d; \tag{9.63}$$

$$f(x) \leqslant 4c \qquad\qquad 0 \leqslant x \leqslant c. \tag{9.64}$$

Then, the boundary value problem (9.58)–(9.59) has at least three solutions in $\langle 0, c \rangle$. Moreover, if f is not constant on any subinterval of $[0, c]$, then (9.63) and (9.64) may be replaced by conditions

$$f(x) \geqslant 4ea - (e - 2)m, \quad a \leqslant x \leqslant b;$$

$$f(x) \leqslant 4d, \qquad\qquad 0 \leqslant x \leqslant d.$$

Proof Suppose that the condition (9.61)–(9.64) hold. If $x \in \langle 0, c \rangle_\infty$ and $t_1, t_2 \in \Omega_1$, then

$$|(Fx)(t_1) - (Fx)(t_2)| \leqslant \int_0^1 |G(t_1, s) - G(t_2, s)| f(x(s)) ds$$

$$\leqslant \frac{1}{4e}(4e)(b - a) = b - a.$$

If $x \in \langle 0, c \rangle_\infty$ and $a \leqslant x(s) \leqslant b$, for almost everywhere on Ω_1,

$$(Fx)(t) = \int_0^{e^{-1/2}} G(t, s) f(x(s)) ds + \int_{e^{-1/2}}^0 G(t, s) f(x(s)) ds$$

$$\geqslant \frac{1}{4e}(4ea - (e - 2)m) + \left(\frac{e - 2}{4e} \right) m = a, \ t \in \Omega_1.$$

Also, if $x \in \langle 0, d \rangle_\infty$, then

$$(Fx)(t) < \frac{1}{4}(4d) = d, \ 0 \leqslant t \leqslant 1,$$

and if $x \in \langle 0, c \rangle_\infty$, then

$$(Fx)(t) \leqslant \frac{1}{4}(4c) = c, \ 0 \leqslant t \leqslant 1.$$

Therefore, F has at least three fixed points in $\langle 0, c \rangle$ by Theorem 5.160. For the proof of the last statement of the above theorem, refer to Williams and Leggett [607].

We present an application of Kakutani's fixed point theorem to discuss the existence of solution of ordinary differential equation investigated by Lasota and Opial [355].

Let the Euclidean norm be denoted by $|x|$ for $x \in \mathbb{R}^n$. Let $C(\Omega)$ be the space of all continuous mappings of a fixed compact interval $\Omega \subset \mathbb{R}$ into $\mathbb{R}^n$ with supremum norm. $(L_p(\Omega))^n$ denotes its nth cartesian product of $L_p(\Omega)$.

Let $CF(X)$ denote the set of all closed convex nonempty subsets of X.

Strongly upper semicontinuous mapping – A mapping $F : X \to CF(Y)$ where X and Y are normed linear spaces is called strongly upper semicontinuous if for all sequence $\{x_i\} \subset X, \{y_i\} \subset Y$, the conditions $x_i \to x, y_i \to y$ and $y_i \in Fx_i$ imply $y \in Fx$.

Let the function $f : \Omega \times \mathbb{R}^n \to CF(\mathbb{R}^n)$ satisfy the following conditions:

 (i) For every fixed $x \in \mathbb{R}^n$, the function $f(t, x)$ is measurable on Ω.
 (ii) For every fixed $t \in \Omega$, $f(t, x)$ is strongly upper semicontinuous on $\mathbb{R}^n$.
 (iii) There are functions $\alpha, \beta \in L_1(\Omega)$ such that

$$|f(t, x)| \leqslant \alpha(t) + \beta(t)|x|; \ (t, x, \in \Omega \times \mathbb{R}^n).$$

Given a function $x \in (L_\infty(\Omega))^n$, let $\mathfrak{F}(x)$ denote the set of all measurable functions $y : \Omega \to \mathbb{R}^n$ such that $y \in f(t, x(t))$ a.e. on Ω.

Consider a differential equation of the following type

$$x'(t) \in A(t)x(t) + f(t, x(t)) \tag{9.65}$$

with a condition $Lx = r$, where $A(t)$ is a $n \times n$ matrix summable on $\Omega, r \in \mathbb{R}^n, L$ a linear continuous mapping of $C(\Omega)$ into $\mathbb{R}^n$ and f a mapping of $\Omega \times \mathbb{R}^n$ into $CF(\mathbb{R}^n)$ satisfying conditions (i), (ii) and (iii). An absolutely continuous function $x \in C^{(n)}(\Omega)$ is said to be a solution of (9.54) if it is satisfied on Ω a.e. by x.

Theorem 9.20 *If $x(t) = 0$ is a unique solution of the linear homogeneous problem $x'(t) = A(t)x(t), Lx = 0$, then there exists a positive number β_0 depending on $A(t)$ and L such that for any function $f(t, x)$ with $\displaystyle\int_\Omega \beta(t)dt \leqslant \beta_0$, the system (9.54) has at least one solution for every $r \in \mathbb{R}^n$.*

The following example is given by Lasota and Opial [355].

Example 9.2 Consider the problem

$$x'(t) \in f(t, x(t)), \ x_i(t_i) = r_i \ (t_i \in \Omega, \ r_i \in \mathbb{R}, i = 1, 2, \ldots, n)$$

The corresponding homogeneous problem

$$x'(t) = 0, \ x_i(t_i) = 0, i = 1, 2, \ldots, n$$

has a unique solution $x(t) = 0$. Hence, by Theorem 9.20

$$Lx = (x_1(t_1), \ldots, x_n(t_n))$$

gives the existence of at least one solution of this problem for arbitrary sequence $(r_1, \ldots, r_n)$, provided f satisfies the required condition with $\beta_0 = \int_\Omega \beta(t)dt$ sufficiently small $(\beta_0 < 1)$.

We briefly dwell upon the Dirichlet boundary value problem, to indicate the use of fixed point method or degree theoretic method in partial differential equations.

Let Ω be a bounded domain in $\mathbb{R}^n$ with smooth boundary $\partial\Omega$. Let $f : \overline{\Omega} \times \mathbb{R} \to \mathbb{R}$ be a locally Hölder continuous function. Consider the nonlinear elliptic boundary value problem:

$$- \Delta u = f(x, u) \text{ in } \Omega \qquad\qquad (9.66)$$
$$u = 0 \qquad \text{on } \partial\Omega.$$

By a solution, we mean a classical solution $u \in C^2(\Omega) \cap C(\overline{\Omega})$ satisfying (9.66).

There is a standard method of reducing problem (9.66) to an equivalent fixed point equation in an appropriate function space by using the Green's function for the differential operator subject to given boundary condition. More explicitly, (9.66) is transformed into a nonlinear integral equation of Hammerstein type

$$u(x) = \int_\Omega G(x, y) f(y, u(y)) \, dy, x \in \overline{\Omega}. \qquad\qquad (9.67)$$

This integral equation can be considered as a fixed point equation in various function spaces like $L_p(\Omega)$, $1 < p < \infty$, $C(\overline{\Omega})$ or $C^1(\overline{\Omega})$. Due to regularity properties of Green's function, (9.67) can be shown to be equivalent to (9.66). Also, the nonlinear operator defined as

$$[F(u)](x) = \int_\Omega G(x, y) f(y, u(y)) \, dy \qquad\qquad (9.68)$$

is a completely continuous mapping of some Banach space of functions on $\overline{\Omega}$. Consequently, the boundary value problem (9.66) is equivalent to the fixed point equation $u = Fu$, and hence, nontrivial existence results can be derived by fixed point theory or by degree theoretic methods. As an illustration, consider the following example given by Berger [52].

Example 9.3 Let us consider the existence of the positive solutions of the following Dirichlet problem defined on a bounded domain $\Omega \subset \mathbb{R}^n$:

$$\Delta u + \lambda^2 f(x, u) = 0, \quad f(x, u) \geqslant a > 0 \text{ for } u \geqslant 0 \qquad (9.69)$$
$$u = 0 \text{ on } \partial\Omega$$

and $f(x, u)$ is nondecreasing in u for fixed x and that $f(x, u) \geqslant g(x)u$ for $u \geqslant 0$. It is proved that there is a finite number $\lambda_0 > 0$ such that for $\lambda < \lambda_0$, (9.69) has at least one positive solution, while for $\lambda > \lambda_0$, it has no positive solution.

First, one can show that (9.69) has a positive solution for some λ. The equivalent integral equation is given by

$$u(x) = \lambda^2 \int_\Omega G(x, y) f(y, u(y)) \, dy.$$

Since the Green's function $G(x, y) > 0$ in Ω and is integrable over Ω, for $u \geqslant 0$, there is a constant $c > 0$ with

$$Fu = \int_\Omega G(x, y) f(y, u(y)) \, dy \geqslant a \int_\Omega G(x, y) dy = c.$$

Then, F is a continuous and compact mapping of the positive cone $C(\Omega)$ into itself. Define $Tu = Fu/\|Fu\|$. Then, T is a continuous compact mapping from the closed bounded and convex set $B^+ = \{u : u \geqslant 0, \|u\| \leqslant 1\}$ into $\partial B^+ = \{u : u \geqslant 0, \|u\| = 1\}$. By Schauder's fixed point theorem, T has a fixed point $u^* \in \partial B^+$ and this u^* is a solution of (9.69) with $\lambda^2 = \frac{1}{\|Fu^*\|}$.

Next, it is shown that if (9.69) has a positive solution u_0 for $\lambda_0 > 0$, then (9.69) has a positive solution for all $\lambda \in (0, \lambda_0]$. Let

$$F_\lambda u(x) = \lambda^2 \int_\Omega G(x, y) f(y, u(y)) dy.$$

Then, for any $\lambda \in (0, \lambda_0]$, F_λ maps the closed and convex bounded set $\Sigma_0 = \{u : 0 \leqslant u \leqslant u_0, u \in C(\Omega)\}$ into itself. Since $f(x, u)$ is nondecreasing in u and $G(x, y) > 0$ in Ω, for $u \in \Sigma_0$, we have

$$f(x, 0) \leqslant f(x, u) \leqslant f(x, u_0)$$

and
$$\int_\Omega Gf(x, 0) \leqslant \int_\Omega Gf(x, u) \leqslant \int_\Omega Gf(x, u_0).$$

Hence, for $\lambda \in (0, \lambda_0]$ and $u \in \Sigma_0$, $0 < F_\lambda(0) \leqslant F_\lambda(u) \leqslant F_\lambda(u_0) = u_0$, i.e. $F_\lambda(u) \in \Sigma_0$. As F_λ is continuous and compact, Schauder's fixed point theorem (Theorem 5.55) implies that F_λ has a fixed point u_λ in Σ_0; that is, u_λ is a solution of (9.69). It can be shown that, for sufficiently large λ, (9.69) has no positive solution.

By means of pointwise definition of the ordering $\leq$, the Banach space $C(\overline\Omega)$ becomes an ordered Banach space and the increasing map F in (9.68) preserves the ordering. This additional information leads to fixed point theorems which are better than the results valid for completely continuous operators. It is shown by Amann [17] that in the case of an increasing or isotone mapping, it is sufficient to verify only that F maps two points of a closed bounded and convex set C into itself in order to guarantee the existence of a fixed point. For application of Schauder's fixed point theorem, it is necessary to verify that F maps all of C into itself and this is definitely not so easy. Though the condition that $f(x, \cdot) : \mathbb{R} \to \mathbb{R}$ is increasing is rather restrictive, we can replace it by a regularity hypothesis.

For an abstract formulation, consider the ordered Banach space $C_0^1(\overline{\Omega}) = \{u \in C^1(\overline{\Omega}) : u = 0 \text{ on } \partial\Omega\}$. The positive cone of $C_0^1(\overline{\Omega})$ has nonempty interior. Because F is a strongly increasing mapping, one can work in the interior of the cone; that is, one can consider the fixed point equation (9.67) in an open set. Hence, one can employ the method of calculus. However, $C_0^1(\overline{\Omega})$ has serious drawbacks because the ordering and the topology of this space are not well related. More precisely, the positive cone of this space is not normal and this implies that the order intervals $\langle u, v \rangle$ in this space are not bounded.

But there is another choice of the underlying ordered Banach space for tackling fixed point equations (9.67) which combines the good properties of the natural ordering of $C(\overline{\Omega})$ with the good properties of the mapping F in $C_0^1(\overline{\Omega})$. For details, refer to Amann [17]. For another approach to partial differential equations, see Cesari [137].

9.4.6 Approximation of Solution of Nonlinear Hybrid Ordinary Differential Equations

Suppose a closed and bounded interval $J = [t_0, t_0 + a]$, of the real line $\mathbb{R}$ for some $t_0, a \in \mathbb{R}$ with $t_0 \geq 0, a > 0$ be given. Consider in the function space $C(J, \mathbb{R})$ of continuous real-valued functions defined on J. Let us define a norm $\| \cdot \|$ and the order relation $\leqslant$ in $C(J, \mathbb{R})$ by $\|x\| = \sup_{t \in J} |x(t)|$ and $x \leqslant y \Leftrightarrow x(t) \leqslant y(t)$ for all $t \in J$, respectively. Then, we see that $C(J, \mathbb{R})$ is a Banach space with respect to above supremum norm and also partially ordered w.r.t. the above partially order relation $\leqslant$. Furthermore, it is known that the partially ordered Banach space $C(J, \mathbb{R})$ has some nice properties w.r.t. the above order relation in it.

Consider the following initial value problem (in short IVP) of first-order ordinary nonlinear hybrid differential equation, (in short HDE)

$$x'(t) = f(t, x(t)) + g(t, x(t)),$$
$$x(t_0) = x_0 \in \mathbb{R}, \tag{9.70}$$

for all $t \in J$, where $f, g : J \times \mathbb{R} \to \mathbb{R}$ are continuous functions.

The HDE (9.70) is considered in the function space $C(J, \mathbb{R})$ and is well known in the literature and discussed at length for existence as well as other aspects of the solutions (see, for instance, Krasnoselskii [344], Burton [126], Dhage [174] and the references therein). The HDE (9.70) is a hybrid differential equation with a linear perturbation of first type and can be tackled with the hybrid fixed point theory (cf. Dhage [174]). The existence theorems proved via classical fixed point theorems on the lines of Krasnoselskii [344] require the condition that the nonlinearities involved in (9.70) satisfy strong Lipschitz and compactness type conditions and do not yield any algorithm to determine the numerical solutions. Recently, Dhage et al. [180] proved the existence of the solutions of HDE (9.70) under weaker partially continuity and partially compactness type conditions.

In what follows, suppose E denotes a partially ordered real normed linear space with an order relation $\preceq$ and the norm $\| \cdot \|$. Recall that E is regular if $\{x_n\}_{n \in \mathbb{N}}$ is a nondecreasing (resp. nonincreasing) sequence in E such that $x_n \to x^*$ as $n \to \infty$, then $x_n \preceq x^*$ (resp. $x_n \succeq x^*$) for all $n \in \mathbb{N}$.

Observation

- The partially ordered Banach space $C(J, \mathbb{R})$ is regular,
- The conditions guaranteeing the regularity of any partially ordered normed linear space E may be found in Nieto and Lopez [423] and Heikkili and Lakshmikantham [269] and the references therein.

Assume the following set of assumptions in what follows:

(H1) Functions $f, g : J \times \mathbb{R} \to \mathbb{R}$ are continuous.
(A1) There exist a $\mathcal{D}$-function φ and a constants $\lambda > 0$ such that

$$0 \leqslant [p(t, x) - p(t, y)] \leqslant e^{\lambda t} \varphi(x - y), \text{ for all } t \in J \text{ and } x, y \in \mathbb{R}, x \geq y.$$

(A2) There exists a constant $K > 0$ such that $|p(t, x)| \leqslant K$ for all $t \in J$ and $x \in \mathbb{R}$.
(B1) There exist constants $L, M > 0$ such that $|\tilde{f}(t, x)| \leqslant L$ and $|\tilde{g}(t, x)| \leqslant M$ for all $t \in J$ and $x \in \mathbb{R}$.
(B2) Functions $\tilde{f}(t, x)$ and $\tilde{g}(t, x)$ are nondecreasing in x for all $t \in J$.
(B3) The HDE (9.70) has a lower solution $u \in C(J, \mathbb{R})$.

Consider the IVP of the HDE

$$x'(t) + \lambda x(t) = \mu e^{-\lambda t} p(t, x(t)) + \tilde{f}(t, x(t)) + \tilde{g}(t, x(t))$$
$$x(t_0) = x_0 \in \mathbb{R}, \tag{9.71}$$

for all $t \in J$, where $\tilde{f}, \tilde{g} : J \times \mathbb{R} \to \mathbb{R}$, $\tilde{f}(t, x) = f(t, x) + \lambda x$ and $\tilde{g}(t, x) = g(t, x) - \mu e^{-\lambda t} p(t, x)$, $\lambda > 0$, $\mu > 0$ with $\mu \leqslant \frac{\lambda}{1 - e^{-a}}$.

Lemma 9.3 *Suppose that hypotheses $(H1)$, $(A2)$ and $(B1)$ hold. Then, a function $u \in C(J, \mathbb{R})$ is a solution of the HDE (9.71) if and only if it is a solution of the nonlinear integral equation,*

$$x(t) = x_0 e^{-\lambda(t - t_0)} + \mu e^{-\lambda t} \int_{t_0}^{t} p(s, x(s)) \, ds + e^{-\lambda t} \int_{t_0}^{t} e^{\lambda s} [\tilde{f}(s, x(s)) + \tilde{g}(s, x(s))] \, ds$$

for all $t \in J$.

We now prove the existence of solutions of HDE (9.70) under new hypotheses for its redesigned structure.

Theorem 9.21 *Assume that hypotheses* $(H1), (A1) - (A2)$ *and* $(B1) - (B3)$ *hold. Then, the HDE (9.71) has a solution* x^* *defined on* J *and the sequence* $\{x_n\}, n = 0, 1, 2, \ldots$ *of successive approximations defined by*

$$x_{n+1}(t) = x_0 e^{-\lambda(t-t_0)} + \mu e^{-\lambda t} \int_{t_0}^{t} p(s, x_n(s))\, ds + e^{-\lambda t} \int_{t_0}^{t} e^{\lambda s} [\tilde{f}(s, x_n(s)) + \tilde{g}(s, x_n(s))]\, ds$$

$$(9.72)$$

where $x_0 = u$ *converges monotonically to* x^*.

Proof Set $E = C(J, \mathbb{R})$ and define two operators $\mathcal{P}$ and $\mathcal{Q}$ on E by

$$\mathcal{P}(x(t)) = x_0 e^{-\lambda(t-t_0)} + \mu e^{-\lambda t} \int_{t_0}^{t} p(s, x(s))\, ds, \ t \in J, \qquad (9.73)$$

and

$$\mathcal{Q}(x(t)) = e^{-\lambda t} \int_{t_0}^{t} e^{\lambda s} [\tilde{f}(s, x_n(s)) + \tilde{g}(s, x_n(s))]\, ds, \ t \in J. \qquad (9.74)$$

From the continuity of the integral, it follows that $\mathcal{P}$ and $\mathcal{Q}$ define the maps $\mathcal{P}, \mathcal{Q} : E \to E$. Now by Lemma 9.3, the HDE (9.71) is equivalent to the operator equation $\mathcal{P}x(t) + \mathcal{Q}x(t) = x(t), t \in J$. We shall show that the operators $\mathcal{P}$ and $\mathcal{Q}$ satisfy all the conditions of Theorem 5.168. This is achieved in the series of following steps.

Step I: $\mathcal{P}$ and $\mathcal{Q}$ are nondecreasing on E. Let $x, y \in E$ be such that $x \geq y$. Then, by hypothesis $(A1)$, we obtain

$$\mathcal{P}(x(t)) = x_0 e^{-\lambda(t-t_0)} + \mu e^{-\lambda t} \int_{t_0}^{t} p(s, x(s))\, ds$$

$$= x_0 e^{-\lambda(t-t_0)} + \mu e^{-\lambda t} \int_{t_0}^{t} [p(s, x(s)) - p(s, y(s)]\, ds + \mu e^{-\lambda t} \int_{t_0}^{t} p(s, y(s))\, ds$$

$$\geq x_0 e^{-\lambda(t-t_0)} + \mu e^{-\lambda t} \int_{t_0}^{t} p(s, y(s))\, ds = \mathcal{P}(y(t)), \qquad (9.75)$$

for all $t \in J$. This shows that $\mathcal{P}$ is nondecreasing operator on E into E. Similarly, using hypothesis $(B2)$, it is shown that $\mathcal{Q}$ is also nondecreasing on E into itself. Thus, $\mathcal{P}$ and $\mathcal{Q}$ are nondecreasing operators on E into itself.

Step II: $\mathcal{P}$ is partially bounded and partially contraction on E. Let $x \in E$ be arbitrary. Then, by $(A2)$,

$$|\mathcal{P}(x(t))| \leqslant |x_0 e^{-\lambda(t-t_0)}| + \mu |e^{-\lambda t}| \int_{t_0}^{t} |p(s, x(s))|\, ds$$

$$\leqslant |x_0| + \mu \int_{t_0}^{t_0+a} K\, ds = |x_0| + \mu K a,$$

for all $t \in J$. Taking supremum over t, we obtain $\|\mathcal{P}x\| \leq |x_0| + \mu K a$, and so, $\mathcal{P}$ is bounded. This further implies that $\mathcal{P}$ is partially bounded on E. Next, let $x, y \in E$ be such that $x \geq y$. Then,

$$
\begin{aligned}
|\mathcal{P}(x(t)) - \mathcal{P}(y(t))| &= \mu \left| e^{-\lambda t} \int_{t_0}^{t} [p(s, x(s)) - p(s, y(s))] \, ds \right| \\
&\leq \frac{e^{-\lambda t}}{1 - e^{-a}} \int_{t_0}^{t} \lambda e^{\lambda s} \varphi(x(s) - y(s)) \, ds \\
&\leq \frac{e^{-\lambda t}}{1 - e^{-a}} \int_{t_0}^{t} \frac{d}{ds}(e^{\lambda s}) \varphi(|x(s) - y(s)|) \, ds \\
&\leq \varphi(\|x - y\|),
\end{aligned}
$$

for all $t \in J$. Taking supremum over t, we obtain $\|\mathcal{P}x - \mathcal{P}y\| \leq \varphi(\|x - y\|)$, for all $x, y \in E$ with $x \geq y$. Hence, $\mathcal{P}$ is a partially nonlinear $\mathcal{D}$-contraction on E which further implies that $\mathcal{P}$ is a partially continuous on E.

Step III: $\mathcal{Q}$ is partially continuous on E. Let $\{x_n\}_{n\in\mathbb{N}}$ be a sequence in a chain C such that $x_n \to x$ as $n \to \infty$. Then, by dominated convergence theorem, we have

$$
\begin{aligned}
\lim_{n\to\infty} \mathcal{Q}x_n(t) &= \lim_{n\to\infty} e^{-\lambda t} \int_{t_0}^{t} e^{\lambda s} \left[\tilde{f}(s, x_n(s)) + \tilde{g}(s, x_n(s)) \right] ds \\
&= e^{-\lambda t} \int_{t_0}^{t} e^{\lambda s} \lim_{n\to\infty} \left[\tilde{f}(s, x_n(s)) + \tilde{g}(s, x_n(s)) \right] ds \\
&= e^{-\lambda t} \int_{t_0}^{t} e^{\lambda s} \left[\lim_{n\to\infty} \tilde{f}(s, x_n(s)) + \lim_{n\to\infty} \tilde{g}(s, x_n(s)) \right] ds \\
&= e^{-\lambda t} \int_{t_0}^{t} e^{\lambda s} \left[\tilde{f}(s, x(s)) + \tilde{g}(s, x(s)) \right] ds = \mathcal{Q}x(t),
\end{aligned}
$$

for all $t \in J$. This shows that $\mathcal{Q}x_n$ converges monotonically to $\mathcal{Q}x$ pointwise on J.

Next, we will show that $\{\mathcal{Q}x_n\}_{n\in\mathbb{N}}$ is an equicontinuous sequence of functions in E. Let $t_1, t_2 \in J$ with $t_1 < t_2$. Then,

$$
\begin{aligned}
|\mathcal{Q}x_n(t_2) - \mathcal{Q}x_n(t_1)| &= \left| e^{-\lambda t_2} \int_{t_0}^{t_2} e^{\lambda s} [\tilde{f}(s, x_n(s)) + \tilde{g}(s, x_n(s))] \, ds \right. \\
&\quad \left. - e^{-\lambda t_1} \int_{t_0}^{t_1} e^{\lambda s} [\tilde{f}(s, x_n(s)) + \tilde{g}(s, x_n(s))] \, ds \right| \\
&\leq \left| (e^{-\lambda t_2} - e^{-\lambda t_1}) \int_{t_0}^{t_1} e^{\lambda s} [\tilde{f}(s, x_n(s)) + \tilde{g}(s, x_n(s))] \, ds \right| \\
&\quad + \left| e^{-\lambda t_2} \int_{t_1}^{t_2} e^{\lambda s} [\tilde{f}(s, x_n(s)) + \tilde{g}(s, x_n(s))] \, ds \right| \\
&\to 0 \text{ as } t_2 - t_1 \to 0
\end{aligned}
$$

uniformly for all $n \in \mathbb{N}$. This shows that the convergence $\mathcal{Q}x_n \to \mathcal{Q}x$ is uniform, and hence, $\mathcal{Q}$ is partially continuous on E.

Step IV: $\mathcal{Q}$ is partially compact operator on E. Let C be an arbitrary chain in E. We show that $\mathcal{Q}(C)$ is a uniformly bounded and equicontinuous set in E. First, we show that $\mathcal{Q}(C)$ is uniformly bounded. Let $y \in \mathcal{Q}(C)$ be any element. Then, there is an element $x \in C$ be such that $y = \mathcal{Q}x$. Now, by hypothesis (B1),

$$
\begin{aligned}
|y(t)| &= |\mathcal{Q}x(t)| \\
&= \left| e^{-\lambda t} \int_{t_0}^{t} e^{\lambda s} [\tilde{f}(s, x(s)) + \tilde{g}(s, x(s))] \, ds \right| \\
&\leqslant \int_{t_0}^{t} e^{\lambda s} [|\tilde{f}(s, x(s))| + |\tilde{g}(s, x(s))|] \, ds \\
&\leqslant \int_{t_0}^{t_0 + a} e^{\lambda(t_0 + a)} [L + M] \, ds \leqslant e^{\lambda(t_0 + a)} [L + M] a
\end{aligned}
$$

for all $t \in J$. Taking supremum over t, we obtain $\|y\| = \|\mathcal{Q}x\| \leqslant r$ for all $y \in \mathcal{Q}(C)$. Hence, $\mathcal{Q}(C)$ is a uniformly bounded subset of E. Next, we will show that $\mathcal{Q}(C)$ is an equicontinuous set in E. Let $t_1, t_2 \in J$ with $t_1 < t_2$. Then, for any $y \in \mathcal{Q}(C)$, one has

$$
\begin{aligned}
|y(t_2) - y(t_1)| &= |\mathcal{Q}x(t_2) - \mathcal{Q}x(t_1)| \\
&= \left| e^{-\lambda t_2} \int_{t_0}^{t_2} e^{\lambda s} [\tilde{f}(s, x(s)) + \tilde{g}(s, x(s))] \, ds \right. \\
&\quad \left. - e^{-\lambda t_1} \int_{t_0}^{t_1} e^{\lambda s} [\tilde{f}(s, x(s)) + \tilde{g}(s, x(s))] \, ds \right| \\
&\leqslant \left| \left(e^{-\lambda t_2} - e^{-\lambda t_1} \right) \int_{t_0}^{t_1} e^{\lambda s} [\tilde{f}(s, x(s)) + \tilde{g}(s, x(s))] \, ds \right| \\
&\quad + \left| e^{-\lambda t_2} \int_{t_1}^{t_2} e^{\lambda s} [\tilde{f}(s, x(s)) + \tilde{g}(s, x(s))] \, ds \right| \\
&\to 0 \text{ as } t_2 - t_1 \to 0
\end{aligned}
$$

uniformly for all $y \in \mathcal{Q}(C)$. Hence, $\mathcal{Q}(C)$ is an equicontinuous subset of E. Now, $\mathcal{Q}(C)$ is a uniformly bounded and equicontinuous set of functions in E, so it is compact. Consequently, $\mathcal{Q}$ is a partially compact operator on E into itself.

Step V: u satisfies the operator inequality $u \leqslant \mathcal{P}u + \mathcal{Q}u$. By hypothesis (B3), the HDE (9.70) has a lower solution u defined on J. Then, we have

$$
\begin{aligned}
u'(t) &\leqslant f(t, u(t)) + g(t, u(t)) \\
u(t_0) &\leqslant u_0 \in \mathbb{R},
\end{aligned}
\tag{9.76}
$$

for all $t \in J$. Adding $\lambda u(t)$ on both sides of the first inequality in (9.76), whereas the term $e^{-\lambda t} p(t, u(t))$ is added and subtracted in the right-hand side of (9.76), we obtain

$$u'(t) + \lambda u(t) \leqslant \mu e^{-\lambda t} p(t, u(t)) + f(t, u(t)) + \lambda u(t)$$
$$+ g(t, u(t)) - \mu e^{-\lambda t} p(t, u(t)), \ t \in J$$

i.e.

$$u'(t) + \lambda u(t) \leqslant \mu e^{-\lambda t} p(t, u(t)) + \tilde{f}(t, u(t)) + \tilde{g}(t, u(t)), \ t \in J \quad (9.77)$$

Again, multiplying the above inequality (9.77) by $e^{\lambda t}$, we obtain

$$\left(e^{\lambda t} u(t) \right)' \leqslant \mu p(t, u(t)) + e^{\lambda t} [\tilde{f}(t, u(t)) + \tilde{g}(t, u(t))], \ t \in J. \quad (9.78)$$

A direct integration of (9.78) from t_0 to t yields

$$u(t) \leqslant u_0 e^{-\lambda(t-t_0)} + \mu e^{-\lambda t} \int_{t_0}^{t} p(s, u(s)) \, ds$$
$$+ e^{-\lambda t} \int_{t_0}^{t} e^{\lambda s} [\tilde{f}(s, u(s)) + \tilde{g}(s, u(s))] \, ds, \quad (9.79)$$

for all $t \in J$. From definitions of the operators $\mathcal{P}$ and $\mathcal{Q}$, it follows that $u(t) \leqslant \mathcal{P}u(t) + \mathcal{Q}u(t)$, for all $t \in J$. Hence, $u \leqslant \mathcal{P}u + \mathcal{Q}u$. Thus, $\mathcal{P}$ and $\mathcal{Q}$ satisfy all the conditions of Theorem 5.168, and we apply it to conclude that the operator equation $\mathcal{P}x + \mathcal{Q}x = x$ has a solution. Consequently, the integral equation and the HDE (9.70) has a solution x^* defined on J. Furthermore, the sequence $\{x_n\}_{n=0}^{\infty}$ of successive approximations defined by (9.71) converges monotonically to x^*. This completes the proof.

Example 9.4 Given a closed and bounded interval $J = [0, 1]$, consider the IVP of HDE,

$$\left. \begin{array}{l} x'(t) = \frac{1}{1-e^{-a}} \frac{|x(t)|}{1+|x(t)|} + \tan^{-1} \frac{|x(t)|}{2} - x(t) + \tan^{-1} x(t), \\ x(0) = 1 \in \mathbb{R}, \end{array} \right\} \quad (9.80)$$

for all $t \in J$. Here, $f(t, x) = \tan^{-1} \frac{|x|}{2} - x$, $\tilde{f}(t, x) = \frac{|x|}{1+|x|}$, $g(t, x) = \tan^{-1} x + \frac{1}{1-e^{-a}} \frac{|x|}{1+|x|} = \tan^{-1} x + \frac{e^{-t}}{1-e^{-a}} e^t \frac{|x|}{1+|x|}$, $p(t, x) = e^t \frac{|x|}{1+|x|}$, $\lambda = 1$, $\mu = \frac{1}{1-e^{-a}}$ and $\tilde{g}(t, x) = \tan^{-1} x$.

We observe the following:

(i) The hypotheses $(H1)$ satisfied, since f and g are real-valued continuous functions on $J \times \mathbb{R}$.

(ii) There exists an exponentially continuous mapping $p : J \times \mathbb{R} \to \mathbb{R}$ defined by $p(t, x) = e^t \frac{|x|}{1+|x|}$ and a $\mathcal{D}$-function $\varphi : \mathbb{R}_+ \to \mathbb{R}_+$ defined by $\varphi(t) = \frac{t}{1+t}$ such that

$$0 \leqslant [p(t, x) - p(t, y)] = e^t \left(\frac{|x|}{1+|x|} - \frac{|y|}{1+|y|} \right) = e^t \left(\frac{|x| - |y|}{1+|x|+|y|+|x||y|} \right)$$
$$\leqslant e^t \left(\frac{|x| - |y|}{1+||x|-|y||} \right) \leqslant e^t \left(\frac{|x - y|}{1+|x-y|} \right) = e^t \varphi(t).$$

Hence, the function f satisfies the hypothesis $(A1)$ with $\lambda = 1$.

(iii) There exists a constant $K = e^1 > 0$ such that $|p(t, x)| = e^t \frac{|x|}{1+|x|} \leqslant K$ for all $t \in J = [0, 1]$ and $x \in \mathbb{R}$, and so the hypotheses $(A2)$ is satisfied.

(iv) The functions $\tilde{f}(t, x) = \tan^{-1} \frac{|x|}{2}$ and $\tilde{g}(t, x) = \tan^{-1} x$ are bounded on $J \times \mathbb{R}$ with bounds L and M, respectively, with $L = M = \pi$. Thus, the hypotheses $(B1)$ is satisfied.

(v) Clearly, the functions $\tilde{f}(t, x)$ and $\tilde{g}(t, x)$ are nondecreasing in x for all $t \in J = [0, 1]$, and thus, hypotheses $(B2)$ is satisfied.

(vi) Finally, the HDE (9.70) has a lower solution $u = e^{-1}$ defined on $J = [0, 1]$. Thus, all hypotheses of theorem (9.50) are satisfied. Hence, we apply Theorem 9.21 and conclude that the HDE (9.71) has a solution x^* defined on J and the sequence $\{x_n\}_{n=0}^{\infty}$ defined by

$$x_{n+1}(t) = x_0 e^{-t} + \mu e^{-t} \int_0^t p(s, x_n(s)) \, ds + e^{-t} \int_0^t e^s [\tilde{f}(s, x_n(s)) + \tilde{g}(s, x_n(s))] \, ds \tag{9.81}$$

where $x_0 = e^{-1}$ converges monotonically to x^*.

9.5 Application to Nonlinear Integral Equations

In this section, to illustrate our Theorem 5.183, we consider the following example of nonlinear integral equation.

Example 9.5 Given a closed and bounded interval $J = [0, 1]$ in $\mathbb{R}$, the set of all real numbers, consider the nonlinear integral equation (in short NIE)

$$x(t) = p(t, x(t)) + q(t, x(t)) \left(\lambda(t) + \int_0^t f(s, x(s)) ds \right), \tag{9.82}$$

for all $t \in J$, where $\lambda : J \longrightarrow \mathbb{R}$, $f : J \times \mathbb{R} \longrightarrow \mathbb{R}$ are continuous and $p : J \times \mathbb{R} \longrightarrow \mathbb{R}$ is given by

$$p(t, x) = \frac{2}{3}x,$$

$q : J \times \mathbb{R} \longrightarrow \mathbb{R}$ is given by

$$q(t, x) = \begin{cases} \frac{1}{3(1+x)}, & if \ x \geq 0; \\ \frac{1}{6}, & if \ x < 0. \end{cases}$$

By the solution of the integral equation (9.82), we mean a continuous function $x :$ $J \longrightarrow \mathbb{R}$ that satisfies (9.82) on J. Let $X = C(J, \mathbb{R})$ be a Banach algebra of all continuous real-valued functions on J with the norm

$$\|x\| \doteq \sup_{t \in J} |x(t)|. \tag{9.83}$$

We now intend to obtain the solution of (9.82). To effect this, under some suitable conditions, suppose that the function f satisfies the following condition:

$$|f(t, x)| \leqslant 1 - \|\lambda\|, \qquad \|\lambda\| < 1, \tag{9.84}$$

for all $t \in J$ and $x \in \mathbb{R}$.

Define a subset S of X by

$$S = \{y \in X : \|y\| \leqslant 1\}. \tag{9.85}$$

Consider two mappings $A, B : X \longrightarrow X$ defined by

$$Ax(t) = q(t, x(t)), \qquad t \in J. \tag{9.86}$$

$$Bx(t) = \lambda(t) + \int_0^t f(s, x(s))ds, \qquad t \in J. \tag{9.87}$$

$$Cx(t) = p(t, x(t)), \qquad t \in J. \tag{9.88}$$

We shall show that operators A, B and C satisfies all the conditions of Theorem 5.183.

First, we show that A is a Lipschitzian map on X. Let $x, y \in X$. Then,

$$|Ax(t) - Ay(t)| = |q(t, x(t)) - q(t, y(t))|$$
$$\leqslant \frac{1}{3}\|x - y\|,$$

which shows that A is a Lipschitzian map. Similarly, C is a Lipschitzian map.

Now, it is an easy exercise to prove that B is completely continuous on S. We show that $B : S \longrightarrow S$. Let $x \in S$. Then, by (9.84) and (9.87),

$$|Bx(t)| = \left| \lambda(t) + \int_0^t f(t, x(t))ds \right|$$

$$\leqslant |\lambda(t)| + \int_0^t |f(t, x(t))|ds$$

$$< |\lambda(t)| + \int_0^t (1 - \|\lambda\|)ds.$$

Since $Bx \in C(J, \mathbb{R})$, there is a point $t^* \in J$ such that

$$\|Bx\| = |Bx(t^*)| = \max_{t \in J} |Bx(t)|$$

Therefore, we have

$$\|Bx\| = |Bx(t^*)| < |\lambda(t^*)| + \int_0^{t^*} (1 - \|\lambda\|)ds \leqslant \|\lambda\| + \int_0^1 (1 - \|\lambda\|)ds = 1,$$

i.e. $\|Bx\| < 1$. As a result, $B : S \longrightarrow S$. Finally, we show that condition (5.174) of Proposition 5.14 holds. Now, for any $x \in X$, we have

$$\left\| \left(\frac{I - C}{A} \right)(x) \right\| = \sup_{t \in J} \left| \frac{x(t) - Cx(t)}{Ax(t)} \right| \geq \|x\|.$$

Thus, operators A, B and C satisfy all the conditions of Theorem 5.183. Hence, the integral equation (9.82) has a solution on J.

9.5.1 Nonlinear Integral Equations Involving Urysohn's Operator and Hammerstein Integral Operator

Next, we discuss some applications to nonlinear integral equations involving Urysohn's operator or its particular case, Hammerstein integral operator. For an exhaustive treatment, refer to Krasnoselskii [344, 345]. Our selection is motivated by the desire to show the applications of fixed point theorems. We have discussed some recent results obtained by Joshi and Srikanth [294], Leggett and Williams [363] and Gatica and Smith [241].

Let I be an interval of the real axis, and let X denote the vector apace of bounded, continuous and complex-valued functions on I. X is a complete metric space induced by the norm $\|x\| = \sup_{t \in I} |x(t)|$. Let $f : I \times I \times C \to C$ be a given function. Assumed that function $s \mapsto f(t, s, x(s))$ is integrable over I, while the function

$$t \mapsto \int_I f(t, s, x(s))ds$$

is bounded and continuous on I. Choosing any element y of X, define a map U of X into itself as

$$[U(x)](t) = \int_I f(t, s, x(s))ds + y(t). \tag{9.89}$$

A fixed point of U is the solution $x \in X$ of the nonlinear integral equation

$$x(t) = \int_I f(t, s, x(s))ds + y(t). \tag{9.90}$$

Under suitable conditions, U or U^n (for suitable n) will be a contraction map on X.

The existence of a unique solution in X will follow from the contraction mapping theorem or its corollary.

Example 9.6 Suppose f satisfies an inequality of the form

$$|f(t, s, \xi) - f(t, s, \eta)| \leqslant F(t, s)|\xi - \eta|$$

Then, for $x_1, x_2 \in X$, we have

$$\|Ux_1 - Ux_2\| \leqslant k\|x_1 - x_2\|$$

where $k = \sup\limits_{t \in I} \left\{ \int_I F(t, s)\, ds \right\}$.

If $k < 1$, the operator U is a contraction map.

Similarly, the existence and uniqueness of a solution in X of the equation

$$x(t) = \lambda \int_I f(t, x, x(s))ds + y(t)$$

for any given $y \in X$ are assured provided λ is small enough to make

$$k\lambda < 1.$$

If f takes the form $f(t, x, \xi) = K(t, x)\xi$, then we get linear nonhomogeneous Fredholm integral equation. Volterra integral equation is a particular case of Fredholm integral equation.

Example 9.7 Let us consider the nonlinear integral equation

$$x(t) = \lambda \int_a^t f(t, s, x(s))ds + y(t) \tag{9.91}$$

where y is continuous on $[a, b]$ and $f(t, s, \xi)$ is continuous in the region $[a, b] \times [a, b] \times [a, b]$ and satisfies the Lipschitz condition

$$|f(t, s, \xi) - f(i, s, \eta)| \leqslant M|\xi - \eta|$$

The classical Volterra integral equation is obtained by taking $f(t, s, \xi) = K(t, s)\xi$, with K continuous in $[a, b]$. Let $X = C[a, b]$, and U be the mapping of X into itself defined by

$$(Ux)(t) = \lambda \int_a^t f(t, s, x(s))ds + y(t), x \in X, a \leqslant t \leqslant b.$$

Given $x_1, x_2 \in X$, we can prove by induction that

$$|(U^n x_1)(t) - (U^n x_2)(t)| \leqslant \frac{1}{n!}|\lambda|^n M^n \|x_1 - x_2\|(t - a)^n, \quad a \leqslant t \leqslant b.$$

Then,

$$\|U^n x_1 - U^n x_2\| \leqslant \frac{1}{n!}|\lambda|^n M^n (b - a)^n \|x_1 - x_2\|.$$

Thus, all U^n and in particular U are continuous, and for n sufficiently large, we see that $\frac{1}{n!}|\lambda|^n M^n (b - a)^n < 1$. Moreover, U^n is a contraction mapping for n large. Hence, by Corollary 5.1, we have a unique $x \in X$ such that $Ux = x$, i.e. x is the required unique solution of the Eq. (9.91).

Consider the Urysohn operator

$$[Ux](t) = \int_\Omega f(t, s, x(s)) \, ds \tag{9.92}$$

discussed earlier. There are suitable criteria for the existence of Fréchet derivative of the Urysohn operator and the positivity of the operator when it acts on the cone K of nonnegative functions in the space L_p $(1 < p < \infty)$. We assume that U is continuous as an operator acting on the space L_p, and it maps bounded subsets into compact subsets. Let the function

$$g(t, s, \xi) = \frac{1}{\xi} f(t, s, \xi)$$

approach uniformly with respect to $t, s \in \Omega$ to some function $V(t, s)$ $(t, s \in \Omega)$ as $\xi \to \infty$, where the integral operator

$$(Vx)(t) = \int_\Omega V(t, s)x(s) \, ds$$

acts in a corresponding space. This operator V turns out to be a strong asymptotic derivative with respect to the cone K of the Urysohn operator.

The following theorem deals with the existence of nonnegative solution of a nonlinear homogeneous integral equation involving Urysohn operator.

Theorem 9.22 *Let the Urysohn operator U be defined and continuous on the cone K of nonnegative function of L_p and maps bounded subsets into compact subsets. Let the operator U have a strong asymptotic derivative $U'(\infty)$ with respect to a cone where $[U'(\infty)x](t) = \int_\Omega V(t,s)x(s)\, ds$. Also assume that $f(t,s,\xi) \geqslant 0 (t, s \in \Omega, \xi \geqslant 0)$. Then, the nonlinear integral equation*

$$x(t) = \int_\Omega f(t,s,x(s))\, ds$$

has at least one nonnegative solution if the operator V (generated by $V(t,s)$) has positive characteristic values greater than or equal to 1.

Proof The cone K is invariant under the mapping U. In view of the fact that $f(t,s,\xi) \geqslant 0$, it follows that the kernel $V(t,s)$ is nonnegative and the operator $U'(\infty)$ is a linear, positive, continuous and compact operator acting on L_p. Now the result follows by Theorem 5.146. ∎

Remark 9.5 A simpler condition for the existence of a nonnegative solution for the above integral equation is as follows:

$$0 \leqslant f(t,s,\xi) \leqslant a + b\xi; \; t, s, \in \Omega, \xi \geqslant 0,$$

where $a \geqslant 0$ and $b \times \mu(\Omega) < 1$.

Theorem 9.23 *Let $f(t,s,0) = 0, (t, s \in \Omega), f(t,s,\xi) \geqslant 0 (t, s \in \Omega, \xi \geqslant 0)$. Let U be continuous on the cone K of nonnegative functions of L^p and maps bounded subsets into compact subsets and U has a strong asymptotic derivative*

$$U'(\infty)x(t) = \int_\Omega V(t,s)x(s)\, ds$$

with respect to a cone and a Fréchet derivative $U'(0)$ with respect to a cone at the point 0, where $U'(0)x(t) = \int_\Omega W(t,x)x(s)\, ds$ where $W(t,s) = f_\xi(t,s,0)$. The functions $V(t,s)$ and $W(t,s)$ are nonnegative and assume that the linear integral operators $U'(\infty)$ and $U'(0)$ have a unique normalized eigenvector each in the cone K. Let λ_0 and λ_∞ denote the greatest positive eigenvalue of the operators $U'(0)$ and $U'(\infty)$, respectively. Then, the nonlinear integral equation

$$x(t) = \int_\Omega f(t,sx(s))\, ds$$

has at least one nonnegative nonzero solution if either $\lambda_0 < 1 < \lambda_\infty$ or $\lambda_\infty < 1 < \lambda_0$.

Proof If follows from Theorems 5.147 and 5.148. ∎

Now we discuss the solution of the equation of the type

$$x = K_1 N_1 x + K_2 N_2 x + f$$

that is, more specifically the existence and uniqueness of solution of integral equations of the following type:

$$x(t) = f(t) + \int_0^t k_1(t, s) g(s, x(s))\, ds + \int_0^\infty k_2(t, s) h(s, x(s))\, ds, \qquad (9.93)$$

where x, f, g and h are vectors with n components and k_1 and k_2 are $n \times n$ matrices.

Notice that equations of the type (9.93) arise naturally in the study of boundary value problems on the infinite half line. The following theorems were proved by Miller, Nohel and Wong [398].

For simplicity, all functions considered are assumed to be continuous. Let $|\cdot|$ denote any appropriate matrix norm, and let $BC[0, \infty)$ be the collection of all bounded continuous functions from $[0, \infty)$ into a finite dimensional Euclidean space, with sup norm defined by $\|\phi\| = \sup_{t \geq 0} |\phi(t)|$.

Theorem 9.24 *Let the following conditions hold:*

(i) $\sup_{t \geq 0} \int_0^t |k_1(t, s)| ds \leq K_1 < \infty,$

(ii) $\sup_{t \geq 0} \int_0^\infty |k_2(t, s)| ds \leq K_2 < \infty;$

(iii) $g(t, 0) = h(t, 0) = 0;$

(iv) *for each $r > 0$, there exists $\delta > 0$ such that*

$$|g(t, x) - g(t, y)| \leq r|x - y| \qquad (9.94)$$

for all $|x|, |y| \leq \delta$ and uniformly in t;

(v) *for each $\xi > 0$, there exists $\eta > 0$ such that*

$$|h(t, x) - h(t, y)| \leq \xi|x - y| \qquad (9.95)$$

for all $|x|, |y| \leq \eta$ and uniformly in t; and

(vi) *for all t,* $\int_0^\infty |k_2(t, s) - k_2(t + \alpha, s)| ds \to 0$ *as* $|\alpha| \to 0$.

Then, there exists a positive number r_0 such that to any $r \in [0, r_0]$, there corresponds a $\delta > 0$ such that whenever $\|f\| < \delta$, there exists a unique solution ϕ of (9.93) on $0 \leq t < \infty$ satisfying $\|\phi\| \leq r$.

Proof Fix $\xi \geq 0$ such that $\xi K_2 < 1$. By (iii) and (v) choose $\eta > 0$ such that for $|x| \leq \eta$, we have $|h(t, x)| \leq \xi|x|$ uniformly in t. Let $r = (1 - \xi K_2)/2K_1$ and choose $\delta > 0$ such that (9.94) holds for all $|x|, |y| \leq \delta$ uniformly in t. Let $r_0 = \min(\eta, \delta)$.

Define operators S and T on B_r as follows:

$$S\phi(t) = f(t) + \int_0^t k_1(t, s)g(s, \phi(s))ds,$$

$$T\phi(t) = \int_0^\infty k_2(t, s)h(s, \phi(s))ds.$$

For any $r \in (0, r_0]$, we first show that there exists $\delta > 0$ such that $Sx + Ty \in B_r$ for all $x, y \in B_r$ provided $\|f\| \leqslant \delta$. For all $t \geqslant 0$, we have

$$|Sx(t) + Ty(t)| \leqslant |f(t)| + r\|x\| \int_0^t |K_1(t, s)|ds + \|y\| \int_0^\infty |k_2(t, s)|ds,$$

$$\leqslant \delta + \frac{r}{2}(1 - \xi K_2) + r\xi K_2 \leqslant r$$

provided that $\delta \leqslant (1 - \xi K_2)r/2$. By (9.94) it can be shown that S is a contraction on B_r. Similarly, by (ii) and (vi), it can be shown that T is compact and continuous. Hence, by Krasnoselskii's theorem (Theorem 5.61), equation (9.93) has a solution in B_r.

To prove the uniqueness of the solution in B_r, let $x(t)$ and $y(t)$ be two distinct solution of (9.93) and let $w(t) = x(t) - y(t)$. By Eqs. (9.94) and (9.95), we have

$$|w(t)| \leqslant r \int_0^t |k_1(t, s)||w(s)|ds + \eta \int_0^\infty |k_2(t, s)||w(s)|ds$$

$$\leqslant rK_1\|w\| + \eta K_2\|w\|$$

$$= \frac{1}{2}(1 + \xi K_2)\|w\| < \|w\|.$$

Taking the supremum over all t in the above estimate, we obtain a contradiction. ∎

Remark 9.6 In Eq. (9.93), instead of $f(t)$, we can take $f(t, x(t))$. Then, the proof will go through with the following assumption in addition to a few modified conditions:

For each $\mu > 0$, there exists a λ such that $|f(t, x(t)) - f(t, y(t))| \leqslant \mu|x(t) - y(t)|$ for all $|x|, |y| \leqslant \lambda$ and uniformly in t.

Theorem 9.25 *Consider the integral equation*

$$x(t) = f(t, x(t)) + \int_0^t k_1(t, s)g(s, x(s))ds + \int_0^\infty k_2(t, s)h(s, x(s))ds. \quad (9.96)$$

Let S and T be operators from $BC[0, \infty)$ into itself as defined below:

$$Sx(t) = \int_0^t k_1(t, s)x(s)ds \quad and \quad Tx(t) = \int_0^\infty k_2(t, s)x(s)ds$$

Let the following conditions hold:

(i) $|g(t, x(t)) - g(t, y(t))| \leqslant \lambda |x(t) - y(t)|$ *for all* $x(t), y(t) \in B_r$ *and* $\lambda \geqslant 0$ *a constant;*

(ii) $|h(t, x(t)) - h(t, y(t))| \leqslant \xi |x(t) - y(t)|$ *for all* $x(t), y(t) \in B_r$

(iii) $|f(t, x(t)) - f(t, y(t))| \leqslant r |x(t) - y(t)|$ *for all* $x(t), y(t) \in B_r$ *and* $r \geqslant 0$ *a constant.*

(iv) $k_1(t, s)$ *and* $k_2(t, s)$ *are such that* S *and* T *are continuous operators from* $BC[0, \infty)$ *into itself.*

Then, there exists a unique solution of (9.96) *provided* $K_1 \lambda + K_2 \xi + r < 1$ *and* $|f(t, x(t0) + K_1 |g(t, 0)| + K_2 |h(t, o)| \leqslant r(1 - \lambda K_1 - \xi K_2)$ *where* K_1 *and* K_2 *are norms of* S *and* T, *respectively.*

Proof Let us define operators U and V from B_r into $BC[0, \infty)$ as follows:

$$(Ux)(t) = f(t, x(t)) + \int_0^t k_1(t, x)g(s, x(s))$$

and

$$(Vx)(t) = \int_0^\infty k_2(t, s)h(s, x(s))ds \quad t \geqslant 0.$$

$BC[0, \infty)$ is a Banach space. We show that $(U + V)$ maps B_r into itself and is a contraction.

$$|(Ux)(t) + (Vx)(t)| \leqslant |f(t, x(t))| + K_1 |g(t, x(t))| + K_2 |h(t, x(t))|. \quad (9.97)$$

We have the following inequalities:

$$|g(t, x(t))| \leqslant |g(t, x(t)) - g(t, 0)| + |g(t, 0)|$$

$$\leqslant \lambda |x(t, 0| + |g(t, 0)|. \quad (9.98)$$

Similarly,

$$|h(t, x(t))| \leqslant \xi |x(t)| + |h(t, 0)|. \quad (9.99)$$

From (9.97)–(9.99), we have

$$|(Ux)(t) + (Vx)(t)| \leqslant f(t, x(t))| + \lambda K_1 |x(t)| + K_1 |g(t, 0)| + K_2 |x(t)| + K_2 |h(t, 0)|.$$

Since by assumption

$$|f(t, x(t))| + K_1 |g(t, 0)| + K_2 |h(t, 0)| \leqslant (1 - \lambda K_1 - \xi K_2),$$

we have

$$|(Ux)(t) + (Vx)(t)| \leqslant r(1 - \lambda K_1 - \xi K_2) + K_1 \lambda r + K_2 \xi r = r.$$

Thus, $(Ux)(t) + (Vx)(t) \in B_r$. Also,

$$
\begin{aligned}
&|(Ux)(t) + (Vx)(t) - (Ux)(t) - (Vy)(t)| \\
&\leqslant r|x(t) - y(t)| + \lambda K_1 |x(t) - y(t)| + \xi K_2 |x(t) - y(t)| \\
&= (r + \lambda K_1 + \xi K_2)|x(t) - y(t)|.
\end{aligned}
$$

Since $(r + \lambda K_1 + \xi K_2) < 1$, we have that $U + V$ is a contraction on B_r, and therefore by Banach's contraction Theorem (Theorem 5.1), there exists a unique solution of (9.96). ∎

In Chap. 8, we have discussed the solvability of generalized Hammerstein equation

$$x + K_x N x = y, \tag{9.100}$$

in a Banach space X, where for each $x \in X$, $K_x : X \to X^*$ is a linear operator and $N : X^* \to X$ is a nonlinear operator. In this section, we consider equation of the type

$$x + K_x x = y, \tag{9.101}$$

which is a special case of (9.100) for $N = I$. The operator Eq. (9.101) includes nonlinear integral equation of form

$$x(s) + x(s) \int_0^1 K(s, t) x(t) dt = y(t). \tag{9.102}$$

Here, for each $x \in X$, the linear operator $K_x : X \to X$ is defined as

$$[K_x u](s) = u(s) \int_0^1 K(s, t) x(t) dt,$$

with suitable assumptions on the kernel $K(s, t)$. These type of equations have been considered by Stuart [579, 580] and Leggett [361, 362] and more recently by Joshi and Srikanth [294].

We shall mainly present the results of Joshi and Srikanth wherein an abstract result is obtained in operator theoretic setting, and then, it is applied to get an existence theorem for nonlinear integral equation of the form (9.102). This result is then subsequently used to get the solvability of Chandrasekhar's H-equation

$$1 + x(s) \int_0^1 \frac{s}{s+t} x(t) \Psi(t) dt = x(s). \tag{9.103}$$

Notice that equations of the type (9.103) occur in the theory of radiative that transfer in semi-infinite atmosphere (refer Chandrasekhar [139]) and hence its importance.

We now state and prove the result which uses a fixed point theorem of Himmelberg for multivalued mappings.

Theorem 9.26 *Let X be a separable reflexive Banach space and C a closed convex subset of X. Suppose that for each $x \in C$, we have a map $\varphi_x : X \to Y$, Y a normed space, such that the following hold:*

(i) *$x_n \rightharpoonup x$ in C and $u_n \rightharpoonup u$ in X implies that $\varphi_{x_n}(u_n) \rightharpoonup \varphi_x(u)$.*
(ii) *For each $x \in C$, the set $C_x = \{u \in X : \varphi_x(u) = 0\}$ is a nonempty, convex set and intersects $C \cap B_r$ (for some fixed r) in a nonempty set. Then, the equation $\varphi_x(x) = 0$ has a solution in C.*

Proof Define a map $F : C \cap B_r \to 2^{C \cap B_r}$ by

$$Fx = \{u \in C \cap B_r : \varphi_x(u) = 0\} \text{ for each } x \in X.$$

It may be observe that for each $x \in C \cap B_r$, Fx is a nonempty convex set. Further, for each x, Fx is a close set in the weak topology of X. For, if $u_n \rightharpoonup u$, with $u_n \in Fx$, then $\phi_x(u_n) \rightharpoonup \varphi(u)$ implies that $\varphi_x(u) = 0$ since each $\varphi_x(u_n) = 0$. Hence, $u \in Fx$. We now show that F is upper semicontinuous when $C \cap B_r$ is equipped with the weak topology.

To show this, we shall require a lemma of Himmelberg [273] which states as follows:

Himmelberg's Lemma. A mapping $F : K \to 2^{K_1}$, where K and K_1 are compact sets, is upper semicontinuous iff $G(F)$ is closed in $K \times K_1$.

Because $A \cap B_r$ is weakly compact in X, it suffices to show that if $[x_n, u_n] \in G(F)$ and $[x_n, u_n] \rightharpoonup [x, u]$, then $[x, u] \in G(F)$. Since $x_n \rightharpoonup x$ and $u_n \rightharpoonup u$, we have from (i) that $\varphi_{x_n}(u_n) \rightharpoonup \phi_x(u)$. Since $u_n \in Fs_n$, $\varphi_{x_n}(u_n) = 0$, and hence, $\varphi_x(u) = 0$ or $u \in Fx$. Thus, $F : C \cap B_r \rightharpoonup C \cap B_r$ is an upper semicontinuous multifunction such that Fx is weakly closed and convex for each $x \in C \cap B_r$. Since F satisfies all the conditions of Theorem 5.71, our result follows. ∎

As an application, we get the following result for nonlinear integral equation of the form (9.101):

Theorem 9.27 *Suppose that*

(i) *$K(s, t) \geqslant 0$ a.e. on $[0, 1] \times [0, 1]$*
(ii) *$ess \sup \int_0^1 K^2(s, t) dt < \infty$*
(iii) *$y(s) \geqslant 0$ a.e. on $[0, 1]$.*

Then, (9.101) has a solution $x \in L_2[0, 1]$ such that $0 \leqslant x(s) \leqslant y(s)$ a.e. on $[0, 1]$.

Proof Let $C = \{x \in L_2 : x(s) \geq 0 \text{ a.e. on } [0, 1]\}$, and let

$$[K_x u](s) = u(s) \int_0^1 K(s, t) x(t) dt.$$

Then, for each $x \in C$, K_x is a bounded linear operator on L_2 (by condition (ii)). We now prove that if $x_n \rightharpoonup x$ in C and $u_n \rightharpoonup u$ in L_2, then $K_{x_n}(u) \to K_x(u)$ in $L_2[0, 1]$. First, we note that if $x_n \rightharpoonup x$ in C, then $K_{x_n}(u) \to K_x(u)$. For

$$[Kx_n(u)](s) = u(s) \int_0^1 K(s, t)x_n(t)dt = \int_0^1 K_u(s, t)x_n(t)dt$$

where $K_u(s, t) = u(s)K(s, t)$. Since $K_u(s, t)$ is a Hilbert–Schmidt kernel, we have the corresponding integral operator completely continuous and hence $K_{x_n}(u) \to K_x(u)$. Consider $(K_{x_n}(u)n) - K_x(u), w)$ for some arbitrary $w \in L_2$.

$$(K_{x_n}(u_n) - K_x(u), w) = (K_{x_n}(u_n) - K_x(u_n) - K_x(u_n)w) + (K_x(u_n) - K_x(u), w)$$

$$= \int_0^1 w(s)u_n(s)\left(\int_0^1 K(s, t)(x_n(t) - x(t))dt\right)ds$$

$$+ \int_0^1 w(s)(u_n(s) - u(s))\left(\int_0^1 K(s, t)x(t)dt\right)ds$$

$$= \int_0^1 (u_n(s)\left(\int_0^1 K_w(s, t)(s_n(t) - s(t))dt\right)ds$$

$$+ \int_0^1 (u_n(s) - u(s))\left(\int_0^1 K_w(s, t)x(t)dt\right)ds$$

Since $\int_0^1 K_w(s, t)x_n(t)dt \to \int_0^1 K_w(s, t)x(t)dt$ and $\{u_n\}$ is a bounded sequence, we have that

$$\int_0^1 u_n(s)\left(\int_0^1 K_w(s, t)(x_n(t) - x(t))dt\right)ds \to 0.$$

Also, since

$$\int_0^1 K_w(s, t)x(t)\, dt \in L_2 \text{ and } u_n \rightharpoonup u$$

wc have

$$\int_0^1 (u_n(s) - u(s))\left(\int_0^1 K_w(s, t)x(t)dt\right)ds \to 0.$$

Hence, $(Kx_n(u_n) - K_x(u), w) \to 0$ which implies that $Kx_n(u_n) \rightharpoonup K_x(u)$ as $x_n \rightharpoonup x$ and $u_n \rightharpoonup u$.

Now set $\varphi_x(u) = K_x(u) + u - y$ for $x \in C$. For each $x \in C$, $\int_0^1 K(s, t)x(t)dt \geqslant 0$ and so

$$u(s) = \frac{y(s)}{1 + \int_0^1 K(s, t)x(t)dt}$$

is the only solution of $\varphi_x(u) = 0$ and $0 \leqslant u(s) \leqslant y(s)$. By setting $r = \|y\|$, we see that the hypothesis of Theorem 9.26 is satisfied and hence the result. ∎

As a corollary, we obtain the following solvability result for Chandrasekhar's H-equation (9.103).

Theorem 9.28 *Equation* (9.102) *has a solution in* $L_2[0, 1]$ *iff* $\int_0^1 \lambda(s)ds \leqslant \frac{1}{2}$. *Moreover, all solutions of* (9.102) *are greater than or equal to unity a.e. on* $[0, 1]$.

One can now extend the theory to the following nonlinear integral equation

$$x(s) - x(s) \int_0^1 K(s, t) f(t, x(t))dt = w(s). \tag{9.104}$$

This has earlier been considered by Leggett [361]. By application of the fixed point theorem of Krasnoselskii (Theorem 5.149) in a cone, we get the positive solutions of (9.104) for $w(s) \geqslant 0$.

9.5.2   Mathematical Modelling of Spread of Certain Infectious Diseases and Single Species Population Growth

I. Mathematical Modelling of Spread of Certain Infectious Diseases

We begin with the following definitions.

Species– A group of closely related organisms that are very similar to each other and are usually capable of interbreeding and producing fertile offspring is called species.
Population– A population is the number of all the organisms of the same group or species, which live in a particular geographical area and have the capability of interbreeding.
Infective– A member of the host population is classified as infective if the member has been infected and is infectious.

Leggett and Williams [363] applied Theorem 5.158 to the nonlinear integral equation

$$x(t) = Ax(t) = \int_{t-\tau}^t f(s, x(s))ds \tag{9.105}$$

to prove the existence of nonzero solution.

This equation can be interpreted as a model for the spread of certain infectious diseases whose periodic contract rate varies seasonally. Here,

- $x(t)$ represents the proportion of infectives in the population at time t,

- $f(t, x(t))$ is the proportion of infectives per unit time ($f(t, 0) = 0$), and
- τ is the length of time an individual remains infectious.

Let $\mathbb{R} = (-\infty, \infty)$, $\mathbb{R}^+ = [0, \infty)$, τ and w be positive constants, and let the following conditions concerning f and a be satisfied:

(1) The function $f(t, x)$ is continuous from $\mathbb{R} \times \mathbb{R}^+$ into $\mathbb{R}^+$.
(2) For each $t \in \mathbb{R}$ and $x \geq 0$, $f(t, x) = f(t + w, x)$ and $f(t, 0) = 0$.
(3) There exists $R > 0$ such that $f(t, x) \leqslant \frac{R}{\tau}$ for all $(t, x) \in [0, w] \times [0, R]$.
(4) For each t, $a(t) = \lim_{x \to 0} \frac{f(t, x)}{x}$, and for each $k \in (0, 1)$, there exists $\varepsilon_k > 0$ such that $f(t, x) \geqslant ka(t)x, t \in \mathbb{R}, 0 \leqslant x \leqslant \varepsilon_k$,

Theorem 9.29 *Suppose conditions* (1)–(4) *hold good. Let N be the smallest integer such that $\frac{w}{N} \leqslant \frac{\tau}{2}$. For $j = 0, 1, \ldots, N$, set*

$$I_j = \left[\frac{j-1}{N} w, \frac{j}{N} w \right].$$

If $\prod_{j=1}^{N} \left[\int_{I_j} a(s)ds \right] > 1$, then Eq. (9.105) has a nonzero solution.

Proof Let X be the Banach space of continuous real-valued functions on $\mathbb{R}$ with supremum norm which are w-periodic. Let K be the cone of nonnegative functions in X. Let A be the operator from K into itself defined by

$$Ax(t) = \int_{t-\tau}^{t} f(x, x(s))ds.$$

By Arzela–Ascoli theorem, A is completely continuous. For $x \in K_R$

$$Ax(t) \leqslant \int_{t-\tau}^{t} \frac{R}{\tau} ds = R.$$

This shows that A maps K_R into itself. Let $u(t) = 1$, and choose $k < 1$ such that

$$k^N \left(\prod_{j=1}^{N} \int_{I_j} a(s)ds \right) > 1.$$

Choose $r < R$ sufficiently small such that $f(t, x) \geq ka(t)x, t \in \mathbb{R}, 0 \leqslant x \leqslant r$. Assume that for some $x \in K(u)$ with $\|x\| = r$, $Ax \leqslant x$. For $1 \leqslant j \leqslant N$ and $t \in I_j$, I_{j_1} is a subset of $[t - \tau, t]$. Hence,

$$\int_{I_j} a(t)x(t) \geq \int_{I_j} a(t)Ax(t)dt \geq \int_{I_j} a(t)\left(\int_{I_{j-1}} f(s,x(s))ds\right)dt$$

$$> k\left(\int_{I_j} a(t)dt\right)\left(\int_{I_{j-1}} a(s)x(s)ds\right).$$

Thus,

$$\int_{I_N} a(t)x(t)dt \geq k^N\left(\prod_{j=1}^N \int_{I_j} a(s)ds\right)\left(\int_{I_0} a(t)x(t)dt\right)$$

$$= k^N\left(\prod_{j=1}^N \int_{I_j} a(s)ds\right)\left(\int_{I_N} a(t)x(t)dt\right).$$

We arrive at a contradiction since

$$k^N\prod_{j=1}^N \int_{I_j} a(s)ds > 1 \text{ and } \int_{I_N} a(t)x(t)dt \neq 0.$$

Therefore, $Ax \leqslant x$. Since all the conditions of Theorem 5.158 are satisfied, A has a nonzero fixed point $y \in K_R$. ∎

II. Mathematical Modelling of Single Species Population Growth

We now discuss below two applications given by Gatica and Smith involving the existence of nonzero solutions of integral equations. But first, we have the following definition.

Species– Species is a set of animals or plants in which the members have similar characteristics to each other and can breed with each other.

Notice that a population becomes a species when it accumulates so many genetic differences that it becomes reproductively isolated from its sister populations.

The following nonlinear integral equation modelling a single species population growth was proposed by Swick [582]

$$x(t) = \int_0^L K(L-s)h(t, x(t-L-\xi+s))ds. \tag{9.106}$$

The assumptions made on functions appearing in (9.106) are the following:

(1) $L > 0, L \geqslant \zeta \geqslant 0$. In Swick's model, L is the maximum lifetime of an individual and ζ is the delay between conception and birth.
(2) $h : \mathbb{R} \times \mathbb{R}^+$ is continuous, $h(t, 0) = 0$, and $h(t, x) > 0$ for $x > 0$.
(3) $K : [0, L] \to (0, \infty)$ is continuous.
(4) $h(t + p, x) = h(t, x)$ for all $t, x \in \mathbb{R} \times \mathbb{R}^+$, for some positive number p.

(5) There exists a positive number r such that $h(t, x) \geq \frac{x}{k_0}$ for $0 \leq x \leq r$ and $0 \leq$
$t \leq p$ where $k_0 = \int_0^L K(u)du$.

(6) Either (a) there exists $R > 0$ such that

$$h(t, x) \leq \frac{R}{k_0} \text{ for } (t, x) \in [0, p] \times [0, R] \text{ or}$$

(b) $\lim\limits_{x \to \infty} \frac{h(t,x)}{x} = 0$ uniformly in t.

Theorem 9.30 *Suppose conditions (1)–(6) hold, then the nonlinear integral equation (9.106) has a positive p-periodic solution defined on* $\mathbb{R}$.

Proof Let X be the Banach space of p-periodic continuous function with supremum norm: $\|x\| = \sup\limits_{0 \leq t \leq p} |x(t)|$. Let K denote the cone of nonnegative functions in X. Let the mapping F on K be defined by

$$(Fx)(t) = \int_0^L K(L - s)h(t, x(t - L - \zeta + s))ds.$$

The mapping F maps K into itself and is continuous and maps bounded sets into precompact set. It is shown that conditions (i) and (ii) of Theorem 5.164 are satisfied and hence the assertion of the theorem follows by Theorem 5.164. We show further that the solution is positive.

Let r be the positive number given in (5), and let

$$B : \{x \in K : \|x\| = r\} \to K$$

be given by $(Bx)(t) = 1$ for all $t \in \mathbb{R}$. Suppose y is a solution of $y = Fy + \lambda By$, $0 < \lambda < \infty$ with $\|y\| = r$. Then, by (5), we have

$$y(t) \geq \int_0^L K(L - s)(y(t - L - \zeta + s))/k_0 ds + \lambda \geq \inf_{t \in R} y(t) + \lambda.$$

This is a contradiction as $\lambda > 0$. So $y = Fy + \lambda By$ has no solution in B. This implies that conditions (i) of Theorem 5.164 is satisfied.

For condition (ii) of Theorem 5.164, let R be as given in (6)(a). Suppose $z \in K$ with $\|z\| = R$ and z satisfies $Fz = \lambda z$ for some positive λ, then by (6)(a) we have

$$\lambda z(t) = \int_0^L K(L - s)h(t, x(t - L - \zeta + s))ds \int_0^L K(L - s)\frac{R}{k_0}ds \leq R.$$

For some $t \in R$, $z(t) = R$, so $\lambda \leq 1$, thereby proving condition *(ii)* of Theorem 5.164.

To show that $x(t) > 0$ for $t \in \mathbb{R}$, assume that $x(t_0) = 0$ for some $t_0 \in \mathbb{R}$. Then,

$$\int_0^i K(L - s)h(t_0, x(t_0 - L - \zeta + s))ds = 0.$$

By (2) and (3), $x(t) = 0$ for $t \in [t_0 - L - \zeta, t_0 - \zeta]$. As $L \geq \zeta, t_0 - L \in [t_0 - L - \zeta, t_0 - \zeta]$, we have $x(t_0 - L) = 0$. This implies that $x(t) = 0$ for $t \in [t_0 - 2L - \zeta, t_0 - L - \zeta]$ as before. Continuing this way, it can be shown that $x(t) = 0$ for $t \leqslant t_0 - \zeta$. Then, by periodicity we get $x(t) = 0$, a contradiction.   ∎

9.5.3 Local Attractivity Results for Generalized Nonlinear Functional-Integral Equations

Assume that $E = BC(\mathbb{R}^+, \mathbb{R})$, and let Ω be a subset of E. Let $\mathscr{Q} : E \to E$ be an operator, and consider the operator equation in E,

$$\mathscr{Q}x(t) = x(t) \text{ for all } t \in \mathbb{R}^+. \tag{9.107}$$

In this section, we consider the following generalized nonlinear functional-integral equation (in short GNFIE)

$$x(t) = F\left(t, x(\theta(t)), u(t, x(\alpha(t))), \int_0^{\beta(t)} f(t, s, x(\gamma(s)))ds, \int_0^{\sigma(t)} g(t, s, x(\eta(s)))ds\right),$$
$$\tag{9.108}$$

for $t \in \mathbb{R}^+$, where $u : \mathbb{R}^+ \times \mathbb{R} \to \mathbb{R}$, $f, g : \mathbb{R}^+ \times \mathbb{R}^+ \times \mathbb{R} \to \mathbb{R}$, $F : \mathbb{R}^+ \times \mathbb{R} \times \mathbb{R} \times \mathbb{R} \times \mathbb{R} \to \mathbb{R}$ and $\alpha, \beta, \gamma, \theta, \sigma, \eta : \mathbb{R}^+ \to \mathbb{R}^+$ are continuous functions.

Definition 9.6 We say that solutions of the Eq. (9.107) are locally attractive if there exists a closed ball $B[x_0, r_0]$ in the space $BC(\mathbb{R}^+, \mathbb{R})$ for some $x_0 \in BC(\mathbb{R}^+, \mathbb{R})$ such that for arbitrary solutions $x = x(t)$ and $y = y(t)$ of Eq. (9.107) belonging to $B[x_0, r_0] \cap \Omega$, we have

$$\lim_{t \to \infty} (x(t) - y(t)) = 0. \tag{9.109}$$

The functional-integral equation (9.108) is "general" in the sense that it includes several classes of known integral equations discussed in the literature. See Dhage and Lakshmikantham [182], Dhage et al. [179], Dhage and Ntouyas [183], Krasnoselskii [345], Väth [602], Dhage [174, 175] and the references therein. We now intend to obtain the solutions of GNFIE (9.105) in the space $BC(\mathbb{R}^+, \mathbb{R})$ of all bounded and continuous realvalued functions on $\mathbb{R}^+$.

Let $X = BC(\mathbb{R}^+, \mathbb{R})$ be the space of all continuous and bounded functions on $\mathbb{R}^+$ and define a norm $\| \cdot \|$ in X by

$$\|x\| = \sup\{|x(t)| : t \geq 0\}.$$

Clearly, X is a Banach space with this supremum norm. Let us fix a bounded subset A of X and a positive real number L. For any $x \in A$ and $\varepsilon \geq 0$, denote by $\omega^L(x, \varepsilon)$, the modulus of continuity of x on the interval $[0, L]$ defined by

$$\omega^L(x, \varepsilon) = \sup\{|x(t) - x(s)| : t, s \in [0, L], |t - s| \leqslant \varepsilon\}.$$

Moreover, let

$$\omega^L(A, \varepsilon) = \sup\{\omega^L(x, \varepsilon) : x \in A\},$$
$$\omega_0^L(A) = \lim_{\varepsilon \longrightarrow 0} \omega^L(A, \varepsilon),$$
$$\omega_0(A) = \lim_{L \longrightarrow \infty} \omega_0^L(A).$$

By $A(t)$, we mean a set in $\mathbb{R}$ defined by $A(t) = \{x(t) | x \in A\}$ for $t \in \mathbb{R}^+$. We denote diam $(A(t)) = \sup\{|x(t) - y(t)| : x, y \in A\}$. Finally, we define a function μ on $\mathcal{P}_{bd}(X)$ by the formula

$$\mu(A) = \omega_0(A) + \limsup_{t \longrightarrow \infty} \mathrm{diam}(A(t)). \tag{9.110}$$

It has been shown in [35] that μ is a sublinear measure of noncompactness in X.

Let $X = BC(\mathbb{R}^+, \mathbb{R})$ be the space of all continuous and bounded functions on $\mathbb{R}^+$ and define a norm $\|.\|$ in X by

$$\|x\| = \sup\{|x(t)| : t \geq 0\}.$$

Clearly X is a Banach space with this norm. We define a measure of non compactness in X as follows. Let us fix a bounded subset A of X and a positive real number L. For any $x \in A$ and $\varepsilon \geq 0$, denote by $\omega^L(x, \varepsilon)$, the modulus of continuity of x on the interval $[0, L]$ defined by

$$\omega^L(x, \varepsilon) = \sup\{|x(t) - x(s)| : t, s \in [0, L], |t - s| \leqslant \varepsilon\}.$$

Moreover, let

$$\omega^L(A, \varepsilon) = \sup\{\omega^T(x, \varepsilon) : x \in A\},$$
$$\omega_0^L(A) = \lim_{\varepsilon \longrightarrow 0} \omega^L(A, \varepsilon),$$
$$\omega_0(A) = \lim_{L \longrightarrow \infty} \omega_0^L(A).$$

By $A(t)$ we mean a set in $\mathbb{R}$ defined by $A(t) = \{x(t) | x \in A\}$ for $t \in \mathbb{R}^+$. We denote diam $(A(t)) = \sup\{|x(t) - y(t)| : x, y \in A\}$. Let us denote the set of all bounded subsets of X by $B(X)$. Finally we define a function μ on $B(X)$ by the formula

$$\mu(A) = \omega_0(A) + \limsup_{t \longrightarrow \infty} \mathrm{diam}(A(t)) \tag{9.111}$$

We consider with the following set of assumptions in what follows.

(H_0) The functions $\alpha, \beta, \gamma, \theta, \sigma, \eta : \mathbb{R}^+ \longrightarrow \mathbb{R}^+$ are continuous functions and $\theta(t) \geq t$ and $\alpha(t) \geq t$ for all $t \in \mathbb{R}^+$.

(H_1) There exists a function $\mathcal{D}$-function φ and the constant $L_i, i = 1, 2, 3$ such that

$$|F(t, x, x_1, x_2, x_3) - F(t, y, y_1, y_2, y_3)| \leqslant \varphi(|x - y|) + \sum_{i=1}^{3} L_i |x - i - y_i|$$

for all $t \in \mathbb{R}^+$ and $x, y, x_i, y_i \in \mathbb{R}, i = 1, 2, 3$. Moreover, the map $t \to F(t, 0, 0, 0, 0)$ is bounded with $F_0 = \sup_{t \geq 0} |F(t, 0, 0, 0, 0)|$.

(H_2) There exists a function $\mathcal{D}$-function φ_1 and a continuous function $k_1 \in BC(\mathbb{R}^+, \mathbb{R}^+)$ such that

$$|u(t, x) - u(t, y)| \leqslant k_1(t)\varphi_1(|x - y|)$$

for all $t \in \mathbb{R}^+$. Moreover, $\sup_{t \geq 0} k_1(t) = K_1$.

(H_3) The function $t \to u(t, 0) = u_0(t)$ is bounded, and $C_0 = \sup_{t \geq 0} |u(t, 0)|$.

(H_4) The function $f : \mathbb{R}^+ \times \mathbb{R}^+ \times \mathbb{R} \longrightarrow \mathbb{R}$ is continuous, and there exist a continuous function $q : \mathbb{R}^+ \times \mathbb{R}^+ \longrightarrow \mathbb{R}^+$ and a $\mathcal{D}$-function φ_2 as defined in (H_1) such that

$$|f(t, s, x) - f(t, s, y)| \leqslant q(t, s))\varphi_2(|x - y|)$$

for any $t, s \in \mathbb{R}^+, x, y \in \mathbb{R}$. Moreover, $\lim_{t \to \infty} \int_0^{\beta(t)} q(t, s)\, ds = 0$.

(H_5) The function $f_0 : \mathbb{R}^+ \longrightarrow \mathbb{R}^+$ defined by $f_0(t) = \int_0^{\beta(t)} |f(t, s, 0)| ds$ is bounded with $C_1 = \sup_{t \geq 0} f_0(t)$.

(H_6) $g : \mathbb{R}^+ \times \mathbb{R}^+ \times \mathbb{R} \to \mathbb{R}$ is a continuous function, and there exist continuous functions $a, b : \mathbb{R}^+ \to \mathbb{R}^+$ such that

$$|g(t, s, x)| \leqslant a(t)b(s)$$

for $t, s \in \mathbb{R}^+$ such that $s \leqslant t$, and $x \in \mathbb{R}$. Moreover, $\lim_{t \to \infty} a(t) \int_0^{\sigma(t)} b(s) ds = 0$.

Remark 9.7 Because the hypotheses (H_4) and (H_6) are held, we have that the functions

$$k_2(t) = \int_0^{\beta(t)} q(t, s) ds \quad \text{and} \quad v(t) = a(t) \int_0^{\sigma(t)} b(s) ds$$

are bounded on $\mathbb{R}^+$ and the positive numbers

$$K_2 = \sup_{t \geq 0} \int_0^{\beta(t)} q(t, s)ds \text{ and } V = \sup_{t \geq 0} v(t)$$

exist.

Theorem 9.31 *Assume that the hypotheses* (H_0) *through* (H_6) *hold. Suppose that*

$$\varphi(r) + L_1 K_1 \varphi_1(r) < r, r > 0, \tag{9.112}$$

and there exists a positive solution r_0 *of the inequality*

$$\varphi(r) + L_1 K_1 \varphi_1(r) + L_2 K_2 \varphi_2(r) + q \leqslant r, \tag{9.113}$$

where q *is the constant defined by the equality*

$$q = \{L_1 C_0 + L_2 C_1 + L_3 V + F_0\}.$$

Then, the functional nonlinear integral equation (9.108) *has a solution, and the solutions are uniformly locally attractive on* $\mathbb{R}^+$.

Proof Consider the operator T defined on the space $BC(\mathbb{R}^+, \mathbb{R})$ by the formula

$$(Tx)(t) == F\left(t, x(\theta(t)), u(t, x(\alpha(t))), \int_0^{\beta(t)} f(t, s, x(\gamma(s)))ds, \int_0^{\sigma(t)} g(t, s, x(\eta(s)))ds\right),$$

for $t \in \mathbb{R}^+$.

We shall show that the map T satisfies all the conditions of Theorem 9.31 on X.
Step I: First, we show that T defines a mapping $T : X \longrightarrow X$. Since all the functions involved in (9.108) are continuous, Tx is continuous on $\mathbb{R}^+$ for each $x \in X$. Hence, Tx is mapping from X into itself. As $\theta(\mathbb{R}^+) \subseteq \mathbb{R}^+$, we have $\max_{t \geq 0} |x(\theta(t))| \leqslant \max_{t \geq 0} |x(t)|$. On the other hand, hypotheses (H_0)–(H_5) imply that

$$|Tx(t)|$$

$$= \left| F\left(t, x(\theta(t)), u(t, x(\alpha(t))), \int_0^{\beta(t)} f(t, s, x(\gamma(s)))ds, \int_0^{\sigma(t)} g(t, s, x(\eta(s)))ds\right) \right|$$

$$\leqslant \left| F\left(t, x(\theta(t)), u(t, x(\alpha(t))), \int_0^{\beta(t)} f(t, s, x(\gamma(s)))ds, \int_0^{\sigma(t)} g(t, s, x(\eta(s)))ds\right) \right.$$

$$\left. - F(t, 0, 0, 0, 0) \right| + |F(t, 0, 0, 0, 0)|$$

$$\leqslant \varphi(|x(\theta(t))|) + L_1|u(t, x(\alpha(t)))| + L_2 \left| \int_0^{\beta(t)} f(t, s, x(\gamma(s)))ds \right|$$

$$+ L_3 \left| \int_0^{\sigma(t)} g(t, s, x(\eta(s)))ds \right| + |F(t, 0, 0, 0, 0)|$$

$$\leqslant \varphi(|x(\theta(t))|) + L_1|u(t, x(\alpha(t))) - u(t, 0)| + L_1|u(t, 0)|$$

$$+ L_2 \int_0^{\beta(t)} |f(t, s, x(\gamma(s))) - f(t, s, 0)| ds + L_2 \int_0^{\beta(t)} |f(t, s, 0)| ds$$

$$+ L_3 \left| \int_0^{\sigma(t)} g(t, s, x(\eta(s))) ds \right| + F_0$$

$$\leqslant \varphi(|x(\theta(t))|) + L_1 k_1(t) \varphi_1(|x(\alpha(t))|) + L_1 |u(t, 0)|$$

$$+ L_2 \int_0^{\beta(t)} q(t, s) \varphi_2(|x(\gamma(s))|) ds + L_2 C_1 + L_3 a(t) \int_0^{\sigma(t)} b(s) ds + F_0$$

$$\leqslant \varphi(\|x\|) + L_1 k_1(t) \varphi_1(\|x\|) + L_1 C_0 + L_2 \int_0^{\beta(t)} q(t, s) \varphi_2(\|x\|) ds + L_2 C_1 + L_3 v(t) + F_0$$

$$\varphi(\|x\|) + L_1 K_1 \varphi_1(\|x\|) + L_2 K_2 \varphi_2(\|x\|) + L_1 C_0 + L_2 C_1 + L_3 V + F_0 \tag{9.114}$$

for all $t \in \mathbb{R}^+$. From (9.114), we deduce that $Tx \in X$.

Step II: From (9.114), it follows that

$$\|Tx\| \leqslant \varphi(r) + L_1 K_1 \varphi_1(r) + + L_2 K_2 \varphi_2(r) + q \leqslant r. \tag{9.115}$$

Now consider the closed ball $B[0, r_0] \subset C[0, L]$ in X centered at origin of radius r_0. Then, T defines a mapping $T : B[0, r_0] \to B[0, r_0]$. We show that T is continuous on $B[0, r_0]$. Let $\varepsilon > 0$ be given, and let $x, y \in B[0, r_0]$ be such that $\|x - y\| \leqslant \varepsilon$. Then, by hypotheses (H_0)–(H_5)

$$|Tx(t) - Ty(t)|$$

$$\leqslant \left| F\left(t, x(\theta(t)), u(t, x(\alpha(t))), \int_0^{\beta(t)} f(t, s, x(\gamma(s))) ds, \int_0^{\sigma(t)} g(t, s, x(\eta(s))) ds \right) \right.$$

$$\left. - F\left(t, y(\theta(t)), u(t, y(\alpha(t))), \int_0^{\beta(t)} f(t, s, y(\gamma(s))) ds, \int_0^{\sigma(t)} g(t, s, y(\eta(s))) ds \right) \right|$$

$$\leqslant \varphi(|x(\theta(t)) - y(\theta(t))|) + L_1 k_1(t) \varphi_1(|x(\alpha(t)) - y(\alpha(t))|)$$

$$+ L_2 \left| \int_0^{\beta(t)} f(t, s, x(\gamma(s))) ds - \int_0^{\beta(t)} f(t, s, y(\gamma(s))) ds \right|$$

$$+ L_3 \left| \int_0^{\sigma(t)} g(t, s, x(\eta(s))) ds - \int_0^{\sigma(t)} g(t, s, y(\eta(s))) ds \right|$$

$$\leqslant \varphi(|x(\theta(t)) - y(\theta(t))|) + L_1 k_1(t) \varphi_1(|x(\alpha(t)) - y(\alpha(t))|)$$

$$+ L_2 \int_0^{\beta(t)} |f(t, s, x(\gamma(s))) - f(t, s, y(\gamma(s)))| ds$$

$$+ L_3 \int_0^{\sigma(t)} |g(t, s, x(\eta(s))) - g(t, s, y(\eta(s)))| ds$$

$$\leqslant \varphi(|x(\theta(t)) - y(\theta(t))|) + L_1 k_1(t) \varphi_1(|x(\alpha(t)) - y(\alpha(t))|)$$

$$+ L_2 \int_0^{\beta(t)} q(t, s) \varphi_2(|x(\gamma(s)) - y(\gamma(s))|) ds + L_3 a(t) \int_0^{\sigma(t)} b(s) ds$$

$$\leqslant \varphi(\|x - y\|) + L_1 K_1 \varphi_1(\|x - y\|) + L_2 K_2 \varphi_2(\|x - y\|) + L_3 v(t)$$

$$\leqslant \varphi(\varepsilon) + L_1 K_1 \varphi_1(\varepsilon) + L_2 K_2 \varphi_2(\varepsilon) + L_3 v(t)$$

$$\leqslant (1 + L_1 K_1 + L_2 K_2) \varepsilon + L_3 v(t). \tag{9.116}$$

Since $v(t) \to 0$ as $t \to \infty$, there exists $L > 0$ such that $v(t) \leqslant \varepsilon$, $\forall t > L$. Thus, if $t > L$, then from (9.116) we have that

$$|Tx(t) - Ty(t)| \leqslant (1 + L_1 K_1 + L_2 K_2 + L_3)\varepsilon. \tag{9.117}$$

If $t < L$, then define a function $\omega = \omega(\varepsilon)$ by the formula

$$\omega(\varepsilon) = \sup\{|g(t, s, x) - g(t, s, y)| : t, s \in [0, L], x, y \in [-r_0, r_0], |x - y| \leqslant \varepsilon\}. \tag{9.118}$$

Now $g(t, s, x)$ is continuous and hence uniformly continuous on $[0, L] \times [0, L] \times [-r_0, r_0]$. As a result, we have $\omega(\varepsilon) \to 0$ as $\varepsilon \to 0$. Therefore, from (9.116),

$$|Tx(t) - Ty(t)| \leqslant (1 + L_1 K_1 + L_2 K_2)\varepsilon + L_3 \omega(\varepsilon)$$

for all $t \in \mathbb{R}^+$. Hence, it follows that

$$\|Tx - Ty\| \leqslant \max\{(1 + L_1 K_1 + L_2 K_2 + L_3)\varepsilon, (1 + L_1 K_1 + L_2 K_2)\varepsilon + L_3 \omega(\varepsilon)\}$$
$$\to 0 \text{ as } \varepsilon \to 0.$$

Hence, T is a continuous mapping from $B[0, r_0]$ into itself.

Step III: Here, we show that T is a nonlinear set contraction on $B[0, r_0]$ in the sense of inequality (9.116). This will be done in the following two cases:

Case I: Let $A \subset B[0, r_0]$ be nonempty. Further, fix the number $L > 0$ and $\varepsilon > 0$. Since the functions f and g are continuous on compact $[0, L] \times [0, L] \times [-r_0, r_0]$, there are constants $D_2 > 0$ and $D_3 > 0$ such that $|f(t, s, x)| \leqslant D_2$ and $|g(t, s, x)| \leqslant D_3$ for all $t, s \in [0, L]$ and $x \in [-r_0, r_0]$. Then, choosing $t, \tau \in [0, L]$ such that $|t - \tau| \leqslant \varepsilon$ and taking into account our hypotheses, we obtain

$$|Tx(t) - Tx(\tau)|$$
$$\leqslant \left| F\left(t, x(\theta(t)), u(t, x(\alpha(t))), \int_0^{\beta(t)} f(t, s, x(\gamma(s)))ds, \int_0^{\sigma(t)} g(t, s, x(\eta(s)))ds\right) \right.$$
$$\left. - F\left(\tau, x(\theta(\tau)), u(\tau, x(\alpha(\tau))), \int_0^{\beta(\tau)} f(\tau, s, y(\gamma(s)))ds, \int_0^{\sigma(\tau)} g(\tau, s, y(\eta(s)))ds\right) \right|$$
$$\leqslant \varphi(|x(\theta(t)) - x(\theta(\tau))|) + L_1|u(t, x(\alpha(t))) - u(t, x(\alpha(\tau)))|$$
$$+ L_2\left| \int_0^{\beta(t)} f(t, s, x(\gamma(s)))ds - \int_0^{\beta(\tau)} f(\tau, s, x(\gamma(s)))ds \right|$$
$$+ L_3\left| \int_0^{\sigma(t)} g(t, s, x(\eta(s)))ds - \int_0^{\sigma(\tau)} g(\tau, s, x(\eta(s)))ds \right|$$
$$\leqslant \varphi(|x(\theta(t)) - x(\theta(\tau))|) + L_1|u(t, x(\alpha(t))) - u(t, x(\alpha(\tau)))|$$
$$+ L_2\left| \int_0^{\beta(t)} f(t, s, x(\gamma(s)))ds - \int_0^{\beta(t)} f(\tau, s, x(\gamma(s)))ds \right|$$
$$+ L_2\left| \int_0^{\beta(t)} f(\tau, s, x(\gamma(s)))ds - \int_0^{\beta(\tau)} f(\tau, s, x(\gamma(s)))ds \right|$$

$$+ L_3 \left| \int_0^{\sigma(t)} g(t, s, x(\eta(s))) ds - \int_0^{\sigma(t)} g(\tau, s, x(\eta(s))) ds \right|$$

$$+ L_3 \left| \int_0^{\sigma(t)} g(\tau, s, x(\eta(s))) ds - \int_0^{\sigma(\tau)} g(\tau, s, x(\eta(s))) ds \right|$$

$$\leqslant \varphi(|x(\theta(t)) - x(\theta(\tau))|) + L_1 |u(t, x(\alpha(t))) - u(t, x(\alpha(\tau)))|$$

$$+ L_2 \int_0^{\beta(t)} |f(t, s, x(\gamma(s))) - f(\tau, s, x(\gamma(s)))| ds + L_2 \left| \int_{\beta(\tau)}^{\beta(t)} f(\tau, s, x(\gamma(s))) ds \right|$$

$$+ L_3 \int_0^{\sigma(t)} |g(t, s, x(\eta(s))) - g(\tau, s, x(\eta(s)))| ds + L_3 \left| \int_{\sigma(\tau)}^{\sigma(t)} g(\tau, s, x(\eta(s))) ds \right|$$

$$\leqslant \varphi(|x(\theta(t)) - x(\theta(\tau))|) + L_1 \omega^L(u, \varepsilon) + L_2 \beta_L \omega^L(f, \varepsilon) + L_2 D_2 \omega^L(\beta, \varepsilon)$$

$$+ L_3 \sigma_L \omega^L(g, \varepsilon) + L_3 D_3 \omega^L(\sigma, \varepsilon),$$

where

$$\beta_L = \sup\{\beta(t) : t \in [0, L]\},$$
$$\sigma_L = \sup\{\sigma(t) : t \in [0, L]\},$$
$$\omega^L(\beta, \varepsilon) = \sup\{|\beta(t) - \beta(\tau)| : t, \tau \in [0, L], |t - \tau| \leqslant \varepsilon\},$$
$$\omega^L(\sigma, \varepsilon) = \sup\{|\sigma(t) - \sigma(\tau)| : t, \tau \in [0, L], |t - \tau| \leqslant \varepsilon\},$$

and ʹ

$$\omega^L(u, \varepsilon) = \sup\{|u(t, x) - u(\tau, x)| : t, \tau \in [0, L], |t - \tau| \leqslant \varepsilon, |x| \leqslant r_0\},$$
$$\omega^L(f, \varepsilon) = \sup\{|f(t, s, x) - f(\tau, s, x)| : t, \tau \in [0, L], |t - \tau| \leqslant \varepsilon, |x| \leqslant r_0\},$$
$$\omega^L(g, \varepsilon) = \sup\{|g(t, s, x) - g(\tau, s, x)| : t, \tau \in [0, L], |t - \tau| \leqslant \varepsilon, |x| \leqslant r_0\}.$$

The above inequality further implies that

$$\omega^L(Tx, \varepsilon) \leqslant \varphi(\omega^L(x, \varepsilon)) + L_1 \omega^L(u, \varepsilon) + L_2 \beta_L \omega^L(f, \varepsilon) + L_2 D_2 \omega^L(\beta, \varepsilon)$$
$$+ L_3 \sigma_L \omega^L(g, \varepsilon) + L_3 D_3 \omega^L(\sigma, \varepsilon). \tag{9.119}$$

Since by hypotheses, the functions β, σ, φ, u and f, g are continuous, respectively, on $[0, L]$, $[0, L] \times [-r_0, r_0]$ and $[0, L] \times [0, L] \times [-r_0, r_0]$, we infer that they are uniformly continuous there. Hence, we deduce that $\varphi(\omega^L(x, \varepsilon)) \to 0$, $\omega^L(u, \varepsilon) \to 0$, $\omega^L(\beta, \varepsilon) \to 0$, $\omega^L(\sigma, \varepsilon) \to 0$, $\omega^L(f, \varepsilon) \to 0$, $\omega^L(g, \varepsilon) \to 0$ as $\varepsilon \to 0$. Hence, from the above estimate (9.119), we obtain

$$\omega_0^L(T(A)) = 0,$$

and consequently,

$$\omega_0(T(A)) = 0. \tag{9.120}$$

Case II: Now for any $x, y \in A$, one has

$$|Tx(t) - Ty(t)|$$

$$\leqslant \left| F\left(t, x(\theta(t)), u(t, x(\alpha(t))), \int_0^{\beta(t)} f(t, s, x(\gamma(s)))ds, \int_0^{\sigma(t)} g(t, s, x(\eta(s)))ds\right) \right.$$

$$\left. - F\left(t, y(\theta(t)), u(t, y(\alpha(t))), \int_0^{\beta(t)} f(t, s, y(\gamma(s)))ds, \int_0^{\sigma(t)} g(t, s, y(\eta(s)))ds\right) \right|$$

$$\leqslant \varphi(|x(\theta(t)) - y(\theta(t))|) + L_1 k_1(t)\varphi_1(|x(\alpha(t)) - y(\alpha(t))|)$$

$$+ L_2 \left| \int_0^{\beta(t)} f(t, s, x(\gamma(s)))ds - \int_0^{\beta(t)} f(t, s, y(\gamma(s)))ds \right|$$

$$+ L_3 \left| \int_0^{\sigma(t)} g(t, s, x(\eta(s)))ds - \int_0^{\sigma(t)} g(t, s, y(\eta(s)))ds \right|$$

$$\leqslant \varphi(|x(\theta(t)) - y(\theta(t))|) + L_1 k_1(t)\varphi_1(|x(\alpha(t)) - y(\alpha(t))|)$$

$$+ L_2 \int_0^{\beta(t)} |f(t, s, x(\gamma(s))) - f(t, s, y(\gamma(s)))|ds$$

$$+ L_3 \int_0^{\sigma(t)} |g(t, s, x(\eta(s))) - g(t, s, y(\eta(s)))|ds$$

$$\leqslant \varphi(|x(\theta(t)) - y(\theta(t))|) + L_1 k_1(t)\varphi_1(|x(\alpha(t)) - y(\alpha(t))|)$$

$$+ L_2 \int_0^{\beta(t)} q(t, s)\varphi_2(|x(\gamma(s)) - y(\gamma(s))|)ds + 2L_3 a(t) \int_0^{\sigma(t)} b(s)ds$$

$$\leqslant \varphi(\text{diam}(A(\theta(t)))) + L_1 k_1(t)\varphi_1(\text{diam}A(\alpha(t)))$$

$$+ L_2 \int_0^{\beta(t)} q(t, s)\varphi_2(\text{diam}(A(\gamma(\beta(s)))))ds + 2L_3 v(t)$$

$$\leqslant \varphi(\text{diam}(A(\theta(t)))) + L_1 k_1(t)\varphi_1(\text{diam}A(\alpha(t)))$$

$$+ L_2 \int_0^{\beta(t)} q(t, s)\varphi_2(\text{diam}(A))ds + 2L_3 v(t)$$

for all $t \in \mathbb{R}^+$. Further, we also notice that $A \subset B[0, r_0]$ implies $\text{diam}(A) \leqslant 2r_0$. Again, since $\theta(t) \geq t$ and $\alpha(t) \geq t$, we have that $\text{diam}(A(\theta(t)) \leqslant \text{diam}(A(t))$ for all $t \in \mathbb{R}^+$. Therefore, as a result of above inequality, we obtain

$$\text{diam}(T(A(t))) \leqslant \varphi(\text{diam}(A(t)) + L_1 k_1(t)\varphi_1(\text{diam}A((t)))$$

$$+ L_2 \int_0^{\beta(t)} q(t, s)\varphi_2(2r_0)ds + 2L_3 v(t) \qquad (9.121)$$

for all $t \in \mathbb{R}^+$. Taking the limit superior in the above inequality yields

$$\limsup_{t\to\infty} \operatorname{diam}(T(A(t))) \leqslant \limsup_{t\to\infty} \varphi(\operatorname{diam}(A(t))) + L_1 K_1 \limsup_{t\to\infty} \varphi_1(\operatorname{diam}(A(t)))$$

$$+ L_2 \varphi_2(2r_0) \limsup_{t\to\infty} \int_0^{\beta(t)} q(t,s)ds + 2L_3 \limsup_{t\to\infty} \upsilon(t)$$

$$\leqslant \limsup_{t\to\infty} \varphi(\operatorname{diam}(A(t))) + L_1 K_1 \limsup_{t\to\infty} \varphi_1(\operatorname{diam}(A(t)))$$

$$\leqslant \varphi(\limsup_{t\to\infty} \operatorname{diam}(A(t))) + L_1 K_1 \varphi_1(\operatorname{diam}(A(t)))$$

$$\leqslant \psi(\limsup_{t\to\infty} \operatorname{diam}(A(t))) \tag{9.122}$$

where ψ is again a $\mathcal{D}$-function in view of Remark 6.5 defined by $\psi(r) = \varphi(r) + L_1 K_1 \varphi_1(r)$ and $\psi(r) < r$ for $r > 0$.

Now from inequality (9.120) and (9.122), it follows that

$$\mu(T(A)) = \omega_0(T(A)) + \limsup_{t\to\infty} \operatorname{diam}(T(A(t)))$$

$$\leqslant \psi(0 + \limsup_{t\to\infty} \operatorname{diam}(A(t)))$$

$$\leqslant \psi(\omega_0(A) + \limsup_{t\to\infty} \operatorname{diam}(A(t)))$$

$$\leqslant \psi(\mu(A)), \tag{9.123}$$

where μ is the measure of noncompactness defined in the space $BC(\mathbb{R}^+, \mathbb{R})$. This shows that T is a nonlinear $\mathcal{D}$-set contraction on $B[0, r]$ in the sense of Definition 6.18. Thus, the map T satisfies all the conditions of Theorem 6.29 with $C = B[0, r]$, and an application of it yields that T has a fixed point in $B[0, r]$. This further by definition of T which implies that the GNFIE (9.108) has a solution in $B[0, r]$. Moreover, taking into account that the image of $B[0, r]$ under the operator T is again contained in the ball $B[0, r]$, we infer that the set $\mathfrak{F}(T)$ of all fixed points of T is contained in $B[0, r]$. If the set $\mathfrak{F}(T)$ contains all solutions of the Eq. (9.108), then we conclude from Remark 9.7 that the set $\mathfrak{F}(T)$ belongs to the family ker μ.

Now, taking into account the description of sets belonging to ker μ, we deduce that all solutions of the Eq. (9.108) are uniformly locally attractive on $\mathbb{R}^+$. This completes the proof.   ∎

Special Cases

As mentioned earlier, the GNFIE (9.108) is more general in the literature on the theory of nonlinear integral equations and includes other several classes of well-known nonlinear integral equations. Below, we list some of our main observations in this direction.

1. Let us define the function F as

$$F(t, x_1, x_2, x_3, x_4) = q(t) + x_3 + x_4,$$

then the GNFIE (9.108) reduces to the following nonlinear function integral equation (NFIE),

$$x(t) = q(t) + \int_0^{\beta(t)} f(t, s, x(\gamma(s)))ds + \int_0^{\sigma(t)} g(t, s, x(\eta(s)))ds, \quad (9.124)$$

for all $t \in \mathbb{R}^+$. NFIE (9.124) has been studied in Dhage [175] and includes the well-known Volterra, Fredholm as well as integral equations of mixed type as special ases by choosing the functions β and σ appropriately.

2. On taking $\sigma(t) = \infty$ for all $t \in \mathbb{R}^+$ and

$$F(t, x_1, x_2, x_3, x_4) = f(t, x_1, x_3, x_4),$$

we obtain the following integral equation studied in Agarwal et al. [3],

$$x(t) = f\left(t, x(\theta(t)), \int_0^{\beta(t)} f(t, s, x(\gamma(s)))ds, \int_0^\infty g(t, s, x(\eta(s)))ds\right),$$
$$(9.125)$$

which again includes other several classes of known integral equations as special cases (cf. Agarwal et al. [2] and the references therein).

3. On taking $F(t, x_1, x_2, x_3, x_4) = f(t, x_2, x_4)$, we obtain the following nonlinear integral equation,

$$x(t) = f\left(t, u(t, x(\alpha(t))), \int_0^{\sigma(t)} g(t, s, x(\eta(s)))ds\right), \quad (9.126)$$

for all $t \in \mathbb{R}^+$. The nonlinear integral equation (9.126) has been studied in Dhage and Lakshmikantham [182] for the global existence and attractivity results for the solutions defined on $\mathbb{R}^+$.

4. When $F(t, x_1, x_2, x_3, x_4) = p(t, x_1) + x_2 x_4$, where $p : \mathbb{R}^+ \times \mathbb{R} \to \mathbb{R}$ is a continuous function, then the GNFIE (9.108) reduces to the following nonlinear quadratic functional-integral equation,

$$x(t) = p(t, x(q\theta(t)) + [u(t, x(\alpha(t)))]\left(\int_0^{\sigma(t)} g(t, s, x(\eta(s)))ds\right) \quad (9.127)$$

for all $t \in \mathbb{R}^+$. The quadratic integral equation (9.127) again includes several other classes of quadratic integral equations as the special cases given in Dhage et al. [179], Dhage and Ntouyas [184] and the references cited therein.

5. On taking $F(t, x_1, x_2, x_3, x_4) = x_1 + p(x_3, x_4)$, where $p : \mathbb{R} \times \mathbb{R} \to \mathbb{R}$ is a continuous function, we obtain the following functional-integral equation recently studied in Dhage et al. [181],

$$x(t)=u(t, x(\alpha(t)))+p\left(\int_0^{\beta(t)} f(t, s, x(\gamma(s)))ds, \int_0^{\sigma(t)} g(t, s, x(\eta(s)))ds\right),$$
$$(9.128)$$

which further yields the functional-integral equation

$$x(t) = u(t, x(\alpha(t))) + \int_0^{\sigma(t)} g(t, s, x(\eta(s)))ds, \qquad (9.129)$$

for all $t \in \mathbb{R}^+$ provided $p(x_3, x_4) = x_4$. The integral equation (9.128) is discussed in Banas and Dhage [34] and Aghajani et al. [5] for existence and asymptotic stability of the solutions.

We now furnish an example to validate all the hypotheses of Theorem 9.31 above.

Example 9.8 We consider the following nonlinear functional-integral equation

$$x(t) = \frac{15}{16} \ln(1 + |x(t^2 + 1)|) + \frac{3}{4}\left(\frac{1+t}{6+7t^2} \ln(1 + \frac{1}{2}|x(t+1)|)\right)$$

$$+ \frac{2}{5}\int_0^{\frac{t}{t^2+1}} \frac{1}{(1+t)}\frac{1}{1+s} \ln\left(1 + \frac{1}{3}|x(s^2+1)|\right) ds$$

$$+ \frac{7}{3}\int_0^{\frac{t}{t^3+1}} t^3 \exp(-t^5) \frac{1}{(1+s^2)}\frac{|\cos x(s^2+3)|}{1 + |\sin x(s^2+3)|} ds \qquad (9.130)$$

for all $t, s \in \mathbb{R}^+$.
Let

$$F(t, x, x_1, x_2, x_3) = \frac{15}{16} \ln(1 + |x|) + \frac{3}{4}x_1 + \frac{2}{5}x_2 + \frac{7}{3}x_3,$$

$$\varphi(t) = \frac{15}{16} \ln(1 + t), \varphi_1(t) = \ln\left(1 + \frac{1}{2}t\right), \varphi_2(t) = \ln\left(1 + \frac{1}{3}t\right),$$

$$\theta(t) = t^2 + 1, \alpha(t) = t + 1, \gamma(s) = s^2 + 1, \eta(s) = s^2 + 3, \beta(t) = \frac{t}{t^2+1}, \sigma(t) = \frac{t}{t^3+1}$$

$$u(t, x) = \frac{1+t}{6+7t^2} \ln\left(1 + \frac{1}{2}|x(t)|\right), v(t) = t^3 \exp(-t^5), b(s) = \frac{1}{1+s^2}$$

for all $t, s \in \mathbb{R}^+$, and

$$f(t, s, x(\gamma(s))) = \frac{1}{(1+t)(1+s)} \ln\left(1 + \frac{1}{3}|x(s^2+1)|\right)$$

$$g(t, s, x(\eta(s))) = t^3 \exp(-t^5) \frac{1}{1+s^2}\frac{|\cos x(s^2+3)|}{1 + |\sin x(s^2+3)|}$$

for all $t, s \in \mathbb{R}^+$ and $x \in \mathbb{R}$. Notice that:

(i) The functions θ, η and u are obviously continuous. Thus, (H_0) is satisfied.

(ii) Since $u(t, x) = \frac{1+t}{6+7t^2} \ln(1 + |x(t)|/3)$, we have that

$$
\begin{aligned}
|u(t, x) - u(t, y)| &= \frac{1+t}{6+7t^2} \ln \frac{1 + |x|/3}{1 + |y|/3} \\
&= \frac{1+t}{6+7t^2} \ln \left(1 + \frac{|x|/3 - |y|/3}{1 + |y|/3} \right) \\
&\leqslant \frac{1+t}{6+7t^2} \ln(1 + |x - y|/3) \\
&= c(t)\varphi(|x(t)| - |y(t)|), \quad \text{where } c(t) = \frac{1+t}{6+7t^2}.
\end{aligned}
$$

Thus, we have $c_0 = \sup_{t\geq 0} c(t) = \sup_{t\geq 0} \frac{1+t}{6+7t^2} = \frac{1}{6} > 0$; i.e., (H_1) is satisfied with $\varphi(r) = \ln(1 + r/3)$, we see that $\varphi(r) < r$ for $r > 0$. Obviously, the function φ is nondecreasing and upper semicontinuous on $\mathbb{R}^+$.

(iii) For arbitrary but fixed $x, y \in \mathbb{R}$ such that $|x| \geq |y|$ and for $t > 0$, we obtain

$$
\begin{aligned}
|f(t, s, x) - f(t, s, y)| &= \frac{1}{3(1+t)} \frac{1}{\arctan(t)} \frac{1}{1+s^2} \ln \frac{1 + |x|/3}{1 + |y|/3} \\
&\leqslant \frac{1}{3(1+t)} \frac{1}{\arctan(t)} \frac{1}{1+s^2} \ln \left(1 + \frac{|x|/3 - |y|/3}{1 + |y|/3} \right) \\
&< \frac{1}{3(1+t)} \frac{1}{\arctan(t)} \frac{1}{1+s^2} \ln(1 + |x - y|/3) \\
&= \frac{1}{3(1+t)} \frac{1}{\arctan(t)} \frac{1}{1+s^2} \varphi(|x - y|) \\
&= q(t, s)\, \varphi(|x - y|), \quad \text{where } q(t, s) = \frac{1}{3(1+t)} \frac{1}{\arctan(t)} \frac{1}{1+s^2}.
\end{aligned}
$$

The case is similar when $|y| \geq |x|$. Furthermore, we obtain

$$
c_1 = \sup_{t\geq 0} \int_0^t q(t, s)ds = \sup_{t\geq 0} \frac{1}{3(1+t)} \frac{1}{\arctan(t)} \int_0^t \frac{1}{1+s^2}ds = \frac{1}{3}.
$$

Hence, (H_2) is satisfied.

(iv) (H_3) is satisfied since $\lim_{t\to\infty} \int_0^t q(t, s)ds = \lim_{t\to\infty} \frac{1}{3(1+t)} = 0$ and $c_2 = \sup_{t\geq 0} F(t) = \sup_{t\geq 0} \int_0^t f(t, s, 0)ds = 0$.

(v) (H_4) is satisfied since the function g acts continuously from the set $\mathbb{R}^+ \times \mathbb{R}^+ \times \mathbb{R}$ into $\mathbb{R}$. Moreover, we have

$$
|g(t, s, x)| \leqslant t^3 \exp(-t^5) \frac{1}{1+s} = v(t)b(s)
$$

for all $t, s \in \mathbb{R}^+$ and $x \in \mathbb{R}$, then we can see that condition (v) is satisfied. Indeed, we have

$$\lim_{t\to\infty} v(t) \int_0^a b(s)ds = \lim_{t\to\infty} t^3 \exp(-t^5) \int_0^a \frac{1}{1+s} ds = \ln(1+a) \lim_{t\to\infty} t^3 \exp(-t^5) = 0.$$

(v) The function $F : \mathbb{R}^+ \longrightarrow \mathbb{R}$ defined by $F(t) = \int_0^t |f(t,s,0)|ds = 0$ is bounded with $c_2 = \sup_{t\geq 0} F(t) = 0$.

(vi) Let us consider the following equality

$$q = \sup \left\{ c_2 + |u(t,0)| + v(t) \int_0^a b(s)ds : t \geq 0 \right\}.$$

Then, we obtain

$$q = \sup \left\{ c_2 + |u(t,0)| + v(t) \int_0^a b(s)ds : t \geq 0 \right\}$$
$$= \ln(1+a) \sup_{t\geq 0} t^3 \exp(-t^5) \approx 0.223 \ln(1+a).$$

(vi) Let us consider the inequality

$$(1 + c_0 + c_1)\varphi(r) + q \leqslant r.$$

For the case $a = 1$, we have

$$\left(1 + \frac{1}{6} + \frac{1}{3} \right) \ln(1+r) + 0.223 \ln(2) \leqslant r \text{ i.e., } (1.5) \ln(1+r) + 0.154 \leqslant r.$$

It is easily seen that each number $r > 0.7$ satisfies the above inequality. Thus, as the number r_0, we can take $r_0 = 0.7$. Note that this estimate of r_0 can be improved.

Thus, the functions $\alpha, \beta, \gamma, \theta, \sigma, \eta, \varphi, \varphi_1, \varphi_2, u, f$ and g involved in (9.108) satisfy all the conditions of Theorem 9.31, and hence, the GNFIE (9.108) has at least one solution in the space $BC(\mathbb{R}^+, \mathbb{R})$, and the solutions are locally uniformly ultimately attractive on $\mathbb{R}^+$ located in the ball $B[0, 0.7]$.

9.5.4 Existence of Solutions for Nonlinear Functional-Integral Equations in Banach Algebra

In this section, using the technique of measure of noncompactness in Banach algebra, we present an existence theorem for a nonlinear integral equation which contains as particular cases a large number of integral and functional equations considered in nonlinear analysis.

It is well known that the differential and integral equations that arise in many physical problems are mostly nonlinear and fixed point theory that provides a powerful tool for obtaining the solutions of such equations which otherwise are difficult to solve by other ordinary methods. In this paper, we consider the solvability of certain functional-integral equation which contains as particular cases a lot of integral and functional-integral equations, those are applicable in many branches on nonlinear analysis. The authors consider the following functional-integral equation:

$$
x(t) = \left(u(t, x(t)) + f\left(t, \int_0^t p(t, s, x(s))ds, x(\alpha(t))\right)\right) \cdot
$$
$$
g\left(t, \int_0^a q(t, s, x(s))ds, x(\beta(t))\right),
$$
(9.131)

for $t \in [0, a]$.

The main tool used in our result is a fixed point theorem which satisfies the Darbo condition with respect to a measure of noncompactness in the Banach algebra of continuous functions in the interval $[0, a]$. First, we introduce some preliminaries and use them to proof Theorem 9.33. Finally, we prove some examples that verify the application of this kind of nonlinear functional-integral equations.

Now let us assume that Ω is a nonempty subset of a Banach space E and $S : \Omega \to E$ is a continuous operator transforming bounded subsets of Ω to bounded ones. Moreover, let μ be a regular measure of noncompactness in E.

Definition 9.7 (*Banas and Goebel* [35]) We say that S satisfies the Darbo condition with a constant k with respect to measure μ provided

$$
\mu(SX) \leqslant k\mu(X),
$$

for each $X \in m_E$ such that $X \subset \Omega$.

If $k < 1$, then S is called a contraction with respect to μ.

In the sequel, we will work in the space $C[a, b]$ consisting of all real functions defined and continuous on the interval $[a, b]$. The space $C[a, b]$ is equipped with the standard norm

$$
\|x\| = \sup\{|x(t)| : t \in [0, a]\}.
$$

Obviously, the space $C[a, b]$ has also the structure of Banach algebra.

In our considerations, we will use a regular measure of noncompactness defined in [36] (cf also [35]).

In order to recall the definitions of that measure let us fix a set $X \in m_{C[a,b]}$. For $x \in X$ and for a given $\varepsilon > 0$ denote by $\omega(x, \varepsilon)$ the modulus of continuity of x, i.e.,

$$
\omega(x, \varepsilon) = \sup\{|x(t) - x(s)| : t, s \in [a, b], |t - s| \leqslant \varepsilon\}.
$$

Further, put

$$
\omega(X, \varepsilon) = \sup\{\omega(x, \varepsilon) : x \in X\},
$$

$$
\omega_0(X) = \lim_{\varepsilon \to 0} w(X, \varepsilon).
$$
(*)

It may be shown in [36] that $\omega_0(X)$ is a regular measure of noncompactness in $C[a, b]$.

For our purpose, we will need the following theorem [36].

Theorem 9.32 *Assume that Ω is a nonempty, bounded, convex and closed subset of $C[a, b]$ and the operators P and T transform continuously the set Ω into $C[a, b]$ in such a way that $P(\Omega)$ and $T(\Omega)$ are bounded. Moreover, assume that the operator $S = P.T$ transform Ω into itself. If the operators P and T satisfy on the set Ω the Darbo condition with the constant k_1 and k_2, respectively, then the operator S satisfies the Darbo condition on Ω with the constant*

$$\|P(\Omega)\|k_2 + \|T(\Omega)\|k_1.$$

Particularly, if $\|P(\Omega)\|k_2 + \|T(\Omega)\|k_1 < 1$, then S is a contraction with respect to the measure ω_0 and has at least one fixed point in the set Ω.

In 2013, Deepmala and Pathak [162] studied the solvability of the nonlinear functional-integral equation (9.124) for $x \in C[a, b]$. Indeed, they formulated the assumptions under which Eq. (9.131) will be investigated. Namely, they assume the following hypothesis.

(H_1)   The function $u : [0, a] \times \mathbb{R} \to \mathbb{R}$, $f, g : [0, a] \times \mathbb{R} \times \mathbb{R} \to \mathbb{R}$ are continuous and there exists constants $l, m \geq 0$ such that

$$|u(t, 0)| \leqslant l, \quad |f(t, 0, x)| \leqslant m, \quad |g(t, 0, x)| \leqslant m.$$

(H_2)   There exists the continuous functions $a_1, a_2, a_3, a_4, a_5 : [0, a] \to [0, a]$ such that

$$|u(t, x_1) - u(t, x_2)| \leqslant a_1(t)|x_1 - x_2|,$$
$$|f(t, y_1, x_1) - f(t, y_2, x_2)| \leqslant a_2(t)|y_1 - y_2| + a_3(t)|x_1 - x_2|,$$
$$|g(t, y_1, x_1) - g(t, y_2, x_2)| \leqslant a_4(t)|y_1 - y_2| + a_5(t)|x_1 - x_2|,$$

for all $t \in [0, a]$ and $x_1, x_2, y_1, y_2 \in \mathbb{R}$.

(H_3)   The functions $p = p(t, s, x)$ and $q = q(t, s, x)$ act continuously form the set $[0, a] \times [0, a] \times \mathbb{R}$ into $\mathbb{R}$ and the functions $\alpha(t)$ and $\beta(t)$ transform continuously the interval $[0, a]$ into itself.

(H_4)   There exists a non negative constant k such that

$$\max\{a_1(t), a_2(t), a_3(t), a_4(t), a_5(t)\} \leqslant k.$$

(H_5)   (sub-linearity condition) There exists constant α and β such that

$$|p(t, s, x)| \leqslant \alpha + \beta|x|, \quad |q(t, s, x)| \leqslant \alpha + \beta|x|.$$

(H_6) $4\gamma\eta < 1$ and $a\beta \geq 1$, for $\gamma = k + ka\beta$ and $\eta = ka\alpha + l + m$.

The following result is obtained by using the above hypothesis.

Theorem 9.33 *Under the assumptions* (H_1)–(H_6) *Eq. (9.131) has at least one solution in the Banach algebra* $C = C[0, a]$.

Proof Let us consider the operators F and G defined on the Banach algebra C by the formula

$$(Fx)(t) = u(t, x(t)) + f\left(t, \int_0^t p(t, s, x(s))ds, x(\alpha(t))\right),$$

$$(Gx)(t) = g\left(t, \int_0^a q(t, s, x(s))ds, x(\beta(t))\right),$$

for $t \in [0, a]$.

From assumptions (H_1) and (H_3), it follows that F and G transform the algebra C into itself.

Further, let us define the operator T on the algebra C by putting

$$Tx = (Fx).(Gx)$$

Obviously, T transform C into itself.

Now, let us fix $x \in C$. Then, using our assumptions for $t \in [0, a]$, we get

$$|(Tx)(t)| = |(Fx)(t)| \cdot |(Gx)(t)|$$

$$= \left| u(t, x(t)) + f\left(t, \int_0^t p(t, s, x(s))ds, x(\alpha(t))\right) \right| \cdot$$

$$\left| g\left(t, \int_0^a q(t, s, x(s))ds, x(\beta(t))\right) \right|$$

$$\leqslant \left(|u(t, x(t)) - u(t, 0)| + |u(t, 0)| \right.$$

$$+ \left| f\left(t, \int_0^t p(t, s, x(s))ds, x(\alpha(t))\right) - f(t, 0, x(\alpha(t))) \right|$$

$$\left. + |f(t, 0, x(\alpha(t)))| \right)$$

$$\cdot \left(\left| g\left(t, \int_0^a q(t, s, x(s))ds, x(\beta(t))\right) - g(t, 0, x(\beta(t))) \right| \right.$$

$$\left. + |g(t, 0, x(\beta(t)))| \right)$$

$$\leqslant \left(a_1(t)|x(t)| + l + a_2(t) \left| \int_0^t p(t, s, x)ds \right| + m \right)$$

$$\left(a_4(t) \left| \int_0^a q(t, s, x)ds \right| + m \right)$$

$$\leqslant (k|x(t)| + l + ka(\alpha + \beta|x(t)|) + m)(ka(\alpha + \beta|x(t)|) + m)$$

$$\leqslant ((k + ka\beta)\|x\| + ka\alpha + l + m)^2$$

Let $\gamma = k + ka\beta$ and $\eta = ka\alpha + l + m$, then from the above estimate, it follows easily that

$$\|Fx\| \leqslant \gamma \|x\| + \eta \tag{9.132}$$

$$\|Gx\| \leqslant \gamma \|x\| + \eta \tag{9.133}$$

$$\|Tx\| \leqslant (\gamma \|x\| + \eta)^2. \tag{9.134}$$

for $x \in C[0, a]$.

From (9.134), we deduce that the operator T maps the ball $B_r \subset C[0, a]$ into itself for $r_1 \leqslant r \leqslant r_2$, where

$$r_1 = \frac{(1 - 2\gamma\eta) - \sqrt{1 - 4\gamma\eta}}{2\gamma^2}, \quad r_2 = \frac{(1 - 2\gamma\eta) + \sqrt{1 - 4\gamma\eta}}{2\gamma^2}.$$

Also, from estimate (9.132) and (9.133), we obtain

$$\|FB_r\| \leqslant \gamma\, r + \eta \tag{9.135}$$

$$\|GB_r\| \leqslant \gamma\, r + \eta. \tag{9.136}$$

Next, we show that the operator F is continuous on the ball B_r. To do this, fix $\varepsilon > 0$ and take arbitrary $x, y \in B_r$ such that $\|x - y\| \leqslant \varepsilon$. Then, for $t \in [0, a]$, we get

$$|(Fx)(t) - (Fy)(t)| = \left| u(t, x(t)) + f\left(t, \int_0^t p(t, s, x(s))ds, x(\alpha(t))\right) \right.$$

$$\left. - u(t, y(t)) - f\left(t, \int_0^t p(t, s, y(s))ds, y(\alpha(t))\right) \right|$$

$$\leqslant a_1(t)|x(t) - y(t)| + \left| f\left(t, \int_0^t p(t, s, x(s))ds, x(\alpha(t))\right) \right.$$

$$- f\left(t, \int_0^t p(t, s, x(s))ds, y(\alpha(t))\right) + f\left(t, \int_0^t p(t, s, x(s))ds, y(\alpha(t))\right)$$

$$\left. - f\left(t, \int_0^t p(t, s, y(s))ds, y(\alpha(t))\right) \right|$$

$$\leqslant a_1(t)|x(t) - y(t)| + a_3(t)|x(\alpha(t)) - y(\alpha(t))| +$$

$$a_2(t) \left| \int_0^t p(t, s, x(s))ds - \int_0^t p(t, s, x(s))ds \right|$$

$$\leqslant 2k\|x - y\| + k\, a\, \omega(p, \varepsilon),$$

where $\omega(p, \varepsilon) = \sup\{|p(t, s, x) - p(t, s, y)| : t, s \in [0, a]; x, y \in [-r, r]; \|x - y\| \leqslant \varepsilon\}$.

Since, we know that $p = p(t, s, x)$ is uniformly continuous on the bounded subset $[0, a] \times [0, a] \times [-r, r]$, we infer that $\omega(p, \varepsilon) \to 0$ as $\varepsilon \to 0$. Thus, the operator F is continuous on B_r. Similarly, one can easily show that G is continuous on B_r, and consequently, we deduce that T is continuous on B_r.

Now, we show that the operators F and G satisfy the Darbo condition with respect to the measure ω_0 as defined in $(*)$, in the ball B_r. Take a nonempty subset X of B_r and $x \in X$, then for a fixed $\varepsilon > 0$ and $t_1, t_2 \in [0, a]$ such that without loss of generality, we may assume that $t_1 \leqslant t_2$ and $t_2 - t_1 \leqslant \varepsilon$, and we obtain

$$
\begin{aligned}
|(Fx)(t_2) - (Fx)(t_1)| &= \left| u(t_2, x(t_2)) + f\left(t_2, \int_0^{t_2} p(t_2, s, x(s))ds, x(\alpha(t_2)) \right) \right. \\
&\quad \left. -u(t_1, x(t_1)) + f\left(t_1, \int_0^{t_1} p(t_1, s, x(s))ds, x(\alpha(t_1)) \right) \right| \\
&\leqslant |u(t_2, x(t_2)) - u(t_2, x(t_1))| + |u(t_2, x(t_1)) - u(t_1, x(t_1))| \\
&\quad + \left| f\left(t_2, \int_0^{t_2} p(t_2, s, x(s))ds, x(\alpha(t_2)) \right) \right. \\
&\quad \left. - f\left(t_2, \int_0^{t_1} p(t_1, s, x(s))ds, x(\alpha(t_2)) \right) \right| \\
&\quad + \left| f\left(t_2, \int_0^{t_1} p(t_1, s, x(s))ds, x(\alpha(t_2)) \right) \right. \\
&\quad \left. - f\left(t_1, \int_0^{t_1} p(t_1, s, x(s))ds, x(\alpha(t_1)) \right) \right| \\
&\leqslant a_1(t)|x(t_2) - x(t_1)| + |u(t_2, x(t_1)) - u(t_1, x(t_1))| \\
&\quad a_2(t)\left| \int_0^{t_2} p(t_2, s, x(s))ds - \int_0^{t_1} p(t_1, s, x(s))ds \right| \\
&\quad + \left| f\left(t_2, \int_0^{t_1} p(t_1, s, x(s))ds, x(\alpha(t_2)) \right) \right. \\
&\quad \left. - f\left(t_1, \int_0^{t_1} p(t_1, s, x(s))ds, x(\alpha(t_2)) \right) \right| \\
&\quad + \left| f\left(t_1, \int_0^{t_1} p(t_1, s, x(s))ds, x(\alpha(t_2)) \right) \right. \\
&\quad \left. - f\left(t_1, \int_0^{t_1} p(t_1, s, x(s))ds, x(\alpha(t_1)) \right) \right| \quad (9.137)
\end{aligned}
$$

For simplicity, we define the following notations

$$\omega_u(\varepsilon, .) = \sup\{|u(t, x) - u(t', x)| : t, t' \in [0, a]; |t - t'| \leqslant \varepsilon; x \in [-r, r]\}$$
$$\omega_p(\varepsilon, ., .) = \sup\{|p(t, s, x) - p(t', s, x)| : t, t' \in [0, a]; |t - t'| \leqslant \varepsilon; x \in [-r, r]\}$$
$$\omega_f(\varepsilon, ., .) = \sup\{|f(t, y, x) - f(t', y, x)| : t, t' \in [0, a]; |t-t'| \leqslant \varepsilon; x \in [-r, r], y \in [-k'a, k'a]\}$$
$$k' = \sup\{|p(t, s, x)| : t, s \in [0, a]; x \in [-r, r]\}$$

Then, using relation (9.137), we obtain the following

$$|(Fx)(t_2) - (Fx)(t_1)| \leqslant 2k|x(\alpha(t_2)) - x(\alpha(t_1))| + \omega_u(\varepsilon, .) + k[\omega_p(\varepsilon, ., .).a + k'\varepsilon] + \omega_f(\varepsilon, ., .)$$

This yields the following estimate

$$\omega(Fx, \varepsilon) \leqslant 2k\omega(x, \omega(\alpha, \varepsilon)) + \omega_u(\varepsilon, .) + k[\omega_p(\varepsilon, ., .).a + k'\varepsilon] + \omega_f(\varepsilon, ., .)$$

In view of our assumption, we infer that the functions $u = u(t, x)$ and $f = f(t, y, x)$ are uniformly continuous on $[0, a] \times \mathbb{R}$ and $[0, a] \times \mathbb{R} \times \mathbb{R}$. Hence, we deduce that $\omega_u(\varepsilon, .) \to 0$, $\omega_p(\varepsilon, ., .) \to 0$, $\omega_f(\varepsilon, ., .) \to 0$ as $\varepsilon \to 0$. Thus,

$$\omega_0(FX) \leqslant 2k\omega_0(X) \tag{9.138}$$

Similarly,

$$\omega_0(GX) \leqslant 2k\omega_0(X) \tag{9.139}$$

Finally, from (9.135), (9.136), (9.138) and (9.139) and Theorem 9.32, we infer that the operator T satisfies the Darbo condition on B_r with respect to the measure ω_0 with constant $(\gamma r + \eta) 2k + (\gamma r + \eta) 2k$. Also, we have

$$(\gamma r + \eta) 2k + (\gamma r + \eta) 2k = 4k(\gamma r + \eta) = 4k(\gamma r_1 + \eta)$$
$$= 4k \left(\gamma \left(\frac{(1 - 2\gamma\eta) - \sqrt{1 - 4\gamma\eta}}{2\gamma^2} \right) + \eta \right)$$
$$= 4k \left(\frac{1 - \sqrt{1 - 4\gamma\eta}}{2\gamma} \right) < 1.$$

Hence, the operator T is a contraction on B_r with respect to ω_0. Thus, by applying Theorem 9.32, we get that T has at least one fixed point in B_r. Consequently, the nonlinear functional-integral equation (9.131) has at least one solution in B_r. This completes the proof. ∎

Special Cases

We now give some examples of classical integral and functional equations considered in applied part of nonlinear analysis which are particular cases of the equation (9.131).

1. If $g(t, y, x) = 1$, then Eq. (9.131) is in the following form which studied in [380].

$$x(t) = \left(u(t, x(t)) + f \left(t, \int_0^t p(t, s, x(s))ds, x(\alpha(t)) \right) \right). \tag{9.140}$$

2. For $u(t, x) = 0$, we obtain the following nonlinear functional-integral equation studied in [38, 379].

$$x(t) = f \left(t, \int_0^t p(t, s, x(s))ds, x(\alpha(t)) \right) g \left(t, \int_0^a q(t, s, x(s))ds, x(\beta(t)) \right). \tag{9.141}$$

3. $f(t, y, x) = y$ and $g(t, y, x) = 1$, then we get the following functional-integral equation studied in [37].

$$x(t) = u(t, x(t)) + \int_0^t p(t, s, x(s))ds. \qquad (9.142)$$

4. If $u(t, x) = 0$, $g(t, y, x) = 1$ and $f(t, y, x) = u(t, x)y$, then Eq. (9.132) has the following form as in the paper [381].

$$x(t) = u(t, x(t)) \int_0^t p(t, s, x(s))ds. \qquad (9.143)$$

5. If $u(t, x) = 0$, $g(t, y, x) = 1$ and $f(t, y, x) = a(t) + y$, then we get following well known nonlinear Volterra integral equation

$$x(t) = a(t) + \int_0^t p(t, s, x(s))ds. \qquad (9.144)$$

6. If $u(t, x) = 0$, $f(t, y, x) = 1$ and $g(t, y, x) = b(t) + y$, then we obtain Urysohn integral equation

$$x(t) = b(t) + \int_0^a q(t, s, x(s))ds. \qquad (9.145)$$

7. If $u(t, x) = 0$, $f(t, y, x) = a(t) + y$ and $g(t, y, x) = y$, then Eq. (9.131) has the form examined in the paper [170].

$$x(t) = a(t) \int_0^a q(t, s, x(s))ds + \left(\int_0^t p(t, s, x(s))ds \right) \left(\int_0^a q(t, s, x(s))ds \right). \qquad (9.146)$$

8. If $u(t, x) = 0$, $f(t, y, x) = 1$, $g(t, y, x) = 1 + xy$ and $q(t, s, x(s)) = \frac{t}{t+s}\varphi(s)$ $x(s)$, $\beta(t) = t$, where $\varphi(s)$ is an even polynomial in s (we may properly call $\varphi(s)$ the characteristic function in terms of which x is defined; however, we notice that in physical applications of FIE (9.147) below certain restrictions are necessary, such as $\int_0^1 \varphi(s)ds \leqslant \frac{1}{2}$, but from the point of view of pure mathematics restrictions of this type are not essential), then Eq. (9.131) has the form

$$x(t) = 1 + x(t) \int_0^a \frac{t}{t+s}\varphi(s)x(s)ds. \qquad (9.147)$$

The above equation is the famous quadratic integral equation [139] of Chandrasekhar type and was considered in many papers.

On the other hand, Eq. (9.131) covers also the well-known functional equation of the first order having the form

$$x(t) = f_1(t, x(\alpha(t))).$$

For this, it is sufficient to put $g(t, y, x) = 1$, $u(t, x) = 0$ and $f(t, y, x) = f_1(t, x)$.

Now, we present an example of functional-integral equation, and consequently, see the existence of its solutions by using Theorem 9.33.

Example 9.9 Consider the following nonlinear functional-integral equation:

$$x(t) = \left(\frac{1}{5}\sin\left(\frac{t}{4}\right) + \frac{1}{4}\int_0^t ts\cos(x(s))ds\right) \cdot \left(\frac{1}{3}\int_0^1 t\sin\left(\frac{sx(s)}{1+x(s)}\right)ds\right),$$
$$(9.148)$$

for $t \in [0, 1]$.

Let us take $u : [0, 1] \to \mathbb{R}$, $f, g : [0, 1] \times \mathbb{R} \times \mathbb{R} \to \mathbb{R}$ and $p, q : [0, 1] \times [0, 1] \times \mathbb{R} \to \mathbb{R}$, and comparing (9.148) with (9.131), we get

$$u(t, x(t)) = \frac{1}{5}\sin\left(\frac{t}{4}\right), f(t, y_1, x) = \frac{1}{4}y_1, g(t, y_2, x) = \frac{t}{3}y_2,$$

$$p(t, s, x) = ts\cos(x(s)), \quad q(t, s, x) = \sin\left(\frac{sx(s)}{1+x(s)}\right).$$

It is easy to prove that these functions are continuous and satisfy the hypothesis (H_2) with $a_1 = a_3 = a_5 = 0$, $a_2 = \frac{1}{4}$, $a_4 = \frac{1}{3}$. In this case, $k = \max\left\{0, \frac{1}{3}, \frac{1}{4}\right\} = \frac{1}{3}$. Moreover,

$$|u(t, 0)| \leqslant \frac{1}{5}, \ |f(t, 0, x)| = 0, \ |g(t, 0, x)| = 0,$$

$$|p(t, s, x)| \leqslant 0 + 1\,|x(t)|, \ |q(t, s, x)| \leqslant 0 + 1\,|x(t)|.$$

It is observed that $\alpha = 0$, $\beta = 1$ and $l = \frac{1}{5}$, $m = 0$, $a = 1$, $a\beta = 1 \geq 1$ and $a\beta = 1$. Also,

$$4\gamma\eta = 4(k + ka\beta)(ka\alpha + l + m) = 4\left(\frac{2}{3}\right) \cdot \left(\frac{1}{5}\right) = \frac{8}{15} < 1.$$

Hence, all the hypothesis from (H_1)–(H_6) are satisfied. Applying the result obtained in Theorem 9.33, we deduce that Eq. (9.131) has at least one solution in Banach algebra $C[0, 1]$.

9.6 Application to Abstract Volterra Integrodifferential Equations

Since the classical work of Volterra, the theory of linear and nonlinear Volterra equations has played an important role in the applications of mathematics in various disciplines. In the last 40 years, interest has also extended to "abstract" Volterra equations. Here, the term "abstract" means that one considers a generalization of the classical equation in one of the following senses:

1. The functions take values in Banach spaces (usually of infinite dimension), or
2. One considers general operator equations in which the operators are assumed to have certain "Volterra-typical" properties.

In this section, we find the common mild solution of abstract Volterra integrodifferential equations of the type (see Theorem 9.34):

$$
u'(t) + Au(t) = f(t, u(t)) + \int_{t_0}^{t} g\left(t, s, u(s), \int_{t_0}^{t} K_1(s, \tau, u(\tau))d\tau\right) ds
$$
$$
+ \int_{t_0}^{t} h\left(t, s, u(s), \int_{t_0}^{\infty} K_2(s, \tau, u(\tau))d\tau\right) ds \qquad (9.149)
$$

$$
u'(t) + Au(t) = \bar{f}(t, u(t)) + \int_{t_0}^{t} \bar{g}\left(t, s, u(s)), \int_{t_0}^{t} \overline{K}_1(s, \tau, u(\tau))d\tau\right) ds
$$
$$
+ \int_{t_0}^{t} \bar{h}\left(t, s, u(s), \int_{t_0}^{\infty} \overline{K}_2(s, \tau, u(\tau))d\tau\right) ds \qquad (9.150)
$$

where $-A$ is the infinitesimal generator of C_0-semigroup $\{T(t) | t \geq 0\}$ of bounded linear operators on a Banach space B with norm $\| \cdot \|$, $f, \bar{f} \in C(\mathbb{R}^+ \times B, B)$, $g, \bar{g}$, $h, \bar{h} \in C(\mathbb{R}^+ \times \mathbb{R}^+ \times B \times B, B)$, $K_1, \overline{K}_1, K_2, \overline{K}_2 \in C(\mathbb{R}^+ \times \mathbb{R}^+ \times B, B)$ and $\mathbb{R}^+ = (0, \infty)$.

In the sequel, we extend Theorem 9.34 to the study of the common mild solution of (9.149) and of the infinite family of integrodifferential equations (see Theorem 9.35):

$$
u'(t) + Au(t) = f_j(t, u(t)) + \int_{t_0}^{t} g_j\left(t, s, u(s)), \int_{t_0}^{t} K_{1j}(s, \tau, u(\tau))d\tau\right) ds
$$
$$
+ \int_{t_0}^{t} h_j\left(t, s, u(s), \int_{t_0}^{\infty} K_{2j}(s, \tau, u(\tau))d\tau\right) ds \qquad (9.151)
$$

where f_j, g_j, K_{1j} and $K_{2j}(j = 1, 2, \ldots)$ play the roles of $\bar{f}, \bar{g}, \overline{K}_1$ and $\overline{K}_2$ of (9.150).

It may be mentioned that equations of the type (9.149) arise naturally in study of initial value problems on the infinite half line. Moreover, many problems arising in various branches of physics and in other areas of mathematical sciences find themselves incorporated in the abstract formulation (9.149) (see, for instance, [44, 388, 397]). Existence, uniqueness and other properties of the solution of several forms of (9.149) using various assumptions and different methods have been studied, among others, by Barbu [39], Fitzgibbon [233], Hussain [282], Martin, Jr. [388], Miller [397], Miller, Nohel and Wong [590], Sinestrari [559], Singh [560], Shuart [580], Travis and Webb [590], Vainberg [599] and Webb [606].

In this section, we study on existence and uniqueness of the common solution of the Eqs. (9.149) and (9.150). To prove the existence of a common solution of Eqs. (9.149) and (9.150), we utilize a fixed point theorem of Yen [613] for two contractive type operators (see Lemma 9.4). On the other hand, the uniqueness of the solution is established using a result of Pachpatte [438] (see Lemma 9.5). In the sequel, some results of Singh [560] and Hussain [282] are obtained as a particular case of our results (see Corollaries 9.3–9.5). The proof of Theorem 9.35 is prefaced by a special case of fixed point theorem of Hussain and Sehgal [283] (see Lemma 9.6) and Lemma 9.5.

Throughout in this section, B stands for a Banach space with norm $\| \cdot \|$ and $-A$ for the infinitesimal generator of C_0 - semigroup of operators $\{T(t), t \geq 0\}$ on B. A family $\{T(t) : t \in \mathbb{R}^+\}$ of bounded operators from B into B is a C_0-semigroup if:

i. $T(0) = I$ the identity operator and $T(t + s) = T(t)T(s)$ for all $t, s \geqslant 0$
ii. $T(\cdot)$ is strongly continuous in $t \in \mathbb{R}^+$
iii. $\| T(t) \| \leq M e^{\omega t}$ for some $M \geq 0$, real ω and $t \in \mathbb{R}^+$ ([398])

For the sake of brevity, we assume that

$$H(T, t_0, f, g, h, K_1, K_2) = T(t - t_0)u_0 + \int_{t_0}^{t} T(t - s)f(s, u(s))ds$$

$$+ \int_{t_0}^{t} T(t - s) \left(\int_{t_0}^{s} g\big(s, \tau, u(\tau), \int_{t_0}^{t} K_1(\tau, \xi, u(\xi))d\xi\big)d\tau \right) ds$$

$$+ \int_{t_0}^{t} T(t - s) \left(\int_{t_0}^{s} h\big(s, \tau, u(\tau), \int_{t_0}^{\infty} K_2(\tau, \xi, u(\xi))d\xi\big)d\tau \right) ds.$$

A continuous $u(t)$ is a mild solution of (9.149) if:

$$u(t) = H(T, t_0, f, h, K_1, K_2).$$

Moreover, a continuous $u(t)$ is a common mild solution of (9.149) and of (9.150) if

$$H(T, t_0, f, h, K_1, K_2) = u(t) = H(T, t_0, \bar{f}, \bar{h}, \overline{K}_1, \overline{K}_2).$$

We use the following assumptions in our first theorem. For all t, s in (t_0, ∞) and x_i, y_i in $B, i = 1, 2$, let nonnegative numbers $L_i, i = 1, 2, 3, 4, 5, 6, 7$ exist, such that

A_1. $\| K_1(t, s, y_1) - \overline{K}_1(t, s, y_2) \| \leqslant L_1 \| y_1 - y_2 \|$

A_2. $\sup\limits_{t \geq t_0} \int_{t_0}^{\infty} \| K_2(t, s, y_1) - \overline{K}_2(t, s, y_2) \| \leqslant L_2 \| y_1 - y_2 \|$

A_3. $\| g(t, s, x_1, y_1) - \bar{g}(t, s, x_2, y_2) \| \leqslant L_3 \| x_1 - x_2 \| + L_4 \| y_1 - y_2 \|$

A_4. $\| h(t, s, x_1, y_1) - \bar{h}(t, s, x_2, y_2) \| \leqslant L_5 \| x_1 - x_2 \| + L_6 \| y_1 - y_2 \|$

A_5. $\| f(t, x_1) - \bar{f}(t, x_2) \| \leqslant L_7 \| x_1 - x_2 \|$

In the proof of Theorem 9.34, we require the lemmas stated below. The following fixed point theorem is essentially given by Yen [613] (see also Hussain and Sehgal [283] Corollary 2 and Rhoades [519] Theorem 14).

Lemma 9.4 *Let T_1 and T_2 be maps on a complete metric space X. If there exist a positive integer m and positive number $k < 1$ such that*

$$d(T_1^m x, T_2^m y) \leqslant kd(x, y) \text{ for all } x, y \in X,$$

then T_1 and T_2 have a unique common fixed point.

The following lemma, due to Pachpatte [438], plays the key role in proving the uniqueness of the common mild solution of (9.149) and (9.150) and the Lipschitz continuity of certain maps.

Lemma 9.5 *Let $x(t), a(t), b(t)$ and $c(t)$ be real-valued nonnegative continuous functions defined on $\mathbb{R}$, for which the inequality*

$$x(t) \leqslant x_0 + \int_0^t a(s)x(s)ds$$

$$+ \int_0^t a(s)\left(\int_0^s b(r)x(r)dr\right)ds + \int_0^s a(s)\left(\int_0^s b(r)\left(\int_0^r c(z)x(z)dz\right)dr\right)ds$$

holds for all t in $\mathbb{R}^+$, where x_0 is a nonnegative constant. Then,

$$x(t) \leqslant x_0\left(1 + \int_0^t a(s)\exp\left(\int_0^s a(r)dr\right)\left\{1 + \int_0^s b(r)\exp\left(\int_0^r (b(z) + c(z))dz\right)dr\right\}ds\right).$$

Theorem 9.34 (Pathak [448]) *Suppose that A_1–A_5 are satisfied. Then, for u_0 in B, the initial value problems (9.149) and (9.150) have a unique common mild solution $u(t)$ in $C([t_0, t_1))$ for $t \geqslant t_0$, where t_1 is arbitrarily fixed, with $t_1 > t_0$. Moreover, the map $u_0 \to u$ from B into $C([t_0, t_1], B)$ is Lipschitz continuous.*

Proof Let $C = C([t_0, t_1], B)$. Define the norm $\| \cdot \|_C$ in C as $\| u \|_C = \max_{p \in [t_0, t_1]} \| u(t) \|$. Then, $\| \cdot \|_C$ is a Banach space. Let $F, \overline{F} : C \to C$ be such that

$$(Fu)(t) = H(T, t_0, f, g, h, K_1, K_2) \qquad t_0 \leqslant t < \infty \qquad (9.152)$$

$$(\overline{F}u)(t) = H(T, t_0, \bar{f}, \bar{g}, \bar{h}, \overline{K}_1, \overline{K}_2) \qquad t_0 \leqslant t < \infty \qquad (9.153)$$

Evidently, a common solution of Eqs. (9.149) and (9.150) is also a common fixed point of the operators F and $\overline{F}$.

Let $\overline{M}$ be an upper bound of $\| T(t - s) \|$ on $[t_0, t_1]$. Evidently, $\overline{M} = M$ if $\omega \leqslant 0$ and $\overline{M} = M \exp(\omega t_1)$ if $\omega > 0$ (See iii.).

$$\| (Fu)(t) - (\overline{F}v)(t) \| \leqslant \int_{t_0}^t \overline{M} L_7 \| u(s) - v(s) \| \, ds$$

$$+ \int_{t_0}^t \overline{M} \int_{t_0}^s (L_5 + L_6 L_2) \| u(\tau) - v(\tau) \| \, d\tau ds$$

$$+ \int_{t_0}^t \overline{M} \int_{t_0}^s \left[L_3 \| u(\tau) - v(\tau) \| + L_4 \int_{t_0}^\tau L_1 \| u(\xi) - v(\xi) \| \, d\xi \right] d\tau ds$$

$$\leqslant \overline{M} L_7 \| u - v \|_C (t - t_0)$$

$$+ \overline{M}(L_3 + L_5 + L_6 L_2) \| u - v \|_C \frac{(t - t_0)^2}{2} + \overline{M} L_4 L_1 \| u - v \|_C \frac{(t - t_0)^3}{6}$$

$$= \overline{M}(t - t_0) \left[\left\{ L_7 + L_3 \frac{(t - t_0)}{2} + L_4 L_1 \frac{(t - t_0)^2}{6} \right\} + (L_5 + L_6 L_2) \frac{(t - t_0)}{2} \right] \| u - v \|_C$$

$$= \overline{M} \alpha \left[\left\{ L_7 + L_3 \frac{\alpha}{2} + L_4 L_1 \frac{\alpha^2}{6} \right\} + (L_5 + L_6 L_2) \frac{\alpha}{2} \right] \| u - v \|_C$$

where $\alpha = t - t_0$.

Repeating this process again, we have

$$\| (F^2 u)(t) - (\overline{F}^2 u)(t) \| \leqslant \int_{t_0}^t \overline{M} L_7 \| (Fu)(s) - (\overline{F}u)(s) \| \, ds$$

$$+ \int_{t_0}^t \overline{M} \int_{t_0}^s \left[L_3 \| (Fu)(\tau) - (\overline{F}u)(\tau) \| + L_4 \int_{t_0}^\tau L_1 \| (Fu)(\xi) - (\overline{F}v)(\xi) \| \, d\xi \right] d\tau ds$$

$$+ \int_{t_0}^t \overline{M} \int_{t_0}^s (L_5 + L_6 L_2) \| (Fu)(\tau) - (\overline{F}v)(\tau) \| \, d\tau ds$$

$$\leqslant \frac{\overline{M}^2}{2} \left[L_7 \int_{t_0}^t \left\{ 2 L_7 (s - t_0) + (L_3 + L_5 + L_6 L_2)(s - t_0)^2 + L_4 L_1 \frac{(s - t_0)^3}{3} \right\} ds \right.$$

$$+ (L_3 + L_5 + L_6 L_2) \int_{t_0}^t \int_{t_0}^s \left\{ 2 L_7 (\tau - t_0) + (L_3 + L_5 + L_6 L_2)(\tau - t_0)^2 + L_4 L_1 \frac{(\tau - t_0)^3}{3} \right\} d\tau ds$$

$$\left. + L_4 L_1 \int_{t_0}^t \int_{t_0}^s \int_{t_0}^\tau \left\{ 2 L_7 (\xi - t_0) + (L_3 + L_5 + L_6 L_2)(\xi - t_0)^2 + L_4 L_1 \frac{(\xi - t_0)^3}{3} \right\} d\xi d\tau ds \right] \| u - v \|_C$$

$$= \frac{\overline{M}^2}{2} \left[L_7 \left\{ L_7 (t - t_0)^2 + (L_3 + L_5 + L_6 L_2) \frac{(t - t_0)^3}{3} + L_4 L_1 \frac{(t - t_0)^4}{12} \right\} \right.$$

$$+ (L_3 + L_5 + L_6 L_2) \left\{ L_7 \frac{(t - t_0)^3}{3} + (L_3 + L_5 + L_6 L_2) \frac{(t - t_0)^4}{12} + L_4 L_1 \frac{(t - t_0)^5}{60} \right\}$$

$$\left. + L_4 L_1 \left\{ L_7 \frac{(t - t_0)^4}{12} + (L_3 + L_5 + L_6 L_2) \frac{(t - t_0)^5}{60} + L_4 L_1 \frac{(t - t_0)^6}{360} \right\} \right] \| u - v \|_C$$

$$= \frac{\alpha^2 \overline{M}^2}{2} \left[L_7 \left\{ L_7 + (L_3 + L_5 + L_6 L_2) \frac{\alpha}{3} + L_4 L_1 \frac{\alpha^2}{12} + L_4 L_1 \frac{\alpha^3}{60} \right\} \right.$$

$$\left. + L_4 L_1 \left\{ L_7 \frac{\alpha^2}{12} + (L_3 + L_5 + L_6 L_2) \frac{\alpha^3}{60} + L_4 L_1 \frac{\alpha^4}{360} \right\} \right] \| u - v \|_C$$

$$= \frac{\alpha^2 \overline{M}^2}{2} \left[\left\{ L_7^2 + L_3^2 \frac{\alpha^2}{12} + L_4^2 L_1^2 \frac{\alpha^4}{360} + L_7 L_3 \frac{\alpha}{3} + L_7 L_4 L_1 \frac{\alpha^2}{6} + L_3 L_4 L_1 \frac{\alpha^3}{60} \right\} \right.$$

$$\left. + \left(L_7 \frac{1}{3} + L_3 \frac{\alpha}{12} + L_4 L_1 \frac{\alpha^2}{60} \right) (L_5 + L_6 L_2) \alpha + (L_5 + L_6 L_2)^2 \frac{\alpha^2}{12} \right] \| u - v \|_C$$

$$\leqslant \frac{\alpha^2 \overline{M}^2}{2} \left[\left\{ L_7^2 + L_3^2 \frac{\alpha^2}{4} + L_4^2 L_1^2 \frac{\alpha^4}{360} + L_7 L_3 \alpha + L_7 L_4 L_1 \frac{\alpha^2}{3} + L_3 L_4 L_1 \frac{\alpha^3}{6} \right\} \right.$$

$$\left. + 2 \left(L_7 + L_3 \frac{\alpha}{2} + L_4 L_1 \frac{\alpha^2}{6} \right) (L_5 + L_6 L_2) \frac{\alpha}{2} + (L_5 + L_6 L_2)^2 \frac{\alpha^2}{4} \right] \| u - v \|_C$$

$$= \frac{\alpha^2}{2!} \overline{M}^2 \left[\left\{ L_7 + L_3 \frac{\alpha}{2} + L_4 L_1 \frac{\alpha^2}{6} \right\} + (L_5 + L_6 L_2) \frac{\alpha}{2} \right]^2 \| u - v \|_C .$$

Further, continuing this process $(n - 2)$ times, it can be seen that

$$\| (F^n u)(t) - (\overline{F}^n v)(t) \| \leqslant \left(\frac{\alpha^n}{n!} \right) \overline{M}^n \left[L_7 + L_3 \frac{\alpha}{2} + L_4 L_1 \frac{\alpha^2}{6} + (L_5 + L_6 L_2) \frac{\alpha}{2} \right]^n \| u - v \|_C .$$

Therefore, $\| F^n u - \overline{F}^n v \|_C \leq k \| u - v \|_C$ where

$$k = \frac{1}{n!} (\bar{\alpha} \overline{M})^n \left[L_7 + L_3 \frac{\bar{\alpha}}{2} + L_4 L_1 \frac{\bar{\alpha}^2}{6} + (L_5 + L_6 L_2) \frac{\bar{\alpha}}{2} \right]^n$$

since when we take the maximum with respect to t, $\alpha = t - t_0$ becomes $\bar{\alpha} = t_1 - t_0$

For sufficiently large n, we can make $k < 1$, and so all the hypothesis of Lemma 9.4 are satisfied. Consequently, there exists a unique u in C such that

$$(Fu)(t) = u(t) = (\overline{F}u)(t).$$

Moreover, this unique common fixed point is a common mild solution of (9.149) and (9.150).

In order to show that (9.149) and (9.150) have exactly one common mild solution, we assume that v is another common mild solution of (9.149) and (9.150) with $v(t_0) = v_0$.

Suppose $L = \max\{L_i \mid i = 1, 2, 3, 4, 5, 6, 7\}$. Then,

$$\| u(t) - v(t) \| = \| (Fu)(t) - (Fv)(t) \| \leq \overline{M} \| u_0 - v_0 \| + \int_{t_0}^t \overline{M} L \| u(s) - v(s) \| \, ds$$

$$+ \int_{t_0}^t \int_{t_0}^s \overline{M} L \left[\| u(\tau) - v(\tau) \| + \int_{t_0}^\tau L \| u(\xi) - v(\xi) \| \, d\xi \right] d\tau \, ds$$

$$+ \int_{t_0}^t \int_{t_0}^s \overline{M} L \left[\| u(\tau) + v(\tau) \| + \int_{t_0}^\infty L \| u(\xi) - v(\xi) \| \, d\xi \right] d\tau \, ds.$$

Now Lemma 9.5 yields $\| u(t) - v(t) \| \leq \overline{M} \| u_0 - v_0 \| R(t)$ where:

$$2R(t) = \left[1 + \int_{t_0}^t \overline{M} L \exp \left(\int_{t_0}^s \overline{M} L d\tau \right) \left\{ 1 + \int_{t_0}^s \exp \left(\int_{t_0}^\tau 2[1 + L] d\xi \right) d\tau \right\} ds \right]$$

$$+ \left[1 + \int_{t_0}^t \overline{M} L \exp \left(\int_{t_0}^s \bar{M} L d\tau \right) \left\{ 1 + \int_{t_0}^s \exp[2(1 + L) d\tau] \, ds \right\} \right]$$

Hence, $\| u - v \|_c \le \overline{M} \| u_0 - v_0 \| \overline{R}$, where $\overline{R} = \max\{R(t), t \in [t_0, t_1]\}$.

By induction, we can prove that

$$\| u - v \|_c \le \frac{\left(\overline{M}\,\overline{R}\right)^n}{n!} \| u_0 - v_0 \|$$

which tends to zero as $n \to \infty$. This yields the unique common mild solution, and the Lipschitz continuity of the map $u_0 \to u$. This completes the proof.

The following results easily follow from Theorem 9.34.

Corollary 9.3 ([560]) *Suppose that A_1, A_3, A_5 and $h = \bar{h} = 0$ with $K_2 = \overline{K}_2$ are satisfied. Then, for u_0 in B, the initial value problems (9.149) and (9.150) have a unique common mild solution u in $C([t_0, t_1], B)$ is Lipschitz continuous.*

Corollary 9.4 ([44]) *Suppose that A_1 with $K_1 = \overline{K}_1$, A_3 with $g = \bar{g}$ and $L_3 = L_4$, A_5 with $f = \bar{f}$ and $h = \bar{h} = 0$ with $K_2 = \overline{K}_2$ are satisfied. Then, for $u_0 \in B$, the initial value problem (9.149) has a unique mild solution $u \in C([t_0, t_1], B)$ for $t \geqslant t_0$ such that $t_0 \leqslant t \leqslant t_1$. Moreover, the map $u_0 \to u$ from B into $C([t_0, t_1], B)$ is Lipschitz continuous.*

Remark 9.8 Fitzgibbon [233] and Webb [606] have studied certain special forms of (9.149) by using different assumptions and methods.

Our next theorem is prefaced by the following result, which is a special case of the fixed point theorem of Hussain and Sehgal [283] (Corollary 2) and Rhoades [519] (Theorem 20).

Lemma 9.6 *Let T and T_i, $i = 1, 2, \ldots$ be maps on a metric space X. If there exist positive integers m_i and positive numbers $k_i < 1$, $i = 1, 2, \ldots$ such that*

$$d(T^m x, T_i^{m_i} y) \leqslant k_i d(x, y) \ \text{for all } x, y \in X,$$

then there exists a unique element u in X such that $T u = u = T_i u$.

We use the following assumptions in Theorem 9.35 (see also A_1–A_5).

For all t, s in $[t_0, \infty)$ and x_q, y_q in B, $q = 1, 2$, let nonnegative numbers L_{ij}, $i = 1, 2, 3, 4, 5, 6, 7$, $j = 1, 2, 3, \ldots$ exist, such that

B_1. $\| K_1(t, s, y_1) - K_{ij}(t, s, y_2) \| \leqslant L_{ij} \| y_1 - y_2 \|$

B_2. $\displaystyle\sup_{t \geqslant t_0} \int_{t_0}^{\infty} \| K_2(t, s, y_1) - K_{2j}(t, s, y_2) \| \, ds \leqslant L_{2j} \| y_1 - y_2 \|$

B_3. $\| g(t, s, x_1, y_1) - g_j(t, s, x_1, y_2) \| \leqslant L_{3j} \| x_1 - x_2 \| + L_{4j} \| y_1 - y_2 \|$

B_4. $\| h(t, s, x_1, y_1) - h_j(t, s, x_2, y_2) \| \leqslant L_{5j} \| x_1 - x_2 \| + L_{6j} \| y_1 - y_2 \|$

B_5. $\| f(t, x_1) - f_j(t, x_2) \| \leqslant L_{7j} \| x_1 - x_2 \|$.

Theorem 9.35 *Suppose that B_1–B_5 are satisfied for each $j = 1, 2, \ldots$. Then, for u_0 in B, the initial value problems (9.149) and (9.151) have a unique common mild solution u in $C([t_0, t_1], B)$. Moreover, the map $u_0 \to u$ from B into $C([t_0, t_1], B)$ is Lipschitz continuous.*

Proof Let C and $F_j : C \to C$, $j = 1, 2, \ldots$ be such that

$$(F_j u)(t) = H(T, t_0, f_j, g_j, h_j, \widetilde{K_{1j}}, K_{2j}) \qquad t_0 \le t \le \infty. \qquad (9.154)$$

Then, a common solution of (9.152) and (9.154) is a common fixed point of the operators F and F_j, $j = 1, 2, \ldots$. It can be seen that for each $j \in \{1, 2, \ldots\}$, we get

$$(F_j u)(t) = H(T, t_0, f_j, g_j, h_j, K_{1j}, K_{2j}) \qquad t_0 \le t \le \infty.$$

Then, a common solution of (9.152) and (9.154) is a common fixed point of the operators F and f_j, $j = 1, 2, \ldots$. It can be seen that for each $j \in \{1, 2, \ldots\}$, we get

$$\| F^n u - F_j^n v \|_C \le k_j \| u - v \|_C$$

where

$$k_j = \frac{1}{n!} (\bar{\alpha}\overline{M})^n \left[L_{7j} + L_{3j} \frac{\bar{\alpha}}{2} + L_{4j} L_{1j} \frac{\bar{\alpha}^2}{6} + (L_{5j} + L_{6j} L_{2j}) \frac{\bar{\alpha}}{2} \right]^n.$$

For sufficiently large n, we can make $k_j < 1$. Consequently, Lemma 9.6 guarantees the existence of a unique u in C such that $Fu = u = F_j u$ for any $j = 1, 2, \ldots$.

The rest of the proof is similar to that given in the proof of Theorem 9.34.

Remark 9.9 Since Lemma 9.6 is true for F and an uncountable family of maps, it follows that Theorem 9.35 is true for (9.149) and an uncountable family of integrodifferential equations of type (9.151).

It may be considered that the following result easily follows from Theorem 9.35.

Corollary 9.5 ([560]) *Suppose that B_1, B_3, B_5 and $h = h_j = 0$, $K_2 = K_{2j}$ are satisfied for each $j = 1, 2, \ldots$. Then, for u_0 in B, the initial value problems (9.149) and (9.151) have a unique common mild solution u in $C([t_0, t_1], B)$ for $t \geqslant t_0$ such that $t_0 \leqslant t \leqslant t_1$. Moreover, the map $u_0 \to u$ from B into $C([t_0, t_1], B)$ is Lipschitz continuous.*

9.7 Application to Surjectivity Theorems

In this section, we introduce a new class of mappings which is known as locally λ-strongly ϕ-accretive mappings, where λ and ϕ have special meanings. Notice that this class of mappings constitutes a generalization of the well-known monotone mappings, accretive mappings and strongly ϕ-accretive mappings. In the sequel, using Caristi–Kirk's fixed point theorem, we extend the results of Park and Park [443], Browder [114] and Ray [502] to locally λ-strongly ϕ-accretive mappings. Finally, we seek to present the notion of generalized directional contractor which generalizes the classical notion of directional contractor introduced by Altman in his fundamental paper [12] and prove a surjectivity theorem which is used to solve certain functional equations in Banach spaces.

Let X and Y be Banach spaces. Throughout, $B(x, r) = \{w \in E : \|w - x\| \leqslant r\}$ will denote a closed ball in a Banach space E (where $E = X$ or $E = Y$ in this case).

Recall that the duality mapping J from X into 2^{X^*} is given by

$$J(x) = \left\{ j \in X^* : \langle x, j \rangle = \|x\|^2 = \|j\|^2 \right\} \quad (x \in X)$$

where $\langle \cdot, \cdot \rangle$ denotes the duality pairing. It is well known that J is single valued in case X is strictly convex, and it is uniformly continuous on bounded subsets of X whenever X^* is uniformly convex.

As a generalization of ϕ-accretive operator, the following definition was introduced by Pathak and Mishra [461].

Definition 9.8 A Lipschitzian mapping $P : X \to Y$ with Lipschitzian constant M is said to be locally λ-strongly ϕ-accretive if for each $y \in Y$ and $r > 0$, there exist constants λ, c with $c/2M > \lambda \geq 0$ such that: if $\|Px - y\| \leqslant r$ and $j \in J$, the duality map on Y, then for all $u \in X$ sufficiently near to x,

$$\langle Pu - Px, \phi(u - x) + \lambda M^{-1} j(Pu - Px) \rangle \geq c\|u - x\|^2. \qquad (9.155)$$

Observation

1. A 0-strongly ϕ-accretive mapping $P : X \to Y$ is strongly ϕ-accretive as defined in [1].

The proof of the next result is prefaced by the following Lemma of Park and Park [8].

Lemma 9.7 *For any $y \in Y$, $y^* \in J(y)$, and $\varepsilon > 0$, there exists an $h \in X$ such that $\|h\| \geq 1$ and $\left\| \phi(h) - y^* \|y\|^{-1} \right\| < \varepsilon$.*

Notice that for any $y^* \in J(y) \in 2^{Y^*}$, we have

$$\|y\|^2 \leqslant \|z\|^2 - 2\langle z - y, y^* \rangle \quad \text{for any } z \in Y. \qquad (9.156)$$

Theorem 9.36 *Let X and Y be Banach spaces and $P : X \to Y$ a locally Lipschitzian and locally λ-strongly ϕ-accretive mapping. If the duality mapping J of Y is strongly upper semicontinuous and $P(X)$ is closed, then $P(X) = Y$.*

Proof As $P(X)$ is closed, to prove the theorem it is just sufficient to show that $P(X)$ is open. It is well known that $J(y) \neq \varnothing$ for each $y \in Y$, so we can choose $y^* \in J(y)$. For a given $x_0 \in X$, choose $\varepsilon_1 > 0$ so small that P is Lipschitzian with constant M on $B(x_0, 2\varepsilon_1)$. Choose $\lambda > 0$ and $\varepsilon_2 > 0$ so that (9.155) holds on $B(Px_0, 2M\varepsilon_1)$ whenever $\|u - x_0\| \leqslant 2\varepsilon_2$; set

$$\varepsilon = \min\{\varepsilon_1, \varepsilon_2\} \text{ and } r = \min\{c\varepsilon/(1 + \sqrt{1 + 4c\lambda M^{-1}}), M\varepsilon\}.$$

Now it suffices to show that $B(Px_0, r) \subset P(X)$. To this end, suppose $y \in B(Px_0, r)$ and $y \notin P(X)$. It follows that $dist(y, P(X)) > 0$. Let $d = dist(y, P(X))$, and let $D = \{x \in B(x_0, \varepsilon) : \|y - Px\| \leqslant r\}$. Clearly, $x_0 \in D$ so that D is nonempty. Moreover, D is closed. Therefore, D is complete. For any $x \in D$, by Lemma 9.7, there exists $h \in X$ such that $\|h\| \geq 1$ and

$$\langle \phi(h), (y - Px)^* \|y - Px\|^{-1} \rangle \leqslant (c/2M - \lambda). \tag{9.157}$$

Set $x_t = x + th, t > 0$. By (9.155), for t sufficiently small, we have

$$\langle Px_t - Px, \phi(x_t - x) + \lambda M^{-1} j(Px_t - Px) \rangle \geq c\|x_t - x\|^2.$$

Thus,

$$\begin{aligned}
\langle Px_t - Px, \phi(x_t - x) \rangle &\geq c\|x_t - x\|^2 - \lambda M^{-1} \langle Px_t - Px, j(Px_t - Px) \rangle \\
&\geq c\|x_t - x\|^2 - \lambda M^{-1} \|Px_t - Px\| \|j(Px_t - Px)\| \\
\text{or} \quad \langle Px_t - Px, \phi(h) \rangle &\geq ct\|h\|^2 - \lambda M^{-1} t^{-1} \|Px_t - Px\| \|j(Px_t - Px)\| \\
&\geq ct\|h\|^2 - \lambda \|h\| \|Px_t - Px\| \\
&\geq ct\|h\| - \lambda \|Px_t - Px\| \\
&\geq (c/M - \lambda) \|Px_t - Px\|
\end{aligned}$$

As P is locally Lipschitzian, we have for $x, x_t \in B(x_0, 2\varepsilon_1)$

$$\|Px_t - Px\| \leqslant M\|x_t - x\|.$$

Applying (9.157), we have

$$\begin{aligned}
\langle Px_t - Px, (y - Px)^* \rangle &= \langle Px_t - Px, \|y - Px\| \phi(h) - \|y - Px\| \phi(h), (y - Px)^* \rangle \\
&\geq (c/M - \lambda) \|Px_t - Px\| \|y - Px\| \\
&\quad - (c/2M - \lambda) \|Px_t - Px\| \|y - Px\| \\
&\geq (c/2M) \|Px_t - Px\| \|y - Px\|. \tag{9.158}
\end{aligned}$$

From (9.156) and (9.158), we have

$$\|y - Px_t\|^2 \leqslant \|y - Px\|^2 - 2\langle Px_t - Px, (y - Px_t)^* \rangle$$
$$= \|y - Px\|^2 - 2\langle Px_t - Px, (y - Px)^* - (y - Px)^* + (y - Px_t)^* \rangle$$
$$= \|y - Px\|^2 - 2\langle Px_t - Px, (y - Px)^* \rangle$$
$$+ 2\langle Px_t - Px, (y - Px)^* - (y - Px_t)^* \rangle$$
$$\leqslant \|y - Px\|^2 - 2\langle Px_t - Px, (y - Px)^* \rangle$$
$$+ 2\|Px_t - Px\| \|(y - Px)^* - (y - Px_t)^*\|$$
$$\leqslant \|y - Px\|^2 - (cd/M)\|Px_t - Px\|$$
$$+ 2\|Px_t - Px\| \|(y - Px)^* - (y - Px_t)^*\|$$

Since $y - Px_t \to y - Px$ as $t \to 0$ and J is strongly u.s.c., we may select $t > 0$ so small that $\|(y - Px)^* - (y - Px_t)^*\| \leqslant (cd/2M)$. Then, it follows that

$$\|y - Px_t\|^2 \leqslant \|y - Px\|^2 - (cd/2M)\|Px_t - Px\|.$$

Recall that for sufficiently small t, we have

$$\langle Px_t - Px, \phi(x_t - x) \rangle \geq c\|x_t - x\|^2 - \lambda M^{-1}\|Px_t - Px\| \|j(Px_t - Px)\|.$$

This yields

$$\|x_t - x\| \leqslant \frac{1 + \sqrt{1 + 4c\lambda M^{-1}}}{2c} \|Px_t - Px\|.$$

So

$$[c^2 d / M(1 + \sqrt{1 + 4c\lambda M^{-1}})]\|x_t - x\| \leqslant \|y - Px\|^2 - \|y - Px_t\|^2.$$

Thus, we find that $\|y - Px\|^2 - \|y - Px_t\|^2 \geq 0$. Hence, $\|y - Px_t\| \leqslant \|y - Px\| \leqslant r$ and $x_t \in B(x_0, 2\varepsilon)$.

Notice that $x_t \in B(x_0, 2\varepsilon)$ and $\|y - Px_t\| \leqslant r$ imply

$$\|x_t - x\| \leqslant \frac{1 + \sqrt{1 + 4c\lambda M^{-1}}}{2c} \|Px_t - Px\|$$
$$\leqslant \frac{1 + \sqrt{1 + 4c\lambda M^{-1}}}{2c} (\|Px_t - y\| + \|y - Px_0\|)$$
$$\leqslant r\frac{1 + \sqrt{1 + 4c\lambda M^{-1}}}{c} \leqslant \varepsilon.$$

Let $\psi(x) = [M(1 + \sqrt{1 + 4c\lambda M^{-1}})/c^2 d]\|y - Px\|^2$ and define $g : D \to D$ such that $gx = x_t$. Then,

$$\|x - gx\| \leqslant \psi(x) - \psi(gx).$$

Observe that D being closed subset of X, it is complete. Since ψ is the continuous map from the complete metric space D into nonnegative reals, by Caristi–Kirk's fixed point theorem g has a fixed point in D. Note that $\|x_t - x\| = t\|h\| \neq 0$; this is a contradiction. ∎

Observation

- Theorem 9.36 generalizes results of Park and Park [443] and hence those of Browder [114] and Ray [502].
- Geometrical structures of Y^* in Theorem 9.34 are not required as opposed to [114] and [502].

Example 9.10 Let $X = Y = \mathbb{R}$. Then, $Y^* = \mathbb{R}^* = \mathbb{R}$. Define $\phi : X \to Y^*$ implicitly which satisfy conditions

(i) $\phi(X)$ is dense in Y^*,
(ii) for each $x \in X$ and each $\alpha \geq 0$,

$$\|\phi(x)\| \leqslant \|x\|, \quad \|\phi(\alpha x)\| = \alpha\|\phi(x)\|.$$

and $P : X \to Y$ explicitly by

$$Px = \frac{2c}{1 + \sqrt{1 + 4c\lambda M^{-1}}}x + \beta \ \forall x \in X, \ \beta \in \mathbb{R}.$$

Notice that the condition

$$\langle Pu - Px, \phi(u - x) + \lambda M^{-1} j(Pu - Px)\rangle \geq c\|u - x\|^2$$

for all $u \in X$ sufficiently near to x yields

$$\|u - x\| \leqslant \frac{1 + \sqrt{1 + 4c\lambda M^{-1}}}{2c} \|Pu - Px\|. \tag{9.159}$$

Indeed, for all $u \in X$ sufficiently near to x,

$$\langle Pu - Px, \phi(u - x) + \lambda M^{-1} j(Pu - Px)\rangle \geq c\|u - x\|^2$$

implies that

$$\langle Pu - Px, \phi(u - x)\rangle + \langle Pu - Px, \lambda M^{-1} j(Pu - Px)\rangle \geq c\|u - x\|^2,$$

i.e.,

$$c\|u - x\|^2 \leqslant \|Pu - Px\|\|\phi(u - x)\| + \lambda M^{-1}\|Pu - Px\|\|j(Pu - Px)\|,$$

i.e.,

$$c\|u - x\|^2 \leqslant \|Pu - Px\|\|u - x\| + \lambda M^{-1}\|Pu - Px\|^2.$$

By solving the above quadratic in $d = \frac{\|u-x\|}{\|Pu-Px\|}$, we can easily find (9.158). Clearly, P satisfies the above condition for all $x \in X$ and all $u \in X$ sufficiently near to x. By definition of P, it is evident that $P(X) = Y$.

9.7.1 Generalized Directional Contractor and Its Application

In this section, we establish a surjectivity theorem for some nonlinear operators by using the notion of generalized directional contractor. In the sequel, we apply our result to obtain solution of certain functional equations.

Altman's fundamental paper [12] contains very useful notion of directional contractor as given below:

Let X and Y be two Banach spaces. Let $P : \mathscr{D}(P) \subseteq X \to Y$ be a nonlinear operator from a linear subspace $\mathscr{D}(P)$ of X to Y, $\Gamma(x) : Y \to \mathscr{D}(P)$ a bounded linear operator associated with $x \in \mathscr{D}(P)$. Suppose there exists a positive number $q = q(P) < 1$ such that for any $x \in \mathscr{D}(P)$ and $y \in Y$, there exist $\varepsilon = \varepsilon(x, y) \in (0, 1]$ satisfying

$$\|P(x + \varepsilon\Gamma(x)y) - Px - \varepsilon y\| \leqslant q\,\varepsilon\,\|y\|. \tag{9.160}$$

Then, $\Gamma(x)$ is called a directional contractor for P at $x \in \mathscr{D}(P)$ and $\Gamma : \mathscr{D}(P) \subset X \to \mathcal{L}(Y, X)$ is called a directional contractor for P, where $\mathcal{L}(Y, X)$ denotes the set of all linear continuous maps of Y into X. If there exists a constant $B(> 0)$ such that $\|\Gamma(x)\| \leqslant B$ for all $x \in \mathscr{D}(P)$, then Γ is called a bounded directional contractor for P.

We now introduce the concept of generalized contractor as follows:

Definition 9.9 Let X and Y be two Banach spaces. Let $P : \mathscr{D}(P) \subset X \to Y$ be a nonlinear operator from a linear subspace $\mathscr{D}(P)$ of X to Y, $\Gamma(x) : Y \to \mathscr{D}(P)$ a bounded linear operator associated with $x \in \mathscr{D}(P)$. Suppose there exists a positive number $q = q(P) < 1$ such that for any $x \in \mathscr{D}(P)$ and $y \in Y$, there exist $\varepsilon = \varepsilon(x, y) \in (0, 1]$ and a nonincreasing function $c : [0, \infty) \to (0, q^{-1/2})$ satisfying

$$\|P(x + \varepsilon\Gamma(x)y) - Px - \varepsilon y\| \leqslant q\,\varepsilon\,c(\|x\|)\|y\|. \tag{9.161}$$

Then, $\Gamma(x)$ is called a generalized directional contractor for P at $x \in \mathscr{D}(P)$, and $\Gamma : \mathscr{D}(P) \subset X \to L(Y, X)$ is called a generalized directional contractor for P, where $L(Y, X)$ denotes the set of all linear continuous maps of Y into X. If there exists a constant $B(> 0)$ such that $\|\Gamma(x)\| \leqslant B$ for all $x \in \mathscr{D}(P)$, then Γ is called a bounded generalized directional contractor for P. It follows from the above definition that $\Gamma(x)y = 0$ implies $y = 0$; i.e., $\Gamma(x)$ is injective.

Notice that every generalized directional contractor is a directional contractor and an inverse Gâuteaux derivative is a directional contractor. Moreover, $P : \mathscr{D}(P) \subseteq$

$X \to Y$ is said to have closed graph if $x_n \to x$, $x_n \in \mathscr{D}(P)$ and $P x_n \to y$ imply $x \in \mathscr{D}(P)$ and $y = Px$.

By applying the ideas of Ray and Walker [503], we are now ready to prove a surjectivity theorem for generalized directional contractor.

Theorem 9.37 *Let X and Y be two Banach spaces. A nonlinear map $P : \mathscr{D}(P) \subset X \to Y$ which has closed graph and a bounded generalized directional contractor Γ is surjective.*

Proof Define a metric ρ on $\mathscr{D}(P)$ by

$$\rho(x, y) = \max\{\|x - y\|, (1 + q^{1/2})^{-1}\|Px - Py\|\}.$$

As $\mathscr{D}(P)$ has closed graph, $(\mathscr{D}(P), \rho)$ is a complete metric space. Suppose $w \notin \mathscr{R}(P)$ (the range of P). For any $x \in \mathscr{D}(P)$, we set $y = w - Px$. Since P has a bounded generalized directional contractor Γ, we have, for some $0 < \varepsilon(x, y) \leqslant 1$,

$$\|P(x + \varepsilon\Gamma(x)y) - Px - \varepsilon y\| \leqslant q\,\varepsilon\,c(\|x\|)\|y\|. \tag{9.162}$$

Set $\varepsilon\Gamma(x)y = h$. Then, we have

$$\|h\| = \|\varepsilon\Gamma(x)y\| \leqslant \varepsilon B\|y\| = \varepsilon B\|w - Px\|. \tag{9.163}$$

From (9.162), we have

$$\|P(x + h) - w + (1 - \varepsilon)(w - Px)\| \leqslant q\,\varepsilon\,c(\|x\|)\|w - Px\|$$

which yields

$$\|P(x + h) - w\| - (1 - \varepsilon)\|w - Px\| \leqslant q\,\varepsilon\,c(\|x\|)\|w - Px\|.$$

Therefore, we have

$$\varepsilon\|w - Px\| - q\,\varepsilon\,c(\|x\|)\|w - Px\| \leqslant \|w - Px\| - \|w - P(x + h)\|$$

i.e.,

$$\varepsilon(1 - q\,c(\|x\|))\|w - Px\| \leqslant \|w - Px\| - \|w - P(x + h)\|. \tag{9.164}$$

Again, from (9.162), we have

$$\|P(x + h) - Px\| - \varepsilon\|w - Px\| \leqslant q\,\varepsilon\,c(\|x\|)\|w - Px\|$$

which yields

$$\|P(x + h) - Px\| \leqslant \varepsilon(1 + q\,c(\|x\|))\|w - Px\|. \tag{9.165}$$

From (9.164) and (9.165), we have

$$\|P(x+h) - Px\| \leqslant \Big(1 + q\,c(\|x\|)\Big)\Big(1 - q\,c(\|x\|)\Big)^{-1}(\|w - Px\| - \|w - P(x+h)\|)$$
$$\leqslant \Big(1 + q^{1/2}\Big)\Big(1 - q^{1/2}\Big)^{-1}(\|w - Px\| - \|w - P(x+h)\|).$$

Using (9.163) again, we have

$$\|h\| \leqslant \varepsilon B \|w - Px\|$$
$$\leqslant B\Big(1 - q\,c(\|x\|)\Big)^{-1}(\|w - Px\| - \|w - P(x+h)\|)$$
$$\leqslant B\Big(1 - q^{1/2}\Big)^{-1}(\|w - Px\| - \|w - P(x+h)\|).$$

Let $a = \max(B, 1)$ and $\varphi(x) = a\Big(1 - q^{1/2}\Big)^{-1}\|w - Px\|$. Then, φ is continuous with respect to metric ρ. Therefore, if we set $fx = x + h$, then $fx \neq x$. Indeed, if $h = 0$, then from (9.161), we have

$$\varepsilon \|y\| \leqslant q\varepsilon\, c(\|x\|)\|y\| \leqslant q^{1/2}\varepsilon\, \|y\| < \varepsilon\, \|y\|.$$

But since $w \notin \mathscr{R}(P), y = Px - w \neq 0$. Therefore, $fx \neq x$ and $\rho(x, fx) \leqslant \varphi(x) - \varphi(fx)$. This is a contradiction to Caristi–Kirk's fixed point theorem. Hence we conclude that $w \in \mathscr{R}(P)$.   ∎

Let X and Y be two Banach spaces. Let $P : \mathscr{D}(P) \subset X \to Y$, and let $x \in X$. We now consider a special class of generalized directional contractors. Let $\Gamma(x)(P)$ be a set of generalized directional contractors for P at $x \in \mathscr{D}(P)$ called class (S) if there exist a positive number $q = q(P) < 1$, a constant $B > 0$ and a nonincreasing function $c : [0, \infty) \to (0, q^{-1/2})$ with the following property:

For each $y \in \Gamma(x)(P)$, there exist a positive number $\varepsilon = \varepsilon(x, y) \leqslant 1$ and an element $\bar{x} \in \mathscr{D}(P)$ such that

(S_1) $\|P\bar{x} - Px - \varepsilon y\| \leqslant q\varepsilon c(\|x\|)\|y\|$ and
(S_2) $\|\bar{x} - x\| \leqslant \varepsilon B\|y\|$.

Now we apply the above results to obtain the solution of certain functional equations.

Theorem 9.38 *Let X and Y be two Banach spaces. Let $P : \mathscr{D}(P) \subset X \to Y$ has closed graph. For $x \in \mathscr{D}(P)$, let $\Gamma(x)(P)$ denote the class (S). Suppose that y_0 is such that for each $x \in \mathscr{D}(P)$, the element $y_0 - Px$ belongs to the closure of the set $\Gamma(x)(P)$ defined by (S_1) and (S_2). Then, the equation $Px - y_0 = 0, x \in \mathscr{D}(P)$ has a solution.*

Proof Suppose, if possible, $Px - y_0 = 0, x \in \mathscr{D}(P)$ has no solution. Set $y = y_0 - Px \neq 0$, then by hypothesis $y \in \overline{\Gamma(x)(P)}$. So we can choose y' in $\Gamma(x)(P)$ and a $\alpha > 0$ such that $\|y - y'\| \leqslant \alpha\|y\|$. Note that $\alpha < 1$ and does not depend on x.

Since $y' \in \Gamma(x)(P)$, there exists $\bar{x}$ such that

$$\|P\bar{x} - Px - \varepsilon y'\| \leqslant q\varepsilon\, c(\|x\|)\|y'\|. \tag{9.166}$$

From the above inequality (9.166), we have

$$\|P\bar{x} - y_0 + y_0 - Px - \varepsilon y'\| \leqslant q\varepsilon\, c(\|x\|)\|y'\|.$$

As $y = y_0 - Px$, we obtain

$$\|P\bar{x} - y_0 + y - \varepsilon y'\| \leqslant q\varepsilon\, c(\|x\|)\|y'\|. \tag{9.167}$$

Choose $q' > 0$ such that $q < q' < 1$. After having chosen q', we may choose $\alpha > 0$ sufficiently small such that $(\alpha + 1) \leq qq'$. Since $\|y - y'\| \leq \alpha\|y\|$, we have $\|y'\| \leqslant (1 + \alpha)\|y\|$. From this and (9.167), we have

$$\|P\bar{x} - y_0 + y - \varepsilon y'\| \leqslant q'\varepsilon\, c(\|x\|)\|y\|.$$

On the other hand, we have

$$(\|P\bar{x} - y_0 + (1 - \varepsilon)y\| - \|P\bar{x} - y_0 + y - \varepsilon y'\|) \leqslant \varepsilon\, \|y - y'\| \leqslant \varepsilon\, \alpha\, \|y\|.$$

Hence from (9.167) and the above inequalities, we have

$$\|P\bar{x} - y_0 + (1 - \varepsilon)y\| - \varepsilon\, \alpha\, \|y\| \leqslant q'\varepsilon\, c(\|x\|)\|y\|.$$

From this, we have

$$\|P\bar{x} - y_0\| - (1 - \varepsilon)\|y\| - \varepsilon\, \alpha\, \|y\| \leqslant q'\varepsilon\, c(\|x\|)\|y\|.$$

Therefore, we obtain

$$\varepsilon(1 - q'c(\|x\|) - \alpha)\|y\| \leqslant \|y\| - \|P\bar{x} - y_0\|$$

which implies

$$\varepsilon(1 - q'q^{-1/2} - \alpha)\|y\| \leqslant \|y\| - \|P\bar{x} - y_0\|.$$

If we choose $\alpha > 0$ so that $\beta = 1 - q'q^{-1/2} - \alpha > 0$, then we obtain

$$\varepsilon\, \beta\|y_0 - Px\| \leqslant \|Px - y_0\| - \|P\bar{x} - y_0\|. \tag{9.168}$$

But from (9.166), we have

$$\|P\bar{x} - Px\| \leqslant \varepsilon(qc(\|x\|) + 1)\|y'\|$$
$$\leqslant \varepsilon(qc(\|x\|) + 1)(\alpha + 1)\|y\|$$

or $\qquad \|P\bar{x} - Px\| \leqslant \varepsilon(q^{1/2} + 1)(\alpha + 1)\|y_0 - Px\|.$

Using (9.168), the above inequality yields

$$\|P\bar{x} - Px\| \leqslant (q^{1/2} + 1)(\alpha + 1)\beta^{-1}(\|Px - y_0\| - \|P\bar{x} - y_0\|).$$

Since $\|\bar{x} - x\| \leqslant \varepsilon B\|y'\| \leqslant \varepsilon B(\alpha + 1)\|y\| = \varepsilon B(\alpha + 1)\|y_0 - Px\|$ we have

$$\|\bar{x} - x\| \leqslant B(\alpha + 1)\beta^{-1}(\|Px - y_0\| - \|P\bar{x} - y_0\|).$$

We now define a metric ρ on $\mathscr{D}(P)$ by

$$\rho(x, y) = \max\{\|x - y\|, (1 + q^{1/2})^{-1}\|Px - Py\|\}.$$

Set $fx = \bar{x}$. Since $y' \neq 0 (\alpha < 1)$, we have $x \neq \bar{x}$. Take $a = \{B, 1\}$ and set $\varphi(x) = a(\alpha + 1)\beta^{-1}\|Px - y_0\|$. Then,

$$\rho(x, fx) \leqslant \varphi(x) - \varphi(fx).$$

This is a contradiction to Caristi–Kirk's fixed point theorem. Hence, we conclude that $Px - y_0 = 0, x \in \mathscr{D}(P)$ has a solution. ∎

9.8   Application to Simultaneous Complementarity Problems

The study of complementarity problems came into existence in the early sixties of twentieth century. Since then, a variety of problems, in particular the explicit complementarity problems and the implicit complementarity problems, were discussed and studied by many researchers. It is fairly well known fact that the complementarity problems have got a wide range of applications in the areas such as optimization theory, engineering, structural mechanics, theory of elasticity, lubrication theory, economics, variational calculus, equilibrium theory on networks, stochastic optimal control. For more details of these applications, one may refer to [149, 150, 153, 193, 194, 284, 316, 382, 383, 412, 440, 441].

Let $\langle E, E^* \rangle$ be a dual system of locally convex spaces and let K be a closed convex cone in E. Let K^* be the dual of K, i.e.,

$$K^* = \{u \in E^* : \langle x, u \rangle \geqslant 0 \text{ for all } x \in E\}.$$

Let $f : K \to E^*$ and $g : K \to E$ be mappings. The explicit complementarity problem and the implicit complementarity problem are as follows, respectively:
E.C.P: Find x_0 in K such that $f(x_0) \in K^*$ and

$$\langle x_0, f(x_0) \rangle = 0,$$

I.C.P: Find x_0 in K such that $f(x_0) \in K^*$, $g(x_0) \in K$ and

$$\langle g(x_0), f(x_0) \rangle = 0.$$

For more details of these problems, the reader may refer to [49, 50, 193, 194, 316, 412, 440, 441].

For mappings $f_1, f_2 : K \to E^*$ and $g : K \to E$, we consider the simultaneous explicit complementarity problem and implicit complementarity problem as follows:
S.E.C.P(f_1, f_2, K) : Find x_0 in K such that $f_1(x_0) \in K^*$, $f_2(x_0) \in K^*$ and

$$\langle x_0, f_1^n(x_0) \rangle = 0 \text{ and } \langle x_0, f_2^n(x_0) \rangle = 0,$$

S.I.C.P(f_1, f_2, g, K) : Find x_0 in K such that $f_1(x_0) \in K^*$, $f_2(x_0) \in K^*$, $g(x_0) \in K$ and

$$\langle g(x_0), f_1^n(x_0) \rangle = 0 \text{ and } \langle g(x_0), f_2^n(x_0) \rangle = 0.$$

Remark 9.10 For $f_1 = f_2$ and $n = 1$, the problem S.E.C.P(f_1, f_2, K) contains E.C.P as a particular case.

Remark 9.11 For $f_1 = f_2$ and $n = 1$, the problem S.I.C.P(f_1, f_2, K) contains I.C.P as a particular case.

Remark 9.12 If we take $n = 1$ and $g(x_0) = x_0 - m(x_0)$, where m is a point to point mapping from E into itself, then the problem S.I.C.P(f_1, f_2, g, K) reduces to the problem of finding x_0 in K such that $f_1(x_0) \in K^*$, $f_2(x_0) \in K^*$ and

$$\langle x_0 - m(x_0), f_1(x_0) \rangle = 0 \text{ and } \langle x_0 - m(x_0), f_2(x_0) \rangle = 0.$$

We note that the strongly nonlinear quasi-complementarity problem discussed in Siddiqi and Ansari [557] is a particular case of the above problem.

Remark 9.13 If we take $n = 1$ and g is the identity mapping on K, then the problem S.I.C.P(f_1, f_2, g, K) reduces to the problem of finding x_0 in K such that $f_1(x_0) \in K^*$, $f_2(x_0) \in K^*$ and

$$\langle x_0, f_1(x_0) \rangle = 0 \text{ and } \langle x_0, f_2(x_0) \rangle = 0.$$

We note that a particular case of the above problem has been discussed and studied by Noor [424, 425].

In what follows, let $(\mathcal{H}, \langle \cdot, \cdot \rangle)$ be a Hilbert space and K a closed convex cone in $\mathcal{H}$. Suppose D is a subset of $\mathcal{H}$ and $f_1, f_2, g : D \to \mathcal{H}$ are mappings. In 1996, Pathak et al. [457] considered the following simultaneous implicit complementarity problem:

S.I.C.P : Find x_0 in D such that $g(x_0) \in K$, $f_1(x_0) \in K^*$, $f_2(x_0) \in K^*$,

$$\langle g(x_0), f_1^n(x_0) \rangle = 0 \text{ and } \langle g(x_0), f_2^n(x_0) \rangle = 0.$$

Definition 9.10 Let D be a subset of $\mathcal{H}$, and consider the mappings $f_1, f_2, g : D \to \mathcal{H}$. We say that f_1 and f_2 are pairwise n-Lipschitz mappings with respect to g if there exists some $\beta > 0$ such that

$$\|f_1^n(x) - f_2^n(y)\| \leqslant \beta \cdot \|g(x) - g(y)\|$$

for all x, y in D.

In 1996, Pathak et al. [457] introduced the following:

Definition 9.11 Let D be a subset of $\mathcal{H}$ and consider the mappings $f_1, f_2, g : D \to \mathcal{H}$. We say that f_1 and f_2 are pairwise n-strongly monotone mappings with respect to g if there exists some $a > 0$ such that

$$\langle f_1^n(x) - f_2^n(y), g(x) - g(y) \rangle \geqslant a \cdot \|g(x) - g(y)\|$$

for all x, y in D.

Theorem 9.39 (Pathak et al. [457]) *Let $\mathcal{H}$ be a Hilbert space and K a closed convex cone in $\mathcal{H}$. If, for a subset D of $\mathcal{H}$ the mappings $f_1, f_2, g : D \to \mathcal{H}$ satisfy the following:*

(1) *f_1 and f_2 are pairwise n-strongly monotone with respect to g,*
(2) *f_1 and f_2 are pairwise n-Lipschitz with respect to g,*
(3) *there exists a real number $r > 0$ such that*

$$r \cdot \beta^2 < 2a < \frac{1}{r} + r \cdot \beta^2$$

where a and β are as in Definitions 9.10 and 9.11, respectively,
(4) *$K \subseteq g(D)$,*

then the problem S.I.C.P is solvable. Moreover, if g is one to one, then the problem S.I.C.P has a unique solution in D.

Proof Using (4), we consider the mappings $h_1, h_2 : K \to \mathcal{H}$ defined by

$$h_1(u) = f_1(x) \text{ and } h_2(u) = f_2(x),$$

where x is an arbitrary element of $g^{-1}(u)$ and $u \in K$. Since f_1 and f_2 are pairwise n-strongly monotone and n-Lipschitz with respect to g, h_1 and h_2 will have the following properties:

(5) $\langle h_1^n(u) - h_2^n(v), u - v \rangle \geqslant a \cdot \|u - v\|^2$,
(6) $\|h_1^n(u) - h_2^n(v)\| \leqslant \beta \cdot \|u - v\|$.

We now see that the problem S.I.C.P is equivalent to the following complementarity problem:

Find u in K such that $h_1(u) \in K^*, h_2(u) \in K^*$,

$$\langle u, h_1^n(u) \rangle = 0 \text{ and } \langle u, h_2^n(u) \rangle = 0.$$

From Proposition 1.22, we see that the above problem has a solution if and only if the mappings $T_1, T_2 : K \to \mathcal{H}$ defined by

$$T_1^n(u) = P_K(u - r \cdot h_1^n(u)) \text{ and } T_2^n(u) = P_K(u - r \cdot h_2^n(u)),$$

respectively, where r is the real number defined in (3), which has a common fixed point in K. In fact, we have

$$\begin{aligned}
\|T_1^n(u) - T_2^n(v)\|^2 &= \|P_K(u - r \cdot h_1^n(u)) - P_K(v - r \cdot h_2^n(v))\|^2 \\
&\leqslant \|u - r \cdot h_1^n(u) - (v - r \cdot h_2^n(v))\| \\
&= \|u - v - r \cdot (h_1^n(u) - h_2^n(v))\| \\
&\leqslant \|u - v\|^2 - 2r \cdot \langle u - v, h_1^n(u) - h_2^n(v) \rangle + r^2 \cdot \|h_1^n(u) - h_2^n(v)\|^2 \\
&\leqslant \|u - v\|^2 - 2r \cdot a\|u - v\|^2 + r^2 \cdot \beta^2 \cdot \|u - v\|^2 \\
&= (1 - 2r \cdot a + r^2 \cdot \beta^2) \cdot \|u - v\|^2
\end{aligned}$$

or

$$\|T_1^n(u) - T_2^n(v)\| < \theta \cdot \|u - v\|,$$

where $\theta = (1 - 2r \cdot a + r^2 \cdot \beta^2)^{1/2}$. By assumption (3) and noting the facts that every Hilbert space is a completely metrically convex metric space is a complete metrically convex metric space and every metrically convex space is locally convex, we have that T_1 and T_2 satisfy all the assumptions of Lemma 9.4. Hence, T_1 and T_2 have a unique common fixed point in K. This completes the proof.

Exercises

9.1 Use Banach's fixed point theorem to show that

$$x_1 = \frac{1}{3}x_1 - \frac{1}{4}x_2 + \frac{1}{4}x_3 - 1$$

$$x_2 = -\frac{1}{2}x_1 + \frac{1}{5}x_2 + \frac{1}{4}x_3 + 2$$

$$x_3 = \frac{1}{5}x_1 - \frac{1}{3}x_2 + \frac{1}{4}x_3 - 2$$

has a unique solution.

9.2 Show that the initial value problem

$$\begin{cases} \frac{du}{dt} = u^{\frac{1}{2}} & t \geq 0, \\ u(0) = 0 \end{cases}$$

has more than one solution.

9.3 Show that the solutions of the initial value problem

$$\frac{dx}{dt} = |x|^{1/2}, \quad x(0) = 0$$

are $x_1(t) = 0$ and $x_2(t) = \frac{t|t|}{4}$. Does this contradicts Picard's theorem? Find other solutions.

9.4 Show that an initial value problem given by

$$\frac{d^2x}{dt^2} = f(t, x), \quad x(t_0) = x_0, \quad x'(t_0) = x_1$$

involving a second-order differential equation can be transform into a Volterra integral equation.

9.5 Consider the Volterra integral equation

$$f(t) = g(t) + \int_0^t K(t, s) f(s) ds, \quad a \leq t \leq b,$$

where K is continuous on $[a, b] \times [a, b]$. Show that the equation has a unique solution $f \in C[a, b]$ for each fixed $g \in C[a, b]$.

9.6 Consider the Fredholm integral equation

$$f(t) = g(t) + \lambda \int_0^1 K(t, s) f(s) ds,$$

where $g \in L_2[0, 1]$ and $K \in L_2([0, 1] \times [0, 1])$. Prove that if $g = 0$ implies $f = 0$, then there exists a unique solution of the equation for any $g \in L_2[0, 1]$.

9.7 Consider the nonlinear integral equation

$$v(t) = x(t) - \lambda \int_a^b K(t, s, x(s)) ds,$$

where v and K are continuous on $[a, b]$ and $[a, b] \times [a, b] \times \mathbb{R}$, respectively, and K satisfies on $[a, b] \times [a, b]$ a Lipschitz condition of the form

$$|K(t, s, u_1) - K(t, s, u_2)| \leqslant c|w_1 - w_2|.$$

Then, show that

$$|\lambda| < \frac{1}{c(b - a)}.$$

9.8 Approximate numerical solutions of $f(x) = 0$. To find approximate numerical solutions of $f(x) = 0$, we may convert the equation in the form $x = g(x)$, choose an initial value x_0 and compute

$$x_n = g(x_{n-1}) \ , n = 1, 2, \ldots.$$

Suppose that g is continuously differentiable on some interval $J = [x_0 - r, x_0 + r]$ and satisfies $|g'(x)| \leqslant c < 1$ on J and

$$|g(x_0) - x_0| < (1 - c)r.$$

Suppose that $x = g(x)$ has a unique solution x on J, the iterative sequence $\{x_n\}$ converges to that solution and the error estimates are:

(i) $|x - x_n| < c^n r$
(ii) $|x - x_n| \leqslant \frac{c}{1-c}|x_n - x_{n-1}|$.

9.9 Newton's method. Let f be real-valued and twice continuously differentiable on an interval $[a, b]$. Let $\hat{x}$ be a simple zero of f in $[a, b]$. Show that Newton's method defined by

$$x_{n+1} = g(x_n), \ \ g(x_n) = x_n - \frac{f(x_n)}{f'(x_n)}$$

is a contraction in some neighbourhood of $\hat{x}$ so that the iterative sequence converges to $\hat{x}$ for any x_0 sufficiently close to $\hat{x}$.

Chapter 10
Applications of Fixed Point Theorems for Multifunction to Integral Inclusions

The most vitally characteristic fact about mathematics, in my opinion, is its quite peculiar relationship to the natural sciences, or, more generally, to any science which interprets experience on a higher more than on a purely descriptive level....

John von Neumann (1947)
Science will not light the lamp in a person whose soul has no fuel.

Meichel de Montaigne
When I was a boy of 14 my father was so ignorant I could hardly stand to have the old man around. But when I got to be 21, I was astonished at how much the old man had learned in 7 years.

Mark Twain

Dynamical systems described by differential equations with continuous right-hand sides were the areas of vigorous steady in the later half of the twentieth century in applied mathematics, in particular, in the study of viscous fluid motion in a porous medium, propagation of light in an optically nonhomogeneous medium, determining the shape of a solid of revolution moving in a flow of gas with least resistance, etc. Euler's equation plays a key role in dealing with the existence of the solution of such problems. On the other hand, Filippov [232] has developed a solution concept for differential equations with a discontinuous right-hand side. In practice, such dynamical systems do arise and require analysis. Examples of such systems are mechanical systems with Coulomb friction modelled as a force proportional to the sign of a velocity, systems whose control laws have discontinuities.

© Springer Nature Singapore Pte Ltd. 2018
H. K. Pathak, *An Introduction to Nonlinear Analysis and Fixed Point Theory*,
https://doi.org/10.1007/978-981-10-8866-7_10

10.1   Continuous Solution for Nonlinear Integral Inclusions

In this section, we present new results which guarantee that the Fredhlom integral inclusion

$$y(t) \in \int_0^T k(t,s)[a(s)g(s,y(s)) + \tilde{f}(s,y(s),y'(s))]ds \quad a.e. \ t \in [0,T], \quad (10.1)$$

having a finite derivative at each $s \in [0,T]$ has a positive solution $y \in L_p[0,T], 1 \leqslant p < \infty$, or has a nonnegative solution $y \in C[0,T]$.

Throughout this section, $\mathbb{R}^+$ and $\mathbb{R}$ denote the set $[0,\infty)$ and the set of real numbers, respectively, $T > 0$ is fixed, $a : [0,T] \to \mathbb{R}^+, k : [0,T] \times [0,T] \to \mathbb{R}$, $g : [0,T] \times \mathbb{R} \to 2^{\mathbb{R}}, f : [0,T] \times \mathbb{R} \to 2^{\mathbb{R}}$ and $\tilde{f} : [0,T] \times \mathbb{R}^+ \times \mathbb{R} \to 2^{\mathbb{R}}$, here $2^{\mathbb{R}}$ denotes the family of nonempty subsets of $\mathbb{R}$.

Our results in this section are motivated by the multivalued analogue of Krasnoselski's fixed point theorem due to Agarwal and O'Regan [6] in a cone.

Let $E = (E, \| \cdot \|)$ be a Banach space and $C \subseteq E$. For $\rho > 0$ let

$$\Omega_\rho = \{x \in E : \|x\| < \rho\} \text{ and } \partial\Omega_\rho = \{x \in E : \|x\| = \rho\}.$$

10.1.1 $L_p[0, T]$ Solutions

Here, we discuss Fredholm nonlinear integral inclusion

$$y(t) \in \int_0^T k(t,s)[a(s)g(s,y(s)) + \tilde{f}(s,y(s),y'(s))]ds \quad a.e. \ t \in [0,T], \quad (10.2)$$

having a finite derivatives at each $s \in [0,T]$, where $a : [0,T] \to \mathbb{R}^+, k : [0,T] \times [0,T] \to \mathbb{R}, g : [0,T] \times \mathbb{R} \to K(\mathbb{R})$ and $\tilde{f} : [0,T] \times \mathbb{R}^+ \times \mathbb{R} \to K(\mathbb{R})$.

At this juncture, it is desirable to introduce generalized multivalued Nemytskiĭ operator.

Definition 10.1 A multivalued operator $N_{g,\tilde{f}}^{\lambda}$ is called *generalized Nemytskiĭ operator* if $\exists \lambda > 0$ such that for each $p : 1 \leqslant p < \infty$,

$$N_{g,\tilde{f}}^{\lambda} u = \Big\{ y \in L_p[0,T] : y(t) \in a(t)g(t,u(t)) + \tilde{f}(t,u(t),u'(t)) \ a.e. \ t \in [0,T]$$

$$\text{and } \sup_{t \in [0,T]} |u'| \leqslant \lambda \Big\}.$$

Remark 10.1 In particular, when nonlinear function $\tilde{f}(t,u(t),u'(t))$ is independent of derivative, we assume $a(t)g(t,u(t)) + \tilde{f}(t,u(t),u'(t)) = f(t,u(t))$ and denote $N_{g,\tilde{f}}^{\lambda} u$ by $N_f u$. Then

$$N_f u = \{y \in L_p[0, T] : y(t) \in f(t, u(t)) \ a.e. \ t \in [0, T]\}.$$

The multivalued operator N_f is called *Nemytskiǐ operator*.

Now, we intend to know what conditions one requires on a, k, g and $\tilde{f}$ in order that the inclusion (10.1) has a positive solution $y \in L_p[0, T]$, where $1 \leqslant p < \infty$. Here by a positive solution y, we mean $y(t) > 0$ for a.e. $t \in [0, T]$. Throughout this subsection, $\| \cdot \|_q$ denotes the usual norm on L_q for $1 \leqslant q \leqslant \infty$.

Theorem 10.1 (Pathak [450]) *Let* $a : [0, T] \to \mathbb{R}^+$, $k : [0, T] \times [0, T] \to \mathbb{R}$, $g : [0, T] \times \mathbb{R} \to K(\mathbb{R})$ *and* $\tilde{f} : [0, T] \times \mathbb{R}^+ \times \mathbb{R} \to K(\mathbb{R})$ *and assume the following conditions hold:*

the map $u \longmapsto a(t)g(t, u) + \tilde{f}(t, u, u')$ *is upper semicontinuous for a.e.* $t \in [0, T]$;
$$(10.3)$$

the graph of $ag + \tilde{f}$ *belongs to the* σ*-field* $\mathscr{L} \otimes \mathscr{B}(\mathbb{R} \times \mathbb{R})$ $\quad (10.4)$

(*here* $\mathscr{L}$ *denotes the Lebesgue* σ*-field on* $[0, T]$ *and* $\mathscr{B}(\mathbb{R} \times \mathbb{R}) = \mathscr{B}(\mathbb{R}) \otimes \mathscr{B}(\mathbb{R})$ *is the* σ *Borel field in* $\mathbb{R} \times \mathbb{R}$);

$\exists p_2, 1 \leqslant p_2 < \infty, a_1 \in L_{p_2}[0, T]$ *and* $a_2, a_3 > 0$ *are constants, with*

$$|a(t)g(t, y) + \tilde{f}(t, y, y')| = \sup\{|z| : z \in a(t)g(t, y) + \tilde{f}(t, y, y')\}$$
$$\leqslant M'\left(a_1(t) + a_2|y|^{\frac{p}{p_2}} + a_3 T^{-\frac{1}{p_2}}|y'|^{\frac{p}{p_2}}\right),$$

for a.e. $t \in [0, T]$ *for all* $y, y' \in \mathbb{R}$, *where* M' *is as defined in (10.7) below*;
$$(10.5)$$

$$(t, s) \longmapsto k(t, s) \ \text{is measurable}; \quad (10.6)$$

$s \longmapsto a(s)$ *is continuous and satisfy the uniform Hölder's continuity condition in the with the exponent* ρ; *i.e., there exists a positive number* b *such that*

$$|a(s_1) - a(s_2)| \leqslant b(|s_1 - s_2|^\rho) \quad (10.7)$$

for all $s_1, s_2 \in [0, T]$ *and* $|a(s)| \leq 2bT + |a(0)| = M'$ *for all* $s \in [0, T], 0 < \rho \leqslant 1$ *and* $M' > 0$;

$\exists 0 < M \leqslant 1, k_1 \in L_p[0, T], k_2 \in L_{p_1}[0, T]$, *here* $\dfrac{1}{p_1} + \dfrac{1}{p_2} = 1$, *such that*

$0 < k_1(t), k_2(t)$ *and a. e.* $t \in [0, T]$ *and* $Mk_1(t)k_2(s) \leqslant k(t, s) \leqslant k_1(t)k_2(s)$
a.e. $t \in [0, T]$, *a.e.* $s \in [0, T]$;
$$(10.8)$$

$$\text{for a. e. } t \in [0, T] \text{ and all } y \in (0, \infty), \ y' \in [-\lambda, \lambda], \ u > 0$$
$$\text{for all } u \in a(t)g(t, y) + \tilde{f}(t, y, y') \tag{10.9}$$

$\exists q \in L^{p_2}[0, T], \ \psi : \mathbb{R}^+ \to \mathbb{R}^+, \psi(u) > 0 \text{ for } u > 0, \text{ is continuous and non-decreasing with } u \text{ and } \theta : \mathbb{R} \to \mathbb{R}^+, \theta(u) > 0 \text{ for } u > 0, \text{ is continuous and nonincreasing} \tag{10.10}$

$$\text{for a.e. } t \in [0, T] \text{ and } y > 0, \ y' \in [-\lambda, \lambda], \ u \geq b(\tau)q(t)\psi(y)\theta(|y'|)$$
$$\text{for all } u \in a(t)g(t, y) + \tilde{f}(t, y, y'); \tag{10.11}$$

$$\exists \alpha > 0 \text{ with } 1 < \frac{\alpha}{2^{\frac{p_2 - 1}{p_2}} M' \|k_1\|_p \|k_2\|_{p_1} \left(\|a_1\|_{p_2}^{p_2} + 2^{p_2 - 1}([a_2]^{p_2}\alpha^p + [a_3]^{p_2}\lambda^p) \right)^{\frac{1}{p_2}}} \tag{10.12}$$

and

$$\exists \beta > 0, \ \beta \neq \alpha \text{ with with } 1 > \frac{\beta}{M\|b\|_p \|k_1\|_p \int_0^T k_2(s)q(s)\psi(\gamma(s)\beta) \, ds}, \tag{10.13}$$

where

$$\gamma(t) = M \frac{k_1(t)}{\|k_1\|_p}. \tag{10.14}$$

Then, (10.1) has at least one positive solution $y \in L_p[0, T]$ and either
(A) $0 < \alpha < \|y\|_p < \beta$ and $y(t) \geq \gamma(t)\alpha$ a.e. $t \in [0, T]$ if $\alpha < \beta$
or
(B) $0 < \beta < \|y\|_p < \alpha$ and $y(t) \geq \gamma(t)\beta$ a.e. $t \in [0, T]$ if $\beta < \alpha$
holds.

Proof Let $E = (L_p[0, T], \|\cdot\|_p)$ and for some $\lambda > 0$,

$$C = \{y \in L_p[0, T] : y(t) \geq \gamma(t)\|y\|_p \text{ a.e. } t \in [0, T] \}.$$

It can easily be shown that $C \subseteq E$ is a cone. Further, let $A = K \circ N^\lambda_{g,\tilde{f}} : C \to 2^E$, where the linear integral (single-valued) operator K is given by

$$Ky(t) = \int_0^T k(t, s)y(s) \, ds,$$

where $y(s) \in a(s)g(s, u(s)) + \tilde{f}(s, u(s), u'(s))$ a.e. $s \in [0, T]$, $\sup\limits_{s \in [0,T]} |u'| \leqslant \lambda$ and $N^\lambda_{g,\tilde{f}}$ is the generalized Nemytskiĭ operator. Then, the nonlinear integral inclusion (10.1) can be expressed as

$$y \in Ay, \quad \text{for some } y \in L_p[0, T] \tag{10.1'}$$

Thus, the problem of finding the solution of nonlinear integral equation (10.1) is reduced into finding fixed point of multivalued operator A.

Notice that A is well defined since if $x \in C$ then (10.3)–(10.5) and using the technique of [133], it guarantees that $N^\lambda_{g,\tilde{f}} x \neq \varnothing$. Now we show that $A : C \to 2^C$. To see this let $x \in C$ and $y \in Ax$. Then, there exists a $v \in N^\lambda_{g,\tilde{f}} x$ with

$$y(t) = \int_0^T k(t, s)v(s) \, ds \quad \text{for a.e. } t \in [0, T].$$

Now

$$|y(t)|^p \leqslant |k_1(t)|^p \left(\int_0^T k_2(s)v(s) \, ds \right)^p \quad \text{for a.e. } t \in [0, T]$$

so

$$\|y\|_p \leqslant \|k_1\|_p \int_0^T k_2(s)v(s) \, ds. \tag{10.15}$$

Now combining (10.15) with (10.8) gives

$$y(t) \geq M \int_0^T k_1(t)k_2(s)v(s) \, ds \geq M \frac{k_1(t)}{\|k_1\|_p} \|y\|_p = \gamma(t)\|y\|_p \quad \text{for a.e. } t \in [0, T].$$

It follows that $y \in C$. Hence, $A : C \to 2^C$. Also notice [4, 133, 394] guarantees that

$$A : C \to 2^C \text{ is upper semicontinuous.} \tag{10.16}$$

Note also that [168, 235, 271](pp. 47 − 49) implies $K : L_{p_2}[0, T] \to L_p[0, T]$ is completely continuous, and $N_{g,\tilde{f}} : L_p[0, T] \to 2^{L_{p_2}[0,T]}$ maps bounded sets into bounded sets. Consequently,

$$A : C \to K(C) \text{ is completely continuous.} \tag{10.17}$$

Let

$$\Omega_\alpha = \{y \in L_p[0, T] : \|y\|_p < \alpha\} \quad \text{and} \quad \Omega_\beta = \{y \in L_p[0, T] : \|y\|_p < \beta\}.$$

Without loss of generality, we may assume that $\beta < \alpha$ (a similar argument holds if $\alpha < \beta$). It is immediate from (10.16) and (10.17) that

$$A : C \cap \overline{\Omega}_\alpha \to K(C) \ \text{is upper semicontinuous and compact.}$$

Next, if we show that

$$\|y\|_p < \|x\|_p \ \text{for all} \ y \in Ax \ \text{and} \ x \in C \cap \partial\Omega_\alpha. \tag{10.18}$$

and

$$\|y\|_p > \|x\|_p \ \text{for all} \ y \in Ax \ \text{and} \ x \in C \cap \partial\Omega_\beta. \tag{10.19}$$

are true, then Theorem 5.167 guarantees that the operator A has a fixed point in $C \cap (\overline{\Omega}_\alpha \backslash \Omega_\beta)$. This in turn implies that (10.1) has at least one solution $y \in L_p[0, T]$ with $\beta \leqslant \|y\|_p \leqslant \alpha$, $y(t) \geq \alpha(t)\beta$ for a.e. $t \in [0, T]$.

Suppose $x \in C \cap \partial\Omega_\alpha$ such that $\sup_{t\in[0,T]} |x'(t)| \leqslant \lambda$, so $\|x\|_p = \alpha$ and $y \in Ax$. Then, for each $t \in [0, T]$, it is easy to observe that there exists a $v \in N_{g,f}^\lambda x$ with

$$y(t) = \int_0^T k(t, s)v(s) \, ds \ \text{for} \ a.e. \ t \in [0, T]$$

$$\leqslant M' \int_0^T k(t, s)\Big[|a_1(s)| + a_2|x(s)|^{\frac{p}{p_2}} + a_3 T^{-\frac{1}{p_2}} |x'(s)|^{\frac{p}{p_2}}\Big] ds \ \text{for} \ a.e. \ t \in [0, T].$$

Now (10.5) and (10.8) guarantee that

$$|y(t)| \leqslant M' k_1(t) \int_0^T k_2(s)\Big[|a_1(s)| + a_2|x(s)|^{\frac{p}{p_2}} + a_3 T^{-\frac{1}{p_2}} |x'(s)|^{\frac{p}{p_2}}\Big] ds \ \text{for} \ a.e. \ t \in [0, T].$$

This together with (10.11) yields

$$\|y(t)\|_p$$

$$\leqslant M' \|k_1\|_p \|k_2\|_{p_1} \Big(\int_0^T \Big[|a_1(s)| + a_2|x(s)|^{\frac{p}{p_2}} + a_3 T^{-\frac{1}{p_2}} |x'(s)|^{\frac{p}{p_2}} \Big]^{p_2} ds \Big)^{\frac{1}{p_2}}$$

$$\leqslant M' \|k_1\|_p \|k_2\|_{p_1} \Big(\int_0^T 2^{p_2-1}\Big[|a_1(s)|^{p_2} + \Big(a_2|x(s)|^{\frac{p}{p_2}} + a_3 T^{-\frac{1}{p_2}} |x'(s)|^{\frac{p}{p_2}}\Big)^{p_2} \Big] ds \Big)^{\frac{1}{p_2}}$$

$$\leqslant 2^{\frac{p_2-1}{p_2}} M' \|k_1\|_p \|k_2\|_{p_1} \Big(\|a_1\|_{p_2}^{p_2} + 2^{p_2-1} \int_0^T ([a_2]^{p_2}|x(s)|^p + [a_3]^{p_2} T^{-1}|x'(s)|^p) ds \Big)^{\frac{1}{p_2}}$$

$$\leqslant 2^{\frac{p_2-1}{p_2}} M' \|k_1\|_p \|k_2\|_{p_1} \Big(\|a_1\|_{p_2}^{p_2} + 2^{p_2-1}([a_2]^{p_2}\alpha^p + [a_3]^{p_2}\lambda^p) \Big)^{\frac{1}{p_2}}$$

$$< \alpha = \|x\|_p$$

and so (10.18) is satisfied.

Now suppose $x \in C \cap \partial\Omega_\beta$ such that $\sup_{t\in[0,T]} |x'(t)| \leqslant \lambda$, so $\|x\|_p = \beta$ and $x(t) \geq b(t)\beta$ for a.e. $t \in [0, T]$, and $y \in Ax$. Then, there exists a $v \in N_{g,f}^\lambda x$ with

$$y(t) = \int_0^T k(t,s)v(s)\,ds \quad \text{for } a.e. \ t \in [0,T].$$

Notice (10.11) guarantees that for each $t \in [0,T]$, $v(s) \geq q(s)\psi(x(s))\theta(|x'(s)|)$ for a.e. $s \in [0,T]$ and this together with (10.8) yields

$$|y(t)| \geq Mk_1(t)\int_0^T k_2(s)q(s)\psi(x(s))\theta(|x'(s)|)\,ds \quad \text{for } a.e. \ t \in [0,T].$$

This together with (10.13) yields

$$\begin{aligned}
\|y\|_p &\geq M\|k_1\|_p\,\|k_2\|_{p_1}\int_0^T q(s)\psi(x(s))\theta(|x'(s)|)\,ds \\
&\geq M\|k_1\|_p\,\|k_2\|_{p_1}\int_0^T q(s)\psi(b(s)\beta)\theta(\lambda)\,ds \\
&> \beta = \|x\|_p
\end{aligned}$$

and so (10.19) is satisfied. To obtain the required conclusion, we now apply Theorem 5.167.

Remark 10.2 If nonlinear function $\tilde{f}(t,u(t),u'(t))$ is independent of derivative, then in view of Remark 10.1 we observe that Theorem 2.1 of Agarwal et al. [4] can be derived from our Theorem 10.1.

10.1.2 C[0, T] Solutions

In this section, we discuss the Fredholm nonlinear integral inclusion

$$y(t) \in \int_0^T k(t,s)[a(t,s)g(t,y(s)) + \tilde{f}(t,s,y(s))]ds \quad a.e. \ t \in [0,T], \quad (10.20)$$

having finite derivative at each $s \in [0,T]$, where $a : [0,T] \times [0,T] \to \mathbb{R}^+$, $k : [0,T] \times [0,T] \to \mathbb{R}$, $g : [0,T] \times \mathbb{R} \to K(\mathbb{R})$ and $\tilde{f} : [0,T] \times [0,T] \times \mathbb{R} \to K(\mathbb{R})$. We will use Theorem 5.167 to establish the existence of a nonnegative solution $y \in C[0,T]$ to (10.20) under less stringent conditions than the conditions as used in Theorem 10.1. Note that $C[0,T] \subset L_p[0,T]$ for each $p : 1 \leq p < \infty$. Let $|\cdot|_0$ denote the usual norm on $C[0,T]$, i.e. $|u|_0 = \sup_{[0,T]} |u(t)|$ for $u \in C[0,T]$.

Theorem 10.2 (Pathak [450]) *Let $1 \leq p < \infty$ and q, $1 < q \leq \infty$, the conjugate to p, $a : [0,T] \times [0,T] \to \mathbb{R}^+$, $k : [0,T] \times [0,T] \to \mathbb{R}$, $g : [0,T] \times \mathbb{R} \to K(\mathbb{R})$ and $\tilde{f} : [0,T] \times [0,T] \times \mathbb{R} \to K(\mathbb{R})$ and assume the following conditions hold:*

$$\text{for each } t \in [0,T], \text{ the map } s \mapsto k(t,s) \text{ is measurable;} \quad (10.21)$$

$$\sup_{t\in[0,T]} \left(\int_0^T |k(t,s)|^q \, ds \right)^{\frac{1}{q}} < \infty; \tag{10.22}$$

$$\int_0^T |k(t',s) - k(t,s)|^q \, ds \to 0 \ \ as \ \ t \to t', \ \ for \ each \ \ t' \in [0,T]; \tag{10.23}$$

$$for \ each \ \ t \in [0,T], \ k(t,s) \ge 0 \ for \ a.e. \ s \in [0,T]; \tag{10.24}$$

for each measurable $u : [0,T] \to \mathbb{R}$, the map $t \mapsto a(t)g(t,u(t))$
$+ \tilde{f}(t,u(t),u'(t))$ has measurable single-valued selections; $\tag{10.25}$

for a.e. $t \in [0,T]$, the map $u \mapsto a(t)g(t,u)) + \tilde{f}(t,u,u'))$ is upper semicontinuous; $\tag{10.26}$

for each $r > 0$, $\exists \xi_r \in L_p[0,T]$ with $|a(t)g(t,y)) + f(t,y,y'))| \leqslant M'\xi_r$
for a.e. $t \in [0,T]$ and every $y, y' \in \mathbb{R}$, with $|y| \leqslant r$ and $|y'| \leqslant \lambda$ for some $\lambda > 0$; $\tag{10.27}$

for a.e. $t \in [0,T]$ and all $y \in (0,\infty)$, $y' \in [-\lambda,\lambda]$, $u > 0$ for all $u \in a(t)g(t,y) + \tilde{f}(t,y,y')$; $\tag{10.28}$

$$\exists \xi \in L_q[0,T] \ with \ k(t,s) \leqslant \xi(s) \ for \ \ t \in [0,T]; \tag{10.29}$$

$\exists \delta, \varepsilon, 0 \leqslant \delta < \varepsilon \leqslant T$ and $M, 0 < M < 1$, with $k(t,s) \ge M\xi(s)$ for $t \in [\delta,\varepsilon]$; $\tag{10.30}$

$\exists h \in L_p[0,T]$ with $h : [0,T] \to (0,\infty)$, $w \ge 0$ continuous and nondecreasing on $\tag{10.31}$

$(0,\infty)$ and $\eta : [0,\lambda] \to \mathbb{R}^+$ is continuous and nondecreasing with

$$|a(t)g(t,y) + \tilde{f}(t,y,y')| \leqslant M'h(t)w(y)\eta(|y'|) \tag{10.32}$$

for a.e. $t \in [0,T]$ and all $y \in (0,\infty)$, $y' \in [-\lambda,\lambda]$;

$\exists \tau \in L_p[\delta,\varepsilon]$ with $\tau > 0$ a.e. on $[\delta,\varepsilon]$ and with for a.e. $t \in [\delta,\varepsilon]$, $y \in (0,\infty)$ and
$y' \in [-\lambda,\lambda]$, and $\theta : \mathbb{R} \to \mathbb{R}^+$, $\theta(u) > 0$ for $u > 0$, is continuous and nonincreasing
$u \ge \tau(t)w(y)\theta(|y'|)$ for all $u \in a(t)g(t,y) + \tilde{f}(t,y,y')$; $\tag{10.33}$

$$\exists \alpha > 0 \ \ with \ \ 1 < \frac{\alpha}{M'w(\alpha)\eta(\lambda) \sup\limits_{t\in[0,T]} \int_0^T k(t,s)h(s) \, ds} \tag{10.34}$$

and

$$\exists \beta > 0, \ \beta \neq \alpha \ \text{ with with } \ 1 > \frac{\beta}{w(M\beta)\theta(\lambda)\int_\delta^\varepsilon \tau(s)k(\sigma, s)\, ds}, \tag{10.35}$$

where $\sigma \in [0, T]$ *is such that*

$$\int_\delta^\varepsilon \eta(s)k(\sigma, s)\, ds = \sup_{t \in [0,T]} \int_\delta^\varepsilon \tau(s)k(t, s)\, ds. \tag{10.36}$$

Then, (10.20) has at least one positive solution $y \in C[0, T]$ *and either*

(A) $0 < \alpha < |y|_0 < \beta$ *and* $y(t) \geq M''\alpha$ *a.e.* $t \in [0, T]$ *if* $\alpha < \beta$
or
(B) $0 < \beta < |y|_0 < \alpha$ *and* $y(t) \geq M''\beta$ *a.e.* $t \in [\delta, \varepsilon]$ *if* $\beta < \alpha$ *holds.*

Proof Let $E = (C[0, T], \| \cdot \|_0)$ and

$$C = \{y \in C[0, T] : y(t) \geq 0 \ \text{ and } \min_{t \in [\delta,\varepsilon]} y(t) \geq M\|y\|_0\}.$$

Also, let $A = KN_{g,\tilde{f}}^\lambda : C \to 2^E$, where $K : L_p[0, T] \to C[0, T]$ and $N_{g,\tilde{f}}^\lambda : C[0, T] \to 2^{L_p[0,T]}$ are given by

$$Ky(t) = \int_0^T k(t, s)y(s)\, ds,$$

where $y(s) \in a(s)g(s, u(s)) + \tilde{f}(s, u(s), u'(s))$ a.e. $s \in [0, T]$, $\sup_{s \in [0,T]} |u'| \leqslant \lambda$ and $N_{g,\tilde{f}}^\lambda$ is the generalized Nemytskij operator.

Notice that A is well defined since if $x \in C$ then [166, 235] guarantee that $N_{g,\tilde{f}}^\lambda x \neq \varnothing$. Now, we show that $A : C \to 2^C$. To see this let $x \in C$ and $y \in Ax$. Then, for each $t \in [0, T]$, there exists a $v \in N_{g,\tilde{f}}^\lambda x$ with

$$y(t) = \int_0^T k(t, s)v(s)\, ds \ \text{ for } \ t \in [0, T].$$

This together with (10.29) yields

$$|y(t)| \leqslant \int_0^T \xi(s)v(s)\, ds \ \text{ for } \ t \in [0, T]$$

and so

$$|y|_0 \leqslant \int_0^T \xi(s)v(s)\, ds. \tag{10.37}$$

On the other hand, (10.30) and (10.31) yields

$$\min_{t\in[\delta,\varepsilon]} y(t) = \min_{t\in[\delta,\varepsilon]} \int_0^T k(t,s)v(s)\,ds \geq M \int_0^T \xi(s)v(s)\,ds \geq M|y|_0.$$

As a consequence, $y \in C$. Thus, $A : C \to 2^C$. A standard result from the literature [235, 504–506] guarantees that

$$A : C \to K(C) \text{ is upper semicontinuous and completely continuous.} \qquad (10.38)$$

Let

$$\Omega_\alpha = \{u \in C[0,T] : |u|_0 < \alpha\} \text{ and } \Omega_\beta = \{u \in C[0,T] : |u|_0 < \beta\},$$

and, for some $\lambda > 0$, let $\max_{s\in[0,T]} |u'| \leqslant \lambda$.

Assume that $\beta < \alpha$ (a similar argument holds if $\alpha < \beta$). If we show

$$|y|_0 < |x|_0 \text{ for all } y \in Ax \text{ and } x \in C \cap \partial\Omega_\alpha \qquad (10.39)$$

and

$$|y|_0 > |x|_0 \text{ for all } y \in Ax \text{ and } x \in C \cap \partial\Omega_\beta \qquad (10.40)$$

are true, then Theorem 5.167 guarantees the result.

Suppose $x \in C \cap \partial\Omega_\alpha$, so $|x|_0 = \alpha$, $\sup_{t\in[0,T]} |x'(t)| \leqslant \lambda$ and $y \in Ax$. Then there exists $v \in N^\lambda_{g,\tilde{f}} x$ with

$$y(t) = \int_0^T k(t,s)v(s)\,ds \quad for \ t \in [0,T].$$

Now (10.31) implies that for $t \in [0,T]$ we have

$$|y(t)| \leqslant M' \int_0^T k(t,s)h(s)w(x(s))\eta(|x'(s)|)\,ds \leqslant w(|x|_0)\eta(\lambda) \int_0^T k(t,s)h(s)\,ds$$

$$\leqslant M'w(\alpha)\eta(\lambda) \sup_{t\in[0,T]} \int_0^T k(t,s)h(s)\,ds.$$

This together with (10.33) yields

$$|y|_0 \leqslant M'w(\alpha)\eta(\lambda) \sup_{t\in[0,T]} \int_0^T k(t,s)h(s)\,ds < \alpha = |x|_0,$$

so (10.38) holds.

Next suppose that $x \in C \cap \partial\Omega_\beta$, so $|x|_0 = \beta$, $\sup_{t\in[0,T]} |x'(t)| \leqslant \lambda$ and $M\beta \leqslant x(t) \leqslant \beta$ for $t \in [\delta, \varepsilon]$, and $y \in Ax$. Then, there exists $v \in N_f x$ with

$$y(t) = \int_0^T k(t, s) v(s) \, ds \quad for \ \ t \in [0, T].$$

Notice (10.33) and (10.35) imply

$$
\begin{aligned}
y(\sigma) &= \int_0^T k(\sigma, s) v(s) \, ds \geq \int_\delta^\varepsilon k(\sigma, s) v(s) \, ds \\
&\geq \int_\delta^\varepsilon k(\sigma, s) \tau(s) w(x(s)) \theta(|x'(s)|) \, ds \geq w(M\beta) \theta(\lambda) \int_\delta^\varepsilon k(\sigma, s) \tau(s) \, ds \\
&> \beta = |x|_0.
\end{aligned}
$$

Thus, $|y|_0 > |x|_0$, so (10.39) holds. To obtain the required conclusion, we now apply Theorem 5.167.

Remark 10.3 If nonlinear function $\tilde{f}(t, u(t), u'(t))$ is independent of derivative, then in view of Remark 10.1 we observe that Theorem 3.1 of Agarwal et al. [4] can be derived from our Theorem 10.2.

10.2 Existence Theorem for Nonconvex Hammerstein Type Integral Inclusions

Let $0 < T < \infty$, $I := [0, T]$ and $\mathscr{L}(I)$ denote the σ-algebra of all Lebesgue measurable subsets of I. Let E be a real separable Banach space with the norm $\| \cdot \|$. Let $\mathscr{P}(E)$ denote the family of all nonempty subsets of E and $\mathscr{B}(E)$ the family of all Borel subsets of E.

In what follows, as usual, we denote by $C(I, E)$ the Banach space of all continuous functions $x(\cdot) : I \rightarrow E$ endowed with the norm $\|x(\cdot)\|_C = \sup_{t\in I} \|x(t)\|$. Consider the following integral equation

$$x(t) = \lambda(t) + \int_0^T k(t, s) \, g(t, s, u(s)) \, ds \quad \text{on} \ \ [0, T]. \tag{10.41}$$

Here, λ, k and g are given functions, where $\lambda(\cdot) : I \rightarrow E$ is a function with Banach space value, $k : I \times I \rightarrow \mathbb{R}^+ = [0, \infty)$ is a positive real single-valued function, while $g : I \times I \times E \rightarrow E$ is a map. Let $p \in [1, \infty)$, $q \in [1, \infty)$, and let $r \in [1, \infty)$ be the conjugate exponent of q, that is $1/q + 1/r = 1$. Let $\| \cdot \|_p$ denote the p-norm of the space $L_p(I, E)$ and is defined by

$$\|u\|_p = (\int_0^T \|u(s)\|^p ds)^{1/p} \text{ for all } u \in L_p(I, E).$$

Consider the Nemitskiǐ operator associated to g, p, q and $G : L_p(I, E) \to L_q(I, E)$ given by

$$G(u) == g(t, s, u(s)) \, a.e. \text{ on } I.$$

Consider the linear integral operator of kernel k, $S : L_q(I, E) \to L_p(I, E)$ given by

$$S(u) = \lambda(t) + \int_0^T k(t, s)u(s)ds \ a.e. \text{ on } I.$$

Thus, the Hammerstein type integral equation (10.41) is transformed into the form

$$x = SG(u), \quad u \in L^p(I, E) \ a.e. \text{ on } I \tag{10.41'}$$

$$u(t) \in F(t, V(x)(t)) \quad a.e. \ (I := [0, T]), \tag{10.42}$$

where $V : C(I, E) \to C(I, E)$ is a given mapping. In the sequel, we also use the following: for any $x \in E, \lambda \in C(I, E), \sigma \in L_p(I, E)$, we define the set-valued maps $M_{\lambda,\sigma}(t) := F(t, V(x_{\sigma,\lambda})(t)), \quad t \in I, T_\lambda(\sigma) := \{\psi(\cdot) \in L_p(I, E) : \psi(t) \in M_{\lambda,\sigma}(t) \ a.e. \ (I)\}$.

In order to study problem (10.41) and (10.42), we introduce the following assumption.

Hypothesis 10.1 Let $F(\cdot, \cdot) : I \times E \to \mathscr{P}(E)$ be a set-valued map with nonempty closed values that verify:

(H_1) The function $k : I \times I \to \mathbf{R}_+$ satisfies that $k(t, \cdot) \in L_r(I)$, and $t \to \|k(t, \cdot)\|_r \in L_p(I)$.

(H_2) The set-valued map $F(\cdot, \cdot)$ is $\mathscr{L}(I) \otimes \mathscr{B}(E)$ measurable.

(H_3) There exists $L(\cdot) \in L_1(I, \mathbb{R}^+)$ such that, for almost all $t \in I, F(t, \cdot)$ is $L(t)$-Lipschitz in the sense that

$$H^+(F(t, x), F(t, y)) \leqslant L(t) \|x - y\| \tag{C1}$$

for all x, y in E, and $T_\lambda(\cdot)$ satisfies the condition: for any $\sigma \in L_p(I, E), \sigma_1 \in T_\lambda(\sigma)$ and any given $\varepsilon > 0$, there exists $\sigma_2 \in T_\lambda(\sigma_1)$ such that

$$\|\sigma_1 - \sigma_2\|_p \leqslant H^+(T_\lambda(\sigma), T_\lambda(\sigma_1)) + \varepsilon. \tag{C2}$$

(H_4) The mappings $k : I \times I \to \mathbb{R}^+, g : I \times I \times E \to E$ are continuous, $V : C(I, E) \to C(I, E)$

and there exist the constants $M_1, M_2, M_3 > 0$ such that

$$\|g(t, s, u_1) - g(t, s, u_2)\|_q \leqslant M_1 \|u_1 - u_2\|^p, \ \forall u_1, u_2 \in E,$$

$$\|V(x_1)(t) - V(x_2)(t)\| \leqslant M_2 \|x_1(t) - x_2(t)\|, \quad \forall t \in I, \forall x_1, x_2 \in C(I, E),$$

and
$$\|k(t, s)\|_r \leqslant M_3 \ \forall t, s \in I.$$

Note that the system (10.41) and (10.42) encompasses a large variety of differential inclusions and control systems including those defined by partial differential equations.

Assume that U be an open bounded subset of $\mathbb{R}^n$ (or Y, a subset of E homeomorphic to $\mathbb{R}^n$) and $U_T = (0, T] \times U$ for some fixed $T > 0$. We say that the partial differential operator $\frac{\partial}{\partial t} + L$ is parabolic if there exists a constant $\theta > 0$ such that

$$\sum_{i,j=1}^n a^{ij}(t, x)\xi_i\xi_j \geq \theta|\xi|^2 \ \text{ for all } (t, x) \in U_T, \xi \in \mathbb{R}^n.$$

The letter L denotes for each time t a second-order partial differential operator, having either the divergence form

$$Lu = -\sum_{i,j=1}^n (a^{ij}(t, x)u_{x_i})_{x_j} + \sum_{i=1}^n b^i(t, x)u_{x_i} + c(t, x)u$$

or else the nondivergence form

$$Lu = -\sum_{i,j=1}^n a^{ij}(t, x)u_{x_i x_j} + \sum_{i=1}^n b^i(t, x)u_{x_i} + c(t, x)u,$$

for given coefficients a^{ij}, b^i, c $(i, j = 1, 2, \ldots, n)$.

A family $\{G(t) : t \in \mathbb{R}^+\}$ of bounded linear operators from X into E is a C_0-semigroup (also called linear semigroup of class (C_0)) on X if

(i) $G(0) = $ the identity operator, and $G(t + s) = G(t)G(s) \ \forall t, s \geq 0$;
(ii) $G(\cdot)$ is strongly continuous in $t \in \mathbb{R}^+$;
(iii) $\|G(t)\| \leqslant Me^{\omega t}$ for some $M > 0$, real ω and $t \in \mathbb{R}^+$.

Example 10.1 Set $k(t, \tau)g(t, \tau, u) = G(t - \tau)u$, $\Phi(x) = x$, $\lambda(t) = G(t)x_0$ where $\{G(t)\}_{t \geq 0}$ is a C_0-semigroup with an infinitesimal generator A. Then, a solution of system (10.41) and (10.42) represents a mild solution of

$$x'(t) \in Ax(t) + F(t, x(t)), \qquad x(0) = x_0. \tag{10.43}$$

In particular, this problem includes control systems governed by parabolic partial differential equations as a special case. When $A = 0$, the relation (10.43) reduces to classical differential inclusions

$$x'(t) \in F(t, x(t)), \qquad x(0) = x_0. \tag{10.44}$$

Denote

$$\Phi(u)(t) = \int_0^T k(t, \tau) g(t, \tau, u(\tau)) \, d\tau, \ t \in I. \tag{10.45}$$

Then, the integral inclusion system (10.41) and (10.42) reduces to the form

$$x(t) = \lambda(t) + \Phi(u)(t) \qquad a.e. \ (I), \tag{S}$$

which may be written in more "compact" form as

$$u(t) \in F(t, V(\lambda + \Phi(u))(t)) \qquad a.e. \ (I).$$

Now, we recall the following:

Definition 10.2 A pair of functions (x, u) is called a solution pair of integral inclusion system (S), if $x(\cdot) \in C(I, E), u(\cdot) \in L_p(I, E)$ and satisfy relation (S).

For our further discussion, we denote by $S(\lambda)$ the solution set of (10.41) and (10.42). Notice that the integral operator in (10.45) plays a key role in the proofs of our main results.

For given $\alpha \in \mathbb{R}$, we denote by $L_p(I, E)$ the Banach space of all Bochner integrable functions $u(\cdot) : I \to E$ endowed with the norm

$$\|u(\cdot)\|_p = \left(\int_0^T e^{-\alpha M_1 M_2 M_3 m(t)} \|u(t)\|^p \, dt \right)^{\frac{1}{p}},$$

where $m(t) = \int_0^t L(s) \, ds, \ t \in I$. For our further discussion, we denote $L = m(T)$.

Theorem 10.3 (Pathak and Shahzad [465]) *Let Hypothesis 10.1 be satisfied, let* $\lambda(\cdot, \cdot), \mu(\cdot, \cdot) \in C(I \times E, E)$ *and let* $v(\cdot) \in L_p(I, E)$ *be such that*

$$d(v(t), F(t, \Phi(y)(t))) \leqslant p(t) \qquad a.e. \qquad (I),$$

where $p(\cdot) \in L_p(I, \mathbb{R}^+)$ *and* $y(t) = \mu(t, v(t)) + \Phi(v)(t), \ \forall t \in I$.
Then, for every $\alpha > 1$, *there exists* $x(\cdot) \in S(\lambda)$ *such that for every* $t \in I$

$$\|x(t) - y(t)\| \leqslant \|\lambda - \mu\|_C + M_1 M_3 e^{\alpha M_1 M_2 M_3 L} \left[\frac{1}{\alpha^{\frac{1}{2p}} (\alpha^{\frac{1}{2p}} - 1) M_1^{\frac{1}{p}} M_3^{\frac{1}{p}}} \|\lambda - \mu\|_C \right.$$

$$\left. + \frac{\alpha^{\frac{1}{2p}}}{\alpha^{\frac{1}{2p}} - 1} \left(\int_0^T e^{-\alpha M_1 M_2 M_3 m(t)} p(t) dt \right)^{\frac{1}{p}} \right]^p.$$

Proof For $\lambda \in C(I, E)$ and $u \in L_p(I, E)$, define

$$x_{u, \lambda}(t) = \lambda(t) + \int_0^T k(t, s) \, g(t, s, u(s)) \, ds, \ t \in I.$$

Let us consider that $\lambda \in C(I, E), \sigma \in L_p(I, E)$ and define the set-valued maps

$$M_{\lambda,\sigma}(t) := F(t, V(x_{\sigma,\lambda})(t)), \ t \in I, \tag{10.46}$$

$$T_\lambda(\sigma) := \{\psi(\cdot) \in L_p(I, E) : \psi(t) \in M_{\lambda,\sigma}(t) \ a.e. \ (I)\}. \tag{10.47}$$

Further, in view of condition (C2) of Hypothesis 10.1 (H_3), $T_\lambda(\cdot)$ satisfies the condition: for any $\sigma \in L_p(I, E), \sigma_1 \in T_\lambda(\sigma)$ and any given $\varepsilon > 0$, there exists $\sigma_2 \in T_\lambda(\sigma_1)$ such that

$$\|\sigma_1 - \sigma_2\|_p \leqslant H^+(T_\lambda(\sigma), T_\lambda(\sigma_1)) + \varepsilon. \tag{10.48}$$

Now, we claim that $T_\lambda(\sigma)$ is nonempty, bounded and closed for every $\sigma \in L_p(I, E)$.

It is well known that the set-valued map $M_{\lambda,\sigma}(\cdot)$ is measurable. For example the map $t \to M_{\lambda,\sigma}(t)$ can be approximated by step functions and so we can apply Theorem III. 40 in [131]. As the values of F are closed, with the measurable selection theorem we infer that $M_{\lambda,\sigma}(\cdot)$ is nonempty.

Further, we note that the set $T_\lambda(\sigma)$ is bounded and closed. Indeed, if $\psi_n \in T_\lambda(\cdot)$ and $\|\psi_n - \psi\|_p \to 0$, then there exists a subsequence ψ_{n_k} such that $\psi_{n_k}(t) \to \psi(t)$ for a.e. $t \in I$ and we find that $\psi \in T_\lambda(\sigma)$.

Let $\sigma_1, \sigma_2 \in L_p(I, E)$ be given. Let $\psi_1 \in T_\lambda(\sigma_1)$ and let $\delta > 0$. Consider the following set-valued map:

$$\mathscr{G}(t) := M_{\lambda,\sigma_2(t)} \cap \left\{ z \in E : \|\psi_1(t) - z\|^p \leqslant M_1 M_2 M_3 L(t) \int_0^T \|\sigma_1(s) - \sigma_2(s)\|^p \, ds + \delta \right\}.$$

By (10.48), it follows that

$$\begin{aligned}
d(\psi_1(t), M_{\lambda,\sigma_2}(t)) &\leqslant H^+\left(F(t, V(x_{\sigma_1,\lambda})(t)), F(t, V(x_{\sigma_2,\lambda})(t)) \right) + \varepsilon \\
&\leqslant L(t)\|V(x_{\sigma_1,\lambda})(t) - V(x_{\sigma_2,\lambda})(t)\| + \varepsilon \\
&\leqslant M_2 L(t)\|x_{\sigma_1,\lambda}(t) - x_{\sigma_2,\lambda}(t)\| + \varepsilon \\
&\leqslant M_2 L(t) \int_0^T \|k(t, s)\|_r \|g(t, s, x_1(s)) \, ds - g(t, s, x_2(s))\|_q \, ds + \varepsilon \\
&\leqslant M_1 M_2 M_3 L(t) \int_0^T \|\sigma_1(s) - \sigma_2(s)\|^p \, ds + \varepsilon.
\end{aligned}$$

Since ε is arbitrary, letting $\varepsilon \to 0$, we deduce that $\mathscr{G}(\cdot)$ is nonempty bounded and has closed values. Further, according to Proposition III.4 in [131], $\mathscr{G}(\cdot)$ is measurable. Let $\psi_2(\cdot)$ be a measurable selector of $\mathscr{G}(\cdot)$. It follows that $\psi_2 \in T_\lambda(\sigma_2)$ and

$$\begin{aligned}
\|\psi_1 - \psi_2\|_p^p &= \int_0^T e^{-\alpha M_1 M_2 M_3 m(t)} \|\psi_1(t) - \psi_2(t)\|^p \, dt \\
&\leqslant \int_0^T e^{-\alpha M_1 M_2 M_3 m(t)} \left(M_1 M_2 M_3 L(t) \int_0^T \|\sigma_1(s) - \sigma_2(s)\|^p \, ds \right) dt
\end{aligned}$$

$$+ \delta \int_0^T e^{-\alpha M_1 M_2 M_3 m(t)} dt$$

$$\leqslant \frac{1}{\alpha} \|\sigma_1 - \sigma_2\|_p^p + \delta \int_0^T e^{-\alpha M_1 M_2 M_3 m(t)} dt.$$

Since δ is arbitrary, so letting $\delta \to 0$ we deduce from the above inequality that

$$\|\psi_1 - \psi_2\|_p^p \leqslant \frac{1}{\alpha} \|\sigma_1 - \sigma_2\|_p^p$$

i.e.

$$\|\psi_1 - \psi_2\|_p \leqslant \frac{1}{\alpha^{\frac{1}{p}}} \|\sigma_1 - \sigma_2\|_p.$$

This yields

$$d(\psi_1, T_\lambda(\sigma_2)) \leqslant \frac{1}{\alpha^{\frac{1}{p}}} \|\sigma_1 - \sigma_2\|_p.$$

Thus, we have

$$\rho(T_\lambda(\sigma_1), T_\lambda(\sigma_2)) = \sup_{\psi_1 \in T_\lambda(\sigma_1)} d(\psi_1, T_\lambda(\sigma_2)) \leqslant \frac{1}{\alpha^{\frac{1}{p}}} \|\sigma_1 - \sigma_2\|_p. \tag{10.49}$$

Now replacing $\sigma_1(\cdot)$ with $\sigma_2(\cdot)$ and arguing as above, we obtain

$$\rho(T_\lambda(\sigma_2), T_\lambda(\sigma_1)) \leqslant \frac{1}{\alpha^{\frac{1}{p}}} \|\sigma_1 - \sigma_2\|_p. \tag{10.50}$$

Now adding (10.49) and (10.50) and dividing by 2, we obtain

$$H^+(T_\lambda(\sigma_1), T_\lambda(\sigma_2)) \leqslant \frac{1}{\alpha^{\frac{1}{p}}} \|\sigma_1 - \sigma_2\|_p$$

$$\leqslant \frac{1}{\alpha^{\frac{1}{p}}} \max\{\|\sigma_1 - \sigma_2\|_p, d(\sigma_1, T_\lambda(\sigma_1)), d(\sigma_2, T_\lambda(\sigma_2)),$$

$$[d(\sigma_1, T_\lambda(\sigma_2)) + d(\sigma_2, T_\lambda(\sigma_1))]/2\}.$$

Hence, we conclude that $T_\lambda(\cdot)$ is an H^+-type multivalued weak contractive mapping on $L_p(I, E)$. Next, we consider the following set-valued maps

$$\tilde{F}(t, x) := F(t, x) + p(t),$$

$$\tilde{M}_{\lambda,\sigma}(t) := \tilde{F}(t, \phi(x_{\sigma,\lambda})(t)), \qquad t \in I,$$

$$\tilde{T}_\lambda(\sigma) := \{\psi(\cdot) \in L_p(I, E); \psi(t) \in \tilde{M}_{\lambda,\sigma}(t) \ a.e. (I)\}.$$

It is obvious that $\tilde{F}(\cdot, \cdot)$ satisfies Hypothesis 10.1.

Let $\phi \in T_\lambda(\sigma), \delta > 0$ and define

$$\mathcal{G}_1(t) := \tilde{M}_{\lambda,\sigma(t)} \cap \left\{ z \in X : \|\phi(t) - z\|^p \leqslant M_2 L(t) \|\lambda - \mu\|_C^p + p(t) + \delta \right\}.$$

Using the same argument as used for the set-valued map $\mathcal{G}(\cdot)$, we deduce that $\mathcal{G}_1(\cdot)$ is measurable with nonempty closed values.

Next, we prove the following estimate:

$$H^+(T_\lambda(\sigma), \tilde{T}_\mu(\sigma)) \leqslant \frac{1}{\alpha^{\frac{1}{p}} M_1^{\frac{1}{p}} M_3^{\frac{1}{p}}} \|\lambda - \mu\|_C + \left(\int_0^T e^{-\alpha M_1 M_2 M_3 m(t)} p(t) dt \right)^{\frac{1}{p}}.$$

$$(10.51)$$

Let $\psi(\cdot) \in T_\mu(\sigma)$. Then

$$\|\phi - \psi\|_p^p \leqslant \int_0^T e^{-\alpha M_1 M_2 M_3 m(t)} \|\phi(t) - \psi(t)\|^p dt$$

$$\leqslant \int_0^T e^{-\alpha M_1 M_2 M_3 m(t)} [M_2 L(t) \|\lambda - \mu\|_C^p + p(t) + \delta] dt$$

$$\leqslant \|\lambda - \mu\|_C^p \int_0^T e^{-\alpha M_1 M_2 M_3 m(t)} M_2 L(t) dt$$

$$+ \int_0^T e^{-\alpha M_1 M_2 M_3 m(t)} p(t) dt + \delta \int_0^T e^{-\alpha M_1 M_2 M_3 m(t)} dt$$

$$\leqslant \frac{1}{\alpha M_1 M_3} \|\lambda - \mu\|_C^p + \int_0^T e^{-\alpha M_1 M_2 M_3 m(t)} p(t) dt$$

$$+ \delta \int_0^T e^{-\alpha M_1 M_2 M_3 m(t)} dt.$$

Since δ is arbitrary, so letting $\delta \to 0$ we deduce from the above inequality that

$$\|\phi - \psi\|_p^p \leqslant \frac{1}{\alpha M_1 M_3} \|\lambda - \mu\|_C^p + \int_0^T e^{-\alpha M_1 M_2 M_3 m(t)} p(t) dt.$$

Thus, by taking $\frac{1}{p}$th power on both sides of the above inequality breaking the right-hand side, one obtains (9.51).
Now applying Proposition 5.6 we obtain

$$H^+(Fix(T_\lambda), Fix(\tilde{T}_\mu)) \leqslant \frac{1}{\alpha^{\frac{1}{2p}} (\alpha^{\frac{1}{2p}} - 1) M_1^{\frac{1}{p}} M_3^{\frac{1}{p}}} \|\lambda - \mu\|_C$$

$$+ \frac{\alpha^{\frac{1}{2p}}}{\alpha^{\frac{1}{2p}} - 1} \left(\int_0^T e^{-\alpha M_1 M_2 M_3 m(t)} p(t) dt \right)^{\frac{1}{p}}.$$

Since $v(\cdot) \in Fix(\tilde{T}_\mu)$, it follows that there exists $u(\cdot) \in Fix(T_\mu)$ such that

$$\|v - u\|_p \leqslant \frac{1}{\alpha^{\frac{1}{2p}}(\alpha^{\frac{1}{2p}} - 1)M_1^{\frac{1}{p}}M_3^{\frac{1}{p}}}\|\lambda - \mu\|_C + \frac{\alpha^{\frac{1}{2p}}}{\alpha^{\frac{1}{2p}} - 1}\Big(\int_0^T e^{-\alpha M_1 M_2 M_3 m(t)}p(t)dt\Big)^{\frac{1}{p}}.$$

(10.52)

We define

$$x(t) = \lambda(t) + \int_0^T k(t, s)\, g(t, s, u(s))\, ds.$$

Then one has the following inequality:

$$\|x(t) - y(t)\| \leqslant \|\lambda(t) - \mu(t)\| + M_1 M_3 \int_0^T \|u(s) - v(s)\|^p\, ds$$

$$\leqslant \|\lambda - \mu\|_C + M_1 M_3 e^{\alpha M_1 M_2 M_3 L}\|u - v\|_p^p.$$

Combining the last inequality with (10.52), we obtain

$$\|x(t) - y(t)\| \leqslant \|\lambda - \mu\|_C + M_1 M_3 e^{\alpha M_1 M_2 M_3 L}\Big[\frac{1}{\alpha^{\frac{1}{2p}}(\alpha^{\frac{1}{2p}} - 1)M_1^{\frac{1}{p}}M_3^{\frac{1}{p}}}\|\lambda - \mu\|_C$$

$$+ \frac{\alpha^{\frac{1}{2p}}}{\alpha^{\frac{1}{2p}} - 1}\Big(\int_0^T e^{-\alpha M_1 M_2 M_3 m(t)}p(t)dt\Big)^{\frac{1}{p}}\Big]^p.$$

This completes the proof.

10.3 Filippov Type Existence Theorem for Integral Inclusions

In this section, we shall consider a nonconvex integral inclusion and prove a Filippov type existence theorem by using an appropriate norm on the space of selection of the multifunction and a H^+-type contraction for set-valued maps.

Let $I := [0, T]$, $T > 0$ and $\mathscr{L}(I)$ denote the σ-algebra of all Lebesgue measurable subsets of I. Let X be a real separable Banach space with the norm $\|\cdot\|$. Let $\mathscr{P}(X)$ denote the family of all nonempty subsets of X and $\mathscr{B}(X)$ the family of all Borel subsets of X.

Throughout this section, let $C(I, X)$ denote the Banach space of all continuous functions $x(\cdot): I \to X$ endowed with the norm $\|x(\cdot)\|_C = \sup_{t \in I} \|x(t)\|$. Consider the following integral inclusion

$$x(t) = \lambda(t) + \int_0^t [a(t, s)\, g(t, u(s)) + f(t, s, u(s))]\, ds$$

(10.53)

$$u(t) \in F(t, V(x)(t)) \text{ a.e. } (I := [0, T]), \tag{10.54}$$

where $\lambda(\cdot): I \to X, g(\cdot, \cdot): I \times X \to X, F(\cdot, \cdot): I \times X \to \mathscr{P}(X), f(\cdot, \cdot, \cdot)$ $: I \times I \times X \to X, V: C(I, X) \to C(I, X), a(\cdot, \cdot): I \times I \to \mathbb{R} = (-\infty, \infty)$ are given mappings.

In order to study problem (10.53) and (10.54), we introduce the following assumption.

Hypothesis 10.2 Let $F(\cdot, \cdot): I \times X \to \mathscr{P}(X)$ be a set-valued map with nonempty closed values that verify:

(H_1) The set-valued map $F(\cdot, \cdot)$ is $\mathscr{L}(I) \otimes \mathscr{B}(X)$ measurable.

(H_2) There exists $L(\cdot) \in L_1(I, \mathbb{R}^+)$ such that, for almost all $t \in I, F(t, \cdot)$ is $L(t)$-Lipschitz in the sense that

(C1) $H^+(F(t, x), F(t, y)) \leqslant L(t) \|x - y\|$ for all x, y in X
and for any $x, y \in X, w \in F(t, x)$ and $\varepsilon > 0$, there exists $z \in F(t, y)$ such that:

(C2) $\|w - z\| \leqslant H^+(F(t, x), F(t, y)) + \varepsilon$
and $T_\lambda(\cdot)$ satisfies the condition: for any $\sigma \in L_1(I, E), \sigma_1 \in T_\lambda(\sigma)$ and any given $\varepsilon > 0$ there exists $\sigma_2 \in T_\lambda(\sigma_1)$ such that:

(C3) $\|\sigma_1 - \sigma_2\|_1 \leqslant H^+(T_(\sigma), T_\lambda(\sigma_1)) + \varepsilon$ for almost all $t \in I$.

(H_3) The mappings $f: I \times I \times X \to X, g, \lambda: I \times X \to X$ are continuous, $V: C(I, X) \to C(I, X)$ and there exist the constants $M_1, M_2, M_3, M_4 > 0$ such that

$$\|f(t, s, u_1) - f(t, s, u_2)\| \leqslant M_1 \|u_1 - u_2\|, \text{ for all } u_1, u_2 \in X,$$
$$\|g(s, u_1) - g(s, u_2)\| \leqslant M_2 \|u_1 - u_2\|, \text{ for all } u_1, u_2 \in X,$$
$$\|V(x_1)(t) - V(x_2)(t)\| \leqslant M_3 \|x_1(t) - x_2(t)\|, \text{ for all } t \in I, \text{ and all } x_1, x_2 \in C(I, X).$$

(H_4) Let $a: I \times I \to \mathbb{R}$ be continuous and satisfy the uniform Hölder's continuity condition in the first and second arguments with the exponent ρ; i.e. there exists a positive number b such that

$$|a(t_1, s_1) - a(t_2, s_2)| \leqslant b(|t_1 - t_2|^\rho + |s_1 - s_2|^\rho)$$

for all $t_1, t_2, s_1, s_2 \in I$ and $|a(t, s)| \leqslant 2bT + |a(0, 0)| = M_4$ for all $t, s \in I$ and $0 < \rho \leqslant 1$.

Note that the system (10.53) and (10.54) encompasses a large variety of differential inclusions and control systems including those defined by partial differential equations.

Assume that U be an open bounded subset of $\mathbb{R}^n$ (or Y, a subset of X homeomorphic to $\mathbb{R}^n$) and $U_T = U \times (0, T]$ for some fixed $T > 0$. We say that the partial differential operator $\frac{\partial}{\partial t} + L$ is parabolic if there exists a constant $\theta > 0$ such that

$$\sum_{i,j=1}^{n} a^{ij}(x,t)\xi_i\xi_j \geq \theta|\xi|^2$$

for all $(x,t) \in U_T, \xi \in \mathbb{R}^n$. The letter L denotes for each time t a second-order partial differential operator, having either the divergence form

$$Lu = -\sum_{i,j=1}^{n} (a^{ij}(x,t)u_{x_i})_{x_j} + \sum_{i=1}^{n} b^i(x,t)u_{x_i} + C(x,t)u$$

or else the nondivergence form

$$Lu = -\sum_{i,j=1}^{n} a^{ij}(x,t)u_{x_ix_j} + \sum_{i=1}^{n} b^i(x,t)u_{x_i} + C(x,t)u,$$

for given coefficients a^{ij}, b^i, c $(i,j = 1, \ldots, n)$.

Example 10.2 Set $f(t, \tau, u) = G(t-\tau)u$, $g(\tau, u(\tau)) = 0$, $V(x) = x$, $\lambda(t) = G(t)x_0$ where $\{G(t)\}_{t \geq 0}$ is a C_0-semigroup with an infinitesimal generator A. Then, a solution of system (10.53) and (10.54) represents a mild solution of

$$x'(t) \in Ax(t) + F(t, x(t)), \quad x(0) = x_0. \tag{10.55}$$

In particular, this problem includes control systems governed by parabolic partial differential equations as a special case. When $A = 0$, the relation (10.55) reduces to classical differential inclusions

$$x'(t) \in F(t, x(t)), \quad x(0) = x_0. \tag{10.56}$$

Denote

$$\Phi(u)(t) = \int_0^t [a(t,\tau)g(\tau, u(\tau)) + f(t, \tau, u(\tau))] d\tau, \quad t \in I. \tag{10.57}$$

Then, the integral inclusion system (10.53) and (10.54) reduces to the form

$$x(t) = \lambda(t) + \Phi(u)(t), u(t) \in F(t, V(x)(t)) \quad \text{a.e. } (I), \tag{10.58}$$

which may be written in more compact form as

$$u(t) \in F(t, V(\lambda + \Phi(u))(t)) \quad \text{a.e.}(I).$$

Now, we recall the following:

Definition 10.3 A pair of functions (x, u) is called a solution pair of integral inclusion system (10.58), if $x(\cdot) \in C(I, X), u(\cdot) \in L_1(I, X)$ and satisfy relation (10.46).

For our further discussion, we denote by $S(\lambda)$ the solution set of (10.53) and (10.54).

Notice that the integral operator in (10.57) plays a key role in the proofs of our main results.

For given $\alpha \in \mathbb{R}$ we denote by $L_1(I, X)$ the Banach space of all Bochner integrable functions $u(\cdot): I \to X$ endowed with the norm

$$\|u(\cdot)\|_1 = \int_0^T e^{-\alpha(M_4 M_2 + M_1) M_3 m(t)} \|u(t)\| \, dt,$$

where $m(t) = \int_0^t L(s) \, ds, t \in I$.

Theorem 10.4 (Pathak and Shahzad [466]) *Let Hypothesis (10.2) be satisfied,* $\lambda(\cdot, \cdot), \mu(\cdot, \cdot) \in C(I \times X, X)$ *and let* $u(\cdot) \in L_1(I, X)$ *be such that*

$$d(v(t), F(t, V(y)(t))) \leqslant p(t) \quad a.e.(I),$$

where $p(\cdot) \in L_1(I, \mathbb{R}^+)$ *and* $y(t) = \mu(t, u(t)) + \Phi(u)(t)$, *for all* $t \in I$. *Then for every* $\alpha > 1$, $0 < h < 1$, *there exist* $v \in \mathbb{N}$ *and* $x(\cdot) \in S(\lambda)$ *such that, for every* $t \in I$,

$$\|x(t) - y(t)\| \leqslant \|\lambda - \mu\|_C \left[1 + \frac{e^{\alpha(M_4 M_2 + M_1) M_3 m(T)}}{\sqrt{\alpha}(\sqrt{\alpha} - 1)} \right]$$
$$+ \frac{\sqrt{\alpha}}{(\sqrt{\alpha} - 1)} (M_4 M_2 + M_1) e^{\alpha(M_4 M_2 + M_1) M_3 m(T)} \int_0^T e^{-\alpha(M_4 M_2 + M_1) M_3 m(t)} p(t) \, dt.$$

Proof For $\lambda \in C(I, X)$ and $u \in L_1(I, X)$, define

$$x_{u,\lambda}(t) = \lambda(t) + \int_0^t [a(t, s) \, g(t, u(s)) + f(t, s, u(s))] \, ds.$$

Let us consider that $\lambda \in C(I, X), \sigma \in L_1(I, X)$ and define the set-valued maps

$$M_{\lambda,\sigma}(t) := F(t, V(x_{\sigma,\lambda})(t)), \quad t \in I, \tag{10.59}$$
$$T_\lambda(\sigma) := \{\psi(\cdot) \in L_1(I, X); \ \psi(t) \in M_{\lambda, \sigma}(t) \text{ a.e. } (I)\}. \tag{10.60}$$

Now, we claim that $T_\lambda(\sigma)$ is nonempty and closed for every $\sigma \in L_1(I, X)$.

It is well known that the set-valued map $M_{\lambda,\sigma}(\cdot)$ is measurable. For example, the map $t \to F(t, V(x_{\sigma,\lambda})(t)$ can be approximated by step functions and so we can apply Theorem III.40 in [131]. As the values of F are closed, with the measurable selection theorem we infer that $M_{\lambda,\sigma}(\cdot)$ is nonempty.

Further, we note that the set $T_\lambda(\sigma)$ is closed. Indeed, if $\psi_n \in T_\lambda(\cdot)$ and $\|\psi_n - \psi\|_1 \to 0$, then there exists a subsequence ψ_{n_k} such that $\psi_{n_k}(t) \to \psi(t)$ for almost every $t \in I$ and we find that $\psi \in T_\lambda(\sigma)$.

Let $\sigma_1, \sigma_2 \in L_1(I, X)$ be given. Assume, without loss of generality, that

$$0 < \rho(F(t, V(x_{\sigma_1, \lambda})(t)), F(t, V(x_{\sigma_2, \lambda})(t)))$$
$$\leqslant \rho(F(t, V(x_{\sigma_2, \lambda})(t)), F(t, V(x_{\sigma_1, \lambda})(t))).$$

Let $w_1 \in T_\lambda(\sigma_1)$ and let $\delta > 0$. Consider the following set-valued map:

$$\mathscr{G}(t) := M_{\lambda, \sigma_2(t)} \bigcap \left\{ z \in X : \|\psi_1(t) - z\| \leqslant M_1 M_3 L(t) \int_0^t \|\sigma_1(s) - \sigma_2(s)\| ds + \delta \right\}.$$

Since

$$d(\psi_1, M_{\lambda, \sigma_2}(t)) \leqslant \rho(F(t, V(x_{\sigma_1, \lambda})(t)), F(t, V(x_{\sigma_2, \lambda})(t)))$$
$$\leqslant \frac{1}{2}[\rho(F(t, V(x_{\sigma_1, \lambda})(t)), F(t, V(x_{\sigma_2, \lambda})(t)))$$
$$+ \rho(F(t, V(x_{\sigma_2, \lambda})(t)), F(t, V(x_{\sigma_1, \lambda})(t)))]$$
$$= H^+(F(t, V(x_{\sigma_1, \lambda})(t)), F(t, V(x_{\sigma_2, \lambda})(t)))$$
$$\leqslant L(t)\|V(x_{\sigma_1, \lambda})(t)) - V(x_{\sigma_2, \lambda})(t))\|$$
$$\leqslant M_3 L(t)\|x_{\sigma_1, \lambda}(t) - x_{\sigma_2, \lambda}(t)\|$$
$$\leqslant M_3 L(t)[\int_0^t |a(t, s)| \|g(t, \sigma_1(s)) - g(t, \sigma_2(s))\| ds$$
$$+ \int_0^t \|f(t, s, \sigma_1(s)) - f(t, s, \sigma_2(s))\| ds]$$
$$\leqslant M_3 L(t) \left[(M_4 M_2 + M_1) \int_0^t \|\sigma_1(s) - \sigma_2(s)\| ds \right],$$

we deduce that $\mathscr{G}(\cdot)$ has nonempty closed values. Further, according to Proposition III.4 in [131], $\mathscr{G}(\cdot)$ is measurable.

Let $\psi_2(\cdot)$ be a measurable selector of $\mathscr{G}(\cdot)$. It follows that $\psi_2 \in T_\lambda(\sigma_2)$ and

$$\|\psi_1 - \psi_2\|_1 = \int_0^T e^{-\sqrt{\alpha}(M_4 M_2 + M_1) M_3 m(t)} \|\psi_1(t) - \psi_2(t)\| dt$$
$$\leqslant \int_0^T e^{-\sqrt{\alpha}(M_4 M_2 + M_1) M_3 m(t)} M_3 L(t) \left[(M_4 M_2 + M_1) \int_0^t \|\sigma_1(s) - \sigma_2(s)\| ds \right] dt$$
$$+ \delta \int_0^T e^{-\sqrt{\alpha}(M_4 M_2 + M_1) M_3 m(t)} dt$$
$$\leqslant \frac{1}{\sqrt{\alpha}} \|\sigma_1 - \sigma_2\|_1 + \delta \int_0^T e^{-\sqrt{\alpha}(M_4 M_2 + M_1) M_3 m(t)} dt.$$

Since δ is arbitrary, so letting $\delta \to 0$ we deduce from the above inequality that

$$d(\psi_1, T_\lambda(\sigma_2)) \leqslant \frac{1}{\sqrt{\alpha}} \|\sigma_1 - \sigma_2\|_1.$$

Now replacing $\sigma_1(\cdot)$ with $\sigma_2(\cdot)$, we obtain

$$H^+(T_\lambda(\sigma_1), T_\lambda(\sigma_2)) \leqslant \frac{1}{\sqrt{\alpha}} \|\sigma_1 - \sigma_2\|_1.$$

Hence, we conclude that $T_\lambda(\cdot)$ is a contraction on $L_1(I, X)$.

Next, we consider the following set-valued maps

$$\widetilde{F}(t, x) := F(t, x) + p(t),$$
$$\widetilde{M}_{\lambda,\sigma}(t) := \widetilde{F}(t, V(x_{\sigma,\lambda})(t)), \quad t \in I,$$
$$\widetilde{T}_\lambda(\sigma) := \{\psi(\cdot) \in L_1(I, X); \ \psi(t) \in \widetilde{M}_{\lambda,\sigma}(t) \text{a.e. } (I)\}.$$

It is obvious that $\widetilde{F}(\cdot, \cdot)$ satisfies Hypothesis 10.2.

Let $\phi \in T_\lambda(\sigma)$, $\delta > 0$ and define

$$\mathscr{G}_1(t) := \widetilde{M}_{\lambda,\sigma(t)} \cap \{z \in X : \|\phi(t) - z\| \leqslant M_3 L(t) \|\lambda - \mu\|_c + p(t) + \delta\}.$$

Using the same argument as used for the set-valued map $\mathscr{G}(\cdot)$, we deduce that $\mathscr{G}_1(\cdot)$ is measurable with nonempty closed values.

Next, we prove the following estimate:

$$H^+(T_\lambda(\sigma), \widetilde{T}_\mu(\sigma)) \leqslant \frac{1}{\sqrt{\alpha}(M_4 M_2 + M_1)} \|\lambda - \mu\|_c$$
$$+ \int_0^T e^{-\sqrt{\alpha}(M_4 M_2 + M_1) M_3 m(t)} p(t) \, dt.$$

Let $\psi(\cdot) \in T_\mu(\sigma)$. Then

$$\|\phi - \psi\|_1 \leqslant \int_0^T e^{-\sqrt{\alpha}(M_4 M_2 + M_1) M_3 m(t)} \|\phi(t) - \psi(t)\| \, dt$$

$$\leqslant \int_0^T e^{-\sqrt{\alpha}(M_4 M_2 + M_1) M_3 m(t)} [M_3 L(t) \|\lambda - \mu\|_c + p(t) + \delta] \, dt$$

$$= \|\lambda - \mu\|_c \int_0^T e^{-\sqrt{\alpha}(M_4 M_2 + M_1) M_3 m(t)} M_3 L(t) \, dt$$

$$+ \int_0^T e^{-\sqrt{\alpha}(M_4 M_2 + M_1) M_3 m(t)} p(t) \, dt$$

$$+ \delta \int_0^T e^{-\sqrt{\alpha}(M_4 M_2 + M_1) M_3 m(t)} \, dt$$

$$\leqslant \frac{1}{\sqrt{\alpha}(M_4 M_2 + M_1)} \|\lambda - \mu\|_C + \int_0^T e^{-\sqrt{\alpha}(M_4 M_2 + M_1) M_3 m(t)} \, p(t) \, dt$$

$$+ \delta \int_0^T e^{-\sqrt{\alpha}(M_4 M_2 + M_1) M_3 m(t)} \, dt.$$

As δ is arbitrary, so letting $\delta \to 0$ we deduce from the above inequality that

$$\|\phi - \psi\|_1 \leqslant \frac{1}{\sqrt{\alpha}(M_4 M_2 + M_1)} \|\lambda - \mu\|_C + \int_0^T e^{-\sqrt{\alpha}(M_4 M_2 + M_1) M_3 m(t)} \, p(t) \, dt.$$

Now applying Proposition 5.6, we obtain

$$H^+(\text{Fix}(T_\lambda), \text{Fix}(\widetilde{T}_\mu)) \leqslant \frac{1}{(\sqrt{\alpha} - 1)(M_4 M_2 + M_1)} \|\lambda - \mu\|_C$$

$$+ \frac{\sqrt{\alpha}}{\sqrt{\alpha} - 1} \int_0^T e^{-\alpha(M_4 M_2 + M_1) M_3 m(t)} \, p(t) \, dt.$$

Since $v(\cdot) \in \text{Fix}(\widetilde{T}_\mu)$, there exists $u(\cdot) \in \text{Fix}(T_\mu)$ such that

$$\|v - u\|_1 \leqslant \frac{1}{\sqrt{\alpha}(\sqrt{\alpha} - 1)(M_4 M_2 + M_1)} \|\lambda - \mu\|_C$$

$$+ \frac{\sqrt{\alpha}}{\sqrt{\alpha} - 1} \int_0^T e^{-\alpha(M_4 M_2 + M_1) M_3 m(t)} \, p(t) \, dt.$$

We define

$$x = \lambda(t) + \int_0^t [a(t, s) g(t, u(s)) + f(t, s, u(s))] \, ds.$$

Then one has the following inequality:

$$\|x(t) - y(t)\| \leqslant \|\lambda(t) - \mu(t)\| + (M_4 M_2 + M_1) \int_0^t \|u(s) - v(s)\| \, ds$$

$$\leqslant \|\lambda - \mu\|_C + (M_4 M_2 + M_1) e^{\alpha(M_4 M_2 + M_1) M_3 m(T)} \|u - v\|_1.$$

Combining the last inequality with (10.59), we obtain

$$\|x(t) - y(t)\|$$

$$\leqslant \|\lambda - \mu\|_C \left[1 + \frac{e^{\alpha(M_4 M_2 + M_1) M_3 m(T)}}{\sqrt{\alpha}(\sqrt{\alpha} - 1)} \right]$$

$$+ \frac{\sqrt{\alpha}}{(\sqrt{\alpha} - 1)} (M_4 M_2 + M_1) e^{\alpha(M_4 M_2 + M_1) M_3 m(T)} \int_0^T e^{-\alpha(M_4 M_2 + M_1) M_3 m(t)} p(t) \, dt.$$

This completes the proof.

Remark 10.4 (*a*) If $a(t, \tau) \equiv 0$, Theorem 10.4 yields the result in [136] obtained for mild solutions of the semilinear differential inclusion (10.55).
(*b*)   If $a(t, \tau) = 0$, $f(t, \tau, u) = \mathscr{G}(t - \tau)u$, $V(x) = x$, $\lambda(t) = \mathscr{G}(t)x_0$ where $\{\mathscr{G}(t)\}_{t \geq 0}$ is a C_0-semigroup with an infinitesimal generator A, Theorem 10.4 yields the result in [135] obtained for mild solutions of the semilinear differential inclusion (10.55).

Exercises

10.1 Let the following integral inclusion be given:

$$x(t) = \lambda(t) + \int_0^t f(t, s, u(s)) ds, \tag{1}$$

$$u(t) \in F(t, V(x)(t)), \quad \text{a.e. } (I := [0, T]), \tag{2}$$

where $T > 0, I : [0, T]$, $\lambda(\cdot) : I \to \mathbb{R}^n$, $F(\cdot, \cdot) : I \times X \to \mathscr{P}(X)$, $f(\cdot, \cdot, \cdot) : I \times I \times X \to X$, $V : C(I, X) \to C(I, X)$ are given mappings, X is a separable Banach space and $\mathscr{P}(X)$ is the class of all nonempty subsets of X.

Let $F(\cdot, \cdot) : I \times X \to \mathscr{P}(X)$ be a set-valued map with nonempty closed values that verify:

(*i*)   The set-valued map $F(\cdot, \cdot)$ is $\mathscr{L}(I) \bigoplus \mathscr{B}(X)$ measurable.
(*ii*)   There exists $L(\cdot) \in L_1(I, \mathbb{R}^+)$ such that, for almost all $t \in I$, $F(t, \cdot)$ is $L(t)$-Lipschitz in the sense that

$$H(F(t, x), F(t, y)) \leqslant L(t) \|x - y\| \; \forall x, y \in X,$$

where H is the Hausdorff generalized metric on $\mathscr{P}(X)$ defined by

$$H(A, B) = \max\{\rho(A, B), \rho(B, A)\}, \; \rho(A, B) = \sup\{d(a, B); a \in A\}.$$

(*iii*)   The mapping $f : I \times I \times X \to X$ is continuous, $V : C(I, X) \to C(I, X)$ and there exist the constants $M_1, M_2 > 0$ such that

$$\|f(t, s, u_1) - f(t, s, u_2)\| = M_1 \|u_1 - u_2\|, \;\; \forall u_1, u_2 \in X,$$

$$\|V(x_1)(t) - V(x_2)(t)\| = M_2 \|x_1(t) - x_2(t)\|, \;\; \forall t \in I, \; \forall x_1, x_2 \in C(I, X).$$

Let $\lambda(\cdot), \mu(\cdot) \in C(I, X)$ and let $v(\cdot) \in L_1(I, X)$ be such that

$$d(v(t), F(t, V(y)(t))) \leqslant p(t) \text{ a.e. } (I),$$

where $p(\cdot) \in L_1(I, \mathbb{R}^+)$ and $y(t) = \mu(t) + \Phi(v)(t), \forall t \in I$, where

$$\Phi(u)(t) = \int_0^t f(t, \tau, u(\tau)d\tau, \ \ t \in I.$$

Then prove that for every $\alpha > 1$ and for every $\varepsilon > 0$, there exists $x(\cdot) \in S(\lambda)$, the solution set of (1)-(2) such that for every $t \in I$

$$\|x(t) - y(t)\| \leqslant \frac{\alpha}{\alpha - 1} e^{\alpha M_1 M_2 m(T)} \left[\|\lambda - \mu\|_C + M_1 \int_0^T e^{-\alpha M_1 M_2 m(t)} p(t) dt \right] + \varepsilon.$$

10.2 Let the following boundary value problem for fourth-order differential inclusion be given:

$$L_4 x(t) + a(t)x(t) \in F(t, x(t)) \ \text{a.e.} \ ([0, T]), \tag{3}$$

$$L_i x(0) = L_i x(T), \ i = 0, 1, 2, 3, \tag{4}$$

where

$$L_0 x(t) = a_0(t)x(t), L_i x(t) = a_i(t)(L_{i-1}x(t))', i = 1, 2, 3,$$
$$L_4 x(t) = (a_3(t)(a_2(t)(a_1(t)(a_0(t)x(t))')')')',$$

$a(\cdot), a_i(\cdot) : [0, T] \to \mathbb{R}$ are continuous mappings such that

$$a_0(t) \equiv 1, a(t) \geqslant 0, \ a_i(t) \overset{.}{>} 0, \ i = 1, 2, \ t \in [0, T], \ a_1(t) \equiv a_3(t)$$

and $F(\cdot, \cdot) : [0, T] \times \mathbb{R} \to \mathscr{P}(\mathbb{R})$ is a set-valued map. Suppose also that the following conditions hold:

(*i*) $F(\cdot, \cdot) : I \times \mathbb{R} \to \mathscr{P}(\mathbb{R})$ has nonempty closed values and for every $x \in \mathbb{R}$, $F(\cdot, x)$ is measurable.

(*ii*)   There exists $L(\cdot) \in L_1(I, \mathbb{R}^+)$ such that for almost all $t \in I$, $F(t, \cdot)$ is $L(t)$-Lipschitz in the sense that

$$H(F(t, x), F(t, y)) \leqslant L(t)|x - y| \ \forall x, y \in \mathbb{R}.$$

(*iii*) $d(0, F(t, 0)) \leqslant L(t)$ a.e. (*I*).

Denote $L_0 := \int_0^1 L(s)ds$.
Assume that $G_0 L_0 < 1$. Let $y(\cdot) \in AC^3(I, \mathbb{R})$ be such that $L_i y(0) = L_i y(T), i = 0, 1, 2, 3$ and there exists $q(\cdot) \in L_1(I, \mathbb{R}^+)$ with

$$d(L_4 y(t) + a(t)y(t), F(t, y(t))) \leqslant q(t), \ \text{a.e.} \ (I).$$

Then prove that for every $\varepsilon > 0$, there exists $x(.)$ a solution of problem (10.3) and (10.4) satisfying for all $t \in I$

$$|x(t) - y(t)| = \frac{G_0}{1 - G_0 L_0} \int_0^T q(t)dt + \varepsilon.$$

10.3 Assume that conditions (i), (ii) of 10.2 are satisfied and $G_0 L_0 < 1$. Let $y(\cdot) \in AC^3(I, \mathbb{R})$ be such that $L_i y(0) = L_i y(T), i = 0, 1, 2, 3$ and there exists $q(\cdot) \in L_1(I, \mathbb{R}^+)$ with $d(L_4 y(t) + a(t)y(t), F(t, y(t))) = q(t)$,   a.e. (I). Then prove that there exists $x(.)$ a solution of the BVP for fourth-order differential inclusion (10.3) and (10.4) satisfying for all $t \in I$

$$|x(t) - y(t)| = \frac{G_0}{1 - G_0 L_0} \int_0^T q(t)dt.$$

Appendix A
Basic Definitions

A.1 Basic Inequalities

Lemma A.1.1 *Let* $a, b \in \mathbb{R}^+$ *and* $0 < p \leqslant 1$. *Then,*

$$(a+b)^p \leqslant a^p + b^p.$$

Lemma A.1.2 *Let* $a, b \in \mathbb{R}^+$ *and* $2 \leqslant p < \infty$. *Then, we have the following inequalities:*

(a) $a^p + b^p \leqslant (a^2 + b^2)^{p/2}$,

(b) $(a^2 + b^2)^{p/2} \leqslant 2^{(p-1)/2}(a^p + b^p)$.

Proof It is easy to observe that both the inequalities hold if either a or b is zero. So, we prove the Lemma for $a \neq 0$ and $b \neq 0$.

(a) For $a, b > 0$, we have

$$\frac{a^2}{a^2 + b^2} \leqslant 1 \text{ and } \frac{b^2}{a^2 + b^2} \leqslant 1.$$

It follows that

$$\frac{a^p}{(a^2 + b^2)^{p/2}} + \frac{b^p}{(a^2 + b^2)^{p/2}} = \left(\frac{a^2}{a^2 + b^2}\right)^{p/2} + \left(\frac{b^2}{a^2 + b^2}\right)^{p/2}$$

$$\leqslant \frac{a^2}{a^2 + b^2} + \frac{b^2}{a^2 + b^2} \quad (\text{since } p/2 \geq 1)$$

$$= 1.$$

(b) For $p = 2$, the inequality is obvious. So, we assume that $p > 2$, i.e., $p/2 > 1$. Set $\alpha = p/2$ and $\beta = \alpha/(\alpha - 1) = p/(p-2)$. Then, $\frac{1}{\alpha} + \frac{1}{\beta} = 1$. By Hölder's

© Springer Nature Singapore Pte Ltd. 2018

H. K. Pathak, *An Introduction to Nonlinear Analysis and Fixed Point Theory*,
https://doi.org/10.1007/978-981-10-8866-7

inequality, we obtain

$$a^2 + b^2 = a^2 \cdot 1 + b^2 \cdot 1 \leqslant \left((a^2)^\alpha + (b^2)^\alpha \right)^{1/\alpha} \left((1)^\beta + (1)^\beta \right)^{1/\beta}$$
$$= 2^{(p-2)/p}(a^p + b^q)^{2/p},$$

which implies that

$$(a^2 + b^2)^{p/2} \leqslant 2^{(p-1)/2}(a^p + b^p).$$

Lemma A.1.3 (Minkowski inequality) *If $p \geq 1$, then for any complex numbers $a_1, a_2, \ldots, a_n$; $b_1, b_2, \ldots, b_n$ we have the following inequality:*

$$\left(\sum_{i=1}^{n} |a_i + b_i| \right)^{1/p} \leqslant \left(\sum_{i=1}^{n} |a_i|^p \right)^{1/p} + \left(\sum_{i=1}^{n} |b_i|^p \right)^{1/p}.$$

Lemma A.1.4 (Hölder inequality) *If $p > 1$ and q is defined by $\frac{1}{p} + \frac{1}{q} = 1$, then for any complex numbers $a_1, a_2, \ldots, a_n$; $b_1, b_2, \ldots, b_n$ we have the following inequality:*

$$\sum_{i=1}^{n} |a_i b_i| \leqslant \left(\sum_{i=1}^{n} |a_i|^p \right)^{1/p} \left(\sum_{i=1}^{n} |b_i|^q \right)^{1/q}.$$

An extension of Minkowski's inequality:

Lemma A.1.5 (Minkowski inequality) *If $1 \leqslant p < \infty$ and if $x = \{x_i\}, y = \{y_i\} \in \ell_p$, then $x + y = \{x_i + y_i\} \in \ell_p$ and*

$$\left(\sum_{i=1}^{\infty} |x_i + y_i|^p \right)^{1/p} \leqslant \left(\sum_{i=1}^{\infty} |x_i|^p \right)^{1/p} + \left(\sum_{i=1}^{\infty} |y_i|^p \right)^{1/p}.$$

An extension of Hölder's inequality:

Lemma A.1.6 (Hölder inequality) *If $1 \leqslant p < \infty$, q is conjugate to p, $x = \{x_i\} \in \ell_p$ and $y = \{y_i\} \in \ell_q$, then $\{x_1 y_1, x_2 y_2, \ldots\} \in \ell_1$ and*

$$\sum_{i=1}^{\infty} |x_i y_i| \leqslant \left(\sum_{i=1}^{\infty} |x_i|^p \right)^{1/p} \left(\sum_{i=1}^{\infty} |y_i|^q \right)^{1/q}.$$

The above inequality with $p = q = 2$ gives the Cauchy–Schwarz inequality:

$$\left(\sum_{i=1}^{\infty} |x_i y_i| \right)^2 \leqslant \left(\sum_{i=1}^{\infty} |x_i|^2 \right) \left(\sum_{i=1}^{\infty} |y_i|^2 \right).$$

A.2 Integral Formula

$$\int_a^t \int_a^t \cdots \int_a^t f(x)\,(dx)^n = \frac{1}{(n-1)!}\int_a^t (t-x)^{(n-1)} f(x)\,dx.$$

A.3 Green's Theorem

If ϕ, ψ are both single-valued and continuously differentiable scalar point functions such that $\nabla\phi$ and $\nabla\psi$ are also continuously differentiable, then

$$\int_V \nabla\phi \cdot \nabla\psi\,dV = -\int_V \phi\nabla^2\psi\,dV - \int_S \phi\frac{\partial\psi}{\partial n}dS$$
$$= -\int_V \psi\nabla^2\phi\,dV - \int_S \psi\frac{\partial\phi}{\partial n}dS$$

where S is a closed surface bounding any simple connected region, δn is an element of the normal at any point on the boundary drawn into region considered, and V is the volume enclosed by S.

Appendix B
Basic Definitions

n-simplex $-$ In $\mathbb{R}^n$, a (non-degenerate) n-simplex is the convex hull K of $n + 1$ points $a_j = (a_{1j}, a_{2j}, \ldots, a_{nj}) \in \mathbb{R}^n$, which are known as the vertices of the n-simplex such that the matrix

$$A = \begin{pmatrix} a_{11} & a_{12} & \cdots & a_{1\ n+1} \\ a_{21} & a_{22} & \cdots & a_{2\ n+1} \\ \cdots & \cdots & \cdots & \cdots \\ a_{n1} & a_{n2} & \cdots & a_{n\ n+1} \\ 1 & 1 & \cdots & 1 \end{pmatrix}$$

is regular; i.e., $n + 1$ points a_j are not contained in a hyperplane. In other words, K is said to be an n-simplex if

$$K = \left\{ x = \sum_{j=1}^{n+1} \lambda_j a_j \mid 0 \leqslant \lambda_j \leqslant 1, \ \sum_{j=1}^{n+1} \lambda_j = 1 \right\}.$$

Particular cases:
- a 2-simplex is a triangle.
- a 3-simplex is a tetrahedron.

Positive semidefinite matrix $-$ A symmetric matrix $A \in S\mathbb{R}^{n \times n}$ is called positive semidefinite if $x^T A x \geqslant 0$ for all $x \in \mathbb{R}^n$ and is called positive definite if $x^T A x > 0$ for all nonzero $x \in \mathbb{R}^n$. The set of positive semidefinite matrices is denoted S_+^n, and the set of positive definite matrices is denoted by S_{++}^n.

Some characterizations of positive semidefinite matrices.
The following statements are equivalent:

- The symmetric matrix A is positive semidefinite.

© Springer Nature Singapore Pte Ltd. 2018
H. K. Pathak, *An Introduction to Nonlinear Analysis and Fixed Point Theory*,
https://doi.org/10.1007/978-981-10-8866-7

- All eigenvalues of A are nonnegative.
- All the principal minors of A are nonnegative.
- There exists B such that $A = B^T B$.

Let (X, d) be a metric space.

Definition B.1 *The point p belongs to the lower limit $Li_{n \to \infty} A_n$ of a sequence (A_n) of subsets of (X, d), if every open ball with center p intersects all the A_n from a sufficiently great index n onward.*

Definition B.2 *The point p belongs to the upper limit $Ls_{n \to \infty} A_n$ of a sequence (A_n) of subsets of (X, d), if every open ball with center p intersects an infinite set of the terms A_n.*

Definition B.3 *The sequence (A_n) of subsets of (X, d) is said to be convergent to A, if $Li_{n \to \infty} A_n = A = Ls_{n \to \infty} A_n$. We then write $A = Lim_{n \to \infty} A_n$.*

Appendix C
Solutions

Problems of Chapter 7

7.3 $y = x^2$.

7.4 $y(x) = \frac{\sinh x}{\sinh 1}$.

7.5 $y(x) = C_1 \cos(e^x) + C_2 \sin(e^x)$.

7.7 $y(x) = \sin \frac{\pi x}{2}$.

7.8 $y(x) = C_1 \cosh \frac{x - C_2}{C_1}$, where the constants C_1 and C_2 are determined from the given boundary conditions that the given curve passes through the points (x_1, y_1) and (x_2, y_2).

7.10 $(x - a)^2 + (y - b)^2 = \lambda^2$ which is a circle.

7.11 $y = \pm 2 \sin m\pi x$, where m is an integer.

7.12 $y = \lambda x^2 + ax + b$, where λ, a, b determined from the isometric and boundary conditions.

7.15 Total resistance experienced by the body is $F = 4\pi\rho v^2 \int_0^l y'^3 y \, dx$. From this, constitute a variational problem with conditions $y(0) = 0, y(l) = R$. The shape of the solid of revolution is $y(x) = R(\frac{x}{l})^{3/4}$.

7.16 $P_0 = \frac{\pi^2}{l^2} EI$, where $l = $ effective length, $E = $ modulus of elasticity and $I = $ minimum area moment of inertia of the cross section of the rod.

7.17 $\lambda_1 = 6$ with $z_1 = \alpha(x^2 + y^2 - 1)$.

7.18 $\lambda_1 = 0.493$.

© Springer Nature Singapore Pte Ltd. 2018
H. K. Pathak, *An Introduction to Nonlinear Analysis and Fixed Point Theory*,
https://doi.org/10.1007/978-981-10-8866-7

7.19 Seeking the solution of the form

$$y_2(x) = x(x-1)(\alpha_1 + \alpha_2 x), \quad y_3(x) = x(x_1)(\beta x + \beta_3 x^2),$$

we finds that $\alpha_1 = 0.1708, \alpha_2 = 0.1744; \beta_1 = 0.1705, \beta_2 = 0.1760, \beta_2 = 0.0018$. The values of y_2 and y_3 agree at the specified points to within 0.0001.

7.20 Seek the solution of the form $z = \alpha(x^2 - l^2)(y^2 - l^2)$. Then, it is found that $\alpha = \frac{5}{16l^2}$.

References

1. A. Agadi, J.-P. Penot, *Asymptotic approximation of sets with application in mathematical programming* (University of Pau, Preprint, 1996)
2. R.P. Agarwal, J. Banas, B.C. Dhage, T. Gnana Bhaskar, Local and global attractivity results for quadratic functional integral equations. Funct. Diff. Equ. **17**, 3–19 (2010)
3. R.P. Agarwal, J. Banas, B.C. Dhage, S.D. Sarkate, Attractivity results for a nonlinear functional integral equation. Georgian Math. J. **18**, 1–19 (2011)
4. R.P. Agarwal, M. Meehan, D. O'Regan, Positive L_p and continuous solutions for Fredholm integral inclusions. SIMAA **4**, 1–9 (2002)
5. A. Aghajani, J. Banas, N. Sabzali, Some generalizations of Darbo fixed point theorem and applications. Bull. Belg. Math. Soc. Simon Stevin **20**, 345–358 (2013)
6. R.P. Agarwal, D. O'Regan, A note on the existence of multiple fixed points for multivalued maps with applications. J. Differ. Equ. **160**, 389–403 (2000)
7. S. Agman, *Lectures on Elliptic Boundary Value Problems* (Von Nostrand, New York, 1965)
8. Ya I. Alber, S. Guerre-Delabriere, Principles of weakly contractive maps, in *Hilbert spaces, in New result in Operator Theory and Applications, vol. 98, Operator Theory: Advances and Applications*, ed. by I. Gohberg, Yu. Lyubich (Birkhauser, Basel, 1997), pp. 7–22
9. P. Alexandroff, H. Hopf, *Topologie* (Springer, Berlin, 1935), reprinted by Chelsa, New York (1965)
10. M. Altman, A fixed point theorem in Hilbert space. Bull. Acad. Polon. Sci. Cl. **III**(5), 89–92 (1957)
11. M. Altman, A fixed point theorem in Banach space. Bull. Acad. Polon. Sci. Cl. **III**(5), 17–22 (1957)
12. M. Altman, Contractor directions, directional contractors and directional contractions for solving equations. Pac. J. Math. **62**, 1–18 (1976)
13. H. Amann, Uber die Existenz and iterative Berechnung linear einer Losung der Hammtien-schen Gleichung. Aequ. Math. **1**, 242–266 (1968)
14. H. Amann, Ein Existenz ana Eindeutigkeits satz fur die Hammersteinsche Gleichung in Banach-Raumen. Math. Z. **111**, 179–190 (1969)
15. H. Amann, Uber die Naherungsweise Losung nichtlinearer integral gleichungen. Num. Math
16. H. Amann, Ein Existenz and Eindwutigkeits satz fur die Hammerstein type. Appl. Anal. **2**, 385–397 (1972)
17. H. Amann, Fixed point equations and non-linear eigenvalue problems in ordered Banach spaces. SIAM Rev. **18**(4), 620–709 (1976)
18. H. Amann, On the number of solutions of nonlinear equations in ordered Banach spaces. J. Funct. Anal. **11**, 346–384 (1972)

© Springer Nature Singapore Pte Ltd. 2018
H. K. Pathak, *An Introduction to Nonlinear Analysis and Fixed Point Theory*,
https://doi.org/10.1007/978-981-10-8866-7

19. H. Amann, Fixed points of asymptotically linear maps in ordered Banach spaces. J. Funct. Anal. **14**, 162–171 (1972)
20. A. Ambrosetti, Un teorema di esistenza per le equations differenzialinegli spazi di Banach. Rend. Sem. Math. Univ. Padova **39**, 349–361 (1967)
21. A. Ambrosetti, Proprieta spectrali di certi operatori lineari non-compactti. Rend. Sem. Math. Univ. Padova **42**, 189–200 (1969)
22. A.M. Arthur, *Complementary Variational Principles* (Clarendon Press, Oxford, 1970)
23. E. Asplund, Frechet differentiability of convex functions. Acta Mathematica **121**, 31–48 (1968)
24. E. Asplund, Positivity and duality mappings. Bull. Am. Math. Soc. **73**, 200–203 (1967)
25. N.A. Assad, W.A. Kirk, Fixed point theorems for set valued mappings of contractive type. Pac. J. Math. **43**, 553–562 (1972)
26. J.P. Aubin, H. Frankowska, *Set-valued Analysis* (Birkhäuser, Boston, Basel, Berlin, 1990)
27. J.P. Aubin, J. Siegel, Fixed points and stationary points of dissipative multivalued maps. Proc. Am. Math. Soc. **78**, 391–398 (1980)
28. M. Backwinkel, M. Schillings, Verallgemeinerte Hammersteinsche Gleichungen und einge Anvendungen, Dissertation, Bochum (1974)
29. M. Backwinkel, M. Schillings, Existence theorems for generalized Hammerstein equations. J. Funct. Anal. **23**, 177–194 (1976)
30. J.B. Bailon, These, *University de Paris* (1978)
31. I.A. Bakhtin, The contraction mapping principle in quasimetric spaces. Funct. Anal. Unianowsk Gos. Ped. Inst. **30**, 26–37 (1989)
32. S. Banach, Sur les operations dans les ensembles abstraits et leur application aux equtaions untegrales. Fund. Math. **3**, 133–181 (1922)
33. S. Banach, *Theorie des Operations Linearries* (Monografge Matematyezne, Warsaw, 1932)
34. J. Banas, B.C. Dhage, Global asymptotic stability of solutions of a functional integral equation. Nonlinear Anal. **69**, 1945–1952 (2008)
35. J. Banas, K. Goebel, *Measures of Noncompactness in Banach Spaces*, vol. 60 (Lecture Notes in Pure and Applied Mathematics (Marcel Dekker, New York, 1980)
36. J. Banas, M. Lecko, Fixed points of the product of operators in Banach algebra. Panamer. Math. J. **12**, 101–109 (2002)
37. J. Banas, B. Rzepka, On existence and asymptotic stability of solutions of a nonlinear integral equation. J. Math. Anal. Appl. **284**, 165–173 (2003)
38. J. Banas, K. Sadarangani, Solutions of some functional-integral equations in Banach algebra. Math. Comput. Modell. **38**, 245–250 (2003)
39. V. Barbu, *Nonlinear semigroups and differential equations in Banach spaces (Noordhoff, Leiden* (Editura Academiei R. S. R, Bucutesti, 1976)
40. R. Bass, Contributions to the of nonlinear oscillations ed. by S. Leffschetz, vol. IV (Princiton University Press, 1958), pp. 201–211 (On nonlinear repulsive forces)
41. I. Beg, A. Azam, Fixed point theorems for Kannan mappings. Indian J. Pure Appl. Math. **17**(11), 1270–1275 (1986)
42. I. Beg, A. Azam, Fixed points of assymptotically regular multivalued mappings. J. Aust. Math. Soc. (Ser. A) **53**, 313–326 (1992)
43. S.A. Belbas, I.D. Mayergoyz, Applications of fixed-point methods to discrete variational and quasi-variational inequalities. Numer. Math. **51**, 631–654 (1987)
44. A. Belent-Morante, An integro-differential equation arising from the theory of heat conduction in rigid materials with memory. Boll. Un. Mat. Ital. B **15**, 470–482 (1978)
45. L.P. Belluec, W.A. Kirk, Nonexpensive mappings and fixed points in Banach spaces. III. J. Math. **11**, 471–479 (1967)
46. L.P. Belluce, W.A. Kirk, Fixed point theorems for families of contraction mappings. Pac. J. Math. **18**, 213–217 (1966)
47. L.P. Belluce, W.A. Kirk, E.F. Steiner, Normal structure in Banach spaces Fixed point theorems for certain classes of nonexpensive mappings. Pac. J. Math. **26**, 433–440 (1968)

48. T.D. Benavides, A renorming of some nonseparable Banach spaces with the fixed point property. J. Math. Anal. Appl. **350**(2), 525–530 (2009)
49. A. Bensoussan, M. Gourset, J.L. Lions, Controle impulssionel et applications. C. R. Acad. Sci. Paris Ser. **I**(276), 1279–1284 (1973)
50. A. Bensoussan, J.L. Lions, Nouvelle formulation de problems de controle impulssionnel et applications. C. R. Acad. Sci. Paris Ser. **I**(276), 1189–1192 (1973)
51. A. Bensoussan, J.L. Lions, *Applications des inequations variationnel en control stochastique* (Dunod, Paris, 1978)
52. M.S. Berger, *Nonlinearity and Functional Analysis, Lectures on Nonlinear Problems in Mathematical Analysis* (Academic Press, New york, 1977)
53. M.S. Berger, M. Berger, *Perspective in Nonlinearity* (W. A. Banjamin Inc, New York, 1968)
54. V. Berinde, Approximating fixed points of weak contractions using the Picard iteration. Nonlinear Anal. Forum **9**, 45–53 (2004)
55. M. Berinde, V. Berinde, On a general class of multi-valued weakly Picard mappings. J. Math. Anal. Appl. **326**, 772–782 (2007)
56. C. Bessaga, On the converse of the Banach fixed point principle. Colloq. Math. **7**, 41–43 (1959)
57. A. Beurling, A.E. Livingston, A theorem on duality mappings in Banach spaces. Ark. Fur. Math. **4**, 405–411 (1962)
58. G. Birkhoff, *Lattice Theory*, vol. 25 (Revised edn (American Mathematical Society Colloquium Publication, New York, 1948)
59. C.D. Birkoff, O.D. Kellog, Invariant points in afunction space. Tran. Am. Math. Soc. **23**, 96–115 (1922)
60. E. Bishop, R.R. Phelps, *The support functionals of a convex set, in Proceeding of Symposia in Pure Mathematics* (Convexity (American Mathematical Society, VII, 1963), pp. 27–36
61. E. Blum, W. Oettli, From optimization and variational inequalities to equilibrium problems. Math. Stud. **63**(1–4), 123–145 (1994)
62. P. Bohl, Uber die Bewwging eines mechanischen systems in der Nahe einer Gleichgewicht stage. J. Reine Angew. Math. **127**, 279–286 (1904)
63. A. Bollenbacher, T.L. Hicks, A fixed point theorem revisited. Proc. Am. Math. Soc. **102**(4), 898–900 (1988)
64. F.F. Bonsal, *Lectures on Some Fixed Point Theorems of Functional Analysis* (Tata institute of Fundamental Research, Bombay, India, 1962)
65. K. Borsuk, Sur les rétractes. Fund. Math. **17**, 152–170 (1931)
66. R.K. Bose, R.N. Mukharjee, Comman fixed points of multivalued mapping. Tamukang J. Math. **8**, 245–249 (1977)
67. R.K. Bose, R.N. Mukharjee, Stability of fixed point sets and comman fixed points of families of mappings. Indian J. Pure Appl. Math. **11**(9), 1130–1138 (1980)
68. R.K. Bose, R.N. Mukharjee, On fixed points of nonexpansive set-valued mappings. Proc. Am. Math. Soc. **72**, 97–98 (1978)
69. R.K. Bose, R.N. Mukharjee, Approximating fixed points of some mappings. Proc. Am. Math. Soc. **82**, 603–606 (1981)
70. R.K. Bose, D. Sahani, Weak convergence and comman fixed points of nonexpansive mappings by iteration. Ind. J. Pure Appl. Math. **15**(2), 123–126 (1984)
71. S.C. Bose, Weak convergence to the fixed points of asymptotically nonexapansive map. Pro. Am. Math. Soc. **68**, 304–308 (1978)
72. S.C. Bose, Comman fixed points of mappings in a uniformly convex spaces. J. Lond. Math. Soc. **18**, 151–156 (1978)
73. N. Bourbaki, *Espaces Vectoriels Topologiques* (Herman, Paris, 1955)
74. W.M. Boyce, Commuting functions with no common fixed point. Trans. Am. Math. Soc. **137**, 77–92 (1969)
75. D. Boyd, J.S.W. Wong, On nonlinear contractions. Proc. Am. Math. Soc. **20**, 458–464 (1969)
76. H. Brezis, *Operateurs maximaux momtones et semi-groupes de contractions dans les epsaces de Hilbert* (American Elsevier, Holland, 1973)

77. H. Brezis, Equations et inequations nonlineaires dans les espaces vectoriels on duality. Ann. Inst. Fourier **18**, 115–175 (1968)

78. H. Brezis, Pertubations nonlinearies d'operateures maximaux monotones. C.R. Acad. Sci. **260**, 566–569 (1969)

79. H. Brezis, Inequations variationnelles assocees a des poerateures d'evolution, Theory and applications of monotone operators, in *Proceedings of NATO Institute Venice* (Oderisi Gubbio, 1968)

80. H. Brezis, F.E. Browder, Nonlinear integral equations and system of Hammerstien type. Adv. Math. **18**, 115–147 (1975)

81. H. Brezis, P.L. Lions, Produits infinis de resolvantes. Isr. J. Math. **29**, 329–345 (1978)

82. M.S. Brodskii, D.P. Milman, On the centre of a convex set. Dokl. Akad. Nauk SSSR(NS) **59**, 837–840 (1948)

83. A. Brønstedt, On a lemma of Bishop and Phelps. Pac. J. Math. **55**, 335–341 (1974)

84. A. Brønstedt, Fixed points and partial orders. Proc. Am. Math. Soc. **60**, 365–366 (1976)

85. A. Brønstedt, R.T. Rockafer, On the subdifferentiabity of convex functions. Proc. Am. Math. Soc. **16**, 605–611 (1965)

86. L.E.J. Brouwer, Über Abbildungen Von Mannigfaltigen. Math. Ann. **71**, 97–115 (1912)

87. L.E.J. Brouwer, An intuitionist correction of the fixed point theorem on the sphere. Proc. R. Soc. Lond. (A) **213**, 1–2 (1952)

88. F.E. Browder, An intuitionist correction of the fixed point theorem on the sphere. Proc. R. Soc. Lond. (A) **213**, 1–2 (1952)

89. F.E. Browder, Nonlinear elliptic boundary value problems. Bull. Am. Math. Soc. **69**, 862–874 (1963)

90. F.E. Browder, The solvability of nonlinear functional equations. On Theorem Beurling and Livingston. Canad. J. Math. **17**, 367–372 (1965)

91. F.E. Browder, Multivalued monotone nonlinear mapping and duality mappings in Banach spaces. Trans. Am. Math. Soc. **118**, 338–351 (1965)

92. F.E. Browder, The solvability of nonlinear functional equations. Duke Math. J. **30**, 557–566 (1963)

93. F.E. Browder, On the constructive solution of nonlinear functional equation. J. Funct. Anal. **25**, 345–356 (1977)

94. F.E. Browder, Nonlinear operators and nonlinear equations of evolution in Banach spaces, Nonlinear Functional Analysis, in *Proceeding of Symposium Pure Mathematics* vol. 18, Part II (American Mathematical Society, Providence, R.I., 1976), pp. 862–874

95. F.E. Browder, Nonlinear functional analysis and nonlinear integral equations of Hammerstein and Uryshon type, in *Contributions to Nonlinear Functional Analysis*, ed. by C.H. Zarantonello (Academic Press, Cambridge, 1971), pp. 99–154

96. F.E. Browder, Nonexpansive nonlinear operators in a Banach space. Proc. Natl. Sci. U.S.A. **54**, 1041–1044 (1965)

97. F.E. Browder, Semiconrtactive and semiaccretive nonlinear mappings in Banach spaces. Bull. Am. Math. Soc. **74**, 660–665 (1968)

98. F.E. Browder, The fixed point theory of multi-valued mappings in topological vector spaces. Math. Ann. **177**, 283–301 (1968)

99. F.E. Browder, Existence of periodic solutions for nonlinear equations of evolution. Proc. Natl. Acad. Sci. **53**, 1100–1103 (1965)

100. F.E. Browder, Nonexpansive nonlinear operators in a Banach space. Proc. Natl. Acad. Sci. U.S.A. **54**, 1041–1044 (1965)

101. F.E. Browder, Fixed point theorems for nonlinear semi-contractive mappings in Banach spaces. Arch. Rat. Mech. Anal. **21**, 259–269 (1966)

102. F.E. Browder, Convergence theorems for sequences of nonlinear operators in Banach spaces. Math. Z. **100**, 201–225 (1967)

103. F.E. Browder, Nonlinear mappings of nonexapansive and accretive type in Banach spaces. Bull. Am. Math. Soc. **73**, 875–881 (1967)

104. F.E. Browder, Continuity properties of monotone nonolinear poerators in Banach spaces. Bull. Am. Math. Soc. **70**, 551–553 (1964)

105. F.E. Browder, On a generalization of the Schauder fixed point theorem. Duke Math. J. **26**, 291–303 (1959)

106. F.E. Browder, Fixed point theorem for noncompact mappings in Hilbert spaces. Proc. Natl. Acad. Sci. U.S.A. **53**, 1272–1276 (1965)

107. F.E. Browder, *Normal solvability and existence theorems for nonlinear mappings in Banach spaces, Problems in Nonlinear Analysis* (Rome, Italy, Edizioni Cremones, 1971), pp. 19–35

108. F.E. Browder, *Normal solvability for nonlinear mappings and the geometry of Banach spaces, Problems in Nonlinear Analysis* (Roma, Italy, Edizioni Cremonese, 1971), pp. 37–66

109. F.E. Browder, Normally solvable nonlinear mappings in Banach spaces and their homotopies. J. Funct. Anal. **17**, 441–446 (1974)

110. F.E. Browder, *Problems Nonlinears* (Les Presses de L'Universite de Montreal, Montreal, 1966)

111. F.E. Browder, On a theorem of Caristi an dKirk, in *Proceedings of Seminar on Fixed Point and its Applications*, Dalhausie University, June 1975 (Academic Press, 1976), pp. 23–27

112. F.E. Browder, Nonlinear equations of evolutions. Ann. Math. **80**, 485–523 (1964)

113. F.E. Browder, Normal solvability for nonlinear mappings in Banach spaces. Bull. Am. Math. Soc. **77**, 73–77 (1971)

114. F.E. Browder, Normal solvability and ϕ-accertive mappings in Banach spaces. Bull. Am. Math. Soc. **78**, 186–192 (1972)

115. F.E. Browder, Fixed point theory and nonlinear problems. Bull. Am. Math. Soc. **9**, 1–40 (1983)

116. F.E. Browder, D. De Figueiredo, J-monotone nonlinear operators in Banach spaces. Proc. Kon. Neder. Akad. Amsterdam **28**, 412–420 (1966)

117. F.E. Browder, C.P. Gupta, Nonlinear monotone opertors and integral equations of Hammerstien type. Bull. Am. Math. Soc. **75**, 1347–1353 (1969)

118. F.E. Browder, P. Hess, Nonlinear mappings of monotone type in Banach spaces. J. Funct. Anal. **11**, 251–294 (1972)

119. F.E. Browder, D. Nussbaum, The topological degree for noncompact nonlinear mappings in Banach spaces. Bull. Am. Math. Soc. **74**, 571–676 (1968)

120. F.E. Browder, W.V. Petryshyn, The solution by iteration of nonlinear functional equations in Banach spaces. Bull. Am. Math. Soc. **72**, 571–575 (1966)

121. F.E. Browder, W.V. Petryshyn, Construction of fixed points of nonlinear mappings in Hilbert spaces. J. Math. Anal. Appl. **20**, 197–228 (1967)

122. R.E. Bruck Jr., On the iterative solution of $y \in x + Tx$ for a bounded monotone operator T in A Hilbert space. Bull. Am. Math. Soc. **79**(6), 1258–1261 (1973)

123. R.E. Bruck Jr., A strongly convergent iterative solution of $0 \in Ux$ for maximal monotone operator U in a Hilbert space. J. Math. Anal. Appl. **48**, 114–126 (1974)

124. R.E. Bruck Jr., A simple proof of the mean ergodic theorem for nonlinear contractions in Banach spaces. Isr. J. Math. **32**, 107–116 (1979)

125. R.E. Bruck Jr., An iterative solution of a variational inequality for certain monotone operators in Hilbert space. Bull. Am. Math. Soc. **81**, 890–892 (1975)

126. T. Burton, A fixed point theorem of Krasnoselskii. Appl. Math. Lett. **11**(1), 85–88 (1998)

127. N.P. Cac, J.A. Gatica, Fixed point theorems for mappings in ordered Banach spaces. J. Math. Anal. Appl. **71**, 547–557 (1979)

128. J.V. Caristi, Ph.D. Thesis, *University of Iowa* (1975)

129. J.V. Caristi, Fixed point theorems for mappings satisfying inward conditions. Trans. Am. Math. Soc. **215**, 241–251 (1976)

130. J. V. Caristi and W. A. Kirk, Geometric fixed point theory and inward condiotions, Proceeding of the conference on geometry of metric and linear spaces, *Michigan State University, 1974, Lecture notes in Mathematics 490, Springer-Verlag 1975*

131. C. Castaing, M. Valadier, *Convex Analysis and Measurable Multifunctions, LNM 580* (Springer, Berlin, 1977)

132. C. Castiang, Sur les equations differentials multivouges. C.R. Acad. Sci. (Paris), **263**, 63–66 (1966)

133. A. Cellina, A. Fryszkowski, T. Rzezuchowski, Upper semicontinuity of Nemytskij operators. Annali di Matematica Pured Applicata **160**, 321–330 (1991)

134. L.-C. Ceng, J.-C., Yao, A hybrid iterative scheme for mixed equilibrium problems and fixed point problems. J. Comput. Appl. Math. **214**, 186–201 (2008)

135. A. Cernea, A Filippov type existence theorem for infinite horizon operational differential inclusions. Stud. Cerc. Mat. **50**, 15–22 (1998)

136. A. Cernea, Existance for nonconvex integral inclusions via fixed points. Archi. Math. (Brno) **39**, 293–298 (2003)

137. L. Cesary, Alternative methods in nonlinear analysis, in *International Conference on Differential Equations*, ed. by H.A. Antosiewicz (Academic Press, 1975) , pp. 95–149

138. L. Cesary, R. Klannan, J.D. Schuur, *Nonlinear Functional Analysis and Differntial Equations (Edited)* (Marcel Dekker Inc, New York, 1976)

139. S. Chandrasekhar, *Radiative Heat Transfer* (Dover, New York, 1960)

140. S.K. Chaterjea, Fixed point theorems. C. R. Acad. Bulgare Sci. **25**, 727–730 (1972)

141. P.G. Ciarlet, M.H. Schultz, R.S. Varga, Numerical methods of high order accuracy for nonlinear boundary value problems vs. monotone operators theory. Num. Math. **13**, 51–77 (1969)

142. B. Lj, Ćirić, Generalized contractions and fixed-point theorems. Publ. Inst. Math. (Belgr.) **12**(26), 19–26 (1971)

143. B. Lj, Čirič. On contraction type mappings. Math. Balk. **1**, 52–57 (1971)

144. B. Lj, Čirič. Proc. Am. Math. Soc. **45**, 267–273 (1974)

145. B. Lj, S.B. Ciric, Presic, On Presic type generalisation of Banach contraction principle. Acta Math. Univ. Comen. LXXV **I**(2), 143–147 (2007)

146. J.A. Clarkson, Uniformly convex spaces. Trans. Am. Math. Soc. **40**, 396–414 (1936)

147. K.L. Cooke, J.L. Kalpan, A periodicity threshold theorem for epidemics and population growth. Math. Bio. Sci. **31**, 87–104 (1976)

148. R.W. Cottle, F. Giannessi, J.-L. Lions, *Variational Inequalities and Complementarity Problems: Theory and Applications* (Wiley, New York, 1980)

149. R.W. Cottle, Numerical methods for complementarity problems in engineering and applied sciences, Computing Mathematics in Applied Science and Engineering, Lecture Notes in Mathematics, vol. 704 (Springer, 1979), pp. 37–52

150. R.W. Cottle, J.S. Pang, A least element theory of solving linear complementarity problems as linear programs. Math. Op. Res. **3**, 155–170 (1978)

151. N.G. Crandal, A. Pazy, Semigroups of nonlinear contractions and dissipative sets. J. Funct. Anal. **3**, 367–418 (1969)

152. J. Cronin, *Fixed Points and Topological Degree in Nonlinear Analysis* (American Mathematical Society, 1964). Mathematical Survey No. 11

153. C.W. Cryer, M.A.H. Dempster, Equivalence of linear complementarity problems and linear programs in vector lattice Hilbert spaces. SIAM J. Control Optim. **18**, 621–623 (1980)

154. P.Z. Daffer, H. Kaneko, Fixed points of generalized contrative multi-valued mappings. J. Math. Anal. Appl. **192**, 655–666 (1995)

155. G. Darbo, Punti uniti in transformations a condionio non compactie. Rend Sem. Math. Univ. Padova **24**, 84–92 (1955)

156. J.P. Dauer, A controllability technique for nonlinear systems. J. Math. Anal. Appl. **37**, 442–451 (1972)

157. A.C. Davis, A characterization of complete lattices. Pac. J. Math. **5**, 311–319 (1955)

158. M.M. Day, Reflexive Banach spaces not isomorphic to uniformly convex spaces. Bull. Am. Soc. **47**, 313–317 (1941)

159. M.M. Day, *Normed Linear Spaces* (Springer, Berlin, 1958)

160. M.M. Day, Fixed point theorems for compact convex sets. III. J. Math. **5**, 585–590 (1961)

161. J.P. Dedieu, Cone asymptotic d'un ensemble non convex. Application à l'optimisation, C.R. Acad. Sci. Paris **287**, 501–503 (1977)

162. H.K. Deepmala, Pathak, Study on existence of solutions for some nonlinear functional-integral equations with applications. Math. Commun. **18**, 97–107 (2013)
163. D.G. de Figueiredo, C.P. Gupta, Nonlinear integral equations of Hammerstein type involving unbounded monotone linear mappings. J. Math. Anal. Appl. **39**, 37–48 (1972)
164. B.C. Dhage, S. Dhage, H.K. Pathak, A generalization of Darbo's fixed point theorem and local attractivity of generalized nonlinear functional integral equations. Differ. Equ. Appl. **7**(1), 57–77 (2015)
165. R. de Marr, Comman fixed points for commuting contractions mappings. Pac. J. Math. **13**, 1139–41 (1963)
166. K. Deimling, *Multivalued Differential Equations* (Walter De Gruyter, Berlin, 1992)
167. K. Deimling, Zeros of accretive operators, Manuscript Math. 365–394 (1974)
168. K. Deimling, *Multivalued Differential Equations* (Walter De Gruyter, Berlin, 1992)
169. B.C. Dhage, On some variants of Schauder's fixed point principle and applications to nonlinear integral equations. J. Maths. Phy. Sci. **25**, 603–611 (1988)
170. B.C. Dhage, On α-condensing mappings in Banah algebras. Math. Stud. **63**, 146–152 (1994)
171. B.C. Dhage, Remarks on two fixed point theorems involving the sum and the product of two operators. Comput. Math. Appl. **46**, 1779–1785 (2003)
172. B.C. Dhage, A fixed point theorem in Banach algebras with applications to functional integral equations. Kyungpook Math. J. **44**, 145–155 (2004)
173. B.C. Dhage, On a fixed point theorem in Banach algebras with applications. Appl. Math. Lett. **18**(3), 273–280 (2005)
174. B.C. Dhage, A nonlinear alternative with applications to nonlinear perturbed differential equations. Nonlinear Stud. **13**(4), 343–354 (2006)
175. B.C. Dhage, Asymptotic stability of the solution of certain nonlinear functional integral equations via measures of noncompactness. Commun. Appl. Nonlinear Anal. **152**, 89–101 (2008)
176. B.C. Dhage, Attractivity and positivity results for nonlinear functional integral equations via measures of noncompactness. Differ. Equ. Appl. **2**(1), 299–318 (2010)
177. B.C. Dhage, Hybrid fixed point theory in partially ordered normed linear spaces and applications to fractional integral equations. Differ. Equ. Appl. **5**, 155–184 (2013)
178. B.C. Dhage, Partially condensing mappings in ordered normed linear spaces and applications to functional integral equations. Tamkang J. Math. **45**(4), 397–426 (2014)
179. B.C. Dhage, A.V. Deshmukh, J.R. Graef, On asymptotic behavior of a nonlinear functional integral equation. Commun. Appl. Nonlinear Anal. **14**, 1–12 (2007)
180. B.C. Dhage, S.B. Dhage, S.K. Ntouyas, Approximating solutions of nonlinear hybrid differential equations. Appl. Math. Lett. **34**, 76–80 (2014)
181. B.C. Dhage, S.B. Dhage, S.K. Ntouyas, H.K. Pathak, On local attractivity of nonlinear functional integral equations via measures of noncompactness. Malaya J. Math. **3**(1), 191–201 (2015)
182. B.C. Dhage, V. Lakshmikantham, On global existence and attractivity results for nonlinear functional integral equations. Nonlinear Anal. **72**, 2219–227 (2010)
183. B.C. Dhage, S.K. Ntouyas, Existence results for nonlinear functional integral equations via a fixed point theorem of Krasnoselskii-Schaefer type. Nonlinear Stud. **9**(3), 307–317 (2002)
184. B.C. Dhage, S.K. Ntouyas, Existence of positive monotonic solutions of functional hybrid fractional integral equations of quadratic type. Fixed Point Theory **16**(2), 273–284 (2015)
185. J.B. Diaj, B. Margolis, A fixed point theorem of the alternative for contraction on a generalized complete metric spaces. Bull. Am. Math. **74**, 305–309 (1968)
186. J.B. Diaz, F.T. Metcalf, On the set of subsequential limit points of successive approximations. Trans. Am. Math. Soc. **135**, 459–485 (1969)
187. J.B. Diaz, F.T. Metcalf, On the structure of the set of subsequential limits of successsive approximations. Bull. Am. Math. Soc. **73**, 516–519 (1967)
188. B. Djafari, Rouhani. Asymptotic behavior of unbounded trajectories for some nonautonomous systems in a Hilbert space. Nonlinear Anal. **19**, 741–751 (1992)
189. B. Djafari, Rouhani, Asymptotic behavior of unbounded nonexpansive sequences in Banach spaces. Proc. Am. Math. Soc. **117**, 951–956 (1993)

190. B. Djafari, Rouhani, On the unbounded behaviour for some non-autonomous systems in Banach spaces. J. Differ. Equ. **110**, 276–288 (1994)

191. Dj, S. Djukic, Trans. ASME (USA), p. 822 (1974)

192. P.G. Dodds, C.J. Lennard, Normality in trace ideals, J. Operator Theory **16**, 127–145 (1986)

193. I.C. Dolcetta, M. Lorenzani, F. Spizzichino, A degenerate complementarity system and application to the optimal stopping of Morkov chains. Boll. Un Mat. Italy **17 - B**(5), 692–703 (1980)

194. I C. Dolcetta, V. Mosco, Implicit complementarity problem and quasi-variational ineqaulities, in *Variational Inequalities and Comlementarity Problem* ed. by R.W. Cottle, F. Giannessi, J.L. Lions. Theory and Application (Willey, 1980), pp. 75–87

195. C.L. Dolph, C.J. Minty, On nonlinear integral equations of Hammerstein type, in *Nonlinear Integral Equations* ed. by P.M. Anselone (University of Wise. Press. Madison, Wisconsin, 1964), pp. 99–154

196. W.G. Dotson Jr., On the mann iteration process. Trans. Am. Math. Soc. **149**, 65–73 (1970)

197. W.G. Dotson Jr., On fixed points of nonexpansive mappings in nonconvex sets. Proc. Am. Math. Soc. **38**, 155–156 (1973)

198. W.G. Dotson Jr., Fixed points of quasi-nonexpansive mappings. J. Aust. Math. Soc. **13**, 167–170 (1972)

199. P.N. Dowling, W.B. Johnson, C.J. Lennard, B. Turett, The optimality of James's distortion theorems, Proc. Amer. Math. Soc. **125**(1), 167–174 (1997)

200. P.N. Dowling, C.J. Lennard, B. Turett, Some fixed point results in ℓ_1 and c_0. Nonlinear Anal. **39**, 929–936 (2000)

201. D. Downing, W.A. Kirk, Fixed point theorems for set-valued mappings in metric and Banach spaces. Math. Jpn. **22**, 99–112 (1977)

202. D. Downing, W.A. Kirk, A generalization of Caristi's theorem with applications to nonlinear mapping theory. Pac. J. Math. **69**, 339–346 (1977)

203. D.J. Downing, W.O. Ray, Some remarks on set valued mappings. Nonlinear Anal. Theory, Methods Appl. **5**(12), 1367–1377 (1981)

204. P. Drabek, Semilinear boundary value problems at resonance with general nonlinearities. Differ. Integral Equ. **5**, 339–55 (1992)

205. W.-S. Du, Nonlinear contractive conditions for coupled cone fixed point theorems. Fixed Point Theory Appl. **2010**, 16 (2010). https://doi.org/10.1155/2010/190606. Article ID 190606

206. W-S. Du, On coincidence point and fixed point theorems for nonlinear multivalued maps. Topol. Appl. **159**(1), 49–56 (2012)

207. J. Dugundji, Absolute neighbourhood retracts and local connectedness in arbitrary metric spaces. Campos. Math. **13**, 229–246 (1958)

208. D. van Dulst, Equivalent norms and the fixed point property for nonexpansive mappings. J. London Math. Soc. **25**(2), 139–144 (1982)

209. N. Dunford, J. Schwartz, *Linear Operators, vol* (I and II (Willey Interscience, New York, 1958)

210. G. Duvuat, J.L. Lions, *Les inequations en mechniqùe et en physique* (Dunod, Paris, 1971)

211. G. Duvaut, J.L. Lions, *Inequalities in Mechanics and Physics* (Springer, Berlin, 1976)

212. M. Edelstien, An extension of Banach contraction principle. Proc. Am. Math. Soc. **12**, 7–10 (1961)

213. M. Edelstien, On fixed points and periodic points under contractive mappings. J. Lond. Math. Soc. **37**, 74–79 (1962)

214. M. Edelstien, The construction of an asymptotic centre with a fixed point property. Bull. Am. Math. Soc. **78**, 206–208 (1972)

215. M. Edelstien, Fixed point theorems in uniformly convex Banach spaces. Proc. Am. Math. Soc. **44**, 369–374 (1974)

216. M. Edelstien, A remark on a theorem of M.A. Krasnoselskii. Am. Math. Mon. **73**, 509–510 (1966)

217. M. Edelstien, On nonexpansive mappings of Banach spaces. Proc. Cambridge Philos. Soc. **60**, 439–447 (1964)

218. R.E. Edwards, *Functional Analysis, Theory and Application* (Holt, Reinehart and Winston, 1965)
219. J. Eisenfeld, V. Lakshmikantham, Remarks on nonlinear contraction and comparison principle in abstract cones. J. Math. Anal. Appl. **61**, 116–121 (1977)
220. J. Eisenfeld, V. Lakshmikantham, Comparison principle and nonlinear contractions in abstract spaces. J. Math. Anal. Appl. **49**, 504–511 (1975)
221. I. Ekeland, Sur les problems variationnels. C.R. Acad. Sci., Paris **275**, 1057–1059 (1972)
222. L. Elsgolts, *Differential Equations and Calculus of Variations* (Mir Publishers, Moscow, 1973)
223. P. Enflo, A counter-example to the approximation problem in Banach spaces. Acta Math. **103**, 309–317 (1973)
224. L.C. Evans, *Partial Differential Equations*, vol. 19 (American Mathematical Society, 1998)
225. K. Fan, Fixed points and min-max theorems in locally convex topological linear spaces. Proc. Natl. Acad. Sci. U.S.A. **38**, 121–126 (1952)
226. K. Fan, A generalization of Tychonoff's fixed point theorem. Math. Ann. **142**, 305–310 (1960-61)
227. K. Fan, Extensions of two fixed point theorems of F.E. Browder. Math. Z. **112**, 234–240 (1969)
228. K. Fan, A minimax inequality and applications, in *by O*, ed. by I.I.I. Inequality (New York, Shisha (Academic Press, 1972), pp. 103–113
229. W. Fenchel, *Convex Cones* (Mimeograph Lecture Notes (Princeton University, Sets and Functions, 1971)
230. W. Fenchel, On conjugate convex functions. J. Math. **1**, 73–77 (1949)
231. M.C. Ferris, J.-S. Pang, *Complementarity and Variational Problems: State of the Art* (SIAM Publications, 1997)
232. A.F. Filippov, Differential equations with second members discontinuous on intersecting surfaces. Differentsial'nye Uravneniya (English translation) **15**, 1814–1832 (1979)
233. W.E. Fitzgibbon, Semilinearintegrodifferential equations in Banach space. Nonlinear Anal. **4**, 745–760 (1980)
234. P.M. Fitzpatrick, W.V. Petryshyn, Fixed point theorems for multivalued noncompact acyclic mappings. Pac. J. Math. **54**, 121–126 (1974)
235. M. Frigon, Thoremes d'existence de solutions d'inclusions diffrentilles, in *Topological Methods in Differential Equations and Inclusions* ed. by A. Granas, M. Frigon, NATO ASI Series C, vol. 472 (Kluwer Academic Publishers, Dordrecht, 1995), pp. 51–87
236. M. Furi, A. Vignoli, A fixed point theorem in complete metric spaces. Boll. Un. Mat. Italy Ser. **4–5**, 505–509 (1969)
237. M. Furi, A. Vignoli, On α-nonexpansive mappings and fixed points. Rend. Acad. Naz. Lincoi **18**(2), 195–198 (1970)
238. D. Gabay, Applications of the method of multipliers to variational inequalities, in *Augmented Lagrangian Methods: Application to Numerical Solution of Boundary Value Problems*, ed. by M. Fortin, R. Glowinski (North-Holland, Amsterdam, 1983), pp. 229–332
239. A.L. Garkavi, On the Chebyshev centre of a set in a normed space. Investigation of Contemporary Problems in the Constructive Theory of Functions (Moscow, 1961), pp. 328–331
240. R. George, K.P. Reshma, R. Rajagopalan, A generalised fixed point theorem of Presic type in cone metric spaces and application to Markov process. Fixed Point Theory Appl. (2011). https://doi.org/10.1186/1687-1812-2011-85
241. J.A. Gatica, H.L. Smith, Fixed point techniques in a cone with applications. J. Math. Anal. Appl. **61**, 58–71 (1977)
242. R. George, K.P. Reshma, R. Rajagopalan, A generalised fixed point theorem of Presic type in cone metric spaces and application to Markov process. Fixed Point Theory Appl. (2011). https://doi.org/10.1186/1687-1812-2011-85
243. I.L. Glicksberg, A further generalization of the Kakutani fixed point theorem with application to Nash equilibrium points. Porc. Am. Math. Soc. **3**, 170–174 (1952)
244. R. Glowinski, J. -L. Lions, R. Trémolières, *Numerical Analysis of Variational Inequalities* (North-Holland, Amsterdam, 1981)

245. K. Goebel, On a fixed point theorem for multivalued nonexpansive mappings. Annaales Univ. Maria Curie-Sklodowska

246. K. Goebel, W.A. Kirk, A fixed point theorem for asymptotically nonexpansive mappings. Am. Math. Soc. **35**, 171–174 (1972)

247. K. Goebel, W.A. Kirk, *Topics in Metric Fixed Point Theory* (Cambride University Press, Cambridge, 1990)

248. K. Goebel, W.A. Kirk, T.N. Shimi, A fixed point theorem in uniformly convex space. Boll. Un. Mat. Italy **7**, 65–67 (1973)

249. K. Goebel, S. Riech, *Uniform Convexity Hyperbolic Geometry and Nonexpansive Mappings* (Dekker, New York and Basel, 1984)

250. I.T. Gokhberg, L.S. Goldenstein, A.S. Markus, Investigation of some properties of bounded linear operators in connection with their q-norms, Uch. zap. Kikishinevsk, In ta **29**, 29–36 (1957)

251. D. Göhde, Zum Prinzip der Kontraktiven Abbildung. Math. Nachr. **30**, 251–258 (1965)

252. L.S. Goldenstein, A.S. Markus, On a measure of noncopmpactness of bounded sets and linear operators, in *Studies in Algebra and Mathematical Analysis* (Russian), *Izdat, karta moldovenjaski kishmev* (1965), pp. 45–54

253. A.J. Goldman, Resolution and seperation theorems for polyhedral convex sets. Linear inequalities and related systems. Ann. Math. Stud. **38**, 41–51 (1956)

254. M. Golomb, On the theory of nonlinear integral equations, integral systems and general functional equations. Math. Z. **39**, 45–75 (1935)

255. M.S. Gowda, M.A. Tawhid, Existence and limiting behavior of trajectories associated with P_0 equations. Comput. Optim. Appl. **12**, 229–251 (1999)

256. F.P. Greenleaf, *Invariace Means Topological Groups and Their Applications* (Van Nostrand, New York, 1969)

257. O. Güler, On the convergence of the proximal point algorithm for convex minimization. SIAM J. Control Optim. **29**, 403–419 (1991)

258. C.P. Gupta, On a class of nonlinear integral equations of Uryshon's type in Banach space. Comment Math. Univ. Carol. **16**(2), 377–386 (1975)

259. L.F. Guseman Jr., Fixed point theorems for mappings with a contractive iterate at a point, proc. Am. Math. Soc. **26**, 615–617 (1970)

260. G.B. Gustafson, K. Schmitt, Nonzero solutions of boundary value problems for second order ordinary and delay-differential equations. J. Differ. Equ. **12**, 129–147 (1972)

261. G.B. Gustafson, K. Schmitt, Methods of Nonlinear Analysis in the theory of Differential Equations, Lecture Notes, Department of Math, University of Utah

262. G.B. Gustafson, Fixed point methods for nodal problem in differential equations, in *Proceedings of Seminar on Fixed Point Theory and its Applications*, Dalhausie University June 1975 (Academic press, 1976)

263. J. Hadamard, Sur queles application de l'indice de Kronecker in *Introduction a'la Theorie des Functions d'une variable* ed. by J. Tannery (Herman, Paris, 1910), pp. 437–477

264. B. Halpern, *Fixed point theorems for outward maps, Doctoral Thesis* (University of California, Los Angeles, 1965)

265. G.H. Hardy, J.E. Littlewood, G. Polya, *Inequalities*, Paperback edn. (Cambridge University Press, London, 1988)

266. G.E. Hardy, T.D. Rogers, A generalization of a fixed point thorem of Riech. Bull. Cand. Math. Soc. **16**, 201–206 (1963)

267. J. Harjani, K. Sadarangani, Fixed point theorems for weakly contractive mappings in partially ordered sets. Nonlinear Anal. TMA **71**, 3403–3410 (2009)

268. P. Hartman, G. Stampacchia, On some nonlinear elliptic differential functional equations. Acta Math. **115**, 271–310 (1966)

269. S. Heikkili, V. Lakshmikantham, *Monotone Iterative Techniques for Discontinuous Nonlinear Differential Equations* (Marcel Dekker Inc., New York, 1994)

270. E. Heinz, An elemantary analytical theory of degree. J. Math. Mech. **8**, 231–247 (1959)

271. T.L. Hicks, B.E. Rhoades, A Banach type fixed-point theorem. Math. Jpn.**24**(3), 327–330 (1979-1980)

272. E. Hille, R.H. Phillips, *Funtional Analysis and Semigroups*, vol. 31 (American Mathematical Society Colloquium Publications, New York, 1957)

273. S.C.J. Himmelberg, Fixed points of compact multifunctions. J. Math. Anal. Appl. **38**, 205–207 (1972)

274. S.C.J. Himmelberg, Measurable relations. Fund. Math. **87**, 53–72 (1975)

275. S.C.J. Himmelberg, J.R. Portor, F.S. Van Vleck, Fixed point theorem for condensing multi-functions. Proc. Am. Math. Soc. **23**, 634–641 (1969)

276. H. Hopf, Vektorfelder in n-dimensionalen Mannigfaltigkeiten. Math. Ann. **96**, 225–250 (1926)

277. H. Hopf, Abbildungsklassen *n*-dimensionaler Mannigfaltigkeiten. Math. Ann. **97**, 209–224 (1927)

278. H. Hopf, Zur Topologie der Abbildungen von Manigfaltigkeiten. Math Ann. **100–102**, 562–623 (1929)

279. T. Hu, Fixed point theorems for multivalued mappings. Canad. Math. Bull. **23**, 193–197 (1980)

280. R.E. Huff, Existence and uniqueness of fixed points for semi-group of affine maps. Trans. Am. Math. Soc. **152**, 99–106 (1970)

281. J.P. Huneka, On comman fixed points of commuting continuous functions on an interval. Trans. Am. Math. Soc. **139**, 371–81 (1969)

282. M.A. Hussain, On a nonlinear integrodifferetial equation in Banach space. Indian J. Pure. Appl. Math. **19**, 516–529 (1980)

283. S.A. Hussain, V.M. Sehgal, On common fixed points for a family of mappings. Bull. Aust. Math. Soc. **13**, 261–267 (1975)

284. G. Isac, Un theroem de point fixe. Application au problem d'optimisation d'ersov, Seminari dell' instituto di mat. aaplicata, (Universita de Firenze, 1980), pp. 1–23

285. J.R. Isbal, Commuting mappings of tree. Bull. Am. Math. Soc. **63**, 419 (1957)

286. K. Iseki, On common fixed points of mappings. Bull. Aust. Math. Soc. **10**, 365–370 (1975)

287. S. Ishikawa, Comman fixed points and iteration of commuting nonexpansive mappings. Pac. J. Math. **80**, 493–501 (1979)

288. V.I. Istratescu, *An Introduction Fixed Point Theory* (D. Reidel Publishing company, Holland, 1981)

289. S. Itoh, W. Takahashi, Single valued mappings, multivalued mappings and fixed point theorems. J. Math. Anal. Appl. **59**, 514–521 (1977)

290. J. Jachymski, Equivalent conditions and the Meir-Keeler type theorems. J. Math. Anal. Appl. **194**, 293–303 (1995)

291. R. C. James, Uniformly non-square Banach spaces, Ann. Math. **80**, 542–550 (1964)

292. M. Joshi, Existense theorems for a generalized Hammerstein type equation. Comment Math. Univ. Carol. **15**(2), 283–291 (1974)

293. M. Joshi, Existense theorem for Uryshons integral equation. Proc. Am. Math. Soc. **49**(2), 387–392 (1975)

294. M. Joshi, P.N. Srikanth, On a class of nonlinear integral equations. Proc. Ind. Acad. Sci. **87A**, 169–175 (1978)

295. J.S. Jung, J.S. Park, Asymptotic behavior of nonexpansive sequences and mean points. Proc. Am. Math. Soc. **124**, 475–480 (1996)

296. G. Jungck, Commuting mappings and fixed points. Am. Math. Mon. **83**, 261–263 (1976)

297. A. Jürgen, P.P. Zabreĭko, *Nonlinear Superposition Operators* (Cambridge University Press, Cambridge, 1990)

298. A. Jürgen and M. Väth, The space of Carathiodory functions, Trudy Inst. Mat. NAN Belarus. Minsk. 2 (A999), 39-43

299. R.I. Kachurovskiĭ, On monotone operators and convex functionals (russian). Akademija Nauk SSSR i Moskovskoe Matematičeskoe Obščestvo. Uspehi Matematičeskih Nauk **15**(4), 213–215 (1960)

300. R.I. Kachurovskiĭ, Nonlinear monotone operators on Banach spaces. Russian Math. Surv. **23**, 117–165 (1968)

301. S. Kakutani, A generalization of Brouwer fixed point theorem. Duke Math. J. **8**, 457–459 (1941)

302. S. Kakutani, Weak topology and regularity of Banach spaces. Proc. Imp. Acad. Tokyo **15**, 169–173 (1939)

303. S. Kamimura, W. Takahashi, Approximating solutions of maximal monotone operators in Hilbert spaces. J. Approx. Theory **106**, 226–240 (2000)

304. S. Kamimura, W. Takahashi, Weak and strong convergence of solutions to accretive operator inclusions and applications. Set-valued Anal. **8**, 361–374 (2000)

305. S. Kamimura, W. Takahashi, Strong convergence of a proximal-type algorithm in a Banach space. SIAM J. Optim. **13**(3), 938–945 (2003)

306. H. Kaneko, Single-valued and multi-valued f-contractions. Boll. Un. Mat. Ital. **4-A**, 29–33 (1985), 29-33

307. H. Kaneko, A common fixed point of weakly commuting multi-valued mappings. Math. Jpn. **33**(5), 741–744 (1988)

308. R. Kannan, Some results on fixed points. Bull. Cal. Math. Soc. **60**, 71–76 (1968)

309. R. Kannan, Some results on fixed points II. Amer. Math. Mon. **76**, 71–76 (1970)

310. R. Kannan, Some results on fixed points III. Fund. Math. Mon. **70**, 169–177 (1971)

311. R. Kannan, Some results on fixed points IV. Fund. Math. Mon. **74**, 181–187 (1972)

312. R. Kannan, Existense of periodic solutions of nonlinear diffrential equations. Trans. Am. Math. Soc. **217**, 225–236 (1976)

313. B. Knaster, Un theoreme sur les fonctions d'ensembles. Ann. Soc. Polon. Math. **6**, 133–134 (1928)

314. S. Karamardian, Complementarity problems over cones with monotone or pseudomonotone maps. J. Optim. Theory Appl. **18**, 445–454 (1976)

315. S. Karamardian, *Fixed Points, Algorithms and Applications(Edited)* (Academic Press, New York, 1977)

316. S. Karamardian, The nonlinear complemtarity problem with applications. J. Optim. Theory Appl. **4**, 87–98 (1969)

317. K. Kartàk, A generalization of the Carathéodory theory of differential equations. Czech. Math. J. **17**(92), 482–514 (1967)

318. S. Kashara, On fixed points in partially ordered sets and Kirk-Caristi theorem. Math. Sem. Notes **3**(35), 229–232 (1975)

319. T. Kato, Math. Ann. **126**, 253 (1953)

320. T. Kato, Accretive operators and nonlinear evolution equations in Banach spaces, Nonlinear Functional Analysis, in *Proceedings of Symposium Pure Mathematics*, ed. by F. Browder vol. 18 Part I (American Mathematical Society, 1970), pp. 138–161

321. T. Kato, Demicontinuity, hemicontinuity and monotonicity. Bull. Am. Math. Soc. **70**, 548–550 (1964); **73**, 886–889 (1967)

322. T. Kato, Nonlinear semigroups and evolution equations. J. Math. Soc. Jpn. **19**, 508–520 (1967)

323. J.L. Kelley, *General Topology* (VanNostrand-Reinholt, Princeton, NJ, 1955)

324. P. Kenderov, Multivalued monotone mappings are almost everywhere single valued. Stud. Math. **56**, 199–203 (1976)

325. J.A. Khan, R.M.A. Zahoor, I.M. Qureshi, Swarm intelligence for the problem of non-linear ordinary differential equations and its application to well known Wessinger's equation. Eur. J. Sci. Res. **34**(4), 514–525 (2009)

326. W.A. Kirk, A fixed point theorem for mappings which do not increase distance. Am. Math. Mon. **72**, 1004–1006 (1963)

327. W.A. Kirk, Fixed point theorems for nonexpansive mappings, in *Nonlinear Function Proceedings of Sysposiun Pure Mathematics, PartI* vol. 18 (A.M.S., 1970), pp. 162–168

328. W.A. Kirk, Caristi's fixed point theorem and theory of normal solvability, in *Proceeding Conference on Fixed Point Theory and Its Applications* (Dalhàusie University, 1975) (Academic Press, 1976), pp. 109–120

329. W.A. Kirk, Caristi's fixed point theorem and metric convexity. Colloquium Math. **36**, 81–86 (1976)
330. W.A. Kirk, Remarks on approximationand approximate fixed points in metric fixed point theory. Annales Universitatis Mariae Curie-Skodowska, Section A **51**(2), 167–178 (1997)
331. W.A. Kirk, Nonexpansive mappings and asymptotic regularity. Nonlinear Anal. TMA **40A**, 323–332 (2000)
332. W.A. Kirk, The fixed point property and mappings which are eventually nonexpansive, in *Theory and Applications of Nonlinear Operators of Accertive and Monotone Type* ed. by A.G. Kartsatos. Lecture Notes in Pure and Applied Mathematics, vol. 178 (Marcel Dekker, New York, 1996), 141–147
333. W.A. Kirk, J. Caristi, Mappings theorems in metric and Banach spaces. Bull. Acad. Polon. Sci. Math. Astr. Phy. **23**, 891–894 (1979)
334. W.A. Kirk, W.O. Ray, Fixed points theorems for mappings defind on unbounded sets in Banach spaces. Stud. Math. **114**, 127–138 (1979)
335. W.A. Kirk, R. Schoneberg, Some results on Pseudo contractive mappings. Pac. J. Math. **71**, 89–100 (1977)
336. M.D. Kirzbraun, Uber die zusammen zichend und Lipschitzsche transformation. Fund. Math. **22**, 77–108 (1934)
337. K. Klein, A.C. Thompson, *Theory of Correspondence: Including Applications to Mathematical Economics* (Wiley, 1984)
338. H.M. Ko, Fixed point theorems for point-to-set mappings and the set of fixed points. Pac. J. Math. **42**, 369–379 (1972)
339. A.N. Kolmogoroff, Beitrige zur Maftheorie. Math. Ann. **107**, 351–366 (1933)
340. J. Kolomy, The solvability of nonlinear integral equations. Comment. Math. Univ. Carolinae **8**, 273–289 (1967)
341. V. Komkov, *Variational Principles of Continuum Mechanics with Engineering Applications*, vol. I (D. Reidel Publishing Co., Dordrecht Holland, 1985)
342. M.A. Krasnoselski, The continuity of a certain operator. Dokl. Akad. Nauk SSSR **77**, 2 (1951)
343. M.A. Krasnoselski, Two comments on the method of successive approximations. Uspehi Mat. Nauk **10**, 123–127 (1955)
344. M.A. Krasnoselski, *Topological Methods in the Theory of Nonlinear Integral Equations* (Pergamon Press, Oxford, 1964)
345. M.A. Krasnoselski, *Positive Solutions of Operator Equations* (P. Noordhoff Ltd., Netherlands, 1964)
346. M.A. Krasnoselski, Some problems of nonlinear analysis. Am. Math. Soc. Trans. **10**(2), 345–409 (1958)
347. M.A. Krasnoselski, L.A. Ladyzehenskii, Conditiond for complete convergence for P.S. Uryshon's operator. Trud. Mosk. Mat. Obsch. **3** (1954)
348. D. Kravvaritis, Continuity properties of monotone nonlinear operators in lcss. Proc. Am. Math. Soc. **72**(1), 46–48 (1978)
349. L. Kronecker, Uber systems von functionen mebrerer variablen. Monatsb-Deutsch Akad-Wiss. Berlin (1869), pp. 159–193, 685–698
350. P.K.F. Kuhfittig, comman fixed points of non-expansive mappings by iteration. Pac. J. Math. **97**, 137–139 (1980)
351. C. Kuratowski, Sur les espaces complete. Fund. Math. **15**, 301–309 (1930)
352. K. Kuratowskii, *Topology*, vol. I (Academic Press, New York and London, 1966)
353. K. Kuratowskii, C. Ryll-Nardzewski, A general theorem on selectors. Bull. Acad. Polonss. Sci. Ser. Math. Sci. Astr. Phys. **13**, 397–403 (1965)
354. L. Ladyzhenskii, Conditions for complete continuity of P.S. Uryshon's inetegral operator in the spaces of continuous functions. Dok. Akad. Nauk SSSR, **97**(5) (1954)
355. A. Lasota, Z. Opial, An application of the Kakutani-Fan theorem in the theory of ordinary differential equations. Bull. Acad. Pol. Sci. Ser. Sci. Math. Astr. Phys. **13**, 781–786 (1965)
356. E. Lami, Dozo, Multivalued nonexpansive mappings and Opial's condition. Proc. Am. Math. Soc. **38**, 286–292 (1973)

357. E. Lami Dozo, Centres Asymptotiques dans certain F-espaces *(preprint)*
358. E. Lami Dozo, Operators nonexpansifs P-compact et propacrtie geometriques de la norme. *These Univ. Libre le Bruxelles* (1970)
359. E. Lami, J.P.Gossez Dozo, Some gemetric properties related to the fixed point thoery of nonexpansive mappings. Pac. J. Math. **40**, 565–573 (1972)
360. P.D. Lax, A.N. Milgram, Ann. Math. Stud. **33**, 167 (1954). Princeton
361. R.W. Legett, On certain nonlinear integral equations. J. Math. Anal. Appl. **57**, 462–468 (1977)
362. R.W. Legett, A new approach to the H-equations of Chandrashekhar. SIAM J. Math. Anal **7**(4), 542–550 (1976)
363. R.W. Legett, L.R. Williams, A fixed point theorem with applications to an infectious disease model. J. Math. Anal. Appl. **76**, 91–97 (1980)
364. C. Lennard, V. Nezir, Reflexivity is equivalent to the perturbed fixed point property for cascading nonexpansive maps in Banach lattices, Nonlinear Analysis **95**, 414–420 (2014)
365. J. Leray, J. Schauder, Topologie et equations functionelles. Ann. Sci. Ecole Norm. Sup. **3**(51), 45–78 (1934)
366. T.C. Lim, On asymptotiques centres and fixed points of nonexpansive mappings. Can. J. Math. **32**, 421–430 (1980)
367. T.C. Lim, A fixed point theorem for multivalued nonexpansive mappings in a uniformly convex Banach space. Bull. Am. Math. Sci. **80**, 1083–1126 (1974)
368. T.C. Lim, Remarks on some fixed point theorems. Proc. Am. Math. Soc. **60**, 179–182 (1976)
369. Y. Lim, Solving the nonlinear matrix equation $X = Q + \sum_{i=1}^{m} M_i^* X^{\delta_i} M_i^*$, via a contraction principle. Linear Algebra Appl. **430**, 1380–1383 (2009)
370. P.K. Lin, There is an equivalent norm on ℓ_1 that has the fixed point property. Nonlinear Anal. **68**, 2303–2308 (2008)
371. J. Lindenstrauss and L. Tzafriri, Classical Banach spaces I: Sequence Spaces, Ergebnisse der Mathematik und ihrer Grenzgebiete, vol. 92, Springer- Verlag, 1977
372. J. Lindenstrauss and L. Tzafriri, Classical Banach spaces II: Function Spaces, Ergebnisse der Mathematik und ihrer Grenzgebiete, vol. 97, Springer-Verlag, 1979
373. P.L. Lions, Une méthode itérative de résolution d'une inéquation variationnelle. Isr. J. Math. **31**, 204–208 (1978)
374. A.R. Lovaglia, Locally uniformly convex Banach spaces. Trans. Am. Math. Soc. **78**, 225–238 (1955)
375. D.T. Luc, Recession maps and applications. Optimization **27**, 1–15 (1993)
376. D.T. Luc, Recessively compact sets: uses and properties. Set-Valued Anal. **10**, 15–35 (2002)
377. D.T. Luc, J.-P. Penot, Convergence of asymptotic directions. Trans. Am. Math. Soc. **352**, 4095–4121 (2001)
378. L.A. Lusternik, V.J. Sobolev, *Elements of Functional Analysis* (Hindustan Publ. Co., Delhi, 1961)
379. K. Maleknejad, R. Mollapourasl, K. Nouri, Study on existence of solutions for some nonlinear functional-integral equations. Nonlinear Anal. **69**, 2582–2588 (2008)
380. K. Maleknejad, K. Nouri, R. Mollapourasl, Existence of solutions for some nonlinear integral equations. Commun. Nonlinear Sci. Numer. Simul. **14**, 2559–2564 (2009)
381. K. Maleknejad, K. Nouri, R. Mollapourasl, Investigation on the existence of solutions for some nonlinear functional-integral equations. Nonlinear Anal. **71**, 1575–1578 (2009)
382. O.L. Mangasaran, Equivalence of complementarity problem to a system of nonlinear equations. SIAM J. Appl. Math. **31**, 89–92 (1976)
383. O.L. Mangasaran, Iterative solutions of linear programs. SIAM J. Numer. Anal. **18**(4), 606–614 (1981)
384. W.R. Mann, Mean value methods in iteration. Proc. Am. Math. Soc. **4**, 506–510 (1953)
385. J.T. Markin, Continuous dependence of fixed point set. Proc. Am. Math. Soc. **38**, 545–547 (1973)
386. J.T. Markin, A fixed point stability theorem for nonexpansive set valued mappings. J. Math. Anal. Appl. **54**, 441–443 (1978)

387. A. Markov, Quelques theorems sur les ensembles abelinans. Dokl. Akad. Nauk SSSR **10**, 311–314 (1936)
388. R.H. Martin, *Nonlinear Operators and Differential Equations in Banach Spaces* (Willey, New York, 1976)
389. B. Martinet, Régularisation d'inéquations variationnelles par approximatios successives. Rev. Franc. Inf. Rech. Op. **4**, 154–159 (1970)
390. J. Matkowski, Integral solutions of functional equations. Diss. Math. **77** (1975)
391. J. Matkowski, J. Mis, Examples and remarks to a fixed point theorem. Facta Univ. Ser. Math. Inform. **1**, 53–56 (1986)
392. W.S. Massey, *A Basic Course in Algebraic Topology* (Springer, New York, 1991)
393. A. Meir, E. Keeler, A theorem on contraction mappings. J. Math. Anal. Appl. **28**, 326–329 (1969)
394. M. Meehan, D. O'Regan, Positive L^p solutions of Fredholm integral equations. Archiv der Mathematik **76**, 366–376 (2001)
395. K. Menger, L.M. Blumenthal, Theory and application of distance geometry. *The Clarendon Press* (Oxford, 1953), p. 41
396. S.G. Mikhlin, *Variational Methods in Mathematical Physics* (Macmillan, New York, 1964)
397. R.K. Miller, Volterra integral equation in a Banach spaces. Funkeial Ekvac. **18**, 163–193 (1975)
398. R.K. Miller, T.A. Nohel, J.S.W. Wong, A stability theorem fo nonlinar mixed integral euqations. J. Math. Anal. Appl. **25**, 446–449 (1965)
399. D. Milman, On some criteria for the equality os spaces of the type(B). Doklady, New Ser. **20**, 243–246 (1938)
400. G. Minty, Monotone (nonlinear) operator in a Hilbert spaces. Duke Math. J. **29**, 341–346 (1962)
401. G. Minty, On the maximal domain of a monotone function. Mich. Math. J. **8**, 135–137 (1961)
402. G. Minty, On the simultaneous solution of a certain system of linear inequalities. Proc. Am. Math. Soc. **13**, 11–12 (1962)
403. G. Minty, On a 'Monotonicity' method for the solution of nonlinear equations in Banach spaces. Proc. Nat. Acad. Sci., U.S.A. **50**, 1038–1041 (1963)
404. C. Miranda, Partial differential equations of Elliptic Type. Grenzeb. der Math. Grenzeb. vol. 2 (Springer New York, 1970)
405. T. Mitchell, Topological semigroup and fixed point theorems. III. J. Math. **14**, 630–641 (1970)
406. N. Mizoguchi, W. Takahashi, Fixed point theorems for multi-valued mappings on complete metric spaces. J. Math. Anal. Appl. **141**, 177–188 (1989)
407. J.J. Moreau, Functionneless sous-differentiables. C.R. Acad. Sci. Paris **257**, 4117–4119 (1963)
408. J.J. Moreau, Functionneless convex, Multilith notes, (Colleg de France, 1966–67), p. 108
409. J.J. Moreau, Sous-differentiabilite, in *Proceedings of Colloquiumon convexity Compenhagen, 1985*, ed. by W. Fenenel, (Copenhagen Mathematics Institute, 1967), pp. 185–201
410. J.J. Moreau, Fonctions convex en duality, *Faculte des Sciences de Montipellier seminaries de Mathematiques* (1962)
411. P.M. Morse, H. Feshbach, *Methods of Theoretical Physics, Part I* (McGraw-Hill, New York, 1953)
412. V. Mosco, On some non-linear quasi-variational inequalities and implicit complementarity problems in stochastic control theory, in *Variational Inequalities and Complemetarity Problems* ed. by R.W. Cottle, F. Giannnessi, J.L. Lions. Theory and Appl. (Willey, 1980), pp. 271–283
413. A. Mukherjee, K. Pathoven, *Real and Functional Analysis* (Plenum Press, New York, 1978)
414. J.R. Munkres, *Topology* (Prentice Hall Inc, Upper Saddle River, 2000)
415. A.D. Myskis, *Advanced Mathematics for Engineers* (Mir Publisher, Moscow, 1975)
416. S.B. Nadler Jr., Multivalued contraction mappings. Pac. J. Math. **30**, 475–488 (1968)
417. S.B. Nadler Jr., Sequences of contractions and fixed points. Pac. J. Math. **27**, 579–585 (1968)
418. N. Nagumo, A theory of degree of mappings based on infinitesimal analysis. Am. J. Math. **73**, 485–496 (1951)

419. N. Nagumo, Degree of mappings in convex linear topological spaces Amer. J. Math. **73**, 497–511 (1951)

420. S. Naimpally, S. Singh, J.H.M. Whitefield, JHM, Coincidence theorems for hybrid contractions. Math. Nachr. **127**, 177–180 (1986)

421. V.V. Nemytskiĭ, Existence and uniqueness theorems for nonlinear integral equations. Math. Sb. **41**(3) (1934)

422. Nevalinna, Global iteration schemes for monotone operators. Nonlinear Anal. Theory, Methods Appl. **3**(4), 505–514 (1979)

423. J.J. Nieto, R. Rodriguez-Lopez, Contractive mappings theorems in partially ordered sets and applications to ordinary differential equations. Order **22**, 223–239 (2005)

424. M.A. Noor, On the non-linear complementarity problem. J. Math. Anal. Appl. **123**, 455–460 (1987)

425. M.A. Noor, Mildly nonlinear variational inequalities. Math. Anal. Numer. Theory Approx. **24**, 99–110 (1982)

426. R.D. Nussbaum, Hilbert's projective metric and iterated nonlinear maps. Mem. Am. Math. Soc. **14**, 438–443 (1963)

427. R.D. Nussbaum, The fixed point index and asymptotic fixed point theorem for k-set contractions. Bull. Am. Soc. **75**, 490–495 (1969)

428. R.D. Nussbaum, Degree theory for local condensing maps. J. Math. Anal. Appl. **37**, 741–767 (1972)

429. R.D. Nussbaum, The fixed point index and fixed point theorems for k-set contractios, Ph. D. dissertation (Chicago, 1969)

430. R.D. Nussbaum, *The Fixed Point Theorem for Local-Condensing Mappings* (Anali de Matem, Pura et Appl, 1972)

431. R.D. Nussbaum, Asymptotic fixed point theorems for local condensing maps. Math. Ann. **101**, 181–195 (1971)

432. R.D. Nussbaum, The radius of essential spectrum. Duke Math. J. **38**, 473–478 (1970)

433. R.D. Nussbaum, Periodic solution of nonlinear autonomous funtional differential equations, in *Functional Differetial Equations and Approximation of Fixed Points* ed. by H.O. Peitgen, H.O. Walther (Lecture notes Math. 730) (Springer, 1970), pp. 283–325

434. R.D. Nussbaum, Periodic solution of some nonlinear autonomous fnctional equations. Ann. Math. Pure. Appl. **101**, 263–306 (1974)

435. G.O. Okikiolu, *Aspects of the Theory of Bounded Integral Operators in L_p-Spaces* (Academic Press, 1971)

436. Z. Opial, Weak convergence of the seuence of successive approximations for nonexpansive mappings. Bull. Am. Math. Soc. **73**, 591–597 (1967)

437. J.M. Ortega, W.C. Rheinboldt, *Iterative Solution of Nonlinear Equations in Several Variables* (Academic Press, New York, 1970)

438. B.G. Pachpatte, On some integral inequalities and their applications to integridifferential equations. Indian J. Pure. Appl. Math. **8**, 1157–1175 (1977)

439. M. Pacurar, A multi-step iterative method for approximating common fixed points of Presic-Rus type operators on metric spaces. Stud. Univ. "Babes-Bolyai", Math. **LV**(1), 149–162 (2010)

440. J.S. Pang, The implicit comlementarity problem, in *Nonlinear Programing* vol. 4 ed. by O.L. Mangasarian, R.R. Meyer, S.M. Roinson (1981), pp. 487–518

441. J.S. Pang, On the convergence of a basic iterative method for the imlicit complementarity problem. J. Optim. Theory Appl. **37**, 149–162 (1982)

442. S. Park, Equivalent formulations of Ekeland's variational principle for approximate solutions of minimization problems and their applications, in *Operator Equations and Fixed Point Theorems* ed. by S.P. Singh, V.M. Sehgal, J.H.W. Burry (MSRI-Korea Publishers 1, Soul, 1986), pp. 55–68

443. J.A. Park, S. Park, Surjectivity of ϕ-accretive operators. Proc. Am. Math. Soc. **90**(2), 289–292 (1984)

444. S.V. Parter, Solutions of a differential equations arising in chemical reactor processes. SIAM J. Appl. Math. **26**, 687–716 (1974)
445. D. Pascali, S. Sburlan, Nonlinear mappings of monotone type. Sythoff and Noordhoff, álphen Aan den Rijn (The Netherland, 1978)
446. G.B. Passty, Ergodic convergence to a zero of the sum of monotone operators in Hilbert space. J. Math. Anal. Appl. **72**, 383–390 (1979)
447. G.B. Passty, Construction of fixed point for asymptotically nonexpansive mappings. Proc. Am. Math. Soc. **84**, 212–216 (1982)
448. H.K. Pathak, Application of fixed point theorems to abstract Volterra integrodifferential equations. Rev. Mat. Univ. Parma **5**(3), 193–202 (1994)
449. H.K. Pathak, Fixed point theorems for weak compatible multi-valued and single valued mappings. Acta. Math. Hungar. **67**(1–2), 69–78 (1995)
450. H.K. Pathak, Integral Φ-type contractions and existence of continuous solutions for nonlinear integral inclusions. Nonlinear Anal. **71**, e2577–e2591 (2009)
451. H.K. Pathak, A note on convergence of a sequence and its applications to geometry of banach spaces. Adv. Pure Math. **1**, 33–41 (2011)
452. H.K. Pathak, Monotoniof NFP mappings associated with variational inequality its application. Applied Mathematics in Electrical and Computer Engineering, in *Proceedings of the Conference on Applied Mathematics (AMERICAN-MATH'12)* (Harvard, Cambridge, USA (2012), pp. 83–91
453. H.K. Pathak, Y.J. Cho, Strong convergence of a proximal-type algorithm for an occasionally pseudomonotone operator in Banach spaces. Fixed Point Theory Appl. **2012**(190), 13 (2012)
454. H.K. Pathak, R.P. Agarwal, Y.J. Cho, Coincidence and fixed points for multi-valued mappings and its application to nonconvex integral inclusions. J. Comput. Appl. Math. **283**, 201–217 (2015)
455. H.K. Pathak, Y.J. Cho, S.M. Kang, Remarks on R-weakly commuting mappings and common fixed point theorems. Bull. Korean Math. Soc. **34**, 247–257 (1997)
456. H.K. Pathak, Deepmala, Remarks on some fixed point theorems of Dhage. Appl. Math. Lett. **25**(11), 1969–1975 (2012)
457. H.K. Pathak, R. George, S.M. Kang, Y.J. Cho, Simultaneous complementarity problems and their solutions via fixed points. Pure Math. Appl. **7**, 175–182 (1996)
458. H.K. Pathak, R. George, H.A. Nabwey, S. El-Paoumy Mahdy, K.P. Reshma, Some generalized fixed point results in a b-metric space and application to matrix equations. Fixed Point Theory Appl. **2015:101**, 17 (2015)
459. H.K. Pathak, S.M. Kang, Y.J. Cho, Coincidence and fixed point theorems for nonlinear hybrid generalized contractions. Czechoslovak. Math. J. **48**, 341–357 (1998)
460. H.K. Pathak, M.S. Khan, Fixed and coincidence points of hybrid mappings. Archi. Math. (Brno) **38**, 201–208 (2002)
461. H.K. Pathak, S.N. Mishra, Some surjectivity theorems with applications. Arch. Math. (BRNO) **49**, 17–27 (2013)
462. H.K. Pathak, B.E. Rhoades, M.S. Khan, Common fixed point theorems for asymptotically I-contractive maapings without convexity. Fixed Point Theory **8**(2), 285–296 (2007)
463. H.K. Pathak, R. Rodríguez-López, Noncommutativity of mappings in hybrid fixed point results. Boundary Value Prob. **2013**, 1–21 (2013)
464. H.K. Pathak, N. Shahzad, Fixed points for generalized contractions and applications to control theory. Nonlinear Anal. **68**, 2181–2193 (2008)
465. H.K. Pathak, N. Shahzad, A new fixed point result and its application to existence theorem for nonconvex hammerstein type integral inclusions. Electron. J. Qual. Theory Differ. Equ. **1**, 1–12 (2012)
466. H.K. Pathak, N. Shahzad, A generalization of Nadler's fixed point theorem and its application to nonconvex integral inclusions. Topol. Methods Nonlinear Anal. **41**(1), 207–227 (2013)
467. H.K. Pathak, D. O'Regan, M.S. Khan, R.P. Agarwal, Asymptotic behavior of generalized nonexpansive sequences and mean points. Georgian Math. J. **11**(3), 539–548 (2004)

468. A. Pazy, *Semigroups of Linear Operators and Applications to Partial Differential Equations* (Springer, New York, 1983)

469. J.-P. Penot, Fixed point theorems without convexity. Memoire Soc. Math. de France **60**, 129–152 (1977)

470. J.-P. Penot, Compact nets, filters and relations. J. Math. Anal. Appl. **93**, 400–417 (1983)

471. J.-P. Penot, A metric approach to asymptotic analysis. Bull. Sci. Math. **127**, 815–833 (2003)

472. J.-P. Penot, A fixed point theorem for asymptotically contractive mappings. Proc. Am. Math. Soc. **131**, 2371–2377 (2003)

473. J.-P. Penot, C. Zalinescu, Continuity of usual operations and variational convergences, Preprint, University of Pau (2000 and 2001)

474. W. Petry, Ein Leray-Schaududer Satz mit Anwendung duf verallemeinerte Hammersteinsche Gleichungen. Math. Nachr. **48**, 49–68 (1971)

475. W. Petry, Verallgeinerte Hammersteinsche Gleichungen and quasilinear Randwert problems. Math. Nachr. **50**, 150–166 (1971)

476. W. Petry, Generalized Hammerstein equation and integral equations of Hammerstein type. Math. Nachr. **59**, 51–62 (1974)

477. W.V. Petryshyn, Projection methods in nonlinear numerical functional analysis. J. Math. Mech. **17**, 353–372 (1967)

478. W.V. Petryshyn, On the approximate solvebility of nonlinear equations. Math. Ann. **177**, 156–164 (1968)

479. W.V. Petryshyn, Remarks on condensing and k-set conrtactive mappings. J. Math. Anal. **39**, 717–741 (1972)

480. W.V. Petryshyn, On the approximation-solvability of equations involving A-proper and pseudo A-proper mapping. Bull. Am. Math. Soc. **81**, 223–312 (1975)

481. W.V. Petryshyn, A characterization of strict convexity of Banach spaces and other uses of duality mapping. J. Funct. Anal. **6**, 282–291 (1970)

482. W.V. Petryshyn, *Approximation-Solvability of Nonlinear Functional and Differential Equations (Monographs and Textbooks in Pure and Applied Math)*, vol. 171 (Marcel Dekker, New York, 1993)

483. W.V. Petryshyn, P.M. Fitzpatrik, New existence theorems for nonlinear equations of Hammmerstein type. Trans. Am. Math. Soc. **160**, 39–63 (1970)

484. W.V. Petryshyn, P.M. Fitzpatrik, Fixed point theorems for multivalued noncompact inward maps. J. Math. Anal. Appl. **46**, 756–767 (1974)

485. W.V. Petryshyn, T.E. Williamson, A necessary and suuficient condition for the convergence of iterates for quasi-nonexpansive mappings. Bull. Am. Math. Soc. **78**, 1027–1031 (1972)

486. W.V. Petryshyn, T.E. Williamson, Strong and weak convergence of the sequence of successive approximation for quasi-nonexpansive mappings. J. Math. Anal. Appl. **43**, 459–497 (1973)

487. B.J. Pettis, A proof that every uniformly convex space is reflexive. Duke Math. J. **5**, 249–253 (1939)

488. R.L. Pluket, A fixed point theorem for continuous multi-valued transformations. Proc. Am. Math. Soc. **79**, 160–163 (1956)

489. H. Poincare, Sur un theoreme les courtes definis par les equations differentielles. J. de Math. **2** (1886)

490. L. Pontrjagin, L. Schnirelmann, Sur une propriété de la dimension. Ann. Math. **33**, 156–162 (1932)

491. A.J.B. Potter, An elementary version of Leary-Schaude theorem. J. Lond. Math. Soc. **2**(5), 414–426 (1972)

492. A.J.B. Potter, A fixed point theorem for positive set contractions. Proc. Edinb. Soc. Ser. I **I**(19), 93–102 (1974)

493. B.L.S. Prakasa, Stocastic integral equations of mixed type. J. Math. Phys. Sci. Ser. **7**, 245–260 (1971)

494. S. B Presic, Sur la convergence des suites. Comptes. Rendus. de l'Acad. de Sci. de Paris, **260**, 3828–3830 (1965)

495. P.M. Prenter, *Splines and Variational Methods* (Wiley, New York, 1975)

496. S.B. Presic, Sur une classe d'inequations aux differences finite et sur la convergence de certaines suites. Publ. de l'Inst. Math. Belgrade **5**(19), 75–78 (1965)

497. E. Rakotch, A note on contractive mappings. Proc. Am. Math. Soc. **13**, 459–465 (1962)

498. E. Rakoch, On ε-contractive mappings. Bull. Res. Counc. Isr. **10F**, 53–58 (1962)

499. L.B. Rall, J. Math. Anal. Appl. **14**, 174 (1966)

500. G. Ravindran, M.S. Gowda, Regularization of P_0-function in box variational inequalities problems. SIAM J. Optim. **11**, 748–760 (2000)

501. B.K. Ray, A fixed point theorem in Banach space. Ind. J. Pure Appl. Math. **8**, 903–907 (1977)

502. W.O. Ray, Phi-accretive operators and Ekeland's theorem. J. Math. Anal. Appl. **88**, 566–571 (1982)

503. W.O. ray, A.M. Walker, Mapping theorems for Gâteaux differetiable and accretive operators. Nonlinear Anal. TMA **6**, 423–433 (1982)

504. D. O'Regan, Integral inclusions of upper semicontinuous or lower semicontinuous type. Proc. Am. Math. Soc. **124**, 2391–2399 (1996)

505. D. O'Regan, A topological approach to integral inclusions. Proc. R. Irish Acad. **97A**, 101–111 (1997)

506. D. O'Regan, M. Meehan, *Existence Theory for Nonlinear Integral and Integrodifferential Equations* (Kluwer Academic Publishers, Dordrecht, 1998)

507. M. Reed, B. Simon, *Methods of Nodern Mathematical Physics*, vol. I. Functional Analysis (Academic Press, 1973)

508. S. Reich, Some remarks concerning contraction mapping. Canad. Math. Bull. **14**, 121–124 (1971)

509. S. Reich, Kannan's fixed point theorem. Bolletino U.M.I. **4** (1971)

510. S. Reich, Fixed points of contractive functions. Boll. Unione Mat. Ital. **5**, 26–42 (1972)

511. S. Reich, Some fixed point problems. Atti Acad. Naz. Lincei **57**, 194–198 (1974)

512. S. Reich, Weak convergence theorems for nonexpansive mappings in Banach spaces. J. Math. Anal. Appl. **67**, 274–276 (1976)

513. S. Reich, Approximate selections best approximations, fixed points and invariant sets. J. Math. Appl. **62**, 104–113 (1968)

514. S. Reich, Fixed point theorem for setvalued mappings. J. Math. Anal. Appl. **69**, 353–358 (1979)

515. S. Reich, On the asymptotic behaviour of nonlinear semigroups and the range of accretive operators II. J. Math. Anal. **87**, 134–146 (1982)

516. S. Reich, Some problems and results in fixed point theory. Contemp. Math. **21**, 179–187 (1983)

517. J. Reinermann, R. Schoneberg, Some results and problems in the fixed theory for nonexpansive and pseudo contractive mappings, in Hilbert space, *Fixed Point Theory and its Applications*, ed. by S. Swaminathan (Academic Press, 1976), pp. 187–196

518. J. Reinermann, R. Schoneberg, Some results in the fixed point theory of nonexpansive mappings and generalized contractions, in *Fixed Point Theory and its Applications* ed. by S. Swaminathan (Academic Press, 1976), pp. 175–186

519. B.E. Rhoades, A comparison of various definitions of contractive mappings. Trans. Am. Math. Soc. **226**, 257–290 (1977)

520. B.E. Rhoades, Contractive definitions and continuity. Contemp. Math. **72**, 233–245 (1988)

521. B.E. Rhoades, Some theorems on weakly contractive maps. Nonlinear Anal. TMA **47**, 2683–2693 (2001)

522. B.E. Rhoades, H.K. Pathak, S.N. Mishra, Some weakly contractive mapping theorems in partially ordered spaces and applications. Demonstratio Math. (2010)

523. R.T. Rockfellar, Convex functions and dual extremum problems. Doctoral dissertation, Multilith Notes, Harvard (1963)

524. R.T. Rockfellar, *Convex Analysis* (Princeton University Press, Princeton, 1970)

525. R.T. Rockfellar, Level sets and continuity of convex conjugate functions. Trans. Am. Math. Soc. **1239**, 46–63 (1966)

526. R.T. Rockfellar, Local boundedness of nonlinear monotone operators. Mich. Math. J. **16**, 397–407 (1969)

527. R.T. Rockfellar, On the virtual convexity of the domain and range of nonlinear maximal monotone operator. Math. Ann. **185**, 81–90 (1970)

528. R.T. Rockfellar, On the maximality of sums of nonlinear monotone operators. Trans. A.M.S. **149**, 75–88 (1970)

529. R.T. Rockfellar, Characterization of the subdifferetials of convex functions. Pac. J. Math. **17**, 497–510 (1966)

530. E. Rothe, Zur theorie der topologischen ordnung una der vectoefelder in Bnachschen Raumen. Composito Math. **5**, 177–197 (1937)

531. B.D. Rouhani, W.A. Kirk, Asymptotic properties of nonexpansive iterations in reflexive spaces. J. Math. Anal. Appl. **236**, 281–289 (1999)

532. C. Ryll-Nardzewski, On fixed points of semigroups of endomorphisms of linear spaces, Proc. Fifth Berkeley Symp. **2**, 55–61(1967), Part 1, 9

533. W. Rudin, *Principles of Mathematical Analysis*, 3rd edn. (McGraw-Hill Book Company, New York, 1976)

534. W. Rudin, *Functional Analysis* (McGraw-Hill International Editions, 1991)

535. V.N. Sadovski, On a fixed point principle. Funct. Anal. Appl. **1**, 74–76 (1967)

536. V.N. Sadovski, Limit-compact and condensing operators. Russian Math. Surv. **27**, 85–155 (1972)

537. V.N. Sadovski, Application of topological methods in the theory of periodic solution of nonlinear differential operator equations of neutral type. Dokl. Akad. Nauk S.S.S.R. **200**, 1037–1040 (1971); Sov. Phys. Dokkl. **12** (1971)

538. I.W. Sandberg, On the L_2-boundedness of solutions of nonlinear function equations. Bell. Syst. Tech. J. **43**(4), 1581–1599 (1964)

539. I.W. Sandberg, A frequency-domain condition for the stability of feedback systems containing a single time-varying nonlinear element. Bell. Syst. Tech. J. **43**(4), 1601–1608 (1964)

540. I.W. Sandberg, Some results on the theory of physical systems governed by nonlinear functional equations. Bell. Syst. Tech. J. **44**(15), 871–898 (1965)

541. A. Sard, The measyres of the critical values of differentiable maps. Bull. Am. Math. Soc. **48**, 883–890 (1942)

542. J. Schauder, Der Fixpunksatz Funktional raumen. Stud. Math. **2**, 171–180 (1930)

543. I.J. Schoenberg, On a theorem of Kirzbraun and valentine. Am. Math. Mon. **60**, 620–622 (1953)

544. Z.P. Schulman, B.M. Berkovsky, Nauka i Tehnika (in Russian), Minsk (1966), p. 100

545. J. Schwartz, *Nonlinear Functional Analysis* (Gorden and Breach Science Publishers, New York, 1969)

546. J. Scarf, The approximation of fixed points of a continuous mappings. SIAM J. Appl. Math. **15**, 1328 (1967)

547. V.M. Sehgal, A fixed point theorem for mapping with a contractive iterate. Proc. Am. Math. Soc. **23**, 631–644 (1969)

548. V.M. Sehgal, On fixed and periodic points of a class of mappings. J. Lond. Math. Soc. **5**(2), 571–576 (1972)

549. V.M. Sehgal, E. Morrison, A fixed point theorem for multifunctions in a lcss. Proc. Am. Math. Soc. **38**, 643–646 (1973)

550. S. Sengupta, A.S. Gupta, Publications de l'institute Mathematique. Bergrade, Nouvelle Series **19**, 147 (1975)

551. H.F. Senter, W.G. Doston Jr., Approximating fixed points of nonexpansive mappings. Proc. Am. Math. Soc. **44**, 375–379 (1974)

552. T.N. Shimi, Approximations of fixed points of certain nonlinear mappings. J. Math. Anal. Appl. **65**, 565–571 (1978)

553. C. Shiau, K.K. Tan, C.S. Wong, A class of quasi-nonexpansive multi-valued maps. Canad. Math. Bull. **18**, 707–714 (1975)

554. R.E. Showalter, R.E. Continuity, Continuity of maximal monotone sets in Banach Space. Proc. Am. Math. Soc. **42**(2), 543–546 (1974)

555. M. Sibony, Methods iteratives pour les equations et inequalities aus derivees partielles non-linearies de type monotone. Calcolo **14**, 65–183 (1970)

556. A.H. Siddiqi, *Functional Analysis with Applications* (Tata McGraw-Hill, New Delhi, 1986)

557. A.H. Siddiqi, Q.H. Ansari, Strongly nonlinear quasi varitional inequality. J. Math. Anal. Appl. **149**(2) (1990)

558. J. Siegel, A new proof of Caristi's fixed point theorem. Proc. Am. Math. Soc. **66**, 54–56 (1977)

559. F. Sinestrari, Time-dependent integrodifferential equations in Banach space, in *Nonlinear Phenomena in Mathematical Sciences* ed. by V. Lakshmikantham (Academic Press, New York, 1982)

560. S.L. Singh, Applications of fixed point theorems to nonlinear integrodifferential equations. Riv. Mat. Univ. Parma **16**, 205–212 (1990)

561. S.L. Singh, K.S. Ha, Y.J. Cho, Coincidence and fixed points of nonlinear hybrid contractions. Internat. J. Math. Math. Sci. **12**(2), 247–256 (1989)

562. S.P. Singh, B.A. Meade, Fixed point theorems in complete metric spaces. Bull. Austr. Math. Soc. **16**, 49–53 (9177)

563. I.V. Skrypnik, Methods for analysis of nonlinear elliptic boundary value problems. Am. Math. Soc. **139** (1994)

564. I.V. Skrypnik, *Nonlinear Elliptic Equations of Higher Order (Russian)* (Naukova Dumka, Kiev, 1973)

565. M.V. Solodov, B.F. Svaiter, A globally convergent inexact Newton method for systems of monotone equations, in *Reformulation: Nonsmooth, Piecewise Smooth, Semismooth and Smoothing Methods*, ed. by M. Fukushima, L. Qi (Kluwer Academic Publisher, Dordecht, The Netherlands, 1999), pp. 355–369

566. M.V. Solodov, B.F. Svaiter, A new projection method for variational inequality problems. SIAM J. Control Optim. **37**, 765–776 (1999)

567. M.V. Solodov, B.F. Svaiter, A hybrid projection proximal point algorithm. J. Convex. Anal. **6**, 59–70 (1999)

568. M.V. Solodov, B.F. Svaiter, Forcing strong convergence of proximal point iterations in a Hilbert space. Math. Progr. **87**, 189–202 (2000)

569. M.V. Solodov, B.F. Svaiter, A truly globally convergent Newton-type method for the monotone nonlinear complementarity problem. SIAM J. Optim. **10**, 605–625 (2000)

570. S. Smale, An infinite dimensional generalization of Sards theorem. Am. J. Math. **87**, 861–890 (1965)

571. D.R. Smart, *Fixed Point Theorems, Paper*, Back edn. (Cambridge University Press, Cambridge, 1980)

572. R.E. Smithson, Fixed points for contractive multifunctions. Proc. Am. Math. Soc. **27**, 192–194 (1971)

573. V. Smulian, Sur la derivabilite de la norme dans l'espace de Banach, C.R. (Doklady). Acad. Sci., USSR **27**, 643–648 (1940)

574. V. Smulian, On the principle of inclusion in the spaces of type (B). Mat. Sb. (N.S.) **5**, 327–328 (1939)

575. I. Stakgold, *Boundary Value Problems of Mathematical Physics*, vol. I & II (Macmillan, New York, 1968)

576. P.N. Srikanth, M.C. Joshi, Existence theorems for generalized Hammerstein equations. Proc. Am. Math. Soc. **783**, 369–374 (1980)

577. G. Stampachhia, *Variational inequalities, Theory and Applications of Monotone Operators*, ed. by G. Oderisi (1969), pp. 101–192

578. G. Stampachhia, C.R. Acad, Sc. (Paris) **258**, 4413 (1964)

579. C.A. Stuart, Positive solutions of a nonlinear integral equation. Math. Ann. **192**, 119–124 (1971)

580. C.A. Stuart, Existence theorems for a class of nonlinear integral equations. Math. Z. **137**, 49–66 (1974)

581. T. Suzuki, Mizoguchi-Takahashi's fixed point theorem is a real generalization of Nadlers. J. Math. Anal. Appl. **340**, 752–755 (2008)

582. K.E. Swick, A model of single species population growth. SIAM J. Math. Anal. **7**, 565–576 (1976)

583. S. Szufla, Some remarks on ordinary differential equations in Banach spaces, bull. Acad. Polon Sci. **16**, 795–800 (1968)

584. W. Takahashi, The asymptotic behaviour of nonlinear semigroups and invariant means. J. Math. Anal. **109**, 130–139 (1985)

585. W. Takahashi, Existence theorems generalizing fixed point theorems for multivalued mappings, in *FixedPoint Theory and Applications* ed. by J.B. Baillon, M. Thera, Pitman Research Notes in Mathematics, vol. 252 (Longman, Harlow, 1991), pp. 397–406

586. S. Takahashi, W. Takahashi, Strong convergence theorem for a generalized equilibrium problem and a nonexpansive mapping in a Hilbert space. Nonlinear Anal. **69**, 1025–1033 (2008)

587. E. Tarafdar, X.-Z. Yuan, Set-valued topological contrations. Appl. Math. Lett. **8**(6), 79–81 (1995)

588. A. Taraski, A Lattice-theoretical fixed point point theorem and its applications. Pac. J. Math. **5**, 285–309 (1955)

589. A.C. Thompson, On certain contraction mappings in a partially ordered vector space. Proc. Am. Math. Soc. **14**, 438–443 (1963)

590. C.C. Travis, G.F. Webb, Existence and stability for partial functional differential equations. Trans. Am. Math. Soc. **200**, 395–418 (1974)

591. S.C. Troyanskii, On locally uniformly convex and differetiable norms in certain nonseparable Banach spaces. Stud. Math. **37**, 173–180 (1971)

592. E.L. Turner, Transverasality and cone maps. Arch. Ration. Mech. Anal. **58**, 151–179 (1975)

593. A. Tychonoff, Ein flxpunktsatz. Math. Ann. **111**, 767–776 (1935)

594. P.S. Uryshon, On a type of nonlinear integral equation. Math. Sb. **31**, 236–255 (1974)

595. M.M. Vainberg, Existence theorems for systems of nonlinear integral equations. Uchen. Zap. Mosk. Obl. Pedinst. 18, Trud, Kafedr. Fizmata **2** (1951)

596. M.M. Vainberg, A new principle in the theory of nonlinear equations. Usphehi Mat. Nauk SSS **15**, 243–244 (1960)

597. M.M. Vainberg, *Variaional Methods for the Steady of Nonlinear Operators* (Holden-Day, San Francisco, 1964)

598. M.M. Vainberg, Nonlinear problems for potential and monotone operators. Soviet Math. Dokl. **9**, 1427–1430 (1968)

599. M.M. Vainberg, Existence theorems for system of nonlinear integral equations. Uchen. Zap. Mosk. Olb.Pedints 18, Trud. Kafedr. Fizamata. **2**, 56–73 (1951)

600. F.A. Valentine, On the extension of a vector function so as to preserve a Lipschitz condition. Bull. A.M.S. **49**, 100–108 (1943)

601. R.S. Varga, *Matrix Iterative Analysis* (Prentice Hall, Englewood Cliff, 1982)

602. M. Vath, *Voltera and Integral Equations of Vector Functions* (PAM, Marcel Dekker, New York, 2000)

603. I. Vrkoč, The representation of Carathéodery operators. Czech. Math. J. **19**(1), 99–109 (1969)

604. H.O. Walther, On instability, w-limit sets periodic solutions of nonlinear autonomous differential delay equations, in *Functional Differential Equations and Approximations of Fixed Points* ed. by H.O. Peitgen, H.O. Walther (Lecture notes Math. 730) (Springer, 1970), pp. 489–503

605. L.E. Ward Jr., Charaterization of the fixed point property for a class of set value mappings. Fund. Math. **50**, 159–167 (1961)

606. G.F. Webb, An abstract semilinear Volterra integro-differential equation. Am. Math. Soc. **69**, 155–260 (1978)

607. L.R. William, R.W. Leggett, Multiple fixed point theorems for problem on chemical reactor theory. J. Math. Anal. Appl. **69**, 180–193 (1979)

608. C.S. Wong, Comman fixed points of two mappings. Pac. J. Math. **48**, 299–312 (1973)

609. C.S. Wong, On a fixed point theorem of contractive type. Proc. Am. Math. Soc. **57**, 283–284 (1976)
610. C.S. Wong, *Maps of Contractive Type in Fixed Point Theory and Its Applications* ed. by (S.Swaminathan) (Academic Press, New York, 1976), pp. 197–206
611. Z.B. Xu, G.F. Roach, Characteristic inequalities in uniformly convex and uniformly smooth Banach spaces. J. Math. Anal. Appl. **157**, 189–210 (1991)
612. K. Yanagi, On some fixed point theorems for multivalued mappings. Pac. J. Math. **87**, 233–240 (1980)
613. C.L. Yen, Remark on common fixed points. Tamkang J. Math. **3**, 95–96 (1972)
614. K. Yosida, Functional Analysis (Springer, 1974)
615. C. Zalinescu, Recession cones and asymptotically compact sets. J. Optim. Theory Appl. **77**, 209–220 (1993)
616. T. Zamfirescu, Fixed point theorems in metric spaces. Arch. Math., (Basel) **23**, 292–298 (1972)
617. E.H. Zarantonello, Solving functional equations by contractive mappings, Math. Res. Center Re No. 160, Madison, Wisconsin (1960)
618. E.H. Zarantonello, Projections on convex sets in Hilbert space and spectral theory, in *Contribution to Nonlinear Functional Analysis*, ed. by E.H. Zarantoned (Academic Press, New York, 1971), pp. 237–424
619. E. Zeidler, *Nonlinear Functional Analysis and Applications, Part 1: Fixed -Point Theorems* (Springer, New York, 1986)
620. E. Zeidler, *Nonlinear Functional Analysis and Its Applications, II: Nonlinear Monotone Operators* (Springer, Berlin, 1990)
621. Y.B. Zhao, D. Li, Strict feasibility condition in nonlinear complementarity problems. J. Optim. Theory Appl. **107**, 641–664 (2000)
622. Y.B. Zhao, D. Li, On a new homotopy continuation trajectory for nonlinear complementarity problems. Math. Op. Res. **26**, 119–146 (2001)
623. Y.B. Zhao, D. Li, Monotonicity of fixed point and normal mappings associated with variational inequality and its application. SIAM J. Optim. **4**, 962–973 (2001)
624. D. Zhu, P. Marcotte, New classes of generalized monotonicity. J. Optim. Theory Appl. **87**, 457–471 (1995)
625. D. Zhu, P. Marcotte, Co-coercivity and its role in the convergence of iterative schemes for solving variational inequalities. SIAM J. Optim. **6**, 714–726 (1996)

Index

© Springer Nature Singapore Pte Ltd. 2018
H. K. Pathak, *An Introduction to Nonlinear Analysis and Fixed Point Theory*,
https://doi.org/10.1007/978-981-10-8866-7

Printed in the United States
By Bookmasters